Topological Geometrodynamics

Topological Geometrodynamics

Matti Pitkänen

Department of Physical Sciences
University of Helsinki, Finland

Luniver Press

Published in 2006 by Luniver Press

Cover picture: "Wormhole" by Mark Thornally, `http://pweb.netcom.com/~mthorn/intrnmsn.htm`

Luniver Press
Beckington, Frome BA11 6TT, United Kingdom
`www.luniver.com`

British Library Cataloguing in Publication Data
A catalogue record for this book is available from the British Library

Library of Congress Cataloguing in Publication Data
A catalogue record is available from the Library of Congress

ISBN-13:   978-0-9551170-8-4
ISBN-10:   0-9551170-8-9

Preface

This book represents the recent state of a unified theory of fundamental interactions for which I have devoted last 27 years of my life. I remember having got the basic idea of Topological Geometrodynamics (TGD) during autumn 1978, perhaps it was October. What I realized was that the representability of physical space-times as 4-dimensional surfaces of some higher-dimensional space-time obtained by replacing the points of Minkowski space with some very small compact internal space could resolve the conceptual difficulties of general relativity related to the definition of the notion of energy.

It soon became clear that the approach leads to a generalization of the notion of space-time with particles being represented by space-time surfaces with finite size so that TGD could be also seen as a generalization of the string model. Much later it became clear that this generalization is consistent with conformal invariance only if space-time is 4-dimensional surfaces and the Minkowski space factor of imbedding space is 4-dimensional.

It took some time to discover that also the geometrization of also gauge interactions and elementary particle quantum numbers could be possible in this framework: it required two years to find the unique internal space providing this geometrization involving also the realization that family replication phenomenon for fermions has a natural topological explanation in TGD framework and that the symmetries of the standard model symmetries are much more profound than pragmatic TOE builders have believed them to be. If TGD is correct, main stream particle physics chose the wrong track leading to the recent deep crisis when people decided that quarks and leptons belong to same multiplet of the gauge group implying instability of proton.

There were however also longstanding problems. Gravitational energy is well-defined in cosmological models but is not conserved. Hence the conservation of the inertial energy is not consistent with the Equivalence Principle. Furthermore, the imbeddings of Robertson-Walker cosmologies turned out to be vacuum extremals with respect to the inertial energy. It took about 25 years to realize that the sign of the inertial energy can be also negative and in cosmological scales the density of inertial energy vanishes: physically acceptable universes are creatable from vacuum. If gravitational energy is the absolute value of the inertial energy at elementary particle level, the breaking of Equivalence Principle is minimal.

From the beginning it was clear that the theory predicts the presence of long ranged classical electro-weak and color gauge fields and that these fields necessarily accompany classical electromagnetic fields. It took about 26 years to gain the maturity to admit the obvious: these fields are classical correlates for long range color and weak interactions assignable to dark matter. TGD physics is a fractal consisting of an entire hierarchy of fractal copies of standard model physics.

I started the attempts to construct quantum TGD after my thesis around 1982. It took some years to discover that the only working approach is based on the generalization of Einstein's program. Quantum physics involves the geometrization of the infinite-dimensional world of classical worlds identified as 3-dimensional surfaces. Still few years had to pass before I understood that general coordinate invariance leads to a more or less unique solution of the problem and implies that space-time surfaces are analogous to Bohr orbits. Still a coupled of years and I discovered that quantum states of the Universe can be identified as classical spinor fields in the world of classical worlds. Only quantum jump is the genuinely quantal aspect of quantum physics.

During these years TGD led to a rather profound generalization of the space-time concept. Quite general properties of the theory led to the notion of many-sheeted space-time with sheets representing physical subsystems of various sizes. At the beginning of 90s I became dimly aware of the importance of p-adic number fields and soon ended up with the idea that p-adic thermodynamics for a conformally invariant system allows to understand elementary particle massivation with amazingly few input assumptions. The attempts to understand p-adicity from basic principles led gradually to the vision about physics as a generalized number theory. One of its elements was a generalization of the number concept obtained by fusing real numbers and various p-adic numbers along common rationals. The vision involves also quaternions and octonions and the notion of infinite prime.

TGD inspired theory of consciousness entered the scheme after 1995 as I started to write book about consciousness. Gradually it became difficult to say where physics and and consciousness theory begins since consciousness theory could be seen as a generalization of quantum measurement theory by identifying quantum jump as a moment of consciousness and by replacing the observer with the notion of self identified as a system which is conscious as long as it can avoid entanglement with environment. "Everything is conscious and consciousness can be only lost" summarizes the basic philosophy neatly.

The idea about p-adic physics as physics of cognition and intentionality emerged also rather naturally and implies perhaps the most dramatic generalization of the space-time concept in which most points of p-adic space-time sheets are infinite in real sense and the projection to the real imbedding space consists of discrete set of points.

The last thread in the evolution of ideas is only slightly more than one year old. Learning about the paper of Laurent Nottale about the possibility to identify planetary orbits as Bohr orbits with a gigantic value of gravitational Planck constant made once again possible to see the obvious. Dynamical quantized Planck constant is actually implied by quantum classical correspondence and the fact that space-time sheets identifiable as quantum coherence regions can have arbitrarily large sizes.

In this book my goal is to provide a bird's eye of view about TGD as it is now. My hope is that the book would stimulate the reader to familiarize with the 15 online books at my home page providing a more detail documentation of TGD and TGD inspired theory of consciousness.

Matti Pitkänen

Hanko,
March 23

Acknowledgements

Neither TGD nor this book would have been possible without the help and encouragement of many people. The friendship with Heikki and Raija Haila and their family have been kept me in contact with the everyday world and without this friendship I would not have survived through these lonely 27 years most of which I have remained unemployed as a scientific dissident. I am happy that my children have understood my difficult position and like my friends have believed that what I am doing is something valuable although I have not received any official recognition for it. During last years Tapio Tammi has helped me quite concretely by providing the necessary computer facilities and being one of the few persons in Finland with whom to discuss about my work.

The collaboration with Lian Sidoroff was extremely fruitful and she also helped my to survive financially through the hardest years. The participation to CASYS conferences in Liege has been an important window to the academic world and I am grateful for Daniel Dubois and Peter Marcer for making this participation possible. The recent discussions and collaboration with Eduardo de Luna and Istvan Dienes stimulated the hope that the communication of new vision might not be a mission impossible after all. Andrew Adamatsky proposed the writing of this book when I had already got used to the thought that my work would not be published during my life time. He also helped me with the otherwise time consuming practical problems related to the transformation of the texts to latex book format. I am grateful for Mark Thornally for allowing to use "Wormhole", a really fascinating piece of electronic art, in the book cover. During these years I have received innumerable email contacts from people around the world. These contacts have helped me to avoid the depressive feeling of being some kind of Don Quijote of Science and helped me to widen my views: I am grateful for all these people.

Contents

III   ALGEBRAIC PHYSICS   237

7   Equivalence of Loop and Tree Diagrams and Divergence Cancellation   239

IV APPLICATIONS TO ELEMENTARY PARTICLE PHYSICS 361

9 General View About Physics in Many-Sheeted Space-Time 363

List of Figures

Chapter 1

Introduction

1.1 Background

T(opological) G(eometro)D(ynamics) is one of the many attempts to find a unified description of basic interactions. The development of the basic ideas of TGD to a relatively stable form took time of about half decade [n1]. The great challenge is to construct a mathematical theory around these physically very attractive ideas and I have devoted the last twenty-three years for the realization of this dream and this has resulted in seven online books [TGDview, TGDgeom, TGDquant, TGDnumber, TGDclass, TGDpad, TGDfree] about TGD and eight online books about TGD inspired theory of consciousness and of quantum biology [TGDconsc, TGDselforg, TGDware, TGDholo, TGDgeme, TGDeeg, TGDmagn, TGDmathc].

Quantum T(opological)D(ynamics) as a classical spinor geometry for infinite-dimensional configuration space, p-adic numbers and quantum TGD, and TGD inspired theory of consciousness have been for last decade of the second millenium the basic three strongly interacting threads in the tapestry of quantum TGD.

For few yeas ago the discussions with Tony Smith generated a fourth thread which deserves the name 'TGD as a generalized number theory'. The work with Riemann hypothesis made time ripe for realization that the notion of infinite primes could provide, not only a reformulation, but a deep generalization of quantum TGD. This led to a thorough and extremely fruitful revision of the basic views about what the final form and physical content of quantum TGD might be.

The fifth thread came with the realization that by quantum classical correspondence TGD predicts an infinite hierarchy of macroscopic quantum systems with increasing sizes, that it is not at all clear whether standard quantum mechanics can accommodate this hierarchy, and that a dynamical quantized Planck constant might be necessary and certainly possible in TGD framework. The identification of hierarchy of Planck constants whose values TGD "predicts" in terms of dark matter hierarchy would be natural. This also led to a solution of a long standing puzzle: what is the proper interpretation of the predicted fractal hierarchy of long ranged classical electro-weak and color gauge fields. Quantum classical correspondences allows only single answer: there is infinite hierarchy of p-adically scaled up variants of standard model physics and for each of them also dark hierarchy. Thus TGD Universe would be fractal in very abstract and deep sense.

TGD forces the generalization of physics to a quantum theory of consciousness, and represent TGD as a generalized number theory vision leads naturally to the emergence of p-adic physics as physics of cognitive representations. The seven online books [TGDview, TGDgeom, TGDquant, TGDnumber, TGDclass, TGDpad, TGDfree] about TGD and eight online books about TGD inspired theory of consciousness and of quantum biology [TGDconsc, TGDselforg, TGDware, TGDholo, TGDgeme, TGDeeg, TGDmagn, TGDmathc] are warmly recommended to the interested reader.

1.2 Basic Ideas of TGD

The basic physical picture behind TGD was formed as a fusion of two rather disparate approaches: namely TGD is as a Poincare invariant theory of gravitation and TGD as a generalization of the old-fashioned string model.

1.2.1   TGD as a Poincare invariant theory of gravitation

The first approach was born as an attempt to construct a Poincare invariant theory of gravitation. Space-time, rather than being an abstract manifold endowed with a pseudo-Riemannian structure, is regarded as a surface in the 8-dimensional space $H = M^4_+ \times CP_2$, where M^4_+ denotes the interior of the future light cone of the Minkowski space (to be referred as light cone in the sequel) and $CP_2 = SU(3)/U(2)$ is the complex projective space of two complex dimensions [bc3, bc2, bc1, bc4]. The identification of the space-time as a sub-manifold [ba1, ba3] of $M^4 \times CP_2$ leads to an exact Poincare invariance and solves the conceptual difficulties related to the definition of the energy-momentum in General Relativity [ia4]. The actual choice $H = M^4_+ \times CP_2$ implies the breaking of the Poincare invariance in the cosmological scales but only at the quantum level. It soon however turned out that sub-manifold geometry, being considerably richer in structure than the abstract manifold geometry, leads to a geometrization of all basic interactions. First, the geometrization of the elementary particle quantum numbers is achieved. The geometry of CP_2 explains electro-weak and color quantum numbers. The different H-chiralities of H-spinors correspond to the conserved baryon and lepton numbers. Secondly, the geometrization of the field concept results. The projections of the CP_2 spinor connection, Killing vector fields of CP_2 and of H-metric to four-surface define classical electro-weak, color gauge fields and metric in X^4.

1.2.2   TGD as a generalization of the hadronic string model

The second approach was based on the generalization of the mesonic string model describing mesons as strings with quarks attached to the ends of the string. In the 3-dimensional generalization 3-surfaces correspond to free particles and the boundaries of the 3- surface correspond to partons in the sense that the quantum numbers of the elementary particles reside on the boundaries. Various boundary topologies (number of handles) correspond to various fermion families so that one obtains an explanation for the known elementary particle quantum numbers. This approach leads also to a natural topological description of the particle reactions as topology changes: for instance, two-particle decay corresponds to a decay of a 3-surface to two disjoint 3-surfaces.

1.2.3   Fusion of the two approaches via a generalization of the space-time concept

The problem is that the two approaches seem to be mutually exclusive since the orbit of a particle like 3-surface defines 4-dimensional surface, which differs drastically from the topologically trivial macroscopic space-time of General Relativity. The unification of these approaches forces a considerable generalization of the conventional space-time concept. First, the topologically trivial 3-space of General Relativity is replaced with a "topological condensate" containing matter as particle like 3-surfaces "glued" to the topologically trivial background 3-space by connected sum operation. Secondly, the assumption about connectedness of the 3-space is given up. Besides the "topological condensate" there is "vapor phase" that is a "gas" of particle like 3-surfaces (counterpart of the "baby universes" of GRT) and the nonconservation of energy in GRT corresponds to the transfer of energy between the topological condensate and vapor phase.

1.3   The five threads in the development of quantum TGD

The development of TGD has involved four strongly interacting threads: physics as infinite-dimensional geometry; p-adic physics; TGD inspired theory of consciousness and TGD as a generalized number theory. In the following these five threads are briefly described.

1.3.1   Quantum TGD as configuration space spinor geometry

A turning point in the attempts to formulate a mathematical theory was reached after seven years from the birth of TGD. The great insight was "Do not quantize". The basic ingredients to the new approach have served as the basic philosophy for the attempt to construct Quantum TGD since then and are the following ones:

a) Quantum theory for extended particles is free(!), classical(!) field theory for a generalized Schrödinger amplitude in the configuration space CH consisting of all possible 3-surfaces in H. "All possible" means that surfaces with arbitrary many disjoint components and with arbitrary internal topology and also singular surfaces topologically intermediate between two different manifold topologies are included. Particle reactions are identified as topology changes [aa2, aa1, aa3]. For instance, the decay of a 3-surface to two 3-surfaces corresponds to the decay $A \to B + C$. Classically this corresponds to a path of configuration space leading from 1-particle sector to 2-particle sector. At quantum level this corresponds to the dispersion of the generalized Schrödinger amplitude localized to 1-particle sector to two-particle sector. All coupling constants should result as predictions of the theory since no nonlinearities are introduced.

b) Configuration space is endowed with the metric and spinor structure so that one can define various metric related differential operators, say Dirac operator, appearing in the field equations of the theory.

1.3.2 p-Adic TGD

The p-adic thread emerged for roughly ten years ago as a dim hunch that p-adic numbers might be important for TGD. Experimentation with p-adic numbers led to the notion of canonical identification mapping reals to p-adics and vice versa. The breakthrough came with the successful p-adic mass calculations using p-adic thermodynamics for Super-Virasoro representations with the super-Kac-Moody algebra associated with a Lie-group containing standard model gauge group. Although the details of the calculations have varied from year to year, it was clear that p-adic physics reduces not only the ratio of proton and Planck mass, the great mystery number of physics, but all elementary particle mass scales, to number theory if one assumes that primes near prime powers of two are in a physically favored position. Why this is the case, became one of the key puzzless and led to a number of arguments with a common gist: evolution is present already at the elementary particle level and the primes allowed by the p-adic length scale hypothesis are the fittest ones.

It became very soon clear that p-adic topology is not something emerging in Planck length scale as often believed, but that there is an infinite hierarchy of p-adic physics characterized by p-adic length scales varying to even cosmological length scales. The idea about the connection of p-adics with cognition motivated already the first attempts to understand the role of the p-adics and inspired 'Universe as Computer' vision but time was not ripe to develop this idea to anything concrete (p-adic numbers are however in a central role in TGD inspired theory of consciousness). It became however obvious that the p-adic length scale hierarchy somehow corresponds to a hierarchy of intelligences and that p-adic prime serves as a kind of intelligence quotient. Ironically, the almost obvious idea about p-adic regions as cognitive regions of space-time providing cognitive representations for real regions had to wait for almost a decade for the access into my consciousness.

There were many interpretational and technical questions crying for a definite answer. What is the relationship of p-adic non-determinism to the classical non-determinism of the basic field equations of TGD? Are the p-adic space-time region genuinely p-adic or does p-adic topology only serve as an effective topology? If p-adic physics is direct image of real physics, how the mapping relating them is constructed so that it respects various symmetries? Is the basic physics p-adic or real (also real TGD seems to be free of divergences) or both? If it is both, how should one glue the physics in different number field together to get *The Physics*? Should one perform p-adicization also at the level of the configuration space of 3-surfaces? Certainly the p-adicization at the level of super-conformal representation is necessary for the p-adic mass calculations. Perhaps the most basic and most irritating technical problem was how to precisely define p-adic definite integral which is a crucial element of any variational principle based formulation of the field equations. Here the frustration was not due to the lack of solution but due to the too large number of solutions to the problem, a clear symptom for the sad fact that clever inventions rather than real discoveries might be in question.

Despite these frustrating uncertainties, the number of the applications of the poorly defined p-adic physics growed steadily and the applications turned out to be relatively stable so that it was clear that the solution to these problems must exist. It became only gradually clear that the solution of the problems might require going down to a deeper level than that represented by reals and p-adics.

1.3.3   TGD as a generalization of physics to a theory consciousness

General coordinate invariance forces the identification of quantum jump as quantum jump between entire deterministic quantum histories rather than time=constant snapshots of single history. The new view about quantum jump forces a generalization of quantum measurement theory such that observer becomes part of the physical system. Thus a general theory of consciousness is unavoidable outcome. This theory is developed in detail in the books [TGDconsc, TGDselforg, TGDware, TGDholo, TGDgeme, TGDeeg, TGDmagn, TGDmathc].

Quantum jump as a moment of consciousness

The identification of quantum jump between deterministic quantum histories (configuration space spinor fields) as a moment of consciousness defines microscopic theory of consciousness. Quantum jump involves the steps

$$\Psi_i \to U\Psi_i \to \Psi_f \ ,$$

where U is informational "time development" operator, which is unitary like the S-matrix characterizing the unitary time evolution of quantum mechanics. U is however only formally analogous to Schrödinger time evolution of infinite duration although there is *no* real time evolution involved. It is not however clear whether one should regard U-matrix and S-matrix as two different things or not: U-matrix is a completely universal object characterizing the dynamics of evolution by self-organization whereas S-matrix is a highly context dependent concept in wave mechanics and in quantum field theories where it at least formally represents unitary time translation operator at the limit of an infinitely long interaction time. The S-matrix understood in the spirit of superstring models is however something very different and could correspond to U-matrix.

The requirement that quantum jump corresponds to a measurement in the sense of quantum field theories implies that each quantum jump involves localization in zero modes which parameterize also the possible choices of the quantization axes. Thus the selection of the quantization axes performed by the Cartesian outsider becomes now a part of quantum theory. Together these requirements imply that the final states of quantum jump correspond to quantum superpositions of space-time surfaces which are macroscopically equivalent. Hence the world of conscious experience looks classical. At least formally quantum jump can be interpreted also as a quantum computation in which matrix U represents unitary quantum computation which is however not identifiable as unitary translation in time direction and cannot be 'engineered'.

The notion of self

The concept of self is absolutely essential for the understanding of the macroscopic and macro-temporal aspects of consciousness. Self corresponds to a subsystem able to remain un-entangled under the sequential informational 'time evolutions' U. Exactly vanishing entanglement is practically impossible in ordinary quantum mechanics and it might be that 'vanishing entanglement' in the condition for self-property should be replaced with 'subcritical entanglement'. On the other hand, if space-time decomposes into p-adic and real regions, and if entanglement between regions representing physics in different number fields vanishes, space-time indeed decomposes into selves in a natural manner.

It is assumed that the experiences of the self after the last 'wake-up' sum up to single average experience. This means that subjective memory is identifiable as conscious, immediate short term memory. Selves form an infinite hierarchy with the entire Universe at the top. Self can be also interpreted as mental images: our mental images are selves having mental images and also we represent mental images of a higher level self. A natural hypothesis is that self S experiences the experiences of its subselves as kind of abstracted experience: the experiences of subselves S_i are not experienced as such but represent kind of averages $\langle S_{ij}\rangle$ of sub-subselves S_{ij}. Entanglement between selves, most naturally realized by the formation of join along boundaries bonds between cognitive or material space-time sheets, provides a possible a mechanism for the fusion of selves to larger selves (for instance, the fusion of the mental images representing separate right and left visual fields to single visual field) and forms wholes from parts at the level of mental images.

Relationship to quantum measurement theory

The third basic element relates TGD inspired theory of consciousness to quantum measurement theory. The assumption that localization occurs in zero modes in each quantum jump implies that the world of conscious experience looks classical. It also implies the state function reduction of the standard quantum measurement theory as the following arguments demonstrate (it took incredibly long time to realize this almost obvious fact!).

a) The standard quantum measurement theory a la von Neumann involves the interaction of brain with the measurement apparatus. If this interaction corresponds to entanglement between microscopic degrees of freedom m with the macroscopic effectively classical degrees of freedom M characterizing the reading of the measurement apparatus coded to brain state, then the reduction of this entanglement in quantum jump reproduces standard quantum measurement theory provide the unitary time evolution operator U acts as flow in zero mode degrees of freedom and correlates completely some orthonormal basis of configuration space spinor fields in non-zero modes with the values of the zero modes. The flow property guarantees that the localization is consistent with unitarity: it also means 1-1 mapping of quantum state basis to classical variables (say, spin direction of the electron to its orbit in the external magnetic field).

b) Since zero modes represent classical information about the geometry of space-time surface (shape, size, classical Kähler field,...), they have interpretation as effectively classical degrees of freedom and are the TGD counterpart of the degrees of freedom M representing the reading of the measurement apparatus. The entanglement between quantum fluctuating non-zero modes and zero modes is the TGD counterpart for the $m - M$ entanglement. Therefore the localization in zero modes is equivalent with a quantum jump leading to a final state where the measurement apparatus gives a definite reading.

This simple prediction is of utmost theoretical importance since the black box of the quantum measurement theory is reduced to a fundamental quantum theory. This reduction is implied by the replacement of the notion of a point like particle with particle as a 3-surface. Also the infinite-dimensionality of the zero mode sector of the configuration space of 3-surfaces is absolutely essential. Therefore the reduction is a triumph for quantum TGD and favors TGD against string models.

Standard quantum measurement theory involves also the notion of state preparation which reduces to the notion of self measurement. Each localization in zero modes is followed by a cascade of self measurements leading to a product state. This process is obviously equivalent with the state preparation process. Self measurement is governed by the so called Negentropy Maximization Principle (NMP) stating that the information content of conscious experience is maximized. In the self measurement the density matrix of some subsystem of a given self localized in zero modes (after ordinary quantum measurement) is measured. The self measurement takes place for that subsystem of self for which the reduction of the entanglement entropy is maximal in the measurement. In p-adic context NMP can be regarded as the variational principle defining the dynamics of cognition. In real context self measurement could be seen as a repair mechanism allowing the system to fight against quantum thermalization by reducing the entanglement for the subsystem for which it is largest (fill the largest hole first in a leaking boat).

Selves self-organize

The fourth basic element is quantum theory of self-organization based on the identification of quantum jump as the basic step of self-organization [I1]. Quantum entanglement gives rise to the generation of long range order and the emergence of longer p-adic length scales corresponds to the emergence of larger and larger coherent dynamical units and generation of a slaving hierarchy. Energy (and quantum entanglement) feed implying entropy feed is a necessary prerequisite for quantum self-organization. Zero modes represent fundamental order parameters and localization in zero modes implies that the sequence of quantum jumps can be regarded as hopping in the zero modes so that Haken's classical theory of self organization applies almost as such. Spin glass analogy is a further important element: self-organization of self leads to some characteristic pattern selected by dissipation as some valley of the "energy" landscape.

Dissipation can be regarded as the ultimate Darwinian selector of both memes and genes. The mathematically ugly irreversible dissipative dynamics obtained by adding phenomenological dissipation terms to the reversible fundamental dynamical equations derivable from an action principle can be understood as a phenomenological description replacing in a well defined sense the series of reversible quantum histories with its envelope.

Classical non-determinism of Kähler action

The fifth basic element are the concepts of association sequence and cognitive space-time sheet. The huge vacuum degeneracy of the Kähler action suggests strongly that the absolute minimum space-time is not always unique. For instance, a sequence of bifurcations can occur so that a given space-time branch can be fixed only by selecting a finite number of 3-surfaces with time like(!) separations on the orbit of 3-surface. Quantum classical correspondence suggest an alternative formulation. Space-time surface decomposes into maximal deterministic regions and their temporal sequences have interpretation a space-time correlate for a sequence of quantum states defined by the initial (or final) states of quantum jumps. This is consistent with the fact that the variational principle selects preferred extremals of Kähler action as generalized Bohr orbits.

In the case that non-determinism is located to a finite time interval and is microscopic, this sequence of 3-surfaces has interpretation as a simulation of a classical history, a geometric correlate for contents of consciousness. When non-determinism has long lasting and macroscopic effect one can identify it as volitional non-determinism associated with our choices. Association sequences relate closely with the cognitive space-time sheets defined as space-time sheets having finite time duration and psychological time can be identified as a temporal center of mass coordinate of the cognitive space-time sheet. The gradual drift of the cognitive space-time sheets to the direction of future force by the geometry of the future light cone explains the arrow of psychological time.

p-Adic physics as physics of cognition and intentionality

The sixth basic element adds a physical theory of cognition to this vision. TGD space-time decomposes into regions obeying real and p-adic topologies labelled by primes $p = 2, 3, 5, ...$. p-Adic regions obey the same field equations as the real regions but are characterized by p-adic non-determinism since the functions having vanishing p-adic derivative are pseudo constants which are piecewise constant functions. Pseudo constants depend on a finite number of positive pinary digits of arguments just like numerical predictions of any theory always involve decimal cutoff. This means that p-adic space-time regions are obtained by gluing together regions for which integration constants are genuine constants. The natural interpretation of the p-adic regions is as cognitive representations of real physics. The freedom of imagination is due to the p-adic non-determinism. p-Adic regions perform mimicry and make possible for the Universe to form cognitive representations about itself. p-Adic physics space-time sheets serve also as correlates for intentional action.

A more more precise formulation of this vision requires a generalization of the number concept obtained by fusing reals and p-adic number fields along common rationals (in the case of algebraic extensions among common algebraic numbers). This picture is discussed in [E1]. The application this notion at the level of the imbedding space implies that imbedding space has a book like structure with various variants of the imbedding space glued together along common rationals (algebraics). The implication is that genuinely p-adic numbers (non-rationals) are strictly infinite as real numbers so that most points of p-adic space-time sheets are at real infinity, outside the cosmos, and that the projection to the real imbedding space is discrete set of rationals (algebraics). Hence cognition and intentionality are almost completely outside the real cosmos and touch it at a discrete set of points only.

This view implies also that purely local p-adic physics codes for the p-adic fractality characterizing long range real physics and provides an explanation for p-adic length scale hypothesis stating that the primes $p \simeq 2^k$, k integer are especially interesting. It also explains the long range correlations and short term chaos characterizing intentional behavior and explains why the physical realizations of cognition are always discrete (say in the case of numerical computations). Furthermore, a concrete quantum model for how intentions are transformed to actions emerges.

The discrete real projections of p-adic space-time sheets serve also space-time correlate for a logical thought. It is very natural to assign to p-adic pinary digits a p-valued logic but as such this kind of logic does not have any reasonable identification. p-Adic length scale hypothesis suggest that the $p = 2^k - n$ pinary digits represent a Boolean logic B^k with k elementary statements (the points of the

k-element set in the set theoretic realization) with n taboos which are constrained to be identically true.

1.3.4   TGD as a generalized number theory

Quantum T(opological)D(ynamics) as a classical spinor geometry for infinite-dimensional configuration space, p-adic numbers and quantum TGD, and TGD inspired theory of consciousness, have been for last ten years the basic three strongly interacting threads in the tapestry of quantum TGD. For few yeas ago the discussions with Tony Smith generated a fourth thread which deserves the name 'TGD as a generalized number theory'. It relies on the notion of number theoretic compactifiction stating that space-time surfaces can be regarded either as hyper-quaternionic, and thus maximally associative, 4-surfaces in M^8 identifiable as space of hyper-octonions or as surfaces in $M^4 \times CP_2$ [E2].

The discovery of the hierarchy of infinite primes and their correspondence with a hierarchy defined by a repeatedly second quantized arithmetic quantum field theory gave a further boost for the speculations about TGD as a generalized number theory. The work with Riemann hypothesis led to further ideas.

After the realization that infinite primes can be mapped to polynomials representable as surfaces geometrically, it was clear how TGD might be formulated as a generalized number theory with infinite primes forming the bridge between classical and quantum such that real numbers, p-adic numbers, and various generalizations of p-adics emerge dynamically from algebraic physics as various completions of the algebraic extensions of rational (hyper-)quaternions and (hyper-)octonions. Complete algebraic, topological and dimensional democracy would characterize the theory.

What is especially satisfying is that p-adic and real regions of the space-time surface could emerge automatically as solutions of the field equations. In the space-time regions where the solutions of field equations give rise to in-admissible complex values of the imbedding space coordinates, p-adic solution can exist for some values of the p-adic prime. The characteristic non-determinism of the p-adic differential equations suggests strongly that p-adic regions correspond to 'mind stuff', the regions of space-time where cognitive representations reside. This interpretation implies that p-adic physics is physics of cognition. Since Nature is probably extremely brilliant simulator of Nature, the natural idea is to study the p-adic physics of the cognitive representations to derive information about the real physics. This view encouraged by TGD inspired theory of consciousness clarifies difficult interpretational issues and provides a clear interpretation for the predictions of p-adic physics.

1.3.5   Dynamical quantized Planck constant and dark matter hierarchy

By quantum classical correspondence space-time sheets can be identified as quantum coherence regions. Hence the fact that they have all possible size scales more or less unavoidably implies that Planck constant must be quantized and have arbitrarily large values. If one accepts this then also the idea about dark matter as a macroscopic quantum phase characterized by an arbitrarily large value of Planck constant emerges naturally as does also the interpretation for the long ranged classical electroweak and color fields predicted by TGD. Rather seldom the evolution of ideas follows simple linear logic, and this was the case also now. In any case, this vision represents the fifth, relatively new thread in the evolution of TGD and the ideas involved are still evolving.

Dark matter as large $\hbar$ phase

D. Da Rocha and Laurent Nottale [if1] have proposed that Schrödinger equation with Planck constant $\hbar$ replaced with what might be called gravitational Planck constant $\hbar_{gr} = \frac{GmM}{v_0}$ ($\hbar = c = 1$). v_0 is a velocity parameter having the value $v_0 = 144.7 \pm .7$ km/s giving $v_0/c = 4.6 \times 10^{-4}$. This is rather near to the peak orbital velocity of stars in galactic halos. Also subharmonics and harmonics of v_0 seem to appear. The support for the hypothesis coming from empirical data is impressive.

Nottale and Da Rocha believe that their Schrödinger equation results from a fractal hydrodynamics. Many-sheeted space-time however suggests astrophysical systems are not only quantum systems at larger space-time sheets but correspond to a gigantic value of gravitational Planck constant. The gravitational (ordinary) Schrödinger equation would provide a solution of the black hole collapse (IR catastrophe) problem encountered at the classical level. The resolution of the problem inspired by

TGD inspired theory of living matter is that it is the dark matter at larger space-time sheets which is quantum coherent in the required time scale [D6].

Already before learning about Nottale's paper I had proposed the possibility that Planck constant is quantized [O3] and the spectrum is given in terms of logarithms of Beraha numbers: the lowest Beraha number B_3 is completely exceptional in that it predicts infinite value of Planck constant. The inverse of the gravitational Planck constant could correspond a gravitational perturbation of this as $1/\hbar_{gr} = v_0/GMm$. The general philosophy would be that when the quantum system would become non-perturbative, a phase transition increasing the value of $\hbar$ occurs to preserve the perturbative character and at the transition $n = 4 \rightarrow 3$ only the small perturbative correction to $1/\hbar(3) = 0$ remains. This would apply to QCD and to atoms with $Z > 137$ as well.

TGD predicts correctly the value of the parameter v_0 assuming that cosmic strings and their decay remnants are responsible for the dark matter. The harmonics of v_0 can be understood as corresponding to perturbations replacing cosmic strings with their n-branched coverings so that tension becomes n^2-fold: much like the replacement of a closed orbit with an orbit closing only after n turns. $1/n$-sub-harmonic would result when a magnetic flux tube split into n disjoint magnetic flux tubes. Also a model for the formation of planetary system as a condensation of ordinary matter around quantum coherent dark matter emerges [D6].

Dark matter as a source of long ranged weak and color fields

Long ranged classical electro-weak and color gauge fields are unavoidable in TGD framework. The smallness of the parity breaking effects in hadronic, nuclear, and atomic length scales does not however seem to allow long ranged electro-weak gauge fields. The problem disappears if long range classical electro-weak gauge fields are identified as space-time correlates for massless gauge fields created by dark matter. Also scaled up variants of ordinary electro-weak particle spectra are possible. The identification explains chiral selection in living matter and unbroken $U(2)_{ew}$ invariance and free color in bio length scales become characteristics of living matter and of bio-chemistry and bio-nuclear physics. An attractive solution of the matter antimatter asymmetry is based on the identification of also antimatter as dark matter. The fluctuations in the rates of radio-active decays and chemical reactions correlating with astrophysical periods discovered by Shnoll [je3] can be seen as evidence for the presence of long ranged weak fields in astrophysical length scales [G1].

p-Adic and dark matter hierarchies and hierarchy of moments of consciousness

Dark matter hierarchy assigned to a spectrum of Planck constant having arbitrarily large values brings additional elements to the TGD inspired theory of consciousness.

a) Macroscopic quantum coherence can be understood since a particle with a given mass can in principle appear as arbitrarily large scaled up copies (Compton length scales as $\hbar$). The phase transition to this kind of phase implies that space-time sheets of particles overlap and this makes possible macroscopic quantum coherence.

b) The space-time sheets with large Planck constant can be in thermal equilibrium with ordinary ones without the loss of quantum coherence. For instance, the cyclotron energy scale associated with EEG turns out to be above thermal energy at room temperature for the level of dark matter hierarchy corresponding to magnetic flux quanta of the Earth's magnetic field with the size scale of Earth and a successful quantitative model for EEG results [M3]. A strong experimental guideline in the construction of the model have the experimental findings dating back to seventies [la8] suggesting that ELF em fields at frequencies which are multiples of cyclotron frequencies of biologically important ions in Earth's magnetic field, have effects on vertebrate brains. These effects look definitely quantal in a dramatic conflict with the fact that the energies of the ELF photons are ridiculously small as compared to thermal energy meaning that thermal noise should mask these effects completely.

Dark matter hierarchy leads to detailed quantitative view about quantum biology with several testable predictions [M3]. The applications to living matter suggests that the basic hierarchy corresponds to a hierarchy of Planck constants coming as $\hbar(k) = \lambda^k(p)\hbar_0$, $\lambda \simeq 2^{11}$ for $p = 2^{127-1}$, $k = 0, 1, 2, ...$ [M3]. Also integer valued sub-harmonics and integer valued sub-harmonics of λ might be possible. Each p-adic length scale corresponds to this kind of hierarchy and number theoretical arguments suggest a general formula for the allowed values of Planck constant λ depending logarithmically on p-adic prime [O5]. Also the value of $\hbar_0$ has spectrum characterized by Beraha numbers

$B_n = 4cos^2(\pi/n)$, $n \geq 3$, varying by a factor in the range $n > 3$ [O5]. It must be however emphasized that the relation of this picture to the model of quantized gravitational Planck constant h_{gr} appearing in Nottale's model is not yet completely understood.

The general prediction is that Universe is a kind of inverted Mandelbrot fractal for which each bird's eye of view reveals new structures in long length and time scales representing scaled down copies of standard physics and their dark variants. These structures would correspond to higher levels in self hierarchy. This prediction is consistent with the belief that 75 per cent of matter in the universe is dark.

1. Living matter and dark matter

Living matter as ordinary matter quantum controlled by the dark matter hierarchy has turned out to be a particularly successful idea. The hypothesis has led to models for EEG predicting correctly the band structure and even individual resonance bands and also generalizing the notion of EEG [M3]. Also a generalization of the notion of genetic code emerges resolving the paradoxes related to the standard dogma [L2, M3]. A particularly fascinating implication is the possibility to identify great leaps in evolution as phase transitions in which new higher level of dark matter emerges [M3].

It seems safe to conclude that the dark matter hierarchy with levels labelled by the values of Planck constants explains the macroscopic and macro-temporal quantum coherence naturally. That this explanation is consistent with the explanation based on spin glass degeneracy is suggested by following observations. First, the argument supporting spin glass degeneracy as an explanation of the macro-temporal quantum coherence does not involve the value of $\hbar$ at all. Secondly, the failure of the perturbation theory assumed to lead to the increase of Planck constant and formation of macroscopic quantum phases could be precisely due to the emergence of a large number of new degrees of freedom due to spin glass degeneracy. Thirdly, the phase transition increasing Planck constant has concrete topological interpretation in terms of many-sheeted space-time consistent with the spin glass degeneracy.

2. Dark matter hierarchy and the notion of self

The vision about dark matter hierarchy leads to a more refined view about self hierarchy and hierarchy of moments of consciousness [J6, M3]. The larger the value of Planck constant, the longer the subjectively experienced duration and the average geometric duration $T(k) \propto \lambda^k$ of the quantum jump.

Quantum jumps form also a hierarchy with respect to p-adic and dark hierarchies and the geometric durations of quantum jumps scale like $\hbar$. Dark matter hierarchy suggests also a slight modification of the notion of self. Each self involves a hierarchy of dark matter levels, and one is led to ask whether the highest level in this hierarchy corresponds to single quantum jump rather than a sequence of quantum jumps. The averaging of conscious experience over quantum jumps would occur only for sub-selves at lower levels of dark matter hierarchy and these mental images would be ordered, and single moment of consciousness would be experienced as a history of events. The quantum parallel dissipation at the lower levels would give rise to the experience of flow of time. For instance, hadron as a macro-temporal quantum system in the characteristic time scale of hadron is a dissipating system at quark and gluon level corresponding to shorter p-adic time scales. One can ask whether even entire life cycle could be regarded as a single quantum jump at the highest level so that consciousness would not be completely lost even during deep sleep. This would allow to understand why we seem to know directly that this biological body of mine existed yesterday.

The fact that we can remember phone numbers with 5 to 9 digits supports the view that self corresponds at the highest dark matter level to single moment of consciousness. Self would experience the average over the sequence of moments of consciousness associated with each sub-self but there would be no averaging over the separate mental images of this kind, be their parallel or serial. These mental images correspond to sub-selves having shorter wake-up periods than self and would be experienced as being time ordered. Hence the digits in the phone number are experienced as separate mental images and ordered with respect to experienced time.

3. The time span of long term memories as signature for the level of dark matter hierarchy

The simplest dimensional estimate gives for the average increment τ of geometric time in quantum jump $\tau \sim 10^4 \, CP_2$ times so that $2^{127} - 1 \sim 10^{38}$ quantum jumps are experienced during secondary p-adic time scale $T_2(k = 127) \simeq 0.1$ seconds which is the duration of physiological moment and predicted to be fundamental time scale of human consciousness [L1]. A more refined guess is that $\tau_p = \sqrt{p}\tau$ gives the dependence of the duration of quantum jump on p-adic prime p. By multi-p-fractality predicted by TGD and explaining p-adic length scale hypothesis, one expects that at least $p = 2$-adic level is also always present. For the higher levels of dark matter hierarchy τ_p is scaled up by $\hbar/\hbar_0$. One can understand evolutionary leaps as the emergence of higher levels at the level of individual organism making possible intentionality and memory in the time scale defined τ [L2].

Higher levels of dark matter hierarchy provide a neat quantitative view about self hierarchy and its evolution. For instance, EEG time scales corresponds to $k = 4$ level of hierarchy and a time scale of .1 seconds [J6], and EEG frequencies correspond at this level dark photon energies above the thermal threshold so that thermal noise is not a problem anymore. Various levels of dark matter hierarchy would naturally correspond to higher levels in the hierarchy of consciousness and the typical duration of life cycle would give an idea about the level in question.

The level would determine also the time span of long term memories as discussed in [M3]. $k = 7$ would correspond to a duration of moment of conscious of order human lifetime which suggests that $k = 7$ corresponds to the highest dark matter level relevant to our consciousness whereas higher levels would in general correspond to transpersonal consciousness. $k = 5$ would correspond to time scale of short term memories measured in minutes and $k = 6$ to a time scale of memories measured in days.

The emergence of these levels must have meant evolutionary leap since long term memory is also accompanied by ability to anticipate future in the same time scale. This picture would suggest that the basic difference between us and our cousins is not at the level of genome as it is usually understood but at the level of the hierarchy of magnetic bodies [L2, M3]. In fact, higher levels of dark matter hierarchy motivate the introduction of the notions of super-genome and hyper-genome. The genomes of entire organ can join to form super-genome expressing genes coherently. Hyper-genomes would result from the fusion of genomes of different organisms and collective levels of consciousness would express themselves via hyper-genome and make possible social rules and moral.

1.4 Bird's eye of view about the topics of the book

The topics of this book are following.

a) In the first part of the book I will try to give an overall view about TGD and the evolution of TGD to its recent form. I cannot avoid the use of various concepts without detailed definitions and my hope is that reader only gets a bird's eye of view about TGD. Two visions about physics are discussed. According to the first vision physical states of the Universe correspond to classical spinor fields in the world of the classical worlds identified as 3-surfaces or equivalently as corresponding 4-surfaces analogous to Bohr orbits and identified as special extreme of Kähler action. TGD as a generalized number theory vision leading naturally also to the emergence of p-adic physics as physics of cognitive representations is the second vision. Number theoretical vision involves three loosely related approaches: fusion of real and various p-adic physics to a larger whole as algebraic continuations of what might be called rational physics; space-time as a hyper-quaternionic surface of hyper-octonion space, and space-time surfaces as a representations of infinite primes.

b) The second part of the book is devoted to the vision about physics as infinite-dimensional configuration space geometry. The basic idea is that classical spinor fields in infinite-dimensional "world of classical worlds", space of 3-surfaces in $M^4 \times CP_2$ describe the quantum states of the Universe. Quantum jump remains the only purely quantal aspect of quantum theory in this approach since there is no quantization at the level of the configuration space. Space-time surfaces correspond to preferred extremals of the Kähler action analogous to Bohr orbits and define what might be called classical TGD discussed in the first chapter. The construction of the configuration space geometry and spinor structure are discussed in remaining chapters.

c) The third part of the book carries title "Algebraic Physics". Classical physics is an exact part of the configuration space geometry and the identification of space-time surfaces as generalized Bohr orbits is not a mere approximation. Path integral is replaced by a functional integral over 3-surfaces and there are reasons to expect that the decomposition of the configuration space to a union of symmetric spaces allows to carry out the functional integral exactly. This together with quantum classical correspondence and Bohr orbit property suggests a generalization of the duality symmetry of hadron models stating that the addition of loops does not affect the physical amplitude assignable to a given diagram. Hopf algebras are used to formulate this symmetry. In the second chapter the profound implications of the fact that configuration space spinors correspond to von Neumann algebras known as hyper-finite factors of type II_1 are considered.

d) The applications of TGD to elementary particle physics are the topic of the fourth part of the book. In the first chapter a general view about many-sheeted space-time and topological condensation is developed using quantum classical correspondence as a guideline. TGD explains particle families in terms of generation genus correspondence (particle families correspond to 2-dimensional topologies labelled by genus). The notion of elementary particle vacuum functional is developed and an argument based on hyper-ellipticity predicts that the number of visible light fermion families is three whereas the possible higher genera would behave effectively like dark matter. The general theory of particle massivation based on p-adic thermodynamics and TGD variant of Higgs mechanism has now a rather rigorous justification on basis of quantum classical correspondence although the interpretation of the theory is still developing. For instance, the identification of Higgs boson as a wormhole contact carrying left handed weak charge and the improved understanding of the relationship between gravitational and inertial masses emerged during the writing of this book. The general theory is applied calculate elementary particle masses and hadron masses and to construct a model for CKM matrix. Also the predicted new physics effects such as scaled quarks and leptons with scaled up/down masses are discussed.

e) In the fifth part of the book the relationship of TGD with General Relativity is discussed and the cosmological and astrophysical applications of the many-sheeted space-time are summarized.

The seven online books about TGD [TGDview, TGDgeom, TGDquant, TGDnumber, TGDclass, TGDpad, TGDfree] and eight online books about TGD inspired theory of consciousness and quantum biology [TGDconsc, TGDselforg, TGDware, TGDholo, TGDgeme, TGDeeg, TGDmagn, TGDmathc] are warmly recommended for the reader willing to get overall view about what is involved.

1.5 The contents of the book

1.5.1 PART I: General Overview

Overall View About Evolution of TGD

The evolution of quantum TGD involve five threads which have become more and more entangled with each other. The first great vision was the reduction of the entire quantum physics (apart from quantum jump) to the geometry of classical spinor fields of the infinite-dimensional space of 3-surfaces in H, the great idea being that infinite-dimensional Kähler geometric existence and thus physics is unique from the requirement that it is free of infinities. The outcome is geometrization and generalization of the known structures of the quantum field theory and of string models.

The second thread is p-adic physics. p-Adic physics was initiated by more or less accidental observations about reduction of basic mass scale ratios to the ratios of square roots of Mersenne primes and leading to the p-adic thermodynamics explaining elementary particle mass scales and masses with an unexpected success. p-Adic physics turned eventually to be the physics of cognition and intentionality. Consciousness theory based ideas have led to a generalization of the notion of number obtained by gluing real numbers and various p-adic number fields along common rationals to a more general structure and implies that many-sheeted space-time contains also p-adic space-time sheets serving as space-time correlates of cognition and intentionality. The hypothesis that real and p-adic physics can be regarded as algebraic continuation of rational number based physics provides extremely strong constraints on the general structure of quantum TGD.

TGD inspired theory of consciousness can be seen as a generalization of quantum measurement theory replacing the notion of observer as an outsider with the notion of self. The detailed analysis of what happens in quantum jump have brought considerable understanding about the basic structure of quantum TGD itself. It seems that even quantum jump itself could be seen as a number theoretical necessity in the sense that state function reduction and state preparation by self measurements are necessary in order to reduce the generalized quantum state which is a formal superposition over components in different number fields to a state which contains only rational or finitely-extended rational entanglement identifiable as bound state entanglement. The number theoretical information measures generalizing Shannon entropy (always non-negative) are one of the important outcomes of consciousness theory combined with p-adic physics.

Physics as a generalized number theory is the fourth thread. The key idea is that the notion of divisibility could make sense also for literally infinite numbers and perhaps make them useful from the point of view of physicist. The great surprise was that the construction of infinite primes corresponds to the repeated quantization of a super-symmetric arithmetic quantum field theory. This led to the vision about physics as a generalized number theory involving infinite primes, integers, rationals and reals, as well as their quaternionic and octonionic counterparts. A further generalization is based on the generalization of the number concept already mentioned. Space-time surfaces could be regarded in this framework as concrete representations for infinite primes and integers, whereas the dimensions 8 and 4 for imbedding space and space-time surface could be seen as reflecting the dimensions of octonions and quaternions and their hyper counterparts obtained by multiplying imaginary units by $\sqrt{-1}$. Also the dimension 2 emerges naturally as the maximal dimension of commutative sub-number field and relates to the ordinary conformal invariance central also for string models.

By quantum classical correspondence space-time sheets can be identified as quantum coherence regions. Hence the fact that they have all possible size scales more or less unavoidably implies that Planck constant must be quantized and have arbitrarily large values. If one accepts this then also the idea about dark matter as a macroscopic quantum phase characterized by an arbitrarily large value of Planck constant emerges naturally as does also the interpretation for the long ranged classical electro-weak and color fields predicted by TGD. Rather seldom the evolution of ideas follows simple linear logic, and this was the case also now. In any case, this vision represents the fifth, relatively new thread in the evolution of TGD and the ideas involved are still evolving.

This chapter represents a overall view about evolution of classical TGD and of p-adic concepts, a summary of the ideas generated by TGD inspired theory of consciousness, the vision about physics as a generalized number theory, and a discussion of how TGD could predict the spectrum of values for the dynamical Planck constant.

Overall View About Quantum TGD

This chapter provides a summary about quantum TGD. The discussions are based on the general vision that quantum states of the Universe correspond to the modes of classical spinor fields in the "world of the classical worlds" identified as the infinite-dimensional configuration space of 3-surfaces of $H = M^4 \times CP_2$ (more or less-equivalently, the corresponding 4-surfaces defining generalized Bohr orbits). The following topics are discussed on basis this vision.

a) The two basic approaches to the construction of the Kähler geometry based on direct guess of the Kähler function and construction of Kähler metric from the hypothesis that configuration space is a union of infinite-dimensional symmetric spaces with metric possessing maximal isometry group are summarized.

b) The great vision guiding the construction of the configuration space spinor structure is that the second quantization of the induced spinor fields can be understood geometrically in terms of the configuration space spinor structure in the sense that the anti-commutation relations for configuration space gamma matrices require anti-commutation relations for the oscillator operators for free second quantized induced spinor fields.

c) p-Adic mass calculations were carried out for the first time for about one decade ago. The calculations rely p-adic thermodynamics giving the p-adic mass squared essentially as the thermal expectation value of conformal weight. Real mass squared is obtained by mapping the p-adic mass squared to its real counterpart using so called canonical identification. The justification for this picture from quantum TGD and for what massivation corresponds at space-time level is sketched.

d) It is not yet possible to deduce the length scale evolution of gauge coupling constants from Quantum TGD proper. Quantum classical correspondence however encourages the hope that it might be possible to achieve some understanding of the coupling constant evolution by using the classical theory. This turns out to be the case and the earlier speculative picture about gauge coupling constants associated with a given space-time sheet as RG invariants finds support. It remains an open question whether gravitational coupling constant is RG invariant inside given space-time sheet. The discrete p-adic coupling constant evolution replacing in TGD framework the ordinary RG evolution allows also formulation at space-time level as also does the evolution of $\hbar$ associated with the phase resolution.

e) Concerning the construction of the S-matrix the basic philosophy is provided by TGD inspired theory of consciousness which can be seen as a generalization of quantum measurement theory. The huge symmetries of the configuration space geometry pose extremely tight constraints on the S-matrix.

The identification of the classical space-time surface as a generalized Bohr orbit in turn implies that there is no path integral over space-time surfaces. Hence there should be no summation over loops in the generalized Feynman diagrams. Or stated using the language for generalized stringy diagrams: all generalized stringy diagrams differing by loops are equivalent. The formulation of this picture in terms of Hopf algebras is briefly discussed.

1.5.2 PART II: Physics as infinite-dimensional spinor geometry in the world of classical worlds

Classical TGD

In this chapter the classical field equations associated with the Kähler action are studied. The study of the extremals of the Kähler action has turned out to be extremely useful for the development of TGD. Towards the end of year 2003 quite dramatic progress occurred in the understanding of field equations and it seems that field equations might be in well-defined sense exactly solvable.

1. General considerations

The vanishing of Lorentz 4-force for the induced Kähler field means that the vacuum 4-currents are in a mechanical equilibrium. Lorentz 4-force vanishes for all known solutions of field equations which inspires the hypothesis that all extremals or at least the absolute minima of Kähler action satisfy the condition. The vanishing of the Lorentz 4-force in turn implies local conservation of the ordinary energy momentum tensor. The corresponding condition is implied by Einstein's equations in General Relativity. The hypothesis would mean that the solutions of field equations are what might be called generalized Beltrami fields. The condition implies that vacuum currents can be non-vanishing only provided the dimension D_{CP_2} of the CP_2 projection of the space-time surface is less than four so that in the regions with $D_{CP_2} = 4$, Maxwell's vacuum equations are satisfied.

The hypothesis that Kähler current is proportional to a product of an arbitrary function ψ of CP_2 coordinates and of the instanton current generalizes Beltrami condition and reduces to it when electric field vanishes. Kähler current has vanishing divergence for $D_{CP_2} < 4$, and Lorentz 4-force indeed vanishes. The remaining task would be the explicit construction of the imbeddings of these fields and the demonstration that field equations can be satisfied.

Under additional conditions magnetic field reduces to what is known as Beltrami field. Beltrami fields are known to be extremely complex but highly organized structures. The natural conjecture is that topologically quantized many-sheeted magnetic and Z^0 magnetic Beltrami fields and their generalizations serve as templates for the helical molecules populating living matter, and explain both chirality selection, the complex linking and knotting of DNA and protein molecules, and even the extremely complex and self-organized dynamics of biological systems at the molecular level.

Field equations can be reduced to algebraic conditions stating that energy momentum tensor and second fundamental form have no common components (this occurs also for minimal surfaces in string models) and only the conditions stating that Kähler current vanishes, is light-like, or proportional to instanton current, remain and define the remaining field equations. The conditions guaranteing topologization to instanton current can be solved explicitly. Solutions can be found also in the more general case when Kähler current is not proportional to instanton current. On basis of these findings there are strong reasons to believe that classical TGD is exactly solvable.

2. Absolute minimization of Kähler action and second law of thermodynamics

By quantum classical correspondence the non-deterministic space-time dynamics should mimic the dissipative dynamics of the quantum jump sequence. Beltrami fields appear in physical applications as asymptotic self organization patterns for which Lorentz force and dissipation vanish. This suggests that absolute minima of Kähler action correspond to space-time sheets which asymptotically satisfy generalized Beltrami conditions so that one can indeed assign to the final (rather than initial!) 3-surface a unique 4-surface apart from effects related to non-determinism. Absolute minimization abstracted to purely algebraic generalized Beltrami conditions would make sense also in the p-adic context. Also the equivalence of absolute minimization with the second law strongly suggests itself. Of course, one must keep mind open for the possibility that it is the second law of thermodynamics which replaces absolute minimization as the fundamental principle.

3. The dimension of CP_2 projection as classifier for the fundamental phases of matter

The dimension D_{CP_2} of CP_2 projection of the space-time sheet encountered already in p-adic mass calculations classifies the fundamental phases of matter. For $D_{CP_2} = 4$ empty space Maxwell equations hold true. This phase is chaotic and analogous to de-magnetized phase. $D_{CP_2} = 2$ phase is analogous to ferromagnetic phase: highly ordered and relatively simple. $D_{CP_2} = 3$ is the analog of spin glass and liquid crystal phases, extremely complex but highly organized by the properties of the generalized Beltrami fields. This phase is the boundary between chaos and order and corresponds to life emerging in the interaction of magnetic bodies with bio-matter. It is possible only in a finite temperature interval (note however the p-adic hierarchy of critical temperatures) and characterized by chirality just like life.

4. Specific extremals of Kähler action

The study of extremals of Kähler action represents more than decade old layer in the development of TGD.

a) The huge vacuum degeneracy is the most characteristic feature of Kähler action (any 4-surface having CP_2 projection which is Legendre sub-manifold is vacuum extremal, Legendre sub-manifolds of CP_2 are in general 2-dimensional). This vacuum degeneracy is behind the spin glass analogy and leads to the p-adic TGD. As found in the second part of the book, various particle like vacuum extremals also play an important role in the understanding of the quantum TGD.

b) The so called CP_2 type vacuum extremals have finite, negative action and are therefore an excellent candidate for real particles whereas vacuum extremals with vanishing Kähler action are candidates for the virtual particles. These extremals have one dimensional M^4 projection, which is light like curve but not necessarily geodesic and locally the metric of the extremal is that of CP_2: the quantization of this motion leads to Virasoro algebra. Space-times with topology $CP_2 \# CP_2 \# ... CP_2$ are identified as the generalized Feynmann diagrams with lines thickened to 4-manifolds of "thickness" of the order of CP_2 radius. The quantization of the random motion with light velocity associated with the CP_2 type extremals in fact led to the discovery of Super Virasoro invariance, which through the construction of the configuration space geometry, becomes a basic symmetry of quantum TGD.

c) There are also various non-vacuum extremals.
i) String like objects, with string tension of same order of magnitude as possessed by the cosmic strings of GUTs, have a crucial role in TGD inspired model for the galaxy formation and in the TGD based cosmology.
ii) The so called massless extremals describe non-linear plane waves propagating with the velocity of light such that the polarization is fixed in given point of the space-time surface. The feature characteristic for TGD is the light like Kähler current: in the ordinary Maxwell theory vacuum gauge currents are not possible. This current serves as a source of coherent photons, which might play an important role in the quantum model of bio-system as a macroscopic quantum system.
iii) In the so called Maxwell's phase, ordinary Maxwell equations for the induced Kähler field are satisfied in an excellent approximation. A special case is provided by a radially symmetric extremal having an interpretation as the space-time exterior to a topologically condensed particle. The sign of the gravitational mass correlates with that of the Kähler charge and one can understand the generation of the matter antimatter asymmetry from the basic properties of this extremal. The possibility to understand the generation of the matter antimatter asymmetry directly from the basic equations of the theory gives strong support in favor of TGD in comparison to the ordinary EYM theories, where the generation of the matter antimatter asymmetry is still poorly understood.

Construction of configuration space geometry

The topics of this chapter are the purely geometric aspects of the vision about physics as an infinite-dimensional Kähler geometry of the "world of classical worlds", with " classical world" identified either as 3-D surface of the unique Bohr orbit like 4-surface traversing through it. The non-determinism of Kähler action forces to generalize the notion of 3-surfaces so that unions of space-like surfaces with time like separations must be allowed. The considerations are restricted mostly to real context and the problems related to the p-adicization are discussed later.

There are two separate tasks involved.

a) Provide configuration space of 3-surfaces with Kähler geometry which is consistent with 4-dimensional general coordinate invariance so that the metric is Diff^4 degenerate. General coordinate invariance implies that the definition of metric must assign to a give 3-surface X^3 a 4-surface as a kind of Bohr orbit $X^4(X^3)$.

b) Provide the configuration space with a spinor structure. The great idea is to identify configuration space gamma matrices in terms of super algebra generators expressible using second quantized fermionic oscillator operators for induced free spinor fields at the space-time surface assignable to a given 3-surface. The isometry generators and contractions of Killing vectors with gamma matrices would thus form a generalization of Super Kac-Moody algebra.

From the experience with loop spaces one can expect that there is no hope about existence of well-defined Riemann connection unless this space is union of infinite-dimensional symmetric spaces with constant curvature metric and simple considerations requires that Einstein equations are satisfied by each component in the union. The coordinates labelling these symmetric spaces are zero modes having interpretation as genuinely classical variables which do not quantum fluctuate since they do not contribute to the line element of the configuration space.

The construction of the Kähler structure involves also the identification of complex structure. Direct construction of Kähler function as action associated with a preferred Bohr orbit like extremal for some physically motivated action action leads to a unique result. Second approach is group theoretical and is based on a direct guess of isometries of the infinite-dimensional symmetric space formed by 3-surfaces with fixed values of zero modes. The group of isometries is generalization of Kac-Moody group obtained by replacing finite-dimensional Lie group with the group of canonical transformations of $\delta M^4_+ \times CP_2$, where δM^4_+ is the boundary of 4-dimensional future light-cone.

Configuration space spinor structure

In this chapter the recent situation in the construction of the configuration space spinor structure is discussed. The construction involves several deep ideas besides the original idea about geometrization of fermionic anti-commutation relations in terms of infinite-dimensional geometry.

1. Configuration space gamma matrices as super algebra generators

The basic idea is that the space of the configuration space spinors must correspond to the Fock space for the second quantized induced spinor fields. In accordance with this the gamma matrices of the configuration space must be expressible as superpositions of the fermionic oscillator operators for the second quantized induced free spinor fields in X^4 so that they are analogous to spin $3/2$ fields. The Dirac equation is fixed from the requirement of super symmetry and has same vacuum degeneracy as Kähler action. A further assumption is that the contractions of the gamma matrices with isometry currents correspond to super charges of the group of isometries of the configuration space so that the construction reduces to group theory. Also the super Kac Moody algebra associated with light like 3-D CDs defines candidates for gamma matrices defining the components of the metric as anti-commutators and the question is whether the two definitions are mutually consistent.

2. 7-3 duality

The failure of the classical non-determinism forces to introduce two kinds of causal determinants (CDs). 7-D light like CDs are unions of the boundaries of future and past directed light cones in M^4 at arbitrary positions (also more general light like surfaces $X^7 = X^3_l \times CP_2$ might be considered). CH is a union of sectors associated with these 7-D CDs playing in a very rough sense the roles of big bangs and big crunches. The creation of pairs of positive and negative energy space-time sheets occurs at $X^3 \subset X^7$ in the sense that negative and positive energy space-time sheet meet at X^3. Negative and positive energy space-time sheets are space-time correlates for bras and kets and the meeting of negative and positive energy space-time sheets is the space-time correlate for their scalar product.

Also 3-D light like causal determinants $X^3_l \subset X^4$ must be introduced: elementary particle horizons provide a basic example of this kind of CDs. The deformations of the 2-surfaces defining X^3_l define Kac Moody type conformal symmetries.

7-3 duality states that the two kind of CDs play a dual role in the construction of the theory and implies that 3-surfaces are effectively two-dimensional with respect to the CH metric in the sense that all relevant data about CH geometry is contained by the two-dimensional intersections $X^2 = X^3_l \cap X^7$ defining 2-sub-manifolds of $X^3 \subset X^7$.

3. Modified Dirac equation and gamma matrices

The modified Dirac equation is deduced from Kähler action by requiring it to have the same vacuum degeneracy as Kähler action itself. The interpretation of the solutions of the modified Dirac equation is as super gauge symmetry generators whereas physical degrees of freedom corresponds to generalized eigen modes at X_l^3 and at space-like 3-surfaces $X^3 \subset X^7$.

The decisive property of the modified Dirac equation is that it allows shock wave solutions restricted to X_l^3: in terms of field-particle duality these shock waves correspond to the click caused by a particle in a detector. This allows to realize quantum gravitational holography and 7–3 duality in the sense that the induced second quantized spinor fields at the intersections $X^2 = X_l^3 \cap X^7$ determine the super-generators super-canonical and super Kac Moody algebras invariant under the super gauge symmetries generated by the solutions of the modified Dirac equation.

Both the function algebra and Poisson algebra of X^7 allow super-symmetrization and both N-S and Ramond type representations are possible. For Ramond type representation the modified Dirac operators D_+ and D_-^{-1} associated with the positive and negative energy space-time sheets $X_\pm^4$ meeting at X^3 are present in the expressions of the super generators. NS-type representations correspond to the replacement of these operators with projection operators to the space of spinor modes with non-vanishing eigenvalues of $D_\pm$. Both representations are necessary and correspond to leptonic and quark like representations of configuration space gamma matrices. Similar statements apply to super Kac-Moody representations. These two kinds of representations correspond to super and kappa symmetries of super-string models.

4. Expressing Kähler function in terms of Dirac determinants

Although quantum criticality in principle predicts the possible values of Kähler coupling strength, one might hope that there exists even more fundamental approach involving no coupling constants and predicting even quantum criticality and realizing quantum gravitational holography. The ratio of Dirac determinants of the modified Dirac operators D_+ and D_- associated with the space-time regions separated by a 3-dimensional causal determinant is an excellent candidate in this respect and the guess is that it is expressible as an exponent for the difference of Kähler actions in the adjacent regions. This allows to deduce the exponent of Kähler function and one could identify it as a renormalized Dirac determinant.

If the operators D_+ and D_- commute and if spectrum of the operator $D_+ D_-^{-1}$ is invariant under $\lambda \to 1/\lambda$ apart from a finite number of eigenvalues for which the ratio equals to the ratio of exponents of respective Kähler actions, the Dirac determinant in question is well-defined even without zeta function regularization, and allows to deduce Kähler function.

The modified Dirac operator indeed allows to realize quantum gravitational holography since it reduces to an effectively 3-dimensional Dirac operator by boundary conditions but depends on the normal derivatives of the imbedding space coordinates at the causal determinants, which are most naturally light like 3-surfaces. Hermiticity of $D_\pm$ poses conditions on normal derivatives of imbedding space coordinates unless Kähler action density vanishes at X_l^3 but does not fully eliminate the effects of the classical non-determinism. Hence there are good hopes of evaluating the exponent of Kähler function as a Dirac determinant without solving the field equations.

5. The relationship between super-canonical and super Kac-Moody algebras

The conformal weights of the generating elements of super-canonical representations correspond to the zeros of Riemann Zeta and one can identify the Cartan decomposition of the super-canonical algebra crucial for defining configuration space of 3-surfaces as a union of symmetric Kähler manifolds labelled by zero modes. Super-canonical algebra differs dramatically from super Kac Moody algebra. 7-3 duality would suggest that super-canonical and super Kac-Moody algebras as associated with two different tangent space basis for CH and giving rise to different coordinate systems.

TGD inspired quantum measurement theory suggests however that the two super-conformal algebras correspond to each other like classical and quantal degrees of freedom. Super Kac-Moody algebra and super conformal algebra would act as transformations preserving the conformal equivalence class of the partonic 2-surfaces X^2 associated with the maxima of the Kähler function whereas super-canonical algebra in general changes conformal moduli and induces a conformal anomaly in this manner. Hence Kac-Moody algebra seems to act in the zero modes of the configuration space metric. In TGD inspired quantum measurement zero modes correspond to classical non-quantum fluctuating dynamical variables in 1-1 correspondence with quantum fluctuating degrees of freedom like the positions of the pointer of the measurement apparatus with the directions of spin of electron. Hence Kac-Moody algebra would define configuration space coordinates in terms of the map induced by correlation between classical and quantal degrees of freedom induced by entanglement.

7–3 duality allows to generalize the Olive-Goddard-Kent coset construction. By 7-3 duality the differences of the commuting Virasoro generators of super-canonical and super Kac-Moody algebras must annihilate the physical states. For the same reason the central charges of the two Virasoro algebras must be identical so that the net central charge vanishes. This condition leads to a generalization of stringy mass formula involving besides super Kac-Moody algebra also the super-canonical algebra but it seems that continuous mass spectrum is not possible.

The $N = 4$ super symmetries generated by the solutions of the modified Dirac equation are pure super gauge transformations. All CP_2 spinor harmonics except the covariantly constant right handed neutrino spinor carry color quantum numbers and thus a non-vanishing vacuum conformal weight: hence only an $N = 1$ global super symmetry is in principle possible. Since the Ramond type super-generator corresponding to the covariantly constant neutrino vanishes identically even $N = 1$ global super-symmetry is absent and no sparticles are predicted. This means a decisive difference in comparison with super string models and M-theory.

Physical states satisfy both N-S and Ramond type Super Virasoro conditions separately: note that in super-canonical degrees of freedom Ramond/NS representations super generators involve carry quark/lepton number. The most obvious application of the mass formula would be to hadron physics. The effective 2-dimensionality allows to identify partons as 2-dimensional surfaces X_i^2, and the more than decade old notion of elementary particle vacuum functional finds a first principle justification as a functional in the modular degrees of freedom of X^2.

By quantum classical correspondence is that Virasoro algebra associated with super Kac-Moody algebra acts on the conformal weights of the super-canonical representations as conformal transformations and the generators of the super-canonical algebra can be regarded as conformal fields. This dictates the matrix elements of the algebra to a high degree as function of conformal weights. A connection with braid and quantum groups and II_1 sub-factors of type von Neumann algebras associated with the Clifford algebra of the configuration space emerges.

1.5.3 PART III: Algebraic Physics

Equivalence of Loop and Tree Diagrams and Divergence Cancellation

The great dream of a physicist believing in reductionism and TGD would be a formalism generalizing Feynman diagrams allowing any graduate student to compute the predictions of the theory. TGD has forced myself to give up naive reductionism but I believe that TGD allows generalization of Feynman diagram in such a manner that one gets rid of the infinities plaguing practically all existing theories. The purpose of this chapter is to develop general vision about how this might be achieved. The vision is based on generalization of mathematical structures discovered in the construction of topological quantum field theories (TQFT), conformal field theories (CQFT). In particular, the notions of Hopf algebras and quantum groups, and categories are central. The following gives a very concise summary of the basic ideas.

1. Feynman diagrams as generalized braid diagrams

The first key idea is that generalized Feynman diagrams with diagrams which are analogous to knot and link diagrams in the sense that diagrams involving loops are equivalent with tree diagrams. This would be a generalization of duality symmetry of string models.

TGD itself provides general arguments supporting same idea. The identification of absolute minimum of Kähler action or some other preferred extremal as a four-dimensional Feynman diagram characterizing particle reaction means that there is only single Feynman diagram instead of functional integral over 4-surfaces: this diagram is expected to be minimal one. S-matrix element as a representation of a path defining continuation of the configuration space spinor field between different sectors of it corresponding different 3-topologies leads also to the conclusion that all continuations and corresponding Feynman diagrams are equivalent. Universe as a computer metaphor idea allowing a quite concrete realization by generalization of what is meant by space-time point leads to the view that generalized Feynman diagrams characterize equivalent computations.

2. Coupling constant evolution from infinite number of critical values of Kähler coupling strength

The basic objection that this vision does not allow to understand coupling constant evolution involving loops in an essential manner can be circumvented. Quantum criticality requires that Kähler coupling strength α_K is analogous to critical temperature (so that the loops for configuration space integration vanish). The hypothesis motivated by the enormous vacuum degeneracy of Kähler action is that α_K has an infinite number of possible values labelled by p-adic length scales and also probably also by the dimensions of effective tensor factors associated with inclusions of von Neumann algebras known as hyper-finite type II_1 factors (so called Beraha numbers) as found already earlier. The dependence on p-adic length scale L_p corresponds to the usual renormalization group evolution whereas the latter dependence would correspond to angular resolution and finite-dimensional extensions of p-adic number fields R_p. Finite resolution and renormalization group evolution are forced by the algebraic continuation of rational number based physics to real and p-adic number fields since p-adic and real notions of distance between rational points differ dramatically.

3. R-matrices, complex numbers, quaternions, and octonions

A crucial observation is that physically equivalent R-matrices of 6-vertex models are labelled by points of CP_2 whereas maximal space of commuting R-matrices are labelled by points of 2-sphere S^2. CP_2 also labels maximal associative and thus quaternionic subspaces of 8-dimensional octonion space containing a preferred imaginary unit whereas S^2 labels the maximally commutative subspaces of quaternion space, which suggests that R-matrices and these structures correspond to each other.

Number theoretic vision leads to the notion of number theoretic compactification meaning that space-time surfaces could be regarded either as hyper-quaternionic, and thus maximal associative, 4-surfaces in M^8 regarded as the space of hyper-octonions or as surfaces in $M^4 \times CP_2$ (the imaginary units of hyper-octonions are multiplied with $\sqrt{-1}$ so that the number theoretical norm has Minkowski signature). No dynamics is involved and duality might be more appropriate term. Associativity constraint is an essential element of also Yang-Baxter equations.

These observations lead to a concrete proposal how how quantum classical correspondence is realized classically at space-time level. Each point of CP_2 corresponds to R-matrix and 3-surfaces can be identified as preferred sections of space-time surface by requiring that unitary R-matrices are commuting in the section (micro-causality) whereas commutativity fails for R-matrices corresponding to different values of time coordinate defined by this foliation.

4. Ordinary conformal symmetries act on the space of super-canonical conformal weights

TGD predicts two kinds of super-conformal symmetries.

a) The conformal symmetries defining super Kac-Moody representations are realized at light-like 3-surfaces appearing as boundaries of space-time sheets and boundaries between space-time regions with Euclidian and Minkowskian signature of metric. Conformal weights are half integer valued in this case.

b) Super-canonical conformal invariance acts at the level of imbedding space and corresponds to light-like 7-surfaces of form $X_l^3 \times CP_2 \subset M^4 \times CP_2$ believed to appear as causal determinants too. In this case conformal weights are complex numbers of form $\Delta = n/2 + iy$ and for physical states they are expected to reduce to the form $\Delta = 1/2 + iy$.

Quantum classical correspondence suggests that the complex conformal weights of super-canonical algebra generators have space-time counterparts. The proposal is that the weights are mapped to the points of geodesic sphere of CP_2 (and thus also of space-time surface) labelling also mutually commuting R-matrices. The map is completely analogous to the map of momenta of quantum particles to the points of celestial sphere. One can thus regard super-generators as conformal fields in space-time or complex plane in which conformal weights appear as punctures. The action of super-conformal algebra and braid group on these points realizing monodromies of conformal field theories induces by a pull-back a braid group action on the conformal labels of configuration space gamma matrices (super generators) and corresponding isometry generators.

The gamma matrices and isometry generators spanning the super-canonical algebra can be regarded as fields of a conformal field theory in the complex plane containing infinite number of punctures defined by the complex conformal weights. Quaternion conformal Kac Moody and Super Virasoro algebras act as symmetries of this theory and S-matrix of TGD should involve the n-point functions of this conformal field theory.

This picture also justifies the earlier proposal that configuration space Clifford algebra defined by the gamma matrices acting as super generators defines an infinite-dimensional von Neumann algebra possessing hierarchies of type II_1 factors having a close connection with non-trivial representations of braid group and quantum groups. The zeros of Riemann Zeta along the line $Re(s) = 1/2$ in the plane of conformal weights could be regarded a the infinite braid behind the von Neumann algebra. Contrary to the expectations, also trivial zeros seem to be important. Super-canonical algebra predicts that also the superposition for the imaginary parts of non-trivial zeros makes sense so that one would have an entire hierarchy of one-dimensional lattices depending on the number of non-trivial zeros included. The braids defined by sub-lattices defined by subsets of non-trivial zeros could be seen as completely integrable 1-dimensional spin chains leading to a hierarchy of quantum groups and braiding matrices naturally.

It seems that not only Riemann's zeta but also polyzetas could play a fundamental role in TGD Universe. The super-canonical conformal weights of interacting particles, in particular of those forming bound states, are expected to have "off mass shell" values. An attractive hypothesis is that they correspond to the zeros of Riemann's polyzetas. Interaction would allow quite concretely the realization of braiding operations dynamically.

5. Equivalence of loop diagrams with tree diagrams from the axioms of generalized ribbon category

The fourth idea is that Hopf algebra related structures and appropriately generalized ribbon categories could provide a concrete realization of this picture. Generalized Feynman diagrams which are identified as braid diagrams with strands running in both directions of time and containing besides braid operations also boxes representing algebra morphisms with more than one incoming and outgoing strands describing particle reactions (3-particle vertex should be enough). In particular, fusion of 2-particles and decay of particle to two would correspond to generalizations of algebra product μ and co-product Δ to morphisms of the category defined by the super-canonical algebras associated with 3-surfaces with various topologies and conformal structures. The basic axioms for this structure generalizing Hopf algebra axioms state that diagrams with self energy loops, vertex corrections, and box diagrams are equivalent with tree diagrams.

6. Quantum criticality and renormalization group invariance

Quantum criticality means that renormalization group acts like isometry group at a fixed point rather than acting like a gauge symmetry as in the standard quantum field theory context. Despite this difference it is possible to understand how Feynman graph expansion with vanishing loop corrections relates to generalized Feynman graphs and a nice connection with the Hopf- and Lie algebra structures assigned by Connes and Kreimer to Feynman graphs emerges. The condition that loop diagrams are equivalent with tree diagrams gives explicit equations which might fix completely also the p-adic length scale evolution of vertices. Quantum criticality in principle fixes completely the values of the masses and coupling constants as a function of p-adic length scale.

7. The spectrum of zeros of Zeta and quantum TGD

The question whether the imaginary parts of the Riemann Zeta are linearly independent (as assumed in the previous work) or not is of crucial physical significance. Linear independence implies that the spectrum of the super-canonical weights is essentially an infinite-dimensional lattice. Otherwise a more complex structure results.

The hypothesis that p^{iy} is an algebraic phase for every prime implies that p^{iy} is expressible as a product of a Pythagorean phase and a root of unity for every prime p. The numerical evidence supporting the translational invariance of the correlations for the spectrum of zeros together with p-adic considerations leads to the working hypothesis that for any prime p one can express the spectrum of zeros as the product of a subset of Pythagorean prime phases and of a fixed subset U of roots of unity. The spectrum of zeros could be expressed as a union over the translates of the same basic spectrum defined by the roots of unity translated by the phase angles associated with a subset of Pythagorean phases: this is consistent with what the spectral correlations strongly suggest. That decompositions defined by different primes p yield the same spectrum would mean a powerful number theoretical symmetry realizing p-adicities at the level of the spectrum of Zeta.

The model for the scalar propagator as a partition function for the super-canonical algebra supports this view. The approximation that the zeros are linearly independent implies for any subalgebra defined by a finite set of zeros a universal spectrum of singularities for real values of mass squared, and at the limit of entire algebra the propagator is not mathematically defined since the singularities are dense in real axis. Linear independence however shifts the poles to complex values of mass and genuine resonances result and the propagator is expected to be well-defined for the entire algebra.

The theory predicts universal momentum and p-adic length scale dependence of propagators in this picture and the option based on super-canonical algebra predicts very simple and elegant expression for the propagator with all the desired properties.

8. Quantum field theory formulation of quantum TGD

The super-canonical generalization of a 2-dimensional conformal field theory seems to be indispensable for the construction of S-matrix at the fundamental level and defines the vertices as n-point functions of a conformal field theory in turn used to construct S-matrix as tree diagrams. Also a quantum field theory defined for 3-surface X^3 belonging to the light like 7-surfaces $X_l^3 \times CP_2$ defining causal determinants seems to make sense. The QFT in question is determined by the absolute minimum $X^4(X^3)$ associated with the maximum of Kähler function and is consistent with the loop corrections in configuration space. The restriction to X^3 realizes quantum holography and means that a minimum amount of data about $X^4(X^3)$ is needed.

The modified Dirac action for the induced spinor fields at the maximum of the Kähler function with induced spinor fields identified as Grassmann algebra valued fields provides the sought for action. The fermionic propagator is defined by the inverse of the modified Dirac operator. Bosonic kinetic term is Grassmann algebra valued and vanishes for $\psi = 0$ and contributes nothing to the perturbation series in accordance with the effective freezing of the configuration space degrees of freedom. Since the fermionic action is free action, a divergence free quantum field theory is in question. One could also see the action as a fixed point of the map sending action to effective action in accordance with the idea that loop corrections vanish. Different sub-algebras of the super-canonical algebra correspond naturally to various conformally inequivalent configuration space metrics and infinite-dimensional subalgebras correspond to the singular limits when the space-time surface becomes vacuum extremal so that the modified Dirac operator vanishes identically and super-canonical propagator becomes ill-defined.

Was von Neumann right after all?

The work with TGD inspired model for quantum computation led to the realization that von Neumann algebras, in particular hyper-finite factors of type II_1 could provide the mathematics needed to develop a more explicit view about the construction of S-matrix. In this chapter I will discuss various aspects of type II_1 factors and their physical interpretation in TGD framework.

1. Philosophical ideas behind von Neumann algebras

The goal of von Neumann was to generalize the algebra of quantum mechanical observables. The basic ideas behind the von Neumann algebra are dictated by physics. The algebra elements allow Hermitian conjugation and observables correspond to Hermitian operators. A measurable function of operator belongs to the algebra.

The predictions of quantum theory are expressible in terms of traces of observables. The highly non-trivial requirement of von Neumann was that identical a priori probabilities for a detection of states of infinite state system must make sense. Since quantum mechanical expectation values are expressible in terms of operator traces, this requires that unit operator has unit trace.

For finite-dimensional case it is easy to build observables out of minimal projections to 1-dimensional eigen spaces of observables. For infinite-dimensional case the probably of projection to 1-dimensional sub-space vanishes if each state is equally probable. The notion of observable must thus be modified by excluding 1-dimensional minimal projections, and allow only projections for which the trace would be infinite using the straightforward generalization of the matrix algebra trace as dimension of the projection.

The definitions of adopted by von Neumann allow more general algebras than algebras of II_1 for with traces are not larger than one. Type I_n algebras correspond to finite-dimensional matrix algebras with finite traces whereas I_∞ does not allow bounded traces. For algebras of type III traces are always infinite and the notion of trace becomes useless.

2. von Neumann, Dirac, and Feynman

The association of algebras of type I with the standard quantum mechanics allowed to unify matrix mechanism with wave mechanics. Because of the finiteness of traces von Neumann regarded the factors of type II_1 as fundamental and factors of type III as pathological. The highly pragmatic and successful approach of Dirac based on the notion of delta function, plus the emergence of Feynman graphs and functional integral meant that von Neumann approach was forgotten to a large extent.

Algebras of type II_1 have emerged only much later in conformal and topological quantum field theories allowing to deduce invariants of knots, links and 3-manifolds. Also algebraic structures known as bi-algebras, Hopf algebras, and ribbon algebras to type II_1 factors. In topological quantum computation based on braids and corresponding topological S-matrices they play an especially important role.

3. Factors of type II_1 and quantum TGD

There are good reasons to believe that hyper-finite (ideal for numerical approximations) von Neumann algebras of type II_1 are of a direct relevance for TGD.

Equivalence of generalized loop diagrams with tree diagrams.

The work with bi-algebras led to the proposal that the generalized Feynman diagrams of TGD at space-time level satisfy a generalization of the duality of old-fashioned string models. Generalized Feynman diagrams containing loops are equivalent with tree diagrams so that they could be interpreted as representing computations or analytic continuations. This symmetry can be formulated as a condition on algebraic structures generalizing bi-algebras. The new element is the possibility of vacuum lines having natural counterpart at the level of bi-algebras and braid diagrams. At space-time level they correspond to vacuum extremals.

Inclusions of hyper-finite II_1 factors as a basic framework to formulate quantum TGD.

The basic facts about von Neumann factors of II_1 suggest a more concrete view about the general mathematical framework needed.

a) The effective 2-dimensionality of the construction of quantum states and configuration space geometry in quantum TGD framework makes hyper-finite factors of type II_1 very natural as operator algebras of the state space. Indeed, the elements of conformal algebras are labelled by discrete numbers and also the modes of induced spinor fields are labelled by discrete label, which guarantees that the tangent space of the configuration space is a separable Hilbert space and Clifford algebra is thus a hyper-finite type II_1 factor. The same holds true also at the level of configuration space degrees of freedom so that bosonic degrees of freedom correspond to a factor of type I_∞ unless super-symmetry reduces it to a factor of type II_1.

b) Four-momenta relate to the positions of tips of future and past directed light cones appearing naturally in the construction of S-matrix. In fact, configuration space of 3-surfaces can be regarded as union of big-bang/big crunch type configuration spaces obtained as a union of light-cones with parameterized by the positions of their tips. The algebras of observables associated with bounded regions of M^4 are hyper-finite and of type III_1. The algebras of observables in the space spanned by the tips of these light-cones are not needed in the construction of S-matrix so that there are good hopes of avoiding infinities coming from infinite traces.

c) Many-sheeted space-time concept forces to refine the notion of sub-system. Jones inclusions $\mathcal{N} \subset \mathcal{M}$ for factors of type II_1 define in a generic manner imbedding interacting sub-systems to a universal II_1 factor which now corresponds naturally to infinite Clifford algebra of the tangent space of configuration space of 3-surfaces and contains interaction as $\mathcal{M} : \mathcal{N}$-dimensional analog of tensor factor. Topological condensation of space-time sheet to a larger space-time sheet, formation of bound states by the generation of join along boundaries bonds, interaction vertices in which space-time surface branches like a line of Feynman diagram: all these situations could be described by Jones inclusion characterized by the Jones index $\mathcal{M} : \mathcal{N}$ assigning to the inclusion also a minimal conformal field theory and conformal theory with k=1 Kac Moody for $\mathcal{M} : \mathcal{N} = 4$. $\mathcal{M} : \mathcal{N}$=4 option need not be realized physically as quantum field theory but as string like theory whereas the limit $D = 4 - \epsilon \to 4$ could correspond to $\mathcal{M} : \mathcal{N} \to 4$ limit. An entire hierarchy of conformal field theories is thus predicted besides quantum field theory.

d) von Neumann's somewhat artificial idea about identical a priori probabilities for states could replaced with the finiteness requirement of quantum theory. Indeed, it is traces which produce the infinities of quantum field theories. That $\mathcal{M} : \mathcal{N} = 4$ option is not realized physically as quantum field theory (it would rather correspond to string model type theory characterized by a Kac-Moody algebra instead of quantum group), could correspond to the fact that dimensional regularization works only in $D = 4 - \epsilon$. Dimensional regularization with space-time dimension $D = 4 - \epsilon \to 4$ could be interpreted as the limit $\mathcal{M} : \mathcal{N} \to 4$. $\mathcal{M}$ as an $\mathcal{M} : \mathcal{N}$-dimensional $\mathcal{N}$-module would provide a concrete model for a quantum space with non-integral dimension as well as its Clifford algebra. An entire sequence of regularized theories corresponding to the allowed values of $\mathcal{M} : \mathcal{N}$ would be predicted.

Generalized Feynman diagrams are realized at the level of $\mathcal{M}$ as quantum space-time surfaces.

The key idea is that generalized Feynman diagrams realized in terms of space-time sheets have counterparts at the level of $\mathcal{M}$ identifiable as the Clifford algebra associated with the entire space-time surface X^4. 4-D Feynman diagram as part of space-time surface is mapped to its $\beta = \mathcal{M} : \mathcal{N} \leq 4$-dimensional quantum counterpart.

a) von Neumann algebras allow a universal unitary automorphism $A \rightarrow \Delta^{it} A \Delta^{-it}$ fixed apart from inner automorphisms, and the time evolution of partonic 2-surfaces defining 3-D light-like causal determinant corresponds to the automorphism $\mathcal{N}_i \rightarrow \Delta^{it} \mathcal{N}_i \Delta^{-it}$ performing a time dependent unitary rotation for $\mathcal{N}_i$ along the line. At configuration space level however the sum over allowed values of t appear and should gives rise to the TGD counterpart of propagator as the analog of the stringy propagator $\int_0^t exp(iL_0 t) dt$. Number theoretical constraints from p-adicization suggest a quantization of t as $t = \sum_i n_i y_i > 0$, where $z_i = 1/2 + y_i$ are non-trivial zeros of Riemann Zeta.

b) At space-time level the "ends" of orbits of partonic 2-surfaces coincide at vertices so that also their images $\mathcal{N}_i \subset \mathcal{M}$ also coincide. The condition $\mathcal{N}_i = \mathcal{N}_j = ... = \mathcal{N}$, where the sub-factors $\mathcal{N}$ at different vertices differ only by automorphism, poses stringent conditions on the values t_i and Bohr quantization at the level of $\mathcal{M}$ results. Vertices can be obtained as a vacuum expectations of of the operators creating the states associated with the incoming lines (crossing symmetry is automatic).

c) The equivalence of loop diagrams with tree diagrams would be due to the possibility to move the ends of the internal lines along the lines of the diagram so that only diagrams containing 3-vertices and self energy loops remain. Self energy loops are trivial if the product associated with fusion vertex and co-product associated with annihilation compensate each other. The possibility to assign quantum group or Kac Moody group to the diagram gives good hopes of realizing product and co-product. Octonionic triality would be an essential prerequisite for transforming N-vertices to 3-vertices. The equivalence allows to develop an argument proving the unitarity of S-matrix.

d) A formulation using category theoretical language suggests itself. The category of space sheets has as the most important arrow topological condensation via the formation of wormhole contacts. This category is mapped to the category of II_1 sub-factors of configurations space Clifford algebra having inclusion as the basic arrow. Space-time sheets are mapped to the category of Feynman diagrams in $\mathcal{M}$ with lines defined by unitary rotations of $\mathcal{N}_i$ induced by Δ^{it}.

e) The hierarchy of imbeddings for type II_1 factors generalizes to the hierarchy of generalized Feynman diagrams in which the particles of given level correspond to Feynman diagrams of the previous level. These Feynman diagrams provide representations for the projections of S-matrix to subspaces of incoming and outgoing states providing a hierarchy of self representations about the system. The vertices for the interactions of these Feynman diagrams reduce to the interactions at the lowest level apart from the automorphisms $\Delta_{\mathcal{M}_n}$ defining free propagation. The interpretation is as an infinite cognitive hierarchy realizing theory about the material world as cognitive quantum states. This hierarchy corresponds to states with vanishing conserved net quantum numbers but having non-vanishing "gravitational" charges identifiable as classical charges. A fascinating possibility is that dark matter corresponds to thoughts!

Is $\hbar$ dynamical?

The work with topological quantum computation inspired the hypothesis that $\hbar$ might be dynamical, and that its values might relate in a simple manner to the logarithms of Beraha numbers giving Jones indices $\mathcal{M} : \mathcal{N}$. The model for the evolution of $\hbar$ implied that $\hbar$ is infinite for the minimal value $\mathcal{M} : \mathcal{N} = 1$ of Jones index.

The construction of a model explaining the strange finding that planetary orbits seem to correspond to a gigantic value of "gravitational" Planck constant led to the hypothesis that when the system gets non-perturbative so that the perturbative expansion in terms of parameter $k = alpha Q_1 Q_2$ ceases to converge, a phase transition increasing the value of $\hbar$ to $\hbar_s = k \times \hbar / v_0$, where $v_0 = 4.8 \times 10^{-4}$ is the ratio of Planck length to CP_2 length, occurs. This involves also a transition to a macroscopic quantum phase since Compton lengths and times increase dramatically. Dark matter would correspond to ordinary matter with large value of $\hbar$, which is conformally confined in the sense the sum of complex super-canonical conformal weights (related in a simple manner to the complex zeros of Riemann Zeta) is real for the many-particle state behaving like a single quantum coherent unit.

The value of $\hbar$ for $\mathcal{M} : \mathcal{N} = 1$ is large but not infinite, and thus in conflict with the original original proposal. A more refined suggestion is that the evolution of $\hbar$ as a function of $\mathcal{M} : \mathcal{N} = 4cos^2(\pi/n)$ can be interpreted as a renormalization group evolution for the phase resolution. The earlier identification is replaced by a linear renormalization group equation for $1/\hbar$ allowing as its solutions the earlier solution plus an arbitrary integration constant. Hence $1/\hbar$ can approach to a finite value $1/\hbar(3) = v_0/k \times \hbar(n \to \infty)$ at the limit $n \to 3$. The evolution equation gives a concrete view about how various charges should be imbedded in Jones inclusion to the larger algebra so that the value of $\hbar$ appearing in commutators evolves in the required manner.

The dependence of $\hbar$ on the parameters of interacting systems means that it is associated with the interface of the interacting systems. Instead of being an absolute constant of nature $\hbar$ becomes something characterizing the interaction between two systems, the "position" of II_1 factor $\mathcal{N}$ inside $\mathcal{M}$. The interface could correspond to wormhole contacts, join along boundaries bond, light-like causal determinant, etc... This property of $\hbar$ is consistent with the fact that vacuum functional expressible as an exponent of Kähler action does not depend at all on $\hbar$.

1.5.4 PART IV: Applications to Elementary Particle Physics

General View About Physics in Many-Sheeted Space-Time

Topological condensation and evaporation represent the basic new concepts of TGD and an attempt to formulate a general qualitative theory of the topological condensation and evaporation and TGD based space-time concept is made. The topics to be discussed in the sequel will be following.

1. The general structure of topological condensate

The question what 3-space looks like in various scales and end up to a purely topological description for the generation of structures. Topological arguments imply a finite size for non-vacuum 3-surfaces and the conservation of the gauge and gravitational fluxes requires that 3-surface feeds these fluxes to a larger 3-surface via # contacts situated near the boundaries of the 3-surface. Renormalization group invariance (RGI) hypothesis suggests that 3-surfaces with all sizes are important in the functional integral and this leads to the idea of the many-sheeted space-time with hierarchical, fractal like structure such that each level of the hierarchy corresponds to a characteristic length scale.

2. Topological field quantization

The general space-time picture suggested by RGI hypothesis can be justified mathematically. Due to the compactness of CP_2, a general space-time surface representable as a map $M^4 \to CP_2$ decomposes into regions, "topological field quanta", characterized by certain vacuum quantum numbers and 3-surface is in general unstable against the decay to disjoint components along the boundaries of the field quanta.

Topological field quanta have finite size depending on the values of the vacuum quantum numbers: the size increases as the values of the vacuum quantum numbers increase. Topological field quantum is therefore a good candidate for a quantum coherent system provided some Bose Einstein condensate or quantum coherent state is available. The BE condensate or coherent state of the light # contacts near the boundaries of the topological field quantum is a good candidate in this respect. It came as a total surprise that this the generation of vacuum expectation value of Higgs field corresponds to the generation of this kind of macroscopic quantum phase.

The requirement of the gauge charge conservation in turn implies the hierarchical structure of the topological condensate: gauge fluxes must go somewhere from the outer boundaries of the topological field quantum with finite size and this 'somewhere' must be a larger topological field quantum, which in turn feeds its gauge fluxes to a larger topological field quantum,.... Of course, the nonlinearity of the theory could allow vacuum charge densities which can cancel the net charge near boundaries.

Most importantly, topological field quanta allow discrete scalings as a dynamical symmetry. p-Adic length scale hypothesis states that the allowed scaling factors correspond to powers of $\sqrt{p}$, where the prime p satisfies $p \simeq 2^k$, k integer with prime values favored. p-Adic fractality (actually multi-p-fractality) can be justified more rigorously by a precise formulation for the fusion of real and various p-adic physics based on the generalization of the notion of number.

3. General physical consequences of new view about space-time

The physical consequences of the new space-time picture are nontrivial at all length scales.

a) A natural interpretation for the hierarchical structure is in terms of bound state formation. Quarks condense to form hadrons, nucleons condense to form atomic nuclei, nuclei and electrons condense to form atoms, how atoms condense to form molecules, and so on. One ends up with a general picture for the topology of 3-space associated with, say, solid state and with the idea that even the macroscopic bodies of the everyday world correspond to topologically condensed 3-surfaces.

b) The join of 3-surfaces along their boundaries defines a new kind of interaction, which has in fact has been used in phenomenological modelling of chemical reactions. Usually chemical bond is believed to result from Schrödinger equation. At the macroscopic level this interaction is rather familiar to us since it means that two macroscopic bodies just touch each other.

v) In TGD context there are purely topological necessary conditions for quantum coherence and a topological description for dissipative phenomena. The formation of the join along boundaries bonds plays a decisive role in the description and this process provides a universal manner to generate macroscopic quantum systems. There is also a topological description for the formation of the supra phases and the phase of the order parameter of the supra phase ground state contains information about the homotopy of the join along boundaries condensate.

4. Higgs boson as wormhole contact, electro-weak symmetry breaking, the weakening of Equivalence Principle, and color confinement

The proper understanding of the concepts of gauge charges and fluxes and their gravitational counterparts in TGD space-time has taken a lot of efforts. At the fundamental level gauge charges assignable to light-like 3-D elementary particle horizons surrounding a topologically condensed CP_2 type extremals can be identified as the quantum numbers assignable to fermionic oscillator operators generating the state associated with horizon identifiable as a parton. Quantum classical correspondence requires that commuting classical gauge charges are quantized and this is expected to be true by the generalized Bohr orbit property of the space-time surface.

There are however non-trivial questions. Do vacuum charge densities give rise to renormalization effects or imply non-conservation so that weak charges would be screened above intermediate boson length scale? Could one assign the non-conservation of gauge fluxes to the wormhole (#) contacts, which are identifiable as pieces of CP_2 extremals and for which electro-weak gauge currents are not conserved so that weak gauge fluxes would be non-vanishing but more or less random so that long range correlations would be lost?

It indeed turns that one can understand the non-conservation of weak gauge fluxes in terms of wormhole contacts carrying pairs of right/left handed fermion and left/right handed antifermion having interpretation as Higgs bosons. The average non-conserved light-like gravitational four-momentum of wormhole contact representing Higgs boson can be identified as the inertial four-momentum apart from the sign factor so that one can also understand particle massivation at fundamental level and a connection with p-adic thermodynamics based description of Higgs mechanism emerges. Also a detailed understanding about how Equivalence Principle is weakened in TGD framework emerges.

Also color confinement can be understood using only quantum classical correspondence and general properties of classical color gauge field. Spin glass degeneracy allows to understand the generation of macro-temporal quantum coherence and the same mechanism allows also to understand more quantitatively color confinement by applying unitarity conditions.

5. Wormhole contacts, super-conductivity, and biology

Wormhole contacts, feeding gauge fluxes from a given sheet of the 3-space to a larger one, which are a necessary concomitant of the many-sheeted space-time concept. # contacts can be regarded as particles carrying classical charges defined by the gauge fluxes but behaving as extremely tiny dipoles quantum mechanically in the case that gauge charge is conserved. # contacts must be light, which suggests that they can form Bose-Einstein condensates and coherent states. The real surprise (after 27 years of TGD) was that the formation of these rather exotic macroscopic quantum phases could be identified as formation of vacuum expectation value of Higgs field for various scaled up copies of standard model physics. This kind of macroscopic quantum phases could be in a central role in the TGD inspired model for a bio-system as a macroscopic quantum system. Electromagnetically charged # contacts are also possible and would explain the massivation of photons in super-conductors implying that long ranged exotic W boson exchanges play a key role in super-conductivity.

6. Dark matter hierarchy and fractal copies of standard model physics

The most dramatic prediction obvious from the beginning but mis-interpreted for about 26 years is the presence of long ranged classical electro-weak and color gauge fields in the length scale of the space-time sheet. The only interpretation consistent with quantum classical correspondence is in terms of a hierarchy of scaled up copies of standard model physics corresponding to p-adic length scale hierarchy and dark matter hierarchy labelled by arbitrarily large values of dynamical quantized Planck constant. Chirality selection in the bio-systems provides direct experimental evidence for this fractal hierarchy of standard model physics.

7. The interpretation of long range weak and color gauge fields

In TGD gravitational fields are accompanied by long ranged electro-weak and color gauge fields. The only possible interpretation is that there exists a p-adic hierarchy of color and electro-weak physics such that weak bosons are massless below the p-adic length scale determining the mass scale of weak bosons. By quantum classical correspondence classical long ranged gauge fields serve as space-time correlates for gauge bosons below the p-adic length scale in question.

The unavoidable long ranged electro-weak and color gauge fields are created by dark matter and dark particles can screen dark nuclear electro-weak charges below the weak scale. Above this scale vacuum screening occurs as for ordinary weak interactions. Dark gauge bosons are massless below the appropriate p-adic length scale but massive above it and $U(2)_{ew}$ is broken only in the fermionic sector. For dark copies of ordinary fermions masses are essentially identical with those of ordinary fermions.

This interpretation is consistent with the standard elementary particle physics for visible matter apart from predictions such as the possibility of p-adically scaled up versions of ordinary quarks predicted to appear already in ordinary low energy hadron physics. The most interesting implications are seen in longer length scales. Dark variants of ordinary valence quarks and gluons and a scaled up copy of ordinary quarks and gluons are predicted to emerge already in ordinary nuclear physics. Chiral selection in living matter suggests that dark matter is an essential component of living systems so that non-broken $U(2)_{ew}$ symmetry and and free color in bio length scales become characteristics of living matter and of bio-chemistry and bio-nuclear physics. An attractive solution of the matter antimatter asymmetry is based on the identification of also antimatter as dark matter.

8. Renormalization group equations at space-time level

Renormalization group evolution equations for gauge couplings at given space-time sheet are discussed using quantum classical correspondence. For known extremals of Kähler action gauge couplings are RG invariants inside single space-time sheet, which supports the view that discrete p-adic coupling constant evolution replaces the ordinary coupling constant evolution.

Elementary particle vacuum functionals

Genus-generation correspondence is one of the basic ideas of TGD approach. In order to answer various questions concerning the plausibility of the idea, one should know something about the dependence of the elementary particle vacuum functionals on the vibrational degrees of freedom for the boundary component. The construction of the elementary particle vacuum functionals based on Diff invariance, 2-dimensional conformal symmetry, modular invariance plus natural stability requirements indeed leads to an essentially unique form of the vacuum functionals and one can understand why $g > 2$ bosonic families are experimentally absent and why lepton numbers are conserved separately.

An argument suggesting that the number of the light fermion families is three, is developed. The argument goes as follows. Elementary particle vacuum functionals represent bound states of g handles and vanish identically for hyper-elliptic surfaces having $g > 2$. Since all $g \leq 2$ surfaces are hyper-elliptic, $g \leq 2$ and $g > 2$ elementary particles cannot appear in same non-vanishing vertex and therefore decouple. The $g > 2$ vacuum functionals not vanishing for hyper-elliptic surfaces represent many particle states of $g \leq 2$ elementary particle states being thus unstable against the decay to $g \leq 2$ states. The failure of Z_2 conformal symmetry for $g > 2$ elementary particle vacuum functionals would in turn explain why they are heavy: this however not absolutely necessary since these particles would behave like dark matter in any case.

Massless states and particle massivation

The massless sector of the TGD and particle massivation is studied in this chapter. The identification of the spectrum of light particles reduces to two tasks: the construction of massless states and the identification of the states which remain light in p-adic thermodynamics.

1. Physical states as representations of super-canonical and Super Kac-Moody algebras

Physical states belong to the representation of super-canonical algebra and Super Kac-Moody algebra of $E^2 \times SU(3) \times U(2)_{ew}$ associated with the 2-D surfaces X^2 defined by the intersections of 3-D light like causal determinants (CDs) with 7-D CDs $X^7 = X_l^3 \times CP_2$, where X_l^3 is union of future and past directed light cone boundaries. These 2-surfaces have interpretation as partons.

A generalization of the coset construction is in question in the sense that Virasoro generators are differences of those for Super Kac-Moody algebra and super-canonical algebra. The justification comes from the miraculous geometry of the light cone boundary implying that Super Kac-Moody conformal symmetries of X^2 can be compensated by super-canonical local radial scalings so that the differences of corresponding Super Virasoro generators annihilate physical states. What is done is to construct first a state with a non-positive conformal weight using super-canonical generators, and then to apply Super-Kac Moody generators to compensate this conformal weight to get a state with vanishing conformal weight and thus mass.

The conformal weights of super-canonical algebra generators are complex and expressible in terms of zeros of Riemann Zeta. The anti-commutators and commutators of the super-canonical generators in turn are expressible in terms of Riemann Zeta. The conformal weights of physical states must be real. This implies conformal confinement to which color confinement might reduce: what would happen that partons can have complex super-canonical conformal weights but particles have real conformal weights.

The hierarchy of p-adic primes is dual to the hierarchy of zeros of Zeta, and it is possible to express operators O_p creating states for given p-adic prime p as superpositions of products of operators O_z such that the net conformal weight is constant for each term in the superposition. The ansatz for the coefficients of the superposition fixes completely the p-adic coupling constant evolution.

For spinor harmonics of CP_2 the correlation between color and electro-weak quantum numbers is not correct and super-canonical generators provide a natural mechanism allowing to cure the problem. Boson states are identified as bi-local bilinears of fermions and anti-fermions in X^2 characterized by charge matrices and conformally invariant correlation function. $BF\overline{F}$ coupling constants can be identified in terms of normalization factors of the boson states. The small value of gravitational coupling can be understood as resulting by a fractal mechanism reducing its value from the square of p-adic length L_p, and a concrete physical interpretation for the expression of gravitational constant in terms of CP_2 length derived from number theoretic arguments emerges.

The possibility of exotic states poses a serious problem for the proposed scenario. It however turns out that if elementary particles correspond to CP_2 type extremals, all exotic massless particles can be constructed using colored generators and by color confinement cannot induce macroscopic long range interactions. The essential assumption is that the fermionic quantization for the space-time sheets having CP_2 projection of dimension $D(CP_2) < 4$ is non-conventional. This has also direct relevance for the understanding of the matter antimatter asymmetry.

2. Particle massivation

Particle massivation can be regarded as a change of the vanishing parton conformal weights describable as a thermal mixing with higher conformal weights. The interpretation is made possible by the fact that the general mass formula for particle does not fix completely the conformal weights of the partons. The values of the partonic conformal weights can be only deduced from the poles of S-matrix assuming that an approximate decomposition to propagators and vertices occurs. The physical mechanism causing the thermal massivation is hydrodynamical mixing by the braiding flow defined by the normal components of energy momentum tensor of the induced Kähler field at light like 3-D CDs describing the orbits of partons.

This mechanism cannot explain the massivation of electro-weak gauge bosons, which could be caused either by TGD variant of Higgs mechanism (TGD indeed predicts a candidate for a Higgs field) or by the fact that the charge matrices of W boson and left handed component of Z^0 are not covariantly constant, which together with the hydrodynamical mixing leads to a loss of correlations.

The massivation induced by the ergodic hydrodynamical flow can be described in terms of p-adic thermodynamics. The underlying philosophy is that real number based TGD can be algebraically continued to various p-adic number fields. This poses extremely strong conditions on the parameters of the theory. Instead of energy, the Super Kac-Moody Virasoro generator L_0 (essentially mass squared) is thermalized. This guarantees Lorentz invariance. p-Adic thermodynamics forces to conclude that CP_2 radius is essentially the p-adic length scale $R \sim L$ and thus of order $R \simeq 10^4\sqrt{G}$ and therefore 10^4 times larger than the naive guess.

p-Adic temperature is quantized by purely number theoretical constraints (Boltzmann weight $exp(-E/kT)$ is replaced with p^{L_0/T_p}, $1/T_p$ integer) and fermions correspond to $T_p = 1$ whereas $T_p = 1/2$ seems to be the only reasonable choice for bosons.

The interpretation for the modular contribution to the particle mass as resulting from p-adic thermodynamics in super-canonical degrees of freedom is proposed. The p-adic counterparts of elementary particle vacuum functionals are constructed and shown to imply quantization of moduli expected to occur also for the maxima of Kähler function.

The predictions of the general theory are consistent with the earlier mass calculations, and the earlier ad hoc parameters disappear. In particular, optimal lowest order predictions for the charged lepton masses are obtained and photon, gluon and graviton appear as essentially massless particles. The negative conformal weight created by super-canonical generators can have arbitrarily large magnitude so that an infinite hierarchy of exotic massless states is predicted. These states receive mass by the proposed mechanism and they are expected to be unstable but it remains to be shown that they do not appear in the spectrum of light particles.

p-Adic Mass Calculations: Elementary Particle Masses

The calculation of elementary fermion and boson masses using p-adic thermodynamics is carried out. Leptons and quarks are obey almost identical mass formulas. Charged lepton mass ratios are predicted with relative errors of order one cent and QED renormalization corrections provide a plausible explanation for the discrepancies. Neutrino masses and neutrino mixing matrix can be predicted highly uniquely if the existing experimental inputs are taken seriously: the best fit of the mass squared differences requires $k = 13^2 = 169$ so that extended form of the p-adic length scale hypothesis is needed.

The prediction or quark masses is more difficult since even the deduction of even the p-adic length scale determining the masses of u, d, and s is a non-trivial task. Second difficulty is related to the topological mixing of quarks. Somewhat surprisingly, the model for U and D matrices constructed for a decade ago predicts realistic quark mass spectrum although the new mass formula is based on different assumptions and different identification of p-adic mass scales.

Contrary to the long-held belief, p-adic thermodynamics cannot explain Z^0 and W boson masses: thermal masses are completely negligible for the p-adic temperature $T = 1/2$ whereas for $T = 1$ they are 20-30 per cent too high. TGD allows a candidate for a Higgs field with the same quantum numbers as its standard model counterpart. Thus p-adic thermodynamics *resp.* Higgs mechanism would predict in excellent accuracy fermion *resp.* boson masses and allow the Higgs production rate to be about one per cent of the rate predicted by the standard model (the dominating fermionic couplings are now small). The fact that left handed parts of electro-weak charge matrices are not covariantly constant provides a possible mechanism of boson massivation involving no Higgs boson.

The possibility of exotic states poses a serious problem for the proposed scenario. If elementary particles correspond to CP_2 type extremals, all exotic massless particles can be constructed using colored generators and by color confinement cannot induce macroscopic long range interactions. The essential assumption is that the fermionic quantization for the space-time sheets having CP_2 projection of dimension $D(CP_2) < 4$ is non-conventional. This has also direct relevance for the understanding of the matter antimatter asymmetry.

p-Adic Mass Calculations: Hadron Masses

In this chapter the results of the calculation of elementary particle masses will be used to construct a model predicting hadron masses. The new elements are a revised identification for the p-adic length scales of quarks and the realization that number theoretical constraints on topological mixing can be realized by assuming that topological mixing leads to a thermodynamical equilibrium. This gives an upper bound of 1200 for the number of different U and D matrices and the input from top quark mass and $\pi^+ - \pi^0$ mass difference implies that physical U and D matrices can be constructed as small perturbations of matrices expressible as direct sum of essentially unique 2×2 and 1×1 matrices. The maximally entropic solutions can be found numerically by using the fact that only the probabilities p_{11} and p_{21} can be varied freely. The solutions are unique in the accuracy used, which suggests that the system allows only single thermodynamical phase.

The matrices U and D associated with the probability matrices can be deduced straightforwardly in the standard gauge. The U and D matrices derived from the probabilities determined by the entropy maximization turn out to be unitary for most values of n_1 and n_2. This is a highly non-trivial result and means that mass and probability constraints together with entropy maximization define a sub-manifold of $SU(3)$ regarded as a sub-manifold in 9-D complex space. The choice $(n(u), n(c)) = (4, n)$, $n < 9$, does not allow unitary U whereas $(n(u), n(c)) = (5, 6)$ does. This choice is still consistent with top quark mass and together with $n(d) = n(s) = 5$ it leads to a rather reasonable CKM matrix with a value of CP breaking invariant within experimental limits. The elements V_{23} and V_{32}, $i = 1, 2$ are however roughly twice larger than their experimental values deduced assuming standard model. V_{31} is too large by a factor 1.6. The possibility of scaled up variants of light quarks could lead to too small experimental estimates for these matrix elements. The whole parameter space has not been scanned so that better candidates for CKM matrices might well exist.

The assumption about the presence of scaled up variants of light quarks in light hadrons leads to a surprisingly successful model for pseudo scalar meson masses using only quark masses. This conforms with the idea that pseudo scalar mesons are Goldstone bosons in the sense that color Coulombic and magnetic contributions to the mass cancel each other. Also the mass differences between hadrons containing different numbers of strange and heavy quarks can be understood if s, b and c quarks appear as several scaled up versions. The earlier model for the purely hadronic contributions to hadron masses simplifies dramatically and only the color Coulombic and magnetic contributions to the color conformal weight are needed.

Color magnetic spin-spin splittings for $\Delta - N$ and $\rho - \pi$ systems predict with only 23 percent error the value of the nucleon mass assuming that color interaction conformal weight is associated with color flux tubes. If the color Coulombic and magnetic net contribution of quarks to the nucleon conformal weight is multiplied by two, an excellent prediction for the nucleon mass results. The p-adic variant of the equipartition theorem of thermodynamics suggests that the additional contribution relates to sea partons.

p-Adic Mass Calculations: New Physics

Concerning new physics the basic predictions are following. Also bosons have counterpart of family replication but the mixing is maximal. The covariant constancy of the right handed neutrino generating a potential super-symmetry implies that no sparticles are predicted: super-generators with vanishing conformal weight simply vanish. TGD predicts a rich spectrum of massless states for which ground states of negative super-canonical conformal weight are created by colored super-generators. By color confinement these states do not however give rise to macroscopic long range forces. A hierarchy of scaled up copies of electro-weak and color physics is predicted. The status of Higgs in TGD Universe is still unestablished. TGD provides a Higgs candidate but particle massivation might be understood without Higgs. The topics discussed in this chapter are following.

1. Color coupling constant evolution without asymptotic freedom or dark matter hierarchy?

QCD coupling constant evolution is discussed it is found that asymptotic freedom could be lost making possible existence of several scaled up versions of QCD existing only in a finite length scale range. The basic counter arguments against lepto-hadron hypothesis are considered and it is found that the loss of asymptotic freedom could allow lepto-hadron physics.

The discovery of dark matter hierarchy about fifteen years after these argument were developed resolves the problems in much more elegant manner. TGD predicts an infinite hierarchy of electro-weak and color physics physics for which particles couple directly only via gravitons. De-coherence phase transitions can however induce processes allowing the decay of particles of a given physics to particles of another physics.

2. Summary of new physics effects

Various new physics effects are discussed.

a) There is a discussion of higher electro-weak boson and gluon families and the possible physical evidence for them, and an argument forcing the identification of partonic vertices as branchings of partonic 2-surfaces is developed.

b) ALEPH anomaly is interpreted in terms of a fractal copy of b-quark corresponding to k=197.

c) The possible signatures of M_{89} hadron physics in e^+e^- annihilation experiments are discussed using a naive scaling of ordinary hadron physics.

d) It is found that the newly born concept of Pomeron of Regge theory could be identified as the sea of perturbative QCD.

3. Anomalously large direct CP breaking in $K - \overline{K}$ system and exotic gluons

The recently observed anomalously large direct CP breaking in $K_L \to \pi\pi$ decays is explained in terms of loop corrections due to the predicted 2 exotic gluons having masses around 33.6 GeV. It will be also found that the TGD version of the chiral field theory believed to provide a phenomenological low energy description of QCD differs from its standard model version in that quark masses correspond in TGD framework to the shifts of quark masses induced by the vacuum expectation values of the scalar meson fields. This conforms with the TGD view about Higgs mechanism as causing only small mass shifts.

1.5.5   PART V: Applications to Cosmology and Astrophysics

The Relationship Between TGD and GRT

In this chapter the relationship of TGD and GRT is studied. The basic problem is that TGD predicts an exact conservation of the four-momentum at the classical level on one hand and GRT space-time is a well-established concept experimentally on the other hand. Einstein's equations for the space-time surface however imply non-conservation of the energy-momentum. There are several possibilities to circumvent the problem.

a) Energy-momentum conservation is broken quantum mechanically due to the presence of the light cone boundary.

b) In TGD space-time is many-sheeted and there could also be a momentum flow between various sheets of the many-sheeted space-time. Also a momentum flow between topological vapor phase and space-time sheet is possible.

c) It took 25 years to discover the third option. When space-times are 4-surfaces, four-momentum currents are vector currents rather than components of a tensor and the sign of energy depends on the time orientation of the space-time sheet. This led to the identification of phase conjugate photons as negative energy photons and forced to ask whether also phase conjugate fermions are possible and whether matter anti-matter asymmetry could be resolved by assuming that anti-fermions have negative energies. This would also eliminate the only infinity appearing in the formulation of quantum TGD and resulting from the zero point energy of second quantized induced spinor fields.

Gravitational mass can be identified as the absolute value of the inertial mass so that its density corresponds to the difference of inertial mass densities for matter and antimatter and is not conserved. Similar interpretation applies to the gravitational counterparts of gauge charges coupling to the long range electro-weak and color gauge fields. The resulting Eastern view about cosmology is maximally predictive, consistent with the crossing symmetry of elementary particle physics, and also conforms with the materialistic illusion about unique objective reality at the limit when the interaction between negative and positive energy matter can be neglected.

Following topics are discussed in the chapter.

a) The relationship between TGD and GRT is discussed in light of the new view about energy. One of the most fascinating outcomes is the resolution of the most gigantic failure in the art of order magnitude estimates. The naive estimate for the cosmological constant predicted also by TGD is by a factor 10^{120} larger than its value deduced from the accelerated expansion of the Universe. The resolution comes naturally from the p-adic fractality predicting that cosmological constant is reduced by a factor of 2 in a step wise manner in phase transitions occurring at times $T(k) \propto 2^{k/2}$, which correspond to p-adic time scales. On the average $\Lambda(k)$ behaves as $1/a^2$, where a is the light-cone proper time. This predicts correctly the observed value of Λ.

b) The notion of many-sheeted space time interpreted as a hierarchy of smoothed out space-times produced by Nature itself rather than only renormalization group theorist is discussed. space-time are compared. The dynamics of gravitational charges is discussed.

c) The theory is applied to the vacuum extremal embeddings of Reissner-Nordström and Schwartschild metric.

d) A model for the final state of a star, which indicates that Z^0 force might have an important role in the dynamics of the compact objects. During year 2003, more than decade after the formulation of the model, the discovery of the connection between supernovas and gamma ray bursts provided strong support for the predicted axial magnetic and Z^0 magnetic flux tube structures predicted by the model for the final state of a rotating star.

e) Black holes represent a basic failure of GRT. Two TGD based views about the relationship of black holes, topological evaporation, and negative energies are represented with a special attention paid to the technological implications. The first view is based on the earlier identification $E_{gr} = E_{inert}$ allowing negative gravitational energies. Second view relies on the more recent identification $E_{gr} = |E_{inert}|$.

f) There is experimental evidence for gravimagnetic fields in rotating superconductors which are by 20 orders of magnitudes stronger than predicted by general relativity. A TGD based explanation of these observations is discussed.

Cosmic Strings

Cosmic strings belong to the basic extremals of the Kähler action. The string tension of the cosmic strings is $T \simeq .2 \times 10^{-6}/G$ and slightly smaller than the string tension of the GUT strings and this makes them very interesting cosmologically. Concerning the understanding of cosmic strings a decisive breakthrough came through the identification of gravitational four-momentum as the difference of inertial momenta associated with matter and antimatter and the realization that the net inertial energy of the Universe vanishes. This forced to conclude cosmological constant in TGD Universe is non-vanishing. p-Adic length fractality predicts that Λ scales as $1/L^2(k)$ as a function of the p-adic scale characterizing the space-time sheet. The recent value of the cosmological constant comes out correctly. The gravitational energy density described by the cosmological constant is identifiable as that associated with topologically condensed cosmic strings and of magnetic flux tubes to which they are gradually transformed during cosmological evolution.

1. Topological condensation of cosmic strings

The basic question whether one can model the exterior region of the topologically condensed cosmic string using General Relativity.

a) Cosmic strings must be created from vacuum as pairs of strings with opposite time orientation and inertial energy. For tightly bound, perhaps even coiled pairs of cosmic strings radial Kähler fields have opposite directions and tend to cancel each other. There is however a slight matter antimatter asymmetry so that the field is not a pure dipole field. It is possible to satisfy Einstein equations and obtain cancellation of Kähler magnetic action of strings and Kähler electric action of the larger space-time sheet.

b) The two cosmic strings correspond to phase conjugates of zero energy fermionic vacua and reduce their inertial mass by Hawking radiation involving generation of fermion anti-fermion pairs, whose second member remains inside string. This generates at least part of visible matter. The splitting of cosmic strings followed by a "burning" of the string ends provides a second manner to generate visible matter. Fermions and negative energy anti-fermions dominate the energy density at the space-time sheet containing strings and strings themselves contain dominantly negative energy fermions and positive energy anti-fermions. This explains the effective absence of antimatter.

c) The phase conjugacy of two cosmic strings has deep implications for the cosmic and ultimately also for biological evolution (magnetic flux tubes paly a fundamental role in TGD inspired biology and cosmic strings are limiting cases of them). The arrows of geometric time are opposite for the strings and also for positive energy matter and negative energy antimatter. This implies a competition between two dissipative time developments proceeding in different directions of geometric time and looking self-organization and even self-assembly from the point of view of each other. This resolves paradoxes created by gravitational self-organization contra second law of thermodynamics.

2. Cosmic strings and generation of structures

p-Adic fractality and simple quantitative observations lead to the hypothesis that pairs of cosmic strings are responsible for the evolution of astrophysical structures in a very wide length scale range. Large voids with size of order 10^8 light years can be seen as structures containing knotted and linked cosmic string pairs wound around the boundaries of the void. Galaxies correspond to same structure with smaller size and linked around the supra-galactic strings. This conforms with the finding that galaxies tend to be grouped along linear structures. Simple quantitative estimates show that even stars and planets could be seen as structures formed around cosmic strings of appropriate size. Thus Universe could be seen as fractal cosmic necklace consisting of cosmic strings linked like pearls around longer cosmic strings linked like...

3. Cosmic strings, gamma ray bursts, and supernovae

During year 2003 two important findings related to cosmic strings were made.

a) A correlation between supernovae and gamma ray bursts was observed.

b) Evidence that some unknown particles of mass $m \simeq 2m_e$ and decaying to gamma rays and/or electron positron pairs annihilating immediately serve as signatures of dark matter. These findings challenge the identification of cosmic strings and/or their decay products as dark matter, and also the idea that gamma ray bursts correspond to cosmic fire crackers formed by the decaying ends of cosmic strings. This forces the updating of the more than decade old rough vision about topologically condensed cosmic strings and about gamma ray bursts described in this chapter.

According to the updated model, cosmic strings transform in topological condensation to magnetic flux tubes about which they represent a limiting case. Primordial magnetic flux tubes forming ferro-magnet like structures become seeds for gravitational condensation leading to the formation of stars and galaxies. The TGD based model for the asymptotic state of a rotating star as dynamo leads to the identification of the predicted magnetic flux tube at the rotation axis of the star as Z^0 magnetic flux tube of primordial origin. Besides Z^0 magnetic flux tube structure also magnetic flux tube structure exists at different space-time sheet but is in general not parallel to the Z^0 magnetic structure. This structure cannot have primordial origin (the magnetic field of star can even flip its polarity).

The flow of matter along Z^0 magnetic (rotation) axis generates synchrotron radiation, which escapes as a precisely targeted beam along magnetic axis and leaves the star. The identification is as the rotating light beam associated with ordinary neutron stars. During the core collapse leading to the supernova this beam becomes gamma ray burst. The mechanism is very much analogous to the squeezing of the tooth paste from the tube. The fact that all nuclei are fully ionized Z^0 ions, the Z^0 charge unbalance caused by the ejection of neutrinos, and the radial compression make the effect extremely strong so that there are hopes to understand the observed incredibly high polarization of 80 ± 20 per cent.

TGD suggests the identification of particles of mass $m \simeq 2m_e$ accompanying dark matter as lepto-pions formed by color excited leptons, and topologically condensed at magnetic flux tubes having thickness of about lepto-pion Compton length. Lepto-pions would serve as signatures of dark matter whereas dark matter itself would correspond to the magnetic energy of topologically condensed cosmic strings transformed to magnetic flux tubes.

TGD and Cosmology

A proposal for what might be called TGD inspired cosmology is made. The basic ingredient of this cosmology is the TGD counter part of the cosmic string. It is found that "Yin-Yang" principle; many-sheeted space-time concept; the new view about the relationship between inertial and gravitational four-momenta; the basic properties of the paired cosmic strings; the existence of the limiting temperature (as in string model, too); the assumption about the existence of the vapor phase dominated by cosmic strings; and quantum criticality imply a rather detailed picture of the cosmic evolution, which differs from that provided by the standard cosmology in several respects but has also strong resemblances with inflationary scenario.

1. Quantum criticality

TGD Universe is quantum counterpart of a statistical system at critical temperature. As a consequence, topological condensate is expected to possess hierarchical, fractal like structure containing topologically condensed 3-surfaces with all possible sizes. Both Kähler magnetized and Kähler electric 3-surfaces ought to be important and string like objects indeed provide a good example of Kähler magnetic structures important in TGD inspired cosmology. In particular space-time is expected to be many-sheeted even at cosmological scales and ordinary cosmology must be replaced with many-sheeted cosmology. The presence of vapor phase consisting of free cosmic strings and possibly also elementary particles is second crucial aspects of TGD inspired cosmology.

Quantum criticality of TGD Universe supports the view that many-sheeted cosmology is in some sense critical. Criticality in turn suggests fractality. Phase transitions, in particular the topological phase transitions giving rise to new space-time sheets, are (quantum) critical phenomena involving no scales. If the curvature of the 3-space does not vanish, it defines scale: hence the flatness of the cosmic time=constant section of the cosmology implied by the criticality is consistent with the scale invariance of the critical phenomena. This motivates the assumption that the new space-time sheets created in topological phase transitions are in good approximation modellable as critical Robertson-Walker cosmologies for some period of time at least.

2. Only sub-critical cosmologies are globally imbeddable

TGD allows global imbedding of subcritical cosmologies. A partial imbedding of one-parameter families of critical and overcritical cosmologies is possible. The infinite size of the horizon for the imbeddable critical cosmologies is in accordance with the presence of arbitrarily long range fluctuations at criticality and guarantees the average isotropy of the cosmology. Imbedding is possible for some critical duration of time. The parameter labelling these cosmologies is scale factor characterizing the duration of the critical period. These cosmologies have the same optical properties as inflationary cosmologies. Critical cosmology can be regarded as a 'Silent Whisper amplified to Bang' rather than 'Big Bang' and transformed to hyperbolic cosmology before its imbedding fails. Split strings decay to elementary particles in this transition and give rise to seeds of galaxies. In some later stage the hyperbolic cosmology can decompose to disjoint 3-surfaces. Thus each sub-cosmology is analogous to biological growth process leading eventually to death.

3. Fractal many-sheeted cosmology

The critical cosmologies can be used as a building blocks of a fractal cosmology containing cosmologies containing ... cosmologies. p-Adic length scale hypothesis allows a quantitative formulation of the fractality. Fractal cosmology predicts cosmos to have essentially same optic properties as inflationary scenario but avoids the prediction of unknown vacuum energy density. Fractal cosmology explains the paradoxical result that the observed density of the matter is much lower than the critical density associated with the largest space-time sheet of the fractal cosmology. Also the observation that some astrophysical objects seem to be older than the Universe, finds a nice explanation.

4. New view about gravitational and inertial energy

An absolutely essential element of the considerations (and longstanding puzzle of TGD inspired cosmology) is the conservation of energy implied by Poincare invariance which seems to be in conflict with the non-conservation of gravitational energy. It took time to discover the natural resolution of the paradox. In TGD Universe matter and antimatter have opposite energies and gravitational four-momentum is identified as difference of the four momenta of matter and antimatter (or vice versa, so that gravitational energy is positive). The assumption that the net inertial energy density vanishes in cosmological length scales is the proper interpretation for the fact that Robertson-Walker cosmologies correspond to vacuum extremals of Kähler action.

5. Cosmic strings

Tightly bound, possibly coiled pairs of cosmic strings are the basic building block of TGD inspired cosmology and all al structures including large voids, galaxies, stars, and even planets can be seen as pearls in a cosmic fractal necklace consisting of cosmic strings containing smaller cosmic strings linked around them containing... During cosmological evolution the cosmic strings are transformed to magnetic flux tubes and these structures are also key players in TGD inspired quantum biology.

Cosmic strings generate part of ordinary matter as Hawking radiation. Also split cosmic strings can generate visible galactic matter. Generation of negative energy gravitation as "acceleration radiation" in strong gravitational fields could generate gravitational mass during the critical period of string dominated cosmology. The absorption of negative energy gravitons by photons implies gradual redshifting of the microwave background radiation. Negative energy virtual gravitons give also rise to a negative gravitational potential energy. Quite generally, negative energy virtual bosons build up the negative interaction potential energy. An important constraint to TGD inspired cosmology is the requirement that Hagedorn temperature $T_H \sim 1/R$, where R is CP_2 size, is the limiting temperature of radiation dominated phase.

Many-sheeted fractal cosmology containing both hyperbolic and critical space-time sheets based on cosmic strings suggests an explanation for several puzzles of GRT based cosmology such as dark matter problem, origin of matter antimatter asymmetry, the problem of cosmological constant and mechanism of accelerated expansion, the problem of several Hubble constants, and the existence of stars apparently older than the Universe. Under natural assumptions TGD predicts same optical properties of the large scale Universe as inflationary scenario does. The recent balloon experiments however favor TGD inspired cosmology.

TGD Inspired Speculations on Astrophysics and Consciousness

In this chapter some speculative ideas relating TGD based cosmology and astrophysics are discussed. p-Adic length scale hypothesis can be applied in astrophysical length scales, too and some examples of possible applications are discussed. One of the most interesting implications of p-adicity is the possibility of series of phase transitions changing the value of cosmological constant behaving as $\Lambda \propto 1/L^2(k)$ as a function of p-adic length scale characterizing the size of the space-time sheet.

D. Da Rocha and Laurent Nottale have proposed that Schrödinger equation with Planck constant $\hbar$ replaced with what might be called gravitational Planck constant $\hbar_{gr} = \frac{GmM}{v_0}$ ($\hbar = c = 1$). v_0 is a velocity parameter having the value $v_0 = 144.7 \pm .7$ km/s giving $v_0/c = 4.6 \times 10^{-4}$. This is rather near to the peak orbital velocity of stars in galactic halos. Also subharmonics and harmonics of v_0 seem to appear. The support for the hypothesis coming from empirical data is impressive.

Nottale and Da Rocha believe that their Schrödinger equation results from a fractal hydrodynamics. Many-sheeted space-time however suggests astrophysical systems are not only quantum systems at larger space-time sheets but correspond to a gigantic value of gravitational Planck constant. The gravitational (ordinary) Schrödinger equation would provide a solution of the black hole collapse (IR catastrophe) problem encountered at the classical level. The resolution of the problem inspired by TGD inspired theory of living matter is that it is the dark matter at larger space-time sheets which is quantum coherent in the required time scale.

I have proposed already earlier the possibility that Planck constant is quantized and the spectrum is given in terms of logarithms of Beraha numbers: the lowest Beraha number B_3 is completely exceptional in that it predicts infinite value of Planck constant. The inverse of the gravitational Planck constant could correspond a gravitational perturbation of this as $1/\hbar_{gr} = v_0/GMm$. The general philosophy would be that when the quantum system would become non-perturbative, a phase transition increasing the value of $\hbar$ occurs to preserve the perturbative character and at the transition $n = 4 \to 3$ only the small perturbative correction to $1/\hbar(3) = 0$ remains. This would apply to QCD and to atoms with $Z > 137$ as well.

TGD predicts correctly the value of the parameter v_0 assuming that cosmic strings and their decay remnants are responsible for the dark matter. The harmonics of v_0 can be understood as corresponding to perturbations replacing cosmic strings with their n-branched coverings so that tension becomes n^2-fold: much like the replacement of a closed orbit with an orbit closing only after n turns. $1/n$-subharmonic would result when a magnetic flux tube split into n disjoint magnetic flux tubes.

Long ranged classical electro-weak and color gauge fields are unavoidable in TGD framework. The smallness of the parity breaking effects in hadronic, nuclear, and atomic length scales does not however seem to allow long ranged electro-weak gauge fields. The problem disappears if long range classical electro-weak gauge fields are identified as space-time correlates for massless gauge fields created by dark matter. The identification explains chiral selection in living matter and unbroken $U(2)_{ew}$ invariance and free color in bio length scales become characteristics of living matter and of bio-chemistry and bio-nuclear physics. An attractive solution of the matter antimatter asymmetry is based on the identification of also antimatter as dark matter.

Consciousness and cosmology represents a rather weird association from the point of view of materialistically inclined cosmologist. p-Adic physics of cognition however predicts that cognitive consciousness is unavoidably a cosmic phenomenon as far its space-time correlates are considered. Magnetic flux tube hierarchy provides the template for the evolution of conscious, intelligent systems in all length scales in TGD Universe, and bio-systems are predicted to possess magnetic bodies of astrophysical size. Adding to this the enormous spectrum of non-deterministic vacuum extremals (with respect to inertial energy) of field equations allowing interpretation as space-time correlates of intentional action, one has good motivations for a serious consideration of the possibility that intentionality might be realized in astrophysical length scales. There is even some evidence that Sun might act as an intentional system. These speculations are not empty since rather dramatic testable phenomena are predicted. The model explains also the anomalous acceleration of spacecrafts and the finding that the age distribution of stars in a given galaxy does not seem to depend on the age of the galaxy.

Part I

OVERVIEW ABOUT QUANTUM TGD

Chapter 2

Overall View About Evolution of TGD

2.1 Introduction

Topological Geometrodynamics was born for twenty five years ago as an attempt to construct a Poincare invariant theory of gravitation by assuming that physically allowed space-times are representable as surfaces in the space $H = M^4 \times CP_2$, where M^4 denotes Minkowski space and CP_2 is complex projective space having real dimension four (see the appendix of the book).Poincare group was identified as the isometry group of M^4 rather than of the space-time surface itself. The isometries of CP_2 were identified as color group and the geometrization of electro-weak gauge fields and elementary particle quantum numbers was achieved in terms of the spinor structure of CP_2. Rather remarkably, for a quarter century after this discovery one can still say that CP_2 codes the known elementary particle quantum numbers and interactions in its geometry. The construction of quantum theory suggests the replacement of M^4 with M^4_+, the interior of the future light cone of Minkowski space so that Poincare invariance is broken by the global geometry of the light cone but not locally.

It took almost half decade to develop the new view about space-time implied by the basic hypothesis: this is summarized in my PhD thesis [n3]. The construction of a mathematical theory around these physically very attractive ideas became the basic challenge and I have devoted my professional life to the realization of this dream. The great idea was that quantum physics reduces to the construction of Kähler metric and spinor structure for the infinite-dimensional space CH of all possible 3-surfaces of H. Physical states correspond to classical spinor fields in this space and a natural geometrization of fermionic statistics in terms of gamma matrices emerges [B1, B2, B3].

p-Adic number fields R_p [cc3] (one number field for each prime obtained as a completion of the rational numbers) emerged for about ten years ago as a separate thread only loosely related to quantum TGD. What made them so attractive was that, with certain additional assumptions about physically favored p-adic primes, it became possible to understand the basic elementary particle mass scales number theoretically. This led to a successful calculation of the elementary particle masses using p-adic thermodynamics assuming that Super Virasoro algebra and related Kac Moody algebras, which are also basic algebraic structures of string models, act as symmetries of TGD [F2, F3, F4, F5]. The success of the mass calculations in turn forced the attempts to understand how Super Virasoro and related symmetries might emerge from basic TGD. Several trials led finally to the realization that these super algebras (or actually the proper generalizations of them) are the basic symmetries of quantum TGD. One of the most dramatic predictions is the uniqueness of the space H: quantum TGD exists mathematically (cancellation of various infinities occurs) only for the space $M^4_+ \times CP_2$, the choice which is forced also by the cosmological and symmetry considerations. One can say that infinite-dimensional Kähler geometric existence and thus physics is unique.

A third thread to the development emerged when I started systematic development of TGD inspired theory of consciousness [TGDconsc]. This work has led to dramatic increase of understanding also at the level of basic quantum TGD and allowed to develop quantum measurement theory in which conscious observer is not anymore Cartesian outsider but an essential part of quantum physics. The need to understand the mechanism making bio-systems macroscopic quantum systems led to a dramatic progress in the understanding of the new physics implied by the notion of many-sheeted space-time. Dramatic change in views about the relation between subjectively experienced and geometric time of physicist emerges and leads to the solution of the basic paradoxes of quantum physics. It became also clear that p-adic numbers are indeed an absolutely essential element of the mathematical formulation of quantum TGD proper and that the general properties of quantum TGD force the introduction of the p-adic numbers. One can say that physics involves both real and p-adic number fields with real numbers describing the topology of the real world and various p-adic number fields serving as correlates of cognition with the prime p labelling the p-adic topology serving as kind of intelligence quotient.

A further thread into the development of ideas came from the realization that physics might be basically number theory in generalized sense. TGD more or less forces the notion of infinite primes [E3], and it turned out that their construction reduces to a repeated second quantization of arithmetic quantum field theory. Generalization of the concept of integer and real number emerges implying that the configuration space and state space of TGD could be imbedded into the field of generalized reals which is infinite-dimensional algebraic extension of ordinary reals. Physics could be basically theory of generalized reals! The dimensions of space-time *resp.* imbedding space correspond to the dimensions of quaternion *resp.* octonion fields as well as the dimensions of algebraic extensions of $p > 2$- *resp.* 2-adics allowing square root of ordinary p-adic number. The discussions with Tony Smith suggested that one can endow space-time and imbedding space with what might be called local quaternion and octonion structures.

This stimulated a development, which led to the notion of number theoretic compactification. Space-time surfaces can be regarded either as hyper-quaternionic , and thus maximally associative, 4-surfaces in M^8 or as surfaces in $M^4 \times CP_2$ [E2]. What makes this duality possible is that CP_2 parameterizes different quaternionic planes of octonion space containing a fixed imaginary unit. Hyper-quaternions/-octonions form a sub-space of complexified quaternions/-octonions for which imaginary units are multiplied by $\sqrt{-1}$: they are needed in order to have a number theoretic norm with Minkowski signature.

Further important number theoretical ideas emerged from the attempt to construct a model for how intentions are transformed to actions. The process was interpreted as a quantum jump in which p-adic space-time sheet representing intention is transformed to a real one. This model led to a bundle of ideas and conjectures.

a) The core idea is the generalization of the notion of number obtained by gluing all number number fields together along rationals and algebraic numbers common to them. This means a generalization of the notion of manifold. In particular, imbedding space is obtained by gluing real and p-adic imbedding spaces together along rational points. This picture also justifies the decomposition of space-time surface to real and p-adic space-time sheets. Also finite-dimensional algebraic extensions, even extensions involving transcendentals like e are needed.

b) p-Adic space-time sheets are identified as correlates of intentionality and cognition. The differences between real and p-adic topologies (two rationals near to each other as p-adic numbers are very far in real sense) have deep implications concerning the understanding of cognitive consciousness. The evolution of cognition corresponds naturally to the increase of the p-adic prime and dimension of the extension of p-adic numbers.

c) Real physics and various p-adic physics are obtained from finitely extended rational physics by algebraic continuation to p-adic number fields and their extensions analogous to analytic continuation in complex analysis. This algebraic continuation is performed both at space-time level, state space level, and configuration space level. One can also generalize the notion of unitarity and the generalization poses extremely strong conditions on S-matrix.

This chapter represents a overall view of classical TGD, a discussion of the p-adic concepts, a summary of the ideas generated by TGD inspired theory of consciousness, and the vision about physics as generalized number theory.

2.2 Evolution of classical TGD

The TGD based space-time concept means a radical generalization of standard views already in the real context. Many-sheetedness means a hierarchy of space-time sheets of increasing size making possible to understand the emergence of structures in terms of the macroscopic space-time topology.

The non-determinism of the Kähler action forces the notion of the association sequence defined as a union of space-like 3-surfaces with time-like separations: association sequence provides a geometric correlate for thought as simulation of the classical history. Non-determinism forces also the notion of mind like space-time sheet defined as a space-time sheet having finite temporal duration, which is an attractive candidate for the geometric correlate of self. Topological field quantization means that space-time topology provides classical correlates for the basic notions of the quantum field theory. The decomposition of space-time surface into real and p-adic regions brings in besides the matter also cognitive representations of material world.

2.2.1 Quantum classical correspondence and why classical TGD is so important?

In standard quantum physics classical theory is seen as a result of some kind of approximation procedure, say stationary phase approximation. In TGD framework classical physics is an exact part of quantum physics, and even more of configuration space geometry since, apart from the complications caused by the classical non-determinism of the Kähler action, the definition of the Kähler geometry in terms of Kähler action assigns to a given 3-surface X^3 a unique space-time surface $X^4(X^3)$.

The evolution of TGD inspired theory of consciousness has gradually led to the notion of quantum classical correspondence which states that every quantum aspect of existence has space-time correlate. The correspondence is certainly not faithful but rather like the representation of contents of consciousness provided by spoken or written language. Space-time surface can be indeed seen as a symbolic representation, kind of written language. Not only the characteristics of quantum states, but also quantum jumps and their sequences defining the contents of conscious experience, have space-time correlates made possible by the classical determinism of the Kähler action, and the inherent p-adic non-determinism of p-adic counterparts of the field equations. In fact, there are reasons to believe that classical non-determinism of the Kähler action and a p-adic non-determinism have close relationship in the sense that the effective topology of the real space-time sheets is expected to correspond to p-adic topology in some length scale range.

2.2.2 Classical fields

In TGD framework the physics of classical fields are an essential part of the quantum theory and the study of classical fields has provided the easiest manner to get grasp about the physics of TGD Universe.

Geometrization of classical fields and of quantum numbers

The basic motivation for TGD was provided by the finding that known interactions at classical level and quantum number spectrum of known particles could be readily understood from the assumption that space-time is a 4-surface in $H = M^4 \times CP_2$.

The geometrization of classical gauge fields is based on the following identifications.

a) The classical gravitational field is identified as the induced metric. The still open question is whether the classical gravitational fields couple to matter with the gravitational constant $G \simeq kR^2$, $k \simeq 10^{-8}$, where R is CP_2 size (the length of CP_2 geodesic line). There is however an argument leading to a precise and correct prediction for k, and fixing the value of the Kähler coupling strength α_K at electron length scale to a value very near to that of the fine structure constant.

b) The geometrization of electro-weak gauge fields reduces to the curvature of CP_2 just like the geometrization of gravitation reduces to the curvature of the space-time surface. Classical electro-weak fields are identified as components of CP_2 spinor connection projected to the space-time surface. The holonomy group of CP_2 spinor connection is $U(2)$ and naturally identifiable as electro-weak gauge group.

c) Color symmetries correspond to the isometries of CP_2 so that there is deep and unexpected connection between electro-weak and color interactions. Color gauge potentials are identified in the spirit of Kaluza-Klein theory as projections of the Killing vector fields of color isometries to the space-time surface. Color gauge fields are of form $F^A_{\alpha\beta} \propto H^A \times J_{\alpha\beta}$, where H^A is the Hamiltonian of the color isometry and J denotes the induced Kähler form. Therefore the vacuum extremals of Kähler action carry also non-vanishing color gauge fields.

Also elementary particle quantum numbers can be understood in terms of the induced spinor structure and simple 3-topology.

a) CP_2 does not allow ordinary spinor structure and it is necessary to couple CP_2 spinors to the Kähler potential of CP_2. The couplings are different for different H-chiralities identifiable as leptonic and quark like spinors. Baryon and lepton numbers are separately conserved for both the ordinary massless Dirac action and modified Dirac action. The modified Dirac action is fixed uniquely by requiring that it has the vacuum degeneracy of Kähler action. The modified Dirac action allows local super-symmetries generated by the right-handed neutrino.

b) At the fundamental level color quantum numbers are not spin like quantum numbers but can be said to correspond to the color partial waves in CP_2 center of mass degrees of freedom of the 3-surface representing the elementary particle. Ordinary Dirac equation for CP_2 predicts wrong correlations between electro-weak and color quantum numbers of the color partial waves associated with the spinor harmonics. This was a longstanding problem of TGD approach but the construction of physical states as representations of the Super Kac Moody algebra allows to obtain correct correlations and an interpretation in terms of electro-weak symmetry breaking coded already into the CP_2 geometry.

c) The first guess was that the genus of the two-dimensional boundary associated with the 3-surface representing particle explains family replication phenomenon. The identification of the super-conformal symmetries as symmetries associated with light like effectively 2-dimensional 3-surfaces X_l^3 acting as causal determinants suggests a more concrete identification.

Quaternion conformal invariance allows to assign to X_l^3 a highly unique 2-dimensional surface X^2 as a surface at which superconformal structure reduces to ordinary conformal structure and thus becomes Abelian. The genus of this surface telling whether the surface is sphere, torus, etc... determines the particle family. X_l^3 could correspond to either a boundary of 3-surface or to an elementary particle horizon. Elementary particle horizon would surround the wormhole contact connecting CP_2 extremal with an Euclidian signature of the induced metric to a larger space-time sheet with a Minkowskian signature of metric. The induced metric is degenerate at the elementary particle horizon so that this surface is indeed metrically 2-dimensional.

More concretely, sphere, torus, and sphere with two handles would correspond to (e, ν_e), (μ, ν_μ), (τ, ν_τ) in the leptonic sector and and (u, d), (c, s), and (t, b) in the quark sector respectively. The experimental absence of heavier particle families would be most naturally due to the fact that they are extremely heavy. The 3 lowest particle families differ from the higher genera in the sense that 2-surfaces with genus $g < 3$ are always hyper-elliptic, that is they allow always Z_2 conformal symmetry, whereas higher genera generically do not allow any conformal symmetries. Hyper-ellipticity is an excellent candidate for an explanation of the lightness of $g < 3$ genera. The construction of elementary particle functionals as functionals in the conformal equivalence classes of the 2-surface X^2 associated with X_l^3 allows to formulate this argument more precisely.

The explanation of Cabibbo mixing as being due to the mixing of boundary topologies, and number theoretic arguments (complex rationality of CKM matrix) lead to a highly unique CKM matrix for quarks and also leptonic mixings can be fixed highly uniquely. Also bosons are predicted to possess family replication phenomenon.

The new physics associated with classical gauge fields

Long range electro-weak, in particular Z^0, vacuum gauge fields are unavoidable in TGD: this is a necessary outcome of the induced gauge field concept reducing the number of the primary bosonic field variables to four (CP_2 coordinates)! The interpretation of this puzzling prediction has been a long standing challenge of TGD. There are three alternative options to consider.

Option I: Classical gauge fields are space-time correlates for gauge bosons with mass scale determined by the p-adic length scale of the space-time sheet in question. The electro-weak charges of elementary particles are screened by vacuum gauge charges (possible in TGD) in a region of size L_W of order intermediate boson length scale. This option does not explain the presence of long range electro-weak gauge fields unavoidably present if the dimension of CP_2 projection of space-time sheet is higher than 2 nor classical color gauge fields present for non-vacuum extremals.

Option II: Electro-weak gauge charges are not screened in the length scale L_W and the gauge fluxes of elementary particles flow to larger space-time sheets via # throats within region of size L_W and elementary particles have the quantized values of em Z^0 charges. The problem for this option are anomalously large Rutherford cross sections in condensed matter and large parity breaking effects in hadronic, nuclear, and atomic length scales. Despite this I regarded this option as the most realistic one until the realization that the mysterious long ranged weak fields could be assigned to dark matter particles at various space-time sheets.

Option III: There is a hierarchy of color electro-weak physics such that weak bosons are massless below the p-adic length scale determining the mass scale of weak bosons. Classical long range gauge fields serve as space-time correlates for gauge bosons below the p-adic length scale in question.

The unavoidable long ranged electro-weak and color gauge fields are created by dark matter and dark particles can screen dark nuclear electro-weak charges below the weak scale above which vacuum screening occurs as for ordinary weak interactions. Dark gauge bosons are massless below the appropriate p-adic length scale but massive above it and $U(2)_{ew}$ is broken only in the fermionic sector. For dark copies of ordinary fermions masses are essentially identical with those of ordinary fermions.

This option is consistent with the standard elementary particle physics for visible matter apart from predictions such as the possibility of p-adically scaled up versions of ordinary quarks predicted to appear already in ordinary low energy hadron physics. The most interesting implications are seen in longer length scales. Dark quarks and gluons and a scaled up copy of ordinary gluons emerge already in ordinary nuclear physics [F8] and explain some recently discovered anomalies such as neutron halos and tetraneutron. The field bodies associated with are predicted to have sizes of order atom size. Also scaled down versions of weak bosons giving to interactions between exotic quarks with a range of order atomic length scale are predicted.

The new nuclear physics has deep implications for chemistry and condensed matter where color bonds between neighboring atoms might be part of the chemical bonding [F9]. Long ranged repulsive weak force behind exotic quarks compensated by color force would contribute to the repulsive force assumed in van der Waals equations of state for condensed matter. No strong isotopic dependence is predicted.

Classical long range weak and color forces become also key players at the level of molecular physics and biophysics. Chiral selection of bio-molecules can be seen as one direct signature of the long ranged weak force which suggests that non-broken $U(2)_{ew}$ symmetry and and free color in bio length scales become characteristics of living matter and of bio-chemistry and bio-nuclear physics. The central role of the long ranged weak forces in bio-systems and in pre-biotic evolution is discussed in [TGDware, M3, TGDgeme].

Classical em fields and Z^0 fields are not invariant under color rotations acting as exact symmetries and are accompanied by classical color gauge fields. This implies new physics potentially important for TGD inspired theory of consciousness. For instance, in TGD Universe the original joke like term "quark color" inspired by certain algebraic similarities ceases to be a joke since it is possible to reduce the 3+3 primary colors in color vision to the 3+3 different increments of color quantum numbers induced by the absorption or emission of color octet gluon.

2.2.3 Many-sheeted space-time concept

The detailed study of TGD led to a further generalization of the space-time concept and the end result is what I have used to call topological condensate or many-sheeted space-time. The 3-space is many-sheeted such that the sheets of 3-space have finite size and outer boundary. The physical interpretation of a given space-time sheet of a finite size is as a 'particle'. Depending on their size, these particles correspond to elementary particles, nucleons, atomic nuclei, atoms, molecules, cells, ourselves, stars, galaxies, etc. For instance, my skin corresponds to the outer boundary of a 3-surface glued to a larger 3-surface identifiable as the room in which I sit! I am a small Universe glued to a larger one, the 3-space associated with me literally ends on my skin just as string ends at its end! The surface of earth, the outer surfaces of trees, etc...: everywhere I can see nontrivial 3-topology.

Important new physics is associated with the extremely tiny wormholes contacts with size of order CP_2 length needed to perform the gluing operation. Join along boundaries bonds serving as space-time correlates for the bound state formation is second important notion. The larger sheets of the many-sheeted space-time are ideal for carrying various macroscopic quantum phases. Topological field quantization allows to define precisely the notions of coherence and de-coherence and also means that one can assign to a given material system what might be called field body or magnetic body.

Obviously the outcome is a thorough-going generalization of the space-time concept and means that TGD has highly nontrivial consequences in all length scales rather than in particle physics only, as one might naively expect.

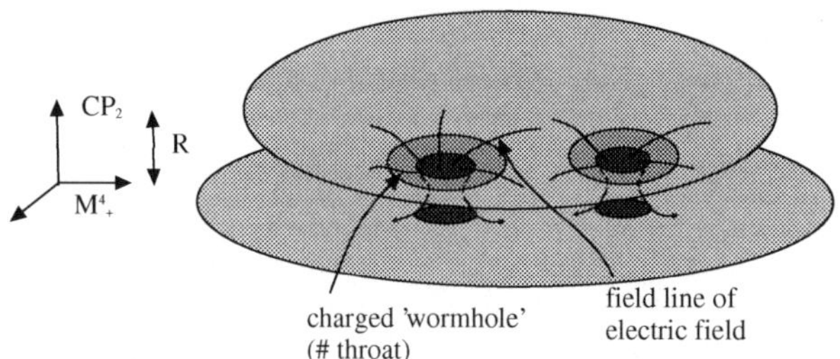

Figure 2.1: Charged wormholes feed the electromagnetic gauge flux to the 'lower' space-time sheet.

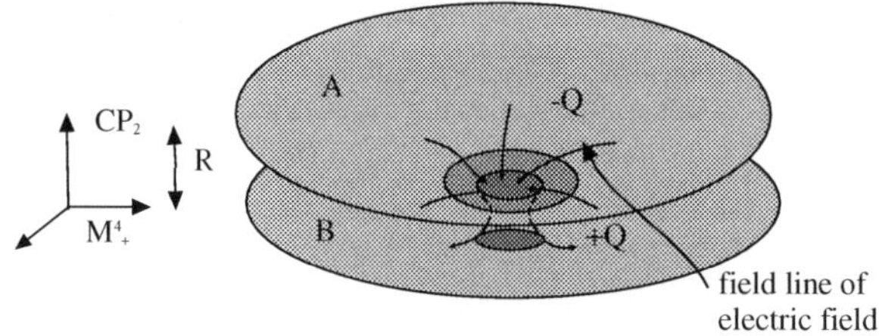

Figure 2.2: The two throats of wormhole behave as classical charges of opposite sign.

Join along boundaries contacts and join along boundaries condensate

The recipe for constructing 3-space is simple. Take 3-surfaces with boundaries, glue them by topological sum to larger 3-surfaces, glue these 3-surfaces in turn on even larger 3-surfaces, etc.. The smallest 3-surfaces correspond to CP_2 type extremals that is elementary particles and they are at the top of hierarchy. In this manner You get quarks, hadrons, nuclei, atoms, molecules,... cells, organs, ..., stars, ..,galaxies, etc...

Besides this, one can also glue different 3-surfaces together by tubes connecting their *boundaries*: this is just connected sum operation for boundaries. Take disks D^2 on the boundaries of two objects and connect these disks by cylinder D^2xD1 having D^2:s as its ends. Or more concretely: let the two 3-surfaces just touch each other.

Depending on the scale join along boundaries bonds are identified as color flux tubes connecting quarks, bonds giving rise to strong binding between nucleons inside nuclei, bonds connecting neutrons inside neutron star, chemical bonds between atoms and molecules, gap junctions connecting cells, the bond which is formed when You touch table with Your finger, etc.

One can construct from a group of nearby disjoint 3-surfaces so called join along boundaries condensate by allowing them to touch each other here and there.

The formation of join along boundaries condensates creates clearly strong correlation between two quantum systems and it is assumed that the formation of join along boundaries condensate is necessary prerequisite for the formation of *macroscopic quantum systems*. Crucially important examples in biology are gap junctions connecting cells and MAPs (micro-tubule associated proteins) connecting micro-tubules.

Quantum classical correspondence inspires the hypothesis that quite generally join along boundaries bonds are space-time correlates for the formation of the bound state entanglement. Since join along boundaries bonds between space-time sheets condensed on larger space-time sheets having no join along boundaries bonds between them is possible, one is forced to conclude that entanglement between sub-systems of un-entangled systems is possible in the many-sheeted space-time. The paradox disappears when entanglement is understood as a length scale dependent notion so that the bound

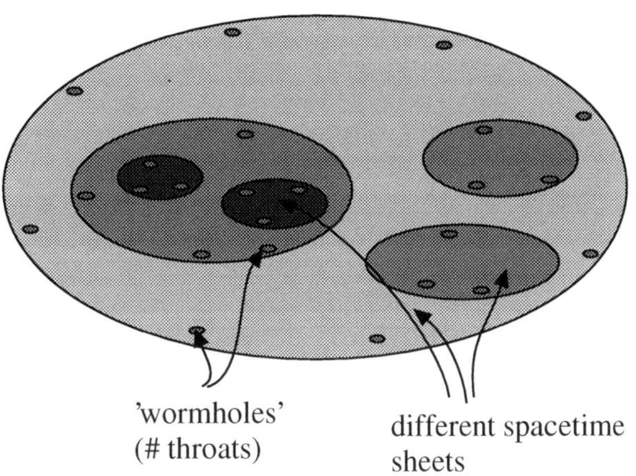

Figure 2.3: Many-sheeted space-time structure results from the requirement of gauge flux conservation.

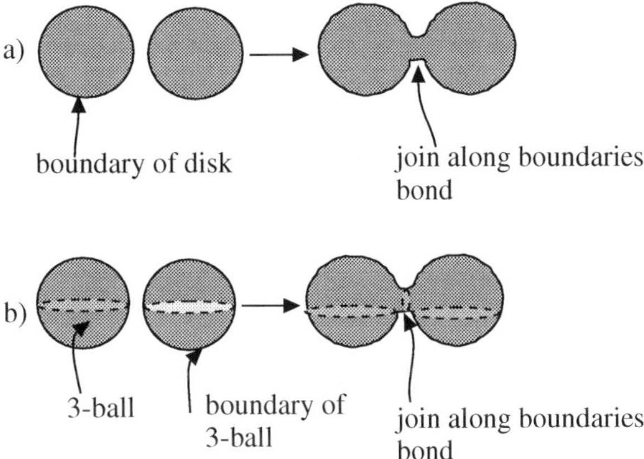

Figure 2.4: Join along boundaries bond a): in two dimensions and b): in 3-dimensions for solid balls.

state entanglement of sub-systems is not visible in the length and time scales of the systems.

Wormhole contacts

The gauge and gravitational fluxes at the boundary of a given space-time sheet must go somewhere by gauge flux conservation. This forces the existence of a larger space-time sheet and of tiny wormhole contacts connecting the two space-time sheets and feeding the gauge fluxes from the smaller sheet to the larger one. Wormhole contacts (# contacts) are elementary particle like objects (actually deformed pieces of so called CP_2 type extremals) having size of order CP_2 size about 10^4 Planck lengths and, being sources and sinks of gauge field lines, wormhole throats effectively like classical charges, the charges of throats at the two space-time sheets being of opposite sign. Hence wormhole contacts look like dipoles and couple to the difference of the classical gauge potentials associated with the two space-time sheets. Also the coupling to the difference of the gauge potentials serving as order parameters for the coherent states of photons is possible.

The crucial experiment would be the one demonstrating the existence of the wormholes.

a) There are good reasons to expect that wormhole gauge flux is quantized. The reason for quantization would be the absolute minimization of Kähler action (or its variant discussed in [E2]), which is mathematically a condition very similar to the Bohr's quantization condition. In the usual initial value problem one would fix only the imbedding space coordinates of 4-surface for given value of time and allow their time derivatives be arbitrary. Now absolute minimization fixes the values of the time derivatives just like Bohr's quantization rules fix the momenta. The most aesthetic possibility is that the unit of wormhole em charge is the smallest possible elementary particle charge 1/3 associated with d quarks but also integer charge could be considered.

b) If wormhole charge is quantized then the gauge flux of an external em field running from a larger space-time sheet to a smaller one is quantized. The experimental arrangement should demonstrate that this flux indeed can change by a multiple of the elementary flux only. One could also try to detect wormhole currents. It must be emphasized that wormhole current is a pseudo current in the sense that two space-time sheets carry opposite classical currents. These currents are created, when magnetic field penetrates from space-time sheet to another. The detection of charge 1/3 for the charge carriers of this current would be a triumph.

c) One cannot exclude the possibility that the recently found evidence for 1/3 charge in condensed matter systems (quantum Hall effect) could be interpreted in terms of an em gauge flux quantized in this manner. Electron current flowing inside a planar layer like structure is studied. Strong magnetic field, which could lead to a generation of wormhole currents is present! Evidence for some quasi particles in current flow possessing this charge has been found. The anyon interpretation of quasi particles as bound states of magnetic flux quanta and electrons explains the effect (McLaughlin wave function). The prediction is however that also fluxes of m/5, m/7,... , m integer, should be observed. Only 1/3 has been detected hitherto and it is not understood why higher charges have not been observed. The question is whether the quasi particles are actually wormholes created by the penetration of magnetic field and flowing along the boundaries of the arrangement.

One application of the new space-time concept is a model of brain. The basic idea is that brain can be regarded as a macroscopic quantum system and that our experiences of free will correspond to quantum jumps which are unpredictable as also is the end result of a free choice. The idea that quantum theory might provide some light in the problem of consciousness has become popular during the last years and a serious building of quantum theories of consciousness has begun. The bottleneck problem is how the brain can be a macroscopic quantum system. Some kind of super conductivity looks a promising idea but standard physics does not provide promising candidates for a super conductor like system. Womholes might provide one such system besides high T_c electronic and protonic superconductors and Bose-Einstein condensates of bosonic ions.

To see what is involved, consider in more precise manner how many sheeted 3-space is constructed. When one glues a sheet of 3-space to a larger sheet of 3-space one does it by constructing extremely tiny elementary particle sized wormholes connecting the two sheets of 3-space.

These wormholes serve important function. For instance, the flux of the electric field (usually it is unlucky space traveller) flows to this kind of wormhole on the smaller sheet of 3-space and and comes back from it to the larger sheet of 3-space. Since the field lines of the electric field flow to the wormhole on the smaller sheet of 3-space, the wormhole looks like a charge since it acts as a sink of field lines. Same applies on the larger sheet of 3-space except that the sign of the charge is opposite. Hence, on both space-time sheets wormhole looks classically like a charged particle. Shortly, wormholes behave like particles and represent a new exotic form of matter. More generally, it seems that many-sheeted nature of the space-time is crucial for the understanding of a bio-system as a macroscopic quantum system.

The interaction between space-time sheets is mediated by these wormholes having size of order CP_2 radius R and located near the boundaries of the smaller space-time sheet. Wormholes feed various gauge fluxes from the smaller space-time sheet to the larger one (say from the atomic sheet to some molecular sheet). p-Adic considerations suggest that wormholes are light having mass of order $1/L_p$: this implies that they suffer Bose-Einstein condensation on the ground state. One could even say that space-time sheets "perceive" the external world and act on it with the help of the charged wormhole BE condensates near their boundaries. Wormholes provide a very general mechanism making possible the transfer of classical electromagnetic fields and various quantum numbers such as energy, momentum and angular momentum, between different space-time sheets and bio-systems are especially promising as far as applications are considered.

Topological field quantization

Topological field quantization [D7] implies that various notions of quantum field theory have rather precise classical analogies. Topological field quantization is basically implied by the compactness of

CP_2, which typically implies that a given Maxwell field allows only a partial imbedding as a space-time surface in H. One can say that magnetic fields, electric fields and radiation fields decompose into field quanta.

The energies and other classical charges of the topological field quanta are quantized by the absolute minimization of the Kähler action making classical space-time surfaces the counterparts of the Bohr orbits. Feynman diagrams become classical space-time surfaces with lines thickened to 4-manifolds. For instance, "massless extremals" representing topologically quantized classical radiation fields are the classical counterparts of gravitons and photons. Topologically quantized non-radiative nearby fields give rise to various geometric structures such as magnetic and electric flux tubes.

Topological field quantization provides the correspondence between the abstract Fock space description of elementary particles and the description of the elementary particles as concrete geometric objects detected in the laboratory. In standard quantum field theory this kind of correspondence is lacking since classical fields are regarded as a phenomenological concept only.

Topological field quanta define coherence regions for the classical gauge fields and induced spinor fields and classical coherence is the prerequisite of the quantum coherence. Whether and how macroscopic and macro-temporal quantum coherence are possible in living matter is the basic question of quantum consciousness theories and quantum biology. In TGD this question is even more difficult since the first estimate for de-coherence time is CP_2 time which is about 10^4 Planck times. The length scale hierarchy of space-time sheets allows immediately to understand at the level of space-time correlates how macroscopic and macro-temporal quantum coherence are possible. A good order of magnitude guess for the zero point energy of a particle at a space-time sheet of size L is given by $E = \pi^2/2mL^2$. $T \leq \pi^2/2mL^2$ gives an estimate for the temperature of the space-time sheet populated by particles of mass m: the larger the size of the space-time sheet, the lower the temperature. Superconductivity and various macroscopic phenomena become thus possible at larger space-time sheets. TGD based model of living matter is based on the hypothesis that large space-time sheets are responsible for quantum control.

The virtual particles of quantum field theory have also classical counterparts. In particular, the virtual particles of quantum field theory can have negative energies: this is true also for the TGD counterparts of the virtual particles. The fundamental difference between TGD and GRT is that in TGD the sign of energy depends on the time orientation of the space-time sheet: this is due to the fact that in TGD energy current is vector field rather than part of tensor field. Therefore space-time sheets with negative energies are possible.

One can criticize the notion of time orientation. A more precise definition of the time orientation requires the realization that configuration space of 3-surfaces, call it CH, can be understood as a union of corresponding configuration spaces associated with unions of arbitrary many light cones, both future and past light cones with positive/negative energies assignable to to future/past lightcones. This brings in in a natural manner also the super-canonical symmetries associated with the boundaries of the lightcones.

Negative energies would have quite dramatic technological consequences: consider only the possibility of generating energy from vacuum and classical signalling backwards in time along negative energy space-time sheets [G1]. Also bio-systems might have invented negative energy space-time sheets: in fact, they define the basic mechanism for the realization of intentional action, long term memory, and metabolism [I5].

Quantum classical correspondence suggests that quantum entanglement has the formation of the join along boundaries bonds as its geometric correlate. The superposition of the topologically quantized space-time surfaces in the state $U\Psi$ could be regarded as a geometric correlate for quantum fields: creation/annihilation operators would correspond to positive/negative energy space-time sheets. This hypothesis, together with the expansion of the interacting quantum field in terms of creation and annihilation operators, would make it possible to make quantitative estimates about the fraction of energy density carried by the negative energy space-time sheets, in particular, about the energy density associated with the massless extremals.

In TGD Universe topological field quanta serve as templates for the formation of the bio-structures. Thus topologically quantized classical electromagnetic fields associated with the material objects, field bodies or more concretely, magnetic bodies, could be equally important for the functioning of the living systems as the structures formed by the visible bio-matter and the visible part of bio-system might represent only a dip of an ice berg. For instance, in [L1] the implications of the notion of field body for the understanding of bio-systems and pre-biotic evolution are discussed in detail.

Negative energy space-time sheets and new view about energy

Negative energy space-time sheets represents an important distinction between TGD and standard physics. They are possible because energy momentum tensor is replaced by a collection of conserved currents associated with various components of four momentum. This resolves the energy problem of general relativity but, since the sign of the conserved charged depends on the time orientation of the space-time sheet, the sign of energy is not positive definite anymore.

Quantum classical correspondence implies that also elementary particles can have negative energies and this means a new kind of physics. It seems that this physics has been already discovered: the strange properties of phase conjugate laser waves can be understood if they consist of negative energy photons.

Negative energy space-time sheets have far reaching implications for TGD inspired theory of consciousness. The so called time mirror mechanism involves the reflection of negative energy signals sent to the geometric past from population inverted lasers as amplified positive energy signals propagating to the geometric future. Time mirror mechanism provides the holy grail to the understanding of the mechanisms of brain functioning and also of the workings of the living matter. There are obvious implications for communication and energy technologies since negative energy signals could make possible instantaneous remote sensing and quantum control over arbitrarily long distances so that light velocity would cease to be a restriction forcing us to be habitants of 3-space instead of space-time.

If Kähler action were strictly deterministic, the only possible choice for H would be $M_+^4 \times CP_2$. Together with negative energies the classical non-determinism of the Kähler action it is possible to assume that imbedding space is $M^4 \times CP_2$ meaning exact Poincare invariance. The point is that generation of pairs of positive and negative energy space-time sheet at light-like 7-surfaces $X_l^3 \times CP_2$ means emergence of new kind of causal determinants generalizing the light cone boundary $\delta M_+^4 \times CP_2$ as a fundamental causal determinant. All states of the Universe have vanishing net quantum numbers and everything in the Universe would have been pair-created from vacuum. Future light cones containing positive energy could also be created when negative energy radiation (in particular gravitons) is generated and propagates to the geometric past and leaks from the future light cone. This vision can be applied also to the second quantization of fermions by giving fermions and anti-fermions opposite energies. Depending on time orientation either fermions or anti-fermions have negative energy.

By crossing symmetry the assumption that the net quantum numbers of the Universe vanish is not in conflict with elementary particle physics. In macroscopic length scales the identification of the gravitational energy as the difference of inertial (Poincare) energies of positive and negative energy matter plus the possibility that negative and positive energy matter interact weakly allows to understand why western view about objective reality with conserved positive total energy is so good an approximation. The non-conservation of the gravitational energy can be understood, and vacuum extremals, of which Robertson-Walker cosmologies, are most important examples find interpretation. The non-determinism of Kähler action explains naturally the fact that Universe is to some extended an outcome of engineering. The notion of gravitational energy generalizes to that of gravitational quantum numbers and the inertial-gravitational dichotomy is a direct correlate for the geometric-subjective dichotomy for time discovered while developing TGD inspired theory of consciousness. Indeed, positive and negative energy space-time sheets correspond to initial and final states of quantum jump so that gravitational quantum numbers characterize changes.

This vision would resolve the unpleasant philosophical questions like "What is the total fermion number of the Universe". One could see entire universe as a result of intentional actions in which intentions represented by p-adic space-time sheets are transformed to actions represented by real space-time sheets. Everyone knows the anecdotes about yogis and gurus creating material objects from nothing and very few "scientifically thinking" westerner can take these stories really seriously. Whether or not these stories are true, they might however express a deep truth about reality.

More precise view about topological condensate

The challenge is to define precisely the concepts like classical gauge charge, gauge flux, wormhole contacts, join along boundaries bonds, topological condensation and evaporation, etc... Number theoretical vision allows to achieve this goal [F6, D2].

The crucial ingredients in the model are so called CP_2 type vacuum extremals. The realization that

\# contacts (topological sum contacts and $\#_B$ contacts (join along boundaries bonds) are accompanied by causal horizons which carry quantum numbers and allow identification as partons leads to a more detailed articulation of these notions.

The partons associated with topologically condensed CP_2 type extremals carry elementary particle vacuum numbers whereas the parton pairs associated with \# contacts connecting two space-time sheets with Minkowskian signature of induced metric define parton pairs. These parton pairs do not correspond to ordinary elementary particles. Gauge fluxes through \# contacts can be identified as gauge charges of the partons. Gauge fluxes between space-time sheets can be transferred through \# and $\#_B$ contacts concentrated near the boundaries of the smaller space-time sheet. The dynamics of topological condensation and evaporation can be formulated in terms of gauge interactions of partons and splitting and fusion of CP_2 type extremals. This picture generalizes to the case of gravitational flux which need not be well-defined purely classically.

Number theoretical vision and p-adic length scale hypothesis allow to quantify this picture and lead to an overall view about interactions of particles in many-sheeted space-time. A far reaching generalization of standard physics results predicting an infinite hierarchy of dark matters besides ordinary elementary particles of standard model. In particular, the partons associated with \# and $\#_B$ contacts represent dark matter.

2.2.4 Classical non-determinism of Kähler action

The classical non-determinism of Kähler action has been deep source of inspiration and challenges and guided the evolution of TGD inspired theory of consciousness and finally also of quantum TGD proper. In nut-shell, classical non-determinism makes possible quantum-classical correspondence in the sense that space-time surface becomes a symbolic representation for the quantum states and quantum jump sequences defining conscious experience.

Matter-mind duality geometrically

The non-determinism of Kähler action implies huge vacuum degeneracy: any 4-surface whose projection belongs to $M_+^4 \times Y^2$, where Y^2 is so called Lagrange manifold of CP_2 (has vanishing induced Kähler form), is a vacuum extremal. This suggests that one must radically generalize the concept of space-time. It seems that the correct picture is roughly like follows. Space-time is many-sheeted. Each sheet can be regarded as a slightly deformed piece of M^4 in H containing smaller sheets glued to it and being itself glued to a larger space-time sheet. Gluing means the formation of topological sum contacts between the space-time sheets. There are reasons to believe that topological sum contacts, "wormhole contacts" are located near the boundaries of the smaller space-time sheet.

Material space-time sheets have infinitely long time duration if they possess non-vanishing energy (and provided that they do feed their energy to some other space-time sheets). "Mind like" space-time sheets can be regarded as obtained by gluing space-time sheets with finite time duration to material space-time sheets. The gluing operation implies that tiny amounts of energy and momenta and other conserved quantities flow to the mind like space-time sheet when it begins and back to the material space-time sheets when mind like space-time sheet ends. Mind like space-time sheets are space-time correlates for contents of consciousness. In particular, they form symbolic representations for material space-time sheets. For instances, the frequencies of various oscillatory processes are mapped also to frequencies of processes occurring in mind like space-time sheets. The possibility of mind like space-time sheets implies that the absolute minima of Kähler action (or more general preferred extremals defining analogs of Bohr orbits [E2]) are degenerate: one can glue mind like space-time sheets to given absolute minimum to get new absolute minima. This conforms with the fact that contents of consciousness are defined by a sequence of non-deterministic quantum jumps.

This picture must of course be taken with strong reservations, and one should actually state more precisely what "mind like" means. The interpretation of p-adic space-time sheets as correlates of intentions and cognitions gives some ideas about what aspects of consciousness mind like space-time sheets correlate with. The model for how intentions are realized as actions in quantum jump assumes that p-adic "topological light rays" representing intentions are transformed to real topological light rays with negative energy serving as correlates of desires, which in turn induce the action initiated in the geometric past. Thus it would seem that real "mind like" space-time sheets with negative energy would serve as correlates for desires.

The precise definition of p-adic space-time sheets is a separate question and requires a precise vision about how real and various p-adic physics integrate to a coherent whole. This requires a generalization of the number concept based on the fusion of real and p-adic number number fields to a larger book like structure along common rationals. The precise definition of p-adic space-time sheets is discussed in [E2]. The surprising outcome, basically due to the difference between real and p-adic notions of distance, is that most points of p-adic space-time sheets can be said to reside at infinity of the real imbedding space and the projection to real space-time consists of a discrete set of rational points. Thus cognition can be said to look material cosmos from outside.

Association sequence concept and a mind like space-time sheets

The vacuum degeneracy of the Kähler action defining quantum TGD solves the difficulty. The vacuum degeneracy implies spin glass analogy and strongly suggests that the Bohr orbit like space-time surface defined as a preferred extremal of Kähler action and going through a given space-like 3-surface, cannot be unique in general. To achieve uniqueness one must generalize the concept of 3-surface to what might be called association sequence . In order to specify uniquely one of the degenerate absolute minimum space-times going through a given 3-surface one must fix some minimum number, say N, of 3-surfaces on a given preferred extremal. These sequences of disjoint 3-surfaces with time-like separations can be regarded as a simulations of the classical time development and hence as a geometric correlate of conscious experience localized temporally. It seems that in real case geometric correlates of sensory experiences are in question whereas in p-adic case correlates of thoughts are in question.

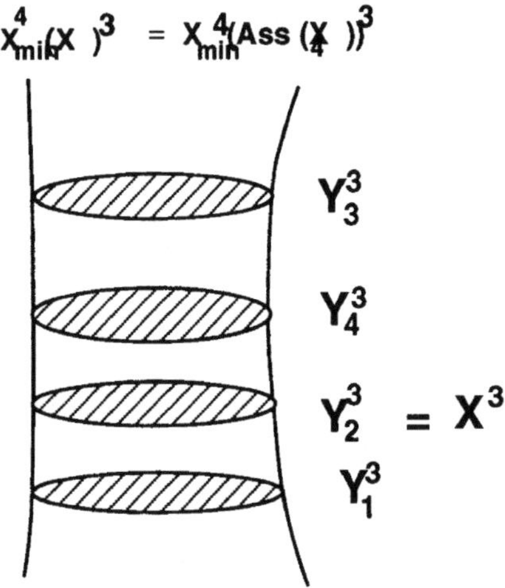

Figure 2.5: 'Association sequence': a geometric model for thought as a sequence of disjoint 3-surfaces with time-like separations.

Association sequences are very probably not all that is needed to overcome the complications caused by the non-determinism of Kähler action. The enormous vacuum degeneracy of Kähler action suggests strongly that the classical non-determinism does not reduce to simple sequences of bifurcations. Hence it seems that must give up the idea of identifying space-like 3-surfaces given value of geometric time as causal determinants which are possibly degenerate because of the bifurcations.

Vacuum degeneracy and spin glass analogy

Kähler action determines configuration space geometry and is hence a cornerstone of quantum TGD. Kähler action can be regarded as a Maxwell action for the Kähler form of CP_2 induced to space-time surface and defining nonlinear Maxwell field. Kähler action possesses enormous vacuum degeneracy. Any space-time surface in $M^4_+ \times CP_2$, where Y^2 is so called Lagrange sub-manifold of CP_2 having by definition vanishing induced Kähler form, is vacuum extremal. In canonical coordinates (P_i, Q_i) for CP_2 Lagrange sub-manifolds correspond to functions

$$P_i = \nabla_i f(Q_j) \ .$$

This means that there is infinite number of vacuum sectors since all 4-surfaces in any six-dimensional space $M_+^4 \times Y^2$ are vacua.

Also non-vacuum configurations are almost degenerate. Only the gravitational effects caused by the presence of the induced metric in the Maxwell action for the induced Kähler form of CP_2 on space-time surface breaks the canonical invariance of the Kähler action. Canonical transformations of CP_2 act as $U(1)$ gauge transformations and in the absence of gravitation one would have ordinary $U(1)$ gauge invariance. Gravitation however changes the situation. Various canonically related configurations are physically *non-equivalent*. This means a characteristic degeneracy analogous to the degeneracy of the states for spin glass rather than to the physically uninteresting gauge degeneracy. The effective breaking of $U(1)$ gauge invariance makes possible vacuum charge densities, scalar wave pulses propagating with light velocity and carrying longitudinal electric field parallel to the propagation direction, and topological light rays carrying light like vacuum current and transversal electric and magnetic fields are predicted.

Contrary to the original beliefs, p-adic physics does not seem to follow from vacuum degeneracy alone. Rather, p-adic space-time topology is a genuine rather than only effective space-time topology and emerges independently from the vacuum degeneracy. p-Adic topology seems however to serve as effective topology for the real space-time sheets in the sense that the non-determinism implied by the vacuum degeneracy mimics the inherent non-determinism of p-adic field equations for some value of p so that one can indeed assign a definite p-adic prime to a given real space-time sheet. Vacuum degeneracy has a wide spectrum of implications. For instance, the spin glass degeneracy implied by it allows to understand at quantum level generation of macroscopic and macro-temporal quantum coherence. The same mechanism explains also color confinement.

The p-adic fractality of real space-time sheets is in turn implied by the fact that p-adic and real space-time sheets have common rational points which implies that the purely local p-adic physics sets constraints on the long ranged real physics because rational points close to each other p-adically are very distant in real sense.

Connection with catastrophe theory and Haken's theory of self-organization for spin glass

If the effects related to the induced metric (classical gravitation) are neglected, canonical transformations of CP_2 act as $U(1)$ gauge symmetries and all canonically related surfaces are physically equivalent. Classical gravitation however breaks this gauge invariance but due to the extreme weakness of the gravitational interaction one has good reasons to expect that the maxima of Kähler function for given values of the zero modes are highly degenerate. The hypothesis that single maximum of Kähler function with respect to fiber degrees of freedom is selected in quantum jump, means huge simplification of the mathematical theory.

Besides the degeneracy resulting from the non-determinism, there is also the spin glass degeneracy related to zero modes. The nonphysical $U(1)$ gauge degeneracy is transformed to physical spin glass degeneracy. The energies of various absolute minima differ only by the classical gravitational energy. Zero modes serve as coordinates for the "energy" landscape of quantum spin glass and the energy landscape of non-equilibrium thermodynamics is fractal containing valleys inside valleys...inside valleys.

One naturally ends up with a generalization of the catastrophe theory [aa1] to the infinite-dimensional configuration space context. Zero modes play the role of the control parameters forming master slave-hierarchy and non-zero modes characterizing various degenerate absolute minima of Kähler action correspond to the state variables [I1]. There is natural connection with the non-equilibrium thermodynamics of Haken [ka1]. Since time development by quantum jumps means hopping in the zero modes characterizing the macroscopic space-time surfaces associated with the final states of the quantum jumps, Haken's classical theory applies almost as such. Asymptotically the self-organizing quantum jumping system (self) ends up to a fixed point, limiting cycle, strange attractor, etc. near the bottom of some valley of the energy landscape. The bottom of a valley corresponds to a maximum of the Kähler function rather than minimum of free energy as in thermodynamics since vacuum functional is exponent of Kähler function. Self-organization in spin glass energy landscape by quantum jumps is extremely powerful notion allowing to understand general features of living systems.

2.2.5 Quantum classical correspondence as an interpretational guide

The overall view about interpretation of TGD can be deduced from the general properties of space-time surfaces, the notion of induced gauge field, the general properties of Kähler action, and the known extremals using quantum classical correspondence. The most dramatic predictions follow without even considering field equations in detail by using quantum classical correspondence.

The implications deriving from the topology of space-time surface and from the properties of induced gauge fields

The notions of many-sheeted space-time, topological field quantization and the notion of field/magnetic body, follow from simple topological considerations. The observation that space-time sheets can have arbitrarily large sizes and their interpretation as quantum coherence regions forces to conclude that in TGD Universe macroscopic and macro-temporal quantum coherence are possible in arbitrarily long scales. It took relatively long time to realize that perhaps the only manner to understand this is a generalization of the quantum theory itself by allowing Planck constant to be dynamical and quantized. TGD leads indeed to a "prediction" for the spectrum of Planck constants and macroscopic quantum phases with large value of Planck constant allow an identification as a dark matter hierarchy.

Also long ranged classical color and electro-weak fields are an unavoidable prediction and it took a considerable time to make the obvious conclusion: TGD Universe is fractal containing fractal copies of standard model physics at various space-time sheets and labelled by the collection of p-adic primes assignable to elementary particles and by the level of dark matter hierarchy defines as $\hbar = \lambda^k \hbar_0$, $k_d = 0, 1, \ldots$. λ depends logarithmically on p-adic length scale $L(k)$ and satisfies $\lambda \simeq 2^{11}$ in atomic length scale $L(k = 137)$. Dark space-time sheets are identifiable as space-time sheets defining locally λ^k-fold covering of M^4 factor of imbedding space.

The new view about energy and time means that the sign of inertial energy depends on the time orientation of the space-time sheet and that negative energy space-time sheets serve as correlates for communications to the geometric past. This alone leads to profoundly new views about metabolism, long term memory, and realization of intentional action.

A further important fact is that the holonomy group of induced color gauge field is Abelian. Together with quantum classical correspondences this suggests a weak form of color confinement in the sense that only color neutral states of color multiplets are realized as physical states. This would mean a weak form of color confinement.

2.3 Evolution of p-adic ideas

It took quite a long time to end up with the recent picture how p-adic numbers emerge as a basic aspect of quantum TGD and what p-adicization of TGD might mean. Of course, recent picture need not be the final yet and there are several unsolved problems. In the following the basic properties of the p-adic numbers are described shortly and then it is demonstrated how p-adic numbers might emerge from TGD and how one should formulate p-adic version of quantum TGD formalism.

2.3.1 p-Adic numbers

Like real numbers, p-adic numbers can be regarded as completions of the rational numbers to a larger number field allowing the generalization of differential calculus. Each prime p defines a p-adic number field allowing the counterparts of the usual arithmetic operations. The basic difference between real and p-adic numbers is that p-adic topology is ultra-metric. Ultrametricity means that the distance function $d(x, y)$ (the counterpart of $|x - y|$ in the real context) satisfies the inequality

$$d(x, z) \leq Max\{d(x, y), d(y, z)\} \ ,$$

(Max(a,b) denotes maximum of a and b) rather than the usual triangle inequality

$$d(x, z) \leq d(x, y) + d(y, z) \ .$$

p-Adic numbers have expansion in powers of p analogous to the decimal expansion

$$x = \sum_{n \geq 0} x_n p^n \ ,$$

and the number of terms in the expansion can be infinite so that p-adic number need not be finite as a real number. The norm of the p-adic number (counterpart of $|x|$ for real numbers) is defined as

$$N_p(x = \sum_{n \geq 0} x_n p^n) = p^{-n_0} \ ,$$

and depends only very weakly on p-adic number. The ultra-metric distance function can be defined as $d_p(x, y) = N_p(x - y)$.

p-Adic numbers allow the generalization of the differential calculus and of the concept of analytic function $f(x) = \sum f_n x^n$. The basic rules of the p-adic differential calculus are the same as those of the ordinary differential calculus. There is however one important new element: the set of the functions having vanishing p-adic derivative consists of so called pseudo constants, which depend on a finite number of positive pinary digits of x only so that one has

$$f_N(x = \sum_n x_n p^n) = f(x_N = \sum_{n < N} x_n p^n) \ .$$

In the real case only constant functions have vanishing derivative. This implies that p-adic differential equations are non-deterministic.

An essential element is the map of the p-adic numbers to the positive real numbers by the so called canonical identification I:

$$I : \sum x_n p^n \in R_p \to \sum_n x_n p^{-n} \in R \ .$$

Canonical identification has inverse, which is single valued for the real numbers having infinite number of pinary digits but two-valued for real numbers having finite number of pinary digits (the reason is that real number with finite number or pinary digits has two equivalent pinary expansions: ($x = 1 = .999999...$ in case of decimal expansion and $x = 1 = 0yyyy...$, $y = p - 1$, in the case of pinary expansion).

Canonical identification in its basic form cannot map real space-time surface to p-adic ones or vice versa because it is not a general coordinate invariant notion. A variant of canonical identification, call it I_Q, maps defined only for rationals is given by $I(q = m/n) = I(m)/I(n)$, where $q = m/n$ is the unique representation of rational q in terms of integers [E1].

I_Q can be applied to map rational points of p-adic CP_2 to their real counterparts whereas the points of p-adic M^4 are mapped as such to real points as such [E1]. General coordinate invariance is not lost since the projection of p-adic space-time sheet to real imbedding space is discrete and genuinely p-adic points are at infinite real distance, "outside the real cosmos". This means a deep number theoretic difference between M^4 and CP_2 and gives one reason for the product decomposition of the imbedding space. O_Q makes it also possible to map the predictions of the p-adic probability theory and thermodynamics to real numbers so that probability is conserved.

2.3.2 Evolution of physical ideas

In the sequel the evolution of physical ideas related to p-adic numbers is summarized.

p-Adic length scale hypothesis

p-Adic length scale hypothesis [TGDpad] states that to a given p-adic prime p there corresponds a primary p-adic length scale $L_p = \sqrt{p}l$, $l \simeq 1.288 \times 10^4 \sqrt{G}$ ($\sqrt{G}$ denotes Planck length) and that physically favored primes correspond to $p \simeq 2^k$, k power of prime. The corresponding p-adic time scale is obtained as $T_p = L_p/c$. The justification for the first part of the hypothesis comes from Uncertainty Principle and from the p-adic mass calculations [TGDpad] predicting that the mass of elementary particle, resulting from the mixing of massless states with $10^{-4} m_{Planck}$ mass states described by p-adic thermodynamics, is of order $1/L_p$ for the light states.

The first principle explanation for p-adic length scale hypothesis derives from the fusion of real and p-adic physics to a single larger framework. The fact that real and p-adic space-time sheets can have common points implies that local p-adic physics give rise to p-adic fractality of real physics. Also multi-p p-adic fractality is possible. $p \simeq 2^k$ would reflect the presence of 2-adic fractality besides $p > 2$-adic fractality.

A heuristic justification for the preferred values of p comes from elementary particle black hole analogy [E5] generalizing the Bekenstein-Hawking area-entropy law to apply to the elementary particle horizon defined as the surface at which the Euclidian signature for the so called CP_2 type extremal describing elementary particle changes to the Minkowskian signature of the background space-time at which elementary particle has suffered topological condensation.

The hypothesis is especially interesting above the elementary particle length scales $p > M_{127}$ and has testable implications in nuclear physics, atomic physics and condensed matter length scales. The most convincing support for this hypothesis are provided by the elementary particle mass calculations: if one assumes that the p-adic primes associated with elementary particles are primes near prime powers of two, one can predict lepton and gauge boson masses with accuracy better than one per cent. Also quark masses can be predicted but the calculation of the hadron masses requires some modelling (CKM matrix, color force, etc...). The existing empirical information about neutrino mass squared differences suggests that the allowed values of k are indeed *powers* of prime rather than primes.

It is natural to postulate that space-time sheets form a hierarchy with respect to p in the sense that the lower bound for the size of the space-time sheets at level p is of order L_p and that $p_1 < p_2$ sheets condensed on p_2 sheets behave like particles on sheet p_2.

The following table lists the p-adic length scales L_p, $p \simeq 2^k$, k power or prime, which might be interesting as far as condensed matter is considered (the notation $L(k)$ will be used instead of L_p). It must be emphasized that the definition of the length scale is bound to contain some unknown numerical factor K: the requirement that the thickness of cell membrane corresponds to $L(151)$ fixes the proportionality coefficient K to $K \simeq 1.1$.

k	127	131	137	139	149
$L_p/10^{-10}m$	.025	.1	.8	1.6	50

k	151	157	163	167	169
$L_p/10^{-8}m$	1	8	64	256	512

k	173	179	181	191	193
$L_p/10^{-4}m$	.2	1.6	3.2	100	200

k	197	199	211	223	227
L_p/m	.08	.16	10	640	2560

Table 1. Primary p-adic length scales $L_p = 2^{k-151}L_{151}$, $p \simeq 2^k$, k prime, possibly relevant to bio physics. The last 3 scales are included in order to show that twin pairs are very frequent in the biologically interesting range of length scales. The length scale $L(151)$ is take to be thickness of cell scale, which is 10^{-8} meters in good approximation.

The assumption that p-adic space-time regions provide cognitive representations of the real space-time regions forces to conclude that cognition is present in all length scales and that the properties of the p-adic space-time regions reflect those of the real space-time regions. p-adic–real phase transitions and identifiable as transformation of intentions to actions [H8] occurring even at elementary particle length scales would explain this elegantly.

Besides primary p-adic length scales also n-ary p-adic length scales defined as $L_p(n) = p^{(n-1)/2}L_p$ and corresponding time scales are possible and form a fractal hierarchy coming as powers of $\sqrt{p}$. Accepting these scales means that all length scales $L(n)$ coming as powers of $2^{n/2}$, n a positive integer, should have a preferred physical role. The TGD inspired model for living matter lends support for the hypothesis that biologically important length and time scales indeed appear as half octaves. A possible explanation for this is the existence of a hierarchy of cognitive codes associated with the time scales $T(n)$. Any prime power factor k^i in the decomposition of the integer n to a product of prime power factors defines a candidate for a cognitive code. The duration of code word would be $T(n)$ and the number of bits would be k^i. For prime values of n the information content of the code word is maximal so that one could understand why prime values of n are especially important.

CP_2 type extremals and elementary particle black hole analogy

CP_2 type extremals are vacuum extremals having a finite negative action so that one can lower the action of the ordinary vacuum extremals by gluing CP_2 type extremals to them. CP_2 type extremals have one-dimensional M_+^4 projection which is light like random curve. Light likeness condition leads to classical Virasoro algebra constraints. $M^4 \times SO(3,1) \times SU(3) \times SU(2)_{ew}$ Super-Kac-Moody algebra acts as symmetries and the spectrum of elementary particles is precisely known. The obvious interpretation of the CP_2 type extremals is as a model of elementary particle.

CP_2 type extremals are much like black holes in the sense that they possess elementary particle horizon: this is the surface at which the Euclidian signature of the metric of the CP_2 type extremal changes to the Minkowskian signature of the background space-time. One can indeed generalize Bekenstein-Hawking law to a statement saying that the real counterpart of the p-adic entropy predicted by the p-adic thermodynamics is proportional to the surface area of the elementary particle horizon. In particular, for primes $p \sim 2^k$, where k is power of prime, the radius of the elementary particle horizon is itself a p-adic length scale. This suggests a double p-adicization associated with p and k and an additional cognitive degeneracy due to the k-adic non-determinism, and hence also the dominance of the final states of quantum jump for which $p \simeq 2^k$ holds true: there would be simply very many physically equivalent physical states for these values of p.

p-Adic thermodynamics and particle massivation

The underlying idea of TGD based description of particle massivation is following. Due to the interaction of a topologically condensed 3-surface describing elementary particle with the background space-time, massless ground states are thermally mixed with the excitations with mass of order $m_0 \sim 1/R$ (R is CP_2 length scale, $1/R$ of order 10^{-4} Planck masses) created by the Super Virasoro generators. Instead of energy, the Virasoro generator L_0 (essentially mass squared) is thermalized. This guarantees Lorentz invariance automatically. p-Adic temperature is quantized by purely number theoretical constraints (the Boltzmann weight $exp(-E/kT)$ is replaced with p^{L_0/T_p}, $1/T_p$ integer) and fermions correspond to $T_p = 1$ whereas $T_p = 1/2$ seems to be the only reasonable choice for bosons. That mass squared, rather than energy, is a fundamental quantity at CP_2 length scale is also suggested by a simple dimensional argument (Planck mass squared is proportional to $\hbar$ so that it should correspond to a generator of some Lie-algebra (Virasoro generator L_0 representing scaling!)).

Optimal lowest order predictions for the charged lepton masses are obtained and photon, gluon and graviton appear as essentially massless particles. The calculations support the existence of massless gluons and electro-weak quanta associated with so called massless extremals (MEs). One important prediction is that p-adic thermodynamics cannot explain the masses of the intermediate gauge bosons although the predictions for the fermion masses are excellent. This observation led to the identification of the TGD counterpart of Higgs field whose vacuum expectation provides the dominating contribution to the bosonic masses and only shifts bosonic masses [F2].

p-Adic coupling constant evolution

The original hypothesis was that Kähler coupling strength α_K is completely fixed by quantum criticality implying that α_K is analogous to critical temperature. p-Adic considerations led to the view that there is infinite number of critical values of α_K labelled by p-adic primes. In many-sheeted space-time one can indeed consider the possibility that α_K is not a universal constant. This would mean that space-time sheets joined only by wormhole contacts and surrounded by light like elementary particle horizons would be characterized by different values of Kähler coupling strength.

Since p-adic primes correspond to p-adic length scales this inspires the idea that the ordinary coupling constant evolution is replaced by a discrete coupling constant evolution. This view is also consistent with the criticality of the Kähler coupling constant. The assumption that gravitational constant is invariant under critical temperature-adic coupling constant evolution fixes highly unique the evolution of Kähler coupling strength. This picture makes sense if one can assign to a given 3-surface a unique p-adic prime and there are good reasons to believe that this is indeed the case.

Vacuum degeneracy of the Kähler action and spin glass analogy

The space of minima of free energy for spin glass is known to have ultra-metric topology. p-Adic topology is also ultra-metric and this motivated the hypothesis that quantum average space-time, 'topological condensate', defined as a maximum of Kähler function can be obtained by gluing together regions characterized by various values of the p-adic prime p. It must be emphasized that this hypothesis is just a guess and not even correct as such, and it seems that TGD as a generalized number theory vision gives the real justification for the p-adics. A good guess is however that the ultra-metric topology of the reduced configuration space consisting of the maxima of the Kähler function is induced from the p-adic norm and that there is a close connection between the two p-adicities. The following arguments tries to make this idea more precise.

The unique feature of the Kähler action is its enormous vacuum degeneracy: any space-time surface, whose CP_2 projection is a so called Lagrange manifold (having dimension $D \leq 2$) is vacuum extremal. This is expected to imply a large degeneracy of the absolute minimum space-times: for instance, several absolute minima with the same action are possible for single 3-surface (this forces to a generalization of space-time concept obtained by introducing 'association sequences'). The degeneracy means an obvious analogy with the spin glass phase characterized by 'frustration' implying a large number of degenerate ground states. In the construction of the configuration space geometry the analogy between quantum TGD and spin glass becomes precise.

Spin glass consists of magnetized regions such that the direction of the magnetization varies randomly in the spatial degrees of freedom but is frozen in time. What is peculiar that, although there are large gradients on the boundaries of the regions with a definite direction of magnetization, no large surface energies are generated. An obvious p-adic explanation suggests itself: p-adic magnetization could be pseudo constant and hence piecewise constant with a vanishing derivative on the boundaries of the magnetized regions so that no p-adic surface energy would be generated.

In the description of the spin glass phase also ultra-metricity, which is the basic property of the p-adic topology, emerges in a natural manner. The energy landscape describing the free energy of spin glass as a function of various parameters characterizing spin glass, is fractal like function and there are infinite number of energy minima. In this case there is a standard manner to endow the space of the free energy minima with an ultra-metric topology [cc4].

The counterpart of the energy landscape in TGD can be constructed as follows. The configuration space of TGD (the space of 3-surfaces in H) has fiber-space like structure deriving from the decomposition $CH = \cup_{zeromodes} G/H$. The fiber is the coset space G/H such that G is the group of the canonical transformation of the light cone boundary. In particular, the canonical transformations of CP_2 act in the fiber as isometries. The base space is the infinite-dimensional space of the zero modes characterizing the size and shape as well as the classical Kähler field at the 3-surface.

To calculate S-matrix element, one must form Fock space inner product as a functional of 3-surface X^3 multiplied with the vacuum functional $exp(K)$ and integrate it over the entire configuration space:

$$S_{i \to f} = \int \langle \Psi_f, \Psi_i \rangle (X^3) exp(K(X^3)) \sqrt{G} DX^3 \ .$$

The integration over the fiber degrees of freedom reduces to a Gaussian integration around the maxima of the Kähler function with respect to the fiber coordinates. The equally poorly defined Gaussian and metric determinants cancel each other in this integration and one obtains a well defined end result. Canonical transformations are 'almost gauge symmetries' since only classical gravitational fields destroy canonical symmetries acting as $U(1)$ gauge transformations. This means that the action for several canonically related configurations can be degenerate and several maxima are expected for given values of the zero modes. This means that the subset CH_0 of the configuration space consisting of the maxima of the Kähler function has many sheets parameterized by the zero modes and that generalized catastrophe theory is obtained.

If a localization in the zero modes occurs in the quantum jump, one can circumvent the integration over the zero modes in practice. The exponent for the maximum of the Kähler action is expected to have maxima as a function of the zero modes too. The maxima of $exp(K_{max})$ as function of zero modes define the counterpart of the energy landscape and $exp(K_{max})$ is the counterpart of the energy serving as a height function of the energy landscape. It could quite well be that this height function can be induced from a p-adic norm. If so, the allowed values of p define a decomposition of the space of zero modes to sectors D_p. For 'full' CP_2 type extremals representing virtual gravitons the exponent is indeed proportional to $1/p$ if one takes seriously the argument determining the possible values of the Kähler coupling strength. Thus cognitive p-adicity and spin glass p-adicity would be related to each other. The connection with gravitons is especially interesting since also classical gravitation is closely related to the spin glass degeneracy.

2.3.3 Evolution of mathematical ideas

The evolution of mathematical ideas has been driven by the following frequently asked questions.

a) Is p-adicity realized at space-time level or only at the level of p-adic thermodynamics which was the first application of p-adic numbers? If p-adic space-time regions really make sense, what is their physical interpretation?

b) Physics seems to require correspondence between p-adic and real numbers. What is the role of canonical identification: does it only map p-adic probabilities to their real counterparts or could it be applied also at space-time level despite the obvious difficulties with general coordinate invariance? What about correspondence defined by rational numbers which can be regarded as numbers common to all number fields. Is it possible to assign to a real space-time surface a p-adic counterpart by procedure respecting general coordinate invariance?

c) Does the notion of p-adicization of real physics make sense? How one might achieve the p-adicization in general coordinate invariance manner? What should one p-adicize: only probability calculus and thermodynamics? Or should one include also Hilbert space level? What about p-adicization at space-time level and perhaps even configuration space-level?

d) What is the origin of p-adicity? What is the origin of p-adic length scale hypothesis? How it is possible to assign p-adic prime to a given real space-time sheet as required by the p-adic mass calculations?

e) There have been also technical problems. Besides differential calculus also integral calculus is basic element of classical physics since all variational principles involve integrals over space-time. Also the functional integral over configuration space is needed in order to define S-matrix elements. How one could circumvent the difficulties caused by the non-existence of a p-adic valued define integral based on Riemann sum.

p-Adic physics as physics of cognition and intentionality and generalization of number concept

The identification of p-adic physics as physics of cognition and intention suggests strongly connections between cognition, intentionality, and number theory. The new idea is that also real transcendental numbers can appear in the extensions of p-adic numbers which must be assumed to be finite-dimensional at least in the case of human cognition.

The basic ingredient is the new view about numbers: real and p-adic number fields are glued together like pages of a book along common rationals representing the rim of the book. Also the rational multiples of algebraic numbers existing p-adically are shared in this manner so that the pages of the book can be stuck together along these lines. This generalizes to the extensions of p-adic number fields and the outcome is a complex fractal book like structure containing books within books. This holds true also for manifolds and one ends up to the view about many-sheeted space-time realized as 4-surface in 8-D generalized imbedding space and containing both real and p-adic space-time sheets. The transformation of intention to action corresponds to a quantum jump in which p-adic space-time sheet is replaced with a real one.

One implication is that the rationals having short distance p-adically are very far away in the real sense. This implies that p-adically short temporal and spatial distances correspond to long real distances and that the evolution of cognition proceeds from long to short temporal and spatial scales whereas material evolution proceeds from short to long scales. Together with p-adic non-determinism due the fact that the integration constants of p-adic differential equations are piecewise constant functions this explains the long range temporal correlations and apparent local randomness of intentional behavior. The failure of the real statistics and its replacement by p-adic fractal statistics for time series defined by varying number N of measurements performed during a fixed time interval T allows very general tests for whether the system is intentional and what is the p-adic prime p characterizing the "intelligence quotient" of the system. The replacement of $log(p_n)$ in the formula $S = -\sum_n p_n log(p_n)$ of Shannon entropy with the logarithm of the p-adic norm $|p_n|_p$ of the rational valued probability allows to define a hierarchy of number theoretic information measures which can have both negative and positive values.

Since p-adic numbers represent a highly number theoretical concept one might expect that there are deep connections between number theory and intentionality and cognition. The discussions with Uwe Kämpf in CASYS'2003 conference in Liege indeed stimulated a bundle of ideas allowing to develop a more detailed view about intention-to-action transformation and to disentangle these connections. These discussions made me aware of the fact that my recent views about the role of extensions of p-adic numbers are perhaps too limited. To see this consider the following arguments.

a) Pure p-adic numbers predict only p-adic length scales proportional to $p^{n/2}l$, l CP_2 length scale about 10^4 Planck lengths, $p \simeq 2^k$, k prime or power of prime. As a matter fact, all positive integer values of k are possible. This is however not enough to explain all known scale hierarchies. Fibonacci numbers $F_n : F_n + 1 = F_n + F_{n-1}$ behave asymptotically like $F_n = kF_{n-1}$, k solution of the equation $k^2 = k + 1$ given by $k = \Phi = (1 + \sqrt{5})/2 \simeq 1.6$. Living systems and self-organizing systems represent a lot of examples about scale hierarchies coming in powers of the Golden Mean $\Phi = (1 + \sqrt{5})/2$.

By allowing the extensions of p-adics by algebraic numbers one ends up to the idea that also the length scales coming as powers of x, where x is a unit of algebraic extension analogous to imaginary unit, are possible. One would however expect that the generalization of the p-adic length scale hypothesis alone would predict only the powers $\sqrt{x}p^{n/2}$ rather than $x^kp^{n/2}$, $k = 1, 2, ...$ Perhaps the purely kinematical explanation of these scales is not possible and genuine dynamics is needed. For sinusoidal logarithmic plane waves the harmonics correspond to the scalings of the argument by powers of some scaling factor x. Thus the powers of Golden Mean might be associated with logarithmic sinusoidal plane waves.

b) Physicist Hartmuth Mueller has developed what he calls Global Scaling Theory [ik10] based on the observation that powers of e (Neper number) define preferred length scales. These powers associate naturally with the nodes of logarithmic sinusoidal plane waves and correspond to various harmonics (matter tends to concentrate on the nodes of waves since force vanishes at the nodes). Mueller talks about physics of number line and there is great temptation to assume that deep number theory is indeed involved. What is troubling from TGD point of view that Neper number e is not algebraic. Perhaps a more general approach allowing also transcendentals must be adopted.

c) Classical mathematics, such as the theory of elementary functions, involves few crucially important transcendentals such as e and π. This might reflect the evolution of cognition: these numbers should be cognitively and number theoretically very special. The numbers e and π appear also repeatedly in the basic formulas of physics. They however look p-adically very troublesome since it has been very difficult to imagine a physically acceptable generalization of such simple concepts as exponent function, trigonometric functions, and logarithm resembling its real counterpart by allowing only the extensions of p-adic numbers based on algebraic numbers.

These considerations stimulate the question whether, besides the extensions of p-adics by algebraic numbers, also the extensions of p-adic numbers involving π and e and other transcendentals might be needed. The intuitive expectation motivated by the finiteness of human intelligence is that these extensions should have finite algebraic dimensions, and it indeed turns out that this is possible under some conditions which can be formulated as very general number theoretical conjectures. Since e^p exists p-adically, the powers $e, ..., e^{p-1}$ define a p-dimensional extension as do also the roots of polynomials with coefficients which are in an extension of rationals containing e and its powers. Contrary to the original conjecture, π however cannot belong to a finite-dimensional extension of p-adics. It is an open question whether one should allow infinite-dimensional extension of p-adic numbers containing π.

In any case, the hypothesis that $log(p) \times \pi$ for p prime, and $log(\Phi) \times \pi$ are rational numbers, guarantees the minimum amount of transcendentality required by classical physics and basic calculus using only a finite-dimensional extensions of p-adic numbers. The special role of π however becomes an extremely strong constraint for the p-adicization of quantum TGD by algebraic continuation from the realm of rationals to real and p-adic number fields.

Second question is whether there might be some dynamical mechanism allowing to understand the hierarchy of scalings coming in powers of some preferred transcendentals and algebraic numbers like Golden Mean. Conformal invariance implying that the system is characterized by a universal spectrum of scaling momenta for the logarithmic counterparts of plane waves seems to provide this mechanism. This spectrum is determined by the requirement that it exists for both reals and all p-adic number fields assuming that finite-dimensional extensions are allowed in the latter case. The spectrum corresponds to the zeros of the Riemann Zeta if Zeta is required to exist for all number fields in the proposed sense, and a lot of new understanding related to Riemann hypothesis emerges and allows to develop further the previous TGD inspired ideas about how to prove Riemann hypothesis [n4, jf2].

Algebraic continuation as a basic principle

One general idea which results as an outcome of the generalized notion of number is the idea of a universal function continuable from a function mapping rationals to rationals or to a finite extension of rationals to a function in any number field. This algebraic continuation is analogous to the analytical continuation of a real analytic function to the complex plane. Rational functions with rational coefficients are obviously functions satisfying this constraint. Algebraic functions with rational coefficients satisfy this requirement if appropriate finite-dimensional algebraic extensions of p-adic numbers are allowed. Exponent function is such a function. Logarithm is also such a function provided that the above mentioned number theoretic conjecture holds true.

The definition of a definite integral for p-adic numbers has been the key challenge in attempts to construct p-adic physics and algebraic continuations seems to solve this problem. The first problem is that p-adic numbers are not well ordered and one cannot define what ordered integration interval $[a, b]$ means p-adically. The second problem is that Riemann sum gives identically vanishing p-adic integral if coordinate increments approach zero at the limit. One can however define the definite integral in terms of the integral function:

$$\int_a^b f(x)dx = F(b) - F(a) \ ; \ f(x) = \frac{dF(x)}{dx} \ .$$

Integral function $F(x)$ is obtained using the inverse of the derivation just as in the real context. If integration limits are restricted to be rational numbers or finitely extended rational numbers, they can be ordered using the ordering of real numbers. This would essentially mean that p-adic integration measure is an algebraic continuation of the real integration measure.

Also residy calculus might be generalized so that the value of an integral along the real axis could be calculated by continuing it instead of the complex plane to any number field via its values in the subset of rational numbers forming the rim of the book like structure having number fields as its pages. If the poles of the continued function in the finitely extended number field allow interpretation as real numbers it might be possible to generalize the residy formula. One can also imagine of extending residy calculus to any algebraic extension. An interesting situation arises when the poles correspond to extended p-adic rationals common to different pages of the "great book". This could mean that the integral could be calculated at any page having the pole common. In particular, could a p-adic residy integral be calculated in the ordinary complex plane by utilizing the fact that in this case numerical approach makes sense.

Gaussian integration as a purely algebraic process gives hopes to define p-adic variants of configuration space integrals but only in the case that the integral over the configuration space reduces effectively to the Gaussian integral of a free quantum field theory. If configuration space is indeed a union of symmetric spaces, there are good hopes for achieving this (Duistermaat-Hecke theorem).

p-Adic integration is not necessarily needed to define the p-adic counterpart for the field equations associated with Kähler action but the continuation of the physics from real configuration space to the p-adic variants of the configuration spaces requires the existence of the p-adic valued Kähler action. If it is possible to assign to a given real space-time surface a p-adic counterpart uniquely in a given resolution for rational numbers, one can define the p-adic Kähler action as the real action interpreted as p-adic number in case that the real action belongs to a finite extension of rationals. This would also take care of the absolute minimization of the p-adic Kähler action which does not make sense as a genuinely p-adic concept.

2.3.4 Generalized Quantum Mechanics

One can consider two generalizations of quantum mechanics to a fusion of p-adic and real quantum mechanics.

a) For the first generalization the guiding principle for the generalization of quantum mechanics is that quantum mechanics in a given number field is obtained as an algebraic continuation of the quantum mechanics in the field of rational numbers common to all number fields or in finite-dimensional extensions of rational numbers. This means that U-matrices U_F for transitions from H_Q to H_F, where F refers to various completions of rationals, are obtained as algebraic continuations of the unitary U-matrix U_Q for H_Q. The generalization means enormously strong algebraic constraints on the form of the U-matrix.

b) A more radical option is that transitions from rational Hilbert space H_Q to the Hilbert spaces H_F associated with different number fields occur. This requires that U-process is followed by a process analogous to a state function reduction and preparation takes care that the resulting states become states in H_Q: this is what makes this generalization of a special interest. In this case one can speak about total scattering probability from H_Q to H_F. The U-matrices U_F are not anymore mere analytic continuations of U_Q. A possible interpretation of the unitary process $H_Q \to H_F$ is as generation of intention whereas the reduction and preparation means the transformation of the intention to action.

The assumption that H_Q allows an algebraic continuation to the spaces H_F is probably too strong an idealization in p-adic and even in the real case. For instance, one cannot allow all rational valued momenta in p-adic case for the simple reason that the continuation to the p-adic case involves always some momentum cutoff if the extension of p-adics remains finite. Even in the real case the summation over all rational momenta in the unitarity conditions of U-matrix fails to make sense and cutoff is needed. A hierarchy of cutoffs suggests itself and has a natural interpretation as number theoretical hierarchy of extensions of p-adics.

In order to avoid un-necessary complications the following formal discussion however uses H_Q as a universal Hilbert space contained by the various state spaces H_F.

Quantum mechanics in H_F as a algebraic continuation of quantum mechanics in H_Q

The rational Hilbert space H_Q is representable as the set of sequences of real or complex rationals of which only finite number are non-vanishing. Real and p-adic Hilbert spaces are obtained as the numbers in the sequences to become real or p-adic numbers and no limitations are posed to the number of non-vanishing elements. All these Hilbert spaces have rational Hilbert space H_Q as a common sub-space. Also momenta and other continuous quantum numbers are replaced by a discrete value set. Superposition principle holds true only in a restricted sense, and state function reduction and preparation leads always to a final state which corresponds to a state in H_Q. This picture differs from the earlier one in which p-adic and real Hilbert spaces were assumed to form a direct sum.

The notion of unitarity generalizes. Contrary to the earlier beliefs, U-matrix does not possess matrix elements between different number fields but between rational Hilbert space and Hilbert spaces associated with various completions of rationals. This makes sense since the final state of the quantum jump (and thus the initial state of the unitary process, is always in H_Q.

The U-matrix is a collection of matrices U_F having matrix elements in the number field F. U_F maps H_Q to H_F. Each of these U-matrices is unitary. Also U_Q is unitary and U_F is obtained by algebraic continuation in the quantum numbers labelling the states of U_Q to U_F.

Hermitian conjugation makes sense since the defining condition

$$\langle \alpha_F | U n_Q \rangle \;\; = \;\; \langle U^\dagger \alpha_F | n_Q \rangle \; . \tag{2.3.1}$$

allows to interpret $|n_Q\rangle$ also as an element of H_F. If U would map different completed number fields to each other, hermiticity conditions would not make sense.

The hermitian conjugate of U-matrix maps H_F to H_Q so that $UU^\dagger$ *resp.* $U^\dagger U$ maps H_F *resp.* H_Q to itself. This means that there are two independent unitarity conditions

$$\begin{aligned} U_F U_F^\dagger &= Id_F \; , \\ U_F^\dagger U_F &= Id_Q \; . \end{aligned} \tag{2.3.2}$$

One can write $U = P_Q + T_F$ and $U^\dagger = P_Q + T_F^\dagger$, where P_Q refers to the projection operator to H_Q. This gives

$$\begin{aligned} T_F + T_F^\dagger &= -T_F T_F^\dagger \ , \\ P_Q T_F + T_F^\dagger P_Q &= -T_F^\dagger T_F \ . \end{aligned} \qquad (2.3.3)$$

It is convenient to introduce the notations $T_Q = P_Q T_F$ and $T_Q^\dagger = T_F^\dagger P_Q$ with analogous notations for U and $U^\dagger$. The first condition, when multiplied from both sides by P_Q, gives together with the second equation unitarity conditions for T_Q

$$\begin{aligned} T_Q + T_Q^\dagger &= -T_Q T_Q^\dagger \ , \\ T_Q + T_Q^\dagger &= -T_F^\dagger T_F \ . \end{aligned} \qquad (2.3.4)$$

This means that the restriction of the U-matrix to H_Q is unitary.

The difference between the right hand sides of the equation should vanish. The understanding of how this happens requires more delicate considerations. For instance, in the case of $F = C$ continuous sum over indices appears at the right hand side coming from four-momenta labelling the states. The restrictions of quantum numbers to Q and its subsets could be a process analogous to the momentum cutoff of quantum field theories. The continuation from discrete integer valued labels of, say discrete momenta, to continuous values is performed routinely in various physical models routinely, and it would seem that this process has cognitive and physical counterparts. This picture conforms with the vision that the rational (or extended rational) U-matrix U_Q gives the U-matrices U_F by an algebraic continuation in the quantum numbers labelling the states (say 4-momenta).

Could U_F describe dispersion from H_Q to the spaces H_F?

One can also consider a more general situation in which the states in H_Q can be said to disperse to the sectors H_F. In this case one can write

$$T = "\sum_F" T_F \ . \qquad (2.3.5)$$

Here the sum has only a symbolic meaning since different number fields are in question and an actual summation is not possible. The T-matrix T_Q is the sum of the restrictions of T_F to H_Q and is the sum of rational valued T-matrices: $T_Q = \sum_F P_Q T_F$.

The T-matrices T_F are not anymore obtainable by algebraic continuation from same T-matrix T_Q. The unitarity conditions

$$\sum_F (P_Q T_F + T_F^\dagger P_Q) = -\sum_F T_F^\dagger T_F \qquad (2.3.6)$$

make sense only if they are satisfied separately for each T_F, exactly as in the previous case. T

The diagonal elements

$$T_F^{mm} + \overline{T}_F^{mm} = \sum_\alpha T_F^{m\alpha} \overline{T}_F^{m\alpha} = \sum_r T_F^{mr} \overline{T}_F^{mr}$$

give essentially total scattering probabilities from the state $|m\rangle$ of H_Q to the sector H_F, and must be rational (or extended rational) numbers. One can therefore say that each U-process leads with a definite probability to a particular sector of the state space.

The fact that states which are superpositions of states in different spaces H_F does not make sense mathematically, forces the occurrence of a process, which might be regarded as a number theoretical counterpart of state function reduction and preparation. First a sector H_F is selected with probability p_F. Then F-valued (in particular complex valued) entanglement in H_F is reduced by state reduction and preparation type processes to a rational or extended rational entanglement having interpretation as bound state entanglement. It would be natural to assume that Negentropy Maximization Principle governs this process. Obviously the possibility to reduce state function reduction to number theory forces to consider quite seriously the proposed option.

2.3.5 Do state function reduction and state-preparation have number theoretical origin?

The foregoing considerations support the view that state function reduction and state preparation are number theoretical necessities so that there would be a deep connection between number theory and free will. One could even say that free will is a number theoretic necessity. The resulting more unified view provides the reason why for state function reduction, and preparation and allows to generalize previous views developed gradually by physics and consciousness inspired educated guess work.

Negentropy Maximization Principle as variational principle of cognition

It is useful to discuss the original view about Negentropy Maximization Principle (NMP) before considering the possible generalization of NMP inspired by the number theoretic vision.

NMP was originally motivated by the need to construct a TGD based quantum measurement theory. Gradually it however became clear that standard quantum measurement theory more or less follows from the assumption that the world of conscious experience is classical: this meant that NMP became a principle governing only state preparation.

State function reduction is achieved if a localization in zero modes occurs in each quantum jump, and if U matrix in zero modes corresponds to a flow in some orthogonal basis for the configuration space spinor fields in the quantum fluctuating fiber degrees of freedom of the configuration space. The requirement that U-matrix induces effectively a flow in zero modes is consistent with the effective classicality of the zero modes requiring that quantum evolution causes no dispersion. The one-one correlation between preferred quantum state basis in quantum fluctuating degrees of freedom and zero modes implies nothing but a one-one correspondence between quantum states and classical variables crucial for the interpretation of quantum theory. It seems that number theoretical vision forces to generalize this view, and to raise NMP to a completely general principle applying also to the state function reduction as the original proposal indeed was.

In its original form NMP governs the dynamics of self measurements and thus applies to the quantum jumps reducing the entanglement between quantum fluctuating degrees of freedom for given values of zero modes. Self measurements reduce the entanglement only between sub-systems in quantum fluctuating degrees of freedom since they occur after the localization in the zero modes. Self measurement is repeated again and again for the unentangled sub-systems resulting in each self measurement. This cascade of self measurements leads to a state possessing only extended rational entanglement identifiable as bound state entanglement and having negative number theoretic entanglement entropy. This process should be equivalent with the state preparation process assumed to be performed by a conscious observer in standard quantum measurement theory.

NMP states that the self measurement can be regarded as a quantum measurement of the sub-system's density matrix reducing the counterpart of the entanglement entropy of some sub-system to a smaller value, and that this occurs for the sub-system for which the reduction of the entanglement entropy is largest among all sub-systems of the p-adic self. Inside each self NMP fixes some sub-system which is quantum measured in the quantum jump. One could perhaps say that self measurements make possible quantum level self repair since they allow the system in self state to fight against thermalization which results from the generation of unbound entanglement between sub-system-complement pairs.

NMP and number theory

The requirement the universe of conscious experience is classical is one manner to justify the notion of quantum jump. This hypothesis could be replaced by a postulate that state function reduction and preparation project quantum states to a definite number field and that only extended rational entanglement identifiable as bound state entanglement is stable. This is consistent with NMP since it is possible to assign to an extended rational entanglement a non-negative number theoretic negentropy as the maximum over entropies defined by various p-adic entropies $S_p = -\sum p_k log(|p_k|_p)$.

The unitary process U would thus start from a product ofbound states for which entanglement coefficient are extended rationals, and would lead to a formal superposition of states belonging to different number fields. Both state function reduction and state preparation would begin with a localization to a definite number field. This localization would be followed by a self measurement cascade reducing the entanglement to extended rational entanglement.

This vision forces to challenge the earlier views about state function reduction.

a) There is no good reason for why NMP could not be applied to both state function reduction and preparation.

b) If the entanglement between zero modes and quantum fluctuating degrees of freedom involves only discrete values of zero modes, the problems caused by the fact that no well-defined functional integral measure over zero modes exists, find an automatic resolution. Since extended rational entanglement possesses negative entanglement entropy, it is stable also against reduction if NMP applies completely generally. A discrete entanglement involving transcendentals not contained to any *finite* extension of any p-adic number field is unstable and reduced.

c) The quantum measurement lasts for a time determined by the life-time of the bound state entanglement between zero modes and quantum fluctuating degrees of freedom. Physical considerations of course support the view that it takes more than single quantum jump (10^{-39} seconds of psychological time) for the state function reduction to take place. The notion of zero mode-zero mode bound state entanglement seems however to be self-contradictory. If join along boundaries bonds are space-time correlates for the bound state entanglement, their formation should transform roughly half of the zero modes associated with the two space-time sheets to quantum fluctuating degrees of freedom.

d) If p-adic length scale hierarchy has as its counterpart a hierarchy of state function reduction and preparation cascades, one must accept the quantum parallel occurrence of state function reduction and preparation processes in the parallel quantum universes corresponding to different p-adic length scales. This picture provides a justification for the modelling of hadron as a quantum system in long length and time scales and as a dissipative system consisting of quarks and gluons in shorter length and time scales. The bound state entanglement between sub-systems of entangled systems having as a space-time correlate join along boundaries bonds connecting sub-system space-time sheets, is a second important implication of the new sub-system concept, and plays a central role in TGD inspired theory of consciousness.

2.4 The boost from TGD inspired theory of consciousness

Quite generally, TGD inspired theory of consciousness can be seen as a generalization of quantum measurement theory. The identification of quantum jump as a moment of consciousness is analogous to the identification of elementary particles as basic building blocks of matter. The observer is an outsider in standard quantum measurement theory and is replaced by the notion of self in TGD inspired theory of consciousness. Selves identified as systems able to avoid bound state entanglement and identifiable as ensembles of quantum jumps, are analogous to many-particle states. The sensory and other qualia of self are determined as statistical averages over quantum number and zero mode increments for the increasing sequence of quantum jumps defining self. Especially important are selves, which are in a state of macro-temporal quantum coherence since for these selves the entropy of the ensemble defined by the quantum jumps does not increase and the qualia stay sharp. These selves are analogous to bound states of elementary particles and their formation actually corresponds to the generation of bound state entanglement.

2.4.1 The anatomy of the quantum jump

In TGD framework quantum transitions correspond to a quantum jump between two different quantum histories rather than to a non-deterministic behavior of a single quantum history. Therefore U-matrix relates to each other two quantum histories rather than the initial and final states of a single quantum history.

To understand the philosophy behind the construction of U-matrix it is useful to notice that in TGD framework there is actually a 'holy trinity' of time developments instead of single time development encountered in ordinary quantum field theories.

a) The classical time development determined by the absolute minimization of Kähler action or some more general principle selecting preferred extremals as generalized Bohr orbits [E2].

b) The unitary "time development" defined by U associated with each quantum jump

$$\Psi_i \rightarrow U\Psi_i \rightarrow \Psi_f \ ,$$

and defining U-matrix. One cannot however assign to the U-matrix an interpretation as a unitary time-translation operator and this means that one must leave open the identification of U-matrix with S-matrix.

c) The time development of subjective experiences by quantum jumps identified as moments of consciousness. The value of psychological time associated with a given quantum jump is determined by the contents of consciousness of the observer. The understanding of psychological time and its arrow and of the dynamics of subjective time development requires the construction of theory of consciousness. A crucial role is played by the classical non-determinism of Kähler action implying that the non-determinism of quantum jump and hence also the contents of conscious experience can be concentrated into a finite volume of the imbedding space.

U is informational "time development" operator, which is unitary like the S-matrix characterizing the unitary time evolution of quantum mechanics. U is however only formally analogous to Schrödinger time evolution of infinite duration since there is *no* real time evolution or translation involved. It is not clear whether one should regard U-matrix and S-matrix as two different things or not: U-matrix is a completely universal object characterizing the dynamics of evolution by self-organization whereas S-matrix is a highly context dependent concept in wave mechanics and in quantum field theories where it at least formally represents unitary time translation operator at the limit of an infinitely long interaction time. The S-matrix understood in the spirit of superstring models is however something very different and could correspond to U-matrix.

The requirement that quantum jump corresponds to a measurement in the sense of quantum field theories implies that each quantum jump involves localization in zero modes which parameterize also the possible choices of the quantization axes. Thus the selection of the quantization axes performed by the Cartesian outsider becomes now a part of quantum theory. Together these requirements imply that the final states of quantum jump correspond to quantum superpositions of space-time surfaces which are macroscopically equivalent. Hence the world of conscious experience looks classical. Physically it seems obvious that U matrix should decompose to a cosmological U-matrix representing dispersion in configuration space and U-matrix representing local dynamics: this indeed occurs thanks to the classical non-determinism of the Kähler action. At least formally quantum jump can be interpreted also as a quantum computation in which matrix U represents unitary quantum computation. An important exception are the zero modes characterizing center of mass degrees of freedom of 3-surface which correspond to the isometries of $M^4_+ \times CP_2$. In these degrees of freedom localization does not occur. At the limit when 3-surfaces are regarded as pointlike objects theory should obviously reduce to quantum field theory.

The three non-determinisms

Besides the non-determinism of quantum jump, TGD allows two other kinds of non-determinisms: the classical non-determinism basically due the vacuum degeneracy of the Kähler action and p-adic non-determinism of p-adic differential equations due to the fact that functions with vanishing p-adic derivative correspond to piecewise constant functions.

To achieve classical determinism in a generalized sense, one must generalize the definition of the 3-surfaces Y^3 (belonging to light cone boundary) by allowing also "association sequences", that is 3-surfaces which have, besides the component belonging to the light cone boundary, also disjoint components which do not belong to the light cone boundary and have mutual *time-like separations*. This means the introduction of additional, one might hope typically discrete, degrees of freedom (consider non-determinism based on bifurcations as an example). It is even possible to have quantum entanglement between the states corresponding to different values of time.

Without the classical and p-adic non-determinisms general coordinate invariance would reduce the theory to the light cone boundary and this would mean essentially the loss of time which occurs also in the quantization of general relativity as a consequence of general coordinate invariance. Classical and p-adic non-determinisms imply that one can have quantum jumps with non-determinism (in conventional sense) located to a finite time interval. If quantum jumps correspond to moments of consciousness, and if the contents of consciousness are determined by the locus of the non-determinism, then these quantum jumps must give rise to a conscious experience with contents located in a finite time interval.

Also p-adic space-time sheets obey their own quantum physics and are identifiable as seats of cognitive representations. p-Adic non-determinism is the basic prerequisite for imagination and simulation. The notion of cognitive space-time sheet as a space-time sheet having finite time duration is one aspect of the p-adic non-determinism and allows to understand how the notion of psychological time emerges. Cognitive space-time sheets simply drift quantum jump by quantum to the direction of geometric future since there is much more room there in the light cone cosmology.

The classical non-determinism is maximal for CP_2 type extremals for which the M^4_+ projection of the space-time surface is random lightlike curve. In this case, basic objects are essentially four- rather than 3-dimensional. The basic implication of the classical non-determinism is that quantum theory does not reduce to the light cone boundary. Secondly, U-matrix reduces to a tensor product of a cosmologcal U-matrix and local U-matrices relevant for particle physics. As a matter fact, an entire hierarchy of U-matrices defined in various p-adic time scales is expected to appear in the hierarchy. Thirdly, the classical non-determinism of CP_2 type extremals allows a topologization of the Feynman diagrammatics of quantum field theories and string models. Although localization in zero modes characterizing zitterbewegung orbit occur in quantum jump, there is integral over the positions of vertices which correspond to cm degrees of freedom for imbedding space, and this gives rise to a sum over various Feynman diagrams.

How psychological time and its arrow emerge?

How psychological time and its arrow emerge is the basic challenge for the hypothesis that quantum jumps occur between quantum histories and are identifiable as moments of consciousness. Mind like space-time sheets provide a geometric model of unconscious mind in TGD framework and make it possible to solve the puzzle of psychological time. The first argument is following.

Mind like space-time sheets have well center of mass time coordinate and this coordinate is zero mode identifiable as psychological time. Localization in zero modes means that final states of quantum jumps correspond to quantum superpositions of space-time surfaces having same number of mind like space-time sheets such that given mind like space-time sheet possesses same value of psychological time for all space-time surfaces appearing in the superposition. The arrow of psychological time follows from the gradual drift of the mind like space-time sheets in future direction occurring quantum jump by quantum jump and is implied by the geometry of future light cone (there is more volume in the future of a given light cone point than in its future). The simplest assumption is that the average increment of psychological time in single quantum jump is of order CP_2 time, which is about 10^4 Planck times.

Besides classical non-determinism there is also p-adic non-determinism and one should keep mind open in the attempts to identify the roles of these two non-determinisms. The interpretation taken as a working hypothesis in the recent version of TGD inspired theory of consciousness is that p-adic space-time regions provide cognitive representations of the real regions and serve as correlates for intentions. Real regions are in turn symbolic representations for the material world in TGD sense of the word. This means that besides ordinary matter also higher level physical states associated with the real space-time sheets of a finite duration and having vanishing net energy are possible. The zero energy states representing pairs of incoming and outgoing states could make possible self-referential real physics representing the laws of physics in the structure of the higher level physical states. Real space-time sheets of finite temporal duration might be interpreted also as correlates of pure sensory experience as opposed to p-adic space-time sheets which can be identified as correlates of thoughts. Also volition could be assigned to the quantum jumps involving selection between various branches of multifurcations implied by the classical non-determinism.

A more refined argument explaining the arrow of psychological time is based on the idea that psychological time correspond to the moment of geometric time which gives the dominant contribution to the conscious experience, and that it is the transformation of intentions to actions which provides this contribution. The transformation of intentions to actions corresponds to the transformation of p-adic space-time sheets to real ones, and one can identify psychological time as characterizing the position of the intention-to action phase transition front. In order to have consistency with the basic facts about everyday conscious experience one must assume that the geometric past remains unable to express intentions for a period of time longer than the life cycle since otherwise the decisions made in say my geometric youth subjectively now could induce dramatic changes in my recent life. This dead time would be analogous to the recovery time of neuron after the generation of nerve pulse.

Macro-temporal quantum coherence and spin glass degeneracy

At the space-time level the generation of macroscopic quantum coherence is easy to understand if one accepts the identification of the space-time sheets as coherence regions. Quantum criticality and the closely related spin glass degeneracy are essential for the fractal hierarchy of space-time sheets. The problem of understanding macro-temporal and macroscopic quantum coherence at the level of configuration space (of 3-surfaces) is a more tricky challenge although quantum-classical correspondence strongly suggests that this is possible.

Concerning macro-temporal quantum coherence, the situation in quantum TGD seems at the first glance to be even worse than in standard physics. The problem is that simplest estimate for the increment in psychological time in single quantum jump is about 10^{-39} seconds derived from the idea that single quantum jump represent a kind of elementary particle of consciousness and thus corresponds to CP_2 time of about 10^{-39} seconds. If this time interval defines coherence time one ends up to a definite contradiction with the standard physics. Of course, the average increment of the geometric time during single quantum jump could vary and correspond to the de-coherence time. The idea of quantum jump as an elementary particle of consciousness does not support this assumption.

To understand how this naive conclusion is wrong, one must look more precisely the anatomy of quantum jump. The unitary process $\Psi_i \rightarrow U\Psi_i$, where Ψ_i is a prepared maximally unentangled state, corresponds to the quantum computation producing maximally entangled multi-verse state. Then follows the state function reduction and after this the state preparation involving a sequence of self measurements and given rise to a new maximally unentangled state Ψ_f.

a) What happens in the state function reduction is a localization in zero modes, which do not contribute to the line element of the configuration space metric. They are non-quantum fluctuating degrees of freedom and TGD counterparts of the macroscopic, classical degrees of freedom. There are however also quantum-fluctuating degrees of freedom and the assumption that zero modes and quantum fluctuating degrees of freedom are correlated like the direction of a pointer of a measurement apparatus and quantum numbers of the quantum system, implies standard quantum measurement theory.

b) Bound state entanglement is assumed to be stable against state function reduction and preparation. Bound state formation has as a geometric correlate formation of join along boundaries bonds between space-time sheets representing free systems. Thus the members of a pair of disjoint space-time sheets are joined to single space-time sheet. Half of the zero modes is transformed to quantum fluctuating degrees of freedom and only overall center of mass zero modes remain zero modes. These new quantum fluctuating degrees of freedom represent macroscopic quantum fluctuating degrees of freedom. In these degrees of freedom localization does not occur since bound states are in question.

Both state function reduction and state preparation stages leave this bound state entanglement intact, and in these degrees of freedom the system behaves effectively as a quantum coherent system. One can say that a sequence of quantum jumps binds to form a single long-lasting quantum jump effectively. This is in complete accordance with the fractality of consciousness. Quantum jumps represent moments of consciousness which are "elementary particles of consciousness" and in macro-temporal quantum coherent state these elementary particles bind to form atoms, molecules, etc. of consciousness.

c) The properties of the bound state plus its interaction with the environment allow to estimate the typical duration of the bound state. This time takes the role of coherence time. This suggests a connection with the standard approach to quantum computation. An essential element is spin glass degeneracy. The generation of join along boundaries bonds connecting the space-time sheets of the composite systems is the space-time correlate for the formation of the bound states. Spin glass degeneracy is much higher for the bound states because of the presence of the join along boundaries bonds. This together with the fact that these degenerate states are almost identical so that transition amplitudes between them are also almost identical, implies that the life-time of the majority of bound states is much longer than one might expect otherwise. The detailed argument is carried out in [C1] and can be applied to show that spin glass degeneracy for the color flux tubes explains color confinement [D2].

e) The number theoretic notion of information relies on Shannon entropy in which the logarithms of probabilities are replaced by logarithms of their p-adic norms. This requires that the probabilities are rational or belong to an finite-dimensional extension of rationals. What is so important is that this entropy can have also negative values. If one assumes that bound states form a hierarchy such that the entanglement coefficients belong always to a finite-dimensional extension of rationals, one can define the entanglement entropy as a number theoretic entropy associated with some prime p. In p-adic context the prime is unique whereas in the real context the value of the prime can be selected in such a manner that the entropy is maximally negative. This prime would be naturally a maximal prime factor of the integer N defining the number of strictly deterministic regions of the space-time sheet in question. If this assumption is made, NMP alone implies the stability of bound states against state preparation by self measurements. This generalization of the information concept has far reaching implications in TGD inspired theory consciousness.

2.4.2 Negentropy Maximization Principle and new information measures

TGD inspired theory of consciousness, in particular the formulation of Negentropy Maximization Principle (NMP) in p-adic context, has forced to rethink the notion of the information concept. In TGD state preparation process is realized as a sequence of self measurements. Each self measurement means a decomposition of the sub-system involved to two unentangled parts. The decomposition is fixed highly uniquely from the requirement that the reduction of the entanglement entropy is maximal.

The additional assumption is that bound state entanglement is stable against self measurement. This assumption is somewhat ad hoc and it would be nice to get rid of it. The only manner to achieve this seems to be a generalized definition of entanglement entropy allowing to assign a negative value of entanglement entropy to the bound state entanglement, so that bound state entanglement would actually carry information, in fact conscious information (experience of understanding). This would be very natural since macro-temporal quantum coherence corresponds to a generation of bound state entanglement, and is indeed crucial for ability to have long lasting non-entropic mental images.

The generalization of the notion of number concept leads immediately to the basic problem. How to generalize the notion of entanglement entropy that it makes sense for a genuinely p-adic entanglement? What about the number-theoretically universal entanglement with entanglement probabilities, which correspond to finite extension of rational numbers? One can also ask whether the generalized notion of information could make sense at the level of the space-time as suggested by quantum-classical correspondence.

In the real context Shannon entropy is defined for an ensemble with probabilities p_n as

$$S = -\sum_n p_n log(p_n) \ . \tag{2.4.1}$$

As far as theory of consciousness is considered, the basic problem is that Shannon entropy is always non-negative so that as such it does not define a genuine information measure. One could define information as a change of Shannon entropy and this definition is indeed attractive in the sense that quantum jump is the basic element of conscious experience and involves a change. One can however argue that the mere ability to transfer entropy to environment (say by aggressive behavior) is not all that is involved with conscious information, and even less so with the experience of understanding or moment of heureka. One should somehow generalize the Shannon entropy without losing the fundamental additivity property.

p-Adic entropies

The key observation is that in the p-adic context the logarithm function $log(x)$ appearing in the Shannon entropy is not defined if the argument of logarithm has p-adic norm different from 1. Situation changes if one uses an extension of p-adic numbers containing $log(p)$: the conjecture is that this extension is finite-dimensional. One might however argue that Shannon entropy should be well defined even without the extension.

p-Adic thermodynamics inspires a manner to achieve this. One can replace $log(x)$ with the logarithm $log_p(|x|_p)$ of the p-adic norm of x, where log_p denotes p-based logarithm. This logarithm is integer valued ($log_p(p^n) = n$), and is interpreted as a p-adic integer. The resulting p-adic entropy

$$S_p = \sum_n p_n k(p_n) \ ,$$
$$k(p_n) = -log_p(|p_n|) \ . \tag{2.4.2}$$

is additive: that is the entropy for two non-interacting systems is the sum of the entropies of composites. Note that this definition differs from Shannon's entropy by the factor $log(p)$. This entropy vanishes identically in the case that the p-adic norms of the probabilities are equal to one. This means that it is possible to have non-entropic entanglement for this entropy.

One can consider a modification of S_p using p-adic logarithm if the extension of the p-adic numbers contains $log(p)$. In this case the entropy is formally identical with the Shannon entropy:

$$S_p \quad = \quad -\sum_n p_n log(p_n) = -\sum_n p_n \left[-k(p_n)log(p) + p^{k_n} log(p_n/p^{k_n})\right] \quad . \tag{2.4.3}$$

It seems that this entropy cannot vanish.

One must map the p-adic value entropy to a real number and here canonical identification can be used:

$$\begin{aligned} S_{p,R} \quad &= \quad (S_p)_R \times log(p)) \ , \\ (\sum_n x_n p^n)_R \quad &= \quad \sum_n x_n p^{-n} \ . \end{aligned} \tag{2.4.4}$$

The real counterpart of the p-adic entropy is non-negative.

Number theoretic entropies and bound states

In the case that the probabilities are rational or belong to a finite-dimensional extension of rationals, it is possible to regard them as real numbers or p-adic numbers in some extension of p-adic numbers for any p. The visions that rationals and their finite extensions correspond to islands of order in the seas of chaos of real and p-adic transcendentals suggests that states having entanglement coefficients in finite-dimensional extensions of rational numbers are somehow very special. This is indeed the case. The p-adic entropy entropy $S_p = -\sum_n p_n log_p(|p_n|)log(p)$ can be interpreted in this case as an ordinary rational number in an extension containing $log(p)$.

What makes this entropy so interesting is that it can have also negative values in which case the interpretation as an information measure is natural. In the real context one can fix the value of the value of the prime p by requiring that S_p is maximally negative, so that the information content of the ensemble could be defined as

$$I \quad \equiv \quad Max\{-S_p, \ p \ \text{prime}\} \ . \tag{2.4.5}$$

This information measure is positive when the entanglement probabilities belong to a finite-dimensional extension of rational numbers. Thus kind of entanglement is stable against NMP, and has a natural interpretation as bound state entanglement. The prediction would be that the bound states of real systems form a number theoretical hierarchy according to the prime p and and dimension of algebraic extension characterizing the entanglement.

Number theoretically state function reduction and state preparation could be seen as information generating processes projecting the physical states from either real or p-adic sectors of the state space to their intersection. Later an argument that these processes have a purely number theoretical interpretation will be developed based on the generalized notion of unitarity allowing the U-matrix to have matrix elements between the sectors of the state space corresponding to different number fields.

Number theoretic information measures at the space-time level

Quantum classical correspondence suggests that the notion of entropy should have also space-time counterpart. Entropy requires ensemble and both the p-adic non-determinism and the non-determinism of Kähler action allow to define the required ensemble as the ensemble of strictly deterministic regions of the space-time sheet. One can measure various observables at these space-time regions, and the frequencies for the outcomes are rational numbers of form $p_k = n(k)/N$, where N is the number of strictly deterministic regions of the space-time sheet. The number theoretic entropies are well defined and negative if p divides the integer N. Maximum is expected to result for the largest prime power factor of N. This would mean the possibility to assign a unique prime to a given real space-time sheet and thus the solve the basic problem created already by p-adic mass calculations.

The classical non-determinism resembles p-adic non-determinism in the sense that the space-time sheet obeys effective p-adic topology in some length and time scale range is consistent with this idea since p-adic fractality suggests that N is power of p.

2.5 TGD as a generalized number theory

The vision about a number theoretic formulation of quantum TGD is based on the gradual accumulation of wisdom coming from different sources. The attempts to find a formulation allowing to understand real and p-adic physics as aspects of some more general scenario have been an important stimulus and generated a lot of, not necessarily mutually consistent ideas, some of which might serve as building blocks of the final formulation. The original chapter representing the number theoretic vision as a consistent narrative grew so massive that I decided to divide it to three parts.

The first part is devoted to the p-adicization program attempting to construct physics in various number fields as an algebraic continuation of physics in the field of rationals (or appropriate extension of rationals). The program involves in essential manner the generalization of number concept obtained by fusing reals and p-adic number fields to a larger structure by gluing them together along common rationals. Highly non-trivial number theoretic conjectures are an i outcome of the program.

Second part focuses on the idea that the tangent spaces of space-time and imbedding space can be regarded as 4- *resp.* 8-dimensional algebras such that space-time tangent space defines sub-algebra of imbedding space. The basic candidates for the pair of algebras are hyper-quaternions and hyper-octonions. The problems are caused by the Euclidian signature of the Euclidian norm.

The great idea is that space-time surfaces X^4 correspond to hyper-quaternionic or co-hyper-quaternionic sub-manifolds of $HO = M^8$. The possibility to assign to X^4 a surface in $M^4 \times CP_2$ means a number theoretic analog for spontaneous compactification. Of course, nothing dynamical is involved: a dual relation between totally different descriptions of the physical world are in question. In the spirit of generalized algebraic geometry one can ask whether hyper-quaternionic space-time surfaces and their duals could be somehow assigned to hyper-octonion analytic maps $HO \to HO$, and there are good arguments suggesting that this is the case.

The third part is devoted to infinite primes. Infinite primes are in one-one correspondence with the states of super-symmetric arithmetic quantum field theories. The infinite-primes associated with hyper-quaternionic and hyper-octonionic numbers are the most natural ones physically because of the underlying Lorentz invariance, and the possibility to interpret them as momenta with mass squared equal to prime. Most importantly, the polynomials associated with hyper-octonionic infinite primes have automatically space-time surfaces as representatives so that space-time geometry becomes a representative for the quantum states.

2.5.1 The painting is the landscape

The work with TGD inspired theory of consciousness has led to a vision about the relationship of mathematics and physics. Physics is not in this view a model of reality but objective reality itself: painting is the landscape. One can also equate mathematics and physics in a well defined sense and the often implicitly assumed Cartesian theory-world division disappears. Physical realities are mathematical ideas represented by configuration space spinor fields (quantum histories) and quantum jumps between quantum histories give rise to consciousness and to the subjective existence of mathematician.

The concrete realization for the notion algebraic hologram based on the notion of infinite prime is a second new element. The notion of infinite rationals leads to the generalization of also the notion of finite number since infinite-dimensional space of real units obtained from finite rational valued ratios q of infinite integers divided by q. These units are not units in p-adic sense. The generalization to the (hyper-)quaternionic and (hyper-)octonionic context means that ordinary space-time points become infinitely structured and space-time point is able to represent even the quantum physical state of the Universe in its algebraic structure. Single space-time point becomes the Platonia not visible at the level of real physics but essential for mathematical cognition.

In this view evolution becomes also evolution of mathematical structures, which become more and more self-conscious quantum jump by quantum jump. The notion of p-adic evolution is indeed a basic prediction of quantum TGD but even this vision might be generalized by allowing rational-adic topologies for which topology is defined by a ring with unit rather than number field.

2.5.2　p-Adic physics as physics of cognition

Real and p-adic regions of the space-time as geometric correlates of matter and mind

The solutions of the equations determining space-time surfaces are restricted by the requirement that imbedding space-coordinates are real. When this is not the case, one might apply instead of a real completion with some rational-adic or p-adic completion: this is how rational-adic p-adic physics could emerge from the basic equations of the theory. One could interpret the resulting rational-adic or p-adic regions as geometrical correlates for 'mind stuff'.

p-Adic non-determinism implies extreme flexibility and therefore makes the identification of the p-adic regions as seats of cognitive representations very natural. Unlike real completion, p-adic completions preserve the information about the algebraic extension of rationals and algebraic coding of quantum numbers must be associated with 'mind like' regions of space-time. p-Adics and reals are in the same relationship as map and territory.

The implications are far-reaching and consistent with TGD inspired theory of consciousness: p-adic regions are present even at elementary particle level and provide some kind of model of 'self' and external world. In fact, p-adic physics must model the p-adic cognitive regions representing real elementary particle regions rather than elementary particles themselves!

The generalization of the notion of number and p-adicization program

The unification of real physics of material work and p-adic physics of cognition and intentionality leads to the generalization of the notion of number field. Reals and various p-adic number fields are glued along their common rationals (and common algebraic numbers too) to form a fractal book like structure. Allowing all possible finite-dimensional extensions of p-adic numbers brings additional pages to this "Big Book".

At space-time level the book like structure corresponds to the decomposition of space-time surface to real and p-adic space-time sheets. This has deep implications for the view about cognition. For instance, two points infinitesimally near p-adically are infinitely distant in real sense so that cognition becomes a cosmic phenomenon.

One general idea which results as an outcome of the generalized notion of number is the idea of a universal function continuable from a function mapping rationals to rationals or to a finite extension of rationals to a function in any number field. This algebraic continuation is analogous to the analytical continuation of a real analytic function to the complex plane. Rational functions with rational coefficients are obviously functions satisfying this constraint. Algebraic functions with rational coefficients satisfy this requirement if appropriate finite-dimensional algebraic extensions of p-adic numbers are allowed. Exponent function is such a function.

For instance, residue calculus might be generalized so that the value of an integral along the real axis could be calculated by continuing it instead of the complex plane to any number field via its values in the subset of rational numbers forming the rim of the book like structure having number fields as its pages. If the poles of the continued function in the finitely extended number field allow interpretation as real numbers it might be possible to generalize the residue formula. One can also imagine of extending residue calculus to any algebraic extension. An interesting situation arises when the poles correspond to extended p-adic rationals common to different pages of the "great book". Could this mean that the integral could be calculated at any page having the pole common. In particular, could a p-adic residue integral be calculated in the ordinary complex plane by utilizing the fact that in this case numerical approach makes sense.

Algebraic continuation is the basic tool of p-adicization program. Entire physics of the TGD Universe should be algebraically continuable to various number fields. Real number based physics would define the physics of matter and p-adic physics would describe correlates of cognition and intentionality. The basic stumbling block of this program is integration and algebraic continuation should allow to circumvent this difficulty. Needless to say, the requirement that the continuation exists must pose immensely tight constraints on the physics.

Due to the fact that real and p-adic topologies are fundamentally different, ultraviolet and infrared cutoffs in the set of rationals are unavoidable notions and correspond to a hierarchy of different physical phases on one hand and different levels of cognition on the other hand. Two types of cutoffs are predicted: p-adic length scale cutoff and a cutoff due to phase resolution. The latter cutoff seems to correspond naturally to the hierarchy of algebraic extensions of p-adic numbers and Beraha numbers $B_n = 4cos^2(\pi/n)$, $n \geq 3$ related closely to the hierarchy of quantum groups, braid groups, and II_1 factors of von Neumann algebra [O4]. This cutoff hierarchy seems to relate closely to the hierarchy of cutoffs defined by the hierarchy of subalgebras of the super-canonical algebra defined by the hierarchy of sets $(z_1, ...z_n)$, where z_i are the first n non-trivial zeros of Riemann Zeta [C5]. Hence there are good hopes that the p-adicization program might unify apparently unrelated branches of mathematics.

2.5.3 Space-time-surface as a hyper-quaternionic sub-manifold of hyper-octonionic imbedding space?

Second thread in the development of ideas has been present for only few years ideas inspired by the possibility that quaternions and octonions might allow a deeper understanding of TGD. This thread emerged from the discussions with Tony Smith which stimulated very general ideas about space-time surface as associative, quaternionic sub-manifold of octonionic 8-space. Also the observation that quaternionic and octonionic primes have norm squared equal to prime in complete accordance with p-adic length scale hypothesis, led to suspect that the notion of primeness for quaternions, and perhaps even for octonions, might be fundamental for the formulation of quantum TGD [E2]. It turned out that, much in spirit with transition from Riemannian to pseudo-Riemannian geometry, hyper-quaternins and hyper-octonions are forced by physical considerations.

Transition from string models to TGD as replacement of real/complex numbers with quaternions/octonions

One can fairly say, that quantum TGD results from string model with the pair of real and complex numbers replaced with the pair of hyper-quaternions and hyper-octonions. Hyper is necessary in order to take into the Minkowskian signature of the metric.

Space-time identified as a hyper-quaternionic sub-manifold of the hyper-octonionic space in the sense that the tangent space of the space-time surface defines a hyper-quaternionic sub-algebra of the hyper-octonionic tangent space of H at each space-time point, looks an attractive idea. Second possibility is that the tangent space-algebra of the space-time surface is either associative or co-associative at each point. One can also consider possibility that the dynamics of the space-time surface is determined from the requirement that space-time surface is algebraically closed in the sense that tangent space at each point has this property. Also the possibility that the property in question is associated with the normal space at each point of X^4 can be considered.

Some delicacies are caused by the question whether the induced algebra at X^4 is just the hyper-octonionic product or whether the algebra product is projected to the space-time surface. If the normal part of the product is projected out, the space-time algebra closes automatically.

The first guess would be that space-time surfaces are hyper-quaternionic sub-manifolds of hyper-octonionic space $HO = M^8$ with the property that complex structure is fixed and same at all points of space-time surface. This corresponds to a global selection of a preferred octonionic imaginary unit. The automorphisms leaving this selection invariant form group $SU(3)$ identifiable as color group. The selections of hyper-quaternionic sub-space under this condition are parameterized by CP_2. This means that each 4-surface in HO defines a 4-surface in $M^4 \times CP_2$ and one can speak about number-theoretic analog of spontaneous compactification having of course nothing to do with dynamics. It would be possible to make physics in two radically different geometric pictures: HO picture and $H = M^4 \times CP_2$ picture.

For a theoretical physicists of my generation it is easy to guess that the next step is to realize that it is possible to fix the preferred octonionic imaginary at each point of HO separately so that local $S^6 = G_2/SU(3)$, or equivalently the local group G_2 subject to $SU(3)$ gauge invariance, characterizes the possible choices of hyper-quaternionic structure with a preferred imaginary unit. $G_2 \subset SO(7)$ is the automorphism group of octonions, and appears also in M-theory. This local choice has interpretation as a fixing of the plane of non-physical polarizations and rise to degeneracy which is a good candidate for the ground state degeneracy caused by the vacuum extremals.

$OH - -M^4 \times CP_2$ duality allows to construct a foliation of HO by hyper-quaternionic space-time surfaces in terms of maps $HO \rightarrow SU(3)$ satisfying certain integrability conditions guaranteeing that the distribution of hyper-quaternionic planes integrates to a foliation by 4-surfaces. In fact, the freedom to fix the preferred imaginary unit locally extends the maps to $HO \to G_2$ reducing to maps $HO \to SU(3) \times S^6$ in the local trivialization of G_2. This foliation defines a four-parameter family of 4-surfaces in $M^4 \times CP_2$ for each local choice of the preferred imaginary unit. The dual of this foliation defines a 4-parameter familiy co-hyper-quaternionic space-time surfaces.

Hyper-octonion analytic functions $HO \to HO$ with real Taylor coefficients provide a physically motivated ansatz satisfying the integrability conditions. The basic reason is that hyper-octonion analyticity is not plagued by the complications due to non-commutativity and non-associativity. Indeed, this notion results also if the product is Abelianized by assuming that different octonionic imaginary units multiply to zero. A good candidate for the HO dynamics is free massless Dirac action with Weyl condition for an octonion valued spinor field using octonionic representation of gamma matrices and coupled to the G_2 gauge potential defined by the tensor 7×7 tensor product of the imaginary parts of spinor fields.

The basic conjecture is that the absolute minima of Kähler action in $H = M^4 \times CP_2$ correspond to the hyper-quaternion analytic surfaces in HO. The map $f : HO \to S^6$ would probably satisfy some constraints posed by the requirement that the resulting surfaces define solutions of field equations in $M^4 \times CP_2$ picture. This conjecture has several variants. It could be that only the asymptotic behavior corresponds to hyper-quaternion analytic function but that hyper-quaternionicity is a general property of absolute minima. It could also be that maxima of Kähler function correspond to this kind of 4-surfaces. The encouraging hint is the fact that Hamilton-Jacobi coordinates coding for the local selection of the plane of non-physical polarizations, appear naturally also in the construction of general solutions of field equations [D1].

Physics as a generalized algebraic number theory and Universe as algebraic hologram

The third stimulus encouraging to think that TGD might be reduced to algebraic number theory and algebraic geometry in some generalized sense, came from the work with Riemann hypothesis [E8]. One can assign to Riemann Zeta a super-conformal quantum field theory and identify Zeta as a Hermitian form in the state space possibly defining a Hilbert space metric. The proposed form of the Riemann hypothesis implies that the zeros of ζ code for infinite primes which in turn have interpretation as Fock states of a super-symmetric quantum field theory if the proposed vision is correct.

A further stimulus came from the realization that algebraic extensions of rationals, which make possible a generalization of the notion of prime, could provide enormous representative and information storage power in arithmetic quantum field theory. Algebraic symmetries defined as transformations preserving the algebraic norm represent new kind of symmetries commuting with ordinary quantum numbers. Fractal scalings and discrete symmetries are in question so that the notion of fractality emerges to the fundamental physics in this manner.

The basic observation, completely consistent with fractality, is that these symmetries make possible what might be called *algebraic hologram*. The algebraic quantum numbers associated with elementary particle depend on the environment of the particle. The only possible conclusion seems to be that these fractal quantum numbers provide some kind of 'cognitive representation' about external world. This kind of an algebraic hologram would be in complete accordance with fractality and would provide first principle realization for fractality observed everywhere in Nature but not properly understood in standard physics framework. A further basic idea which emerged was the principle of *algebraic democracy*: all possible algebraic extensions of rational (hyper-)quaternions and (hyper-)octonions are possible and emerge dynamically as properties of physical systems in algebraic physics.

2.5.4 Infinite primes and physics in TGD Universe

The notion of infinite primes emerged originally from TGD inspired theory of consciousness [TGDmathc] but it soon turned out that the notion could be used to build a number theoretic interpretation of quantum TGD and relate quantum to classical. Also the notion of infinite-P p-adicity emerges naturally and could replaces real topology with something more refined and appropriate for description of the space-time correlates of cognition.

Infinite primes and infinite hierarchy of second quantizations

The discovery of infinite primes was one important step in the development suggesting strongly the possibility to reduce physics to number theory. The construction of infinite primes can be regarded as a repeated second quantization of a super-symmetric arithmetic quantum field theory. Later it became clear that the process generalizes so that it applies even in the case of hyper-quaternionic and hyper-octonionic primes. This hierarchy of second quantizations means enormous generalization of physics to what might be regarded a physical counterpart for a hierarchy of abstractions about abstractions about.... The ordinary second quantized quantum physics corresponds only to the lowest level infinite primes.

What is remarkable is that one has quite realistic possibilities to understand the quantum numbers of physical particles in terms of hyper-octonionic infinite primes. Also the TGD inspired model for $1/f$ noise [I5] based on thermal arithmetic quantum field theory encouraged also to consider the idea about hyper-quaternionic or hyper-octonionic arithmetic quantum field theory as an essential element of quantum TGD.

Infinite primes as a bridge between quantum and classical

The final stimulus came from the observation stimulated by algebraic number theory [ca5]. Infinite primes can be mapped to polynomial primes and this observation allows to identify completely generally the spectrum of infinite primes whereas hitherto it was possible to construct explicitly only what might be called generating infinite primes. Infinite primes allow nice interpretation as Fock states of a second quantized super-symmetric quantum field theory. Also bound states are included.

This in turn led to the observation that one can represent infinite primes (integers) geometrically as surfaces related to the polynomials associated with infinite primes (integers). Thus infinite primes would serve as a bridge between Fock-space descriptions and geometric descriptions of physics: quantum and classical. Geometric objects could be seen as concrete representations of infinite numbers providing amplification of infinitesimals to macroscopic deformations of space-time surface. We see the infinitesimals as concrete geometric shapes!

The original mapping to 4-surfaces inspired by algebraic geometry was essentially as zeros of polynomials. It however turned out that the mapping is more delicate and based on the idea that space-time surfaces correspond to hyper-quaternionic or co-hyper-quaternionic sub-manifolds of imbedding space with hyper-octonionic structure. Also the attribute maximally associative or co-associate could be used. The assignment of a space-time surface to an infinite prime boils down to an assignment of a hyper-octonion analytic polynomial to infinite prime, which in turn defines a foliation of $M^4 \times CP_2$ by hyper-quaternionic space-time surfaces. The procedure generalizes also to the higher levels of the hierarchy and the natural interpretation is in terms of the hierarchical structure of the many-sheeted space-time.

The connection with the basic ideas of algebraic geometry from the possibility to order space-time surfaces according to the complexity of the polynomial involved (at higher levels rational coefficients of the polynomial are replaced with rational polynomials). In particular, the notions of degree and genus make sense for space-time surface.

Various equivalent characterizations of space-times as surfaces

The idea about space-times as associative, hyper-quaternionic surfaces of a hyper-octonionic imbedding space M^8 and the notion of infinite prime serving as a bridge between classical and quantum are the two basic tenets of the algebraic approach. This vision leads to an equivalence of quite different views about space-time: space-time as an associative/hyper-quaternionic or co-associative/co-hyperquaterionic surface of an hyper-octonionic imbedding space $HO = M^8$; space-time as a surface in $H = M^4 \times CP_2$; space-time as a geometric counterpart of an infinite prime representing also Fock state identifiable as a particular ground state of super-canonical representation; and finally, space-time surface as an absolute minimum of the Kähler action. The great challenge is to prove that the last characterization is equivalent with the others.

Infinite primes and quantum gravitational holography

Infinite primes emerge naturally in the realization of the quantum gravitational holography in terms of the modified Dirac operator and provide a deeper understanding of the basic aspects of the configuration space geometry.

a) Two types of infinite primes are predicted corresponding to the two types of fermionic vacua $X \pm 1$, where X is the product of all finite primes. The physical interpretation for the two types of infinite primes $X \pm 1$ is in terms of two quantizations for which creation and oscillator operators change role and which correspond to the two signs of inertial energy in TGD Universe. In particular, phase conjugate photons would be negative energy photons erratically believed to reduce to standard physics.

b) The new view about gravitational and inertial masses forced by TGD leads also the view that positive and negative energy space-time sheets are created pairwise at space-like 3-surfaces located at 7-D light-like causal determinants $X^7_{\pm} = \delta M^4_{\pm} \times CP_2$. The conjecture is that the ratio of Dirac determinants associated with the positive and negative energy space-time sheets, which is finite, equals to the exponent of Kähler function which would be thus determined completely by the data at 3-dimensional causal determinants and realizing quantum gravitational holography.

c) The spectra associated with the space-time sheets X^4_+ and X^4_- meeting at X^3 would correspond to the infinite primes built from the vacua corresponding to the infinite primes $X \pm 1$. The close analogy of the product of all finite hyper-octonionic primes with Dirac determinant suggest that the ratio of the determinants corresponds to the ratio of infinite primes defining X^4_+ and X^4_-. The theory predicts the dependence of the eigenvalues of the modified Dirac operator on the value of the Kähler action. Both Kähler coupling strength and gravitational coupling strength are expressible in terms of the finite primes characterizing the ratio of the infinite primes and this ratio depends on the p-adic prime characterizing X^4_+ and X^4_-.

d) Some modes of the spectrum of the modified Dirac operator at $X^4_{\pm}$ become zero modes, and by the resulting spectral asymmetry the ratio of the determinants differs from unity. Thus the spectral asymmetry or the infinite primes defining the space-time sheets X^4_+ and X^4_- is all that would be needed to deduce the value of the vacuum functional once causal determinants are known.

2.5.5 Infinite primes and more precise view about p-adic length scale hypothesis

Number theoretical considerations allow to develop more quantitative vision about the how p-adic length scale hypothesis relates to the ideas just described.

How to define the notion of elementary particle?

p-Adic length scale hierarchy forces to reconsider carefully also the notion of elementary particle. p-Adic mass calculations led to the idea that particle can be characterized uniquely by single p-adic prime characterizing its mass squared. It however turned out that the situation is probably not so simple.

The work with modelling dark matter suggests that particle could be characterized by a collection of p-adic primes to which one can assign weak, color, em, gravitational interactions, and possibly also other interactions. It would also seem that only the space-time sheets containing common primes in this collection can interact. This leads to the notions of relative and partial darkness. An entire hierarchy of weak and color physics such that weak bosons and gluons of given physics are characterized by a given p-adic prime p and also the fermions of this physics contain space-time sheet characterized by same p-adic prime, say M_{89} as in case of weak interactions. In this picture the decay widths of weak bosons do not pose limitations on the number of light particles if weak interactions for them are characterized by p-adic prime $p \neq M_{89}$. Same applies to color interactions.

The p-adic prime characterizing the mass of the particle would perhaps correspond to the largest p-adic prime associated with the particle. Graviton which corresponds to infinitely long ranged interactions, could correspond to the same p-adic prime or collection of them common to all particles. This might apply also to photons. Infinite range might mean that the join along boundaries bonds mediating these interactions can be arbitrarily long but their transversal sizes are characterized by the p-adic length scale in question.

The natural question is what this collection of p-adic primes characterizing particle means? The hint about the correct answer comes from the number theoretical vision, which suggests that at fundamental level the branching of boundary components to two or more components, completely analogous to the branching of line in Feynman diagram, defines vertices [C2, C5, E3].

a) If space-time sheets correspond holographically to multi-p p-adic topology such that largest p determines the mass scale, the description of particle reactions in terms of branchings indeed makes sense. This picture allows also to understand the existence of different scaled up copies of QCD and weak physics. Multi-p p-adicity could number theoretically correspond to q-adic topology for $q = m/n$ a rational number consistent with p-adic topologies associated with prime factors of m and n ($1/p$-adic topology is homeomorphic with p-adic topology).

b) One could also imagine that different p-adic primes in the collection correspond to different space-time sheets condensed at a larger space-time sheet or boundary components of a given space-time sheet. If the boundary topologies for gauge bosons are completely mixed, as the model of hadrons forces to conclude, this picture is consistent with the topological explanation of the family replication phenomenon and the fact that only charged weak currents involve mixing of quark families. The problem is how to understand the existence of different copies of say QCD. The second difficult question is why the branching leads always to an emission of gauge boson characterized by a particular p-adic prime, say M_{89}, if this p-adic prime does not somehow characterize also the particle itself.

What effective p-adic topology really means?

The need to characterize elementary particle p-adically leads to the question what p-adic effective topology really means. p-Adic mass calculations leave actually a lot of room concerning the answer to this question.

a) The naivest option is that each space-time sheet corresponds to single p-adic prime. A more general possibility is that the boundary components of space-time sheet correspond to different p-adic primes. This view is not favored by the view that each particle corresponds to a collection of p-adic primes each characterizing one particular interaction that the particle in question participates.

b) A more abstract possibility is that a given space-time sheet or boundary component can correspond to several p-adic primes. Indeed, a power series in powers of given integer n gives rise to a well-defined power series with respect to all prime factors of n and effective multi-p-adicity could emerge at the level of field equations in this manner.

One could say that space-time sheet or boundary component corresponds to several p-adic primes through its effective p-adic topology in a hologram like manner. This option is the most flexible one as far as physical interpretation is considered. It is also supported by the number theoretical considerations predicting the value of gravitational coupling constant [E3].

An attractive hypothesis is that only space-time sheets characterized by integers n_i having common prime factors can be connected by join along boundaries bonds and can interact by particle exchanges and that each prime p in the decomposition corresponds to a particular interaction mediated by an elementary boson characterized by this prime.

Do infinite primes code for q-adic effective space-time topologies?

Besides the hierarchy of space-time sheets, TGD predicts, or at least suggests, several hierarchies such as the hierarchy of infinite primes [E3], hierarchy of Jones inclusions [O5], hierarchy of dark matters with increasing values of $\hbar$ [F9, J6], the hierarchy of extensions of given p-adic number field, and the hierarchy of selves and quantum jumps with increasing duration with respect to geometric time. There are good reasons to expect that these hierarchies are closely related.

1. Some facts about infinite primes

The hierarchy of infinite primes can be interpreted in terms of an infinite hierarchy of second quantized TGD super-symmetric arithmetic quantum field theories allowing a generalization to quaternionic or perhaps even octonionic context [E3]. Infinite primes, integers, and rationals have decomposition to primes of lower level.

Infinite prime has fermionic and bosonic parts having no common primes. Fermionic part is finite and corresponds to an integer containing and bosonic part is an integer multiplying the product of all primes with fermionic prime divided away. The infinite prime at the first level of hierarchy corresponds in a well defined sense a rational number $q = m/n$ defined by bosonic and fermionic integers m and n having no common prime factors.

2. Do infinite primes code for effective q-adic space-time topologies?

The most obvious question concerns the space-time interpretation of this rational number. Also the question arises about the possible relation with the integers characterizing space-time sheets having interpretation in terms of multi-p-adicity. On can assign to any rational number $q = m/n$ so called q-adic topology. This topology is not consistent with number field property like p-adic topologies. Hence the rational number q assignable to infinite prime could correspond to an effective q-adic topology.

If this interpretation is correct, arithmetic fermion and boson numbers could be coded into effective q-adic topology of the space-time sheets characterizing the non-determinism of Kähler action in the relevant length scale range. For instance, the power series of $q > 1$ in positive powers with integer coefficients in the range $[0, q)$ define q-adically converging series, which also converges with respect to the prime factors of m and can be regarded as a p-adic power series. The power series of q in negative powers define in similar converging series with respect to the prime factors of n.

I have proposed earlier that the integers defining infinite rationals and thus also the integers m and n characterizing finite rational could correspond at space-time level to particles with positive *resp.* negative time orientation with positive *resp.* negative energies. Phase conjugate laser beams would represent one example of negative energy states. With this interpretation super-symmetry exchanging the roles of m and n and thus the role of fermionic and bosonic lower level primes would correspond to a time reversal.

a) The first interpretation is that there is single q-adic space-time sheet and that positive and negative energy states correspond to primes associated with m and n respectively. Positive (negative) energy space-time sheets would thus correspond to p-adicity ($1/p$-adicity) for the field modes describing the states.

b) Second interpretation is that particle (in extremely general sense that entire universe can be regarded as a particle) corresponds to a pair of positive and negative energy space-time sheets labelled by m and n characterizing the p-adic topologies consistent with $m-$ and n-adicities. This looks natural since Universe has necessary vanishing net quantum numbers. Unless one allows the non-uniqueness due to $m/n = mr/nr$, positive and negative energy space-time sheets can be connected only by $\#$ contacts so that positive and negative energy space-time sheets cannot interact via the formation of $\#_B$ contacts and would be therefore dark matter with respect to each other.

Positive energy particles and negative energy antiparticles would also have different mass scales. If the rate for the creation of $\#$ contacts and their CP conjugates are slightly different, say due to the presence of electric components of gauge fields, matter antimatter asymmetry could be generated primordially.

These interpretations generalize to higher levels of the hierarchy. There is a homomorphism from infinite rationals to finite rationals. One can assign to a product of infinite primes the product of the corresponding rationals at the lower level and to a sum of products of infinite primes the sum of the corresponding rationals at the lower level and continue the process until one ends up with a finite rational. Same applies to infinite rationals. The resulting rational $q = m/n$ is finite and defines q-adic effective topology, which is consistent with all the effective p-adic topologies corresponding to the primes appearing in factorizations of m and n. This homomorphism is of course not 1-1.

If this picture is correct, effective p-adic topologies would appear at all levels but would be dictated by the infinite-p p-adic topology which itself could refine infinite-P p-adic topology [E3] coding information too subtle to be catched by ordinary physical measurements [O4].

Obviously, one could assign to each elementary particle infinite prime, integer, or even rational to this a rational number $q = m/n$. q would associate with the particle q-adic topology consistent with a collection of p-adic topologies corresponding to the prime factors of m and n and characterizing the interactions that the particle can participate directly. In a very precise sense particles would represent both infinite and finite numbers.

Under what conditions space-time sheets can be connected by $\#_B$ contact?

Assume that particles are characterized by a p-adic prime determining it mass scale plus p-adic primes characterizing the gauge bosons to which they couple and assume that $\#_B$ contacts mediate gauge interactions. The question is what kind of space-time sheets can be connected by $\#_B$ contacts.

a) The first working hypothesis that comes in mind is that the p-adic primes associated with the two space-time sheets connected by $\#_B$ contact must be identical. This would require that particle is many-sheeted structure with no other than gravitational interactions between various sheets. The problem of the multi-sheeted option is that the characterization of events like electron-positron annihilation to a weak boson looks rather clumsy.

b) If the notion of multi-p p-adicity is accepted, space-time sheets are characterized by integers and the largest prime dividing the integer might characterize the mass of the particle. In this case a common prime factor p for the integers characterizing the two space-time sheets could be enough for the possibility of $\#_B$ contact and this contact would be characterized by this prime. If no common prime factors exist, only $\#$ contacts could connect the space-time sheets. This option conforms with the number theoretical vision. This option would predict that the transition to large $\hbar$ phase occurs simultaneously for all interactions.

What about the integer characterizing graviton?

If one accepts the hypothesis that graviton couples to both visible and dark matter, graviton should be characterized by an integer dividing the integers characterizing all particles. This leaves two options.

Option I: gravitational constant characterizes graviton number theoretically

The argument leading to an expression for gravitational constant in terms of CP_2 length scale led to the proposal that the product of primes $p \leq 23$ are common to all particles and one interpretation was in terms of multi-fractality. If so, graviton would be characterized by a product of some or all primes $p \leq 23$ and would thus correspond to a very small p-adic length scale. This might be also the case for photon although it would seem that photon cannot couple to dark matter always. $p = 23$ might characterize the transversal size of the massless extremal associated with the space-time sheet of graviton.

Option II: graviton behaves as a unit with respect to multiplication

One can also argue that if the largest prime assignable to a particle characterizes the size of the particle space-time sheet it does not make sense to assign any finite prime to a massless particle like graviton. Perhaps graviton corresponds to simplest possible infinite prime $P = X \pm 1$, X the product of all primes.

As found, one can assign to any infinite prime, integer, and rational a rational number $q = m/n$ to which one can assign a q-adic topology as effective space-time topology and as a special case effective p-adic topologies corresponding to prime factors of m and n.

In the case of $P = X \pm 1$ the rational number would be equal to ± 1. Graviton could thus correspond to $p = 1$-adic effective topology. The "prime" $p = 1$ indeed appears as a factor of any integer so that graviton would couple to any particle. Formally the 1-adic norm of any number would be 1 or 0 which would suggest that a discrete topology is in question.

The following observations help in attempts to interpret this.

a) CP_2 type extremals having interpretation as gravitational instantons are non-deterministic in the sense that M^4 projection is random light-like curve. This condition implies Virasoro conditions which suggests interpretation in terms topological quantum theory limit of gravitation involving vanishing four-momenta but non-vanishing color charges. This theory would represent gravitation at the ultimate CP_2 length scale limit without the effects of topological condensation. In longer length scales a hierarchy of effective theories of gravitation corresponds to the coupling of space-time sheets by join along boundaries bonds would emerge and could give rise to "strong gravities" with strong gravitational constant proportional to L_p^2. It is quite possible that the M-theory based vision about duality between gravitation and gauge interactions applies to electro-weak interactions and in these "strong gravities".

b) p-Adic length scale hypothesis $p \simeq 2^k$, k integer, implies that $L_k \propto \sqrt{k}$ corresponds to the size scale of causal horizon associated with $\#$ contact. For $p = 1$ k would be zero and the causal horizon would contract to a point which would leave only generalized Feynman diagrams consisting of CP_2 type vacuum extremals moving along random light-like orbits and obeying Virasoro conditions so that interpretation as a kind of topological gravity suggests itself.

c) $p = 1$ effective topology can make marginally sense for vacuum extremals with vanishing Kähler form and carrying only gravitational charges. The induced Kähler form vanishes identically by the mere assumption that X^4, be it continuous or discontinuous, belongs to $M^4 \times Y^2$, Y^2 a Lagrange sub-manifold of CP_2.

Why topological graviton, or whatever the particle represented by CP_2 type vacuum extremals should be called, should correspond to the weakest possible notion of continuity? The most plausible answer is that discrete topology is *consistent* with any other topology, in particular with any p-adic topology. This would express the fact that CP_2 type extremals can couple to any p-adic prime. The vacuum property of CP_2 type extremals implies that the splitting off of CP_2 type extremal leaves the physical state invariant and means effectively multiplying integer by $p = 1$.

It seems that Option I suggested by the deduction of the value of gravitational constant looks more plausible as far as the interpretation of gravitation is considered. This does not however mean that CP_2 type vacuum extremals carrying color quantum numbers could not describe gravitational interactions in CP_2 length scale.

2.5.6 Complete algebraic, topological, and dimensional democracy?

Without the notion of Platonia allowing realization of all imaginable algebraic structures cognitively but leaving no trace on the physics of matter, the idea about dimensional democracy would look almost compelling despite the fact that it might well be in conflict with the special role of the dimensions associated with the classical number fields. One can imagine several realizations of this idea.

a) The most (if not the only) plausible realization for the dimensional hierarchy would be following. Both fractal cosmology, non-determinism of Kähler action, and Poincare invariance favor the option in which configuration space is a union of sectors characterized by unions of future and past light cones $M^4_\pm(a)$ where a characterizes the position a of the dip of the light-cone in M^4. Future/past dichotomy would correspond to positive/negative energy dichotomy and to the two kinds of infinite primes constructed from $X \pm 1$, X the product of all finite primes. Hence the cm degrees of freedom for the sectors of the configuration space would correspond to the union of the spaces $(M^4)^m \times (M^4)^n$ of dimension $D = 4(m+n)$, and the dimensional democracy would conform with the 8-dimensionality of the imbedding space.

b) The most plausible identification consistent with the p-adic length scale hierarchy is as unions of n disjoint 4-surfaces of H. This correspondence is completely analogous to that involved when the configuration space of n point-like particles is identified as $(E^3)^n$ in wave mechanics.

c) One might also consider of assigning with hyper-octonionic infinite primes of level n $4n$-dimensional surfaces in $8n$-dimensional space $H^n = (M^4_+ \times CP_2)^n$. This would suggests a dimensional hierarchy of space-time surfaces and a complete dimensional and algebraic democracy: quite a considerable generalization of quantum TGD from its original formulation. This option does not however look physically plausible since it is not consistent with the hierarchical "abstractions about abstractions" structure of infinite primes and corresponding space-time representations.

Since quantum field theories are based on the notion of point like particles, the hierarchy of arithmetic quantum field theories associated with infinite primes cannot code entire quantum TGD but only the ground states of the super-canonical representations. This might however be the crucial element needed to understand the construction S-matrix of quantum TGD at the general level.

One can imagine also a topological democracy and an evolution of algebraic topological structures. At the lowest, primordial level there are just algebraic surfaces allowing no completion to smooth ...-adic or real surfaces, and defined only in algebraic extensions of rationals by algebraic field equations. At higher levels rational-adic, p-adic and even infinite-P p-adic completions of infinite primes could appear and provide natural completions of function spaces. Of course, all these generalizations might make sense only as cognitive structures in Platonia and it is comforting to know that there is room in just a single point of TGD Universe for all this richness of imaginable structures!

The reader not familiar with the basic algebra of quaternions and octonions is encourated to study some background material: the homepage of Tony Smith provides among other things an excellent introduction to quaternions and octonions [cb5]. String model builders are beginning to grasp the potential importance of octonions and quaternions and the articles about possible applications of octonions [cb2, cb3, bb1] provide an introduction to octonions using the language of physicist.

Personally I found quite frustrating to realize that I had neglected totally learning of the basic ideas of algebraic geometry, despite its obvious potential importance for TGD and its applications in string models. This kind of losses are the price one must pay for working outside the scientific community. It is not easy for a physicist to find readable texts about algebraic geometry and algebraic number theory from the bookshelves of mathematical libraries. The book "Algebraic Geometry for Scientists and Engineers" by Abhyankar [ce1], which is not so elementary as the name would suggest, introduces in enjoyable manner the basic concepts of algebraic geometry and binds the basic ideas with the more recent developments in the field. "Problems in Algebraic Number Theory" by Esmonde and Murty [ca5] in turn teaches algebraic number theory through exercises which concretize the abstract ideas. The book "Invitation to Algebraic Geometry" by K. E. Smith. L. Kahanpää, P. Kekäläinen and W. Traves is perhaps the easiest and most enjoyable introduction to the topic for a novice. It also contains references to the latest physics inspired work in the field.

Chapter 3

Overall view about Quantum TGD

3.1 Introduction

This chapter provides a summary about quantum TGD. The discussions are based on the general vision that the quantum states of the Universe correspond to the modes of classical spinor fields in the "world of the classical worlds" identified as the infinite-dimensional configuration space of 3-surfaces of $H = M^4 \times CP_2$ (more or less-equivalently, the corresponding 4-surfaces defining generalized Bohr orbits) [B2, B3]. The following topics are discussed on basis this vision.

a) The two basic approaches to the construction of the Kähler geometry based on direct guess of the Kähler function and construction of Kähler metric from the hypothesis that configuration space is a union of infinite-dimensional symmetric spaces with metric possessing maximal isometry group are summarized.

b) The great vision guiding the construction of the configuration space spinor structure is that the second quantization of the induced spinor fields can be understood geometrically in terms of the configuration space spinor structure in the sense that the anti-commutation relations for configuration space gamma matrices require anti-commutation relations for the oscillator operators for free second quantized induced spinor fields.

c) p-Adic mass calculations were carried out for the first time for about one decade ago. The calculations rely p-adic thermodynamics giving the p-adic mass squared essentially as the thermal expectation value of conformal weight. Real mass squared is obtained by mapping the p-adic mass squared to its real counterpart using so called canonical identification. The justification for this picture from quantum TGD and for what massivation corresponds at space-time level is sketched.

d) It is not yet possible to deduce the length scale evolution of gauge coupling constants from Quantum TGD proper. Quantum classical correspondence however encourages the hope that it might be possible to achieve some understanding of the coupling constant evolution by using the classical theory. This turns out to be the case and the earlier speculative picture about gauge coupling constants associated with a given space-time sheet as R(enormalization)G(roup) invariants finds support. It remains an open question whether gravitational coupling constant is RG invariant inside given space-time sheet. The discrete p-adic coupling constant evolution replacing in TGD framework the ordinary RG evolution allows also formulation at space-time level as also does the evolution of $\hbar$ associated with the phase resolution.

e) Concerning the construction of the S-matrix the basic philosophy is provided by TGD inspired theory of consciousness which can be seen as a generalization of quantum measurement theory. The huge symmetries of the configuration space geometry pose extremely tight constraints on the S-matrix. The identification of the classical space-time surface as a generalized Bohr orbit in turn implies that there is no path integral over space-time surfaces. Hence there should be no summation over loops in the generalized Feynman diagrams. Or stated using the language for generalized stringy diagrams: all generalized stringy diagrams differing by loops are equivalent. The formulation of this picture in terms of Hopf algebras is briefly discussed.

3.2   Physics as geometry of configuration space spinor fields

The construction of the configuration space geometry has proceeded rather slowly. The experimentation with various ideas has however led to the identification of the basic constraints on the configuration space geometry.

3.2.1   Reduction of quantum physics to the Kähler geometry and spinor structure of configuration space of 3-surfaces

The basic philosophical motivation for the hypothesis that quantum physics could reduce to the construction of configuration space Kähler metric and spinor structure, is that infinite-dimensional Kähler geometric existence could be unique not only in the sense that the geometry of the space of 3-surfaces could be unique but that also the dimension of the space-time is fixed to $D = 4$ by this requirement and $M_+^4 \times CP_2$ is the only possible choice of imbedding space. This optimistic vision derives from the work of Dan Freed with loops spaces demonstrating that they possess unique Kähler geometry and from the fact that in $D > 1$ case the existence of Riemann connection, finiteness of Ricci tensor, and general coordinate invariance poses even stronger constraints.

3.2.2   Constraints on configuration space geometry

The detailed considerations of the constraints on configuration space geometry suggests that it should possess at least the following properties.

a) Metric should be Kähler metric. This property is necessary if one wants to geometrize the oscillator algebra used in the construction of the physical states and to obtain a well defined divergence free functional integration in the configuration space.

b) Metric should allow Riemann connection, which, together with the Kähler property, very probably implies the existence of an infinite dimensional isometry group as the construction of Kähler geometry for the loop spaces demonstrates [df6].

c) The so called symmetric spaces classified by Cartan [ba2] are Cartesian products of the coset spaces G/H with maximal isometry group G. Symmetric spaces possess G invariant metric and curvature tensor is constant so that all points of the symmetric space are metrically equivalent. Symmetric space structure means that the Lie-algebra of G decomposes as

$$g = h \oplus t \ ,$$
$$[h,h] \subset h \ , \quad [h,t] \subset t \ , \quad [t,t] \subset h \ ,$$

where g and h denote the Lie-algebras of G and H respectively and t denotes the complement of h in g. The existence of the $g = t + h$ decomposition poses an extremely strong constraint on the symmetry group G.

In the infinite-dimensional context symmetric space property would mean a drastic calculational simplification. The most one can hope is that configuration space is expressible as a union $\cup_i (G/H)_i$ of symmetric spaces. Reduction to a union of G/H is the best one can hope since 3-surface of Planck size cannot be metrically equivalent with a 3-surface having the size of galaxy! The coordinates labelling the symmetric spaces in this union do not appear as differentials in the line element of configuration space and are thus zero modes. They correspond to non-quantum fluctuating degrees of freedom in a well defined sense and are identifiable as classical variables of quantum measurement theory.

d) Metric should be Diff^4 (not only $Diff^3$!) invariant and degenerate and the definition of the metric should associate a unique space-time surface $X^4(X^3)$ to a given 3-surface X^3 to act on. This requirement is absolutely crucial for all developments.

e) Divergence cancellation requirement for the functional integral over the configuration space requires that the metric is Ricci flat and thus satisfies vacuum Einstein equations.

3.2.3   Configuration space as a union of symmetric spaces

In the finite-dimensional context, globally symmetric spaces are of form G/H and connection and curvature are independent of the metric, provided it is left invariant under G. Good guess is that same holds true in the infinite-dimensional context. The task is to identify the infinite-dimensional groups G and H. Only quite recently, more than seven years after the discovery of the candidate for Kähler function defining the metric, it became finally clear that these identifications follow quite nicely from Diff^4 invariance and Diff^4 degeneracy.

The crux of the matter is Diff^4 degeneracy : all 3-surfaces on the orbit of 3-surface X^3 must be physically equivalent so that one can effectively replace all 3-surfaces Z^3 on the orbit of X^3 with a suitably chosen surface Y^3 on the orbit of X^3. The Lorentz and Diff^4 invariant choice of Y^3 is as the intersection of the 4-surface with the set $\delta M_+^4 \times CP_2$, where δM_+^4 denotes the boundary of the light cone: effectively the imbedding space can be replaced with the product $\delta M_+^4 \times CP_2$ as far as vibrational degrees of freedom are considered. More precisely: configuration space has a fiber structure: the 3-surfaces $Y^3 \subset \delta M_+^4 \times CP_2$ correspond to the base space and the 3-surfaces on the orbit of given Y^3 and diffeomorphic with Y^3 correspond to the fiber and are separated by a zero distance from each other in the configuration space metric.

These observations lead to the identification of the isometry group as some subgroup G of the group of the diffeomorphisms of $\delta H = \delta M_+^4 \times CP_2$. These diffeomorphisms indeed act in a natural manner in δCH, the space of the 3-surfaces in δH. Therefore one can identify the configuration space as the union of the coset spaces G/H, where H corresponds to the subgroup of G acting as diffeomorphisms for a given X^3. H depends on the topology of X^3 and since G does not change the topology of the 3-surface, each 3-topology defines a separate orbit of G. Therefore, the union involves the sum over all topologies of X^3 plus possibly other 'zero modes'.

The task is to identify correctly G as a sub-algebra of the diffeomorphisms of δH. The only possibility seems to be that the canonical transformations of δH generated by the function algebra of δH act as isometries of the configuration space. The canonical transformations act nontrivially also in δM_+^4 since δM_+^4 allows Kähler structure and thus also symplectic structure.

The magic properties of the light like 3-surfaces

In case of the Kähler metric, G- and H Lie-algebras must allow a complexification so that the isometries can act as holomorphic transformations. The unique feature of the lightcone boundary δM_+^4, realized already seven years ago, is its metric degeneracy: the boundary of the light cone is metrically 2-dimensional sphere although it is topologically 3-dimensional! This implies that light cone boundary allows an infinite-dimensional group of conformal symmetries generated by an algebra, which is a generalization of the ordinary Virasoro algebra! There is actually also an infinite-dimensional group of isometries (!) isomorphic with the group of the conformal transformations! Even more, in case of δH the groups of the conformal symmetries and isometries are local with respect to CP_2. Furthermore, light cone boundary allows infinite dimensional group of canonical transformations as the symmetries of the symplectic structure automatically associated with the Kähler structure. Therefore 4-dimensional Minkowski space is in a unique position in TGD approach. δM_+^4 allows also complexification and Kähler structure unlike the boundaries of the higher-dimensional light cones so that it becomes possible to define a complexification in the tangent space of the configuration space, too.

The space of the vector fields on $\delta H = \delta M_+^4 \times CP_2$ inherits the complex structure of the light cone boundary and CP_2. The complexification can be induced from the complex conjugation for the functions depending on the radial coordinate of the light cone boundary playing the same role as the time coordinate associated with string space-time sheet. In M_+^4 degrees of freedom complexification works only provided that the radial vector fields posses zero norm as configuration space vector fields (they have also zero norm as vector fields).

The effective two-dimensionality of the light cone boundary allows also to circumvent the no-go theorems associated with the higher-dimensional Abelian extensions. First, in the dimensions $D > 2$ Abelian extensions of the gauge algebra are extensions by an infinite dimensional Abelian group rather than central extensions by the group $U(1)$. In the present case the extension is a symplectic extension analogous to the extension defined by the Poisson bracket $\{p, q\} = 1$ rather than the standard central extension but is indeed 1- dimensional and well defined provided that the configuration space metric is Kähler. Secondly, $D > 2$ extensions possess no unitary faithful representations (satisfying certain well motivated physical constraints) [dg1]. The point is that light cone boundary is metrically and conformally 2-sphere and therefore the gauge algebra is effectively the algebra associated with the 2-sphere and, as a consequence, also the configuration space metric is Kähler.

There is counter argument against complexifixation. The Kähler structure of the light cone boundary is not unique: various complex structures are parameterized by $SO(3,1)/SO(3)$ (Lobatchewski space). The definition of the Kähler function as absolute minimum of Kähler action however makes it possible to assign unique space-time surface $X^4(Y^3)$ to any Y^3 on the light cone boundary and the requirement that the group $SO(3)$ specifying the Kähler structure is isotropy group of the classical four-momentum associated with $X^4(Y^3)$, fixes the complex structure uniquely as a function of Y^3. Thus it seems that Kähler action is necessary ingredient of the group theoretical approach.

Symmetric space property reduces to conformal and canonical invariance

The idea about symmetric space is extremely beautiful but it millenium had to change before I was ripe to identify the precise form of the Cartan decomposition. The solution of the puzzle turned out to be amazingly simple.

The inspiration came from the finding that quantum TGD leads naturally to an extension of Super Algebras by combining Ramond and Neveu-Schwartz algebras into single algebra. This led to the introduction Virasoro generators and generators of canonical algebra of CP_2 localized with respect to the light cone boundary and carrying conformal weights with a half integer valued real part. Soon came the realization that the conformal weights $h = -1/2 - i\sum_i y_i$, where $z_i = 1/2 + y_i$ are non-trivial zeros of Riemann Zeta, are excellent candidates for the conformal weights. It took some time to answer affirmatively the question whether also the negatives of the trivial zeros $z = -2n$, $n > 0$ should be included. Thus the conjecture inspired by the work with Riemann hypothesis stating that the zeros of Riemann Zeta appear at the level of basic quantum TGD turned out be correct.

The generators whose commutators define the basis of the entire algebra have conformal weights given by the negatives of the zeros of Rieman Zeta. The algebra is a direct sum $g = g_1 \oplus g_2$ such that g_1 has $h = n$ as conformal weights and g_2 $h = n - 1/2 + iy$, where y is sum over imaginary parts y_i of non-trivial zeros of Zeta. Only $h = 2n$, $n > 1$, and $h = -1/2 - iy + n$, such that n is even (odd) if y is sum of odd (even) number of y_i correspond to the weights labelling the generators of t in the Cartan decomposition $g = h + t$. The resulting super-canonical algebra would quite well be christened as Riemann algebra.

The requirement that ordinary Virasoro and Kac Moody generators annihilate physical states corresponds now to the fact that the generators of h vanish at the point of configuration space, which remains invariant under the action of h. The maximum of Kähler function corresponds naturally to this point and plays also an essential role in the integration over configuration space by generalizing the Gaussian integration of free quantum field theories.

The light cone conformal invariance differs in many respects from the conformal invariance of string theories. Finite-dimensional group defining Kac-Moody group is replaced by an infinite-dimensional canonical group. Conformal weights correspond to zeros of Riemann zeta and suitable superpositions of them in case of trivial zeros, and physical states can have non-vanishing conformal weights just as the representations of color group in CP_2 can have non-vanishing color isospin and hyper charge. The conformal weights have also interpretation as quantum numbers associated with unitary representations of Lorentz group: thus there is no conflict between conformal invariance and Lorentz invariance in TGD framework.

3.2.4   An educated guess for the Kähler function

The turning point in the attempts to construct configuration space geometry was the realization that four-dimensional $Diff$ invariance (not only 3-dimensional $Diff$ invariance!) of General Relativity must have a counterpart in TGD. In order to realize this symmetry in the space of 3-surfaces, the definition of the configuration space metric should somehow associate to a given 3-surface X^3 a unique space-time surface $X^4(X^3)$ for $Diff^4$ to act on. Physical considerations require that the metric should be, not only $Diff^4$ invariant, but also $Diff^4$ degenerate so that infinitesimal $Diff^4$ transformations should correspond to zero norm vector fields of the configuration space.

Since Kähler function determines Kähler geometry, the definition of the Kähler function should associate a unique space-time surface $X^4(X^3)$ to a given 3-surface X^3. The natural physical interpretation for this space-time surface is as the classical space-time associated with X^3 so that in TGD classical physics ($X^4(X^3)$) becomes a part of the configuration space geometry and of the quantum theory.

One could try to construct the configuration space geometry by finding the metric for a single representative 3-surface at each orbit of G and extending it by left translations to the entire orbit of G. The metric for this representative should be $Diff^3$ invariant and somehow it should associate a unique space-time surface to the 3-surface in question. The original attempt was however more indirect and based on the realization that the construction of the Kähler geometry reduces to that of finding Kähler function $K(X^3)$ with the property that it associates a unique space-time surface $X^4(X^3)$ to a given 3-surface X^3 and possesses mathematically and physically acceptable properties. The guess for the Kähler function is the following one.

By Diff^4 invariance one can restrict the consideration on the set of 3-surfaces Y^3 on the 'light cone boundary' $\delta H = \delta M_+^4 \times CP_2$ since one can define the space-time surface associated with $X^3 \subset X^4(Y^3)$ to be $X^4(X^3) = X^4(Y^3)$ in case that the initial value problem for X^3 has $X^4(Y^3)$ as its solution. This implies $K(X^3) = K(Y^3)$.

The value of the Kähler function K for a given 3-surface Y^3 on light cone boundary is obtained in the following manner.

a) Consider all possible 4-surfaces $X^4 \subset M_+^4 \times CP_2$ having Y^3 as its sub-manifold: $Y^3 \subset X^4$. If Y^3 has boundary then it belongs to the boundary of X^4: $\delta Y^3 \subset \delta X^4$.

b) Associate to each four surface Kähler action as the Maxwell action for the Abelian gauge field defined by the projection of the CP_2 Kähler form to the four-surface. For a Minkowskian signature of the induced metric Kähler electric field gives a negative contribution to the action density whereas for an Euclidian signature the action density is always non-positive.

c) Define the value of the Kähler function K for Y^3 as the absolute minimum of the Kähler action S_K over all possible four-surfaces having Y^3 as its sub-manifold: $K(Y^3) = Min\{S_K(X^4)|X^4 \supset Y^3\}$.

This definition of the Kähler function has several physically appealing features.

a) Kähler geometry associates with each X^3 a unique four-surface, which will be interpreted as the classical space-time associated with X^3. This means that the so called classical space time (and physics!) in TGD approach is not defined via some approximation procedure (stationary phase approximation of the functional integral) but is an essential part of not only quantum theory, but also of the configuration space geometry, which in turn might be determined by a mere mathematical consistency! Since quantum states are superpositions over these classical space-times, it is clear that the observed classical space-time is some kind of effective, quantum average space-time, presumably defined as an absolute minimum for the effective action of the theory.

b) The space-time surface associated with a given 3-surface is analogous to a Bohr orbit of the old fashioned quantum theory. The point is that the initial value problem in question differs from the ordinary initial value problem in that although the values of the H coordinates h^k as functions $h^k(x)$ of X^3 coordinates can be chosen arbitrarily, the time derivatives $\partial_t h^k(x)$ at X^3 are uniquely fixed by the principle selecting preferred extremals as generalized Bohr orbits (absolute minimization or something more delicate [E2]) unlike in the ordinary variational problems encountered in the classical physics. This implies something closely analogous to the quantization of the canonical momenta so that the space-time surface can be regarded as a generalized Bohr orbit. The classical quantization of electric charge and mass are possible consequences of the Bohr orbit property.

c) Kähler function is Diff^4 invariant in the sense that the value of the Kähler function is same for all 3-surfaces belonging to the orbit of a given 3-surface. As a consequence, configuration space metric is Diff^4 degenerate. The implications of the Diff^4 invariance have turned out to be decisive, not only for the geometrization of the configuration space, but also for the construction of the quantum theory. For instance, time like vibrational modes tangential to the 4-surface imply tachyonic mass spectrum unless they correspond to the zero modes of the configuration space metric. Diff^4 invariance however guarantees the required kind of degeneracy of the metric.

d) The non-determinism of Kähler action means that the complete reduction to the light cone boundary is not possible. This means a mathematical challenge but is physically a highly desirable feature since otherwise time would be lost as it is lost in the canonically quantized general relativity.

The most general expectation is that configuration space can be regarded as a union of coset spaces: $C(H) = \cup_i G/H(i)$. Index i labels 3-topology and zero modes. The group G, which can depend on 3-surface, can be identified as a subgroup of diffeomorphisms of $\delta M_+^4 \times CP_2$ and H must contain as its subgroup a group, whose action reduces to $Diff(X^3)$ so that these transformations leave 3-surface invariant.

The task is to identify plausible candidate for G and to show that the tangent space of the configuration space allows Kähler structure, in other words that the Lie-algebras of G and $H(i)$ allow complexification. One must also identify the zero modes and construct integration measure for the functional integral in these degrees of freedom. Besides this one must deduce information about the explicit form of configuration space metric from symmetry considerations combined with the hypothesis that Kähler function is determined as absolute minimum of Kähler action.

It will be found that in the case of $M_+^4 \times CP_2$ Kähler geometry, or strictly speaking contact Kähler geometry, characterized by a degenerate Kähler form (Diff^4 degeneracy and plus possible other degeneracies) seems possible. Although it seems that this construction must be generalized by allowing all light like 7-surfaces $X_l^3 \times CP_2$, at least those for which X_l^3 is boundary of light-cone inside M_+^4 or M^4, with the physical interpretation differing dramatically from the original one, the original construction discussed in the sequel involves the most essential aspects of the problem.

3.2.5 An alternative for the absolute minimization of Kähler action

One can criticize the assumption that extremals correspond to absolute minima, and the number theoretical vision discussed in [E2] indeed favors the separate minimization of magnitudes of positive and negative contributions to the Kähler action.

For this option Universe would do its best to save energy, being as near as possible to vacuum. Also vacuum extremals would become physically relevant: note that they would be only inertial vacua and carry non-vanishing density gravitational energy. The non-determinism of the vacuum extremals would have an interpretation in terms of the ability of Universe to engineer itself.

The 3-surfaces for which CP_2 projection is at least 2-dimensional and not a Lagrange manifold would correspond to non-vacua since conservation laws do not leave any other option. The variational principle would favor equally magnetic and electric configurations whereas absolute minimization of action based on S_K would favor electric configurations. The positive and negative contributions would be minimized for 4-surfaces in relative homology class since the boundary of X^4 defined by the intersections with 7-D light-like causal determinants would be fixed. Without this constraint only vacuum bubbles would result.

The attractiveness of the number theoretical variational principle from the point of calculability of TGD would be that the initial values for the time derivatives of the imbedding space coordinates at X^3 at light-like 7-D causal determinant could be computed by requiring that the energy of the solution is minimized. This could mean a computerizable solution to the construction of Kähler function.

It should be noticed that the considerations of this chapter relate only to the extremals of Kähler action which need not be absolute minima nor more general preferred extremals discussed in [E2] although this is suggested by the high symmetries. The number theoretic approach based on the properties of quaternions and octonions discussed in the chapter [E2] leads to a proposal for the general solution of field equations based on the generalization of the notion of calibration [bb1] providing absolute minima of volume to that of Kähler calibration. This approach will not be discussed in this chapter.

3.2.6 The construction of the configuration space geometry from symmetry principles

The gigantic size of the isometry group suggests that it might be possible to deduce very detailed information about the metric of the configuration space by group theoretical arguments. This turns out to be the case. In order to have a Kähler structure, one must define a complexification of the configuration space. Also one should identify the Lie algebra of the isometry group and try to derive explicit form of the Kähler metric using this information. One can indeed construct the metric in this manner but a rigorous proof that the corresponding Kähler function is the one defined by Kähler action does not exist yet although both approaches predict the same general qualitative properties for the metric. The argument stating the equivalence of the two approaches reduces to the hypothesis stating electric-magnetic duality of the theory. For the Bohr orbit like preferred extremals of Kähler action magnetic configuration space Hamiltonians derivable from group theoretical approach are essentially identical with electric configuration space Hamiltonians derivable from Kähler action.

General Coordinate Invariance and generalized quantum gravitational holography

The basic motivation for the construction of configuration space geometry is the vision that physics reduces to the geometry of classical spinor fields in the infinite-dimensional configuration space of 3-surfaces of $M_+^4 \times CP_2$ or of $M^4 \times CP_2$. Hermitian conjugation is the basic operation in quantum theory and its geometrization requires that configuration space possesses Kähler geometry. Kähler geometry is coded into Kähler function.

The original belief was that the four-dimensional general coordinate invariance of Kähler function reduces the construction of the geometry to that for the boundary of configuration space consisting of 3-surfaces on $\delta M_+^4 \times CP_2$, the moment of big bang. The proposal was that Kähler function $K(Y^3)$ could be defined as absolute minimum of so called Kähler action for the unique space-time surface $X^4(Y^3)$ going through given 3-surface Y^3 at $\delta M_+^4 \times CP_2$. For Diff^4 transforms of Y^3 at $X^4(Y^3)$ Kähler function would have the same value so that Diff^4 invariance and degeneracy would be the outcome.

This picture is however too simple.

a) The degeneracy of the absolute minima caused by the classical non-determinism of Kähler action however brings in additional delicacies, and it seems that the reduction to the light cone boundary which in fact corresponds to what has become known as quantum gravitational holography must be replaced with a construction involving more general light like 7-surfaces $X_l^3 \times CP_2$.

b) It has also become obvious that the gigantic symmetries associated with $\delta M_+^4 \times CP_2$ manifest themselves as the properties of propagators and vertices, and that M^4 is favored over M_+^4. Cosmological considerations, Poincare invariance, and the new view about energy favor the decomposition of the configuration space to a union of configuration spaces associated with various 7-D causal determinants. The minimum assumption is that all possible unions of future and past light cone boundaries $\delta M_\pm^4 \times CP_2 \subset M^4 \times CP_2$ label the sectors of CH: the nice feature of this option is that the considerations of this chapter restricted to $\delta M_+^3 \times CP_2$ generalize almost trivially. This option is beautiful because the center of mass degrees of freedom associated with the different sectors of CH would correspond to M^4 itself and its Cartesian powers. One cannot exclude the possibility that even more general light like surfaces $X_l^3 \times CP_2$ of M^4 are important as causal determinants.

The definition of the Kähler function requires that the many-to-one correspondence $X^3 \to X^4(X^3)$ must be replaced by a bijective correspondence in the sense that X^3 is unique among all its Diff^4 translates. This also allows physically preferred "gauge fixing" allowing to get rid of the mathematical complications due to Diff4 degeneracy. The internal geometry of the space-time sheet $X^4(X^3)$ must define the preferred 3-surface X^3 and also a preferred light like 7-surface $X_l^3 \times CP_2$.

This is indeed possible. The possibility of negative values of Poincare energy(or equivalently inertial energy) inspires the hypothesis that the total quantum numbers and classical conserved quantities of the Universe vanish. This view is consistent with experimental facts if gravitational energy is defined as a difference of Poincare energies of positive and negative energy matter. Space-time surface consists of pairs of positive and negative energy space-time sheets created at some moment from vacuum and branching at that moment. This allows to select X^3 uniquely and define $X^4(X^3)$ as the absolute minimum of Kähler action in the set of 4-surfaces going through X^3. These space-time sheets should also define uniquely the light like 7-surface $X_l^3 \times CP_2$, most naturally as the "earliest" surface of this kind. Note that this means that it become possible to assign a unique value of geometric time to the space-time sheet.

The realization of this vision means a considerable mathematical challenge. The effective metric 2-dimensionality of 3-dimensional light-like surfaces X_l^3 of M^4 implies generalized conformal and canonical invariances allowing to generalize quantum gravitational holography from light like boundary so that the complexities due to the non-determinism can be taken into account properly.

Light like 3-D causal determinants, 7-3 duality, and effective 2-dimensionality

Thanks to the non-determinism of Kähler action, also light like 3-surfaces X_l^3 of space-time surface appear as causal determinants (CDs). Examples are boundaries and elementary particle horizons at which Minkowskian signature of the induced metric transforms to Euclidian one. This brings in a second conformal symmetry related to the metric 2-dimensionality of the 3-D CD. This symmetry is identifiable as TGD counterpart of the Kac Moody symmetry of string models. The challenge is to understand the relationship of this symmetry to configuration space geometry and the interaction between the two conformal symmetries.

The possibility of spinorial shock waves at X_l^3 leads to the hypothesis that they correspond to particle aspect of field particle duality whereas the physics in the interior of space-time corresponds to field aspect. More generally, field particle duality in TGD framework states that 3-D light like CDs and 7-D CDs are dual to each other. In particular, super-canonical and Super Kac Moody symmetries are also dually related.

The underlying reason for 7–3 duality be understood from a simple geometric picture in which 3-D light like CDs X_l^3 intersect 7-D CDs X^7 along 2-D surfaces X^2 and thus form 2-sub-manifolds of the space-like 3-surface $X^3 \subset X^7$. One can regard either canonical deformations of X^7 or Kac-Moody deformations of X^2 as defining the tangent space of configuration space so that 7–3 duality would relate two different coordinate choices for CH.

The assumption that the data at either X^3 or X_l^3 are enough to determine configuration space geometry implies that the relevant data is contained to their intersection X^2. This is the case if the deformations of X_l^3 not affecting X^2 and preserving light likeness corresponding to zero modes or gauge degrees of freedom and induce deformations of X^3 also acting as zero modes. The outcome is effective 2-dimensionality. One cannot over-emphasize the importance of this conclusion. It indeed stream lines dramatically the earlier formulas for configuration space metric involving 3-dimensional integrals over $X^3 \subset M_+^4 \times CP_2$ reducing now to 2-dimensional integrals. Most importantly, no data about absolute minima of Kähler are needed to construct the configuration space metric so that the construction is also practical.

The reduction of data to that associated with 2-D surfaces conforms with the number theoretic vision about imbedding space as having hyper-octonionic structure [E2]: the commutative sub-manifolds of $OH = M^8$ have dimension not larger than two and for them tangent space is complex sub-space of hyper-octonion tangent space. Number theoretic counterpart of quantum measurement theory forces the reduction of relevant data to 2-D commutative sub-manifolds of X^3. These points are discussed in more detail in the next chapter whereas in this chapter the consideration will be restricted to $X_l^3 = \delta M_+^4$ case which involves all essential aspects of the problem.

Magnetic Hamiltonians

Assuming that the elements of the radial Virasoro algebra of δM_+^4 have zero norm, one ends up with an explicit identification of the symplectic structure of the configuration space. There is almost unique identification for the symplectic structure. Configuration space counterparts of $\delta M^4 \times CP_2$ Hamiltonians are defined by the generalized signed and and unsigned Kähler magnetic fluxes

$$Q_m(H_A, X^2) = Z \int_{X^2} H_A J \sqrt{g_2} d^2 x \ ,$$

$$Q_m^+(H_A, r_M) = Z \int_{X^2} H_A |J| \sqrt{g_2} d^2 x \ ,$$

$$J \equiv \epsilon^{\alpha\beta} J_{\alpha\beta} \ .$$

H_A is CP_2 Hamiltonian multiplied by a function of coordinates of light cone boundary belonging to a unitary representation of the Lorentz group. Z is a conformal factor depending on canonical invariants. The symplectic structure is induced by the symplectic structure of CP_2.

The most general flux is superposition of signed and unsigned fluxes Q_m and Q_m^+.

$$Q_m^{\alpha,\beta}(H_A, X^2) = \alpha Q_m(H_A, X^2) + \beta Q_m^+(H_A, X^2) \ .$$

Thus it seems that symmetry arguments fix the form of the configuration space metric apart from the presence of a conformal factor Z multiplying the magnetic flux and the degeneracy related to the signed and unsigned fluxes.

The notion of 7–3-duality described in the introduction implies that the relevant data about configuration space geometry is contained by 2-D surfaces X^2 at the intersections of 3-D light like CDS and 7-D CDs such as $M_+^4 \times CP_2$. In this case the entire Hamiltonian could be defined as the sum of magnetic fluxes over surfaces $X_i^2 \subset X^3$. The maximally optimistic guess would be that it is possible to fix both X_i^2 and 7-D CDs freely with X_i^2 possibly identified as commutative sub-manifold of octonionic H.

Electric Hamiltonians and electric-magnetic duality

Absolute minimization of Kähler action in turn suggests that one can identify configuration space Hamiltonians as classical charges $Q_e(H_A)$ associated with the Hamiltonians of the canonical transformations of the light cone boundary, that is as variational derivatives of the Kähler action with respect to the infinitesimal deformations induced by $\delta M_+^4 \times CP_2$ Hamiltonians. Alternatively, one might simply replace Kähler magnetic field J with Kähler electric field defined by space-time dual $*J$

in the formulas of previous section. These Hamiltonians are analogous to Kähler electric charge and the hypothesis motivated by the experience with the instantons of the Euclidian Yang Mills theories and 'Yin-Yang' principle, as well as by the duality of CP_2 geometry, is that for the absolute minima of the Kähler action these Hamiltonians are affinely related:

$$Q_e(H_A) = Z\left[Q_m(H_A) + q_e(H_A)\right] \ .$$

Here Z and q_e are constants depending on canonical invariants only. Thus the equivalence of the two approaches to the construction of configuration space geometry boils down to the hypothesis of a physically well motivated electric-magnetic duality.

The crucial technical idea is to regard configuration space metric as a quadratic form in the entire Lie-algebra of the isometry group G such that the matrix elements of the metric vanish in the sub-algebra H of G acting as $Diff^3(X^3)$. The Lie-algebra of G with degenerate metric in the sense that H vector fields possess zero norm, can be regarded as a tangent space basis for the configuration space at point X^3 at which H acts as an isotropy group: at other points of the configuration space H is different. For given values of zero modes the maximum of Kähler function is the best candidate for X^3. This picture applies also in symplectic degrees of freedom.

Complexification and explicit form of the metric and Kähler form

The identification of the Kähler form and Kähler metric in canonical degrees of freedom follows trivially from the identification of the symplectic form and definition of complexification. The requirement that Hamiltonians are eigen states angular momentum (and possibly also of Lorentz boost), isospin and hypercharge implies physically natural complexification. In order to fix the complexification completely one must introduce some convention fixing which states correspond to 'positive' frequencies and which to 'negative frequencies' and which to zero frequencies that is to decompose the generators of the canonical algebra to three sets Can_+, Can_- and Can_0. One must distinguish between Can_0 and zero modes, which are not considered here at all. For instance, CP_2 Hamiltonians correspond to zero modes.

The natural complexification relies on the imaginary part of the radial conformal weight whereas the real part defines the $g = t + h$ decomposition naturally. The wave vector associated with the radial logarithmic plane wave corresponds to the angular momentum quantum number associated with a wave in S^1 in the case of Kac Moody algebra. One can imagine three options.

a) It is quite possible that the spectrum of k_2 does not contain $k_2 = 0$ at all so that the sector Can_0 could be empty. This complexification is physically very natural since it is manifestly invariant under $SU(3)$ and $SO(3)$ defining the preferred spherical coordinates. The choice of $SO(3)$ is unique if the classical four-momentum associated with the 3-surface is time like so that there are no problems with Lorentz invariance.

b) If $k_2 = 0$ is possible one could have

$$\begin{aligned}
Can_+ &= \{H^a_{m,n,k=k1+ik_2}, k_2 > 0\} \ , \\
Can_- &= \{H^a_{m,n,k}, k_2 < 0\} \ , \\
Can_0 &= \{H^a_{m,n,k}, k_2 = 0\} \ .
\end{aligned} \tag{3.2.1}$$

c) If it is possible to $n_2 \neq 0$ for $k_2 = 0$, one could define the decomposition as

$$\begin{aligned}
Can_+ &= \{H^a_{m,n,k}, k_2 > 0 \ \text{or} \ k_2 = 0, n_2 > 0\} \ , \\
Can_- &= \{H^a_{m,n,k}, k_2 < 0 \ \text{or} k_2 = 0, n_2 < 0\} \ , \\
Can_0 &= \{H^a_{m,n,k}, k_2 = n_2 = 0\} \ .
\end{aligned} \tag{3.2.2}$$

In this case the complexification is unique and Lorentz invariance guaranteed if one can fix the $SO(2)$ subgroup uniquely. The quantization axis of angular momentum could be chosen to be the direction of the classical angular momentum associated with the 3-surface in its rest system.

The only thing needed to get Kähler form and Kähler metric is to use the "half Poisson bracket"

$$\begin{aligned} J_f(X(H_A), X(H_B)) &= 2Im\left(iQ_f(\{H_A, H_B\}_{-+})\right) \ , \\ G_f(X(H_A), X(H_B)) &= 2Re\left(iQ_f(\{H_A, H_B\}_{-+})\right) \ . \end{aligned} \qquad (3.2.3)$$

Here the subscript $+$ and $-$ refer to complex isometry current and its complex conjugate in terms of which the "half Poisson bracket" can be expressed.

Symplectic form, and thus also Kähler form and Kähler metric, could contain a conformal factor depending on the isometry invariants characterizing the size and shape of the 3-surface. At this stage one cannot say much about the functional form of this factor.

3.2.7 Configuration space spinor structure

Quantum TGD should be reducible to the classical spinor geometry of the configuration space. In particular, physical states should correspond to the modes of the configuration space spinor fields. The immediate consequence is that configuration space spinor fields cannot, as one might naively expect, be carriers of a definite spin and unit fermion number. Concerning the construction of the configuration space spinor structure there are some important clues.

a) The classical bosonic physics is coded into the definition of the configuration space metric; therefore the classical physics associated with the spinors of the imbedding space should be coded into the definition of the configuration space spinor structure. This means that the generalized massless Dirac equation for the induced spinor fields on $X^4(X^3)$ should be closely related to the definition of the configuration space gamma matrices.

b) Complex probability amplitudes (scalar fields) in the configuration space correspond to the second quantized boson fields in X^4. Hence the spinor fields of the configuration space should correspond to the second quantized, free, induced spinor fields on X^4. The space of the configuration space spinors should be just the Fock space of the second quantized fermions on X^4!

c) Canonical algebra might generalize to a super canonical algebra and that super generators should be linearly related to the gamma matrices of the configuration space. If this indeed is the case then the construction of the configuration space spinor structure becomes a purely group theoretical problem.

The realization of these ideas is simple in principle. Perform a second quantization for the free induced spinor field in X^4. Express configuration space gamma matrices and canonical super generators as superpositions of the fermionic oscillator operators. This means that configuration space gamma matrices are analogous to spin 3/2 fields and can be regarded as a superpartner of the gravitational field of the configuration space. Deduce the anti-commutation relations of the spinor fields from the requirement of super canonical invariance. Generalize the flux representation for the configuration space Hamiltonians to a spinorial flux representation for their super partners.

Configuration space gamma matrices as super algebra generators

The basic idea is that the space of the configuration space spinors must correspond to the Fock space for the second quantized induced spinor fields. In accordance with this the gamma matrices of the configuration space must be expressible as superpositions of the fermionic oscillator operators for the second quantized induced free spinor fields in X^4 so that they are analogous to spin 3/2 fields. The Dirac equation is fixed from the requirement of super symmetry and has same vacuum degeneracy as Kähler action. A further assumption is that the contractions of the gamma matrices with isometry currents correspond to super charges of the group of isometries of the configuration space so that the construction reduces to group theory. Also the super Kac Moody algebra associated with light like 3-D CDs defines candidates for gamma matrices defining the components of the metric as anti-commutators and the question is whether the two definitions are mutually consistent.

7–3 duality

The failure of the classical non-determinism forces to introduce two kinds of causal determinants (CDs). 7-D light like CDs are unions of the boundaries of future and past directed light cones in M^4 at arbitrary positions (also more general light like surfaces $X^7 = X_l^3 \times CP_2$ might be considered). CH is a union of sectors associated with these 7-D CDs playing in a very rough sense the roles of big bangs and big crunches. The creation of pairs of positive and negative energy space-time sheets occurs at $X^3 \subset X^7$ in the sense that negative and positive energy space-time sheet meet at X^3. Negative and positive energy space-time sheets are space-time correlates for bras and kets and the meeting of negative and positive energy space-time sheets is the space-time correlate for their scalar product.

Also 3-D light like causal determinants $X_l^3 \subset X^4$ must be introduced: elementary particle horizons provide a basic example of this kind of CDs. The deformations of the 2-surfaces defining X_l^3 define Kac Moody type conformal symmetries.

7–3 duality states that the two kind of CDs play a dual role in the construction of the theory and implies that 3-surfaces are effectively two-dimensional with respect to the CH metric in the sense that all relevant data about CH geometry is contained by the two-dimensional intersections $X^2 = X_l^3 \cap X^7$ defining 2-sub-manifolds of $X^3 \subset X^7$.

The modified Dirac equation and gamma matrices

The modified Dirac equation is deduced from Kähler action by requiring it to have the same vacuum degeneracy as Kähler action itself. The interpretation of the solutions of the modified Dirac equation is as super gauge symmetry generators whereas physical degrees of freedom corresponds to generalized eigen modes at X_l^3 and at space-like 3-surfaces $X^3 \subset X^7$.

The decisive property of the modified Dirac equation is that it allows shock wave solutions restricted to X_l^3: in terms of field-particle duality these shock waves correspond to the click caused by a particle in a detector. This allows to realize quantum gravitational holography and 7–3 duality in the sense that the induced second quantized spinor fields at the intersections $X^2 = X_l^3 \cap X^7$ determine the super-generators super-canonical and super Kac Moody algebras invariant under the super gauge symmetries generated by the solutions of the modified Dirac equation.

Both the function algebra and Poisson algebra of X^7 allow super-symmetrization and both N-S and Ramond type representations are possible. For Ramond type representation the modified Dirac operators D_+ and D_-^{-1} associated with the positive and negative energy space-time sheets $X_\pm^4$ meeting at X^3 are present in the expressions of the super generators. NS-type representations correspond to the replacement of these operators with projection operators to the space of spinor modes with non-vanishing eigenvalues of $D_\pm$. Both representations are necessary and correspond to leptonic and quark like representations of configuration space gamma matrices. Similar statements apply to super Kac-Moody representations. These two kinds of representations correspond to super and kappa symmetries of super-string models.

Expressing Kähler function in terms of Dirac determinants

Although quantum criticality in principle predicts the possible values of Kähler coupling strength, one might hope that there exists even more fundamental approach involving no coupling constants and predicting even quantum criticality and realizing quantum gravitational holography. The ratio of Dirac determinants of the modified Dirac operators D_+ and D_- associated with the space-time regions separated by a 3-dimensional causal determinant is an excellent candidate in this respect and the guess is that it is expressible as an exponent for the difference of Kähler actions in the adjacent regions. This allows to deduce the exponent of Kähler function and one could identify it as a renormalized Dirac determinant.

If the operators D_+ and D_- commute and if spectrum of the operator $D_+ D_-^{-1}$ is invariant under $\lambda \to 1/\lambda$ apart from a finite number of eigenvalues for which the ratio equals to the ratio of exponents of respective Kähler actions, the Dirac determinant in question is well-defined even without zeta function regularization, and allows to deduce Kähler function.

The modified Dirac operator indeed allows to realize quantum gravitational holography since it reduces to an effectively 3-dimensional Dirac operator by boundary conditions but depends on the normal derivatives of the imbedding space coordinates at the causal determinants, which are most naturally light like 3-surfaces. Hermiticity of $D_\pm$ poses conditions on normal derivatives of imbedding space coordinates unless Kähler action density vanishes at X_l^3 but does not fully eliminate the effects of the classical non-determinism. Hence there are good hopes of evaluating the exponent of Kähler function as a Dirac determinant without solving the field equations.

The relationship between super-canonical and super Kac-Moody algebras

The conformal weights of the generating elements of super-canonical representations correspond to the zeros of Riemann Zeta and one can identify the Cartan decomposition of the super-canonical algebra crucial for defining configuration space of 3-surfaces as a union of symmetric Kähler manifolds labelled by zero modes. Super-canonical algebra differs dramatically from super Kac Moody algebra. 7-3 duality however allows to see super-canonical and super Kac-Moody algebras as associated with two different tangent space basis for CH and giving rise to different coordinate systems. Hence both super algebras could give rise to a gamma matrix algebra of CH.

7–3 duality allows to generalize the Olive-Goddard-Kent coset construction. By 7-3 duality the differences of the commuting Virasoro generators of super-canonical and super Kac-Moody algebras must annihilate the physical states. For the same reason the central charges of the two Virasoro algebras must be identical so that the net central charge vanishes. This condition leads to a generalization of stringy mass formula involving besides super Kac-Moody algebra also the super-canonical algebra and allowing continuum mass spectrum for many particle states.

The $N = 4$ super symmetries generated by the solutions of the modified Dirac equation are pure super gauge transformations. All CP_2 spinor harmonics except the covariantly constant right handed neutrino spinor carry color quantum numbers and thus a non-vanishing vacuum conformal weight: hence only an $N = 1$ global super symmetry is in principle possible. Since the Ramond type super-generator corresponding to the covariantly constant neutrino vanishes identically even $N = 1$ global super-symmetry is absent and no sparticles are predicted. This means a decisive difference in comparison with super string models and M-theory.

Physical states satisfy both N-S and Ramond type Super Virasoro conditions separately: note that in super-canonical degrees of freedom Ramond/NS representations super generators involve carry quark/lepton number. The most obvious application of the mass formula would be to hadron physics. The effective 2-dimensionality allows to identify partons as 2-dimensional surfaces X_i^2, and the more than decade old notion of elementary particle vacuum functional finds a first principle justification as a functional in the modular degrees of freedom of X^2.

By quantum classical correspondence is that Virasoro algebra associated with super Kac-Moody algebra acts on the conformal weights of the super-canonical representations as conformal transformations and the generators of the super-canonical algebra can be regarded as conformal fields. This dictates the matrix elements of the algebra to a high degree as function of conformal weights. A connection with braid and quantum groups and II_1 sub-factors of type von Neumann algebras associated with the Clifford algebra of the configuration space emerges.

3.2.8 What about infinities?

The construction of a divergence free and unitary inner product for the configuration space spinor fields is one of the major challenges. In the sequel constraints on the geometry of the configuration space posed by the finiteness of the inner product are analyzed.

Inner product from divergence cancellation

Forgetting the delicacies related to the non-determinism of the Kähler action, the inner product is given by integrating the usual Fock space inner product defined at each point of the configuration space over the reduced configuration space containing only the 3-surfaces Y^3 belonging to $\delta H = \delta M_+^4 \times CP_2$ ('light cone boundary') using the exponent $exp(K)$ as a weight factor:

$$\langle \Psi_1 | \Psi_2 \rangle = \int \overline{\Psi}_1(Y^3) \Psi_2(Y^3) exp(K) \sqrt{G} dY^3 \ ,$$

$$\overline{\Psi_1(Y^3)} \Psi_2(Y^3) \equiv \langle \Psi_1(Y^3) | \Psi_2(Y^3) \rangle_{Fock} \ . \tag{3.2.4}$$

The degeneracy for the absolute minima of Kähler action implies additional summation over the degenerate minima associated with Y^3. The restriction of the integration on light cone boundary is Diff^4 invariant procedure and resolves in elegant manner the problems related to the integration over Diff^4 degrees of freedom. A variant of the inner product is obtained dropping the bosonic vacuum functional $exp(K)$ from the definition of the inner product and by assuming that it is included into the spinor fields themselves. Probably it is just a matter of taste how the necessary bosonic vacuum functional is included into the inner product: what is essential that the vacuum functional $exp(K)$ is somehow present in the inner product.

The unitarity of the inner product follows from the unitary of the Fock space inner product and from the unitarity of the standard L^2 inner product defined by configuration space integration in the set of the L^2 integrable scalar functions. It could well occur that $Diff^4$ invariance implies the reduction of the configuration space integration to $C(\delta H)$.

Consider next the bosonic integration in more detail. The exponent of the Kähler function appears in the inner product also in the context of the finite dimensional group representations. For the representations of the noncompact groups (say $SL(2, R)$) in coset spaces (now $SL(2, R)/U(1)$ endowed with Kähler metric) the exponent of Kähler function is necessary in order to get square integrable representations [df5]. The scalar product for two complex valued representation functions is defined as

$$(f, g) = \int \overline{f} g\, exp(nK) \sqrt{g} dV \ . \tag{3.2.5}$$

By unitarity, the exponent is an integer multiple of the Kähler function. In the present case only the possibility $n = 1$ is realized if one requires a complete cancellation of the determinants. In finite dimensional case this corresponds to the restriction to single unitary representation of the group in question.

The sign of the action appearing in the exponent is of decisive importance in order to make theory stable. The point is that the theory must be well defined at the limit of infinitely large system. Minimization of action is expected to imply that the action of infinitely large system is bound from above: the generation of electric Kähler fields gives negative contributions to the action. This implies that at the limit of the infinite system the average action per volume is non-positive. For systems having negative average density of action vacuum functional $exp(K)$ vanishes so that only configurations with vanishing average action per volume have significant probability. On the other hand, the choice $exp(-K)$ would make theory unstable: probability amplitude would be infinite for all configurations having negative average action per volume. In the fourth part of the book it will be shown that the requirement that average Kähler action per volume cancels has important cosmological consequences.

Consider now the divergence cancellation in the bosonic integration. One can develop the Kähler function as a Taylor series around maximum of Kähler function and use the contravariant Kähler metric as a propagator. Gaussian and metric determinants cancel each other for a unique vacuum functional. Ricci flatness guarantees that metric determinant is constant in complex coordinates so that one avoids divergences coming from it. The non-locality of the Kähler function as a functional of the 3-surface serves as an additional regulating mechanism: if $K(X^3)$ were a local functional of X^3 one would encounter divergences in the perturbative expansion.

The requirement that quantum jump corresponds to a quantum measurement in the sense of quantum field theories implies that quantum jump involves localization in zero modes. Localization in the zero modes implies automatically p-adic evolution since the decomposition of the configuration space into sectors D_P labelled by the infinite primes P is determined by the corresponding decomposition in zero modes. Localization in zero modes would suggest that the calculation of the physical predictions does not involve integration over zero modes: this would dramatically simplify the calculational apparatus of the theory. Probably this simplification occurs at the level of practical calculations if U-matrix separates into a product of matrices associated with zero modes and fiber degrees of freedom.

One must also calculate the predictions for the ratios of the rates of quantum transitions to different values of zero modes and here one cannot actually avoid integrals over zero modes. To achieve this one is forced to define the transition probabilities for quantum jumps involving a localization in zero modes as

$$P(x, \alpha \to y, \beta) = \sum_{r,s} |S(r, \alpha \to s, \beta)|^2 |\Psi_r(x)|^2 |\Psi_s(y)|^2 \ ,$$

where x and y correspond to the zero mode coordinates and r and s label a complete state functional basis in zero modes and $S(r, m \to s, n)$ involves integration over zero modes. In fact, only in this manner the notion of the localization in the zero modes makes mathematically sense at the level of S-matrix. In this case also unitarity conditions are well-defined. In zero modes state function basis can be freely constructed so that divergence difficulties could be avoided. An open question is whether this construction is indeed possible.

Some comments about the actual evaluation of the bosonic functional integral are in order.

a) Since configuration space metric is degenerate and the bosonic propagator is essentially the contravariant metric, bosonic integration is expected to reduce to an integration over the zero modes. For instance, isometry invariants are variables of this kind. These modes are analogous to the parameters describing the conformal equivalence class of the orbit of the string in string models.

b) α_K is a natural small expansion parameter in configuration space integration. It should be noticed that α_K, when defined by the criticality condition, could also depend on the coordinates parameterizing the zero modes.

c) Semiclassical approximation, which means the expansion of the functional integral as a sum over the extrema of the Kähler function, is a natural approach to the calculation of the bosonic integral. Symmetric space property suggests that for the given values of the zero modes there is only single extremum and corresponds to the maximum of the Kähler function. There are theorems stating that semiclassical approximation is exact for certain systems (for example for integrable systems (Duistermaat-Hecke theorem [db3]). Symmetric space property suggests that Kähler function might possess the properties guaranteing the exactness of the semiclassical approximation. This would mean that the calculation of the integral $\int exp(K)\sqrt{G}dY^3$ and even more complex integrals involving configuration space spinor fields would be completely analogous to a Gaussian integration of free quantum field theory. This kind of reduction actually occurs in string models and is consistent with the criticality of the Kähler coupling constant suggesting that all loop integrals contributing to the renormalization of the Kähler action should vanish. Also the condition that configuration space integrals are continuable to p-adic number fields requires this kind of reduction.

Divergence cancellation, Ricci flatness, and symmetric space and Hyper Kähler properties

In the case of the loop spaces left invariance implies that Ricci tensor is a multiple of the metric tensor so that Ricci scalar has an infinite value. Mathematical consistency (essentially the absence of the divergences in the integration over the configuration space) forces the geometry to be Ricci flat: in other words, vacuum Einstein's equations are satisfied. It can be shown that Hyper Kähler property guarantees Ricci flatness. The reason is that the contractions of the curvature tensor appearing in the components of the Ricci tensor transform to traces over Lie algebra generators, which are $SU(\infty)$ generators instead of $U(\infty)$ generators as in case of loop spaces, so that the traces vanish.

Hyper Kähler property requires a quaternionic structure in the tangent space of the configuration space. Since any direction on the sphere S^2 defined by the linear combinations of quaternionic imaginary units with unit norm defines a particular complexification physically, Hyper-Kähler property means the possibility to perform complexification in S^2-fold manners. An interesting possibility raised by the notion of number theoretical compactification [E2] is that hyper Kähler structure could be replaced with what might be called "hyper-hyper-Kähler structure" resulting when quaternionic tangent space is replaced with its hyper-quaternionic variant. This would conform with the Minkowski signature of the space-time surface. In this framework also hyper-octonionic structure might be considered. An interesting question not yet even touched, is whether the conjectured $M^8 - -M^4 \times CP_2$ duality is realized also at the level of the configuration space of 3-surfaces.

Consider now the arguments in favor of Ricci flatness of the configuration space.

a) The canonical algebra of δM^4_+ takes effectively the role of the $U(1)$ extension of the loop algebra. More concretely, the $SO(2)$ group of the rotation group $SO(3)$ takes the role of $U(1)$ algebra. Since volume preserving transformations are in question, the traces of the canonical generators vanish identically and in finite-dimensional this should be enough for Ricci flatness even if Hyper Kähler property is not achieved.

b) The comparison with CP_2 allows to link Ricci flatness with conformal invariance. The elements of the Ricci tensor are expressible in terms of traces of the generators of the holonomy group $U(2)$ at the origin of CP_2, and since $U(1)$ generator is non-vanishing at origin, the Ricci tensor is non-vanishing. In recent case the origin of CP_2 is replaced with the maximum of Kähler function and holonomy group corresponds to super-canonical generators labelled by integer valued real parts k_1 of the conformal weights $k = k_1 + i\rho$. If generators with $k_1 = n$ vanish at the maximum of the Kähler function, the curvature scalar should vanish at the maximum and by the symmetric space property everywhere. These conditions correspond to Virasoro conditions in super string models.

A possible source of difficulties are the generators having $k_1 = 0$ and resulting as commutators of generators with opposite real parts of the conformal weights. It might be possible to assume that only the conformal weights $k = k_1 + i\rho$, $k_1 = 0, 1, \ldots$ are possible since it is the imaginary part of the conformal weight which defines the complexification in the recent case. This would mean that the commutators involve only positive values of k_1.

c) In the infinite-dimensional case the Ricci tensor involves also terms which are non-vanishing even when the holonomy algebra does not contain $U(1)$ factor. It will be found that symmetric space property guarantees Ricci flatness even in this case and the reason is essentially the vanishing of the generators having $k_1 = n$ at the maximum of Kähler function.

There are also arguments in favor of the Hyper Kähler property. In the following argument reader can well consider replacing the attribute "quaternionic" with "hyper-quaternionic".

a) The dimensions of the imbedding space and space-time are 8 and 4 respectively so that the dimension of configuration space in vibrational modes is indeed multiple of four as required by Hyper Kähler property. Hyper Kähler property requires a quaternionic structure in the tangent space of the configuration space. Since any direction on the sphere S^2 defined by the linear combinations of quaternionic imaginary units with unit norm defines a particular complexification physically, Hyper Kähler property means the possibility to perform complexification in S^2-fold manners.

b) S^2-fold degeneracy is indeed associated with the definition of the complex structure of the configuration space. First of all, the direction of the quantization axis for the spherical harmonics or for the eigen states of Lorentz Cartan algebra at $X_+^2 \times CP_2$ can be chosen in S^2-fold manners. Quaternion conformal invariance means Hyper Kähler property almost by definition and the S^2-fold degeneracy for the complexification is obvious in this case.

c) One can see the super-canonical conformal weights as points in a particular complex plane of the quaternionic space and the choice of this plane corresponds to a selection of one configuration space Kähler structure which are parameterized by S^2. The necessity to restrict the conformal weights to a complex plane brings in mind the commutativity constraint on simultaneously measurable quantum observables.

If these naive arguments survive a more critical inspection, the conclusion would be that the effective 2-dimensionality of light like 3-surfaces implying generalized conformal and canonical symmetries would also imply Hyper Kähler property of the configuration space and make the theory well-defined mathematically. This obviously fixes the dimension of space-time surfaces as well as the dimension of Minkowski space factor of the imbedding space.

3.3 Particle massivation from the first principles

It took a decade to understand how p-adic thermodynamics description of the particle massivation emerges from the fundamental TGD. The understanding of 7–3 duality [B4, C2, F2] played a key role in the development of ideas leading to the understanding of what is involved. The thermodynamics for Virasoro generator L_0 makes sense also at the real context, and the requirement that there exists an algebraic continuation to the p-adic context implies the crucial quantization of the temperature parameter implying the emergence of the p-adic mass scales.

Basically configuration space Dirac equation corresponds to Super Virasoro conditions giving rise to Virasoro conditions determining the mass spectrum of partons. There are two kinds causal determinants (CDs): 3-dimensional light like surfaces $X_l^3 \subset X^4$ representing generalized Feynman diagrams and 7-D CDs of form $X_l^3 \times CP_2$, where X_l^3 in the general case is a union of boundaries of future and past directed light cones. X_l^4 have ends at future and past directed 7-D CDs somewhat like the ends of string connecting branes.

The two kinds of causal determinants give rise to super-canonical and Super Kac-Moody Dirac operators. In both cases quark-lepton degeneracy gives rise to two different Dirac operators corresponding to N-S (quarks) and Ramond (leptons) representations of Super Virasoro algebras. Dirac operators are identifiable as super-generators G of the Super Virasoro algebra and Dirac equation states that Super Virasoro algebra acts as gauge transformations.

a) Supercanonical Dirac operators $D(SC)$ are associated with 7-D CDs $X^7 = X_l^3 \times CP_2$, X_l^3 light like surface. The interpretation is in terms of creation of pairs of space-time sheets branching from X^3 and with vanishing total classical energy. In the "vibrational" part of the $L_0(SC)$ appears the quadratic Casimir operator of Lorentz group, which can have complex values in accordance with the fact that super-canonical conformal weights are in general complex. The corresponding excitations have thus spin, color, electro-weak quantum numbers, and fermion numbers.

b) The light like 3-surfaces $Y_l^3 \subset X^4$ defining light like CDs event horizons and light like boundaries of space-time sheets define Super-Kac Moody Dirac operators. 7–3 duality reduces these operators as well as super-canonical Dirac operators to the intersections X_i^2 of Y_l^3 and X^3. These operators correspond to $P \times SU(3) \times U(2)_{ew}$ (P denotes Poincare group) localized with respect to Y_l^3 in the general case whereas the restriction of X_l^3 to surfaces $\delta M_{\pm}^4$ would reduce P to M^4. The counterpart of the stringy mass formula would result from the vanishing of $L_0(SKM)$.

The Dirac equations associated with these Dirac equations two kinds of Dirac operators can be regarded as independent, and the correlation comes from the assumption that super-conformal algebra generates super-canonical spectral flows acting as local gauge symmetries and braiding operations at global level. 7–3 duality indeed realizes this correlation concretely. In the sequel super symmetry based arguments are used to guess the explicit form of these operators and a general solution of configuration space Dirac equation is proposed by exploiting the analog with the Dirac equation in Minkowski space.

3.3.1   The analog of coset construction for super generators

The analog of Olive-Goddard-Kent coset construction for super Kac-Moody and super-canonical algebras provides a highly attractive manner to obtain a Virasoro algebra with a vanishing central extension term. A generalization of the coset construction is in question in the sense that Virasoro generators are differences of those for Super Kac-Moody algebra and super-canonical algebra. The justification comes from the miraculous geometry of the light cone boundary implying that Super Kac-Moody conformal symmetries of X^2 can be compensated by super-canonical local radial scalings so that the differences of corresponding Super Virasoro generators annihilate physical states. What is done is to construct first a state with a non-positive conformal weight using super-canonical generators, and then to apply Super-Kac Moody generators to compensate this conformal weight to get a state with vanishing conformal weight and thus mass.

The coset construction giving $L_n = L_n(SKM) - L_n(SC)$ means at the level of super generators G_n the expression

$$
\begin{aligned}
G_n^+ &= G_n^+(SKM) - G_n^+(SC) \ , \\
G_n^- &= G_n^-(SKM) - G_n^-(SC) \ .
\end{aligned}
\tag{3.3.1}
$$

Here $\pm$ refers to the positive and negative energy space-time sheets meeting at X^2 since "hermitian conjugation" (ket $\leftrightarrow$ bra) corresponds geometrically to a replacement of the space-time sheet with a positive time orientation with that having a negative time orientation. Note also that the super generators possess fermion number so that creation operators are replaced by annihilation operators in the conjugation.

If the conditions

$$
\{G^+(SKM), G^-(SC)\} = \{G^+(SKM), G^-(SKM)\}
\tag{3.3.2}
$$

hold true, the anti-commutators give just the differences $L(SKM) - L(SC)$. If 7–3 duality corresponds to a coordinate change in CH, the central extension charges for SKM and SC Super Virasoro algebras are same and resulting Super Virasoro algebra has vanishing central extension terms.

This construction works perfectly for quark sector. Leptonic super-canonical generators defined assuming the presence of operators D_+ and D_-^{-1} however anti-commute to products of configuration space Hamiltonians rather than components of Kähler form. Hence the formula for super generators $G(SC)$ does not work. The conclusion can be avoided if there exists also a leptonic representation CH gamma matrices anti-commuting to Kähler form. As already found, this kind of representation is obtained by replacing the operators D^+ and D_1^{-1} in the definitions of the super generators by projectors to the space of spinor modes with non-vanishing generalized eigen value of $D_\pm$. Also quark like super generators anti-commuting to a function algebra exist with this assumption.

The Ramond-NS degeneracy is a further important factor which must be taken into account. Possible sparticles are associated with Ramond type representation and generated by the super generators defined by the covariantly constant right handed neutrino. The helicity for neutrino is fixed by requiring that the neutrino spinor is proportional to the operator $n^k\gamma_k$, where n^k corresponds to the radial light like vector for $M_\pm^4$. If the super generators containing D and D^{-1} correspond to a Ramond type representation then covariantly right handed neutrino spinor defines an identically vanishing super generator and spartners with degenerate masses are absent. In NS sector the conformal weights of the generators generating ground states are $\pm 1/2$ so that there is no problem with the ground state degeneracy. The conclusion would be that quark/lepton like super generators G correspond to Ramond/N-S type representation. Note that leptonic super generators $G(SC)$ must carry half odd integer conformal weight.

3.3.2 General mass formula

The vanishing condition for $L_0(SKM) - L_0(SC)$ leads to a general mass formula in quark and lepton sectors separately. One of the blessings of 7–3 duality is that one can treat different 2-surfaces X_i^2 as almost independent degrees of freedom. In the case of translations this does not seem to be true since independent translations lead the surfaces X^2 outside $X_l^3 \times CP_2$. Therefore one must consider two options.

a) If one neglects the correlation between the translations and assigns to each X_i^2 independent translational degrees of freedom a separate mass formula for each X_i^2 would result:

$$M_i^2 \;=\; -\sum_i L_{0i}(SKM) + \sum_i L_{0i}(SC) \; . \tag{3.3.3}$$

Here $L_{0i}(SKM)$ contains a CP_2 cm term giving the CP_2 contribution to the mass squared known once the spinorial partial waves associated with super generators used to construct the state are known.

b) Perhaps the only internally consistent option is based on the assignment of the mass squared with the total cm. This would give

$$M^2 \;=\; \left(\sum_i p_i\right)^2 = \sum_i M_i^2 + 2\sum_{i\neq j} p_i \cdot p_j = -\sum_i L_{0i}(SKM) + \sum_i L_{0i}(SC) \; .$$

$$\tag{3.3.4}$$

Here $L_{0i}(SKM)$ contains a CP_2 cm term giving the CP_2 contribution to the mass squared known once the spinorial partial waves associated with super generators used to construct the state are known. $L_0(SC)$ term contains only leptonic or quark oscillator operators unless one allows both the lepto-quark type gamma matrices involving both D^+ and D_-^{-1} and leptonic gamma matrices involving instead of $D_\pm^{\pm 1}$ the projector P to the spinor modes with a non-vanishing eigenvalue of D.

The decomposition of the net four momentum to a sum of individual momenta can be regarded as subjective unless there is a manner to measure the individual masses. It might be that there is no unique assignment of momenta to individual partons and that this non-uniqueness is part of the gauge symmetry implied by 7–3 duality.

Mass formula before massivation

If interactions can be neglected one could assume that the free particle mass formulas

$$M_i^2 \;=\; L_{0i}(SKM) + h_i(vac) \; , \tag{3.3.5}$$

where $h_i(vac)$ corresponds to vacuum conformal weight are satisfied and what remains is the condition

$$2\sum_{i\neq j} p_i \cdot p_j = \sum_i L_{0i}(SC) - \sum_i h_i(vac) \;=\; 0 \; . \tag{3.3.6}$$

In the case of single Kac Moody algebra this condition would be replaced with

$$2 \sum_{i \neq j} p_i \cdot p_j \;=\; 0 \;. \tag{3.3.7}$$

The presence of the super-canonical algebra obviously allows much more flexibility but it seems that continuum mass spectrum is not possible.

The conformal weights of the super-canonical algebra related closely to the zeros of Riemann Zeta are complex and for the physical states the sum of the imaginary parts of the conformal weights must vanish: this is satisfied if the zeros and their conjugates appear as conformal weights of generators creating the state. The presence of super-canonical half integer contribution from super-canonical sector could explain the negative values of the ground state conformal weights forced by p-adic mass calculations.

The above formula does not contain all that is needed: also a contribution from modular degrees of freedom associated with the complex structures of X_i^2 is necessary in order to understand the dependence of the mass of fermion families on the genus of X_i^2.

Hadron physics offers the most obvious application for the mass formula. The often used metaphor for the hadronic collision as a mini big bang would have a precise meaning in TGD framework, and the effective 2-dimensionality would provide a precise realization for the parton model of hadrons. The presence of both algebras could be essential for understanding the relationship between hadron and quark masses, and the presence of super-canonical spin could allow insights to the problem of proton spin.

Particle massivation

There is an objection against the proposed mass formula. The individual contributions to the mass squared eigenvalues are not fixed uniquely. The proposed identification of parton masses is only the simplest possibility, and one could add to the individual parton conformal weights contributions compensated by the super-canonical conformal weights. This might however be a blessing rather than a curse since it suggests a manner to understand particle massivation at the fundamental level.

a) A natural requirement is that parton masses are consistent with the poles of the S-matrix elements [C2]. This assumption is quite general and certainly makes sense if the S-matrix elements allow a decomposition to vertices and propagators for a tree diagram.

b) The time evolution for the quantum states of individual partons, which are always on mass shell (that is generalized eigen states of the modified Dirac operator) is a unitary process, and corresponds to a braiding for an N-puncture system defined by the product of N local operators creating the parton state. The basic requirement is that the flow contains information about the presence of other partons and thus about the normal derivatives of the imbedding space coordinates at X_l^3. Hence the S-matrix indeed contains information about the interior of the space-time surface and the effective two-dimensionality is indeed only effective. The condition that the S-matrix elements remain unchanged in conformal transformations obviously poses explicit conditions on the normal derivatives and can be regarded as conditions stating the vanishing of the corresponding beta function.

The best candidate for the flow is as the hydrodynamic flow defined by the discontinuity of the energy momentum tensor associated with the Kähler action at X_l^3 representing what might be regarded as a hydrodynamical shock wave. This flow is in general not integrable in the sense that one could assign a global coordinate varying along the flow lines. By identifying the points of X_i^2 and X_f^2 having the same value of the complex coordinate z, the flow $X_i^2 \to X_f^2$ defines a map $w : (x,y) \to (u(x,y), v(x,y))$ mixing the points of X_i^2. The inner products of the states created by the local operator $\Psi_n(x,y)$ creating one-parton state at X_i^2 with the state created by the transformed operator $\Psi_n(u(x,y), v(x,y)$ define correlation functions, which vanish above some length scale determining the mass of the parton. Massivation occurs if this map fails to be a conformal transformation.

b) By quantum classical correspondence this braiding can be also regarded as a braiding for the points of X_i^2, which correspond to the super-canonical conformal weights just like the points of the celestial sphere correspond to momenta. If the number of operators creating the parton state is larger than two, the minimal number three of threads in the braid is present and the conformal weights become "off mass shell" conformal weights. The massivation however can in principle occur always.

In [O3, C5] the proposal was made that the bound state conformal weights could correspond to the zeros of poly-zetas: obviously this is a very strong prediction. Riemann Zeta would correspond to Higgs zero phase with the minimum of 2 operators with conjugate super-canonical conformal weights creating the state.

c) The successful description of particle massivation in terms of p-adic thermodynamics for the Virasoro generator L_0 (with p^2 not included) of the partons encourages to think that the change of the parton conformal weight could be understood as a generation of a thermal conformal weight by the flow induced by an ergodic braiding flow. This interpretation would allow to circumvent the problem created by the fact that the thermalization for mass squared is not consistent with the Lorentz invariance.

3.3.3   Particle massivation and anyonic shydrodynamics

It took quite a long time to understand how the strange hydrodynamic character of the field equations for both Kähler action and modified Dirac action relates to the particle massivation and S-matrix. It was the interaction with M-theory, which led to the discovery of 7–3 duality and effective 2-dimensionality, which in turn made possible the breakthrough in the attempts to understand how p-adic thermodynamics follows from the fundamental theory.

What is the flow defining the braiding?

The basic condition on the braiding is that it contains information about the interior of X^4 and thus about interactions with other partons. Second constraint is that the braiding flow is trivial for massless particles such as photon for which the space-time correlate should correspond to X_l^3 carrying a light-like energy momentum current.

The components $X^{n\alpha}$ of some tensor field with α restricted to X_l^3 define the most natural candidate for the braiding flow. The existence of the preferred light-like normal coordinate x^n constant at X_l^3 (in the case of δM_+^4 the light like coordinates would be $x^{\pm} = t \pm r$) is essential to achieve general coordinate invariance.

The discontinuity of the normal component $T^{n\alpha}$ of the energy momentum tensor associated with the Kähler action is a good candidate. At light like CDs the discontinuity of $T^{n\alpha}$ could be non-vanishing if allows light like CD to carry a shock wave also in imbedding space degrees of freedom as suggested by the super-symmetry. The discontinuity of $T^{n\alpha}$ would have an interpretation as a shock wave like hydrodynamic flow at the boundary. For massless particles the energy momentum current would have only a longitudinal component, the braiding would be trivial and particle would remain massless. The appearance of the energy momentum tensor in the definition of the S-matrix conforms with the hydrodynamic character of field equations and with the fact that the theory must be also a quantum theory of gravitation.

This guess is supported by the modified Dirac equation. By multiplying the modified Dirac equation at X_l^3 for shock waves localized at X_l^3 with the o_t defined by the light like gamma matrix along X_l^3 and doing the anti-commutations with the modified Dirac operator D, one finds

$$T^{\alpha n} D_\alpha \Psi = 0 \ . \tag{3.3.8}$$

The equation states that Ψ is covariantly constant along the flow lines of the flow defined by $T^{\alpha n}$. The equation can be written as

$$D_t \Psi + v^i D_i \Psi = 0 \ ,$$
$$v_i = \frac{T^{ni}}{T^{nt}} \ . \tag{3.3.9}$$

This differs from a standard flow equation for a quantity Ψ moving along field lines only by the fact that ordinary derivatives ∂_α are replaced by covariant derivatives D_α. This means that Ψ suffers a braiding transformation in spin and electro-weak degrees of freedom. Obviously, this equation states super-conformal invariance in the sense that it is not possible to poses the values of Ψ arbitrarily in entire X_l^3 but only at X^2. One could regard TGD as anyonic shydrodynamics at the space-time level.

The usual dispersion characterizing Schödinger equation emerges only at the level of imbedding space when one assigns wave equations to propagators defined by S-matrix elements.

If spinor connection is continuous at X_l^3, the discontinuity for this equation is the ordinary hydrodynamic equation

$$\partial_t \Psi + w^i \partial_i \Psi \;=\; 0 \;,$$
$$w_i \;=\; \Delta \left[\frac{T^{ni}}{kT^{nt}} \right] \;. \tag{3.3.10}$$

The correspondence with Higgs mechanism

A priori one can imagine three sources of mass.

a) The mixing caused by the flow reduces correlations unless it induces a conformal transformation leaving the physical state invariant. This mechanism could obviously relate to the massivation described in terms of p-adic thermodynamics.

b) The lacking covariant constancy of the charge matrices appearing in the definition of electroweak gauge bosons except photon can induce a loss of correlations, and is a good candidate for TGD counterpart of Higgs coupling. Also fermionic states involve this kind of dependences inducing a Higgs type contribution to the mass squared.

c) In contrast to the first guess, the holonomy transformation associated with braiding does not lead to a loss of correlations. The 2-dimensionality of X^2 implies that the holonomy group defining the braiding is Abelian. In spin degrees of freedom spinor connection vanishes by the flatness of M^4 so that the braiding cannot make neither graviton nor gluons massive. The vanishing of photon mass requires that the braiding in elecro-weak spin degrees of freedom is generated by an element of the holonomy algebra commuting with the electromagnetic charge. This however means that for charge eigen states the braiding brings in only a phase factor and would not lead to a loss of correlations.

These observations lead to a view about how TGD description relates to the standard description in terms of the phases defined by the vacuum expectation values of the Higgs field.

The model for the modular contribution to the mass squared to be discussed suggests that the p-adic prime p characterizes the size and effective p-adic topology of X^2. This raises the question why the states created by super-canonical operators O_q, $q \neq p$ do not appear in the spectrum. The intuitive guess is that the states corresponding to "wrong" p-adic topology are very massive. For instance, for $q \neq p$ the states could have inverse p-adic temperature $T_q = n > 1$ possible if roots of q are allowed in the extension of R_q used. Various phases would be characterized by the primes labelling various operator basis O_q.

Higgs zero phase would correspond to CP_2 type extremals for which the operators O_p for various primes reduce to same operator apart from the normalization constant since the projection of X^2 to $\delta M_\pm^4$ becomes zero-dimensional. This phase would be unstable but the assumption that elementary particles are near to this phase allows to identify a mechanism eliminating light exotics.

One can consider also an alternative interpretation based on the notion of conformal homotopy class implying that p-adic prime p characterizes X_l^3 rather than X^2.

a) Homotopy theory could generalize in the sense that there are several conformal equivalence classes of space-time sheets connecting given partonic 2-surfaces X_i^2 at various 7-D CDs representing initial and final states. The derivatives of the imbedding space coordinates with respect to the light like coordinate t at X_i^2 are same for these surfaces. These conformal homotopy classes could define the TGD counterparts of the "phases" of gauge theories labelled by vacuum expectation values of Higgs fields.

b) In the trivial conformal equivalence class braiding flow reduces to a mere conformal transformation: photon, graviton, and gluons remain massless. The thermal massivation of photon might be describable in terms of a non-trivial conformal homotopy class.

c) The success of the p-adic mass calculations would suggest that p-adic primes label conformal homotopy classes and thus phases and characterize the mass scale as well as the effective p-adic topology for which p-adic non-determinism corresponds to the classical non-determinism of Kähler action. p-Adic prime would characterize X_l^3 rather than X^2.

There are however objections against this interpretation. First, the notion of conformal homotopy class is not consistent with the notion of generalized Feynman diagram requiring that even topologically different surfaces X_l^3 are equivalent. Secondly, the model for the modular contribution to the mass squared forces to assign p-adic prime p to X^2 rather than X_l^3.

3.3.4 The identification of Higgs as a weakly charged wormhole contact

Quantum classical correspondence suggests that electro-weak massivation should have simple space-time description allowing also to identify Higgs if it exists. This description indeed exists and allows also to understand the precise relationship between gravitational and inertial masses and how Equivalence Principle is weakened in TGD framework.

The basic observation is that gauge and gravitational fluxes flow to larger space-time sheets through # (wormhole) contacts. If gravitational energy can be regarded in the Newtonian limit as a gauge charge, the contacts feed the gravitational energy regarded as a gauge flux to the lower condensate levels. The non-conservation of gravitational gauge flux means that # contacts can carry gravitational four-momentum. Since CP_2 type vacuum extremals are the natural candidates for # contacts, the natural hypothesis is that the non-vanishing light-like gravitational four-momentum of # contacts is responsible for the non-conservation of gravitational four-momentum flux. The non-conservation of the light-like gravitational four-momentum of CP_2 type extremals is in turn responsible for the non-conservation of the net gravitational four-momentum.

contacts can be also carriers of inertial four-momentum which must be conserved in absence of four-momentum exchange between environment and wormhole contact. Therefore Equivalence Principle cannot hold true in strict sense. Equivalence Principle is satisfied in a weak sense if the inertial four-momentum is equal to the average four-momentum associated with the zitterbewegung motion and corresponds to the center of mass motion for the # contact.

The non-conservation of weak gauge currents for CP_2 type extremals implies a non-conservation of weak charges and the finite range of weak forces. If wormhole contacts correspond to pieces of CP_2 type vacuum extremal, electro-weak gauge currents are not conserved classically unlike color and Kähler current. The non-conservation of weak isospin corresponds to the presence of pairs of right/left handed fermion and left/right handed antifermion at wormhole contacts. These wormhole contacts are excellent candidates for the TGD counterpart of Higgs boson providing the most natural mechanism for the massivation of weak bosons.

The dominant contribution to fermion masses would be due to p-adic thermodynamics. If weak form of Equivalence Principle holds true, inertial mass would result simply as the average of non-conserved light-like gravitational four-momentum. The part of the inertial mass is generated in the topological condensation of CP_2 type extremal representing elementary particle involving only single light like elementary particle horizon, say fermion, corresponds naturally to the contribution given by p-adic thermodynamics.

For gauge bosons this contribution should be very small or vanishing if the radius of zitterbewegung orbit is larger than the size of the space-time sheet containing the topologically condensed boson so that the motion is along a light-like geodesic in a good approximation. The space-time sheet representing massless state suffered secondary topologically condensation at a larger space-time sheet and viewed as a particle can develop mass via Higgs mechanism since the wormhole contacts cannot be regarded as moving along light like geodesics in the length and time scale involved.

p-Adic thermodynamics and couplings to Higgs boson as the source of elementary particle masses emerges from this picture naturally.

a) The identification of the inertial four-momentum as average gravitational four-momentum conforms with the identification of inertial mass squared as p-adic thermal mass squared.

b) # contacts carrying net weak isospin have interpretation as TGD counterparts of neutral Higgs bosons and the formation of a coherent state involving superposition of states with varying number of wormhole contacts corresponds to the generation of a vacuum expectation value of Higgs field.

c) The finding that that CP_2 parts of the induced gamma matrices connect different M^4 chiralities of induced spinor fields provided the original motivation for the belief that Higgs mechanism is realized in some manner in TGD Universe. This coupling must be crucial for the formation of weakly charged wormhole contacts.

3.4 Coupling constant evolution at space-time level

It is not yet possible to deduce the length scale evolution gauge coupling constants from Quantum TGD proper. Quantum classical correspondence however encourages the hope that it might be possible to achieve some understanding of the coupling constant evolution by using the classical theory.

This turns out to be the case and the earlier speculative picture about gauge coupling constants associated with a given space-time sheet as RG invariants finds support [C3]. It remains an open question whether gravitational coupling constant is RG invariant inside give space-time sheet. The discrete p-adic coupling constant evolution replacing in TGD framework the ordinary RG evolution allows also formulation at space-time level as also does the evolution of $\hbar$ associated with the phase resolution.

3.4.1 The evolution of gauge couplings at single space-time sheet

The renormalization group equations of gauge coupling constants g_i follow from the following idea. The basic observation is that gauge currents have vanishing covariant divergences whereas ordinary divergence does not vanish except in the Abelian case. The classical gauge currents are however proportional to $1/g_i^2$ and if g_i^2 is allowed to depend on the space-time point, the divergences of currents can be made vanishing and the resulting flow equations are essentially renormalization group equations. The physical motivation for the hypothesis is that gauge charges are assumed to be conserved in perturbative QFT. The space-time dependence of coupling constants takes care of the conservation of charges.

A surprisingly detailed view about RG evolution emerges.

a) The UV fixed points of RG evolution correspond to CP_2 type extremals (elementary particles).

b) The Abelianity of the induced Kähler field means that Kähler coupling strength is RG invariant which has indeed been the basic postulate of quantum TGD. The only possible interpretation is that the coupling constant evolution in sense of QFT:s corresponds to the discrete p-adic coupling constant evolution.

c) IR fixed points correspond to space-time sheets with a 2-dimensional CP_2 projection for which the induced gauge fields are Abelian so that covariant divergence reduces to ordinary divergence. Examples are cosmic strings (, which could be also seen as UV fixed points), vacuum extremals, solutions of a sub-theory defined by $M^4 \times S^2$, S^2 a homologically non-trivial geodesic sphere, and "massless extremals".

d) At the light-like boundaries of the space-time sheet gauge couplings are predicted to be constant by conformal invariance and by effective two-dimensionality implying Abelianity: note that the 4-dimensionality of the space-time surface is absolutely essential here.

e) In fact, all known extremals of Kähler action correspond to RG fixed points since gauge currents are light-like so that coupling constants are constant at a given space-time sheet. This is consistent with the earlier hypothesis that gauge couplings are renormalization group invariants and coupling constant evolution reduces to a discrete p-adic evolution. As a consequence also Weinberg angle, being determined by a ratio of $SU(2)$ and $U(1)$ couplings, is predicted to be RG invariant. A natural condition fixing its value would be the requirement that the net vacuum em charge of the space-time sheet vanishes. This would state that em charge is not screened like weak charges.

f) When the flow determined by the gauge current is not integrable in the sense that flow lines are identifiable as coordinate curves, the situation changes. If gauge currents are divergenceless for all solutions of field equations, one can assume that gauge couplings are constant at a given space-time sheet and thus continuous also in this case. Otherwise a natural guess is that the coupling constants obtained by integrating the renormalization group equations are continuous in the relevant p-adic topology below the p-adic length scale. Thus the effective p-adic topology would emerge directly from the hydrodynamics defined by gauge currents.

3.4.2 RG evolution of gravitational constant at single space-time sheet

Similar considerations apply in the case of gravitational and cosmological constants.

a) In this case the conservation of gravitational mass determines the RG equation (gravitational energy and momentum are not conserved in general).

b) The assumption that coupling cosmological Λ constant is proportional to $1/L_p^2$ (L_p denotes the relevant p-adic length scale) explains the mysterious smallness of the cosmological constant and leads to a RG equation which is of the same form as in the case of gauge couplings.

c) Asymptotic cosmologies for which gravitational four momentum is conserved correspond to the fixed points of coupling constant evolution now but there are much more general solutions satisfying the constraint that gravitational mass is conserved.

d) It seems that gravitational constant cannot be RG invariant in the general case and that effective p-adicity can be avoided only by a smoothing out procedure replacing the mass current with its average over a four-volume 4-volume of size of order p-adic length scale.

3.4.3   p-Adic evolution of gauge couplings

If RG invariance at given space-time sheet holds true, the question arises whether it is possible to understand p-adic coupling constant evolution at space-time level.

a) Simple considerations lead to the idea that M^4 scalings of the intersections of 3-surfaces defined by the intersections of space-time surfaces with light-cone boundary induce transformations of space-time surface identifiable as RG transformations. If sufficiently small they leave gauge charges invariant: this seems to be the case for known extremals which form scaling invariant families. When the scaling corresponds to a ratio p_2/p_1, $p_2 > p_1$, bifurcation would become possible replacing p_1-adic effective topology with p_2-adic one.

b) Stability considerations determine whether p_2-adic topology is actually realized and could explain why primes near powers of 2 are favored. The renormalization of coupling constant would be dictated by the requirement that Q_i/g_i^2 remains invariant.

3.4.4   p-Adic evolution in angular resolution and dynamical $\hbar$

For a given p-adic topology algebraic extensions of p-adic numbers define also a hierarchy ordered by the dimension of the extension and this hierarchy naturally corresponds to an increasing angular resolution so that RG flow would be associated also with it.

a) A characterization of angular scalings consistent with the identification of $\hbar$ as a characterizer of the topological condensation of 3-surface X^3 to a larger 3-surface Y^3 is that angular scalings correspond to the transformations $\Phi \to r\Phi$, $r = m/n$ in the case of X^3 and $\Phi \to \Phi$ in case of Y^3 so that X^3 becomes analogous to an m-fold covering of Y^3. Rational coverings could also correspond to m-fold scalings for X^3 and n-fold scalings for Y^3.

b) The formation of these stable multiple coverings could be seen as an analog for a transition in chaos via a process in which a closed Bohr orbit regarded as a particle itself becomes an orbit closing only after m turns. TGD predicts a hierarchy of higher level zero energy states representing S-matrix of lower level as entanglement coefficients. Particles identified as "tracks" of particles at orbits closing after m turns might serve as space-time correlates for this kind of states. There is a direct connection with the fractional quantum numbers, anyon physics and quantum groups.

c) The simplest generalization from the p-adic length scale evolution consistent with the proposed role of Beraha numbers $B_n = 4cos^2(\pi/n)$ is that bifurcations can occur for integer values of r=m and change the value of $\hbar$. The interpretation would be that single 2π rotation in δM_+^4 corresponds to the angular resolution with respect to the angular coordinate ϕ of space-time surface varying in the range $(0, 2\pi)$ and is given by $\Delta\phi = 2\pi/m$.

d) The evidence for a gigantic but finite value of "gravitational" Planck constant suggests that large values of $\hbar$ corresponding to $3 < n < 4$ and defining a "generalized" Beraha number B_r are possible. For $n = 3$ corresponding to the minimal resolution of $\Delta\phi = 2\pi/3$ $\hbar$ would be infinite. This would allow to keep the formula for $\hbar(n)$ in its original form by replacing n with a rational number. This would mean that also rational values of r correspond to bifurcations in the range $3 < r < 4$ at least.

3.5   About the construction of S- and U-matrices

The enormous symmetries of quantum are bound to lead to a highly unique S-matrix but the practical construction of S-matrix is a formidable challenge and necessitates deep grasp about the physics involved so that one can make the needed approximations. The evolution of the ideas related to S-matrix involves several side-tracks and strange twists characteristic for a mathematical problem solving when a direct contact with the experimental reality is lacking. The work with S-matrix has taught me that principles are more important than formulas and that the only manner to proceed is from top to bottom by gradually solving the philosophical problems, identifying all the relevant symmetries and understanding the horribly nonlinear dynamics defined by the Kähler action.

The poor understanding of the philosophical issues has led to frustratingly many candidates for S-matrix. TGD inspired theory of consciousness has however gradually led to a clarification of various issues and it seems that it is safer to distinguish between two matrices: U-matrix and S-matrix. Moreover, it seems that one must talk in plural: there is entire hierarchy of U-matrices and S-matrices correspond to higher levels of the hierarchy.

a) U-matrix is much more fundamental object than S-matrix conventionally defined as time-translation operator and characterizes what happens in single quantum jump $\Psi_i \rightarrow U\Psi_i \rightarrow \Psi_f$. A good candidate for U-matrix is as Glebsch-Gordan coefficients relating free and interacting Super Virasoro representations.

b) U-matrix decomposes into a tensor product of a U-matrix characterizing dispersion in zero modes and a U-matrix characterizing the dynamics in fiber degrees of freedom. The U-matrix associated with the fiber degrees of freedom in turn decomposes into a tensor product of the local U-matrices associated with various space-time sheets. The tensor factor of U describing dynamics in conformal degrees of freedom for a given space-time sheet could correspond to the TGD counterpart of the stringy S-matrix.

c) The new view about sub-system forced by the many-sheeted space-time and the integration of quantum jump sequences to single quantum as far as conscious experience is considered, suggests that one must introduce entire hierarchy of U-matrices corresponding to p-adic length scale hierarchy and hierarchy of durations for sequences of quantum jumps. The identification as S-matrices is natural since the duration of quantum jump sequence allows identification as counterpart for the duration of time evolution associated with S-matrix in standard physics. Actually subjective time duration is in question and corresponds only in a statistical sense to a definite duration of geometric time. In particular, one expects that these higher level U-matrices do not provide only an approximate description of something more fundamental but express all that can be said and tested by conscious observer.

d) The hierarchy of p-adic number fields and their extensions of increasing dimension should correspond to the hierarchy of U-matrices. This means that the matric elements of S-matrix should be in finite-dimensional extensions of rationals possibly involving transcendentals. This would mean that S-matrix theory becomes number theory at the deepest and most challenging level one can imagine. A typical problem can be formulated as a question whether some finite set of transcendentals is a closed system under the process of taking logarithms.

The lack of explicit formulas for S-matrix elements have been the basic weakness of quantum TGD approach as compared to the concrete perturbative formulas provided by super-string approach. Fortunately, the new number theoretic vision leads to concrete Feynman rules for S-matrix in the approximation that elementary particles can be regarded as CP_2 type extremals. Of course, this is only small piece of quantum TGD but certainly the most important one as far as the empirical testing of the theory is considered.

3.5.1 Inner product from divergence cancellation

The construction of a divergence free and unitary inner product for the configuration space spinor fields has been already considered in the first part of the book. The inner product is given by integrating the usual Fock space inner product defined at each point of the configuration space over the reduced configuration space containing only the 3-surfaces Y^3 belonging to $\delta H = \delta M_+^4 \times CP_2$ ('light cone boundary') using the exponent $exp(K)$ as a weight factor:

$$\langle \Psi_1 | \Psi_2 \rangle = \int \overline{\Psi}_1(Y^3)\Psi_2(Y^3)exp(K)\sqrt{G}dY^3 \ ,$$

$$\overline{\Psi}_1(Y^3)\Psi_2(Y^3) \equiv \langle \Psi_1(Y^3)|\Psi_2(Y^3)\rangle_{Fock} \ . \tag{3.5.1}$$

The degeneracy for the absolute minima of Kähler action implies additional summation over the degenerate minima associated with Y^3. The restriction of the integration on light cone boundary is Diff^4 invariant procedure and resolves in elegant manner the problems related to the integration over Diff^4 degrees of freedom. A variant of the inner product is obtained dropping the bosonic vacuum functional $exp(K)$ from the definition of the inner product and by assuming that it is included into the spinor fields themselves. Probably it is just a matter of taste how the necessary bosonic vacuum functional is included into the inner product: what is essential that the vacuum functional $exp(K)$ is somehow present in the inner product.

The unitarity of the inner product follows from the unitary of the Fock space inner product and from the unitarity of the standard L^2 inner product defined by configuration space integration in the set of the L^2 integrable scalar functions. It could well occur that $Diff^4$ invariance implies the reduction of the configuration space integration to $C(\delta H)$.

Consider next the bosonic integration in more detail. The exponent of the Kähler function appears in the inner product also in the context of the finite dimensional group representations. For the representations of the noncompact groups (say $SL(2,R)$) in coset spaces (now $SL(2,R)/U(1)$ endowed with Kähler metric) the exponent of Kähler function is necessary in order to get square integrable representations [df5]. The scalar product for two complex valued representation functions is defined as

$$(f,g) \;\; = \;\; \int \overline{f} g exp(nK) \sqrt{g} dV \; . \tag{3.5.2}$$

By unitarity, the exponent is an integer multiple of the Kähler function. In the present case only the possibility $n = 1$ is realized if one requires a complete cancellation of the determinants. In finite dimensional case this corresponds to the restriction to single unitary representation of the group in question.

The sign of the action appearing in the exponent is of decisive importance in order to make theory stable. The point is that the theory must be well defined at the limit of infinitely large system. Minimization of action is expected to imply that the action of infinitely large system is bound from above: the generation of electric Kähler fields gives negative contributions to the action. This implies that at the limit of the infinite system the average action per volume is nonpositive. For systems having negative average density of action vacuum functional $exp(K)$ vanishes so that only configurations with vanishing average action per volume have significant probability. On the other hand, the choice $exp(-K)$ would make theory unstable: probability amplitude would be infinite for all configurations having negative average action per volume. In the fourth part of the book it will be shown that the requirement that average Kähler action per volume cancels has important cosmological consequences.

Divergence cancellation in the bosonic integration has been already demonstrated in the first part of the book. One can develop the Kähler function as a Taylor series around maximimum of Kähler function and use the contravariant Kähler metric as a propagator. Gaussian and metric determinants cancel each other for a unique vacuum functional. Ricci flatness guarantees that metric determinant is constant in complex coordinates so that one avoids divergences coming from it. The non-locality of the Kähler function as a functional of the 3-surface serves as an additional regulating mechanism: if $K(X^3)$ were a local functional of X^3 one would encounter divergences in the perturbative expansion.

The requirement that quantum jump corresponds to a quantum measurement in the sense of quantum field theories implies that quantum jump involves localization in zero modes. Localization in the zero modes implies automatically p-adic evolution since the decomposition of the configuration space into sectors D_P labelled by the infinite primes P is determined by the corresponding decomposition in zero modes. Localization in zero modes would suggest that the calculation of the physical predictions does not involve integration over zero modes: this would dramatically simplify the calculational apparatus of the theory. Probably this simplification occurs at the level of practical calculations if U-matrix separates into a product of matrices associated with zero modes and fiber degrees of freedom.

One must also calculate the predictions for the ratios of the rates of quantum transitions to different values of zero modes and here one cannot actually avoid integrals over zero modes. To achieve this one is forced to define the transition probabilities for quantum jumps involving a localization in zero modes as

$$P(x, \alpha \to y, \beta) = \sum_{r,s} |S(r, \alpha \to s, \beta)|^2 |\Psi_r(x)|^2 |\Psi_s(y)|^2 \ ,$$

where x and y correspond to the zero mode coordinates and r and s label a complete state functional basis in zero modes and $S(r, m \to s, n)$ involves integration over zero modes. In fact, only in this manner the notion of the localization in the zero modes makes mathematically sense at the level of S-matrix. In this case also unitarity conditions are well-defined. In zero modes state function basis can be freely constructed so that divergence difficulties could be avoided. An open question is whether this construction is indeed possible.

Some comments about the actual evaluation of the bosonic functional integral are in order.

a) Since configuration space metric is degenerate and the bosonic propagator is essentially the contravariant metric, bosonic integration is expected to reduce to an integration over the zero modes. For instance, isometry invariants are variables of this kind. These modes are analogous to the parameters describing the conformal equivalence class of the orbit of the string in string models. The identification of all isometry invariants and corresponding integration measure in the configuration space integral was considered in the first part of the book.

b) α_K is a natural small expansion parameter in configuration space integration. It should be noticed that α_K, when defined by the criticality condition, could also depend on the coordinates parametrizing the zero modes.

c) Semiclassical approximation, which means the expansion of the functional integral as a sum over the extrema of the Kähler function, is a natural approach to the calculation of the bosonic integral. There are theorems stating that semiclassical approximation is exact for certain systems (for example for integrable systems [db3]). An interesting question is whether the Kähler function might possess the properties guaranteing the exactness of the semiclassical approximation. This would mean that the calculation of the integral $\int exp(K)\sqrt{G}dY^3$ would reduce to an integration over the zero modes: this kind of reduction actually occurs in string models.

3.5.2 The fundamental identification of U- and S-matrices

Single quantum jump corresponds to the sequence

$$\Psi_i \to U\Psi_i \to ....\Psi_f \ .$$

U does not certainly correspond to a genuine time-development in the sense of a unitary time translation operator. A good guess is that U has interpretation as Glebsch-Gordan coefficients between free an interacting representations of Super Virasoro algebra associated with surfaces $\cup_i X^4(Y_i^3)$ and $X^4(\cup_i Y_i^3)$. For a given unentangled sub-system (sub-system in self-organizing self-state) the eigen states of the density matrix of the sub-system becoming unentangled in quantum jump determines what are the final states of the quantum jump. Negentropy Maximization Principle states that the sub-system of unentangled sub-systems whose measurement gives rise to maximal entanglement negentropy gain, is quantum measured. U is much more fundamental object than S-matrix. If sub-system is entangled then in a reasonable approximation nothing happens to it during quantum jump sequence and the sub-quantum history remains unchanged.

What could then be the interpretation of S-matrix in this framework?

a) The first glance to the problem is based on macro-temporal quantum coherence. If the dissipative effects caused by the state function reductions and state preparations are absent completely or are not visible in the time resolution defined by the duration of macro-temporal quantum coherence, one might expect that S-matrix is product of U-matrices occurred during the period of the macro-temporal quantum coherence. The system in self state would indeed effectively behave like its own Universe. One could say that time-evolution is discretized with CP_2 time defining the duration of the chronon.

b) Observer is represented by a cognitive space-time sheet drifting towards the geometric future quantum jump by quantum jump along material space-time sheet. Observer is basically interested in the unitary time development induced by the time-translation operator P_0 associated with the modified Dirac operator. This time development can have any duration T and defines time evolution operator $exp(iP_0T)$ if sub-system develops as essentially free system. The measurement of the scattering probabilities defined by S-matrix corresponds to a construction of an empirical arrangement guaranteing that quantum measurement observed by (the sufficiently intelligent!) cognitive space-time sheet at time $t = T$ reduces the quantum state to some of the state of the initial state basis at time $t = 0$. In the ideal situation the measured system would develop unitarily during this interval and stay thus entangled so that it cannot self-organize by quantum jumps.

This picture would suggest that the S-matrix could be defined as a generalization of the exponential $exp(iP_0 T)$ of the second quantized Poincare energy operator P_0 associated with the modified Dirac action for the interacting space-time surface $X^4(\cup_i Y_i^3)$ and acts on the tensor product of the state spaces associated with $X^4(Y_i^3)$. This picture might make sense at the limit when wave mechanics is a good approximation but does not work in elementary particle length scales where CP_2 type extremals whose M_+^4 projection is a random light like curve, are expected to dominated.

Interacting space-time surfaces defined by a connected sum of CP_2 type extremals can be regarded as a Feynman graph with lines thickened to 4-manifolds. This suggests the assignment of the exponent with the internal lines of the generalized Feynman graph acting as translation operators whereas vertices where lines join give rise to vertex operators which can be regarded as Glebch-Gordan coefficients for super-Kac-Moody representations.

Since the Hamiltonian in question is quadratic in oscillator operators, the theory is free in the standard sense of the word, and it is only the dynamics of Kähler action which induces interactions and makes the theory nontrivial. Feynman diagram structure has purely topological origin. For instance, topological sums of CP_2 type extremals can be regarded as an example of Feynman diagrams with lines thickened to 4-manifolds. The absence of interaction terms guarantees that there are no sources of divergences. For the CP_2 type extremals one can develop rather detailed form of the Feynman rules.

TGD inspired theory of consciousness suggests that one should not take too dogmatic view about S-matrix as a summary for the predictions of quantum theory. Even the idealizations about experimental situation involved with the S-matrix formalism might be quite too strong since experimental observations are always about sub-systems.

3.5.3   7–3 duality and construction of S-matrix

The notion of 7–3 duality emerged from the (one might say violent) interaction between TGD and M-theory [A3]. The attempts to construct quantum TGD have gradually led to the conclusion that the geometry of the configuration space ("world of classical worlds") involves both 7-D and 3-D light like surfaces as causal determinants. 7-D light like surfaces X^7 are unions of future and past light cone boundaries and play a role somewhat resembling that of branes. Each 7-D CD determines one particular sector of the configuration space. 3-D light like surfaces X_l^3 can correspond to boundaries of space-time sheets, regions separating two maximally deterministic space-time regions, and elementary particle horizons at which the signature of the induced metric changes.

7–3 duality states that it is possible to formulate the theory using either the data at 3-D space-like 3-surfaces resulting as intersections of the space-time surface with 7-D CDs or the data at 3-D light like CDs [B4]. This results if the data needed is actually contained by 2-D intersections $X^2 = X_l^3 \cap X^7$. This effective 2-dimensionality has far-reaching implications. The notion simplifies dramatically the basic formulas related to the configuration space geometry and spinor structure and allows to understand the relationship between various conformal symmetries of TGD.

Effective 2-dimensionality also leads to an explicit identification of the generalized Feynman diagrams at space-time level as light like 3-D CDs. The basic philosophy is that quantum-classical correspondence stating that space-time sheets provide a description for the physics associated with the configuration space spin degrees of freedom (fermionic degrees of freedom). This means kind of self-referentiality making it possible for quantum states to represent the dynamics of quantum jumps.

The generalized Feynman diagrammatics is simple. The propagator associated with 3-D CD is simply the inverse of the modified Dirac operator D, and vertices the inner products at X^2 for the positive energy states and negative energy states entering to the vertex and in principle computable. The equivalence of generalized Feynman diagrams with tree diagrams is expected on basis of the effective 2-dimensionality and gives very strong additional constraints to the vertices. It implies also unitarity. Since no loop summations are involved, the theory does not differ from effective theories defined by effective action.

The counterparts of loop sums are absent in TGD framework and p-adic number fields and their extensions defining an infinite hierarchy of fixed point values of Kähler coupling strength and thus of gauge coupling constants. The question is whether this discrete coupling constant evolution can mimic a QFT type coupling constant evolution (or vice versa). Is it possible to have renormalization without renormalization? The construction of quantum state using generalization of coset construction for super-canonical and super Kac-Moody algebra allows to answer this question. The counterparts of bare states are non-orthogonal and have a natural multi-grading. Gram-Schmidt orthogonalization procedure makes the bare states dressed and brings in TGD counterpart of loop corrections to the S-matrix. The counterparts of renormalization group equations result by formally regarding p-adic prime p as a continuous variable. Quantum field theory approximation results when the inner products defining simplest particle decays are described as coupling constants.

Certainly this generalized Feynman diagrammatics could be seen as the fulfillment of a long held dream and forms the basic for various approximate theories. This does not however exclude alternative approaches to S-matrix assuming that they are consistent with the fundamental equivalence of the generalized Feynman diagrams with tree diagrams.

3.5.4 Equivalence of loop diagrams with tree diagrams and cancellation of infinities in Quantum TGD

In [C5] a vision about how dual diagrams generalize in TGD context is developed. The vision is based on generalization of mathematical structures discovered in the construction of topological quantum field theories (TQFT) and conformal field theories (CQFT). In particular, the notions of Hopf algebras and quantum groups, and categories are central. The following gives a very concise summary of the basic ideas.

Feynman diagrams as generalized braid diagrams

The first key idea is that generalized Feynman diagrams are analogous to knot and link diagrams in the sense that they allow also "moves" allowing to identify classes of diagrams and that the diagrams containing loops are equivalent with tree diagrams, so that there would be no summation over diagrams. This would be a generalization of duality symmetry of string models.

TGD itself provides general arguments supporting same idea. The identification of absolute minimum of Kähler action as a four-dimensional Feynman diagram characterizing particle reaction means that there is only single Feynman diagram instead of functional integral over 4-surfaces: this diagram is expected to be minimal one. At quantum level S-matrix element can be seen as a representation of a path defining continuation of configuration space (CH) spinor field between different sectors of CH corresponding to different 3-topologies. All continuations and corresponding Feynman diagrams are equivalent. The idea about Universe as a computer and algebraic hologram allows a concrete realization based on the notion of infinite primes, and space-time points become infinitely structured monads [O4]. The generalized Feynman diagrams differing only by loops are equivalent since they characterize equivalent computations.

Coupling constant evolution from infinite number of critical values of Kähler coupling strength

The basic objection against the new view about Feynman diagrams is that it is not consistent with the notion of coupling constant evolution involving loops in an essential manner. The objection can be circumvented. Quantum criticality requires that Kähler coupling constant α_K is analogous to critical temperature (so that the loops for configuration space integration vanish). The hypothesis motivated by the enormous vacuum degeneracy of Kähler action is that α_K has an infinite number of possible values labelled by p-adic length scales and also also by the dimensions of effective tensor factors defined hierarchy of II_1 factors (so called Beraha numbers) as found in [O4]. The dependence on p-adic length scale L_p corresponds to the usual renormalization group evolution whereas the latter dependence would correspond to a finite angular resolution and to a hierarchy of finite-dimensional extensions of p-adic number fields R_p. The finiteness of the resolution is forced by the algebraic continuation of rational number based physics to real and p-adic number fields since p-adic and real notions of distance between rational points differ dramatically. The higher the algebraic dimension of the extension and the higher the value of p-adic prime the better the angular (or phase) resolution and nearer the p-adic topology to that for real numbers.

R-matrices, complex numbers, quaternions, and octonions

A crucial observation is that physically equivalent R-matrices of 6-vertex models are labelled by points of CP_2 whereas maximal space of commuting R-matrices are labelled by points of 2-sphere S^2. CP_2 also labels maximal associative and thus quaternionic subspaces of 8-dimensional octonion space whereas S^2 labels the maximally commutative subspaces of quaternion space, which suggests that R-matrices and these structures correspond to each other.

Number theoretic vision leads to a number theoretic variant of spontaneous compactification meaning that space-time surfaces could be regarded either as hyper-quaternionic, and thus maximal associative, 4-surfaces in M^8 regarded as the space of hyper-octonions or as surfaces in $M^4 \times CP_2$ (the imaginary units of hyper-octonions are multiplied with $\sqrt{-1}$ so that the number theoretical norm has Minkowski signature). Associativity constraint is an essential element of also Yang-Baxter equations [ec4, ee2].

These observations lead to a concrete proposal how quantum classical correspondence is realized classically at the space-time level. Each point of CP_2 corresponds to R-matrix and 3-surfaces can be identified as preferred sections of the space-time surface by requiring that unitary R-matrices are commuting in the section (micro-causality) whereas commutativity fails for R-matrices corresponding to different values of time coordinate defined by this foliation.

Ordinary conformal symmetries act on the space of super-canonical conformal weights

TGD predicts two kinds of super-conformal symmetries when space-time surfaces are regarded as sub-manifolds of $H = M^4 \times CP_2$ (in $OH = M^8$ picture hyper-quaternionic and hyper-octonionic conformal symmetries are relevant).

a) The ordinary super-conformal symmetries realized at the space-time level are associated with super Kac-Moody representations realized at light-like 3-surfaces appearing as boundaries of space-time sheets and boundaries between space-time regions with Euclidian and Minkowskian signature of metric. Conformal weights are half-integer valued in this case.

b) Super-canonical conformal invariance acts at the level of imbedding space and corresponds to light-like 7-surfaces of form $X_l^3 \times CP_2 \subset M^4 \times CP_2$ believed to appear as causal determinants too. In this case conformal weights are complex numbers of form $\Delta = n/2 + iy$ for the generators if super algebra, and I have proposed that the conformal weights of the physical states correspond are of form $1/2 + iy$, where y is zero of Riemann Zeta or possibly even superposition of them. In the latter case the imaginary parts of zeros would define a basis of an infinite-dimensional Abelian group spanning the weights.

The basic question concerns the interaction of these conformal symmetries.

a) Quantum classical correspondence suggests that the complex conformal weights of super-canonical algebra generators have space-time counterparts. The proposal is that the weights are mapped to the points of geodesic sphere of CP_2 (and thus also of space-time surface) labelling also mutually commutating R-matrices. The map is completely analogous to the map of momenta of quantum particles to the points of celestial sphere. One can thus regard super-generators as conformal fields in space-time or complex plane having conformal weights as punctures. The action of super-conformal algebra and braid group on these points realizing monodromies of conformal field theories [ee2] induces by a pull-back a braid group action on the conformal labels of configuration space gamma matrices (super generators) and corresponding isometry generators.

b) The gamma matrices and isometry generators spanning the super-canonical algebra can be regarded as fields of a conformal field theory in the complex plane containing infinite number of punctures defined by the complex conformal weights. Quaternion conformal Super Virasoro algebra and Kac Moody algebra would act as symmetries of this theory and the S-matrix of TGD would involve the n-point functions of this conformal field theory.

This picture also justifies the earlier proposal that configuration space Clifford algebra defined by the gamma matrices acting as super generators defines an infinite-dimensional von Neumann algebra possessing hierarchies of type II_1 factors [eb1] having a close connection with the non-trivial representations of braid group and quantum groups. The sequence of non-trivial zeros of Riemann Zeta along the line $Re(s) = 1/2$ in the plane of conformal weights could be regarded an an infinite braid behind the von Neumann algebra [eb1]. Contrary to the expectations, also trivial zeros seem to be important. Purely algebraic considerations support the view that superposition for the imaginary parts of non-trivial zeros makes sense so that one would have entire hierarchy of one-dimensional lattices depending on the number of zeros included. The finite braids defined by subsets of zeros could be seen as a hierarchy of completely integrable 1-dimensional spin chains leading to quantum groups and braid groups [ec4, ee2] naturally. In conformal field theories it is possible to construct explicitly the generators of quantum group in terms of operators creating screening charges [ee2]: interestingly, the charges are located going along a line parallel to imaginary axis to infinity.

It seems that not only Riemann's zeta but also polyzetas [cf3, cf4, ed4, ed5] could play a fundamental role in TGD Universe. The super-canonical conformal weights of interacting particles, in particular of those forming bound states, are expected to have "off mass shell" values. An attractive hypothesis is that they correspond to zeros of Riemann's polyzetas. Interaction would allow quite concretely the realization of braiding operations dynamically. The physical justification for the hypothesis would be quantum criticality. Indeed, it has been found that the loop corrections of quantum field theory are expressible in terms of polyzetas [ee9]. If the arguments of polyzetas correspond to conformal weights of particles of many-particle bound state, loop corrections vanish when the super-canonical conformal weights correspond to the zeros of polyzetas including zeta. This argument does not allow superpositions of imaginary parts of zeros.

Equivalence of loop diagrams with tree diagrams from the axioms of generalized ribbon category

The fourth idea is that Hopf algebra related structures and appropriately generalized ribbon categories [ec4, ee2] could provide a concrete realization of this picture. Generalized Feynman diagrams which are identified as braid diagrams with strands running in both directions of time and containing besides braid operations also boxes representing algebra morphisms with more than one incoming and outgoing strands. 3-particle vertex should be enough, and the fusion of 2-particles and $1 \to 2$ particle decay would correspond to generalizations of the algebra product μ and co-product Δ to morphisms of the category defined by the super-canonical algebras associated with 3-surfaces with various topologies and conformal structures. The basic axioms for this structure generalizing ribbon algebra axioms [ec4] would state that diagrams with self energy loops, vertex corrections, and box diagrams are equivalent with tree diagrams.

Tensor categories might provide a deeper understanding of p-adic length scale hypothesis. Tensor primes can be identified as vector/Hilbert spaces, whose real or complex dimension is prime. They serve as "elementary particles" of tensor category since they do not allow a decomposition to a tensor product of lower-dimensional vector spaces. The unit I of the tensor category would have an interpretation as a one-dimensional Hilbert space or as the number field associated with the Hilbert space and would act like identity with respect to tensor product. Quantum jump cannot decompose tensor prime system to an unentangled product of sub-systems. This elementary particle like aspect of tensor primes might directly relate to the origin of p-adicity. Also infinite primes are possible and could distinguish between different infinite-dimensional state spaces.

For quantum dimensions $[n]_q \equiv (q^n - q^{-n})/(q - q^{-1})$ [ec4] no decomposition into a product of prime quantum dimensions exist and one can say that all non-vanishing quantum integers $[n]_q$ are primes. For q an nth root of unity, quantum integers form a finite set containing only the elements $0, 1, ..., [n-1]$ so that quantum dimension is always finite. The numbers $[2]_q^2$, for $q = exp(i\pi/n)$ define a hierarchy of Beraha numbers having an interpretation as a renormalized dimension $[2]_q^2 \leq 4$ for the spinor space of 4-dimensional space and appearing as effective dimensions of type II_1 sub-factors of von Neumann algebras.

Quantum criticality and renormalization group invariance

Quantum criticality means that renormalization group acts like isometry group at a fixed point rather than acting like a gauge symmetry as in the standard quantum field theory context. Despite this difference it is possible to understand how Feynman graph expansion with vanishing loop corrections relates to generalized Feynman graphs and a nice connection with the Hopf- and Lie algebra structures assigned by Connes and Kreimer to Feynman graphs emerges. It is possible to deduce an explicit representation for the universal momentum and p-adic length scale dependence of propagators in this picture. The condition that loop diagrams are equivalent with tree diagrams gives explicit equations which might fix completely also the p-adic length scale evolution of vertices. Quantum criticality in principle fixes completely the values of the masses and coupling constants as a function of p-adic length scale.

Modified Dirac action for the induced spinor fields as the QFT description of quantum TGD?

Quantum-classical correspondence suggests that quantum TGD should allow a quantum field theory formulation of some kind allowing to relate the abstract physics at the level of configuration space to the space-time physics defined by Kähler action. QFT formulation indeed seems to exist and the constraints satisfied by the QFT formulation of quantum TGD are so strong that the formulation is essentially unique.

a) The QFT in question should be determined by the absolute minimum $X^4(X^3)$ of Kähler action corresponding to a given causal determinant (7-dimensional light like surface $X_l^3 \times CP_2$) rather than in M^4. The averaging over all Poincare and color translates of this space-time surface would give rise to an S-matrix respecting the basic symmetries. Note that the theory would be 3-D quantum field theory at X^3. The only information needed about $X^4(X^3)$ would be the time derivatives of the imbedding space coordinates at X^3. Also the value of Kähler action seems to be needed but even the exponent of Kähler function might disappear from the Greens functions in normalization just as the exponent e^G of the generating functional G of connected Green's functions disappears in the quantum field theory.

The effective 3-dimensionality has an interesting connection to unresolved difficulties encountered in the attempt to formulate bound state problems in quantum field theory context. Non-relativistic, essentially 3-dimensional, Schrödinger equation works and yield correct predictions whereas Bethe-Salpeter equation in Minkowski space fails. The TGD based explanation of this failure is that bound state formation means that 3-surfaces of particles involved form a join along boundaries condensate. This means that non-relativistic 3-dimensional formulation is necessary in order to catch the essential aspects of the physics involved. In quantum field theory context point-likeness of the particles allows only the modelling of those aspects of particle interactions which do not involve bound states.

b) To calculate correlation functions one should expand the super-canonical generators as functional Taylor series around the maximum of the Kähler function at the 3-surface X^3. If super-canonical generators can be regarded as functionals of the second quantized induced spinor field ψ, also the functional series with respect to ψ is needed. This would make it possible to evaluate the correlation functions perturbatively in terms of the tree diagrams defined by the effective action using bosonic and fermionic propagators defined by it. One would calculate M^4 Fourier transforms of the correlation functions and integration over Poincare translates would give Poincare invariant correlation functions.

c) The complete localization at configuration space level means the vanishing of the bosonic loops and effective freezing of configuration space degrees of freedom. This is achieved if the action is such that the bosonic part vanishes when the induced spinor fields vanish. I have represented in [B4] arguments that the action for the induced spinor fields treated as Grassman variables is all that is needed to define quantum physics. Kähler action would be the effective action associated with the Dirac action and its absolute minimization (or some more general variational principle selecting the preferred extremals as generalized Bohr orbits [E2]) would be thus natural. The approach would also predict the possible values of Kähler coupling strength as part of data characterizing the effective action. This approach also conforms with the fact that elementary bosons are predicted to be bound states of fermion-anti-fermion pairs.

This approach would also bring induced electro-weak gauge potentials into play so that quantum-classical correspondence would be realized. The absence of quark color as spin like quantum number of induced spinor fields would not be a problem. Also the topologization of the family replication phenomenon in terms of the genus of 2-surface would result without representation as an additional spin like degeneracy of fermion fields.

d) The action would be the modified Dirac action for the induced spinor fields at the maximum of the Kähler function. Modified Dirac action is defined by replacing the induced gamma matrices $\Gamma_\alpha = \partial_\alpha h^k \Gamma_k$ by the modified gamma matrices

$$\hat{\Gamma}^\alpha = \frac{\partial(L\sqrt{|g|})}{\partial(\partial_\alpha h^k)} \Gamma^k \quad . \tag{3.5.3}$$

Here L denotes the action density of Kähler action and Γ^k denotes gamma matrices of the imbedding space H. Modified Dirac action is supersymmetric and shares the vacuum degeneracy of Kähler action [B4].

The fermionic propagator would be defined by the inverse of the modified Dirac operator. Bosonic kinetic term would be Grassmann algebra valued and vanish for $\psi = 0$ and would contribute nothing to the perturbation series. Since the fermionic action is free action, a divergence free quantum field theory would be in question irrespective of whether the fermionic action is interpreted as an action or an effective action. One could also see the action as a fixed point of the map sending action to effective action in a complete accordance with the idea that loop corrections vanish.

e) The nice feature of this approach is that the information needed about the absolute minimum would be minimal since one can restrict the consideration to 3-surface X^3 which can be selected arbitrarily. In fact, the outcome is a 3-dimensional free field theory in the fermionic degrees of freedom and the integral over Poincare and color translates guarantees isometry symmetries. Grassmannian functional integral would give exponent of Kähler function as the analog of the generating functional G for connected Green functions. The functional derivatives of the configuration space spinor field with respect to the induced spinor field are very simple by the conservation of fermion number. This approach could be seen as an alternative approach to calculate n-point functions by treating fermionic fields as Grassmann fields whereas in the super-algebra approach fermionic fields would be second quantized.

3.5.5 Various approaches to the construction of S-matrix

The gigantic symmetries of quantum TGD are bound to lead to a highly unique U-matrix but the practical construction of U-matrix remains still a formidable challenge. Despite this one can write Feynman rules for the S-matrix in the approximation that the consideration is restricted to elementary particles modelled as CP_2 type extremals. This approximation might well be all that is needed for practical purposes and leads to precise predictions.

7–3 duality as a key to the construction of S-matrix

The notion of 7–3 duality emerged from the (one might say violent) interaction between TGD and M-theory [A3]. The attempts to construct quantum TGD have gradually led to the conclusion that the geometry of the configuration space ("world of classical worlds") involves both 7-D and 3-D light like surfaces as causal determinants. 7-D light like surfaces X^7 are unions of future and past light cone boundaries and play a role somewhat resembling that of branes. 3-D light like surfaces X_l^3 can correspond to boundaries of space-time sheets, regions separating two maximally deterministic space-time regions, and elementary particle horizons at which the signature of the induced metric changes.

7–3 duality states that it is possible to formulate the theory using either the data at 3-D space-like 3-surfaces resulting as intersections of the space-time surface with 7-D CDs or the data at 3-D light like CDs [B4]. This results if the data needed is actually contained by 2-D intersections $X^2 = X_l^3 \cap X^7$. This effective 2-dimensionality has far-reaching implications. It simplifies dramatically the basic formulas related to the configuration space geometry and spinor structure, it leads to the explicit identification of the generalized Feynman diagrams at space-time level as light like 3-D CDs. The basic philosophy · is that quantum-classical correspondence stating that space-time sheets provide a description for the physics associated with the configuration space spin degrees of freedom (fermionic degrees of freedom).

The generalized Feynman diagrammatics is simple. The fermions do not carry four-momenta but are on mass shell particles characterized by the eigenvalues of the modified Dirac operator D. There is no propagator associated with 3-D CDs: only a unitary transformation U_λ representing braiding in spin and electro-weak spin degrees of freedom can be present. Vertices are the inner products at X^2 for the positive energy states and negative energy states entering to the vertex, finite, and in principle computable. The equivalence of generalized Feynman diagrams with tree diagrams is expected on basis of the effective 2-dimensionality, and indeed follows from on mass shell property directly. Unitarity follows trivially. No loop summations are thus involved.

Quantum criticality and Hopf algebra approach to S-matrix

Quantum criticality leads to a generalization of duality symmetry of string models stating that the generalized Feynman diagrams with loops are equivalent with diagrams having no loops. This means that each S-matrix element corresponds to a unique tree diagram. Note however that photon-photon scattering for which the lowest order contribution is loop diagram, forces to consider also a more general formulation. The lowest diagram contributing to a given scattering amplitude can have loops, and can be classified by the minimal genus of the Riemann surface at which it is imbeddable as a diagram having no intersecting lines and minimum number of non-contractible loops. All diagrams are equivalent to this diagram.

The conditions for this equivalence can be formulated as algebraic conditions characterizing a Hopf algebra like structure, and, using the language of ordinary Feynman diagrams, correspond to the vanishing of the loop corrections in the configuration space integral crucial for the p-adicization. This symmetry is expected to be of crucial importance for practical evaluation of S-matrix elements as should be also the reduction of the matrix elements of generators of the enveloping algebra of super-canonical algebra to n-point functions of conformal field theory in the complex plane of super-canonical conformal weights.

von Neumann algebras and S-matrix

The work with TGD inspired model for quantum computation led to the realization that von Neumann algebras, in particular hyper-finite factors of type II_1 could provide the mathematics needed to develop a more explicit view about the construction of S-matrix [O5].

1. Inclusions of hyper-finite II_1 factors as a basic framework to formulate quantum TGD

a) The effective 2-dimensionality of the construction of quantum states and configuration space geometry in quantum TGD framework makes hyper-finite factors of type II_1 very natural as operator algebras of the state space. Indeed, the elements of conformal algebras are labelled by discrete numbers and also the modes of induced spinor fields are labelled by discrete label, which guarantees that the tangent space of the configuration space is a separable Hilbert space and Clifford algebra is thus a hyper-finite type II_1 factor. The same holds true also at the level of configuration space degrees of freedom so that bosonic degrees of freedom correspond to a factor of type I_∞ unless super-symmetry reduces it to a factor of type II_1.

b) Four-momenta relate to the positions of tips of future and past directed light cones appearing naturally in the construction of S-matrix. In fact, configuration space of 3-surfaces can be regarded as union of big-bang/big crunch type configuration spaces obtained as a union of light-cones with parameterized by the positions of their tips. The algebras of observables associated with bounded regions of M^4 are hyper-finite and of type III_1. The algebras of observables in the space spanned by the tips of these light-cones are not needed in the construction of S-matrix so that there are good hopes of avoiding infinities coming from infinite traces.

c) Many-sheeted space-time concept forces to refine the notion of sub-system. Jones inclusions $\mathcal{N} \subset \mathcal{M}$ for factors of type II_1 define in a generic manner imbedding interacting sub-systems to a universal II_1 factor which now corresponds naturally to infinite Clifford algebra of the tangent space of configuration space of 3-surfaces and contains interaction as $\mathcal{M} : \mathcal{N}$-dimensional analog of tensor factor. Topological condensation of space-time sheet to a larger space-time sheet, formation of bound states by the generation of join along boundaries bonds, interaction vertices in which space-time surface branches like a line of Feynman diagram: all these situations could be described by Jones inclusion characterized by the Jones index $\mathcal{M} : \mathcal{N}$ assigning to the inclusion also a minimal conformal field theory and conformal theory with k=1 Kac Moody for $\mathcal{M} : \mathcal{N} = 4$. $\mathcal{M} : \mathcal{N}=4$ option need not be realized physically as quantum field theory but as string like theory whereas the limit $D = 4 - \epsilon \rightarrow 4$ could correspond to $\mathcal{M} : \mathcal{N} \rightarrow 4$ limit. An entire hierarchy of conformal field theories is thus predicted besides quantum field theory.

d) Von Neumann's somewhat artificial idea about identical a priori probabilities for states could replaced with the finiteness requirement of quantum theory. Indeed, it is traces which produce the infinities of quantum field theories. That $\mathcal{M} : \mathcal{N} = 4$ option is not realized physically as quantum field theory (it would rather correspond to string model type theory characterized by a Kac-Moody algebra instead of quantum group), could correspond to the fact that dimensional regularization works only in $D = 4 - \epsilon$. Dimensional regularization with space-time dimension $D = 4 - \epsilon \rightarrow 4$ could be interpreted as the limit $\mathcal{M} : \mathcal{N} \to 4$. $\mathcal{M}$ as an $\mathcal{M} : \mathcal{N}$-dimensional $\mathcal{N}$-module would provide a concrete model for a quantum space with non-integral dimension as well as its Clifford algebra. An entire sequence of regularized theories corresponding to the allowed values of $\mathcal{M} : \mathcal{N}$ would be predicted.

2. Generalized Feynman diagrams are realized at the level of $\mathcal{M}$ as quantum space-time surfaces

The key idea is that generalized Feynman diagrams realized in terms of space-time sheets have counterparts at the level of $\mathcal{M}$ identifiable as the Clifford algebra associated with the entire space-time surface X^4. 4-D Feynman diagram as part of space-time surface is mapped to its $\beta = \mathcal{M} : \mathcal{N} \leq 4$-dimensional quantum counterpart.

a) Von Neumann algebras allow a universal unitary automorphism $A \to \Delta^{it} A \Delta^{-it}$ fixed apart from inner automorphisms, and the time evolution of partonic 2-surfaces defining 3-D light-like causal determinant corresponds to the automorphism $\mathcal{N}_i \to \Delta^{it} \mathcal{N}_i \Delta^{-it}$ performing a time dependent unitary rotation for $\mathcal{N}_i$ along the line. At configuration space level however the sum over allowed values of t appear and should gives rise to the TGD counterpart of propagator as the analog of the stringy propagator $\int_0^t exp(iL_0 t)dt$. Number theoretical constraints from p-adicization suggest a quantization of t as $t = \sum_i n_i y_i > 0$, where $z_i = 1/2 + y_i$ are non-trivial zeros of Riemann Zeta.

b) At space-time level the "ends" of orbits of partonic 2-surfaces coincide at vertices so that also their images $\mathcal{N}_i \subset \mathcal{M}$ also coincide. The condition $\mathcal{N}_i = \mathcal{N}_j = ... = \mathcal{N}$, where the sub-factors $\mathcal{N}$ at different vertices differ only by automorphism, poses stringent conditions on the values t_i and Bohr quantization at the level of $\mathcal{M}$ results. Vertices can be obtained as a vacuum expectations of the operators creating the states associated with the incoming lines (crossing symmetry is automatic).

c) The equivalence of loop diagrams with tree diagrams (or a minimal genus g diagram with loops) would be due to the possibility to move the ends of the internal lines along the lines of the diagram so that only diagrams containing 3-vertices and self energy loops remain. Self energy loops are trivial if the product associated with fusion vertex and co-product associated with annihilation compensate each other. The possibility to assign quantum group or Kac Moody group to the diagram gives good hopes of realizing product and co-product. Octonionic triality would be an essential prerequisite for transforming N-vertices to 3-vertices. The equivalence allows to develop an argument proving the unitarity of S-matrix.

d) A formulation using category theoretical language suggests itself. The category of space sheets has as the most important arrow topological condensation via the formation of wormhole contacts. This category is mapped to the category of II_1 sub-factors of configurations space Clifford algebra having inclusion as the basic arrow. Space-time sheets are mapped to the category of Feynman diagrams in $\mathcal{M}$ with lines defined by unitary rotations of $\mathcal{N}_i$ induced by Δ^{it}.

e) The hierarchy of imbeddings for type II_1 factors generalizes to the hierarchy of generalized Feynman diagrams in which the particles of given level correspond to Feynman diagrams of the previous level. These Feynman diagrams provide representations for the projections of S-matrix to subspaces of incoming and outgoing states providing a hierarchy of self representations about the system. By crossing symmetry the entanglement defined by S-matrix corresponds to the so called Connes tensor product [ed1]. At space-time level it corresponds to time-like entanglement made possible by the failure of strict determinism. That the failure is only partial explains why the entanglement is so special.

The vertices for the interactions of these Feynman diagrams reduce to the interactions at the lowest level apart from the automorphisms $\Delta_{\mathcal{M}_n}$ defining free propagation. Also transitions between between different levels are possible. The interpretation is as an infinite cognitive hierarchy realizing theory about the material world as cognitive quantum states. This hierarchy corresponds to states with vanishing conserved net quantum numbers but having non-vanishing "gravitational" charges identifiable as classical charges.

Crossing symmetry and configuration space spinor fields as representations of S-matrix elements

In case of CP_2 type extremals the picture seems to be more complicated than this since the classical space-time surface is not unique. There is summation over the positions of the vertices of a Feynman diagram represented as a topological sum over CP_2 type extremals as well as sum over the Feynman diagrams. One could perhaps see these summations as an integral over the fiber of freedom and possible also zero modes of the configuration space surfaces $X^4(\cup_i Y_i^3)$: the same infinite prime indeed defines space-time surface for each values of the fiber coordinates and this would give rise to Feynman diagrams labelled by the positions of the vertices.

There is also a further problem involved: classical space-time surface represents *both* the incoming and outgoing states. The correspondence between Fock-space and classical descriptions of particle reactions suggests that also the outgoing space-time $X^4(\cup_i Y_i^3(out))$ surfaces appear as asymptotic space-time regions of $X^4(\cup_i Y_i^3)$. This might be possible thanks to the non-determinism of Kähler action.

The simple Glebch-Gordan vision fails in the case of CP_2 type extremals, since it neglects the non-determinism of the classical time evolution, which in case of CP_2 type extremals plays a crucial role in the construction of the theory. The construction of the S-matrix for CP_2 type extremals relies on the correspondence between Feynman diagrams as space-time surfaces constructed from the topological sums of CP_2 type extremals. Stringy Feynman rules are however not consistent with the Glebch-Gordan philosophy for the simple reason that space-time surface represents both the incoming and outgoing states rather than only the initial states.

The following interpretation inspired by crossing symmetry and the idea that configuration space spinor fields represent, not only physical states, but also S-matrix geometrically, however allows to understand why the Feynman rules make sense.

a) Feynman graphs with lines thickened to CP_2 type extremals represent classically quantum jumps between initial and final states characterized by incoming and outgoing CP_2 type extremals. Space-time surface containing topologically condensed CP_2 type extremals does not represent quantum state but a pair of quantum states and a quantum transition between these states. Incoming and outgoing states are represented as positive and negative energy particles and the net quantum numbers of the space-time sheet containing particles vanishes.

b) The initial state for quantum transition in question is vacuum and the final state $\Psi_f(m \to n)$ is represented by a which is superposition of all possible Feynman graphs for the transition in question. Thus S-matrix is coded into a physical state itself and the amplitudes $\langle 0|\Psi_f(m \to n)\rangle$ are proportional to S-matrix elements $S(m \to n)$. The amplitudes are indeed equal to S-matrix elements if crossing symmetry holds true.

Thus one ends up with the view that classical non-determinism and crossing symmetry make it possible for the configuration space spinor fields to represent not only physical states but also the U-matrix between them. This is possible also for other particles than those represented by CP_2 type extremals since one can construct space-time sheets containing arbitrary pairs of positive and negative energy states such that the net quantum numbers of the state pair cancel. The beauty of this representation is that it allows also to derive S-matrix easily as overlap integrals of vacuum state and zero energy states. Obviously this means huge simplification and gives a connection with ordinary Feynman diagrammatics.

The representation of entire sequences of quantum transitions as zero energy states of form $|\Psi_f(m_1 \to m_2... \to m_n)\rangle$ becomes possible so that classical non-determinism makes possible self-referential Universe able to represent the laws of physics in the structure of its own states. Also this idea supports the view that cognition is present already in elementary particle length scales.

U-matrix as Glebsch-Gordan coefficients

U-matrix relates 'free' and 'interacting' representations of the super-canonical and super Kac-Moody algebras acting as symmetries of quantum TGD. The construction is based on the association of 3-surfaces Y_i^3 and corresponding absolute minima $X^4(Y_i^3)$ to incoming states as well as the interacting four-surface $X^4(\cup_i Y_i^3)$ describing the interactions classically. The generators for various super-algebras associated with $X^4(\cup_i Y_i^3)$ are modified by interactions so that the generator basis is not just a union of the generator basis associated with $X^4(Y_i^3)$. U-matrix relates the tensor product for the representations associated with the incoming 'free' space-time surfaces $X^3(Y_i^3)$ and the interaction representation associated with $X^4(\cup Y_i^3)$: generalized Glebch-Gordan coefficients are clearly in question and unitarity is obvious.

Number theoretic approach to U-matrix

The task of assigning to the surfaces Y_i^3 the free space-time surfaces $X^4(Y_i^3)$ and interacting space-time surface $X^4(\cup_i Y_i^3)$ is the basic stumbling block for the construction of U-matrix. The super-algebra generators creating the excitations of the incoming ground states are super-algebra generators associated with $\cup X^4(Y_i^3)$ whereas the outgoing states are created by the super-algebra generators associated with $X^4(\cup_i Y_i^3)$. The surfaces $X^4(Y_i^3)$ correspond to the space-time surfaces associated with infinite primes P_i representing ground states of super-conformal representations whereas $X^4(\cup_i Y_i^3)$ corresponds to the space-time surface associated with the infinite integer $N = \prod_i P_i^{k_i}$. This means that the worst part of the problem is solved. The remaining challenge is to relate the super-algebra basis to each other.

Perturbation theoretic approach to U-matrix

This formal approach starts from the identification of U-matrix elements as Glebsch-Gordan coefficients relating free and interacting states and tries to construct U-matrix perturbatively by reducing it to stringy perturbation theory. The starting point is that U-matrix must follow from Super Virasoro invariance alone and that the condition $L_0(tot)\Psi = 0$ (plus the corresponding conditions for other super-Virasoro generators) must determine U-matrix. Here $L_0(tot)$ corresponds to the Virasoro generators associated with the interacting space-time surface $X^4(\cup_i Y_i^3)$ whereas $L_0(free, i)$ correspond to the free generators associated with $X^3(Y_i^3)$. It is however not at all obvious whether the generators $L_0(tot)$ are perturbatively related to the generators $L_0(free, i)$ and whether U-matrix allows perturbative expansion.

Construction of the S-matrix at high energy limit

It is possible to write Feynman rules for the S-matrix in the approximation that only CP_2 type extremals appear as virtual and real particles. All CP_2 type extremals are locally isometric with CP_2 itself and only the random lightlike curve is dynamical. The classical dynamics is actually isomorphic with stringy dynamics since classical Virasoro conditions are satisfied. Fermions belong to the representations of Super-Kac-Moody algebra of $M^4 \times SO(3,1) \times SU(3) \times U(2)_{ew}$. The classical nondeterminism of the dynamics implies that Feynman graph expansion is topologized. This saves from the troubles caused by fermionic divergences since the exponent of the momentum generator effecting translation along the line of the Feynman graph corresponds to that associated with the modified Dirac action and thus to a free quantum theory for fermions.

Vertex operators $V(a,b,c)$ are generalizations of the vertex operators of string theory: instead of strings 3-surface inside CP_2 type extremal fuse together. Propagator factors are products of the exponent of the Kähler action for CP_2 type extremal proportional to the volume of the CP_2 type extremal; the 'stringy' $1/(L_0 + i\epsilon)$ factor, which comes from the vertices; and a unitary translation operator (counterpart of S-matrix as time translation operator) along the geodesic representing average cm motion.

The theory has some features which are characteristic for quantum TGD.

a) One can assume that each quantum jump involves localization in zitterbewegung degrees of freedom. The resulting S-matrix is independent of the choice of the representative for the zitterbewegung orbit as long as the cm motion connects the lines of the vertices. The predictions depend however on an arbitrary function of U of CP_2 coordinates giving rise to a decomposition of CP_2 to 'time slices'. The dependence of the propagator is only through the volume of CP_2 type extremal determined by U whereas coupling constants have more complicated, but presumably very mild dependence on U. The dependence on the function U means that one must average the scattering rates over the allowed spectrum of functions U. This dependence of the fundamental coupling constants on U is in accordance with spin glass analogy and means that fundamental coupling constants are not strictly speaking constants.

b) The volume of the internal line, which is a fraction of CP_2 volume determines the value of the exponent of Kähler action and provides thus a suppression factor serving as an infrared cutoff. A constraint to the allowed functions U results from the topological condensation of CP_2 in particle like space-time sheet (for instance, massless extremal), which implies that CP_2 type extremals cannot extend outside the region with size of order p-adic length scale L_p. The only plausible interpretation seems to be that the information about the infrared cutoff length scale is coded into the structure of particle: particle in the box is quite not the same as free particle. This suggests new view about color confinement: quarks and gluons correspond to CP_2 type extremals which cannot exist too long time as free particles and therefore cannot leave hadron.

3.6 Does TGD predict the values of Planck constant?

The idea that $\hbar$ is dynamical and can have arbitrarily large values is about one and half year old as a write this. A lot of progress has occurred during the last year but I have not yet been able to seriously pose the question whether and how TGD could predict the values of the Planck constant. In the following a proposal for how TGD predicts the value spectrum of $\hbar$ as one aspect of quantum criticality is discussed and number theoretical arguments are used to make a guess about the spectrum of $\hbar$. In very concise form the argument goes as follows.

a) The freedom to choose the value of $\hbar$ corresponds to the freedom to choose the overall scaling $\lambda = \hbar/\hbar_0$ of M^4 metric associated with various copies of M^4 obtained identified as various algebraic extensions of rational M^4 glued together along common set of rationals consistent with the isometric identification.

b) The dependence of λ on the algebraic extension for a given p-adic prime p is fixed by the quantum criticality condition stating that the critical Kähler coupling strength is same for various algebraic extensions associated with given p-adic prime p.

c) Number theoretic ideas allow to make good guesses concerning the dependence of λ on the algebraic extension. Simplest guess is that λ for union of linearly independent extensions is product or $1/\lambda$ a sum of λ:s for composites. It turns out that sum is the most plausible guess.

3.6.1 The basic ideas

In order to build a more coherent theoretical framework for the quantization of $\hbar$ let us summarize the basic vision as it exists now.

Four-momentum is invariant in the transition changing $\hbar$

The basic constraint is that four-momentum and other quantum numbers of the state are conserved in the scaling $\hbar \to \lambda\hbar$ whereas various quantum lengths and times are scaled up by λ so that one obtains the basic predictions such as macroscopic and macro-temporal quantum coherence in arbitrarily long scales. It is however not at all obvious how to realize this condition.

The identification of the value of the parameter v_0 in terms of Kähler coupling strength

The parameter $v_0 \simeq 2^{-11}$, which has has actually dimension of velocity unless on puts $c = 1$, and its harmonics and sub-harmonics appear in the scaling of $\hbar$. v_0 corresponds to the velocity of distant stars in the model of galactic dark matter. TGD allows to identify this parameter as the parameter

$$
\begin{aligned}
v_0 &= 2\sqrt{TG} = \sqrt{\frac{1}{2\alpha_K}}\sqrt{\frac{G}{R^2}} \;, \\
T &= \frac{1}{8\alpha_K}\frac{\hbar_0}{R^2} \;.
\end{aligned}
\tag{3.6.1}
$$

Here T is the string tension of cosmic strings, R denotes the "radius" of CP_2 ($2R$ is the radius of geodesic sphere of CP_2). $\hbar_0$ corresponds to Beraha number B_∞. α_K is Kähler coupling strength whose evolution is dictated by the condition that G is invariant in coupling constant evolution and by number theoretical arguments to

$$
\begin{aligned}
\alpha_K &= k\frac{1}{log(p) + log(K)} \;, \\
K &= \frac{R^2}{\hbar_0 G} = 2 \times 3 \times 5 \times 7 \times 11 \times 13 \times 17 \times 19 \times 23 = 223,092,870 \;, \\
k &\simeq \pi/4 \;.
\end{aligned}
\tag{3.6.2}
$$

Equivalence Principle requires that $\hbar$ in the formula for K corresponds to the value $\hbar_0 \equiv 1$ [D6] so that gravitational coupling constant is invariant also with respect to coupling constant evolution associated with Beraha numbers. One can define "dynamical" Planck length as

$$L_d = \sqrt{G\hbar_{gr}} = \sqrt{M_2/M_1}\,\frac{1}{v_0}GM_1 \ .$$

The order of magnitude is not too far from Schwartschild radius.

Number theoretic constraints are expected to pose strong constraints on the value of k. A discrete version of a typical logarithmic evolution of U(1) coupling constant strength as a function of length scale is in question and for $k = 127$ (M_{127}) the value of α_K is very near to fine structure constant $\alpha \simeq 1/137$. Note that the ratio R^2/G is predicted to be constant so that G would be renormalization group invariant even in the sense that it would not depend on the p-adic length scale.

The resulting prediction for v_0 reads as

$$v_0 = \sqrt{\frac{1}{2\,[k \times log(2) + log(K)]\,K}} \quad \text{for} \ \ p \simeq 2^k \ ,$$

$$v_0 \simeq .54412 \times 10^{-3} \ \text{for} \ \ p = M_{127} \ . \tag{3.6.3}$$

v_0 approaches to zero at logarithmic rate for large values of k. For $k = 127$ it is rather near to $v_0 = 2^{-11}$. In a good approximation the value 2^{-11} is reached for $k = 135$ which corresponds to the p-adic length scale $L(k_{eff} = 113+22)$ associated with dark variants of $k = 113$ weak bosons appearing in the TGD based model of atomic nucleus. The mean experimental value of $1/v_0$ is $1/v_0 = 2174$ with an accuracy of 1 per cent.

TGD allows also to develop arguments explaining the harmonics and sub-harmonics of v_0. Higher harmonics would be due to wrapped magnetic flux tubes for which CP_2 coordinates are n-valued functions of M^4 coordinates defining local coordinates of the flux tube space-time sheet. Sub-harmonics would result when M^4 coordinates become n-valued functions of CP_2 coordinates.

The challenge is to understand why v_0 appears in the basic formula expressing the change of $\hbar$ in the transition increasing the value of $\hbar$. It would seem that $\hbar$ characterizes the magnetic flux tubes (join along boundaries bonds) connecting the interacting systems and serving as space-time correlates for the interaction giving rise to bound state.

The value of v_0 deduced for cosmic strings does not make sense in astrophysical or condensed matter context, where cosmic strings are replaced with magnetic flux tubes. v_0 remains invariant in this scaling down if R^2 is replaced by the p-adic length scale L_p^2 apart from a multiplicative factor in the formulas for G and T so that the product TG remains invariant. $T_m \propto 1/L_p^2$ characterizes the magnetic energy density of the magnetic flux tube and $G_m \to L_p^2$ is identifiable as a "strong" gravitational coupling strength characterizing the interactions of magnetic flux tubes behaving like string like objects.

The criterion for the occurrence of the phase transition increasing the value of $\hbar$

In the case of planetary orbits the large value of $\hbar = 2GM/v_0$ makes possible to apply Bohr quantization to planetary orbits. This leads to a more general idea that the phase transition increasing $\hbar$ occurs when the system consisting of interacting units with charges Q_i becomes non-perturbative in the sense that the perturbation series in the coupling strength $\alpha Q_i Q_j$, where α is the appropriate coupling strength and $Q_i Q_j$ represents the maximum value for products of gauge charges, ceases to converge. Thus Mother Nature would resolve the problems of theoretician.

A primitive formulation for this criterion is the condition $\alpha Q_i Q_j \geq 1$ and predicts the existence of dark matter hierarchies with $\hbar = \lambda^k \hbar_0$, $k = 0, 1, ...$, $\lambda = n/v_0$ or $\lambda = 1/nv_0$, $v_0 \simeq 2^{-11}$. This rule of thumb has now been applied with success the interpretation of hadronic mass calculations and to build models for systems like atomic nucleus and high T_c superconductor and seems to work. Of course, the criterion for transition is primitively formulated and the understanding what really happens in the transition to large $\hbar$ phase behaving like dark matter.

What are the allowed values of $\hbar$?

From the beginning I had strong gut feeling that the allowed values of $\hbar$ are expressible in terms of Beraha numbers $B_n = 4cos^2(\pi/n)$, $n \geq 3$ related to Jones inclusion hierarchies of hyperfinite factors of type II_1, which correspond to von Neumann algebra naturally associated with configuration space spinors.

Consider the inclusion $N \subset M$ of these factors as von Neumann algebras. A deep result is that one can express M as $N : M$-dimensional module over N with fractal dimension $N : M = B_n$. $\sqrt{B_n}$ represents the dimension of a space of spinor space renormalized from the value 2 corresponding to $n = \infty$ down to $\sqrt{B_n} = 2cos(\pi/n)$ varying thus in the range $[1, 2]$. B_n in turn would represent the dimension of the corresponding Clifford algebra.

This observation leads to the idea about how one could predict the value spectrum of $\hbar$. There are two arguments based on similar reasoning but leading to quite different predictions.

Option I: Consider the trace of $\sigma_z^2 = \hbar^2 \times 1$ for $d = 2$-component spinors. Assume that the trace of σ_z^2 does not depend on $\hbar$ so that one has

$$Tr(\sigma_z^2) = \sqrt{M : N} \hbar_n^2 = 2\hbar_\infty^2 .$$

By comparing the two sides one obtains

$$\frac{\hbar_n}{\hbar_\infty} = \sqrt{\frac{1}{cos(\pi/n)}} , \quad n \geq 3 . \tag{3.6.4}$$

This would predict that the values of $\hbar_n/\hbar_\infty$ vary in the range $[1, \sqrt{2}]$. This is not enough to explain the large values of $\hbar$ but the arguments represented in the sequel allow to consider this option seriously.

Option II: In order to understand the large values of $\hbar$ I developed an argument the outcome of which seems to be correct although the proposed explanation of large values of $\hbar$ need not be correct. The idea is that the dimension of the spinor space is given by $d = 2^{D/2}$, where D is the dimension of the space-time. By applying this formula in the recent fractal context one obtains

$$D_n = 2log_2(\sqrt{B_n}) = 2log_2(2cos(\pi/n)) .$$

In recent case the invariance for the trace of the scaling operator $L_0 = \hbar m^k d/dm^k$ acting in the representation defined by M^D coordinates implies that the trace $Trace(L_0) = D\hbar$ equals to $2\hbar_\infty$.

$$D_n \hbar_n = 2\hbar_\infty \tag{3.6.5}$$

allowing to identify the renormalized value of $\hbar$ as

$$\frac{\hbar_n}{\hbar_\infty} = \frac{2}{D_n} = \frac{1}{log_2(2cos(\pi/n))} . \tag{3.6.6}$$

This would give $D_3 = 1$ and $D_\infty = 2$ and $\hbar_3 = \infty$. The idea was that the gigantic value of $\hbar$ in astrophysical length scales would result as a small perturbation of $x = 1/\hbar$ at $x = 0$. The infinite value of $\hbar$ for $n = 3$ would have a nice interpretation in terms of vacuum degeneracy. $n = 3$ would correspond to vacuum extremals for which perturbation theory does not make sense. The situation would be saved by the vanishing value of gauge coupling strengths meaning that all quantum corrections to the classical prediction vanish. It will turn out that option II is indeed more natural in the TGD framework.

Do Jones inclusions correspond to inclusions of rationals to their algebraic extensions?

Jones inclusions for type II_1 factors have several interpretations and applications in TGD framework. The most obvious interpretation Jones inclusion is as sub-system-system inclusion. Jones inclusion could be also accompany the inclusions of rationals to algebraic extensions of rationals at the level of configuration space spinor fields. One can also consider the inclusions of p-adic number fields to their complex extensions.

The inclusion $Q \subset Q[exp(i\pi/n)]$ would correspond to Jones inclusion characterized by n at the level of configuration space spinors. In the framework of rational physics the hierarchy of algebraic extensions of rationals would define a hierarchy of Jones inclusions and these extensions would give rise to a hierarchy of Planck constants.

The problem is that the large values of $\hbar$ do not seem to be natural in either case except possibly for option II by allowing perturbative corrections to $x = 1/\hbar_3 = 0$. This need not to be the case. $M : N$ characterizes configuration space Clifford algebra assignable with an algebraic extension of rationals as N-module where N corresponds to CH Clifford algebra assignable to rationals. What comes in mind that M could be much higher-dimensional as N-module than in the case of standard Jones inclusions defined using reals or complex numbers as a coefficient field of the Clifford algebra. The simplest guess is that M is tensor power of $M : N$-dimensional spinor base as an N module indeed allowed by option II as will be found.

3.6.2 Mathematical constraints

In order to gain insight to how one might understand the hierarchy of Planck constants in TGD framework it is good to start from good old Schrödinger equation.

What is behind minimal substitution?

The key property of Schrödinger equation is that kinetic energy term depends on $\hbar$ whereas the potential energy term does not depend on it. This makes the scaling of $\hbar$ a non-trivial transformation. In the case of Dirac equation same conclusion applies and corresponds to the minimal substitution $p - eA \to i\hbar\nabla - eA$. Consider next the situation in TGD framework.

1. Minimal substitution does not make sense in CP_2 degrees of freedom

The first crucial observation is that the minimal substitution $p - eA \to i\hbar\nabla - eA$ does not make sense in the case of CP_2 Dirac operator since, by the non-triviality of spinor connection, one cannot choose the value of $\hbar$ freely. In fact, spinor connection of CP_2 is defined naturally in such a manner that spinor connection corresponds to the quantity $eQ/\hbar$ and there is no natural manner to separate $e/\hbar$ from it.

The only reasonable conclusion is that the Dirac operator in $M^4 \times CP_2$ depends on $\lambda = \hbar/\hbar_0$ only via its M^4 part and that Dirac equation gives the eigenvalues of wave vector squared $k^2 = k^i k_i$ rather than four-momentum squared $p^2 = p^i p_i$. The values of k^2 are proportional to $1/\lambda^2$ so that p^2 does not depend on it for $p^i = \hbar k^i$. This gives rise to the invariance of mass squared and the desired scaling of wave vector when $\hbar$ changes.

2. The dynamical character of $\hbar$ can be only due to the freedom to select the scale of M^4 metric

The second crucial observation is that the freedom to vary λ can be only due to the freedom to vary the overall scaling of M^4 metric which is indeed possible. Whether the freedom to choose the scaling of M^4 metric freely corresponds to a genuine dynamical degree of freedom has been a continual source of worried thoughts now and then. If $\hbar$ has only single value this freedom does not seem to have meaning but if it has spectrum the situation changes.

The question is how to realize this freedom to choose the scale of the M^4 metric realized as replacements $m_{kl} \to m_{kl}/\lambda$, where allowed values of λ define the spectrum of Planck constant. The number theoretic vision suggests an answer to the question.

TGD leads to a generalization of the notion of number by gluing reals and various extensions of p-adic number fields to a larger structure. The gluing is carried out along common rationals. On the other hand, the rational physics approach suggests that the physics in various number fields satisfy the enormously powerful constraint that they are obtained by an algebraic continuation from rational physics. Hence rationals and the algebraic extensions of rationals should play key role in TGD.

The obvious idea encouraged also by the observation about the role of Beraha numbers is that different algebraic extensions of rationals (or p-adic numbers) correspond to different scalings of M^4 metric. This would mean that the identification along common rationals would be replaced with the identification $y = \lambda x$, where y is rational and x is arbitrary number in extension, and λ characterizing the scaling of M^4 metric belongs to extension. If lambda is not rational, the values of x are obtained by multiplying rationals by $1/\lambda$. Rationality would give a strong constraint on λ, which might be however too strong as the guesses for the form of λ suggest. This identification would generalize the original $y = x$ identification and guarantee that the distances of points y and x from the origin of M^4 would be same in respective M^4 metrics.

3. Other Dirac operators and dynamical scale of M^4 metric

The number theoretically realized spectrum of scales for M^4 metric fits nicely with the properties of the Dirac operator of imbedding space, with the properties of the modified Dirac operator defined for induced metric and spinor structure, and with the properties super-conformal Dirac operators of configuration space since both these operators involve natural separation of M^4 and CP_2 degrees of freedom. Induced metric and spinor structure depend non-linearly on λ. Super-conformal mass squared formula is replaced by a formula for wave vector squared involving λ as a scaling factor and the oscillator operator algebra used depends on the value of λ since it is expressed in terms fermionic oscillator operators associated with second quantized induced spinor fields.

4. Quantum classical correspondence and the values of λ

Quantum classical correspondence suggests that the values of λ should be represented also at space-time level. The variational principle making space-time sheets counterparts of Bohr orbits indeed implies the quantization of Kähler magnetic flux and the quantum need not be the standard flux quantum. The generalized quantization condition would be $\int BdS = n\lambda\hbar_0$ and in principle it is possible to deduce the values of λ from the classical theory. The flux integral does not involve the induced metric so that there is no explicit dependence on λ. There is however an implicit dependence via field equations which involve λ via the induced metric. This should be the case since the information about algebraic extension must be coded to the classical theory somehow.

The invariance of angular momentum under the scaling of $\hbar$

The assumption that four-momentum is invariant in the scaling of M^4 metric by λ combined with Poincare invariance implies that also angular momentum is invariant under the scaling by λ. Hence the analog of wave vector defined as $L_z/\hbar$ would scale down by a factor $1/\lambda$ giving $L_z/\hbar = m/\lambda$. The one-valuedness of the wave function however requires $L_z/\hbar = m$.

A possible resolution of the paradox is based on following argument.

a) Consider first a situation in which λ is integer. Effective 2-dimensionality means that 2-dimensional partonic surfaces are basic structures to be considered. By conformal invariance scalings and rotations are generated by L_0 and iL_0. Hence it would not be surprising if the radial scaling by λ^k would be accompanied by a similar angular scaling. If M^4 projection of the partonic 2-surface has dimension $D \geq 1$, this scaling would at space-time level mean that partonic 2-surface becomes analogous to λ^k-sheeted Riemann surface defining λ^k-fold covering of $E^2 \subset M^4$. Using M^4 coordinates for the space-time sheet as local coordinates, one finds that induced spinor fields have fractional angular momentum eigenvalues $L_z = m/\lambda^k$.

b) In this picture the approach to quantum criticality would correspond to the emergence of classical chaos at space-time level by a step-wise process in which the step $\hbar \to \lambda\hbar$ can be regarded as generalization of the period doubling bifurcation. As $\hbar$ increases by a factor λ^k, the space-time sheet representing an orbit of particle closing after one turn transforms to an orbit closing only after λ^k turns. Note that the volume of space-time sheet remains finite only if the orbit closes after finite number of turns. The step $k \to k + 1$ would correspond to a local fractal operation making each sheet of the λ^k sheeted surface λ-sheeted so that λ^{k+1} sheeted surface would result. Instead of period doubling one would have period λ-folding with the value of λ depending on p-adic prime $p \simeq 2^k$.

c) The model of Nottale for planetary orbits requires besides the harmonics of $\lambda_0 \simeq 2^{11}$ also some of its sub-harmonics, at least third and fifth one. This conflicts with the assumption that λ is integer unless λ_0 is divisible by 3 and 5. This would be the case for $\lambda_0 = 2175 = 3 \times 5^2 \times 29$ consistent with the mean value $\lambda = 2174$ deduced with one per cent accuracy from the model for planetary orbits [if1]. Only integer factors of $\lambda_0(p \simeq 2^k)$ would be allowed as sub-harmonics for given p for this option and integer valuedness of λ_0 would also pose strong conditions on the values of p if the proposed formula for λ_0 is accepted.

d) If one allows the quantization of angular momentum using integer multiple of $\hbar_0$ as a unit, λ need not be an integer. In this case one could have $\lambda = (r/s) \times \lambda_0$, where $\lambda_0 \simeq 2^{11}$ is integer. The orbit represented by the space-time sheet would close after $(r\lambda_0)^k$ turns and angular momentum would be quantized as $L_z = ms^k \hbar_0$. Even in this case the proposed formula for $\lambda(p)$ would give strong constraints on the values of p.

Kähler function codes for a perturbative expansion in powers of λ

Suppose that one accepts the number theoretical interpretation for the spectrum of $\hbar$ in terms of a hierarchy of overall scalings of M^4 metric. The first implication of this picture is that the modified Dirac operator determined by the induced metric and spinor structure depends on the λ in a highly nonlinear manner. This in turn implies that the fermionic oscillator algebra used to define configuration space spinor structure and metric depends on the value of λ. Same is true also for Kähler action and configuration space Kähler function. Hence Kähler function is analogous to an effective action expressible as infinite powers series in powers of $\hbar$ replaced now with λ.

This interpretation allows to overcome the paradox caused by the hypothesis that loop corrections to the functional integral over configuration space defined by the exponent of Kähler function serving as vacuum functional vanish so that tree approximation is exact. This would imply that all higher order corrections usually interpreted in terms of perturbative series in powers of $1/\hbar$ vanish. The paradox would result from the fact that scattering amplitudes would not receive higher order corrections and classical approximation would be exact. This certainly cannot be the case always: consider only the photon-photon scattering. This paradox can be also regarded as an objection against the proposal that generalized Feynman diagrams are equivalent with tree diagrams or more generally, that each diagram is equivalent with a minimal loopy diagram allowing homologically non-trivial imbedding with non-intersecting lines to a higher genus Riemann surface.

The dependence of both states created by Super Kac-Moody algebra and the Kähler function and corresponding propagator identifiable as contravariant configuration space metric would mean that the expressions for scattering amplitudes indeed allow an expression in powers of λ. What is so remarkable is that the TGD approach would be non-perturbative from the beginning and "semiclassical" approximation, which might be actually exact, automatically would give a full expansion in powers of $\hbar$. This is in a sharp contrast to the usual quantization approach.

How the spectrum of $\hbar$ could be predicted by quantum TGD?

Number theoretical vision combined quantum criticality could allow to determine the allowed values of λ to a high degree. Quantum criticality means that different values for $\hbar$ corresponding to algebraic extensions of rationals correspond to same form of Kähler function, that is same value of Kähler coupling constant. Number theoretical vision in turn has led to the proposal that the exponent of Kähler function is expressible as a Dirac determinant for the modified Dirac operator. As a matter fact, the ratio of Dirac determinants for space-time regions separated by a light-like causal determinant acting as causal horizon would give the exponent for the difference of Kähler functions of the two regions [B4].

The condition that the value of g_K^2, defined as the analog of critical temperature, is same for all algebraic extensions would fix the value of λ as a function of algebraic extension apart from possible multi-valuedness.

This condition does not require the restriction of the modified Dirac operator to the set of rationals or their algebraic extensions. The only thing that is required is that the subset of allowed eigenvalues of the modified Dirac operator belongs to the extension considered. In this manner one can calculate a Dirac determinant for each extension as a function of λ. Note that λ could belong to the algebraic extension considered. By requiring that g_K^2 corresponds to its value for rationals, one can fix the value of λ. Even better, it is quite possible that the value of Dirac determinant, rather than only the ratio of Dirac determinants, is finite since the number theoretic restriction might be satisfied only by a finite number of eigenvalues. Hence nature would also take care of regularization of Dirac determinants by using the number theoretic hierarchy.

Part II

PHYSICS AS SPINOR GEOMETRY IN THE WORLD OF CLASSICAL WORLDS

Chapter 4

Classical TGD

4.1 Introduction

A brief summary of what might be called basic principles is in order to facilitate the reader to assimilate the basic tools and rules of intuitive thinking involved.

4.1.1 Quantum-classical correspondence

The fundamental meta level guiding principle is quantum-classical correspondence (classical physics is an exact part of quantum TGD). The principle states that all quantum aspects of the theory, which means also various aspects of consciousness such as volition, cognition, and intentionality, should have space-time correlates [TGDconsc]. Real space-time sheets provide kind of symbolic representations whereas p-adic space-time sheets provide correlates for cognition and intentions. All that we can symbolically communicate about conscious experience relies on quantal space-time engineering to build these representations.

4.1.2 Classical physics as exact part of quantum theory

Classical physics corresponds to the dynamics of space-time surfaces determined by the absolute minimization of Kähler action or some other principle selecting preferred extremals of Kähler action [E2]. This dynamics have several unconventional features basically due to the possibility to interpret the Kähler action as a Maxwell action expressible in terms of the induced metric defining classical gravitational field and induced Kähler form defining a non-linear Maxwell field not as such identifiable as electromagnetic field however.

Classical long ranged weak and color fields as signature for a fractal hierarchy of copies standard model physics

The geometrization of classical fields means that various classical fields are expressible in terms of imbedding space-coordinates and are thus not primary dynamical variables. This predicts the presence of long range weak and color (gluon) fields not possible in standard physics context. It took 26 years to end up with a convincing interpretation for this puzzling prediction.

What seems to be the correct interpretation is in terms of an infinite fractal hierarchy of copies of standard models physics with appropriately scaled down mass spectra for quarks, leptons, and gauge bosons. Both p-adic length scales and the values of Planck constant predicted by TGD [O5] label various physics in this hierarchy. Also other quantum numbers are predicted as labels. This means that universe would be analogous to an inverted Mandelbrot fractal with each bird's eye of view revealing new long length scale structures serving also as correlates for higher levels of self hierarchy.

Exotic dark weak forces and their dark variants are consistent with the experimental widths for ordinary weak gauge bosons since the particles belonging to different levels of the hierarchy do not have direct couplings at Feynman diagram level although they have indirect classical interactions and also the de-coherence reducing the value of $\hbar$ is possible. Classical long ranged weak fields play a key role in quantum control and communications in living matter [M3, N4]. Long ranged classical color force in turn is the backbone in the model of color vision [K3]: colors correspond to the increments of color quantum numbers in this model. The increments of weak isospin in turn could define the basic color like quale associated with hearing (black-white $\leftrightarrow$ to silence-sound [K3, M5, M6]).

Topological field quantization and the notion of many-sheeted space-time

The compactness of CP_2 implies the notions of many-sheeted space-time and topological field quantization. Topological field quantization means that various classical field configurations decompose into topological field quanta. One can see space-time as a gigantic Feynman diagram with lines thickened to 4-surfaces. Absolute minimization of Kähler action (or any other analogous selection principle) implies that only selected field configurations analogous to Bohr's orbits are realized physically so that quantum-classical correspondence becomes very predictive. An interpretation as a 4-D quantum hologram is a further very useful picture [K2] but will not be discussed in this chapter in any detail.

Topological field quantization implies that the field patterns associated with material objects form extremely complex topological structures which can be said to belong to the material objects. The notion of field body, in particular magnetic body, typically much larger than the material system, differentiates between TGD and Maxwell's electrodynamics, and has turned out to be of fundamental importance in the TGD inspired theory of consciousness. One can say that field body provides an abstract representation of the material body.

One implication of many-sheetedness is the possibility of macroscopic quantum coherence. By quantum classical correspondence large space-time sheets as quantum coherence regions are macroscopic quantum systems and therefore ideal sites of the quantum control in living matter.

a) The original argument was that each space-time sheet carrying matter has a temperature determined by its size and the mass of the particles residing at it via de Broglie wave length $\lambda_{dB} = \sqrt{2mE}$ assumed to define the p-adic length scale by the condition $L(k) < \lambda_{dB} < L(k_>)$. This would give very low temperatures when the size of the space-time sheet becomes large enough. The original belief indeed was that the large space-time sheets can be very cold because they are not in thermal equilibrium with the smaller space-time sheets at higher temperature.

b) The assumption about thermal isolation is not needed if one accepts the possibility that Planck constant is dynamical and quantized and that dark matter corresponds to a hierarchy of phases characterized by increasing values of Planck constant [O5, J6]. From $E = hf$ relationship it is clear that arbitrarily low frequency dark photons (say EEG photons) can have energies above thermal energy which would explain the correlation of EEG with consciousness. This vision allows to formulate more precisely the basic notions of TGD inspired theory of consciousness and leads to a model of living matter giving precise quantitative predictions. Also the ability of this vision to generate new insights to quantum biology provides strong support for it [M3].

Many-sheeted space-time predicts also fundamental mechanisms of metabolism based on the dropping of particles between space-time sheets with an ensuing liberation of the quantized zero point kinetic energy. Also the notion of many-sheeted laser follows naturally and population inverted many-sheeted lasers serve as storages of metabolic energy [K6].

Space-time sheets topologically condense to larger space-time sheets by wormhole contacts which have Euclidian signature of metric. This implies causal horizon (or elementary particle horizon) at which the signature of the induced metric changes from Minkowskian to Euclidian. This forces to modify the notion of sub-system. What is new is that two systems represented by space-time sheets can be unentangled although their sub-systems bound state entangle with the mediation of the join along boundaries bonds connecting the boundaries of sub-system space-time sheets. This is not allowed by the notion of sub-system in ordinary quantum mechanics. This notion in turn implies the central concept of fusion and sharing of mental images by entanglement [TGDconsc].

The possibility of negative energies

A further prediction derives from the fact that space-time is 4-surface rather than an abstract manifold. The energy momentum tensor of general relativity is replaced by a collection of conserved energy and momentum currents, which are 4-vector fields. This makes the notions of energy and momentum precisely defined but also implies that the sign of energy and momentum depend on the time-orientation of the space-time sheet. Negative energies become therefore possible somewhat like in the lines of a Feynman diagram. Negative energy topological light rays have phase conjugate laser waves [jb1] as the most plausible standard physics counterparts, and play a fundamental role in quantum metabolism as a kind of quantum credit card [K6]. They generate also time like entanglement which corresponds to a formation of new kind of bound states.

Negative energies might be possible even for ordinary particles and could mean dramatic deviation from the standard quantum theory. The roles of annihilation and creation operators have changed for negative energy space-time sheets. This would mean that operator combinations involving both annihilation and creation operators would generate states involving positive and negative energy space-time sheets. One can even imagine that a intentional action could create states with vanishing net quantum numbers and that positive and negative energy particles could be separated from each other.

TGD Universe is quantum spin glass

Since Kähler action is Maxwell action with Maxwell field and induced metric expressed in terms of $M_+^4 \times CP_2$ coordinates, the gauge invariance of Maxwell action as a symmetry of the vacuum extremals (this implies is a gigantic vacuum degeneracy) but not of non-vacuum extremals. Gauge symmetry related space-time surfaces are not physically equivalent and gauge degeneracy transforms to a huge spin glass degeneracy. Spin glass degeneracy provides a universal mechanism of macro-temporal quantum coherence and predicts degrees of freedom called zero modes not possible in quantum field theories describing particles as point-like objects. Zero modes not contributing to the configuration space line element are identifiable as effectively classical variables characterizing the size and shape of the 3-surface as well as the induced Kähler field. Spin glass degeneracy as mechanism of macroscopic quantum coherence should be equivalent with dark matter hierarchy as a source of the coherence [K2].

Classical and p-adic non-determinism

The vacuum degeneracy of Kähler action implies classical non-determinism, which means that space-like 3-surface is not enough to fix the space-time surface associated with it uniquely as an absolute minimum of action, and one must generalize the notion of 3-surface by allowing sequences of 3-surfaces with time like separations to achieve determinism in a generalized sense. These "association sequences" can be seen as symbolic representations for the sequences of quantum jumps defining selves and thus for contents of consciousness. Not only speech and written language define symbolic representations but all real space-time sheets of the space-time surfaces can be seen in a very general sense as symbolic representations of not only quantum states but also of quantum jump sequences. An important implication of the classical non-determinism is the possibility to have conscious experiences with contents localized with respect to geometric time. Without this non-determinism conscious experience would have no correlates localized at space-time surface, and there would be no psychological time.

p-Adic non-determinism follows from the inherent non-determinism of p-adic differential equations for any action principle and is due to the fact that integration constants, which by definition are functions with vanishing derivatives, are not constants but functions of the pinary cutoffs x_N defined as $x = \sum_k x_k p^k \rightarrow x_N = \sum_{k<N} x_k p^k$ of the arguments of the function. In p-adic topology one can therefore fix the behavior of the space-time surface at discrete set of space-time points *above* some length scale defined by p-adic concept of nearness by fixing the integration constants. In the real context this corresponds to the fixing the behavior *below* some time/length scales since points p-adically near to each other are in real sense faraway. This is a natural correlate for the possibility to plan the behavior and p-adic non-determinism is assumed to be a classical correlate for the non-determinism of intentionality, and perhaps also imagination and cognition.

These two non-determinisms allow to understand the self-referentiality of consciousness at a very general level. In a given quantum jump a space-time surface can be created with the property that it represents symbolically or cognitively something about the contents of consciousness before the quantum jump. Thus it becomes possible to become conscious about being conscious of something. This is very much like mathematician expressing her thoughts as symbol sequences which provides feedback to go the next abstraction level.

Classical and p-adic non-determinisms force also the generalization of the notion of quantum entanglement. Time-like entanglement, crucial for understanding long term memory and precognition becomes possible. The notion of many-sheeted space-time forces also to modify the notion of sub-system, which implies that unentangled systems can have entangled sub-systems. One can partially understand this in terms of length scale dependent notion of entanglement (the entanglement of sub-systems is not seen in the length scale resolution defined by the size of unentangled systems) but only partially. The formation of join along boundaries bonds between sub-system space-time sheets and the fact that topologically condensed space-time sheets are separated by elementary particle horizons from larger space-time sheets, provide the deeper topological motivation for the generalization of sub-system concept.

p-Adic fractality of life and consciousness

p-Adic fractality of biology and consciousness has become an increasingly important guide line in the construction of the theory. This notion allows to relate phenomena occurring in the molecular level to phenomena like remote viewing and psychokinesis and it leads also to the view that topological field quanta of various fields of astrophysical size are crucial for the functioning of bio-systems. If one accepts p-adic fractality, the theory can be tested in unexpected manners, in particular in molecular and cellular length scales where the systems are much simpler. Sensory perception, long term memory, remote mental interactions, metabolism: all these phenomena rely on the same basic mechanisms. p-Adic length scale hypothesis allows to quantify the hypothesis with testable quantitative predictions.

Double slit experiment and classical non-determinism

Bohr's complementarity principle is the basic element of Copenhagen interpretation and at the same time one of the most poorly defined aspects of this interpretation. If the possibility of macroscopic quantum entanglement between measurement instrument and quantum system is accepted, complementary principle becomes un-necessary. This is however not all that is needed. If classical non-determinism makes it possible to represent quantum jump sequences at space-time level, a revision of space-time description of quantum measurement is necessary. This sounds very logical but to be honest, I write these lines only after having learned about the remarkable experiment done by Shahriar Afshar [je2].

The variant of double slit experiment by Shahriar Afshar seems to contradict the Copenhagen interpretation which states that the particle and field aspects are complementarity and thus mutually exclusive. In the case of double slit experiment complementarity predicts that the measurement of whether the photon came to the detector through slit 1 or 2 should destroy the interference pattern of electromagnetic fields in the region behind the screen.

The experimental arrangement of Afshar differs from the standard double slit experiment in that a lens was added behind the screen. The lens transmitted the photons coming from slits 1 and 2 via mirrors to detectors A and B so that in particle picture a photon detected by A (B) could be regarded as coming from slit 1 (2). In the first step both slits were open and the detectors represented interference patterns representing diffraction through single slit. The other slit was then closed and metal wires at the positions of dark interference rings were added. These wires degraded somewhat the image in the second detector. After this the second slit was opened again. Surprisingly, the resulting interference pattern was the original one.

The measurement certainly measures the particle aspect of photons. On the other hand, the preservation of the detected patterns means that no photons did enter in the regions containing the wires so that also interference pattern is there. Hence wave and particle aspects seem to be mutually consistent.

This finding is difficult to understand in Copenhagen interpretation and also in the many-worlds interpretation of quantum mechanics. Afshar himself suggest that the very notion of photon must be questioned. It is however difficult to accept this view since the photon absorption quite concretely corresponds to a click in the detector and also because the mathematical formalism of second quantization works so fantastically.

The conclusion can be criticized. What is primarily measured is not basically through which slit the photons came but whether the direction of the momentum of the photon emerging from the lens was in the angle range characterizing the detector or not. One can however argue that in deterministic physics for fields the two measurements are equivalent so that the problem remains.

In TGD framework the classical physics is not completely deterministic and this has led to a generalization of the notion of quantum classical correspondence. Space-time surface provides a classical (unfaithful) representation not only for quantum states but for quantum jump sequences or equivalently, for sequences of quantum states. The most obvious identification for the quantum states is as the maximal non-deterministic regions of a given space-time sheet.

In the recent context this would mean that the fields in the region between the screen and lens represent the state before the state function reduction and thus the interference pattern, whereas the fields in the region between lens and detectors represent the situation after the state function reduction. The interaction with lens involves classical non-determinism.

This picture conforms also with the notion of topological field quantization. The space-time decomposes into space-time sheets interpreted, topological field quanta (topological light rays containing photons, flux quanta of magnetic field, etc..). Topological field quanta correspond to the coherence regions for classical fields with spinor fields included. De-coherence corresponds to the splitting of space-time sheet to smaller, possibly parallel space-time sheets. Topological field quantum carries classical fields inside it but behaves as a whole like particle. Hence particle and wave aspects are consistent in the sense that below the size scale L of the topological field quantum (say the thickness of a magnetic flux tube or topological light ray) the description as a wave applies and above L particle description makes sense. In the recent case the coherence is lost at the lens space-time sheet where the space-time sheet representing interference pattern decomposes to two sheets representing photon beams going to the two detectors.

4.1.3 Some basic ideas of TGD inspired theory of consciousness and quantum biology

The following ideas of TGD inspired theory of consciousness and of quantum biology are the most relevant ones for what will follow.

a) "Everything is conscious and consciousness can be only lost" is the briefest manner to summarize TGD inspired theory of consciousness. Quantum jump as moment of consciousness and the notion of self are key concepts of the theory. Self is a system able to avoid bound state entanglement with environment and can be formally seen as an ensemble of quantum jumps. The contents of consciousness of self are defined by the averaged increments of quantum numbers and zero modes (sensory and geometric qualia). Moment of consciousness can be said to be the counterpart of elementary particle and self the counterpart of many-particle state, either bound and free. The selves formed by macro-temporal quantum coherence are in turn the counterparts of atoms, molecules and larger structures. Macro-temporal quantum coherence effectively binds a sequence of quantum jumps to a single quantum jump as far as conscious experience is considered. The idea that conscious experience is about changes amplified to macroscopic quantum phase transitions, is the key philosophical guideline in the construction of various models, such as the model of qualia, the capacitor model of sensory receptor, the model of cognitive representations, and declarative memories.

b) Macro-temporal quantum coherence is a second consequence of the spin glass degeneracy [K2]. It is essentially due to the formation of bound states and has as a topological correlate the formation of join along boundaries bonds connecting the boundaries of the component systems. During macro-temporal coherence quantum jumps integrate effectively to single long-lasting quantum jump and one can say that system is in a state of oneness, eternal now, outside time. Macro-temporal quantum coherence makes possible stable non-entropic mental images. Negative energy MEs are one particular mechanism making possible macro-temporal quantum coherence via the formation of bound states, and remote metabolism and sharing of mental images are other facets of this mechanism. The real understanding of the origin of macroscopic quantum coherence requires the generalization of quantum theory allowing dynamical and quantized Planck constant [J6, M3].

d) p-Adic physics as physics of intentionality and of cognition is a further key idea of TGD inspired theory of consciousness. p-Adic space-time sheets as correlates for intentions and p-adic-to-real transformations of them as correlates for the transformation of intentions to actions allow deeper understanding of also psychological time as a front of p-adic-to-real transition propagating to the direction of the geometric future. Negative energy MEs are absolutely essential for the understanding of how precisely targeted intentionality is realized.

4.2 Many-sheeted space-time, magnetic flux quanta, electrets and MEs

TGD inspired theory of consciousness and of living matter relies on space-time sheets carrying ordinary matter, topological light rays (massless extremals, MEs), and magnetic and electric flux quanta. There are some new results which motivate a separate discussion of them.

4.2.1 Dynamical quantized Planck constant and dark matter hierarchy

By quantum classical correspondence space-time sheets can be identified as quantum coherence regions. Hence the fact that they have all possible size scales more or less unavoidably implies that Planck constant must be quantized and have arbitrarily large values. If one accepts this then also the idea about dark matter as a macroscopic quantum phase characterized by an arbitrarily large value of Planck constant emerges naturally as does also the interpretation for the long ranged classical electroweak and color fields predicted by TGD. Rather seldom the evolution of ideas follows simple linear logic, and this was the case also now. In any case, this vision represents the fifth, relatively new thread in the evolution of TGD and the ideas involved are still evolving.

Dark matter as large $\hbar$ phase

D. Da Rocha and Laurent Nottale have proposed that Schrödinger equation with Planck constant $\hbar$ replaced with what might be called gravitational Planck constant $\hbar_{gr} = \frac{GmM}{v_0}$ ($\hbar = c = 1$). v_0 is a velocity parameter having the value $v_0 = 144.7 \pm .7$ km/s giving $v_0/c = 4.6 \times 10^{-4}$. This is rather near to the peak orbital velocity of stars in galactic halos. Also sub-harmonics and harmonics of v_0 seem to appear. The support for the hypothesis coming from empirical data is impressive.

Nottale and Da Rocha believe that their Schrödinger equation results from a fractal hydrodynamics. Many-sheeted space-time however suggests astrophysical systems are not only quantum systems at larger space-time sheets but correspond to a gigantic value of gravitational Planck constant. The gravitational (ordinary) Schrödinger equation would provide a solution of the black hole collapse (IR catastrophe) problem encountered at the classical level. The resolution of the problem inspired by TGD inspired theory of living matter is that it is the dark matter at larger space-time sheets which is quantum coherent in the required time scale [D6].

Dark matter as a source of long ranged weak and color fields

Long ranged classical electro-weak and color gauge fields are unavoidable in TGD framework. The smallness of the parity breaking effects in hadronic, nuclear, and atomic length scales does not however seem to allow long ranged electro-weak gauge fields. The problem disappears if long ranged classical electro-weak gauge fields are identified as space-time correlates for massless gauge fields created by dark matter. Also scaled up variants of ordinary electro-weak particle spectra are possible. The identification explains chiral selection in living matter and unbroken $U(2)_{ew}$ invariance and free color in bio length scales become characteristics of living matter and of bio-chemistry and bio-nuclear physics. An attractive solution of the matter antimatter asymmetry is based on the identification of also antimatter as dark matter.

Dark matter hierarchy and consciousness

The emergence of the vision about dark matter hierarchy has meant a revolution in TGD inspired theory of consciousness. Dark matter hierarchy means also a hierarchy of long term memories with the span of the memory identifiable as a typical geometric duration of moment of consciousness at the highest level of dark matter hierarchy associated with given self so that even human life cycle represents at this highest level single moment of consciousness.

Dark matter hierarchy leads to detailed quantitative view about quantum biology with several testable predictions [M3]. The applications to living matter suggests that the basic hierarchy corresponds to a hierarchy of Planck constants coming as $\hbar(k) = \lambda^k(p)\hbar_0$, $\lambda \simeq 2^{11}$ for $p = 2^{127-1}$, $k = 0, 1, 2, \ldots$ [M3]. Also integer valued sub-harmonics and integer valued sub-harmonics of λ might be possible. Each p-adic length scale corresponds to this kind of hierarchy and number theoretical arguments suggest a general formula for the allowed values of Planck constant λ depending logarithmically on p-adic prime [O5]. Also the value of $\hbar_0$ has spectrum characterized by Beraha numbers $B_n = 4cos^2(\pi/n)$, $n \geq 3$, varying by a factor in the range $n > 3$ [O5].

The general prediction is that Universe is a kind of inverted Mandelbrot fractal for which each bird's eye of view reveals new structures in long length and time scales representing scaled down copies of standard physics and their dark variants. These structures would correspond to higher levels in self hierarchy. This prediction is consistent with the belief that 75 per cent of matter in the universe is dark.

1. Living matter and dark matter

Living matter as ordinary matter quantum controlled by the dark matter hierarchy has turned out to be a particularly successful idea. The hypothesis has led to models for EEG predicting correctly the band structure and even individual resonance bands and also generalizing the notion of EEG [M3]. Also a generalization of the notion of genetic code emerges resolving the paradoxes related to the standard dogma [L2, M3]. A particularly fascinating implication is the possibility to identify great leaps in evolution as phase transitions in which new higher level of dark matter emerges [M3].

It seems safe to conclude that the dark matter hierarchy with levels labelled by the values of Planck constants explains the macroscopic and macro-temporal quantum coherence naturally. That this explanation is consistent with the explanation based on spin glass degeneracy is suggested by following observations. First, the argument supporting spin glass degeneracy as an explanation of the macro-temporal quantum coherence does not involve the value of $\hbar$ at all. Secondly, the failure of the perturbation theory assumed to lead to the increase of Planck constant and formation of macroscopic quantum phases could be precisely due to the emergence of a large number of new degrees of freedom due to spin glass degeneracy. Thirdly, the phase transition increasing Planck constant has concrete topological interpretation in terms of many-sheeted space-time consistent with the spin glass degeneracy.

2. Dark matter hierarchy and the notion of self

The vision about dark matter hierarchy leads to a more refined view about self hierarchy and hierarchy of moments of consciousness [J6, M3]. The larger the value of Planck constant, the longer the subjectively experienced duration and the average geometric duration $T(k) \propto \lambda^k$ of the quantum jump.

Dark matter hierarchy suggests also a slight modification of the notion of self. Each self involves a hierarchy of dark matter levels, and one is led to ask whether the highest level in this hierarchy corresponds to a single quantum jump rather than a sequence of quantum jumps. The averaging of conscious experience over quantum jumps would occur only for sub-selves at lower levels of dark matter hierarchy and these mental images would be ordered, and single moment of consciousness would be experienced as a history of events. One can ask whether even entire life cycle could be regarded as a single quantum jump at the highest level so that consciousness would not be completely lost even during deep sleep. This would allow to understand why we seem to know directly that this biological body of mine existed yesterday.

The fact that we can remember phone numbers with 5 to 9 digits supports the view that self corresponds at the highest dark matter level to single moment of consciousness. Self would experience the average over the sequence of moments of consciousness associated with each sub-self but there would be no averaging over the separate mental images of this kind, be their parallel or serial. These mental images correspond to sub-selves having shorter wake-up periods than self and would be experienced as being time ordered. Hence the digits in the phone number are experienced as separate mental images and ordered with respect to experienced time.

4.2.2 p-Adic length scale hypothesis and the connection between thermal de Broglie wave length and size of the space-time sheet

Also real space-time sheets are assumed to be characterized by p-adic prime p and assumed to have a size determined by primary p-adic length scale L_p or possibly n-ary p-adic length scale $L_p(n)$. Since multi-p-fractality is allowed [E1], one cannot exclude even the possibility that each space-time dimension might correspond to its own p-adic length scale and even several p-adic primes could be associated with single dimension.

The possibility to assign a p-adic prime to the real space-time sheets is required by the success of the elementary particle mass calculations and various applications of the p-adic length scale hypothesis. Rational numbers are common to reals and all p-adic number fields. The p-adic-to-real transition transforming intentions to actions is made possible by a large number of common rational points between p-adic and real space-time surfaces, which supports the view that real space-time sheets obeys effective p-adic topology as an approximate topology in some resolution and below some length scale. p-Adic prime thus characterizes the classical non-determinism of the Kähler action.

Parallel space-time sheets with distance about 10^4 Planck lengths form a hierarchy. Each material object (...,atom, molecule, ..., cell,...) would correspond to this kind of space-time sheet. The p-adic primes $p \simeq 2^k$, k prime or power of prime, characterize the size scales of the space-time sheets in the hierarchy. The p-adic length scale $L(k)$ can be expressed in terms of cell membrane thickness as

$$L(k) = 2^{(k-151)/2} \times L(151) \ , \tag{4.2.1}$$

$L(151) \simeq 10$ nm. These are so called primary p-adic length scales but there are also n-ary p-adic length scales related by a scaling of power of $\sqrt{p}$ to the primary p-adic length scale. Quite recent model for photosynthesis [K6] gives additional support for the importance of also n-ary p-adic length scales so that the relevant p-adic length scales would come as half-octaves in a good approximation but prime and power of prime values of k would be especially important.

4.2.3 Topological light rays (massless extremals, MEs)

I have described MEs, or "topological light rays", in detail in [n7] and in [J4, J7], and describe here only very briefly the basic characteristics of MEs and concentrate on new idea about their possible role for consciousness and life.

What MEs are?

MEs (massless extremals, topological light rays) can be regarded as topological field quanta of classical radiation fields [J4, J7]. They are typically tubular space-time sheets inside which radiation fields propagate with light velocity in single direction without dispersion. The simplest case corresponds to a straight cylindrical ME but also curved MEs, kind of curved light rays, are possible. The initial values for a given moment of time are arbitrary by light likeness. Therefore MEs are ideal for precisely targeted communications. What distinguishes MEs from Maxwellian radiation fields in empty space is that light like vacuum 4-current is possible: ordinary Maxwell's equations would state that this current vanishes. Quite generally, purely geometric vacuum charge densities and 3-currents are purely TGD based prediction and could be seen as a classical correlate of the vacuum polarization predicted by quantum field theories.

MEs are fractal structures containing MEs within MEs. The so called scaling law of homeopathy predicts that the high frequency MEs inside low frequency MEs are in a ratio having discrete values [K5]. One can indeed justify this relationship. As ions drop from smaller space-time sheets to magnetic flux tubes, zero point kinetic energy is liberated as high frequency MEs, and the ions dropped to magnetic flux tubes generate cyclotron radiation, and the ratio of the fundamental frequencies is constant not depending on particle mass and being determined solely by p-adic length scale hypothesis. The model for the radio waves induced by the irradiation of DNA by laser light [la6] gives support for this picture [K2].

Two basic types of MEs

MEs have 2-dimensional CP_2 projection which means that electro-weak holonomy group is Abelian (color holonomy is always Abelian which suggests that physical states in TGD Universe correspond to states of color multiplets with vanishing color hypercharge and isospin rather than color singlets). If CP_2 projection belongs to a homologically non-trivial geodesic sphere, only em and Z^0 fields and Abelian color gauge fields are present. In the homologically trivial case only classical W fields are non-vanishing.

a) Neutral MEs can be assigned to various kinds of communications from biological body to the magnetic body and fractal hierarchy of EEGs and ZEGs represent the basic example in this respect [M3].

b) Dark W MEs serving as correlate for dark W exchanges induce an exotic ionization of atomic nuclei [F8, F9, M3]. This induces charge entanglement between magnetic body and biological body generating dark plasma oscillation patterns inducing nerve pulse patterns and ion waves at the space-time sheets occupied by the ordinary matter. The mechanism is based on many-sheeted Faraday law inducing electromagnetic fields at ordinary space-time sheet in turn giving rise to ohmic currents. State function reduction selects one of the exotically ionized configurations. This mechanism is the most plausible candidate for how magnetic body as an intentional agent controls biological body.

Negative energy MEs

MEs can have either positive or negative energy depending on the time orientation. The understanding of negative energy MEs has increased considerably. Phase conjugate laser beams [jb1] are the most plausible standard physics counterparts of negative energy MEs since they can be interpreted as time reversed laser beams and do not possess direct Maxwellian analog. By quantum-classical correspondence one can interpret the frequencies associated with negative energy MEs as energies. One can also assume that the Bose-Einstein condensed photons associated with negative energy MEs and with the coherent light generated by the light like vacuum current have negative energies.

For frequencies for which energy is above the thermal energy there is no system which could interact with negative energy MEs or absorb negative energy photons. Therefore negative energy MEs and corresponding photons should propagate through matter practically without any interaction. Feinberg has demonstrated that phase conjugate laser beams behave similarly: for instance, one can see through chickens using these laser beams [jb2]. This means that negative energy MEs do not respect Faraday cages and thus represent an attractive candidate for the hypothetical Psi field.

Negative energy MEs have many applications.

a) Negative energy MEs ideal for generating time like entanglement. Since negative energies are involved, this entanglement can be seen as a correlate for the bound state entanglement leading to a macro-temporal quantum coherence. Negative energy MEs make thus possible telepathic sharing of mental images. Negative energy MEs are involved with both sensory perception, long term memory, and motor action. In the model for living matter [M3] The charge entanglement generated by W MEs inducing exotic weak charge and electromagnetic charge is assumed to be responsible for bio-control whereas neutral MEs in general carrying both em and Z^0 fields are responsible for communications.

b) Negative energy MEs are ideal for a precisely targeted realization of intentions. p-Adic ME having a large number of common rational points with negative energy ME is generated and transformed to a real ME in quantum jump. The system receives positive energy and momentum as a recoil effect and the transition is not masked by ordinary spontaneously occurring quantum transitions since the energy of the system increases. One can say that negative energy ME represents the desires communicated to the geometric past and inducing as a reaction the desired action realized as say neuronal activity and generation of positive energy MEs.

c) The generation of negative energy MEs is also in a key role in remote metabolism and MEs serve as quantum credit cards implying an extreme flexibility of the metabolism. If the system receiving negative energy MEs is a population inverted laser or its many-sheeted counterpart, then quite a small field intensity associated with negative energy MEs (intensity of negative energy photons) can lead to the amplification of the time reflected positive energy signal. The reason is that the rate for the induced emission is proportional to the number of particles dropped to the ground state from the excited state. Therefore even negative energy bio-photons might serve as quantum controllers of metabolism and induce much more intense beams of positive energy photons, say when interacting with mitochondria.

4.2.4 Magnetic flux quanta and electrets

Magnetic flux tubes and electrets are extremals of Kähler action dual to each other. Also layer like magnetic flux quanta and their electric counterparts are possible. The magnetic/electric field is in a good approximation of constant magnitude but has varying direction.

Magnetic fields and life

The magnetic field associated with any material system is topologically quantized, and one can assign to any system a magnetic body. An attractive idea is that the relationship of the magnetic body to the material system is to some degree that of the manual to an electronic instrument. Quantitative arguments related to the dark matter hierarchy assuming that magnetic bodies are dark suggest that cognitions and emotions are regarded as somatosensory qualia of the magnetic body [M3, K3]. Magnetic body would in this case serve as a kind of computer screen at which the data items processes in say brain are communicated either classically (positive energy MEs) or by sharing of mental images (negative energy MEs).

Magnetic body is also an active intentional agent: motor actions are controlled from magnetic body and proceed as cascade like processes from long to short length and time scales as quantum communications of desires at various levels of hierarchy of magnetic bodies. Communication occurs backwards in geometric time by negative energy MEs. Motor action as a response to these desires occurs by classical communications by positive energy MEs and as neural activities. This explains the coherence and synchrony of motor actions difficult to understand in neuroscience framework. The sizes of flux quanta are astrophysical: for instance, EEG frequency of 7.8 Hz corresponds to a wave length defined by Earth's circumference. The non-locality in the length scale of magnetosphere, and even in length scales up to light life, is forced by Uncertainty Principle alone, if taken seriously in macroscopic length scales.

The leakage of supra currents of ions and their Cooper pairs from magnetic flux tubes of the Earth's magnetic field to smaller space-time sheets and their dropping back involving liberation of the zero point kinetic energy defines one particular metabolic "Karma's cycle". The dropping of protons from $k = 137$ atomic space-time sheet involved with the utilization of ATP molecules is only a special instance of the general mechanism involving an entire hierarchy of zero point kinetic energies defining universal metabolic currencies. This leads to the idea that the topologically quantized magnetic field of Earth defines the analog of central nervous system and blood circulation present already during the pre-biotic evolution and making possible primitive metabolism. This has far reaching implications for the understanding of how pre-biotic evolution led to living matter as we understand it [N4].

For instance, it has recently become clear that the dropping of atoms and molecules from space-time sheets labelled by p-adic prime $p \simeq 2^k$, $k = 131$, liberates photons at visible and near infrared wave lengths. The hot $k = 131$ space-time sheets (with temperatures above 1000 K) could have served as a source of metabolic energy for life-forms at cool $k = 137$ sheets. Photosynthesis could have developed in the circumstances where solar radiation was replaced with these photons. The correct prediction is that chlorophylls should be especially sensitive to these wave lengths. In particular, it is predicted that also IR wave lengths 700-1000 nm should have been utilized. There indeed are bacteria using only this portion of solar radiation. This leads to a scenario making sense only in TGD universe. Pre-biotic life could have developed at the cool space-time sheets in the hot interior of Earth below crust, where $k = 131$ space-time sheets are possible and this life could still be there [N4]. Also the life as we know it, could involve hot spots generated by the cavitation of water inside cell. The classical repulsive Z^0 force causes a strong acceleration during final stages of bubble collapse creating high temperatures, and could explain also sono-luminescence [jc1] as suggested in [F9].

Magnetic Mother Gaia could also form sensory and other representations receiving input from several brains via negative energy EEG MEs entangling magnetosphere with brains. The multi-brained magnetospheric selves could be responsible for the third person aspect of consciousness and for the evolution of social structures. For instance, the successful healing by prayer and meditation groups [ma1], and the experiments of Mark Germine [ma2] provide support for the notion of multi-brained magnetospheric selves are involved. Magnetic flux tubes could function as wave guides for MEs and this aspect is crucial in the model of long term memory.

Electrets and bio-systems

Bio-systems are known to be full of electrets and liquid crystals [la7]. Perhaps the most fundamental electret structure is cell membrane. In particular, the water inside cells tends to be in gel phase which is liquid crystal phase. There are many good reasons for why water should be in ordered phase. One very fundamental reason is that bio-polymers are stable in liquid crystal/ordered water phase since there are no free water molecules available for the depolymerization by hydration. In fact, only a couple of years ago it was experimentally discovered that bio-polymers can be stabilized around ice.

The capacitor model for sensory receptor is one very important application of the electret concept [n6, K3]. Sensory qualia result in the flow of particles with given quantum numbers from the plate to another one in quantum discharge. This kind of amplification of quantum number *resp.* zero mode increments would give rise to both geometric *resp.* non-geometric qualia [K3].

Also micro-tubuli are electrets. Sol-gel transition, as any phase transition, is an good candidate for the representation of a conscious bit and controlled local sol-gel transitions between ordinary and liquid crystal water could be a basic control tool making possible cellular locomotion, changes of protein conformations, etc... The tubulin dimers of micro-tubuli could induce sol-gel transformations by generating negative energy MEs, and micro-tubular surface could provide bit maps of their environment somewhat like sensory areas of brain provide maps of body. If gel→sol transition around tubulin inducing conformational change induces sol→gel transformation in some point of environment as would be the case for the seesaw mechanism to be discussed below, a one-one correspondence would result. By this one-one correspondence micro-tubules would automatically generate kind of conscious log files about the control activities which could have evolved to micro-tubular declarative memory representations about what happens inside cell [K6].

4.3 General considerations

In this section field equations and their physical interpretation are discussed. Quantum classical correspondence suggests that the non-deterministic dynamics of Kähler action makes possible self-referential dynamics in the sense that larger space-time sheet perform smoothed out mimicry of the dynamics at smaller space-time sheets. The fact that the divergence of the energy momentum tensor, Lorentz 4-force, does not vanish in general makes possible the mimicry of even dissipation and of the second law. For asymptotic self organization patterns for which dissipation is absent the Lorentz 4-force must vanish. This condition is guaranteed if Kähler current is proportional to the instanton current in the case that CP_2 projection of the space-time sheet is smaller than four and vanishes otherwise. An attractive identification for the vanishing of Lorentz 4-force is as a condition equivalent with the principle selecting preferred extremals of Kähler action so that this principle would be essentially equivalent with the second law of thermodynamics.

4.3.1 Long range classical weak and color gauge fields as correlates for dark massless weak bosons

Long ranged classical electro-weak gauge fields are unavoidably present when the dimension D of the CP_2 projection of the space-time sheet is larger than 2. Classical color gauge fields are non-vanishing for all non-vacuum extremals. This poses deep interpretational problems. If ordinary quarks and leptons are assumed to carry weak charges feeded to larger space-time sheets within electro-weak length scale, large hadronic, nuclear, and atomic parity breaking effects, large contributions of the classical Z^0 force to Rutherford scattering, and strong isotopic effects, are expected. If weak charges are screened within electro-weak length scale, the question about the interpretation of long ranged classical weak fields remains.

Various interpretations for the long ranged classical electro-weak fields

During years I have discussed several solutions to the problems listed above.

Option I: The trivial solution of the constraints is that Z^0 charges are neutralized at electro-weak length scale. The problem is that this option leaves open the interpretation of classical long ranged electro-weak gauge fields unavoidably present in all length scales when the dimension for the CP_2 projection of the space-time surface satisfies $D > 2$.

Option II: Second option involves several variants but the basic assumption is that nuclei or even quarks feed their Z^0 charges to a space-time sheet with size of order neutrino Compton length. The large parity breaking effects in hadronic, atomic, and nuclear length scales is not the only difficulty. The scattering of electrons, neutrons and protons in the classical long range Z^0 force contributes to the Rutherford cross section and it is very difficult to see how neutrino screening could make these effects small enough. Strong isotopic effects in condensed matter due to the classical Z^0 interaction energy are expected. It is far from clear whether all these constraints can be satisfied by any assumptions about the structure of topological condensate.

Option III: During 2005 (27 years after the birth of TGD!) third option solving the problems emerged based on the progress in the understanding of the basic mathematics behind TGD.

In ordinary phase the Z^0 charges of elementary particles are indeed neutralized in intermediate boson length scale so that the problems related to the parity breaking, the large contributions of classical Z^0 force to Rutherford scattering, and large isotopic effects in condensed matter, trivialize.

Classical electro-weak gauge fields in macroscopic length scales are identified as space-time correlates for the gauge fields created by dark matter, which corresponds to a macroscopically quantum coherent phase for which elementary particles possess complex conformal weights such that the net conformal weight of the system is real.

In this phase $U(2)_{ew}$ symmetry is not broken below the scaled up weak scale except for fermions so that gauge bosons are massless below this length scale whereas fermion masses are essentially the same as for ordinary matter. By charge screening gauge bosons look massive in length scales much longer than the relevant p-adic length scale. The large parity breaking effects in living matter (chirality selection for bio-molecules) support the view that dark matter is what makes living matter living.

Classical color gauge fields

Classical long ranged color gauge fields always present for non-vacuum extremals are interpreted as space-time correlates of gluon fields associated with dark copies of hadron physics. It seems that this picture is indeed what TGD predicts. A very special feature of classical color fields is that the holonomy group is Abelian. This follows directly from the expression $g^A_{\alpha\beta} \propto H^A J_{\alpha\beta}$ of induced gluon fields in terms of Hamiltonians H^A of color isometries and induced Kähler form $J_{\alpha\beta}$. This means that classical color magnetic and electric fluxes reduce to the analogs of ordinary magnetic fluxes appearing in the construction of configuration space geometry [B2, B3].

By a local color rotation the color field can be rotated to a fixed direction so that genuinely Abelian field would be in question apart from the possible presence of gauge singularities making impossible a global selection of color direction. These singularities could be present since Kähler form defines a magnetic monopole field. An interesting question inspired by quantum classical correspondence is what the Abelian holonomy tells about the sources of color gauge fields and color confinement.

For instance, could Abelian color holonomy mean that colored gluons (and presumably also other colored particles) do not propagate in the p-adic length scale considered? Color neutral gluons would propagate but since also their sources must be color neutral, they should have vanishing net color electric fluxes. This form of confinement would allow those states of color multiplets which have vanishing color charges and obviously symmetry breaking down to $U(1) \times U(1)$ would be in question. This would serve as a signal for monopole confinement which would not exclude higher multipoles for the Abelian color fields. This kind of fields appear in the TGD based model for nuclei as nuclear strings bound together by color flux tubes [F8].

4.3.2 Is absolute minimization the correct variational principle?

One can criticize the original assumption that extremals correspond to absolute minima, and the number theoretical vision discussed in [E2] indeed favors the separate minimization of magnitudes of positive and negative contributions to the Kähler action.

For this option Universe would do its best to save energy, being as near as possible to vacuum. Also vacuum extremals would become physically relevant: note that they would be only inertial vacua and carry non-vanishing density gravitational energy. The non-determinism of the vacuum extremals would have an interpretation in terms of the ability of Universe to engineer itself.

The 3-surfaces for which CP_2 projection is at least 2-dimensional and not a Lagrange manifold would correspond to non-vacua since conservation laws do not leave any other option. The variational principle would favor equally magnetic and electric configurations whereas absolute minimization of action based on S_K would favor electric configurations. The positive and negative contributions would be minimized for 4-surfaces in relative homology class since the boundary of X^4 defined by the intersections with 7-D light-like causal determinants would be fixed. Without this constraint only vacuum bubbles would result.

The attractiveness of the number theoretical variational principle from the point of calculability of TGD would be that the initial values for the time derivatives of the imbedding space coordinates at X^3 at light-like 7-D causal determinant could be computed by requiring that the energy of the solution is minimized. This could mean a computerizable solution to the construction of Kähler function.

Since the considerations of this chapter relate only to the extremals of Kähler action, I have not bothered to replace "absolute minimum" with "preferred extremal" in the text. The number theoretic approach based on the properties of quaternions and octonions discussed in the chapter [E2] leads to a proposal for the general solution of field equations based on the generalization of the notion of calibration [bb1] providing absolute minima of volume to that of Kähler calibration. This approach will not be discussed in this chapter.

4.3.3   Field equations

The requirement that Kähler action is stationary leads to the following field equations in the interior of the four-surface

$$
\begin{aligned}
D_\beta(T^{\alpha\beta}h^k_\alpha) \quad - \quad & j^\alpha J^k_l \partial_\alpha h^l = 0 \ , \\
T^{\alpha\beta} \quad = \quad & J^{\nu\alpha} J_\nu{}^\beta - \frac{1}{4} g^{\alpha\beta} J^{\mu\nu} J_{\mu\nu} \ .
\end{aligned}
\tag{4.3.1}
$$

Here $T^{\alpha\beta}$ denotes the traceless canonical energy momentum tensor associated with the Kähler action.

An equivalent form for the first equation is

$$
\begin{aligned}
T^{\alpha\beta} H^k_{\alpha\beta} \quad - \quad & j^\alpha(J_\alpha{}^\beta h^k_\beta + J^k_l \partial_\alpha h^l) = 0 \ . \\
H^k_{\alpha\beta} \quad = \quad & D_\beta \partial_\alpha h^k \ .
\end{aligned}
\tag{4.3.2}
$$

$H^k_{\alpha\beta}$ denotes the components of the second fundamental form and $j^\alpha = D_\beta J^{\alpha\beta}$ is the gauge current associated with the Kähler field.

On the boundaries of X^4 the field equations are given by the expression

$$
T^{n\beta} \partial_\beta h^k - J^{n\alpha}(J_\alpha{}^\beta \partial_\beta h^k + J^k_l)\partial_\alpha h^k) = 0 \ .
\tag{4.3.3}
$$

A general manner manner to solve the field equations on the boundaries is to assume that the induced Kähler field associated with the boundaries vanishes:

$$
J_{\alpha\beta}(\delta) \quad = \quad 0 \ .
\tag{4.3.4}
$$

In this case the energy-momentum tensor vanishes identically on the boundary component. On the outer boundaries of the 3-surface this solution ansatz makes sense only provided the gauge fluxes and gravitational flux (defined by Newtonian potential in the non-relativistic limit) associated with the matter in the interior go somewhere. The only possibility seems to be that 3-surface is topologically condensed on a larger 3-surface and feeds its gauge fluxes to the larger 3-surface via # contacts (topological sum). This assumption forces the concept of topological condensate defined as a hierarchical structure of 3-surfaces condensed on each other and thus giving rise to the many-sheeted space-time.

An important thing to notice is that the boundary conditions do not force the normal components of the gauge fields to zero even if the Kähler electric field vanishes near the boundaries. This makes in principle possible gauge charge renormalization classically resulting from the hierarchical structure of the topological condensation.

4.3.4   Topologization and light-likeness of the Kähler current as alternative manners to guarantee vanishing of Lorentz 4-force

The general solution of 4-dimensional Einstein-Yang Mills equations in Euclidian 4-metric relies on self-duality of the gauge field, which topologizes gauge charge. This topologization can be achieved by a weaker condition, which can be regarded as a dynamical generalization of the Beltrami condition. An alternative manner to achieve vanishing of the Lorentz 4-force is light-likeness of the Kähler 4-current. This does not require topologization.

Topologization of the Kähler current for $D_{CP_2} = 3$: covariant formulation

The condition states that Kähler 4-current is proportional to the instanton current whose divergence is instanton density and vanishes when the dimension of CP_2 projection is smaller than four: $D_{CP_2} < 4$. For $D_{CP_2} = 2$ the instanton 4-current vanishes identically and topologization is equivalent with the vanishing of the Kähler current.

$$j^\alpha \equiv D_\beta J^{\alpha\beta} \quad = \quad \psi \times j_I^\alpha = \psi \times \epsilon^{\alpha\beta\gamma\delta} J_{\beta\gamma} A_\delta \ . \tag{4.3.5}$$

Here the function ψ is an arbitrary function $\psi(s^k)$ of CP_2 coordinates s^k regarded as functions of space-time coordinates. It is essential that ψ depends on the space-time coordinates through the CP_2 coordinates only. Hence the representation as an imbedded gauge field is crucial element of the solution ansatz.

The field equations state the vanishing of the divergence of the 4-current. This is trivially true for instanton current for $D_{CP_2} < 4$. Also the contraction of $\nabla\psi$ (depending on space-time coordinates through CP_2 coordinates only) with the instanton current is proportional to the winding number density and therefore vanishes for $D_{CP_2} < 4$.

The topologization of the Kähler current guarantees the vanishing of the Lorentz 4-force. Indeed, using the self-duality condition for the current, the expression for the Lorentz 4-force reduces to a term proportional to the instanton density:

$$\begin{aligned} j^\alpha J_{\alpha\beta} \quad &= \quad \psi \times j_I^\alpha J_{\alpha\beta} \\ &= \quad \psi \times \epsilon^{\alpha\mu\nu\delta} J_{\mu\nu} A_\delta J_{\alpha\beta} \ . \end{aligned} \tag{4.3.6}$$

Since all vector quantities appearing in the contraction with the four-dimensional permutation tensor are proportional to the gradients of CP_2 coordinates, the expression is proportional to the instanton density, and thus winding number density, and vanishes for $D_{CP_2} < 4$.

Remarkably, the topologization of the Kähler current guarantees also the vanishing of the term $j^\alpha J^{kl}\partial_\alpha s^k$ in the field equations for CP_2 coordinates. This means that field equations reduce in both M_+^4 and CP_2 degrees of freedom to

$$T^{\alpha\beta} H_{\alpha\beta}^k \quad = \quad 0 \ . \tag{4.3.7}$$

These equations differ from the equations of minimal surface only by the replacement of the metric tensor with energy momentum tensor. The earlier proposal that quaternion conformal invariance in a suitable sense might provide a general solution of the field equations could be seen as a generalization of the ordinary conformal invariance of string models. If the topologization of the Kähler current implying effective dimensional reduction in CP_2 degrees of freedom is consistent with quaternion conformal invariance, the quaternion conformal structures must differ for the different dimensions of CP_2 projection.

Topologization of the Kähler current for $D_{CP_2} = 3$: non-covariant formulation

In order to gain a concrete understanding about what is involved it is useful to repeat these arguments using the 3-dimensional notation. The components of the instanton 4-current read in three-dimensional notation as

$$\bar{j}_I = \overline{E} \times \overline{A} + \phi\overline{B} \ , \quad \rho_I = \overline{B} \cdot \overline{A} \ . \tag{4.3.8}$$

The self duality conditions for the current can be written explicitly using 3-dimensional notation and read

$$\begin{aligned} \nabla \times \overline{B} - \partial_t \overline{E} \quad &= \quad \bar{j} = \psi \bar{j}_I = \psi \left(\phi\overline{B} + \overline{E} \times \overline{A} \right) \ , \\ \nabla \cdot E \quad &= \quad \rho = \psi \rho_I \ . \end{aligned} \tag{4.3.9}$$

For a vanishing electric field the self-duality condition for Kähler current reduces to the Beltrami condition

$$\nabla \times \overline{B} = \alpha \overline{B} \ , \quad \alpha = \psi \phi \ . \tag{4.3.10}$$

The vanishing of the divergence of the magnetic field implies that α is constant along the field lines of the flow. When ϕ is constant and $\overline{A}$ is time independent, the condition reduces to the Beltrami condition with $\alpha = \phi = constant$, which allows an explicit solution [ab4].

One can check also the vanishing of the Lorentz 4-force by using 3-dimensional notation. Lorentz 3-force can be written as

$$\rho_I \overline{E} + \overline{j} \times \overline{B} = \psi \overline{B} \cdot \overline{AE} + \psi \left(\overline{E} \times \overline{A} + \phi \overline{B} \right) \times \overline{B} = 0 \ . \tag{4.3.11}$$

The fourth component of the Lorentz force reads as

$$\overline{j} \cdot \overline{E} = \psi \overline{B} \cdot \overline{E} + \psi \left(\overline{E} \times \overline{A} + \phi \overline{B} \right) \cdot \overline{E} = 0 \ . \tag{4.3.12}$$

The remaining conditions come from the induction law of Faraday and could be guaranteed by expressing $\overline{E}$ and $\overline{B}$ in terms of scalar and vector potentials.

The density of the Kähler electric charge of the vacuum is proportional to the helicity density of the so called helicity charge $\rho = \psi \rho_I = \psi B \cdot A$. This charge is topological charge in the sense that it does not depend on the induced metric at all. Note the presence of arbitrary function ψ of CP_2 coordinates.

Further conditions on the functions appearing in the solution ansatz come from the 3 independent field equations for CP_2 coordinates. What is remarkable that the generalized self-duality condition for the Kähler current allows to understand the general features of the solution ansatz to very high degree without any detailed knowledge about the detailed solution. The question whether field equations allow solutions consistent with the self duality conditions of the current will be dealt later. The optimistic guess is that the field equations and topologization of the Kähler current relate to each other very intimately.

Vanishing or light likeness of the Kähler current guarantees vanishing of the Lorentz 4-force for $D_{CP_2} = 2$

For $D_{CP_2} = 2$ one can always take two CP_2 coordinates as space-time coordinates and from this it is clear that instanton current vanishes so that topologization gives a vanishing Kähler current. In particular, the Beltrami condition $\nabla \times \overline{B} = \alpha \overline{B}$ is not consistent with the topologization of the instanton current for $D_{CP_2} = 2$.

$D_{CP_2} = 2$ case can be treated in a coordinate invariant manner by using the two coordinates of CP_2 projection as space-time coordinates so that only a magnetic or electric field is present depending on whether the gauge current is time-like or space-like. Light-likeness of the gauge current provides a second manner to achieve the vanishing of the Lorentz force and is realized in case of massless extremals having $D_{CP_2} = 2$: this current is in the direction of propagation whereas magnetic and electric fields are orthogonal to it so that Beltrami conditions is certainly not satisfied.

Under what conditions topologization of Kähler current yields Beltrami conditions?

Topologization of the Kähler 4-current gives rise to magnetic Beltrami fields if either of the following conditions is satisfied.

a) The $\overline{E} \times \overline{A}$ term contributing besides $\phi \overline{B}$ term to the topological current vanishes. This requires that $\overline{E}$ and $\overline{A}$ are parallel to each other

$$\overline{E} \ = \ \nabla \Phi - \partial_t \overline{A} = \beta \overline{A} \tag{4.3.13}$$

This condition is analogous to the Beltrami condition. Now only the 3-space has as its coordinates time coordinate and two spatial coordinates and and $\overline{B}$ is replaced with $\overline{A}$. Since E and B are orthogonal, this condition implies $\overline{B} \cdot \overline{A} = 0$ so that Kähler charge density is vanishing.

b) The vector $\overline{E} \times \overline{A}$ is parallel to $\overline{B}$.

$$\overline{E} \times \overline{A} = \beta \overline{B} \qquad (4.3.14)$$

The condition is consistent with the orthogonality of $\overline{E}$ and $\overline{B}$ but implies the orthogonality of $\overline{A}$ and $\overline{B}$ so that electric charge density vanishes

In both cases vector potential fails to define a contact structure since $B \cdot A$ vanishes (contact structures are discussed briefly below), and there exists a global coordinate along the field lines of $\overline{A}$ and the full contact structure is lost again. Note however that the Beltrami condition for magnetic field means that magnetic field defines a contact structure irrespective of whether $\overline{B} \cdot \overline{A}$ vanishes or not. The transition from the general case to Beltrami field would thus involve the replacement

$$(\overline{A}, \overline{B}) \rightarrow_{\nabla \times} (\overline{B}, \overline{j})$$

induced by the rotor.

One must of course take these considerations somewhat cautiously since the inner product depends on the induced 4-metric and it might be that induced metric could allow small vacuum charge density and make possible genuine contact structure.

Hydrodynamic analogy

The field equations of TGD are basically hydrodynamic equations stating the local conservation of the currents associated with the isometries of the imbedding space. Therefore it is intriguing that Beltrami fields appear also as solutions of ideal magnetohydrodynamics equations and as steady solutions of non-viscous incompressible flow described by Euler equations [ab3].

In hydrodynamics the role of the magnetic field is taken by the velocity field. TGD based models for nuclei [F6] and condensed matter [F9] involve in an essential manner valence quarks having large $\hbar$ and exotic quarks giving nucleons anomalous color and weak charges creating long ranged color and weak forces. Weak forces have a range of order atomic radius and could be responsible for the repulsive core in van der Waals potential.

This raises the idea that the incompressible flow could occur along the field lines of the Z^0 magnetic field so that the velocity field would be proportional to the Z^0 magnetic field and the Beltrami condition for the velocity field would reduce to that for Z^0 magnetic field. Thus the flow lines of hydrodynamic flow would directly correspond to those of Z^0 magnetic field. The generalized Beltrami flow based on the topologization of the Z^0 current would allow to model also the non-stationary incompressible non-viscous hydrodynamical flows.

It would seem that one cannot describe viscous flows using flows satisfying generalized Beltrami conditions since the vanishing of the Lorentz 4-force says that there is no local dissipation of the classical field energy. One might claim that this is not a problem since in TGD framework viscous flow could be seen as a practical description of a quantum jump sequence by replacing the corresponding sequence of space-time surfaces with a single space-time surface.

One the other hand, quantum classical correspondence requires that also dissipative effects have space-time correlates. Kähler fields, which are dissipative, and thus correspond to a non-vanishing Lorentz 4-force, represent one candidate for correlates of this kind. If this is the case, then the fields satisfying the generalized Beltrami condition provide space-time correlates only for the asymptotic self organization patterns for which the viscous effects are negligible, and also the solutions of field equations describing effects of viscosity should be possible.

One must however take this argument with a grain of salt. Dissipation, that is the transfer of conserved quantities to degrees of freedom corresponding to shorter scales, could correspond to a transfer of these quantities between different space-time sheets of the many-sheeted space-time. Here the opponent could however argue that larger space-time sheets mimic the dissipative dynamics in shorter scales and that classical currents represent "symbolically" averaged currents in shorter length scales, and that the local non-conservation of energy momentum tensor consistent with local conservation of isometry currents provides a unique manner to mimic the dissipative dynamics. This view will be developed in more detail below.

The stability of generalized Beltrami fields

The stability of generalized Beltrami fields is of high interest since unstable points of space-time sheets are those around which macroscopic changes induced by quantum jumps are expected to be localized.

1. Contact forms and contact structures

The stability of Beltrami flows has been studied using the theory of contact forms in three-dimensional Riemann manifolds [ab5]. Contact form is a one-form A (that is covariant vector field A_α) with the property $A \wedge dA \neq 0$. In the recent case the induced Kähler gauge potential A_α and corresponding induced Kähler form $J_{\alpha\beta}$ for any 3-sub-manifold of space-time surface define a contact form so that the vector field $A^\alpha = g^{\alpha\beta}A_\beta$ is not orthogonal with the magnetic field $B^\alpha = \epsilon^{\alpha\beta\delta}J_{\beta\gamma}$. This requires that magnetic field has a helical structure. Induced metric in turn defines the Riemann structure.

If the vector potential defines a contact form, the charge density associated with the topologized Kähler current must be non-vanishing. This can be seen as follows.

a) The requirement that the flow lines of a one-form X_μ defined by the vector field X^μ as its dual allows to define a global coordinate x varying along the flow lines implies that there is an integrating factor ϕ such that $\phi X = dx$ and therefore $d(\phi X) = 0$. This implies $dlog(\phi) \wedge X = -dX$. From this the necessary condition for the existence of the coordinate x is $X \wedge dX = 0$. In the three-dimensional case this gives $\overline{X} \cdot (\nabla \times \overline{X}) = 0$.

b) This condition is by definition not satisfied by the vector potential defining a contact form so that one cannot identify a global coordinate varying along the flow lines of the vector potential. The condition $\overline{B} \cdot \overline{A} \neq 0$ states that the charge density for the topologized Kähler current is non-vanishing. The condition that the field lines of the magnetic field allow a global coordinate requires $\overline{B} \cdot \nabla \times \overline{B} = 0$. The condition is not satisfied by Beltrami fields with $\alpha \neq 0$. Note that in this case magnetic field defines a contact structure.

Contact structure requires the existence of a vector ξ satisfying the condition $A(\xi) = 0$. The vector field ξ defines a plane field, which is orthogonal to the vector field A^α. Reeb field in turn is a vector field for which $A(X) = 1$ and $dA(X;) = 0$ hold true. The latter condition states the vanishing of the cross product $X \times B$ so that X is parallel to the Kähler magnetic field B^α and has unit projection in the direction of the vector field A^α. Any Beltrami field defines a Reeb field irrespective of the Riemannian structure.

2. Stability of the Beltrami flow and contact structures

Contact structures are used in the study of the topology and stability of the hydrodynamical flows [ab5], and one might expect that the notion of contact structure and its proper generalization to the four-dimensional context could be useful in TGD framework also. An example giving some idea about the complexity of the flows defined by Beltrami fields is the Beltrami field in R^3 possessing closed orbits with all possible knot and link types simultaneously [ab5]!

Beltrami flows associated with Euler equations are known to be unstable [ab5]. Since the flow is volume preserving, the stationary points of the Beltrami flow are saddle points at which also vorticity vanishes and linear instabilities of Navier-Stokes equations can develop. From the point of view of biology it is interesting that the flow is stabilized by vorticity which implies also helical structures. The stationary points of the Beltrami flow correspond in TGD framework to points at which the induced Kähler magnetic field vanishes. They can be unstable by the vacuum degeneracy of Kähler action implying classical non-determinism. For generalized Beltrami fields velocity and vorticity (both divergence free) are replaced by Kähler current and instanton current.

More generally, the points at which the Kähler 4-current vanishes are expected to represent potential instabilities. The instanton current is linear in Kähler field and can vanish in a gauge invariant manner only if the induced Kähler field vanishes so that the instability would be due to the vacuum degeneracy also now. Note that the vanishing of the Kähler current allows also the generation of region with $D_{CP_2} = 4$. The instability of the points at which induce Kähler field vanish is manifested in quantum jumps replacing the generalized Beltrami field with a new one such that something new is generated around unstable points. Thus the regions in which induced Kähler field becomes weak are the most interesting ones. For example, unwinding of DNA could be initiated by an instability of this kind.

4.3.5　How to satisfy field equations?

The topologization of the Kähler current guarantees also the vanishing of the term $j^\alpha J^{k}{}_l \partial_\alpha s^k$ in the field equations for CP_2 coordinates. This means that field equations reduce in both M^4_+ and CP_2 degrees of freedom to

$$T^{\alpha\beta} H^k_{\alpha\beta} \;=\; 0 \;. \tag{4.3.15}$$

These equations differ from the equations of minimal surface only by the replacement of the metric tensor with energy momentum tensor. The earlier proposal that quaternion conformal invariance in a suitable sense might provide a general solution of the field equations could be seen as a generalization of the ordinary conformal invariance of string models. If the topologization of the Kähler current implying effective dimensional reduction in CP_2 degrees of freedom is consistent with quaternion conformal invariance, the quaternion conformal structures must differ for the different dimensions of CP_2 projection. In the following somewhat different approach is however considered utilizing the properties of Hamilton Jacobi structures of M^4_+ introduced in the study of massless extremals and contact structures of CP_2 emerging naturally in the case of generalized Beltrami fields.

String model as a starting point

String model serves as a starting point.

a) In the case of Minkowskian minimal surfaces representing string orbit the field equations reduce to purely algebraic conditions in light cone coordinates (u, v) since the induced metric has only the component g_{uv}, whereas the second fundamental form has only diagonal components H^k_{uu} and H^k_{vv}.

b) For Euclidian minimal surfaces (u, v) is replaced by complex coordinates $(w, \overline{w})$ and field equations are satisfied because the metric has only the component $g^{w\overline{w}}$ and second fundamental form has only components of type H^k_{ww} and $H^{\overline{k}}_{\overline{w}\overline{w}}$. The mechanism should generalize to the recent case.

The general form of energy momentum tensor as a guideline for the choice of coordinates

Any 3-dimensional Riemann manifold allows always a orthogonal coordinate system for which the metric is diagonal. Any 4-dimensional Riemann manifold in turn allows a coordinate system for which 3-metric is diagonal and the only non-diagonal components of the metric are of form g^{ti}. This kind of coordinates might be natural also now. When $\overline{E}$ and $\overline{B}$ are orthogonal, energy momentum tensor has the form

$$T = \begin{pmatrix} \frac{E^2+B^2}{2} & 0 & 0 & EB \\ 0 & \frac{E^2+B^2}{2} & 0 & 0 \\ 0 & 0 & \frac{-E^2+B^2}{2} & 0 \\ EB & 0 & 0 & \frac{E^2-B^2}{2} \end{pmatrix} \tag{4.3.16}$$

in the tangent space basis defined by time direction and longitudinal direction $\overline{E} \times \overline{B}$, and transversal directions $\overline{E}$ and $\overline{B}$. Note that T is traceless.

The optimistic guess would be that the directions defined by these vectors integrate to three orthogonal coordinates of X^4 and together with time coordinate define a coordinate system containing only g^{ti} as non-diagonal components of the metric. This however requires that the fields in question allow an integrating factor and, as already found, this requires $\nabla \times X \cdot X = 0$ and this is not the case in general.

Physical intuition suggests however that X^4 coordinates allow a decomposition into longitudinal and transversal degrees freedom. This would mean the existence of a time coordinate t and longitudinal coordinate z the plane defined by time coordinate and vector $\overline{E} \times \overline{B}$ such that the coordinates $u = t - z$ and $v = t + z$ are light like coordinates so that the induced metric would have only the component g^{uv} whereas g^{vv} and g^{uu} would vanish in these coordinates. In the transversal space-time directions complex space-time coordinate coordinate w could be introduced. Metric could have also non-diagonal components besides the components $g^{w\overline{w}}$ and g^{uv}.

Hamilton Jacobi structures in M_+^4

Hamilton Jacobi structure in M_+^4 can understood as a generalized complex structure combing transversal complex structure and longitudinal hyper-complex structure so that notion of holomorphy and Kähler structure generalize.

a) Denote by m^i the linear Minkowski coordinates of M^4. Let (S^+, S^-, E^1, E^2) denote local coordinates of M_+^4 defining a *local* decomposition of the tangent space M^4 of M_+^4 into a direct, not necessarily orthogonal, sum $M^4 = M^2 \oplus E^2$ of spaces M^2 and E^2. This decomposition has an interpretation in terms of the longitudinal and transversal degrees of freedom defined by local light-like four-velocities $v_\pm = \nabla S_\pm$ and polarization vectors $\epsilon_i = \nabla E^i$ assignable to light ray. Assume that E^2 allows complex coordinates $w = E^1 + iE^2$ and $\overline{w} = E^1 - iE^2$. The simplest decomposition of this kind corresponds to the decomposition $(S^+ \equiv u = t + z, S^- \equiv v = t - z, w = x + iy, \overline{w} = x - iy)$.

b) In accordance with this physical picture, S^+ and S^- define light-like curves which are normals to light-like surfaces and thus satisfy the equation:

$$(\nabla S_\pm)^2 = 0 \quad .$$

The gradients of $S_\pm$ are obviously analogous to local light like velocity vectors $v = (1, \overline{v})$ and $\tilde{v} = (1, -\overline{v})$. These equations are also obtained in geometric optics from Hamilton Jacobi equation by replacing photon's four-velocity with the gradient ∇S: this is consistent with the interpretation of massless extremals as Bohr orbits of em field. $S_\pm = constant$ surfaces can be interpreted as expanding light fronts. The interpretation of $S_\pm$ as Hamilton Jacobi functions justifies the term Hamilton Jacobi structure.

The simplest surfaces of this kind correspond to $t = z$ and $t = -z$ light fronts which are planes. They are dual to each other by hyper complex conjugation $u = t - z \to v = t + z$. One should somehow generalize this conjugation operation. The simplest candidate for the conjugation $S^+ \to S^-$ is as a conjugation induced by the conjugation for the arguments: $S^+(t - z, t + z, x, y) \to S^-(t - z, t + z, x, y) = S^+(t + z, t - z, x, -y)$ so that a dual pair is mapped to a dual pair. In transversal degrees of freedom complex conjugation would be involved.

c) The coordinates $(S_\pm, w, \overline{w})$ define local light cone coordinates with the line element having the form

$$
\begin{aligned}
ds^2 &= g_{+-} dS^+ dS^- + g_{w\overline{w}} dw d\overline{w} \\
&+ g_{+w} dS^+ dw + g_{+\overline{w}} dS^+ d\overline{w} \\
&+ g_{-w} dS^- dw + g_{-\overline{w}} dS^- d\overline{w} \quad .
\end{aligned}
\tag{4.3.17}
$$

Conformal transformations of M_+^4 leave the general form of this decomposition invariant. Also the transformations which reduces to analytic transformations $w \to f(w)$ in transversal degrees of freedom and hyper-analytic transformations $S^+ \to f(S^+), S^- \to f(S^-)$ in longitudinal degrees of freedom preserve this structure.

d) The basic idea is that of generalized Kähler structure meaning that the notion of Kähler function generalizes so that the non-vanishing components of metric are expressible as

$$
\begin{aligned}
g_{w\overline{w}} = \partial_w \partial_{\overline{w}} K \quad , \quad g_{+-} = \partial_{S^+} \partial_{S^-} K \quad , \\
g_{w\pm} = \partial_w \partial_{S^\pm} K \quad , \quad g_{\overline{w}\pm} = \partial_{\overline{w}} \partial_{S^\pm} K \quad .
\end{aligned}
\tag{4.3.18}
$$

for the components of the metric. The expression in terms of Kähler function is coordinate invariant for the same reason as in case of ordinary Kähler metric. In the standard lightcone coordinates the Kähler function is given by

$$K = w_0 \overline{w}_0 + uv \quad , \quad w_0 = x + iy \quad , \quad u = t - z \quad , \quad v = t + z \quad .
\tag{4.3.19}$$

The Christoffel symbols satisfy the conditions

$$\{{}^{\,k}_{w\,\overline{w}}\} = 0 \quad , \quad \{{}^{\,k}_{+-}\} = 0 \quad .
\tag{4.3.20}$$

If energy momentum tensor has only the components $T^{w\overline{w}}$ and T^{+-}, field equations are satisfied in M_+^4 degrees of freedom.

e) The Hamilton Jacobi structures related by these transformations can be regarded as being equivalent. Since light-like 3- surface is, as the dynamical evolution defined by the light front, fixed by the 2-surface serving as the light source, these structures should be in one-one correspondence with 2-dimensional surfaces with two surfaces regarded as equivalent if they correspond to different time=constant snapshots of the same light front, or are related by a conformal transformation of M_+^4. Obviously there should be quite large number of them. Note that the generating two-dimensional surfaces relate also naturally to generalized conformal invariance and corresponding Kac Moody invariance for which deformations defined by the M^4 coordinates as functions of the light-cone coordinates of the light front evolution define Kac Moody algebra, which thus seems to appear naturally also at the level of solutions of field equations.

The task is to find all possible local light cone coordinates defining one-parameter families 2-surfaces defined by the condition $S_i = constant$, $i = +$ or $= -$, dual to each other and expanding with light velocity. The basic open questions are whether the generalized Kähler function indeed makes sense and whether the physical intuition about 2-surfaces as light sources parameterizing the set of all possible Hamilton Jacobi structures makes sense.

Contact structure and generalized Kähler structure of CP_2 projection

In the case of 3-dimensional CP_2 projection it is assumed that one can introduce complex coordinates $(\xi, \overline{\xi})$ and the third coordinate s. These coordinates would correspond to a contact structure in 3-dimensional CP_2 projection defining transversal symplectic and Kähler structures. In these coordinates the transversal parts of the induced CP_2 Kähler form and metric would contain only components of type $g_{w\overline{w}}$ and $J_{w\overline{w}}$. The transversal Kähler field $J_{w\overline{w}}$ would induce the Kähler magnetic field and the components J_{sw} and $J_{s\overline{w}}$ the Kähler electric field.

It must be emphasized that the non-integrability of the contact structure implies that J cannot be parallel to the tangent planes of $s = constant$ surfaces, s cannot be parallel to neither A nor the dual of J, and ξ cannot vary in the tangent plane defined by J. A further important conclusion is that for the solutions with 3-dimensional CP_2 projection topologized Kähler charge density is necessarily non-vanishing by $A \wedge J \neq 0$ whereas for the solutions with $D_{CP_2} = 2$ topologized Kähler current vanishes.

b) Also the CP_2 projection is assumed to possess a generalized Kähler structure in the sense that all components of the metric except s_{ss} are derivable from a Kähler function by formulas similar to M_+^4 case.

$$s_{w\overline{w}} = \partial_w \partial_{\overline{w}} K \quad , \quad s_{ws} = \partial_w \partial_s K \quad , \quad s_{\overline{w}s} = \partial_{\overline{w}} \partial_s K \quad . \tag{4.3.21}$$

Generalized Kähler property guarantees that the vanishing of the Christoffel symbols of CP_2 (rather than those of 3-dimensional projection), which are of type $\{{}^{k}_{\xi\,\overline{\xi}}\}$.

$$\{{}^{k}_{\xi\,\overline{\xi}}\} = 0 \quad . \tag{4.3.22}$$

Here the coordinates of CP_2 have been chosen in such a manner that three of them correspond to the coordinates of the projection and fourth coordinate is constant at the projection. The upper index k refers also to the CP_2 coordinate, which is constant for the CP_2 projection. If energy momentum tensor has only components of type T^{+-} and $T^{w\overline{w}}$, field equations are satisfied even when if non-diagonal Christoffel symbols of CP_2 are present. The challenge is to discover solution ansatz, which guarantees this property of the energy momentum tensor.

A stronger variant of Kähler property would be that also s_{ss} vanishes so that the coordinate lines defined by s would define light like curves in CP_2. The topologization of the Kähler current however implies that CP_2 projection is a projection of a 3-surface with strong Kähler property. Using $(s, \xi, \overline{\xi}, S^-)$ as coordinates for the space-time surface defined by the ansatz $(w = w(\xi, s), S^+ = S^+(s))$ one finds that g_{ss} must be vanishing so that stronger variant of the Kähler property holds true for $S^- = constant$ 3-surfaces.

The topologization condition for the Kähler current can be solved completely generally in terms of the induced metric using $(\xi, \bar{\xi}, s)$ and some coordinate of M_+^4, call it x^4, as space-time coordinates. Topologization boils down to the conditions

$$\begin{aligned} \partial_\beta(J^{\alpha\beta}\sqrt{g}) &= 0 \text{ for } \alpha \in \{\xi, \bar{\xi}, s\} \ , \\ g^{4i} &\neq 0 \ . \end{aligned} \tag{4.3.23}$$

Thus 3-dimensional empty space Maxwell equations and the non-orthogonality of X^4 coordinate lines and the 3-surfaces defined by the lift of the CP_2 projection.

A solution ansatz yielding light-like current in $D_{CP_2} = 3$ case

The basic idea is that of generalized Kähler structure and solutions of field equations as maps or deformations of canonically imbedded M_+^4 respecting this structure and guaranteing that the only non-vanishing components of the energy momentum tensor are $T^{\xi\xi}$ and T^{s-} in the coordinates $(\xi, \bar{\xi}, s, S^-)$.

a) The coordinates (w, S^+) are assumed to holomorphic functions of the CP_2 coordinates (s, ξ)

$$S^+ = S^+(s) \ , \quad w = w(\xi, s) \ . \tag{4.3.24}$$

Obviously S^+ could be replaced with S^-. The ansatz is completely symmetric with respect to the exchange of the roles of (s, w) and (S^+, ξ) since it maps longitudinal degrees of freedom to longitudinal ones and transverse degrees of freedom to transverse ones.

b) Field equations are satisfied if the only non-vanishing components of the energy momentum tensor are of type $T^{\xi\bar{\xi}}$ and T^{s-}. The reason is that the CP_2 Christoffel symbols for projection and projections of M_+^4 Christoffel symbols are vanishing for these lower index pairs.

c) By a straightforward calculation one can verify that the only manner to achieve the required structure of energy momentum tensor is to assume that the induced metric in the coordinates $(\xi, \bar{\xi}, s, S^-)$ has as non-vanishing components only $g_{\xi\bar{\xi}}$ and g_{s-}

$$g_{ss} = 0 \ , \quad g_{\xi s} = 0 \ , \quad g_{\bar{\xi}s} = 0 \ . \tag{4.3.25}$$

Obviously the space-time surface must factorize into an orthogonal product of longitudinal and transversal spaces.

d) The condition guaranteing the product structure of the metric is

$$\begin{aligned} s_{ss} &= m_{+w}\partial_s w(\xi, s)\partial_s S^+(s) + m_{+\bar{w}}\partial_s \bar{w}(\xi, s)\partial_s S^+(s) \ , \\ s_{s\xi} &= m_{+w}\partial_\xi w(\xi)\partial_s S^+(s) \ , \\ s_{s\bar{\xi}} &= m_{+w}\partial_{\bar{\xi}} w(\bar{\xi})\partial_s S^+(s) \ . \end{aligned} \tag{4.3.26}$$

Thus the function of dynamics is to diagonalize the metric and provide it with strong Kähler property. Obviously the CP_2 projection corresponds to a light-like surface for all values of S^- so that space-time surface is foliated by light-like surfaces and the notion of generalized conformal invariance makes sense for the entire space-time surface rather than only for its boundary or elementary particle horizons.

e) The requirement that the Kähler current is proportional to the instanton current means that only the j^- component of the current is non-vanishing. This gives the following conditions

$$j^\xi \sqrt{g} = \partial_\beta(J^{\xi\beta}\sqrt{g}) = 0 \ , \quad j^{\bar{\xi}}\sqrt{g} = \partial_\beta(J^{\bar{\xi}\beta}\sqrt{g}) = 0 \ ,$$

$$j^+\sqrt{g} = \partial_\beta(J^{+\beta}\sqrt{g}) = 0 \ . \tag{4.3.27}$$

Since $J^{+\beta}$ vanishes, the condition

$$\sqrt{g}j^+ = \partial_\beta(J^{+\beta}\sqrt{g}) = 0 \tag{4.3.28}$$

is identically satisfied. Therefore the number of field equations reduces to three.

The physical interpretation of the solution ansatz deserves some comments.

a) The light-like character of the Kähler current brings in mind CP_2 extremals for which CP_2 projection is light like. This suggests that the topological condensation of CP_2 type extremal occurs on $D_{CP_2} = 3$ helical space-time sheet representing zitterbewegung. In the case of many-body system light-likeness of the current does not require that particles are massless if particles of opposite charges can be present. Field tensor has the form $(J^{\xi\bar{\xi}}, J^{\xi-}, J^{\bar{\xi}-})$. Both helical magnetic field and electric field present as is clear when one replaces the coordinates (S^+, S^-) with time-like and space-like coordinate. Magnetic field dominates but the presence of electric field means that genuine Beltrami field is not in question.

b) Since the induced metric is product metric, 3-surface is metrically product of 2-dimensional surface X^2 and line or circle and obeys product topology. If absolute minima correspond to asymptotic self-organization patterns, the appearance of the product topology and even metric is not so surprising. Thus the solutions can be classified by the genus of X^2. An interesting question is how closely the explanation of family replication phenomenon in terms of the topology of the boundary component of elementary particle like 3-surface relates to this. The heaviness and instability of particles which correspond to genera $g > 2$ (sphere with more than two handles) might have simple explanation as absence of (stable) $D_{CP_2} = 3$ solutions of field equations with genus $g > 2$.

c) The solution ansatz need not be the most general. Kähler current is light-like and already this is enough to reduce the field equations to the form involving only energy momentum tensor. One might hope of finding also solution ansätze for which Kähler current is time-like or space-like. Space-likeness of the Kähler current might be achieved if the complex coordinates $(\xi, \bar{\xi})$ and hyper-complex coordinates (S^+, S^-) change the role. For this solution ansatz electric field would dominate. Note that the possibility that Kähler current is always light-like cannot be excluded.

d) Suppose that CP_2 projection quite generally defines a foliation of the space-time surface by light-like 3-surfaces, as is suggested by the conformal invariance. If the induced metric has Minkowskian signature, the fourth coordinate x^4 and thus also Kähler current must be time-like or light-like so that magnetic field dominates. Already the requirement that the metric is non-degenerate implies $g_{s4} \neq 0$ so that the metric for the $\xi = constant$ 2-surfaces has a Minkowskian signature. Thus space-like Kähler current does not allow the lift of the CP_2 projection to be light-like.

Are solutions with time-like or space-like Kähler current possible in $D_{CP_2} = 3$ case?

The following ansatz gives good hopes for obtaining solutions with space-like and time-like Kähler currents.

a) Assign to light-like coordinates coordinates (T, Z) by the formula $T = S^+ + S^-$ and $Z = S^+ - S^-$. Space-time coordinates are taken to be $(\xi, \bar{\xi}, s)$ and coordinate Z. The solution ansatz with time-like Kähler current results when the roles of T and Z are changed. It will however found that same solution ansatz can give rise to both space-like and time-like Kähler current.

b) The solution ansatz giving rise to a space-like Kähler current is defined by the equations

$$T = T(Z, s) \ , \quad w = w(\xi, s) \ . \tag{4.3.29}$$

If T depends strongly on Z, the g_{ZZ} component of the induced metric becomes positive and Kähler current time-like.

c) The components of the induced metric are

$$g_{ZZ} = m_{ZZ} + m_{TT}\partial_Z T\partial_s T \ , \quad g_{Zs} = m_{TT}\partial_Z T\partial_s T \ ,$$

$$g_{ss} = s_{ss} + m_{TT}\partial_s T\partial_s T \ , \qquad g_{w\bar{w}} = s_{w\bar{w}} + m_{w\bar{w}}\partial_\xi w\partial_{\bar{\xi}}\bar{w} \ , \tag{4.3.30}$$

$$g_{s\xi} = s_{s\xi} \ , \qquad\qquad g_{s\bar{\xi}} = s_{s\bar{\xi}} \ .$$

Topologized Kähler current has only Z-component and 3-dimensional empty space Maxwell's equations guarantee the topologization.

In CP_2 degrees of freedom the contractions of the energy momentum tensor with Christoffel symbols vanish if T^{ss}, $T^{\xi s}$ and $T^{\xi\xi}$ vanish as required by internal consistency. This is guaranteed if the condition

$$J^{\xi s} = 0 \tag{4.3.31}$$

holds true. Note however that $J^{\xi Z}$ is non-vanishing. Therefore only the components $T^{\xi\bar{\xi}}$ and $T^{Z\bar{\xi}}$, $T^{Z\bar{\xi}}$ of energy momentum tensor are non-vanishing, and field equations reduce to the conditions

$$\partial_{\bar{\xi}}(J^{\xi\bar{\xi}}\sqrt{g}) + \partial_Z(J^{\xi Z}\sqrt{g}) = 0 \ ,$$
$$\partial_{\xi}(J^{\bar{\xi}\xi}\sqrt{g}) + \partial_Z(J^{\bar{\xi}Z}\sqrt{g}) = 0 \ . \tag{4.3.32}$$

In the special case that the induce metric does not depend on z-coordinate equations reduce to holomorphicity conditions. This is achieve if T depends linearly on Z: $T = aZ$.

The contractions with M_+^4 Christoffel symbols come from the non-vanishing of $T^{Z\xi}$ and vanish if the Hamilton Jacobi structure satisfies the conditions

$$\{{}^k_{T\ w}\} = 0 \ , \quad \{{}^k_{T\ \overline{w}}\} = 0 \ ,$$
$$\{{}^k_{Z\ w}\} = 0 \ , \quad \{{}^k_{Z\ \overline{w}}\} = 0 \tag{4.3.33}$$

hold true. The conditions are equivalent with the conditions

$$\{{}^k_{\pm\ w}\} = 0 \ , \quad \{{}^k_{\pm\ \overline{w}}\} = 0 \ . \tag{4.3.34}$$

These conditions possess solutions (standard light cone coordinates are the simplest example). Also the second derivatives of $T(s, Z)$ contribute to the second fundamental form but they do not give rise to non-vanishing contractions with the energy momentum tensor. The cautious conclusion is that also solutions with time-like or space-like Kähler current are possible.

$D_{CP_2} = 4$ case

The preceding discussion was for $D_{CP_2} = 3$ and one should generalize the discussion to $D_{CP_2} = 4$ case.

a) Hamilton Jacobi structure for M_+^4 is expected to be crucial also now.

b) One might hope that for $D = 4$ the Kähler structure of CP_2 defines a foliation of CP_2 by 3-dimensional contact structures. This requires that there is a coordinate varying along the field lines of the normal vector field X defined as the dual of the three-form $A \wedge dA = A \wedge J$. By the previous considerations the condition for this reads as $dX = d(log\phi) \wedge X$ and implies $X \wedge dX = 0$. Using the self duality of the Kähler form one can express X as $X^k = J^{kl}A_l$. By a brief calculation one finds that $X \wedge dX \propto X$ holds true so that (somewhat disappointingly) a foliation of CP_2 by contact structures does not exist.

For $D_{CP_2} = 4$ case Kähler current vanishes and this case corresponds to what I have called earlier Maxwellian phase since empty space Maxwell's equations are indeed satisfied.

1. Solution ansatz with a 3-dimensional M_+^4 projection

The basic idea is that the complex structure of CP_2 is preserved so that one can use complex coordinates (ξ^1, ξ^2) for CP_2 in which CP_2 Christoffel symbols and energy momentum tensor have automatically the desired properties. This is achieved the second light like coordinate, say v, is non-dynamical so that the induced metric does not receive any contribution from the longitudinal degrees of freedom. In this case one has

$$S^+ = S^+(\xi^1, \xi^2) \ , \quad w = w(\xi^1, \xi^2) \ , \quad S^- = constant \ . \tag{4.3.35}$$

The induced metric does possesses only components of type $g_{i\bar{j}}$ if the conditions

$$g_{+w} = 0 \ , \quad g_{+\overline{w}} = 0 \ . \tag{4.3.36}$$

This guarantees that energy momentum tensor has only components of type $T^{i\bar{j}}$ in coordinates (ξ^1, ξ^2) and their contractions with the Christoffel symbols of CP_2 vanish identically. In M_+^4 degrees of freedom one must pose the conditions

$$\{^{\,k}_{w+}\} = 0 \ , \quad \{^{\,k}_{\overline{w}+}\} = 0 \ , \quad \{^{\,k}_{++}\} = 0 \ . \tag{4.3.37}$$

on Christoffel symbols. These conditions are satisfied if the M_+^4 metric does not depend on S^+:

$$\partial_+ m_{kl} \quad = \quad 0 \ . \tag{4.3.38}$$

This means that m_{-w} and $m_{-\overline{w}}$ can be non-vanishing but like m_{+-} they cannot depend on S^+.

The second derivatives of S^+ appearing in the second fundamental form are also a source of trouble unless they vanish. Hence S^+ must be a linear function of the coordinates ξ^k:

$$S^+ \quad = \quad a_k \xi^k + \overline{a}_k \overline{\xi}^k \ . \tag{4.3.39}$$

Field equations are the counterparts of empty space Maxwell equations $j^\alpha = 0$ but with M_+^4 coordinates (u, w) appearing as dynamical variables and entering only through the induced metric. By holomorphy the field equations can be written as

$$\partial_j (J^{j\bar{i}} \sqrt{g}) = 0 \ , \quad \partial_{\bar{j}} (J^{\bar{j}i} \sqrt{g}) = 0 \ , \tag{4.3.40}$$

and can be interpreted as conditions stating the holomorphy of the contravariant Kähler form.

What is remarkable is that the M_+^4 projection of the solution is 3-dimensional light like surface and that the induced metric has Euclidian signature. Light front would become a concrete geometric object with one compactified dimension rather than being a mere conceptualization. One could see this as topological quantization for the notion of light front or of electromagnetic shock wave, or perhaps even as the realization of the particle aspect of gauge fields at classical level.

If the latter interpretation is correct, quantum classical correspondence would be realized very concretely. Wave and particle aspects would both be present. One could understand the interactions of charged particles with electromagnetic fields both in terms of absorption and emission of topological field quanta and in terms of the interaction with a classical field as particle topologically condenses at the photonic light front.

For CP_2 type extremals for which M_+^4 projection is a light like curve correspond to a special case of this solution ansatz: transversal M_+^4 coordinates are constant and S^+ is now arbitrary function of CP_2 coordinates. This is possible since M_+^4 projection is 1-dimensional.

2. Are solutions with a 4-dimensional M_+^4 projection possible?

The most natural solution ansatz is the one for which CP_2 complex structure is preserved so that energy momentum tensor has desired properties. For four-dimensional M_+^4 projection this ansatz does not seem to make promising since the contribution of the longitudinal degrees of freedom implies that the induced metric is not anymore of desired form since the components $g_{ij} = m_{+-}(\partial_{\xi^i} S^+ \partial_{\xi^j} S^- + m_{+-} \partial_{\xi^i} S^- \partial_{\xi^j} S^+)$ are non-vanishing.

a) The natural dynamical variables are still Minkowski coordinates $(w, \overline{w}, S^+, S^-)$ for some Hamilton Jacobi structure. Since the complex structure of CP_2 must be given up, CP_2 coordinates can be written as (ξ, s, r) to stress the fact that only "one half" of the Kähler structure of CP_2 is respected by the solution ansatz.

b) The solution ansatz has the same general form as in $D = 3$ case and must be symmetric with respect to the exchange of M_+^4 and CP_2 coordinates. Transverse coordinates are mapped to transverse ones and longitudinal coordinates to longitudinal ones:

$$(S^+, S^-) = (S^+(s,r), S^-(s,r)) \quad , \quad w = w(\xi) \ . \tag{4.3.41}$$

This ansatz would describe ordinary Maxwell field in M^4_+ since the roles of M^4_+ coordinates and CP_2 coordinates are interchangeable.

It is however far from obvious whether there are any solutions with a 4-dimensional M^4_+ projection. That empty space Maxwell's equations would allow only the topologically quantized light fronts as its solutions would realize quantum classical correspondence very concretely.

$D_{CP_2} = 2$ case

Hamilton Jacobi structure for M^4_+ is assumed also for $D_{CP_2} = 2$, whereas the contact structure for CP_2 is in $D = 2$ case replaced by the induced Kähler structure. Topologization yields vanishing Kähler current. Light-likeness provides a second manner to achieve vanishing Lorentz force but one cannot exclude the possibility of time- and space-like Kähler current.

1. Solutions with vanishing Kähler current

a) String like objects, which are products $X^2 \times Y^2 \subset M^4_+ \times CP_2$ of minimal surfaces Y^2 of M^4_+ with geodesic spheres S^2 of CP_2 and carry vanishing gauge current. String like objects allow considerable generalization from simple Cartesian products of $X^2 \times Y^2 \subset M^4 \times S^2$. Let $(w, \overline{w}, S^+, S^-)$ define the Hamilton Jacobi structure for M^4_+. $w = constant$ surfaces define minimal surfaces X^2 of M^4_+. Let ξ denote complex coordinate for a sub-manifold of CP_2 such that the imbedding to CP_2 is holomorphic: $(\xi^1, \xi^2) = (f^1(\xi), f^2(\xi))$. The resulting surface $Y^2 \subset CP_2$ is a minimal surface and field equations reduce to the requirement that the Kähler current vanishes: $\partial_{\overline{\xi}}(J^{\xi \overline{\xi}} \sqrt{g_2}) = 0$. One-dimensional strings are deformed to 3-dimensional cylinders representing magnetic flux tubes. The oscillations of string correspond to waves moving along string with light velocity, and for more general solutions they become TGD counterparts of Alfwen waves associated with magnetic flux tubes regarded as oscillations of magnetic flux lines behaving effectively like strings. It must be emphasized that Alfwen waves are a phenomenological notion not really justified by the properties of Maxwell's equations.

b) Also electret type solutions with the role of the magnetic field taken by the electric field are possible. $(\xi, \overline{\xi}, u, v)$ would provide the natural coordinates and the solution ansatz would be of the form

$$(s, r) = (s(u,v), r(u,v)) \quad , \quad \xi = constant \ , \tag{4.3.42}$$

and corresponds to a vanishing Kähler current.

c) Both magnetic and electric fields are necessarily present only for the solutions carrying non-vanishing electric charge density (proportional to $\overline{B} \cdot \overline{A}$). Thus one can ask whether more general solutions carrying both magnetic and electric field are possible. As a matter fact, one must first answer the question what one really means with the magnetic field. By choosing the coordinates of 2-dimensional CP_2 projection as space-time coordinates one can define what one means with magnetic and electric field in a coordinate invariant manner. Since the CP_2 Kähler form for the CP_2 projection with $D_{CP_2} = 2$ can be regarded as a pure Kähler magnetic field, the induced Kähler field is either magnetic field or electric field.

The form of the ansatz would be

$$(s, r) = (s, r)(u, v, w, \overline{w}) \quad , \quad \xi = constant \ . \tag{4.3.43}$$

As a matter fact, CP_2 coordinates depend on two properly chosen M^4 coordinates only.

1. Solutions with light-like Kähler current

There are large classes of solutions of field equations with a light-like Kähler current and 2-dimensional CP_2 projection.

a) Massless extremals for which CP_2 coordinates are arbitrary functions of one transversal coordinate $e = f(w, \overline{w})$ defining local polarization direction and light like coordinate u of M^4_+ and carrying in the general case a light like current. In this case the holomorphy does not play any role.

b) The string like solutions thickened to magnetic flux tubes carrying TGD counterparts of Alfwen waves generalize to solutions allowing also light-like Kähler current. Also now Kähler metric is allowed to develop a component between longitudinal and transversal degrees of freedom so that Kähler current develops a light-like component. The ansatz is of the form

$$\xi^i = f^i(\xi) \ , \ \ w = w(\xi) \ , \ \ S^- = s^- \ , \ \ S^+ = s^+ + f(\xi,\overline{\xi}) \ .$$

Only the components $g_{+\xi}$ and $g_{+\overline{\xi}}$ of the induced metric receive contributions from the modification of the solution ansatz. The contravariant metric receives contributions to $g^{-\xi}$ and $g^{-\overline{\xi}}$ whereas $g^{+\xi}$ and $g^{+\overline{\xi}}$ remain zero. Since the partial derivatives $\partial_\xi\partial_+ h^k$ and $\partial_{\overline{\xi}}\partial_+ h^k$ and corresponding projections of Christoffel symbols vanish, field equations are satisfied. Kähler current develops a non-vanishing component j^-. Apart from the presence of the electric field, these solutions are highly analogous to Beltrami fields.

3. Do scalar wave pulses represent a solution type with non-vanishing but not light-like Kähler current?

Since longitudinal polarizations are possible only for off mass shell virtual photons, physical intuition suggests that scalar wave pulse solutions describing the propagation of longitudinal electric field with light velocity cannot appear as asymptotic field patterns. This is also consistent with the claim that scalar wave pulses are associated with the transients involved with sudden switching of electric voltage on or off. Let $M^4 = M^2 \oplus E^2$ be the standard decomposition of M^4 to flat longitudinal and transversal spaces, and S^2 a homologically non-trivial geodesic sphere of CP_2. The simplest solution ansatz corresponds to a surface $X^2 \times Y^2$, $X^2 \subset E^2$, such that Y^2 is a surface defined by a map $S^2 \to M^2$ (or vice versa).

Energy momentum tensor is in both longitudinal and transversal degrees of freedom proportional to the corresponding part of the induced metric. Field equations are trivially true in the transversal degrees of freedom. The calculation of the divergence of energy momentum tensor demonstrates that Kähler current can be regarded as a vector field

$$j^\alpha = \frac{1}{4}J^{\alpha\beta}\partial_\beta L$$

defined by the Kähler action density acting as Hamiltonian. Poisson bracket is defined by the pseudo-symplectic form associated with the induced Kähler form with respect to the induced metric rather that that of S^2 (using S^2-coordinates as coordinates for Y^2, the square of this pseudo-symplectic form is equal to metric multiplied by the ratio $det(g(Y^2))/det(g(S^2))$).

In longitudinal degrees of freedom field equations are minimal surface equations with a source term proportional to the Kähler current divided by the Kähler action density. The vanishing of the Kähler current is possible only if Kähler action density is constant. This condition is true in the approximation that the induced metric for Y^2 is flat, that is at the limit when M^4 projection has size larger than size of CP_2 projection and that induced metric has Minkowskian signature). It is not clear whether the minimal surface property of Y^2 in $M^2 \times S^2$ is consistent with the constancy of the Kähler action density. This would suggest that classical gravitational interactions eliminate scalar wave pulses as asymptotic field patterns and cause the deviation from the minimal surface property and the non-vanishing of the Kähler current. The fact that solution becomes "instanton" like Euclidian solution when S^+ and S^- become constant suggests that the M^4 projection of the solution quite generally has a finite extension in time direction.

Could $D_{CP_2} = 2 \to 3$ transition occur in rotating magnetic systems?

I have studied the imbeddings of simple cylindrical and helical magnetic fields in various applications of TGD to condensed matter systems, in particular in attempts to understand the strange findings about rotating magnetic systems [G2].

Let S^2 be the homologically non-trivial geodesic sphere of CP_2 with standard spherical coordinates $(U \equiv cos(\theta), \Phi)$ and let (t, ρ, ϕ, z) denote cylindrical coordinates for a cylindrical space-time sheet. The simplest possible space-time surfaces $X^4 \subset M^4_+ \times S^2$ carrying helical Kähler magnetic field depending on the radial cylindrical coordinate ρ, are given by:

$$U = U(\rho) \ , \qquad \Phi = n\phi + kz \ ,$$
$$J_{\rho\phi} = n\partial_\rho U \ , \quad J_{\rho z} = k\partial_\rho U \ . \tag{4.3.44}$$

This helical field is not Beltrami field as one can easily find. A more general ansatz corresponding defined by

$$\Phi = \omega t + kz + n\phi$$

would in cylindrical coordinates give rise to both helical magnetic field and radial electric field depending on ρ only. This field can be obtained by simply replacing the vector potential with its rotated version and provides the natural first approximation for the fields associated with rotating magnetic systems.

A non-vanishing vacuum charge density is however generated when a constant magnetic field is put into rotation and is implied by the condition $\overline{E} = \overline{v} \times \overline{B}$ stating vanishing of the Lorentz force. This condition does not follow from the induction law of Faraday although Faraday observed this effect first. This is also clear from the fact that the sign of the charge density depends on the direction of rotation.

The non-vanishing charge density is not consistent with the vanishing of the Kähler 4-current and requires a 3-dimensional CP_2 projection and topologization of the Kähler current. Beltrami condition cannot hold true exactly for the rotating system. The conclusion is that rotation induces a phase transition $D_{CP_2} = 2 \to 3$. This could help to understand various strange effects related to the rotating magnetic systems [G2]. For instance, the increase of the dimension of CP_2 projection could generate join along boundaries contacts and wormhole contacts leading to the transfer of charge between different space-time sheets. The possibly resulting flow of gravitational flux to larger space-time sheets might help to explain the claimed antigravity effects.

4.3.6 $D = 3$ phase allows infinite number of topological charges characterizing the linking of magnetic field lines

When space-time sheet possesses a $D = 3$-dimensional CP_2 projection, one can assign to it a non-vanishing and conserved topological charge characterizing the linking of the magnetic field lines defined by Chern-Simons action density $A \wedge dA/4\pi$ for induced Kähler form. This charge can be seen as classical topological invariant of the linked structure formed by magnetic field lines.

The topological charge can also vanish for $D = 3$ space-time sheets. In Darboux coordinates for which Kähler gauge potential reads as $A = P_k dQ^k$, the surfaces of this kind result if one has $Q^2 = f(Q^1)$ implying $A = fdQ^1$, $f = P_1 + P_2 \partial_{Q_1} Q^2$, which implies the condition $A \wedge dA = 0$. For these space-time sheets one can introduce Q^1 as a global coordinate along field lines of A and define the phase factor $exp(i \int A_\mu dx^\mu)$ as a wave function defined for the entire space-time sheet. This function could be interpreted as a phase of an order order parameter of super-conductor like state and there is a high temptation to assume that quantum coherence in this sense is lost for more general $D = 3$ solutions.

Chern-Simons action is known as helicity in electrodynamics [ab6]. Helicity indeed describes the linking of magnetic flux lines as is easy to see by interpreting magnetic field as incompressible fluid flow having A as vector potential: $B = \nabla \times A$. One can write A using the inverse of $\nabla \times$ as $A = (1/\nabla \times)B$. The inverse is non-local operator expressible as

$$\frac{1}{\nabla \times} B(r) = \int dV' \frac{(r - r')}{|r - r'|^3} \times B(r') \ ,$$

as a little calculation shows. This allows to write $\int A \cdot B$ as

$$\int dV A \cdot B = \int dV dV' B(r) \cdot \left(\frac{(r - r')}{|r - r'|^3} \times B(r') \right) \ ,$$

which is completely analogous to the Gauss formula for linking number when linked curves are replaced by a distribution of linked curves and an average is taken.

For $D = 3$ field equations imply that Kähler current is proportional to the helicity current by a factor which depends on CP_2 coordinates, which implies that the current is automatically divergence free and defines a conserved charge for $D = 3$-dimensional CP_2 projection for which the instanton density vanishes identically. Kähler charge is not equal to the helicity defined by the inner product of magnetic field and vector potential but to a more general topological charge.

The number of conserved topological charges is infinite since the product of any function of CP_2 coordinates with the helicity current has vanishing divergence and defines a topological charge. A very natural function basis is provided by the scalar spherical harmonics of $SU(3)$ defining Hamiltonians of CP_2 canonical transformations and possessing well defined color quantum numbers. These functions define and infinite number of conserved charges which are also classical knot invariants in the sense that they are not affected at all when the 3-surface interpreted as a map from CP_2 projection to M_+^4 is deformed in M_+^4 degrees of freedom. Also canonical transformations induced by Hamiltonians in irreducible representations of color group affect these invariants via Poisson bracket action when the $U(1)$ gauge transformation induced by the canonical transformation corresponds to a single valued scalar function. These link invariants are additive in union whereas the quantum invariants defined by topological quantum field theories are multiplicative.

Also non-Abelian topological charges are well-defined. One can generalize the topological current associated with the Kähler form to a corresponding current associated with the induced electro-weak gauge fields whereas for classical color gauge fields the Chern-Simons form vanishes identically. Also in this case one can multiply the current by CP_2 color harmonics to obtain an infinite number of invariants in $D = 3$ case. The only difference is that $A \wedge dA$ is replaced by $Tr(A \wedge (dA + 2A \wedge A/3))$.

There is a strong temptation to assume that these conserved charges characterize colored quantum states of the conformally invariant quantum theory as a functional of the light-like 3-surface defining boundary of space-time sheet or elementary particle horizon surrounding wormhole contacts. They would be TGD analogs of the states of the topological quantum field theory defined by Chern-Simons action as highest weight states associated with corresponding Wess-Zumino-Witten theory. These charges could be interpreted as topological counterparts of the isometry charges of configuration space of 3-surfaces defined by the algebra of canonical transformations of CP_2.

The interpretation of these charges as contributions of light-like boundaries to configuration space Hamiltonians would be natural. The dynamics of the induced second quantized spinor fields relates to that of Kähler action by a super-symmetry, so that it should define super-symmetric counterparts of these knot invariants. The anti-commutators of these super charges cannot however contribute to configuration space Kähler metric so that topological zero modes are in question. These Hamiltonians and their super-charge counterparts would be responsible for the topological sector of quantum TGD.

4.3.7 Is the principle selecting preferred extremals of Kähler action equivalent with the topologization/light-likeness of Kähler current and with second law?

The basic question is whether the Kähler current is either topologized or light-like for all extremals or only for the preferred extremals of Kähler action analogous to Bohr orbits in some sense, presumably asymptotically as suggested by the fact that generalized Beltrami fields correspond to asymptotic self-organization patterns, when dissipation has become insignificant.

a) The generalized Beltrami conditions or light-likeness can hold true only asymptotically. First of all, generic non-asymptotic field configurations have $D_{CP_2} = 4$, and would thus carry a vanishing Kähler four-current if Beltrami conditions were satisfied universally rather than only asymptotically. $j^\alpha = 0$ would obviously hold true also for the asymptotic configurations, in particular those with $D_{CP_2} < 4$ so that empty space Maxwell's field equations would be universally satisfied for asymptotic field configurations with $D_{CP_2} < 4$.

b) The failure of the generalized Beltrami conditions would mean that Kähler field is completely analogous to a dissipative Maxwell field since $\overline{j} \cdot \overline{E}$ is non-vanishing (note that isometry currents are conserved although energy momentum tensor is not). Quantum classical correspondence states that classical space-time dynamics is by its classical non-determinism able to mimic the non-deterministic sequence of quantum jumps at space-time level, in particular dissipation in various length scales defined by the hierarchy of space-time sheets. Classical fields could represent "symbolically" the average dynamics, in particular dissipation, in shorter length scales. For instance, vacuum 4-current would be a symbolic representation for the average of the currents consisting of elementary particles.

The obvious objection to the idea is that second law realized as an asymptotic vanishing of Lorentz-

Kähler force implies that all 3-surfaces approaching same asymptotic state have the same value of Kähler function. This is actually not a problem since it means an additional symmetry extending general coordinate invariance. The exponent of Kähler function would be highly analogous to a partition function defined as an exponent of Hamiltonian with Kähler coupling strength playing the role of temperature. The principle selecting preferred extremals of Kähler action would guarantee that the predicted both signs of Kähler coupling strength and depending on sign Kähler magnetic or electric field configurations are stable. The two phases would relate by time reversal to each other. The role of time reversed dissipation manifesting itself as processes like self assembly would involve in essential manner the time reversed negative energy world. The fact that Kähler coupling strength is of opposite sign for the time reversed dynamics is essential for internal consistency. For instance, creation of matter from vacuum in big bang can be seen as time reflection of stable negative energy cosmic strings as positive energy cosmic strings unstable against decay to magnetic flux tubes.

Is the principle selecting preferred extremals equivalent with generalized Beltrami conditions?

Previous findings inspire the hypothesis that generalized Beltrami conditions express algebraically the conditions selecting preferred extremal of Kähler action so that they make sense also in the p-adic case.

a) Generalized Beltrami conditions are satisfied by the asymptotic field configurations representing self-organization patterns. For non-asymptotic fields vacuum Lorentz force is non-vanishing and does work in Maxwellian sense so that $\overline{j} \cdot \overline{E}$ is non-vanishing. This would mean that the dynamics defined by Kähler action could in principle predict even the values of the parameters related to dissipation such as conductivities and viscosities. The space-time sheets of the many-sheeted space-time would be busily modelling its own physics in shorter length scales.

b) Absolute minimization of Kähler action implies that single space-time surface goes through given 3-surface apart from the non-uniqueness caused by the non-determinism of Kähler action. This gives *four* additional local conditions to the initial values of field equations fixing the time derivatives of the four dynamical imbedding space coordinates (conditions are analogous to Bohr conditions).

The topologization of the Kähler current current gives also *four* local conditions:
i) For $D_{CP_2} < 4$ the vanishing of instanton density gives one condition, and the proportionality of the Kähler current to instanton current gives 3 conditions since the proportionality factor is an arbitrary function of CP_2 coordinates. Altogether this makes four conditions.
ii) For $D_{CP_2=4}$ the vanishing of the Kähler current gives four conditions.
This encourages to think that the principle selecting preferred extremals of Kähler action forces the asymptotic behavior (final values instead of initial values) to correspond to dissipation-less state characterized by the generalized Beltrami conditions.

c) Absolute minimization in a strict sense of the word does not make sense in the p-adic context since p-adic numbers are not well-ordered, and one cannot even define the action integral as a p-adic number except perhaps by algebraic continuation procedure described in the first part of the book. The generalized Beltrami conditions are however purely algebraic conditions and make sense also in the p-adic context. Therefore it possible to give a precise content to the notion of the principle selecting preferred extremals also in p-adic context.

Is the principle selecting preferred extremals equivalent with the second law?

The fact that Beltrami conditions are associated with the asymptotic dynamics suggests that absolute minimization (or its variant discussed in [E2]) is equivalent with the second law at space-time level. Or putting it more cautiously: second law at space-time level could be equivalent with the principle selecting preferred extremals of Kähler action. If not, one must give up the principle selecting preferred extremals and replace it with the second law.

For space-time sheets with negative time orientation and negative energy, say "massless extremals" representing phase conjugate laser waves, field configurations would approach non-dissipating ones in the geometric past, and the arrow of geometric time would be opposite to the standard one in this case. This situation is possible for space-time sheets of finite duration, in particular virtual particle like space-time sheets or the negative energy space-time sheets extending down to the boundary of imbedding space (moment of "big bang"). This would explain at the space-time level the change of arrow of time and breaking of the second law observed for the phase conjugate laser waves (used to generate healing and error correction for instance). In TGD framework second law is not a producer of a thermal chaos but Darwinian selector since state function reduction and state preparation by self measurements lead from a state with positive entanglement entropy to that with a negative entanglement entropy (defined number theoretically), and possessing only finitely extended rational entanglement identifiable as a bound state entanglement.

Absolute minimization of Kähler action indeed induces long range correlations since the positive Kähler action of space-time sheets carrying magnetic fields must be compensated by the negative Kähler action of space-time sheets dominated by Kähler electric fields. The resulting non-local long range correlations could serve as correlates for bound state entanglement. More concretely, the stable join along boundaries bonds would be the correlates for bound state entanglement whereas topological light rays analogous to the exchange of virtual photons could serve as classical correlates for unbound entanglement. The closedness (periodicity) of the field lines of Beltrami fields for space-like Kähler current and periodicity of the field pattern for the time like Kähler current could be space-time correlates for the rational entanglement. The pinary expansions of rational numbers which are periodic after finite number of pinary digits indeed represent closed orbits in the set of integers modulo p. Amusingly, the first non-periodic pits of the expansion would in fact be analogous to the dissipative period.

Macro-temporal quantum coherence integrates sequences of quantum jumps to single effective quantum jump so that effectively a fractal hierarchy of quantum jumps emerges having the fractal hierarchy of time scales of dissipation resulting from many-sheetedness as a correlate. Even the anatomy of quantum jump could have space-time correlate. The final state of the quantum jump would correspond to highly negentropic and non-dissipating topologically quantized generalized Beltrami fields. State function reduction and preparation would correspond to the non-deterministic dissipative approach to the non-dissipative Beltrami field configuration. The points of space-time sheets with vanishing Kähler 4-currents would be unstable against quantum jumps generating an instability of the Beltrami field leading to a field configuration with a non-vanishing Lorentz 4-force and emission of topological light rays representing unstable entanglement. Quantum jump would have this kind of instability as a natural space-time correlate.

To sum up, the main lessons would be following.

a) The ability of basically non-dissipative dynamics to mimic dissipative dynamics in terms of energy momentum tensor would be the basic reason for why space-times must be 4-surfaces.

b) If absolute minimization or its variant discussed in [E2] of the Kähler action is correct principle, it must correspond to the second law, which is the Darwinian selector of the most information rich patterns rather than a thermal killer.

4.3.8   Generalized Beltrami fields and biological systems

The following arguments support the view that generalized Beltrami fields play a key role in living systems, and that $D_{CP_2} = 2$ corresponds to ordered phase, $D_{CP_2} = 3$ to spin glass phase and $D_{CP_2} = 4$ to chaos, with $D_{CP_2} = 3$ defining life as a phenomenon at the boundary between order and chaos.

Why generalized Beltrami fields are important for living systems?

Chirality, complexity, and high level of organization make $D_{CP_2} = 3$ generalized Beltrami fields excellent candidates for the magnetic bodies of living systems.

a) Chirality selection is one of the basic signatures of living systems. Beltrami field is characterized by a chirality defined by the relative sign of the current and magnetic field, which means parity breaking. Chirality reduces to the sign of the function ψ appearing in the topologization condition and makes sense also for the generalized Beltrami fields.

b) Although Beltrami fields can be extremely complex, they are also extremely organized. The reason is that the function α is constant along flux lines so that flux lines must in the case of compact Riemann 3-manifold belong to 2-dimensional $\alpha = constant$ closed surfaces, in fact two-dimensional invariant tori [ab3].

For generalized Beltrami fields the function ψ is constant along the flow lines of the Kähler current. Space-time sheets with 3-dimensional CP_2 projection serve as an illustrative example. One can use the coordinates for the CP_2 projection as space-time coordinates so that one space-time coordinate disappears totally from consideration. Hence the situation reduces to a flow in a 3-dimensional sub-manifold of CP_2. One can distinguish between three types of flow lines corresponding to space-like, light-like and time-like topological current. The 2-dimensional $\psi = constant$ invariant manifolds are sub-manifolds of CP_2. Ordinary Beltrami fields are a special case of space-like flow with flow lines belonging to the 2-dimensional invariant tori of CP_2. Time-like and light-like situations are more complex since the flow lines need not be closed so that the 2-dimensional $\psi = constant$ surfaces can have boundaries.

For periodic self-organization patterns flow lines are closed and $\psi = constant$ surfaces of CP_2 must be invariant tori. The dynamics of the periodic flow is obtained from that of a steady flow by replacing one spatial coordinate with effectively periodic time coordinate. Therefore topological notions like helix structure, linking, and knotting have a dynamical meaning at the level of CP_2 projection. The periodic generalized Beltrami fields are highly organized also in the temporal domain despite the potentiality for extreme topological complexity.

For these reasons topologically quantized generalized Beltrami fields provide an excellent candidate for a generic model for the dynamics of biological self-organization patterns. A natural guess is that many-sheeted magnetic and Z^0 magnetic fields and their generalizations serve as templates for the helical molecules populating living matter, and explain both chiral selection, the complex linking and knotting of DNA and protein molecules, and even the extremely complex and self-organized dynamics of biological systems at the molecular level.

The intricate topological structures of DNA, RNA, and protein molecules are known to have a deep significance besides their chemical structure, and they could even define something analogous to the genetic code. Usually the topology and geometry of bio-molecules is believed to reduce to chemistry. TGD suggests that space-like generalized Beltrami fields serve as templates for the formation of bio-molecules and bio-structures in general. The dynamics of bio-systems would in turn utilize the time-like Beltrami fields as templates. There could even exist a mapping from the topology of magnetic flux tube structures serving as templates for bio-molecules to the templates of self-organized dynamics. The helical structures, knotting, and linking of bio-molecules would thus define a symbolic representation, and even coding for the dynamics of the bio-system analogous to written language.

$D_{CP_2} = 3$ systems as boundary between $D_{CP_2} = 2$ order and $D_{CP_2} = 4$ chaos

The dimension of CP_2 projection is basic classifier for the asymptotic self-organization patterns.

1. $D_{CP_2} = 4$ phase, dead matter, and chaos

$D_{CP_2} = 4$ corresponds to the ordinary Maxwellian phase in which Kähler current and charge density vanish and there is no topologization of Kähler current. By its maximal dimension this phase would naturally correspond to disordered phase, ordinary dead matter. If one assumes that Kähler charge corresponds to either em charge or Z^0 charge then the signature of this state of matter would be em neutrality or Z^0 neutrality.

2. $D_{CP_2} = 2$ phase as ordered phase

By the low dimension of CP_2 projection $D_{CP_2} = 2$ phase is the least stable phase possible only at cold space-time sheets. Kähler current is either vanishing or light-like, and Beltrami fields are not possible. This phase is highly ordered and much like a topological quantized version of ferro-magnet. In particular, it is possible to have a global coordinate varying along the field lines of the vector potential also now. The magnetic and Z^0 magnetic body of any system is a candidate for this kind of system. Z^0 field is indeed always present for vacuum extremals having $D = 2$ and the vanishing of em field requires that that $sin^2(\theta_W)$ (θ_W is Weinberg angle) vanishes.

3. $D_{CP_2} = 3$ corresponds to living matter

$D_{CP_2} = 3$ corresponds to highly organized phase characterized in the case of space-like Kähler current by complex helical structures necessarily accompanied by topologized Kähler charge density $\propto \overline{A} \cdot \overline{B} \neq 0$ and Kähler current $\overline{E} \times \overline{A} + \phi\overline{B}$. For time like Kähler currents the helical structures are replaced by periodic oscillation patterns for the state of the system. By the non-maximal dimension of CP_2 projection this phase must be unstable against too strong external perturbations and cannot survive at too high temperatures. Living matter is thus excellent candidate for this phase and it might be that the interaction of the magnetic body with living matter makes possible the transition from $D_{CP_2} = 2$ phase to the self-organizing $D_{CP_2} = 3$ phase.

Living matter which is indeed populated by helical structures providing examples of space-like Kähler current. Strongly charged lipid layers of cell membrane might provide example of time-like Kähler current. Cell membrane, micro-tubuli, DNA, and proteins are known to be electrically charged and Z^0 charge plays key role in TGD based model of catalysis discussed in [N4]. For instance, denaturing of DNA destroying its helical structure could be interpreted as a transition leading from $D = 3$ phase to $D = 4$ phase. The prediction is that the denatured phase should be electromagnetically (or Z^0) neutral.

Beltrami fields result when Kähler charge density vanishes. For these configurations magnetic field and current density take the role of the vector potential and magnetic field as far as the contact structure is considered. For Beltrami fields there exist a global coordinate along the field lines of the vector potential but not along those of the magnetic field. As a consequence, the covariant consistency condition $(\partial_s - qeA_s)\Psi = 0$ frequently appearing in the physics of super conducting systems would make sense along the flow lines of the vector potential for the order parameter of Bose-Einstein condensate. If Beltrami phase is super-conducting, then the state of the system must change in the transition to a more general phase. Since the field lines of the vector potential define chaotic orbits in this phase, the loss of coherence of the order parameter implying the loss of superconductivity by random collisions of particles is what one expects to happen.

The existence of these three phases brings in mind systems allowing chaotic de-magnetized phase above critical temperature T_c, spin glass phase at the critical point, and ferromagnetic phase below T_c. Similar analogy is provided by liquid phase, liquid crystal phase possible in the vicinity of the critical point for liquid to solid transition, and solid phase. Perhaps one could regard $D_{CP_2} = 3$ phase and life as a boundary region between $D_{CP_2} = 2$ order and $D_{CP_2} = 4$ chaos. This would naturally explain why life as it is known is possible in relatively narrow temperature interval.

4.4   Basic extremals of Kähler action

The solutions of field equations can be divided to vacuum extremals and non-vacuum extremal. Vacuum extremals come as two basic types: CP_2 type vacuum extremals for which the induced Kähler field and Kähler action are non-vanishing and the extremals for which the induced Kähler field vanishes. The deformations of both extremals are expected to be of fundamental importance in TGD universe.

4.4.1 CP_2 type vacuum extremals

These extremals correspond to various isometric imbeddings of CP_2 to $M_+^4 \times CP_2$. One can also drill holes to CP_2. Using the coordinates of CP_2 as coordinates for X^4 the imbedding is given by the formula

$$
\begin{aligned}
m^k &= m^k(u) \ , \\
m_{kl}\dot{m}^k\dot{m}^l &= 0 \ ,
\end{aligned}
\tag{4.4.1}
$$

where $u(s^k)$ is an arbitrary function of CP_2 coordinates. The latter condition tells that the curve representing the projection of X^4 to M^4 is light like curve. One can choose the functions $m^i, i = 1, 2, 3$ freely and solve m^0 from the condition expressing light likeness so that the number of this kind of extremals is very large.

The induced metric and Kähler field are just those of CP_2 and energy momentum tensor $T^{\alpha\beta}$ vanishes identically by the self duality of the Kähler form of CP_2. Also the canonical current $j^\alpha = D_\beta J^{\alpha\beta}$ associated with the Kähler form vanishes identically. Therefore the field equations in the interior of X^4 are satisfied. The field equations are also satisfied on the boundary components of CP_2 type extremal because the non-vanishing boundary term is, besides the normal component of Kähler electric field, also proportional to the projection operator to the normal space and vanishes identically since the induced metric and Kähler form are identical with the metric and Kähler form of CP_2.

As a special case one obtains solutions for which M^4 projection is light like geodesic. The projection of $m^0 = constant$ surfaces to CP_2 is $u = constant$ 3-sub-manifold of CP_2. Geometrically these solutions correspond to a propagation of a massless particle. In a more general case the interpretation as an orbit of a massless particle is not the only possibility. For example, one can imagine a situation, where the center of mass of the particle is at rest and motion occurs along a circle at say (m^1, m^2) plane. The interpretation as a massive particle is natural. Amusingly, there is nice analogy with the classical theory of Dirac electron: massive Dirac fermion moves also with the velocity of light (zitterbewegung). The quantization of this random motion with light velocity leads to Virasoro conditions and this led to a breakthrough in the understanding of the symmetries of TGD. Super Virasoro invariance is a general symmetry of the configuration space geometry and quantum TGD.

The action for all extremals is same and given by the Kähler action for the imbedding of CP_2. The value of the action is given by

$$S = -\frac{\pi}{8\alpha_K} . \tag{4.4.2}$$

To derive this expression we have used the result that the value of Lagrangian is constant: $L = 4/R^4$, the volume of CP_2 is $V(CP_2) = \pi^2 R^4/2$ and the definition of the Kähler coupling strength $k_1 = 1/16\pi\alpha_K$ (by definition, πR is the length of CP_2 geodesics). Four-momentum vanishes for these extremals so that they can be regarded as vacuum extremals. The value of the action is negative so that these vacuum extremals are indeed favored by the minimization of the Kähler action. The the principle selecting preferred extremals of Kähler action suggests that ordinary vacuums with vanishing Kähler action density are unstable against the generation of CP_2 type extremals. There are even reasons to expect that CP_2 type extremals are for TGD what black holes are for GRT. Indeed, the nice generalization of the area law for the entropy of black hole [E5] supports this view.

In accordance with the basic ideas of TGD topologically condensed vacuum extremals should somehow correspond to massive particles. The properties of the CP_2 type vacuum extremals are in accordance with this interpretation. Although these objects move with a velocity of light, the motion can be transformed to a mere so that the center of mass motion is trivial. Even the generation of the rest mass could might be understood classically as a consequence of the minimization of action. Long range Kähler fields generate negative action for the topologically condensed vacuum extremal (momentum zero massless particle) and Kähler field energy in turn is identifiable as the rest mass of the topologically condensed particle.

An interesting feature of these objects is that they can be regarded as gravitational instantons [bc1]. A further interesting feature of CP_2 type extremals is that they carry nontrivial classical color charges. The possible relationship of this feature to color confinement raises interesting questions. Could one model classically the formation of the color singlets to take place through the emission of "colorons": states with zero momentum but non-vanishing color? Could these peculiar states reflect the infrared properties of the color interactions?

4.4.2 Vacuum extremals with vanishing Kähler field

Vacuum extremals correspond to 4-surfaces with vanishing Kähler field and therefore to gauge field zero configurations of gauge field theory. These surfaces have CP_2 projection, which is Lagrange manifold. The condition expressing Lagrange manifold property is obtained in the following manner. Kähler potential of CP_2 can be expressed in terms of the canonical coordinates (P_i, Q_i) for CP_2 as

$$A = \sum_k P_k dQ^k . \tag{4.4.3}$$

The conditions

$$P_k = \partial_{Q^k} f(Q^i) , \tag{4.4.4}$$

where $f(Q^i)$ is arbitrary function of its arguments, guarantee that Kähler potential is pure gauge. It is clear that canonical transformations, which act as local $U(1)$ gauge transformations, transform different vacuum configurations to each other so that vacuum degeneracy is enormous. Also M_+^4 diffeomorphisms act as the dynamical symmetries of the vacuum extremals. Some sub-group of these symmetries extends to the isometry group of the configuration space in the proposed construction of the configuration space metric. The vacuum degeneracy is still enhanced by the fact that the topology of the four-surface is practically free.

Vacuum extremals are certainly not absolute minima of the action. For the induced metric having Minkowski signature the generation of Kähler electric fields lowers the action. For Euclidian signature both electric and magnetic fields tend to reduce the action. Therefore the generation of Euclidian regions of space-time is expected to occur. CP_2 type extremals, identifiable as real (as contrast to virtual) elementary particles, can be indeed regarded as these Euclidian regions.

Particle like vacuum extremals can be classified roughly by the number of the compactified dimensions D having size given by CP_2 length. Thus one has $D = 3$ for CP_2 type extremals, $D = 2$ for string like objects, $D = 1$ for membranes and $D = 0$ for pieces of M^4. As already mentioned, the rule $h_{vac} = -D$ relating the vacuum weight of the Super Virasoro representation to the number of compactified dimensions of the vacuum extremal is very suggestive. $D < 3$ vacuum extremals would correspond in this picture to virtual particles, whose contribution to the generalized Feynmann diagram is not suppressed by the exponential of Kähler action unlike that associated with the virtual CP_2 type lines.

M^4 type vacuum extremals (representable as maps $M^4_+ \to CP_2$ by definition) are also expected to be natural idealizations of the space-time at long length scales obtained by smoothing out small scale topological inhomogenities (particles) and therefore they should correspond to space-time of GRT in a reasonable approximation.

The reason would be "Yin-Yang principle" discussed in [D1].

a) Consider first the option for which Kähler function corresponds to an absolute minimum of Kähler action. Vacuum functional as an exponent of Kähler function is expected to concentrate on those 3-surfaces for which the Kähler action is non-negative. On the other hand, the requirement that Kähler action is absolute minimum for the space-time associated with a given 3-surface, tends to make the action negative. Therefore the vacuum functional is expected to differ considerably from zero only for 3-surfaces with a vanishing Kähler action per volume. It could also occur that the degeneracy of 3-surfaces with same large negative action compensates the exponent of Kähler function.

b) If preferred extrema correspond to Kähler calibrations or their duals [E2], Yin-Yang principle is modified to a more local principle. For Kähler calibrations (their duals) the absolute value of action in given region is minimized (maximized). A given region with a positive (negative sign) of action density favors Kähler electric (magnetic) fields. In long length scales the average density of Kähler action per four-volume tends to vanish so that Kähler function of the entire universe is expected to be very nearly zero. This regularizes the theory automatically and implies that average Kähler action per volume vanishes. Positive and finite values of Kähler function are of course favored.

In both cases the vanishing of Kähler action per volume in long length scales makes vacuum extremals excellent idealizations for the smoothed out space-time surface. Robertson-Walker cosmologies provide a good example in this respect. As a matter fact the smoothed out space-time is not a mere fictive concept since larger space-time sheets realize it as a essential part of the Universe.

Several absolute minima could be possible and the non-determinism of the vacuum extremals is not expected to be reduced completely. The remaining degeneracy could be even infinite. A good example is provided by the vacuum extremals representable as maps $M^4_+ \to D^1$, where D^1 is one-dimensional curve of CP_2. This degeneracy could be interpreted as a space-time correlate for the non-determinism of quantum jumps with maximal deterministic regions representing quantum states in a sequence of quantum jumps.

4.4.3   Cosmic strings

Cosmic strings are extremals of type $X^2 \times S^2$, where X^2 is minimal surface in M^4_+ (analogous to the orbit of a bosonic string) and S^2 is the homologically non-trivial geodesic sphere of CP_2. The action of these extremals is positive and thus absolute minima are certainly not in question. One can however consider the possibility that these extremals are building blocks of the absolute minimum space-time surfaces since the principle selecting preferred extremals of the Kähler action is global rather than a local. Cosmic strings can contain also Kähler charged matter in the form of small holes containing elementary particle quantum numbers on their boundaries and the negative Kähler electric action for a topologically condensed cosmic string could cancel the Kähler magnetic action.

The string tension of the cosmic strings is given by

$$T \;=\; \frac{1}{8\alpha_K R^2} \simeq .2210^{-6}\frac{1}{G} \;,\qquad\qquad (4.4.5)$$

where $\alpha_K \simeq \alpha_{em}$ has been used to get the numerical estimate. The string tension is of the same order of magnitude as the string tension of the cosmic strings of GUTs and this leads to the model of the galaxy formation providing a solution to the dark matter puzzle as well as to a model for large voids as caused by the presence of a strongly Kähler charged cosmic string. Cosmic strings play also fundamental role in the TGD inspired very early cosmology.

4.4.4 Massless extremals

Massless extremals are characterized by massless wave vector p and polarization vector ε orthogonal to this wave vector. Using the coordinates of M^4 as coordinates for X^4 the solution is given as

$$
\begin{aligned}
s^k &= f^k(u,v) \;, \\
u &= p \cdot m \;, \qquad v = \varepsilon \cdot m \;, \\
p \cdot \varepsilon &= 0 \;, \qquad\; p^2 = 0 \;.
\end{aligned}
$$

CP_2 coordinates are arbitrary functions of $p \cdot m$ and $\varepsilon \cdot m$. Clearly these solutions correspond to plane wave solutions of gauge field theories. It is important to notice however that linear super position doesn't hold as it holds in Maxwell phase. Gauge current is proportional to wave vector and its divergence vanishes as a consequence. Also cylindrically symmetric solutions for which the transverse coordinate is replaced with the radial coordinate $\rho = \sqrt{m_1^2 + m_2^2}$ are possible. In fact, v can be *any* function of the coordinates m^1, m^2 transversal to the light like vector p.

Boundary conditions on the boundaries of the massless extremal are satisfied provided the normal component of the energy momentum tensor vanishes. Since energy momentum tensor is of the form $T^{\alpha\beta} \propto p^\alpha p^\beta$ the conditions $T^{n\beta} = 0$ are satisfied if the M^4 projection of the boundary is given by the equations of form

$$
\begin{aligned}
H(p \cdot m, \varepsilon \cdot m, \varepsilon_1 \cdot m) &= 0 \;, \\
\varepsilon \cdot p = 0 \;, \qquad\qquad \varepsilon_1 \cdot p &= 0 \;, \quad \varepsilon \cdot \varepsilon_1 = 0 \;.
\end{aligned}
\qquad\qquad (4.4.6)
$$

where H is arbitrary function of its arguments. Recall that for M^4 type extremals the boundary conditions are also satisfied if Kähler field vanishes identically on the boundary.

The following argument suggests that there are not very many manners to satisfy boundary conditions in case of M^4 type extremals. The boundary conditions, when applied to M^4 coordinates imply the vanishing of the normal component of energy momentum tensor. Using coordinates, where energy momentum tensor is diagonal, the requirement boils down to the condition that at least one of the eigen values of $T^{\alpha\beta}$ vanishes so that the determinant $det(T^{\alpha\beta})$ must vanish on the boundary: this condition defines 3-dimensional surface in X^4. In addition, the normal of this surface must have same direction as the eigen vector associated with the vanishing eigen value: this means that three additional conditions must be satisfied and this is in general true in single point only. The boundary conditions in CP_2 coordinates are satisfied provided that the conditions

$$J^{n\beta} J^k_{\;l} \partial_\beta s^l = 0$$

are satisfied. The identical vanishing of the normal components of Kähler electric and magnetic fields on the boundary of massless extremal property provides a manner to satisfy all boundary conditions but it is not clear whether there are any other manners to satisfy them.

The characteristic feature of the massless extremals is that in general the Kähler gauge current is non-vanishing. In ordinary Maxwell electrodynamics this is not possible. This means that these extremals are accompanied by vacuum current, which contains in general case both weak and electromagnetic terms as well as color part.

A possible interpretation of the solution is as the exterior space-time to a topologically condensed particle with vanishing mass described by massless CP_2 type extremal, say photon or neutrino. In general the surfaces in question have boundaries since the coordinates s^k are are bounded: this is in accordance with the general ideas about topological condensation. The fact that massless plane wave is associated with CP_2 type extremal combines neatly the wave and particle aspects at geometrical level.

The fractal hierarchy of space-time sheets implies that massless extremals should interesting also in long length scales. The presence of a light like electromagnetic vacuum current implies the generation of coherent photons and also coherent gravitons are generated since the Einstein tensor is also non-vanishing and light like (proportional to $k^\alpha k^\beta$). Massless extremals play an important role in the TGD based model of bio-system as a macroscopic quantum system. The possibility of vacuum currents is what makes possible the generation of the highly desired coherent photon states.

4.4.5 Generalization of the solution ansatz defining massless extremals (MEs)

The solution ansatz for MEs has developed gradually to an increasingly general form and the following formulation is the most general one achieved hitherto. Rather remarkably, it rather closely resembles the solution ansatz for the CP_2 type extremals and has direct interpretation in terms of geometric optics. Equally remarkable is that the latest generalization based on the introduction of the local light cone coordinates was inspired by quantum holography principle.

The solution ansatz for MEs has developed gradually to an increasingly general form and the following formulation is the most general one achieved hitherto. Rather remarkably, it rather closely resembles the solution ansatz for the CP_2 type extremals and has direct interpretation in terms of geometric optics. Equally remarkable is that the latest generalization based on the introduction of the local light cone coordinates was inspired by quantum holography principle.

Local light cone coordinates

The solution involves a decomposition of M_+^4 tangent space localizing the decomposition of Minkowski space to an orthogonal direct sum $M^2 \oplus E^2$ defined by light-like wave vector and polarization vector orthogonal to it. This decomposition defines what might be called local light cone coordinates.

a) Denote by m^i the linear Minkowski coordinates of M^4. Let (S^+, S^-, E^1, E^2) denote local coordinates of M_+^4 defining a *local* decomposition of the tangent space M^4 of M_+^4 into a direct *orthogonal* sum $M^4 = M^2 \oplus E^2$ of spaces M^2 and E^2. This decomposition has interpretation in terms of the longitudinal and transversal degrees of freedom defined by local light-like four-velocities $v_\pm = \nabla S_\pm$ and polarization vectors $\epsilon_i = \nabla E^i$ assignable to light ray.

b) With these assumptions the coordinates $(S_\pm, E^i)$ define local light cone coordinates with the metric element having the form

$$ds^2 = 2g_{+-}dS^+dS^- + g_{11}(dE^1)^2 + g_{22}(dE^2)^2 \quad . \tag{4.4.7}$$

If complex coordinates are used in transversal degrees of freedom one has $g_{11} = g_{22}$.

c) This family of light cone coordinates is not the most general family since longitudinal and transversal spaces are orthogonal. One can also consider light-cone coordinates for which one non-diagonal component, say m_{1+}, is non-vanishing if the solution ansatz is such that longitudinal and transversal spaces are orthogonal for the induced metric.

A conformally invariant family of local light cone coordinates

The simplest solutions to the equations defining local light cone coordinates are of form $S_\pm = k \cdot m$ giving as a special case $S_\pm = m^0 \pm m^3$. For more general solutions of from

$$S_\pm = m^0 \pm f(m^1, m^2, m^3) \quad , \quad (\nabla_3 f)^2 = 1 \quad ,$$

where f is an otherwise arbitrary function, this relationship reads as

$$S^+ + S^- = 2m^0 \quad .$$

This condition defines a natural rest frame. One can integrate f from its initial data at some two-dimensional $f = constant$ surface and solution describes curvilinear light rays emanating from this surface and orthogonal to it. The flow velocity field $\overline{v} = \nabla f$ is irrotational so that closed flow lines are not possible in a connected region of space and the condition $\overline{v}^2 = 1$ excludes also closed flow line configuration with singularity at origin such as $v = 1/\rho$ rotational flow around axis.

One can identify E^2 as a local tangent space spanned by polarization vectors and orthogonal to the flow lines of the velocity field $\overline{v} = \nabla f(m^1, m^2, m^3)$. Since the metric tensor of any 3-dimensional space allows always diagonalization in suitable coordinates, one can always find coordinates (E^1, E^2) such that (f, E^1, E^2) form orthogonal coordinates for $m^0 = constant$ hyperplane. Obviously one can select the coordinates E^1 and E^2 in infinitely many manners.

Closer inspection of the conditions defining local light cone coordinates

Whether the conformal transforms of the local light cone coordinates $\{S_\pm = m^0 \pm f(m^1, m^2, m^3), E^i\}$ define the only possible compositions $M^2 \oplus E^2$ with the required properties, remains an open question. The best that one might hope is that any function S^+ defining a family of light-like curves defines a local decomposition $M^4 = M^2 \oplus E^2$ with required properties.

a) Suppose that S^+ and S^- define light-like vector fields which are not orthogonal (proportional to each other). Suppose that the polarization vector fields $\epsilon_i = \nabla E^i$ tangential to local E^2 satisfy the conditions $\epsilon_i \cdot \nabla S^+ = 0$. One can formally integrate the functions E^i from these condition since the initial values of E^i are given at $m^0 = constant$ slice.

b) The solution to the condition $\nabla S_+ \cdot \epsilon_i = 0$ is determined only modulo the replacement

$$\epsilon_i \rightarrow \hat{\epsilon}_i = \epsilon_i + k\nabla S_+ \ ,$$

where k is any function. With the choice

$$k = -\frac{\nabla E^i \cdot \nabla S^-}{\nabla S^+ \cdot \nabla S^-}$$

one can satisfy also the condition $\hat{\epsilon}_i \cdot \nabla S^- = 0$.

c) The requirement that also $\hat{\epsilon}_i$ is gradient is satisfied if the integrability condition

$$k - k(S^+)$$

is satisfied: in this case $\hat{\epsilon}_i$ is obtained by a gauge transformation from ϵ_i. The integrability condition can be regarded as an additional, and obviously very strong, condition for S^- once S^+ and E^i are known.

d) The problem boils down to that of finding local momentum and polarization directions defined by the functions S^+, S^- and E^1 and E^2 satisfying the orthogonality and integrability conditions

$$(\nabla S^+)^2 = (\nabla S^-)^2 = 0 \ , \quad \nabla S^+ \cdot \nabla S^- \neq 0 \ ,$$

$$\nabla S^+ \cdot \nabla E^i = 0 \ , \qquad \frac{\nabla E^i \cdot \nabla S^-}{\nabla S^+ \cdot \nabla S^-} = k_i(S^+) \ .$$

The number of integrability conditions is 3+3 (all derivatives of k_i except the one with respect to S^+ vanish): thus it seems that there are not much hopes of finding a solution unless some discrete symmetry relating S^+ and S^- eliminates the integrability conditions altogether.

A generalization of the spatial reflection $f \rightarrow -f$ working for the separable Hamilton Jacobi function $S_\pm = m^0 \pm f$ ansatz could relate S^+ and S^- to each other and trivialize the integrability conditions. The symmetry transformation of M_+^4 must perform the permutation $S^+ \leftrightarrow S^-$, preserve the light-likeness property, map E^2 to E^2, and multiply the inner products between M^2 and E^2 vectors by a mere conformal factor. This encourages the conjecture that all solutions are obtained by conformal transformations from the solutions $S_\pm = m^0 \pm f$.

General solution ansatz for MEs for given choice of local light cone coordinates

Consider now the general solution ansatz assuming that a local wave-vector-polarization decomposition of M_+^4 tangent space has been found.

a) Let $E(S^+, E^1, E^2)$ be an arbitrary function of its arguments: the gradient ∇E defines at each point of E^2 an S^+-dependent (and thus time dependent) polarization direction orthogonal to the direction of local wave vector defined by ∇S^+. Polarization vector depends on E^2 position only.

b) Quite a general family of MEs corresponds to the solution family of the field equations having the general form

$$s^k = f^k(S^+, E) \ ,$$

where s^k denotes CP_2 coordinates and f^k is an arbitrary function of S^+ and E. The solution represents a wave propagating with light velocity and having definite S^+ dependent polarization in the direction of ∇E. By replacing S^+ with S^- one obtains a dual solution. Field equations are satisfied because energy momentum tensor and Kähler current are light-like so that all tensor contractions involved with the field equations vanish: the orthogonality of M^2 and E^2 is essential for the light-likeness of energy momentum tensor and Kähler current.

c) The simplest solutions of the form $S_\pm = m^0 \pm m^3$, $(E^1, E^2) = (m^1, m^2)$ and correspond to a cylindrical MEs representing waves propagating in the direction of the cylinder axis with light velocity and having polarization which depends on point (E^1, E^2) and S^+ (and thus time). For these solutions four-momentum is light-like: for more general solutions this cannot be the case. Polarization is in general case time dependent so that both linearly and circularly polarized waves are possible. If m^3 varies in a finite range of length L, then 'free' solution represents geometrically a cylinder of length L moving with a light velocity. Of course, ends could be also anchored to the emitting or absorbing space-time surfaces.

d) For the general solution the cylinder is replaced by a three-dimensional family of light like curves and in this case the rectilinear motion of the ends of the cylinder is replaced with a curvilinear motion with light velocity unless the ends are anchored to emitting/absorbing space-time surfaces. The non-rotational character of the velocity flow suggests that the freely moving particle like 3-surface defined by ME cannot remain in a infinite spatial volume. The most general ansatz for MEs should be useful in the intermediate and nearby regions of a radiating object whereas in the far away region radiation solution is excepted to decompose to cylindrical ray like MEs for which the function $f(m^1, m^2, m^2)$ is a linear function of m^i.

e) One can try to generalize the solution ansatz further by allowing the metric of M_+^4 to have components of type g_{i+} or g_{i-} in the light cone coordinates used. The vanishing of T^{11}, T^{+1}, and T^{--} is achieved if $g_{i\pm} = 0$ holds true for the induced metric. For $s^k = s^k(S^+, E^1)$ ansatz neither $g_{2\pm}$ nor g_{1-} is affected by the imbedding so that these components of the metric must vanish for the Hamilton Jacobi structure:

$$ds^2 \quad = \quad 2g_{+-}dS^+dS^- + 2g_{1+}dE^1dS^+ + g_{11}(dE^1)^2 + g_{22}(dE^2)^2 \ . \qquad (4.4.8)$$

$g_{1+} = 0$ can be achieved by an additional condition

$$m_{1+} \quad = \quad s_{kl}\partial_1 s^k \partial_+ s^k \ . \qquad (4.4.9)$$

The diagonalization of the metric seems to be a general aspect of absolute minima. The absence of metric correlations between space-time degrees of freedom for asymptotic self-organization patterns is somewhat analogous to the minimization of non-bound entanglement in the final state of the quantum jump.

Are the boundaries of space-time sheets quite generally light like surfaces with Hamilton Jacobi structure?

Quantum holography principle naturally generalizes to an approximate principle expected to hold true also in non-cosmological length and time scales.

a) The most general ansatz for topological light rays or massless extremals (MEs) inspired by the quantum holographic thinking relies on the introduction of the notion of local light cone coordinates S_+, S_-, E_1, E_2. The gradients ∇S_+ and ∇S_- define two light like directions just like Hamilton Jacobi functions define the direction of propagation of wave in geometric optics. The two polarization vector fields ∇E_1 and ∇E_2 are orthogonal to the direction of propagation defined by either S_+ or S_-. Since also E_1 and E_2 can be chosen to be orthogonal, the metric of M_+^4 can be written locally as $ds^2 = g_{+-}dS_+dS_- + g_{11}dE_1^2 + g_{22}dE_2^2$. In the earlier ansatz S_+ and S_- where restricted to the variables $k \cdot m$ and $\tilde{k} \cdot m$, where k and $\tilde{k}$ correspond to light like momentum and its mirror image and m denotes linear M^4 coordinates: these MEs describe cylindrical structures with constant direction of wave propagation expected to be most important in regions faraway from the source of radiation.

b) Boundary conditions are satisfied if the 3-dimensional boundaries of MEs have one light like direction (S_+ or S_- is constant). This means that the boundary of ME has metric dimension $d = 2$ and is characterized by an infinite-dimensional super-canonical and super-conformal symmetries just like the boundary of the imbedding space $M_+^4 \times CP_2$: The boundaries are like moments for mini big bangs (in TGD based fractal cosmology big bang is replaced with a silent whisper amplified to not necessarily so big bang).

c) These observations inspire the conjecture that boundary conditions for M^4 like space-time sheets fixed by the variational principle selecting preferred extremals of Kähler action quite generally require that space-time boundaries correspond to light like 3-surfaces with metric dimension equal to $d = 2$. This does not yet imply that light like surfaces of imbedding space would take the role of the light cone boundary: these light like surface could be seen only as a special case of causal determinants analogous to event horizons.

Chapter 5

The geometry of the world of the classical worlds

5.1 Introduction

In this chapter a summary about basic ideas related to the construction of the Kähler geometry of infinite-dimensional configuration space of 3-surfaces (more or less-equivalently, the corresponding 4-surfaces defining generalized Bohr orbits).

5.1.1 The quantum states of Universe as modes of classical spinor field in the "world of classical worlds"

The vision behind the construction of configuration space geometry is that physics reduces to the geometry of classical spinor fields in the infinite-dimensional configuration space of 3-surfaces of $M_+^4 \times CP_2$ or $M^4 \times CP_2$, where M^4 and M_+^4 denote Minkowski space and its light cone respectively. This configuration space might be called the "world of classical worlds".

a) Hermitian conjugation is the basic operation in quantum theory and its geometrization requires that configuration space possesses Kähler geometry. One of the basic features of the Kähler geometry is that it is solely determined by the so called , which defines both the J and the components of the g in complex coordinates via the general formulas [bc3]

$$
\begin{aligned}
J &= i\partial_k\partial_{\bar{l}}Kdz^k \wedge d\bar{z}^l \ , \\
ds^2 &= 2\partial_k\partial_{\bar{l}}Kdz^k d\bar{z}^l \ .
\end{aligned}
\tag{5.1.1}
$$

Kähler form is covariantly constant two-form and can be regarded as a representation of imaginary unit in the tangent space of the configuration space

$$
J_{mr}J^{rn} = -g_m{}^n \ .
\tag{5.1.2}
$$

As a consequence Kähler form defines also symplectic structure in configuration space.

5.1.2 Definition of Kähler function

The task of finding Kähler geometry for the configuration space reduces to that of finding Kähler function and identifying the complexification. The main constraints on the Kähler function result from the requirement of Diff^4 symmetry and degeneracy. requires that the definition of the Kähler function assigns to a given 3-surface X^3 a unique space-time surface $X^4(X^3)$, the generalized Bohr orbit defining the classical physics associated with X^3. The natural guess is that Kähler function is defined by what might be called Kähler action, which is essentially Maxwell action with Maxwell field expressible in terms of CP_2 coordinates. Absolute minimization is the first guess for how to fix $X^4(X^3)$ uniquely.

It has however become clear that this option might well imply that Kähler is negative and infinite for the entire Universe so that the vacuum functional would be identically vanishing. I The number theoretical vision [E2] leads to a more attractive option. According to this option the absolute value of the contribution to the Kähler action coming from each region where the action density has a definite sign is minimized separately. As a consequence the preferred extremals are as near to vacuum extremals as possible. This approach has the very attractive feature that for physically allowed space-time surfaces action density would reduce to a closed 4-form (Kähler calibration) so that TGD would be in certain sense a topological quantum field theory.

If Kähler action would define a strictly deterministic variational principle, Diff^4 degeneracy and general coordinate invariance would be achieved by restricting the consideration to 3-surfaces Y^3 at the boundary of M^4_+ and by defining Kähler function for 3-surfaces X^3 at $X^4(Y^3)$ and diffeo-related to Y^3 as $K(X^3) = K(Y^3)$. This reduction might be called . The classical non-determinism of the Kähler action however introduces complications which might be however overcome by generalizing the notion of quantum gravitational holography.

5.1.3   Minkowski space or its light cone?

For a long time I believed that the question "M^4_+ or M^4?" had been settled in favor of M^4_+. The work with the conceptual problems related to energy and time, and with the symmetries of quantum TGD, however led gradually to the realization that there are strong reasons for considering M^4 instead of M^4_+.

a) It has become clear that the gigantic symmetries associated with the $\delta M^4_+ \times CP_2$ and more general surfaces $X^3_l \times CP_2$, X^3_l light like 3-surface of M^4 are also laboratory symmetries visible directly at the level of propagators and vertices [C5]. X^3_l could be restricted to be a union of future and past directed light cone boundaries since arbitrary light like surfaces X^3_l contain singularities such as self intersections. fits very elegantly with the two types of super-conformal symmetries of TGD. The first conformal symmetry corresponds to the light-like surfaces $X^3_l \times CP_2$ of the imbedding space. More general light like 7-surfaces are not favored because they do not possess the huge conformal symmetries of $X^3_l \times CP_2$. Second conformal symmetry corresponds to light like boundaries of X^4 and light-like surfaces separating space-time regions with different signatures of the induced metric and is identifiable as the counterpart of the Kac Moody symmetry of string models.

A rather plausible conclusion is that configuration space is a union of configuration spaces associated with $X^3_l \times CP_2$, with X^3_l identified as unions of future and past directed light cones. Thus the construction reduces to a high degree to a study of a simple special case $\delta M^4_+ \times CP_2$.

b) The replacement of the energy momentum tensor by a collection of conserved currents means that the sign of the energy depends on the time orientation of the space-time sheet. The simplest theory results if one assumes that the net quantum numbers of physical states vanishes. guarantees the consistency with elementary particle physics. A consistency with the macroscopic physics results if gravitational energy is the difference of positive and negative inertial (Poincare) energies of matter. This option allows also M^4 and leads to a fractal cosmology in which light like 7-surfaces of $X^3_l \times CP_2$ serve as causal determinants at the level of the imbedding space.

c) There is however a conceptual hurdle involved. Suppose that X^4 is the absolute minimum (or a more general Bohr orbit like extremal) associated with X^3, and let Y^3 be some other $Diff^4$ related 3-surface at X^4. One cannot require that the absolute minima $X^4(X^3)$ and $X^4(Y^3)$ are same and one certainly cannot assign an absolute minimum to every 3-surface separately since general coordinate invariance for 3-surfaces would imply that Kähler function is infinitely many-valued. X^4 can thus be an absolute minimum only for some preferred 3-surface $X^3(X^4)$ at X^4 and the question is what makes this 3-surface preferred.

d) It seems that the internal geometry of $X^4(X^3)$ must be such that it defines uniquely $X^3(X^4)$, or perhaps even more generally, a light like causal determinant $X^3_l \times CP_2$ of H to which X^3 belongs. If one requires that the net values of the conserved classical quantities are zero, one could regard X^4 as consisting of space-time sheets with opposite time orientation which are created at some moment from vacuum and possibly also disappear to vacuum. If this is the case then the 3-surface at which positive and negative energy space-time sheets are created and begin to evolve as separate branches presents a natural candidate for X^3. Also this view is also consistent with quantum holography and supported strongly by number theoretical considerations (M^4 has an interpretation as quaternion space with Minkowski metric defined as the real part of q^2).

5.1.4   Configuration space metric from symmetries

A complementary approach to the problem of constructing configuration space geometry is based on symmetries. The work of Dan Freed [df6] has demonstrated that the Kähler geometry of loop spaces is unique from the existence of Riemann connection and fixed completely by the Kac Moody symmetries of the space. In 3-dimensional context one has even better reasons to expect uniqueness. The guess is that configuration space is a union of symmetric spaces labelled by zero modes not appearing in the line element as differentials. The generalized conformal invariance of metrically 2-dimensional light like 3-surfaces acting as causal determinants is the corner stone of the construction. The construction works only for 4-dimensional space-time and imbedding space which is a product of four-dimensional Minkowski space or its future light cone with CP_2.

5.1.5   Is absolute minimization the correct variational principle?

One can criticize the assumption that extremals correspond to the absolute minima of Kähler action Any other principle allowing to assign to a given 3-surface a unique space-time surface in principle must in principle be considered as a viable alternative. The number theoretical vision discussed in [E2] indeed favors the separate minimization of magnitudes of positive and negative contributions to the Kähler action.

For this option Universe would do its best to save energy, being as near as possible to vacuum. Also vacuum extremals would become physically relevant: note that they would be only inertial (Poincare) vacua and carry non-vanishing density gravitational energy. The non-determinism of the vacuum extremals would have an interpretation in terms of the ability of Universe to engineer itself.

The 3-surfaces for which CP_2 projection is at least 2-dimensional and not a Lagrange manifold would correspond to non-vacua since conservation laws do not leave any other option. The variational principle would favor equally magnetic and electric configurations whereas absolute minimization of action based on S_K would favor electric configurations. The positive and negative contributions would be minimized for 4-surfaces in relative homology class since the boundary of X^4 defined by the intersections with 7-D light-like causal determinants would be fixed. Without this constraint only vacuum bubbles would result.

The attractiveness of the number theoretical variational principle from the point of calculability of TGD would be that the initial values for the time derivatives of the imbedding space coordinates at X^3 at light-like 7-D causal determinant could be computed by requiring that the energy of the solution is minimized. This could mean a computerizable solution to the construction of Kähler function. The number theoretic approach based on the properties of quaternions and octonions discussed in the chapter [E2] leads to a proposal for the general solution of field equations based on the generalization of the notion of calibration [bb1] providing absolute minima of volume to that of Kähler calibration. This approach will not be discussed in this chapter.

In this chapter I will first consider the basic properties of the configuration space, discuss briefly the various approaches to the geometrization of the configuration space, and introduce the two complementary strategies based on a direct guess of Kähler function and on the group theoretical approach assuming that configuration space can be regarded as a union of symmetric spaces. After these preliminaries a definition of the Kähler function is proposed and various physical and mathematical motivations behind the proposed definition are discussed. The key feature of the Kähler action is classical non-determinism, and various implications of the classical non-determinism are discussed.

5.2   Constraints on the configuration space geometry

The constraints on the configuration space geometry result both from the infinite dimension of the configuration space and from physically motivated symmetry requirements. There are three basic physical requirements on the configuration space geometry: namely four-dimensional Diff invariance, Kähler property and the decomposition of configuration space into a union $\cup_i G/H_i$ of symmetric spaces G/H_i, each coset space allowing G-invariant metric such that G is subgroup of some 'universal group' having natural action on 3-surfaces. Together with the infinite dimensionality of the configuration space these requirements pose extremely strong constraints on the configuration space geometry. In the following we shall consider these requirements in more detail.

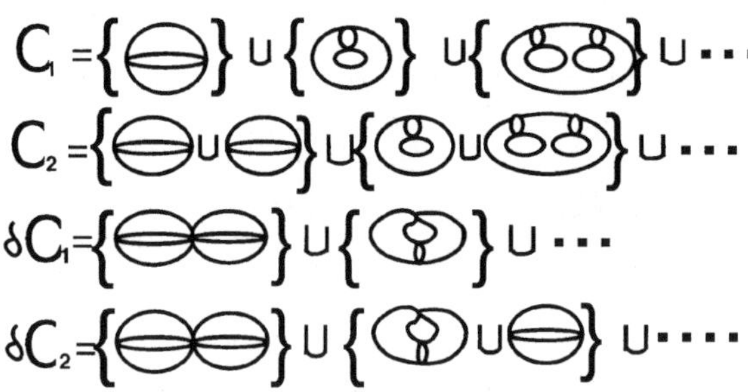

Figure 5.1: Structure of the configuration space: two-dimensional visualization

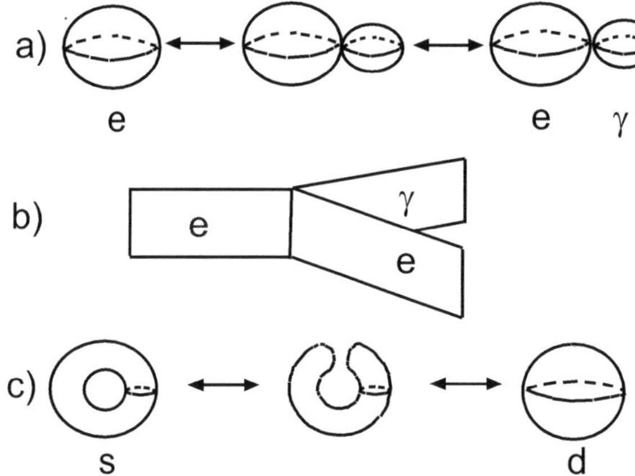

Figure 5.2: Two-dimensional visualization of topological description of particle reactions. a) Generalization of stringy diagram describing particle decay: 4-surface is smooth manifold and vertex a non-unique singular 3-manifold, b) Topological description of particle decay in terms of a singular 4-manifold but smooth and unique 3-manifold at vertex. c) Topological origin of Cabibbo mixing.

5.2.1   Configuration space as "the world of classical worlds"

The configuration space of TGD consists of all 3-surfaces of $M_+^4 \times CP_2$ containing sets of
i) surfaces with all possible manifold topologies and arbitrary numbers of components (N-particle sectors)
ii) singular surfaces topologically intermediate between two manifold topologies (see Fig. 5.2.1)
We shall use the symbol $C(H)$ to denote the set of 3-surfaces $X^3 \subset H$. It should be emphasized that surfaces related by $Diff^3$ transformations will be regarded as different surfaces in the sequel.

These surfaces form a connected(!) space since it is possible to glue various N-particle sectors to each other along their boundaries consisting of sets of singular surfaces topologically intermediate between corresponding manifold topologies. The connectedness of the configuration space is a necessary prerequisite for the description of topology changing particle reactions as continuous paths in configuration space (see Fig. 5.2.1).

5.2.2   Diff^4 invariance and Diff^4 degeneracy

Diff^4 plays fundamental role as the gauge group of General Relativity. In string models $Diff^2$ invariance ($Diff^2$ acts on the orbit of the string) plays central role in making possible the elimination of the time like and longitudinal vibrational degrees of freedom of string. Also in the present case the elimination of the tachyons (time like oscillatory modes of 3-surface) is a physical necessity and Diff^4 invariance provides an obvious manner to do the job.

In the standard functional integral formulation the realization of Diff^4 invariance is an easy task at the formal level. The problem is however that the path integral over four-surfaces is plagued by divergences and doesn't make sense. In the present case the configuration space consists of 3-surfaces and only $Diff^3$ emerges automatically as the group of re-parameterizations of 3-surface. Obviously one should somehow define the action of Diff^4 in the space of 3-surfaces. Whatever the action of Diff^4 is it must leave the configuration space metric invariant. Furthermore, the elimination of tachyons is expected to be possible only provided the time like deformations of the 3-surface correspond to zero norm vector fields of the configuration space so that 3-surface and its Diff^4 image have zero distance. The conclusion is that configuration space metric should be both Diff^4 invariant and Diff^4 degenerate.

The problem is how to define the action of Diff^4 in $C(H)$. Obviously the only manner to achieve Diff^4 invariance is to require that the very definition of the configuration space metric somehow associates a unique space-time surface to a given 3-surface for Diff^4 to act on! The obvious physical interpretation of this space time surface is as "classical space time" so that "Classical Physics" would be contained in configuration space geometry. It is this requirement, which has turned out to be decisive concerning the understanding of the configuration space geometry. Amusingly enough, the historical development was not this: the definition of Diff^4 degenerate Kähler metric was found by a guess and only later it was realized that Diff^4 invariance and degeneracy could have been postulated from beginning!

5.2.3 Decomposition of the configuration space into a union of symmetric spaces G/H

The extremely beautiful theory of finite-dimensional symmetric spaces constructed by Elie Cartan suggests that configuration space should possess a decomposition into a union of coset spaces $CH = \cup_i G/H_i$ such that the metric inside each coset space G/H_i is left invariant under the infinite dimensional isometry group G. The metric equivalence of surfaces inside each coset space G/H_i does not mean that 3-surfaces inside G/H_i are physically equivalent. The reason is that the vacuum functional is exponent of Kähler action which is not isometry invariant so that the 3-surfaces, which correspond to maxima of Kähler function for a given orbit, are in a preferred position physically. For instance, one can calculate functional integral around this maximum perturbatively. The sum of over i means actually integration over the zero modes of the metric (zero modes correspond to coordinates not appearing as coordinate differentials in the metric tensor).

The coset space G/H is a symmetric space only under very special Lie-algebraic conditions. Denoting the Cartan decomposition of the Lie-algebra g of G to the direct sum of H Lie-algebra h and its complement t by $g = h \oplus t$, one has

$$[h,h] \subset h \ , \quad [h,t] \subset t \ , \quad [t,t] \subset h \ .$$

This decomposition turn out to play crucial role in guaranteing that G indeed acts as isometries and that the metric is Ricci flat.

The four-dimensional $Diff$ invariance indeed suggests to a beautiful solution of the problem of identifying G. The point is that any 3-surface X^3 is $Diff^4$ equivalent to the intersection of $X^4(X^3)$ with the light cone boundary. This in turn implies that 3-surfaces in the space $\delta H = \delta M_+^4 \times CP_2$ should be all what is needed to construct configuration space geometry. The group G can be identified as some subgroup of diffeomorphisms of δH and H_i contains that subgroup of G, which acts as diffeomorphisms of the 3-surface X^3. Since G preserves topology, configuration space must decompose into union $\cup_i G/H_i$, where i labels 3-topologies and various zero modes of the metric. For instance, the elements of the Lie-algebra of G invariant under configuration space complexification correspond to zero modes.

The reduction to the light cone boundary, identifiable as the moment of big bang, looks perhaps odd at first. In fact, it turns out that the classical non-determinism of Kähler action forces does not allow the complete reduction to the light cone boundary: physically this is a highly desirable implication but means a considerable mathematical challenge.

5.2.4　Kähler property

Kähler property implies that the tangent space of the configuration space allows complexification and that there exists a covariantly constant two-form J_{kl}, which can be regarded as a representation of the imaginary unit in the tangent space of the configuration space:

$$J_k{}^r J_{rl} = -G_{kl} \ . \tag{5.2.1}$$

There are several physical and mathematical reasons suggesting that configuration space metric should possess Kähler property in some generalized sense.

a) Kähler property turns out to be a necessary prerequisite for defining divergence free configuration space integration. We will leave the demonstration of this fact later although the argument as such is completely general.

b) Kähler property very probably implies an infinite-dimensional isometry group. The study of the loop groups $Map(S^1, G)$ [df6] shows that loop group allows only single Kähler metric with well defined Riemann connection and this metric allows local G as its isometries!

To see this consider the construction of Riemannian connection for $Map(X^3, H)$. The defining formula for the connection is given by the expression

$$
\begin{aligned}
2(\nabla_X Y, Z) \ &= \ X(Y, Z) + Y(Z, X) - Z(X, Y) \\
&+ \ ([X, Y], Z) + ([Z, X], Y) - ([Y, Z], X)
\end{aligned}
\tag{5.2.2}
$$

X, Y, Z are smooth vector fields in $Map(X^3, G)$. This formula defines $\nabla_X Y$ uniquely provided the tangent space of Map is complete with respect to Riemann metric. In the finite-dimensional case completeness means that the inverse of the covariant metric tensor exists so that one can solve the components of connection from the conditions stating the covariant constancy of the metric. In the case of the loop spaces with Kähler metric this is however not the case.

Now the symmetry comes into the game: if X, Y, Z are left (local gauge) invariant vector fields defined by the Lie-algebra of local G then the first three terms drop away since the scalar products of left invariant vector fields are constants. The expression for the covariant derivative is given by

$$\nabla_X Y \ = \ (Ad_X Y - Ad_X^* Y - Ad_Y^* X)/2 \tag{5.2.3}$$

where Ad_X^* is the adjoint of Ad_X with respect to the metric of the loop space.

At this point it is important to realize that Freed's argument does not force the isometry group of the configuration space to be $Map(X^3, M^4 \times SU(3))$! Any symmetry group, whose Lie algebra is complete with respect to the configuration space metric (in the sense that any tangent space vector is expressible as superposition of isometry generators modulo a zero norm tangent vector) is an acceptable alternative.

The Kähler property of the metric is quite essential in one-dimensional case in that it leads to the requirement of left invariance as a mathematical consistency condition and we expect that dimension three makes no exception in this respect. In 3-dimensional case the degeneracy of the metric turns out to be even larger than in 1-dimensional case due to the four-dimensional Diff degeneracy. So we expect that the metric ought to possess some infinite-dimensional isometry group and that the above formula generalizes also to the 3-dimensional case and to the case of local coset space. Note that in M^4 degrees of freedom $Map(X^3, M^4)$ invariance would imply the flatness of the metric in M^4 degrees of freedom.

The physical implications of the above purely mathematical conjecture should not be underestimated. For example, one natural looking manner to construct physical theory would be based on the idea that configuration space geometry is dynamical and this approach is followed in the attempts to construct string theories [gb2]. Various physical considerations (in particular the need to obtain oscillator operator algebra) seem to imply that configuration space geometry is necessarily Kähler. The above result however states that configuration space Kähler geometry cannot be dynamical quantity and is dictated solely by the requirement of internal consistency. This result is extremely nice since it has been already found that the definition of the configuration space metric must somehow associate a unique classical space time and "classical physics" to a given 3-surface: uniqueness of the geometry implies the uniqueness of the "classical physics".

c) The choice of the imbedding space becomes highly unique. In fact, the requirement that configuration space is not only symmetric space but also (contact) Kähler manifold inheriting its (degenerate) Kähler structure from the imbedding space suggests that spaces, which are products of four-dimensional Minkowski space with complex projective spaces CP_n, are perhaps the only possible candidates for H. The reason for the unique position of the four-dimensional Minkowski space turns out to be that the boundary of the light cone of D-dimensional Minkowski space is metrically a sphere S^{D-2} despite its topological dimension $D-1$: for $D=4$ one obtains two-sphere allowing Kähler structure and infinite parameter group of conformal symmetries.

d) It seems possible to understand the basic mathematical structures appearing in string model in terms of the Kähler geometry rather nicely.

i) The projective representations of the infinite-dimensional isometry group (not necessarily Map!) correspond to the ordinary representations of the corresponding centrally extended group [df2]. The representations of Kac Moody group indeed play a central role in string models [ga1, ga2] and configuration space approach would explain their occurrence, not as a result of some quantization procedure, but as a consequence of symmetry of the underlying geometric structure.

ii) The bosonic oscillator operators of string models would correspond to centrally extended Lie-algebra generators of the isometry group acting on spinor fields of the configuration space.

iii) The "fermionic" fields (Ramond fields, [ga1, ga2]) should correspond to gamma matrices of the configuration space. Fermionic oscillator operators would correspond simply to contractions of isometry generators j_A^k with complexified gamma matrices of configuration space

$$\begin{aligned}
\Gamma_A^{\pm} &= j_A^k \Gamma_k^{\pm} \\
\Gamma_k^{\pm} &= (\Gamma^k \pm J_l^k \Gamma^l)/\sqrt{2}
\end{aligned} \qquad (5.2.4)$$

(J_l^k is the Kähler form of the configuration space) and would create various spin excitations of the configuration space spinor field. $\Gamma_k^{\pm}$ are the complexified gamma matrices, complexification made possible by the Kähler structure of the configuration space.

This suggests that some generalization of the so called Super Kac Moody algebra of string models [ga1, ga2] should be regarded as a spectrum generating algebra for the solutions of field equations in configuration space.

Although the Kähler structure seems to be physically well motivated there is a rather heavy counter argument against the whole idea. Kähler structure necessitates complex structure in the tangent space of the configuration space. In CP_2 degrees of freedom no obvious problems of principle are expected: configuration space should inherit in some sense the complex structure of CP_2.

In Minkowski degrees of freedom the signature of the Minkowski metric seems to pose a serious obstacle for complexification: somehow one should get rid of two degrees of freedom so that only two Euclidian degrees of freedom remain. An analogous difficulty is encountered in quantum field theories: only two of the four possible polarizations of gauge boson correspond to physical degrees of freedom: mathematically the wrong polarizations correspond to zero norm states and transverse states span a complex Hilbert space with Euclidian metric. Also in string model analogous situation occurs: in case of D-dimensional Minkowski space only $D-2$ transversal degrees of freedom are physical. The solution to the problem seems therefore obvious: configuration space metric must be degenerate so that each vibrational mode spans effectively a 2-dimensional Euclidian plane allowing complexification.

We shall find that the definition of Kähler function to be proposed indeed provides a solution to this problem and also to the problems listed before.

a) The definition of the metric doesn't differentiate between 1- and N-particle sectors, avoids spin statistics difficulty and has the physically appealing property that one can associate to each 3-surface a unique classical space time: classical physics is described by the geometry of the configuration space! And the geometry of the configuration space is determined uniquely by the requirement of mathematical consistency!

b) Complexification is possible only provided the dimension of the Minkowski space equals to four(!).

c) It is possible to identify a unique candidate for the necessary infinite-dimensional isometry group G. G is subgroup of the diffeomorphism group of $\delta M_+^4 \times CP_2$. Essential role is played by the fact that the boundary of the four-dimensional light cone, which, despite being topologically 3-dimensional, is metrically two-dimensional(!) Euclidian sphere, and therefore allows infinite-parameter groups of isometries as well as conformal and canonical symmetries and also Kähler structure unlike the higher-dimensional light cone boundaries. Therefore configuration space metric is Kähler only in the case of four-dimensional Minkowski space and allows symplectic $U(1)$ central extension without conflict with the no-go theorems about higher dimensional central extensions.

The study of the vacuum degeneracy of Kähler function defined by Kähler action forces to conclude that the isometry group must consist of the canonical transformations of $\delta H = \delta M_+^4 \times CP_2$. The corresponding Lie algebra can be regarded as a loop algebra associated with the canonical group of $S^2 \times CP_2$, where S^2 is $r_M = constant$ sphere of light cone boundary. Thus the finite-dimensional group G defining loop group in case of string models extends to an infinite-dimensional group in TGD context. This group is a real monster! The radial Virasoro localized with respect to $S^2 \times CP_2$ defines naturally complexification for both G and H. The general form of the Kähler metric deduced on basis of this symmetry has same qualitative properties as that deduced from Kähler function identified as the absolute minimum of Kähler action. Also the zero modes, among them isometry invariants, can be identified.

d) The construction of the configuration space spinor structure is based on the identification of the configuration space gamma matrices as linear superpositions of the oscillator operators associated with the second quantized induced spinor fields. The extension of the canonical invariance to super canonical invariance fixes the anti-commutation relations of the induced spinor fields, and configuration space gamma matrices correspond directly to the super generators. Physics as number theory vision suggests strongly that configuration space geometry exists for 8-dimensional imbedding space only and that the choice $M_+^4 \times CP_2$ for the imbedding space is the only possible one.

5.3   Configuration space geometry from Kähler function

The motivation the construction of configuration space geometry is that physics reduces to the geometry of classical spinor fields in the infinite-dimensional configuration space of 3-surfaces of $M_+^4 \times CP_2$ or $M^4 \times CP_2$, where M^4 and M_+^4 denote Minkowski space and its light cone respectively.

a) Hermitian conjugation is the basic operation in quantum theory and its geometrization requires that configuration space possesses Kähler geometry. One of the basic features of the Kähler geometry is that it is solely determined by the so called Kähler function, which defines both the Kähler form J and the components of the Kähler metric g in complex coordinates via the formulas [bc3]

$$
\begin{aligned}
J &= i\partial_k \partial_{\bar{l}} K dz^k \wedge d\bar{z}^l \ , \\
ds^2 &= 2\partial_k \partial_{\bar{l}} K dz^k d\bar{z}^l \ .
\end{aligned}
\tag{5.3.1}
$$

Kähler form is covariantly constant two-form and can be regarded as a representation of imaginary unit in the tangent space of the configuration space

$$
J_{mr} J^{rn} = -g_m{}^n \ .
\tag{5.3.2}
$$

As a consequence Kähler form defines also symplectic structure in configuration space.

5.3.1   Definition of Kähler function

The task of finding Kähler geometry for the configuration space reduces to that of finding Kähler function. The main constraints on the Kähler function result from the requirement of Diff^4 symmetry and degeneracy. General coordinate invariance requires that the definition of the Kähler function assigns to a given 3-surface X^3 a unique space-time surface $X^4(X^3)$, the generalized Bohr orbit defining the classical physics associated with X^3. The natural guess is that Kähler function is defined by what might be called Kähler action, which is essentially Maxwell action with Maxwell field expressible in terms of CP_2 coordinates. Absolute minimization is the first guess for how to fix $X^4(X^3)$ uniquely.

It has however become clear that this option might well imply that Kähler is negative and infinite for the entire Universe so that vacuum functional would be identically vanishing. I ended up with a more attractive option from number theoretical vision [E2]. According to this option the absolute value of the contribution to the Kähler action coming from each region where the action density has definite sign is minimized separately. As a consequence the preferred extremals are as near to vacuum extremals as possible.

If Kähler action would define a strictly deterministic variational principle, Diff^4 degeneracy and invariance would be achieved by restricting the consideration to 3-surfaces Y^3 at the boundary of M_+^4 and by defining Kähler function for 3-surfaces X^3 at $X^4(Y^3)$ and diffeo-related to Y^3 as $K(X^3) = K(Y^3)$. This reduction might be called quantum gravitational holography. The classical non-determinism of the Kähler action however introduces complications which might be however overcome by generalizing the notion of quantum gravitational holography.

5.3.2 Minkowski space or its light cone?

For a long time I believed that the question "M_+^4 or M^4?" had been settled in favor of M_+^4. The work with the conceptual problems related to energy and time, and with the symmetries of quantum TGD, however led gradually to the realization that there are strong reasons for considering M^4 instead of M_+^4.

a) It has become clear that the gigantic symmetries associated with the $\delta M_+^4 \times CP_2$ and more general surfaces $X_l^3 \times CP_2$, X_l^3 light like 3-surface of M^4 are also laboratory symmetries visible directly at the level of propagators and vertices [C5]. X_l^3 could be restricted to be a union of future and past directed light cone boundaries since arbitrary light like surfaces X_l^3 contain singularities such as self intersections. Poincare invariance fits very elegantly with the two types of super-conformal symmetries of TGD. The first conformal symmetry corresponds to the light-like surfaces $X_l^3 \times CP_2$ of the imbedding space. More general light like 7-surfaces are not favored because they do not possess the huge conformal symmetries of $X_l^3 \times CP_2$. Second conformal symmetry corresponds to light like boundaries of X^4 and light-like surfaces separating space-time regions with different signatures of the induced metric and is identifiable as the counterpart of the Kac Moody symmetry of string models. A rather plausible conclusion is that configuration space is a union of configuration spaces associated with $X_l^3 \times CP_2$, with X_l^3 identified as unions of future and past directed light cones. Thus the construction reduces to a high degree to a study of a simple special case $\delta M_1^4 \times CP_2$.

b) The replacement of the energy momentum tensor by a collection of conserved currents means that the sign of the energy depends on the time orientation of the space-time sheet. The simplest theory results if one assumes that the net quantum numbers of physical states vanishes. Crossing symmetry guarantees consistency with elementary particle physics. A consistency with the macroscopic physics results if gravitational energy is the difference of positive and negative inertial (Poincare) energies of matter. This option allows also M^4 and leads to a fractal cosmology in which light like 7-surfaces of $X_l^3 \times CP_2$ serve as causal determinants at the level of the imbedding space.

c) There is however a conceptual hurdle involved. Suppose that X^4 is the absolute minimum associated with X^3, and let Y^3 be some other $Diff^4$ related 3-surface at X^4. One cannot require that the absolute minima $X^4(X^3)$ and $X^4(Y^3)$ are same and one certainly cannot assign an absolute minimum to every 3-surface separately since general coordinate invariance for 3-surfaces would imply that Kähler function is infinitely many-valued. X^4 can thus be an absolute minimum only for some preferred 3-surface $X^3(X^4)$ at X^4 and the question is what makes this 3-surface preferred.

d) It seems that the internal geometry of $X^4(X^3)$ must be such that it defines uniquely $X^3(X^4)$, or perhaps even more generally, a light like causal determinant $X_l^3 \times CP_2$ of H to which X^3 belongs. If one requires that the net values of the conserved classical quantities are zero, one could regard X^4 as consisting of space-time sheets with opposite time orientation which are created at some moment from vacuum and possibly also disappear to vacuum. If this is the case then the 3-surface at which positive and negative energy space-time sheets are created and begin to evolve as separate branches presents a natural candidate for X^3. Also this view is also consistent with quantum holography and supported strongly by number theoretical considerations (M^4 has an interpretation as quaternion space with Minkowski metric defined as the real part of q^2).

5.3.3 Configuration space metric from symmetries

A complementary approach to the problem of constructing configuration space geometry is based on symmetries. The work of Dan Freed has demonstrated that the Kähler geometry of loop spaces is unique from the existence of Riemann connection and fixed completely by the Kac Moody symmetries of the space. In 3-dimensional context one has even better reasons to expect uniqueness. The guess is that configuration space is a union symmetric spaces labelled by zero modes not appearing in the line element as differentials. The generalized conformal invariance of metrically 2-dimensional light like 3-surfaces acting as causal determinants is the corner stone of the construction. The construction works only for 4-dimensional space-time and imbedding space which is a product of four-dimensional Minkowski space or its future light cone with CP_2.

5.3.4   Is absolute minimization the correct variational principle?

One can criticize the assumption that extremals correspond to the absolute minima of Kähler action. Any other principle allowing to assign to a given 3-surface a unique space-time surface in principle must in principle be considered as a viable alternative. The number theoretical vision discussed in [E2] indeed favors the separate minimization of magnitudes of positive and negative contributions to the Kähler action.

For this option Universe would do its best to save energy, being as near as possible to vacuum. Also vacuum extremals would become physically relevant: note that they would be only inertial vacua and carry non-vanishing density gravitational energy. The non-determinism of the vacuum extremals would have an interpretation in terms of the ability of Universe to engineer itself.

The 3-surfaces for which CP_2 projection is at least 2-dimensional and not a Lagrange manifold would correspond to non-vacua since conservation laws do not leave any other option. The variational principle would favor equally magnetic and electric configurations whereas absolute minimization of action based on S_K would favor electric configurations. The positive and negative contributions would be minimized for 4-surfaces in relative homology class since the boundary of X^4 defined by the intersections with 7-D light-like causal determinants would be fixed. Without this constraint only vacuum bubbles would result.

The attractiveness of the number theoretical variational principle from the point of calculability of TGD would be that the initial values for the time derivatives of the imbedding space coordinates at X^3 at light-like 7-D causal determinant could be computed by requiring that the energy of the solution is minimized. This could mean a computerizable solution to the construction of Kähler function. The number theoretic approach based on the properties of quaternions and octonions discussed in the chapter [E2] leads to a proposal for the general solution of field equations based on the generalization of the notion of calibration [bb1] providing absolute minima of volume to that of Kähler calibration. This approach will not be discussed in this chapter.

In this chapter I will first consider the basic properties of the configuration space, discuss briefly the various approaches to the geometrization of the configuration space, and introduce the two complementary strategies based on a direct guess of Kähler function and on the group theoretical approach assuming that configuration space can be regarded as a union of symmetric spaces. After these preliminaries a definition of the Kähler function is proposed and various physical and mathematical motivations behind the proposed definition are discussed. The key feature of the Kähler action is classical non-determinism, and various implications of the classical non-determinism are discussed.

5.4   Configuration space Kähler geometry from symmetry principles

The most general expectation is that configuration space can be regarded as a union of coset spaces: $C(H) = \cup_i G/H(i)$. Index i labels 3-topology and zero modes. The group G, which can depend on 3-surface, can be identified as a subgroup of diffeomorphisms of $\delta M_+^4 \times CP_2$ and H must contain as its subgroup a group, whose action reduces to $Diff(X^3)$ so that these transformations leave 3-surface invariant.

The task is to identify plausible candidate for G and to show that the tangent space of the configuration space allows Kähler structure, in other words that the Lie-algebras of G and $H(i)$ allow complexification. One must also identify the zero modes and construct integration measure for the functional integral in these degrees of freedom. Besides this one must deduce information about the explicit form of configuration space metric from symmetry considerations combined with the hypothesis that Kähler function is determined as absolute minimum of Kähler action or by some more general principle guaranteing Bohr orbit property.

It will be found that in the case of $M_+^4 \times CP_2$ Kähler geometry, or strictly speaking contact Kähler geometry, characterized by a degenerate Kähler form (Diff^4 degeneracy and plus possible other degeneracies) seems possible. Although it seems that this construction must be generalized by allowing all light like 7-surfaces $X_l^3 \times CP_2$, at least those for which X_l^3 is boundary of light-cone inside M_+^4 or M^4, with the physical interpretation differing dramatically from the original one, the original construction discussed in the sequel involves the most essential aspects of the problem.

5.4.1 General Coordinate Invariance and generalized quantum gravitational holography

The basic motivation for the construction of configuration space geometry is the vision that physics reduces to the geometry of classical spinor fields in the infinite-dimensional configuration space of 3-surfaces of $M_+^4 \times CP_2$ or of $M^4 \times CP_2$. Hermitian conjugation is the basic operation in quantum theory and its geometrization requires that configuration space possesses Kähler geometry. Kähler geometry is coded into Kähler function.

The original belief was that the four-dimensional general coordinate invariance of Kähler function reduces the construction of the geometry to that for the boundary of configuration space consisting of 3-surfaces on $\delta M_+^4 \times CP_2$, the moment of big bang. The proposal was that Kähler function $K(Y^3)$ could be defined as absolute minimum of so called Kähler action for the unique space-time surface $X^4(Y^3)$ going through given 3-surface Y^3 at $\delta M_+^4 \times CP_2$. For Diff^4 transforms of Y^3 at $X^4(Y^3)$ Kähler function would have the same value so that Diff^4 invariance and degeneracy would be the outcome.

This picture is however too simple.

a) The degeneracy of the absolute minima caused by the classical non-determinism of Kähler action however brings in additional delicacies, and it seems that the reduction to the light cone boundary which in fact corresponds to what has become known as quantum gravitational holography must be replaced with a construction involving more general light like 7-surfaces $X_l^3 \times CP_2$.

b) It has also become obvious that the gigantic symmetries associated with $\delta M_+^4 \times CP_2$ manifest themselves as the properties of propagators and vertices, and that M^4 is favored over M_+^4. Cosmological considerations, Poincare invariance, and the new view about energy favor the decomposition of the configuration space to a union of configuration spaces associated with various 7-D causal determinants. The minimum assumption is that all possible unions of future and past light cone boundaries $\delta M_\pm^4 \times CP_2 \subset M^4 \times CP_2$ label the sectors of CH: the nice feature of this option is that the considerations of this chapter restricted to $\delta M_+^3 \times CP_2$ generalize almost trivially. This option is beautiful because the center of mass degrees of freedom associated with the different sectors of CH would correspond to M^4 itself and its Cartesian powers. One cannot exclude the possibility that even more general light like surfaces $X_l^3 \times CP_2$ of M^4 are important as causal determinants.

The definition of the Kähler function requires that the many-to-one correspondence $X^3 \to X^4(X^3)$ must be replaced by a bijective correspondence in the sense that X^3 is unique among all its Diff^4 translates. This also allows physically preferred "gauge fixing" allowing to get rid of the mathematical complications due to Diff4 degeneracy. The internal geometry of the space-time sheet $X^4(X^3)$ must define the preferred 3-surface X^3 and also a preferred light like 7-surface $X_l^3 \times CP_2$.

This is indeed possible. The possibility of negative Poincare energies inspires the hypothesis that the total quantum numbers and classical conserved quantities of the Universe vanish. This view is consistent with experimental facts if gravitational energy is defined as a difference of Poincare energies of positive and negative energy matter. Space-time surface consists of pairs of positive and negative energy space-time sheets created at some moment from vacuum and branching at that moment. This allows to select X^3 uniquely and define $X^4(X^3)$ as the absolute minimum of Kähler action in the set of 4-surfaces going through X^3. These space-time sheets should also define uniquely the light like 7-surface $X_l^3 \times CP_2$, most naturally as the "earliest" surface of this kind. Note that this means that it become possible to assign a unique value of geometric time to the space-time sheet.

The realization of this vision means a considerable mathematical challenge. The effective metric 2-dimensionality of 3-dimensional light-like surfaces X_l^3 of M^4 implies generalized conformal and canonical invariances allowing to generalize quantum gravitational holography from light like boundary so that the complexities due to the non-determinism can be taken into account properly.

5.4.2 Light like 3-D causal determinants, 7-3 duality, and effective 2-dimensionality

Thanks to the non-determinism of Kähler action, also light like 3-surfaces X_l^3 of space-time surface appear as causal determinants (CDs). Examples are boundaries and elementary particle horizons at which Minkowskian signature of the induced metric transforms to Euclidian one. This brings in a second conformal symmetry related to the metric 2-dimensionality of the 3-D CD. This symmetry is identifiable as TGD counterpart of the Kac Moody symmetry of string models. The challenge is to understand the relationship of this symmetry to configuration space geometry and the interaction between the two conformal symmetries.

The possibility of spinorial shock waves at X_l^3 leads to the hypothesis that they correspond to particle aspect of field particle duality whereas the physics in the interior of space-time corresponds to field aspect. More generally, field particle duality in TGD framework states that 3-D light like CDs and 7-D CDs are dual to each other. In particular, super-canonical and Super Kac Moody symmetries are also dually related.

The underlying reason for 7–3 duality be understood from a simple geometric picture in which 3-D light like CDs X_l^3 intersect 7-D CDs X^7 along 2-D surfaces X^2 and thus form 2-sub-manifolds of the space-like 3-surface $X^3 \subset X^7$. One can regard either canonical deformations of X^7 or Kac-Moody deformations of X^2 as defining the tangent space of configuration space so that 7–3 duality would relate two different coordinate choices for CH.

The assumption that the data at either X^3 or X_l^3 are enough to determine configuration space geometry implies that the relevant data is contained to their intersection X^2. This is the case if the deformations of X_l^3 not affecting X^2 and preserving light likeness corresponding to zero modes or gauge degrees of freedom and induce deformations of X^3 also acting as zero modes. The outcome is effective 2-dimensionality. One cannot over-emphasize the importance of this conclusion. It indeed stream lines dramatically the earlier formulas for configuration space metric involving 3-dimensional integrals over $X^3 \subset M_+^4 \times CP_2$ reducing now to 2-dimensional integrals. Most importantly, no data about absolute minima of Kähler are needed to construct the configuration space metric so that the construction is also practical.

The reduction of data to that associated with 2-D surfaces conforms with the number theoretic vision about imbedding space as having hyper-octonionic structure [E2]: the commutative sub-manifolds of H have dimension not larger than two and for them tangent space is complex sub-space of hyper-octonion tangent space. Number theoretic counterpart of quantum measurement theory forces the reduction of relevant data to 2-D commutative sub-manifolds of X^3. These points are discussed in more detail in the next chapter whereas in this chapter the consideration will be restricted to $X_l^3 = \delta M_+^4$ case which involves all essential aspects of the problem.

5.4.3 Magic properties of light cone boundary and isometries of configuration space

The special conformal, metric and symplectic properties of the light cone of four-dimensional Minkowski space: δM_+^4, the boundary of four-dimensional light cone is metrically 2-dimensional(!) sphere allowing infinite-dimensional group of conformal transformations and isometries(!) as well as Kähler structure. Kähler structure is not unique: possible Kähler structures of light cone boundary are parameterized by Lobatchevski space $SO(3,1)/SO(3)$. The requirement that the isotropy group $SO(3)$ of S^2 corresponds to the isotropy group of the unique classical 3-momentum assigned to $X^4(Y^3)$ defined as absolute minimum of Kähler action (or some other preferred extremal [E2]), fixes the choice of the complex structure uniquely. Therefore group theoretical approach and the approach based on Kähler action complement each other.

The allowance of an infinite-dimensional group of isometries isomorphic to the group of conformal transformations of 2-sphere is completely unique feature of the 4-dimensional light cone boundary. Even more, in case of $\delta M_+^4 \times CP_2$ the isometry group of δM_+^4 becomes localized with respect to CP_2! Furthermore, the Kähler structure of δM_+^4 defines also symplectic structure.

Hence any function of $\delta M_+^4 \times CP_2$ would serve as a Hamiltonian transformation acting in both CP_2 and δM_+^4 degrees of freedom. These transformations obviously differ from ordinary local gauge transformations. This group leaves the symplectic form of $\delta M_+^4 \times CP_2$, defined as the sum of light cone and CP_2 symplectic forms, invariant. The group of canonical transformations of $\delta M_+^4 \times CP_2$ is a good candidate for the isometry group of the configuration space.

The approximate canonical invariance of Kähler action is broken only by gravitational effects and is exact for vacuum extremals. This suggests that Kähler function is in a good approximation invariant under the canonical transformations of CP_2 would mean that CP_2 canonical transformations correspond to zero modes having zero norm in the Kähler metric of configuration space.

The groups G and H, and thus configuration space itself, should inherit the complex structure of the light cone boundary. The diffeomorphims of M^4 act as dynamical symmetries of vacuum extremals. The radial Virasoro localized with respect to $S^2 \times CP_2$ could in turn act in zero modes perhaps inducing conformal transformations: note that these transformations lead out from the symmetric space associated with given values of zero modes.

5.4.4   Canonical transformations of $\delta M_+^4 \times CP_2$ as isometries of configuration space

The canonical transformations of $\delta M_+^4 \times CP_2$ are excellent candidates for inducing canonical transformations of the configuration space acting as isometries. There are however deep differences with respect to the Kac Moody algebras.

a) The conformal algebra of the configuration space is gigantic as compared with the Virasoro + Kac Moody algebras of string models as is clear from the fact that the Lie-algebra generator of a canonical transformation of $\delta M_+^4 \times CP_2$ corresponding to a Hamiltonian which is product of functions defined in δM_+^4 and CP_2 is sum of generator of δM_+^4-local canonical transformation of CP_2 and CP_2-local canonical transformations of δM_+^4. This means also that the notion of local gauge transformation generalizes.

b) The physical interpretation is also quite different: the relevant quantum numbers label the unitary representations of Lorentz group [de3] and color group, and the four-momentum labelling the states of Kac Moody representations is not present. Physical states carrying no energy and momentum at quantum level are predicted. The appearance of a new kind of angular momentum not assignable to elementary particles might shed some light to the longstanding problem of baryonic spin (quarks are not responsible for the entire spin of proton). The possibility of a new kind of color might have implications even in macroscopic length scales.

c) The central extension induced from the natural central extension associated with $\delta M_+^4 \times CP_2$ Poisson brackets is anti-symmetric with respect to the generators of the canonical algebra rather than symmetric as in the case of Kac Moody algebras associated with loop spaces. At first this seems to mean a dramatic difference. For instance, in the case of CP_2 canonical transformations localized with respect to δM_+^4 the central extension would vanish for Cartan algebra, which means a profound physical difference. For $\delta M_+^4 \times CP_2$ canonical algebra a generalization of the Kac Moody type structure however emerges naturally.

The point is that δM_+^4-local CP_2 canonical transformations are accompanied by CP_2 local δM_+^4 canonical transformations. Therefore the Poisson bracket of two δM_+^4 local CP_2 Hamiltonians involves a term analogous to a central extension term symmetric with respect to CP_2 Hamiltonians, and resulting from the δM_+^4 bracket of functions multiplying the Hamiltonians. This additional term could give the entire bracket of the configuration space Hamiltonians at the maximum of the Kähler function where one expects that CP_2 Hamiltonians vanish and have a form essentially identical with Kac Moody central extension because it is indeed symmetric with respect to indices of the canonical group.

5.4.5   Symmetric space property reduces to conformal and canonical invariance

The idea about symmetric space is extremely beautiful but millenium had to change before I was ripe to identify what seems to be the precise form of the Cartan decomposition. The solution of the puzzle turned out to be amazingly simple.

The inspiration came from the finding that quantum TGD leads naturally to an extension of Super Algebras by combining Ramond and Neveu-Schwartz algebras into single algebra. This led to the introduction Virasoro generators and generators of canonical algebra of CP_2 localized with respect to the light cone boundary and carrying conformal weights with a half integer valued real part. Soon came the realization that the conformal weights $h = -1/2 - i\sum_i y_i$, where $z_i = 1/2 + y_i$ are non-trivial zeros of Riemann Zeta, are excellent candidates for the conformal weights. It took some time to answer affirmatively the question whether also the negatives of the trivial zeros $z = -2n$, $n > 0$ should be included. Thus the conjecture inspired by the work with Riemann hypothesis stating that the zeros of Riemann Zeta appear at the level of basic quantum TGD turned out be correct.

The generators whose commutators define the basis of the entire algebra have conformal weights given by the negatives of the zeros of Rieman Zeta. The algebra is a direct sum $g = g_1 \oplus g_2$ such that g_1 has $h = n$ as conformal weights and g_2 $h = n - 1/2 + iy$, where y is sum over imaginary parts y_i of non-trivial zeros of Zeta. Only $h = 2n$, $n > 1$, and $h = -1/2 - iy + n$, such that n is even (odd) if y is sum of odd (even) number of y_i correspond to the weights labelling the generators of t in the Cartan decomposition $g = h + t$. The resulting super-canonical algebra would quite well be christened as Riemann algebra.

The requirement that ordinary Virasoro and Kac Moody generators annihilate physical states corresponds now to the fact that the generators of h vanish at the point of configuration space, which remains invariant under the action of h. The maximum of Kähler function corresponds naturally to this point and plays also an essential role in the integration over configuration space by generalizing the Gaussian integration of free quantum field theories.

The light cone conformal invariance differs in many respects from the conformal invariance of string theories. Finite-dimensional Kac-Moody group is replaced by infinite-dimensional canonical group. Conformal weights correspond to zeros of Riemann zeta and suitable superpositions of them in case of trivial zeros, and physical states can have non-vanishing conformal weights just as the representations of color group in CP_2 can have non-vanishing color isospin and hyper charge. The conformal weights have also interpretation as quantum numbers associated with unitary representations of Lorentz group: thus there is no conflict between conformal invariance and Lorentz invariance in TGD framework.

5.4.6 Magnetic Hamiltonians

Assuming that the elements of the radial Virasoro algebra of δM^4_+ have zero norm, one ends up with an explicit identification of the symplectic structure of the configuration space. There is an almost unique identification for the symplectic structure. Configuration space counterparts of $\delta M^4 \times CP_2$ Hamiltonians are defined by the generalized signed and and unsigned Kähler magnetic fluxes

$$Q_m(H_A, X^2) = Z \int_{X^2} H_A J \sqrt{g_2} d^2x \ ,$$

$$Q_m^+(H_A, r_M) = Z \int_{X^2} H_A |J| \sqrt{g_2} d^2x \ ,$$

$$J \equiv \epsilon^{\alpha\beta} J_{\alpha\beta} \ .$$

H_A is CP_2 Hamiltonian multiplied by a function of coordinates of light cone boundary belonging to a unitary representation of the Lorentz group. Z is a conformal factor depending on canonical invariants. The symplectic structure is induced by the symplectic structure of CP_2.

The most general flux is superposition of signed and unsigned fluxes Q_m and Q_m^+.

$$Q_m^{\alpha,\beta}(H_A, X^2) = \alpha Q_m(H_A, X^2) + \beta Q_m^+(H_A, X^2) \ .$$

Thus it seems that symmetry arguments fix the form of the configuration space metric apart from the presence of a conformal factor Z multiplying the magnetic flux and the degeneracy related to the signed and unsigned fluxes.

The notion of 7–3-duality described in the introduction implies that the relevant data about configuration space geometry is contained by 2-D surfaces X^2 at the intersections of 3-D light like CDs (causal determinants) and 7-D CDs such as $M^4_+ \times CP_2$. In this case the entire Hamiltonian could be defined as the sum of magnetic fluxes over surfaces $X_i^2 \subset X^3$. The maximally optimistic guess would be that it is possible to fix both X_i^2 and 7-D CDs freely with X_i^2 possibly identified as commutative sub-manifold of hyper-octonionic H [E2].

5.4.7 Electric Hamiltonians and electric-magnetic duality

Absolute minimization of Kähler action in turn suggests that one can identify configuration space Hamiltonians as classical charges $Q_e(H_A)$ associated with the Hamiltonians of the canonical transformations of the light cone boundary, that is as variational derivatives of the Kähler action with respect to the infinitesimal deformations induced by $\delta M_+^4 \times CP_2$ Hamiltonians. Alternatively, one might simply replace Kähler magnetic field J with Kähler electric field defined by space-time dual $*J$ in the formulas of previous section. These Hamiltonians are analogous to Kähler electric charge and the hypothesis motivated by the experience with the instantons of the Euclidian Yang Mills theories and 'Yin-Yang' principle, as well as by the duality of CP_2 geometry, is that for the absolute minima of the Kähler action these Hamiltonians are affinely related:

$$Q_e(H_A) = Z\left[Q_m(H_A) + q_e(H_A)\right] \ .$$

Here Z and q_e are constants depending on canonical invariants only. Thus the equivalence of the two approaches to the construction of configuration space geometry boils down to the hypothesis of a physically well motivated electric-magnetic duality.

The crucial technical idea is to regard configuration space metric as a quadratic form in the entire Lie-algebra of the isometry group G such that the matrix elements of the metric vanish in the sub-algebra H of G acting as $Diff^3(X^3)$. The Lie-algebra of G with degenerate metric in the sense that H vector fields possess zero norm, can be regarded as a tangent space basis for the configuration space at point X^3 at which H acts as an isotropy group: at other points of the configuration space H is different. For given values of zero modes the maximum of Kähler function is the best candidate for X^3. This picture applies also in symplectic degrees of freedom.

5.5 Complications due to the failure of determinism in standard sense of the word

If Kähler action were strictly deterministic, the construction of the configuration space geometry would reduce to $\delta H = \delta M_+^4 \times CP_2$. The classical non-determinism of the Kähler action does not however allow to realize quantum gravitational holography in the simplest sense of the word. One can however consider a generalization of the notion of the quantum holography.

5.5.1 The challenges posed by the non-determinism of Kähler action

The vision discussed in this chapter is far from a complete solution to the problem of constructing the configuration space geometry. The non-determinism of Kähler action means that the reduction of the construction of the configuration space geometry to the light cone boundary fails. Besides degeneracy of the absolute minima of Kähler action, the non-determinism should manifest itself as presence of causal determinants also other than light cone boundary. One can imagine two kinds of causal determinants.

a) Elementary particle horizons and light-like boundaries $X_l^3 \subset X^4$ of 4-surfaces. In this case only quaternion conformal algebra makes sense.

b) Causal determinants could also correspond to light like 7-surfaces of $X_l^3 \times CP_2 \subset H = M^4 \times CP_2$ or perhaps even more general light like 7-surfaces of H. It would be a pity if Nature would have not expressed physically the extremely elegant mathematic associated with the light like surfaces. Hence the intuitive and somewhat irrational expectation is that super-canonical algebra emerges in some natural manner also for these more general light like surfaces.

c) The light-like 7-surfaces $X_l^3 \times CP_2$ allow super-canonical symmetries whereas for completely general light like surfaces $X_l^7 \subset H$ this symmetry is broken. Hence mathematical elegance does not favor these surfaces. The requirement that that $X_l^3 \subset M^4$ is non-singular and allows Lorentz group as isometries, leaves only the option $X_l^4 = \delta M_\pm^4$ with $\pm$ telling whether a boundary of a future or past light cone is in question.

d) An elegant looking manner to take into account the non-determinism is motivated by TGD inspired cosmology: replace the configuration space CH associated with the future light cone with the union of configuration spaces associated with all possible unions of future and past light cones $M_\pm^4 \times CP_2$ with dips at arbitrary points of M^4. This would make the theory Poincare invariant and super-canonical algebra allows string mass formula with translations realized as translations of an entire sector of configuration space. An attractive hypothesis is that this is enough to take into account the non-determinism of Kähler action and that there exists a duality in the sense that the use of 7-D and 3-D light like causal determinants (imbedding space level and space-time level) provide dual approaches to the construction of the configuration space geometry. The dual construction utilizing 3-D light like causal determinants will be discussed in the next chapter.

5.5.2 Category theory and configuration space geometry

Due the effects caused by the classical non-determinism even classical TGD universes are very far from simple Cartesian clockworks, and the understanding of the general structure of the configuration space is a formidable challenge. Category theory is a branch of mathematics which is basically a theory about universal aspects of mathematical structures. Thus category theoretical thinking might help in disentangling the complexities of the configuration space geometry and the basic ideas of category theory are discussed in this spirit. It indeed turns out that the approach makes highly non-trivial predictions. So called ribbon categories discussed in [C5] seem to be tailor made for the formulation of quantum TGD and allow to build bridge to topological and conformal field theories.

5.5.3 Super-conformal symmetries and duality

There are two types of causal determinants (CDs) corresponding to 7-surfaces $X_l^3 \times CP_2$ of imbedding space and light like 3-surfaces of space-time surface. The basic question is whether both of them contribute separately to the configuration space geometry or whether they provide descriptions which are in some sense dual.

Duality would allow to organize various super conformal symmetries to dual pairs.

a) In [B4] it will be found that 3-D light like CDs allow genuinely 3-D solutions of the modified Dirac equation, kind of spinorial shock waves besides four-dimensional solutions. The interpretation of elementary particles as this kind of spinorial shock waves is an extremely attractive option. By the metric 2-dimensionality of light like 3-D CDs a slight generalization of ordinary 2-D super-conformal invariance is involved. Note that the role of light like 3-D CDs are very much like closed string world sheets.

b) Number theoretic argument lead to the hypothesis that also the interior of space-time surface allows conformal invariance. The first identification is as quaternion conformal invariance acting as a gauge invariance. Second, a less plausible, identification is as Abelian 4-D conformal invariance based on the generalization of hyper-complex numbers whose action is realized in terms of sigma matrices. Both super conformal symmetries would act as pure gauge symmetries and give rise to $N = 4$ super gauge symmetries [B4]. A natural interpretation would be in terms of quantum holography: the super-conformal gauge symmetries in the interior and at 3-D causal determinants and correspond to field and particle aspects of field particle duality.

c) One can assign to the conformal gauge symmetries of 7-D CDs what I have used to call super-canonical invariance and to the conformal symmetries of 3-D light like CDs super Kac-Moody algebra crucial for p-adic mass calculations. Duality would mean that the two descriptions of quantum would be dual. This would resolve the difficult question like "Do both 3-D and 7-D causal determinants contribute separately to the configuration space geometry?". In [B4] it will be argued that the Dirac determinant associated with the modified Dirac action at 3-D light like CDs indeed allows to construct configuration space geometry: this description would be dual to the construction of the previous chapter.

5.5.4 Divergence cancellation and configuration space geometry

Divergences, which have plagued quantum field theories since their discovery, are basically due to the micro-locality of quantum field theories. In TGD framework 3-surface becomes the basic dynamical object instead of a point like particle and physics is local only at the level of configuration space whereas Kähler function is a non-local functional of the 3-surface: this eliminates the loop divergences resulting from the local interaction vertices in quantum fluctuating degrees of freedom. The localization occurring in each quantum jump in zero modes and having interpretation as state function reduction saves from the divergences related to the lack of Gaussian integration in zero modes.

This does not however eliminate all sources of divergences. The cancellation of metric and Gaussian determinants in the configuration space functional integral eliminates the TGD counterparts of the standard divergences of quantum field theories. In higher orders divergence cancellation implies Ricci flatness. The conditions guaranteing Ricci flatness are discussed and it is shown that the basic Lie-algebraic properties implied by the symmetric space metric property imply Ricci flatness.

The so called Hyper Kähler property meaning the existence of quaternionic structure in the tangent space of the configuration space would imply Ricci flatness and the quaternion structure of space-time surface forces to take seriously the possibility of Hyper Kähler structure. Contrary to the earlier expectations, it seems that Hyper Kähler property means that sphere S^2 labels the possible complexifications. The choice of the imaginary unit reduces basically to the choice of the quantization axis for the rotation group $SO(3)$ for $r_M = constant$ sphere at the light cone boundary so that S^2 parameterizes the possible choices.

Number theoretic constraints from the p-adicization of the theory sharpen the requirement about divergence cancellation: loops are not only finite but vanish. For generalization Feynman diagrams analogous to tangles with chords this corresponds to the equivalence loop diagrams with tree diagrams: a generalization of the duality symmetry of string models would be in question. In [C5] these ideas are developed in detail.

Chapter 6

Configuration Space Spinor Structure

6.1 Introduction

Quantum TGD should be reducible to the classical spinor geometry of the configuration space. In particular, physical states should correspond to the modes of the configuration space spinor fields. The immediate consequence is that configuration space spinor fields cannot, as one might naively expect, be carriers of a definite spin and unit fermion number. Concerning the construction of the configuration space spinor structure there are some important clues.

6.1.1 Geometrization of fermionic statistics in terms of configuration space spinor structure

The great vision has been that the second quantization of the induced spinor fields can be understood geometrically in terms of the configuration space spinor structure in the sense that the anti-commutation relations for configuration space gamma matrices require anti-commutation relations for the oscillator operators for free second quantized induced spinor fields.

a) One must identify the counterparts of second quantized fermion fields as objects closely related to the configuration space spinor structure. Ramond model [ga1] has as its basic field the anti-commuting field $\Gamma^k(x)$, whose Fourier components are analogous to the gamma matrices of the configuration space and which behaves like a spin 3/2 fermionic field rather than a vector field. This suggests that the complexified gamma matrices of the configuration space are analogous to spin 3/2 fields and therefore expressible in terms of the fermionic oscillator operators so that their anti-commutativity naturally derives from the anti-commutativity of the fermionic oscillator operators.

As a consequence, configuration space spinor fields can have arbitrary fermion number and there would be hopes of describing the whole physics in terms of configuration space spinor field. Clearly, fermionic oscillator operators would act in degrees of freedom analogous to the spin degrees of freedom of the ordinary spinor and bosonic oscillator operators would act in degrees of freedom analogous to the 'orbital' degrees of freedom of the ordinary spinor field.

b) The classical theory for the bosonic fields is an essential part of the configuration space geometry. It would be very nice if the classical theory for the spinor fields would be contained in the definition of the configuration space spinor structure somehow. The properties of the modified massless Dirac operator associated with the induced spinor structure are indeed very physical. The modified massless Dirac equation for the induced spinors predicts a separate conservation of baryon and lepton numbers. Contrary to the long held belief it seems that covariantly constant right handed neutrino does not generate $N = 1$ super symmetry. The differences between quarks and leptons result from the different couplings to the CP_2 Kähler potential. In fact, these properties are shared by the solutions of massless Dirac equation of the imbedding space.

c) Since TGD should have a close relationship to the ordinary quantum field theories it would be highly desirable that the second quantized free induced spinor field would somehow appear in the definition of the configuration space geometry. This is indeed true if the complexified configuration space gamma matrices are linearly related to the oscillator operators associated with the second quantized induced spinor field on the space-time surface and its boundaries. There is actually no deep reason forbidding the gamma matrices of the configuration space to be spin half odd-integer objects whereas in the finite-dimensional case this is not possible in general. In fact, in the finite-dimensional case the equivalence of the spinorial and vectorial vielbeins forces the spinor and vector representations of the vielbein group $SO(D)$ to have same dimension and this is possible for $D = 8$-dimensional Euclidian space only. This coincidence might explain the success of 10-dimensional super string models for which the physical degrees of freedom effectively correspond to an 8-dimensional Euclidian space.

d) It took a long time to realize that the ordinary definition of the gamma matrix algebra in terms of the anti-commutators $\{\gamma_A, \gamma_B\} = 2g_{AB}$ must in TGD context be replaced with

$$\{\gamma_A^\dagger, \gamma_B\} = iJ_{AB} \quad ,$$

where J_{AB} denotes the matrix elements of the Kähler form of the configuration space. The presence of the Hermitian conjugation is necessary because configuration space gamma matrices carry fermion number. This definition is numerically equivalent with the standard one in the complex coordinates. The realization of this delicacy is necessary in order to understand how the square of the configuration space Dirac operator comes out correctly.

e) What second quantization for spinor fields at space-time surface really means is a highly non-trivial problem. The simplest option would be that free spinor fields of the imbedding space are second quantized and simply restricted to the space-time surface. This however leads to several problems. The relationship between color and electro-weak quantum numbers comes out incorrectly. The number of degrees of freedom for second quantized imbedding space spinor fields seems to be too small. The explicit construction of the configuration space gamma matrices as super-charges fails. Thus the only possible option is that induced spinor fields are second quantized at the space-time surface.

TGD as a generalized number theory vision leads to the understanding of how the second quantization of the induced spinor fields should be carried out and space-time conformal symmetries allow to explicitly solve the Dirac equation associated with the modified Dirac action in the interior and at the 3-D light like causal determinants (spinorial shock waves). These solutions represent super gauge degrees of freedom. For 3-surfaces at 7-D CDs the anti-commutation relations for the induced spinor fields are fixed by the anti-commutation relations of the canonical super-charges whereas at light like 3-D CDs they are fixed by the anti-commutation relations of Kac Moody super charges. These two manners to fix the anti-commutation relations must be equivalent.

6.1.2   Dualities and representations of configuration space gamma matrices as super-canonical and super Kac-Moody super-generators

There are several approaches to the construction of the configuration space geometry and dualities provide a powerful framework for unifying these seemingly disparate approaches. One must however be extremely cautious in order to avoid exaggerations here and must honestly admit that the proposed dualities are just interesting hypothesis rather than proven mathematical facts.

a) The construction of the configuration space geometry led to the notion of electric-magnetic dualityfor more than decade ago. One can imagine a variant of this duality by allowing both signs of Kähler coupling strength. This would correspond to the dominance of electric/magnetic fields and to the two different arrows of the geometric time and 7-D causal determinants which are future/past directed. The first guess was that the possibility to identify space-time surfaces as surfaces for which tangent/normal space defines a quaternionic sub-algebra of octonionic tangent space of H could be the geometric counterpart of this duality.

A more promising identification is based on the number theoretic spontaneous compactification conjecture stating that space-time surfaces can be regarded either as hyper-quaternionic or co-hyper-quaternionic 4-surfaces of hyper-quaternionic imbedding space M^8 or as 4-surfaces in $H = M^4 \times CP_2$ [E2]. Here the attribute "hyper" means that the imaginary units of quaternion/octonion algebra are multiplied by $\sqrt{-1}$ to obtain a sub-space of complexified quaternions/octonions and Minkowskian signature of number theoretic norm.

b) A duality between 7-D and 3-D light like CDs would simplify a lot the construction of the theory. The basic idea behind 7-3 duality is that space-time surface is fixed completely once either

3-D light like CDs or space-like 3-surfaces at 7-D CDs are given: one cannot fix both arbitrarily. Also other variants can be considered such as fixing 3-D and 7-D light like CDs. The two choices would roughly correspond to the possibility of fixing either initial values or boundary values to solve field equations.

3-D space-like surfaces at 7-D causal determinants define the Kähler metric of CH. They contribute to Kähler function and the idea would be that they *alone* define Kähler function and thus also Kähler metric indirectly. The data from space like 3-surfaces X^3 at 7-D CDs would not be needed in order to deduce Kähler function. As will be found, that this means that the Kähler actions associated with the maximal deterministic regions $X^4_\pm$ of the positive and negative energy branches of the space-time surface meeting at X^3 have the same value. This is guaranteed if the modified Dirac operators D_+ and D_- have same eigenvalue spectrum at X^3.

The strongest form of 7–3 duality would be quantum gravitational holography in a strong form: light like 3-D CDs would provide a representation for the theory and a very intimate connection with closed super-string models would result. This alternative leads to a remarkable simplification of the basic formulas for configuration space Hamiltonians, Kähler metric, and gamma matrices since one can restrict the integrals in the defining formulas to 2-D intersections X^2 of 3-D CDs with 7-D CDs identifiable as sub-manifolds of space-like 3-surfaces.

Duality would means that configuration space gamma matrices identified as Super Kac Moody generators should also anticommute to configuration space metric just like the super-canonical charges do. This is quite possible: the representations would correspond to two different coordinates for the tangent space of CH determined by the Hamiltonians of $\delta M^4_\pm \times CP_2$ and by Kac Moody Lie-algebra and if the coordinatizations are faithful 7–3 duality corresponds to a change of CH coordinates.

The objection is that super-canonical representations associated with 7-D CDs and Super-Kac Moody algebras associated with 3-D CDs do not seem to be in dual relation. Super-canonical and Super Kac-Moody representations can be realized at the above mentioned 2-D intersections X^2_i, and the action of Kac Moody algebra on super-canonical algebra is well defined and does not lead out of super-canonical algebra. Hence one can hope that same representation space defines representations of both algebras, at least when one allows the representation to consist of several irreducible representations of both algebras.

There are some reasons to consider the possibility that the corresponding algebras are in same relation and Virasoro algebras of group G and its subgroup H in Goddard-Kent construction giving rise to super Virasoro representation with a vanishing central charge using differences of the super Virasoro generators in question. Hence the super Virasoro algebras would be dual in the sense that their actions cancel each other.

6.1.3 Modified Dirac equation for induced classical spinor fields

The earlier approach to the definition of the configuration space spinor structure relied on the second quantized ordinary massless Dirac action for the induced spinors. This action had some anomalous looking features. The first anomaly was the appearance of the effective tachyonic mass term proportional to the trace of the second fundamental form vanishing only for minimal surfaces. The breaking of $N = 2$ super symmetry generated by right-handed neutrinos for other than minimal surfaces was the second anomalous feature. It became also clear that the divergences of the fermionic isometry currents can have a non-vanishing c-number anomaly unless one varies Dirac action also with respect to the configuration space coordinates. This anomaly obviously might destroy the definition of the configuration space spinor structure.

The vision about quantum TGD as a generalized number theory [E1, E2, E3] comes in rescue here. One of its outcomes was the realization that, in order to achieve exact super-symmetry, one must modify Dirac action so that its variation with respect to the imbedding space coordinates gives the field equations derivable from the Kähler action. By taking the modified Dirac action as the fundamental action, one can identify Kähler action as a c-number term resulting from the normal ordering of the modified Dirac action, and predict the value of the Kähler coupling constant and quantum criticality.

The interpretation of the solutions of the modified Dirac equations differs from the conventional one and is motivated by the construction of the Kähler function in terms of Dirac determinants. The solutions of the modified Dirac equation are interpreted as generators of exact $N = 4$ super symmetries in both quark and lepton sectors. These super-symmetries correspond to pure super gauge transformations and no spartners of ordinary particles are predicted: in particular $N = 2$ super-symmetry (more or less equivalent with $N = 1$ supersymmetry since so called short representations were involved) generated by the righthanded neutrino is absent contrary to the earlier beliefs. There is no need to emphasize the experimental implications of this finding.

There are arguments suggesting that the space-time super-symmetries can be regarded as hyper-quaternion conformal symmetries [E2]. This could be due to the additional degeneracy of the configuration space metric caused by 7–3 duality and implying that there is larger number of gauge equivalent space-time surfaces having representatives allowing hyper-quaternion analytic symmetries.

The physics is coded by the eigen-states of the modified Dirac operator at space-like 3-surfaces. By their light likeness of 7-D and 3-D CDs modified Dirac operator is Hermitian and its square vanishes so that in this sense massless particles are in question. The generators of super-canonical and super Kac-Moody algebras are invariant under the super-gauge transformations generated by the solutions of the modified Dirac equation.

In particular, configuration space gamma matrices identified as super generators of super-canonical and Super Kac-Moody algebras (depending on CH coordinates used) are expressible in terms of the oscillator operators associated with the eigen modes of D^+ and D_-^{-1} associated with the two branches of the space-time surface meeting at X^3 along 2-surfaces X^2. By 7–3 duality the different light-like surfaces having a given intersection with 7-D CD correspond to different values of the configuration space zero modes, and the super charges are expressible as two-dimensional integrals over 2-D surface X^2. The resulting situation is highly reminiscent of WZW model and the results imply that at technical level the methods of 2-D conformal field theories should allow to construct quantum TGD.

6.1.4 Could the solutions of the modified Dirac equation define c-number valued space-time correlates for physical states?

The super-positions of the c-number valued solutions of the modified Dirac equation define classical commuting spinor fields. The conventional view about Dirac equation would suggest that these classical spinor fields serve as a classical space-time correlate for quantal fermionic degrees of freedom. If the solutions of the modified Dirac equation can carry non-trivial conserved isometry charges, this interpretation makes sense.

The interpretation implies that the classical isometry currents associated with the Dirac equation must superpose with the isometry currents associated with Kähler action. Therefore the field equations, which state the conservation of the bosonic isometry currents, would be modified to the statement that net isometry currents are conserved. This would conform with the standard interpretation and in fact give a precise meaning to the otherwise fuzzy-at-classical-level notion of fermion-boson coupling. Instead of Kähler action, the sum of Kähler action and Dirac action would be minimized for this option.

This interpretation does not conform with the identification of the solutions of the modified Dirac equation as c-number valued super conformal gauge symmetry generators albeit being inspired by it. Also the assumption that the ratio of Dirac determinants defines the exponent of Kähler action is in conflict with the idea that the sum of Kähler action and Dirac action defines the classical dynamics.

Obviously, both of these interpretations cannot be correct. The following arguments support the new interpretation.

a) Fermionic zeros modes in $Int(X^4)$ can be assigned with space-like 3-surfaces X^3 at 7-D light-like causal determinants X^7, and their number is probably very small because boundary conditions tend to eliminate them. The modes might well have vanishing isometry charges. For instance, only the covariantly constant right handed neutrino solves massless Dirac equation in CP_2. Hence boundary conditions alone could imply that fermionic and bosonic isometry currents are conserved separately in the interior.

b) For the "shock wave" solutions of the modified Dirac equation associated with light-like boundaries and causal determinants the normal component of a bosonic isometry current would be equal to the divergence of the corresponding fermionic isometry current. The proposed boundary conditions however imply that the normal components of bosonic currents vanish so that decoupling occurs. As a matter fact, even the mathematical existence of the fermionic isometry charges becomes questionable since the determinant of the induced metric vanishes and time-like direction is light-like.

6.1.5 The exponent of Kähler function as Dirac determinant for the modified Dirac action?

Although quantum criticality in principle predicts the possible values of Kähler coupling strength, one might hope that there exists even more fundamental approach involving no coupling constants and predicting even quantum criticality and realizing quantum gravitational holography. The ratio of Dirac determinants of the modified Dirac operators D_+ and D_- associated with the space-time regions separated by a 3-dimensional causal determinant is an excellent candidate in this respect and the guess is that it is expressible as an exponent for the difference of Kähler actions in the adjacent regions. This allows to deduce the exponent of Kähler function and one could identify it as a renormalized Dirac determinant.

If the operators D_+ and D_- commute and if spectrum of the operator $D_+ D_-^{-1}$ is invariant under $\lambda \to 1/\lambda$ apart from a finite number of eigenvalues for which the ratio equals to the ratio of exponents of respective Kähler actions, the Dirac determinant in question is well-defined even without zeta function regularization and allows to deduce Kähler function.

The modified Dirac operator indeed allow to realize quantum gravitational holography since it reduces at lightlike 3-D CDs to an effectively 3-dimensional Dirac operator by boundary conditions allowing "shock wave" solutions with particle interpretation but depends on the normal derivatives of the imbedding space coordinates. Internal consistency poses conditions on the normal derivatives are good hopes of evaluating the exponent of Kähler function as a Dirac determinant without solving the field equations. The non-determinism of Kähler actions is however present so that the values of the normal derivatives are not expected to be unique.

6.2 Configuration space spinor structure: general definition

The basic problem in constructing configuration space spinor structure is clearly the construction of the explicit representation for the gamma matrices of the configuration space. One should be able to identify the space, where these gamma matrices act as well as the counterparts of the "free" gamma matrices, in terms of which the gamma matrices would be representable using generalized vielbein coefficients.

6.2.1 Defining relations for gamma matrices

The ordinary definition of the gamma matrix algebra is in terms of the anti-commutators

$$\{\gamma_A, \gamma_B\} = 2g_{AB} \ .$$

This definition served implicitly also as a basic definition of the gamma matrix algebra in TGD context until the difficulties related to the understanding of the configuration space d'Alembertian defined in terms of the square of the Dirac operator forced to reconsider the definition. If configuration space allows Kähler structure, the most general definition allows to replace the metric any covariantly constant Hermitian form. In particular, g_{AB} can be replaced with

$$\{\Gamma_A^\dagger, \Gamma_B\} = iJ_{AB} \ , \tag{6.2.1}$$

where J_{AB} denotes the matrix element of the Kähler form of the configuration space. The reason is that gamma matrices carry fermion number and are non-hermitian in all coordinate systems. This definition is numerically equivalent with the standard one in the complex coordinates but in arbitrary coordinates situation is different since in general coordinates iJ_{kl} is a nontrivial positive square root of g_{kl}. The realization of this delicacy is necessary in order to understand how the square of the configuration space Dirac operator comes out correctly. Obviously, what one must do is the equivalent of replacing $D^2 = (\Gamma^k D_k)^2$ with $D\hat{D}$ with $\hat{D}$ defined as

$$\hat{D} = iJ^{kl}\Gamma_l^\dagger D_k \ .$$

6.2.2 General vielbein representations

There are two ideas, which make the solution of the problem obvious.

a) Since the classical time development in bosonic degrees of freedom (induced gauge fields) is coded into the geometry of the configuration space it seems natural to expect that same applies in the case of the spinor structure. The time development of the induced spinor fields dictated by TGD counterpart of the massless Dirac action should be coded into the definition of the configuration space spinor structure. This leads to the challenge of defining what classical spinor field means.

b) Since classical scalar field in the configuration space corresponds to second quantized boson fields of the imbedding space same correspondence should apply in the case of the fermions, too. The spinor fields of configuration space should correspond to second quantized fermion field of the imbedding space and the space of the configuration space spinors should be more or less identical with the Fock space of the second quantized fermion field of imbedding space or $X^4(X^3)$. Since classical spinor fields at space-time surface are obtained by restricting the spinor structure to the space-time surface, one might consider the possibility that life is really simple: the second quantized spinor field corresponds to the free spinor field of the imbedding space satisfying the counterpart of the massless Dirac equation and more or less standard anti-commutation relations. Unfortunately life is not *so* simple as the construction of configuration space spinor structure demonstrates: second quantization must be performed for induced spinor fields.

It is relatively simple to fill in the details once these basic ideas are accepted.

a) The only natural candidate for the second quantized spinor field is just the induced spinor field on X^4. Since this field is free field, one can indeed perform second quantization and construct fermionic oscillator operator algebra with unique anti-commutation relations. The space of the configuration space spinors can be identified as the Fock space associated with these oscillator operators. This space depends on 3-surface and strictly speaking one should speak of the Fock bundle having configuration space as its base space.

b) The gamma matrices of the configuration space (or rather fermionic Kac Moody generators) are representable as super positions of the fermionic oscillator algebra generators:

$$\begin{aligned}
\Gamma_A^+ &= E_A^n a_n^\dagger \\
\Gamma_A^- &= \bar{E}_A^n a_n \\
iJ_{A\bar{B}} &= \sum_n E_A^n \bar{E}_B^n
\end{aligned} \qquad (6.2.2)$$

where E_A^n are the vielbein coefficients. Induced spinor fields can possess zero modes and there is no oscillator operators associated with these modes. Since oscillator operators are spin 1/2 objects, configuration space gamma matrices are analogous to spin 3/2 spinor fields (in a very general sense). Therefore the generalized vielbein and configuration space metric is analogous to the pair of spin 3/2 and spin 2 fields encountered in super gravitation! Notice that the contractions $j^{Ak}\Gamma_k$ of the complexified gamma matrices with the isometry generators are genuine spin 1/2 objects labelled by the quantum numbers labelling isometry generators. In particular, in CP_2 degrees of freedom these fermions are color octets.

c) A further great idea inspired by the symplectic and Kähler structures of the configuration space is that configuration gamma matrices are actually generators of super-canonical symmetries. This simplifies enormously the construction allows to deduce explicit formulas for the gamma matrices.

6.2.3 Inner product for configuration space spinor fields

The conjugation operation for configuration space spinors corresponds to the standard *ket* $\rightarrow$ *bra* operation for the states of the Fock space:

$$\begin{aligned}
\Psi &\leftrightarrow |\Psi\rangle \\
\bar{\Psi} &\leftrightarrow \langle\Psi|
\end{aligned} \qquad (6.2.3)$$

The inner product for configuration space spinors at a given point of the configuration space is just the standard Fock space inner product, which is unitary.

$$\bar{\Psi}_1(X^3)\Psi_2(X^3) \;=\; \langle\Psi_1|\Psi_2\rangle_{|X^3} \tag{6.2.4}$$

Configuration space inner product for two configuration space spinor fields is obtained as the integral of the Fock space inner product over the whole configuration space using the vacuum functional $exp(K)$ as a weight factor

$$\langle\Psi_1|\Psi_2\rangle \;=\; \int \langle\Psi_1|\Psi_2\rangle_{|X^3}\, exp(K)\sqrt{G}dX^3 \tag{6.2.5}$$

This inner product is obviously unitary. A modified form of the inner product is obtained by including the factor $exp(K/2)$ in the definition of the spinor field. In fact, the construction of the central extension for the isometry algebra leads automatically to the appearance of this factor in vacuum spinor field.

The inner product differs from the standard inner product for, say, Minkowski space spinors in that integration is over the entire configuration space rather than over a time= constant slice of the configuration space. Also the presence of the vacuum functional makes it different from the finite dimensional inner product. These are not un-physical features. The point is that (apart from classical non-determinism forcing to generalized the concept of 3-surface) Diff^4 invariance dictates the behavior of the configuration space spinor field completely: it is determined form its values at the moment of the big bang. Therefore there is no need to postulate any Dirac equation to determine the behavior and therefore no need to use the inner product derived from dynamics.

6.2.4   Holonomy group of the vielbein connection

Generalized vielbein allows huge gauge symmetry. An important constraint on physical observables is that they do not depend at all on the gauge chosen to represent the gamma matrices. This is indeed achieved using vielbein connection, which is now quadratic in fermionic oscillator operators. The holonomy group of the vielbein connection is the configuration space counterpart of the electro weak gauge group and its algebra is expected to have same general structure as the algebra of the configuration space isometries. In particular, the generators of this algebra should be labelled by conformal weights like the elements of Kac Moody algebras. In present case however conformal weights are complex as the construction of the configuration space geometry demonstrates.

6.2.5   Realization of configuration space gamma matrices in terms of super symmetry generators

In string models super symmetry generators behave effectively as gamma matrices and it is very tempting to assume that configuration space gamma matrices can be regarded as generators of the canonical algebra extended to super-canonical Kac Moody type algebra. The experience with string models suggests also that radial Virasoro algebra extends to Super Virasoro algebra. There are good reasons to expect that configuration space Dirac operator and its square give automatically a realization of this algebra. It this is indeed the case, then configuration space spinor structure as well as Dirac equation reduces to mere group theory.

One can actually guess the general form of the super-canonical algebra. The form is a direct generalization of the ordinary super Kac Moody algebra. The complexified super generators S_A are identifiable as configuration space gamma matrices:

$$\Gamma_A \;=\; S_A \;. \tag{6.2.6}$$

The anti-commutators $\{\Gamma_A^\dagger, \Gamma_B\}_+ = i2J_{A,B}$ define a Hermitian matrix, which is proportional to the Kähler form of the configuration space rather than metric as usually. Only in complex coordinates the anti-commutators equal to the metric numerically. This is, apart from the multiplicative constant n, is expressible as the Poisson bracket of the configuration space Hamiltonians H_A and H_B. Therefore one should be able to identify super generators $S_A(r_M)$ for each values of r_M as the counterparts of fluxes. The anti-commutators between the super generators S_A and their Hermitian conjugates should read as

$$\{S_A, S_B^\dagger\}_+ \;\; = \;\; iQ_m(H_{[A,B]}) \; . \tag{6.2.7}$$

and should be induced directly from the anti-commutation relations of free second quantized spinor fields of the imbedding space restricted to the light cone boundary.

The commutation relations between Hamiltonians and super generators follow solely from the transformation properties of the super generators under canonical transformations, which are same as for the Hamiltonians themselves

$$\{H_{Am}, S_{Bn}\}_- \;\; = \;\; S_{[Am,Bn]} \; , \tag{6.2.8}$$

and are of the same form as in the case of Super-Kac-Moody algebra.

The task is to derive an explicit representation for the super generators S_A in both cases. For obvious reason the spinor fields restricted to the 3-surfaces on the light cone boundary $\delta M_+^4 \times CP_2$ can be used. Leptonic/quark like oscillator operators are used to construct Ramond/NS type algebra.

What is then the strategy that one should follow?

a) Configuration space Hamiltonians correspond to either magnetic or electric flux Hamiltonians and the conjecture is that these representations are equivalent. It turns out that this electric-magnetic duality generalizes to the level of super charges. It also turns out that quark representation is the only possible option whereas leptonic super charges super-symmetrize the ordinary function algebra of the light cone boundary.

b) The simplest option would be that second quantized imbedding space spinors could be used in the definition of super charges. This turns out to not work and one must second quantize the induced spinor fields.

c) The task is to identify a super-symmetric variational principle for the induced spinors: ordinary Dirac action does not work. It turns out that in the most plausible scenario the modified Dirac action varied with respect to *both* imbedding space coordinates and spinor fields is the fundamental action principle. The c-number parts of the conserved canonical charges associated with this action give rise to bosonic conserved charges defining configuration space Hamiltonians. The second quantization of the spinor fields reduces to the requirement that super charges and Hamiltonians generate super-canonical algebra determining the anti-commutation relations for the induced spinor fields.

6.2.6   Configuration space Clifford algebra as a hyper-finite factor of type II_1

The naive expectation is that the trace of the unit matrix associated with the Clifford algebra spanned by configuration space sigma matrices is infinite and thus defines an excellent candidate for a source of divergences in perturbation theory. This potential source of infinities remained un-noticed until it became clear that there is a connection with von Neumann algebras [ea4]. In fact, infinite-dimensional Clifford algebra for a separable Hilbert space defines a standard representation for so called hyper-finite factor of type II_1 [eb3]. This guarantees that the trace of the unit matrix equals to unity and there is no danger about divergences.

Philosophical ideas behind von Neumann algebras

The goal of von Neumann was to generalize the algebra of quantum mechanical observables. The basic ideas behind the von Neumann algebra are dictated by physics. The algebra elements allow Hermitian conjugation * and observables correspond to Hermitian operators. Any measurable function $f(A)$ of operator A belongs to the algebra and one can say that non-commutative measure theory is in question.

The predictions of quantum theory are expressible in terms of traces of observables. Density matrix defining expectations of observables in ensemble is the basic example. The highly non-trivial requirement of von Neumann was that identical a priori probabilities for a detection of states of infinite state system must make sense. Since quantum mechanical expectation values are expressible in terms of operator traces, this requires that unit operator has unit trace: $tr(Id) = 1$.

In the finite-dimensional case it is easy to build observables out of minimal projections to 1-dimensional eigen spaces of observables. For infinite-dimensional case the probably of projection to 1-dimensional sub-space vanishes if each state is equally probable. The notion of observable must thus be modified by excluding 1-dimensional minimal projections, and allow only projections for which the trace would be infinite using the straightforward generalization of the matrix algebra trace as the dimension of the projection.

The non-trivial implication of the fact that traces of projections are never larger than one is that the eigen spaces of the density matrix must be infinite-dimensional for non-vanishing projection probabilities. Quantum measurements can lead with a finite probability only to mixed states with a density matrix which is projection operator to infinite-dimensional subspace. The simple von Neumann algebras for which unit operator has unit trace are known as factors of type II_1 [eb3].

The definitions of adopted by von Neumann allow however more general algebras. Type I_n algebras correspond to finite-dimensional matrix algebras with finite traces whereas I_∞ associated with a separable infinite-dimensional Hilbert space does not allow bounded traces. For algebras of type III non-trivial traces are always infinite and the notion of trace becomes useless.

von Neumann, Dirac, and Feynman

The association of algebras of type I with the standard quantum mechanics allowed to unify matrix mechanism with wave mechanics. Note however that the assumption about continuous momentum state basis is in conflict with separability but the particle-in-box idealization allows to circumvent this problem (the notion of space-time sheet brings the box in physics as something completely real).

Because of the finiteness of traces von Neumann regarded the factors of type II_1 as fundamental and factors of type III as pathological. The highly pragmatic and successful approach of Dirac based on the notion of delta function, plus the emergence of Feynman graphs, the possibility to formulate the notion of delta function rigorously in terms of distributions, and the emergence of path integral approach meant that von Neumann approach was forgotten by particle physicists.

Algebras of type II_1 have emerged only much later in conformal and topological quantum field theories [ec3, ec2] allowing to deduce invariants of knots, links and 3-manifolds. Also algebraic structures known as bi-algebras, Hopf algebras, and ribbon algebras [ec4, ee2] relate closely to type II_1 factors. In topological quantum computation [ef1] based on braid groups [eb2] modular S-matrices they play an especially important role.

Clifford algebra of configuration space as von Neumann algebra

The Clifford algebra of the configuration space provides a school example of a hyper-finite factor of type II_1, which means that fermionic sector does not produce divergence problems. Super-symmetry means that also "orbital" degrees of freedom corresponding to the deformations of 3-surface define similar factor. The general theory of hyper-finite factors of type II_1 is very rich and leads to rather detailed understanding of the general structure of S-matrix in TGD framework. For instance, there is a unitary evolution operator intrinsic to the von Neumann algebra defining in a natural manner single particle time evolution. Also a connection with 3-dimensional topological quantum field theories and knot theory, conformal field theories, braid groups, quantum groups, and quantum counterparts of quaternionic and octonionic division algebras emerges naturally. These aspects are discussed in detail in [O5].

6.3   Dualities and conformal symmetries in TGD framework

The reason for discussing the rather speculative notion of dualities before considering the definition of the modified Dirac action and discussing the proposal how to define Kähler function in terms of Dirac determinants, is that the duality thinking gives the necessary overall view about the complex situation: even wrong vision is better than no vision at all.

The first candidate for a duality in TGD is electric-magnetic duality appearing in the construction of configuration space geometry. Also the duality between 7-D and 3-D CDs relating closely to quantum gravitational holography and YM-gravity duality and representing basically field-particle duality suggests itself. In this case strict duality seems however too strong an assumption.

6.3.1 Electric-magnetic duality

Electric-magnetic duality for the induced Kähler induced field is present also in TGD (CP_2 Kähler form is self-dual). My original belief was that it corresponds to a self duality leaving Kähler coupling constant invariant as an analog of critical temperature: $\alpha_K \to \alpha_K$ in this transformation [B2, B3]. This duality would allow to construct configuration space Kähler metric in terms of Kähler electric or magnetic fluxes.

It is however possible to imagine a second variant of electric-magnetic duality, not in fact a genuine duality, could correspond to $\alpha_K \to -\alpha_K$ [B2, B3, E3], which would be a logarithmic version of $g \to 1/g$ duality and makes sense in TGD framework since g_K does not appear as a coupling constant. This "duality" would have nothing to do with whether the configuration space Hamiltonians are defined in terms of Kähler magnetic or electric fluxes. The two directions of geometric and subjective time and two possible signs of inertial energy would correspond to the two signs (phase conjugate photons would provide example of negative energy particles propagating in the direction of geometric past). Electric *resp.* magnetic flux tubes would be favored in the two phases and a beautiful consistency with TGD inspired cosmology results [D5]. It is questionable whether one can speak of duality now since the two phases seem to be physically different so that alternative descriptions would not be in question. In any case, if a genuine duality is in question it is enough to use either future directed 7-D CDs or past directed CDs. If not, both are required.

This duality relates in an interesting manner to the conjecture about number theoretic spontaneous compactification [E2] in the sense that space-time surfaces in $M^4 \times CP_2$ could be equivalently regarded as hyper-quaternionic 4-surfaces in M^8 possessing hyper-octonionic structure. The point is that one can consider also the dual definition for which the 4-D normal space defines 4-D subalgebra of 8-D algebra at each point of the space-time surface. Future-past duality could basically reduce to this purely geometric duality and would basically reflect bra-ket duality.

Unfortunately, the construction of S-matrix [C2] suggests strongly that the change of sign of α_K breaks unitarity (this is not related to the naive expectation that g_K becomes imaginary since the absolute minima would be different and tend to be dominated by Kähler magnetic fields). In pure quantum context the cosmological argument does not actually bite since one can simply regard the moment of "big bang" as a creation of pairs of positive and negative energy cosmic strings. Whether the time development is dissipative in ordinary or reversed time direction for negative energy cosmic strings is not relevant. Negative energy space-time sheets could be final points of dissipative evolution from the beginning somewhere in the geometric future. It is also possible that the dissipative time evolution occurs in the reversed time direction only in the fermionic degrees of freedom.

6.3.2 Duality of 3-D and 7-D causal determinants as particle-field duality

As already described, TGD predicts two kinds of super-conformal symmetries corresponding to 7-D and 3-D causal determinants and that their duality would generalize the age-old field-particle particle duality so that quantum gravitational holography and YM-gravitational duality could be seen as particular aspects of field particle duality. The two dual super symmetry algebras defined by super-canonical and Super Kac-Moody algebras at configuration space level define spectrum generating algebras whereas at space-time level they define pure super gauge symmetry algebras eliminating half of the helicities of the induced Dirac spinor fields at each point.

1. The conformal symmetries associated with 7-D CDs and space-time interior

The super-canonical conformal invariance is associated with 7-dimensional light like CDs $\delta M^4_\pm(a) \times CP_2$ and their unions at the level of the imbedding space. The GRT counterpart is the moment of local "big bang". The string model counterpart is a Kaluza-Klein type representation of quantum numbers used in super string models relying on closed strings. Obviously this corresponds to the field aspect of the duality.

At space-time level these dynamical super-conformal symmetries have gauge super-symmetries as their counterpart. The number theoretic spontaneous compactification conjecture [E2] suggests that a possibly equivalent identification is in terms of deformations of the boundaries of the space-time surface induced by deformations of $X^4 \subset M^8$ caused by hyper-quaternion analytic maps of $X^4 \to X^4$ inducing no deformation in the interior but moving the boundaries of X^4 [E2]. These deformations would correspond to deformations of the corresponding surface $X^4 \subset M^4 \times CP_2$.

This symmetry could be equivalent with $N = 4$ local gauge super-symmetry ($N = 4$ super-symmetric YM theory has been proposed to be closely linked with string models). There would be no global super symmetry. Since the hyper-quaternion conformal symmetry can make sense only in interior, and since only the induced spinor field in the interior of space-time surface contributes to the configuration space super charges, hyper-quaternion conformal symmetry would indeed correspond to the field aspect of the field-particle duality.

2. Conformal symmetries associated with 3-D light like CDs and quantum gravitational holography

The Super-Kac Moody symmetry at the 3-D light like causal determinants and super-canonical symmetry at 7-D causal determinants define configuration space super-symmetries. These super-symmetries are dynamical and contrary to the original beliefs do not imply the existence of sparticles.

The conformal symmetry associated with the 3-dimensional light like CDs reduces to a generalization of the ordinary super-conformal symmetry. The derivative of the normal coordinate disappears from the modified Dirac operator and solutions are 3-dimensional spinorial shock waves having a very natural interpretation as representations of elementary particles. The GRT analog is black hole horizon. $N = 4$ superconformal symmetry in an almost ordinary sense is in question.

The induced spinor fields carry electro-weak quantum numbers as YM type quantum numbers, Poincare and color isometry charges, but not color as spin like degrees of freedom. Hence color degrees of freedom are analogous to rigid body rotational degrees of freedom of the 3-D causal determinant and genuine configuration space degrees of freedom having no counterpart at space-time level although conserved classical color charges make sense: obviously Kaluza-Klein type quantum numbers are in question.

7–3 duality however suggests that it is possible to code the information about configuration space color partial wave to the induced spinor field at X_l^3 (for 7-D CDs this is not necessary) as a functional of X_l^3. The guess is that the shock wave solutions of the modified Dirac equation at 3-D CDs can be constructed by taking imbedding space spinor harmonics, operating on them by appropriate color Kac Moody generators to get a correct correlation between electro-weak and color quantum numbers, and applying the modified Dirac operator D to get a spinor basis $D\Psi_m$. If the spinor basis obtained in this manner satisfies $D\Psi_m = c_{mn}o\Psi_n$, where o is the contraction of the light like normal vector of CD with the induced gamma matrices appearing in the eigenvalue equation $D\Psi = \lambda o\Psi$ and defining boundary states for the induced spinor fields and Dirac determinant, the construction works. The fact that quark color does not have a direct space-time counterpart (imbedding space spinors allow color partial waves but induced spinors do not) might correlate with color confinement and with the impossibility to detect free quarks.

3. 7–3 duality and effective 2-dimensionality of 3-surfaces

Whether super-canonical and Kac-Moody algebras are dual is not at all obvious. The assumption that the situation reduces to the intersections X_i^2 of the 3-D CDs X_l^3 with 7-D CDs defining 2-sub-manifolds of X^3 concretizes the idea about duality. Duality would imply effective 2-dimensionality of 3-surfaces and the task is to understand what this could mean.

a) By duality both X_i^2-local H-isometries and the Hamiltonians of $\delta M_+^4 \times CP_2$ restricted to X_i^2 span the tangent space of CH. A highly non-trivial implication would be a dramatic simplification of the construction of the configuration space Hamiltonians, Kähler metric, and gamma matrices since one could just sum only over the flux integrals over the sub-manifolds X_i^2. The best that one might hope is that it is possible to fix both 3-D light like CDs and their sub-manifolds X_i^2 and 7-D CDs freely. There would be good hopes about achieving the p-adicization of the basic definitions.

One could assign unique modular degrees of freedom to X_i^2: this would be crucial for the unique definition of the elementary particle vacuum functionals [F1]. This would give rather good hopes of achieving a better understanding of why particle families corresponding to genera $g > 2$ are effectively absent from the spectrum. Elementary particle vacuum functionals vanish when $g > 2$ is hyper-elliptic, that is allows Z_2 conformal symmetry. The requirement that the tangent space 2-surface defines at each point a commutative sub-space of the octonionic tangent space might force $g > 2$ surfaces to be hyper-elliptic.

b) The reduction to dimension 2 could be understood in terms of the impossibility to choose X^3 freely once light like 3-D CDs are fixed but this does not remove the air of paradox. The resolution of the paradox comes from the following observation. The light likeness condition for 3-D CD can be written in the coordinates for which the induced metric is diagonal as a vanishing of one of the diagonal components of the induced metric, say g_{11}:

$$g_{1i}h_{kl}\partial_1 h^k \partial_i h^l \quad = \quad 0 \ , \quad i = 1, 2, 3 \ . \tag{6.3.1}$$

The condition $g_{11} = 0$ is exactly like the light likeness condition for the otherwise random M^4 projection of CP_2 type extremals [D1]. When written in terms of the Fourier expansion this conditions gives nothing but classical Virasoro conditions. This analog of the conformal invariance is different from the conformal invariance associated with transversal degrees of freedom and and from hyper-quaternion conformal invariance. This symmetry conforms nicely with the duality idea since also the boundary of the light cone allows conformal invariance in both light like direction and transversal degrees of freedom.

One can consider two interpretations of this symmetry.

i) The degrees of freedom generating different light like 3-D CDs X^3_l with a given intersection X^2 with 7-D CD correspond to zero modes. Physically this would mean that in each quantum jump a complete localization occurs in these degrees of freedom so that particles behave effectively classically. With this interpretation these degrees of freedom could perhaps be seen as dual for the zero mode degrees of freedom associated with the space-like 3-surfaces X^3 at 7-D CDs: deformation of X^3_l would induce deformation of X^3.

ii) Gauge degrees of freedom could be in question so that one can make a gauge choice fixing the orbits within certain limits. If ones assumes a suitable correlation between the longitudinal and transversal conformal degrees of freedom, hyper-quaternion conformal invariance could be the outcome. The two symmetries could correspond to two different choices of gauge reflected as a choice of different space-time sheets. This would mean additional flexibility in the interpretation of this symmetry at the level of solutions of the modified Dirac equation.

At the level of configuration space geometry the result would mean that one can indeed code all data using only two-dimensional surfaces X^2_i of X^3. This brings in mind a number theoretic realization for the quantum measurement theory. That only mutually commuting observables can be measured simultaneously would correspond to the assumption that all data about configuration space geometry and quantum physics must be given at 2-dimensional surfaces of H for which the tangent space at each point corresponds to an Abelian sub-algebra of octonions. Quantum TGD would reduce to something having very high resemblance with WZW model. One cannot deny the resemblance with M-theories with M interpreted as a membrane.

4. 7–3 duality and the equivalence of loop diagrams with tree diagrams

The 3-D light like CDs are expected to define analogs of Feynman diagrams. In the simplest case there would be past of future and past directed 7-D CDs $X^7_\pm = \delta M^4_\pm \times CP_2$, and the lines of the generalized Feynman diagram would begin from X^7_+ and terminate to X^7_-. In [C5] the generalization of duality symmetry of string models stating that generalized Feynman diagrams with loops are equivalent with tree diagrams is discussed. By quantum-classical correspondence this would mean that the conformal equivalence for Feynman diagrams defined by 3-D light like CDs generalizes to a topological equivalence. This is indeed as it should be since it is the intersections X^2_i with $X^7_\pm$ which should code for physics and these intersections do not contain information about loops.

Interesting questions relate to the interpretation of the negative energy branches of the space-time surface. It would seem that also the surfaces X^2_i are accompanied by negative energy branch. The branching brings in mind a space-time correlate for bra-ket dichotomy. The two branches would represent Feynman diagrams which are equivalent but correspond to different sign of Kähler coupling strength if the generalization of electric-magnetic duality is accepted.

5. 7–3 duality and quantum measurement theory

The action of Super Kac-Moody generators on configuration space Hamiltonians is well defined and one might hope that as a functional of 2-surface it could give rise to a unique superposition of super-canonical Hamiltonians. Same should apply to the action of super-canonical algebra on Kac Moody algebra. At the level of gamma matrices the question is whether the configuration space metric can be defined equivalently in terms of anti-commutators of super-canonical and Super Kac-Moody generators. If the answer is affirmative, then 7–3 duality would be nothing but a transformation between two preferred coordinates of the configuration space.

TGD inspired quantum measurement theory suggests however that the two super-conformal algebras correspond to each other like classical and quantal degrees of freedom. Super Kac-Moody algebra and super conformal algebra would act as transformations preserving the conformal equivalence class of the partonic 2-surfaces X^2 associated with the maxima of the Kähler function whereas super-canonical algebra in general changes conformal moduli and induces a conformal anomaly in this manner. Hence Kac-Moody algebra seems to act in the zero modes of the configuration space metric. In TGD inspired quantum measurement zero modes correspond to classical non-quantum fluctuating dynamical variables in 1-1 correspondence with quantum fluctuating degrees of freedom like the positions of the pointer of the measurement apparatus with the directions of spin of electron. Hence Kac-Moody algebra would define configuration space coordinates in terms of the map induced by correlation between classical and quantal degrees of freedom induced by entanglement.

Duality would be also realized in a well-defined sense at the level of configuration space conformal symmetries. The idea inspired by Olive-Goddard-Kent coset construction is that the generators of Super Virasoro algebra corresponds to the differences of those associated with Super Kac-Moody and super-canonical algebras. The justification comes from the miraculous geometry of the light cone boundary implying that Super Kac-Moody conformal symmetries of X^2 can be compensated by super-canonical local radial scalings so that the differences of corresponding Super Virasoro generators annihilate physical states. If the central extension parameters are same, the resulting central extension is trivial. What is done is to construct first a state with a non-positive conformal weight using super-canonical generators, and then to apply Super-Kac Moody generators to compensate this conformal weight to get a state with vanishing conformal weight and thus mass.

6.3.3 Quantum gravitational holography

The so called AdS/CFT of Maldacena[gb5] correspondence relates to quantum-gravitational holography states roughly that the gravitational theory in 10-dimensional $AdS_{10-n} \times S^n$ manifold is equivalent with the conformal field theory at the boundary of AdS_D factor, which is $D-1$-dimensional Minkowski space. This duality has been seen as a manifestation of a duality between super-gravity with Kaluza-Klein quantum numbers (closed strings) and super Yang-Mills theories (open strings with quantum numbers at the ends of string).

In TGD quantum gravitational holography is realized in terms of the modified Dirac action at light like 3-D CDs, which by their metric 2-dimensionality allow superconformal invariance and are very much like world sheets of closed super string or the ends of an open string.

It is possible to deduce the values of Kähler action at maximally deterministic regions of space-time sheets from the Dirac determinants at the CDs [E3] so that the enormously difficult solution of the problem selecting preferred extremals of Kähler action as generalized Bohr orbits would be reduced to local data stating that CDs are light like 3-surfaces which are also minimal surfaces in the case that Kähler action density is non-vanishing at them. This reduction has enormous importance for the calculability of the theory. Also the values of Kähler coupling strength and gravitational constant are predicted [E3].

Here a word of warning is in order: I do not know how to prove that the minimal surface property of the CDs implies the absolute minimization of Kähler action or the more general variational principle discussed in [E2].

Perhaps the most practical form of the quantum gravitational holography would be that the correlation functions of $N = 4$ super-conformal field theory at the light like 3-D CDs allow to construct the vertices needed to construct S-matrix of quantum TGD. Computationally TGD would reduce to almost string model since light like CD:s are analogous to closed string word sheets on one hand, and to the ends of open string on the other hand. There is also an analogy with the Wess-Zumino-Witten model: light like CDs would correspond to the 2-D space of WZW model and 4-surface to the associated 3-D space defining the central extension of the Kac-Moody algebra.

Quantum gravitational holography could also mean that light like CDs define what might be called fundamental central nervous systems able to represent and process conscious information about the interior of the space-time surface in terms of its own quantum states which have interpretation either as a time evolution or state (duality again!). Topological quantum computation might be one of the activities associated with the light like CDs as proposed in [O3].

6.3.4 Super-symmetry at the space-time level

The interpretation of the bosonic Kac Moody symmetries is as deformations preserving the light likeness of the light like 3-D CD X_l^3. Gauge symmetries are in question when the intersections of X_l^3 with 7-D CDs X^7 are not changed. Since general coordinate invariance corresponds to gauge degeneracy of the metric it is possible to consider reduced configuration space consisting of the light like 3-D CDs. The conformal symmetries in question imply a further degeneracy of the configuration space metric and effective metric 2-dimensionality of 3-surfaces as a consequence. These conformal symmetries are accompanied by $N = 4$ local super conformal symmetries defined by the solutions of the induced spinor fields.

Contrary to the original beliefs these conformal symmetries do not seem to be continuable to quaternion conformal super symmetries in the interior of the space-time surface realized as real analytic power series of a quaternionic space-time coordinate. The reason is that these symmetries involve both transversal complex coordinate and light like coordinate as independent variables whereas quaternion conformal symmetries are algebraically one-dimensional. If the number theoretic spontaneous compactification conjecture [E2] holds true, hyper-quaternionic symmetries make however sense for the representation of the space-time surface as a 4-surface in M^8 possessing hyper-octonion structure.

The solutions of the modified Dirac equation $D\Psi = 0$, define the modes which do not contribute to the Dirac determinant of the modified Dirac operator in terms of which the vacuum functional assumed to correspond to the exponent of the Kähler action is defined. Thus they define gauge supersymmetries. Usually D selects the physical helicities by the requirement that it annihilates physical states: now the situation is just the opposite. D^2 annihilates the generalized eigen states both at space-like and light like 3-surfaces. Hence the roles of the physical and non-physical helicities are switched. It is the generalized eigen modes of D with non-vanishing eigenvalues λ, which code for the physics whereas the solutions of the modified Dirac equation define super gauge symmetries.

At the space-like 3-surfaces associated with 7-D CDs the spinor harmonics of the configuration space satisfy the $M^4 \times CP_2$ counterpart of the massless Dirac equation so that non-physical helicities are eliminated in the standard sense at the imbedding space level. The righthanded neutrino does not generate an $N = 1$ space-time super-symmetry contrary to the long held belief.

6.3.5 Super-symmetry at the level of configuration space

The gamma matrices of the configuration space are defined as matrix elements of properly chosen operators between right-handed neutrino and second quantized induced spinor field at space-like boundaries X^3. These generators define the fermionic generators of what I call super-canonical algebra. The right handed neutrino can be replaced with any spinor harmonic of the imbedding space to obtain an extended super-algebra, which can be used to construct the physical states.

The requirement that super-generators vanish for the vacuum extremals requires that the modified Dirac operator D_+ or the inverse of D_- appearing in the matrix element of the "Hermitian conjugate" $S^- = (S^+)^\dagger$ of the super charge S^+. Here $\pm$ refers to the negative and positive energy space-time sheets meeting at X^3 or to the two maximally deterministic space-time regions separated by the causal determinant. The operators D_+ and D_-^{-1} are restricted to the spinor modes not annihilated by $D_\pm$. The super-generator generated by the covariantly constant right handed neutrino vanishes identically: a more rigorous argument showing that $N = 1$ global super symmetry is indeed absent.

If the configuration space decomposes into a union of sectors labelled by unions of light cones having dips at arbitrary points of M^4, the spinor harmonics can be assumed to define plane waves in M^4 and even possess well-defined four-momenta and mass squared values. Same applies to the super-canonical generators defined by their commutators. This means that the generators of the super-canonical algebra generated in this manner would possess well defined four-momenta and thus their action would change the mass of the state. Space-time super-symmetries would be absent. Similar argument applies to the Kac Moody algebras associated with the light like 3-D CDs if super-canonical Super Kac-Moody algebras provide dual representations of quantum states.

If the gist of these admittedly heuristic arguments is correct, they force to modify drastically the existing view about space-time super-symmetries. The problem how to break super-symmetry disappears since there is no space-time symmetry be broken down. Super-symmetries are realized as a spectrum generating algebra rather than symmetries in the standard sense.

I hasten to admit that I have myself believed that right handed neutrino defines a global super-symmetry and proposed that the topological condensation of sparticles and particles at space-time sheets with different p-adic primes would provide an elegant model for super-symmetry breaking using same general mass formulas but only a different mass scale. Giving up this assumption causes however only a sigh of relief. The predicted spectrum of massless states is reduced dramatically [F3]. p-Adic mass calculations based on p-adic thermodynamics and representations of super-conformal algebra are not affected since the global $N = 1$ super-symmetry implies only an additional vacuum degeneracy. Most predictions of TGD remain intact. The speculation that sneutrinos might be light and play a role in TGD based condensed matter physics is the only possible exception [F9]. One can however consider the possibility of light colored sneutrinos obtained by applying to a neutrino state a colored and thus non-vanishing super-canonical generator defined by right handed antineutrino.

It deserves to be noticed that the notion super-symmetry in configuration space sense was discovered with the advent of super string models and generalized to a space-time super-symmetry when gauge theories made their breakthrough. The notion of spontaneous compactification (we meet our friend again and again!) inspired then the hypothesis that this super-symmetry has a space-time counterpart and everyone believed. There is now an entire industry making similar purely formal out of context applications and generalizations of quantum groups, which originally emerged naturally in knot and braid theory and in the theory of von Neumann algebras [O4, O3].

6.4 Modified Dirac action as the fundamental action principle?

Although quantum criticality in principle predicts the possible values of Kähler coupling strength, one might hope that there exists even more fundamental approach involving no coupling constants and predicting even quantum criticality and realizing quantum gravitational holography. The Dirac determinant associated with the modified Dirac action is and excellent candidate in this respect.

6.4.1 The exponent of Kähler function as Dirac determinant for the modified Dirac action?

There is quite fundamental and elegant interpretation of the modified Dirac action as a fundamental action principle. In this approach vacuum functional can be defined as the Grassmannian functional integral associated with the exponent of the modified Dirac action. This definition is invariant with respect to the scalings of the Dirac action so that theory contains no free parameters.

There are good reasons to expect that this functional integral is nothing but absolute minimum of the Kähler action with the value of Kähler coupling strength coming out as a prediction. Massless Dirac equation is satisfied in the sense that the vacuum expectation value of the modified Dirac action vanishes and implies Euler-Lagrange equations for Kähler action. The conservation of the canonical currents in turn implies the selection of preferred extremal of Kähler action and c-number parts of the currents would correspond to the conserved currents associated with the Kähler action.

In second quantized formalism, which is more natural in TGD context, massless Dirac equation and Euler-Lagrange equations for the Kähler action follow from the variation of the modified Dirac action with respect to Ψ and imbedding space coordinates. This approach predicts non-vanishing conserved classical vacuum charges in accordance with the physical intuitions. The exponent of the Dirac action operating on vacuum state gives exponent of a term coming from the normal ordering expected to be identical with Kähler action.

The knowledge of the canonical currents and super-currents, together with the anti-commutation relations stating that the fermionic super-currents S_A and S_B associated with Hamiltonians H_A and H_B anti-commute to a bosonic current $H_{[A,B]}$, allows in principle to deduce the anti-commutation relations satisfied by the induced spinor field. In fact, these conditions replace the usual anti-commutation relations used to quantize free spinor field. Since the normal ordering of the Dirac action would give Kähler action,

Kähler coupling strength would be determined completely by the anti-commutation relations of the super-canonical algebra. Kähler coupling strength would be dynamical and the selection of preferred extremals of Kähler action as generalized Bohr orbits would be more or less equivalent with quantum criticality because criticality corresponds to conformal invariance and the hyper-quaternionic version of the super-conformal invariance would results only for the extrema of Kähler action. p-Adic (or possibly more general) coupling constant evolution and quantum criticality would come out as a prediction whereas in the case that Kähler action is introduced as primary object, the value of Kähler coupling strength must be fixed by quantum criticality hypothesis.

The mixing of the M^4 chiralities of the imbedding space spinors serves as a signal for particle massivation and breaking of super-symmetry. The induced gamma matrices for the space-time surfaces which are deformations of M^4 indeed contain a small contribution from CP_2 gamma matrices: this implies a mixing of M^4 chiralities even for the modified Dirac action so that there is no need to introduce this mixing by hand.

6.4.2   Definition of the Dirac determinant

The integral over Grassmann variables gives a determinant of the modified Dirac operator which should reduce to the exponent of Kähler function in some sense perhaps involving TGD counterpart of regularization and renormalization procedures. The attempt to define the Dirac determinant rigorously leads to unexpected insights giving real hopes that one can calculate the exponent of the Kähler function from local data.

The definition of Dirac determinant as a key to the understanding of the configuration space geometry

In TGD framework spinor modes have a definite H-chirality. As such the chiral symmetry is not a problem since a given matrix element of the modified Dirac operator is between spinor modes of the chirality and inner product is defined by the "time" component of conserved fermion current. The fermionic determinant by using zeta function regularization however requires the expression of the determinant as a product of the eigenvalues of the Dirac operator. The problem is that by the chirality condition the solution basis cannot correspond to the eigen modes of the modified Dirac operator in the ordinary sense ($D\Psi = \lambda\Psi$).

Usually this problem is avoided by defining the determinant as the square root of the determinant of the square of the Dirac operator. In TGD context the matrix defined by the square of the Dirac operator would transform quarks to leptons and this makes the trick not only ugly but intolerably un-physical. There is also a regularization corresponding to the division of the Dirac determinant by the determinant associated with the free Dirac operator.

TGD approach suggests and elegant solution of the problems. Consider first the constraints which the definition of the Dirac determinant should satisfy.

a) The definition should be general coordinate invariant. It should also realize quantum gravitational holography. Perhaps even in the sense that the value of the Kähler function should be in principle deducible from the data at the 3-dimensional causal determinants X_l^3 for which light-like 3-surfaces defining elementary particle horizons are excellent candidates. The hope is that the consistency conditions associated with definition of the Dirac determinant fix the normal derivatives $\partial_n h^k$ of the imbedding space-coordinates and hence the spectrum of the modified Dirac operator at X_l^3 and hence also the value of Kähler function.

b) The determinant should carry non-trivial information about the absolute minimum of Kähler action. Ordinary 3-dimensional Dirac operator for the induced spinors would not carry this information but the modified Dirac operator indeed does since the normal derivatives $\partial_n h^k$ of the imbedding space coordinates appear in the definition of energy momentum tensor appearing in the definition of the modified gamma matrices. Furthermore, the boundary condition states that the generalized energy momentum currents $T^{\alpha k}$ have a vanishing normal component at the causal determinant X^3:

$$T^{nk} \;=\; 0 \;. \tag{6.4.1}$$

This condition makes sense for light like boundaries (or more generally light like CDs) and states that there is no flow of momentum through the CD.

For space-like boundaries the condition does not make sense unless they are vacuum extremals. This is not the case: rather, positive and negative energy space-time sheets carrying opposite sign conserved quantities meet at the space-like boundaries. This means a profound difference in the treatment of the two situations.

Rather remarkably, by this boundary condition the modified Dirac equation at light like CDs involves only derivatives with respect to the coordinates x^i of X^3 and only the modified gamma matrices contain information about normal derivatives. This means that 3-D shock wave like spinor fields restricted to the light like CDs are possible. The idea of quantum gravitational holography suggests itself strongly. 3-D light like CDs could be interpreted in terms of the field particle duality: the click in the particle detector would be like the bang caused by an acoustic shock wave in my ears.

c) The eigenvalue equation for the modified Dirac operator can be understood in the following sense at light like 3-D CD X_l^3:

$$
\begin{aligned}
D\Psi \equiv \hat{\Gamma}^i D_i \Psi &= \lambda o \Psi \ , \\
o &= n^\alpha \Gamma_\alpha \ , \ o^2 = 0 \ ,
\end{aligned}
\tag{6.4.2}
$$

where n^α is some vector field. It is natural to require that n^α defines a light like direction. n_α naturally corresponds either to the gradient of the light like coordinate x^T associated with X_l^3 so that by its light likeness n^α formally defines a normal direction for X_l^3:

$$
n_\alpha = \partial_\alpha x^T \ .
\tag{6.4.3}
$$

For a space-like 3-surface X^3 at 7-D light like CD of form $X_l^3 \times CP_2$, where X_l^3 is union of future and past light cone boundaries, the operator o appearing in the condition must be expressed in terms of the imbedding space gamma matrices:

$$
\begin{aligned}
D\Psi \equiv \hat{\Gamma}^i D_i \Psi &= \lambda o \Psi \ , \\
o &= n^k \Gamma_k \ , \ o^2 = 0 \ ,
\end{aligned}
\tag{6.4.4}
$$

and n^k corresponds naturally to the radial light like vector field associated with δM_+^4.

In both cases the consistency condition

$$
[D, o] = 0 \ ,
\tag{6.4.5}
$$

is necessary for Hermiticity and the possibility to orthonormalize the eigenmodes using the natural inner product. The orthonormalized inner product in turns allows to fix the anticommutators of the fermionic oscillator operators.

By operating to the modified Dirac equation by D one finds

$$
D^2 \Psi = \lambda^2 o^2 \Psi = 0 \ .
\tag{6.4.6}
$$

The spinor fields $Do\Psi$ define solutions of the modified Dirac equation $D\Psi = 0$ and there is a direct correspondence with the spectrum of Dirac operator and solutions of Dirac operator. This is undeniably something really nice. A further elegant feature is that zero eigenvalues drop automatically from the determinant. This plus quantum gravitational holography makes the hypothesis that all causal determinants are actually light like 3-surfaces very attractive.

The physical interpretation for the solutions of the modified Dirac equation is in the recent case in terms of non-physical helicities representing fermionic gauge degrees of freedom. Usually the interpretation is just the opposite. The solutions of $D\Psi = 0$ define $N = 4$ super conformal symmetry algebra in both leptonic and quark sectors and the definition of configuration space spinor structure must remain invariant under the super symmetries $\Psi \to \Psi + \epsilon \Psi_1$, $D\Psi_1 = 0$. Also the covariantly constant right handed neutrino corresponds to a pure gauge super symmetry so that $N = 1$ global super symmetries would not be predicted.

d) For 3-D CDs the commutator condition decomposes into two terms corresponding to the commutator and anti-commutator of the matrices $T^{ik}\Gamma_k$ and Γ_α. The commutator term is expressible in terms of sigma matrices and proportional to $D_\alpha n_\beta - D_\beta n_\alpha$ and vanishes if the vector field n_α is a gradient of some scalar function Φ, most naturally the tangential coordinate x^T.

For light like 3-D CDs the vanishing of the anti-commutator term in turn boils down to the vanishing of the term proportional to $g^{\alpha\beta}L_K$ in the energy momentum tensor:

$$L_K D_\alpha n^\alpha \;\; = \;\; 0 \;\; . \tag{6.4.7}$$

Unless the Kähler action density L_K vanishes at X^3, Φ must satisfy 4-D d'Alembert equation at X^3. If Φ corresponds to x^T, this sharpens the minimal surface condition satisfied automatically since the volume vanishes automatically for light like CDs. The equation poses condition on normal derivatives of imbedding space coordinates at X_l^4 and the hope is that it could code purely locally the absolute minimization selection of preferred extremals of Kähler action. Certainly it gives constraint to the initial values although the classical non-determinism of Kähler action is not expected to fix the initial values uniquely.

e) For 7-D CDs the commutator condition decomposes into two terms corresponding to the commutator and anti-commutator of the matrices $T^{\alpha k}\Gamma_k$ and Γ_l. For space-like X^3 at 7-D CD the vanishing of these terms follows from the covariant constancy condition

$$D_l n_k \;\; = \;\; 0 \tag{6.4.8}$$

for the vector defined by the gradient of the radial light like coordinate of $\delta M_{\pm}^4$. There are no conditions on the initial values of the imbedding space coordinates. This conforms with the duality in the sense that light like 3-D CDs fix the normal derivatives of the imbedding space coordinates at 7-D CDs. A stronger form of the duality would be that also 7-D CDs from which 3-D CDs emanate are fixed.

These conditions might give the desired conditions fixing the values of the normal derivatives of the imbedding space coordinates and therefore allow to deduce the value of the exponent of Kähler function from the information provided by the Dirac determinants associated with the light like causal determinants associated with the space-time sheets.

Is the regularization of the Dirac determinant needed?

The definition of the Dirac determinant involves potential problems. In the case of ordinary Dirac operator the formal product of eigenvalues diverges and regularization is needed.

a) The first step consists of dividing the determinant with the determinant of the free Dirac operator D_0.

b) The next step is to express the resulting determinant D_A as an exponent of the trace of the logarithm of D_A, or effectively of the operator D_A/D_0 . If D is diagonalizable in some sense the trace reduces to the sum of the logarithms of the eigenvalues: $Tr(D) = \sum_n log(\lambda_n)$. This in turn can be expressed as a logarithmic derivative of the zeta function $\zeta(D) = \sum_n \lambda_n^{-s}$ defined by the eigenvalues of D at $s = 0$. This function can be defined by analytically continuing $\zeta(D)$ to the complex plane.

The modified Dirac operator is expected to deviate strongly from the ordinary Dirac operator and this might make its definition easier. For vacuum extremals all eigenvalues vanish and in this case Dirac determinant equals to unity even without regularization. This is consistent with the vanishing of the Kähler function.

One can imagine several manners of having a finite Dirac determinant.

1. The option based on hydrogen atom like bound state spectrum

If the allowed spinor modes are analogous to bound states localized inside light like boundary components, then even the possibility that the eigenvalues are bounded from above by a constant as in the case of hydrogen atom must be considered. In this case the asymptotic eigenvalue would be renormalizable to unity by multiplying the Dirac operator by a properly chosen constant. This could fix the value of the gravitational constant. Unfortunately, this option does not conform with the number theoretical vision.

2. The option based on a partially broken $\lambda \rightarrow \lambda$ symmetry

One might hope that time translations are replaced by scalings in TGD framework and the eigenvalues of the modified Dirac operator can be interpreted as scaling momenta. The symmetry $\lambda \leftrightarrow 1/\lambda$

would replace the $E \leftrightarrow -E$ symmetry of the ordinary Dirac operator transforming positive energy solutions to their negative energy counterparts. Exact symmetry would predict unit Dirac determinant and indeed a vanishing Kähler function. The non-vanishing value of Kähler function would correspond to a breaking of the conformal invariance so that $\lambda \leftrightarrow 1/\lambda$ symmetry would fail for a finite number of eigenvalues. Unfortunately, there is no real justification for the belief on partially broken $\lambda \leftrightarrow 1/\lambda$ symmetry.

3. The option inspired by the regularization of the ordinary Dirac determinant

The third and most realistic option is inspired by the regularization of the ordinary Dirac determinant by dividing it by the Dirac determinant of the free Dirac operator. Causal determinants involve always pairs of maximal strictly deterministic space-time regions, and the natural hypothesis is that the ratio of the Dirac determinants of the two adjacent deterministic space-time regions contains information about their Kähler actions.

The assumption the spectra of the modified Dirac operators of two regions coincide at the causal determinant apart from a finite number of eigenvalues whose ratio differs from unity is a strong statement about the character of the classical non-determinism. If the eigenvalue ratios are proportional to the difference of Kähler actions, Dirac determinant gives an exponent for the difference of Kähler actions of the two regions, and one can identify the result as the ratio of exponents of Kähler actions for the two regions. Nothing hinders from defining the regularized Dirac determinant for a given region as a corresponding finite exponent throwing away the common eigen values. This would give Dirac determinant as a global section of a determinant bundle.

This approach has several nice features.

a) The construction brings in mind the difference bundle construction giving rise to a non-trivial gerbe in turn defining a regularized Dirac determinant [dg1] as the multiplicative analog of gauge potential for which curvature form corresponds formation of the ratio of determinants and gives the ratios of determinants correctly. One could see this approach as the deeper one and Dirac determinants as a mere calculational trick to deduce the vacuum functional and Kähler action.

b) The value of the Dirac determinant does not depend on the normalization of the modified Dirac operator if the non-vanishing eigenvalues are in one-one correspondence.

c) There is also a consistency with the number theoretical conjectures about the relationship between Kähler coupling strength and gravitational constant discussed in [E3, O5].

d) If the ratio of the Dirac determinants reduces to a product of ratios for a finite number of eigenvalues of the modified Dirac operator changing at the causal determinant, the number theoretic properties and hence the p-adicization of the Dirac determinant and of Kähler function are in control.

4. Could G/R^2 be fixed from finiteness conditions?

The reason why options a) and b) looked so nice at first is following. Since the energy momentum tensor replaces metric tensor in the modification of the gamma matrices, contravariant modified gamma matrices have dimension $1/L^5$ instead of $1/L$ of the ordinary gamma matrices. Hence the modified Dirac action must be multiplied by a factor nR^4, n a numerical factor.

a) For option $a)$ this constant should be such that the eigen values approach to unity. For option $b)$ n could be determined by the requirement that almost exact conformal symmetry $\lambda \leftrightarrow 1/\lambda$ holds true, one can hope that n corresponds essentially to the ratio G^2/R^4, $\sqrt{G}/R \sim 10^{-4}$. In both cases the value of the gravitational constant would be fixed by the finiteness of the theory. In particular, the hypothesis is related to the number theoretical hypothesis relating Kähler coupling strength and gravitational constant to each other.

b) For option $c)$ one could think (although reluctantly) of giving up the assumption about the one-one correspondence of eigen values the ratio of Dirac determinants so that some eigenvalues can become vanishing at the causal determinant. Hence the ratio of Dirac determinants would involve besides the product of ratios $\lambda_i^{1)}/\lambda_i^{2)}$ also "lonely" eigenvalues, say $\lambda_i^{1)}$ depending on the normalization of the modified Dirac operator. The ratio G/R^2 naturally appearing in the normalization could be fixed by the requirement that the "loners" are proportional to the exponent of Kähler action and for the deterministic regions with a vanishing Kähler action reduce to unity.

Could generalized index theorems provide information about the spectrum?

The presence of only a finite number of "active" eigen values for a given causal determinant enhances the hopes that information about the exponent of Kähler function could be deduced by using a generalization of index theorems [ea8]. Ordinary index theorems typically give the number of solutions $D\Psi = 0$ of the modified Dirac operator not expressible in the form $\Psi = D\Psi_1$ (covariantly constant right handed neutrino spinor with two spin directions is one example) in terms of topological invariants of the manifold.

For the option c) the conservative view would be that the generalized index theorem expresses the number of the eigenvalues which become vanishing in the transition between the adjacent regions. A more radical interpretation is that index theorem expresses the number of those solutions of the modified Dirac operators in the adjacent deterministic regions which correspond to different eigen values in terms of some natural topological invariant associated with the causal determinant.

The Chern-Simons action associated with the induced Kähler form defines a 3-connection form of 2-gerbe (see the chapter "Basic Extremals of the Kähler Action"). The more than obvious guess is that its value for a given causal determinant gives the number of "active" eigen values with sign telling the sign of asymmetry. Presumably the integer value of C-S action corresponds to a surface for which the projection of the causal determinant to CP_2 is many-to-one map.

The vanishing of the net C-S charge does not imply the vanishing of the Kähler function since only the value of the Kähler action for a particular maximal strictly non-deterministic region follows from the spectral asymmetry. If the value of the entire Kähler function were in question, Kähler function would be non-vanishing only when the dimension D of CP_2 projection would be $D = 4$ in some region in the interior of the space-time sheet.

If the CP_2 projection of the causal determinant is 2-dimensional, the spectral asymmetry is predicted to be trivial. The Kähler function could be non-vanishing for the space-time sheets having $D = 3$ everywhere. The regions of the space-time sheet with $D = 4$ (such as CP_2 extremals representing elementary particles and non-asymptotic regions of sheets having $D = 3$ asymptotically) would certainly contribute to the Kähler function.

About the conditions satisfied by the modified Dirac operators at the causal determinants

The discussion of the chapter "TGD as a Generalized Number Theory" inspires the idea that, at least for the causal determinants representing creation of matter from vacuum occurs, two space-time sheets branching from single 3-sheet are involved. This means that negative energy space-time sheets are time reflected as positive energy space-time sheets at the moment of "big bang". There are no initial nor final singularities nor not any big bang but just creation or annihilation of gravitational mass, and even living systems are excellent candidates sub-cosmologies consisting mainly of topological field quanta of classical fields.

Let the modified Dirac operator $D_\pm$ correspond to a space-time sheet possessing positive *resp.* negative time orientation. A similar association could hold true at the level of imbedding space for future *resp.* past directed light cones, which seem to play a key role in the construction of the configuration space geometry.

In accordance with option c), consider the ratio as the ratio of the determinants D_+ and D_-. The conditions making zeta function regularization un-necessary would be

$$
\begin{aligned}
[D_+, D_-] &= 0 \ , \\
(D_+ D_-^{-1})_{ij} &= \frac{\lambda_{+i}}{\lambda_{-i}} \delta i_{ij} \ .
\end{aligned}
\tag{6.4.9}
$$

The condition implies the anti-commutator condition

$$
\{D_+, D_-^{-1}\}_{ij} = 2\frac{\lambda_{+i}}{\lambda_{-i}} \delta i_{ij} \ .
\tag{6.4.10}
$$

D_+ (D_-^{-1}) is associated with the creation of positive (negative) energy space-time sheet would effectively act like a fermionic creation (annihilation operator). Also the interpretation as a super-symmetry might be considered.

Commutativity conditions allow also to consider the operator $D_+ D_-^{-1}$ as defining the "regularized" Dirac determinant giving the ratio of the exponents of Kähler action. Most eigenvalues of this operator would be equal to unity and the possible zero eigenvalues of D_+ would not contribute to the determinant.

Consider now the interpretation of these conditions for various kinds of causal determinants.

1. "Big bangs" and "big crushes" associated with sub-cosmologies

The first kind of causal determinants correspond to surfaces $\delta M_{\pm}^4(a) \times CP_2$ at which matter is created from vacuum. Here a labels the position of the light cone in M^4. The unions of these light cones form a category with the inclusion defining the arrow of geometric time and they define sectors $CH(\cup_i M_{\epsilon_i}^4(a_i))$, $\epsilon_i = \pm$, of the configuration space. In TGD based fractal cosmology $CH(M_{\epsilon_i}^4(a_i))$ defines a sub-cosmology in an extremely general sense.

A negative energy space-time sheet (D_-) would be time reflected as a positive energy space-time sheet (D_+). Besides $\lambda_+ = 1/\lambda_-$ satisfied apart from a finite number of eigenvalues, also the condition $o_+ o_- = 1$ for the contractions of normal vectors with induced gamma matrices would be satisfied. Lifted to vectors of the imbedding space the normal vectors are expected to be light like vectors with temporal components of opposite sign.

2. Elementary particle horizons and regions separating maximal strictly deterministic regions of space-time sheet

The wormhole contacts with Euclidian signature of metric are separated from the space-time sheets with a Minkowskian signature of induced metric by two elementary particle horizons with a distance of order of CP_2 length $R \sim 10^4$ Planck lengths. D_+ *resp.* D_- could be associated with the Minkowskian *resp.* Euclidian sides of the elementary particle horizon. For causal determinants separating maximal deterministic regions inside space-time sheet and having same signature of the induced metric an analogous treatment applies.

The global existence of Dirac determinant is not obvious

In gauge theories with chirality conservation there are also problems related to the global existence of the Dirac determinant as a functional of background gauge field A when one induces the determinant bundle to the space of gauge equivalence classes of gauge potentials. Since chirality condition is involved also now and since the signs of eigenvalues can be also negative, this problem might be present also now.

In the recent case the space of connections divided by gauge group is replaced with the configuration space, and the task is to demonstrate that the fermionic determinant expected to reduce to an exponent of the Kähler function is indeed single-valued. The gauge theory situation is discussed in excellent manner in [dg1] and the following discussion represents basic ideas plus a straightforward translation to recent case without any attempt to really prove anything.

1. Spectral flow through origin as the basic problem

The difficulties are basically due to the fact that the spectrum of the Dirac operator contains both negative and positive part, and it can happen that some eigenvalues change sign as a functional of vector potential. This is not as such a catastrophe since the resulting Fock spaces are canonically isomorphic but there can be obstructions for the global existence of the Dirac determinant.

The naive expectation is that the exponent of Kähler function changes sign when an eigenvalue changes sign. This is however not possible if Kähler function is real. Hence there is a temptation to conclude that the determinant is trivially globally defined for given values of zero modes. Also the selection of preferred extremals of Kähler action as generalized Bohr orbits suggests that the determinant is negative for all 3-surfaces (or rather space-time sheets) and cannot change sign. Regularization by analytic continuation might spoil this argument.

2. Global existence of the Dirac determinant from the triviality of bundle gerbe

The conditions for the global existence of the Dirac determinant can be formulated as a condition that the so called bundle gerbe [dg1]. Gerbe is a generalization of connection 1-form to a connection 2-form, and would now be defined in the configuration space. The global existence of the Dirac determinant states that that gerge is trivial ("pure gauge"), that is expressible as an exterior derivative of 1-form [dg1]. Gerbe is non-trivial only if the third cohomology group $H^3(CH)$ of the configuration space is trivial just as the non-triviality of second cohomology group is necessary for the existence of magnetic monopole. This is the case if the configuration space is Kähler manifold.

3. About rigorous definition of Dirac determinant

The rigorous definition of the Dirac determinant [dg1] forces to consider the subspaces $H_+(X^3, \lambda)$ consisting of eigen modes of Dirac operator with eigenvalues of Dirac operator strictly larger than λ, as a functional of 3-surface having fixed values of zero modes.

Second key notion is the Grassmann manifold $Gr_p(H_+)$ consisting of closed sub-spaces $W \subset H$ for which $P_W - P_{H_+}$ is in the Schatten ideal L_p of operators T with $|T|^p$ a trace class operator. The spaces $H_+(\lambda)$ belong to $Gr_p(H_+)$ for $p > dim(M)$, where M is the dimension of the manifold in which Dirac operator is defined.

By quantum-gravitational holography one has $dim(M) = 3$ now: this is essentially due to the special properties of the modified Dirac operator. What makes this so interesting is that the type II_1 factors of von Neumann algebras which appear in the construction of quantum TGD, have in a well defined sense a discrete valued fractal dimension $D < 4$. Gr_p allows an extension to a complex line bundle defined by single element of second cohomology group which can be regarded as Dirac determinant.

One can cover X^3 with the open sets $U_\lambda = \{X^3 | \lambda \notin Spec(D(X^3))\}$ (λ does not belong to the spectrum of D). In each U_λ the map $X^3 \to H_+(X^3, \lambda)$ allows to define a determinant bundle in fixed zero modes sector of the configuration space via pull back.

Furthermore, one can define difference bundles in the intersections $U_{\lambda\lambda'} = U_\lambda \cap U_{\lambda'}$ and the determinant in the intersection is essentially the ratio $DET(X^3, \lambda)/DET(X^3, \lambda')$ and thus the product of eigenvalues in the open interval (λ, λ') and well-defined even without any regularization.

This system of difference bundles satisfies in the triple intersections

$$U_{\lambda\lambda'\lambda''} = U_\lambda \cap U_{\lambda'} U_{\lambda''}$$

the cocycle conditions

$$DET_{p,\lambda\lambda'} DET_{p,\lambda'\lambda''} \quad = \quad DET_{p,\lambda\lambda''} \tag{6.4.11}$$

guaranteing that the differences are single valued.

This cocycle property defines what is known as a trivial gerbe. The triviality follows from the decomposition to the difference of bundles and corresponds to the ability to express connection 2-form in the configuration space in terms of 1-form. A good guess is that this connection two-form is just the Kähler form of configuration space. The triviality of course requires the existence of $DET_{p\lambda}$ and the this might require a regularization procedure.

The Fock spaces $\mathcal{F}(X^3, \lambda)$ are canonically isomorphic for different values of λ. Hence the change of the sign of eigenvalue is as such not a catastrophe since one can map the fibers to each other. Using physics terminology, the Dirac sea filled up to λ is equivalent with the Dirac sea filled up to λ'. The isomorphism mediating this filling brings in the Dirac determinant a determinant R of a unitary transformation mixing the eigen modes with eigenvalues in the range (λ, λ'). For a trivial bundle gerbe the emergence of this determinant does not lead to non-uniqueness.

6.4.3   About the p-adicization of quantum TGD

The p-adicization of quantum TGD looks a formidable challenge. One should achieve nothing less than the p-adicization of the S-matrix elements between configuration space spinor fields by an algebraic continuation of rational physics. This is not an easy task. The definition of Kähler action as a space-time integral is out of consideration, to say nothing about p-adic configuration of the functional integral over configuration space. The only hope is the algebraic continuation of matrix elements of configuration space spinor fields and the first question is what this procedure might mean. Here the proposed definition of Dirac determinant might come in rescue.

Suppose that the calculation of S-matrix elements reduces to a perturbative functional integral over quantum fluctuations around maximum of the Kähler function for given values of zero modes and that this can be done exactly by the symmetric space property. If the values of the maxima of the exponent of the Kähler function for the maximal deterministic regions are rational or belong to some algebraic extension of rationals, then vacuum functionals satisfy the basic constraint. This assumption looks internally consistent.

Suppose that the Dirac determinant indeed gives the ratio of exponents of Kähler function for two maximal deterministic regions separated by a light-like causal determinant. In this case the algebraic number property of the vacuum functional in both regions is guaranteed if the Dirac determinant belongs to an algebraic extension of rational numbers associated with the algebraic extension of the p-adic number field considered. This can be guaranteed if the Dirac determinant reduces to a product of a finite number of eigenvalues in the algebraic extension of rational numbers considered provided that the number of eigenvalues involved is finite: this would also guarantee finiteness. This condition should fix the value of Kähler coupling strength at quantum criticality.

Thus the restriction of the allowed eigenvalues of the modified Dirac operator to a hierarchy of algebraic extensions of rationals would give hopes of continuing the vacuum functional to various p-adic number fields at the maxima of Kähler function. This would also mean a prediction of an infinite hierarchy of physics ordered by the dimension of the algebraic extension: this hierarchy conforms nicely with the many-sheeted space-time concept and also with the idea about dynamical Planck constant. The symmetric space property gives also hopes of defining the matrix elements of the configuration space spinors at maxima as numbers belonging to the algebraic extension.

6.5 Representations for configuration space gamma matrices

A priori one can consider two representations for the gamma matrices and canonical supercharges in terms of the leptonic (Ramond) and quark like (N-S) oscillator operators. This degeneracy might relate to the degeneracy associated with the magnetic and electric representations of the configuration space Hamiltonians and also to the ordinary and kappa super-symmetries. The representations differ only by the replacement of Kähler magnetic flux with Kähler electric flux in the defining formulas. In the following it will be found that quark like N-S super generators indeed provide simple representations of configuration space gamma matrices.

6.5.1 Magnetic flux representation of the canonical algebra

In order to derive representation of the configuration space gamma matrices and super charges it is good to restate the basic facts about the magnetic flux representation of the configuration space gamma matrices.

Kähler magnetic invariants

The Kähler magnetic fluxes defined both the normal component of the Kähler magnetic field and by its absolute value

$$
\begin{aligned}
Q_m(X^2) &= \int_{X^2} J_{CP_2} = J_{\alpha\beta}\epsilon^{\alpha\beta}\sqrt{g_2}d^2x \ , \\
Q_m^+(X^2) &= \int_{X^2} |J_{CP_2}| \equiv \int_{X^2} |J_{\alpha\beta}\epsilon^{\alpha\beta}|\sqrt{g_2}d^2x \ ,
\end{aligned}
\tag{6.5.1}
$$

over two-surfaces X^2 are invariants under both Lorentz isometries and the canonical transformations of CP_2 and can be calculated once X^3 is given. Assuming 7–3 duality, the surfaces X^2 corresponds to an intersections of the 3-D light like CDs with the $X_l^3 \times CP_2$ and 2-sub-manifolds of the space-like 3-surface X^3. In this case no additional data about the extremal of Kähler action is needed.

For a closed surface $Q_m(X^2)$ vanishes unless the homology equivalence class of the surface is nontrivial in CP_2 degrees of freedom. In this case the flux is quantized. $Q_M^+(X^2)$ is non-vanishing for closed surfaces, too. Signed magnetic fluxes over non-closed surfaces depend on the boundary of X^2 only:

$$
\begin{aligned}
\int_{X^2} J &= \int_{\delta X^2} A \ . \\
J &= dA \ .
\end{aligned}
$$

Un-signed fluxes can be written as sum of similar contributions over the boundaries of regions of X^2 in which the sign of J remains fixed.

$$
\begin{aligned}
Q_m(X^2) &= \int_{X^2} J_{CP_2} = J_{\alpha\beta}\epsilon^{\alpha\beta}\sqrt{g_2}d^2x \;, \\
Q_m^+(X^2) &= \int_{X^2} |J_{CP_2}| \equiv \int_{X^2} |J_{\alpha\beta}\epsilon^{\alpha\beta}|\sqrt{g_2}d^2x \;,
\end{aligned}
\tag{6.5.2}
$$

There are also canonical invariants, which are Lorentz covariants and defined as

$$
\begin{aligned}
Q_m(K,X^2) &= \int_{X^2} f_K J_{CP_2} \;, \\
Q_m^+(K,X^2) &= \int_{X^2} f_K |J_{CP_2}| \;, \\
f_{K\equiv(s,n,k)} &= e^{is\phi} \times \frac{\rho^{n-k}}{(1+\rho^2)^k} \times \left(\frac{r_M}{r_0}\right)^k
\end{aligned}
\tag{6.5.3}
$$

These canonical invariants transform like an infinite-dimensional unitary representation of Lorentz group.

Generalized magnetic fluxes

Isometry invariants are just a special case of fluxes defining natural coordinate variables for the configuration space. Canonical transformations of CP_2 act as $U(1)$ gauge transformations on the Kähler potential of CP_2 (similar conclusion holds at the level of $\delta M_+^4 \times CP_2$).

One can generalize these transformations to local canonical transformations by allowing the Hamiltonians to be products of the CP_2 Hamiltonians with the real and imaginary parts of the functions $f_{s,n,k}$ defining the Lorentz covariant function basis H_A, $A \equiv (a,s,n,k)$ at the light cone boundary: $H_A = H_a \times f(s,n,k)$, where a labels the Hamiltonians of CP_2.

One can associate to any Hamiltonian H^A of this kind both signed and unsigned magnetic flux via the following formulas:

$$
\begin{aligned}
Q_m(H_A|X^2) &= \int_{X^2} H_A J \;, \\
Q_m^+(H_A|X^2) &= \int_{X^2} H_A |J| \;.
\end{aligned}
\tag{6.5.4}
$$

Both signed and unsigned magnetic fluxes and their superpositions

$$
Q_m^{\alpha,\beta}(H_A|X^2) = \alpha Q_m(H_A|X^2) + \beta Q_m^+(H_A|X^2) \;, \quad A \equiv (a,s,n,k)
\tag{6.5.5}
$$

provide representations of Hamiltonians. Note that canonical invariants $Q_m^{\alpha,\beta}$ correspond to $H^A = 1$ and $H^A = f_{s,n,k}$. $H^A = 1$ can be regarded as a natural central term for the Poisson bracket algebra. Therefore, the isometry invariance of Kähler magnetic and electric gauge fluxes follows as a natural consequence.

6.5.2　Expressions for the canonical supercharges

The natural hypothesis generalizing the identification of the configuration space Hamiltonians as classical canonical charges, is that quark like super-canonical generators S_q^A are obtained as the supercharges, which are in a well-defined sense square roots of flux Hamiltonians. There are two possibilities.

a) The quantities J or $|J|$ appearing in the definition of the flux Hamiltonians are replaced with their square roots. The requirement that super charge is real favors the use of unsigned flux $|J|$.

b) The anti-commutation relations of the induced fermion fields at Y^3 are proportional to J or to $|J|$. This conforms with the fact that modified Dirac equation is related to Kähler action by super-symmetry. In the following only this option is discussed.

This guarantees that the definitions of the flux Hamiltonians and corresponding super-charges is consistent with the basic properties of the Kähler action. For 3-surfaces which reduce to vacuum extremals at light cone boundary the super charges should vanish identically and this is indeed the case since the Kähler magnetic field is identically vanishing. Thus one can say that although Kähler action need not appear explicitly in the definition of configuration space Hamiltonians, only the dynamics defined by the Kähler action conforms with the basic properties of the super-canonical algebra.

These supercharges do not correspond to the symmetries of the Kähler action but to the isometries of the configuration space. The requirement that these currents are conserved on the light cone boundary would presumably mean that Kähler function is stationary with respect to the canonical variations: obviously this does not make sense.

The representation of super-canonical generators are derived from following requirements.

a) Super generators and their "hermitian conjugates" correspond to the positive and negative energy space-time sheets meeting at space-like 3-surface at $X_l^3 \times CP_2$.

b) The solutions of the modified Dirac equation generate gauge super symmetries and thus the addition of a solution of the modified Dirac equation to $\Psi_\pm$ does not affect the configuration space super generator and gamma matrix. This is guaranteed D_+ *resp.* D_-^{-1} appears in the definition of the super generator $S^{A,+}$ and its "hermitian conjugate" $S^{A,-}$.

c) A further assumption crucial for anticommutation relations of gamma matrices is the condition $D_+D_-^{-1} = 1$. Hence D_+ and D_- act only as projectors eliminating the super gauge degrees of freedom. This implies that Dirac determinant defining the exponent for the difference of the Kähler action for the maximal deterministic space-time regions meeting at X^3 equals to one. This means that space like 3-surfaces at 7-D CDs do not code information about configuration space Kähler function, which is thus determined by the data at 3-D light like CDs.

A priori there are two candidates for the super-canonical sub-algebra defining gamma matrices.

a) A candidate for a quark representation of the super-canonical generators is obtained as

$$
\begin{aligned}
S_q'^{A+} &= \frac{1}{2i} \int_{Y^3} \overline{\nu}_R j_{\bar{k}}^{A} \Gamma^{\bar{k}} D_+ \Psi_+ dV \ , \\
S_q^{A-} &= -\frac{1}{2i} \int_{Y^3} \overline{\Psi}_- \bar{D}_-^{-1} j_{\bar{k}}^{A} \Gamma^k \nu_R dV \ , \\
dV &= f(h) \times \sqrt{g} d^2 x per, \\
j^{Ak} &= J^{kl} \partial_l H_A \ .
\end{aligned}
\tag{6.5.6}
$$

The effective two-dimensionality implied by 7-3 duality allows to drop from the earlier formula for dV the factor dr_M/r_M.

It is essential that complex coordinates and holomorphic half of the canonical current $j_{\bar{k}}^A$ of CP_2 are used. For the Hermitian conjugate S^{A+} j_k^A appears in the formula. Ψ denotes the second quantized free induced quark field at $X^4(Y^3)$. The function $f(h)$ is some function of the coordinates of $\delta M_+^4 \times CP_2$ needed to take care that the anti-commutations are precisely correct. It is important to notice that j_k^A are restricted to the light cone boundary in these formulas and the selection of preferred extremals as generalized Bohr orbits defines the continuation of the canonical deformations to the entire space-time surface $X^4(Y^3)$. One can extend this algebra by allowing arbitrary H-spinors harmonics instead of only ν_R so that generators carrying both quark and lepton number become possible.

b) One might criticize the super-algebra in which contraction of second quantized quark field and classical lepton spinor harmonics occur, and argue that only vector currents make sense. One can indeed define a set of leptonic vector currents by replacing D_+ and D_-^{-1} in the defining formulas by the projector P into spinor modes with non-vanishing eigenvalue so that the local $N = 4$ super symmetry is not lost. If D^+ and D^- have same eigen value spectrum, the resulting super generators anti-commute to same expressions as the super generators already defined. If this argument is accepted then only the leptonic super-algebra with D replaced by a projector P can be used to define the configuration space gamma matrices. It must be re-emphasized that the covariantly constant right-handed neutrino can be replaced by any spinor harmonic of H so that the algebra extends to a much larger super algebra which can be used to construct physical states.

These algebras can be assumed to be Ramond type algebras. The reason is that constant Hamiltonian does not contribute to the algebra and one thus avoids the un-desired global $N = 1$ super-symmetry generated by the right handed neutrino.

It is also possible to define super-algebras which anti-commute to the function algebra defined by the super-canonical generators.

a) A vectorial leptonic super-algebra is obtained by generalizing the super-symmetry $\Psi \to \Psi + \epsilon \nu_R$ to canonical super-symmetries $\Psi \to \Psi + \epsilon H_A \nu_R$. This would give rise to super-canonical generators

$$
\begin{aligned}
S_l^{A,+} &= \frac{1}{2i} \int_{Y^3} \overline{\nu}_R H_A D_+ \Psi_+ dV \ , \\
S_l^{A,-} &= -\frac{1}{2i} \int_{Y^3} \overline{\Psi}_- \overleftarrow{D}_-^{-1} H_A \nu_R dV
\end{aligned}
\tag{6.5.7}
$$

depending on the leptonic oscillator operators only. If the representation is of Ramond type, constant Hamiltonian has a non-vanishing super-counterpart whereas for the quark representation it vanishes. This phenomenon also occurs for the electric and magnetic flux representation of the configuration space Hamiltonians but need not have any special significance. If one assumes that this algebra is N-S type algebra with Hamiltonians possessing half-odd integer conformal weight with respect to the radial coordinate, one can avoid non-vanishing generator associated with a constant Hamiltonian and generating a global super-symmetry. Also this algebra extends by allowing all spinor harmonics.

b) The replacement of $D_\pm$ with P gives a set of generators which have quantum numbers of lepto-quark. This option is subject to the same criticism as the quark like algebra for $D^\pm$. The expression for super-Virasoro algebra seems however to require both lepto-quark like and vectorial representations of the configuration space gamma matrices.

6.5.3   Is second quantization performed for imbedding space spinor fields or for induced spinors?

Whether the second quantization should be performed for imbedding space spinors or for induced spinor fields is key question concerning the construction of the configuration space spinor structure. The simplest option would certainly be that the restriction of the second quantized imbedding space spinor field to Y^3 appears in the definition of super-canonical charges. An alternative option is that the second quantization is done for the induced spinor field. It seem that the latter option is the correct one.

The anti-commutators of super-canonical charges can give canonical super charges only if the anti-commutators of leptonic and and quark fields are proportional to a 2-dimensional delta function

$$
\begin{aligned}
\delta_3 &= \frac{\delta_2(x,y)}{X} \sqrt{g_2} \ , \\
X &= J \ \text{or} \ |J| \ .
\end{aligned}
$$

where x and y refer to the coordinates of X_i^2 (intersection of X_l^3 with X^3). This is what second quantized induced spinor fields should give as anti-commutation relations with normal direction of X_i^2 playing the role of the time coordinate in the ordinary quantization. The appearance of J or —J— is strongly favored by the fact that induced spinor fields should represent vacuum degrees of freedom when 3-surface becomes vacuum extremal.

On the other hand, the anti-commutation relations for the imbedding space spinor fields are expected to produce delta function of $\delta M_+^4 \times CP_2$ rather than that of light cone boundary:

$$
\{\Psi^\dagger(h^1), \Psi(h^2)\} = \gamma^0 \delta_7(h_1, h_2) = \gamma^0 \delta^3(m_1, m_2) \delta^4(s_1, s_2) \ .
\tag{6.5.8}
$$

The problem is that the anti-commutator would be proportional to δ_7 rather than δ_3. The formal introduction of $f(h) = \sqrt{\delta_2/\delta_7}$ together with the assumption that $\sqrt{|J|}$ is included in the definition of S_q and S_l certainly extremely non-aesthetic and artificial manner to resolve the problem.

This result supports the number theoretical vision encouraging to think that space-time surfaces as hyper-quaternionic manifolds of M^8 (recall the conjecture about number theoretic spontaneous compactification [E2]) induce ”polarizations” of the imbedding space $M^4 \times CP_2$ such that anti-commutation relations for the imbedding space spinor fields can be given only inside that polarization.

The associativity requirement would allow only quantization in the hyper-quaternionic sub-manifold of $X^4 \subset M^8$ just as Uncertainty Principle forces to restrict Schrödinger amplitudes to n-dimensional Legendre sub-manifolds of $2n$-dimensional phase space. What is nice that the anti-commutation relations are fixed uniquely for induced spinor fields from the anti-commutations of super-canonical algebra and are very nearly what one might naively expect.

6.5.4 Anti-commutation relations for super-canonical charges

Assuming that the induced spinor fields at the surfaces X_i^2 satisfy the 2-dimensional anti-commutation relations

$$
\begin{aligned}
\{\Psi^\dagger(x), \Psi(y)\} &= \frac{\delta_2(x,y)X}{\sqrt{g_2}} \ , \\
X &= \alpha J + \beta |vertJ| \ \text{ or } \ |J|
\end{aligned}
\tag{6.5.9}
$$

the anti-commutator $\{S^{A+}, S^{B-}\}$ gives the term

$$
\begin{aligned}
\{S^{A-}, S^{B+}\} &= \int XY\sqrt{g_2}d^2x \ , \\
Y &= \{j_{A\bar{k}}\Gamma^{\bar{k}}, j_{Bl}\Gamma^l\} = J^{\bar{k}l}j_{A\bar{k}}j_{Bl} \\
&= J^{\bar{k}l}\partial_{\bar{k}}H_A\partial_l H_B \equiv \{H_A, H_B\}_{-+} \ .
\end{aligned}
\tag{6.5.10}
$$

The result is just the half commutator of the imbedding space Hamiltonians with respect to CP_2 symplectic structure:

$$
\{S_q^{A-}, S_q^{B+}\} = Q_m^{\alpha,\beta}(\{H_A, H_B\}_|) \ .
\tag{6.5.11}
$$

This means that the real and imaginary parts of the resulting anti-commutator give the metric and Kähler form of the configuration space and that quark super charges correspond to configuration space gamma matrices.

In the case of leptonic super-canonical charges one obtains the product $H_A H_B$ of the Hamiltonians rather than their Poisson bracket.

$$
\{S_l^{A-}, S_l^{B+}\} = Q_m(H_A H_B) \ .
\tag{6.5.12}
$$

Therefore it seems that in the case of leptons it is not possible to obtain the required kind of super-algebra and that leptons cannot define the representation of configuration space gamma matrices. This resolves the problem whether quarks (N-S representation) or leptons (Ramond representation) should define configuration space gamma matrices. Also the coset space structure of the configuration space based on the generalization of conformal algebras to contain also generators with half-odd integer conformal charge allows only N-S type representation for gamma matrices.

This does not of course mean that leptonic super algebra would be something artificial: leptonic super algebra can be seen as defining the super-symmetrization of the function algebra of $\delta M_+^4 \times CP_2$ so that both symplectic and function algebras are super-symmetrized. In string model context this degeneracy corresponds to the ordinary and kappa super-symmetries. One can also assign to leptons Ramond type super-conformal algebra and corresponding Super Virasoro conditions define leptonic variant of configuration space Dirac equation so that two different Super Virasoro conditions emerge.

Besides magnetic flux Hamiltonians one can also define electric flux Hamiltonians and the conjecture is that the selection of preferred extremals of Kähler action as genearalized Bohr orbits implies that these two representations for the configuration space Hamiltonians are equivalent. This conjecture is nothing but electric-magnetic duality at the level of the configuration space geometry. It seems that the generalization of the electric magnetic duality to the case of super-charges requires only the trivial replacement of magnetic flux J with the electric flux in the defining formulas.

6.6 Super-symmetries at space-time and configuration space level

The first difference between TGD and standard conformal field theories and string models is that super-symmetry generators acting as configuration space gamma matrices acting as super generators carry either lepton or quark number. This corresponds to the situation in $N = 2$ super-conformal field theories allowing an Abelian charge having now interpretation as a fermion number. Only the anti-commutators of quark like generators expressible in terms of Hamiltonians H_A of $X_l^3 \times CP_2$ can contribute to the super-symmetrization of the Poisson algebra and thus to CH metric via Poisson central extension, whereas leptonic generators, which are proportional to $j^{Ak}\Gamma_k$ can contribute to the super-symmetrization of the function algebra of CH. Quarks correspond to N-S type representations and kappa symmetry of string models whereas leptons correspond to Ramond type representations and ordinary super-symmetry.

Also Super Kac-Moody invariance allows lepton-quark dichotomy. What forces to assign leptons with Ramond representation is that covariantly constant neutrino must correspond to one conformal mode $(z^n, n = 0)$. The p-adic mass calculations [TGDpad] carried for more than decade ago led to the same assignment on physical grounds: p-adic mass calculations also forced to include $SO(3,1)$ besides M^4 a tensor factor to super-conformal representations, which in recent context suggests that causal determinants $X_l^3 \times CP_2$, $X_l^3 \subset M^4$ an arbitrary light like 3-surface rather than just a translate of δM_+^4, must be allowed. Also now the lepton-quark, Ramond-NS and SUSY-kappa dichotomies correspond to one and same dichotomy so that the general structure looks quite satisfactory although it must be admitted that it is based on heuristic guess work.

Second deep difference is the appearance of the zeros of Riemann Zeta as conformal weights of the generating elements of the super-canonical algebra and the expected action of conformal algebra associated with 3-D CDS as a spectral flow in the space of super-canonical conformal weights inducing a mere gauge transformation infinitesimally and a braiding action in topological degrees of freedom.

In this section the relationship of Super Kac-Moody invariance to ordinary super-conformal symmetry and the interaction between Super-Kac Moody and super-canonical symmetries are discussed. For years the role of quaternions and octonions in TGD has been under an active speculation. These aspects are considered in [E2], where the number theoretic equivalent of spontaneous compactification is proposed. The conjecture states that space-time surfaces can be regarded either as 4-surfaces in $M^4 \times CP_2$ or as hyper-quaternionic 4-surfaces in the space $HO = M^8$ possessing hyper-octonionic structure (the attribute 'hyper' means that imaginary units are multiplied by $\sqrt{-1}$ in order to achieve number theoretic norm with Minkowskian signature).

6.6.1 Modified Dirac equation

In the following the problems of the ordinary Dirac action are discussed and the notion of modified Dirac action is introduced.

Problems associated with the ordinary Dirac action

Minimal 2-surface represents a situation in which the representation of surface reduces to a complex-analytic map. This implies that induced metric is hermitian so that it has no diagonal components in complex coordinates $(z, \overline{z})$ and the second fundamental form has only diagonal components of type H_{zz}^k. This implies that minimal surface is in question since the trace of the second fundamental form vanishes. At first it seems that the same must happen also in the more general case with the consequence that the space-time surface is a minimal surface. Although many basic extremals of Kähler action are minimal surfaces, it seems difficult to believe that minimal surface property plus extremization of Kähler action could really boil down to the absolute minimization of Kähler action or a more general principle selecting preferred extremals as Bohr orbits [E2].

This brings in mind a similar long-standing problem associated with the Dirac equation for the induced spinors. The problem is that right-handed neutrino generates super-symmetry only provided that space-time surface and its boundary are minimal surfaces. Although one could interpret this as a geometric symmetry breaking, there is a strong feeling that something goes wrong. Induced Dirac equation and super-symmetry fix the variational principle but this variational principle is not consistent with Kähler action.

One can also question the implicit assumption that Dirac equation for the induced spinors is consistent with the super-symmetry of the configuration space geometry. Super-symmetry would obviously require that for vacuum extremals of Kähler action also induced spinor fields represent vacua. This is however not the case. This super-symmetry is however assumed in the construction of the configuration space geometry so that there is internal inconsistency.

Super-symmetry forces modified Dirac equation

The above described three problems have a common solution. Nothing prevents from starting directly from the hypothesis of a super-symmetry generated by covariantly constant right-handed neutrino and finding a Dirac action which is consistent with this super-symmetry. Field equations can be written as

$$
\begin{aligned}
D_\alpha T_k^\alpha &= 0 \ , \\
T_k^\alpha &= \frac{\partial}{\partial h_\alpha^k} L_K \ .
\end{aligned}
\tag{6.6.1}
$$

If super-symmetry is present one can assign to this current its super-symmetric counterpart

$$
\begin{aligned}
J^{\alpha k} &= \overline{\nu_R}\Gamma^k T_l^\alpha \Gamma^l \Psi \ , \\
D_\alpha J^{\alpha k} &= 0 \ .
\end{aligned}
\tag{6.6.2}
$$

having a vanishing covariant divergence. The isometry currents currents and super-currents are obtained by contracting $T^{\alpha k}$ and $J^{\alpha k}$ with the Killing vector fields of super-symmetries. Note also that the super current

$$
J^\alpha = \overline{\nu_R} T_l^\alpha \Gamma^l \Psi
\tag{6.6.3}
$$

has a vanishing divergence.

By using the covariant constancy of the right-handed neutrino spinor, one finds that the divergence of the super current reduces to

$$
D_\alpha J^{\alpha k} = \overline{\nu_R}\Gamma^k T_l^\alpha \Gamma^l D_\alpha \Psi \ .
\tag{6.6.4}
$$

The requirement that this current vanishes is guaranteed if one assumes that modified Dirac equation

$$
\begin{aligned}
\hat{\Gamma}^\alpha D_\alpha \Psi &= 0 \ , \\
\hat{\Gamma}^\alpha &= T_l^\alpha \Gamma^l \ .
\end{aligned}
\tag{6.6.5}
$$

This equation must be derivable from a modified Dirac action. It indeed is. The action is given by

$$
L = \overline{\Psi}\hat{\Gamma}^\alpha D_\alpha \Psi \ .
\tag{6.6.6}
$$

Thus the variational principle exists. For this variational principle induced gamma matrices are replaced with effective induced gamma matrices and the requirement

$$
D_\mu \hat{\Gamma}^\mu = 0
\tag{6.6.7}
$$

guaranteing that super-symmetry is identically satisfied if the bosonic field equations are satisfied. For the ordinary Dirac action this condition would lead to the minimal surface property. What sounds strange that the essentially hydrodynamical equations defined by Kähler action have fermionic counterpart: this is very far from intuitive expectations raised by ordinary Dirac equation and something which one might not guess without taking super-symmetry very seriously.

How can one avoid minimal surface property?

These observations suggest how to avoid the emergence of the minimal surface property as a consequence of field equations. It is not induced metric which appears in field equations. Rather, the effective n metric appearing in the field equations is defined by the anti-commutators of $\hat{\gamma}_\mu$

$$\hat{g}_{\mu\nu} \;=\; \{\hat{\Gamma}_\mu, \hat{\Gamma}_\nu\} = 2T_\mu^k T_{\nu k} \;. \tag{6.6.8}$$

Here the index raising and lowering is however performed by using the induced metric so that the problems resulting from the non-invertibility of the effective metric are avoided. It is this dynamically generated effective metric which must appear in the number theoretic formulation of the theory.

Field equations state that space-time surface is minimal surface with respect to the effective metric. Note that a priori the choice of the bosonic action principle is arbitrary. The requirement that effective metric defined by energy momentum tensor has only non-diagonal components except in the case of non-light-like coordinates, is satisfied for the known solutions of field equations.

Does the modified Dirac action define the fundamental action principle?

There is quite fundamental and elegant interpretation of the modified Dirac action as a fundamental action principle discussed also in [E2]. In this approach vacuum functional can be defined as the Grassmannian functional integral associated with the exponent of the modified Dirac action. This definition is invariant with respect to the scalings of the Dirac action so that theory contains no free parameters.

There are good reasons to expect that this functional integral is nothing but absolute minimum of the Kähler action with the value of Kähler coupling strength coming out as a prediction. Massless Dirac equation would satisfied in the sense that the vacuum expectation value of the modified Dirac action vanishes and implies Euler-Lagrange equations for Kähler action. The conservation of the canonical currents in turn implies the selection of preferred extremals of Kähler action as generalized Bohr orbits and c-number parts of the currents would correspond to the conserved currents associated with the Kähler action.

In second quantized formalism, which is more natural in TGD context, massless Dirac equation and Euler-Lagrange equations for the Kähler action follow from the variation of the modified Dirac action with respect to Ψ and imbedding space coordinates. This approach predicts non-vanishing conserved classical vacuum charges in accordance with the physical intuitions. The exponent of the Dirac action operating on vacuum state gives exponent of a term coming from the normal ordering expected to be identical with Kähler action.

The knowledge of the canonical currents and super-currents, together with the anti-commutation relations stating that the fermionic super-currents S_A and S_B associated with Hamiltonians H_A and H_B anti-commute to a bosonic current $H_{[A,B]}$, allows in principle to deduce the anti-commutation relations satisfied by the induced spinor field. In fact, these conditions replace the usual anti-commutation relations used to quantize free spinor field. Since the normal ordering of the Dirac action would give Kähler action,

Kähler coupling strength would be determined completely by the anti-commutation relations of the super-canonical algebra. Kähler coupling strength would be dynamical and the selection of preferred extremals of Kähler action would be more or less equivalent with quantum criticality because criticality corresponds to conformal invariance and the hyper-quaternionic version of the super-conformal invariance results only for the extrema of Kähler action. p-Adic (or possibly more general) coupling constant evolution and quantum criticality would come out as a prediction whereas in the case that Kähler action is introduced as primary object, the value of Kähler coupling strength must be fixed by quantum criticality hypothesis.

The mixing of the M^4 chiralities of the imbedding space spinors serves as a signal for particle massivation and breaking of super-symmetry. The induced gamma matrices for the space-time surfaces which are deformations of M^4 indeed contain a small contribution from CP_2 gamma matrices: this implies a mixing of M^4 chiralities even for the modified Dirac action so that there is no need to introduce this mixing by hand.

The detailed construction of the Dirac determinant shows that the modified Dirac operator indeed allows to realize quantum gravitational holography since it reduces to an effectively 3-dimensional

Dirac operator by boundary conditions but depends on the normal derivatives of the imbedding space coordinates at the causal determinants, which are most naturally light like 3-surfaces.

Internal consistency requires that the causal determinants are minimal hyper surfaces so that there are good hopes of evaluating the exponent of Kähler function as a Dirac determinant without solving the field equations.

The hypothesis that the spectrum is apart from a finite number of eigenvalues invariant under $\lambda \to 1/\lambda$ symmetry implies that Dirac determinant requires no regularization and is determined by a finite number of eigenvalues. The symmetry condition however makes sense only if one replaces the modified Dirac operator D with $D_+ D_-^{-1}$ where D_+ and D_- are commuting Dirac operators associated with adjacent maximal deterministic space-time regions. This is the analog for the regularization of the Dirac determinant by dividing it with the Dirac determinant of the free Dirac operator. These points will be discussed [E3], when explicit formulas for the Kähler coupling strength and gravitational coupling strength will be deduced.

Super-symmetries

The modified form of Dirac action allows a large number of super-symmetries. These super-symmetries actually extend the super-symmetries which the theory was originally believed to have: not only right-handed neutrino but all solutions of the induced Dirac equation generate super-symmetries.

The first group of super-symmetries derives from the conserved fermion number current

$$\overline{\Psi}\hat{\Gamma}^\alpha \Psi \tag{6.6.9}$$

by replacing the conjugate of Ψ by a conjugate of any c-number solution of Dirac equation. This gives the super-currents

$$J_i \;\; = \;\; \overline{\Psi}_i \hat{\Gamma}^\alpha \Psi \;\; . \tag{6.6.10}$$

The conserved charges associated with these generators define supercharges which are both quark and lepton like. There is great temptation to assume that they correspond to N-S and/or Ramond representations of super algebra. Of course, also the Hermitian conjugates of these currents are conserved and define supercharges.

Second group of super-currents can be derived from the vanishing of the Dirac action,

$$L \;\; = \;\; \overline{\Psi}\hat{\Gamma}^\alpha D_\alpha \Psi = 0 \;\; , \tag{6.6.11}$$

in the set of the space-time surfaces obtained from each other by symmetry transformations. The variational equation stating the invariance of Dirac action reads as

$$\Delta L \;\; = \;\; \delta \left[\overline{\Psi}\hat{\Gamma}^\alpha D_\alpha \Psi \right] = 0 \;\; . \tag{6.6.12}$$

The variational principle gives rise to the currents

$$J^{A\mu} \;\; = \;\; \overline{\Psi} j^{Ak} \frac{\partial}{\partial h_\mu^k} \left[\hat{\Gamma}^\alpha \right] D_\alpha \Psi \tag{6.6.13}$$

having a vanishing divergence. Here j^{Ak} denotes the vector field representing the infinitesimal symmetry in question. Isometries are certainly symmetries of the theory. If canonical transformations are indeed dynamical symmetries for the absolute minima of Kähler action, then canonical currents give rise to solutions of the bosonic field equations so that fermionic current is also conserved. If the modified Dirac action is the fundamental action containing Kähler action as c-number part resulting from the normal ordering, then these currents represent the action of the configuration space isometries on configurations space spinors with the generalized spin-rotation terms automatically included. The c-number parts of these currents represent configuration space Hamiltonians.

The canonical super-currents result by replacing the conjugate of Ψ with any c-number solution Ψ_i of the induced Dirac equation in the basis used in second quantization.

$$ J_i^{A\mu} \;=\; \overline{\Psi}_i j^{Ak} \frac{\partial}{\partial h_\mu^k} \left[\hat{\Gamma}^\alpha \right] D_\alpha \Psi \;. \tag{6.6.14} $$

For a covariantly constant right handed neutrino one certainly obtains conserved super-currents in this manner. In the general case the conservation of the super current is plausible because it guarantees that the canonical current is conserved. This is because each term in the superposition of terms bilinear in oscillator operators is separately conserved, very natural in a free theory.

6.6.2   Super-canonical and Super Kac-Moody symmetries

The proper understanding of super symmetries has turned out to be crucial for the understanding of quantum TGD and it seems that the mis-interpreted super-symmetries are one of the basic reasons for the difficulties of super string models too. At this moment one can fairly say that the construction of the configuration space spinor structure reduces to a purely group theoretical problem of constructing representations for the super generators of the super-canonical algebra of CP_2 localized with respect to $\delta M_\pm^4$ in terms of second quantized induced spinor fields.

Super canonical symmetries

One can imagine two kinds of causal determinants besides $\delta M_+^4 \times CP_2$. In principle all surfaces $X_l^3 \times CP_2$, where X_l^3 is a light like 3-surface of M^4, could act as effective causal determinants: the reason is that the creation of pairs of positive and negative energy space-time sheets is possible at these surfaces. There are good hopes that the super-canonical and super-conformal symmetries associated with δX_l^3 allow to generalize the construction of the configuration space geometry performed at $\delta M_+^4 \times CP_2$. If X_l^3 can be restricted to be unions of future and past light cone boundaries, the generalization is more or less trivial: one just forms a union of configuration spaces associated with unions of translates of δM_+^4 and δM_-^4.

As explained in the previous chapter, one can understand how the causal determinants $X_l^3 \times CP_2$ emerge from the facts that space-time sheets with negative time orientation carry negative energy and that the most elegant theory results when the net quantum numbers and conserved classical quantities vanish for the entire Universe. Crossing symmetry allows consistency with elementary particle physics and the identification of gravitational 4-momentum as difference of conserved inertial (Poincare) 4-momenta for positive and negative energy matter provides consistency with macroscopic physics.

The emergence of these additional causal determinants means that super-canonical symmetries become macroscopic, rather than only cosmological, symmetries commuting with Poincare transformations exactly for $M^4 \times CP_2$ and apart from small quantum gravitational effects for $M_+^4 \times CP_2$. Super-canonical symmetry differs in many respects from Kac-Moody symmetries of particle physics, which in fact correspond to the conformal invariance associated with the modified Dirac action and correspond to the product of Poincare, electro-weak and color groups. It seems that these symmetries are dually related.

Super Kac-Moody symmetries associated with the light like causal determinants

Also the light like 3-surfaces X_l^3 of H defining elementary particle horizons at which Minkowskian signature of the metric is changed to Euclidian and boundaries of space-time sheets can act as causal determinants, and thus contribute to the configuration space metric. In this case the symmetries correspond to the isometries of the imbedding space localized with respect to the complex coordinate of the 2-surface X^2 determining the light like 3-surface X_l^3 so that Kac-Moody type symmetry results. Also the condition $\sqrt{(g_3)} = 0$ for the determinant of the induced metric seems to define a conformal symmetry associated with the light like direction. This conforms with duality since also the 7-D causal determinants $X_l^3 \times CP_2$ allow both radial and transversal conformal symmetry.

Good candidates for the counterpart of this symmetry in the interior of space-time surface are hyper-quaternion conformal invariance [E2]. All that is needed for these symmetries to be equivalent that the spaces of super-gauge degrees of freedom defined by them are equivalent. Kac Moody generators and their super counterparts can be associated with the 3-D light like CDs.

If is enough to localize only the H-isometries with respect to X_l^3, the purely bosonic part of the Kac-Moody algebra corresponds to the isometry group $M^4 \times SO(3,1) \times SU(3)$. The physical interpretation of these symmetries is not so obvious as one might think. The point is that one can generalize the formulas characterizing the action of infinitesimal isometries on spinor fields of finite-dimensional Kähler manifold to the level of the configuration space. This gives rise to bosonic generators containing also a sigma-matrix term bilinear in fermionic oscillator operators. This representation is not equivalent with the purely fermionic representations provided by induced Dirac action. Thus one has two groups of local color charges and the challenge is to find a physical interpretation for them. The following arguments fix the identification.

a) The hint comes from the fact that $U(2)$ in the decomposition $CP_2 = SU(3)/U(2)$ corresponds in a well-defined sense electro-weak algebra identified as a holonomy algebra of the spinor connection. Hence one could argue that the $U(2)$ generators of either $SU(3)$ algebra might be identifiable as generators of local $U(2)_{ew}$ gauge transformations whereas non-diagonal generators would correspond to Higgs field. This interpretation would conform with the idea that Higgs field is a genuine scalar field rather than a composite of fermions.

b) Since X_l^3-local $SU(3)$ transformations represented by fermionic currents are characterized by central extension they would naturally correspond to the electro-weak gauge algebra and Higgs bosons. This is also consistent with the fact that both leptons and quarks define fermionic Kac Moody currents.

c) The fact that only quarks appear in the gamma matrices of the configuration space supports the view that action of the generators of X_l^3-local color transformations on configuration space spinor fields represents local color transformations. If the action of X_l^3-local $SU(3)$ transformations on configuration space spinor fields has trivial central extension term the identification as a representation of local color symmetries is possible.

The topological explanation of the family replication phenomenon is based on an assignment of 2-dimensional boundary to a 3-surface characterizing the elementary particle. The precise identification of this surface has remained open and one possibility is that the 2-surface X^2 defining the light light-like surface associated with an elementary particle horizon is in question. This assumption would conform with the notion of elementary particle vacuum functionals defined in the zero modes characterizing different conformal equivalences classes for X^2.

The relationship of the Super-Kac Moody symmetry to the standard super-conformal invariance

Super-Kac Moody symmetry can be regarded as $N = 4$ complex super-symmetry ($N = 8$ real super-symmetry) with complex H-spinor modes of H representing the 4 physical helicities of 8-component leptonic and quark like spinors acting as generators of complex dynamical super-symmetries. The super-symmetries generated by the covariantly constant right handed neutrino appear with *both* M^4 helicities: it however seems that covariantly constant neutrino does not generate any global super-symmetry in the sense of particle-sparticle mass degeneracy. Only righthanded neutrino spinor modes (apart from covariantly constant mode) appear in the expressions of configuration space gamma matrices forming a subalgebra of the full super-algebra.

$N = 2$ real super-conformal algebra is generated by the energy momentum tensor $T(z)$, $U(1)$ current $J(z)$, and super generators $G^{\pm}(z)$ carrying $U(1)$ charge. Now $U(1)$ current would correspond to right-handed neutrino number and super generators would involve contraction of covariantly constant neutrino spinor with second quantized induced spinor field. The further facts that $N = 2$ algebra is associated naturally with Kähler geometry, that the partition functions associated with $N = 2$ super-conformal representations are modular invariant, and that $N = 2$ algebra defines so called chiral ring defining a topological quantum field theory [ee2], lend a further support for the belief that $N = 2$ super-conformal algebra acts in super-canonical degrees of freedom.

The values of c and conformal weights for $N = 2$ super-conformal field theories are given by

$$c \;=\; \frac{3k}{k+2} \;,$$

$$\Delta_{l,m}(NS) \;=\; \frac{l(l+2) - m^2}{4(k+2)} \;,\quad l = 0, 1, ..., k \;,$$

$$q_m \;=\; \frac{m}{k+2} \;,\quad m = -l, -l+2, ..., l-2, l \;. \tag{6.6.15}$$

q_m is the fractional value of the $U(1)$ charge, which would now correspond to a fractional fermion number. For $k = 1$ one would have $q = 0, 1/3, -1/3$, which brings in mind anyons. $\Delta_{l=0,m=0} = 0$ state would correspond to a massless state with a vanishing fermion number. Note that $SU(2)_k$ Wess-Zumino model has the same value of c but different conformal weights. More information about conformal algebras can be found from the appendix of [ee2].

For Ramond representation $L_0 - c/24$ or equivalently G_0 must annihilate the massless states. This occurs for $\Delta = c/24$ giving the condition $k = 2\left[l(l+2) - m^2\right]$ (note that k must be even and that $(k, l, m) = (4, 1, 1)$ is the simplest non-trivial solution to the condition). Note the appearance of a fractional vacuum fermion number $q_{vac} = \pm c/12 = \pm k/4(k+2)$. I have proposed that NS and Ramond algebras could combine to a larger algebra containing also lepto-quark type generators but this not necessary.

The conformal algebra defined as a direct sum of Ramond and NS $N = 4$ complex sub-algebras associated with quarks and leptons might further extend to a larger algebra if lepto-quark generators acting effectively as half odd-integer Virasoro generators can be allowed. The algebra would contain spin and electro-weak spin as fermionic indices. Poincare and color Kac-Moody generators would act as symplectically extended isometry generators on configuration space Hamiltonians expressible in terms of Hamiltonians of $X_l^3 \times CP_2$. Electro-weak and color Kac-Moody currents have conformal weight $h = 1$ whereas T and G have conformal weights $h = 2$ and $h = 3/2$.

The experience with $N = 4$ complex super-conformal invariance suggests that the extended algebra requires the inclusion of also second quantized induced spinor fields with $h = 1/2$ and their super-partners with $h = 0$ and realized as fermion-antifermion bilinears. Since G and Ψ are labelled by 2×4 spinor indices, super-partners would correspond to $2 \times (3 + 1) = 8$ massless electro-weak gauge boson states with polarization included. Their inclusion would make the theory highly predictive since induced spinor and electro-weak fields are the fundamental fields in TGD.

How do conformal symmetries of light like 3-D CDs act on super-canonical degrees of freedom?

An important challenge is to understand the action of super-conformal symmetries associated with the light like 3-D CDs on super-canonical degrees of freedom. The breakthrough in this respect via the algebraic formulation for the vision about vanishing loop corrections of ordinary Feynman diagrams in terms of equivalence of generalized Feynman diagrams with loops with tree diagrams (see the chapter "Equivalence of Loop Diagrams with Tree Diagrams and Cancellation of Infinities in Quantum TGD"). The formulation involves Yang-Baxter equations, braid groups, Hopf algebras, and so called ribbon categories and led to the following vision.

a) Quantum classical correspondence suggests that the complex conformal weights of super-canonical algebra generators have space-time counterparts. The proposal is that the weights are mapped to the points o of the homologically non-trivial geodesic sphere S^2 of CP_2 corresponds to the super-canonical conformal weights, and corresponds to a discrete set of points at the space-time surface. These points would also label mutually commuting R-matrices. The map is completely analogous to the map of momenta of quantum particles to the points of celestial sphere. These points would belong to a "time=constant" section of 2-dimensional "space-time", presumably circle, defining physical states of a two-dimensional conformal field theory for which the scaling operator L_0 takes the role of Hamiltonian.

b) One could thus regard super-generators as conformal fields in space-time or complex plane having super-canonical conformal weights as punctures. The action of super-conformal algebra and braid group on these points realizing monodromies of conformal field theories [ee2] induces by a pullback a braid group action on the super-canonical conformal weights of configuration space gamma matrices (super generators) and corresponding isometry generators.

At the first sight the explicit realization of super-canonical and Kac Moody generators seems however to be in conflict with this vision. The interaction of the conformal algebra of X_l^3 on super-canonical algebra is a pure gauge interaction since the definition of super canonical generators is not changed by the action of conformal transformations of X_l^3. This is however consistent with the assumption that the action defined by the quantum-classical correspondence is also a pure gauge interaction locally. The braiding action would be analogous to the holonomies encountered in the case of non-Abelian gauge fields with a vanishing curvature in spaces possessing non-trivial first homotopy group.

Quantum classical correspondence would allow to map abstract configuration space level to space-time level.

a) The complex argument z of Kac Moody and Virasoro algebra generators $T(z) = \sum T_n z^n$ is discretized so that has values on the set of supercanonical conformal corresponding to the space t in the Cartan decomposition $g = t + h$ of the tangent space of configuration space. These points can be interpreted as punctures of the complex plane restricted to the lines $Re(z) = \pm 1/2$ and positive real axis.

b) The vacuum expectation values of the enveloping algebra of the super-canonical algebra would reduce to n-point functions of a super-conformal quantum field theory in the complex plane containing infinite number of punctures defined by the super-canonical conformal weights, for which primary fields correspond to the representations of $SO(3) \times SU(3)$. These representations would combine to form infinite-dimensional representations of super-canonical algebra. The presence of the gigantic super-canonical symmetries raises the hope that quantum TGD could be solvable to a very high degree.

c) The Super Virasoro algebra and Super Kac Moody algebra associated with 3-D light like CDs would act as symmetries of this theory and the S-matrix of TGD would involve the n-point functions of this field theory. By 7–3 duality this indeed makes sense. The situation would reduce to that encountered in WZW theory in the sense that one would have space-like 3-surfaces X^3 containing two-dimensional closed surfaces carrying representations of Super Kac-Moody algebra.

This picture also justifies the earlier proposal that configuration space Clifford algebra defined by the gamma matrices acting as super generators defines an infinite-dimensional von Neumann algebra possessing hierarchies of type II_1 factors [eb1] having a close connection with the non-trivial representations of braid group and quantum groups. The sequence of non-trivial zeros of Riemann Zeta along the line $Re(s) = 1/2$ in the plane of conformal weights could be regarded an an infinite braid behind the von Neumann algebra [eb1]. Contrary to the expectations, also trivial zeros seem to be important. The finite braids defined by subsets of zeros, and also superpositions of non-trivial zeros of form $1/2 + \sum_i y_i$, could be seen as a hierarchy of completely integrable 1-dimensional spin chains leading to quantum groups and braid groups [ec4, ee2] naturally.

It seems that not only Riemann's zeta but also polyzetas [cf3, cf4, ed4, ed5] could play a fundamental role in TGD Universe. The super-canonical conformal weights of interacting particles, in particular of those forming bound states, are expected to have "off mass shell" values. An attractive hypothesis is that they correspond to zeros of Riemann's polyzetas. Interaction would allow quite concretely the realization of braiding operations dynamically. The physical justification for the hypothesis would be quantum criticality. Indeed, it has been found that the loop corrections of quantum field theory are expressible in terms of polyzetas [ee9]. If the arguments of polyzetas correspond to conformal weights of particles of many-particle bound state, loop corrections vanish when the super-canonical conformal weights correspond to the zeros of polyzetas including zeta.

How Super-Kac Moody algebra acts on super-canonical algebra

If 7–3 duality is assumed, it is clear how Super Kac-Moody algebra acts on super-canonical algebra. The action would be commutation action on $SO(3) \times SU(3)$ labels of the Hamiltonians associated with $X_l^3 \times CP_2$ used in the construction of configuration space metric. The power z^n associated with the Kac Moody generator would simply multiply the configuration space Hamiltonian or super charge.

Also electro-weak quantum numbers associated with ground states would be present and would also appear as indices of the fermionic parts of Kac Moody generators. The first guess was that configuration space Hamiltonians, which in turn correspond to Hamiltonians associated with $X_l^3 \times CP_2$, label highest weight representations of the Super Kac-Moody algebra. This guess is however not correct as becomes clear just by looking the action of the Kac Moody generators on configuration space Hamiltonians: the power z^n, when expressed in terms of $X_l^3 \times CP_2$ Hamiltonians, contains in general infinite number of them. Rather, the two algebras share a common representation space containing several irreducible representations of both algebras: this of course conforms with 7–3 duality. Sugawara construction would provide the expressions for the Virasoro algebra generators of Super Kac-Moody algebra.

Extensions of super-canonical and Super Kac-Moody algebras

Super-canonical algebra allows an extension obtained by adding an integer multiple of Hamiltonian to the vector field of imbedding space inducing the action at configuration space level: $jAk\partial_k \rightarrow j^{Ak}\partial_k + kH_A$. The same extension applies also to the Kac Moody generators whose Lie algebra parts correspond to $SO(3) \times SU(3)$ Killing vector fields of the light cone boundary or more, generally of light like 7-surface $X_l^3 \times CP_2$, allowing representation as Hamiltonians. Therefore super canonical and Kac Moody extensions correspond to each other. For $k \neq 0$ the multiplets defined by Hamiltonians do not remain closed with respect to the action of Lie-algebra generators of H isometries and super-canonical representation gives rise to a highest weight Kac-Moody representation.

Comparison with string models

The super-conformal algebra of 7-D CDs differs in many respects from the corresponding algebra of string models and these differences could be seen also as objections against the proposed construction of the spinor structure.

a) Configuration space gamma matrices must carry definite fermion number and hence cannot be Hermitian. Besides this the fermion number is either leptonic or quark like. It seems that either quarks or leptons are needed to construct vielbein but not both. It indeed turns out that one can construct representations of the configuration space gamma matrices only in terms of quark like oscillator operators and that these representations correspond to N-S representations of super-conformal algebra. This degeneracy corresponds also to the ordinary and kappa super-symmetries of super-string models. Leptonic super generators define super-symmetrization of the function algebra of $\delta M_+^4 \times CP_2$ whereas quark super charges defined super-symmetrization of the Poisson algebra.

b) In string model context Super Virasoro symmetry requires Majorana spinors and the Majorana condition guaranteing the hermiticity of the super generator G satisfying (very schematically) $G^2 = L$ and implying imbedding space dimension $D = 2 \ mod \ 8$. How could TGD avoid this difficulty? The answer is simple: Super Virasoro algebra must be generalized so that the local super generator $G(r)$ is non-hermitian and one has symbolically $\{G, G^\dagger\} = 2L$ instead of the usual $GG = L$. This modification is forced by the fact that configuration space gamma matrices carry fermion number and configuration space Dirac equation reduces to Super Virasoro conditions such that super generators G carries fermion number.

c) TGD framework suggests the possibility that Ramond and N-S type super-conformal algebras could be extended to a larger algebra containing also bosonic generators with half-odd integer conformal weights and carrying lepto-quark quantum numbers.

d) The canonical algebra spanning the tangent space of the configuration space contains vector fields with complex conformal weights such that real parts of the conformal weights are half integers. Configuration space itself is obtained as a coset space with Cartan decomposition $g = t + h$.

t corresponds to conformal weights of form

i) $z = -2n$, $n > 0$,

ii) $z = n - 1/2 - i \sum n_i y_i$, $\sum n_i$ odd for even n and $\sum n_i$ even for odd n. At least for physical states $n = 0, 1$ option is favored over $n \geq 0$ option since in this case an orthogonal basis is certainly obtained. Also the scalar propagator defined as super-canonical partition function is non-singular for this option unlike for $n \geq 0$ option (see the chapter "Equivalence of Loop Diagrams with Tree Diagrams and Cancellation of Infinities in Quantum TGD").

h in turn corresponds to

i) $z = -2n + 1$, $n > 0$,

ii) $z = n - 1/2 - i \sum n_i y_i$, $\sum n_i$ odd for odd n and $\sum n_i$ even for even n.

e) The physical interpretation of super-canonical algebra differs from that of Kac-Moody algebra. The elements of the super-canonical algebra are labelled by complex conformal weights and can be regarded as irreducible representations of Lorentz group at light cone boundary. The quantum numbers involved are thus eigen values of Lorentz boost generator and of rotation generator around given axis.

When one allows decomposition of the configuration space to a union of sectors defined by unions of future and past directed light cones, translations shift the configuration space and also momentum can be assigned to the super-canonical algebra.

By 7–3 duality the super-canonical and super Kac-Moody representations are defined in the same representation space but large number of irreducible representations combined together. A very attractive hypothesis is a generalization of the Goddard-Kent-Olive coset construction so that the net conformal algebra is generated by the differences of Super Virasoro generators associated with the super-canonical and Super-Kac Moody algebras. This could guarantee the vanishing of the central extension of the Virasoro algebra since the central extensions are expected to be same if super-canonical and super Kac-Moody super algebras are related by coordinate change in CH.

A second prediction is that both spin and color quantum numbers not assignable to elementary particles are possible. Super-canonical degrees of freedom represent completely new degrees of freedom not present in string models and are predicted to manifest themselves also in elementary particle physics at the level of propagators and vertices (see the chapter "Equivalence of Loop Diagrams with Tree Diagrams and Cancellation of Infinities in Quantum TGD").

f) The $N = 4$ super-conformal gauge invariance associated with the modified Dirac action at light like 3-D CDs predicts Noether charges and super-charges which span super Kac-Moody algebra associated with Poincare, color, and electro-weak symmetries. The new interpretation of the solutions of Dirac equations as generators of local gauge super symmetries implies that the $N = 4$ supersymmetry at space-time level is pure gauge symmetry so that there is no global super-symmetry and no spartners of ordinary particles are predicted: this means quite a sigh of relief since low mass sparters predicted by super string theories are experimentally absent. At configuration space level the super algebras act as dynamical super-symmetries and no mass degeneracies are predicted. Super-Kac-Moody algebra and corresponding Super Virasoro algebra have a natural action in the super-canonical algebra whose conformal weights take the role of complex coordinate of complex plane.

6.6.3 Representations of super-canonical and Super Kac-Moody algebras

The first belief was that super-canonical and Super Kac-Moody Super Virasoro conditions hold true separately. The assumption that the 3-D light like CDs intersect X^3 at 7-D CDs suggests much more intimate connection.

One can think Super Kac-Moody states as delocalized states in the space of super-canonical conformal weights. The finite-dimensional representations of $SO(3) \times SU(3)$ associated with the Hamiltonians define highest weight- or finite-dimensional representations of Super Kac Moody algebra. Super-canonical transformations in turn transform Super Kac-Moody representations to each other since they mix the irreducible representations of δX^3 Hamiltonians.

Representations of super-canonical conformal algebra

Super-canonical Virasoro algebra acts purely geometrically on the light like radial coordinate of X^3_l. Sugawara construction does not make sense nor is needed in this case. Somewhat loosely one can say that The role of the translation group is taken by the Lorentz group of δM^4_+ or its generalization applying in the case of more general light like surfaces $X^3_l \times CP_2$. Note however that also translational contribution is present in the mass formula involving both super-canonical and super Kac-Moody degrees of freedom.

The conformal weights associated with the sub-space t of the super-canonical algebra do not in general correspond to those allowed by Kac formula for the conformal weights $\Delta_{m,n}$ of the primary fields yielding Verma modules with vectors annihilated by L_n, $n > 0$, encountered in conformal field theories. The presence of complex conformal weights is however allowed also by Kac formula. The complex conformal weights Δ closely related to the zeros of Riemann Zeta are allowed but the sum of the imaginary parts of the super-canonical conformal weights for the physical states must sum up to zero.

Also non-unitary representations of Lorentz group associated with $h = 2n$ states can be allowed. The situation is analogous to that for field theories in M^4: non-unitary representations of Lorentz group defined by field components at a point give rise to unitary representations of Poincare group. Now M^4 is replaced with configuration space and finite non-unitary representations of Lorentz group are replaced with infinite-dimensional non-unitary representations.

The super-canonical Virasoro generators L_n, $n > 0$, must annihilate the physical states is required in order to eliminate the states with conformal weights $n - 1/2 - i \sum_i n_i y_i$, $n \geq 2$, (with $\sum n_i =$ odd/even for n even/odd) generated by the super-canonical algebra. Their elimination is necessary since they are not orthogonal to the states with $n = 0, 1$.

Representations of Super Kac-Moody algebra

The conformal algebra associated with $Y_l^3 \subset X^4$ acts as local deformations on geometrically preferred light like 3-surfaces associated with space-time sheets (say boundaries or elementary particle horizons surrounding wormhole contacts). The assumption is that this action can be lifted to an action in super-canonical algebra. Besides the Kac Moody type action on $SO(3) \times SU(3) \times U(2)_{ew}$ degrees of freedom there is also action in center of mass degrees of freedom defined by the possibility to perform Poincare transformations for $X_l^3 \times CP_2$. If $X_l^3 = \delta M_+^4$ restriction is made, only translational degrees of freedom remain. The p-adic mass calculations necessitated to include Kac Moody algebra of Lorentz group respecting the dip of the light cone. It is not necessary to assume $X_l^3 \times CP_2$ to be more general than $M_\pm^4$ or union of them in order to have local Kac Moody invariance in M_+^4 degrees of freedom.

When the discrete super-canonical conformal weight varies a generator of super-canonical algebra (or of its enveloping algebra) defines what might be regarded as a conformal field in the Super Kac-Moody sense. The eigen states of L_0 of Super Kac-Moody algebra are delocalized in the space of super-canonical conformal weights and thus a superposition over the states belonging in different representations of super-canonical Lorentz group results.

The complex coordinates defined by the super-canonical conformal weights appear as arguments in the coefficients for the products of configuration space gamma matrices and their bosonic counterparts acting as infinitesimal isometries of CH. The differential operator action of Super Kac-Moody Virasoro algebra generators L_n on the super-canonical conformal weights is translated to a corresponding action on the coefficients by partial integration so that the analogs of momentum eigen states $\sum_n exp(ikna)|na\rangle$ is lattice of lattice constant a result with momentum replaced with the scaling momentum L_0. The dependence of Super Kac Moody generators on z means multiplications of these coefficients by functions of conformal weight.

If y_k are linearly independent, the superpositions of zeros of non-trivial zeros of Riemann Zeta define effectively a union of infinite number of superposed 1-dimensional lattices, or equivalently, an infinite-dimensional discrete lattice with each zero $z_k = 1/2 + iy_k$ defining its own algebraic dimension. The question whether the imaginary parts of the Riemann Zeta are linearly independent or not is of crucial physical significance. Linear independence would imply that the spectrum of the super-canonical weights is essentially an infinite-dimensional lattice. Otherwise a more complex structure results. The hypothesis that p^{iy} is an algebraic phase for every prime implies that p^{iy} is expressible as a product of a Pythagorean phase and a root of unity for every prime p.

The numerical evidence supporting the translational invariance of the correlations for the spectrum of zeros [cf5] together with p-adic considerations leads to the working hypothesis that for any prime p one can express the spectrum of zeros as the product of pth powers for a subset of Pythagorean prime phases and p^{th} power U^p of a fixed subset U of roots of unity. The spectrum of zeros could be expressed as a union over the translates of the same basic spectrum defined by the roots of unity translated by the phase angles associated with p^{th} powers of a subset of Pythagorean phases: this is consistent with what the spectral correlations strongly suggest. That decompositions defined by different primes p yield the same spectrum would mean a powerful number theoretical symmetry realizing p-adicities at the level of the spectrum of Zeta.

The model for the scalar propagator as a partition function for the super-canonical algebra supports this view. The approximation that the zeros are linearly independent implies for any subalgebra defined by a finite set of zeros a universal spectrum of singularities for real values of mass squared, and at the limit of entire algebra the propagator is not mathematically defined since the singularities are dense in real axis. Linear independence however shifts the poles to complex values of mass and genuine resonances result and the propagator is expected to be well-defined for the entire algebra. The theory predicts universal momentum and p-adic length scale dependence of propagators in this picture and the option based on super-canonical algebra predicts very simple and elegant expression for the propagator with all the desired properties.

For a sub-algebra generated by any K distinct zeros with $z_{i_k} = 1/2 + y_{i_k}$, $k = 1, .., K$, one has K-dimensional lattice. The eigen states of the operators $L_0 = z_k d/dz_k$ of Super Kac-Moody algebra acting as a differential operator on a function of conformal weights $z_1, ..., z_k$ would be of form

$$|n_1, ..., n_k\rangle = \sum_{z_{i_k}} \prod_i z_{i_k}^{n_k} |n_1 y_{i_1}, n_2 y_{i_2}, ..., n_k y_{i_k}\rangle \ , \quad z_{i_k} = \frac{\epsilon}{2} + i y_{i_k} \ ,$$

$$\sum n_{i_k} = 2N \text{ for } \epsilon = 1 \ , \quad \sum n_k = 2N + 1 \text{ for } \epsilon = -1 \ , \tag{6.6.16}$$

where $h = \pm 1/2 - \sum_i n_i y_i$ corresponds to the conformal weight at the line $Re(h) = \pm 1/2$. Note that the cutoff for the non-trivial zeros of Zeta as a reduction to a finite-dimensional sub-algebra of the super-canonical algebra emerges completely naturally. The linear combinations of zeros are analogous to many particle excitations in accordance with the fact that one could interpret y_k as analogs of harmonic oscillator frequencies.

At the half-line $h = 2n$, $n > 0$ corresponding to trivial zeros one does not obtain strict eigen states of discrete translations although eigen states in the same sense as above are possible

$$|n\rangle = \sum_{z=2n} z^n |n\rangle \ . \tag{6.6.17}$$

Note that the states have also $SU(2) \times SU(3)$ and electro-weak quantum numbers and these quantum numbers give then conformal weight Δ which also contributes to the L_0 eigen values.

The sub-algebras spanned by K zeros of Riemann Zeta naturally correspond to K-particle free states. In this picture the idea about zeros of polyzetas $\zeta(z_1, ..., z_K)$ as conformal weights of K-particle bound states analogous to bound state energies emerges very naturally. This idea is discussed in more detail in [C5].

Constraints from unitarity

For the representations of Virasoro algebra allowing Verma modules containing null states annihilated by Virasoro generators L_n, $n > 0$ conformal weights are quantized [ee2]. The simplest situation would correspond to rational conformal weights Δ and maximally symmetric super-conformal quantum field theories known as rational super-conformal field theories (RCFT) having a finite number of primary fields. In TGD framework super-generators carry fermion number which leaves $N = 2$ super conformal theories (or possibly $N = 4$ theories, situation is still unsettled) as the only physically acceptable option. The conformal weights are predicted to be rational numbers in this case.

For Sugawara construction [ee2] the conformal weights associated with the Hamiltonians are expressible in terms of the value of Kac-Moody central extension parameter k and Casimir invariant of the highest weight representation. For instance, in the case of $SU(2)$ one has $\Delta(j) = \frac{j(j+1)}{k+2}$. Unitarity poses strong constraints on the conformal weights.

In Wess-Zumino-Witten model also the analogs of Verma modules having finite number of non-zero norm states for Kac Moody algebras emerge and relate very closely to the corresponding phenomenon for quantum groups [ee2]. Thus only a finite number of highest weight representations of Kac Moody algebra appear as primary fields in WZW model. The theory predicts the dependence of the conformal weights on the finite-dimensional representations of the Kac Moody algebra. In the recent case Super Kac-Moody representations relate naturally also the finite dimensional representations of $SO(3) \times U_{ew}(2) \times SU(3)$ labelling finite-dimensional or highest weight Kac Moody representations.

For Sugawara construction central extension parameter is given by $c = k\, dimg(G)/(k + h^\star)$, where $h^\star$ is the dual Coxeter number ($h^\star = n$ for $SU(n)$ and is non-vanishing in general case and one faces the question whether Super Kac Moody symmetry is gauge symmetry or not. String model based mass formulas would suggest the Virasoro representations should correspond to a trivial central extension $c = 0$. The best manner to guarantee this is the analog of the coset construction already described: the central charges of super-canonical and super Kac-Moody representations would be identical but otherwise arbitrary.

Hence the most general hypothesis is that all values of k and c consistent with unitarity are possible. The first thing to come into mind is that elementary particles could correspond to $k = 0, c = 0$ and thus to finite-dimensional representations of Super Kac-Moody algebra defined by Hamiltonians belonging to a given representation of $SO(3) \times SU(3)$. This view is consistent with p-adic mass calculations if massivation and corresponding symmetry breaking correspond to a process in which k and c become non-vanishing. Also anyons could correspond to a non-vanishing value of k. At configuration space level this would correspond to an extension of isometry generators obtained by adding an integer multiple of Hamiltonian as a scalar term to the isometry generator so that the action couples together several irreducible representations of $SO(3) \times SU(3)$ corresponding to states with non-vanishing norm.

How to achieve the vanishing of the central charge of the conformal algebra?

The central charge of the Virasoro representation should vanish whereas the central charges of the Kac Moody algebras involved should be non-vanishing since physical states are created by Kac Moody generators and their super counterparts. One can imagine two manners to achieve this.

1. Does the difference of super Kac-Moody and super-canonical Virasoro algebras annihilate physical states?

The non-vanishing value of k means that Virasoro algebra possesses a non-vanishing central extension and there should exist some elegant manner to construct a representation for which the central extension vanishes. A modification of Olive-Goddard-Kent coset construction utilizing the commuting Virasoro representations of group G and its sub-group H suggest itself as a solution of the problem. The essential point is that the two Virasoro representations commute so that the difference representation has central charge which is the difference $c_g - c_h$ of the central charges.

The obvious guess is that the differences of commuting generators of Virasoro algebras accompanying Super Kac-Moody and super-canonical algebras define Virasoro algebra for which central charge is the difference of central charges involved and vanishes if super-canonical and Kac-Moody super algebras indeed correspond to two different coordinates for CH. Physical states would be annihilated by this algebra. The possibility to satisfy the fundamental vanishing of central charge condition would be the deep underlying reason why for two conformal algebras.

2. Less general variants of coset construction

Also less general and not so attractive variants of Olive-Goddard-Kent construction can be considered as means of achieving vanishing central charge [ee2] and deserve mentioning. Consider Kac-Moody algebras associated with G and $H \subset G$. For a semi-simple Lie group Sugawara construction gives representations with central charge

$$c_G = \frac{kdim(G)}{k + h^\star} \ , \tag{6.6.18}$$

where k is the central extension parameter of Kac-Moody algebra and $h^\star$ is the Coxeter number of G. For $SU(n)$ one has $h^\star = n$. The Virasoro representation associated with G commutes with that associated with H and the value of central charge for the representation defined by the differences of Virasoro generators is

$$c_{G/H} = c_G - c_H \ . \tag{6.6.19}$$

This procedure can be applied to color group $SU(3$ and $SU(2)$ and in this case one has

$$c \ = \ \frac{8k_1}{k_1 + 3} - \frac{3k_2}{k_2 + 2} \ . \tag{6.6.20}$$

The maximal reduction of the central charge is obtained for $k_1 = 1$ at the limit $k_2 \to \infty$ and one has $c \to -1$ holds at this limit. $c = 0$ is obtained for $k_1 = 1$ and $k_2 = 4$.

For the diagonal subgroup $SU(2)$ of $SU(2) \otimes SU(2)$, which could correspond to the vectorial part of the electro-weak gauge group and entire electro-weak gauge group, one has

$$c \ = \ \frac{3k_1}{k_1 + 2} + \frac{3k_2}{k_2 + 2} - \frac{3k_3}{k_3 + 2} \ , \tag{6.6.21}$$

Maximal reduction corresponds now to $k_1 = k_2 = 1$ and $k_3 \to \infty$ and also now one has $c \to -1$ at this limit. $c = 0$ is obtained for $k_1 = k_2 = 1$ and $k_3 = 4$.

There are good reasons to believe that the central extension for the Kac Moody algebra associated with the two transversal M^4 degrees of freedom has $k = 1$ and $c = 2$. One could argue that this representation can be equivalently regarded also as a $(k = 1, c = 2)$ representation of $SO(3,1)$, which should result by an analytical continuation from that of $SO(3) \times SO(3)$. Subtracting $SO(3)$ representation with Kac-Moody central charge k_1 one would indeed obtain $c \to -1$ at the limit $k_1 \to \infty$. $c = 0$ would result for $k_1 = 4$.

One can however raise an an objection against $M^4 \leftrightarrow SO(3,1)$ equivalence. p-Adic mass calculations encourage to think that one should treat M^4 and $SO(3,1)$ as separate factors. This holds true for light like 7-surfaces $X_l^3 \times CP_2$ when X_l^3 can be any light like 3-surface of M^4 since in this case the entire Poincare group defines cm coordinate for the causal determinants $X_l^3 \times CP_2$ whereas $X_l^3 = \delta M_{\pm}^4$ corresponds to a degenerate case.

Assume $SO(3,1) \times SU(3) \times SU(2)_L \times SU(2)_R$ representation with $k_1 = 1$ for $SO(3,1)$ and $k_3 = k_4$ for the factors of electro-weak gauge group. The total central charge is given by

$$c(1, k_2, k_3) \quad = \quad c(SO(3,1), 1) + c(SU(3), k_2) + c(SU(2)_L, k_3) + c(SU(2)_R, k_3)$$

$$= \quad 2 + \frac{8k_1}{k_1 + 3} + \frac{3k_2}{k_2 + 2} + 2\frac{3k_3}{k_3 + 2} \geq c(1,1,1) = 6 \ . \tag{6.6.22}$$

This central charge can in principle be reduced to zero since the reduction of the central charge satisfies $|\Delta c| < 3 \times 3 = 9$. Coset construction is however required in all tensor factors. The condition for the vanishing of the central charge reads for $k_1 = k_2 = k_2 = k_4 = 1$ as

$$\frac{3k_a}{k_a + 2} + \frac{3k_b}{k_b + 2} + \frac{3k_c}{k_c + 2} \quad = \quad 6 \ . \tag{6.6.23}$$

Here a, b, c refer to $SO(3,1)/SO(3)$, $SU(3)/SU(2)$ and $SU(2)_L \times SU(2)_R/SU(2)_V$. The condition can be written in terms of $k_i + 2 = n_i$ as

$$\frac{1}{n_a} + \frac{1}{n_b} + \frac{1}{n_c} \quad = \quad \frac{1}{2} \ . \tag{6.6.24}$$

giving the condition $k_i > 0$. For instance, $k_a = k_b = k_c = 4$, $k_a = 1, k_b = k_c = 10$ and $k_a = k_b = 3, k_c = 8$ and the solutions resulting by the permutations of (a, b, c) represent solutions to these conditions.

A more general manner to generate $c = 0$ is to consider tensor products of the representations of $SO(3,1) \times SU(3) \times SU(2)_L \times SU(2)_R$. This procedure should apply to the construction of bound states, say color bound states. In this case one obtains a considerably richer spectrum of solutions. It must be however admitted that this kind of constructions look rather tricky when compared with the analog of the coset construction using super-canonical and super Kac-Moody algebras.

Quaternion conformal braiding and monodromy

Monodromy operations in complex plane related closely to braid groups are an essential element of conformal field theories and and should relate very closely to von Neumann algebra structure of the Clifford algebra defined by the gamma matrices of the configuration space. The interpretation of configuration space gamma matrices as conformal fields restricted to the subset of allowed conformal weights justifies this expectation if it is possible to define braiding in the space of conformal weights.

In the case of Super Kac-Moody algebra complex plane is replaced by a discrete set of punctures of complex plane and one can ask whether the braiding operations make sense any more. One can imagine several manners to realized the braiding operation.

a) By the quantum classical correspondence Super Kac-Moody algebra acting on complex space-time coordinate (coordinate for points of the dual of the commutative sub-manifold X^2 of hyper-quaternionic 4-surface of hyper-octonionic imbedding space M^8 [E2]) could be thought of as acting on the super-canonical conformal weights by a pull-back and induce braiding operations naturally and realize thus quantum classical correspondence. These braiding operations would act as gauge transformations. One might think that braidings are trivial since the definitions of super-canonical generators are invariant with respect to the classically represented conformal transformations of $X^2 = X_l^3 \cap X^7$. The quantum realization of conformal transformations involves however a conformal anomaly induced by the central extension of Kac Moody algebra and making the braiding non-trivial.

b) Right-handed neutrinos define $N = 2$ super-conformal symmetry. $N = 2$ super-conformal algebra is known to allow a continuous spectral flow in the indices associated with super-generators G_n [ee2]. The flow acts as

$$
L_n \rightarrow L_n + \alpha J_n + \frac{c}{6}\alpha^2 \delta_{n,0}
$$
$$
J_n \rightarrow J_n + \frac{c}{3}\alpha \delta_{n,0} \ ,
$$
$$
G_n^{\pm} \rightarrow G_{n \pm \alpha}^{\pm} \ . \tag{6.6.25}
$$

The choice $\alpha = \pm 1/2$ transforms NS representation to Ramond representation.

Quaternion conformal algebra could induce an analogous local spectral flow in the space of super-canonical conformal weights. For K-particle states involving superposition of K imaginary parts y_k of zeros of Zeta this interpretation is perfectly sensible. The resulting spectral flows would generalize complex translations in conformal weight generated by L_{-1} associated with NS type kappa symmetry (super generators containing quarks) to general conformal transformations.

c) For the bound state conformal weights represented as zeros of poly-zetas [cf4] the space of allowed conformal weights is $2K - 2$-dimensional in K-particle sector and braiding operations become possible. At the classical limit one can quite well imagine classical orbits mapped to orbits in $T \times C$ and defining braids. A possible space-time correlate would be braids defined by magnetic flux tubes with light like boundaries so that the longitudinal light-like coordinate along the boundary would have interpretation as a time coordinate.

d) The spectral flow should act as a gauge symmetry and the challenge is to understand in what sense the super-canonical generator $T^{\Delta'}$ with a shifted a conformal weight $\Delta' = \Delta + \epsilon \Delta^n$ is equivalent with that having the original conformal weight Δ. Since Δ appears as the exponent of the power $(r_M/r_0)^{\Delta}$ of the light like coordinate r_M/r_0 in the definition of the configuration space Hamiltonian, and since a given three-surface corresponds to a finite range for r_M/r_0, a natural idea is to write the transformed exponent as

$$
\left(\tfrac{r_M}{r_0}\right)^{\Delta'} = \left(\tfrac{r_M}{r_0}\right)^{\Delta} \times \left(1 + u\right)^{\epsilon \Delta^n} \ , \quad u = \tfrac{r_M - r_0}{r_0} \ .
$$

One can expand the latter factor at the right hand side in power series with respect to u, and in this power series expand the terms $u^n = [(r_M - r_0)/r_0)]^n$ using binomial formula. What one obtains is a series containing super-canonical generators $T^{\Delta + k}$. For $k > 0$ these generators generate gauge transformations so that in physical states only $k = 0$ contribution remains but this contribution corresponds to a binomial series of $(1 - 1)^n = 0$ and vanishes for $n > 0$.

One can therefore say that infinitesimally spectral flows do not affect the physical states at all. Non-trivial topological effects are however possible. In gauge theories a non-trivial holonomy of a flat connection induces non-trivial physical effects and makes topological quantum field theories interesting. In the recent case conformal transformations of X^2 inducing the non-trivial braiding of punctures represented by conformal weights can induce non-trivial effects on the physical states.

6.7   Configuration space Dirac equation as super Virasoro conditions

Basically configuration space Dirac equation corresponds to Super Virasoro conditions. There are several configuration space Dirac operators. The two kinds of causal determinants give rise to super-canonical and Super Kac-Moody Dirac operators. In both cases quark-lepton degeneracy gives rise to two different Dirac operators corresponding to N-S (quarks) and Ramond (leptons) representations of Super Virasoro algebras. Dirac operators are identifiable as super-generators G of the light Super Virasoro algebra and Dirac equation states that Super Virasoro algebra acts as gauge transformations.

a) Supercanonical Dirac operators $D(SC)$ are associated with 7-D CDs $X^7 = X_l^3 \times CP_2$, X_l^3 light like surface. The interpretation is in terms of creation of pairs of space-time sheets branching from X^3 and with vanishing total classical energy. In the "vibrational" part of the $L_0(SC)$ appears the quadratic , which can have complex values in accordance with the fact that super-canonical conformal weights are in general complex. The corresponding excitations have thus spin, color, electro-weak quantum numbers, and fermion numbers.

b) The light like 3-surfaces $Y_l^3 \subset X^4$ defining light like CDs event horizons and light like boundaries of space-time sheets define Super-Kac Moody Dirac operators. 7–3 duality reduces these operators as well as super-canonical Dirac operators to the intersections X_i^2 of Y_l^3 and X^3. These operators correspond to $P \times SU(3) \times U(2)_{ew}$ (P denotes Poincare group) localized with respect to Y_l^3 in the general case whereas the restriction of X_l^3 to surfaces $\delta M_\pm^4$ would reduce P to M^4. The counterpart of the stringy mass formula would result from the vanishing of $L_0(SKM)$.

The Dirac equations associated with these Dirac equations two kinds of Dirac operators can be regarded as independent, and the correlation comes from the assumption that super-conformal algebra generates super-canonical spectral flows acting as local gauge symmetries and braiding operations at global level. 7–3 duality indeed realizes this correlation concretely. In the sequel super symmetry based arguments are used to guess the explicit form of these operators and a general solution of configuration space Dirac equation is proposed by exploiting the analog with the Dirac equation in Minkowski space.

6.7.1   The analog of coset construction for super generators

As already found, the analog of Olive-Goddard-Kent coset construction for super Kac-Moody and super-canonical algebras provides a highly attractive manner to obtain a Virasoro algebra with a vanishing central extension term. The coset construction for groups G and H relies on $H \subset G$ relation. The commutators of the Virasoro generators L_G with generators L_H satisfy $[L_G, L_H] = [L_H, L_H]$ by the subgroup property in Sugawara construction. This relationship generalizes to the anti-commutators of super generators: $\{G_G, G_H\} = \{G_H, G_H\}$.

The following arguments support the view that Kac Moody algebra can be regarded as a sub-algebra of the super-canonical algebra.

a) By 7–3 duality both canonical and Kac-Moody algebras define two basis which are in 1-1 correspondence with each other. They need not however correspond to the tangent space basis for configuration space. Rather, Kac-Moody algebra could define zero modes in one-one correspondence with super-canonical degrees of freedom contributing to the configuration space metric.

b) Essential for the coset construction is that super-canonical conformal algebra acts effectively as Super Kac-Moody conformal algebra on Super Kac-Moody algebra so that one has $[L_G - L_H, L_H] = 0$. This is indeed true in the geometric sense if the induced metric of the partonic 2-surface X^2 suffers only an X^2-conformal scaling in the super-canonical conformal transformations realized as S^2-local radial scalings of the light like coordinate r_M of $M_+^4 = S^2 \times R_+$. This point will be discussed in detail in [F2].

c) One can wonder to what aspects of the coset construction are preserved in the recent case. Super-canonical basis is highly degenerate and this suggests that one can symbolically write $SC = SKM + degeneracy$, where "degeneracy" refers to generators which do not affect the configuration metric at all. If this is the case, Kac-Moody algebra can be regarded as a subalgebra of super-canonical algebra. The geometric argument suggests that the action of the super-canonical degeneracy algebra on super Kac-Moody conformal algebra is trivial.

The coset construction giving $L_n = L_n(SKM) - L_n(SC)$ means at the level of super generators G_n the expression

$$
\begin{aligned}
G_n^+ &= G_n^+(SKM) - G_n^+(SC) \ , \\
G_n^- &= G_n^-(SKM) - G_n^-(SC) \ .
\end{aligned}
\tag{6.7.1}
$$

Here $\pm$ refers to the positive and negative energy space-time sheets meeting at X^2 since "hermitian conjugation" (ket $\leftrightarrow$ bra) corresponds geometrically to a replacement of the space-time sheet with a positive time orientation with that having a negative time orientation. Note also that the super generators possess fermion number so that creation operators are replaced by annihilation operators in the conjugation.

If the conditions

$$\{G^+(SKM), G^-(SC)\} \;\; = \;\; \{G^+(SKM), G^-(SKM)\} \tag{6.7.2}$$

hold true, the anti-commutators give just the differences $L(SKM) - L(SC)$. If 7–3 duality corresponds to a coordinate change in CH, the central extension charges for SKM and SC Super Virasoro algebras are same and resulting Super Virasoro algebra has vanishing central extension terms.

This construction works perfectly for quark sector. Leptonic super-canonical generators defined assuming the presence of operators D_+ and D_-^{-1} however anti-commute to products of configuration space Hamiltonians rather than components of Kähler form. Hence the formula for super generators $G(SC)$ does not work. The conclusion can be avoided if there exists also a leptonic representation CH gamma matrices anti-commuting to Kähler form. As already found, this kind of representation is obtained by replacing the operators D^+ and D_1^{-1} in the definitions of the super generators by projectors to the space of spinor modes with non-vanishing generalized eigen value of $D_\pm$. Also quark like super generators anti-commuting to a function algebra exist with this assumption.

The Ramond-NS degeneracy is a further important factor which must be taken into account. Possible sparticles are associated with Ramond type representation and generated by the super generators defined by the covariantly constant right handed neutrino. The helicity for neutrino is fixed by requiring that the neutrino spinor is proportional to the operator $n^k \gamma_k$, where n^k corresponds to the radial light like vector for $M_\pm^4$. If the super generators containing D and D^{-1} correspond to a Ramond type representation then covariantly right handed neutrino spinor defines an identically vanishing super generator and spartners with degenerate masses are absent. In NS sector the conformal weights of the generators generating ground states are $\pm 1/2$ so that there is no problem with the ground state degeneracy. The conclusion would be that quark/lepton like super generators G correspond to Ramond/N-S type representation. Note that leptonic super generators $G(SC)$ must carry half odd integer conformal weight.

6.7.2   General mass formula

The vanishing condition for $L_0(SKM) - L_0(SC)$ leads to a general mass formula in quark and lepton sectors separately. One of the blessings of 7–3 duality is that one can treat different 2-surfaces X_i^2 as almost independent degrees of freedom. In the case of translations this does not seem to be true since independent translations lead the surfaces X^2 outside $X_l^3 \times CP_2$. Therefore one must consider two options.

a) If one neglects the correlation between the translations and assigns to each X_i^2 independent translational degrees of freedom a separate mass formula for each X_i^2 would result:

$$M_i^2 \;\; = \;\; -\sum_i L_{0i}(SKM) + \sum_i L_{0i}(SC) \; . \tag{6.7.3}$$

Here $L_{0i}(SKM)$ contains a CP_2 cm term giving the CP_2 contribution to the mass squared known once the spinorial partial waves associated with super generators used to construct the state are known.

b) Perhaps the only internally consistent option is based on the assignment of the mass squared with the total cm. This would give

$$M^2 \;\; = \;\; \left(\sum_i p_i\right)^2 = \sum_i M_i^2 + 2\sum_{i \neq j} p_i \cdot p_j = -\sum_i L_{0i}(SKM) + \sum_i L_{0i}(SC) \; .$$

$$\tag{6.7.4}$$

Here $L_{0i}(SKM)$ contains a CP_2 cm term giving the CP_2 contribution to the mass squared known once the spinorial partial waves associated with super generators used to construct the state are known. $L_0(SC)$ term contains only leptonic or quark oscillator operators unless one allows both the lepto-quark type gamma matrices involving both D^+ and D_-^{-1} and leptonic gamma matrices involving instead of $D_\pm^{\pm 1}$ the projector P to the spinor modes with a non-vanishing eigenvalue of D.

The decomposition of the net four momentum to a sum of individual momenta can be regarded as subjective unless there is a manner to measure the individual masses. It might be that there is no unique assignment of momenta to individual partons and that this non-uniqueness is part of the gauge symmetry implied by 7–3 duality.

Mass formula before massivation

If interactions can be neglected one could assume that the free particle mass formulas

$$M_i^2 \;=\; L_{0i}(SKM) + h_i(vac) \;,\tag{6.7.5}$$

where $h_i(vac)$ corresponds to vacuum conformal weight are satisfied and what remains is the condition

$$2\sum_{i\neq j} p_i \cdot p_j = \sum_i L_{0i}(SC) - \sum_i h_i(vac) \;=\; 0 \;.\tag{6.7.6}$$

In the case of single Kac Moody algebra this condition would be replaced with

$$2\sum_{i\neq j} p_i \cdot p_j \;=\; 0 \;.\tag{6.7.7}$$

The presence of the super-canonical algebra obviously allows much more flexibility but it seems that continuum mass spectrum is not possible.

The conformal weights of the super-canonical algebra related closely to the zeros of Riemann Zeta are complex and for the physical states the sum of the imaginary parts of the conformal weights must vanish: this is satisfied if the zeros and their conjugates appear as conformal weights of generators creating the state. The presence of super-canonical half integer contribution from super-canonical sector could explain the negative values of the ground state conformal weights forced by p-adic mass calculations.

The above formula does not contain all that is needed: also a contribution from modular degrees of freedom associated with the complex structures of X_i^2 is necessary in order to understand the dependence of the mass of fermion families on the genus of X_i^2 [F2, F3].

Hadron physics offers the most obvious application for the mass formula. The often used metaphor for the hadronic collision as a mini big bang would have a precise meaning in TGD framework, and the effective 2-dimensionality would provide a precise realization for the parton model of hadrons. The presence of both algebras could be essential for understanding the relationship between hadron and quark masses, and the presence of super-canonical spin could allow insights to the problem of proton spin.

Particle massivation

There is an objection against the proposed mass formula. The individual contributions to the mass squared eigenvalues are not fixed uniquely. The proposed identification of parton masses is only the simplest possibility, and one could add to the individual parton conformal weights contributions compensated by the super-canonical conformal weights. This might however be a blessing rather than a curse since it suggests a manner to understand particle massivation at the fundamental level.

a) A natural requirement is that parton masses are consistent with the poles of the S-matrix elements [C2]. This assumption is quite general and certainly makes sense if the S-matrix elements allow a decomposition to vertices and propagators for a tree diagram.

b) The time evolution for the quantum states of individual partons, which are always on mass shell (that is generalized eigen states of the modified Dirac operator) is a unitary process, and corresponds to a braiding for an N-puncture system defined by the product of N local operators creating the parton state. The basic requirement is that the flow contains information about the presence of other partons and thus about the normal derivatives of the imbedding space coordinates at X_i^3. Hence the

S-matrix indeed contains information about the interior of the space-time surface and the effective two-dimensionality is indeed only effective. The condition that the S-matrix elements remain unchanged in conformal transformations obviously poses explicit conditions on the normal derivatives and can be regarded as conditions stating the vanishing of the corresponding beta function.

The best candidate for the flow is as the hydrodynamic flow defined by the discontinuity of the energy momentum tensor associated with the Kähler action at X_l^3 representing what might be regarded as a hydrodynamical shock wave. This flow is in general not integrable in the sense that one could assign a global coordinate varying along the flow lines. By identifying the points of X_i^2 and X_f^2 having the same value of the complex coordinate z, the flow $X_i^2 \to X_f^2$ defines a map $w : (x, y) \to (u(x, y), v(x, y))$ mixing the points of X_i^2. The inner products of the states created by the local operator $\Psi_n(x, y)$ creating one-parton state at X_i^2 with the state created by the transformed operator $\Psi_n(u(x, y), v(x, y))$ define correlation functions, which vanish above some length scale determining the mass of the parton. Massivation occurs if this map fails to be a conformal transformation.

b) By quantum classical correspondence this braiding can be also regarded as a braiding for the points of X_i^2, which correspond to the super-canonical conformal weights just like the points of the celestial sphere correspond to momenta. If the number of operators creating the parton state is larger than two, the minimal number three of threads in the braid is present and the conformal weights become "off mass shell" conformal weights. The massivation however can in principle occur always. In [O3, C5] the proposal was made that the bound state conformal weights could correspond to the zeros of poly-zetas: obviously this is a very strong prediction. Riemann Zeta would correspond to Higgs zero phase with the minimum of 2 operators with conjugate super-canonical conformal weights creating the state.

c) The successful description of particle massivation in terms of p-adic thermodynamics for the Virasoro generator L_0 (with p^2 not included) of the partons encourages to think that the change of the parton conformal weight could be understood as a generation of a thermal conformal weight by the flow induced by an ergodic braiding flow. This interpretation would allow to circumvent the problem created by the fact that the thermalization for mass squared is not consistent with the Lorentz invariance.

What is the flow defining the braiding?

The basic condition on the braiding is that it contains information about the interior of X^4 and thus about interactions with other partons. Second constraint is that the braiding flow is trivial for massless particles such as photon for which the space-time correlate should correspond to X_l^3 carrying a light-like energy momentum current.

The components $X^{n\alpha}$ of some tensor field with α restricted to X_l^3 define the most natural candidate for the braiding flow. The existence of the preferred light-like normal coordinate x^n constant at X_l^3 (in the case of δM_+^4 the light like coordinates would be $x^\pm = t \pm r$) is essential to achieve general coordinate invariance.

The discontinuity of the normal component $T^{n\alpha}$ of the energy momentum tensor associated with the Kähler action is a good candidate. At light like CDs the discontinuity of $T^{n\alpha}$ could be non-vanishing if allows light like CD to carry a shock wave also in imbedding space degrees of freedom as suggested by the super-symmetry. The discontinuity of $T^{n\alpha}$ would have an interpretation as a shock wave like hydrodynamic flow at the boundary. For massless particles the energy momentum current would have only a longitudinal component, the braiding would be trivial and particle would remain massless. The appearance of the energy momentum tensor in the definition of the S-matrix conforms with the hydrodynamic character of field equations and with the fact that the theory must be also a quantum theory of gravitation.

This guess is supported by the modified Dirac equation. By multiplying the modified Dirac equation at X_l^3 for shock waves localized at X_l^3 with the o_t defined by the light like gamma matrix along X_l^3 and doing the anti-commutations with the modified Dirac operator D, one finds

$$T^{\alpha n} D_\alpha \Psi = 0 . \tag{6.7.8}$$

The equation states that Ψ is covariantly constant along the flow lines of the flow defined by $T^{\alpha n}$. The equation can be written as

$$D_t \Psi + v^i D_i \Psi \;=\; 0 \;,$$
$$v_i \;=\; \frac{T^{ni}}{T^{nt}} \;. \tag{6.7.9}$$

This differs from a standard flow equation for a quantity Ψ moving along field lines only by the fact that ordinary derivatives ∂_α are replaced by covariant derivatives D_α. This means that Ψ suffers a braiding transformation in spin and electro-weak degrees of freedom. Obviously, this equation states super-conformal invariance in the sense that it is not possible to poses the values of Ψ arbitrarily in entire X_l^3 but only at X^2. One could regard TGD as at the space-time level. The usual dispersion characterizing Schödinger equation emerges only at the level of imbedding space when one assigns wave equations to propagators defined by S-matrix elements.

Since the induced spinor connection is continuous at X_l^3, the discontinuity for this equation reads as

$$w^i D_i \Psi \;=\; 0 \;,$$
$$w_i \;=\; \Delta \left[\frac{T^{ni}}{T^{nt}} \right] \;, \tag{6.7.10}$$

thus defines an adiabatic series of braiding flows.

An interesting special case corresponds to the situation when the flow lines allow a longitudinal coordinate varying along them. This requires integrability conditions guaranteing that the flow lines define a gradient flow. The condition that the flow lines correspond to the coordinate lines of a global coordinate Ψ and the corresponding integrability conditions read as

$$\nabla_\alpha \Psi \;=\; \Phi X_\alpha \;. \tag{6.7.11}$$

Ψ and Φ are scalar functions. By $g_{\alpha t} = 0$ (t denotes the light like coordinate) the integrability condition implies that the light like component of X_t is zero so that X^t can be arbitrary, and Ψ can depend only on transversal coordinates. The integrability condition reads as

$$dX \;=\; -dlog(\Phi) \wedge X \;. \tag{6.7.12}$$

A second interesting situation corresponds to a braiding flow that reduces to a conformal transformation. In this case there are excellent reason to expect that the flow does not cause massivation since it only generates a global conformal transformation leaving the physical state invariant.

6.7.3   Super canonical Dirac operators

The definition of configuration space Dirac operators relies on the use of generators B_A of configuration space isometries and of corresponding super-symmetries S_A identifiable as gamma matrices of CH. Both leptonic and quark like super generators are possible and define Ramond and NS type Dirac operators having also interpretation in terms of generators G of Super Virasoro algebra. The Dirac operator is simply the bilinear $g^{AB} B_A S_B$, where g^{AB} defines the contravariant metric of configuration space. The special feature is that only quark like oscillator operators appear in the definition of S_A and B_A so that there is a strong asymmetry between quarks and leptons perhaps reflecting the actual physical asymmetry between these particles. B_A does not however commute with leptonic super charges so that leptons do not decouple from the physics.

Symplectic extension

The Abelian extension of the super-canonical algebra is obtained by an extremely simple trick. Replace the ordinary derivatives appearing in the definition of, say spinorial isometry generator, by the covariant derivatives defined by a coupling to a multiple of the Kähler potential.

$$
\begin{aligned}
j^{Ak}\partial_k &\;\rightarrow\; j^{Ak}D_k \;, \\
D_k &\;=\; \partial_k + ikA_k/2 \;.
\end{aligned}
\tag{6.7.13}
$$

where A_k denotes Kähler potential. The reality of the parameter k is dictated by the Hermiticity requirement and also by the requirement that Abelian extension reduces to the standard form in Cartan algebra. k is expected to be integer also by the requirement that covariant derivative corresponds to connection (quantization of magnetic charge).

The commutation relations for the centrally extended generators J^A read:

$$
[J^A, J^B] \;=\; J^{[A,B]} + ik\,j^{Ak}J_{kl}j^{Bl} \equiv J^{[A,B]} + ikJ_{AB} \;.
\tag{6.7.14}
$$

Since Kähler form defines symplectic structure in configuration space one can express Abelian extension term as a Poisson bracket of two Hamiltonians

$$
J_{AB} \;\equiv\; j^{Ak}J_{kl}j^{Bl} = \{H^A, H^B\} \;.
\tag{6.7.15}
$$

Notice that Poisson bracket is well defined also when Kähler form is degenerate.

The extension indeed has acceptable properties:

a) Jacobi-identities reduce to the form

$$
\sum_{cyclic} H^{[A,[B,C]]} \;=\; 0 \;,
\tag{6.7.16}
$$

and therefore to the Jacobi identities of the original Lie- algebra in Hamiltonian representation.

b) In the Cartan algebra Abelian extension reduces to a constant term since the Poisson bracket for two commuting generators must be a multiple of a unit matrix. This feature is clearly crucial for the non-triviality of the Abelian extension and is encountered already at the level of ordinary (q,p) Poisson algebra: although the differential operators ∂_p and ∂_q commute the Poisson bracket of the corresponding Hamiltonians p and q is nontrivial: $\{p,q\} = 1$. Therefore the extension term commutes with the generators of the Cartan subalgebra. Extension is also local $U(1)$ extension since Poisson algebra differs from the Lie-algebra of the vector fields in that it contains constant Hamiltonian ("1" in the commutator), which commutes with all other Hamiltonians and corresponds to a vanishing vector field.

c) For the generators not belonging to Cartan sub-algebra of CH isometries Abelian extension term is not annihilated by the generators of the original algebra and in this respect the extension differs from the standard central extension for the loop algebras. It must be however emphasized that for the super-canonical algebra generators correspond to products of δM_+^4 and CP_2 Hamiltonians and this means that generators of say δM_+^4-local $SU(3)$ Cartan algebra are non-commuting and the commutator is completely analogous to central extension term since it is symmetric with respect to $SU(3)$ generators.

d) The proposed method yields a trivial extension in the case of Diff^4. The reason is the (four-dimensional!) Diff degeneracy of the Kähler form. Abelian extension term is given by the contraction of the Diff^4 generators with the Kähler potential

$$
j^{Ak}J_{kl}j^{Bl} \;=\; 0 \;,
\tag{6.7.17}
$$

which vanishes identically by the Diff degeneracy of the Kähler form. Therefore neither 3- or 4-dimensional Diff invariance is not expected to cause any difficulties. Recall that 4-dimensional Diff degeneracy is what is needed to eliminate time like vibrational excitations from the spectrum of the theory. By the way, the fact that the loop space metric is not Diff degenerate makes understandable the emergence of Diff anomalies in string models [ga1, ga2].

e) The extension is trivial also for the other zero norm generators of the tangent space algebra, in particular for the $k_2 = Im(k) = 0$ canonical generators possible present so that these generators indeed act as genuine $U(1)$ transformations.

f) Concerning the solution of configuration space Dirac equation the maximum of Kähler function is expected to be special, much like origin of Minkowski space and symmetric space property suggests that the construction of solutions reduces to this point. At this point the generators and Hamiltonians of the algebra h in the defining Cartan decomposition $g = h+t$ should vanish. h corresponds to integer values of $k_1 = Re(k)$ for Cartan algebra of super-canonical algebra and integer valued conformal weights n for Super Kac-Moody algebra. The algebra reduces at the maximum to an exceptionally simple form since only central extension contributes to the metric and Kähler form. In the ideal case the elements of the metric and Kähler form could be even diagonal. The degeneracy of the metric might of course pose additional complications.

Super canonical action on configuration space spinors

The generators of canonical transformations are obtained in the spinor representation of the isometry group of the configuration space by the following formal construction. Take isometry generator in the spinor representation and add to the covariant derivative D_k defined by vielbein connection the coupling to the multiple of the Kähler potential: $D_k \to D_k + ikAk/2$.

$$
\begin{aligned}
J^A &= j^{Ak}D_k + D_l j_k \Sigma^{kl}/2 \ , \\
&\to \hat{J}^A = j^{Ak}(D_k + ikA_k/2) + D_l j_k^A \Sigma^{kl}/2 \ ,
\end{aligned}
$$

$$(6.7.18)$$

This induces the required central term to the commutation relations. Introduce complex coordinates and define bosonic creation and annihilation operators as $(1,0)$ and $(0,1)$ parts of the modified isometry generators

$$
\begin{aligned}
B_A^\dagger &= J_+^A = j^{Ak}(D_k + \ldots \ , \\
B_A &= J_-^A = j^{A\bar{k}}(D_{\bar{k}} + \ldots \ .
\end{aligned}
$$

$$(6.7.19)$$

where "k" refers now to complex coordinates and "$\bar{k}$" to their conjugates.

Fermionic generators are obtained as the contractions of the complexified gamma matrices with the isometry generators

$$
\begin{aligned}
\Gamma_A^\dagger &= j^{Ak}\Gamma_k \ , \\
\Gamma_A &= j^{A\bar{k}}\Gamma_{\bar{k}} \ .
\end{aligned}
$$

$$(6.7.20)$$

Notice that the bosonic Cartan algebra generators obey standard oscillator algebra commutation relations and annihilate fermionic Cartan algebra generators. Hermiticity condition holds in the sense that creation type generators are hermitian conjugates of the annihilation operator type generators. There are two kinds of representations depending on whether one uses leptonic or quark like oscillator operators to construct the gammas. These will be assumed to correspond to Ramond and NS type generators with the radial plane waves being labelled by integer and half odd integer indices respectively.

The non-vanishing commutators between the Cartan algebra bosonic generators are given by the matrix elements of the Kähler form in the basis of formed by the isometry generators

$$
[B_A^\dagger, B_B] = J(j^{A\dagger}, j^B) \equiv J_{\bar{A}B} \ .
$$

$$(6.7.21)$$

and are isometry invariant quantities. The commutators between local $SU(3)$ generators not belonging to Cartan algebra are just those of the local gauge algebra with Abelian extension term added.

The anti-commutators between the fermionic generators are given by the elements of the metric (as opposed to Kähler form in the case of bosonic generators) in the basis formed by the isometry generators

$$\{\Gamma_A\dagger, \Gamma_B\} \;\; = \;\; 2g(j^{A\dagger}, j^B) \equiv 2g_{\bar{A}B} \;\; . \tag{6.7.22}$$

and are invariant under isometries. Numerically the commutators and anti-commutators differ only the presence of the imaginary unit and the scale factor R relating the metric and Kähler form to each other (the factor R is same for CP_2 metric and Kähler form).

The commutators between bosonic and fermionic generators are given by

$$[B_A, \Gamma_B] \;\; = \;\; \Gamma_{[A,B]} \;\; . \tag{6.7.23}$$

The presence of vielbein and rotation terms in the representation of the isometry generators is essential for obtaining these nice commutations relations. The commutators vanish identically for Cartan algebra generators. From the commutation relations it is clear that Super Kac Moody algebra structure is directly related to the Kähler structure of the configuration space: the anti-commutator of fermionic generators is proportional to the metric and the commutator of the bosonic generators is proportional to the Kähler form. It is this algebra, which should generate the solutions of the field equations of the theory.

The vielbein and rotational parts of the bosonic isometry generators are quadratic in the fermionic oscillator operators and this suggests the interpretation as the fermionic contribution to the isometry currents. This means that the action of the bosonic generators is essentially non-perturbative since it creates fermion antifermion pairs besides exciting bosonic degrees of freedom.

Super canonical Dirac operators

The basic difference with respect to super string models is that configuration space gamma matrices carry fermion number and this means some deviations from the string model picture. The super-canonical and Super Kac-Moody Dirac operators are defined by the same general formula

$$\begin{aligned} D \;\; &= \;\; g^{A\bar{B}} B_A \Gamma_B^{\dagger} \;\; , \\ \Gamma_A \;\; &= \;\; j^{AK} \Gamma_{AK} = S_A \;\; . \end{aligned} \tag{6.7.24}$$

B^A are the symplectically extended canonical generators and Γ_A are gamma matrices associated with the isometry generators. Since Γ_A carry fermion number only second half of gamma matrices can be included unless one is willing to give up fermion number conservation. Both D and its Hermitian conjugate $D^{\dagger}$ should annihilate physical states. Since both bosonic commutators and fermionic anti-commutators are proportional to $g_{A\bar{B}}$, the square of the Dirac operator gives the analog of mass squared formula of string models in which mass squared is proportional to a sum of terms equal to the fermionic or bosonic occupation number times the diagonalized element of $g_{A\bar{B}}$. Therefore string model type mass formula basically results.

6.7.4 Construction of solutions of the super-canonical Dirac operators

For a long time I regarded the construction of the solutions of Dirac operator of configuration space as representations of super-canonical algebra as a rather academic exercise unless one is doing quantum cosmology. The creation of pairs of positive and negative energy space-time sheets makes possible more general causal determinants than $\delta M_+^4 \times CP_2$ and this might imply that the exotic new physics predicted by super canonical symmetry could be tested in laboratory.

Some general comments about the construction are in order.

Relation to the symmetric space structure

a) The assumption that configuration space is union of symmetric spaces G/H corresponding to Cartan decomposition $g = t + h$: $[h,h] \subset h$, $[h,t] \subset t$, $[t,t] \subset h$ and labelled by the values of zero modes. Here g corresponds to the entire super-canonical algebra, whose elements are labelled by the conformal weights associated with the radial coordinate of light cone boundary or light-like surface $X_l^3 \subset M^4$ in the more general case. The details of the decomposition have been already explained.

These generators are completely analogous to $U(2)$ generators and corresponding Hamiltonians at the origin of CP_2: both vanish.

b) Super-canonical and Super Kac-Moody generators of h must either vanish or annihilate the physical states at the point which corresponds to the maximum of Kähler function (the origin of symmetric space with vanishing t-coordinates). One can say that the subalgebra h of the entire algebra acts as gauge symmetries. Although only quark like oscillator operators appear in configuration space Dirac operator and configuration space Dirac equation, also leptonic operators appear in the Ramond type conditions stating that leptonic super generators vanish or at least annihilate the physical states at the maximum of Kähler function.

c) Symmetric space structure suggests that the construction of the solutions of configuration space Dirac equation could be reduced to a single point of the configuration space identifiable as the maximum of Kähler function for given values of zero modes. In the case of M^4 plane wave solutions of massless Dirac equation are constructed by taking constant spinor u_0 at say origin of M^4, selecting four momentum p^k satisfying mass shell condition $p^2 = 0$ and forming the spinor $u = p^k \gamma_k u_0$. The Dirac equation at the origin reduces to the condition $p^k p_k = 0$ and the spinor plane wave elsewhere is obtained as a unitary representation of translation group M^4 defined by p^k by translating the value of u from the origin and multiplying it by matrix element $exp(ip_k m^k)$ representing translation.

In the recent case the group of translations should be replaced by some subgroup of G acting effectively in the fiber of configuration space corresponding to constant values of zero modes. In the recent case the group is not commutative and the situation is therefore more complex. The second difference is that complexification is involved now.

Also now one could however start from a spinor field u_0 and apply to it configuration space Dirac operator D to get $u = Du_0$ annihilated automatically by D by not by $D^\dagger$ so that the counterpart of d'Alembert equation giving rise to the analog of mass squared condition of string models would be $D^\dagger D u = 0$. If u can be selected in such a manner that this condition reduces to a simple form, it might be possible to calculate all that is needed for the deduction of mass spectrum and even S-matrix (by the analog of the Gaussian functional integral at the maximum of Kähler function). As far as mass squared condition is considered, the maximum of Kähler function might provide the needed simplification.

e) Concerning the selection of the configuration space spinor u_0 at the maximum of Kähler function, the best that one might hope is that all possible choices correspond to the states obtained from a vacuum state by applying fermionic super generators acting as gamma matrices. This is indeed the case for the solutions of ordinary Dirac equation in Minkowski space. Configuration space Hamiltonians are expected vanish at the maximum of Kähler function just as Hamiltonians of color transformations vanish at the origin of CP_2.

This raises the question whether also the leptonic and quark counterparts of configuration space Hamiltonians vanish identically at the maximum. If this were the case, the action of D would be essentially that of a d'Alembertian containing only ordinary derivatives and the situation would simplify enormously. For instance, one can imagine following procedure for the construction of the solutions.
i) Take a state created by fermionic super generators. Suppose that this state vanishes at the maximum of Kähler function. Also configuration space Hamiltonians could be used. Essentially the element of the super function algebra of the configuration space vanishing at the maximum would be in question. One can also say that physics is effectively classical at the maximum in the sense that Poisson brackets and anti-commutators both vanish at the maximum.
ii) Apply to this state a sufficient number of configuration space Dirac operators D so that it is non-vanishing at the origin. u might also vanish at origin but be such that $D^\dagger u$ is non-vanishing.
iii) Require that $D^\dagger$ annihilates the state. Since covariant derivatives are effectively absent from the action of D, the resulting equation $D^\dagger u$ could reduce to a sum of simple enough equations. If u but not $D^\dagger u$ vanishes at origin the resulting equation might be simple enough to allow to deduce the spectrum.

Properties of the vacuum CH spinor field

Vacuum (CH-) spinor field Ψ_{vac} at the maximum of Kähler function has the following properties.

a) Vacuum spinor field is Diff^4 invariant and is annihilated by all annihilation operator type symplectically extended bosonic and fermionic generators of the canonical algebra in the case that they have even conformal weight:

$$B_A \Psi_{vac} = 0 \; ,$$
$$\Gamma_A \Psi_{vac} = 0 \; . \qquad\qquad (6.7.25)$$

b) Vacuum spinor field is the product of a real exponent of Kähler function and Fock vacuum

$$\Psi_{vac} = exp(kK/2)\xi_{Fock} \; , \qquad\qquad (6.7.26)$$

where ξ_{Fock} is just the Fock vacuum associated with the fermionic oscillator operators (and in general depends on 3- surface being a section of the Fock bundle over the configuration space) and depends on the oscillator basis used. k denotes the central extension parameter associated with the symplectic extension.

The additional conditions to be satisfied by the Fock vacuum can be found from the spinorial representation

$$j^A = j^{Ak}(D_k - k\partial_k K/2) + \frac{1}{2}\partial_l j_k^A \Sigma^{kl} \; , \qquad\qquad (6.7.27)$$

of the complexified bosonic generator (only sum over the holomorphic indices k appears). Notice the disappearance of the imaginary unit resulting from the expression of the Kähler form in complex coordinates, which implies that the exponent of the Kähler function must be real.

The exponent of the Kähler action guarantees the cancellation of the term $j^{Ak}(\partial_k - kK/2)$ in the complexified annihilation operator type bosonic generator. The requirement of divergence cancellation implies that the total exponent appearing in the inner product is just $exp(K)$. The vielbein and rotation terms appearing in the isometry generator annihilate Fock vacuum provided these terms are automatically normal ordered with respect to the fermionic operators. The holomorphy of the bosonic isometry generators and the complexification implies that normal ordering indeed takes place.

Canonical invariants define an important class of zero mode coordinates and wave functions in zero modes representing various vacua would depend on these canonical invariants. If each quantum jump involves a localization occurring in zero modes represented by canonical invariants, as implied by the requirement that quantum measurement is local with respect to these zero modes, the wave function in zero modes is perhaps not a very practical notion.

6.7.5 Definition of super Kac-Moody Dirac operator

The definition of SKMD operator associated with super Kac-Moody degrees of freedom is based on the assumption that SKMD equation is equivalent with the generalized Super Virasoro conditions SKMD operator should thus have the same general form as the super generators G_n and their Hermitian conjugates in the generalized Sugawara construction [df3] and spinor d'Alembertian should correspond to the Virasoro generators L_n.

a) One must express SKMD operator in terms of the uper Kac-Moody generators. This operator decomposes into center of mass term representing Dirac operator of imbedding space and vibrational part which has some coefficient, call it k. This coefficient determines the mass scale of the theory (counterpart of the string tension). p-Adic mass calculations imply that the mass scale should be of order $m_0^2 \sim \frac{10^{-8}}{G}$. For $R \sim \sqrt{G}$ this would require $k \sim 10^{-8}/R^2$ instead of the expected $k \sim 1/R^2$. In any case, the value of k should follow from the basic structure of the theory. It indeed follows. The square of SKMD must contain also CP_2 Laplacian as a cm term and the scale of this term must be the same as the scale of vibrational part. Hence CP_2 radius must be essentially equal to the p-adic length scale $L \simeq 10^4\sqrt{G}$ so that one indeed has $k \sim 1$! This simple result was quite a surprise and led to a solution of several longstanding problems of TGD. The precise value of k for light particles is fixed from the requirement that mass spectrum is integer valued so that also massless states are possible for the lowest color partial waves.

The solutions of SKMD operator are essentially solutions of Super Virasoro conditions for SKM algebra associated with the group $SO(3,1) \times SU(3) \times U(2)_{ew}$. The algebra differs from the standard one in that super generators $G(z)$ carry lepton (Ramond representation with integer conformal weights) and quark (N-S representation with half-odd integer conformal weights) numbers are not Hermitian as in super-string models (Majorana conditions are not satisfied). If only quarks seem to contribute to configuration space metric also in Super Kac-Moody degrees of freedom, only quark like oscillator operators appear in the Dirac equation.

It seems possible to combine the algebra so that it contains also non-Hermitian Virasoro generators having half-odd integer conformal weights. These generators have quantum numbers of lepto-quark and cannot annihilate physical states. Already the p-adic mass calculations [TGDpad] that both Ramond and N-S type Virasoro conditions can be assumed to be satisfied by the physical states. This indeed makes sense since super generators carry quark and lepton number. At the maximum of Kähler function gamma matrices corresponding to the sub-algebra h in Cartan decomposition should vanish or at least annihilate physical states.

The structure of the SKM algebra

The task is to identify the SKM algebra serving as symmetry algebra of the theory. Here physical intuition helps to guess what the result must be. First of all, the conformal invariance of 3-D CDs is the counterpart of the gauge invariance associated with gravitational, color and electro-weak interactions. Hence the bosonic charges of Super-Kac-Moody algebra must correspond to the charges associated with these gauge symmetries. A further guideline is the fact that p-adic mass calculations require $D = 5$ sectors of super-conformal algebra.

a) There are two kinds of fundamental super-generators. The lepton and quark like super-generators $L_{n\alpha}$ and $Q_{r\alpha}$ (α is spin label) associated with the super counterpart of the conserved quark and lepton currents are Abelian charges and do not correspond physically to local gauge symmetries.

b) There are charges associated with the isometries of the imbedding space. The super generators $L^{nA\alpha}$ and $Q^{nA\alpha}$ associated with $M^4 \times SO(3,1) \times SU(3)$, where M^4 denotes here translations in M^4. For M^4 option Poincare transformations are exact symmetries of the modified Dirac action and of configuration space Kähler function: this option conforms with the original vision which I gave up later but found again after realizing what the relationship between gravitational and inertial energies must be in TGD Universe.

c) Also super Kac-Moody symmetries act as symmetries of the Dirac action. Super Kac-Moody symmetries can be identified as conformal localization of the translations of M^4. A physically well-motivated guess is that the entire local isometry group of $H = M^4 \times CP_2$ becomes local. $M^4 \times SO(3,1)$ would act as the local gauge group of gravitational interactions. For the complex-conformal symmetry only the two transversal M^4 degrees of freedom generating physical states would contribute two sectors to the super-conformal algebra.

Number theoretical considerations suggest a kind of an algebraic Uncertainty Principle implying naturally the decomposition of the tangent space of the space-time surface to longitudinal and transverse degrees of freedom. At the level of the configuration space super-canonical algebra would correspond to conformal weights which are complex numbers: these complex numbers can correspond to any sub-space of hyper-quaternions, which means a degeneracy labelled by a sphere S^2 and a Hyper Kähler structure of configuration space implying its Ricci flatness crucial for the cancellation of divergences in the configuration space integration. At the space-time level two-dimensional surfaces, which correspond to Abelian sub-manifolds of the hyper-quaternionic space-time surface, would in some sense define what is quantum perceivable. If so, the 2-dimensionality of strings would catch something very essential about fundamental physics.

d) Physical intuition suggest that the remaining super-conformal sector corresponds to the conserved charges associated with electro-weak symmetries. In fact, $SU(3)$ isometry charges decompose to two separately conserved parts for the solutions of the field equations for which energy momentum tensor term and Kähler current term vanish separately. There are good reasons to believe that this occurs completely generally and means that space-time surfaces are minimal surfaces with respect to the effective metric defined by the energy momentum tensor associated with Kähler action. What is important is that for Poincare charges the corresponding decomposition brings in nothing new since the Kähler current part of charge vanishes identically. Since $SU(3)$ rotations are represented as electro-weak rotations, and since the $U(2)$ sub-group of $SU(3)$ corresponds naturally to the holonomy group of CP_2 spinor connection, one can identify electro-weak SKM algebra that generate by $U(2)$ color charges, and identify color charges as those associated with the energy momentum tensor. This gives 2 sectors of super-conformal algebra so that one has $N = 5$ sectors as required by the p-adic mass calculations.

Vacuum sector and the absence of sparticles

Leptonic and quark like super generators in the vacuum sector have conformal weights 0 and $\pm 1/2$ respectively. Usually the combinations of both these operators generate ground states. As already found, the super-generators with conformal weight 0 vanish identically if the super-generators anti-commuting to the function algebra and correspond to to N-S type representation not allowing constant Hamiltonian. For other spinor harmonics the analogs of Ramond type super-symmetry generators are non-vanishing but the spinor harmonics are proportional to color partial waves giving them mass of order CP_2 mass so that there is no mass degeneracy in Ramond sector. Obviously, the symmetry breaking induced by CP_2 geometry is absolutely essential for the elimination of the $N = 4$ global super-symmetry.

These operators are very much like ordinary oscillator operators creating Fock states in standard quantum theory and states can be labelled by quark and lepton numbers. The number of leptonic creation operator like super-generators L^{0i} super-generators is $2^8 = 2^4 + 2^4$ correspond to fermion and anti-fermion with spin and electro-weak spin. If leptons correspond to NS representation (as they should if sparticles are absent) there are $8 - 2 = 6$ super generators with conformal weights $h = 1/2$ and $h = -1/2$. For quarks the corresponding number would be 8. In absence of other contribution to the conformal weight, the smallest possible conformal weight for ground state would be $h_{vac} = -3$ for leptons and $h_{vac} = -4$ for quarks and would correspond to the presence of 3 *resp.* 4 fermion-antifermion pairs. $h_{vac} = -3$ is indeed required by p-adic mass calculations.

CM degrees of freedom

An important element in the discussion are center of mass degrees of freedom parameterized by imbedding space. One can assign to each eigen state of momentum and color quantum numbers plane wave in these degrees of freedom as well as color partial wave in CP_2 degrees of freedom. Thus color quantum numbers are not spin like quantum numbers in TGD framework. For CP_2 spinor harmonics the correlation between color partial waves and electro-weak quantum numbers is not physical in general: only the covariantly constant right handed neutrino has vanishing color. It is however possible to used operators of conformal weight 0 and $\pm 1/2$ to compensate the anomalous color so that correct correlation results (see the chapter "Massless States and Particle Massivation"). I saw this wrong correlation for a long time as a strong objection against TGD but it turned out to be a blessing since it is just this feature which allows to eliminate the $N = 4$ sparticle degeneracy not favored by the experimental data.

SKMD equation, the mass formula, and condition determining the effective string tension

SKMD operator, which can be written as

$$G_{tot} \quad \equiv \quad D = D_H + \sqrt{k}G \tag{6.7.28}$$

annihilates physical states amongst other super generators. Here D_H is the counterpart of the ordinary Dirac operator of the imbedding space defined by the contraction of the Kac-Moody and super-Kac-Moody generators associated with translations and color isometries. Its conformal weight is 0 for quark like supercharges (Ramond representations) and $\pm 1/2$ for leptonic supercharges (N-S representation).

The square of D determines the possible values of mass squared operator and $L_0(tot)|phys\rangle = 0$ Virasoro condition. Mass squared eigenvalues are given by

$$M^2 \quad = \quad m^2_{CP_2} + kL_0 \ . \tag{6.7.29}$$

The contribution of CP_2 spinor d'Alembertian to the mass squared operator is in general not integer valued. The requirement that massless states are possible fixes the possible values of k determining the effective string tension. For the lowest color partial waves in leptonic and bosonic sector $k = 1$ is the only possible choice. For the lowest lying color partial waves for quarks $k = 2/3$ is the only possible choice. What determines these values of k remains to be understood.

Solutions of the SKMD equation

The construction of the solutions of SKMD equation reduces to the construction of representations for the super Kac-Moody algebra. Apart from the fact that super-generators carry fermion number, this involves no new mathematics and the procedure is familiar from super string models and conformal field theories. The proposed picture is achieved by combining the experience gained from the p-adic mass calculations.

a) Super Virasoro corresponds to $\{G, G^\dagger\} = 2L$ instead of the usual one. Assuming that the states in each sector of the Super Virasoro are created by fermion number zero generators one obtains the same partition function as in the case of the ordinary Super Virasoro. This is absolutely crucial for the p-adic mass calculations in the fermionic sector since a change of the degeneracy by one unit can change light particle to a particle with mass of order $m_0 \sim 1/R$ in p-adic thermodynamics [TGDpad].

b) The basic experimental input from the p-adic mass calculations is that the number of sectors of the Super Virasoro algebra is five. If the bosonic Kac Moody sectors correspond to $SO(3) \times SU(3)$ and $SU(2)_L \times U(1)$, there are only 4 sectors. The construction must be however performed also for the holonomy algebra $U(1)$ of $\delta M_+^4 = S^2 \times R_+$. The covariantly constant sigma matrices of $E^3 \subset M^4$ assignable to the rest frame defined by S^2 in $\delta M_+^4 = S^2 \times R_+$ extend this representation to a representation of $SO(3)$. There are thus 5 sectors altogether.

c) Ramond and NS algebras form a larger algebraic structure consisting of the fermionic oscillator operators and this could make the construction of the vertex operators much simpler task than in string models with both N-S and Ramond type Super Virasoro conditions being satisfied. The representations for Kac Moody generators contain both quark and leptonic contributions unlike the super-canonical generators. This guarantees that the partition functions crucial for the mass calculations are not changed.

d) Kaluza-Klein picture applies on the boundary degrees of freedom. Tachyonic mass term, motivated by the properties of the Dirac equation for the induced spinors, gives the same negative vacuum weight for all representations involving boundary degrees of freedom. Color partial wave gives their own contribution to the effective vacuum weight and implies electro-weak symmetry breaking. The color quantum numbers of the light non-exotic particles result from a color compensation mechanism.

Part III

ALGEBRAIC PHYSICS

Chapter 7

Equivalence of Loop and Tree Diagrams and Divergence Cancellation

7.1 Introduction

The great dream of a physicist believing in reductionism and TGD would be a formalism generalizing Feynman diagrams allowing any graduate student to compute the predictions of the theory. TGD has forced myself to give up naive reductionism but I believe that TGD allows a generalization of Feynman diagrammatics free of the infinities plaguing practically all existing theories. The purpose of this chapter is to develop a general vision about how this might be achieved. The vision is based on generalization of mathematical structures discovered in the construction of topological quantum field theories (TQFT) and conformal field theories (CQFT). In particular, the notions of Hopf algebras and quantum groups, and categories are central. The following gives a very concise summary of the basic ideas.

7.1.1 Feynman diagrams as generalized braid diagrams

The first key idea is that generalized Feynman diagrams are analogous to knot and link diagrams in the sense that they allow also "moves" allowing to identify classes of diagrams and that the diagrams containing loops are equivalent with tree diagrams, so that there would be no summation over diagrams. This would be a generalization of duality symmetry of string models.

TGD itself provides general arguments supporting same idea. The identification of absolute minimum of Kähler action as a four-dimensional Feynman diagram characterizing particle reaction means that there is only single Feynman diagram instead of functional integral over 4-surfaces: this diagram is expected to be minimal one. At quantum level S-matrix element can be seen as a representation of a path defining continuation of configuration space (CH) spinor field between different sectors of CH corresponding to different 3-topologies. All continuations and corresponding Feynman diagrams are equivalent. The idea about Universe as a computer and algebraic hologram allows a concrete realization based on the notion of infinite primes, and space-time points become infinitely structured monads [O4]. The generalized Feynman diagrams differing only by loops are equivalent since they characterize equivalent computations.

7.1.2 Coupling constant evolution from infinite number of critical values of Kähler coupling strength

The basic objection against the new view about Feynman diagrams is that it is not consistent with the notion of coupling constant evolution involving loops in an essential manner. The objection can be circumvented. Quantum criticality requires that Kähler coupling constant α_K is analogous to critical temperature (so that the loops for configuration space integration vanish). The hypothesis motivated by the enormous vacuum degeneracy of Kähler action is that α_K has an infinite number of possible values labelled by p-adic length scales and also also by the dimensions of effective tensor factors defined hierarchy of II_1 factors (so called Beraha numbers) as found in [O4]. The dependence on p-adic length scale L_p corresponds to the usual renormalization group evolution whereas the latter dependence would correspond to a finite angular resolution and to a hierarchy of finite-dimensional extensions of p-adic number fields R_p. The finiteness of the resolution is forced by the algebraic continuation of rational number based physics to real and p-adic number fields since p-adic and real notions of distance between rational points differ dramatically. The higher the algebraic dimension of the extension and the higher the value of p-adic prime the better the angular (or phase) resolution and nearer the p-adic topology to that for real numbers.

7.1.3   R-matrices, complex numbers, quaternions, and octonions

A crucial observation is that physically equivalent R-matrices of 6-vertex models are labelled by points of CP_2 whereas maximal space of commuting R-matrices are labelled by the points of 2-sphere S^2 [ee2]. As found in [O4, E2], CP_2 also labels maximal associative and thus quaternionic sub-spaces of 8-dimensional octonion space. S^2 in turn labels the maximally commutative sub-spaces of quaternion space, which suggests that R-matrices and these structures correspond to each other.

Number theoretic vision leads to a number theoretic variant of spontaneous compactification meaning that space-time surfaces could be regarded either as hyper-quaternionic, and thus maximal associative, 4-surfaces in M^8 regarded as the space of hyper-octonions or as surfaces in $M^4 \times CP_2$ (the imaginary units of hyper-octonions are multiplied with $\sqrt{-1}$ so that the number theoretical norm has Minkowski signature)[E2]. Associativity constraint is an essential element of also Yang-Baxter equations [ec4, ee2].

These observations lead to a concrete proposal how quantum classical correspondence is realized classically at the space-time level. Each point of CP_2 corresponds to R-matrix and 3-surfaces can be identified as preferred sections of the space-time surface by requiring that unitary R-matrices are commuting in the section (micro-causality) whereas commutativity fails for R-matrices corresponding to different values of time coordinate defined by this foliation.

7.1.4   Ordinary conformal symmetries act on the space of super-canonical conformal weights

TGD predicts two kinds of super-conformal symmetries.

a) The ordinary super-conformal symmetries realized at the space-time level are associated with super Kac-Moody representations realized at light-like 3-surfaces appearing as boundaries of space-time sheets and boundaries between space-time regions with Euclidian and Minkowskian signature of metric. Conformal weights are half-integer valued in this case.

b) Super-canonical conformal invariance acts at the level of imbedding space and corresponds to light-like 7-surfaces of form $X_l^3 \times CP_2 \subset M^4 \times CP_2$ believed to appear as causal determinants too. In this case conformal weights are complex numbers of form $\Delta = n/2 + iy$ for the generators if super algebra, and I have proposed that the conformal weights of the physical states correspond are of form $1/2 + iy$, where y is zero of Riemann Zeta or possibly even superposition of them. In the latter case the imaginary parts of zeros would define a basis of an infinite-dimensional Abelian group spanning the weights. Later it will be found that quantum criticality favors zeros but not superpositions of their imaginary parts.

The basic question concerns the interaction of these conformal symmetries.

a) Quantum classical correspondence suggests that the complex conformal weights of super-canonical algebra generators have space-time counterparts. The proposal is that the weights are mapped to the points of geodesic sphere of CP_2 (and thus also of space-time surface) labelling also mutually commuting R-matrices. The map is completely analogous to the map of momenta of quantum particles to the points of celestial sphere. One can thus regard super-generators as conformal fields in space-time or complex plane having conformal weights as punctures. The action of super-conformal algebra and braid group on these points realizing monodromies of conformal field theories [ee2] induces by a pull-back a braid group action on the conformal labels of configuration space gamma matrices (super generators) and corresponding isometry generators.

b) The gamma matrices and isometry generators spanning the super-canonical algebra can be regarded as fields of a conformal field theory in the complex plane containing infinite number of punctures defined by the complex conformal weights. Quaternion conformal Super Virasoro algebra and Kac Moody algebra would act as symmetries of this theory and the S-matrix of TGD would involve the n-point functions of this conformal field theory.

This picture also justifies the earlier proposal that configuration space Clifford algebra defined by the gamma matrices acting as super generators defines an infinite-dimensional von Neumann algebra possessing hierarchies of type II_1 factors [eb1] having a close connection with the non-trivial representations of braid group and quantum groups. The sequence of non-trivial zeros of Riemann Zeta along the line $Re(s) = 1/2$ in the plane of conformal weights could be regarded an an infinite braid behind the von Neumann algebra [eb1]. Contrary to the expectations, also trivial zeros seem to be important. Purely algebraic considerations support the view that superposition for the imaginary parts of non-trivial zeros makes sense so that one would have entire hierarchy of one-dimensional lattices depending on the number of zeros included. The finite braids defined by subsets of zeros could be seen as a hierarchy of completely integrable 1-dimensional spin chains leading to quantum groups and braid groups [ec4, ee2] naturally. In conformal field theories it is possible to construct explicitly the generators of quantum group in terms of operators creating screening charges [ee2]: interestingly, the charges are located going along a line parallel to imaginary axis to infinity.

It seems that not only Riemann's zeta but also polyzetas [cf3, cf4, ed4, ed5] could play a fundamental role in TGD Universe. The super-canonical conformal weights of interacting particles, in particular of those forming bound states, are expected to have "off mass shell" values. An attractive hypothesis is that they correspond to zeros of Riemann's polyzetas. Interaction would allow quite concretely the realization of braiding operations dynamically. The physical justification for the hypothesis would be quantum criticality. Indeed, it has been found that the loop corrections of quantum field theory are expressible in terms of polyzetas [ee9]. If the arguments of polyzetas correspond to conformal weights of particles of many-particle bound state, loop corrections vanish when the super-canonical conformal weights correspond to the zeros of polyzetas including zeta. This argument does not allow superpositions of imaginary parts of zeros.

7.1.5 Equivalence of loop diagrams with tree diagrams from the axioms of generalized ribbon category

The fourth idea is that Hopf algebra related structures and appropriately generalized ribbon categories [ec4, ee2] could provide a concrete realization of this picture. Generalized Feynman diagrams which are identified as braid diagrams with strands running in both directions of time and containing besides braid operations also boxes representing algebra morphisms with more than one incoming and outgoing strands. 3-particle vertex should be enough, and the fusion of 2-particles and $1 \rightarrow 2$ particle decay would correspond to generalizations of the algebra product μ and co-product Δ to morphisms of the category defined by the super-canonical algebras associated with 3-surfaces with various topologies and conformal structures. The basic axioms for this structure generalizing ribbon algebra axioms [ec4] would state that diagrams with self energy loops, vertex corrections, and box diagrams are equivalent with tree diagrams.

Tensor categories might provide a deeper understanding of p-adic length scale hypothesis. Tensor primes can be identified as vector/Hilbert spaces, whose real or complex dimension is prime. They serve as "elementary particles" of tensor category since they do not allow a decomposition to a tensor product of lower-dimensional vector spaces. The unit I of the tensor category would have an interpretation as a one-dimensional Hilbert space or as the number field associated with the Hilbert space and would act like identity with respect to tensor product. Quantum jump cannot decompose tensor prime system to an unentangled product of sub-systems. This elementary particle like aspect of tensor primes might directly relate to the origin of p-adicity. Also infinite primes are possible and could distinguish between different infinite-dimensional state spaces.

For quantum dimensions $[n]_q \equiv (q^n - q^{-n})/(q - q^{-1})$ [ec4] no decomposition into a product of prime quantum dimensions exist and one can say that all non-vanishing quantum integers $[n]_q$ are primes. For q an nth root of unity, quantum integers form a finite set containing only the elements $0, 1, ..., [n-1]$ so that quantum dimension is always finite. The numbers $[2]_q^2$, for $q = exp(i\pi/n)$ define a hierarchy of Beraha numbers having an interpretation as a renormalized dimension $[2]_q^2 \leq 4$ for the spinor space of 4-dimensional space and appearing as effective dimensions of type II_1 sub-factors of von Neumann algebras.

7.1.6   What about loop diagrams with a non-singular homologically non-trivial imbedding to a Riemann surface of minimal genus?

There is an objection against the reduction to tree diagrams. Photon-photon scattering rate vanishes classically (zeroth order in $\hbar$), and the lowest non-vanishing contribution to the scattering amplitude corresponds to a box diagram and diagrams obtained from it by permuting the outgoing photon lines. All diagrams that can have intersecting lines as planar diagrams when outgoing lines are permuted, can be imbedded to a Riemann surface with some minimal genus as diagrams without intersecting lines containing a minimal number of homologically non-trivial loops.

This suggests a more general axiom: diagrams are characterized by genus and only homologically trivial loops can be eliminated. Also the question arises whether a sum over the different genera must be assumed as string model would suggest or whether single value of genus is enough. The discussion of this chapter assumes a reduction to a tree diagram but the generalization allowing all values of genus should be rather straightforward.

7.1.7   Quantum criticality and renormalization group invariance

Quantum criticality means that renormalization group acts like isometry group at a fixed point rather than acting like a gauge symmetry as in the standard quantum field theory context. Despite this difference it is possible to understand how Feynman graph expansion with vanishing loop corrections relates to generalized Feynman graphs and a nice connection with the Hopf- and Lie algebra structures assigned by Connes and Kreimer to Feynman graphs emerges. It is possible to deduce an explicit representation for the universal momentum and p-adic length scale dependence of propagators in this picture. The condition that loop diagrams are equivalent with tree diagrams gives explicit equations which might fix completely also the p-adic length scale evolution of vertices. Quantum criticality in principle fixes completely the values of the masses and coupling constants as a function of p-adic length scale.

To sum up, although these new ideas are rather speculative and my poor algebraic skills restrict severely the attempts to develop them into a more detailed theory, I believe that the new vision makes a big step in concretizing the physics as a generalized number theory vision. There is a lot of work to do. The implications for the construction of configuration space geometry and spinor structure are expected to be highly non-trivial and should tighten the loose web of ideas about the relationship and role of super-canonical and Super-Kac-Moody algebras. The new ideas relate also closely to the p-adicization program being partly inspired by it. Also the fascinating idea that $M^4 \times CP_2$ a emerges naturally from the interpretation of space-time surfaces as maximally associative 4-manifolds of octonionic space should be developed further.

7.2   Generalizing the notion of Feynman diagram

In this section various motivations for generalizing the notion of Feynman diagram such that diagrams with loops are equivalent with tree diagrams are discussed.

7.2.1   Divergence cancellation mechanisms in TGD

TGD provides general mechanisms for the cancellation of ultraviolet divergences.

a) The standard divergence due to the Gaussian determinant is absent in TGD. The Kähler geometry of the configuration space implies a cancellation of the configuration space metric determinant and Gaussian determinant associated with the integral over the quantum fluctuating configuration space degrees of freedom by Gaussian integration around a maximum of Kähler function.

b) The divergences due to the 3-dimensional micro-locality are absent due to the generalization of point particle which means that state functionals are non-local functionals of the 3-surface and local interaction vertices are smoothed out. Locality is realized at the configuration space level but there is now second quantization at this level since physical states correspond to classical configuration space spinor fields.

c) The vacuum energy for induced second quantized spinor fields cancels when fermions and anti-fermions have opposite sign of inertial energy. Crossing symmetry allows vanishing of the net inertial energy at least in cosmological length scales and even shorter length scales assuming that gravitational four-momentum is the difference of 4-momenta associated with matter and antimatter with an appropriate sign factor taking care that gravitational energy is positive.

d) The infinite-dimensional space of zero modes does not allow any natural integration measure. If localization in zero modes occurs in each quantum jump and leads to a discrete subspace of zero modes, the integration over zero modes becomes trivial. The localization can be interpreted in terms of a quantum measurement correlating classical macroscopic degrees of freedom represented by zero modes correlated with quantum fluctuating degrees of freedom. Entanglement need not be reduced completely since discrete subset of zero modes can remain entangled in this manner.

7.2.2 Motivation for generalized Feynman diagrams from topological quantum field theories and generalization of string model duality

There are many manners to end up to the idea that generalized Feynman diagrams are analogous to knot and link diagrams and that diagrams with loops are equivalent with tree diagrams.

Topological quantum field theories

The work with topological quantum computation (TQC, see [O3]) provided very fruitful stimuli and insights about quantum TGD itself. TQC [ef2, ef3, ef4] involves topological notions like links, knots and braids, and topological invariants of 3-manifolds. The pioneering works was doen by Jones with knot polynomials [eb5] who noticed also the connection with von Neumann algebras and so called Beraha numbers [ec6] appearing also in the quantum group context and in representations of braid groups. Topological quantum field theories (TQFT) [ec2] pioneered by Witten [ec3] provide an extremely powerful and abstract approach to the deduction of these invariants. The computations of these invariants involve a diagrammatic approach based on recursion bring in mind Feynman diagrams.

There is close relationship with conformal quantum field theories (CQFT) and related mathematical structures referred to as co-algebras, bi-algebras,Hopf algebras,quantum groups and category theory [ee2]. Since the generalized conformal invariance is a basic element of quantum TGD, this raises the hope that the replacement of "topological" with "conformal" might not not be so drastic a modification after all. If this is the case, one might hope that the generalization of Feynman diagrams might be based on the generalization of braid diagrams by bringing in the interaction in which the strands of braid can decay and fuse.

Even more, one might hope that the notion of equivalence for braid diagrams allowing to transform these diagrams to each other by "elementary moves" could allow to identify loop diagrams with tree diagrams so that divergence problem would disappear.

Fusion rules of conformal field theories and generalization of duality

Conformal invariance is the basic symmetry of also quantum TGD so that it is natural to look what one might learn from conformal quantum field theories, which rely on the notion of field algebra.

In conformal field theories fusion rules [ee2] for operator algebra code the vertices of the theory to numerical coefficients: $\Phi_k(z,\overline{z})\Phi_l(w,\overline{w}) = C_{kl}^m(z-w,\overline{z}-\overline{w})\Phi_m(w,\overline{w})$ modulo derivative terms which vanish in vacuum expectation values. Conformal invariance dictates two-point functions completely and three-point functions apart from numerical coefficients since one can write $C_{kl}^m(z,\overline{z}) = z^{\Delta_m - \Delta_k - \Delta_l}\overline{z}^{\overline{\Delta}_m - \overline{\Delta}_k - \overline{\Delta}_l}C_{kl}^m$. 4-point functions and also higher n-point functions can be constructed once the coefficients C_{kl}^m are known.

In particular, four-point functions can be expressed in terms of three point functions by interpreting them in terms of particle scattering as a sum of s-channel resonance or t-channel exchanges and the coefficients C_{kl}^m appear explicitly in these conditions. Duality corresponds to crossing symmetry which is of special importance in TGD framework where particle reaction can be interpreted as a creation of a state with vanishing conserved quantum numbers from vacuum. Rather remarkably, crossing symmetry is in turn equivalent with the associativity of the field algebra. One can express four-point function in terms of so called conformal blocks and crossing symmetry leads to conditions which can be solved in many cases.

This occurs when the primary field content of the theory is finite. This occurs when the Verma modules V^Δ associated with the primary fields contain zero norm states and thus contain a finite number of descendants (being thus effectively finite-dimensional as representations of conformal algebra) and when the operator products of the primary fields close to a finite algebra. The exceptional weights Δ are algebraic numbers involving only square roots of integers and thus p-adically highly interesting. In the case of rational (super-)conformal field theories the conformal weights are rational numbers and conformal Ward identities allow the construction of the descendants of the primary field Φ_Δ explicitly. The conformal weights for the finite number primary fields can be listed and fusion coefficients written explicitly. Same applies to super-symmetric variants of these theories. Rational conformal field theories are of special interest in TGD framework since there are good hopes that at least they allow a continuation to p-adic number fields. Note however that the allowed conformal weights for theories allowing null vectors are always algebraic numbers.

It would seem that the field algebra of conformal field theories might closely relate to what happens in 3-particle vertex for incoming particles. Fusion corresponds to an algebra product. In conformal field theories also co-algebras for which multiplication is replaced with co-multiplication analogous to a particle decay are important. This encourages to think that a suitable generalization of Hopf algebra with duality involving both multiplication μ and co-multiplication Δ in a well defined sense time reversals of each other by the duality, should be fundamental for the algebraization of Feynman rules.

String model duality which allows to identify diagrams involving resonances in s-channel with diagrams involving exchanges in t- or u-channel. The additional space-dimension brought in by TGD might mean a generalization of duality so that any diagram with loops would be equivalent with tree diagram, or more generally, to a diagram imbeddable without intersections to a Riemann surface with minimal genus and having minimal number of loops (photon photon scattering suggests this). In the following the consideration is restricted to tree diagrams.

The commuting diagrams expressing the fact that multiplication μ is co-algebra morphism and co-multiplication Δ is algebra morphism indeed state that box diagram describing two-particle scattering is equivalent with tree diagram and it turns out that additional natural postulates would state that also self energy diagrams and vertex correction diagrams are equivalent with corresponding tree diagrams.

7.2.3   How to end up with generalized Feynman diagrams in TGD framework?

In TGD framework there are several manners to end up with the proposed vision about Feynman diagrams with generalized moves eliminating loops. The idea that loop integrations giving rise to the renormalization of coupling constants and to anomalous dimensions vanish is consistent with this view and also emerges naturally in TGD framework.

Quantum classical correspondence and the replacement of sum over Feynman diagrams with single diagram

The basic difference between TGD and standard QFT:s is that the construction of configuration space geometry assigns to a given 3-surface X^3 a unique space-time surface $X^4(X^3)$ as absolute minimum of Kähler action or some more general preferred extremal [E2]. Although the classical non-determinism of Kähler action forces to modify this picture, the implication is that the functional integral over all 4-surfaces X^4 going through X^3 is replaced with single 3-surface $X^4(X^3)$. Quantum classical correspondence allows the interpretation of $X^4(X^3)$ as a generalized Feynman diagram. Hence single Feynman diagram without loops, "tree diagram", would characterize particle reaction and correspond the simplest possible generalized Feynman diagram among many equivalent ones. The vacuum degeneracy of Kähler action and the consequent non-determinism might mean that there indeed exists a large, or even infinite, number of equivalent absolute preferred Bohr orbit like extremals of Kähler action [B1, E2] equivalent as far as computation of S-matrix is considered. Deterministic regions of space-time sheets could be seen as diagrams with incoming and outgoing on mass shell particles. Also the possibility that these diagrams are not equivalent can be considered. "Homotopically non-equivalent" diagrams could be perhaps interpreted as time evolutions occurring in different quantum phases.

Quantum criticality requires the vanishing of loop corrections

Configuration space integration over quantum fluctuating degrees of freedom leads to a perturbative expansion similar to the ordinary Feynman diagrammatics. Quantum criticality states that the Kähler coupling strength is analogous to a critical temperature and has only a discrete set of values, at least one for each p-adic number field R_p and each algebraic extension of R_p. Quantum criticality requires loop corrections to the configuration space integral vanish so that it would reduce effectively to a Gaussian integral around a maximum of Kähler function for given values of zero modes.

Coupling constant evolution assignable with loop diagrams would be coded by the dependence of "critical temperature" on p (p-adic length scale) and algebraic extension of R_p. The loop gymnastics could be seen as an extremely tedious manner to model discrete p-adic coupling constant evolution having more elegant description in terms of quantum criticality.

p-Adicization requires the vanishing of loop corrections

Also the hypothesis that S-matrix elements can be algebraically continued to real and various p-adic number fields encourages the hope that configuration space integration must effectively reduce to an effectively Gaussian integral around a maximum of Kähler function so that simple expressions involving only rational, algebraic, and exponential functions requiring a finite-dimensional extension of p-adic numbers result making it possible to continue the expressions to p-adic number fields. The reason is that infinite sums of diagrams are not expected to give a result boiling down something expressible in terms of these functions.

Theory would formally reduce to a free field theory and the non-triviality of the theory would be basically due to the topological description of particle reactions. The symmetric space property of the quantum fluctuating degrees freedom and Duistermaat-Hecke theorem [db3] stating that functional integral can be expressed as an exponential of Kähler function K associated with the maximum of K give indeed good hopes that the dream might be realized.

The construction of S-matrix as analytic continuation between different sectors of configuration space

By quantum classical correspondence particle reactions correspond to processes changing 3-topology. For instance, the decay of a particle to two final state particles in Fock sense corresponds to a decay of a connected 3-surface to two disjoint components. In string models this can be described using smooth string world sheet whereas the string representing the vertex is singular eye glass like configuration. TGD suggests a different approach in which lines are replaced with singular four-manifolds whose ends meet at non-singular 3-manifold representing vertex is more appropriate. As a matter fact, the first proposal to describe S-matrix in terms of CP_2 type extremals led to this picture but was thought to be a mere approximation at that time [C2]. In this picture the stringy branching of the 3-surface would have interpretation as a space-time correlate for what it means that particle traverses simultaneously through several paths as in the case of double slit experiment. This certainly conforms with what happens to the induced second quantized spinor field in the branching.

For both options the generalized Feynman diagrams could be seen as characterizing the paths involved with a continuation of a configuration space spinor field from a given sector to another one. Any path will do and this implies a large number of moves allowing to construct the simplest possible Feynman diagram characterizing the continuation. Each topology change would give rise to a vertex and motion inside given sector would presumably correspond to a stringy propagator. Also other than topological characteristics, say the parameters characterizing conformal structure, might be involved.

At the space-time level the continuation would reduce to a continuation of second quantized free induced spinor fields from a given 3-topology to a new one. The 3-vertices describing decay and fusion provide the basic examples. The task of continuing an induced spinor field and corresponding fermionic Fock space from single-particle sector to two-particle sector means finding of an algebra morphism imbedding a single particle Fock algebra to a tensor product of Fock algebras and is highly analogous to finding of a co-algebra product Δ. The time reversal of Δ corresponds to a fusion vertex at the Fock space level. The construction of S-matrix would reduce to the solution of this continuation problem and one might hope that the process boils down to general algebra axioms.

The notion of Platonia and Feynman diagrams as computations

As found in [O4], the notion of infinite primes [E3] leads to a generalization of real numbers since an infinite algebra of real units becomes possible. These units are not units in the p-adic sense and have a finite p-adic norm which can be differ from one. Infinite primes form an infinite hierarchy so that the points of space-time and imbedding space can be seen as infinitely structured and able to represent all imaginable algebraic structures. Certainly counter-intuitively, single space-time point is even capable of representing the quantum state of the entire physical Universe in its structure. For instance, in real sense surfaces in the space of units correspond to the same real number 1, and single point, which is structure-less in the real sense could represent arbitrarily high-dimensional spaces as unions of real units. For real physics this structure is completely invisible and is relevant only for the physics of cognition. One can say that Universe is an algebraic hologram, and there is an obvious connection both with Brahman=Atman identity of Eastern philosophies and Leibniz's notion of monad.

The along curve makes means that both real and p-adic norms of the unit multiplying the point along curve are constant. The curve allows interpretation as a representation of an energy conserving time evolution of an arithmetic quantum field theory. In a complete analogy with the continuation at the configuration space level and analytic continuation, the time evolution defined by the cobordism does not depend on the path traversed. Each curve represents a discrete sequence of phase transitions changing the unit and the interpretation as a Feynman diagram like structure is very attractive.

The requirement that the configuration space paths defining generalized Feynman graphs are cognitively representable as cobordisms of a point provides a powerful heuristic tool in the attempts to say something non-trivial about the general properties of S-matrix. The cobordism of a point can be generalized in several manners.

In this approach Feynman diagrams could be seen as sequences of computations satisfying some algebraic rules fixed by the model that is mimicked. Computations themselves form a local algebra. The basic super rule satisfied always would be the conservation of guaranteing real and p-adic continuities. The generalized Feynman diagram defining a representation of an algebraic computation would determine an element of algebra since the algebra elements associated with different points of space-time surface form a local algebra structure. The simplest possible Feynman diagram would correspond to a tree diagram.

Many interpretations are possible. The preferred Bohr orbit like extremals of Kähler action could represent the simplest computations leading from an input to the outcome or minimal sequences of steps making up a proof of a theorem. At the level of configuration space, the simplest Feynman diagram would correspond to the simplest path allowing to continue configuration space spinor field from a given sector D_1 with a fixed 3-topology to another sector D_2.

The algebraic interpretation implies powerful symmetries. By replacing the topological invariance with the generalized conformal invariance serving as the basic symmetry of also TGD, it might be possible to generalize the construction of modular invariant S-matrices of topological quantum field theories to that of conformally invariant S-matrices by adding the fusion and decay of braid strands as additional operations to an appropriate braided category. The very intimate relationship between topological and conformal quantum field theories raises the hope about concrete predictions.

7.3   Algebraic physics, the two conformal symmetries, and Yang Baxter equations

TGD predicts two conformal symmetries corresponding to super-canonical symmetries acting at the level of imbedding space $H = M^4 \times CP_2$ and Super-Kac-Moody symmetries acting at the space-time level. If the notion of number theoretic spontaneous compactification makes sense there are also the analogs of conformal symmetries defined by hyper-octonion analytic maps of $OH = M^8$ and hyper-quaternion analytic maps of space-time surface regarded as a hyper-quaternionic 4-surface in M^8 [E2]. These symmetries would induce dynamical symmetries at the level of H.

a) The super-canonical conformal algebra appears naturally in the construction of the configuration space geometry for $H = M^4_+ \times CP_2$ option at the 7-dimensional light-like boundary of $M^4_+ \times CP_2$ and has complex conformal weights, possibly identifiable as zeros of Riemann zeta. The non-determinism of Kähler action and aesthetic considerations encourage the expectation that all light-like surfaces X^3_l of M^4 define 7-dimensional light-like causal determinants $X^3_l \times CP_2$ in this sense. Light-like 3-surfaces form a foliation of M^4 or M^4 by 4-dimensional general coordinate invariance and one can use any light-like 3-surface. The classical non-determinism of course poses restrictions.

b) The super-conformal algebra associated with the 3-dimensional light-likes surfaces X_l^3, which can appear as boundaries of space-time sheets and as elementary particle horizons at which the induced metric is degenerate corresponds to the ordinary super Kac-Moody algebra SKM of elementary particle physics and of string models. Conformal weights are now real and half integer valued.

The challenge is to understand how these symmetries relate to each other and and how they interact. Quantum classical correspondence allows to deduce a rather detailed interpretation of the basis of super-canonical algebra interpreted as super-conformal fields in the space defined by the complex conformal weights labelling them, and SKM conformal algebra having natural infinitesimal action in the space of super-canonical weights. Therefore he powerful machinery of super-conformal theories becomes directly available and allows to make educated guesses about the basic structure of S-matrix.

Yang-Baxter equations characterize the physics of two-dimensional integrable systems [ec4, ee2] and define a core element of braided Hopf algebras, braid group representations appearing in topological and conformal quantum field theories. R-matrices relate very directly to the Clifford algebra of the configuration space gamma matrices defining the fermionic part of super-canonical algebra and they act at the space-level too. Hence the physics of two-dimensional integrable systems and conformal field theories become a quintessential part of quantum TGD.

7.3.1 Space-time sheets as maximal associative sub-manifolds of the imbedding space with hyper-octonion structure

The vision about physics as a generalized number theory stimulated a development, which led to the notion of number theoretic compactification [E2]. Space-time surfaces can be regarded either as hyper-quaternionic, and thus maximal associative, 4-surfaces in M^8 having hyper-octonion structure, or as surfaces in $M^4 \times CP_2$ [E2]. Hyper-quaternions/-octonions form a sub-space of complexified quaternions/-octonions for which imaginary units are multiplied by $\sqrt{-1}$: they are needed in order to have a number theoretic norm with Minkowski signature. What makes this duality possible is that CP_2 parametcrizes different quaternionic planes of octonion space containing a fixed imaginary unit. S^2 can in turn be interpreted as the parameter space labelling the maximal commutative subspaces, i.e. complex planes, of a quaternionic space. The specification of quaternion plane is necessary in order to introduce octonion Hermitian structure whereas the specification of preferred imaginary unit (that is complex plane) is necessary in order to fix Hermitian structure in hyper-quaternionic tangent plane [E2].

Hyper-quaternionic sub-spaces are also maximal associative sub-spaces. The physical counterpart of associativity is crossing symmetry: in conformal field theories crossing symmetry follows from the associativity for the products of fields. The Yang-Baxter equations emerging naturally in 2-dimensional integrable quantum field theories and statistical systems state the associativity of braiding operations in the case of three strand braid. Thus it would not be surprising if there were connection between Yang-Baxter equations and space CP_2 of quaternionic sub-spaces of octonions. Commutativity of R-matrices means commutativity of braiding operations and this suggests that also the sphere labelling complex planes of quaternion space might emerge naturally in the context of Yang-Baxter equations.

Indeed, the Yang-Baxter matrices defining equivalent 6-vertex models are parameterized by CP_2 whereas mutually commuting Yang-Baxter models by sphere S^2. These findings lead to a quite detailed albeit speculative view about quantum classical correspondence.

a) One can select a 4-parameter subset of equivalent R-matrices

$$R = \begin{pmatrix} a & 0 & 0 & 0 \\ 0 & b & c & 0 \\ 0 & b & c & 0 \\ 0 & 0 & 0 & a \end{pmatrix} \tag{7.3.1}$$

in such a manner that they are apart from a real multiplicative constant r unitary matrices labelled by 3-parameters (Θ, Φ, Ψ) and one has

$$R = rexp(i\Phi) \times \begin{pmatrix} exp(i\Psi) & 0 & 0 & 0 \\ 0 & cos(\theta) & isin(\theta) & 0 \\ 0 & isin(\theta) & cos(\theta) & 0 \\ 0 & 0 & 0 & exp(i\Psi) \end{pmatrix} \qquad (7.3.2)$$

r is identifiable as the $U(2)$ invariant radial coordinate of CP_2 and (Θ, Φ, Ψ) are identifiable as spherical coordinates of $r = constant$ 3-sphere of CP_2.

If it is possible to interpret the points of CP_2 as labels of R-matrices, one could assign to a given point of space-time surface a unitary R-matrix labelling the quaternion structure of the tangent space at that point. The R-matrices with the same value of the parameter

$$\Delta = \frac{a^2 + b^2 - c^2}{2ab} \qquad (7.3.3)$$

commute as is easy to verify. With the parametrization used this means that the commuting R-matrices corresponds to the sphere defined by the equation

$$\frac{cos(\theta)}{cos(\psi)} = constant \ . \qquad (7.3.4)$$

b) A classical space-time correlate for equal time commutation relations of quantum field theories suggests itself. In the generic case the space-time surface would decompose into lines along which r varies and the "S-matrix" defined by the R-matrix would be constant along these lines. One could choose the time coordinate in such a manner that time constant 3-surface would foliated this kind of lines. The R-matrices associated with a 3-surface would be labelled by two coordinates, and by a proper selection of time coordinate it might be possible to decompose time=constant 3-surfaces to regions inside which R-matrices commute with each other and correspond to different complex planes in the space of quaternions. The R-matrices corresponding to different values of time coordinate would not commute.

c) Space-time evolution would induce an adiabatic evolution of a conformal quantum field theory. In particular, an evolution of the transfer matrix or S-matrix of a two-dimensional integrable model with respect to the parameters characterizing the matrix, and of n-point functions of a conformal field theory would be induced.

d) Particle decay could be interpreted in terms of a flow in which 3-surface X^3 decomposes to two disjoint components Y^3 and Z^3 such that the R matrices associated with different points y of Y^3 resp. z of Z^3 commute but the commutativity fails for R-matrices associated with y and z. The loss of commutativity and decay to pieces can be interpreted in terms of quantum de-coherence. This flow interpretation might significantly help in the construction of mathematical description of particle reactions.

Of course, here one encounters two possible interpretations for what particle decay means corresponding to the stringy picture and vertex identified as a 3-surfaces which defines the common end of incoming 4-surfaces. If both 3-surfaces decompose to union of one-dimensional curves along which the S-matrix defined by R-matrix is constant, the stringy decay could be illustrated graphically also as a decay of string. For second option a replication of string would be in question.

7.3.2   Quaternion conformal symmetries act on the space of super-canonical conformal weights

TGD predicts to super-conformal algebras. Super-canonical algebra certainly appears in the construction of the configuration space metric and spinor structure. Quaternion conformal algebra which generalizes the Kac Moody algebras of string models and its role is somewhat unclear in this respect.

Single particle super-canonical conformal weights and zeros of Riemann Zeta

Configuration space spinors can be interpreted as systems describing Fock states consisting of fermions with spin, electro-weak and color quantum numbers, plus a complex conformal weight. The work with conformal invariance and Riemann hypothesis (see [E8] and [n4, jf2]) inspires the hypothesis that the complex conformal weights labelling configuration space gamma matrices and isometry generators defining super-canonical algebra could be determined by the zeros of Riemann Zeta.

The conformal weights of the physical states would correspond to $Re(h) = n - 1/2 - i\sum n_i y_i$, $n \geq 0$ or $n = 0, 1$ at least, and to real axis the points $h = 2n$, $n > 0$, corresponding to trivial zeros at $s = -2n$. The superpositions $\sum n_i y_i$ of the imaginary parts of trivial zeros $s = 1/2 + y_i$ must be allowed. These points label the algebra elements obtained by commuting the algebra generators labelled by conformal weights which are the negatives of the zeros of Riemann Zeta and negative integers. If one assumes that Virasoro generators L_n, $n \geq 1$, generate zero norm states, only the $n = 0, 1$ option for the complex conformal weights remains. This option is forced also by the construction of the scalar propagator as a partition function in the super-canonical algebra. $n = 0, 1$ restriction guarantees also the orthogonality of the physical states.

These 1-dimensional lattices could define braid groups possessing an infinite number of generators appearing also in the description of type II_1 factors of von Neumann algebras [eb1] as described in [O4]. The Clifford algebra of the configuration space spinors would be the natural identification for von Neumann algebra in question. The connection between hyperfinite type II_1 factors of von Neumann algebras, braid groups, and quantum groups suggests that the existence of a quantum group structure at least in configuration space spinor degrees of freedom. Also the braiding for ordinary conformal field theories is known to define quantum group structure in a natural manner. The connection between R-matrices and quaternionic and complex structures supports this expectation. The minimal assumption would be that configuration space Clifford algebra for a given three-topology can be regarded as ordinary algebra without co-product and that co-product structure is associated only with the continuation of the Fock space structure from single-particle sector to two-particle sector.

Configuration space gamma matrices defining super generators and corresponding bosonic generators could be interpreted as conformal fields in the complex plane restricted to the punctures so that the vacuum expectation values of Clifford algebra elements would have interpretation as correlation functions of a conformal quantum field theory evaluated at points which define the lattice of super-canonical conformal weights. Also a connection with integrable lattice models suggests itself. In field theory context one possible interpretation is that a particle of given mass has an infinite degeneracy corresponding to the various super-canonical conformal weights and that one must sum in Feynmann diagrams over this degeneracy somehow. Physical intuition suggests wave functions in the set of conformal weights. This point will be discussed in detail later.

Mapping of super-canonical conformal weights to a geodesic sphere of CP_2

By quantum classical correspondence the construction of the S-matrix in configuration space spinor degrees of freedom should reduce to a corresponding construction at the space-time level. The idea that real physics and various p-adic physics result through an algebraic continuation of rational physics does not force discretization at the space-time level but leads in a natural manner to emergence of discrete sets of rational points as intersections of real and p-adic space-time sheets [E1]. Both infrared and ultraviolet cutoffs are involved since p-adically infinitesimal is infinite in the real sense and vice versa (for some values of p for given real infinitesimal) so that most points of p-adic space-time sheets are at real infinity. Each discretization corresponds to a different extension of p-adic numbers such that an improving resolution means increasing dimension of extension implying improved cognitive resolution.

The second quantized free spinor fields with a discretization at the space-time level could be interpreted in terms of a conformal field theory. Quantum classical correspondence would mean the mapping of the super-canonical conformal weights to the space-time points, which correspond to points of geodesic sphere of CP_2 giving rise to a family of commuting R-matrices. The mapping of complex scaling momenta to complex surface S^2 of CP_2 is analogous to the mapping of ordinary momenta to the points of the celestial sphere relating quantum to classical. This mapping is natural because configuration space gamma matrices are linearly related to the second quantized induced spinor fields. The physical states in configuration spin degrees of freedom might be characterized by the states of this system. In the real context this would require taking the limit of vanishing UV cutoff.

Braiding operation characterized by the Yang-Baxter matrix R for a tensor product of identical representations of braid group is the key element of braided Hopf algebras and the considerations of [O4] led to the speculation that braiding operation acts in the Clifford algebra of configuration space. The light like boundaries of space-time surfaces allow generalized conformal structure as metrically 2-dimensional structures. In a light hearted manner I christened this conformal symmetry as quaternion conformal invariance (the punishment has been a colossal find and replace procedure). A better term would have been Super Kac-Moody symmetry. Hyper-quaternionic and -octonionic analogs of conformal symmetry emerge naturally if the notion of the number theoretic spontaneous compactification makes sense.

The braiding operation for "particles" located in 2-dimensional sections could be seen as a space-time correlate for a corresponding operation performed for configuration space gamma matrices. The complex conformal weights of configuration space gamma matrices for which also imaginary part should be discrete in order to allow p-adicization could be interpreted as counterparts for the coordinates of points of a one-dimensional lattice defining an integrable system characterized by R-matrix. Configuration space gamma matrices would become conformal fields as function of super-canonical conformal weight on which super Kac-Moody conformal symmetries act infinitesimally.

Super Kac-Moody conformal algebra acts on super-canonical conformal weights as infinitesimal conformal transformations

Quantum classical correspondence inspires the hypothesis that the super Kac-Moody conformal algebra acts on the complex conformal weights of the super-canonical algebra. The discrete set of points of the sphere S^2 of CP_2 labelling commuting R-matrices would correspond to the super-canonical conformal weights, and would be mapped to a discrete set of points at the space-time surface. These points would belong to a "time=constant" section of 2-dimensional "space-time", presumably circle, defining physical states of a two-dimensional conformal field theory for which the scaling operator L_0 takes the role of Hamiltonian. Quaternion conformal algebra would act on super-canonical conformal weights by a pull-back and induce braiding operations naturally and realize thus quantum classical correspondence.

1. Action of super Kac-Moody conformal algebra on super-canonical algebra

Conformal algebra would act on $SO(3) \times SU(3)$ labels of the Hamiltonians associated with $X_l^3 \times CP_2$ used in the construction of configuration space metric. Also electro-weak quantum numbers associated with ground states would be present and would also appear as indices of the fermionic parts of Kac Moody generators. A good guess is that configuration space Hamiltonians, which in turn correspond to Hamiltonians associated with $X_l^3 \times CP_2$, label highest weight representations of the Kac-Moody algebra. Sugawara construction would provide the expressions for the super Kac-Moody Virasoro algebra.

The complex coordinates defined by the super-canonical conformal weights appear as arguments in the coefficients for the products of configuration space gamma matrices and their bosonic counterparts acting as isometries. The differential operator action of super Kac-Moody Virasoro algebra generators L_n on the super-canonical conformal weights is translated to a corresponding action on the coefficients by partial integration. The dependence of super Kac-Moody generators on z means multiplications of these coefficients by functions of conformal weight.

Super-canonical algebra allows an extension obtained by adding an integer multiple of Hamiltonian to the vector field of imbedding space inducing the action at configuration space level: $jAk\partial_k \rightarrow j^{Ak}\partial_k + kH_A$. The same extension applies also to the Kac Moody generators whose Lie algebra parts correspond to $SO(3) \times SU(3)$ Killing vector fields of the light cone boundary or more, generally of light like 7-surface $X_l^3 \times CP_2$, allowing representation as Hamiltonians. Therefore super canonical and Kac Moody extensions correspond to each other. For $k \neq 0$ the multiplets defined by Hamiltonians do not remain closed with respect to the action of Lie-algebra generators of H isometries and super-canonical representation gives rise to a highest weight Kac-Moody representation.

What differentiates TGD from the standard conformal field theories and string models is that super-symmetry generators acting as configuration space gamma matrices acting as super generators carry either lepton or quark number [B4]. As a matter fact, this corresponds to the situation in $N = 2$ super-conformal field theories allowing an Abelian charge having now interpretation as a fermion number. It seems that only the anti-commutators of quark like generators can contribute to the CH metric and super-symmetrization of Poisson algebra, whereas leptonic generators can contribute to the super-symmetrization of the function algebra of CH. This would mildly suggest that super Kac-Moody algebra is leptonic whereas super-canonical algebra would correspond to quarks.

The transfer matrices and factorizable S-matrices of integrable 2-D quantum field theories would be expressible at the level of configuration space as products of R-matrices acting on configuration super algebra generators regarded as conformal fields. They would act as monodromies in the space defined by complex plane punctured by the lattice of super-canonical conformal weights.

The vacuum expectation values of the enveloping algebra of the super-canonical algebra would reduce to n-point functions of a conformal quantum field theory, for which primary fields correspond to the representations of $SO(3) \times SU(3)$. These representations would combine to form infinite-dimensional representations of super-canonical algebra. The presence of the gigantic super-canonical symmetries raises the hope that quantum TGD could be solvable to a very high degree.

2. Generalization of the coset construction as a manner to satisfy Super Virasoro conditions

Super Virasoro constraints pose an extremely powerful constraint on theories possessing conformal invariance. The recent view about particle massivation [F2] inspires the hypothesis that Olive-Kent-Goddard coset construction [df3]generalizes in such a manner that the sums of super-canonical and Super Kac Moody Super Virasoro generators annihilate the physical states. This would provide a very elegant manner to understand and realized super conformal invariance.

A possible interpretation would be in terms of a duality stating that these two super algebras correspond to two different coordinatizations of the physical degrees of freedom of the configuration space and configuration space spinors using imbedding space vector fields *resp.* imbedding space vector valued fields defined at the space-time surface. In this picture one has also hopes of understanding why the conformal transformations associated with Super-Kac Moody algebra affect the conformal weights of super-canonical algebra.

3. Constraints from unitarity

For the representations of Virasoro algebra allowing Verma modules containing null states annihilated by Virasoro generators L_n, $n > 0$ conformal weights are quantized [ee2]. The simplest situation would correspond to rational conformal weights Δ and maximally symmetric super-conformal quantum field theories known as rational super-conformal field theories (RCFT) having a finite number of primary fields. In TGD framework super-generators carry fermion number which leaves $N = 2$ super conformal theories (or possibly $N = 4$ theories, situation is still unsettled) as the only physically acceptable option. The conformal weights are predicted to be rational numbers in this case.

For Sugawara construction [ee2] the super Kac-Moody conformal weights associated with the Hamiltonians are expressible in terms of the value of Kac-Moody central extension parameter k and Casimir invariant of the highest weight representation. For instance, in the case of $SU(2)$ one has $\Delta(j) = \frac{j(j+1)}{k+2}$. Unitarity poses strong constraints on the super Kac-Moody conformal weights.

In Wess-Zumino-Witten model also the analogs of Verma modules having finite number of non-zero norm states for Kac Moody algebras emerge and relate very closely to the corresponding phenomenon for quantum groups [ee2]. Thus only a finite number of highest weight representations of Kac Moody algebra appear as primary fields in WZW model. The theory predicts the dependence of the conformal weights on the finite-dimensional representations of the Kac Moody algebra. In the case of super Kac-Moody conformal algebra the super Kac-Moody conformal Kac-Moody representations relate naturally also the finite dimensional representations of $SO(3) \times U_{ew}(2) \times SU(3)$ labelling finite-dimensional or highest weight Kac Moody representations.

For Sugawara construction central extension parameter is given by $c = kdimg(G)/(k + h^\star)$, where $h^\star$ is the dual Coxeter number ($h^\star = n$ for $SU(n)$ and is non-vanishing in general case and one faces the question whether super Kac-Moody conformal invariance is gauge symmetry or not. String model based mass formulas would suggest the super Kac-Moody Virasoro representations should correspond to a trivial central extension $c = 0$.

The most general hypothesis is that all values of k and c consistent with unitarity are possible. Elementary particles would correspond to $k = 0, c = 0$ and finite-dimensional representations defined by Hamiltonians belonging to a given representation of $SO(3) \times SU(3)$. Anyons could correspond to a non-vanishing value of k. At configuration space level this would correspond to an extension of isometry generators obtained by adding an integer multiple of Hamiltonian as a scalar term to the isometry generator so that the action couples together several irreducible representations of $SO(3) \times SU(3)$ corresponding to states with non-vanishing norm.

7.3.3   Stringy diagrammatics and quantum classical correspondence

One expects that S-matrix can be constructed by generalizing the ordinary Feynman diagrams in the manner already discussed whereas generalization of stringy tree diagrams would have different interpretation in TGD framework. This expectation combined with quantum classical correspondence has some interesting implications.

Vertex operators and configuration space super algebra

Vertex operators are a fundamental notion in conformal field theories and string models. Quantum classical correspondence suggests that S-matrix should be expressible in terms of the vertex operators which at the configuration space level correspond to the continuations between one- and two-particle sectors of configuration space. By quantum classical correspondence the conformal weights of the generators of super-canonical algebra generators at X^3 are mapped to complex points of a two-sphere of CP_2 in turn defining points of X^3 and interpreted as punctures of a compactified complex plane. This picture conforms with the fact that sewing procedure allows to construct n-point functions for arbitrary 2-topology and with an arbitrary number of punctures using spheres with 3 punctures as a basic building block.

One can understand why the number of punctures appearing in vertex function cannot be larger than three. Three punctures means three complex super-canonical conformal weights and globally defined conformal transformations acting on the complex sphere must map this set of punctures to itself. The group $SL(2,C)$ of Möbius transformations acts as global conformal symmetries of sphere and can indeed map these points to each other. If the number of points is larger than 3 this is not possible unless the points satisfy some special symmetries.

Conformal invariance fixes the dependence of 3-vertices $\langle \Phi(z_1)\Phi(z_2)\Phi(z_3) \rangle$ on the coordinates z_i, which now correspond to super-canonical conformal weights. What remains is the possibility that the fusion coefficients C_{lm}^k appearing in both the product and co-product depend on two light-like coordinates of the space-time sheet complementary to the complex coordinates $(w, \overline{w})$. The dependence on CP_2 coordinates labelling different braiding matrices suggests itself as an additional dependence. One might hope that the construction of vertices might not differ much from that in conformal field theories and string models. The most optimistic expectation is that the theory effectively reduces to a construction of stringy diagrams without loops.

Delocalization in the space of super-canonical conformal weights as localization in the imbedding space

The identification of conformal weights as punctures raises the question how to interpret the propagators $1/L_0$ appearing in stringy diagrams. The propagating states are eigen states of the scaling momentum L_0 associated with super Kac-Moody conformal algebra. Therefore the physical states should be quantum superpositions of states with various complex super-canonical conformal weights defining two one-dimensional sub-lattices along lines $Re(s) = \pm 1/2$ and $(Re(s) < 0, Im(s) = 0)$. Scaling momentum eigen states represent Bloch waves in the lattice of punctures analogous to Bloch waves in a one-dimensional lattice encountered for Bethe ansatz [ee2] for integrable lattice models and spin chain models. Bethe ansatz should apply also now to the construction of states created by configuration space gamma matrices as eigen-states of super Kac-Moody conformal Virasoro generator $exp(iaL_0)$.

The construction of scalar field propagator as a partition function for the physical states of super-canonical algebra demonstrates that the introduction of a hierarchy of y-cutoffs is necessary defined by sub-algebras having as conformal weights the lattice defined by trivial zeros of zeta and n first nontrivial zeros $s = 1/2 + iy_i$, $i = 1, ..., n$, of Riemann Zeta. This cutoff hierarchy is probably related to the hierarchy of type II_1 factors of von Neumann algebras.

In the construction of super-canonical algebra ordinary translations are replaced by scalings acting on the radial light-like coordinate r associated with a light-like 3-surface X_l^3 of M^4.

a) The construction of the super-canonical algebra discussed in [B2] demonstrates also trivial zeros of Zeta correspond naturally to a super-canonical representation and the generators associated with trivial zeros span together with the generators associated with the non-trivial zeros an orthogonal basis. As a matter fact, non-trivial zeros correspond to $SO(3)$- and trivial zeros to $SO(2)$ reduction for the representations of the Lorentz group and the appearance of only even integers corresponds physically to the fact that only even parity excitations can be assigned to a given particle [B2].

Orthogonality allows also superposition for the imaginary parts of non-trivial zeros implied by the algebra structure. The most general conformal weights are of form

i) $s = 2n$, $n > 0$, and

ii) $s = n - 1/2 - \sum_i n_i y_i$, $n \geq 0$, with $\sum_i n_i = 2N + 1$ for even n and $\sum_i n_i = 2N$ for odd n. The states of type ii) with $n = 0, 1$ form an orthogonal basis but it is not clear whether orthogonality allows all values of n. The states of type ii) having $n > 1$ would naturally correspond to gauge degrees of freedom obtained via the action of the Virasoro algebra generators L_n, $n \geq 1$. The detailed properties of the super-canonical generators with weights $s = 2n$, $n > 0$, discussed in [B2, B3] guarantee that they indeed define non-gauge degrees of freedom.

b) The model for the renormalization group evolution of the scalar field propagator to be discussed later requires the inclusion of both kinds of zeros. The least singular propagator results if only the orthogonal $n = 0, 1$ states of type ii) are included.

c) p-Adicization requires a scaling momentum cutoff so that super-canonical operators appearing in a given length scale have conformal weights above some resolution. Hence one could have a finite lattice determined by the selection of a lattice generated by a subset of non-trivial zeros of ζ forced by the necessity of pinary cutoff.

d) The appearance of a one-dimensional lattices would be a direct counterpart for the one-dimensional lattice of strands defining the infinite braid characterizing sub-factors of type II_1 for von Neumann algebra. The infinite braids of super-canonical conformal weights at lines $Re(s) = n - 1/2$ ($n = 0, 1$ at least) defined by the points $s = n - 1/2 - \sum_i n_i y_i$, where y_i denote of the first K non-trivial zeros of Riemann Zeta, would correspond to different sub-factor hierarchies for von Neumann algebras. At the space-time level the hierarchies would correspond to different angular (or phase) resolutions and the appearance of the effective non-integer dimensions given by Beraha numbers labelling unitary representations of Temperley Lieb algebra would correspond to anomalous dimensions also in space-time sense [O4].

Quantum classical correspondence implies a strange duality in which the localization at the level of configuration space implies delocalization at the level of space-time. More explicitly, since the eigen-states of the super-canonical scaling momentum are totally delocalized in the radial coordinate r, the Bloch wave localizes the state with respect to r and thus both in imbedding and configuration configuration space. Physically the localization in the scale of 3-surface containing particles is natural. Inside the 3-surface containing particles however delocalization occurs since for the induced spinor fields plane waves over punctures mean de-localization at the level of 3-surface. This is natural since the modes of induced spinor fields are fixed by the requirement that they are eigen-states of L_0.

The zeros of Riemann polyzetas as super-canonical conformal weights of bound or virtual states?

Drinfeld's associator Φ encountered in monodromy considerations of conformal field theories [ec4] is expressible in terms of Riemann Zeta and polyzetas $\zeta(z_1, ..., z_n)$ [cf3, cf4], and this observation [O4] led to a number theoretical conjecture about the values of ζ and polyzetas at integer valued points (for the role of polyzetas are discussed in a wider context in [ed4]).

1. Definition of Polyzetas

Riemann's polyzeta $\zeta(z_1, ..., z_n)$ is defined via the sum

$$\zeta(z_1, ..., z_k) = \sum_{n_1 > n_2 > ... > n_k \geq 1} \prod_i n_i^{-z_i} \tag{7.3.5}$$

A possible interpretation for the ordering of the integers is in terms of fermion statistics in the sense that symmetrized polyzetas could represent a partition function for a system with states labelled by integers. Already this observation suggests that polyzetas might play some role in physics.

Polyzetas satisfy identities following from their defining representations: how one can deduced these identities from quantum field theory is discussed in [cf4]. For instance, the identity

$$P_2(a,b) \equiv \zeta(a,b) + \zeta(b,a) = \zeta(a)\zeta(b) - \zeta(a+b) \tag{7.3.6}$$

holds true. An example of a more complex identity holding true for 3-zeta is

$$P_3(a_1,a_2,a_3) \equiv \sum_P \zeta(a_{P(1)},a_{P(2)},a_{P(3)})$$
$$=$$
$$2\zeta(a_1+a_2+a_3) + \zeta(a_1)P_2(a_2,a_3) + \zeta(a_2)P_2(a_3,a_1) + \zeta(a_3)P_2(a_1,a_2) \tag{7.3.7}$$
$$-2\zeta(a_1)\zeta(a_2)\zeta(a_3) \ .$$

Here the subscript P refers to permutations of three objects.

2. Polyzetas and divergences of quantum field theories

Polyzetas $\zeta(k_1, ..., k_n)$ with integer values are found to appear as coefficients of counter terms in quantum field theories. Furthermore, Connes and Kreimer who worked with the renormalization in quantum field theory found that the "forest formula" of Zimmermann can be translated into the language of Hopf algebras and that the counter terms are proportional to the values $\zeta(k_1, ..., k_m)$ [ed2].

The question discussed Broadhurst and Kreimer [ee9] concerns the number of polyzeta values $\zeta(k_1, ...k_m)$ for a given weight $\sum_1^m k_i = n$ and depth m in terms of which one can express all polyzeta values of same weight and depth using rational coefficients. These generating polyzeta values are called irreducible. On basis of a numerical work they ended up to several conjectures, one of them being that the number M_n of irreducible polyzeta values of weight n is coded by the generating function $1 - x^2 - x^3 = \prod_n (1-x^n)^{M_n}$. They also demonstrated that for $3 \leq n \leq 9$ M_n enumerates so called positive knots with n crossings. The results emerged from the study of divergences of Feynman diagrams of ϕ^4 theory. A rule for assigning knots to Feynman diagrams allows to predict the level of transcendentality characterized by M_n associated with the counter term coefficients of a given UV divergent Feynman diagram.

p-Adicization philosophy inspires the following conjecture about the values of polyzetas. From $\zeta(2) = \pi^2/6$, and $\zeta(4) = \pi^4/90$, and more general result $\zeta(2n) \propto \pi^{2n}$, one could guess that that the values of polyzetas at the level $K = n$ are proportional to $x\pi^n$, where x is a number which belongs to an extension of rationals defining a finite extension of p-adic numbers (say combination of algebraic number and root of e). Hence the identities between polyzetas would make sense for finite extensions of p-adic numbers. Also Drinfeld's associator would make sense for these extensions.

3. Do the zeros of polyzetas correspond to the values of off mass shell super-canonical conformal weights?

The following considerations lend support to the speculation that Riemann Zeta and polyzetas are deeply related to the basic structure of quantum TGD so that Riemann Hypothesis might have much deeper physical content than previously thought.

a) The complex conformal weights associated with the super-canonical algebra could correspond to the zeros of Riemann Zeta or of their subset.

b) The intermediate particles appearing in stringy tree diagrams could have off mass shell super canonical conformal weights. This would make possible braiding during off mass-shell propagation and the interpretation as a time evolution for a braid would make sense. The zeros of polyzetas forming $2(n-1)$-dimensional surfaces might correspond to conformal weights of intermediate n-particle states appearing in the generalized Feynman diagrams involving n incoming particles in interaction. Also conformal weights of particles forming bound states could satisfy this condition, and one generalize the notion of bound state energy to that of "bound state conformal weight". According to the proposal of [O4], the braiding of join along boundaries bonds is the space-time correlate for bound state entanglement. At the level of configuration the braiding in the space of conformal weights would be in question.

i) The symmetrized 2-zeta $P_2(a_1,a_2)$, which might be relevant at least in the case of identical bosons resulting in the decay $a \to b + c$ or its reversal. The pairs of integers (m,n), $m + n = -2k$, $k > 0$ are its trivial zeros as one might expect from the addition of conformal weights. Since symmetrized n-zeta is expressible in terms of lower symmetrized poly-zetas the same holds true for arbitrary number of particles.

ii) In 2-particle intermediate (or bound) state the conformal weights of interacting particles must belong to a 2-dimensional surface of non-trivial zeros of $P_2(a_1, a_2)$. Let $z_i = 1/2 + y_i$, $i = 1, 2$ be two non-trivial zeros of ζ. The pairs (a_1, a_2), where $a_1 = 1/2 + iy_1$, $a_2 = z_2 - z_1$ so that $a_1 + a_2$ is zero of ζ, define nontrivial zeros of $P_2(a_1, a_2)$. A possible interpretation for a_2 would be as a conformal weight of an off mass shell virtual particle resulting when the intermediate state decays to an on mass shell particle and virtual particle. The on mass shell decay of the intermediate state $a \to b + c$ is forbidden by Riemann hypothesis unless one allows trivial zeros as conformal weights. The conservation of imaginary part of conformal weight for $a + b \to c + d$ by exchange in t-channel yields $(y_a, y_c) = (y_b, y_d)$.

iii) For the symmetrized 3-zeta $P_3(a_1, a_2, a_3)$ the zeros define a 4-dimensional surface. The study of the expression of P_3 given above shows that in this case there are no zeros expressible as combinations of non-trivial zeros of ζ. The emission of a_1 as on-mass shell particle from the 3-particle intermediate state would imply that a_1 and $a_1 + a_2 + a_3$ correspond to zeros of ζ. $a_2 + a_3$ would be a difference for zeros of ζ. The remaining condition would be $\zeta(a_2)P_2(a_1, a_3) + \zeta(a_3)P_2(a_1, a_2) = 0$, which allows only discrete points as solutions for given values of a_1 and $a_2 + a_3$. $a_2 + a_3$ has interpretation as an intermediate state and could decay to two on mass shell particles.

4. Does quantum criticality imply the identification of conformal weights as zeros of polyzetas?

The physical justification for the identification of zeros of polyzetas as allowed super-canonical conformal weights comes from quantum criticality.

a) Riemann Zeta has interpretation as a partition function [E8]. Also polyzetas might have a similar interpretation. Partition functions vanish at critical points representing phase transitions so that the vanishing of Riemann Zeta for allowed free particle conformal weights is consistent with quantum criticality, and allowed conformal weights would be analogous to critical temperature.

b) Quantum criticality means the vanishing of loop corrections. The brave guess is that for given values of super-canonical conformal weights the loop corrections predicted by TGD are proportional to polyzeta values just as they are in ordinary quantum field theories. The difference would be that conformal weights can be complex in the case of super-canonical algebra. The vanishing of loop corrections at criticality would force the identification of the super-canonical conformal weights as zeros of polyzetas.

c) The integer valued arguments of polyzetas correlate strongly with the number of loops correlating in turn with the superficial divergence of the Feynman diagram. A multi-loop specialist could probably immediately tell whether also in quantum field context the integers appearing as arguments of polyzeta could have an interpretation as conformal weights. In [cf4] a simple quantum field theoretical model yielding the values of polyzetas as vacuum diagrams is constructed, and vacuum diagrams are given by $\lambda^n \times (g\overline{g})^{2m+1}\zeta(k_1, ..., k_m)$. λ has dimension of inverse length so that the weight $n = \sum_i k_i$ indeed appears as a scaling dimension of the vacuum diagram. The depth m defines the powers of the coupling constants g and $\overline{g}$ associated with the vacuum diagram. Unfortunately this model is purely formal so that one cannot draw strong conclusions from it.

7.4 Hopf algebras and ribbon categories as basic structures

In this section the basic notions related to Hopf algebras and categories are discussed from TGD point of view. Examples are left to appendix. The new element is the graphical representation of the axioms leading to the idea about the equivalent of loop diagrams and tree diagrams based on general algebraic axioms.

7.4.1 Hopf algebras and ribbon categories very briefly

An algebraic formulation generalizing braided Hopf algebras and related structures to what might be called quantum category would involve the replacement of the co-product of Hopf algebras with morphism of quantum category having as its objects the Clifford algebras associated with configuration space spinor structure for various 3-topologies. The corresponding Fock spaces would would define algebra modules and the objects of the category would consists of pairs of algebras and corresponding modules. The underlying primary structure would be second quantized free induced spinor fields associated with 3-surfaces with various 3-topologies and generalized conformal structures.

1. Bi-algebras

Bi-algebras have two algebraic operations. Besides ordinary multiplication $\mu : H \otimes H \rightarrow H$ there is also co-multiplication $\Delta : H \rightarrow H \otimes H$. Algebra satisfies the associativity axiom (Ass): $a(bc) = (ab)c$, or more formally, $\mu(id \otimes \mu) = \mu(\mu \otimes id)$, and the unit axiom (Un) stating that there is morphism $\eta : k \rightarrow A$ mapping the unit of A to the unit of field k. Commutativity axiom (Co) $ab = ba$ translates to $\mu \otimes \tau \equiv \mu^{op} = \mu$, where τ permutes factors in tensor product $A \otimes A$.

Δ satisfies mirror images of these axioms. Co-associativity axiom (Coass) reads as $(\Delta \otimes id)\Delta = (id \otimes \Delta)\Delta$, co-unit axiom (Coun) states existence of morphism $\epsilon : k \rightarrow C$ mapping the unit of A to that of k, and co-commutativity (Coco) reads as $\tau \circ \Delta \equiv \Delta^{op} = \Delta$. For a bi-algebra H also additional axioms are satisfied: in particular, Δ (μ) acts as algebra (bi-algebra) morphism. When represented graphically, this constraint states that a box diagram is equivalent to a tree diagram as will be found and served as the stimulus for the idea that loop diagrams might be equivalent with tree diagrams.

Left and right algebra modules and algebra representations are defined in an obvious manner and satisfy associativity and unit axioms. A left co-module corresponds a pair (V, Δ_V) where the co-action $\Delta_N : V \rightarrow A \otimes V$ satisfies co-associativity and co-unit axioms. Right co-module is defined in an analogous manner.

Particle fusion $A \otimes B \rightarrow C$ corresponds to μ: $A \otimes B \rightarrow C = AB$. Co-multiplication Δ corresponds time reversal $C \rightarrow A \otimes B$ of this process, which is kind of a time-reversal for multiplication. The generalization would mean that μ and Δ become morphisms $\mu : B \otimes C \rightarrow A$ and $\Delta : A \rightarrow B \otimes C$, where A, B, C are objects of the quantum category. They could be either representations of same algebra or even different algebras.

2. Drinfeld's quantum double

Drinfeld's quantum double [ec4, ee2] is a braided Hopf algebra obtained by combining Hopf algebra $(H, \mu, \Delta, \eta, \epsilon, S, R)$ and its dual $H^\star$ to a larger Hopf algebra known as quasi-triangular Hopf algebra satisfying $\Delta = R\Delta^{op}R^{-1}$, where $\Delta^{op}(a)$ is obtained by permuting the two tensor factors. Duality means existence of a scalar product and the two algebras correspond to Hermitian conjugates of each other.

In TGD framework the physical states associated with these algebras have opposite energies since in TGD framework antimatter (or matter depending on the phase of matter) corresponds to negative energy states. The states of the Universe would correspond to states with vanishing conserved quantum numbers, and in concordance with crossing symmetry, particle reactions could be interpreted as transitions generating zero energy states from vacuum.

The notion of duality [ec4] is needed to define an inner product and S-matrix. Essentially Dirac's bra-ket formalism is in question. The so called evaluation map $ev : V \otimes V^\star \rightarrow k$ defined as $ev(v^i \otimes v_j) = \langle v^i, v_j \rangle = \delta_{ij}$ defines an inner product in any Hopf algebra module. The inverse of this map is the linear map $k \rightarrow V$ defined by $\delta_v(1) = v_i \otimes v^i$. For a tensor category with unit I, field k is replaced with unit I, and left duality these maps are replaced with maps $b_V : I \rightarrow V \otimes V^\star$ and $d_V = V \otimes V^\star \rightarrow I$. Right duality is defined in an analogous manner. The map d_V assigns to a given zero energy state S-matrix element. Algebra morphism property $b_V(ab) = b_V(a)b_V(b)$ would mean that the outcome is essentially the counterpart of free field theory Feynman diagram. This diagram is convoluted with the S-matrix element coded to the entanglement coefficients between positive and negative energy particles of zero energy state.

3. Ribbon algebras and ribbon categories

The so called ribbon algebra [ec4] is obtained by replacing one-dimensional strands with ribbons and adding to the algebra the so called twist operation θ acting as a morphism in algebra and in any algebra module. Twist allows to introduce the notion of trace, in particular quantum trace.

The thickening of one-dimensional strands to 2-dimensional ribbons is especially natural in TGD framework, and corresponds to a replacement of points of time=constant section of 4-surface with one-dimensional curves along which the S-matrix defined by R-matrix is constant. Ribbon category is defined in an obvious manner. There is also a more general definition of ribbon category with objects identified as representations of a given algebra and allowing morphisms with arbitrary number of incoming and outgoing strands having interpretation as many-particle vertices in TGD framework. The notion of quantum category defined as a generalization of a ribbon category involving the generalization of algebra product and co-product as morphisms between different objects of the category and allowing objects to correspond different algebras might catch the essentials of the physics of TGD Universe.

7.4.2 Algebras, co-algebras, bi-algebras, and related structures

It is useful to formulate the notions of algebra, co-algebra, bi-algebra, and Hopf algebra in order to understand how they might help in attempt to formulate more precisely the view about what generalized Feynman diagrams could mean. Since I am a novice in the field of quantum groups, the definitions to be represented are more or less as such from the book "Quantum Groups" of Christian Kassel [ec4] with some material (such as the construction of Drinfeld double) taken from [ee2]. What is new is a graphical representation of algebra axioms and the proposal that algebra and co-algebra operations have interpretation in terms of generalized Feynman diagrams.

In the following considerations the notation id_k for the isomorphism $k \to k \otimes k$ defined by $x \to x \otimes x$ and its inverse will be used.

Algebras

Algebra can be defined as a triple (A, μ, η), where A is a vector space over field k and $\mu : A \otimes A \to A$ and $\eta : k \to A$ are linear maps satisfying the following axioms (Ass) and (Un).

(Ass): The square

$$
\begin{array}{ccc}
A \otimes A \otimes A & \xrightarrow{\mu \otimes id} & A \otimes A \\
\downarrow{id \otimes \mu} & & \downarrow{\mu} \\
A \otimes A & \xrightarrow{\mu} & A
\end{array}
\tag{7.4.1}
$$

commutes.

(Un): The diagram

$$
\begin{array}{ccccc}
k \otimes A & \xrightarrow{\eta \otimes id} & A \otimes A & \xleftarrow{id \otimes \eta} & A \otimes k \\
 & \searrow{\cong} & \downarrow{\mu} & \swarrow{\cong} & \\
 & & A & &
\end{array}
\tag{7.4.2}
$$

commutes. Note that η imbeds field k to A.

(Comm) If algebra is commutative, the triangle

$$
\begin{array}{ccc}
A \otimes A & \xrightarrow{\tau_{A,A}} & A \otimes A \\
 & \searrow{\mu} \quad \swarrow{\mu} & \\
 & A &
\end{array}
\tag{7.4.3}
$$

commutes. Here $\tau_{A,A}$ is the flip switching the factors: $\tau_{A,A}(a \otimes a') = a' \otimes a$.

A morphism of algebras $f : (A, \mu, \eta) \to (A', \mu', \eta')$ is a linear map $A \to A'$ such that

$$
\mu' \circ (f \otimes f) = f \circ \mu, \ \text{ and } \ f \circ \eta = \eta' \ .
$$

A graphical representation of the algebra axioms is obtained by assigning to the field k a dashed line to be referred as a vacuum line in the sequel and to A a full line, to η a vertex $\times$ at which k-line changes to A-line. The product μ can be represented as 3-particle vertex in which algebra lines fuse together. The three axioms (Ass), (Un) and (Comm) can are expressed graphically in figure 7.4.2. Note that associativity axiom implies that two tree diagrams not equivalent as Feynman diagrams are equivalent in the algebraic sense.

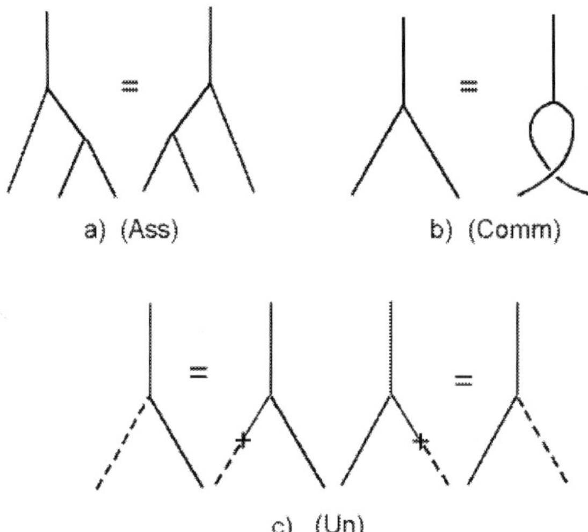

Figure 7.1: Graphical representation for the axioms of algebra. a) $a(bc) = (ab)c$, b) $ab = ba$, c) $ka = \mu(\eta(k), a)$ and $ak = \mu(a, \eta(k))$.

Co-algebras

The definition of co-algebra is obtained by systematically reversing the directions of arrows in the previous diagrams.

A co-algebra is a triple (C, Δ, ϵ), where C is a vector space over field k and $\Delta : C \to C \otimes C$ and $\epsilon : C \to k$ are linear maps satisfying the following axioms (Coass) and (Coun).

(Coass): The square

$$
\begin{array}{ccc}
C & \xrightarrow{\Delta} & C \otimes C \\
\downarrow{\Delta} & & \downarrow{id \otimes \Delta} \\
C \otimes C & \xrightarrow{\Delta \otimes id} & C \otimes C \otimes C
\end{array}
\tag{7.4.4}
$$

commutes.

(Coun): The diagram

$$
\begin{array}{ccccc}
k \otimes C & \xleftarrow{\epsilon \otimes id} & C \otimes C & \xrightarrow{id \otimes \epsilon} & C \otimes k \\
& {\scriptstyle\cong}\nwarrow & \uparrow{\Delta} & \nearrow{\scriptstyle\cong} & \\
& & C & &
\end{array}
\tag{7.4.5}
$$

commutes. The map Δ is called co-product or co-multiplication whereas ϵ is called the counit. The commutative diagram state that the co-product is co-associative and that co-unit commutes with co-product.

(Cocomm) If co-algebra is commutative, the triangle

$$
\begin{array}{ccc}
& C & \\
{\scriptstyle\Delta}\swarrow & & \searrow{\scriptstyle\Delta} \\
C \otimes C & \xrightarrow{\tau_{C,C}} & C \otimes C
\end{array}
\tag{7.4.6}
$$

commutes. Here $\tau_{C,C}$ is the flip switching the factors: $\tau_{C,C}(c \otimes c') = c' \otimes c$.

A morphism of co-algebras $f : (C, \Delta, \epsilon) \to (C', \Delta', \epsilon')$ is a linear map $C \to C'$ such that

$$(f \otimes f) \circ \Delta = \Delta' \circ f \ , \quad \text{and} \quad \epsilon = \epsilon' \circ f \ .$$

It is straightforward to define notions like co-ideal and co-factor algebra by starting from the notions of ideal and factor algebra. A very useful notation is Sweedler's sigma notation for $\Delta(x)$, $x \in C$ as element of $C \otimes C$:

$$\Delta(x) = \sum_i x'_i \otimes x''_i \equiv \sum_{\{x\}} x' \otimes x'' \ .$$

Also co-algebra axioms allow graphical representation. One assigns to ϵ a vertex $\times$ at which C-line changes to k-line: the interpretation is as an absorption of a particle by vacuum. The co-product Δ can be represented as 3-particle vertex in which C-line decays to two C-lines. The graphical representation of the three axioms (Coass), (Coun), and (Cocomm) is related to the representation of algebra axioms by "time reversal", that is turning the diagrams for the algebra axioms upside down (see figure 7.4.2).

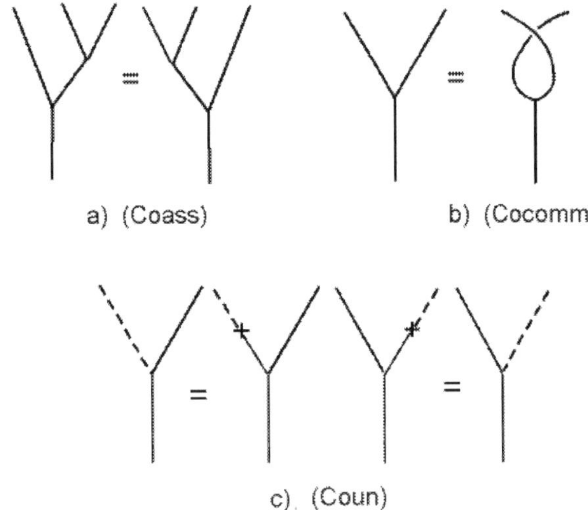

Figure 7.2: Graphical representation for the axioms of co-algebra is obtained by turning the representation for algebra axioms upside down. a) $(id \otimes \Delta)\Delta = (\Delta \otimes id)\Delta$, b) $\Delta = \Delta^{op}$, c) $(\epsilon \otimes id) \circ \Delta = (id \otimes \epsilon) \circ \Delta = id$.

Bi-algebras

Consider next a vector space H equipped simultaneously with an algebra structure (H, μ, η) and a co-algebra structure (H, Δ, ϵ). There are some compatibility conditions between these two structures. $H \otimes H$ can be given the induced structures of a tensor product of algebras and of co-algebras.

The following two statements are equivalent.

a) The maps μ and η are morphisms of co-algebras. For μ this means that the diagrams

$$
\begin{array}{ccc}
H \otimes H & \xrightarrow{\ \mu\ } & H \\[2pt]
\Big\downarrow{\scriptstyle (id \otimes \tau \otimes id) \otimes (\Delta \otimes \Delta)} & & \Big\downarrow{\scriptstyle \Delta} \\[2pt]
(H \otimes H) \otimes (H \otimes H) & \xrightarrow{\ \mu \otimes \mu\ } & H \otimes H
\end{array}
\qquad (7.4.7)
$$

and

$$
\begin{array}{ccc}
H \otimes H & \xrightarrow{\ \epsilon \otimes \epsilon\ } & k \otimes k \\
\downarrow{\scriptstyle \mu} & & \downarrow{\scriptstyle id} \\
H & \xrightarrow{\ \epsilon\ } & k
\end{array}
\tag{7.4.8}
$$

commute. For η this means that the diagrams

$$
\begin{array}{ccc}
k & \xrightarrow{\ \eta\ } & H \\
\downarrow{\scriptstyle id} & & \downarrow{\scriptstyle \Delta} \\
k \otimes k & \xrightarrow{\ \eta \otimes \eta\ } & H \otimes H
\end{array}
\qquad\qquad
\begin{array}{c}
k \xrightarrow{\ \eta\ } H \\
{\scriptstyle id}\searrow \ \ \swarrow{\scriptstyle \epsilon} \\
k
\end{array}
\tag{7.4.9}
$$

commute.

   b) The maps Δ and ϵ are morphisms of algebras.
For Δ this means that diagrams

$$
\begin{array}{ccc}
H \otimes H & \xrightarrow{\ \Delta \otimes \Delta\ } & (H \otimes H) \otimes (H \otimes H) \\
\downarrow{\scriptstyle \mu} & & \downarrow{\scriptstyle (\mu \otimes \mu)(id \otimes \tau \otimes id)} \\
H & \xrightarrow{\ \Delta\ } & H \otimes H
\end{array}
\tag{7.4.10}
$$

and

$$
\begin{array}{ccc}
k & \xrightarrow{\ \eta\ } & H \\
\downarrow{\scriptstyle id} & & \downarrow{\scriptstyle \Delta} \\
k \otimes k & \xrightarrow{\ \eta \otimes \eta\ } & H \otimes H
\end{array}
\tag{7.4.11}
$$

commute.
   For ϵ this means that the diagrams

$$
\begin{array}{ccc}
H \otimes H & \xrightarrow{\ \epsilon \otimes \epsilon\ } & k \otimes k \\
\downarrow{\scriptstyle \mu} & & \downarrow{\scriptstyle id} \\
H & \xrightarrow{\ \epsilon\ } & k
\end{array}
\qquad\qquad
\begin{array}{c}
k \xrightarrow{\ \eta\ } H \\
{\scriptstyle id}\searrow \ \ \swarrow{\scriptstyle \epsilon} \\
k
\end{array}
\tag{7.4.12}
$$

commute. The proof of the theorem involves the comparison of the commutative diagrams expressing both statements to see that they are equivalent.

   The theorem inspires the following definition.
   <u>Definition</u>: A bi-algebra is a quintuple $(H, \mu, \eta, \Delta, \epsilon)$, where (H, μ, η) is an algebra and (H, Δ, ϵ) is co-algebra satisfying the mutually equivalent conditions of the previous theorem. A morphisms of bi-algebras is a morphism for the underlying algebra and bi-algebra structures.

   An element $x \in H$ is known as primitive if one has $\Delta(x) = 1 \otimes x + x \otimes 1$ and have $\epsilon(x) = 0$. The subspace of primitive elements is closed with respect to the commutator $[x, y] = xy - yx$. Note that for primitive elements $\mu \circ \Delta = 2id_H$ holds true so that $\mu/2$ acts as the left inverse of Δ.

Given a vector space V, there exists a unique bi-algebra structure on the tensor algebra $T(V)$ such that $\Delta(v) = 1 \otimes v + v \otimes 1$ and $\epsilon(v) = 0$ for any element v of V. By the symmetry of Δ this bi-algebra structure is co-commutative and corresponds to the "classical limit". Also the Grassmann algebra associated with V allows bi-algebra structure defined in the same manner.

Figure 7.4.2 provides a representation for the axioms of bi-algebra stating that Δ and ϵ act as algebra morphisms of algebra and or equivalent that μ and η act as co-algebra morphisms. The axiom stating that Δ (μ) is algebra (co-algebra) morphism implies that scattering diagrams differing by a box loop are equivalent. The statement that μ is co-algebra morphism reads $(id \otimes \mu \otimes id)(\Delta \otimes \Delta) = \Delta \circ \mu$ whereas the mirror statement $\Delta(ab) = \Delta(a)\Delta(b)$ for Δ reads as $\Delta \circ \mu = \mu(\Delta \otimes \Delta)$ and gives rise to the same graph.

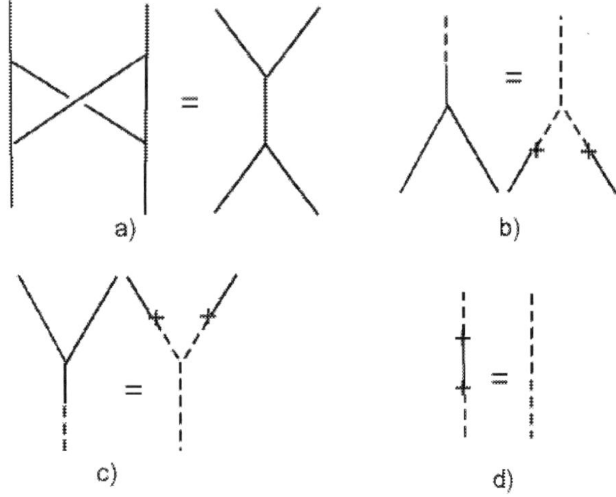

Figure 7.3: Graphical representation for the conditions guaranteing that μ and η ($\wedge$ and ϵ) act as homomorphisms of co-algebra (algebra). a)$(id \otimes \mu \otimes id)(\Delta \otimes \Delta) = \Delta \circ \mu$, b) $\epsilon \circ \mu = id \circ (\epsilon \otimes \epsilon)$, c) $\Delta \circ \eta = \mu \otimes \circ id_k$, d) $\epsilon \circ \eta = id_k$.

Hopf algebras

Given an algebra (A, μ, η) and co-algebra (C, Δ, ϵ), one can define a bilinear map, the convolution on the vector space $Hom(C, A)$ of linear maps from C to A. By definition, if f and g are such linear maps, then the convolution $f \star g$ is the composition of the maps

$$C \;\xrightarrow{\Delta}\; C \otimes C \;\xrightarrow{f \otimes g}\; A \otimes A \;\xrightarrow{\mu}\; A \tag{7.4.13}$$

Using Sweedler's sigma notion one has

$$f \star g(x) = \sum_{\{x\}} f(x')g(x'') \ . \tag{7.4.14}$$

It can be shown that the triple $(Hom(C, A), \star, \Delta, \eta \circ \epsilon)$ is an algebra and that the map $\Lambda_{C,A} : A \otimes C^\star \to Hom(C, A)$ defined as

$$\Lambda_{C,A}(a \otimes \gamma)(c) = \gamma(c)a$$

is a morphism of algebras, where $C^\star$ is the dual of the finite-dimensional co-algebra C.

For $A = C$ the result gives a mathematical justification for the crossing symmetry inspired re-interpretation of the unitary S-matrix interpreted usually as an element of $Hom(A, A)$ as a state generated by element of $A \otimes A^\star$ from the vacuum $|vac\rangle = |vac_A\rangle \otimes |vac_{A^\star}\rangle$. This corresponds to the interpretation of the reaction $a_i|vac_A\rangle \to a_f|vac_A\rangle$ as a transition creating state $a_i \otimes a_f^\star|vac\rangle$ with vanishing conserved quantum numbers from vacuum.

With these prerequisites one can introduce the notion of Hopf algebra. Let $(H, \mu, \eta, \Delta, \epsilon)$ be a bi-algebra. An endomorphism S of H is called an antipode for the bi-algebra H if

$$S \star id_H = id_H \star S = \eta \circ \epsilon \ .$$

A Hopf algebra is a bi-algebra with an antipode. A morphism of a Hopf algebra is a morphism between the underlying bi-algebras commuting with the antipodes.

The graphical representation of the antipode axiom is given in the figure below.

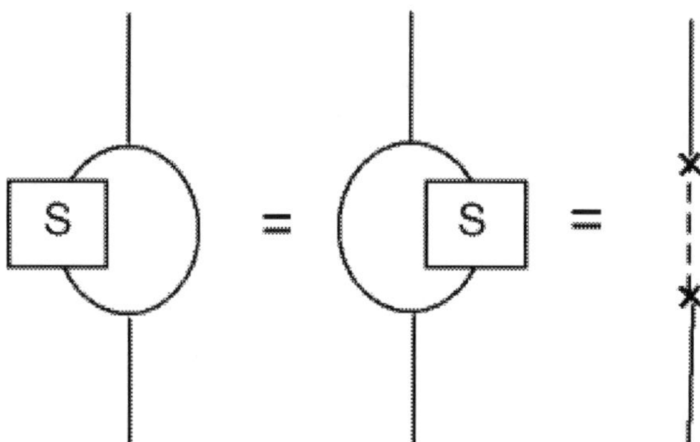

Figure 7.4: Graphical representation of antipode axiom $S \star id_H = id_H \star S = \eta \circ \epsilon$.

The notion of scalar product central for physical applications boils down to the notion of duality. Duality between Hopf algebras U and H means the existence of a morphism $x \to \Psi(x)$: $H \to U^\star$ defined by a bilinear form $\langle u, x \rangle = \Psi(x)(u)$ on $U \times H$, which is a bi-algebra morphism. This means that the conditions

$$\langle uv, x \rangle = \langle u \otimes v, \Delta(x) \rangle \ , \quad \langle u, xy \rangle = \langle \Delta(u), x \otimes y \rangle \ ,$$

$$\langle 1, x \rangle = \epsilon(x) \ , \qquad \langle u, 1 \rangle = \epsilon(u) \ , \tag{7.4.15}$$

$$\langle S(u), x \rangle = \langle u, S(x) \rangle$$

are satisfied. The first condition on multiplication and co-multiplication, when expressed graphically, states that the decay $x \to u \otimes v$ can be regarded as time reversal for the fusion of $u \otimes v \to x$. Second condition has analogous interpretation.

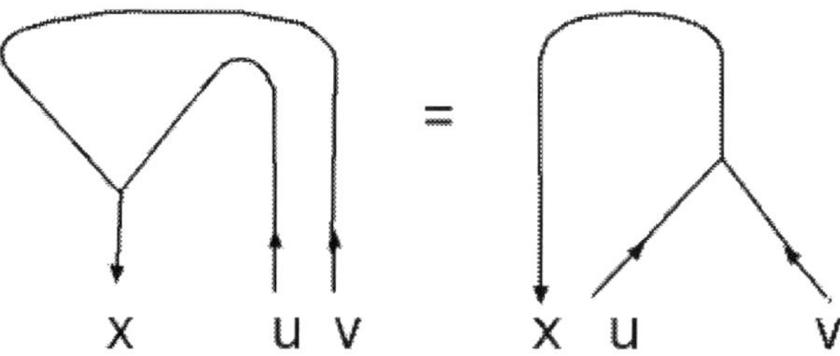

Figure 7.5: Graphical representation of the duality condition $\langle uv, x \rangle = \langle u \otimes v, \Delta(x) \rangle$.

Modules and comodules

Left and right algebra modules and algebra representations are defined in an obvious manner and satisfy associativity and unit axioms having diagrammatic representation similar to that for corresponding algebra axioms.

A left co-module corresponds a pair (V, Δ_V), where the co-action $\Delta_N \colon V \to C \otimes V$ satisfies co-associativity axiom $(id_C \otimes \Delta_N) \circ \Delta_N = (\Delta \otimes id_N) \circ \Delta_N$ and co-unit axiom $(\epsilon \otimes id) \circ \Delta_N = id_N$. A right co-module is defined in an analogous manner. It is convenient to introduce Sweedlers's notation for Δ_N as $\Delta_N = \sum_{\{c\}} x_C \otimes x_N$.

One can define module and comodule morphisms and tensor product of modules and co-modules in a rather obvious manner. The module N could be also algebra, call it A, in which case μ_A and η_A are assumed to act as H-comodule morphisms.

The standard example is quantum plane $A = M(2)_q$ is the free algebra generated variables x, y subject to to relations $yx = qxy$ and having coefficients in k. The action of Δ_A reads as

$$\Delta_A \begin{pmatrix} x \\ y \end{pmatrix} = \begin{pmatrix} a & b \\ c & d \end{pmatrix} \otimes \begin{pmatrix} x \\ y \end{pmatrix} \ .$$

Δ_A defines algebra morphism from A to $SL(2)_q \otimes A$: $\Delta_a(yx) = \Delta_A(y)\Delta_A(x) = q\Delta_A(x)\Delta_A(y) = \Delta(qxy)$.

Braided bi-algebras

$\Delta^{op} = \tau_{H,H} \circ \Delta$ defines the opposite co-algebra H^{op} of H. A braided bi-algebra $(H, \mu, \eta, \Delta, \epsilon)$ is called quasi-co-commutative (or quasi-triangular) if there exists an element R of algebra $H \otimes H$ such that for all $x \in H$ one has

$$\Delta^{op} = R\Delta R^{-1} \ .$$

One can express R in the form

$$R = \sum_i s_i \otimes t_i \ .$$

It is convenient to denote by R_{ij} the R matrix acting in i^{th} and j^{th} tensor factors of nth tensor power of H. More precisely, R_{ij} can be defined as an operator acting in an n-fold tensor power of H by the formula $R_{ij} = y^{(1)} \otimes y^{(2)} \otimes ... \otimes y^{(p)}$, $p \leq n$, $y^{(k_i)} = s_i$ and $y^{(k_j)} = t_j$, $y^{(k)} = 1$ otherwise. For instance, one has $R_{13} = \sum_i s_i \otimes 1 \otimes t_i$.

With these prerequisites one can define a braided bi-algebra as a quasi-commutative bi-algebra $(H, \mu, \eta, \Delta, \epsilon, S, S^{-1}, R)$ as an algebra with a preferred element $R \in H \otimes H$ satisfying the two relations

$$\begin{aligned} (\Delta \otimes id_H)(R) &= R_{13}R_{23} \ , \\ (id_H \otimes \Delta)(R) &= R_{13}R_{12} \ . \end{aligned}$$

$$\tag{7.4.16}$$

Braided bi-algebras, known also as quasi-triangular bi-algebras, are central in the theory of quantum groups, R-matrices, and braid groups. By a direct calculations one can verify the following relations.

a) Yang-Baxter equations

$$R_{12}R_{13}R_{23} = R_{23}R_{13}R_{12} \ , \tag{7.4.17}$$

and the relation

$$(\epsilon \otimes id_H)(R) = 1 \tag{7.4.18}$$

hold true.

b) Since H has an invertible antipode S, one has

$$\begin{aligned}
(S \otimes id_H)(R) &= R^{-1} = (id_H \otimes S^{-1})(R) \ , \\
(S \otimes S)(R) &= R \ .
\end{aligned}$$
(7.4.19)

The graphical representation of the Yang-Baxter equation in terms of the relations of braid group generators is given in the figure 7.4.2.

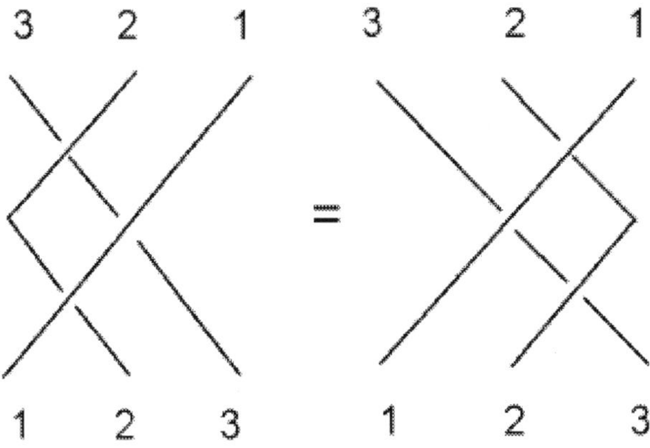

Figure 7.6: Graphical representation of Yang-Baxter equation $R_{12}R_{13}R_{23} = R_{23}R_{13}R_{12}$.

Ribbon algebras

Let H be a braided Hopf algebra with a universal matrix $R = \sum_i s_i \otimes t_i$ and set $u = \sum_i S(t_i)s_i$. It can be shown that u is invertible with the inverse $u^{-1} = \sum s_i S^2(t_i)$ and that $uS(u) = S(u)u$ is central element in H. Furthermore, one has $\epsilon(u) = 1$ and $\Delta(u) = (R_{21}R)^{-1}(u \otimes u)$, and the antipode is given for any $x \in H$ by $S^2(x) = uxu^{-1}$.

Ribbon algebra has besides $R \in H \otimes H$ also a second preferred element called θ. A braided Hopf algebra is called ribbon algebra if there exists a central element θ of H satisfying the relations

$$\Delta(\theta) = (R_{21}R)^{-1}(\theta \otimes \theta) \ , \quad \epsilon(\theta) = 1 \ , \quad S(\theta) = \theta \ .$$
(7.4.20)

It can be shown that θ^2 acts like $S(u)u$ on any finite-dimensional module [ec4].

Drinfeld's quantum double

Drinfeld's quantum double construction allows to build a quasi-triangular Hopf algebra by starting from any Hopf algebra H and its dual $H^\star$, which exists in a finite-dimensional case always, and as a vector space is isomorphic with H. Besides duality normal ordering is second ingredient of the construction. Physically the generators of the algebra and its dual correspond to creation and annihilation operator type operators. Drinfeld's quantum double construction is represented in a very general manner in [ec4]. A construction easier to understand by a physicist is discussed in [ee2]. For this reason this representation is summarized here although the style differs from the representation of [ec4] followed in the other parts of appendices.

Consider first what is known.

a) Duality means the existence of basis $\{e_a\}$ for H and $\{e^a\}$ for $H^\star$ and inner product (or evaluation as it is called in [ec4]) $ev : H^\star \otimes H \rightarrow k$ defined as $ev(e^a e_b) \equiv \langle e^a, e_b \rangle = \delta^a_b$ and its inverse $\delta : k \rightarrow H^\star \otimes H$ defined by $\delta(1) = e^a e_a$. One can extend the inner product to an inner product in the tensor product $(H^\star \otimes H^\star) \otimes (H \otimes H)$ in an obvious manner.

b) The product (co-product) in H ($H^\star$) coincides with the co-product (product) in $H^\star$ (H) in the sense that one has

$$\langle e^c, e_a e_b \rangle = m^c_{ab} = \langle \Delta(e^c), e_b \otimes e_a \rangle \ ,$$
$$\langle e^a e^b, e_c \rangle = \mu^{ab}_c = \langle e^a \otimes e^b, \Delta(e_c) \rangle \ , \tag{7.4.21}$$

These equations are quite general expressions for the duality expressed graphically in figure 7.4.2.

c) The antipodes S for H and $H^\star$ can be represented as matrices

$$S_H(e_a) = S_a{}^b e_b \ , \quad S_{H^\star}(e^a) = (S^{-1})^a{}_b e^b \ . \tag{7.4.22}$$

The task is to construct algebra product μ and co-algebra product Δ, unit η and co-unit ϵ, antipode, and R-matrix R for for $H \otimes H^\star$. The natural basis for $H \otimes H^\star$ consists of $e_a \otimes e^b$.

a) Co-product Δ is simply the product of co-products

$$\Delta(e_a e^b) = \Delta(e_a) \Delta(e^b) = m^b_{vu} \mu^{cd}_a e_c e^u \otimes e_d e^v \ . \tag{7.4.23}$$

b) Product μ involves normal ordering prescription allowing to transform products $e^a e_b$ (elements of $H^\star \otimes H$) to combinations of basis elements $e_a e^b$ (elements of $H \otimes H^\star$. This map must be consistent with the requirement that co-product acts as an algebra morphism. Drinfeld's normal ordering prescription, or rather a map $c_{H^\star,H}\colon H^\star \otimes H \to H \otimes H^\star$ is given by

$$c_{H^\star,H}(e^a e_b) = R^{ac}_{bd} e_c e^d \ , \quad R^{ac}_{bd} = m^x_{kd} m^a_{xu} \mu^{vy}_b \mu^{ck}_y (S^{-1})^u_v e_c e^d \ . \tag{7.4.24}$$

The details of the formula are far from being obvious: the axioms of tensor category with duality to be discussed later might allow to relate $R_{H^\star,H}$ to $R_{H,H}$ and this might help to understand the origin of the expression. Normal ordering map can be interpreted as braid operation exchanging H and $H^\star$ and the matrix defining the map could be regarded as R-matrix $R_{H \otimes H^\star}$.

c) The universal R-matrix is given by

$$R \ = \ (e_a \otimes id_{H^\star}) \otimes (id_H \otimes e^a) \ , \tag{7.4.25}$$

where the summation convention is applied. One can show that $R\Delta = \Delta^{op} R$ by a direct calculation.

d) The antipode $S_{H \otimes H^\star}$ follows from the product of antipodes for H and $H^\star$ using the fact that antipode is antihomomorphism using the normal ordering prescription

$$S_{H \otimes H^\star}(e_a e^b) = c_{H^\star,H}(S(e^b) S(e_a)) \ . \tag{7.4.26}$$

Quasi-Hopf algebras and Drinfeld associator

Braided Hopf algebras are quasi-commutative in the sense that one has $\Delta^{op} = R \Delta R^{-1}$. Also the strict co-associativity can be given up and this means that one has

$$(\Delta \otimes id)\Delta = \Phi(id \otimes \Delta)\Phi^{-1} \ , \tag{7.4.27}$$

where $\Phi \in H \otimes H \otimes H$ is known as Drinfeld's associator and appears in the of conformal fields theories. If the resulting structure satisfies also the so called Pentagon Axiom (to be discussed later, see Eq. 7.4.36 and figure 7.4.3), it is called quasi-Hopf algebra. Pentagon Axiom boils down to the condition

$$(id \otimes id \otimes \Delta)(\Phi)(\Delta \otimes id \otimes id)(\Phi) = (id \otimes \Phi)(id \otimes id \otimes \Delta)(\Phi)(\Phi \otimes id) \ . \tag{7.4.28}$$

The Yang-Baxter equation for quasi-Hopf algebra reads as

$$R_{12} \Phi_{312} R_{13} \Phi^{-1}_{1322} R_{23} \Phi_{123} = \Phi_{321} R_{23} \Phi^{-1}_{231} R_{13} \Phi_{213} R_{12} \Phi_{123} \ . \tag{7.4.29}$$

The left-hand side arises from a sequence of transformations

$$(12)3 \xrightarrow{\Phi_{123}} 1(23) \xrightarrow{R_{23}} 1(32) \xrightarrow{\Phi_{132}^{-1}} (31)2 \xrightarrow{R_{13}} 3(12) \xrightarrow{\Phi_{312}} 3(12) \xrightarrow{R_{12}} 3(21) \ .$$

The right-hand side arises from the sequence

$$(12)3 \xrightarrow{R_{12}} (21)3 \xrightarrow{\Phi_{213}} 2(13) \xrightarrow{R_{13}} 2(31) \xrightarrow{\Phi_{231}^{-1}} (23)1 \xrightarrow{R_{23}} (32)1 \xrightarrow{\Phi_{321}} 3(21) \ .$$

One can produce new quasi-Hopf algebras by gauge (or twist) transformations using invertible element $\Omega \in H \otimes H$ called twist operator

$$
\begin{aligned}
\Delta(a) \ &\rightarrow \ \Omega \Delta(a) \Omega^{-1} \ , \\
\Phi \ &\rightarrow \ \Omega_{23}(id \otimes \Delta)(\Omega)\Phi(\Delta \otimes id)(\Omega^{-1})\Omega_{12}^{-1} \ , \\
R \ &\rightarrow \ \Omega R \Omega^{-1} \ .
\end{aligned}
\tag{7.4.30}
$$

Quasi-Hopf algebras appear in conformal field theories and correspond quantum universal enveloping algebras divided by their centralizer. Consider as an example the R-matrix R^{j_1, j_2} relating $j_1 \otimes j_2$ and $j_2 \otimes j_1$ representations $\Delta^{j_1, j_2}(a)$ and $\Delta^{j_2, j_1}(a)$ of the co-product Δ of $U(sl(2))_q$. $\Delta^{j,j}(a)$ commutes with R^{jj} for all elements of the quantum group. The action of $g_i = qR^{jj}$ acting in i^{th} and $(i+1)^{th}$ tensor factors extends to the representation $(V_j)^{\times n}$ in an obvious manner. From the Yang-Baxter equation it follows that the operators g_i define a representation of braid group B_n:

$$
\begin{aligned}
g_i g_{i+1} g_i \ &= \ g_{i+1} g_i g_{i+1} \ , \\
g_i g_j \ &= \ g_j g_i \ , \quad \text{for } |j - k| \geq 2 \ .
\end{aligned}
\tag{7.4.31}
$$

Under certain conditions the braid group generators generate the whole centralizer C_q^n for the representation of quantum group. For instance, this occurs for $j = 1/2$. In this case the additional condition

$$g_i^2 \ = \ (q^2 - 1)g_i + q^2 \times 1 \ , \tag{7.4.32}$$

so that the centralizer is isomorphic with the Hecke algebra $H_n(q)$, which can be regarded as a q-deformation of permutation group S_n.

The result generalizes. In Wess-Zumino-Witten model based on group G the relevant algebraic structure is $U(G_q)/C^n(q)$. This is quasi-Hopf algebra and the so called Drinfeld associator characterizes the quasi-associativity.

7.4.3 Tensor categories

Hopf algebras and related structures do not seem to be quite enough in order to formulate elegantly the construction of S-matrix in TGD framework. A more general structure known as a braided tensor category with left duality and twist operation making the category to a ribbon category is needed. The algebra product μ and co-product Δ must be generalized so that they appear as morphisms $\mu_{A \otimes B \to C}$ and $\Delta_{A \to B \otimes C}$: this gives hopes of describing 3-vertices algebraically. It is not clear whether one can assume single underlying algebra so that objects would correspond to different representations of this algebra or whether one allow even non-isomorphic algebras.

In the tensor category the tensor products of objects and corresponding morphisms belong to the category. In a braided category the objects $U \otimes V$ and $V \otimes U$ are related by a braiding morphism. The notion of braided tensor category appears naturally in topological and conformal quantum field theories and seems to be an appropriate tool also in TGD context. The basic category theoretical notions are discussed in [ec4] and I have already earlier considered category theory as a possible tool in the construction of quantum TGD and TGD inspired theory of consciousness [O1].

In braided tensor categories one introduces the braiding morphism $c_{V,W} : V \otimes W \to W \otimes V$, which is closely related to R-matrix. In categories allowing duality arrows with both directions are allowed ad diagrams analogous to pair creation from vacuum are possible. In ribbon categories one introduces also the twist operation θ_V as a morphism of object and the Θ_W satisfies the axiom: $\theta_{V \otimes W} = (\theta_V \otimes \theta_W) c_{W,V} c_{V,W}$. One can also introduce morphisms with arbitrary number of incoming lines and outgoing lines and visualize them as boxes, coupons. Isotopy principle, originally related to link and knot diagrams provides a powerful tool allowing to interpret the basic axioms of ribbon categories in terms of isotopy invariance of the diagrams and to invent theorems by just isotoping.

Categories, functors, natural transformations

Categories [dh1, dh2, dh3, ec4] are roughly collections of objects A, B, C... and morphisms $f(A \to B)$ between objects A and B such that decomposition of two morphisms is always defined. Identity morphisms map objects to objects. Examples of categories are open sets of some topological spaces with continuous maps between them acting as morphisms, linear spaces with linear maps between them acting as morphisms, groups with group homomorphisms taking the role of morphisms. Practically any collection of mathematical structures can be regarded as a category. Morphisms can be very general: for instance, partial ordering $a \leq b$ can define a morphism $f(A \to B)$.

Functors between categories map objects to objects and morphisms to morphisms so that a product of morphisms is mapped to the product of the images and identity morphism is mapped to identity morphism. Functor $F : \mathcal{C} \to \mathcal{D}$ commutes also with the maps s and b assigning to a morphism $f : V \to W$ its source $s(f) = V$ and target $b(f) = W$.

A natural transformation between functors F and G from $\mathcal{C} \to \mathcal{C}'$ is a family of morphisms $\eta(V) : F(V) \to G(V)$ in $\mathcal{C}'$ indexed by objects V of $\mathcal{C}$ such that for any morphisms $f : V \to W$ in $\mathcal{C}$, the square

$$
\begin{array}{ccc}
F(V) & \xrightarrow{\eta(V)} & G(V) \\
\downarrow{F(f)} & & \downarrow{G(f)} \\
F(W) & \xrightarrow{\eta(W)} & G(W)
\end{array}
\tag{7.4.33}
$$

commutes.

The functor $F : \mathcal{C} \to \mathcal{D}$ is said to be equivalence of categories if there exists a functor $G : \mathcal{D} \to \mathcal{C}$ such and natural isomorphisms

$$
\eta : id_{\mathcal{D}} \to FG \quad \text{and} \quad \theta : GF \; \text{›} \; id_{\mathcal{C}} FG \; .
$$

The notion of adjoint functor is a more general notion than equivalence of categories. In this case η and θ are natural transformations but not necessary natural isomorphisms in such a manner that the composite maps

$$
F(V) \; \xrightarrow{\eta(F(V))} \; (FGF)(V) \; \xrightarrow{F(\theta(V))} \; F(V)
$$

$$
G(W) \; \xrightarrow{G(\eta(W))} \; (GFG)(W) \; \xrightarrow{\theta(G(W)))} \; G(W)
\tag{7.4.34}
$$

are identify morphisms for all objects V in $\mathcal{C}$ and W in $\mathcal{D}$.

The product $C = AB$ for objects of categories is defined by the requirement that there exist projection morphisms π_A and π_B from C to A and B and that for any object D and pair of morphisms $f(D \to A)$ and $g(D \to B)$ there exist morphism $h(D \to C)$ such that one has $f = \pi_A h$ and $g = \pi_B h$. Graphically this corresponds to a square diagram in which pairs A,B and C,D correspond to the pairs formed by opposite vertices of the square and arrows DA and DB correspond to morphisms f and g, arrows CA and CB to the morphisms π_A and π_B and the arrow h to the diagonal DC. Examples of product categories are Cartesian products of topological spaces, linear spaces, differentiable manifolds, groups, etc. The tensor products of linear spaces and algebras provides an especially interesting example of product in the recent case. One can define also more advanced concepts such as limits and inverse limits. Also the notions of sheafs, presheafs, and topos are important.

Tensor categories

Let $\mathcal{C}$ be a category. Tensor product $\otimes$ is a functor from $\mathcal{C} \times \mathcal{C}$ to $\mathcal{C}$ if

a) there is an object $V \otimes W$ associated with any pair (V, W) of objects of $\mathcal{C}$

b) there is an morphism $f \otimes g$ associated with any pair (f, g) of morphisms of $\mathcal{C}$ such that $s(f \otimes g) = s(f) \otimes s(g)$ and $b(f \otimes g) = b(f) \otimes b(g)$,

c) if f' and g' are morphisms such that $s(f') = b(f)$ and $s(g') = b(g)$ then $(f' \otimes g') \circ (f \otimes g) = (f' \circ f) \otimes (g' \circ g)$,

d) $id_{V \otimes W} = id_{W \otimes V}$.

Any functor with these properties is called tensor product. The tensor product of vector spaces provides the most familiar example of a tensor product functor.

In figure 7.4.3 the general rules for graphical representations of morphisms are given.

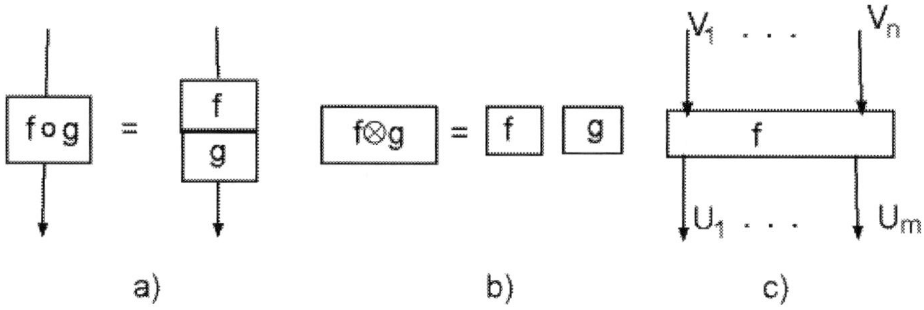

Figure 7.7: The graphical representation of morphisms. a) $g \circ f$: $V \to W$, b) $f \otimes g$, c) $f : U_1 \otimes ... \otimes U_m \to V_1 \otimes .... \otimes V_n$.

An associativity constraint for the tensor product is a natural isomorphism

$$a : \otimes(\otimes \times id) \to \otimes(id \times \otimes) \ .$$

On basis of general definition of natural isomorphisms (see Eq. 7.4.33) one can conclude that for any triple (U, V, W) of objects of $\mathcal{C}$ there exists an isomorphism

$$\begin{array}{ccc} (U \otimes V) \otimes W & \overset{a_{U,V,W}}{\rightarrow} & U \otimes (V \otimes W) \\ \big\downarrow (f \otimes g) \otimes h & & \big\downarrow f \otimes (g \otimes h) \\ (U' \otimes V') \otimes W' & \xrightarrow{a_{U',V',W'}} & U' \otimes (V' \otimes W') \end{array} \tag{7.4.35}$$

Associativity constraints satisfies Pentagon Axiom [ec4] if the following diagrams commutes.

$$\begin{array}{ccc} U \otimes (V \otimes W) \otimes X & \underset{a_{U,V,W} \otimes id_X}{\longleftarrow} & ((U \otimes V) \otimes W)) \otimes X \\ & & \big\downarrow a_{U \otimes V,W,X} \\ \big\downarrow a_{U,V \otimes W,X} & & (U \otimes V) \otimes (W \otimes X) \\ & & \big\downarrow a_{U,V,W \otimes X} \\ U \otimes ((V \otimes W) \otimes X) & \xrightarrow{id_U \otimes a_{V,W,X}} & U \otimes (V \otimes (W \otimes X)) \end{array} \tag{7.4.36}$$

Pentagon axiom has been already mentioned while discussing the definition of quasi-Hopf algebras. In figure 7.4.3 are graphical illustrations of associativity morphism $a(U, V, W)$, Triangle Axiom, and Pentagon Axiom are given.

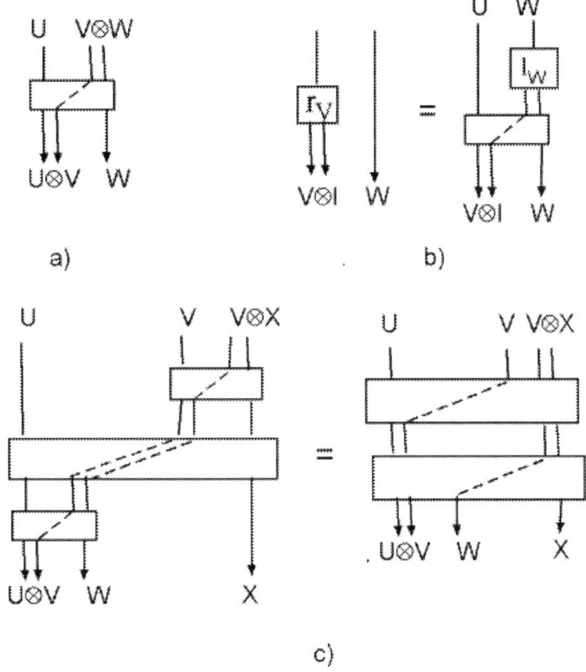

Figure 7.8: Graphical representations of a) the associativity isomorphism $a_{U,V,W}$, b) Triangle Axiom, c) Pentagon Axiom.

Assume that an object I is fixed in the category. A left unit constraint with respect to I is a natural isomorphism

$$l : \otimes(I \times id) \to id$$

By Eq. 7.4.33 this means that for any object V of C there exists an isomorphism

$$l_V : I \otimes V \to V \tag{7.4.37}$$

such that

$$\begin{array}{ccc} I \otimes V & \xrightarrow{\ l_V\ } & V \\ \downarrow id_I \otimes f & & \downarrow f \\ I \otimes V' & \xrightarrow{\ l_{V'}\ } & V' \end{array} \tag{7.4.38}$$

The right unit constraint $r : \otimes(id \times I) \to id$ can be defined in a completely analogous manner.

Given an associativity constraint a, and left and right unit constraints l, r with respect to an object I, one can say that the Triangle Axiom is satisfies if the triangle

$$\begin{array}{ccc} (V \otimes I) \otimes W & \xrightarrow{\ a_{V,I,W}\ } & V \otimes (I \otimes W) \\ & \searrow{\scriptstyle r_V \otimes id_W} \quad \swarrow{\scriptstyle id_W \otimes l_W} & \\ & V \otimes W & \end{array} \tag{7.4.39}$$

commutes (see figure 7.4.3).

These ingredients lead allow to define tensor category $(\mathcal{C}, I, a, l, r)$ as a category $\mathcal{C}$ which is equipped with a tensor product $\otimes : \mathcal{C} \times \mathcal{C} \to \mathcal{C}$ satisfying associativity constraint a, left unit constraint l and right unit constraint r with respect to I, such that Pentagon Axiom and Triangle Axiom are satisfied.

The definition of a tensor functor $F : \mathcal{C} \to \mathcal{D}$ involves also additional isomorphisms. $\phi_0 : I \to F(I)$ satisfies commutative diagrams involving right and left unit constraints l and r. The family of isomorphisms

$$\phi_2(U, V) : F(U) \otimes F(V) \to F(U \otimes V)$$

satisfies a commutative diagram stating that ϕ_2 commutes with associativity constraints. The interested reader can consult [ec4] for details. One can also define the notions of natural tensor transformation, natural tensor isomorphism, and tensor equivalence between tensor categories by applying the general category theoretical tools.

Keeping track of associativity isomorphisms is obviously a rather heavy burden. Fortunately, it can be shown that one can assign to a tensor category $\mathcal{C}$ a strictly associative (or briefly, strict) tensor category which is tensor equivalent of $\mathcal{C}$.

Braided tensor categories

Braided tensor categories satisfy also commutativity constraint c besides associativity constraint a. Denote by $\tau : \mathcal{C} \times \mathcal{C} \to \mathcal{C} \times \mathcal{C}$ the flip functor defined by $\tau(V, W) = (W, V)$. Commutativity constraint is a natural isomorphism

$$c : \otimes \to \otimes \tau \ .$$

This means that for any pair (V, W) of objects there exists isomorphism

$$c_{V,W} : V \otimes W \to W \otimes V$$

such that the square

$$
\begin{array}{ccc}
V \otimes W & \xrightarrow{c_{V,W}} & W \otimes V \\
\downarrow{f \otimes g} & & \downarrow{g \otimes f} \\
V' \otimes W' & \xrightarrow{c_{V',W'}} & W' \otimes V'
\end{array}
\tag{7.4.40}
$$

commutes.

The commutativity constraint satisfies Hexagon Axiom if the two hexagonal diagrams

(H1)

$$
\begin{array}{ccc}
U \otimes (V \otimes W) & \xrightarrow{c_{U,V \otimes W}} & (V \otimes W) \otimes U \\
\uparrow{a_{U,V,W}} & & \downarrow{a_{V,W,U}} \\
(U \otimes V) \otimes W & & V \otimes (W \otimes U) \\
\downarrow{c_{U,V} \otimes id_W} & & \uparrow{id_V \otimes c_{U,W}} \\
(V \otimes U) \otimes W & \xrightarrow{a_{V,U,W}} & V \otimes (U \otimes W)
\end{array}
\tag{7.4.41}
$$

and (H2)

$$\begin{array}{ccc}
(U \otimes V) \otimes W & \xrightarrow{\;c_{U \otimes V, W}\;} & W \otimes (U \otimes V) \\
\big\uparrow a^{-1}_{U,V,W} & & \big\downarrow a^{-1}_{W,U,V} \\
U \otimes (V \otimes W) & & (W \otimes U) \otimes V \\
\big\downarrow id_U \otimes c_{V,W} & c_{U,W} \otimes id_V \big\uparrow & \\
U \otimes (W \otimes V) & \xrightarrow{\;a^{-1}_{U,W,V}\;} & (U \otimes W) \otimes V
\end{array}$$

$$(7.4.42)$$

commute.

The braiding operation $c_{V,W}$ and the association operation $a(U, V, W)$, and pentagon and hexagon axioms are illustrated in the figure 7.4.3 below.

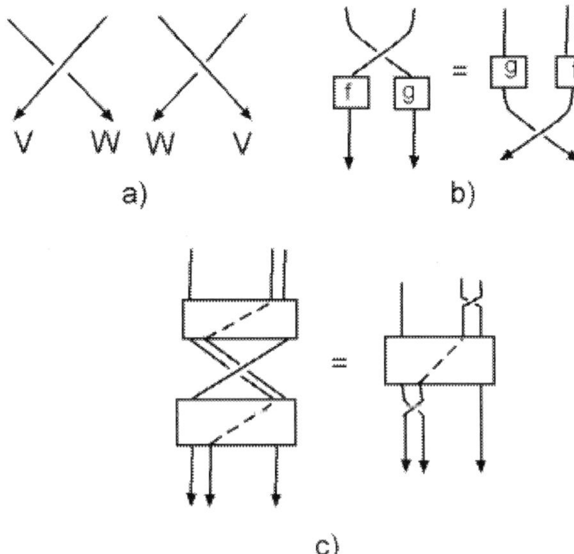

Figure 7.9: Graphical representations a) of the braiding morphism $c_{V,W}$ and its inverse $c^{-1}_{V,W}$, b) of naturality of $c_{V,W}$, c) of First Hexagon Axiom.

Duality and tensor categories

The notion of a dual of the finite-dimensional vector space as a space of linear maps from V to field k can lifted to a concept applying for tensor category. A strict (strictly associative) tensor category $(\mathcal{C}, \otimes, I)$ with unit object I is said to possess left duality if for each object V of $\mathcal{C}$ there exists an object $V^\star$ and morphisms

$$b_V : I \to V \otimes V^\star \quad \text{and} \quad d_V : V^\star \otimes V \to I$$

such that

$$(id \otimes d_V)(b_V \otimes id_V) = id_V \quad \text{and} \quad (d_V \otimes id_{V^\star})(id_{V^\star} \otimes b_V) = id_{V^\star} \; . \qquad (7.4.43)$$

One can define the transpose of f in terms of b_V and d_V. The idea how this is achieved is obvious from figure 7.4.3.

$$f^* = (d_V \otimes id_{U^*})(id_{V^*} \otimes f \otimes id_{U^*})(id_{V^*} \otimes b_U) \ . \tag{7.4.44}$$

Also the braiding operation $c_{V^*,W}$ can be expressed in terms of $c_{V,W}^{-1}$, b_V and d_V by using the isotopy of Fig. 7.4.3:

$$c_{V^*,W} = (d_V \otimes id_{W \otimes V^*})(id_{V^*} \otimes c_{V,W}^{-1} \otimes id_{V^*})(id_{V^* \otimes W} \otimes b_V) \ . \tag{7.4.45}$$

Drinfeld quantum double can be regarded as a tensor product of Hopf algebra and its dual and in this case one can introduce morphisms $ev_H : H \otimes H^* \to k$ defined as $e^i \otimes e_j \to \delta_j^i$ defining inner product and its inverse $\delta : k \to H \otimes H$ defined as $1 \to e^i e_i$, where summation over i is understood. For categories these morphisms are generalized to morphism d_V from objects V of category to unit object I and b_V from I to object of category. The elements of H and H^* are described as strands with opposite directions, whereas d_V and b_V correspond to annihilation and creation of strand–anti-strand pair as show in figure 7.4.3.

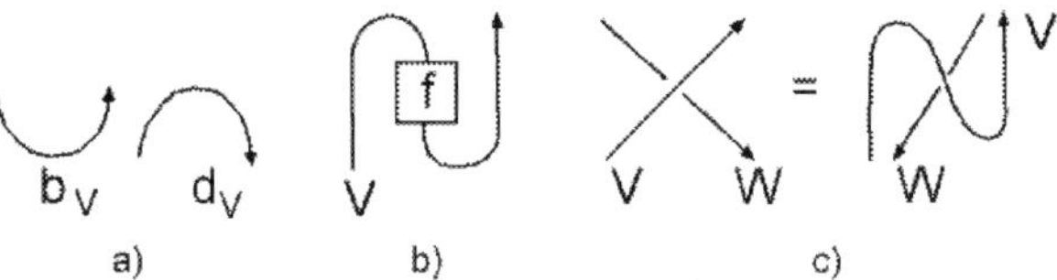

Figure 7.10: Graphical representations a) of the morphisms b_V and d_V, b) of the transpose f^*, c) of braiding operation $c_{V^*,W}$ expressed in terms of $c_{V,W}$.

Ribbon categories

According to the definition of [ec4] ribbon category is a strict braided tensor category $(\mathcal{C}, \otimes, I)$ with a left duality with a family of natural morphisms $\theta_V : V \to V$ indexed by the objects V of $\mathcal{C}$ satisfying the conditions

$$\begin{aligned}
\theta_{V \otimes W} &= \theta_V \otimes \theta_W c_{W,V} c_{V,W} \ , \\
\theta_{V^*} &= (\theta_V)^*
\end{aligned} \tag{7.4.46}$$

for all objects V, W of $\mathcal{C}$. The naturality of twist means for for any morphisms $f : V \to W$ one has $\theta_W f = f \theta_V$. The graphical representation for the axioms and is in Fig. 7.4.3.
The existence of the twist operation provides $\mathcal{C}$ with right duality necessary in order to define trace (see Fig. 7.4.3).

$$\begin{aligned}
d_V' &= (id_{V^*} \otimes \theta_V) c_{V,V^*} b_V \ , \\
b_V' &= d_V c_{V,V^*} (\theta_V \otimes id_{V^*}) \ .
\end{aligned} \tag{7.4.47}$$

One can define quantum trace for any endomorphisms f of ribbon category:

$$tr_q(f) = d_V'(f \otimes id_{V^*}) b_V = d_V c_{V,V^*}(\theta_V f \otimes id_{V^*}) b_V \ . \tag{7.4.48}$$

Again the graphical representation is the best manner to understand the definition, see figure 7.4.3. Quantum trace has the basic properties of trace: $tr_q(fg) = tr_q(gf)$, $tr_q(f \otimes g) = tr_q(f) tr_q(g)$, $tr_q(f) = tr_q(f^*)$. The proof of these properties is easiest using isotropy principle.

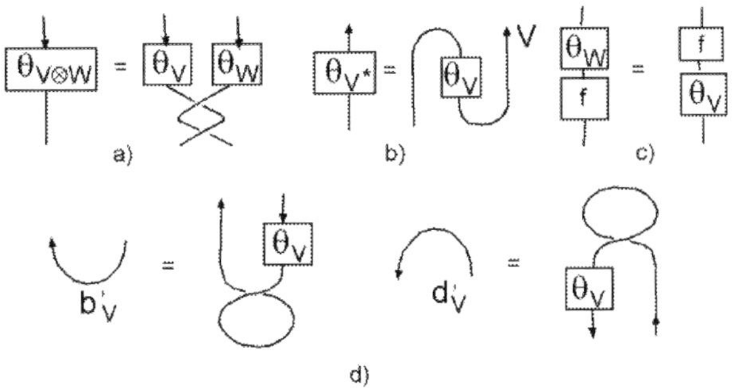

Figure 7.11: Graphical representations a) of $\theta_{V\otimes W} = \theta_V \otimes \theta_W c_{W,V} c_{V,W}$, b) of $\theta_{V^*} = (\theta_V)^*$, c) of $\theta_W f = f\theta_V$, d) of right duality for a ribbon category.

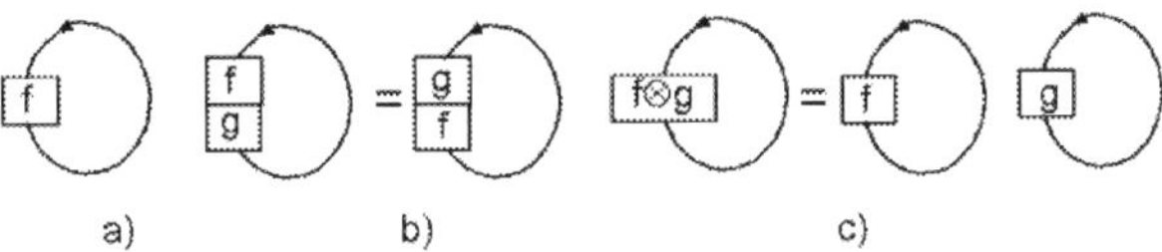

Figure 7.12: Graphical representations of a) $tr_q(f)$, b) of $tr_q(fg) = tr_q(gf)$, c) of $tr(f\otimes g) = tr(f)tr(g)$.

The quantum dimension of an object V of ribbon category can be defined as the quantum trace for the identity morphism of V: $dim_q(V) = tr_q(id_V) = d'_V b_V$. Quantum dimension is represented as a vacuum bubble. Quantum dimension satisfies the conditions $dim_q(V \otimes W) = dim_q(V)dim_q(W)$ and $dim_q(V) = dim_q(V^*)$.

A more general definition of ribbon category inspired by the considerations of [ee2] is obtained by allowing the generalization of morphisms μ and Δ so that they become morphisms $\mu_{A\otimes B\to C}$ and $\Delta_{C\to A\otimes B}$ of ribbon category. Graphically the general morphism with arbitrary number of incoming outgoing strands can be represented as a box or "coupon". An important special case of ribbon categories consists of modules over braided Hopf algebras allowing ribbon algebra structure.

7.5 Axiomatic approach to S-matrix based on the notion of quantum category

This section can be regarded as an attempt of a physicists with some good intuitions and intentions but rather poor algebraic skills to formulate basic axioms about S-matrix in terms of what might be called quantum category. The basic result is an interpretation for the equivalence of loop diagrams with tree diagrams as a consequence of basic algebra and co-algebra axioms generalized to the level of tensor category. The notion of quantum category emerges naturally as a generalization of ribbon category, when algebra product and co-algebra product are interpreted as morphisms between different objects of the ribbon category.

The general picture suggest that the operations Δ and μ generalized to algebra homomorphisms $A \to B\otimes C$ and $A\otimes B \to C$ in a tensor category whose objects are either representations of an algebra or even algebras might provide an appropriate mathematical tool for saying something interesting about S-matrix in TGD Universe. These algebras need not necessarily be bi-algebras. In the following it is demonstrated that the equivalence of loop diagrams to tree diagrams follows from suitably generalized bi-algebra axioms. Also the interpretation of various morphisms involved with Hopf algebra structure is discussed.

7.5.1 Δ and μ and the axioms eliminating loops

The first task is to find a physical interpretation for the basic algebraic operations and how the basic algebra axioms might allow to eliminate loops. The physical interpretation of morphisms Δ and μ as algebra or category morphisms has been already discussed. As already found, the condition that Δ (μ) acts as an algebra (co-algebra) morphism leads to a condition stating that a box graph for 2-particle scattering is equivalent with tree graph. It is interesting to identify the corresponding conditions in the case of self energy loops and vertex corrections.

The condition

$$\mu_{B\otimes C\to A} \circ \Delta_{A\to B\otimes C} \;=\; K\times id_A \;, \tag{7.5.1}$$

where K is a numerical factor, is a natural additional condition stating that a line with a self energy loop is equivalent with a line without the loop. The condition is illustrate in figure 7.5.1. For the co-commutative tensor algebra $T(V)$ of vector space with $\Delta(x) = 1\otimes x + x\otimes 1$ one would have $K = 2$ for the generators of $T(V)$. For a product of n generators one has $K = 2^n$.

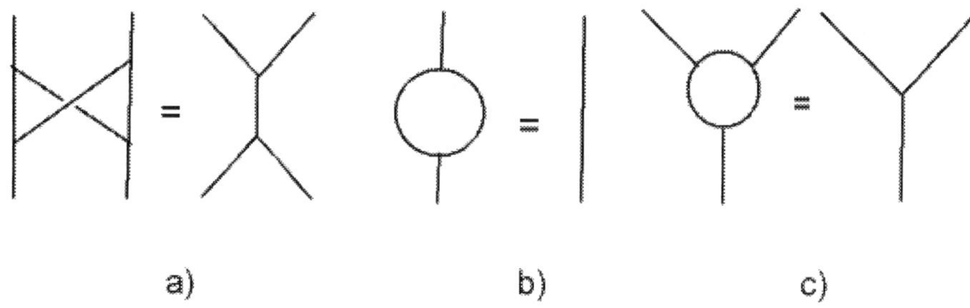

a)   b)   c)

Figure 7.13:   Graphical representations for the conditions a) $(id \otimes \mu \otimes id)(\Delta \otimes \Delta) = \Delta \circ \mu$, b) $\mu_{B\otimes C\to A} \circ \Delta_{A\to B\otimes C} = K \times id_A$, and c) $(\mu \otimes id) \circ (\Delta \otimes id) \circ \Delta = K \times \Delta$.

The condition $\Delta_{A\to B\otimes C} \circ \mu_{B\otimes C\to A} = K \times id_A$ cannot hold true since multiplication is not an irreversible process. If this were the case one could reduce tree diagrams to collections of free propagator lines.

In quantum field theories also vertex corrections are a source of divergences. The requirement that the graph representing a vertex correction is equivalent with a simple tree graph representing a decay gives an additional algebraic condition. For bi-algebras the condition would read

$$(\mu \otimes id) \circ (\Delta \otimes id) \circ \Delta = K\Delta \;, \tag{7.5.2}$$

where K is a simple multiplicative factor. In fact, for the co-commutative tensor algebra $T(V)$ of vector space the left hand side would be $3 \times \Delta(x)$ giving $K = 3$ for generators $T(V)$. The condition is illustrated in figure 7.5.1.

Using the standard formulas of appendix for quantum groups one finds that in the case of $U_q(sl(2))$ the condition $\mu \circ \Delta(X) = K_X X$, K_X constant, is not true in general. Rather, one has $\mu \circ \Delta(X) = X K_X(q^{H/2} + q^{-H/2}, q^{1/2}, q^{-1/2})$. The action on the vacuum state is however proportional to that of X, being given by $K_X(2,1,1)X$. The function K_X for a given X can be deduced from $\mu \circ \Delta(X_\pm) = q^{H/2}X_\pm + X_\pm q^{-H/2} = X_\pm(q^{\pm 1/2} + q^{H/2} + q^{-H/2})$. The eigen states of Cartan algebra generators are expected to be eigen states of $\mu \circ \Delta$ also in the case of a general quantum group. $\mu \circ \Delta$ is analogous to a single particle operator like kinetic energy and its action on multi-particle state is a sum over all tensor factors with $\mu \circ \Delta$ applied to each of them. For eigen states of $\mu \circ \Delta$ the projective equivalence of loop diagrams with tree diagrams would make sense.

Since self energy loops, vertex corrections, and box diagrams represent the basic divergences of renormalizable quantum field theories, these axioms raise the hope that the basic infinities of quantum field theories could be eliminated by the basic axioms for the morphisms of quantum category.

There are also morphisms related to the topology changes in which the 3-surface remains connected. For instance, processes in which the number of boundary components can change could be of special relevance if the family replication phenomenon reduces to the boundary topology. Also 3-topology can change. The experience with topological quantum field theories [ec3], stimulates the hope that the braid group representations of the topological invariants of 3-topology might be of help in the construction of S-matrix.

The equivalence of loop diagrams with tree diagrams must have algebraic formulation using the language of standard quantum field theory. In the third section it was indeed found that thanks to the presence of the emission of vacuons, the equivalence of loop diagrams with tree diagrams corresponds to the vanishing of loop corrections in the standard quantum field theory framework. Furthermore, the non-cocommutative Hopf algebra of Feynman diagrams discussed in [ed2] becomes co-commutative when the loop corrections vanish so that TGD program indeed has an elegant algebraic formulation also in the standard framework.

7.5.2   The physical interpretation of non-trivial braiding and quasi-associativity

The exchange of the tensor factors by braiding could also correspond to a physically non-trivial but unitary operation as it indeed does in anyon physics [ee6, ee5]. What would differentiate between elementary particles and anyons would be the non-triviality of the super-canonical and Super Kac-Moody conformal central extensions which have the same origin (addition of a multiplication by a multiple of the Hamiltonian of a canonical transformation to the action of isometry generator). The proposed interpretation of braiding acting in the complex plane in which the conformal weights of the elements of the super-canonical algebra represent punctures justifies the non-triviality. Hexagon Axioms would state that two generalized Feynman diagrams involving exchanges, dissociations and re-associations are equivalent.

An interesting question is whether the association $(A, B) \rightarrow (A \otimes B)$ could be interpreted as a formation of bound state entanglement between A and B. A possible space-time correlate for association is topological condensation of A and B to the same space-time sheet. Association would be trivial if all particles are at same space-time sheet X^4 but non-trivial if some subset of particles condense at an intermediate space-time sheet Y^4 condensing in turn at X^4.

Be as it may, association isomorphisms $a_{A,B,C}$ would state that the state space obtained by binding A with bound bound states $(B \otimes C)$ is unitarily related with the state space obtained by binding $(A \otimes B)$ bound states with C. With this interpretation Pentagon axiom would state that two generalized Feynman diagrams depicted in figure 7.4.3 leading from initial to final to final state by dissociation and reassociation are equivalent.

7.5.3   Generalizing the notion of bi-algebra structures at the level of configuration space

Configuration space of 3-surfaces decomposes into sectors corresponding to different 3-topologies. Also other signatures might be involved and I have proposed that the sectors are characterized by the collection of p-adic primes labelling space-time sheets of the 3-surface and that a given space-time surface could be characterized by an infinite prime or integer. The general problem is to continue various geometric structures from a given sector A of configuration space to other sector B.

An especially interesting special case corresponds to a continuation from 1-particle sector to two-particle sector or vice versa and corresponds to TGD variant of 3-vertex. All these continuations involve the imbedding of a structure associated with the sector A to a structure associated with sector B. For the continuation from 1-particle sector to 2-particle sector the map is analogous to co-algebra homomorphism Δ. For the reverse continuation it is analogous to the algebra product μ. Now however one does not have maps $\Delta : A \rightarrow A \otimes A$ and $\mu : A \otimes A \rightarrow A$ but $\Delta : A \rightarrow B \otimes C$ and $\mu : B \otimes C \rightarrow A$ unless the algebras are isomorphic. $\mu \circ \Delta = id$ should hold true as an additional condition but $\Delta \circ \mu = id$ cannot hold true since product maps many pairs to the same element.

Continuation of the configuration space spinor structure

The basic example of a structure to be continued is configuration space spinor structure. Configuration space spinor fields in different sectors should be related to each other. The isometry generators and gamma matrices of configuration space span a super-canonical algebra. The continuation requires that the super algebra basis of different sectors are related. Also vacua must be related. Isometry generators correspond to bosonic generators of the super-canonical algebra. There is also a natural extension of the super-canonical algebra defined by the Poisson structure of the configuration space.

This view suggests that in the first approximation one could see the construction of S-matrix as following process.

a) Incoming/outgoing states correspond to positive/negative energy states localized to the sectors of configuration space with fixed 3-topologies.

b) In order to construct an S-matrix matrix element between two states localized in sectors A and B, one must continue the state localized in A to B or vice versa and calculate overlap. The continuation involves a sequence of morphisms mapping various structures between sectors. In particular, topological transformations describing particle decay and fusion are possible so that the analogs of product μ and co-product Δ are involved. The construction of three-manifold topological invariants [ec3] in topological quantum field theories provides concrete ideas about how to proceed.

c) The S-matrix element describing a particular transition can be expressed as any path leading from the sector A to B or vice versa. There is a huge symmetry very much analogous to the independence of the final result of the analytic continuation on the path chosen since generalized Feynman graphs allow all moves changing intermediate topologies so that initial and final 3-topologies are same. Generalized conformal invariance probably also poses restrictions on possible paths of continuation. In the path integral approach one would have simply sum over all these equivalent paths and thus encounter the fundamental difficulties related to the infinite-dimensional integration.

d) Quantum classical correspondence suggests that the continuation operation has a space-time correlate. That is, the absolute minimum of Kähler action going through the initial and final 3-sheets defines a sequence of transitions changing the topology of 3-sheet. The localization to a particular sector of course selects particular absolute minimum. There are two possible interpretations. Either the continuation from A is not possible to all possible sectors but only to those with 3-topologies appearing in X^4, or the absolute minimum represents some kind of minimal continuation involving minimum amount of calculational labor.

e) Quantum classical correspondence and the possibility to represent the rows of S-matrix as zero energy quantum states suggests that the paths for continuation can be also represented at the space-time level, perhaps in terms of braided join along boundaries bonds connecting two light like 3-surfaces representing the initial and final states of particle reaction. Since light like 3-surfaces are metrically two-dimensional and allow conformal invariance, this suggests a connection with braid diagrams in the sense that it should be possible to regard the paths connecting sectors of configuration space consisting of unions of disjoint 3-surfaces (corresponding interacting 4-surfaces are connected) as generalized braids for which also decay and fusion for the strands of braid are possible. Quantum algebra structure and effective metric 2-dimensionality of the light like 3-surfaces suggests different braidings for join along boundaries bonds connecting boundaries of 3-surfaces define non-equivalent 3-surfaces.

Co-multiplication and second quantized induced spinor fields

At the microscopic level the construction of S-matrix reduces to understanding what happens for the classical spinor fields in a vertex, which corresponds to an incoming 3-surface A decaying to two outgoing 3-surfaces B and C. At the classical level incoming spinor field A develops into a spinor fields B and C expressible as linear combinations of appropriate spinor basis. At quantum level one must understand how the Fock space defined by the incoming spinor fields of A is mapped to the tensor product of Fock spaces of B and C. The idea about the possible importance of co-algebras came with the realization that this mapping is obviously is very much like a co-product. Co-algebras and bi-algebras possessing both algebra and co-algebra structure indeed suggest a general approach giving hopes of understanding how Feynman diagrammatics generalizes to TGD framework.

The first guess is that fermionic oscillator operators are mapped by the imbedding Δ to a superposition of operators $a^\dagger_{Bn} \otimes Id_C$ and $Id_B \otimes a^\dagger_{Cn}$ with obvious formulas for Hermitian conjugates. Δ induces the mapping of higher Fock states and the construction of S-matrix should reduce to the construction of this map.

Δ is analogous to the definition for co-product operation although there is also an obvious difference due to the fact that Δ imbeds algebra A to $B \otimes C$ rather than to $A \otimes A$. Only in the case that the algebras are isomorphic, the situation reduces to that for Hopf algebras. Category theoretical approach however allows to consider a more general situation in which Δ is a morphism in the category of Fock algebras associated with 3-surfaces.

Δ preserves fermion number and should respect Fock algebra structure, in particular commute with the anti-commutation relations of fermionic oscillator operators. The basis of fermionic oscillator operators would naturally correspond to fermionic super-canonical generators in turn defining configuration space gamma matrices.

Since any leg can be regarded as incoming leg, strong consistency conditions result on the coefficients in the expression

$$\Delta(a^{\dagger}_{An}) \;=\; C(A,B)_n{}^m a^{\dagger}_{Bm} \otimes Id_C + C(A,C)_n{}^m Id_B \otimes a^{\dagger}_{Cm} \tag{7.5.3}$$

by forming the cyclic permutations in A, B, C. This option corresponds to the co-commutative situation and quantum group structure. If identity matrices are replaced with something more general, co-product becomes non-cocommutative.

7.5.4 Ribbon category as a fundamental structure?

There exists a generalization of the braided tensor category inspired by the axiomatic approach to topological quantum field theories which seems to almost catch the proposed mathematical requirements. This category is also called ribbon category [ec2] but in more general sense than it is defined in [ec4].

One adds to the tangle diagrams (braid diagrams with both directions of strands and possibility of strand–anti-strand annihilation) also "coupons", which are boxes representing morphisms with arbitrary numbers of incoming and outgoing strands. As a special case 3-particle vertices are obtained. The strands correspond to representations of a fixed Hopf algebra H.

In the recent case it would seem safest to postulate that strands correspond to algebras, which can be different because of the potential dependence of the details of Fock algebra on 3-topology and other properties of 3-surface. For instance, configuration space metric defined by anti-commutators of the gamma matrices is degenerate for vacuum extremals so that the infinite Clifford algebra is definitely "smaller" than for surfaces with $D \geq 3$-dimensional CP_2 projection.

One might feel that the full ribbon algebra is an un-necessary luxury since only 3-particle vertices are needed since higher vertices describing decays of 3-surfaces can be decomposed to 3-vertices in the generic case. On the other hand, many-sheeted space-time and p-adic fractality suggest that coupons with arbitrary number of incoming and outgoing strands are needed in order to obtain the p-adic hierarchy of length scale dependent theories.

The situation would be the same as in the effective quantum field theories involving arbitrarily high vertices and would require what might be called universal algebra allowing n-ary multiplications and co-multiplications rather than only binary ones. Also strands within strands hierarchy is strongly suggestive and would require a fractal generalization of the ribbon algebra. Note that associativity and commutativity conditions for morphisms which more than three incoming and outgoing lines would force to generalize the notion of R-matrix and would bring in conditions stating that more complex loop diagrams are equivalent with tree diagrams.

7.5.5 Minimal models and TGD

Quaternion conformal invariance with non-vanishing c and k for anyons is highly attractive option and minimal super-conformal field theories attractive candidate since they describe critical systems and TGD Universe is indeed a quantum critical system.

Rational conformal field theories and TGD

The highest weight representations of Virasoro algebra are known as Verma modules containing besides the ground state with conformal weight Δ the states generated by Virasoro generators L_n, $n \geq 0$. For some values of Δ Verma module contains states with conformal weight $\Delta + l$ annihilated by Virasoro generators L_n, $n \geq 1$. In this case the number of primary fields is reduced since Virasoro algebra acts as a gauge algebra. The conformal weights Δ of the Verma modules allowing null states are given by the Kac formula

$$\Delta_{mm'} = \Delta_0 + \frac{1}{4}(\alpha_+ m + \alpha_- m')^2 \ , \quad m, m' \in \{1, 2...\} \ , \tag{7.5.4}$$

$$\Delta_0 = \frac{1}{24}(c - 1) \ ,$$

$$\alpha_\pm = \frac{\sqrt{1 - c} \pm \sqrt{25 - c}}{\sqrt{24}} \ . \tag{7.5.5}$$

The descendants $\prod_{n \geq 1} L_n^{k_n} |\Delta\rangle$ annihilated by L_n, $n > 0$, have conformal weights at level $l = \sum_n n k_n = mm'$.

In the general case the operator products of primary fields satisfying these conditions form an algebra spanned by infinitely many primary fields. The situation changes if the central charge c satisfies the condition

$$c = 1 - \frac{6(p' - p)^2}{pp'} \ , \tag{7.5.6}$$

where p and p' are mutually prime positive integers satisfying $p < p'$. In this case the Kac weights are rational

$$\Delta_{m,m'} = \frac{(mp' - m'p)^2 - (p' - p)^2)}{4pp'} \ , \quad 0 < m < p \ , \quad 0 < m' < p' \ . \tag{7.5.7}$$

Obviously, the number of primary fields is finite. This option does not seem to be realistic in TGD framework were super-conformal invariance is realized.

For $N = 1$ super-conformal invariance the unitary representations have central extension and conformal weights given by

$$c = \frac{3}{2}\left(1 - \frac{8}{m(m + 2)}\right) \ ,$$

$$\Delta_{p,q}(NS) = \frac{[(m + 2)p) - mq)]^2 - 4}{8m(m + 2)} \ , \quad 0 \leq p \leq m \ , \quad 1 \leq q \leq m + 2 \ . \tag{7.5.8}$$

For Ramond representations the conformal weights are

$$\Delta_{p,q}(R) = \Delta(NS) + \frac{1}{16} \ . \tag{7.5.9}$$

The states with vanishing conformal weights correspond to light elementary particles and the states with $p = q$ have vanishing conformal weight in NS sector. Also this option is non-realistic since in TGD framework super-generators carry fermion number so that G cannot be a Hermitian operator.

$N = 2$ super-conformal algebra is the most interesting one from TGD point of view since it involves also a bosonic $U(1)$ charge identifiable as fermion number and $G^\pm(z)$ indeed carry $U(1)$ charge[1]. Hence one has $N = 2$ super-conformal algebra is generated by the energy momentum tensor $T(z)$, $U(1)$ current $J(z)$, and super generators $G^\pm(z)$. $U(1)$ current would correspond to fermion number and super generators would involve contraction of covariantly constant neutrino spinor with second quantized induced spinor field. The further facts that $N = 2$ algebra is associated naturally with Kähler geometry, that the partition functions associated with $N = 2$ super-conformal representations

---

[1]I realized that TGD super-conformal algebra corresponds to $N = 2$ algebra while writing this and proposed it earlier as a generalization of super-conformal algebra!

are modular invariant, and that $N = 2$ algebra defines so called chiral ring defining a topological quantum field theory [ee2], lend further support for the belief that $N = 2$ super-conformal algebra acts in super-canonical degrees of freedom.

The values of c and conformal weights for $N = 2$ super-conformal field theories are given by

$$
\begin{aligned}
c &= \frac{3k}{k+2} \ , \\
\Delta_{l,m}(NS) &= \frac{l(l+2) - m^2}{4(k+2)} \ , \quad l = 0, 1, ..., k \ , \\
q_m &= \frac{m}{k+2} \ , \quad m = -l, -l+2, ..., l-2, l \ .
\end{aligned}
\qquad (7.5.10)
$$

q_m is the fractional value of the $U(1)$ charge, which would now correspond to a fractional fermion number. For $k = 1$ one would have $q = 0, 1/3, -1/3$, which brings in mind anyons. $\Delta_{l=0,m=0} = 0$ state would correspond to a massless state with a vanishing fermion number. Note that $SU(2)_k$ Wess-Zumino model has the same value of c but different conformal weights. More information about conformal algebras can be found from the appendix of [ee2].

For Ramond representation $L_0 - c/24$ or equivalently G_0 must annihilate the massless states. This occurs for $\Delta = c/24$ giving the condition $k = 2 \left[l(l+2) - m^2 \right]$ (note that k must be even and that $(k, l, m) = (4, 1, 1)$ is the simplest non-trivial solution to the condition). Note the appearance of a fractional vacuum fermion number $q_{vac} = \pm c/12 = \pm k/4(k+2)$. I have proposed that NS and Ramond algebras could combine to a larger algebra containing also lepto-quark type generators.

Quaternion conformal invariance [B4] encourages to consider the possibility of super-symmetrizing also spin and electro-weak spin of fermions. In this case the conformal algebra would extend to a direct sum of Ramond and NS $N = 8$ algebras associated with quarks and leptons. This algebra in turn extends to a larger algebra if lepto-quark generators acting as half odd-integer Virasoro generators are allowed. The algebra would contain spin and electro-weak spin as fermionic indices. Poincare and color Kac-Moody generators would act as symplectically extended isometry generators on configuration space Hamiltonians expressible in terms of Hamiltonians of $X_l^3 \times CP_2$. Electro-weak and color Kac-Moody currents have conformal weight $h = 1$ whereas T and G have conformal weights $h = 2$ and $h = 3/2$.

The experience with $N = 4$ super-conformal invariance suggests that the extended algebra requires the inclusion of also second quantized induced spinor fields with $h = 1/2$ and their super-partners with $h = 0$ and realized as fermion-antifermion bilinears. Since G and Ψ are labelled by 2×4 spinor indices, super-partners would correspond to $2 \times (3+1) = 8$ massless electro-weak gauge boson states with polarization included. Their inclusion would make the theory highly predictive since induced spinor and electro-weak fields are the fundamental fields in TGD.

In TGD framework both quark and lepton numbers correspond to NS and Ramond type representations, which in conformal field theories can be assigned to the topologies of complex plane and cylinder. This would suggest that a given three-surface allows either NS or Ramond representation and is either leptonic or quark like but one must be very cautious with this kind of conclusion. Interestingly, NS and Ramond type representations allow a symmetry acting as a spectral flow in the indices of the generators and transforming NS and Ramond type representations continuously to each other [ee2]. The flow acts as

$$
\begin{aligned}
L_n &\rightarrow L_n + \alpha J_n + \frac{c}{6} \alpha^2 \delta_{n,0} \\
J_n &\rightarrow J_n + \frac{c}{3} \alpha \delta_{n,0} \ , \\
G_n^{\pm} &\rightarrow G_{n \pm \alpha}^{\pm} \ .
\end{aligned}
\qquad (7.5.11)
$$

The choice $\alpha = \pm 1/2$ transforms NS representation to Ramond representation. The idea that leptons could be transformed to quarks in a continuous manner does not sound attractive in TGD framework. Note that the action of Super Kac-Moody Virasoro algebra in the space of super-canonical conformal weights can be interpreted as a spectral flow.

Co-product for Super Kac-Moody and Super Virasoro algebras

By the previous considerations the quantized conformal weights z_1, z_2, z_3 of super-canonical generators defining punctures of 2-surface should correspond to line punctures of 3-surface. One cannot avoid the thought that these line punctures should meet at single point so that three-vertex would have also quantum field theoretical interpretation.

Each point z_k corresponds to its own Virasoro algebra $V_k = \{L_n^{z_k)}\}$ and Kac-Moody algebra $J_k = \{J_n^{z_k)}\}$ defined by Laurent series of $T(z)$ and $J(z)$ at z_k. Also super-generators are involved. To minimize notational labor denote by $X_n^{z_k)}$, $k = 1, 2, 3$ the generators in question.

The co-algebra product for Super-Virasoro and Super-Kac-Moody involves in the case of fusion $A_1 \otimes A_2 \to A_3$ a co-algebra product assigning to the generators $X_n^{z_3)}$ direct sum of generators of $X_k^{z_1)}$ and $X_l^{z_2)}$. The most straightforward approach is to express the generators $X_n^{z_3)}$ in terms of generators $X_k^{z_1)}$ and $X_l^{z_2)}$. This is achieved by using the expressions for generators as residy integrals of energy momentum tensor and Kac Moody currents. For Virasoro generators this is carried out explicitly in [ee2]. The resulting co-product conserves the value of central extension whereas for the naive co-product this would not be the case. Obviously, the geometric co-product does not conserve conformal weight.

7.6   RGI: gauge symmetry or symmetry at fixed point?

The notion of renormalization group invariance (RGI) has a slightly different content in quantum field theories and in TGD framework. This allows to understand what goes wrong with ordinary Feynman diagrammatics and why the renormalization procedure combined with experimental input still works.

7.6.1   How renormalization group invariance and p-adic topology might relate?

Renormalization group invariance in the standard sense would suggest that propagators, vertices, and in fact all generalized Feynman diagrams D, have a vanishing logarithmic derivative with respect to the parameter λ:

$$\lambda \frac{d}{d\lambda} D \;=\; 0 \;. \tag{7.6.1}$$

All parameters are functions of λ and the interpretation would be as a gauge symmetry. The equation holds true for on mass shell external momenta. In TGD context this equation is not expected to hold true generally and only determines the values of λ, which correspond to the fixed points of renormalization group evolution at which effective gauge symmetry property holds true. The critical of λ should correspond to p-adic length scales. This interpretation allows to assign different values of λ to different particles.

p-Adicization suggests another manner to see the situation. The requirement that $X \equiv \lambda^2$ is well defined both in the real and p-adic sense implies that λ^2 is a rational number. This means a discretization. p-Adicization poses a cutoff so that X must be integer. The prediction that these values correspond to stationary points of $\lambda d/d\lambda$ in the real context is a strong constraint on the algebraic structure. It is also expected to lead to the quantization of masses and various dimensionless coupling constants. Formally D as a function of a discrete variable $X = N$ is pseudo constant in the sense that the derivative vanishes everywhere unlike in the real context and quantum criticality holds true in a stronger sense.

The primes p dividing $X = N$ are in a preferred position since the p-adic norm of X is in this case smaller than one. This is in fact necessary for the series $\sum_{n>1} D^{-2n}$, $D = (p^2 - m^2)/\lambda^2$ appearing in the expression for the propagator to converge p-adically when $p^2 - m^2$ has p-adic norm equal to unity. Thus p-adic topologies R_p for which p divides X are favored. Unless this is the case $p^2 - m^2$ must have p-adic norm smaller than one and the particle would be non-relativistic in the p-adic sense.

7.6.2   How generalized Feynman diagrams relate to tangles with chords?

The work of Vassilev, Kontsevich and other mathematicians related to the universal knot invariant of finite type [ec4] generalizes the notion of knot invariant so that it makes sense also for singular knots having double points as self-intersections. The invariant of the singular knot is obtained by replacing singular points with differences of two possible crossings resulting in de-singularization so that ordinary knots result.

1. Tangles with chords as a tool to produce universal knot invariant

The construction of knot invariants for singular knots involves chord diagrams as an auxiliary tool. Chord diagrams are circles containing m pairs of points connected by chords. The immersions of circles defining chord diagrams to R^3 define singular knots resulting when the chords are contracted effectively to points by mapping their end points to a single point of R^3. Vacuum Feynman diagrams can be seen as the physical analogs of the chord diagrams associated with circle.

Chords can be attached also to a tangles, whose endpoints defining their boundaries are fixed and instead of circle the strands of the tangle contain the 3-vertices. The physical analogy of chords would be as particle exchanges allowing only 3-vertices at strands of the tangle but not elsewhere. Tensor product and concatenation of the tangles are well defined operations also in the presence of chords. Chord diagrams associated with tangles allow Hopf algebra structure graded by the number c of chords (c relates to the number L of loops by $c = L - 1$ for chord diagrams associated with circles). The tangles A and B such that the source of A is target of B can be fused to tangle $A \circ B$ and this induces a product $\mu : A \otimes B \to A \circ B$ of the chord diagrams. The co-commutative co-product $\Delta(D)$ of chorded tangle D is obtained by forming the sum of the tensor products $D' \otimes D''$ of all sub-diagrams $D' \subset D$ and their complements D''.

2. How tangles with chords relate to generalized Feynman diagrams?

It is interesting to see how tangles with chords could relate to generalized Feynman diagrams whose symmetries should code the basic axioms of the underlying algebraic structure expressing in turn quantum criticality.

a) The structure of generalized Feynman graphs differs from that of tangles with chords in the sense that also internal 3-vertices are allowed. The imbedding of the tangles to singular braids does not make sense since the mapping of the end points of the chords of a tree diagram to single point would transform tree diagrams to diagrams having no 3-vertices. Neither does the difference $K_+ - K_-$ of knot diagrams have any obvious interpretation in terms of Feynman diagrams. This does not however exclude the possibility of mapping of the diagrams to *non-singular* braids in R^3.

b) The expression for quantum criticality as a symmetry is the free motion of the end points of the chords along the lines of the tangle. The end points of the chord can be made to co-incide in the final situation if the tangle contains a path connecting the points. In this case a self energy loop with coinciding end points results. It is possible to continue the motion of the second end of the loop so that a 3-vertex with a line containing at its end a bubble necessarily representing vacuum state results, and must correspond to emission of vacuon and can thus be eliminated. Also higher vertices can be decomposed to 3-vertices using the same trick. When the end points of the chord cannot be connected by a path contained by the tangle, chord cannot be eliminated. All loops can be eliminated in this manner so that the outcome is a tree diagram.

c) If knotting would matter, one would obtain an infinite number of non-equivalent diagrams since both internal and external lines could be knotted. It seems that only braiding of the external lines can be allowed. Note that braiding is basically linking and thus represents a "many-particle phenomenon" whereas knotting is "single particle phenomenon" (sub-manifolds X and Y with dimensions $n > 0$ and $D - n - 1$ can become linked whereas knotting occurs for sub-manifolds of dimension $D - 2$).

d) The possibility to move freely the end points of the chords along lines means that the moves are more general than those induced by orientation preserving maps acting on the lines preserving the ordering of the points associated with the lines so that it is not possible to get through vertices. For instance, a non-planar diagram describing a vertex correction can be transformed to a corresponding planar diagram. This might relate to the fact that p-adic numbers are not well-ordered and that the diagrams should make sense also in the p-adic context.

7.6.3   Do standard Feynman diagrammatics and TGD inspired diagrammatics express the same symmetry?

The computational formalism of quantum field theories has enjoyed an enormous success, and this forces to ask whether the Feynman rules of renormalized quantum field theory provide a very cumbersome manner to express essentially the same great idea than TGD view about Feynman diagrams tries to do. If so, the equivalence of loop diagrams with tree diagrams must have an algebraic formulation using the language of the standard quantum field theory. In will be indeed found that, thanks to the presence of the emission of vacuons, the equivalence of loop diagrams with tree diagrams corresponds to the vanishing of loop corrections in the standard quantum field theory framework (vacuum lines have vacuum extremals as classical space-time correlates). Thus there is a transition from TGD based notion of Feynman diagram to that of ordinary Feynman diagram. One might say that generalized diagrams relate to Feynman diagrams like an integral to the integrand which is constant.

The following considerations support the view that there is a delicate difference between the philosophies behind the two approaches. In the standard quantum field theory approach scalings act as gauge symmetries whereas in TGD they are analogous to isometries at a fixed point by quantum criticality. p-Adic length scales define the preferred length scales associated with the fixed points. An effective reduction to gauge symmetry however occurs at the fixed points.

1. The Hopf and Lie algebras associated with Feynman diagrams

The work of Connes and Kreimer [ed2] provides support for the vision declared above. Connes and Kreimer Connes and Kreiman accept quantum field theory as such and generalize the Hopf algebra and Lie algebra structures associated with diagrams defined by tangles with chords so that they apply to Feynman diagrams. The enveloping algebra of Lie-algebra of Feynman diagrams can be identified as the dual of the corresponding Hopf algebra and assigns numerical expressions to Feynman graphs. The Lie algebra exponentiates to an infinite-dimensional group G. TGD approach can be seen as a diametrical opposite of this approach: the idea is to replace Feynman diagrammatics with chorded tangle diagrammatics with quantum criticality as the fundamental symmetry allowing to eliminate loops.

The outcome is a systematic and very elegant representation of the dimensional regularization procedure. The complex sphere S^2 defined by a complexified space-time dimension D emerges naturally since the dimensional regularization involves a small complex curve $D = 4 + \epsilon(z)$ surrounding the point $D = 4$. The curve divides S^2 to two disjoint parts C_+ and C_-. The dependence of the Feynman graph on complexified dimension defines a map $z \to \gamma(z)$ from S^2 to the group G. The Birkhoff decomposition $\gamma(z) = \gamma_-^{-1}(z)\gamma_+(z)$ such that $\gamma_\pm(z)$ is holomorphic in $C_\pm$ allows to extract the finite part of the Feynmann diagram as the value of $\gamma_+(z)$ at $z = D = 4$.

Only one-particle irreducible (1PI) Feynman graphs (not decomposing to two disconnected parts when single line is removed) relevant for the effective action are considered. The commutative product for Feynman graphs is simply the union of graphs whereas co-product involves the decomposition of the graph in various manners to a sub-graph D_1 and its complement D_1^c and summing over all possible tensor products $D \otimes D_1^c$. Graph and complement are allowed to meet only through two or three lines, which relates to the assumption that no higher vertices are allowed in the model considered. A somewhat analogous decomposition appears in Bogoliubov's recursion formula and in Zimmermann's general solution to it using the "forest formula" [fa3] and giving renormalized quantities without any need for regularization. Co-product is not co-commutative.

The Lie-algebra action in the space of 1PI Feynman diagrams has interpretation in terms of insertions and eliminations for Feynman diagrams, the Lie bracket being computable from the insertions of one graph to another one and vice versa. Lie-algebra generators corresponds to linear complex valued functionals $f(X)$ in the space of 1PI Feynman diagrams with obvious product and $\star$ operation $f \star g(X) = \mu \circ (f \otimes g)\Delta(X))$ for the dual algebra is used to define associative product in turn defining the Lie algebra commutator. Elimination means the replacement of a sub-graph with a single vertex graph having the same external lines as the original graph. Insertion is the inverse of this procedure. For instance, a bare vertex can be replaced with a vertex containing some radiative corrections.

2. The Lie-algebra of Feynman diagrams annihilates Green's functions at quantum criticality in TGD Universe

In TGD framework the non-cocommutative Hopf algebra of Feynman diagrams becomes effectively co-commutative at quantum critical point since loop corrections vanish. Also the action of the scaling transformation realized infinitesimally as $\lambda d/d\lambda$ operation leaves the expressions associated with Feynman diagrams invariant only for the *critical* values of λ. In standard renormalization theory the action of the scaling transformation would be a gauge symmetry for *all* values of renormalization parameter λ.

Quite generally, there are two basic manners to realize gauge invariance. One can sum over all gauge equivalent configurations and form an average or one can perform a gauge fixing. The standard renormalization group invariant quantum field theory based on the functional integral and summation over Feynman diagrams could be seen as an attempt to produce renormalization group gauge invariance by simply taking an average over all gauge equivalent Feynman diagrams by using a gauge invariant integration measure. Divergence difficulties could be seen as being due to the fact that this integration measure is un-normalized and gives infinite values so that one must introduce infinite renormalization constants. In TGD based diagrammatics the effective gauge invariance would be realized by picking just single tree diagram from the equivalence class of physically equivalent diagrams, or if standard Feynmann diagrammatics is used, by the vanishing of loop corrections.

7.6.4 How p-adic coupling constant evolution is implied by the vanishing of loops?

Consider next how the equivalence of loop diagrams with tree diagrams can be consistent with the vanishing of loop corrections and how this equivalence could fix the coupling constant evolution.

a) What is new as compared to quantum field theory is that algebraic approach predicts the possibility of vacuum lines to which one assigns an identity operator and vanishing momentum and super-canonical conformal weight. This means that the amplitude associated with a given loop diagram contains terms associated with simpler diagrams with some internal propagator lines eliminated. The equivalence of loop diagrams with tree diagrams would mean that the conservation of conformal weight in vertices allows only identity operator in the loop propagator lines describing exchanged particles so that the reduction to tree diagrams occurs. The contributions with non-trivial propagators should vanish by the requirement that the allowed conformal weights correspond to the zeros of relevant polyzetas. The reduction of loop diagrams to tree diagrams uniquely fix evolution of the vertices as a function λ and thus of p-adic length scale.

b) Consider first the conditions from the reduction of loops with vacuum exchanges to tree diagrams. In the case of a self energy bubble defined by 3-vertex the reduced part of the diagram would correspond to an insertion of a term $V_{vac}GV_{vac}$, where V_{vac} is the vertex for the emission of the vacuum line and G is the propagator. The reduction to a tree diagram gives the condition

$$V_{vac}GV_{vac} \;=\; 1 \; . \tag{7.6.2}$$

The reduction of 3-vertex with triangle to tree-vertex gives

$$(V_{vac}G \otimes V_{vac}G)V = V \; . \tag{7.6.3}$$

The reduction of the box diagram to tree diagram gives

$$(1 \otimes G \otimes 1)(V_{vac}G \otimes 1)V \otimes (1 \otimes V_{vac}G)V = (1 \otimes G \otimes 1)V \otimes V \; . \tag{7.6.4}$$

Note that V is analogous to the co-multiplication Δ and $V_{vac}GV_{vac}$ is analogous to co-multiplication $H \to H \otimes k$ defined as the inverse of the multiplication of algebra element by the field k with respect to which algebra is a linear space.

c) The vertices depend on the conformal weights of the particles emanating from the vertex (besides momenta and other quantum numbers) so that for three vertex one has $V_3 = V(z_1, z_2, z_3)$. The natural guess is that conformal weights or at least their imaginary parts, which are highly analogous to ordinary momenta, are conserved in the vertices and that the conformal weights $z_1, z_2, ..., z_k$ emanating from the vertex correspond to zeros of $\zeta(z_1, z_2, ...z_k)$. As already found, the conservation of the entire conformal weight allows non-vanishing tree diagrams in the case of three-vertices so that this condition looks sensible.

d) Loop corrections associated with the vertices should vanish. This is achieved if each external 3-vertex associated with the loop gives in the loop integration a factor proportional to $\zeta(z_1, z_2)$, where z_i are the conformal weights associated with the internal lines of the vertex. For n-vertices $\zeta(z_1, z_2, ..., z_{n-1})$ should result in the loop integration.

A stronger hypothesis is that all loop integrals, say an arbitrary vacuum bubble containing an arbitrary number of propagator lines emanating from the points of a circle and connected to the points inside the circle assigns a factor $\zeta(z_1, z_2, ..z_k)$ with each external k-vertex. Effectively this would mean a separation of variables in the conformal degrees of freedom associated with the external vertices. Since essentially topological degrees of freedom are in question, one might hope that this is possible.

The conditions are very stringent and might allow only three-vertices in accordance with the assumption that the Hopf algebra based category describing vertices relies on binary algebraic operations. The condition would relate propagators to vertices and fix their coupling constant evolution. These considerations encourage the hopes that the proposed program might work.

7.6.5 Hopf algebra formulation of unitarity and failure of perturbative unitarity in TGD framework

Unitarity is the basic property of the S-matrix and this raises the question whether also unitarity might be expressed using Hopf algebra language. Since perturbative unitarity giving infinite number of conditions in various orders of coupling constant is the basic aspect of quantum field theories, one is led to ask whether the generalized Feynmann diagrams could allow to identify the counterpart of perturbative unitarity. It turns out that this is not the case and is expected to be so on basis of quantum criticality.

That perturbative unitarity allows Hopf algebra formulation has been demonstrated to be the case by Yong Zhang [ee8] in the case of Φ^4 theory and the proof seems to generalize in a straightforward manner. The Hopf algebra formulation for unitarity generalizes at the formal level to TGD framework although the perturbative unitarity is lost.

Cutkosky rules

Scalar field propagator Δ_F can be decomposed into a sum of positive and negative cutting propagators Δ_+ and Δ_-.

$$
\begin{aligned}
\Delta_F &= \theta(x^0 - y^0)\Delta_+(x - y) + \theta(y^0 - x^0)\Delta_-(x - y) \ , \\
\Delta_\pm(x - y) &= \int \frac{d^4k}{(2\pi)^4}\theta(\pm k^0)2\pi\delta(k^2 - m^2)exp(ik \cdot (x - y)) \ .
\end{aligned}
\tag{7.6.5}
$$

Perturbative unitarity can be formulated using Cutkosky rules [ee4] in the following manner.

a) Consider the decompositions of the vertices of the Feynman diagram to ordinary and circled vertices by a cut line dividing the diagram to left and right sub-diagrams. Perform Hermitian conjugation for the circled vertices and for the propagators connecting circled vertices. If the propagator connects ordinary (circled) vertex to circled (ordinary) vertex, replace it by Δ_+ (Δ_-). This operation obviously puts the particles at the cut lines on mass shell. Conservation laws give constraints on admissible cuts.

b) Perturbative unitarity states that the sum of Feynman diagrams obtained by dividing the Feynman diagram Γ in all possible manners to ordinary and circled vertices using cut line and performing some modifications vanishes. Also included are the decomposition to the union of Γ and empty diagram Φ and to the union of Φ and conjugate of Γ: this obviously corresponds to the imaginary part $i(T - T^\dagger)$ of the scattering amplitude whereas the remaining terms correspond to the $TT^\dagger$ term in a given order of the perturbation theory defined by the number of vertices in the case of Φ^4 theory.

Largest time equation Hopf algebraically

Consider next the Hopf algebra formulation of the largest time equation (not equivalent with perturbative unitarity).

a) The set H defining the Hopf algebra is generated by connected Feynman diagrams. The number field is C. Algebra structure is obtained in a trivial manner. The addition of diagrams is defined by the formal linear combination $a\Gamma_1 + b\Gamma_2$. The multiplication m corresponds to disjoint union. The unit map η specifies unit e as empty set Φ with $\eta(1) = e$.

b) To define co-algebra structure some notational conventions are necessary. Let Γ denote arbitrary Feynman graph, $\mathcal{V}_N(\Gamma)$ denote the set of vertices of N-vertex Feynman diagram, γ a sub-diagram constructed using a subset $\mathcal{V}_c(\Gamma)$ of vertices of Γ and all lines connecting them. A reduced diagram Γ/γ is what remains when γ is cut out of Γ. The cut internal lines of Γ appear as external lines of γ and Γ/γ. Γ and empty set Φ are called trivial diagrams.

c) The co-product Δ is defined by

$$\Delta(\Gamma) \;=\; e \otimes \Gamma + \Gamma \otimes e + \sum_{1 \leq c < N} \gamma(\mathcal{V}_c) \otimes \Gamma/\gamma \;. \tag{7.6.6}$$

Here the sum includes all possible divisions of the vertices to ordinary and circled vertices.

The co-unit ϵ vanishes except for $\epsilon(e) = 1$. Co-associativity is ensured by the fact that two subsequent divisions of a vertex set commute. Antipode is defined recursively by

$$S(\Gamma) \;=\; -\Gamma - e \otimes \Gamma + \Gamma \otimes e + \sum_{1 \leq c < N} S(\gamma(\mathcal{V}_c)) \otimes \Gamma/\gamma \tag{7.6.7}$$

with $S(e) = e$.

Define a new multiplication m of diagram and conjugate diagram by reconnecting the cut lines ending at the same point and assigning to the resulting internal line Δ_+. Let ϕ *resp.* ϕ_c be the Hopf algebra homomorphisms assigning to the Feynman diagram Feynman integral *resp.* its conjugate. m is in a well-defined sense the reverse of Δ and hence very natural.

The convolution $\phi \star \phi_c$ can be defined using the general formula involving only Δ and m

$$\phi \star \phi_c(\Gamma) \;=\; m(\phi \otimes \phi_c)\Delta(\Gamma) \;. \tag{7.6.8}$$

Largest time equation corresponds to the condition

$$\phi \star \phi_c(\Gamma) \;=\; \phi_c \star \phi(\Gamma) = 0 \tag{7.6.9}$$

The formulation of perturbative unitarity using cutting equation

Cutting equation stating perturbative unitary is obtained when only so called admissible cuts are allowed in the definition of the co-product Δ. An admissible cut is defined by a line dividing the diagram into two parts, the left and the right part, in a manner consistent with conservation laws. The left part γ connects at least one incoming line whereas the right part Γ/γ connects at least one outgoing line. For γ *resp.* Γ/γ ordinary *resp.* conjugate Feynman rules are applied.

Cutting equations are of the same Hopf-algebraic form as the largest time equation with the only difference coming from the different definition of the co-product:

$$\phi \star \phi_c(\Gamma) \;\equiv\; m(\phi \otimes \phi_c)\Delta(\Gamma) = 0 \;. \tag{7.6.10}$$

The convolution operation $\star$ is defined by the general formula with m defined by reconnecting the cut lines with the same quantum numbers so that Δ_- connects a circled vertex to an un-circled one. A sum over connected diagrams expressing unitarity condition in a given order of coupling constant defined by the number of vertices results in the operation. By writing explicitly the formula one has

$$\phi_c(\Gamma) + \phi(\Gamma) + \sum{}^{''} m\left[\phi(\gamma) \otimes \phi_c(\Gamma/\gamma)\right] \;=\; 0 \; . \tag{7.6.11}$$

Here the super-script $''$ denotes sum over all admissible cuttings. The first terms obviously correspond to $i(T - T^\dagger)$ term and latter terms to the sum of intermediate states appearing in $TT^\dagger$ term.

Hopf algebra formulation of unitarity conditions in TGD context

The equivalence of the generalized tree diagrams with an infinite number of generalized loop diagrams implied by quantum criticality can be regarded as something highly non-perturbative, so that the generalization of the perturbative unitarity is not expected to make sense.

This indeed seems to be the case. If perturbative unitarity would make sense, each loopy version of a given tree diagram would define a particular perturbative unitarity condition. The absorbtive part $i(T - T^\dagger)$ of a given tree amplitude would have infinitely many different expressions of form $TT^\dagger$ with the product of T and $T^\dagger$ defined such that only a sum over a subspace of intermediate states occurs in it. Unless the different subspaces of intermediate states are gauge equivalent, unitarity requires that absorbtive part equals to the sum over all these subsets.

The Hopf algebra formulation for unitarity could however makes sense also now.

a) There are good hopes that the stringy propagators G, G_+, and G_- are well defined. Since four-momentum appears as an argument of the propagator G, G_+ can be obtained as the absorptive part of G behaving as $G \propto 1/(p^2 - m^2 + i\epsilon)$ near mass shell and becomes thus restricted on mass shell. G_- can be defined as a Hermitian conjugate of G_+.

b) The co-product Δ for the tree diagram Γ must be defined as a sum over all possible loopy diagrams equivalent with Γ by assigning to each loopy diagram a sum over all admissible cuttings. Antipode S is defined recursively and summing over all loopy diagrams equivalent with Γ. m is essentially the inverse of Δ and defined in the same manner as in quantum field theory for each sub-diagram in the unitarity sum defined by $TT^\dagger$. Also the convolution $\star$ would be defined as before. Unitarity conditions formulated as cutting rules would state the vanishing of $\phi \star \phi_c(\Gamma)$.

7.7   The spectrum of the zeros of Riemann Zeta and physics

The properties of the spectrum of the zeros of Riemann Zeta have crucial implications for the quantum TGD. For instance, the question whether the imaginary parts of the zeros are linearly independent or not has deep physical meaning as the model for the scalar field propagator as a partition function for the super-canonical algebra demonstrates. Combining the existing intriguing numerical findings about the correlation functions of the non-trivial zeros allows to make a precise guess about the structure of the zeros of Zeta satisfying also the most obvious physical constraints.

7.7.1   Are the imaginary parts of the zeros of Zeta linearly independent of not?

Concerning the structure of the weight space of super-canonical algebra the crucial question is whether the imaginary parts of the zeros of Zeta are linearly independent or not. If they are independent, the space of conformal weights is infinite-dimensional lattice. Otherwise points of this lattice must be identified. The model of the scalar propagator identified as a suitable partition function in the super-canonical algebra for which the generators have zeros of Riemann Zeta as conformal weights demonstrates that the assumption of linear independence leads to physically unrealistic results and the propagator does not exist mathematically for the entire super-canonical algebra. Also the findings about the distribution of zeros of Zeta favor a hypothesis about the structure of zeros implying a linear dependence.

Imaginary parts of non-trivial zeros as additive counterparts of primes?

The natural looking (and probably wrong) working hypothesis is that the imaginary parts y_i of the nontrivial zeros $z_i = 1/2 + y_i$, $y_i > 0$, of Riemann Zeta are linearly independent. This would mean that y_i define play the role of primes but with respect to addition instead of multiplication. If there exists no relationship of form $y_i = n2\pi + y_j$, the exponents e^{iy_i} define a multiplicative representation of the additive group, and these factors satisfy the defining condition for primeness in the conventional sense. The inverses e^{-iy_i} are analogous to the inverses of ordinary primes, and the products of the phases are analogous to rational numbers.

There would exist an algebra homomorphism from $\{y_i\}$ to ordinary primes ordered in the obvious manner and defined as the map as $y_i \leftrightarrow p_i$. The beauty of this identification would be that the hierarchies of p-adic cutoffs identifiable in terms of the p-adic length scale hierarchy and y-cutoffs identifiable in terms p-adic phase resolution (the higher the p-adic phase resolution, the higher-dimensional extension of p-adic numbers is needed) would be closely related. The identification would allow to see Riemann Zeta as a function relating two kinds of primes to each other.

A rather general assumption is that the phases p^{iy_i} are expressible as products of roots of unity and Pythagorean phases:

$$
\begin{aligned}
p^{iy} &= e^{i\phi_P(p,y)} \times e^{i\phi(p,y)} \ , \\
e^{i\phi_P(p,y)} &= \frac{r^2 - s^2 + i2rs}{r^2 + s^2} \ , \quad r = r(p,y) \ , \quad s = s(p,y) \ , \\
e^{i\phi(p,y)} &= e^{i\frac{2\pi m}{n}} \ , \quad m = m(p,y) \ , \quad n = n(p,y) \ .
\end{aligned}
\tag{7.7.1}
$$

If the Pythagorean phases associated with two different zeros of zeta are different a linear independence over integers follows as a consequence.

Pythagorean phases form a multiplicative group having "prime" phases, which are in one-one correspondence with the squares of Gaussian primes, as its generators and Gaussian primes which are in many-to-one correspondence with primes $p_1 \ mod \ 4 = 1$. If p^{iy} is a product of algebraic phase and Pythagorean phase for any prime p, one should be able to decompose any zero y into two parts $y = y_1(p) + y_P(p)$ such that one has

$$
log(p)y_1(p) = \frac{m2\pi}{n} \ , \quad log(p)y_P(p) = \Phi_P = arctan\left[\frac{2rs}{r^2 + s^2}\right] \ .
\tag{7.7.2}
$$

Note that the decomposition is not unique without additional conditions. The integers appearing in the formula of course depend on p.

Does the space of zeros factorize to a direct sum of multiples Pythagorean prime phase angles and algebraic phase angles?

As already noticed, the linear independence of the y_i follows if the Pythagorean prime phases associated with different zeros are different. The reverse of this implication holds also true. Suppose that there are two zeros $log(p)y_{1i} = \Phi_{P_1} + q_{1i}2\pi$, $i = a, b$ and two zeros $log(p)y_{2i} = \Phi_{P_2} + q_{2i}2\pi$, $i = a, b$, where q_{ij} are rational numbers. Then the linear combinations $n_1 y_{1a} + n_2 y_{2a}$ and $n_1 y_{1b} + n_2 y_{2b}$ represent same zeros if one has $n_1/n_2 = (q_{2a} - q_{2b})/(q_{1b} - q_{1a})$.

One can of course consider the possibility that linear independence holds true only in the weaker sense that one cannot express any zero of zeta as a linear combination of other zeros. For instance, this guarantees that the super-canonical algebra generated by generators labelled by the zeros has indeed these generates as a minimal set of generating elements.

For instance, one can imagine the possibility that for any prime p a given Pythagorean phase angle $log(p)y_{P_k}$ corresponds to a set of zeros by adding to $\Phi_{P_k} = log(p)y_{P_k}$ rational multiples $q_{k,i}2\pi$ of 2π, where $Q_p(k) = \{q_{k,i}|i = 1, 2, ..\}$ is a subset of rationals so that one obtains subset $\{\Phi_{P_k} + q_{k,i}2\pi|q_{k,i} \in Q_p(k)\}$. Note that the definition of y_P involves an integer multiple of 2π which must be chosen judiciously: for instance, if y_P is taken to be minimal possible (that is in the range $(0, \pi/2)$, one obviously ends up with a contradiction. The same is true if $q_{k,i} < 1$ is assumed. Needless to say, the existence of this kind of decomposition for every prime p is extremely strong number theoretic condition.

The facts that Pythagorean phases are linearly independent and not expressible as a rational multiple of 2π imply that no zero is expressible as a linear combination of other zeros whereas the linear independence fails in a more general sense as already found. An especially interesting situation results if the set $Q_p(k)$ for given p does not depend on the Pythagorean phase so that one can write

$Q_p(k) = Q_p$. In this case the set of zeros of Zeta would be obtained as a union of translates of the set Q_p by a subset of Pythagorean phase angles and approximate translational invariance realized in a statistical sense would result. Note that the Pythagorean phases need not correspond to Pythagorean prime phases: what is needed is that a multiple of the same prime phase appears only once.

An attractive interpretation for the existence of this decomposition to Pythagorean and algebraic phases factors for every prime is in terms of the p-adic length scale evolution. The possibility to express the zeros of Zeta in an infinite number of manners labelled by primes could be seen as a number theoretic realization of the renormalization group symmetry of quantum field theories. Primes p define kind of length scale resolution and in each length scale resolution the decomposition of the phases makes sense. This assumption implies the following relationship between the phases associated with y:

$$\frac{\left[\Phi_{P(p_1)} + q(p_1)2\pi\right]}{log(p_1)} = \frac{\left[\Phi_{P(p_2)} + q(p_2)2\pi\right]}{log(p_2)} \ . \tag{7.7.3}$$

In accordance with earlier number theoretical speculations, assume that $log(p_2)/log(p_1) \equiv Q(p_2, p_1)$ is rational. This condition allows to deduce how the phases p_1^{iy} transform in $p_1 \to p_2$ transformation. Let $p_1^{iy} = U_{P,p_1,y}U_{q,p_1,y}$ be the representation of p_1^{iy} as a product of Pythagorean and algebraic phases. Using the previous equation, one can write

$$p_2^{iy} = U_{P,p_2,y}U_{q,p_2,y} = U_{P,p_1,y}^{Q(p_2,p_1)}U_{q,p_1,y}^{Q(p_2,p_1)} \ . \tag{7.7.4}$$

This means that the phases are mapped to rational powers of phases. In the case of Pythagorean phases this means that Pythagorean phase becomes a product of some Pythagorean and an algebraic phase whereas algebraic phases are mapped to algebraic phases. The requirement that the set of phases p_2^{iy} is same as the set of phases p_1^{iy} implies that the rational power $U_{P,p_1,y}^{Q(p_2,p_1)}$ is proportional to some Pythagorean phase U_{P,p_1,y_1} times algebraic phase U_q such that the product of $U_q U_{q,p_1,y}^{Q(p_2,p_1)}$ gives an allowed algebraic phase. The map $U_{P,p_1,y} \to U_{P,p_1,y_1}$ from Pythagorean phases to Pythagorean phases induced in this manner must be one-to one must be the map between algebraic phases. Thus it seems that in principle the hypothesis might make sense.

The basic question is why the phases q^{iy} should exist p-adically in some finite-dimensional extension of R_p for every p. Obviously some function coding for the zeros of Zeta should exist p-adically. The factors $G_q = 1/(1 - q^{-iy-1/2})$ of the product representation of Zeta obviously exist if this assumption is made for every prime p but the product is not expected to converge p-adically.

Also the logarithmic derivative of Zeta codes for the zeros and can be written as

$$\frac{\zeta'}{\zeta} = -\sum_q log(q)\frac{q^{-1/2-iy}}{1 - q^{-1/2-iy}} \ . \tag{7.7.5}$$

As such this function does not exist p-adically but dividing by $log(p)$ one obtains

$$\frac{1}{log(p)}\frac{\zeta'}{\zeta} = -\sum_q Q(q,p)\frac{q^{-1/2-iy}}{1 - q^{-1/2-iy}} \ . \tag{7.7.6}$$

This function exists if the p-adic norms rational numbers $Q(q,p)$ approach to zero for $q \to \infty$: $|Q(q,p)|_p \to 0$ for $q \to \infty$. The p-adic existence of the logarithmic derivative would thus give hopes of universal coding for the zeros of Zeta and also give strong constraints to the behavior of the factors $Q(q,p)$. The simplest guess would be $Q(q,p) \propto p^q$ for $q \to \infty$.

Correlation functions for the spectrum of zeros favor the factorization of the space of zeros

The idea that the imaginary parts of the zeros of Zeta are linearly independent is a very attractive but must be tested against what is known about the distribution of the zeros of Zeta.

There exists numerical evidence for the linear independence of y_i as well as for the hypothesis that the zeros correspond to a union of translates of a basic set Q_1 by subset of Pythagorean phase angles. Lu and Sridhar have studied the correlation among the zeros of ζ [cf5]. They consider the correlation functions for the fluctuating part of the spectral function of zeros smoothed out from a sum of delta functions to a sum of Lorentzian peaks. The correlation function between two zeros with a constant distance $K_2 - K_1 + s$ with the first zero in the interval $[K_1, K_1 + \Delta]$ and second zero in the interval $[K_2, K_2 + \Delta]$ is studied. The choice $K_1 = K_2$ assigns a correlation function for single interval at K_1 as a function of distance s between the zeros.

a) The first interesting finding, made already by Berry and Keating, is that the peaks for the negative values of the correlation function correspond to the lowest zeros of Riemann Zeta (only those contained in the interval Δ can appear as minima of correlation function). This phenomenon observed already by Berry and Keating is known as resurgence. That the anti-correlation is maximal when the distance of two zeros corresponds to a low lying zero of zeta can be understood if linear combinations of the zeros of Zeta are the least probable candidates for zeros. Stating it differently, large zeros tend to avoid the points which represent linear combinations of the smaller zeros.

b) Direct numerical support the hypothesis that the correlation function is approximately translationally invariant, which means that it depends on $K_2 - K_1 + s$ only. Correlation function is also independent of the width of the spectral window Δ. In the special $K_1 = K_2$ the finding means that correlation function does not depend at all on the position K_1 of the window and depends only on the variable s. Prophecy means that the correlation function between the interval $[K, K + \Delta]$ and its mirror image $[-K - \Delta, -K]$ is the correlation function for the interval $[2K + \Delta]$ and depends only on the variable $2K + s$ allowing to allows to deduce information about the distribution of zeros outside the range $[-K, K]$. This property obviously follows from the proposed hypothesis implying that the spectral function is a sum of translates of a basic distribution by a subset of Pythagorean prime phase angles.

This hypothesis is consistent with the properties of the smoothed out spectral density for the zeros given by

$$\langle \rho(k) \rangle = \frac{1}{2\pi} log(\frac{k}{2\pi}) \ . \tag{7.7.7}$$

This implies that the smoothed out number of zeros y smaller than Y is given by

$$N(Y) = \frac{Y}{2\pi}(log(\frac{Y}{2\pi}) - 1) \ . \tag{7.7.8}$$

$N(Y)$ increases faster than linearly, which is consistent with the assumption that the distribution of zeros with positive imaginary part is sum over translates of a single spectral function ρ_{Q_0} for the rational multiples $q_i X_p$, $X_p = 2\pi/log(p)$, $q_i \in Q_p$, for every prime p.

If the smoothed out spectral function for $q_i \in Q_p$ is constant:

$$\rho_{Q_p} = \frac{1}{K_p 2\pi} \ , \quad K_p > 0 \ , \tag{7.7.9}$$

the number $N_P(Y, p)$ of Pythagorean prime phases increases as

$$N_P(Y|p) = K_p(log(\frac{Y}{2\pi}) - 1) \ , \tag{7.7.10}$$

so that the smoothed out spectral function associated with $N_P(Y|p)$ is given by the function

$$\rho_P(k|p) = \frac{K_p}{k} \tag{7.7.11}$$

for sufficiently large values of k. Therefore the distances between subsequent zeros could quite well correspond to the same Pythagorean phase for a given p and thus should allow to deduce information about the spectral function ρ_{Q_0}. A convenient parametrization of K_p is as $K = K_{p,0}/4\pi^2$ since the points of Q_p are of form $q_i 2\pi = (n(q_i) + q_1(q_i))2\pi$, $q_1 < 1$, and $n(q_i)$ must in the average sense form an evenly spaced subset of reals.

Physical considerations favor the linear dependence of the zeros

The numerical evidence is at best suggestive and one can always argue that by an arbitrary small deformation of the linearly dependent zeros one obtains linearly independent zeros. This would however require that each zero of form $y_{P_i} + q2\pi$, $q \in Q_p$ is very near to a zero $\Phi_{P_{k(i,q)}} + q_{k(i,q)}2\pi$. In other words, the union of the translates of Q_p by a subset of Pythagorean phases would approximate the zeros in one-one correspondence with a larger subset of Pythagorean phases (given prime phase appears only once). This should hold for every prime and this seems rather implausible.

On the other hand, the linear dependence between zeros has deep physical implications for the basic quantum TGD, and as the following arguments demonstrate, is physically highly desirable. The precise arguments are developed later and here only the skeleton of the argument is given.

a) The zeros label the generating elements of the super-canonical algebra and the failure of the linear independence means that the weight system is not just the infinite-dimensional lattice spanned by the zeros but can be regarded as a kind of bundle like structure such that the linear combinations $log(p)y_b = \sum_{i=1}^{N} n_i\Phi_{P_{k_i}}$ form N-dimensional lattice and the fiber at a given point of this lattice consists of the points $log(p)y_f = \sum_i n_i q_i 2\pi$. The set of these points is the lattice $n_1 Q_p \times n_2 Q_p \times ...$ divided by the equivalence defined by $y_{f,1} = y_{f,2}$ and for given values of n_i a discrete analog of the one-dimensional space of parallel hyper-planes of an N-dimensional defined by the equation $\sum_{i=1}^{N} n_i x^i = y$ space parameterized by the values of y. What is essential that the space of the planes is different for each point $y_b = \sum_{i=1}^{N} n_i y_{P_{k_i}}$.

b) The calculation of the scalar propagator as a partition function for the super-canonical algebra assuming linear independence gives without any restrictions to the super-canonical weights an infinite number of delta-function resonances of form $\delta(p^2 - m_n^2)$, and at the limit when all zeros of the Riemann Zeta are included in the sub-algebra of super-canonical algebra the set of delta function resonances defines a dense set on real axis. If only the super-canonical conformal weights generated by the positive zeros of Zeta are included, delta function resonances become ordinary poles of form $1/(p^2 - m_k^2)$. The resonances are infinitely narrow and form also now a dense set of real axis.

c) This result, which can be claimed to be non-physical, can be avoided if the zeros are not linearly independent. Although the partition function cannot be calculated explicitly in this case, one can expect that the linear independence gives a reasonable first approximation and that the failure of the approximation is due to the multiple counting caused by the neglect of the fact that the planes of the fiber space can contain several equivalent points. If the zeros are linearly dependent, resonances get a finite width and singularities are avoided for real values of the masses and there are good hopes that the partition function is well-defined for the entire super-canonical algebra.

d) A further argument favoring the proposed form of zeros relates to the two hierarchies strongly suggested by quantum TGD. The first hierarchy corresponds to ordinary primes labelling p-adic length scales and corresponds to length scale resolution. The second hierarchy corresponds to a hierarchy of algebraic extensions of p-adic numbers and there is strong feeling that this hierarchy should correspond to the hierarchy of Beraha numbers $B_n = 4cos^2(\pi/n)$ associated with the phases $exp(i2\pi/n)$. The phases $exp(i\pi/p)$ or their non-trivial powers, for p prime, are even more interesting because of the structure of finite field $G(p, 1)$.

One could consider the possibility that the rationals $q \in Q_p$ for any p can be ordered by their size in such a manner that this ordering corresponds to the ordering of primes with respect to size. Obviously the condition $Q_p = Q_1$ must hold true. This would imply that the products of the powers of the phases $exp(iq2\pi)$ for the lowest N values of q_i would give the Beraha phases corresponding to square free integers having corresponding primes p_i, $i = 1, ..., N$, as factors. All Beraha phases are obtained if the phases $exp(i2\pi/p^n)$, $n = 1, 2, ..$ or their non-trivial powers, are also present. If this waves the case the full p-adic length scale hierarchy with powers of p would correspond to the hierarchy of Beraha phases. This would mean that the addition of new super-canonical conformal weights of increasing size to the sub-algebra of the super-canonical algebra would mean the increase of the dimension of the extension of p-adic numbers needed to represent the resulting phases p-adically as well as an increasing phase resolution.

e) With the assumptions about the structure of zeros of Zeta, the hierarchies defined by the subset y_{P_i} of multiples of Pythagorean prime phase angles and algebraic phases would neatly factorize and the latter would correspond to the p-adic length scale hierarchy. Pythagorean phases correspond to phases of the squares of Gaussian integers $r + is$ and the squares of Gaussian primes define naturally Pythagorean primes. The norm squared of the Gaussian prime is obviously prime: $r^2 + s^2 = p_1$, and satisfies $p_1 \ mod \ 4 = 1$. Hence there is a natural correspondence between Pythagorean prime phases and primes $p \ mod \ 4 = 1$. One can wonder whether also Pythagorean prime phase angles could be mapped to a subset of primes such that that size ordering for y_{P_i} would correspond to the size ordering for the subset of primes. As already noticed, the primeness property is actually an un-necessary strong requirement for Pythagorean phases: it is enough that only single power of a given Pythagorean prime phase appears.

Needless to emphasize, these speculative assumptions which could make possible to realize the p-adicization program and understand that origin of also effective p-adicity would pose very strong constraints on the spectrum of zeros and are certainly testable numerically.

The notion of dual Zeta

These considerations lead to the idea that Riemann Zeta has a dual for which the role of multiplicative primes is taken by the additive primes. This function, call it $\zeta_d(u)$ should either vanish or diverge at points $u = p$. The partition functions for super-canonical conformal weights define analogs of Riemann Zeta involving analog of restriction of summation to integers which are products of even and odd integers and these functions indeed are singular at powers $u = p^{kx}$, $x = 2\pi k/y$, $k = 1, 2, ...,$ where the transcendental values x do not depend on p. That the singularities do not occur for rational values of u is physically very satisfactory since this would mean that the scattering rates could become infinite.

The precise dual ζ_d of ζ would be the function

$$\zeta_d(u) \quad = \sum_{\sum n(y)y, y \in Y} u^{i \sum n(y)y} = \prod_{y>0, y \in Y} \frac{1}{1 - u^{iy}} \ , \tag{7.7.12}$$

where the summation is over all possible formal linear combinations of positive imaginary parts y of zeros or subset of them with non-negative coefficients $n(y)$. In the case that the zeros of Riemann Zeta are linearly independent, the set Y corresponds to all zeros. If the zeros are of the form $y = y_{P_i} + q2\pi$, $q \in Q_0$, one can restrict the consideration to a subset Y of zeros obtains by selecting only single value of $q \in Q_0$ for each y_{P_i}. The simplest option is that q is same for all values of y_{P_i}.

The interpretation as a product of bosonic partition functions defined by the zeros of ζ or subset of them, obviously makes sense, and the form of the partition function is the same as that of Riemann Zeta in the product representation. By writing $u - \rho exp(i\phi)$, $\phi \geq 0$ one finds that all terms in the product converge if the term corresponding to the smallest value $y_{min} \simeq 14.124725$ of y converges. This gives the condition $\phi > 1/y_{min} \sim 2\pi/14$. One can however extract arbitrary number of the lowest terms in the product as a separate well-defined factor and obtain a convergence above arbitrarily small $\phi_{min} = \epsilon > 0$. Thus the product is well-defined arbitrary near to real axis above it.

The limit $\phi \to 2\pi$ is well-defined and at $z = \rho e^{i2\pi}$, $\rho > 0$ the product can be written as

$$\zeta_d(\rho e^{i2\pi}) \quad = \prod_{y \in Y} \frac{1}{1 - \rho^{-2\pi y} \rho^{iy}} \ . \tag{7.7.13}$$

This expression converges to a finite result at the real axis and pole is not possible. This expression is not consistent with the requirement that $u \to 1/u$ induces a complex conjugation of ζ_d at the real axis.

The conjecture is that the limit $\phi \to 0_+$ limit of ζ_d vanishes or diverges for $u = p^{\pm 1}$. Also now the powers of $u_p = p^{kx}$ define poles of the individual factors in the product at real axis. For $u = p$ one can write

$$\zeta_d(p)\overline{\zeta}_d(p) \quad = \prod_{y>0, y \in Y} \frac{1}{4 sin^2 \left[\frac{\phi(p,y)+\phi_P(y)}{2} \right]} \ . \tag{7.7.14}$$

Here U refers to the subset of zeros of Zeta. This expansion diverges for $sin^2[(\phi(p,y)+\phi_P(y))/2] < 1/4$ for sufficiently many values of y. An interesting possibility inspired by the connection with braid groups and Beraha numbers $B_n = 4cos^2(\pi/n)$ is that the numbers $4cos^2[\phi(p,y)]$ are Beraha numbers so that one would have $\phi(p,y) = \pi/n(p,y)$, $n(p,y) \geq 3$. For $n(p,y) \geq 3$ and $\phi_P(y) = 0$, all factors in the product would be larger than or equal to one so that the product would diverge. The vanishing would be thus due the Pythagorean phases. Of course, these arguments cannot be however taken completely seriously since the product expansion does not converge at the real axis.

Also the zeros $z_i = 1/2 + iy_i$, $y_i > 0$, are generators of an Abelian algebra with integers $n/2 + \sum_i n_i y_i$, $\sum n_i = n > 0$. The corresponding zeta function is

$$\zeta_d(u) \quad = \quad \prod_{y \in Y} \frac{1}{1 - u^{-\frac{1}{2} - iy}} \quad . \tag{7.7.15}$$

This function has even nearer resemblance to the ordinary ζ. Interestingly, the product $\prod_d \zeta_d(p)$ satisfies the identity

$$\prod_p \zeta_d(p) \quad = \quad \prod_{y \in Y} \zeta(\frac{1}{2} + y) \ , \tag{7.7.16}$$

if one exchanges freely the order of producting. The fact that all factors on the right hand side vanish would suggest that also $\zeta_d(p)$ vanishes for all values of p.

7.7.2   Why the zeros of Zeta should correspond to number theoretically allowed values of conformal weights?

The following argument provides support for the belief that the conformal weights $s = 1/2 + iy$ for which $p^{1/2+iy}$ exist in a finite-dimensional extension of rationals for all values of prime p, indeed correspond to the non-trivial zeros of Zeta.

a) The basic idea of the number theoretical approach is that the conformal weights $1/2 + iy$ are such that the radial waves $r^{-1/2-iy}$ exist for all rational (and thus for integer) values of r in some finite-dimensional extension of rationals. The logarithms $log(n)$ of integers can be interpreted as quantum numbers of a system defined by an arithmetic quantum field theory and Zeta function $\zeta = \sum_n n^{-iy-1/2}$ with $s = 1/2 + iy$ interpreted as an inverse temperature, defines the partition function of this system.

b) On the other hand, so called Selberg's Zeta function characterizes the eigen values of the Laplacian in 2-dimensional quantum billiard systems defined in the fundamental domain of some hyperbolic subgroup G of $SL(2, Z)$ acting in the hyperbolic plane $SL(2, R)/SO(2)$ [cf7]. The fundamental domain is analogous to a box containing the particle. At quantum level the boundary conditions are satisfied by summing over all the G translates of $SL(2, R)$ invariant Green function with respect to the second argument. Physically this is analogous to putting to all copies of the fundamental domain an image charge. The confinement to the fundamental domain selects from the continuous energy spectrum a discrete sub-spectrum. Selberg's Zeta (its logarithmic derivative) has the allowed energy eigen values as its zeros (poles). Furthermore, the energy eigen values of Laplacian are of form $E = -l(l+1)$, where $l = -1/2 - iy$ is identifiable as the counterpart of conformal weight and has the same form as the zeros of Zeta. y has discrete spectrum of values characterized by the choice of G. The density of the energy eigenvalues is amazingly similar to that of Zeta.

c) On basis of above resemblances one can argue that Riemann Zeta (its logarithmic derivative) characterizes the purely number theoretical spectrum as its zeros (poles). If this is the case, the zeros of Zeta would coincide with the number theoretically allowed conformal weights $1/2 + iy$.

The p-adically existing conformal weights are zeros of Zeta for 1-dimensional systems allowing discrete scaling invariance

The obvious question is whether one could reduce number theory to symmetry. The following considerations suggests that $D \geq 2$-dimensional spaces do not allow a system having zeros of Zeta as its spectrum.

a) The density of states of the Selberg Zeta function differs in some aspects from that of Zeta so that Riemann Zeta probably has no interpretation as a Selberg Zeta function of a number theoretical system. For instance, the average density of states with respect to y grows linearly rather than logarithmically although the fluctuating part of the density of states is formally very similar to that of Zeta.

b) Lobatchevski space (the hyperboloid of the 4-dimensional future light cone) has $SL(2, C)$ as its isometry group. The energy spectrum of Laplacian in this case is of the form $E = -l(l + 2) = 1 + y^2$ with $l = -1 - iy$ and thus different from the spectrum of 2-dimensional case and of Riemann Zeta. Due to the higher dimension of the system the mean density of states grows even faster than in the 2-dimensional case so that there seems to be no hope of getting the density of states of Riemann Zeta.

Only one-dimensional systems give hopes of the required logarithmically varying mean density of states. The simplest candidate one can imagine is a system with discrete scaling invariance.

a) Instead of Laplacian, and in complete accordance with the view that conformal invariance is the key to the understanding of Riemann Zeta, one can consider the scaling operator $L_0 = xd/dx$ acting at the half line R_+ so that the Green functions defined by the equation

$$(L_0 + z)G(x, x_1) = (xd/dx + z)G(x, x_1) = \delta(\frac{x}{x_1} - 1) \tag{7.7.17}$$

become the object of interest. The solution can be written as

$$G(x, x_1|z) \;=\; (\frac{x}{x_1})^z \times \theta(\frac{x}{x_1} - 1) \;. \tag{7.7.18}$$

Here $\theta(x)$ denotes the step function. The requirement that the integrals

$$\int \overline{G}(x, x_1|z_1)G(x, x_1|z_2)dx$$

reduce to the inner products of ordinary plane waves when $ln(x/y)$ is taken as an integration variable forces the condition $z = 1/2 + iy$. In fact, this might be seen as the physicist's "proof" of the Riemann hypothesis.

b) Following the construction of the automorphic Green functions in the hyperbolic plane described in [cf7], the next step is to form a sum over the $x-$ scaling transforms of $G(x, x_1|z)$ by summing over the integer scaled values nx of x to form a well defined Green function in the fundamental domain associated with the semigroup of integer scalings. Any interval $[n, 2n]$ forms a fundamental domain. This gives

$$\begin{aligned}
G_I(x, x_1|\frac{1}{2} + iy) \;&=\; \sum_n G(nx, x_1|\frac{1}{2} + iy) = \sum_n (\frac{nx}{x_1})^{\frac{1}{2}+iy} \\
&=\; \zeta(\frac{1}{2} + iy) \times (\frac{x}{x_1})^{\frac{1}{2}+iy} \;.
\end{aligned} \tag{7.7.19}$$

The resulting Green function is proportional to Riemann Zeta at the critical line and vanishes for the zeros of Zeta. Note that the logarithmic derivative of ζ divided by $log(p)$ exists in a finite-dimensional extension of R_p for $x = n/2 + i\sum_k m_k y_k$ if the basic number theoretical requirements on the phases p^{iy} defined by the zeros of Zeta are satisfied: in particular $log(p_1)/log(p)$ must have R_p norm which approaches zero for larger values of p_1. Hence the logarithmic derivative of Zeta could codes the number theoretical physics universally.

c) In the usual approach [cf7] the integral of G_I over the fundamental domain would give the density of states $d(E)$. In the recent case the integration over the fundamental domain $[1, 2]$ gives just ζ function

$$\int_1^2 G_I(x, x| - \frac{1}{2} + iy)dx \;=\; \sum_n n^{-\frac{1}{2}-iy} = \zeta(\frac{1}{2} + iy) \;. \tag{7.7.20}$$

The interpretation as a density of states is obviously not possible. The proof for the Riemann hypothesis to be discussed later allows to interpret the vanishing of Riemann Zeta as as orthogonality of physical states labelled by zeros of Zeta with a tachyonic vacuum state with a vanishing conformal weight. The vanishing of Green function could also now have an interpretation stating that the physical states labelled by non-trivial zeros are orthogonal to the scaling invariant tachyonic vacuum.

d) Quite generally, the imaginary part of the logarithmic derivative of any real function $f(E)$ for which energy eigenvalues E_n correspond to zeros of unit multiplicity, defines the density of states as a sum over delta functions. $G(y) = \zeta(1/2 + iy)$ is real at the critical line as is also its logarithmic derivative apart from delta function singularities of the imaginary part at the zeros of Zeta so that its logarithmic derivative indeed gives the density of zeros of Zeta:

$$d(y) \quad = \quad \frac{1}{\pi} Im \left[i \frac{dlog\left[\zeta(\frac{1}{2} + iy)\right]}{dy} \right] = \sum_n \delta(y - y_n) \ . \tag{7.7.21}$$

This ultra simple model realizes the idea that the logarithmic derivative of Green function naturally associated with a system invariant under the semi-group of integer scalings codes as its poles the zeros of Zeta. The p-adic existence of the Green function in turn is equivalent with the requirement that the spectrum corresponds to the zeros of Zeta.

Realization of discrete scaling invariance as discrete 2-dimensional Lorentz invariance

Both the role of the hyperbolic groups and the fact that in quantum TGD zeros of Zeta label representations of Lorentz group, encourage to think that the 1-dimensional hyperbolic subspace $t^2 - x^2 = constant$ of 2-dimensional Minkowski space having Lorentz group $SO(1,1)$ as its symmetries realizes the above described system physically. The counterpart of the hyperbolic subgroup G of $SL(2,R)$ would the semigroup of Lorentz transformations defining integer scalings of the second light like coordinate:

$$u \equiv t + z \to nu \ , \quad v \equiv t - z \to \tfrac{1}{n}v \ .$$

This semigroup corresponds to the diagonal semi-subgroup of $SL(2,Q)$ consisting of matrices $diag(\lambda, 1/lambda) = diag(n, 1/n)$. The reduction to semigroup is natural by the presence of the p-adic length scale cutoff unavoidable in p-adicization.

Taking $u = t + z$ as the coordinate of the hyperboloid, the situation reduces to that already considered. Infinitesimal Lorentz boost acts as a scaling operator and its eigenvalues correspond to the zeros of Zeta by number theoretic existence requirements. The matrices $diag(p, 1/p)$, p prime, are completely analogous to the group elements g_0 defining primitive periodic orbits in the higher-dimensional case so that prime numbers are naturally realized as discrete Lorentz transformations. Prime Lorentz transformations and their inverses generate rational Lorentz group. The length of the primitive periodic orbit corresponds to the scaling parameter $log(p)$ defining the scaling by p as an exponentiated scaling transformation $u \to exp(log(p))u = pu$.

The statistics of generic continued fractions and the rationality of $log(p)/log(q)$

The hypothesis that $log(p)/log(q)$ is rational number is not a mere metaphysical statement since its plausibility can be tested numerically. The reason is that the regular continued fraction expansion $[a_0, a_1, ...a_n, ..]$ defined for arbitrary real number a as

$$a \quad = \quad a_0 + \cfrac{1}{a_1 + \cfrac{1}{a_2 + \cfrac{1}{a_3 + ...}}} \tag{7.7.22}$$

of any rational number has only finite number of non-vanishing coefficients a_n.

For irrational numbers in any quadratic extension of rationals the continued fraction expansion is periodic. In the case of algebraic numbers the expansion does not possess any obvious symmetry. In the case of certain transcendental numbers like e there are symmetries allowing to predict the expansion whereas the expansion of π does not seem to be predictable. Perhaps this relates to the fact that e defines a finite-dimensional extension of p-adic numbers.

What is interesting is that by Kuzmin's theorem [cd1] the statistics of the continued fraction expansion in a dense set of real numbers (with at least rationals and numbers in quadratic extension of rationals excluded) involves the ratios of $log(p)/log(q)$. Namely, the probability $p(k)$ for the appearance of integers k as the value of a_n at the limit $n \to \infty$ is given by

$$p_k \equiv p(k, n \to \infty) = \frac{log\left[\frac{(k+1)^2}{k(k+2)}\right]}{log(2)} \; . \tag{7.7.23}$$

If $log(p)/log(q)$ hypothesis holds true, the ensemble defined by a generic real number is not only universal but the ensemble probabilities are rational.

This ensemble can be identified as an ensemble associated with an arithmetic quantum field theory with energy spectrum $E_k = log(k)$ with the probability of the energy level E_k being given by p_k. For large values of k the probabilities p_k indeed behave as

$$p_k \simeq \frac{e^{-2log(k)}}{log(2)} \; , \tag{7.7.24}$$

which has the form of Boltzmann weight $e^{-E(k)/T}$ with temperature $T = 1/2$. One could say, that the typical real number is analogous to a system in a thermal equilibrium, and rationality of $log/p)/log(2)$ would fit nicely with the basic philosophy behind arithmetic quantum field theories.

From this one can deduce the average of $log(n)$, that is the average of the number theoretical energy, for an ensemble defined by a generic real number as

$$\frac{1}{n} \lim_{n \to \infty} \left\langle \sum_{i=1}^{n} log(a_i) \right\rangle \;\; = \;\; \sum_k p(k)log(k) = \sum_k log(k) \frac{log\left[\frac{(k+1)^2}{k(k+2)}\right]}{log(2)} = log(K) \; ,$$

$$K \;\; = \;\; \prod_k \left[log\left[\frac{(k+1)^2}{k(k+2)}\right]\right]^{log(k)/log(2)} \; . \tag{7.7.25}$$

The value of K identifiable as the geometric average of a_n is approximately $K = 2.684545$, a result deduced first by Khinchin [cd2]. From $p_1 \simeq .415$ and $p_2 \simeq .17$ it is clear that $a_n = 1$ representing the ground state dominates the ensemble.

The notion of ensemble suggests an statistical approach to the complexity of the real numbers. The ensemble could be also seen as a statistical description of the orbit defined by the continued fraction expansion. One-dimensional spin lattice with spin projection having all possible positive integer values provides a physical analog system for real numbers.

a) Rationals correspond to non-chaotic orbits ending to a fixed point $a_n = 0$ in finite number of steps and thus to finite spin lattice. Algebraic numbers in quadratic extensions could be seen as periodic orbits or as completely ordered one-dimensional lattice like systems. Golden Mean could be seen either as a fixed point equivalent with a zero temperature ferromagnet for which only $a_n = 1$ appears. Also $a_n = k > 1$ is possible and corresponds to $X_k = (k + \sqrt{k^2 + 4})/2$. The numbers in the quadratic extensions of rationals would correspond to zero temperature "ferrimagnets" with a periodic magnetization and consisting of finite number of sub-lattices representing ferromagnets with magnetization $a_n = k$.

Perhaps Golden Mean and its generalizations X_k could be seen as representing the simplest irrationals from which other irrationals result as thermal excitations. For instance, a sufficiently small irrational change $x \to x + \Delta x$ of a number with a periodic continued fraction affects only the coefficients a_n for which n is high enough. Perhaps the deviation from ferrimagnet could in this case be seen as being due to a temperature gradient along the one-dimensional lattice gradually spoiling the periodicity. The fact that the continued fraction expansion of the real is unique evidently means that there is more or less unique minimal algebraic number which defines the ferrimagnet characterizing an irrational number.

b) Neper number corresponds to an expansion

$$e = [2; 1, 1, 4, 1, 1, 6, 1, 1, ...2n, 1, 1, ...] \; .$$

$k = 2$ and $k = 1$ appear with non-vanishing probabilities $1/3$ and $2/3$ whereas the multiples of $2n$ appear only once: the interpretation in terms of a predictable orbit is natural. Also an interpretation in terms of a symmetry broken ferrimagnet is possible. Note that the probability of 1 is $2/3$ whereas for generic real number it is .415.

d) These examples encourage the hope of deriving a measure of complexity for irrationals based on the properties of the ensemble defined by the continued fraction expansion. The statistics of algebraic numbers is of special interest. For instance, one could ask whether only finite number of integers have vanishing probabilities $p(k)$ for these ensembles, and whether the average number theoretical energy $log(k)$ could serve as a characteristic of the ensemble. Also correlation functions $\langle log(a_n)log(a_{n+k})\rangle$ could be used to characterize the ensemble. Perhaps more refined notions of statistical physics such as ferro- and ferrimagnets, and even spin glass could be useful for the statistical characterization of the complexity of reals.

7.7.3   Riemann Zeta and particle propagators

TGD should possess a well-defined and divergence free quantum field theory limit. The preceding considerations suggest a manner to get rid of the divergences of quantum field theories. Super-canonical conformal weights, or more generally, zeros of Riemann Zeta, should leave a trace in quantum field theory and this trace must be essential for the cancellation of infinities. In the sequel an attempt to understand how the zeros of Zeta might be visible at single particle level is made.

Is the p-adic length scale evolution of propagators universal and determined by the zeros of zeta

The quantum field theory model for polyzetas [cf4] gives some hints about what generalized Feynman diagrams could be.

a) In the 1-dimensional quantum field theory model for polyzetas discussed in [cf4] the propagator factors correspond to all possible integer powers G^k of a propagator G defined as inverse of the derivative operator $D = d/dt$. The resulting diagrams give results which are proportional to polyzetas. The integer k is completely analogous to a conformal weight. For a single loop approximation of the scalar field theory only the power D^2 would appear as a propagator and this would give only polyzetas with even weight.

b) The work of Kreimer and Broadhurst [ee9] leads to the conclusion that one can assign to each vacuum Feynman diagram a closure of a unique braid with chords connecting the strands of the braid: this is nothing but the TGD counterpart of vacuum diagram defined in terms of a braid with chords representing propagators. Probably this correspondence holds for non-vacuum diagrams also. These braids give rise to combinations of polyzetas such that the number n of strands corresponds to the depth $m = n + 1$ of the polyzeta and the number of crossings equals to the weight $k = \sum_i k_i$ of the polyzeta.

c) The polyzeta associated with the Feynman diagram vanishes if the allowed conformal weights k_i satisfying $\sum k_i = k$ are not positive integers but correspond to the zeros of polyzeta in question. Hence the replacement of the integers k_i identified as conformal weights with the zeros of the polyzeta in question gives hopes that one could understand the p-adic length scale evolution of the propagator and the cancellation of loop corrections. The physical interpretation for the modified conformal weights would as "bound state conformal weights", and the necessity to use them would be something which field theory alone cannot explain. The form of the propagator to be discussed is completely universal and applies also to stringy diagrams when one replaces propagator with the Super Kac-Moody conformal scaling generator L_0.

Various models for scalar propagator

The first thing that comes in mind is to construct a model for the scalar field propagator based based on zeros of ζ. The are several options.

Option I: This model is inspired by the quantum field theoretical model for polyzetas [cf4]. Scalar field propagator is regarded as a kind of partition function in an ensemble whose states are labelled by spectrum of super-canonical weights expressible in terms of zeros of Zeta. This assumption is admittedly somewhat adhoc. A sum over both the non-trivial and trivial zeros is needed and trivial zeros give the usual free propagator contribution. The details of this option are left to Appendix B.

Option II: The inspiration comes now from the formula $\int_0^\infty exp(i(p^2 - m^2 - i\epsilon)t)dt$ for the scalar propagator. The integral is replaced with a discrete sum over allowed values of $t > 0$ which on basis of number theoretical considerations are argued to correspond to linear combinations for the imaginary parts of non-trivial zeros of Zeta with coefficients which are positive integers to guarantee $t > 0$.

Option III: This model is inspired by the observation that the defining integral can be transformed to an integral over positive imaginary axis in t-plane and one can wonder whether it could be replaced with a sum over the trivial zeros of Zeta. This would give a rather realistic looking propagator.

Option IV: One can criticize the notion of the momentum space scalar propagator as being a too phenomenological concept to allow the application of notions of super-canonical invariance and quantum criticality.

a) In TGD the usual momentum space propagators can have only a phenomenological role since 4-momentum does not flow in the lines of the generalized Feynman diagrams. The counterpart of the propagator originates from the unitary evolution for the induced spinor fields at 3-D light-like causal determinants representing orbits of 2-D partons along which generalized eigen modes of the modified Dirac operator propagate. Momentum squared as the argument of the propagator is replaced by the square of the eigen value of the modified Dirac operator D.

b) There are reasons to suspect [O5] that the unitary evolution associated with the lines of generalized Feynman diagrams corresponds to the unitary evolution operator Δ^{it} determined by the positive Hermitian "Hamiltonian" operator Δ inherent for von Neumann algebras. Positivity suggests that Δ corresponds to the square D^2 of the modified Dirac operator. Conformal invariance would in turn suggest that the eigenvalues of Δ and possibly also D^2 might relate in a simple manner to the positive integer valued spectrum of Virasoro generator L_0. In this case the propagator would result through a summation over the number theoretically allowed discrete values of t as for option II. The propagator reduces to a function which can be regarded as a dual of Riemann Zeta.

General conditions on scalar propagator

Consider as an example a scalar field theory with the momentum space inverse propagator $D = p^2 - m^2$.

a) Introduce a mass scale λ playing the role of an ultraviolet cutoff and replace the inverse propagator $D_0 \equiv \lambda^{-2}D$ with a conformal weight $k = -2$ with the sum of propagators with conformal weights z which correspond to the zeros of ζ. In the quantum field theory context the parameter λ would be an ad hoc parameter but in TGD framework its values naturally corresponds to the inverses of p-adic length scales so that one would have entire hierarchy of propagators with $\lambda \propto 1/\sqrt{p}$, p prime and discretized renormalization group evolution.

b) In the standard renormalization theory one poses the conditions

$$G_R^{-1}(p^2 = m^2) = 0 \ , \qquad \frac{dG_R^{-1}}{dp^2}(p^2 = m^2) = 1 \ . \tag{7.7.26}$$

These conditions are posed also now. In the standard renormalization theory the propagator has a cut along real axis above $p^2 = m^2$ and the cut is due to the logarithmic terms $log[(p^2 - m^2)/\lambda^2]$ whose imaginary part is discontinuous. One might expect this behavior also now.

For options II and III the definition of scalar propagator is unique. Concerning the precise definition of the propagator for option I, one can consider several options.

a) The Cartan decomposition $g = h + t$ of the super-canonical algebra discussed in [B2] determines the conformal weights in question and they correspond to the conformal weights of t. This option has a good theoretical justification since the generators of t indeed corresponding to non-gauge degrees of freedom. In this case also conformal weights of form $z = n - 1/2 - \sum_i y_i$, where y_i are imaginary parts of the non-trivial zeros of Riemann zeta must be included. The resulting propagator has an interpretation as a propagator with ultraviolet cutoff. The restriction $n = 0, 1$ emerges from the requirement that the propagator is non-singular on mass shell. This option is implied also by the requirement that Virasoro generators L_n, $n \geq 1$, act as gauge symmetries in the super-canonical algebra.

b) One can also consider the restriction to the conformal sub-algebra of the super-canonical algebra so that only the linear combinations $\sum n(y)y$, $n(y) > 0$, are involved.

c) The minimal option is that only the conformal weights corresponding to the zeros of Riemann Zeta contribute to the scalar propagator.

The character of the non-trivial zeros of Riemann Zeta is of crucial physical significance.

a) The assumption that the imaginary parts of the zeros are linearly independent is attractive from the calculational point of view. It however turns that the resulting propagator has either delta function like poles or infinitely narrow resonance poles ro real mass mass squared values and this is not a physically realistic result although it could be a good approximation. Furthermore the propagator is totally ill defined at the limit when all zeros of Zeta contribute since the singularities are dense on real axis.

b) The distribution of the zeros of Zeta favors strongly the option for which zeros are not linearly independent and that the phases p^{iy} for each prime p correspond to sums of subset of angles of Pythagorean prime phases Φ_P defined by the squares of Gaussian primes and a fixed set of rational fractions $m2\pi$. This implies that the generating super-canonical conformal weights are not linearly independent. With this assumption the singularities of the propagator are not anymore at real values of mass squared and the propagator is well-defined also at the limit when all zeros of Zeta contribute to it.

In the following calculations the assumption that y_i are linearly independent is made although this assumption is probably wrong. The motivation is that it allows to understand the behavior of the propagator approximately and see why the linear independence implied by the ansatz inspired by the properties of the spectral function of Riemann zeros is physically highly desirable.

The hypothesis that p^{iy} can be expressed in terms of Pythagorean prime phase angle and rational multiple of 2π for any prime, if true, is bound to be a very powerful number theoretic symmetry. In particular, it gives infinite number of representations of the scalar field propagator, one for each prime. In the sequel this aspect is not discussed at all.

Propagators for various options

1. Scalar field propagator for option II

In this case the expression for scalar field propagator is easy to derive if linear independence of y_i is assumed. The expression for the propagator reads as

$$G \;=\; \sum_{n_i>0} u^{i\sum_i n_i y_i} \;, \tag{7.7.27}$$

$$u \;=\; exp(\frac{p^2-m^2}{\lambda^2}) \;. \tag{7.7.28}$$

Here the sum over the combinations of zeros with positive integer valued coefficients is assumed. One could assume that only the condition $\sum n_i y_i > 0$ holds true. The restriction is consistent with the hypothesis about the p-adic existence of basic building blocks at zeros of Riemann Zeta if $exp((p^2-m^2)/\lambda^2)$ is restricted to be rational.

Performing the sums, one obtains

$$G(p) \;=\; \prod_i \frac{1}{1-u^{iy_i}} \;,$$

$$u \;=\; exp(\frac{p^2-m^2+i\epsilon}{\lambda^2}) \;. \tag{7.7.29}$$

This expression can be regarded as a dual of Riemann Zeta in the sense that the product over partition functions associated with primes is replaced with the product over partition functions labelled by non-trivial zeros. The terms of product approach to unity with exponential rate for $\epsilon > 0$ so that the product converges.

The propagator has poles at

$$p^2 \;=\; m^2 + k\lambda^2\frac{2\pi}{y_i} \;,\quad k \in Z \;. \tag{7.7.30}$$

$k = 0$ corresponds to $p^2 = m^2$ so that scalar propagator pole is obtained. The mass formula is stringy mass formula with string tension proportional to $1/y_i$.

If u corresponds stringy propagator $L_0^{tot} = (p^2 - L_0)/\lambda^2$ instead of scalar propagator one obtains modification of the stringy mass formula

$$m^2 = (n + k\frac{2\pi}{y_i})\lambda^2 \ , \quad k \in Z \ . \tag{7.7.31}$$

If all non-trivial zeros of ζ are allowed, the poles are dense in the entire m^2-axis so that also tachyons are obtained. One can consider several cures the problem.

a) Assuming that the basic building blocks of Riemann Zeta exist p-adically for its zeros, the poles with $k > 0$ do not correspond to rational values of u so that the poles other than $p^2 = m^2$ ($p^2 = n\lambda^2$ in the stringy case) are not at all visible if the theory is required to satisfy p-adicizability constraint.

b) A natural hierarchy of cutoffs is provided by the subalgebras of super-canonical algebra defined by y-cutoff. In this case poles are not dense anymore but tachyons remain.

c) If y_i are not linearly independent, it is not possible to sum freely over integers n_i and poles are smoothed out and are expected to develop imaginary parts. If the condition $n_i > 0$ is replaced with the weaker condition $\sum_i n_i y_i > 0$ more correlates between y_i emerge and further smoothing out of poles is expected to happen.

2. Scalar field propagator for option III

In this case the propagator is given by as sum over trivial zeros of ζ at $z = -2n$.

$$G(u) \ = \ iu^2/(1 - u^2) = i\frac{exp(\frac{2(p^2-m^2)}{\lambda^2})}{1 - exp(\frac{2(p^2-m^2)}{\lambda^2})} \ . \tag{7.7.32}$$

The pole at $p^2 = m^2$ is not shifted and for large values of $p^2 > 0$ the propagator approaches -1 and vanishes exponentially for space-like momenta.

3. Scalar propagator for option I

There are several alternatives concerning definition of the scalar propagator as a partition function. These are discussed in detail in Appendix B. What makes this option questionable is that the pole at m^2 is shifted to $m^2 + \lambda^2$. This would require that massless particles result from tachyons. The other problems and possible cures are same as those for option II. One can say that the number theoretic discretization of the integral formula for propagator is favored over the partition function based model.

4. Propagator for option IV

In this case the interpretation as momentum space propagator is given up. The general expression of propagator is same as for option II: only the argument $u = exp((p^2 - m^2)/\lambda^2)$ is replaced by a new one. A good guess is that u corresponds to the squares for the eigenvalues of the modified Dirac operator required to be rational by p-adicization. An alternative guess is that u corresponds to the non-negative integer valued spectrum of Virasoro generator L_0.

If y_i are linearly independent and $n_i > 0$ condition is assumed one obtains poles also now but they do not correspond rational values of u. The values of poles would be given by

$$u_{k,y_i} \ = \ exp(\frac{2\pi k}{y_i}) \ . \tag{7.7.33}$$

Rationality requirement, the linear independence of y_i, the weakened condition $\sum_i n_i y_i > 0$, and already the fact that the spectrum of D^2 is pre-determined, excludes poles. It seems that option III favored also by the general physical picture is the only realistic one.

7.8   Quantum TGD as a quantum field theory of some kind?

The super-canonical generalization of a 2-dimensional conformal field theory seems to be indispensable for the construction of S-matrix at the fundamental level and defines the vertices as n-point functions of a conformal field theory in turn used to construct S-matrix as tree diagrams. It is however not at all obvious whether QFT like formulation in a more general sense really makes sense or that it is even needed.

Certainly it is clear that standard quantum field theory starting from an action principle and defining perturbation theory is out of question since it contradicts the basic assumption that S-matrix elements reduce to tree diagrams. This leaves two options.

a) The action of the possibly existing field theory limit must correspond to an effective action for one-particle irreducible (1PI) Green's functions from which Green's functions and S-matrix elements are obtained as tree diagrams. 1PI Greens function must correspond to the vertices identifiable as n-point functions of a conformal field theory for which super-canonical algebra defines the conformal fields.

b) If quantum TGD allows a formulation as a QFT, the formulation must be such that it effectively reduces to a free field theory. This is required by the vanishing of loops implying localization in the configuration space also necessary for the p-adicization of the theory.

7.8.1   Could one formulate quantum TGD as a quantum field theory at the absolute minimum space-time surface?

The original naive belief when I started to develop TGD for 25 years ago was that something like functional integral using EYM action coupled to induced spinor fields would define the quantum theory. The belief turned out to be wrong and the conclusion was that this kind of approach might make sense only as a quantum field theory limit of TGD. The philosophy just described and the view that space-time physics represents only classical correlates for the underlying configuration space physics however makes even the idea about the existence of quantum field theory limit questionable.

On the other hand, one could defend the existence of some kind of QFT type formulation by quantum classical correspondence and by the fact that the phenomenology based on the notion of classical induced gauge fields has been extremely fruitful. One could even hope that the theory would allow formulation as some kind of field theory at $X^4(X^3)$ or even better, at X^3 so that a minimum amount of information about the absolute minimum would be needed in accordance with the gravitational hologram principle.

Does the modified Dirac action for the induced spinor fields define the QFT description of quantum TGD?

The constraints satisfied by the QFT formulation of quantum TGD are so strong that the formulation is essentially unique.

a) The QFT in question should be determined by the absolute minimum $X^4(X^3)$ of Kähler action corresponding to a given causal determinant (7-dimensional light like surface $X^3_l \times CP_2$) rather than in M^4. The averaging over all Poincare and color translates of this space-time surface would give rise to an S-matrix respecting the basic symmetries. Note that the theory would be 3-D quantum field theory at X^3. The only information needed about $X^4(X^3)$ would be the time derivatives of the imbedding space coordinates at X^3. Also the value of Kähler action seems to be needed but even the exponent of Kähler function might disappear from the Greens functions in normalization just as the exponent e^G of the generating functional G of connected Green's functions disappears in the quantum field theory.

The effective 3-dimensionality has an interesting connection to unresolved difficulties encountered in the attempt to formulate bound state problems in quantum field theory context. Non-relativistic, essentially 3-dimensional, Schrödinger equation works and yield correct predictions whereas Bethe-Salpeter equation in Minkowski space fails. The TGD based explanation of this failure is that bound state formation means that 3-surfaces of particles involved form a join along boundaries condensate. This means that non-relativistic 3-dimensional formulation is necessary in order to catch the essential aspects of the physics involved. In quantum field theory context point-likeness of the particles allows only the modelling of those aspects of particle interactions which do not involve bound states.

b) To calculate correlation functions one should expand the super-canonical generators as functional Taylor series around the maximum of the Kähler function at the 3-surface X^3. If super-canonical generators can be regarded as functionals of the second quantized induced spinor field ψ, also the functional series with respect to ψ is needed. This would make it possible to evaluate the correlation functions perturbatively in terms of the tree diagrams defined by the effective action using bosonic and fermionic propagators defined by it. One would calculate M^4 Fourier transforms of the correlation functions and integration over Poincare translates would give Poincare invariant correlation functions.

c) The complete localization at configuration space level means the vanishing of the bosonic loops and effective freezing of configuration space degrees of freedom. This is achieved if the action is such that the bosonic part vanishes when the induced spinor fields vanish. I have represented in [B4] arguments that the action for the induced spinor fields treated as Grassman variables is all that is needed to define quantum physics. Kähler action would be the effective action associated with the Dirac action. The approach would also predict the possible values of Kähler coupling strength as part of data characterizing the effective action. This approach also conforms with the fact that elementary bosons are predicted to be bound states of fermion-anti-fermion pairs.

This approach would also bring induced electro-weak gauge potentials into play so that quantum-classical correspondence would be realized. The absence of quark color as spin like quantum number of induced spinor fields would not be a problem. Also the topologization of the family replication phenomenon in terms of the genus of 2-surface would result without representation as an additional spin like degeneracy of fermion fields. The super-canonical and super Kac-Moody conformal algebras would be however an essential element of the picture.

d) The action would be the modified Dirac action for the induced spinor fields at the maximum of the Kähler function. Modified Dirac action is defined by replacing the induced gamma matrices $\Gamma_\alpha = \partial_\alpha h^k \Gamma_k$ by the modified gamma matrices

$$\hat{\Gamma}^\alpha = \frac{\partial(L\sqrt{|g|})}{\partial(\partial_\alpha h^k)}\Gamma^k \quad . \tag{7.8.1}$$

Here L denotes the action density of Kähler action and Γ^k denotes gamma matrices of the imbedding space H. Modified Dirac action is supersymmetric and shares the vacuum degeneracy of Kähler action [B4].

The fermionic propagator would be defined by the inverse of the modified Dirac operator. Bosonic kinetic term would be Grassmann algebra valued and vanish for $\psi = 0$ and would contribute nothing to the perturbation series. Since the fermionic action is free action, a divergence free quantum field theory would be in question irrespective of whether the fermionic action is interpreted as an action or an effective action. One could also see the action as a fixed point of the map sending action to effective action in a complete accordance with the idea that loop corrections vanish.

e) The nice feature of this approach is that the information needed about the absolute minimum would be minimal since one can restrict the consideration to 3-surface X^3 which can be selected arbitrarily. In fact, the outcome is a 3-dimensional free field theory in the fermionic degrees of freedom and the integral over Poincare and color translates guarantees isometry symmetries. Grassmannian functional integral would give exponent of Kähler function as the analog of the generating functional G for connected Green functions. The functional derivatives of the configuration space spinor field with respect to the induced spinor field are very simple by the conservation of fermion number. This approach could be seen as an alternative approach to calculate n-point functions by treating fermionic fields as Grassmann fields whereas in the super-algebra approach fermionic fields would be second quantized.

Vacuum extremals and the fractality of super-canonical propagator

For vacuum extremals the propagator defined by the modified Dirac operator becomes singular. The interpretation as a space-time correlate for the singular behavior of the scalar propagator defined as a partition function in super-canonical algebra is suggestive.

Two different models for the scalar propagator were considered.

a) In the first case all physical super-canonical conformal weights were assumed to contribute and the resulting propagator has besides the ordinary pole infinite number of delta function type resonances. The real inverse of the propagator is poorly defined at the continuum limit. The mass squared

values of the resonances are transcendental numbers and do not contribute at all to propagation assuming that the virtual momentum square values are rational or algebraic numbers. Therefore all
p-adicizations of the propagator are trivial.

b) For the second option only the "holomorphic" sub-algebra of conformal weights contributes to
the partition function. In this case a very complex spectrum of resonances, which are now poles,
emerges. Also now singularities correspond to transcendental values of mass squared so that singular
S-matrix elements are avoided when rational cutoff is used. The inverse of the propagator is well-
defined but at the limit when all non-trivial zeros of Riemann Zeta are included both the propagator
and itse inverse extremely singular mathematically. Obviously the kinetic part of the local field theory
action for M^4 field theory would be very awkward at this limit and would not provide anything new.

The only manner to avoid filling the entire phase space with resonances, is to assume a physical
(rather than only fictive) hierarchy of p-adic length scale and phase resolution cutoffs defined by the
hierarchy of sub-algebras of the super-canonical algebra defined by the first n non-trivial zeros of
Riemann Zeta.

As noticed, the ordinary perturbation theory fails for the vacuum extremals of Kähler action since
the modified Dirac operator vanishes in this case identically. One might expect that the propagator
develops a lot of singularities in the vicinity of vacuum extremals. Caustics would be the analog in
the classical Maxwell's theory so that these singularities are expected to be very important physically.
This singular behavior could correspond to the predicted presence of a dense set of singularities
(delta function or pole singularities depending on the option) for the super-canonical propagator at
the limit when all zeros of Zeta are included. If this were the case the limit as the cutoff N for
the number of non-trivial zeros of Riemann Zeta included goes to infinity, would correspond to the
approach to a vacuum extremal. Physically this would mean that zero modes would approach to a
limit at which configuration space metric becomes degenerate. In particular, the dimension of CP_2
projection would approach $D \leq 2$ since Lagrange sub-manifold is in question. The sub-algebras of
super-canonical algebra would classify the phases represented by space-time sheets and also conformal
equivalence classes of the configuration space metric. Super-canonical algebra has infinite number of
infinite-dimensional sub-algebras and each could correspond to a vacuum extremals. Note that the
partition function hypothesis would effectively mean the explicit calculation of the n-point functions
from symmetry considerations so that QFT limit would only provide the interpretation in terms of
space-time physics rather than being a practical tool.

The zeros of Riemann Zeta are known to be associated with chaotic systems and vacuum extremals
indeed correspond to chaotic systems by their non-deterministic behavior. The interpretation of the
vacuum extremals as space-time correlates for engineered world (recall the non-determinism) conforms
with this picture. The freedom to engineer would apply even to the mass spectrum of particles realized
as a freedom to choose the sub-algebra of the super-canonical algebra.

7.8.2   Could field theory limit defined in M^4 or H be useful?

The general vision behind quantum TGD suggests that the field theory limit should not contain
any dependence on the details of space-time surfaces. Thus it should be defined in either M^4 or
$M^4 \times CP_2$. The theory would be defined by an effective action defining the vertices and propagators
and tree diagrams would give the n-point functions and S-matrix elements. No problems with loop
divergences would be encountered.

At least formally one could start from vertices and propagators defined by n-point functions of
the conformal field theory as function of four-momenta and other quantum numbers of the incoming
particles and interpret them as local vertices involving differential operators in Minkowski space.
Assuming that the proper 2-point functions $\Gamma^{2)}$ identifiable as the inverse of the connected two point
function $G_c^{2)}$ exist, one could interpret them as analogs of the operators defining the differential
operators defining the kinetic part of the effective action. The construction of the scalar propagator
as a partition function for the super-canonical algebra however suggests that the $\Gamma^{2)}$ fails to exist
at the continuum limit without cutoff for the number of the zeros. As already described, an entire
hierarchy of these functions involving physical length scale and phase resolution cutoffs is involved
and has interpretation in terms of approach to vacuum extremal.

When the propagators defined as super-canonical are defined for finite sub-algebras of a holomorphic super-canonical algebra, the propagator and its inverse are well defined in momentum space. It

is not obvious whether the construction of a local field theory in M^4 by using the inverse brings in anything interesting.

The cautious conclusion would be that effective field theory limit makes sense in M^4 only by taking the propagators and vertices defined as n-point functions of the conformal field theory and using momentum space Feynman rules to build tree diagrams. The QFT limit could be seen as low energy approximation obtained by taking into account the vertices involving light particles.

7.9 Appendix A: Some examples of bi-algebras and quantum groups

The appendix summarizes briefly the simplest bi- and Hopf algebras and some basic constructions related to quantum groups.

7.9.1 Simplest bi-algebras

Let $k(x_1, .., x_n)$ denote the free algebra of polynomials in variables x_i with coefficients in field k. x_i can be regarded as points of a set. The algebra $Hom(k(x_1, ..., x_n), A)$ of algebra homomorphisms $k(x_1, ..., x_n) \to A$ can be identified as A^n since by the homomorphism property the images $f(x_i)$ of the generators $x_1, ...x_n$ determined the homomorphism completely. Any commutative algebra A can be identified as the $Hom(k[x], A)$ with a particular homomorphism corresponding to a line in A determined uniquely by an element of A.

The matrix algebra $M(2)$ can be defined as the polynomial algebra $k(a, b, c, d)$. Matrix multiplication can be represented universally as an algebra morphism Δ from from $M_2 = k(a, b, c, d)$ to $M_2^{\otimes 2} = k(a', a'', b', b'', c', c'', d', d'')$ to $k(a, b, c, d)$ in matrix form as

$$\Delta \begin{pmatrix} a & b \\ c & d \end{pmatrix} = \begin{pmatrix} a' & b' \\ c' & d' \end{pmatrix} \begin{pmatrix} a'' & b'' \\ c'' & d'' \end{pmatrix} .$$

This morphism induces algebra multiplication in the matrix algebra $M_2(A)$ for any commutative algebra A.

$M(2)$, $GL(2)$ and $SL(2)$ provide standard examples about bi-algebras. $SL(2)$ can be defined as a commutative algebra by dividing free polynomial algebra $k(a, b, c, d)$ spanned by the generators a, b, c, d by the ideal $det - 1 = ad - bc - 1 = 0$ expressing that the determinant of the matrix is one. In the matrix representation μ and η are defined in obvious manner and μ gives powers of the matrix

$$A = \begin{pmatrix} a & b \\ c & d \end{pmatrix} .$$

Δ, counit ϵ, and antipode S can be written in case of $SL(2)$ as

$$\begin{pmatrix} \Delta(a) & \Delta(b) \\ \Delta(c) & \Delta(d) \end{pmatrix} = \begin{pmatrix} a & b \\ c & d \end{pmatrix} \otimes \begin{pmatrix} a & b \\ c & d \end{pmatrix} ,$$

$$\begin{pmatrix} \epsilon(a) & \epsilon(b) \\ \epsilon(c) & \epsilon(d) \end{pmatrix} = \begin{pmatrix} 1 & 0 \\ 0 & 1 \end{pmatrix} .$$

$$S \begin{pmatrix} a & b \\ c & d \end{pmatrix} = (ad - bc)^{-1} \begin{pmatrix} d & -b \\ -c & a \end{pmatrix} .$$

Note that matrix representation is only an economical manner to summarize the action of Δ on the generators a, b, c, d of the algebra. For instance, one has $\Delta(a) = a \to a \otimes a + b \otimes c$. The resulting algebra is both commutative and co-commutative.

$SL(2)_q$ can be defined as a Hopf algebra by dividing the free algebra generated by elements a, b, c, d by the relations

$$\begin{aligned} ba &= qab , & db &= qbd , \\ ca &= qac , & dc &= qcd , \\ bc &= cb , & ad - da &= (q^{-1} - 1)bc , \end{aligned}$$

and the relation

$$det_q = ad - q^{-1}bc = 1$$

stating that the quantum determinant of $SL(2)_q$ matrix is one.

$\mu, \eta, \Delta, \epsilon$ are defined as in the case of $SL(2)$. Antipode S is defined by

$$S \begin{pmatrix} a & b \\ c & d \end{pmatrix} = det_q^{-1} \begin{pmatrix} d & -qb \\ -q^{-1}c & a \end{pmatrix} \ .$$

The relations above guarantee that it defines quantum inverse of A. For q an nth root of unity, $S^{2n} = id$ holds true which signals that these parameter values are somehow exceptional. This result is completely general.

Given an algebra, the R point of $SL_q(2)$ is defined as a four-tuple (A, B, C, D) in R^4 satisfying the relations defining the point of $SL_q(2)$. One can say that R-points provide representations of the universal quantum algebra $SL_q(2)$.

7.9.2   Quantum group $U_q(sl(2))$

Quantum group $U_q(sl(2))$ or rather, quantum enveloping algebra of $sl(2)$, can be constructed by applying Drinfeld's quantum double construction (to avoid confusion note that the quantum Hopf algebra associated with $SL(2)$ is the quantum analog of a commutative algebra generated by powers of a 2×2 matrix of unit determinant).

The commutation relations of $sl(2)$ read as

$$[X_+, X_-] = H \ , \quad [H, X_\pm] = \pm 2X_\pm \ . \tag{7.9.1}$$

$U_q(sl(2))$ allows co-algebra structure given by

$$\Delta(J) = J \otimes 1 + 1 \otimes J \ , \quad S(J) = -J \ , \quad \epsilon(J) = 0 \ , \quad J = X_\pm, H \ ,$$
$$S(1) = 1 \ , \qquad\qquad \epsilon(1) = 1 \ . \tag{7.9.2}$$

The enveloping algebras of Borel algebras $U(B_\pm)$ generated by $\{1, X'_+, H\}$ $\{1, X_-, hH\}$ define the Hopf algebra H and its dual $H^\star$ in Drinfeld's construction. h could be called Planck's constant vanishes at the classical limit. Note that $H^\star$ reduces to $\{1, X_-\}$ at this limit. Quantum deformation parameter q is given by $exp(2h)$. The duality map $\star : H \to H^\star$ reads as

$$a \to a^\star \ , \quad ab = (ab)^\star = b^\star a^\star \ ,$$
$$1 \to 1 \ , \quad H \to H^\star = hH \ , \quad X_+ \to (X_+)^\star = hX_- \ . \tag{7.9.3}$$

The commutation relations of $U_q(sl(2)$ read as

$$[X_+, X_-] = \frac{q^H - q^{-H}}{q - q^{-1}} \ , \quad [H, X_\pm] = \pm 2X_\pm \ . \tag{7.9.4}$$

Co-product Δ, antipode S, and co-unit ϵ differ from those $U(sl(2))$ only in the case of $X_\pm$:

$$\Delta(X_\pm) = X_\pm \otimes q^{H/2} + q^{-H/2} \otimes X_\pm \ ,$$
$$S(X_\pm) = -q^{\pm 1}X_\pm \ . \tag{7.9.5}$$

When q is not a root of unity, the universal R-matrix is given by

$$R = q^{\frac{H \otimes H}{2}} \sum_{n=0}^{\infty} \frac{(1-q^{-2})^n}{[n]_q!} q^{\frac{n(1-n)}{2}} q^{\frac{nH}{2}} X_+^n \otimes q^{-\frac{nH}{2}} X_-^n \ . \tag{7.9.6}$$

When q is m:th root of unity the q-factorial $[n]_q!$ vanishes for $n \geq m$ and the expansion does not make sense.

For q not a root of unity the representation theory of quantum groups is essentially the same as of ordinary groups. When q is m^{th} root of unity, the situation changes. For $l = m = 2n$ n^{th} powers of generators span together with the Casimir operator a sub-algebra commuting with the whole algebra providing additional numbers characterizing the representations. For $l = m = 2n + 1$ same happens for m^{th} powers of Lie-algebra generators. The generic representations are not fully reducible anymore. In the case of $U_q(sl(2))$ irreducibility occurs for spins $n < l$ only. Under certain conditions on q it is possible to decouple the higher representations from the theory. Physically the reduction of the number of representations to a finite number means a symmetry analogous to a gauge symmetry. The phenomenon resembles the occurrence of null vectors in the case of Virasoro and Kac Moody representations and there indeed is a deep connection between quantum groups and Kac-Moody algebras [ee2].

One can wonder what is the precise relationship between $U_q(sl(2)$ and $SL_q(2)$ which both are quantum groups using loose terminology. The relationship is duality. This means the existence of a morphism $x \rightarrow \Psi(x)$ $M_q(2) \rightarrow U_q^\star$ defined by a bilinear form $\langle u, x \rangle = \Psi(x)(u)$ on $U_q \times M_q(2)$, which is bi-algebra morphism. This means that the conditions

$$\langle uv, x \rangle = \langle u \otimes v, \Delta(x) \rangle \ , \quad \langle u, xy \rangle = \langle \Delta(u), x \otimes y \rangle \ ,$$

$$\langle 1, x \rangle = \epsilon(x) \ , \qquad \langle u, 1 \rangle = \epsilon(u)$$

are satisfied. It is enough to find $\Psi(x)$ for the generators $x = A, B, C, D$ of $M_q(2)$ and show that the duality conditions are satisfied. The representation

$$\rho(E) = \begin{pmatrix} 0 & 1 \\ 0 & 0 \end{pmatrix} \ , \quad \rho(F) = \begin{pmatrix} 0 & 0 \\ 1 & 0 \end{pmatrix} \ , \quad \rho(K = q^H) = \begin{pmatrix} q & 0 \\ 0 & q^{-1} \end{pmatrix} \ ,$$

extended to a representation

$$\rho(u) = \begin{pmatrix} A(u) & B(u) \\ C(u) & D(u) \end{pmatrix}$$

of arbitrary element u of $U_q(sl(2)$ defines for elements in $U_q^\star$. It is easy to guess that $A(u), B(u), C(u), D(u)$, which can be regarded as elements of $U_q^\star$, can be regarded also as R points that is images of the generators a, b, c, d of $SL_q(2)$ under an algebra morphism $SL_q(2) \rightarrow U_q^\star$.

7.9.3 General semisimple quantum group

The Drinfeld's construction of quantum groups applies to arbitrary semi-simple Lie algebra and is discussed in detail in [ee2]. The construction relies on the use of Cartan matrix.

Quite generally, Cartan matrix $A = \{a_{ij}\}$ is $n \times n$ matrix satisfying the following conditions:
i) A is indecomposable, that is does not reduce to a direct sum of matrices.
ii) $a_{ij} \leq 0$ holds true for $i < j$.
iii) $a_{ij} = 0$ is equivalent with $a_{ij} = 0$.
A can be normalized so that the diagonal components satisfy $a_{ii} = 2$.
The generators e_i, f_i, k_i satisfying the commutations relations

$$\begin{aligned} k_i k_j &= k_j k_i \ , & k_i e_j &= q_i^{a_{ij}} e_j k_i \ , \\ k_i f_j &= q_i^{-a_{ij}} e_j k_i \ , & e_i f_j - f_j e_i &= \delta_{ij} \frac{k_i - k_i^{-1}}{q_i - q_i^{-1}} \ , \end{aligned} \qquad (7.9.7)$$

and so called Serre relations

$$\begin{aligned} \sum_{l=0}^{1-a_{ij}} (-1)^l \begin{bmatrix} 1 - a_{ij} \\ l \end{bmatrix}_{q_i} e_i^{1-a_{ij}-l} e_j e_i^l &= 0, \ i \neq j \ , \\ \sum_{l=0}^{1-a_{ij}} (-1)^l \begin{bmatrix} 1 - a_{ij} \\ l \end{bmatrix}_{q_i} f_i^{1-a_{ij}-l} f_j f_i^l &= 0 \ , \ i \neq j \ . \end{aligned} \qquad (7.9.8)$$

Here $q_i = q^{D_i}$ where one has $D_i a_{ij} = a_{ij} D_i$. $D_i = 1$ is the simplest choice in this case.
Comultiplication is given by

$$
\begin{aligned}
\Delta(k_i) &= k_i \otimes k_i \ , & (7.9.9)\\
\Delta(e_i) &= e_i \otimes k_i + 1 \otimes e_i \ , & (7.9.10)\\
\Delta(f_i) &= f_i \otimes 1 + k_i^{-1} \otimes 1 \ . & (7.9.11)\\
& & (7.9.12)
\end{aligned}
$$

The action of antipode S is defined as

$$
S(e_i) = -e_i k_i^{-1} \ , \quad S(f_i) = -k_i f_i \ , \quad S(k_i) = -k_i^{-1} \ . \tag{7.9.13}
$$

7.9.4   Quantum affine algebras

The construction of Drinfeld and Jimbo generalizes also to the case of untwisted affine Lie algebras, which are in one-one correspondence with semisimple Lie algebras. The representations of quantum deformed affine algebras define corresponding deformations of Kac-Moody algebras. In the following only the basic formulas are summarized and the reader not familiar with the formalism can consult a more detailed treatment can be found in [ee2].

1. Affine algebras

The Cartan matrix A is said to be of affine type if the conditions $det(A) = 0$ and $a_{ij} a_{ji} \geq 4$ (no summation) hold true. There always exists a diagonal matrix D such that $B = DA$ is symmetric and defines symmetric bilinear degenerate metric on the affine Lie algebra.

The Dynkin diagrams of affine algebra of rank l have $l+1$ vertices (so that Cartan matrix has one null eigenvector). The diagrams of semisimple Lie-algebras are sub-diagrams of affine algebras. From the $(l+1) \times (l+1)$ Cartan matrix of an untwisted affine algebra $\hat{A}$ one can recover the $l \times l$ Cartan matrix of A by dropping away 0:th row and column.

For instance, the algebra A_1^1, which is affine counterpart of $SL(2)$, has Cartan matrix a_{ij}

$$
A = \begin{pmatrix} 2 & -2 \\ -2 & 2 \end{pmatrix}
$$

with a vanishing determinant.

Quite generally, in untwisted case quantum algebra $U_q(\hat{G}_l)$ as $3(l+1)$ generators e_i, f_i, k_i ($i = 0, 1, .., l$) satisfying the relations of Eq. 7.9.8 for Cartan matrix of $\mathcal{G}^{(1)}$. Affine quantum group is obtained by adding to $U_q(\hat{G}_l)$ a derivation d satisfying the relations

$$
[d, e_i] = \delta_{i0} e_i \ , \quad [d, f_i] = \delta_{i0} f_i, \quad [d, k_i] = 0 \ . \tag{7.9.14}
$$

with comultiplication $\Delta(d) = d \otimes 1 + 1 \otimes d$.

2. Kac Moody algebras

The undeformed extension $\hat{\mathcal{G}}_l$ associated with the affine Cartan matrix $\mathcal{G}_l^{(1)}$ is the Kac Moody algebra associated with the group G obtained as the central extension of the corresponding loop algebra. The loop algebra is defined as

$$
L(\mathcal{G}) = \mathcal{G} \otimes C\left[t, t^{-1}\right] \ , \tag{7.9.15}
$$

where $C\left[t, t^{-1}\right]$ is the algebra of Laurent polynomials with complex coefficients. The Lie bracket is

$$
[x \times P, y \otimes Q] = [x, y] \otimes PQ \ . \tag{7.9.16}
$$

The non-degenerate bilinear symmetric form $(,)$ in $\mathcal{G}_l$ induces corresponding form in $L(\mathcal{G}_l)$ as $(x \otimes P, y \otimes Q) = (x, y)PQ$.

A two-cocycle on $L(\mathcal{G}_l)$ is defined as

$$\Psi(a, b) = Res(\frac{da}{dt}, b) \ , \tag{7.9.17}$$

where the residue of a Laurent is defined as $Res(\sum_n a_n t^n) = a_{-1}$. The two-cocycle satisfies the conditions

$$\Psi(a, b) = -\Psi(b, a) \ ,$$
$$\Psi([a, b], c) + \Psi([b, c], a) + \Psi([c, a], b) = 0 \ . \tag{7.9.18}$$

The two-cocycle defines the central extension of loop algebra $L(\mathcal{G}_l)$ to Kac Moody algebra $L(\mathcal{G}_l) \otimes Cc$, where c is a new central element commuting with the loop algebra. The new bracket is defined as $[,] + \Psi(,)c$. The algebra $\tilde{L}(\mathcal{G}_l)$ is defined by adding the derivation d which acts as td/dt measuring the conformal weight.

The standard basis for Kac Moody algebra and corresponding commutation relations are given by

$$\begin{aligned} J_n^x &= x \otimes t^n \ , \\ [J_n^x, J_m^y] &= J_{n+m}^{[x,y]} + n\delta_{m+n,0}c \ . \end{aligned} \tag{7.9.19}$$

The finite dimensional irreducible representations of G defined representations of Kac Moody algebra with a vanishing central extension $c = 0$. The highest weight representations are characterized by highest weight vector $|v\rangle$ such that

$$\begin{aligned} J_n^x|v\rangle &= 0, \ n > 0 \ , \\ c|v\rangle &= k|v\rangle \ . \end{aligned} \tag{7.9.20}$$

3. Quantum affine algebras

Drinfeld has constructed the quantum affine extension $U_q(\mathcal{G}_l)$ using quantum double construction. The construction of generators uses almost the same basic formulas as the construction of semi-simple algebras. The construction involves the automorphism $D_t : U_q(\tilde{\mathcal{G}}_l) \otimes C\left[t, t^{-1}\right] \to U_q(\tilde{\mathcal{G}}_l) \otimes C\left[t, t^{-1}\right]$ given by

$$\begin{aligned} D_t(e_i) &= t^{\delta_{i0}}e_i \ , \quad D_t(f_i) = t^{\delta_{i0}}f_i \ , \\ D_t(k_i) &= k_i \quad\quad\ \ D_t(d) = d \ , \end{aligned} \tag{7.9.21}$$

and the co-product

$$\Delta_t(a) = (D_t \otimes 1)\Delta(a) \ , \quad \Delta_t^{op}(a) = (D_t \otimes 1)\Delta^{op}(a) \ , \tag{7.9.22}$$

where the $\Delta(a)$ is the co-product defined by the same general formula as applying in the case of semi-simple Lie algebras. The universal R-matrix is given by

$$\mathcal{R}(t) = (D_t \otimes 1)\mathcal{R} \ , \tag{7.9.23}$$

and satisfies the equations

$$\mathcal{R}(t)\Delta_t(a) = \Delta_t^{op}(a)\mathcal{R} \ ,$$

$$(\Delta_z \otimes id)\mathcal{R}(u) = \mathcal{R}_{13}(zu)\mathcal{R}_{23}(u) \ ,$$

$$(id \otimes \Delta_u)\mathcal{R}(zu) = \mathcal{R}_{13}(z)\mathcal{R}_{12}(zu) \ ,$$

$$\mathcal{R}_{12}(t)\mathcal{R}_{13}(tw)\mathcal{R}_{23}(w) = \mathcal{R}_{23}(w)\mathcal{R}_{13}(tw)\mathcal{R}_{12}(t) \ .$$

(7.9.24)

The infinite-dimensional representations of affine algebra give representations of Kac-Moody algebra when one restricts the consideration to generations $e_i, f_i, k_i,\ i > 0$.

7.10   Appendix B: Scalar field propagator for option I

In the following various options for scalar field propagator defined as a partition function are studied in detail.

7.10.1   Propagator assuming all conformal weights predicted by super-canonical algebra

The super-canonical algebra defining the algebra of isometries of configuration space has conformal weights expressible in terms of both trivial and non-trivial zeros of Riemann Zeta and this leads to the idea that scalar field propagator could be regarded as a partition function in super-canonical algebra. The construction turned out to be extremely useful by providing the physical input making it easier to grasp the structure of the super-canonical algebra. In the following the assumption that y_i are linearly independent is made in the hope that it is a good approximation although the facts about the spectral function favor the ansatz predicting linear dependence.

1. Scalar propagator as a partition function for Cartan decomposition of the super-canonical algebra

The idea is to construct propagator as a partition function for the super-canonical algebra and require that it is faithful to the Cartan decomposition of the algebra defining configuration space of 3-surfaces as a symmetric space. As shown in [B2], the super-canonical algebra decomposes to two parts $g = g_t + g_{nt}$ corresponding to trivial and non-trivial zeros of Zeta.

a) The Cartan decomposition $g = t + h$, $[t, t] \subset h$, $[h, h] \subset h$, $[t, h] \subset t$ means that only the generators of t correspond to physical degrees of freedom because of coset space-structure meaning that h corresponds to gauge degrees of freedom. Thus only the conformal weights of t contribute to the partition function defining the propagator. The conformal weights associated with t_n are $h_t = 2n$ whereas the conformal weights associated with t_{nt} are

$$h_{nt,1} = 2n - \tfrac{1}{2} - i\sum_{y_i>0} n_i y_i \ , \quad \sum_i n_i = 2N+1 \ , \quad n > 0 \ ,$$

$$h_{nt,2} = 2n + \tfrac{1}{2} - i\sum_{y_i>0} n_i y_i \ , \quad \sum_i n_i = 2N \ , \quad n > 0 \ .$$

(7.10.1)

The sub-spaces defined by $n = 0$ defined together with trivial zeros an orthogonal basis. It is however unclear whether $n > 0$ states can satisfy the orthogonality conditions and whether they should be excluded as states generated by Virasoro generators L_{2n}, $n > 0$.

b) The propagator is defined as a partition function

$$\frac{G_R}{\lambda^2} = \sum_{z \in t} u^{f(z)} \ ,$$

$$u = \frac{p^2 - m^2}{\lambda^2} \ ,$$

(7.10.2)

where $f(z)$ is a linear function of the conformal weight $z = h_t, ht_{tn}$.

c) Propagator can be expressed as a sum of three parts

$$\frac{G_R}{\lambda^2} = G_{R,1} + G_{R,2} + G_{R,3}$$
$$= \epsilon_1 \sum_{z=-2n} G_0^{f(z)} + \epsilon_2 \sum_{z=h_{nt,1}} G_0^{f(z)} + \epsilon_3 \sum_{z=h_{nt,2}} G_0^{f(z)} \ . \tag{7.10.3}$$

The sign factors must be chosen in such a manner that the resulting expression for the propagator is physically acceptable. The sign factor should define a Lie algebra homomorphism: $\epsilon([A,B]) = \epsilon(A)\epsilon B$. For the choice $\epsilon_1 = \epsilon_3 = 1, \epsilon_2 = \pm 1$ this condition is satisfied. In the following $\epsilon = \epsilon_3 = 1$ is assumed and the value of ϵ_2 is left free: it turns out that $\epsilon_2 = -1$ is the only possible consistent choice if one allows all values of n for conformal weights hnt. If only $n = 0, 1$ is allowed then $\epsilon_2 = 1$ is possible.

d) The requirement that $1/(p^2 - m^2)$ behavior results as the lowest order contribution from $G_{R,1}$, implies $f(z) = -z/2$, and one has

$$\frac{G_{R,1}}{\lambda^2} = \sum_{n>0} G_0^n \ ,$$
$$\frac{G_{R,2}}{\lambda^2} = \epsilon_2 \sum_n G_0^{2n-1/2} \times \sum_{\sum n(y)y} G_0^{[-i\sum n(y)y]/2} \ | \sum_y n(y) \ odd \ ,$$
$$\frac{G_{R,3}}{\lambda^2} = \sum_n G_0^{2n+1/2} \times \sum_{\sum n(y)y} G_0^{[-i\sum n(y)y]/2} \ | \sum_y n(y) \ even \ . \tag{7.10.4}$$

Here $\sum n(y)y$ appearing as summation index can be regarded a formal superposition over imaginary parts of arbitrary subset of non-trivial zeros of Zeta (note that the linear independence over zeros is an essential although probably only approximate assumption).

Consider the expressions for the various parts of the propagator.

a) The expression for $G_{R,1}$ reads as

$$G_{R,1} = \frac{1}{p^2 - m^2 - \lambda^2} \ . \tag{7.10.5}$$

The pole is shifted unless $m^2(\lambda)$ depends on λ as $m^2 = m_0^2 - \lambda^2$.

b) The expressions for $G_{R,2}$ and $G_{R,3}$ are sums of terms proportional to sums

$$\sum_{\sum_i n(y_i)y_i} exp\left[-ilog(u)\sum_i n(y_i)y_i/2\right]$$

where $n(y)$ is odd or even. For $u > 0$ the conditions guaranteing the vanishing of these sums can be derived by assuming that the imaginary parts y of the zeros of Riemann Zeta are linearly independent do not satisfy any identify of form

$$\sum_i n_i y_i = 0 \ , \ n_i \in Z \ ,$$

so that they generate an Abelian group with an infinite number of generators defined by y. With this assumption the sum over each y gives a delta function $\delta(log(u) - 4\pi k/y)$, and if the exponent contains more than one y_i, the result is zero. The sums give delta functions allowing p^2 to have only discrete values

$$p^2 = m_{y,k}^2 = m^2 + \lambda^2 \times e^{\frac{2\pi k}{y}} \ , \ k \in Z \ , \ y > 0 \ . \tag{7.10.6}$$

The expressions for $G_{R,2}$ and $G_{R,3}$ are

$$\frac{G_{R,2}}{\lambda^2} = \epsilon_2 \sum_{k,y>0} \delta(u - u_{k,y})(-1)^k u_{k,y}^{-1/4} f_i(u_{k,y}) \ ,$$

$$\frac{G_{R,3}}{\lambda^2} = \sum_{k,y>0} \delta(u - u_{k,y}) u_{k,y}^{1/4} f_i(u_{k,y}) \ ,$$

$$u_{k,y} = \frac{m_{k,y}^2 - m^2}{\lambda^2} = e^{\frac{2\pi k}{y_i}} \ . \tag{7.10.7}$$

The function $f_i(u_{k,y})$, $i = 1,2$ results from the summation over powers u^n, and $i = 1,2$ labels $n = 0,1$ and $n = 0,1,...$ options for the conformal weights h_{nt}. One has

$$f_1(u_{k,y}) = 1 + u(k,y) \ ,$$

$$f_2(u_{k,y}) = \frac{1}{1 - u(k,y)} \ . \tag{7.10.8}$$

Except for the dependence coming from the delta functions expressing the fact that resonances masses scale as λ^2, these parts of propagator are independent of λ.

c) The overall expression for the propagator in the region $p^2 - m^2 \geq 0$ reads as

$$\frac{G_R}{\lambda^2} = \frac{1}{p^2 - m^2 - \lambda^2} + \sum_{k,y>0} \delta(p^2 - m_{k,y}^2) \left[u_{k,y}^{1/4} + \epsilon_2(-1)^k u_k^{-1/4} \right] f_i(u_{k,y}) \ . \tag{7.10.9}$$

d) The guess that the propagator in the space-like region $p^2 - m^2 < 0$ is obtained by an analytical continuation turns out to be correct. The explicit summation gives zero always irrespective of the value of whether u is positive or negative unless the resonance conditions are satisfied. This means that resonance contributions vanish for $p^2 - m^2 < 0$, that is for $p^2 < m_0^2 - \lambda^2$ for $m^2 = m_0^2 - \lambda^2$ option implied by renormalization group invariance.

2. Pole and resonance structure of the propagator

The pole of the propagator is shifted to $m^2 + \lambda^2$ and renormalization conditions are satisfied at $p^2 = m^2 + \lambda^2$. The renormalization group invariance condition applied on mass shell gives $\lambda dG_R/d\lambda = 0$. Only $G_{R,1}$ contributes to the condition for on mass shell and one has

$$m^2(\lambda) = m_0^2 - \lambda^2 \ . \tag{7.10.10}$$

Obviously $m^2 = m_0^2$ becomes renormalization group invariant.

The propagator is thus a sum of free propagator with additional delta-function contributions at $p^2 = m_{y_i,k}^2$ for $k \neq 0,$. There appears no pole (that is propagating physical particle) at these masses because one has $[p^2 - m_{y_i,k}^2]\delta(p^2 - m_{y_i,k}^2) = 0$.

Consider now the choice of the parameter ϵ_2 and the choice between $n = 0,1$ and $n < 0$ options corresponding to $f_1(u) = 1 + u$ and $f_2(u) = 1/(1 - u)$.

a) For $n = 0,1$ option corresponding to $f_1(u) = 1 + u$ no pole is generated for $k = 0$ and $\epsilon_2 = 1$ is possible. The choice $\epsilon_2 = -1$ however cancels the resonance at $p^2 = m_0^2$ corresponding to $k = 0$. This is highly desirable since at the limit when all zeros of Zeta are allowed the resonance strength of this resonance becomes infinite. The result conforms with the view that the generators L_n, $n \geq 1$, indeed generate zeros states.

b) For $n > 0$ option corresponding the situation is more delicate assuming that the condition $m^2(\lambda^2) = m_0^2 - \lambda^2$ is satisfied. The factor

$$\frac{\left[u_{k,y}^{1/4} + \epsilon_2(-1)^k u_{k,y}^{-1/4} \right]}{1 - u_{k,y}} = \frac{e^{\frac{\pi k}{2y}} + \epsilon_2(-1)^k e^{-\frac{\pi k}{2y}}}{1 - e^{\frac{2\pi k}{y}}}$$

would have a pole for $\epsilon_2 = 1$ because of vanishing of the denominator $1 - u_{k=0,y}$ and the residue of the propagator would be infinite at $p^2 = m^2$ rather than being equal to 1. Thus the only possible choice is $\epsilon_2 = -1$ in this case. One has however still the problem with the infinite value of the resonance strength when all zeros of Zeta are included.

One can say that the virtual mass squared possesses a discrete resonance spectrum superposed to the free propagation. The interpretation as a kind of fractal noise might make sense. The propagator contains the resonance contributions also in the space-like region for $p^2 - m^2 = m_0^2 - \lambda^2 < 0$.

The mass spectrum for the resonances can be written as

$$m_{k,y}^2 \;=\; m_0^2 + \lambda^2 \left[e^{\frac{2\pi k}{y}} - 1 \right] \;, \tag{7.10.11}$$

and depends on λ. The spectrum contains both s-channel resonances corresponding to $k > 0$ and tachyonic exchange resonances corresponding to $k < 0$. The symmetry between the two spectra brings in mind string model duality. For tree diagrams the mass quantization of exchanged resonances would mean that discretization occurs in the momentum space for resonances. For instance, in the scattering of two particles in cm frame the space-like resonances appear at discrete values of the scattering angle θ as a kind of a fractal rainbow effect. The couplings to the resonances are reduced exponentially at the time-like limit $k \to \infty$ and increase exponentially at the space-like limit $k \to -\infty$ ($m^2 \to m_0^2 - \lambda^2$).

The inverse $D = 1/G$ of the propagator does not make sense at the momenta corresponding to the resonances. The reason is that the geometric series expansion of

$$G_R^{-1} = \frac{1}{G_{R,3}[1 + G_{R,3}^{-1}(G_{R,1} + G_{R,2})]}$$

involves ill-defined powers of delta functions $\delta(p^2 - m_{k,y_i}^2)$. The interpretation is that the propagator does not allow a non-singular field theory limit.

3. Why a cutoff in the number of non-trivial zeros of Riemanm Zeta is necessary?

Any subset $\{y > 0\}$ of non-trivial zeros of Riemann Zeta defines a sub-algebra of the super-canonical algebra. This suggests a hierarchy of sub-algebras defined by the set $Y_n = \{y_1, ..., y_n\}$ of n lowest zeros defining a hierarchy of propagators containing only sum over conformal weights of a sub-algebra of super-canonical algebra. The restriction to sub-algebra defined by Y_n could be interpreted as a cutoff in y-space. This cutoff would not be a mere calculational trick but would represent infinite hierarchy of "phases" with a finite y-cutoff meaning increasing density of points in the lattice defined by the set Y_n. One could even think that the full super-canonical algebra as well as propagators, vertices and S-matrices are *defined* by a nested hierarchy of sub-algebras defined by the subsets $Y_n \subset Y_{n+1}$.

There are several reasons making this kind of cutoff necessary.

a) Without the cutoff the number of resonances in a mass squared interval $[m^2, m^2 + \Delta m^2]$ is infinite. The condition for the resonances can be written as

$$\frac{m^2 - m_0^2}{\lambda^2} + 1 = exp\left[\frac{2\pi k}{y}\right] \;.$$

For sufficiently large values of y one can always find at least one $k(y)$ such that the resulting mass squared is contained in the interval $[m^2, m^2 + \Delta m^2]$. Without the y-cutoff rates for the scattering with final state momenta restricted in a finite volume elements of the phase space would be infinite. The situation changes for rational cutoff since the resonances turn out to correspond to the values of u which are not rational or do not even belong to a finite-dimensional extension of rationals. In this case the delta singularities disappear.

b) The p-adicization of the theory requires y-cutoff. Consider p-adicization in a given p-adic number field R_p. Suppose that the values of $u = (p^2 - m^2)/\lambda^2$ appearing in the propagator are rational. The p-adic existence of the phase factors u^{iy} is required by the definition of the partition function as a function making sense also p-adically. If rationals containing primes smaller than some cutoff prime p_c are allowed, this condition boils down to the existence of the phases q^{iy}, $q \leq p_c$ prime, as a number in a finite-dimensional extension of the p-adic number field R_p.

Suppose that the phase factors q^{iy}, q prime belong to a finite-dimensional algebraic extension of any p-adic number field R_p for any prime q and any non-trivial zero $z = 1/2 + y$ of Zeta. This is achieved if the phases p^{iy_i} are expressible as products of roots of unity and Pythagorean phases:

$$
\begin{aligned}
p^{iy} &= e^{i\phi_P(p,y)} \times e^{i\phi(p,y)} \;, \\
e^{i\phi_P(p,y)} &= \frac{r^2 - s^2 + i2rs}{r^2 + s^2} \;, \quad r = r(p,y) \;, \quad s = s(p,y) \;, \\
e^{i\phi(p,y)} &= e^{i\frac{2\pi m}{n}} \;, \quad m = m(p,y) \;, \quad n = n(p,y) \;.
\end{aligned}
\tag{7.10.12}
$$

The Pythagorean phases associated with two different zeros of zeta must be different in order to have linear independence y_i over integers. Of course, linear independence is just a working hypothesis to be tested and it will be found that there is support for the view that linear independence leads to non-physical results and is not favored by the facts known from the distribution of zeros of Zeta. Pythagorean phases form a multiplicative group having the phases of squares of Gaussian primes as its generators. The squares of Gaussian primes are primes $p \bmod 4 = 1$.

One can decompose for any p any zero y_i into two parts $y_i = y_{ia} + y_{iP}$ such that one has

$$
log(p)y_{ia}(p) = \frac{m2\pi}{n} \;, \quad log(p)y_{iP} = arctan\left[\frac{2rs}{r^2 + s^2}\right] \;,
\tag{7.10.13}
$$

where m, n, r, s of course depend on y and p. The decomposition is unique if one requires $m < n$.

c) The exponents $exp(\pi k/2y)$ appearing in the formula for the propagator can be expressed as

$$
e^{\frac{\pi k}{2y}} = q^{\frac{k}{2[log(p)y_1(p)+y_P(p)]}} \;.
\tag{7.10.14}
$$

Due to the presence of y_P these numbers are not rational nor even algebraic unless additional conditions are posed. Scalar propagator would reduce to a free propagator for rational values of u propagator and the p-adicization of the propagator would always give just the free propagator.

d) The propagator could be formally interpreted as a partition function with the conformal weights of t being in the role of energy spectrum and $2log(u)$ in the role of the inverse temperature. In the real context physical high energy limit corresponds to a formal low temperature limit. The p-adic length scale hierarchy with decreasing p-adic length scales and increasing value of λ presumably corresponds to a series of phase transitions increasing λ and the temperature in a discontinuous manner occur unless one can achieve overcooling. p-Adically the increase of the upper bound for u means that the set of rational values of u increases in size. This means the increase of p_c and thus also of the p-adic prime p.

e) The propagator has several non-physical features. The reduction to single particle sector in the sense that all contributions involving sums of at least two different zeros y vanish, is definitely a pathological feature. The presence of delta function singularities for real mass squared values is a second un-physical feature. The limit when all zeros contribute to the propagator gives real propagator for which delta function singularities fill the real axis. The reduction to a free propagator in p-adic context for all rational cutoffs is a further questionable feature. All these features follow from the assumption that the zeros are linearly independent. The actual linear dependence implied by the ansatz explaining the correlations between zeros implies that this approximation leads to multiple counting of some super-canonical weights and there are good hopes that the singularities are due to this over counting.

7.10.2 Propagator as a partition function associated with the holomorphic sub-algebra of the super-canonical algebra

The interpretation of the conformal weights of super-canonical algebra as punctures of complex plane at which Super Kac-Moody conformal Virasoro algebra acts as conformal transformations encourages to consider also the subalgebra generated by allowing only the generators with conformal weights $z = -1/2 - iy$, $y > 0$ and $z = n$, $n > 0$. In this case the sums appearing in the definition of the propagator as a partition function would be only over sums $\sum_n (y)n(y)y$ involving only positive values of y and one would have $N > 0$ and $n(y) \geq 0$ in the sums of Eq. 7.10.4. Also now the working hypothesis is that y_i are linearly independent.

The outcome is the following general formula

$$\frac{G_{R,2}}{\lambda^2} = \epsilon_2 u^{-1/4} f_i(u) \times \sum_{N \geq 0} \sum_{\sum n(y)y} X\left[\sum n(y)y | 2N + 1\right] ,$$

$$\frac{G_{R,3}}{\lambda^2} = u^{1/4} f_i(u) \times \sum_{N \geq 1} \sum_{\sum n(y)y} X\left[\sum n(y)y | 2N\right] ,$$

$$X\left[\sum n(y)y | M\right] = u^{[-i\sum n(y)y]/2} , \quad \sum n(y) = M. \tag{7.10.15}$$

Here $\sum_y n(y)y$ refers to formal sum of non-trivial zeros of Zeta with $n(y) > 0$. The terms in the sum can be arranged according to how many different zeros contribute to the sum $\sum n(y)y$, denote this number by K. The sum can be decomposed to a sum over all possible different subsets $U_K = \{y_{i_1}, ..., y_{i_K}\}$ of non-trivial zeros of Zeta. Hence one can write

$$\frac{G_{R,2}}{\lambda^2} = \epsilon_2 u^{-1/4} f_i(u) \times \sum_{K \geq 1} \sum_{U_K} \sum_{N \geq K-1} \sum_{\sum n_k = 2N+1} e^{-i\sum_{k=1}^{K} n_k y_{i_k} log(u)} ,$$

$$\frac{G_{R,3}}{\lambda^2} = u^{1/4} f_i(u) \times \sum_{K \geq 1} \sum_{U_K} \sum_{N \geq K-1} \sum_{\sum n_k = 2N} e^{-i\sum_{k=1}^{K} n_k y_{i_k} log(u)} . \tag{7.10.16}$$

Here $n_k \geq 1$ holds true for every y_{i_k}. The evaluation of the sums in terms of geometric sums is in principle straightforward for arbitrary value of K but the expressions get complicated as K increases.

Apart from factors $u^{\pm 1/4} f_i(u)$, $G_{R,2}$ resp. $G_{R,3}$ could be interpreted as a partition function for a system consisting of harmonic oscillators with fundamental frequencies y_i with the restriction that the total number of quanta is odd *resp.* even for the allowed states.

This modification alters dramatically the structure of the propagator.

a) Terms involving sums $\sum_k n_k y_{i_k}$ in the exponent contribute to the propagator and the outcome is terms depending on arbitrary many different y:s instead of only single particle terms as in the previous case.

b) Single particle contributions are transformed from resonances to genuine poles. For instance, in $G_{R,2}$ the sum

$$\sum_{n(y) \in Z} e^{ilog(u)(2n_y + 1)/2}$$

giving single particle contribution is replaced with the sum

$$\sum_{n(y) \geq 0} e^{ilog(u)(2n(y)+1)/2} = \frac{e^{ilog(u)y/2}}{1 - e^{ilog(u)y}} .$$

The delta functions at $u = exp(k2\pi/y)$ transform to poles. This predicts infinite number of particles.

$n = 0, 1$ option with $\epsilon_2 = -1$ seems to be the only sensible choice since in this case there is no singularity at $p^2 = m_0^2$. This option guarantees the vanishing of $G_{R,2} + G_{R,3}$ at $u = 1$ quite generally. For other options the pole contribution at $u = 1$ ($p^2 = m_0^2$) is present, and implies that the propagator normalization is changed and without y-cutoff the residue of the propagator at $p^2 = m_0^2$ is infinite.

Tachyons result unless the condition

$$m_0^2 > \lambda^2$$

holds true so that λ^2 would play the role of infrared- rather than ultraviolet cutoff. This raises the possibility that the restriction to holomorphic sub-algebra makes sense for infrared cutoff whereas full algebra would make sense for ultraviolet cutoff. Ultraviolet cutoff means finite time resolution $\tau \sim 1/\sqrt{\lambda}$ and the interpretation would be that the lifetimes of resonances are shorter than τ. Ultraviolet cutoff would also mean that multiple y states are not seen at all in the spectrum. By previous arguments delta function type singularities disappear completely form the p-adicized theory whereas poles are visible but for rational values of u they do not produce diverging S-matrix elements.

c) Two-particle contributions read as

$$\frac{1}{e^{iy_2x} - e^{iy_1x}} \times \left[\frac{e^{i4y_2x}}{1 - e^{i2y_2x}} - \frac{e^{i4y_1x}}{1 - e^{i2y_1x}} \right] \ , \quad x = log(u) \ .$$

Besides new poles poles at $u = e^{k\pi/y_i}$ there are also poles at $u = e^{k2\pi/(y_1 - y_2)}$.

d) The study of three-particle contributions suggests that the term containing K different y:s gives rise to poles at $u = e^{k2\pi/x}$, for $x = 2y_i$ and $x = y_{ij} = y_i - y_j$ and $xi \neq j \neq k = y_{ij} + y_{kj}$, etc.

e) In order to not fill the entire phase space densely with propagator poles one must assume a hierarchy of y-cutoffs. The most general hierarchy is obtained by restricting the consideration to a set of finitely generated subalgebras defined by the sets U_K, $K < K_{max}$. In this case the convergence is not guaranteed, in particular in p-adic context there could be problems since infinite number of terms are present. Number-theoretical considerations suggest a simpler filtered hierarchy of subalgebras defined by the sets $\{y_1, ... y_K\}$.

f) Also this propagator has non-physical features. The presence of singularities are real values of mass squared is physically unrealistic feature since resonance widths are infinitely narrow in this kind of situation. The propagator is ill-define at the limit when all zeros contribute to it. These features are solely due to the assumed linear independence of y_i.

7.10.3 Propagator assuming that only zeros of ζ contribute to the spectrum of conformal weights

Without any idea about the structure of the super-canonical algebra the most probable guess would be that the partition function defining the scalar propagator contains only the contribution of trivial and non-trivial zeros but not the contributions corresponding to the commutators of these generators. This option could be also defended by saying that the conformal weights $\pm 1/2 + \sum_i n_i y_i$ correspond in some sense to many particle states.

With this assumption the propagator would reduce to a sum of two terms analogous to $G_{R,1}$ and $G_{R,2}$ in the general formula with the summations restricted to $n(y) = 1$. The term corresponding to $G_{R,3}$ would not be present since $G_{R,3}$ can involve only even values of $\sum n(y)$. This would lead to a propagator with a resonance type on mass shell singularity. In the following the option allowing $n(y) = 2$ at the line $Re(h) = 1/2$ is also considered since it gives a non-singular propagator at $p^2 = m_0^2$.

From the general formula of Eq. 7.10.4 for the propagator one finds that the expression for the propagator in this case reads as

$$G_R \ = \ \frac{1}{p^2 - m^2 - \lambda^2} + \epsilon_2 u^{-1/4} \sum_{y>0} cos\left[ilog(u)\frac{y}{2} \right] + \delta \times u^{1/4} \sum_{y>0} cos\left[ilog(u)y \right] \ .$$

$$(7.10.17)$$

$\delta = 0$ characterizes the strictly "single particle" case $n(y) = 1$.

The requirement of on mass shell renormalization group invariance implies $m^2 = m_0^2 - \lambda^2$. ($\delta = 1, \epsilon_2 = -1$) option allows a propagator having only a pole singularity at $p^2 = m_0^2$ and satisfying renormalization conditions. For $\delta = 0$ the additional resonance singularity is unavoidable but renormalization conditions are still satisfied. The propagator reads as

$$G_R \ = \ \frac{1}{p^2 - m_0^2} - (1 + \frac{p^2 - m_0^2}{\lambda^2})^{-1/4} \sum_{y>0} cos\left[ilog(1 + \frac{p^2 - m_0^2}{\lambda^2})\frac{y}{2} \right]$$

$$+ \ \delta \times (1 + \frac{p^2 - m_0^2}{\lambda^2})^{1/4} \sum_{y>0} cos\left[ilog(1 + \frac{p^2 - m_0^2}{\lambda^2})y \right] \ .$$

$$(7.10.18)$$

The powers of the logarithm imply a cut for $p^2 < m_0^2 - \lambda^2$ whereas in the region $p^2 > m_0^2 - \lambda^2$ logarithm and $1/4^{th}$ root have power series expansion. It is not clear whether the sum over zeros converges or not in the general case.

The sum over infinite number of phase factors u^{iy} is problematic p-adically.

i) The basic hypothesis is that the phase factors p^{iy} belong to finite-dimensional extensions of p-adic numbers for all primes p and zeros of Zeta. This would mean that the u^{iy} exists for all rational values of u but the sum $\sum_y u^{iy}$ fails to exist as an infinite sum of numbers possessing unit p-adic norm. y-cutoff would be necessary.

ii) If y would exist in an extension of R_p, cosines exist provided the conditions $|log\left[1 + (p^2 - m^2)/\lambda^2\right])y|_p < 1$ hold true. In the p-adic context the logarithm exists for $|(p^2 - m^2)/\lambda^2|_p < 1$ and has the same p-adic norm as $(p^2 - m^2)/\lambda^2$. In 2-adic case also the condition $|(p^2 - m^2)y/2\lambda^2|_2 < 1$ must be satisfied.

It is interesting to look what happens if one assumes that the hypothesis explaining the correlations between the zeros holds true. In this case the sum over zeros decomposes into a separate sum over Pythagorean phases y_P and over rational phases assumed to be same for each y_P. One can write the term

$$A \equiv \sum_{y>0} cos\left[ilog(1 + \frac{p^2 - m_0^2}{\lambda^2})\frac{y}{x}\right] \;\; ,$$

where one has either $x = 1$ or $x = 2$ as a real part of a product two sums:

$$q = Re(X) \;\; ,$$
$$X = \sum_{y_P>0} exp\left[ilog(1 + \frac{p^2-m_0^2}{\lambda^2})\frac{\Phi_P}{xlog(p)}\right] \times \sum_{q\in Q_0} exp\left[ilog(1 + \frac{p^2-m_0^2}{\lambda^2})\frac{q2\pi}{xlog(p)}\right] \;\; .$$

The first sum is over phase angles associated with a subset of Pythagorean prime phases (squares of Gaussian primes) and second over rational phase angles. Note that the non-uniqueness due to the decomposition to Pythagorean and algebraic phase angle does not affect the value of the X.

The conclusion is that the proposed form of the propagator is not favored by neither physical nor number theoretic conditions.

7.10.4 Do primes and their inverses correspond to the zeros of the propagators $G_{R,...}$?

The previous considerations rely on the probably wrong working hypothesis that the imaginary parts y_i of the nontrivial zeros $z_i = 1/2 + y_i$, $y_i > 0$, of Riemann Zeta are linearly independent.

The scalar propagator is defined as a partition function in the super-canonical algebra and one can decompose it into a sum of three parts with $G_{R,2}$ and $G_{R,3}$ being directly related to the non-trivial zeros of ζ. Also Riemann Zeta has an interpretation as a partition function. This raises interesting questions. Could Riemann Zeta and the building blocks $G_{R,2}$ and/or $G_{R,3}$ or some closely related partition functions be somehow duals of each other? Could Riemann Zeta be for ordinary primes what $G_{R,2}$ and/or $G_{R,3}$ are to y_i? What could be the counterpart of Riemann Zeta for the primes $z_i = 1/2 + y_i$?

If the correspondences makes sense, G_R or $G_{R,2}$ and/or $G_{R,3}$ should vanish for $u = p$ and $u = 1/p$ in an analogy with the fact that $1/2 \pm y_i$ and its conjugate are zeros of Zeta. This would predict the mass formula $M^2(p) = m_0^2 - \lambda^2 + p^{\pm1}\lambda^2$. It is interesting to look whether this hypothesis might make sense in the approximation that the zeros are linearly independent.

a) In the case of the scalar propagator defined by all zeros of ζ the function $G_{R,2} + G_{R,3}$ has delta function peaks at transcendental values of u and vanishes trivially for $u = p^{\pm1}$. $G_{R,1}$ has a pole for $u = 1$. This propagator would have as its analog the rational Zeta function $\zeta_q = \sum_q q^{-z}$, which is ill defined unless one poses cutoff by allowing only primes $p < p_c$. This cutoff would be analogous to the y-cutoff. One can however define ζ_q by noticing that it can be formally written as

$$\hat{\zeta}_q = \frac{1}{N} \sum_{m>0,n>0} (m/n)^{-z} = \frac{1}{N}\zeta(z)\zeta(-z) \;\; ,$$

where N is the number of all integers n. The expression results by writing $q = r/s = nr/ns$ and summing over all values of all positive integers n and dividing by the number N of all integers. By an infinite re-normalization one obtains a well-defined function $\zeta_q \equiv N \times \hat{\zeta}_q = \zeta(z)\zeta(-z)$. Interestingly, the inverse of this function has poles at the lines $Re(z) = \pm 1/2$ containing the super-canonical conformal weights.

b) The fact that Riemann Zeta is a partition function for the states labelled by integers suggests that the dual partition function should be a partition function for the states corresponding to "integers" defined by Riemann Zeta and thus correspond to the "holomorphic" partition function. It is clear from the expressions of $G_{R,2}$ and $G_{R,3}$ that if they vanish for u they vanish also for $1/u$. Also now the transcendental poles at $u = p^{k2\pi/y}$ are present as are also poles at $u = p^{k\pi/y}$: these poles should become complex poles when the linear dependence is taken into account. The points $u = p$ could be zeros but it is difficult to prove this. Note however that $G_{R,2}$ and $G_{R,3}$ do not directly correspond to ζ since only the sums of odd and even number of y_i:s are involved.

Chapter 8

Was von Neumann Right After All?

8.1 Introduction

The work with TGD inspired model [O3] for topological quantum computation [ef1] led to the realization that von Neumann algebras [ea1, ea2, ea3, ea4], in particular so called hyper-finite factors of type II_1 [eb3], seem to provide the mathematics needed to develop a more explicit view about the construction of S-matrix. I have already discussed a vision for how to achieve this [C5]. In this chapter I will discuss various aspects of type II_1 factors and their physical interpretation in TGD framework. The lecture notes of R. Longo [dc6] give a concise and readable summary about the basic definitions and results related to von Neumann algebras and I have used this material freely in this chapter.

8.1.1 Philosophical ideas behind von Neumann algebras

The goal of von Neumann was to generalize the algebra of quantum mechanical observables. The basic ideas behind the von Neumann algebra are dictated by physics. The algebra elements allow Hermitian conjugation * and observables correspond to Hermitian operators. Any measurable function $f(A)$ of operator A belongs to the algebra and one can say that non-commutative measure theory is in question.

The predictions of quantum theory are expressible in terms of traces of observables. Density matrix defining expectations of observables in ensemble is the basic example. The highly non-trivial requirement of von Neumann was that identical a priori probabilities for a detection of states of infinite state system must make sense. Since quantum mechanical expectation values are expressible in terms of operator traces, this requires that unit operator has unit trace: $tr(Id) = 1$.

In the finite-dimensional case it is easy to build observables out of minimal projections to 1-dimensional eigen spaces of observables. For infinite-dimensional case the probably of projection to 1-dimensional sub-space vanishes if each state is equally probable. The notion of observable must thus be modified by excluding 1-dimensional minimal projections, and allow only projections for which the trace would be infinite using the straightforward generalization of the matrix algebra trace as the dimension of the projection.

The non-trivial implication of the fact that traces of projections are never larger than one is that the eigen spaces of the density matrix must be infinite-dimensional for non-vanishing projection probabilities. Quantum measurements can lead with a finite probability only to mixed states with a density matrix which is projection operator to infinite-dimensional subspace. The simple von Neumann algebras for which unit operator has unit trace are known as factors of type II_1 [eb3].

The definitions of adopted by von Neumann allow however more general algebras. Type I_n algebras correspond to finite-dimensional matrix algebras with finite traces whereas I_∞ associated with a separable infinite-dimensional Hilbert space does not allow bounded traces. For algebras of type III non-trivial traces are always infinite and the notion of trace becomes useless.

8.1.2 von Neumann, Dirac, and Feynman

The association of algebras of type I with the standard quantum mechanics allowed to unify matrix mechanism with wave mechanics. Note however that the assumption about continuous momentum state basis is in conflict with separability but the particle-in-box idealization allows to circumvent this problem (the notion of space-time sheet brings the box in physics as something completely real).

Because of the finiteness of traces von Neumann regarded the factors of type II_1 as fundamental and factors of type III as pathological. The highly pragmatic and successful approach of Dirac [da1] based on the notion of delta function, plus the emergence of s [db1], the possibility to formulate the notion of delta function rigorously in terms of distributions [da2, da5], and the emergence of path integral approach [db2] meant that von Neumann approach was forgotten by particle physicists.

Algebras of type II_1 have emerged only much later in conformal and topological quantum field theories [ec3, ec2] allowing to deduce invariants of knots, links and 3-manifolds. Also algebraic structures known as bi-algebras, Hopf algebras, and ribbon algebras [ec4, ee2] relate closely to type II_1 factors. In topological quantum computation [ef1] based on braid groups [eb2] modular S-matrices they play an especially important role.

In algebraic quantum field theory [dc7] defined in Minkowski space the algebras of observables associated with bounded space-time regions correspond quite generally to the type III_1 hyper-finite factor [dc8, dc9].

8.1.3   Factors of type II_1 and quantum TGD

For me personally the realization that TGD Universe is tailored for topological quantum computation [O3] led also to the realization that hyper-finite (ideal for numerical approximations) von Neumann algebras of type II_1 have a direct relevance for TGD.

Equivalence of generalized loop diagrams with tree diagrams

The work with bi-algebras [C5] led to the proposal that the generalized Feynman diagrams of TGD at space-time level satisfy a generalization of the duality of old-fashioned string models. Generalized Feynman diagrams containing loops are equivalent with tree diagrams so that they could be interpreted as representing computations or analytic continuations. This symmetry can be formulated as a condition on algebraic structures generalizing bi-algebras. The new element are the vacuum lines having a natural counterpart at the level of bi-algebras and braid diagrams. At space-time level they correspond to vacuum extremals.

Inclusions of hyper-finite II_1 factors as a basic framework to formulate quantum TGD

The basic facts about hyper-finite von Neumann factors of type II_1 suggest a more concrete view about the general mathematical framework needed.

a) The effective 2-dimensionality of the construction of quantum states and configuration space geometry in quantum TGD framework makes hyper-finite factors of type II_1 very natural as operator algebras of the state space. Indeed, the generators of conformal algebras, the gamma matrices of the configuration space, and the modes of the induced spinor fields are labelled by discrete labels. Hence the tangent space of the configuration space is a separable Hilbert space and its Clifford algebra is a hyper-finite type II_1 factor. Super-symmetry requires that the bosonic algebra generated by configuration space Hamiltonians and the Clifford algebra of configuration space both correspond to hyper-finite type II_1 factors.

b) Four-momenta relate to the positions of tips of future and past directed light cones appearing naturally in the construction of S-matrix. In fact, configuration space of 3-surfaces can be regarded as union of big-bang/big crunch type configuration spaces obtained as a union of light-cones parameterized by the positions of their tips. The algebras of observables associated with bounded regions of M^4 are hyper-finite and of type III_1 in algebraic quantum field theory [dc8]. The algebras of observables in the space spanned by the tips of these light-cones are not needed in the construction of S-matrix so that there are good hopes of avoiding infinities coming from infinite traces.

c) Many-sheeted space-time concept forces to refine the notion of sub-system. Jones inclusions $\mathcal{N} \subset \mathcal{M}$ for factors of type II_1 define in a generic manner to imbed interacting sub-systems to a universal II_1 factor which now naturally corresponds to the infinite Clifford algebra of the tangent space of configuration space of 3-surfaces and contains interaction as $\mathcal{M} : \mathcal{N}$-dimensional analog of tensor factor. Topological condensation of space-time sheet to a larger space-time sheet, the formation of bound states by the generation of join along boundaries bonds, interaction vertices in which space-time surface branches like a line of Feynman diagram: all these situations might be described by Jones inclusion [ea8, ea5] characterized by the Jones index $\mathcal{M} : \mathcal{N}$ assigning to the inclusion also a minimal conformal field theory and quantum group in case of $\mathcal{M} : \mathcal{N} < 4$ and conformal theory with $k = 1$ Kac Moody for $\mathcal{M} : \mathcal{N} = 4$ [ea9].

d) von Neumann's somewhat artificial idea about identical a priori probabilities for states could replaced with the finiteness requirement of quantum theory. Indeed, it is traces which produce the infinities of quantum field theories. That $\mathcal{M} : \mathcal{N} = 4$ option is not realized physically as quantum field theory (it would rather correspond to string model type theory characterized by a Kac-Moody algebra instead of quantum group), could correspond to the fact that dimensional regularization works only in $D = 4 - \epsilon$. Dimensional regularization with space-time dimension $D = 4 - \epsilon \to 4$ could be interpreted as the limit $\mathcal{M} : \mathcal{N} \to 4$. $\mathcal{M}$ as an $\mathcal{M} : \mathcal{N}$-dimensional $\mathcal{N}$-module would provide a concrete model for a quantum space with non-integral dimension as well as its Clifford algebra. An entire sequence of regularized theories corresponding to the allowed values of $\mathcal{M} : \mathcal{N}$ would be predicted.

Generalized Feynman diagrams are realized at the level of $\mathcal{M}$ as quantum space-time surfaces

The key idea is that generalized Feynman diagrams realized in terms of space-time sheets have counterparts at the level of $\mathcal{M}$ identifiable as the Clifford algebra associated with the entire space-time surface X^4. 4-D Feynman diagram as part of space-time surface is mapped to its $\beta = \mathcal{M} : \mathcal{N} \leq 4$-dimensional quantum counterpart.

a) von Neumann algebras allow a universal unitary automorphism $A \to \Delta^{it} A \Delta^{-it}$ fixed apart from inner automorphisms, and the time evolution of partonic 2-surfaces defining 3-D light-like causal determinant corresponds to the automorphism $\mathcal{N}_i \to \Delta^{it} \mathcal{N}_i \Delta^{-it}$ performing a time dependent unitary rotation for $\mathcal{N}_i$ along the line. At configuration space level however the sum over allowed values of t appear and should gives rise to the TGD counterpart of propagator as the analog of the stringy propagator $\int_0^t exp(iL_0 t)dt$. Number theoretical constraints from p-adicization suggest a quantization of t as $t = \sum_i n_i y_i > 0$, where $z_i = 1/2 + y_i$ are non-trivial zeros of Riemann Zeta.

b) At space-time level the "ends" of orbits of partonic 2-surfaces coincide at vertices so that also their images $\mathcal{N}_i \subset \mathcal{M}$ also coincide. The condition $\mathcal{N}_i = \mathcal{N}_j = ... = \mathcal{N}$, where the sub-factors $\mathcal{N}$ at different vertices differ only by automorphism, poses stringent conditions on the values t_i and Bohr quantization at the level of $\mathcal{M}$ results. Vertices can be obtained as a vacuum expectations of the operators creating the states associated with the incoming lines (crossing symmetry is automatic).

c) The equivalence of loop diagrams with tree diagrams would be due to the possibility to move the ends of the internal lines along the lines of the diagram so that only diagrams containing 3-vertices and self energy loops remain [C5]. Self energy loops are trivial if the product associated with fusion vertex and co-product associated with annihilation compensate each other. The possibility to assign quantum group or Kac Moody group to the diagram gives good hopes of realizing product and co-product. Octonionic triality would be an essential prerequisite for transforming N-vertices to 3-vertices [E2]. The equivalence allows to develop an argument proving the unitarity of S-matrix.

d) A formulation using category theoretical language suggests itself. The category of space sheets has as the most important arrow topological condensation via the formation of wormhole contacts. This category is mapped to the category of II_1 sub-factors of configurations space Clifford algebra having inclusion as the basic arrow. Space-time sheets are mapped to the category of Feynman diagrams in $\mathcal{M}$ with lines defined by unitary rotations of $\mathcal{N}_i$ induced by Δ^{it}.

e) The hierarchy of imbeddings for type II_1 factors generalizes to the hierarchy of generalized Feynman diagrams in which the particles of given level correspond to Feynman diagrams of the previous level. These Feynman diagrams provide representations for the projections of S-matrix to subspaces of incoming and outgoing states providing a hierarchy of self representations about the system. The vertices for the interactions of these Feynman diagrams reduce to the interactions at the lowest level apart from the automorphisms $\Delta_{\mathcal{M}_n}$ defining free propagation. The interpretation is as an infinite cognitive hierarchy realizing theory about the material world as cognitive quantum states. This hierarchy corresponds to states with vanishing conserved net quantum numbers but having non-vanishing "gravitational" charges identifiable as classical charges. A fascinating possibility is that dark matter corresponds to thoughts!

e) Algebraic hologram idea is realized in the sense that the operator algebras of all other lines are imbeddable to the operator algebra of any line: whether this has some practical significance remains an open question. Alternatively, the same fundamental entity looks different depending on its relationship with environment characterized by $\mathcal{M} : \mathcal{N}$. The triviality of loops for generalized Feynman diagrams gives strong conditions on vertices.

Is $\hbar$ dynamical?

The work with topological quantum computation [O3] inspired the hypothesis that $\hbar$ might be dynamical, and that its values might relate in a simple manner to the logarithms of Beraha numbers $B_n = 4cos^2(\pi/n)$ giving Jones indices $\mathcal{M} : \mathcal{N}$. The model for the evolution of $\hbar$ implied that $\hbar$ is infinite for the minimal value $\mathcal{M} : \mathcal{N} = 1$ of Jones index.

The construction of a model explaining the strange finding [if1, D6, J6] that planetary orbits seem to correspond to a gigantic value of "gravitational" Planck constant led to the hypothesis that when the perturbative expansion in terms of parameter $k = \alpha Q_1 Q_2$ ceases to converge, a phase transition increasing the value of $\hbar$ to $\hbar_s = k\hbar/v_0$, where v_0 the ratio of Planck length to CP_2 length, occurs. This involves also a transition to a macroscopic quantum phase since Compton lengths and times increase dramatically. Dark matter would correspond to ordinary matter, which corresponds to larger $\hbar$ and is conformally confined in the sense the sum of complex super-canonical conformal weights (related in a simple manner to the complex zeros of Riemann Zeta [B2, B3, B4, E8]) is real is for a block of dark matter behaving thus like a single quantum coherent unit [J6, D5].

The value of $\hbar$ for $\mathcal{M} : \mathcal{N} = 1$ is large but not infinite, and thus in conflict with the original proposal of [O3, C5]. A more refined proposal is that the evolution of $\hbar$ as a function of $\mathcal{M} : \mathcal{N} = 4cos^2(\pi/n)$ can be interpreted as a renormalization group evolution for the phase resolution. The earlier identification is replaced by a linear renormalization group difference equation for $1/\hbar$ allowing as its solutions the earlier solution modified by an additive integration constant. Hence $1/\hbar$ can approach to a finite value $1/\hbar(3) = v_0/k\hbar(n \to \infty)$ at the limit $n \to 3$. The evolution equation gives a concrete view about how various charges should be imbedded in Jones inclusion to the larger algebra.

The dependence of $\hbar$ on the parameters of interacting systems means that it is associated with the interface of the interacting systems. Instead of being an absolute constant of nature $\hbar$ becomes something characterizing the interaction between two systems in terms of the "position" of II_1 factor $\mathcal{N}$ inside $\mathcal{M}$. The interface could correspond to wormhole contacts, join along boundaries bond, lightlike causal determinant, etc... This property of $\hbar$ is consistent with the fact that vacuum functional expressible as an exponent of Kähler action does not depend at all on $\hbar$.

8.2 von Neumann algebras

In this section basic facts about von Neumann algebras are summarized using as a background material the concise summary given in the lecture notes of Longo [dc6].

8.2.1 Basic definitions

A formal definition of von Neumann algebra [ea2, ea3, ea4] is as a *-subalgebra of the set of bounded operators $\mathcal{B}(\mathcal{H})$ on a Hilbert space $\mathcal{H}$ closed under weak operator topology, stable under the conjugation $J =^*: x \to x^*$, and containing identity operator Id. This definition allows also von Neumann algebras for which the trace of the unit operator is not finite.

Identity operator is the only operator commuting with a simple von Neumann algebra. A general von Neumann algebra allows a decomposition as a direct integral of simple algebras, which von Neumann called factors. Classification of von Neumann algebras reduces to that for factors.

$\mathcal{B}(\mathcal{H})$ has involution * and is thus a *-algebra. $\mathcal{B}(\mathcal{H})$ has order order structure $A \geq 0 : (Ax, x) \geq 0$. This is equivalent to $A = BB^*$ so that order structure is determined by algebraic structure. $\mathcal{B}(\mathcal{H})$ has metric structure in the sense that norm defined as supremum of $||Ax||$, $||x|| \leq 1$ defines the notion of continuity. $||A||^2 = inf\{\lambda > 0 : AA^* \leq \lambda I\}$ so that algebraic structure determines metric structure.

There are also other topologies for $\mathcal{B}(\mathcal{H})$ besides norm topology.

a) $A_i \to A$ strongly if $||Ax - A_i x|| \to 0$ for all x. This topology defines the topology of C^* algebra. $\mathcal{B}(\mathcal{H})$ is a Banach algebra that is $||AB|| \leq ||A|| \times ||B||$ (inner product is not necessary) and also C^* algebra that is $||AA^*|| = ||A||^2$.

b) $A_i \to A$ weakly if $(A_i x, y) \to (Ax, y)$ for all pairs (x, y) (inner product is necessary). This topology defines the topology of von Neumann algebra as a sub-algebra of $\mathcal{B}(\mathcal{H})$.

Denote by M' the commutant of $\mathcal{M}$ which is also algebra. von Neumann's bicommutant theorem says that $\mathcal{M}$ equals to its own bi-commutant. Depending on whether the identity operator has a finite trace or not, one distinguishes between algebras of type II_1 and type II_∞. II_1 factor allow trace with properties $tr(Id) = 1$, $tr(xy) = tr(yx)$, and $tr(x^*x) > 0$, for all $x \neq 0$. Let $L^2(\mathcal{M})$ be the Hilbert space obtained by completing $\mathcal{M}$ respect to the inner product defined $\langle x|y \rangle = tr(x^*y)$ defines inner product in $\mathcal{M}$ interpreted as Hilbert space. The normalized trace induces a trace in M', natural trace $Tr_{M'}$, which is however not necessarily normalized. JxJ defines an element of M': if $\mathcal{H} = L^2(\mathcal{M})$, the natural trace is given by $Tr_{M'}(JxJ) = tr_M(x)$ for all $x \in M$ and bounded.

8.2.2 Basic classification of von Neumann algebras

Consider first some definitions. First of all, Hermitian operators with positive trace expressible as products xx^* are of special interest since their sums with positive coefficients are also positive.

In quantum mechanics Hermitian operators can be expressed in terms of projectors to the eigen states. There is a natural partial order in the set of isomorphism classes of projectors by inclusion: $E < F$ if the image of $\mathcal{H}$ by E is contained to the image of $\mathcal{H}$ by a suitable isomorph of F. Projectors are said to be metrically equivalent if there exist a partial isometry which maps the images $\mathcal{H}$ by them to each other. In the finite-dimensional case metric equivalence means that isomorphism classes are identical $E = F$.

The algebras possessing a minimal projection E_0 satisfying $E_0 \leq F$ for any F are called type I algebras. Bounded operators of n-dimensional Hilbert space define algebras I_n whereas the bounded operators of infinite-dimensional separable Hilbert space define the algebra I_∞. I_n and I_∞ correspond to the operator algebras of quantum mechanics. The states of harmonic oscillator correspond to a factor of type I.

The projection F is said to be finite if $F < E$ and $F \equiv E$ implies $F = E$. Hence metric equivalence means identity. Simple von Neumann algebras possessing finite projections but no minimal projections so that any projection E can be further decomposed as $E = F + G$, are called factors of type II.

Hyper-finiteness means that any finite set of elements can be approximated arbitrary well with the elements of a finite-dimensional sub-algebra. The hyper-finite II_∞ algebra can be regarded as a tensor product of hyper-finite II_1 and I_∞ algebras. Hyper-finite II_1 algebra can be regarded as a Clifford algebra of an infinite-dimensional separable Hilbert space sub-algebra of I_∞.

Hyper-finite II_1 algebra can be constructed using Clifford algebras $C(2n)$ of $2n$-dimensional spaces and identifying the element x of $2^n \times 2^n$ dimensional $C(n)$ as the element $diag(x, x)/2$ of $2^{n+1} \times 2^{n+1}$-dimensional $C(n+1)$. The union of algebras $C(n)$ is formed and completed in the weak operator topology to give a hyper-finite II_1 factor. This algebra defines the Clifford algebra of infinite-dimensional separable Hilbert space and is thus a sub-algebra of I_∞ so that hyper-finite II_1 algebra is more regular than I_∞.

von Neumann algebras possessing no finite projections (all traces are infinite or zero) are called algebras of type III. It was later shown by Connes [ea7] that these algebras are labelled by a parameter varying in the range $[0, 1]$, and referred to as algebras of type III_x. III_1 category contains a unique hyper-finite algebra. It has been found that the algebras of observables associated with bounded regions of 4-dimensional Minkowski space in quantum field theories correspond to hyper-finite factors of type III_1 [dc6]. Also statistical systems at finite temperature correspond to factors of type III and temperature parameterizes one-parameter set of automorphisms of this algebra [dc8]. Zero temperature limit correspond to I_∞ factor and infinite temperature limit to II_1 factor.

8.2.3 Non-commutative measure theory and non-commutative topologies and geometries

von Neumann algebras and C^* algebras give rise to non-commutative generalizations of ordinary measure theory (integration), topology, and geometry. It must be emphasized that these structures are completely natural aspects of quantum theory. In particular, for the hyper-finite type II_1 factors quantum groups and Kac Moody algebras [ea9] emerge quite naturally without any need for ad hoc modifications such as making space-time coordinates non-commutative. The effective 2-dimensionality of quantum TGD (partonic or stringy 2-surfaces code for states) means that these structures appear completely naturally in TGD framework.

Non-commutative measure theory

von Neumann algebras define what might be a non-commutative generalization of measure theory and probability theory [dc6].

a) Consider first the commutative case. Measure theory is something more general than topology since the existence of measure (integral) does not necessitate topology. Any measurable function f in the space $L^\infty(X, \mu)$ in measure space (X, μ) defines a bounded operator M_f in the space $\mathcal{B}(L^2(X, \mu))$ of bounded operators in the space $L^2(X, \mu)$ of square integrable functions with action of M_f defined as $M_f g = fg$.

b) Integral over $\mathcal{M}$ is very much like trace of an operator $f_{x,y} = f(x)\delta(x, y)$. Thus trace is a natural non-commutative generalization of integral (measure) to the non-commutative case and defined for von Neumann algebras. In particular, generalization of probability measure results if the case $tr(Id) = 1$ and algebras of type I_n and II_1 are thus very natural from the point of view of non-commutative probability theory.

The trace can be expressed in terms of a cyclic vector Ω or vacuum/ground state in physicist's terminology. Ω is said to be cyclic if the completion $\overline{M\Omega} = H$ and separating if $x\Omega$ vanishes only for $x = 0$. Ω is cyclic for $\mathcal{M}$ if and only if it is separating for M'. The expression for the trace given by

$$Tr(ab) \quad = \quad \left(\frac{(ab + ba)}{2}, \Omega \right) \tag{8.2.1}$$

is symmetric and allows to defined also inner product as $(a, b) = Tr(a^*b)$ in $\mathcal{M}$. If Ω has unit norm $(\Omega, \Omega) = 1$, unit operator has unit norm and the algebra is of type II_1. Fermionic oscillator operator algebra with discrete index labelling the oscillators defines II_1 factor. Group algebra is second example of II_1 factor.

The notion of probability measure can be abstracted using the notion of state. State ω on a C^* algebra with unit is a positive linear functional on $\mathcal{U}$, $\omega(1) = 1$. By so called KMS construction [dc6] any state ω in C^* algebra $\mathcal{U}$ can be expressed as $\omega(x) = (\pi(x)\Omega, \Omega)$ for some cyclic vector Ω and π is a homomorphism $\mathcal{U} \to \mathcal{B}(\mathcal{H})$.

Non-commutative topology and geometry

C^* algebras generalize in a well-defined sense ordinary topology to non-commutative topology.

a) In the Abelian case Gelfand Naimark theorem [dc6] states that there exists a contravariant functor F from the category of unital abelian C^* algebras and category of compact topological spaces. The inverse of this functor assigns to space X the continuous functions f on X with norm defined by the maximum of f. The functor assigns to these functions having interpretation as eigen states of mutually commuting observables defined by the function algebra. These eigen states are delta functions localized at single point of X. The points of X label the eigenfunctions and thus define the spectrum and obviously span X. The connection with topology comes from the fact that continuous map $Y \to X$ corresponds to homomorphism $C(X) \to C(Y)$.

b) In non-commutative topology the function algebra $C(X)$ is replaced with a general C^* algebra. Spectrum is identified as labels of simultaneous eigen states of the Cartan algebra of C^* and defines what can be observed about non-commutative space X.

c) Non-commutative geometry can be very roughly said to correspond to *-subalgebras of C^* algebras plus additional structure such as symmetries. The non-commutative geometry of Connes [ed1] is a basic example here.

8.2.4 Modular automorphisms

von Neumann algebras allow a canonical unitary evolution associated with any state ω fixed by the selection of the vacuum state Ω [dc6]. This unitary evolution is an automorphism fixed apart form unitary automorphisms $A \to UAU^*$ related with the choice of Ω.

Let ω be a normal faithful state: $\omega(x^*x) > 0$ for any x. One can map $\mathcal{M}$ to $L^2(\mathcal{M})$ defined as a completion of $\mathcal{M}$ by $x \to x\Omega$. The conjugation $*$ in $\mathcal{M}$ has image at Hilbert space level as a map $S_0 : x\Omega \to x^*\Omega$. The closure of S_0 is an anti-linear operator and has polar decomposition $S = J\Delta^{1/2}$, $\Delta = SS^*$. Δ is positive self-adjoint operator and J anti-unitary involution. The following conditions are satisfied

$$\begin{aligned}
\Delta^{it} \mathcal{M} \Delta^{-it} &= \mathcal{M} \ , \\
J\mathcal{M}J &= \mathcal{M}' \ .
\end{aligned} \tag{8.2.2}$$

Δ^{it} is obviously analogous to the time evolution induced by positive definite Hamiltonian and induces also the evolution of the expectation ω as $\pi \to \Delta^{it}\pi\Delta^{-it}$.

8.2.5 Joint modular structure and sectors

Let $\mathcal{N} \subset \mathcal{M}$ be an inclusion. The unitary operator $\gamma = J_N J_M$ defines a canonical endomorphisms $M \to N$ in the sense that it depends only up to inner automorphism on $\mathcal{N}$, γ defines a sector of $\mathcal{M}$. The sectors of $\mathcal{M}$ are defined as $Sect(\mathcal{M}) = End(\mathcal{M})/Inn(\mathcal{M})$ and form a semi-ring with respected to direct sum and composition by the usual operator product. It allows also conjugation.

$L^2(\mathcal{M})$ is a normal bi-module in the sense that it allows commuting left and right multiplications. For $a, b \in M$ and $x \in L^2(\mathcal{M})$ these multiplications are defined as $axb = aJb^*Jx$ and it is easy to verify the commutativity using the factor $Jy^*J \in \mathcal{M}'$. Connes [ed1] has shown that all normal bi-modules arise in this way up to unitary equivalence so that representation concepts make sense. It is possible to assign to any endomorphism ρ index $Ind(\rho) \equiv M : \rho(\mathcal{M})$. This means that the sectors are in 1-1 correspondence with inclusions. For instance, in the case of hyper-finite II_1 they are labelled by Jones index. Furthermore, the objects with non-integral dimension $\sqrt{[\mathcal{M} : \rho(\mathcal{M})]}$ can be identified as quantum groups, loop groups, infinite-dimensional Lie algebras, etc...

8.3 Inclusions of II_1 and III_1 factors

Inclusions $\mathcal{N} \subset \mathcal{M}$ of von Neumann algebras have physical interpretation as a mathematical description for sub-system-system relation. For type I algebras the inclusions are trivial and tensor product description applies as such. For factors of II_1 and III the inclusions are highly non-trivial. The inclusion of type II_1 factors were understood by Vaughan Jones [ea8]and those of factors of type III by Alain Connes [ea7].

Sub-factor $\mathcal{N}$ of $\mathcal{M}$ is defined as a closed $*$-stable C-subalgebra of $\mathcal{M}$. Let $\mathcal{N}$ be a sub-factor of type II_1 factor $\mathcal{M}$. Jones index $\mathcal{M} : \mathcal{N}$ for the inclusion $\mathcal{N} \subset \mathcal{M}$ can be defined as $\mathcal{M} : \mathcal{N} = dim_N(L^2(\mathcal{M})) = Tr_{N'}(id_{L^2(\mathcal{M})})$. One can say that the dimension of completion of $\mathcal{M}$ as $\mathcal{N}$ module is in question.

8.3.1 Basic findings about inclusions

What makes the inclusions non-trivial is that the position of $\mathcal{N}$ in $\mathcal{M}$ matters. This position is characterized in case of hyper-finite II_1 factors by index $\mathcal{M} : \mathcal{N}$ which can be said to the dimension of $\mathcal{M}$ as $\mathcal{N}$ module and also as the inverse of the dimension defined by the trace of the projector from $\mathcal{M}$ to $\mathcal{N}$. It is important to notice that $\mathcal{M} : \mathcal{N}$ does not characterize either $\mathcal{M}$ or $\mathcal{M}$, only the imbedding.

The basic facts proved by Jones are following [ea8].
a) For pairs $\mathcal{N} \subset \mathcal{M}$ with a finite principal graph the values of $\mathcal{M} : \mathcal{N}$ are given by

$$a) \ \ \mathcal{M} : \mathcal{N} = 4cos^2(\pi/h) \ , \quad h \geq 3 \ ,$$

$$\tag{8.3.1}$$

$$b) \ \ \mathcal{M} : \mathcal{N} \geq 4 \ .$$

the numbers at right hand side are known as Beraha numbers [ec6]. The comments below give a rough idea about what finiteness of principal graph means.

b) As explained in [ea9], for $\mathcal{M} : \mathcal{N} < 4$ one can assign to the inclusion Dynkin graph of ADE type (quantum) Lie-algebra g with h equal to the Coxeter number h of the Lie algebra given in terms of its dimension and dimension r of Cartan algebra r as $h = (dimg(g) - r)/r$. The Lie algebras of E_7 and D_{2n+1} are however not allowed. For $\mathcal{M} : \mathcal{N} = 4$ one can assign to the inclusion an extended Dynkin graph of type ADE characterizing Kac Moody algebra.

8.3.2   The fundamental construction and Temperley-Lieb algebras

It was shown by Jones [ea5] that for a given Jones inclusion with $\beta = \mathcal{M} : \mathcal{N} < \infty$ there exists a tower of finite II_1 factors $\mathcal{M}_k$ for $k = 0, 1, 2, ...$ such that
   a) $\mathcal{M}_0 = \mathcal{N}$, $\mathcal{M}_1 = \mathcal{M}$,

   b) $\mathcal{M}_{k+1} = End_{\mathcal{M}_{k-1}}\mathcal{M}_k$ is the von Neumann algebra of operators on $L^2(\mathcal{M}_k)$ generated by $\mathcal{M}_k$ and an orthogonal projection $e_k : L^2(\mathcal{M}_k) \to L^2(\mathcal{M}_{k-1})$ for $k \geq 1$, where $\mathcal{M}_k$ is regarded as a subalgebra of $\mathcal{M}_{k+1}$ under right multiplication.

   It can be shown that $\mathcal{M}_{k+1}$ is a finite factor. The sequence of projections on $\mathcal{M}_\infty = \cup_{k \geq 0}\mathcal{M}_k$ satisfies the relations

$$\begin{aligned}
e_i^2 &= e_i \ , \qquad e_i^= e_i \ , \\
e_i &= \beta e_i e_j e_i \quad \text{for } |i - j| = 1 \ , \\
e_i e_j &= e_j e_i \quad \text{for } |i - j| \geq 2 \ .
\end{aligned}$$
(8.3.2)

The construction of hyper-finite II_1 factor using Clifford algebra $C(2)$ represented by 2×2 matrices allows to understand the theorem in $\beta = 4$ case in a straightforward manner. In particular, the second formula involving β follows from the identification of x at $(k - 1)^{th}$ level with $(1/\beta)diag(x, x)$ at k^{th} level.

   By replacing 2×2 matrices with $\sqrt{\beta} \times \sqrt{\beta}$ matrices one can understand heuristically what is involved in the more general case. $\mathcal{M}_k$ is $\mathcal{M}_{k-1}$ module with dimension $\sqrt{\beta}$ and $\mathcal{M}_{k+1}$ is the space of $\sqrt{\beta} \times \sqrt{\beta}$ matrices $\mathcal{M}_{k-1}$ valued entries acting in $\mathcal{M}_k$. The transition from $\mathcal{M}_k$ to $\mathcal{M}_{k-1}$ linear maps of $\mathcal{M}_k$ happens in the transition to the next level. x at $(k-1)^{th}$ level is identified as $(x/\beta) \times Id_{\sqrt{\beta} \times \sqrt{\beta}}$ at the next level. The projection e_k picks up the projection of the matrix with $\mathcal{M}_{k-1}$ valued entries in the direction of the $Id_{\sqrt{\beta} \times \sqrt{\beta}}$.

   The union of algebras $A_{\beta,k}$ generated by $1, e_1, ..., e_k$ defines Temperley-Lieb algebra A_β [ee1]. This algebra is naturally associated with braids. Addition of one strand to a braid adds one generator to this algebra and the representations of the Temperley Lieb algebra provide link, knot, and 3-manifold invariants [eb2]. There is also a connection with systems of statistical physics and with Yang-Baxter algebras [ee3].

   A further interesting fact about the inclusion hierarchy is that the elements in $\mathcal{M}_i$ belonging to the commutator $\mathcal{N}'$ of $\mathcal{N}$ form finite-dimensional spaces. Presumably the dimension approaches infinity for $n \to \infty$.

8.3.3   Connection with Dynkin diagrams

The possibility to assign Dynkin diagrams ($\beta < 4$) and extended Dynkin diagrams ($\beta = 4$ to Jones inclusions can be understood heuristically by considering a characterization of so called bipartite graphs [ea6, ea9] by the norm of the adjacency matrix of the graph.

   Bipartite graphs Γ is a finite, connected graph with multiple edges and black and white vertices such that any edge connects white and black vertex and starts from a white one. Denote by $w(\Gamma)$ $(b(\Gamma))$ the number of white (black) vertices. Define the adjacency matrix $\Lambda = \Lambda(\Gamma)$ of size $b(\Gamma) \times w(\Gamma)$ by

$$w_{b,w} = \begin{cases} m(e) & \text{if there exists } e \text{ such that } \delta e = b - w \ , \\ 0 & \text{otherwise } . \end{cases}$$
(8.3.3)

Here $m(e)$ is the multiplicity of the edge e.
   Define norm $||\Gamma||$ as

$$\|X\| = max\{\|X\|;\ \|x\| \leq 1\}\ ,$$

$$\|\Gamma\| = \|\Lambda(\Gamma)\| = \left\|\ \begin{matrix} 0 & \Lambda(\Gamma) \\ \Lambda(\Gamma)^t & 0 \end{matrix}\ \right\| . \tag{8.3.4}$$

Note that the matrix appearing in the formula is $(m+n) \times (m+n)$ symmetric square matrix so that the norm is the eigenvalue with largest absolute value.

Suppose that Γ is a connected finite graph with multiple edges (sequences of edges are regarded as edges). Then

a) If $\|\Gamma\| \leq 2$ and if Γ has a multiple edge, $\|\Gamma\| = 2$ and $\Gamma = \tilde{A}_1$, the extended Dynkin diagram for $SU(2)$ Kac Moody algebra.

b) $\|\Gamma\| < 2$ if and only Γ is one of the Dynkin diagrams of A,D,E. In this case $\|\Gamma\| = 2cos(\pi/h)$, where h is the Coxeter number of Γ.

c)$\|\Gamma\| = 2$ if and only if Γ is one of the extended Dynkin diagrams $\tilde{A}, \tilde{D}, \tilde{E}$.

This result suggests that one can indeed assign to the Jones inclusions Dynkin diagrams. To really understand how the inclusions can be characterized in terms bipartite diagrams would require a deeper understanding of von Neumann algebras. The following argument only demonstrates that bipartite graphs naturally describe inclusions of algebras.

a) Consider a bipartite graph. Assign to each white vertex linear space $W(w)$ and to each edge of a linear space $W(b,w)$. Assign to a given black vertex the vector space $\oplus_{\delta e=b-w} W(b,w) \otimes W(w)$ where (b,w) corresponds to an edge ending to b.

b) Define $\mathcal{N}$ as the direct sum of algebras $End(W(w))$ associated with white vertices and $\mathcal{M}$ as direct sum of algebras $\oplus_{\delta e=b-w} End(W(b,w)) \otimes End(W(w))$ associated with black vertices.

c) There is homomorphism $N \to M$ defined by imbedding direct sum of white endomorphisms x to direct sum of tensor products x with the identity endomorphisms associated with the edges starting from x.

It is possible to show that Jones inclusions correspond to the Dynkin diagrams of A_n, D_{2n}, and E_6, E_8 and extended Dynkin diagrams of ADE type. In particular, the dual of the bi-partite graph associated with $\mathcal{M}_{n-1} \subset \mathcal{M}_n$ obtained by exchanging the roles of white and black vertices describes the inclusion $\mathcal{M}_n \subset \mathcal{M}_{n+1}$ so that two subsequent Jones inclusions might define something fundamental (the corresponding space-time dimension is $2 \times log_2(\mathcal{M} : \mathcal{N}) \leq 4$.

8.3.4 Indices for the inclusions of type III_1 factors

Type III_1 factors appear in relativistic quantum field theory defined in 4-dimensional Minkowski space [dc8]. An overall summary of basic results discovered in algebraic quantum field theory is described in the lectures of Longo [dc6]. In this case the inclusions for algebras of observables are induced by the inclusions for bounded regions of M^4 in axiomatic quantum field theory. Tomita's theory of modular Hilbert algebras [dc2, dc9] forms the mathematical corner stone of the theory.

The basic notion is Haag-Kastler net [dc3] consisting of bounded regions of M^4. Double cone serves as a representative example. The von Neumann algebra $\mathcal{A}(O)$ is generated by observables localized in bounded region O. The net satisfies the conditions implied by local causality:

a) Isotony: $O_1 \subset O_2$ implies $\mathcal{A}(O_1) \subset \mathcal{A}(O_2)$.

b) Locality: $O_1 \subset O_2'$ implies $\mathcal{A}(O_1) \subset \mathcal{A}(O_2)'$ with O' defined as $\{x : \langle x,y \rangle < 0$ for all $y \in O\}$.

c) Haag duality $\mathcal{A}(O')' = \mathcal{A}(O)$.

Besides this Poincare covariance, positive energy condition, and the existence of vacuum state is assumed.

DHR (Doplicher-Haag-Roberts) [dc4] theory allows to deduce the values of Jones index and they are squares of integers in dimensions $D > 2$ so that the situation is rather trivial. The 2-dimensional case is distinguished from higher dimensional situations in that braid group replaces permutation group since the paths representing the flows permuting identical particles can be linked in $X^2 \times T$ and anyonic statistics [ee5, ee7] becomes possible. In the case of 2-D Minkowski space M^2 Jones inclusions with $\mathcal{M} : \mathcal{N} < 4$ plus a set of discrete values of $\mathcal{M} : \mathcal{N}$ in the range $(4,6)$ are possible. In [dc6] some values are given ($\mathcal{M} : \mathcal{N} = 5, 5.5049..., 5.236...., 5.828...$).

At least intersections of future and past light cones seem to appear naturally in TGD framework such that the boundaries of future/past directed light cones serve as seats for incoming/outgoing states defined as intersections of space-time surface with these light cones. III_1 sectors cannot thus be excluded as factors in TGD framework. On the other hand, the construction of S-matrix at space-time level is reduced to II_1 case by effective 2-dimensionality.

8.4 TGD and hyper-finite factors of type II_1

By effective 2-dimensionality of the construction of quantum states the hyper-finite factors of type II_1 fit naturally to TGD framework. In the following a general vision about how space-time surfaces and generalized Feynman diagrams at space-time level correspond to their quantum counterparts constructed in terms of Jones inclusions and the orbits of sub-factors under canonical unitary automorphism induced by Δ^{it} is formulated.

8.4.1 Problems associated with the physical interpretation of III_1 factors

Algebraic quantum field theory approach [dc7, dc8] has led to a considerable understanding of relativistic quantum field theories in terms of hyper-finite III_1 factors. There are however several reasons to suspect that the resulting picture is in conflict with physical intuition. Also the infinities of non-trivial relativistic QFTs suggest that something goes wrong.

Do the infinities of quantum field theories due the wrong type of von Neumann algebra?

The infinities of quantum field theories involve basically infinite traces and it is now known that the algebras of observables for relativistic quantum field theories for bounded regions of Minkowski space correspond to hyper-finite III_1 algebras, for which non-trivial traces are always infinite. This might be the basic cause of the divergence problems of relativistic quantum field theory.

On basis of this observations there is some temptation to think that the finite traces of hyper-finite II_1 algebras might provide a resolution to the problems but not necessarily in QFT context. One can play with the thought that the subtraction of infinities might be actually a process in which III_1 algebra is transformed to II_1 algebra. A more plausible idea suggested by dimensional regularization is that the elimination of infinities actually gives rise to II_1 inclusion at the limit $\mathcal{M} : \mathcal{N} \to 4$. It is indeed known that the dimensional regularization procedure of quantum field theories can be formulated in terms of bi-algebras assignable to Feynman diagrams and [ed2, ed3] and the emergence of bi-algebras suggests that a connection with II_1 factors and critical role of dimension $D = 4$ might exist.

Continuum of inequivalent representations of commutation relations

There is also a second difficulty related to type III algebras. There is a continuum of inequivalent representations for canonical commutation relations [dc1]. In thermodynamics this is blessing since temperature parameterizes these representations. In quantum field theory context situation is however different and this problem has been usually put under the rug.

Entanglement and von Neumann algebras

In quantum field theories where 4-D regions of space-time are assigned to observables. In this case hyper-finite type III_1 von Neumann factors appear. Also now inclusions make sense and has been studied: in fact, the parameters characterizing Jones inclusions appear also now and this due to the very general properties of the inclusions.

The algebras of type III_1 have rather counter-intuitive properties from the point of view of entanglement. For instance, product states between systems having space-like separation are not possible at all so that one can speak of intrinsic entanglement [dc5]. What looks worse is that the decomposition of entangled state to product states is highly non-unique.

Mimicking the steps of von Neumann one could ask what the notion of observables could mean in TGD framework. Effective 2-dimensionality states that quantum states can be constructed using the data given at partonic or stringy 2-surfaces. This data includes also information about normal derivatives so that 3-dimensionality actually lurks in. In any case this would mean that observables are assignable to 2-D surfaces. This would suggest that hyper-finite II_1 factors appear in quantum TGD at least as the contribution of single space-time surface to S-matrix is considered. The contributions for configuration space degrees of freedom meaning functional (not path-) integral over 3-surfaces could of course change the situation.

Also in case of II_1 factors, entanglement shows completely new features which need not however be in conflict with TGD inspired view about entanglement. The eigen values of density matrices are infinitely degenerate and quantum measurement can remove this degeneracy only partially. TGD inspired theory of consciousness has led to the identification of rational (more generally algebraic entanglement) as bound state entanglement stable in state function reduction. When an infinite number of states are entangled, the entanglement would correspond to rational (algebraic number) valued traces for the projections to the eigen states of the density matrix. The canonical transformations of CP_2 are almost $U(1)$ gauge symmetries broken only by classical gravitation. They imply a gigantic spin glass degeneracy which could be behind the infinite degeneracies of eigen states of density matrices in case of II_1 factors.

8.4.2 Bott periodicity, its generalization, and dimension $D = 8$ as an inherent property of the hyper-finite II_1 factor

Hyper-finite II_1 factor can be constructed as infinite-dimensional tensor power of the Clifford a $C(2)$ in dimension $D = 2$. More precisely one forms the union of the Clifford algebras $C(2n) = C(2)^{\otimes n}$ of $2n$-dimensional spaces by identifying the element $x \in C(2n)$ as block diagonal elements $diag(x, x)$ of $C(2(n+1))$. The union of these algebras is completed in weak operator topology and can be regarded as a Clifford algebra of real infinite-dimensional separable Hilbert space and thus as sub-algebra of I_∞.

a) Dimension 8 is an inherent property of the hyper-finite II_1 factor since Bott periodicity theorem states $C(n+8) = C_n(16)$. In other words, the Clifford algebra $C(n+8)$ is equivalent with the algebra of 16×16 matrices with entries in $C(n)$. Or articulating it still differently: $C(n+8)$ can be regarded as 16×16 dimensional module with $C(n)$ valued coefficients. Hence the elements in the union defining hyper-finite II_1 factor are $16^n \times 16^n$ matrices having $C(0)$, $C(2)$, $C(4)$ or $C(6)$ valued valued elements.

b) The notion of Bott periodicity for Clifford algebra generalizes. Instead of 8-cycle one has several cycles corresponding to the values of $\mathcal{M} : \mathcal{N}$. These cycles correspond to a repeated Jones inclusion with a fixed index $\mathcal{M} : \mathcal{N}$. If the $\mathcal{M} : \mathcal{N}$-dimensional space is interpreted as a Clifford algebra, the fractal space-time dimension D in cycle would be from $\mathcal{M} : \mathcal{N} = 2^D$

$$D(n) = log_2(\mathcal{M} : \mathcal{N}) = 2 + log_2(cos^2(\pi/n)) \leq 2 \ . \tag{8.4.1}$$

One has $0 \leq D(n) \leq 2$. This conforms with the fact that hyper-finite II_1 factor allows a construction in tensor powers of $C(2)$. It also conforms with the effective 2-dimensionality of quantum TGD state construction (partonic or stringy 2-surface code for the states). The 8-cycle of Clifford algebras corresponds to a sequence of four Jones inclusions.

It must be however emphasized that also other interpretations are possible since in Jones inclusion the space of operators becomes the spaces in which the operators at next level act. In fact, the connection with conformal field theory supports the interpretation of $\beta = \mathcal{M} : \mathcal{N}$ as the number of states of β-state Potts model varying in the range $[1, ..., 4]$. Furthermore, the natural periodicity of Jones inclusions corresponds to 2 subsequent inclusions with bipartite diagrams which are duals of each other and this corresponds to space-time dimension $D = 2\mathcal{M} : \mathcal{N} \leq 4$.

8.4.3 Is a new kind of Feynman diagrammatics needed?

In light of these arguments, one can ask whether the approach based on Feynman diagrams and path integrals forced by quantum field theory characterized by a heroic and futile attempt to find infinity free quantum field theory, and culminating to string models, might represent a wrong turn in the history of physics.

In TGD classical physics is not assigned to a stationary phase approximation but is an exact part of quantum physics. In particular, the poorly defined path integrals over all possible 4-surfaces do not appear at all, being replaced by a functional integral over configuration space of 3-surfaces with exponent of Kähler function defining a vacuum functional which is a non-local functional of 3-surface so that no local divergences result. Gaussian and metric determinants cancel each other by Kähler property. This suggests a new approach based on the generalization of stringy diagrams. Generalized Feynman diagrams are analogous to representations of computations or analytic continuations so that diagrams with loops are equivalent with tree diagrams. This generalizes the notion of duality of old fashioned string models.

The low-dimensional algebras correspond to II_1 algebras and there is a strong temptation to believe that the effective 2-dimensionality of the state construction in TGD framework implies II_1 algebras. The proposed generalized Feynman diagrammatics at space-time level [C5] is not expected to give rise to quantum field theory limit except as an effective theory. Configuration space spinor fields would naturally correspond to a tensor product of bosonic and fermionic II_∞ factors: actually a tensor product of four II_1 factors is involved by quark-lepton degeneracy. It must be however admitted that the appearance of future and past oriented light cones of M^4 in the basic construction of the theory means that also factors of type III_1 might creep in.

8.4.4 The interpretation of Jones inclusions in TGD framework

By the basic self-referential property of von Neumann algebras one can consider several interpretations of Jones inclusions consistent with sub-system-system relationship, and it is better to start by considering the options that one can imagine.

How Jones inclusions relate to the new view about sub-system?

Jones inclusion characterizes the imbedding of sub-system $\mathcal{N}$ to $\mathcal{M}$and $\mathcal{M}$ as a finite-dimensional $\mathcal{N}$-module is the counterpart for the tensor product in finite-dimensional context. The possibility to express $\mathcal{N}$ as $\mathcal{N}$ module states fractality and can be regarded as a kind of self-referential "Brahman=Atman identity" at the level of infinite-dimensional systems.

Also the mysterious looking almost identity $CH^2 = CH$ for the configuration space of 3-surfaces would fit nicely with the identity $M \oplus M = M$. $M \otimes M \subset M$ in configuration space Clifford algebra degrees of freedom is also implied and the construction of $\mathcal{M}$ as a union of tensor powers of $C(2)$ suggests that $M \otimes M$ allows $\mathcal{M} : \mathcal{N} = 4$ inclusion to $\mathcal{M}$. This paradoxical result conforms with the strange self-referential property of factors of II_1.

The notion of many-sheeted space-time forces a considerable generalization of the notion of sub-system and simple tensor product description is not enough. Topological picture based on the length scale resolution suggests even the possibility of entanglement between sub-systems of un-entangled sub-systems. The hypothesis that hyper-finite II_1-factors describe the physics of TGD also in bosonic degrees of freedom by configuration space super-symmetry, raises the hope that Jones inclusion could allow to realize mathematically this intuitive view.

The most general view is that Jones inclusion describes all kinds of sub-system-system inclusions. The possibility to assign conformal field theory to the inclusion gives hopes of rather detailed view about dynamics of inclusion.

a) The topological condensation of space-time sheet to a larger space-time sheet mediated by wormhole contacts could be regarded as Jones inclusion. $\mathcal{N}$ would correspond to the condensing space-time sheet, $\mathcal{M}$ to the system consisting of both space-time sheets, and $\mathcal{M} : \mathcal{N}$ would characterize the number of degrees of freedom associated with the interaction between space-time sheets. Note that by general results $\mathcal{M} : \mathcal{N}$ characterizes the fractal dimension of quantum group ($\mathcal{M} : \mathcal{N} < 4$) or Kac-Moody algebra ($\mathcal{M} : \mathcal{N} = 4$) [ea9].

b) The branchings of space-time sheets (space-time surface is thus homologically like branching like of Feynman diagram) correspond naturally to n-particle vertices in TGD framework. What is nice is that vertices are nice 2-dimensional surfaces rather than surfaces having typically pinch singularities. Jones inclusion would naturally appear as inclusion of operator spaces $\mathcal{N}_i$ (essentially Fock spaces for fermionic oscillator operators) creating states at various lines as sub-spaces $N_i \subset M$ of operators creating states in common von Neumann factor $\mathcal{M}$. This would allow to construct vertices and vertices in natural manner using quantum groups or Kac-Moody algebras.

c) If 4-surfaces can branch it is difficult to argue that 3-surfaces and partonic/stringy 2-surfaces could not do the same. This allows a new kind of mechanism of topological condensation. 3-space-sheets and partonic 2-surfaces characterized by different p-adic primes could be connected by "joins" representing branchings of 2-surfaces. The structures formed by soap film foam provide a very concrete illustration about what would happen. In the TGD based model of hadrons [F4] it has been assumed join along boundaries bonds (JABs) connecting quark space-time space-time sheets to the hadronic space-time sheet. The problem is that, at least for identical primes, formation of join along boundaries bond fuses two systems to single bound state. If JABs are replaced joins, this objection is circumvented.

d) The space-time correlate for the formation of bound states is the formation of JABs. Standard intuition tells that the number of degrees of freedom associated with the bound state is smaller than the number of degrees of freedom associated with the pair of free systems. Hence the inclusion of the bound state to the tensor product could be regarded as Jones inclusion. On the other hand, one could argue that the JABs carry additional vibrational degrees of freedom so that the idea about reduction of degrees of freedom might be wrong: free system could be regarded as sub-system of bound state by Jones inclusion. The self-referential holographic properties of von Neumann algebras allow both interpretations: any system can be regarded as sub-system of any system in accordance with the bootstrap idea.

e) Maximal deterministic regions inside given space-time sheet bounded by light-like causal determinants define also sub-systems in a natural manner and also their inclusions would naturally correspond to Jones inclusions.

f) The TGD inspired model for topological quantum computation involves the magnetic flux tubes defined by join along boundaries bonds connecting space-time sheets having light-like boundaries. These tubes condensed to background 3-space can become linked and knotted and code for quantum computations in this manner. In this case the addition of new strand to the system corresponds to Jones inclusion in the hierarchy associated with inclusion $\mathcal{N} \subset \mathcal{M}$.

One can regard $\mathcal{M} : \mathcal{N}$ degrees of freedom correspond to quantum group or Kac-Moody degrees of freedom. Quantum group degrees of freedom relate closely to the conformal and topological degrees of freedom as the connection of II_1 factors with topological quantum field theories and braid matrices suggests itself. There are several questions to be answered.

a) Can one assign the $\mathcal{M} : \mathcal{N}$ degrees of freedom to the interface between $\mathcal{N}$ and $\mathcal{M}$. The interface could correspond to the wormhole contacts, joins, JABs, or light-like causal determinants serving as boundary between maximal deterministic regions, etc... In terms of the bipartite diagrams representing the inclusions, joins (say) would correspond to the edges connecting white vertices representing sub-system (the entire system without the joins) to black vertices (entire system).

b) Can one regard the sequence of phase transitions $n \to n + 1$ as a kind of renormalization group evolution for phase resolution? The limit $n = 3$ would mean that the number of these degrees of freedom becomes one so that the interaction between space-time sheets becomes trivial. Does this mean that join along boundaries bonds become vacuum extremals?

About the interpretation $\mathcal{M} : \mathcal{N}$ degrees of freedom

The operator algebra $\mathcal{N}$ associated with a system formed by two space-time sheet can be regarded as $1 \leq \mathcal{M} : \mathcal{N} \leq 4$-dimensional module having $\mathcal{N}$ as its coefficients. It is possible to imagine several interpretations the degrees of freedom labelled by β.

a) The $\beta = \mathcal{M} : \mathcal{N}$ degrees of freedom relate to the interaction of the space-time sheets. Beraha numbers appear in the construction of S-matrices of topological quantum field theories and an interpretation in terms of braids is possible. This would suggest that the interaction between space-time sheets can be described in terms of conformal quantum field theory and the S-matrices associated with braids describe this interaction. Jones inclusions would characterize the effective number of active conformal degrees of freedom. At $n = 3$ limit these degrees of freedom disappear completely since the conformal field theory defined by the Chern-Simons action describing this interaction would become trivial ($c = 0$ as will be found).

b) The interpretation of von Neumann algebra in terms of imbedding space Clifford algebra would suggest that β-dimensional Clifford algebra of $\beta = 2^d$-dimensional space is in question. For $\beta = 4$ the algebra would be the Clifford algebra of 2-dimensional space. This would conform with the effective 2-dimensionality of TGD. For $\beta < 4$ the fractal dimension of partons would be reduced since they would approach to vacua. At $\beta = 1$ limit they would become vacuum extremals.

c) In the hierarchy of inclusions the space of operators typically becomes a state space in which the operators of next level act. This allows to consider also the possibility that $\beta = 4$ defines complex dimension of spinor space in which Clifford algebra acts. This would mean that $\beta = 4$ Clifford algebra is that of a four-dimensional space-time.

d) A further interpretation is based on the identification of the states as $\beta = 4$ physical helicities of 8-dimensional spinors satisfying chirality condition.

8.4.5   The physical interpretation of ADE diagrams in TGD framework

A possible interpretation for the sequence of Jones inclusions labelled by Beraha numbers $B_n = 4cos^2(\pi/n)$, $n \to \infty$, in terms of excitation of additional degrees of freedom as deformations of vacuum extremals take them farther away from vacuum.

Renormalization group flow associated with phase resolution and Jones inclusions

The basic philosophical idea behind renormalization group approach is thinning of degrees of freedom when the scale of resolution is reduced. One can also consider renormalization group evolution associated with angle/phase resolution and quantum group phases associated with Beraha numbers defining Jones indices $\mathcal{M} : \mathcal{N}$ are excellent candidates for defining angular resolution [O3, C5, O4].

In p-adic TGD these angular resolutions would have very concrete interpretation since only algebraic extensions obtained by introducing phases $exp(i\pi/n)$ make it possible to speak about phases in p-adic sense. For a finite-dimensional extension of p-adic numbers the number of existing angles is thus always finite. Each Jones inclusion would define algebraic extension of p-adic giving rise to definite angular resolution defined by $\Delta\phi = \pi/n$ and at the limit $n \to \infty$ the resolution would become ideal.

a) The change $n \to n - 1$ of Jones inclusion could be seen as thinning of degrees of freedom associated with angular resolution leading from resolution $\Delta\phi = \pi/n$ to $\Delta\phi = \pi/(n-1)$. At the limit $n = 3$ only the multiples of $exp(i\pi/3)$ can be resolved.

b) Since angular resolution becomes poorer in $n \to n - 1$ transition and the dimension $\mathcal{M} : \mathcal{N}$ of $\mathcal{M}$ as $\mathcal{N}$-module is reduced in the transition, it seems natural to assign angular resolution to the dynamics to these $\mathcal{M} : \mathcal{N}$ degrees of freedom.

The possibility to assign Dynkin diagrams with the inclusions of II_1 algebras is highly suggestive concerning possible physical interpretations.

a) For $\beta = \mathcal{M} : \mathcal{N} < 4$ Dynkin diagrams code for the inclusions and correspond to simply laced Lie algebras. $SU(2)$, D_{2n+1}, and E_7 are excluded.

b) Extended ADE Dynkin diagrams coding for simoly laced ADE Kac Moody algebras appear at $\beta = 4$. Also $SU(2)$ Kac Moody algebra appears.

Jones inclusions and topological condensation

One can try to interpret the appearance of ADE Dynkin diagrams in terms of the general vision about renormalization group evolutions associated with phase resolution developed in [O3, O4].

a) The basic observation is that $\beta < 4$ and the corresponding ADE quantum group have nothing to do with the structure of $\mathcal{N}$ and $\mathcal{M}$ but characterizes the imbedding of II_1 factor $\mathcal{N}$ to $\mathcal{M}$, the position of $\mathcal{N}$ in $\mathcal{M}$. Suppose that the topological condensation of space-time sheet to a larger space-time sheet can be modelled as imbedding of the state space $\mathcal{N}$ of the smaller space-time sheet to the full state space $\mathcal{M}$. Under this assumption β and ADE group would characterize the interaction between space-time sheets rather than interactions inside either space-time sheet.

b) One could regard the two-sheet system as β-state system with the $\beta = \mathcal{M} : \mathcal{N}$ states have a fine structure described by the II_1 factor $\mathcal{N}$ associated with the smaller space-time sheet and appearing in the role of a multiplicative field associated with a state space. This fine structure would not be visible to the physicist at larger space-time sheet seeing only a p-state system.

c) According to the earlier considerations based on quantum-classical correspondence and properties of solutions of field equations [D1, O3], the dimension β could correspond fractal dimension for the projection of space-time sheet to CP_2. $\beta = 4$ corresponds to maximal dimension and $\beta = 1$ to a vacuum extremal and a trivial theory in vibrational degrees of freedom so that only center of mass degrees of freedom (translational and rotational as well as color-rotational) degrees of freedom would remain intact.

d) For $\beta = 4$ one would have full Kac Moody dynamical symmetry and for $\beta < 4$ Kac Moody symmetry would reduce to a pure gauge symmetry (its quantum group version) so that physical states would correspond to finite-dimensional representations of quantum group. The growth of n in $\beta(n)$ would corresponds to a gradual excitation of dynamical degrees of freedom when space-time surface deforms farther away from vacuum extremal and gauge degrees of freedom become dynamical spin glass degrees of freedom so that the dimension of the gauge group increases.

e) The effective 2-dimensionality of TGD Universe in the sense that partonic or stringy 2-surfaces code for the data relevant to quantum states would suggest that the sequence of Beraha numbers corresponds to a hierarchy of conformal field theories. For $\beta < 4$ they they would have pure gauge symmetry given by ADE diagram and for $\beta = 4$ this symmetry would become Kac-Moody symmetry as the gauge degrees of freedom become physical.

f) The complex fractal spinor space associated with a β-dimensional Clifford algebra has dimension $d = \sqrt{\beta}$ from the generalization of the finite-dimensional formula. This dimension equals to the dimension $d = [2]_q$, $q = exp(i\pi/n)$ of quantum plane given as $d_q = [2]_q = sin(2\pi/n)/sin(\pi/n) = 2cos(\pi/n)$ [ec4]. The Hilbert space associated with M has interpretation as quantum spinor space defined as $\sqrt{\beta}$ dimensional N-module. On basis of the experience from quantum field theory, generalized Feynman rules for TGD are expected to involve traces of products of gamma matrices equal to d_q. The dimension $D_q = 2 \times log_2(d)$ of the space associated with $d = [2]_q$-dimensional spinors would emerge through the dependence of $\hbar$ on it. I do not know whether any formulation for the quantum counterpart of real space-time with dimension D_q exists. Considering a pair of subsequent Jones inclusions (dual to each other as are also partonic and string-like 2-surfaces [E2]) as a basic inclusion, means the replacement $[2]_q \to ([2]_q)^2$.

Quantitative support for the interpretation

A more detailed analysis of the situation gives support for the proposed vision.

a) A given value of quantum group deformation parameter $q = exp(i\pi/n)$ makes sense for any Lie algebra but now a preferred Lie-algebra is assigned to a given value of quantum deformation parameter. At the limit $\beta = 4$ when quantum deformation parameter becomes trivial, the gauge symmetry is replaced by Kac Moody symmetry.

b) The prediction is that Kac-Moody central extension parameter should vanish for $\beta < 4$. There is an intriguing relationship to formula for the quantum phase q_{KM} associated with (possibly trivial) Kac-Moody central extension and the phase defined by ADE diagram

$$q_{KM} = exp(i\phi) \ , \quad \phi_1 = \frac{\pi}{k+h^v} \ ,$$
$$q_{Jones} = exp(i\phi) \ , \quad \phi = \frac{\pi}{h}$$

In the first formula sum of Kac-Moody central extension parameter k and dual Coxeter number h^v appears whereas Coxeter number h appears in the second formula. Internal consistency requires

$$k + h^v = h \ . \tag{8.4.2}$$

It is easy see that the dual Coxeter number h^v and Coxeter number h given by $h = (dim(g) - r)/r$, where r is the dimension of Cartan algebra of g, are identical for ADE algebras so that the Kac-Moody central extension parameter k must indeed vanish. For $SO(2n + 1)$, $Sp(n)$, G_2, and F_4 the condition $h = h^v$ does not hold true but one has $h(n) = 2n = h^v + 1$ for $SO(2n + 1)$, $h(n) = 2n = 2(h^v - 1)$ for $Sp(n)$, $h = 6 = h^v + 2$ for G_2, and $h = 12 = h^v + 3$ for F_4.

What is intriguing that G_2, which seems to play a fundamental role in the dual formulation of quantum TGD based on the identification of space-times as surfaces in hyper-octonionic space M^8 [E2] is not allowed. As a matter of fact, $G_2 \to SU(3)$ reduction occurs also in the dual formulation based on $G_2/SU(3)$ coset model and is required by the separate conservation of quark and lepton numbers predicted by TGD. ADE groups would be associated with the interaction between space-time sheets rather than entire dynamics and need not have anything to do with the Kac-Moody algebra associated with color and electro-weak interactions appearing in the construction of physical states [F2].

c) There seems to be a concrete connection with conformal field theories. This connection would allow to understand the emergence of quantum groups appearing naturally in these theories. Quite generally, the conformal central extension parameter for unitary Virasoro representations resulting by Sugawara construction from Kac Moody representations satisfies either of the conditions

$$
\begin{aligned}
c & \geq \frac{kdim(g)}{k+h^v} + 1 \ , \\
c & = \frac{kdim(g)}{k+h^v} + 1 - \frac{6}{(h-1)h} \ .
\end{aligned}
\tag{8.4.3}
$$

For $k = 0$, which should be interesting for $\beta < 4$, the second formula reduces to

$$
c = 1 - \frac{6}{(h-1)h} \ .
\tag{8.4.4}
$$

The formula gives the values of c for minimal conformal field theories with finite number of conformal fields and real conformal weights. Indeed, h in this formula seems to correspond to the same h as appearing in the expression $\beta \equiv \mathcal{M} : \mathcal{N} = 4cos^2(\pi/h)$.

$\beta = 3, h = 6$ corresponds to three-state Potts model with $c = 4/5$ which should thus have a gauge group for which Coxeter number is 6: the group should be either $SU(6)$ or $SO(8)$. Two-state Potts model, that is Ising model with $\beta = 2, h = 4$ would correspond to $c = 1/2$ and to a gauge group $SU(4)$ or $SO(4)$. For $h = 3$ ("one-state Potts model") with group $SU(3)$ one would have $c = 0$ and vanishing conformal anomaly so that conformal degrees of freedom would become pure gauge degrees of freedom.

These observations give support for the following picture.

a) Quite generally, the number of states of the generalized β-state Potts model has an interpretation as the dimension $\beta = \mathcal{M} : \mathcal{N}$ of $\mathcal{M}$ as $\mathcal{N}$-module. Besides the models with integer number of states there is an infinite number of models for which the number of states is not an integer. The conditions $c \leq 1$ guaranteing real conformal weights and $\beta \leq 4$ correspond to each other for these models.

b) $\beta > 4$ Potts models would be formally obtained by allowing h to be imaginary in the defining formula for $\mathcal{M} : \mathcal{N}$. In this case c would be however complex so that the theory would not be unitary.

c) For minimal models with $(\beta < 4, c < 1)$ Kac-Moody central extension parameter is vanishing so that Kac Moody algebra indeed acts like gauge symmetries and gauge symmetries would be in question. The limiting case $(\beta = 4, c = 1)$ would define a "four-state Potts model" with infinite-dimensional unitary group acting as a gauge group. On the other hand, the appearance of extended ADE Dynkin diagrams suggests strongly that this limit is not realized but that $\beta = \mathcal{M} : \mathcal{N} = 4$ corresponds to $k = 1$ conformal field theory allowing Kac Moody symmetries for any ADE group, which as simply-laced groups allows vertex operator construction. The appearance of $kdim(g)/(k+g)$ in the more general formula would thus code the Kac Moody group whereas for $\beta < 4$ ADE diagram codes for the preferred gauge group characterizing the minimal CFT.

d) The possibility that any ADE gauge group or Kac-Moody group can characterize the interaction between space-time sheets conforms with the idea about Universe as a Topological Quantum Computer able to simulate any conceivable quantum dynamics. Of course, one cannot exclude the possibility that only electro-weak and color symmetries are realized in this manner.

8.4.6 Construction of S-matrix in TGD framework and II_1 factors

The construction of S-matrix for a single space-time surface is based on generalized Feynman diagrams [C5].

a) Lines correspond to 3-D light-like causal determinants (CDs) X_i^3. They can correspond to boundary components of a space-time sheet (such as boundaries of magnetic flux tubes) but can also serve as horizons separating maximal non-deterministic regions within a space-time sheet.

b) Vertices correspond to 2-D partonic surfaces at which the light-like CDs X_i^3 branch like a lines of Feynman diagrams. Hence both X_i^3 and presumably also space-time surfaces are singular as manifolds. The 2-surfaces representing vertices need not however have pinch like singularities appearing in stringy diagrams. Also partonic surfaces can have branchings and a foam formed from soap films seems to be a good visualization of the most general situation.

c) There is rather close analogy with the branes in that the intersection of space-time surfaces 7-D light-like CDs $X_\pm^7 = \delta M_\pm^4 \times CP_2$ of imbedding space provide a natural gauge fixing for 4-D general coordinate invariance. Hence the incoming partons X_i^2 correspond to intersections $X_i^3 \cap X_{\pm}^7$. Future (past) oriented light-cone corresponds to incoming (outgoing) particles.

The basic challenges are following.

a) Construct explicitly the unitary evolution operators associated with the lines defining the analogs of propagators. These operators should be fixed to a high degree by the dynamics of the second quantized induced spinor fields as has been suggested in [C5]. The existence of a universal unitary von Neumann algebra automorphism Δ^{it} suggesting itself as a candidate for this unitary evolution operator.

b) Construct vertices. Vertices are in principle fixed as vacuum expectation values for the product of operators creating the incoming and outgoing states at the vertices. These operators are constructible from oscillator operators associated with the generalized eigen modes of the modified Dirac operator acting on the second quantized induced spinor fields, whose quantization is fixed by the requirement that the super-canonical charges constructed in terms of oscillator operators define configuration space gamma matrices having super-symmetrized symplectic transformations of $\delta M_+^4 \times CP_2$ as isometries. Intuitively it looks obvious that the product of the operators creating states at the lines defines vertex as its vacuum expectation. The challenge is to imbed the fermionic oscillator operator algebras associated with incoming lines to same oscillator algebra and Jones inclusions suggest themselves here.

Free propagation

von Neumann algebras allow a universal unitary automorphism $A \to \Delta^{it} A \Delta^{-it}$ [dc6], which is unique apart from an inner automorphisms $\Delta^{it} \to U\Delta^{it}V$, with V U and V defining a change of basis for the target and domain. This automorphism is a highly attractive candidate for the unitary transformation associated with the lines of generalized Feynman diagrams.

The first problem is how to fix the inner automorphisms U and V. Second problem concerns the possible values of the parameter t. The sum over all allowed values of t is expected to appear when one integrates over configuration space. ne might hope is that it is possible to forget all the details of space-time surfaces and perform the integral explicitly to get the analog of propagator $1/(L_0 + i\epsilon)$ as in the case of string models.

Unitarity requires that the eigenvalues of $log(\Delta)$ are real: this is the case since the spectrum of Δ is non-negative. Note however that the spectrum of $log(\Delta)$ contains also negative eigenvalues. The eigenvalues of Δ would naturally correspond to the squares of the eigenvalues of the modified Dirac operator D at the partonic 2-surface and Hamilton would correspond to $log(D^2)$. It is encouraging that von Neumann algebra approach provides a concrete interpretation for these eigenvalues which has been lacking hitherto.

p-Adicization certainly poses strong constraints and it would not be surprising if the imaginary parts for the zeros of Riemann Zeta would pop up here as they do in the construction of configuration space geometry involving the 7-D light-like CD $\Delta M_+^4 \times CP_2$ (rather than 3-D CD X_l^3). In this case appropriate combinations of the zeros of zeta correspond to complex super-canonical conformal weights whose sum is real for the physical states (conformal confinement).

a) The condition that the spectrum of the positive definite Hamiltonian Δ consists of rationals follows as a constraint if the idea about rational physics is accepted. This requires that $(m/n)^{it}$ (or equivalently p^{it} for any prime) belongs to a finite- dimensional algebraic extension of rational numbers for allowed values of t and thus defines a finite-dimensional extension of p-adic numbers for any p.

b) Both the product and sum representation of Riemann Zeta involve similar exponents. The requirement $\zeta(s)$ exists for all values of $Im(s)$ expressible as linear combinations of imaginary parts $y = Im(s)$ of non-trivial zeros of ζ implies that p^{iy} satisfies the same condition for any prime p. Hence the allowed values of t would correspond to linear combinations of imaginary parts of zeros of Zeta with integer coefficients presumably forming a dense set along imaginary axis.

c) Time evolution operator should commute with super-canonical symmetries. Hence Δ should be a function of the Casimir operator of the super-canonical algebra identifiable as the Virasoro generator L_0. This encourages to ask whether for a space-time sheet characterized by p-adic prime p, Δ^{it} corresponds to $p^{iL_0 t}$ (recall that the reality of eigenvalues of L_0 implies conformal confinement).

d) I have discussed in [C5] a model for a scalar propagator assuming that it is expressible as superposition of Δ^{it}, $\Delta = p^2 - m_0^2$ and t expressible in terms of zeros of Zeta for various sub-algebras of super-canonical conformal algebra. This identification need not be correct: if the eigen values of

Δ correspond to $p^2 - m_0^2 \geq 0$, non-negativity of Δ excludes virtual masses $p^2 < m_0^2$. The generalized eigen modes of the modified Dirac operator D are however direct analogs off mass shell states. The replacement of $p^2 - m_0^2$ with its exponent and restriction to the non-trivial zeros of ζ would guarantee the positive definiteness of Δ. In this case the propagator would become the discrete analog for the integral representation of scalar propagator as the integral $\int_0^\infty e^{i(p^2-m^2)}dt$.

e) A profound implication of the configuration space super-symmetry is that the unitary time evolution for the lines of generalized Feynman graphs in $\mathcal{M}$ generalizes from CH spinorial degrees of freedom to the level of "orbital" CH degrees of freedom (represented as deformations of 3-surface). This strengthens the hopes about a calculable theory.

Interaction vertices and Jones inclusions

Jones inclusion provides a generic manner to describe sub-system-system relation. $\mathcal{M}$ as an $\mathcal{M} : \mathcal{N}$ dimensional $\mathcal{N}$ module defines a fractal-dimensional quantum space of complex dimension $\mathcal{M} : \mathcal{N} \leq 4$, which characterizes the interface between two interacting systems in the sense that it corresponds to the edge of the bi-partite graph connecting $\mathcal{N}$ to $\mathcal{M}$, where the free system corresponds to $\mathcal{N} = \mathcal{M}_1 \otimes \mathcal{M}_2$ and $\mathcal{M}$ to the interacting system. The interaction between the systems would thus be described in a completely universal manner.

1. Self-referentiality of von Neumann algebras and negative energies

The Hilbert space associated with $\mathcal{M}$ could be identifies as $\mathcal{M}$ itself by the basic self-referential property of von Neumann algebras. This identification would mean that also "negative energy states" (or phase conjugate states) for which the roles of annihilation and creation operators are changed become possible. Negative energies are indeed a basic prediction of quantum TGD [D3, D5]. If must be however emphasized that at space-time level, four-momenta do not appear in the construction of vertices or unitary transformations associated with the lines. This is just what makes possible separable Hilbert space.

2. Jones inclusions and quantum analogs of space-time surfaces

The idea about space-time as a 4-surface replicates itself at the level of operator algebra and state space in the sense that Jones inclusion can be seen as a representation of the operator algebra $\mathcal{N}$ as infinite-dimensional linear sub-space (surface) of the operator algebra $\mathcal{M}$. This encourages to think that generalized Feynman diagrams could correspond to image surfaces in II_1 factor having identification as quantum space-time surfaces.

In vertices the II_1 factors creating states associated with lines must be imbedded to common II_1 factor, which by self-referentiality can be any II_1 associated with the lines, or just abstract II_1 factor. The Eastern idea about entities able to have very different appearances (characterized by $\mathcal{M} : calN$ now) but being basically one and same thing comes in mind. Grothendienck's notion of motive based on similar idea and is proposed to have applications in quantum theory [ed3].

If the idea about mapping state spaces of light-like CDs to sub-factors of von Neumann algebra makes sense, then vertex as a branching space-time surface at which X_i^3 meet along partonic 2-surface X^2 might be mappable to an intersection of II_1 factors inside II_1 factor. Many-sheeted space-time would have a kind of quantum image inside II_1 factor consisting of pieces which are $\mathcal{M} : \mathcal{N} \leq 4$-dimensional quantum spaces. Strictly speaking, quantum Clifford algebras are in question and it is not clear whether corresponding real spaces with fractal dimension $log_2(\mathcal{M} : \mathcal{N})$ exist in some sense.

3. Constraints on the inclusions $N_i \subset M$

The inclusions $N_i \subset M$ are fixed a priori only apart from unitary automorphisms U_i and the vertices are expected to depend on the relative positions of $\mathcal{N}_i$ defined by U_i. The basic question is whether the relative positions are fixed by some natural conditions such as the requirement of common vacuum state Ω in the case that the state space is identified as II_1 factor itself satisfied for $N_i = N_j = ... = N$ or that loop diagrams are equivalent with tree diagrams.

Are partonic orbits mapped directly to $\mathcal{N}_i \subset \mathcal{M}$?

The most stringent scenario is based on a direct mapping of the generalized Feynman graphs to surfaces in II_1 factor $\mathcal{M}$ characterized by a common value of $\beta = \mathcal{M} : \mathcal{N}$ so that entire generalized Feynman diagram corresponds to a unique value of $\hbar$.

Imbedding to X^4 can be also replaced with imbedding to $Y^4 \subset X^4$. One can wonder whether the following correspondences could make sense:

i) $1 < \beta(Y^4) < 4$ (rational CFT) $\leftrightarrow$ the imbeddings to $Y^4 \subset X^4$ with $\mathcal{M}(Y^4) \subset \mathcal{M}$;

ii) $\beta(Y^4) \to 1$ ($n \to 3$) $\leftrightarrow$ the imbeddings $X^4 \supset Y^4 \to X^4$: at this limit theory becomes trivial since entire Universe has vanishing conserved quantum numbers [D5];

iii) $\beta = 4$ ("stringy" description at CP_2 length scales) $\leftrightarrow$ the imbeddings to Y^4, which corresponds to very small space-time sheets at which CP_2 type extremals have suffered the first topological condensation.

$\mathcal{M}$ could be also seen as the counterpart of the imbedding space: $\beta = 4$ would be analogous to $dim(H) = 8 = 2 \times 4$ decomposition where "2" is interpreted as the dimension of partonic 2-surface. A possible interpretation is as a decomposition of the tangent space of H as a tensor product of 2-D and 4-D sub-spaces.

The orbits of partons correspond to time dependent sub-factors $\mathcal{N}_i(t)$ suffering a unitary automorphism by $\Delta_i^{it_i}$ along the 3-D CD defining the orbit at space-time level. At vertices the factors $\mathcal{N}_i$ must coincide just just like the partonic 2-surfaces do. This poses strong constraints on the values of $\Delta_i^{it_i}$ and t_i defining the temporal duration of the line and very much analogous to the time parameter associated with string diagrams. Thus Bohr orbits at the level of space-time surface have counterparts at the level of M, which could be interpreted as the II_1 factor associated with sub-system plus environment. Number theoretical constraints from p-adicization suggest a quantization of t as $t = \sum_i n_i y_i > 0$, where $z_i = 1/2 + y_i$ are non-trivial zeros of Riemann Zeta. The result would be a very geometric picture about interactions even at Hilbert space level.

The ordering of the operators associated with the incoming lines affects physical predictions unless symmetrization or anti-symmetrization or quantum group symmetrization using braid group statistics is performed. The construction should indeed yield as a special case modular S-matrix associated with braids and defining knot and link invariants. The vertex itself should be trace of the resulting operator O expressible as vacuum expectation value $(O\Omega, \Omega)$ creating the state at vertex.

<u>Feynman diagrams inside Feynman diagrams and a hierarchy of topological condensations</u>

It is possible to assign to a given Jones inclusion $\mathcal{N} \subset \mathcal{M}$ an entire hierarchy of Jones inclusions $\mathcal{M}_0 \subset \mathcal{M}_1 \subset \mathcal{M}_2..., \mathcal{M}_0 = N, \mathcal{M}_1 = M$. A natural interpretation for these inclusions would be as a sequence of topological condensations.

This sequence also defines a hierarchy of Feynman diagrams inside Feynman diagrams. The factor $\mathcal{M}$ containing the Feynman diagram having as its lines the unitary orbits of $\mathcal{N}$ under $\Delta_{\mathcal{M}}$ becomes a parton in $\mathcal{M}_1$ and its unitary orbits under $\Delta_{\mathcal{M}_1}$ define lines of Feynman diagrams in M_1. The outcome is a hierarchy of Feynman diagrams within Feynman diagrams, a fractal structure for which many particle scattering events at a given level become particles at the next level. The particles at the next level represent dynamics at the lower level: they have the property of "being about" representing perhaps the most crucial element of conscious experience. Since net conserved quantum numbers can vanish for a system in TGD Universe, this kind of hierarchy indeed allows a realization as zero energy states. Crossing symmetry can be understood in terms of this picture and has been applied to construct a model for S-matrix at high energy limit [C2].

The quantum image for the orbit of parton has dimension $\log_2(\mathcal{M} : \mathcal{N}) + 1 \leq 3$. Two subsequent inclusions form a natural basic unit since the bipartite diagrams are duals of each other by black-white duality. In this double inclusion a two-parameter family of deformations of $\mathcal{M}$ counterpart of a partonic 2-surface is formed and has quantum dimension $\log_2(\mathcal{M} : \mathcal{N}) + 2 \leq 4$. One might perhaps say that quantum space-time corresponds to a double inclusion and that further inclusions bring in N-parameter families of space-time surfaces.

<u>Could Jones inclusions be random?</u>

A second model for the Jones inclusions at vertex is based on the assumption that the inclusions are random. This model is not consistent with the mapping $X^4 \subset H$ inclusion to $N_i \subset M$ inclusion. Conceptual problems are caused by the fact that the values of the quantum group parameter and $\hbar$ can be different for the different incoming lines. The following poor man's argument shows that this option leads to an inconsistency unless $\mathcal{M}$ is chosen to be a factor associated with one of the lines of the diagram.

a) One can try to understand the intersections of Jones inclusion by generalizing naively the rules for the intersections of surfaces with dimensions D_i imbedded in finite-dimensional manifold of dimension D. Intersection is non-empty in non-generic case if the condition $\sum D_i > D$ holds true.

b) In the recent case all inclusions are linear sub-spaces of $\mathcal{M}$ through origin and the requirement is that the intersection defines higher- than 0-dimensional Hilbert space. The dimension of II_1 factor $\mathcal{M}$ equals to 1 and the dimensions of the imbedded factors $\mathcal{N}_i$ are equal to $1/\beta_i$, $\beta_i = 1/\mathcal{M} : \mathcal{N}_i$, since the dimension of $\mathcal{M}$ as $\mathcal{N}_i$ module is β_i. Note that the codimension of $\mathcal{N}$ would be $1 - 1/\beta$ which for $n = 4$ equals to the dimension of $\mathcal{N}$ itself ($\beta = 1/2$).

c) Without further conditions, the generalization for the condition of having non-trivial intersection in the generic case is

$$\sum_{i=1}^{N} \frac{1}{\beta_i} - 1 > 0 \ .$$

This condition does not seem sensible. For instance for $\beta_i = 4$ this would allow only 5-vertex as the lowest non-trivial vertex and elastic 2-particle scattering would not be possible. If at least one of the vertices has $\beta_i = 2$ ($n = 2$), 4-vertex becomes possible.

d) The possibility to choose $\beta_{i_0} = 1$ for some i_0 means that $\mathcal{N}_{i_0}$ fills $\mathcal{M}$ and the condition is trivially satisfied:

$$\sum_{i=1}^{N-1} \frac{1}{\beta_i} \ > \ 0 \ . \tag{8.4.5}$$

3. Conformal invariance and field theory and stringy phases

$\mathcal{M} : \mathcal{N} < 4$ assigns a unique minimal conformal field theory to the inclusion and this should give important information about the vertex. A priori the inclusions $\mathcal{N}_i \subset \mathcal{M}$ can have different values of $\mathcal{M} : \mathcal{N}_i$ determining the quantum phases q_i. Both physical intuition and anyonic statistics encourage to think that the values of quantum phases q_i are identical. This is certainly the case for the most string scenario for which the inclusions $N_i \subset M$ coincide.

It seems conceivable that $\mathcal{M} : \mathcal{N} < 4$ vertices correspond physically to the low energy phase symmetry broken phase describable using field theory. Ordinary QFT would corresponds to $\mathcal{M} : \mathcal{N} \to 4$ limit whereas $\mathcal{M} : \mathcal{N} = 4$ phase with Kac-Moody symmetry would correspond to the "stringy" phase of the theory. The low energy limit would transform from an approximate theoretical description to an actual physical phase. In this phase massivation of massless particles would occur by p-adic thermodynamics [F2] whereas ultra-heavy particles would drop from the spectrum.

Dimensional regularization with complex space-time dimension $D = 4 - \epsilon \to 4$ could be interpreted as the limit $\mathcal{M} : \mathcal{N} \to 4$. $\mathcal{M}$ as an $\mathcal{M} : \mathcal{N}$-dimensional $\mathcal{N}$-module would provide a concrete model for the a quantum Clifford algebra. An entire sequence of regularized theories corresponding to the allowed values of $\mathcal{M} : \mathcal{N}$ is predicted.

As will be discussed, the evolution of Jones index could correspond to renormalization group evolution for phase resolution characterized by the value $\hbar$. For instance, in the range $n \in (4, \infty)$ $\hbar$ would vary by a factor of 2 and the theory predicts the possibility of varying fine structure constant.

There is evidence that fine structure constant has grown somewhat in cosmological time scales [ie2, ie3]. In [ie6] the recent situation is described and also possible mechanisms explaining the variation are discussed at popular level. The figure 8.4.6 taken from [ie6] shows that the possible relative variation of fine structure constant is in the range $10^{-5} - 10^{-6}$ over 10 billion years. There are however later experiments [ie5] finding that the relative time variation is below 10^{-6} over the past 6 to 10 billion years so that the situation remains unsettled.

In [D5] I have considered a possible TGD inspired mechanism explaining the variation of α. The variation might be also understood in terms of the gradual growth of the integer n characterizing $\mathcal{M} : \mathcal{N}$ if $\hbar$ depends on Beraha number in the proposed manner. In the p-adic context this evolution means the emergence of higher-dimensional algebraic extensions allowing higher roots of unity. In TGD inspired theory of consciousness this would correspond to evolution of cognition. This leaves open also the possibility that the value of the fine structure constant is to some degree dependent on the local environment.

The order of magnitude for the variation gives a lower bound for the value of n, which must be relatively large. Using the predicted proportionality $\hbar(n) \propto 1/log(B_n)$, $B(n) = 4cos^2(\pi/n)$, the

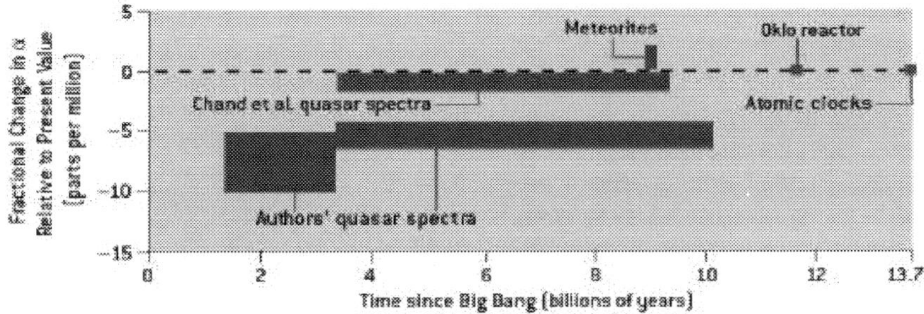

MEASUREMENTS of the fine-structure constant are inconclusive. Some indicate that the constant used to be smaller, and some do not. Perhaps the constant varied earlier in cosmic history and no longer does so. (The boxes represent a range of data.)

Figure 8.1: Data about the time variation of α.

bound $\Delta\alpha/\alpha \leq 10^{-5}$ for the relative variation boils down to the same bound for the relative variation $\Delta log(B_n)/log(B_n) \simeq \Delta cos^2(\pi/n)/log(4)$ giving at large n limit

$$\frac{\Delta\alpha}{\alpha} \simeq \frac{\pi^2\Delta n}{n^3} \leq 10^{-5} \ .$$

This gives $n \geq (log(4) \times \pi^2/10)^{1/3} \times (\Delta n)^{1/3} \times 10^2 \simeq 111$ for $\Delta n = 1$. The upper bound 10^{-6} for the variation would give $n \geq 239$ for $\Delta n = 1$.

In $\mathcal{M} : \mathcal{N} = 4$ case all ADE type $k = 1$ conformal theories are in principle possible. An open question is whether actually the conformal field theory defined by the Kac-Moody algebra characterizing TGD is possible or whether the idea about interfaces as able to emulate any string model is more appropriate.

4. Holonomy of $\mathcal{N} \subset \mathcal{M}$ and unitarity of S-matrix and braid statistics

Assume that the automorphisms induced by $\Delta_{\mathcal{M}}$ defines a closed path of $\mathcal{N}$ in $\mathcal{M}$. The action on the automorphism on individual points of $\mathcal{N}$ could be non-trivial and would have interpretation as a holonomy group defining a unitary action of $\mathcal{M}$ on $\mathcal{N}$. This action generalizes also to the level of Feynman diagrams. Single particle S-matrix would represent this action. This action could relate to the 2-dimensional braid statistics defined by quantum group. For braids the 2π braid rotation of $(k + 1)^{th}$ strand around k^{th} strand induces a non-trivial action on the state and this action could correspond to $\Delta_{\mathcal{M}_{k+1}}$ holonomy on $\mathcal{M}_k$. This interpretation would indeed predict that S-matrix elements for tree diagrams differing by braidings of incoming and outgoing lines are not identical. One could perhaps say that the counterpart of 2π braid rotation (or equivalently, strand exchange) is performed in $d = log_2(\mathcal{M} : \mathcal{N}) \leq 2$-dimensional quantum plane associated with $\mathcal{M}$ interpreted as a $\mathcal{N}$ module and representing Clifford algebra of this space.

The equivalence of loop diagrams with tree diagrams in terms of Jones inclusions

1. Equivalence of loop diagrams with tree diagrams

A strong requirement is that diagrams with loops are equivalent with tree diagrams. Conditions for this to occur have been formulated in algebraic terms for the generalization of ribbon algebras [C5]. The natural hope is that this could be due to the inherent properties of von Neumann algebras and guaranteed by the proper choice of inclusions at the vertices.

Two kinds of loops are possible.

a) The stringy loops in which partonic 2-surface decays temporarily to two partonic 2-surfaces do not seem to correspond in TGD framework to particle decays but to a single particle propagation along two different paths simultaneously. This picture leads to a generalization of the quantum measurement theory and explanation [K1] for the findings of Shahriar Afshar relating to double slit experiment challenging Copenhagen interpretation [je2]. For instance, in double slit experiment the measurement of the particle aspect of photon would reduce the branched photon path in such a manner that second branch corresponds to a vacuum extremal having a vacuum line representing identity operator as its algebraic counterpart. In this case the equivalence of loop diagrams with tree diagrams looks obvious.

b) A stronger condition is that loops obtained by allowing intermediate vertices at which partonic lines are branched, can be eliminated.

The equivalence of loop diagrams with tree diagrams means that one can move the end of any internal line until it becomes a tadpole loop which must represent vacuum line and can be eliminated. This means that all diagrams are equivalent to a simplicial complex representing the homology of planar disk D^2 with the ends of the external lines at the boundary circle of the disk and having Euler characteristic $E + F - L = -2$.

The equivalence implies also that any tree diagram containing N-vertices with arbitrary values of N can be transformed to a single M-vertex, where M is the number of incoming lines (for convenience al lines are regarded as incoming). The diagram can be also transformed to a diagram containing only 3-vertices. Number theoretic vision about the role of classical division algebras in TGD [E2] suggests that this symmetry is closely related to the octonionic triality reflecting itself also as the existence of 3 8-dimensional representations of $SO(8)$.

2. Equivalence of loop diagrams with tree diagrams and Jones inclusions

Consider now how the equivalence could be understood in terms of Jones inclusions. The argument below is a simplification of the algebraic conditions formulated in [C5] guaranteing also the possibility to move the ends of the lines around the graph.

a) If each incoming line is thought of as being imbedded in the same manner to a II_1 factor $\mathcal{M}$ then each incoming line of the vertex can be characterized by the same value of $\mathcal{M} : \mathcal{N}$, and one can assign to each line emanating from a vertex a representation of the same quantum group or Kac Moody group. The notions of product, co-product, and bi-algebra are well-defined [C5].

b) Assume that it is possible to transform the diagram to a diagram containing only 3-vertices by moving around the ends of the lines: this possibility should relate closely to octonionic triality [E2] underlying the vertex construction. As a consequence, all loops reduce to self-energy loops. Assume that the operator in the third line of the vertex is product of the operators associated with other two lines and the operator associated with two lines is a co-product of the operator in the third line. Under this assumption products and co-products in the self energy loops compensate each other and they are trivial.

3. Does the equivalence with tree diagrams imply unitarity?

It would be easier to take seriously the reducibility of generalized Feynman diagrams to tree diagrams if it would guarantee the unitarity of S-matrix. The following heuristics indicates that this is the case.

a) The equivalence with tree diagrams allows to carry out two operations for the generalized Feynman diagrams.

i) It is possible to transform diagrams representing S-matrix elements to diagrams involving single vertex with M incoming lines and N outgoing lines. Incoming lines start from the boundary of a future directed light cone $X^7_+ = \delta M^4_+ CP_2$ and outgoing lines end at the boundary of a past directed light-cone $X^7_- = \delta M^4_+ \times CP_2$ having its tip inside X^7_+.

ii) It is possible to move the position of $M + N$ vertex arbitrarily near to the initial moment. At space-time level this means that the M partonic orbits intersect at the partonic 2-surface already at X^7_+ and decay to N partonic 2-surfaces.

b) Unitary conditions should reduce to the statement that the initial states M_1 and M_2 are orthogonal. The sum over intermediate states in the unitarity relation involves a sum over number N of outgoing lines. Assume that it can be transformed by the completeness of states to a form in which a delta function appears stating that the values t_i are equal for N lines and their conjugates. If this is the possible, the unitary automorphisms Δ^{it_i} and their conjugates compensate each other for each outgoing N line.

c) If only the condition i) is assumed, the unitarity condition reduces to a condition stating the orthogonality of the images of the states $\hat{M}_1$ and $\hat{M}_2$ obtained from M_1 and M_2 by assigning the automorphisms Δ^{it_i} to the lines M_1 and their conjugates Δ^{-it_j} to the lines of M_2. A reaction in which both incoming and outgoing partons belong to the boundary of the same future light-cone X^7_+ is in question. M particles travel to future, react in $M + N$ vertex and produce N particles, which are reflected back to the past.

d) If also the Feynman diagrams characterizing the S-matrix obtained by replacing the automorphisms associated with outgoing lines with their time reversals satisfy the equivalence with tree diagrams, this S-matrix is trivial without further conditions since the lines to future and back can be contracted to points. If also the condition ii) is assumed, $\hat{M}_i = M_i$ and unitarity conditions reduce to the ordinary orthogonality conditions.

8.4.7 Feynman diagrams as higher level particles and their scattering as dynamics of self consciousness

The hierarchy of imbeddings of hyper-finite factors of II_1 as counterpart for many-sheeted space-time lead inevitably to the idea that this hierarchy corresponds to a hierarchy of generalized Feynman diagrams for which Feynman diagrams at a given level become particles at the next level. Accepting this idea, one is led to ask what kind of quantum states these Feynman diagrams correspond, how one could describe interactions of these higher level particles, what is the interpretation for these higher level states, and whether they can be detected.

Higher level Feynman diagrams

The lines of Feynman diagram in $\mathcal{M}_{n+1}$ are geodesic lines representing orbits of $\mathcal{M}_n$ and this kind of lines meet at vertex and scatter. The evolution along lines is determined by $\Delta_{\mathcal{M}_{n+1}}$. These lines contain within themselves $\mathcal{M}_n$ Feynman diagrams with similar structure and the hierarchy continues down to the lowest level at which ordinary elementary particles are encountered.

For instance, the generalized Feynman diagrams at the second level are ribbon diagrams obtained by thickening the ordinary diagrams in the new time direction. The interpretation as ribbon diagrams crucial for topological quantum computation and suggested to be realizable in terms of zero energy states in [O3] is natural. At each level a new time parameter is introduced so that the dimension of the diagram can be arbitrarily high. The dynamics is not that of ordinary surfaces but the dynamics induced by the $\Delta_{\mathcal{M}_n}$.

Quantum states defined by higher level Feynman diagrams

The intuitive picture is that higher level quantum states corresponds to the self reflective aspect of existence and must provide representations for the quantum dynamics of lower levels in their own structure. This dynamics is characterized by S-matrix whose elements have representation in terms of Feynman diagrams.

a) These states correspond to zero energy states in which initial states have "positive energies" and final states have "negative energies". The net conserved quantum numbers of initial and final state partons compensate each other. Gravitational energies, and more generally gravitational quantum numbers defined as absolute values of the net quantum numbers of initial and final states do not vanish. One can say that thoughts have gravitational mass but no inertial mass.

b) States in sub-spaces of positive and negative energy states are entangled with entanglement coefficients given by S-matrix at the level below.

To make this more concrete, consider first the simplest non-trivial case. In this case the particles can be characterized as ordinary Feynman diagrams, or more precisely as scattering events so that the state is characterized by $\hat{S} = P_{in}SP_{out}$, where S is S-matrix and P_{in} *resp.* P_{out} is the projection to a subspace of initial *resp.* final states. An entangled state with the projection of S-matrix giving the entanglement coefficients is in question.

The larger the domains of projectors P_{in} and P_{out}, the higher the representative capacity of the state. The norm of the non-normalized state $\hat{S}$ is $Tr(\hat{S}\hat{S}^\dagger) \leq 1$ for II_1 factors, and at the limit $\hat{S} = S$ the norm equals to 1. Hence, by II_1 property, the state always entangles infinite number of states, and can in principle code the entire S-matrix to entanglement coefficients.

The states in which positive and negative energy states are entangled by a projection of S-matrix might define only a particular instance of states for which conserved quantum numbers vanish. The model for the interaction of Feynman diagrams discussed below applies also to these more general states.

The interaction of $\mathcal{M}_n$ Feynman diagrams at the second level of hierarchy

What constraints can one pose to the higher level reactions? How Feynman diagrams interact? Consider first the scattering at the second level of hierarchy ($\mathcal{M}_1$), the first level $\mathcal{M}_0$ being assigned to the interactions of the ordinary matter.

a) Conservation laws pose constraints on the scattering at level $\mathcal{M}_1$. The Feynman diagrams can transform to new Feynman diagrams only in such a manner that the net quantum numbers are conserved separately for the initial positive energy states and final negative energy states of the diagram. The simplest assumption is that positive energy matter and negative energy matter know nothing about each other and effectively live in separate worlds. The scattering matrix form Feynman diagram like states would thus be simply the tensor product $S \otimes S^\dagger$, where S is the S-matrix characterizing the lowest level interactions. Reductionism would be realized in the sense that, apart from the new elements brought in by $\Delta_{\mathcal{M}_n}$ defining single particle free dynamics, the lowest level would determine in principle everything occurring at the higher level providing representations about representations about... for what occurs at the basic level. The lowest level would represent the physical world and higher levels the theory about it.

b) The description of hadronic reactions in terms of partons serves as a guide line when one tries to understand higher level Feynman diagrams. The fusion of hadronic space-time sheets corresponds to the vertices $\mathcal{M}_1$. In the vertex the analog of parton plasma is formed by a process known as parton fragmentation. This means that the partonic Feynman diagrams belonging to disjoint copies of $\mathcal{M}_0$ find themselves inside the same copy of $\mathcal{M}_0$. The standard description would apply to the scattering of the initial *resp.* final state partons.

c) After the scattering of partons hadronization takes place. The analog of hadronization in the recent case is the organization of the initial and final state partons to groups I_i and F_i such that the net conserved quantum numbers are same for I_i and F_i. These conditions can be satisfied if the interactions in the plasma phase occur only between particles belonging to the clusters labelled by the index i. Otherwise only single particle states in $\mathcal{M}_1$ would be produced in the reactions in the generic case. The cluster decomposition of S-matrix to a direct sum of terms corresponding to partitions of the initial state particles to clusters which do not interact with each other obviously corresponds to the "hadronization". Therefore no new dynamics need to be introduced.

d) One cannot avoid the question whether the parton picture about hadrons indeed corresponds to a higher level physics of this kind. This would require that hadronic space-time sheets carry the net quantum numbers of hadrons. The net quantum numbers associated with the initial state partons would be naturally identical with the net quantum numbers of hadron. Partons and they negative energy conjugates would provide in this picture a representation of hadron about hadron. This kind of interpretation of partons would make understandable why they cannot be observed directly. A possible objection is that the net gravitational mass of hadron would be three times the gravitational mass deduced from the inertial mass of hadron if partons feed their gravitational fluxes to the space-time sheet carrying Earth's gravitational field.

e) This picture could also relate to the suggested duality between string and parton pictures [E2]. In parton picture hadron is formed from partons represented by space-like 2-surfaces X_i^2 connected by join along boundaries bonds. In string picture partonic 2-surfaces are replaced with string orbits. If one puts positive and negative energy particles at the ends of string diagram one indeed obtains a higher level representation of hadron. If these pictures are dual then also in parton picture positive and negative energies should compensate each other. Interestingly, light-like 3-D causal determinants identified as orbits of partons could be interpreted as orbits of light like string word sheets with "time" coordinate varying in space-like direction.

Scattering of Feynman diagrams at the higher levels of hierarchy

This picture generalizes to the description of higher level Feynman diagrams.

a) Assume that higher level vertices have recursive structure allowing to reduce the Feynman diagrams to ordinary Feynman diagrams by a procedure consisting of finite steps.

b) The lines of diagrams are classified as incoming or outgoing lines according to whether the time orientation of the line is positive or negative. The time orientation is associated with the time parameter t_n characterizing the automorphism $\Delta_{\mathcal{M}_\backslash}^{it_n}$. The incoming and outgoing net quantum numbers compensate each other. These quantum numbers are basically the quantum numbers of the state at the lowest level of the hierarchy.

c) In the vertices the $\mathcal{M}_{n+1}$ particles fuse and $\mathcal{M}_n$ particles form the analog of quark gluon plasma. The initial and final state particles of $\mathcal{M}_n$ Feynman diagram scatter independently and the S-matrix S_{n+1} describing the process is tensor product $S_n \otimes S_n^\dagger$. By the clustering property of S-matrix, this scattering occurs only for groups formed by partons formed by the incoming and outgoing particles $\mathcal{M}_n$ particles and each outgoing $\mathcal{M}_{n+1}$ line contains and irreducible $\mathcal{M}_n$ diagram. By continuing the recursion one finally ends down with ordinary Feynman diagrams.

A connection with TGD inspired theory of consciousness

The implications of this picture TGD inspired theory of consciousness are rather breathtaking.

a) The hierarchy of self representations and the reduction of their quantum dynamics to the dynamics of the material world apart from the effects brought in by the automorphisms Δ_{M_n} determining the free propagation of thoughts, would mean a concrete calculable theory for the quantum dynamics of cognition. My sincere hope is however that no one would ever christen these states "particles of self consciousness". These states are not conscious, consciousness would be in the quantum jump between these states.

b) Cognitive representations would possess "gravitational" charges, in particular gravitational mass, so that thoughts could be put into "gravitational scale". I have proposed that "gravitational" charges correspond to classical charges characterizing the systems at space-time level as opposed to quantum charges.

c) As found, even hadrons could form self representations usually assigned with human brain. This is certainly something that neuroscientist would not propose but conforms with the basic prediction of TGD inspired theory of consciousness [TGDconsc] about infinite self hierarchy involving cognitive representations at all levels of the hierarchy [K1].

d) The TGD inspired model of topological quantum computation [O3] in terms of zero energy cognitive states inspired the proposal that the appearance of a representation and its negative energy conjugate could relate very intimately to the fact that DNA appears as double helices of a strand and its conjugate. This could also relate to the fact that binary structures are common in living matter.

e) One is forced to consider a stronger characterization of dark matter [D6, J6] as a matter at higher levels of the hierarchy with vanishing net inertial quantum numbers but with non-vanishing "gravitational" quantum numbers. We would detect dark matter via its "gravitational" charges. We would also experience it directly since our thoughts would be dark matter! The cosmological estimates for the proportion of dark matter and dark energy would give also estimate for the gravitational mass of thoughts in the Universe! This speculation is probably not quite correct since also more general entanglement than that defined by two-sided projections of S-matrix is possible between positive and negative energy states.

Cognitive entanglement as Connes tensor product

In the proposed construction the lowest level $\mathcal{N}$ represents matter and higher levels give cognitive representations. The ordinary tensor product $S \otimes S$ and its tensor powers define a hierarchy of S-matrices and the two-sided projections of these S-matrices in turn define entanglement coefficients for positive and negative energy states at various levels of hierarchy.

The following arguments show that the cognitive tensor product restricted to projections of S-matrix corresponds to the so called Connes tensor product appearing naturally in the hierarchy of Jones inclusions.A slight generalization of earlier scenario predicting matter-mind type transitions is forced by this identification and a beautiful interpretation for these transitions in terms of space-time correlates emerges.

1. Connes tensor product

Connes [ed1, ec5] has introduced a variant of tensor product allowing to express the union $\cup M_i$, where M_i the inclusion hierarchy as infinite tensor product $M \otimes_N M \otimes_N \otimes...$. The Connes tensor product $\otimes_N$ differs from the standard tensor product and is obtained by requiring that in the Connes tensor product of Hilbert spaces $\mathcal{H}_1$ and $\mathcal{H}_2$ the condition $n\xi_1 \otimes_N \xi_2 = \xi_1 \otimes_N n\xi_2$ for all $n \in \mathcal{N}$ holds true. Connes tensor product means forces to replace ordinary statistics with braid statistics. The physical interpretation proposed by Connes is that this tensor product could make sense when

N represents observables common to the Hilbert spaces $\mathcal{H}_1$ and $\mathcal{H}_2$. Later it will be found that TGD suggests quite different interpretation.

Connes tensor product makes sense also for finite-dimensional right and left modules. Consider the spaces $M_{n\times q}$ of $n \times q$-matrices and $M_{p\times n}$ of $p \times n$ matrices for which $n \times n$ matrix algebra $M_{n\times n}$ acts as a left *resp.* right multiplier. The tensor product $\otimes_N$ for these matrices is the ordinary matrix product of $m_{p\times n} \times m_{n\times q}$ and belongs to $M_{p\times q}$ so that the dimension of tensor product space is much lower than $m \times q \times n^2$ and does not depend on n. For Jones inclusion N takes the role of $M_{n\times n}$ and since M can be regarded as β-dimensional $\mathcal{N}$-module, tensor product can be said to give $\sqrt{\beta} \times \sqrt{\beta}$-dimensional matrices with $\mathcal{N}$ valued entries. In particular, the inclusion sequence is an infinite tensor product of $\sqrt{\beta} \times \sqrt{\beta}$-dimensional matrices.

2. Does Connes tensor product generate cognitive entanglement?

One can wonder why the entanglement coefficients between positive and negative energy states should be restricted to the projections of S-matrix. The obvious guess is that it gives rise to an entanglement equivalent with Connes tensor product so that the action of N on initial state is equivalent with its action on the final state. This indeed seems to be the case. The basic symmetry of Connes tensor product translates to the possibility to move an operator creating particles in initial state to final state by conjugating it: this is nothing but crossing symmetry characterizing S-matrix. Thus Connes tensor product generates zero energy states providing a hierarchy of cognitive representations.

3. Do transitions between different levels of cognitive hierarchy occur?

The following arguments suggest that the proposed hierarchy of cognitive representations is not exhaustive.
a) Only tensor powers of S involving $(2^n)^{th}$ powers of S appear in the cognitive hierarchy as it is constructed. Connes tensor product representation of $\cup_i M_i$ would however suggest that all powers of S appear.
b) There is no reason to restrict the states to positive energy states in TGD Universe. In fact, the states of the entire Universe have zero energy. Thus much more general zero energy states are possible in TGD framework than those for which entanglement is given by a projection of S-matrix, and they occur already at the lowest level of the hierarchy.

On basis of these observations there is no reason to exclude transitions between different levels of the cognitive hierarchy transforming ordinary tensor product of positive and negative energy states with vanishing conserved quantum numbers to a Connes tensor product involving only the projection of S-matrix as entanglement coefficients. These transitions would give rise to S-matrices connecting different levels and thus fill the gaps in the spectrum of allowed tensor powers of S.

4. Space-time correlates for the matter-to-mind transitions

Allowing somewhat poetic language, the scatterings in question would represent kind of matter-to-mind transitions, enlightment, or transition to a Buddha state. At space-time level zero energy matter would correspond to positive and negative energy states with a space-like separation whereas cognitive states would correspond to positive and negative energy states with a time-like separation. By the failure of the complete classical determinism time like entanglement makes sense but due to the fact determinism is not completely lost, entanglement could be of a very special kind only, and S-matrix could appear as entanglement coefficients.

Light-like causal determinants (CDs) identifiable as orbits of both space-like partonic 2-surfaces and light-like stringy surfaces, can be said to represent both matter and mind. According to the proposal of [O3], light-like CDs would correspond to both programs and computers for topological quantum computation, and matter-mind transformation would be also involved with the realization of the genetic code both as cognitive and material structures. This would support the view that the stringy 2-surfaces in the foliation of the space-time surface are time-like in the interior of the space-time sheet (or more generally, outside light-like causal determinants) and light-like causal determinants correspond to critical line between matter and mind.

8.4.8 Configuration space, space-time, and imbedding space and hyper-finite type II_1 factors

The preceding considerations have by-passed the question about the relationship of the configuration space tangent space to its Clifford algebra. Also the relationship between space-time and imbedding space and their quantum variants could be better. In particular, one should understand how effective 2-dimensionality can be consistent with the 4-dimensionality of space-time.

Tangent space of CH as a logarithm of Clifford algebra

It would be highly desirable to achieve also a description of the configuration space degrees of freedom using von Neumann algebras. Super-canonical algebra has as its generators configuration space Hamiltonians and their super-counterparts identifiable as CH gamma matrices. Super-symmetry requires that the Clifford algebra of CH and the Hamiltonian vector fields of CH with symplectic central extension both define hyper-finite II_1 factors. The presence of zero modes means direct integral over these factors.

Configuration space gamma matrices anti-commuting to identity operator with unit norm corresponds to the tangent space $T(CH)$ of CH. Thus it would be not be surprising if $T(CH)$ could be imbedded in the sigma matrix algebra as a sub-space of operators defined by the gamma matrices generating this algebra. At least for $\beta = 4$ construction of hyper-finite II_1 factor this definitely makes sense. The dimension of this algebra as N-module obtained by interpolating the formula holding for integer dimensions is equal to $D_q = log_2(\beta)$ appearing in the proposed formula for $\hbar(q)$. Configuration space and its sub-spaces can thus be seen as an $\mathcal{N} \subset \mathcal{M}$-modules with quantum dimension D_q serving as a counterpart for real dimension $D = 2$. By super-symmetry Poisson bracket corresponds to anti-commutator for gamma matrices. The ordinary quantized version of Poisson bracket is obtained as $\{P_i, Q_j\} \rightarrow [P_i, Q_j] = J_{ij} Id$. Finite trace version results by assuming that Id corresponds to the projector CH Clifford algebra having unit norm.

Perhaps the fact that only covariantly constant right handed neutrino contributes to CH gamma matrix algebra [B4] relates to the quantum dimension $D_q \leq 2$ of CH. Since right hand neutrino has two different spin directions, a quantum counterpart for 2×2 Clifford algebra would result naturally.

The dimension of the configuration space defined as the trace of the projection operator to the sub-space spanned by gamma matrices is obviously zero. This follows also from the formula for the space-time dimension in terms of the dimension of $D = 2log_2(Tr(Id)) = 0$, where Id is the identity matrix of Clifford algebra having $Tr(Id) = 1$. Thus configuration space has in this sense the dimensionality of single space-time point. This sounds perhaps absurd but the generalization of the number concept implied by infinite primes indeed leads to the view that single space-time point is infinitely structured in the number theoretical sense although in the real sense all states of the point are equivalent [O4]. The reason is that there is infinitely many numbers expressible as ratios of infinite integers having unit real norm in the real sense but having different p-adic norms.

How to understand the dimensions of space-time and imbedding space?

One should be able to understand the dimensions of 3-space, space-time and imbedding space in a convincing matter in the proposed framework. There is also the question whether space-time and imbedding space emerge uniquely from the mathematics of von Neumann algebras alone.

1. The dimensions of space-time and imbedding space

Two sub-sequent inclusions dual to each other define a special kind of inclusion giving rise to a quantum counterpart of $D = 4$ naturally. This would mean that space-time is something which emerges at the level of cognitive states.

The special role of classical division algebras in the construction of quantum TGD [E2], $D = 8$ Bott periodicity generalized to quantum context, plus self-referential property of type II_1 factors might explain why 8-dimensional imbedding space is the only possibility. Perhaps $M_1^{\otimes N^4} \otimes_N M$ and $\mathcal{M}$ in the basic construction are isomorphic in some special sense by Bott periodicity.

State space has naturally quantum dimension $D \leq 8$ as the following simple argument shows. The space of quantum states has quark and lepton sectors which both are super-symmetric implying $D \leq 4$ for each. Since these sectors correspond to different Hamiltonian algebras (triality one for quarks and triality zero for leptonic sector), the state space has quantum dimension $D \leq 8$.

2. How the lacking two space-time dimensions emerge?

3-surface is the basic dynamical unit in TGD framework. This seems to be in conflict with the effective 2-dimensionality [E2] meaning that partonic 2-surface code for quantum states, and with the fact that hyper-finite II_1 factors have intrinsic quantum dimension 2.

A possible resolution of the problem is that the foliation of 3-surface by partonic two-surfaces defines a one-dimensional direct integral of isomorphic hyper-finite type II_1 factors, and the zero mode labelling the 2-surfaces in the foliation serves as the third spatial coordinate. For a given 3-surface the contribution to the configuration space metric can come only from 2-D partonic surfaces defined as intersections of 3-D light-like CDs with $X^7_{\pm}$ [B2, B3]. Hence the direct integral should somehow relate to the classical non-determinism of Kähler action.

a) The one-parameter family of intersections of light-like CD with $X^7_{\pm}$ inside $X^4 \cap X^7_{\pm}$ could indeed be basically due to the classical non-determinism of Kähler action. The contribution to the metric from the normal light-like direction to $X^3 = X^4 \cap X^7_{\pm}$ can cause the vanishing of the metric determinant $\sqrt{g_4}$ of the space-time metric at $X^2 \subset X^3$ under some conditions on X^2. This would mean that the space-time surface $X^4(X^3)$ is not uniquely determined by the minimization principle defining the value of the Kähler action, and the complete dynamical specification of X^3 requires the specification of partonic 2-surfaces X^2_i with $\sqrt{g_4} = 0$.

b) The known solutions of field equations [D1] define a double foliation of the space-time surface defined by Hamilton-Jacobi coordinates consisting of complex transversal coordinate and two light-like coordinates for M^4 (rather than space-time surface). Number theoretical considerations inspire the hypothesis that this foliation exists always [E2]. Hence a natural hypothesis is that the allowed partonic 2-surfaces correspond to the 2-surfaces in the restriction of the double foliation of the space-time surface by partonic 2-surfaces to X^3, and are thus locally parameterized by single parameter defining the third spatial coordinate.

d) There is however also a second light-like coordinate involved and one might ask whether both light-like coordinates appear in the direct sum decomposition of II_1 factors defining $T(CH)$. The presence of two kinds of light-like CDs would provide the lacking two space-time coordinates and quantum dimension $D = 4$ would emerge at the limit of full non-determinism. Note that the duality of space-like partonic and light-like stringy 2-surfaces conforms with this interpretation since it corresponds to a selection of partonic/stringy 2-surface inside given 3-D CD whereas the dual pairs correspond to different CDs.

e) That the quantum dimension would be $2D_q = \beta < 4$ above CP_2 length scale conforms with the fact that non-determinism is only partial and time direction is dynamically frozen to a high degree. For vacuum extremals there is strong non-determinism but in this case there is no real dynamics. For CP_2 type extremals, which are not vacuum extremals as far action and small perturbations are considered, and which correspond to $\beta = 4$ there is a complete non-determinism in time direction since the M^4 projection of the extremal is a light-like random curve and there is full 4-D dynamics. Light-likeness gives rise to conformal symmetry consistent with the emergence of Kac Moody algebra [D1].

3. Time and cognition

In a completely deterministic physics time dimension is strictly speaking redundant since the information about physical states is coded by the initial values at 3-dimensional slice of space-time. Hence the notion of time should emerge at the level of cognitive representations possible by to the non-determinism of the classical dynamics of TGD.

Since Jones inclusion means the emergence of cognitive representation, the space-time view about physics should correspond to cognitive representations provided by Feynman diagram states with zero energy with entanglement defined by a two-sided projection of the lowest level S-matrix. These states would represent the "laws of quantum physics" cognitively. Also space-time surface serves as a classical correlate for the evolution by quantum jumps with maximal deterministic regions serving as correlates of quantum states. Thus the classical non-determinism making possible cognitive representations would bring in time. The fact that quantum dimension of space-time is smaller than $D = 4$ would reflect the fact that the loss of determinism is not complete.

4. Do space-time and imbedding space emerge from the theory of von Neumann algebras and number theory?

The considerations above force to ask whether the notions of space-time and imbedding space emerge from von Neumann algebras as predictions rather than input. The fact that it seems possible to formulate the S-matrix and its generalization in terms of inherent properties of von Neumann algebras suggest that this might be the case. What is lacking is the proof that the III_1 factors possibly present and related to the positions of the tips of light-cones $M^4_\pm$ characterizing sub-configuration spaces defined by $M^4_\pm CP_2$ force $D = 8$.

8.4.9 Quaternions, octonions, and hyper-finite type II_1 factors

Quaternions and octonions as well as their hyper counterparts obtained by multiplying imaginary units by commuting $\sqrt{-1}$ and forming a sub-space of complexified division algebra, are in in a central role in the number theoretical vision about quantum TGD [E2]. Therefore the question arises whether complexified quaternions and perhaps even octonions could be somehow inherent properties of von Neumann algebras. One can also wonder whether the quantum counterparts of quaternions and octonions could emerge naturally from von Neumann algebras. The following considerations allow to get grasp of the problem.

Quantum quaternions and quantum octonions

Quantum quaternions have been constructed as deformation of quaternions [cb6]. The key observation that the Glebsch Gordan coefficients for the tensor product $3 \otimes 3 = 5 \oplus \oplus 3 \oplus 1$ of spin 1 representation of $SU(2)$ with itself gives the anti-commutative part of quaternionic product as spin 1 part in the decomposition whereas the commutative part giving spin 0 representation is identifiable as the scalar product of the imaginary parts. By combining spin 0 and spin 1 representations, quaternionic product can be expressed in terms of Glebsh-Gordan coefficients. By replacing GGC:s by their quantum group versions for group $sl(2)_q$, one obtains quantum quaternions.

There are two different proposals for the construction of quantum octonions [cb8, cb9]. Also now the idea is to express quaternionic and octonionic multiplication in terms of Glebsch-Gordan coefficients and replace them with their quantum versions.

a) The first proposal [cb8] relies on the observation that for the tensor product of $j = 3$ representations of $SU(2)$ the Glebsch-Gordan coefficients for $7 \otimes 7 \to 7$ in $7 \otimes 7 = 9 \oplus 7 \oplus 5 \oplus 3 \oplus 1$ defines a product, which is equivalent with the antisymmetric part of the product of octonionic imaginary units. As a matter fact, the antisymmetry defines 7-dimensional Malcev algebra defined by the anticommutator of octonion units and satisfying b definition the identity

$$[[x, y, z], x] = [x, y, [x, z]] \quad , \quad [x, y, z] \equiv [x, [y, z]] + [y, [z, x]] + [z, [x, y]] \quad . \tag{8.4.6}$$

7-element Malcev algebra defining derivations of octonionic algebra is the only complex Malcev algebra not reducing to a Lie algebra. The $j = 0$ part of the product corresponds also now to scalar product for imaginary units. Octonions are constructed as sums of $j = 0$ and $j = 3$ parts and quantum Glebsch-Gordan coefficients define the octonionic product.

b) In the second proposal [cb9] the quantum group associated with $SO(8)$ is used. This representation does not allow unit but produces a quantum version of octonionic triality assigning to three octonions a real number.

Quaternionic or octonionic quantum mechanics?

There have been numerous attempts to introduce quaternions and octonions to quantum theory. Quaternionic or octonionic quantum mechanics, which means the replacement of the complex numbers as coefficient field of Hilbert space with quaternions or octonions, is the most obvious approach (for example and references to the literature see for instance [cb4].

In both cases non-commutativity poses serious interpretational problems. In the octonionic case the non-associativity causes even more serious obstacles [cb7, cb4].

a) Assuming that an orthonormalized state basis with respect to an octonion valued inner product has been found, the multiplication of any basis with octonion spoils the orthonormality. The proposal to circumvent this difficulty discussed in [cb7] eliminates non-associativity by assuming that octonions multiply states one by one (rather than multiplying each other before multiplying the state). Effectively this means that octonions are replaced with 8×8-matrices.

b) The definition of the tensor product leads also to difficulties since associativity is lost (recall that Yang-Baxter equation codes for associativity in case of braid statistics [ec4, ee2]).

c) The notion of hermitian conjugation is problematic and forces a selection of a preferred imaginary unit, which does not look nice. Note however that the local selection of a preferred imaginary unit is in a key role in the proposed construction of space-time surfaces as hyper-quaternionic or co-hyper-quaternionic surfaces and allows to interpret space-time surfaces either as surfaces in 8-D Minkowski space M^8 of hyper-octonions or in $M^4 \times CP_2$. This selection turns out to have quite different interpretation in the proposed framework.

Hyper-finite factor II_1 has a natural Hyper-Kähler structure

In the case of hyper-finite factors of type II_1 quaternions a more natural approach is based on the generalization of the Hyper-Kähler structure rather than quaternionic quantum mechanics. The reason is that also configuration space tangent space should and is expected to have this structure [B3]. The Hilbert space remains a complex Hilbert space but the quaternionic units are represented as operators in Hilbert space. The selection of the preferred unit is necessary and natural. The identity operator representing quaternionic real unit has trace equal to one, is expected to give rise to the series of quantum quaternion algebras in terms of inclusions $\mathcal{N} \subset \mathcal{M}$ having interpretation as N-modules.

The representation of the quaternion units is rather explicit in the structure of hyper-finite II_1 factor. The $\mathcal{M} : \mathcal{N} \equiv \beta = 4$ hierarchical construction can be regarded as Connes tensor product of infinite number of 4-D Clifford algebras of Euclidian plane with Euclidian signature of metric $(diag(-1,-1))$. This algebra is nothing but the quaternionic algebra in the representation of quaternionic imaginary units by Pauli spin matrices multiplied by i.

The imaginary unit of the underlying complex Hilbert space must be chosen and there is whole sphere S^2 of choices and in every point of configuration space the choice can be made differently. The space-time correlate for this local choice of preferred hyper-octonionic unit [E2]. At the level of configuration space geometry the quaternion structure of the tangent space means the existence of Hyper-Kähler structure guaranteing that configuration space has a vanishing Einstein tensor. It it would not vanish, curvature scalar would be infinite by symmetric space property (as in case of loop spaces) and induce a divergence in the functional integral over 3-surfaces from the expansion of $\sqrt{g}$ [B3].

The quaternionic units for the II_1 factor, are simply limiting case for the direct sums of 2×2 units normalized to one. Generalizing from $\beta = 4$ to $\beta < 4$, the natural expectation is that the representation of the algebra as $\beta = \mathcal{M} : \mathcal{N}$-dimensional $\mathcal{N}$-module gives rise to quantum quaternions with quaternion units defined as infinite sums of $\sqrt{\beta} \times \sqrt{\beta}$ matrices.

At Hilbert space level one has an infinite Connes tensor product of 2-component spinor spaces on which quaternionic matrices have a natural action. The tensor product of Clifford algebras gives the algebra of 2×2 quaternionic matrices acting on 2-component quaternionic spinors (complex 4-component spinors). Thus double inclusion could correspond to (hyper-)quaternionic structure at space-time level. Note however that the correspondence is not complete since hyper-quaternions appear at space-time level and quaternions at Hilbert space level.

von Neumann algebras and octonions

The octonionic generalization of the Hyper-Kähler manifold does not make sense as such since octonionic units are not representable as linear operators. The allowance of anti-linear operators inherently present in von Neumann algebras could however save the situation. Indeed, the Cayley-Dickson construction for the division algebras (for a nice explanation see [cb1]), which allows to extend any * algebra, and thus also any von Neumann algebra, by adding an imaginary unit it and identified as *, comes in rescue.

The basic idea of the Cayley-Dickson construction is following. The * operator, call it J, representing a conjugation defines an *anti-linear* operator in the original algebra A. One can extend A by adding this operator as a new element to the algebra. The conditions satisfied by J are

$$a(Jb) = J(a^*b) \ , \quad (aJ)b = (ab^*)J \ , \quad (Ja)(bJ^{-1}) = (ab)^* \ . \tag{8.4.7}$$

In the associative case the conditions are equivalent to the first condition.

It is intuitively clear that this addition extends the hyper-Kähler structure to an octonionic structure at the level of the operator algebra. The quantum version of the octonionic algebra is fixed by the quantum quaternion algebra uniquely and is consistent with the Cayley-Dickson construction. It is not clear whether the construction is equivalent with either of the earlier proposals [cb8, cb9]. It would however seem that the proposal is simpler.

Physical interpretation of quantum octonion structure

Without further restrictions the extension by J would mean that vertices contain operators, which are superpositions of linear and anti-linear operators. This would give superpositions of states and their time-reversals and mean that state could be a superposition of states with opposite values of say fermion numbers. The problem disappears if either the linear operators A or anti-linear operators JA can be used to construct physical states from vacuum. The fact, that space-time surfaces are either hyper-quaternionic or co-hyper-quaternionic, is a space-time correlate for this restriction.

The $HQ - coHQ$ duality discussed in [E2] states that the descriptions based on hyper-quaternionic and co-hyper-quaternionic surfaces are dual to each other. The duality can have two meanings.

a) The vacuum is invariant under J so that one can use either complexified quaternionic operators A or their co-counterparts of form JA to create physical states from vacuum.

b) The vacuum is not invariant under J. This could relate to the breaking of CP and T invariance known to occur in meson-antimeson systems. In TGD framework two kinds of vacua are predicted corresponding intuitively to vacua in which either the product of all positive or negative energy fermionic oscillator operators defines the vacuum state, and these two vacua could correspond to a vacuum and its J conjugate, and thus to positive and negative energy states. In this case the two state spaces would not be equivalent although the physics associated with them would be equivalent.

The considerations of [E2] related to the detailed dynamics of $HQ - coHQ$ duality demonstrate that the variational principles defining the dynamics of hyper-quaternionic and co-hyper-quaternionic space-time surfaces are antagonistic and correspond to world as seen by a conscientious book-keeper on one hand and an imaginative artist on the other hand. HQ case is conservative: differences measured by the magnitude of Kähler action tend to be minimized, the dynamics is highly predictive, and minimizes the classical energy of the initial state. $coHQ$ case is radical: differences are maximized (this is what the construction of sensory representations would require). The interpretation proposed in [E2] was that the two space-time dynamics are just different predictions for what would happen (has happened) if no quantum jumps would occur (had occurred). A stronger assumption is that these two views are associated with systems related by time reversal symmetry.

What comes in mind first is that this antagonism follows from the assumption that these dynamics are actually time-reversals of each other with respect to M^4 time (the rapid elimination of differences in the first dynamics would correspond to their rapid enhancement in the second dynamics). This is not the case so that T and CP symmetries are predicted to be broken in accordance with the CP breaking in meson-antimeson systems [F5] and cosmological matter-antimatter asymmetry [D5].

8.4.10 How does the hierarchy of infinite primes relate to the hierarchy of II_1 factors?

The hierarchy of Feynman diagrams accompanying the hierarchy defined by Jones inclusions $\mathcal{M}_0 \subset \mathcal{M}_1 \subset \ldots$ gives a concrete representation for the hierarchy of cognitive dynamics providing a representation for the material world at the lowest level of the hierarchy. This hierarchy seems to relate directly to the hierarchy of space-time sheets.

Also the construction of infinite primes [E3] leads to an infinite hierarchy. Infinite primes at the lowest level correspond to polynomials of single variable x_1 with rational coefficients, next level to polynomials x_1 for which coefficients are rational functions of variable x_2, etc... so that a natural ordering of the variables is involved.

If the variables x_i are hyper-octonions (subs-space of complexified octonions for which elements are of form $x + \sqrt{-1}y$, where x is real number and y imaginary octonion and $\sqrt{-1}$ is commuting imaginary unit, this hierarchy of states could provide a realistic representation of physical states as far as quantum numbers related to imbedding space degrees of freedom are considered in M^8 picture dual to $M^4 \times CP_2$ picture [E2]. Infinite primes are mapped to space-time surfaces in a manner analogous to the mapping of polynomials to the loci of their zeros so that infinite primes, integers, and rationals become concrete geometrical objects.

Infinite primes are also obtained by a repeated second quantization of a super-symmetric arithmetic quantum field theory. Infinite rational numbers correspond in this description to pairs of positive energy and negative energy states of opposite energies having interpretation as pairs of initial and final states so that higher level states indeed represent transitions between the states. For these reasons this hierarchy has been interpreted as a correlate for a cognitive hierarchy coding information about quantum dynamics at lower levels. This hierarchy has also been assigned with the hierarchy of space-time sheets. Just as the hierarchy of generalized Feynman diagrams provides self representations of the lowest matter level and is coded by it, finite primes code the hierarchy of infinite primes.

Infinite primes, integers, and rationals have finite p-adic norms equal to 1, and one can wonder whether a Hilbert space like structure with dimension given by an infinite prime or integer makes sense, and whether it has anything to do with the Hilbert space for which dimension is infinite in the sense of the limiting value for a dimension of sub-space. The Hilbert spaces with dimension equal to infinite prime would define primes for the tensor product of these spaces. The dimension of this kind of space defined as any p-adic norm would be equal to one.

One cannot exclude the possibility that infinite primes could express the infinite dimensions of hyper-finite III_1 factors, which cannot be excluded and correspond to that part of quantum TGD which relates to the imbedding space rather than space-time surface. Indeed, infinite primes code naturally for the quantum numbers associated with the imbedding space. Secondly, the appearance of 7-D light-like causal determinants $X^7_{\pm} = M^4_{\pm} \times CP_2$ forming nested structures in the construction of S-matrix brings in mind similar nested structures of algebraic quantum field theory [dc8]. If this is were the case, the hierarchy of Beraha numbers possibly associated with the phase resolution could correspond to hyper-finite factors of type II_1, and the decomposition of space-time surface to regions labelled by p-adic primes and characterized by infinite primes could correspond to hyper-finite factors of type III_1 and represent imbedding space degrees of freedom.

The state space would in this picture correspond to the tensor products of hyper-finite factors of type II_1 and III_1 (of course, also factors I_n and I_∞ are also possible). III_1 factors could be assigned to the sub-configuration spaces defined by 3-surfaces in regions of M^4 expressible in terms of unions and intersections of $X^7_{\pm} = M^4_{\pm} \times CP_2$. By conservation of four-momentum, bounded regions of this kind are possible only for the states of zero net energy appearing at the higher levels of hierarchy. These sub-configuration spaces would be characterized by the positions of the tips of light cones $M^4_{\pm} \subset M^4$ involved. This indeed brings in continuous spectrum of four-momenta forcing to introduce non-separable Hilbert spaces for momentum eigen states and necessitating III_1 factors. Infinities would be avoided since the dynamics proper would occur at the level of space-time surfaces and involve only II_1 factors.

8.5   Jones inclusions and dynamical $\hbar$

The idea about dynamical $\hbar$ depending on the Jones index was inspired by the considerations related to topological quantum computation, and the modelling of various anomalies has provided considerable support for the notion. In this section an improved formulation for dynamical $\hbar$ in terms of renormalization group evolution associated with phase resolution is discussed.

8.5.1   Are Jones inclusions associated with a renormalization group flow associated with phase resolution?

A kind of generalized renormalization group flow which is associated with phase resolution defined by the quantum phase $exp(i\pi/n)$ would be in question. The change of $\hbar$ in the step $n \to n-1$ would be given by

$$\frac{\hbar(\infty)}{\hbar(n)} - \frac{\hbar(\infty)}{\hbar(n-1)} = 4 \times \left[\frac{1}{log_2(q(n))} - \frac{1}{log_2(q(n-1))} \right] \tag{8.5.1}$$

Since the value of $\hbar(n)$ is determined only up to an additive constant, the equation would allow also solutions for which $1/\hbar(3)$ is non-vanishing. Indeed, the model for the astrophysical findings [if1, D6, J6] suggesting gigantic value of $\hbar$ can be justified if the condition $\hbar(\infty)/\hbar(3) = v_0/GM_1M_2$ is satisfied. At the limit when the gravitational masses M_i approach infinity one would obtain strictly infinite $\hbar$.

At the limit ($n = 3, \beta = 1$) one would obtain a minimal angular resolution with quantum phase equal to $exp(i\pi/3)$. Only three angles values would be discernible. These phases would correspond naturally to the phases assignable to the center of SU(3) whose Dynkin diagram indeed corresponds to $n = 3$ inclusion. Color $SU(3)$ would be the most rigid or minimal symmetry and would not reduce to $SU(2)$. Color degrees of freedom would correspond to the color rotational rigid body degrees of freedom of a topologically condensed space-time sheet.

This picture suggest that $\hbar$ can be assigned with the interaction between two space-time sheets whereas it does not appear explicitly in the exponent of Kähler function defining vacuum functional (this is the case for YM action too). Quite generally, $\hbar$ could be assigned to the interface between sub-system and system.

8.5.2 $n = 3$ case as large $\hbar$ phase

The expression for $\hbar_{gr}$ in the model explaining the Bohr orbits for planets is of form $\hbar = GM_1M_2/v_0$ [D6]. This suggests that the interaction is indeed associated with the interface, in this case most naturally join along boundaries connecting the space-time sheets associated with systems possessing gravitational masses M_1 and M_2. This argument generalizes to the case $\hbar/\hbar(\infty) = Q_1Q_2\alpha/v_0$ in case of generic phase transition to a strongly interacting phase with α describing gauge coupling strength.

There is indeed some experimental evidence for the existence of a phase with a large $\hbar$.

a) I have proposed an explanation of dark matter as a macroscopic quantum phase with a large value of $\hbar$ [D6]. Any interaction, if sufficiently strong, can lead to this kind of phase and conformal confinement is an essential feature of this phase.

b) Living matter could represent a basic example of $n = 3$ phase. Even ordinary condensed matter could be "partially dark" in many-sheeted space-time and this could resolve the age old mystery of why water is transparent [J6].

c) There is claim about a detection in RHIC (Relativistic Heavy Ion Collider in Brookhaven) of states behaving in some respects like mini black holes [hk3]. These states could have explanation as color flux tubes at Hagedorn temperature forming a highly tangled state and identifiable as stringy black holes of strong gravitation. The strings would carry a quantum coherent color glass condensate in which partons are in conformally confined state, and would be characterized by a large value of $\hbar$ naturally resulting in confinement phase with large value of α_s [D5].

d) I have also discussed a model for cold fusion based on the assumption that nucleons can be in large $\hbar$ phase. In this case the relevant strong interaction strength is $Q_1Q_2\alpha_{em}$ for two nucleon clusters inside nucleus which can increase $\hbar$ so large that the Compton length of protons becomes of order atomic size and nuclear protons form a macroscopic quantum phase [J6].

8.5.3 Quantum coherent dark matter and $\hbar$

The argument based on gigantic value of $\hbar_{gr}$ explaining darkness of dark mater is attractive but one should be very cautious.

Consider first ordinary QED: $e = \sqrt{\alpha/4\pi\hbar}$ appears in vertices so that perturbation expansion in powers of $\sqrt{\hbar}$ basically. This would suggest that large $\hbar$ leads to large effects. All predictions are however in powers of alpha and large $\hbar$ means small higher order corrections. What happens can be understood on basis of dimensional analysis. For instance, cross sections are proportional to $(\hbar/m)^2$, where m is the relevant mass and the remaining factor depends on $\alpha = e^2/(4\pi\hbar)$ only. In the more general case tree amplitudes with n vertices are proportional to e^n and thus to $\hbar^{n/2}$ and loop corrections give only powers of α which get smaller when $\hbar$ increases. This must relate to the powers of $1/\hbar$ from the integration measure associated with the momentum loop integrals affected by the change of α.

Consider now the effects of the scaling of $\hbar$. The scaling of Compton lengths and other quantum kinematical parameters is the most obvious effect. An obvious effect is due to the change of $\hbar$ in the commutation relations and in the change of unit of various quantum numbers. In particular, the right hand side of oscillator operator commutation and anti-commutation relations is scaled. A further effect is due to the scaling of the eigenvalues of the modified Dirac operator $\hbar\Gamma^\alpha D_\alpha$.

The exponent $exp(K)$ of Kähler function K defining perturbation series in the configuration space degrees of freedom is proportional to $1/g_K^2$ and does not depend on $\hbar$ at all. The propagator is proportional to g_K^2. This can be achieved also in QED by absorbing e from vertices to e^2 in photon propagator. Hence the dependence on α_K (and $\hbar$) must come from vertices which indeed involve Jones inclusions of the II_1 factors of the incoming and outgoing lines.

8.5.4 What a phase transition increasing $\hbar$ could mean physically?

The general vision is that a phase transition increasing $\hbar$ occurs when perturbation theory ceases to converge. Very roughly, this would occur when the parameter $x = Q_1 Q_2 \alpha$ becomes larger than one. The net quantum numbers for "spontaneously magnetized" regions provide new natural units for quantum numbers. The simplest situation is that conformally confined block of $\mathcal{N}$ particles with identical quantum numbers is formed Compton length scaled up so that $\hbar$ is scaled up by factor $\mathcal{N}$. The assumption that standard quantization rules prevail poses very strong restrictions on allowed physical states and selects a subspace of the original configuration space. One can of course, consider the possibility of giving up these rules at least partially in which case a spectrum of fractionally charged anyon like states would result.

This view about the effect of phase transition increasing the value of $\hbar$ allows an interpretation as a concrete physical realization for the thinning of degrees of freedom associated with the renormalization process. Notice that for phase transition in which $\hbar$ is not scaled by integer cannot be understood from this simple model.

8.5.5 $\hbar$ increasing phase transition as a phase transition changing Jones inclusion

The question is how $\hbar$ increasing phase transition could be formulated concretely as a phase transition changing Jones inclusion. One might start with the following simple picture.

Phase transition changing $n = 4$ inclusion to $n = 3$ inclusion

This phase transition would make $\hbar$ very large. The physically motivated hypothesis is that in $n = 3$ Jones inclusion single particle states are mapped to many-particle states by a formation of kind of Bose-Einstein condensates of identical particles with different super-canonical conformal weights which are complex such that net conformal weight is real.

Consider a phase transition in which N identical particles labelled by index $i = 1, ..., N$ with different conformal weights form an entangled state with real net conformal weight in the Hilbert space associated with factor $\mathcal{N}$. Let $J_{a,i}^{\mathcal{N})}$ be charge operators associated with the particles and assume that the values of these charges are same for all particles. The question is how the direct sum $\oplus_i J_{a,i}^{\mathcal{N})}$ defining charge operators for the "Bose-Einstein condensate" particles is mapped to a single particle charge operator $J_a^{\mathcal{M})}$ acting in the factor $\mathcal{M}$.

The simplest guess is that the mapping is

$$\oplus_i J_{a,i}^{\mathcal{N})} \quad \rightarrow \quad N \times J_a^{\mathcal{M})} = \hat{J}^{\mathcal{M})} \ . \tag{8.5.2}$$

This implies that the commutation relations for the images charges are given by

$$\left[\hat{J}_a^{\mathcal{M})}, \hat{J}_b^{\mathcal{M})}\right] \quad = \quad \hat{\hbar} \times f_{abc}\hat{J}_c^{\mathcal{M})} \ ,$$
$$\hat{\hbar} \quad = \quad N\hbar \ . \tag{8.5.3}$$

Thus $\hbar$ would be scaled up by a factor N in this kind of imbedding. Of course, it is not at all obvious that this kind of map of observables is consistent with a phase transition changing $n = 4$ Jones inclusion to $n = 3$ Jones inclusion. Neither it is obvious that scaling occurs by factor N.

In this kind of situation fractionally charged states result if one allows also n-particle states for which charges are not same. Only rational valued charges of form $q = m/n$ are possible.

Phase transitions between $n > 3$ inclusions

One can consider also phase transitions replacing $n > 4$ Jones inclusion with $n - 1$ Jones inclusion. Since the change of $\hbar$ is now relatively small ($\hbar(\infty)/\hbar(4) = 2$), the most natural assumption would be that single particle charges in $\mathcal{N}$ are mapped to single particle charges in $\mathcal{M}$ with renormalization of charge given by

$$J_a^{\mathcal{N})} \to x \times J_a^{\mathcal{M})} = \hat{J}_a^{\mathcal{M})} \ . \tag{8.5.4}$$

In this case $\hbar$ would be scaled up to $x\hbar$. It clear that the imbedding must determine the value of x and that x could differ from unity. The hypothesis is that renormalization group flow for $\hbar(n)$ determines the value of x. Also now fractionization of charge results but is not rational and could be seen as renormalization in phase resolution.

8.5.6 p-Adic evolution in angular resolution and dynamical $\hbar$

The space-time correlates for both RG invariance at single space-time sheet and for p-adic coupling constant evolution are discussed in [A2, C2]. p-Adic evolution is considered for both length and angular resolution. The model for the space-time correlates of the p-adic evolution of $\hbar$ interpreted in terms of phase resolution provides a concrete geometric view about what is involved, and suggests that the formula for $\hbar(n)$ in terms of Beraha numbers B_n can be generalized by allowing also rational valued argument. This predicts that also phase transitions changing $\hbar$ but not Jones inclusion are possible.

For a given p-adic topology algebraic extensions of p-adic numbers define a hierarchy ordered by the dimension of the extension and this hierarchy naturally corresponds to an increasing angular resolution so that RG flow would be associated also with it.

a) A characterization of angular scalings consistent with the identification of $\hbar$ as a characterizer of the topological condensation of 3-surface X^3 to a larger 3-surface Y^3 is that angular scalings correspond to the transformations $\Phi \to r\Phi$, $r = m/n$ in the case of X^3 and $\Phi \to \Phi$ in case of Y^3 so that X^3 becomes analogous to an m-fold covering of Y^3. Rational coverings could also correspond to m-fold scalings for X^3 and n-fold scalings for Y^3.

b) The formation of these stable multiple coverings could be seen as an analog for a transition in chaos via a process in which a closed Bohr orbit regarded as a particle itself becomes an orbit closing only after m turns. TGD predicts a hierarchy of higher level zero energy states representing S-matrix of lower level as entanglement coefficients. Particles identified as "tracks" of particles at orbits closing after m turns might serve as space-time correlates for this kind of states. There is a direct connection with the fractional quantum numbers, anyon physics and quantum groups.

c) The simplest generalization from the p-adic length scale evolution consistent with the proposed role of Beraha numbers $B_n = 4cos^2(\pi/n)$ is that bifurcations can occur for integer values of r=m and change the value of $\hbar$. The interpretation would be that single 2π rotation in δM_+^4 corresponds to the angular resolution with respect to the angular coordinate ϕ of space-time surface varying in the range $(0, 2\pi)$ and is given by $\Delta\phi = 2\pi/m$.

d) For $n = 3$ corresponding to the minimal resolution of $\Delta\phi = 2\pi/3$ $\hbar$ would be infinite. The evidence for a gigantic but finite value of "gravitational" Planck constant [J6] would mean that the simplest formula

$$\frac{1}{\hbar(n)} = \frac{log(B_n)}{log(4)}$$

for $\hbar$ fails for $n = 3$.

The first cure of the problem would be a replacement of the formula for $\hbar(n)$ by a difference equation

$$\frac{1}{\hbar(n)} - \frac{1}{\hbar(n-1)} = \frac{log(B_n)}{log(4)} - \frac{log(B_{n-1})}{log(4)}$$

having interpretation as RGE difference equation and allowing additive constant in the expression of $1/\hbar(n)$ and thus yielding finite value for $\hbar(3)$.

A slightly more elegant resolution of the problem could be that for a given n characterizing von Neumann inclusion there is spectrum of values for $\hbar(r = n/m)$ expressible in terms of $B_r = 4cos^2(\pi/r)$ as

$$\frac{1}{\hbar(n/m)} = \frac{log(B_{n/m})}{log(4)}$$

such that $m/n < 3$ holds true. This would reflect the presence of an additional degree of freedom related to the Jones inclusion. m could characterize the scaling of Φ for X^3 and n the scaling of Φ for Y^3. For this option also phase transitions change the value of $\hbar$ but not Jones inclusion would be possible and large $\hbar$ phases would be possible for all Jones inclusions. A simple TGD inspired model for dark atoms and dark condensed matter [J6] predicts $\hbar/\hbar_0 = 1/v_0 \simeq 2^{11}$. This would correspond to $r \simeq .3077$.

In the next section it will be found that quantum criticality combined with number theoretical vision leads to a prediction for the possible values of $\hbar$ providing the most convincing explanation for large values of $\hbar$ found hitherto.

8.5.7 Large values of $\hbar$ and classical behavior in macro scales

The necessity of large $\hbar$ phases has been actually highly suggestive since the first days of quantum mechanics. The classical looking behavior of macroscopic quantum systems remains still a poorly understood problem and large $\hbar$ phases provide a natural solution of the problem.

In TGD framework quantum coherence regions correspond to space-time sheets. Since their sizes are arbitrarily large the conclusion is that macroscopic and macro-temporal quantum coherence are possible in all scales. Standard quantum theory definitely fails to predict this and the conclusion is that large $\hbar$ phases for which quantum length and time scales are proportional to $\hbar$ and long are needed.

Somewhat paradoxically, large $\hbar$ phases explain the effective classical behavior in long length and time scales. Quantum perturbation theory is an expansion in terms of gauge coupling strengths inversely proportional to $\hbar$ and thus at the limit of large $\hbar$ classical approximation becomes exact. Also the Coulombic contribution to the binding energies of atoms vanishes at this limit. The fact that we experience world as a classical only tells that large $\hbar$ phase is essential for our sensory perception. Of course, this is not the whole story and the full explanation requires a detailed anatomy of quantum jump.

8.6 Does TGD predict the values of Planck constant?

The idea that $\hbar$ is dynamical and can have arbitrarily large values is about one and half year old as a write this. A lot of progress has occurred during the last year but I have not yet been able to seriously pose the question whether and how TGD could predict the values of the Planck constant. In the following a proposal for how TGD predicts the value spectrum of $\hbar$ as one aspect of quantum criticality is discussed and number theoretical arguments are used to make a guess about the spectrum of $\hbar$. In very concise form the argument goes as follows.

a) The freedom to choose the value of $\hbar$ corresponds to the freedom to choose the overall scaling $\lambda = \hbar/\hbar_0$ of M^4 metric associated with various copies of M^4 obtained identified as various algebraic extensions of rational M^4 glued together along common set of rationals consistent with the isometric identification.

b) The dependence of λ on the algebraic extension for a given p-adic prime p is fixed by the quantum criticality condition stating that the critical Kähler coupling strength is same for various algebraic extensions associated with given p-adic prime p.

c) Number theoretic ideas allow to make good guesses concerning the dependence of λ on the algebraic extension. Simplest guess is that λ for union of linearly independent extensions is product or $1/\lambda$ a sum of λ:s for composites. It turns out that sum is the most plausible guess.

8.6.1 The basic ideas

In order to build a more coherent theoretical framework for the quantization of $\hbar$ let us summarize the basic vision as it exists now.

Four-momentum is invariant in the transition changing $\hbar$

The basic constraint is that four-momentum and other quantum numbers of the state are conserved in the scaling $\hbar \to \lambda\hbar$ whereas various quantum lengths and times are scaled up by λ so that one obtains the basic predictions such as macroscopic and macro-temporal quantum coherence in arbitrarily long scales. It is however not at all obvious how to realize this condition.

The identification of the value of the parameter v_0 in terms of Kähler coupling strength

The parameter $v_0 \simeq 2^{-11}$, which has has actually dimension of velocity unless on puts $c = 1$, and its harmonics and sub-harmonics appear in the scaling of $\hbar$. v_0 corresponds to the velocity of distant stars in the model of galactic dark matter. TGD allows to identify this parameter as the parameter

$$
\begin{aligned}
v_0 &= 2\sqrt{TG} = \sqrt{\frac{1}{2\alpha_K}}\sqrt{\frac{G}{R^2}} \ , \\
T &= \frac{1}{8\alpha_K}\frac{\hbar_0}{R^2} \ .
\end{aligned}
\tag{8.6.1}
$$

Here T is the string tension of cosmic strings, R denotes the "radius" of CP_2 ($2R$ is the radius of geodesic sphere of CP_2). $\hbar_0$ corresponds to Beraha number B_∞. α_K is Kähler coupling strength whose evolution is dictated by the condition that G is invariant in coupling constant evolution and by number theoretical arguments to

$$
\begin{aligned}
\alpha_K &= k\frac{1}{log(p) + log(K)} \ , \\
K &= \frac{R^2}{\hbar_0 G} = 2 \times 3 \times 5 \times 7 \times 11 \times 13 \times 17 \times 19 \times 23 = 223{,}092{,}870 \ , \\
k &\simeq \pi/4 \ .
\end{aligned}
\tag{8.6.2}
$$

Equivalence Principle requires that $\hbar$ in the formula for K corresponds to the value $\hbar_0 \equiv 1$ [D6] so that gravitational coupling constant is invariant also with respect to coupling constant evolution associated with Beraha numbers. One can define "dynamical" Planck length as

$$
L_d = \sqrt{G\hbar_{gr}} = \sqrt{M_2/M_1}\frac{1}{v_0}GM_1 \ .
$$

The order of magnitude is not too far from Schwartschild radius.

Number theoretic constraints are expected to pose strong constraints on the value of k. A discrete version of a typical logarithmic evolution of U(1) coupling constant strength as a function of length scale is in question and for $k = 127$ (M_{127}) the value of α_K is very near to fine structure constant $\alpha \simeq 1/137$. Note that the ratio R^2/G is predicted to be constant so that G would be renormalization group invariant even in the sense that it would not depend on the p-adic length scale.

The resulting prediction for v_0 reads as

$$
\begin{aligned}
v_0 &= \sqrt{\frac{1}{2\left[k \times log(2) + log(K)\right] K}} \quad \text{for} \ \ p \simeq 2^k \ , \\
v_0 &\simeq .54412 \times 10^{-3} \ \text{for} \ \ p = M_{127} \ .
\end{aligned}
\tag{8.6.3}
$$

v_0 approaches to zero at logarithmic rate for large values of k. For $k = 127$ it is rather near to $v_0 = 2^{-11}$. In a good approximation the value 2^{-11} is reached for $k = 135$ which corresponds to the p-adic length scale $L(k_{eff} = 113 + 22)$ associated with dark variants of $k = 113$ weak bosons appearing in the TGD based model of atomic nucleus. The mean experimental value of $1/v_0$ is $1/v_0 = 2174$ with an accuracy of 1 per cent.

TGD allows also to develop arguments explaining the harmonics and sub-harmonics of v_0. Higher harmonics would be due to wrapped magnetic flux tubes for which CP_2 coordinates are n-valued functions of M^4 coordinates defining local coordinates of the flux tube space-time sheet. Sub-harmonics would result when M^4 coordinates become n-valued functions of CP_2 coordinates.

The challenge is to understand why v_0 appears in the basic formula expressing the change of $\hbar$ in the transition increasing the value of $\hbar$. It would seem that $\hbar$ characterizes the magnetic flux tubes (join along boundaries bonds) connecting the interacting systems and serving as space-time correlates for the interaction giving rise to bound state.

The value of v_0 deduced for cosmic strings does not make sense in astrophysical or condensed matter context, where cosmic strings are replaced with magnetic flux tubes. v_0 remains invariant in this scaling down if R^2 is replaced by the p-adic length scale L_p^2 apart from a multiplicative factor in the formulas for G and T so that the product TG remains invariant. $T_m \propto 1/L_p^2$ characterizes the magnetic energy density of the magnetic flux tube and $G_m \to L_p^2$ is identifiable as a "strong" gravitational coupling strength characterizing the interactions of magnetic flux tubes behaving like string like objects.

The criterion for the occurrence of the phase transition increasing the value of $\hbar$

In the case of planetary orbits the large value of $\hbar = 2GM/v_0$ makes possible to apply Bohr quantization to planetary orbits. This leads to a more general idea that the phase transition increasing $\hbar$ occurs when the system consisting of interacting units with charges Q_i becomes non-perturbative in the sense that the perturbation series in the coupling strength $\alpha Q_i Q_j$, where α is the appropriate coupling strength and $Q_i Q_j$ represents the maximum value for products of gauge charges, ceases to converge. Thus Mother Nature would resolve the problems of theoretician.

A primitive formulation for this criterion is the condition $\alpha Q_i Q_j \geq 1$ and predicts the existence of dark matter hierarchies with $\hbar = \lambda^k \hbar_0$, $k = 0, 1, ...$, $\lambda = n/v_0$ or $\lambda = 1/nv_0$, $v_0 \simeq 2^{-11}$. This rule of thumb has now been applied with success the interpretation of hadronic mass calculations and to build models for systems like atomic nucleus and high T_c superconductor and seems to work. Of course, the criterion for transition is primitively formulated and the understanding what really happens in the transition to large $\hbar$ phase behaving like dark matter.

What are the allowed values of $\hbar$?

From the beginning I had strong gut feeling that the allowed values of $\hbar$ are expressible in terms of Beraha numbers $B_n = 4cos^2(\pi/n)$, $n \geq 3$ related to Jones inclusion hierarchies of hyperfinite factors of type II_1, which correspond to von Neumann algebra naturally associated with configuration space spinors.

Consider the inclusion $N \subset M$ of these factors as von Neumann algebras. A deep result is that one can express M as $N : M$-dimensional module over N with fractal dimension $N : M = B_n$. $\sqrt{B_n}$ represents the dimension of a space of spinor space renormalized from the value 2 corresponding to $n = \infty$ down to $\sqrt{B_n} = 2cos(\pi/n)$ varying thus in the range $[1, 2]$. B_n in turn would represent the dimension of the corresponding Clifford algebra.

This observation leads to the idea about how one could predict the value spectrum of $\hbar$. There are two arguments based on similar reasoning but leading to quite different predictions.

Option I: Consider the trace of $\sigma_z^2 = \hbar^2 \times 1$ for $d = 2$-component spinors. Assume that the trace of σ_z^2 does not depend on $\hbar$ so that one has

$$Tr(\sigma_z^2) = \sqrt{M : N} \hbar_n^2 = 2\hbar_\infty^2 \; .$$

By comparing the two sides one obtains

$$\frac{\hbar_n}{\hbar_\infty} = \sqrt{\frac{1}{cos(\pi/n)}} \; , \quad n \geq 3 \; . \tag{8.6.4}$$

This would predict that the values of $\hbar_n/\hbar_\infty$ vary in the range $[1, \sqrt{2}]$. This is not enough to explain the large values of $\hbar$ but the arguments represented in the sequel allow to consider this option seriously.

Option II: In order to understand the large values of $\hbar$ I developed an argument the outcome of which seems to be correct although the proposed explanation of large values of $\hbar$ need not be correct. The idea is that the dimension of the spinor space is given by $d = 2^{D/2}$, where D is the dimension of the space-time. By applying this formula in the recent fractal context one obtains

$$D_n = 2log_2(\sqrt{B_n}) = 2log_2(2cos(\pi/n)) \ .$$

In recent case the invariance for the trace of the scaling operator $L_0 = \hbar m^k d/dm^k$ acting in the representation defined by M^D coordinates implies that the trace $Trace(L_0) = D\hbar$ equals to $2\hbar_\infty$.

$$D_n\hbar_n \;\; = \;\; 2\hbar_\infty \tag{8.6.5}$$

allowing to identify the renormalized value of $\hbar$ as

$$\frac{\hbar_n}{\hbar_\infty} \;\; = \;\; \frac{2}{D_n} = \frac{1}{log_2(2cos(\pi/n))} \ . \tag{8.6.6}$$

This would give $D_3 = 1$ and $D_\infty = 2$ and $\hbar_3 = \infty$. The idea was that the gigantic value of $\hbar$ in astrophysical length scales would result as a small perturbation of $x = 1/\hbar$ at $x = 0$. The infinite value of $\hbar$ for $n = 3$ would have a nice interpretation in terms of vacuum degeneracy. $n = 3$ would correspond to vacuum extremals for which perturbation theory does not make sense. The situation would be saved by the vanishing value of gauge coupling strengths meaning that all quantum corrections to the classical prediction vanish. It will turn out that option II is indeed more natural in the TGD framework.

Do Jones inclusions correspond to inclusions of rationals to their algebraic extensions?

Jones inclusions for type II_1 factors have several interpretations and applications in TGD framework. The most obvious interpretation Jones inclusion is as sub-system-system inclusion. Jones inclusion could be also accompany the inclusions of rationals to algebraic extensions of rationals at the level of configuration space spinor fields. One can also consider the inclusions of p-adic number fields to their complex extensions.

The inclusion $Q \subset Q[exp(i\pi/n)]$ would correspond to Jones inclusion characterized by n at the level of configuration space spinors. In the framework of rational physics the hierarchy of algebraic extensions of rationals would define a hierarchy of Jones inclusions and these extensions would give rise to a hierarchy of Planck constants.

The problem is that the large values of $\hbar$ do not seem to be natural in either case except possibly for option II by allowing perturbative corrections to $x = 1/\hbar_3 = 0$. This need not to be the case. $M : N$ characterizes configuration space Clifford algebra assignable with an algebraic extension of rationals as N-module where N corresponds to CH Clifford algebra assignable to rationals. What comes in mind that M could be much higher-dimensional as N-module than in the case of standard Jones inclusions defined using reals or complex numbers as a coefficient field of the Clifford algebra. The simplest guess is that M is tensor power of $M : N$-dimensional spinor base as an N module indeed allowed by option II as will be found.

8.6.2 Mathematical constraints

In order to gain insight to how one might understand the hierarchy of Planck constants in TGD framework it is good to start from good old Schrödinger equation.

What is behind minimal substitution?

The key property of Schrödinger equation is that kinetic energy term depends on $\hbar$ whereas the potential energy term does not depend on it. This makes the scaling of $\hbar$ a non-trivial transformation.

In the case of Dirac equation same conclusion applies and corresponds to the minimal substitution $p - eA \rightarrow i\hbar\nabla - eA$. Consider next the situation in TGD framework.

1. Minimal substitution does not make sense in CP_2 degrees of freedom

The first crucial observation is that the minimal substitution $p - eA \rightarrow i\hbar\nabla - eA$ does not make sense in the case of CP_2 Dirac operator since, by the non-triviality of spinor connection, one cannot choose the value of $\hbar$ freely. In fact, spinor connection of CP_2 is defined naturally in such a manner that spinor connection corresponds to the quantity $eQ/\hbar$ and there is no natural manner to separate $e/\hbar$ from it.

The only reasonable conclusion is that the Dirac operator in $M^4 \times CP_2$ depends on $\lambda = \hbar/\hbar_0$ only via its M^4 part and that Dirac equation gives the eigenvalues of wave vector squared $k^2 = k^i k_i$ rather than four-momentum squared $p^2 = p^i p_i$. The values of k^2 are proportional to $1/\lambda^2$ so that p^2 does not depend on it for $p^i = \hbar k^i$. This gives rise to the invariance of mass squared and the desired scaling of wave vector when $\hbar$ changes.

2. The dynamical character of $\hbar$ can be only due to the freedom to select the scale of M^4 metric

The second crucial observation is that the freedom to vary λ can be only due to the freedom to vary the overall scaling of M^4 metric which is indeed possible. Whether the freedom to choose the scaling of M^4 metric freely corresponds to a genuine dynamical degree of freedom has been a continual source of worried thoughts now and then. If $\hbar$ has only single value this freedom does not seem to have meaning but if it has spectrum the situation changes.

The question is how to realize this freedom to choose the scale of the M^4 metric realized as replacements $m_{kl} \rightarrow m_{kl}/\lambda$, where allowed values of λ define the spectrum of Planck constant. The number theoretic vision suggests an answer to the question.

TGD leads to a generalization of the notion of number by gluing reals and various extensions of p-adic number fields to a larger structure. The gluing is carried out along common rationals. On the other hand, the rational physics approach suggests that the physics in various number fields satisfy the enormously powerful constraint that they are obtained by an algebraic continuation from rational physics. Hence rationals and the algebraic extensions of rationals should play key role in TGD.

The obvious idea encouraged also by the observation about the role of Beraha numbers is that different algebraic extensions of rationals (or p-adic numbers) correspond to different scalings of M^4 metric. This would mean that the identification along common rationals would be replaced with the identification $y = \lambda x$, where y is rational and x is arbitrary number in extension, and λ characterizing the scaling of M^4 metric belongs to extension. If lambda is not rational, the values of x are obtained by multiplying rationals by $1/\lambda$. Rationality would give a strong constraint on λ, which might be however too strong as the guesses for the form of λ suggest. This identification would generalize the original $y = x$ identification and guarantee that the distances of points y and x from the origin of M^4 would be same in respective M^4 metrics.

3. Other Dirac operators and dynamical scale of M^4 metric

The number theoretically realized spectrum of scales for M^4 metric fits nicely with the properties of the Dirac operator of imbedding space, with the properties of the modified Dirac operator defined for induced metric and spinor structure, and with the properties super-conformal Dirac operators of configuration space since both these operators involve natural separation of M^4 and CP_2 degrees of freedom. Induced metric and spinor structure depend non-linearly on λ. Super-conformal mass squared formula is replaced by a formula for wave vector squared involving λ as a scaling factor and the oscillator operator algebra used depends on the value of λ since it is expressed in terms fermionic oscillator operators associated with second quantized induced spinor fields.

4. Quantum classical correspondence and the values of λ

Quantum classical correspondence suggests that the values of λ should be represented also at space-time level. The variational principle making space-time sheets counterparts of Bohr orbits indeed implies the quantization of Kähler magnetic flux and the quantum need not be the standard flux quantum. The generalized quantization condition would be $\int BdS = n\lambda\hbar_0$ and in principle it is possible to deduce the values of λ from the classical theory. The flux integral does not involve the induced metric so that there is no explicit dependence on λ. There is however an implicit dependence via field equations which involve λ via the induced metric. This should be the case since the information about algebraic extension must be coded to the classical theory somehow.

The invariance of angular momentum under the scaling of $\hbar$

The assumption that four-momentum is invariant in the scaling of M^4 metric by λ combined with Poincare invariance implies that also angular momentum is invariant under the scaling by λ. Hence the analog of wave vector defined as $L_z/\hbar$ would scale down by a factor $1/\lambda$ giving $L_z/\hbar = m/\lambda$. The one-valuedness of the wave function however requires $L_z/\hbar = m$.

A possible resolution of the paradox is based on following argument.

a) Consider first a situation in which λ is integer. Effective 2-dimensionality means that 2-dimensional partonic surfaces are basic structures to be considered. By conformal invariance scalings and rotations are generated by L_0 and iL_0. Hence it would not be surprising if the radial scaling by λ^k would be accompanied by a similar angular scaling. If M^4 projection of the partonic 2-surface has dimension $D \geq 1$, this scaling would at space-time level mean that partonic 2-surface becomes analogous to λ^k-sheeted Riemann surface defining λ^k-fold covering of $E^2 \subset M^4$. Using M^4 coordinates for the space-time sheet as local coordinates, one finds that induced spinor fields have fractional angular momentum eigenvalues $L_z = m/\lambda^k$.

b) In this picture the approach to quantum criticality would correspond to the emergence of classical chaos at space-time level by a step-wise process in which the step $\hbar \to \lambda\hbar$ can be regarded as generalization of the period doubling bifurcation. As $\hbar$ increases by a factor λ^k, the space-time sheet representing an orbit of particle closing after one turn transforms to an orbit closing only after λ^k turns. Note that the volume of space-time sheet remains finite only if the orbit closes after finite number of turns. The step $k \to k+1$ would correspond to a local fractal operation making each sheet of the λ^k sheeted surface λ-sheeted so that λ^{k+1} sheeted surface would result. Instead of period doubling one would have period λ-folding with the value of λ depending on p-adic prime $p \simeq 2^k$.

c) The model of Nottale for planetary orbits requires besides the harmonics of $\lambda_0 \simeq 2^{11}$ also some of its sub-harmonics, at least third and fifth one. This conflicts with the assumption that λ is integer unless λ_0 is divisible by 3 and 5. This would be the case for $\lambda_0 = 2175 = 3 \times 5^2 \times 29$ consistent with the mean value $\lambda = 2174$ deduced with one per cent accuracy from the model for planetary orbits [if1]. Only integer factors of $\lambda_0(p \simeq 2^k)$ would be allowed as sub-harmonics for given p for this option and integer valuedness of λ_0 would also pose strong conditions on the values of p if the proposed formula for λ_0 is accepted.

d) If one allows the quantization of angular momentum using integer multiple of $\hbar_0$ as a unit, λ need not be an integer. In this case one could have $\lambda = (r/s) \times \lambda_0$, where $\lambda_0 \simeq 2^{11}$ is integer. The orbit represented by the space-time sheet would close after $(r\lambda_0)^k$ turns and angular momentum would be quantized as $L_z = ms^k\hbar_0$. Even in this case the proposed formula for $\lambda(p)$ would give strong constraints on the values of p.

Kähler function codes for a perturbative expansion in powers of λ

Suppose that one accepts the number theoretical interpretation for the spectrum of $\hbar$ in terms of a hierarchy of overall scalings of M^4 metric. The first implication of this picture is that the modified Dirac operator determined by the induced metric and spinor structure depends on the λ in a highly nonlinear manner. This in turn implies that the fermionic oscillator algebra used to define configuration space spinor structure and metric depends on the value of λ. Same is true also for Kähler action and configuration space Kähler function. Hence Kähler function is analogous to an effective action expressible as infinite powers series in powers of $\hbar$ replaced now with λ.

This interpretation allows to overcome the paradox caused by the hypothesis that loop corrections to the functional integral over configuration space defined by the exponent of Kähler function serving as vacuum functional vanish so that tree approximation is exact. This would imply that all higher order corrections usually interpreted in terms of perturbative series in powers of $1/\hbar$ vanish. The paradox would result from the fact that scattering amplitudes would not receive higher order corrections and classical approximation would be exact. This certainly cannot be the case always: consider only the photon-photon scattering. This paradox can be also regarded as an objection against the proposal that generalized Feynman diagrams are equivalent with tree diagrams or more generally, that each diagram is equivalent with a minimal loopy diagram allowing homologically non-trivial imbedding with non-intersecting lines to a higher genus Riemann surface.

The dependence of both states created by Super Kac-Moody algebra and the Kähler function and corresponding propagator identifiable as contravariant configuration space metric would mean that the expressions for scattering amplitudes indeed allow an expression in powers of λ. What is so remarkable is that the TGD approach would be non-perturbative from the beginning and "semiclassical" approximation, which might be actually exact, automatically would give a full expansion in powers of $\hbar$. This is in a sharp contrast to the usual quantization approach.

How the spectrum of $\hbar$ could be predicted by quantum TGD?

Number theoretical vision combined quantum criticality could allow to determine the allowed values of λ to a high degree. Quantum criticality means that different values for $\hbar$ corresponding to algebraic extensions of rationals correspond to same form of Kähler function, that is same value of Kähler coupling constant. Number theoretical vision in turn has led to the proposal that the exponent of Kähler function is expressible as a Dirac determinant for the modified Dirac operator. As a matter fact, the ratio of Dirac determinants for space-time regions separated by a light-like causal determinant acting as causal horizon would give the exponent for the difference of Kähler functions of the two regions [B4].

The condition that the value of g_K^2, defined as the analog of critical temperature, is same for all algebraic extensions would fix the value of λ as a function of algebraic extension apart from possible multi-valuedness.

This condition does not require the restriction of the modified Dirac operator to the set of rationals or their algebraic extensions. The only thing that is required is that the subset of allowed eigenvalues of the modified Dirac operator belongs to the extension considered. In this manner one can calculate a Dirac determinant for each extension as a function of λ. Note that λ could belong to the algebraic extension considered. By requiring that g_K^2 corresponds to its value for rationals, one can fix the value of λ. Even better, it is quite possible that the value of Dirac determinant, rather than only the ratio of Dirac determinants, is finite since the number theoretic restriction might be satisfied only by a finite number of eigenvalues. Hence nature would also take care of regularization of Dirac determinants by using the number theoretic hierarchy.

8.6.3 Is it possible to guess the dependence of λ on algebraic extension?

The dependence of λ on the algebraic extension should be something very universal and the notion of algebraic homomorphism is highly suggestive idea. At least one can hope that λ is expressible using functions which define algebraic homomorphisms to additive or multiplicative reals.

1. Does λ define additive or multiplicative homomorphism?

The generating units of the algebraic extension can be decomposed to generating units such that at most one of them is phase factor of form $U_n = exp(i\pi/n)$ and corresponds to spin degrees of freedom whereas the remaining generating units e_i have complex norm different from unit and correspond to scalings. Since scalings and rotations commute, a natural hypothesis is that λ decomposes into a product of factors corresponding to the phase U_n determined by the Beraha number B_n and a multiplicative factor corresponding to scalings and determined by the units e_i. Denoting the extension by $(U_n, E) == (U_n, \{e_i\})$ one would have

$$\lambda(U_n, E) \quad = \quad \lambda_a(U_n) \times \lambda_b(E) \ . \tag{8.6.7}$$

The idea of algebraic homomorphism suggest what happens for λ when two linearly independent extensions E_1 and E_2 are combined together by combining the "imaginary" unit sets $(U_{n_1}, E_1) = (U_{n_1}, \{e_{1,i}\})$ and $(U_{n_2}, E_2) = (U_{n_2}, \{e_{2,i}\})$ to $E_1 \cup E_2$.

a) If n_1 and n_2 have no common factors, the resulting extension has obviously generating phase $U_{n_1 n_2}$ and one has

$$\lambda((U_{n_1}, E_1) \cup (U_{n_2}, E_2)) \quad = \quad \lambda_a(U_{n_1 n_2}) \times \lambda_b(E_1 \cup E_2) \ . \tag{8.6.8}$$

b) Concerning the dependence of λ_b on $\{e_i\}$, multiplicative homomorphism property is most natural and would mean that for linearly independent basis E_1 and E_2 one has

$$\lambda_b(E_1 \cup E_2) \;\; = \;\; \lambda_b(E_1)\lambda_b(E_2) \;. \tag{8.6.9}$$

2. The dependence of λ on Beraha number and homomorphism hypothesis

Let N refer in the sequel to the algebra of configuration space gamma matrices defined by the restriction of fermionic oscillator operators to those corresponding to rational eigenvalues of the modified Dirac operator D. M in turn will denote the configuration space Clifford algebra associated with the spectrum of the modified Dirac operator D allowing eigenvalues in the extension of rationals.

The homomorphism hypothesis gives a powerful tool to deduce the dependence of λ on algebraic extension. Concerning the dependence on U_n one has two options.

$$\lambda_a^I(U_n) \;\; = \;\; \frac{1}{\sqrt{cos(\pi/n)}} \;,$$
$$\frac{1}{\lambda_a^{II}(U_n)} \;\; = \;\; \frac{2}{log_2(B_n)} = \frac{1}{log_2(2cos(\pi/n))} \;. \tag{8.6.10}$$

The interpretation would be that algebraic extension (U_n, E) induces Jones inclusion $N \subset M$ at the level of configuration space Clifford algebra and that this extension is characterized at least partially by Beraha number B_n.

3. The dependence of λ on v_0 and homomorphism hypothesis

The dark matter hierarchy would suggest that powers of v_0 define a hierarchy of values of λ and the question is whether this could be understood in terms of a hierarchy of algebraic extensions. Let $E \equiv \{e_i\}$ be characterized by d generating units e_i, which are not phases. For two linearly independent extensions E_1 and E_2 one has $d(E_1+E_2) = d(E_1)+d(E_2)$ so that d defines an additive homomorphism. The exponent $v_0^{d(E)}$ gives a multiplicative homomorphism. This would suggest the following behavior:

$$\lambda_b(E) \;\; \propto \;\; (\frac{1}{v_0})^{d(E)} \;. \tag{8.6.11}$$

The large value of v_0 does not conform with the idea that the renormalized dimension of 2-D spinor space is smaller than two. A possible interpretation for the large renormalized dimension of configuration space Clifford algebra as an N module is due to the increase of the algebraic dimension of the Clifford algebra as eigenvalues of the modified Dirac operator in extension are allowed. This rapid increase of the dimension of Clifford algebra is expected to bring in a degeneracy which is expected to give a power of M:N instead of M:N for the fractal dimension of M as N module.

a) For option I one would have

$$M : N \rightarrow (\frac{1}{v^0})^d \times M : N \;. \tag{8.6.12}$$

M would be $(\frac{1}{v_0})^d \times M : N$ dimensional N-module. This is not consistent with the idea that the dimension of Clifford algebra as N module should be a power of $M : N$.

b) For option II one would have

$$M : N \rightarrow (M : N)^{(\frac{1}{v_0})^d} \;. \tag{8.6.13}$$

This fits nicely with the idea that $(\frac{1}{v_0})^d$-fold tensor power of $M : N$ module is in question. Hence there are good algebraic reasons to believe that the original proposal for the quantization of $\hbar$ is more plausible option. In both cases the natural expectation is that the allowed values of $1/v_0$ are integers.

To sum up, the most plausible formula for $\lambda(U_n, E)$ when E has $d(E)$ generators reads as

$$\lambda(U_n, E) = (\frac{1}{v_0})^{d(E)} \times log_2(2cos(\pi/n)) = log_2 \left[(\sqrt{M : N})^{(\frac{1}{v_0})^{d(E)}} \right] \;. \tag{8.6.14}$$

8.6.4 Are the possible values of α_K fixed by homomorphism property?

One of the great number theoretical ideas is that α_K assignable to given space-time sheet might be determined by the infinite prime characterizing the space-time sheet and the space-time sheets topologically condensed at it.

There is a natural homomorphism to ordinary rationals from infinite primes, integers, and rationals. For the lowest level infinite primes this homomorphism gives simply the ratio $q = m/n$ for the representation

$$P = m\frac{X}{\prod_i q_i} + n \ ,$$

where m and n have no common prime factors, X is product of all finite primes, and n contains only powers of q_i in its decomposition to prime factors. The integer m characterize the bosonic many particle state consisting of bosons labelled by primes $p \neq q_i$ and integer n characterizing fermionic many-particle state characterized by $\prod q_i$ containing also bosonic excitations labelled by q_i.

The requirement of homomorphism property allows to generalize this homomorphism to higher levels of hierarchy of infinite primes and a recursive construction of $q = m/n$ results based on assignment of lower level infinite rational with a given level and continuing the process down to the lowest level.

The additive homomorphism property for the basic building block functions of α_K raise the hopes of predicting α_K. The proposed expression for $1/\alpha_K$ is indeed consistent with additive homomorphism property

$$\frac{1}{\alpha_K} \;=\; xlog(\frac{m}{n}) \ . \tag{8.6.15}$$

Here the factor $x \simeq 4/\pi$ should be chosen in such a manner that $1/v_0 = \sqrt{2\alpha_K \prod_{q=2,3..}^{23} q}$ is integer. One could require that also $g_K^2 = 4\pi\alpha_K$ is rational. These constraints inspire quite far reaching number theoretical conjectures unless one chooses x by hand so that it takes care of the rationalization.

The beauty of this approach would be that the p-adic coupling constant evolution of α_K, and in principle also of the other coupling constant strengths expressible in terms of α_K, would be fixed by the homomorphism property alone.

Part IV

APPLICATIONS TO ELEMENTARY PARTICLE PHYSICS

Chapter 9

General View About Physics in Many-Sheeted Space-Time

9.1   Introduction

The concept of topological condensation unifies two disparate approaches to TGD, namely TGD as a Poincare invariant theory of gravitation and TGD as a generalization of the string model. The idea is that classical 3- space with matter can be regarded as a 3-surface obtained by "gluing" particle like 3-surfaces to the background 3-surface with possibly macroscopic size: resulting topological inhomogenities correspond to matter. The "gluing" of two n-manifolds together by topological sum means the following operation: drill spherical holes to both n-manifolds and connect the resulting boundary components S^{n-1} with a tube $D^1 \times S^{n-1}$ (see Fig. 9.1.1). Of course, several $\#$ contacts, which are tiny 'wormholes' connecting two parallel space-time sheets, are expected to be present in the general case.

9.1.1   Various types of topological condensation

One can in fact distinguish between three kinds of topological condensation.

a) 3-dimensional topological condensation, which is expected to give rise to the formation of bound states (not necessary all possible bound states).

b) 4-dimensional topological condensation, which results from the properties of the Kähler action: the minimizing four surface associated with a given set of 3-surfaces is in general connected so that long range interactions are generated between the 3-surfaces. This mechanism is in principle all what is needed to generate the so called classical space-time. Although the physical state can consist of arbitrarily many disjoint 3-surfaces, the space-time associated with these surfaces is connected and resembles the "classical" space-time, when topological inhomogenities are smoothed out. It should be noticed that 4-dimensional topological condensation corresponds to unstable 3-dimensional topological condensation. For the visualization purposes, one can consider a simplified example: instead of 3-surfaces consider strings so that space-time is replaced with a two-surface having strings as its boundaries. c) 2-dimensional topological condensation: boundaries of the 3- surfaces are joined together by a tube $D^1 \times D^2$. This process will be referred as a formation of join along boundaries bonds.

There are also reasons to suspect that the actual macroscopic 3-space is not connected but corresponds to a large macroscopic 3-surface, classical 3-space, plus a gas of small particle like 3-surfaces, "Baby Universes". It is to be expected that the effects related to the vapor phase particles are very small. An idealization is obviously needed in order to obtain something resembling the topologically trivial 3-space of the standard theories: topological inhomogenities of size smaller than a given length scale L are smoothed out and their presence is described using various currents, such as energy momentum tensor, gauge currents and particle number currents. To be precise, this works only provided one takes the limit $L \to \infty$ since TGD space-time could well be many sheeted in arbitrarily long length scales.

Figure 9.1: Topological sum of two manifolds

9.1.2   Implications of the topological non-triviality of macroscopic space-time

If one accepts that 3-space is topologically nontrivial, one must sooner or later end up asking following questions. What does 3-space actually look like in various scales? What are the general physical consequences of the new space time concept? Are they seen at elementary particle level only or perhaps at atomic, molecular, etc. levels? What is the 3-topology of the solid/liquid/gas state? What about macroscopic bodies: what do they correspond topologically?

In the following the general ideas about the topological condensation are discussed. These ideas have developed gradually in parallel with the development of the configuration space geometry and Quantum TGD, through the study of the extremals of Kähler action and through the attempts to apply TGD inspired ideas to many not so well understood phenomena like Higgs mechanism or more generally, particle massivation, color confinement, super fluidity, super conductivity, hydrodynamic turbulence, etc.. The ideas to be represented may look rather wild, when encountered outside the context defined by twenty years of personal work with many trials and errors and moments of discovery. It is the internal consistency rather than quantitative details, as well as the radically new approach provided to the problems of even macroscopic physics, which makes the scenario so exciting.

9.1.3   Topics of the chapter

The topics to be discussed in the sequel will be following:

a) The question what 3-space looks like in various scales and end up to a purely topological description for the generation of structures. Topological arguments imply a finite size for non-vacuum 3-surfaces and the conservation of the gauge and gravitational fluxes requires that 3-surface feeds these fluxes to a larger 3-surface via # contacts situated near the boundaries of the 3-surface. Renormalization group invariance (RGI) hypothesis suggests that 3-surfaces with all sizes are important in the functional integral and this leads to the idea of the many-sheeted space-time with hierarchical, fractal like structure such that each level of the hierarchy corresponds to a characteristic length scale.

b) The general space-time picture suggested by RGI hypothesis can be justified mathematically. Due to the compactness of CP_2, a general space-time surface representable as a map $M^4 \to CP_2$ decomposes into regions, "topological field quanta", characterized by certain vacuum quantum numbers and 3-surface is in general unstable against the decay to disjoint components along the boundaries of the field quanta. Topological field quanta have finite size depending on the values of the vacuum quantum numbers: the size increases as the values of the vacuum quantum numbers increase. Topological field quantum is therefore a good candidate for a quantum coherent system provided some Bose Einstein condensate or quantum coherent state is available. The BE condensate or coherent state of the light # contacts near the boundaries of the topological field quantum is a good candidate in this respect. It came as a total surprise that this the generation of vacuum expectation value of Higgs field corresponds to the generation of this kind of macroscopic quantum phase.

The requirement of the gauge charge conservation in turn implies the hierarchical structure of the topological condensate: gauge fluxes must go somewhere from the outer boundaries of the topological field quantum with finite size and this 'somewhere' must be a larger topological field quantum, which in turn feeds its gauge fluxes to a larger topological field quantum,.... Of course, the nonlinearity of the theory could allow vacuum charge densities which can cancel the net charge near boundaries.

Most importantly, topological field quanta allow discrete scalings as a dynamical symmetry. p-Adic length scale hypothesis states that the allowed scaling factors correspond to powers of $\sqrt{p}$, where the prime p satisfies $p \simeq 2^k$, k integer with prime values favored. p-Adic fractality (actually multi-p-fractality) can be justified more rigorously by a precise formulation for the fusion of real and various p-adic physics based on the generalization of the notion of number [E1].

c) The physical consequences of the new space-time picture are nontrivial at all length scales.

i) A natural interpretation for the hierarchical structure is in terms of bound state formation. Quarks condense to form hadrons, nucleons condense to form atomic nuclei, nuclei and electrons condense to form atoms, how atoms condense to form molecules, and so on. One ends up with a general picture for the topology of 3-space associated with, say, solid state and with the idea that even the macroscopic bodies of the everyday world correspond to topologically condensed 3-surfaces.

ii) The join of 3-surfaces along their boundaries defines a new kind of interaction, which in fact has been used in phenomenological modelling of and usually believed to result from Schrödinger equation. At the macroscopic level this interaction is rather familiar to us since it means that two macroscopic bodies just touch each other!

iii) The possibility to understand general qualitative features of the charge renormalization topologically in the proposed scenario for space-time, is considered. This rough vision represents one of the oldest strata in the evolution of TGD: in [C3] the recent view about space-time correlates of gauge charges is developed.

iv) In TGD context there are purely topological necessary conditions for quantum coherence and a topological description for dissipative phenomena. The formation of the join along boundaries bonds plays a decisive role in the description and this process provides a universal manner to generate macroscopic quantum systems.

v) There is also a topological description for the formation of the supra phases and the phase of the order parameter of the supra phase ground state contains information about the homotopy of the join along boundaries condensate.

d) The proper understanding of the concepts of gauge charges and fluxes and their gravitational counterparts in TGD space-time has taken a lot of efforts. At the fundamental level gauge charges assignable to light-like 3-D elementary particle horizons surrounding a topologically condensed CP_2 type extremals can be identified as the quantum numbers assignable to fermionic oscillator operators generating the state associated with horizon identifiable as a parton. Quantum classical correspondence requires that commuting classical gauge charges are quantized and this is expected to be true by the generalized Bohr orbit property of the space-time surface.

There are however non-trivial questions. Do vacuum charge densities give rise to renormalization effects or imply non-conservation so that weak charges would be screened above intermediate boson length scale? Could one assign the non-conservation of gauge fluxes to the wormhole (#) contacts, which are identifiable as pieces of CP_2 extremals and for which electro-weak gauge currents are not conserved so that weak gauge fluxes would be non-vanishing but more or less random so that long range correlations would be lost?

It indeed turns that one can understand the non-conservation of weak gauge fluxes in terms of wormhole contacts carrying pairs of right/left handed fermion and left/right handed antifermion having interpretation as Higgs bosons. The average non-conserved light-like gravitational four-momentum of wormhole contact representing Higgs boson can be identified as the inertial four-momentum apart from the sign factor so that one can also understand particle massivation at fundamental level and a connection with p-adic thermodynamics based description of Higgs mechanism emerges. Also a detailed understanding about how Equivalence Principle is weakened in TGD framework emerges.

e) The most dramatic prediction obvious from the beginning but mis-interpreted for about 26 years is the presence of long ranged classical electro-weak and color gauge fields in the length scale of the space-time sheet. The only interpretation consistent with quantum classical correspondence is in terms of a hierarchy of scaled up copies of standard model physics corresponding to p-adic length scale hierarchy and dark matter hierarchy labelled by arbitrarily large values of dynamical quantized Planck constant. Chirality selection in the bio-systems provides direct experimental evidences for this fractal hierarchy of standard model physics.

f) # contacts, feeding gauge fluxes from a given sheet of the 3-space to a larger one, which are a necessary concomitant of the many-sheeted space-time concept. # contacts can be regarded as particles carrying classical charges defined by the gauge fluxes but behaving as extremely tiny dipoles quantum mechanically in the case that gauge charge is conserved. # contacts must be light, which suggests that they can form Bose-Einstein condensates and coherent states. The real surprise (after 27 years of TGD) was that the formation of these rather exotic macroscopic quantum phases could be identified as formation of vacuum expectation value of Higgs field for various scaled up copies of standard model physics!

This kind of macroscopic quantum phases could be in a central role in the TGD inspired model for a bio-system as a macroscopic quantum system. A related effect is the formation of exotic atoms, when some valence (say) electrons drop from the atomic space-time sheet to a larger space-time sheet. This process is accompanied by the generation of a # contacts. The process leads to the effective lowering of the valence of the original atom and thus to electronic alchemy! Under certain circumstances, the electrons on the nearly empty non-atomic space-time sheets could form a high temperature super conductor.

9.2   What does 3-surface look like?

In the following a general picture of 3-space will be deduced from RGI hypothesis and spin glass analogy, the selection of preferred extremals of the Kähler action as generalized Bohr orbits, and from the special properties of the induced gauge fields implied by the compactness of CP_2.

9.2.1   Renormalization group invariance, quantum criticality and topology of 3-space

Renormalization group invariance, quantum criticality, and spin glass analogy are basic notions of quantum TGD but it is far from clear what these notions really mean at the level of space-time physics.

What quantum criticality means?

RGI (Renormalization group invariance) hypothesis states essentially that TGD Universe is quantum critical meaning that quantum theory is mathematically equivalent with a statistical system at critical point. S-matrix elements are analogous to thermal averages of observables, α_K corresponds to critical temperature and the vacuum functional $exp(K)$ corresponds to $exp(-H/T)$. The physical interpretation of the Kähler function suggests that $\alpha_K(phys)$ might correspond to a critical temperature at which spontaneous Kähler magnetization and formation Kähler electric fields compete.

The analogy with spin glass phase in four-dimensional sense is an additional characteristics feature. This allows the critical value of the α_K to depend on the zero modes of the configuration space metric.

The naive idealized interpretation for the quantum criticality would be that 3-surfaces with all possible sizes contribute to the functional integral. In realistic situations there is some upper bound for the size and duration quantum fluctuations and the size of the largest space-time sheet involved would define the scales in question.

Spin glass analogy leads to the idea that configuration space decomposes into regions D_p characterized by the p-adic prime p such that one can associate a hierarchy of p-adic length scales $L_p(n) = \sqrt{p}^{n-1}l$, $l \sim 10^4\sqrt{G}$ to each value of p [E5]. The critical value of α_K would depend on p. The dependence can be deduced from the requirement that gravitational constant is approximately invariant in the coupling constant evolution associated with the p-adic prime p . These scales define natural upper bounds for the scale of quantum fluctuations associated with the quantum critical space-time sheet. Dark matter hierarchy in turn assigns to each p-adic length scale a hierarchy of further length scales scaled up by the values of $\hbar/\hbar_0$. The typical duration of quantum fluctuation would correspond to the typical geometric duration of maximal deterministic region inside space-time sheet.

What are the competing phases?

Quite generally, critical systems are characterized by long range correlations (correlation length ξ diverges) for the competing phases present in the system. Physically this means the coexistence of arbitrarily large volumes of the two phases. Both Kähler magnetized 3-surfaces and 3-surfaces containing predominantly Kähler electric fields contribute significantly to the functional integral are present. At the infinite volume limit the Kähler action per volume must vanish since otherwise the vacuum functional vanishes: TGD cosmology to be studied later is in accordance with this picture.

If preferred extremals minimize (or maximize) the absolute value of Kähler action in each region with a definite sign of action density, the decomposition into magnetic and electric regions and vanishing of the Kähler action per volume follows automatically [E2]. For the absolute minimization of Kähler action Kähler electric fields dominate and it is not clear whether there are solutions for which the Kähler action of the entire Universe is finite.

How quantum fluctuations and thermal fluctuations relate to each other?

An experimental fact is that quantum critical systems such as high temperature superconductors [J1, J2, J3] exist in a rather narrow parameter range, and one can say that quantum criticality becomes visible only when quantum fluctuations are not masked by thermal fluctuations. One should express this fact using TGD based notions.

p-Adic and dark matter hierarchies correspond also to hierarchies for quantum jumps with time scales given the average geometric duration for quantum jump. This hierarchy means quantum parallel dissipation about which hadrons as quantum systems containing quarks as dissipating subsystem at shorter p-adic length and time scale give a basic example.

At given space-time sheet short scale thermal fluctuations would have interpretation as quantum parallel fluctuations at smaller space-time sheets topologically condensed to the space-time sheet in question whereas the quantum critical fluctuations would correspond to the quantum fluctuations in the scale of the space-time sheet. The duration of maximal deterministic space-time region would correspond to the duration of single quantum state in the sequence of quantum jumps. The interpretation would be that only at quantum criticality the quantal fluctuations in long time scales can mask the thermal fluctuations in shorter scales.

How quantum measurement theory relates to quantum criticality?

A further question is how quantum measurement theory relates to this picture. Configuration space zero modes represent non-quantum fluctuating classical observables correlating with quantum numbers and in quantum measurement a localization in zero modes occurs. Does this mean that the localization in zero modes breaks quantum criticality above the time scale corresponding to the typical geometric time duration of quantum jump by selecting precise values of zero modes?

Formation of join along boundaries condensates and visible-to-dark phase transitions as mechanisms giving rise to quantum critical systems

The phase transition from visible to dark matter, and more generally, the transitions increasing the value of Planck constant define the first mechanism leading to the formation of larger quantum critical system and long range quantum fluctuations can be assigned to dark matter.

The formation of a join along boundaries condensate means also a formation of a quantum critical system. The 3-surfaces with a typical size of order L_p combine together by join along boundaries bonds to form larger surfaces. Above criticality there are no bonds, below criticality all 3-surfaces combine to form larger condensates and at criticality there are join along boundaries condensates with all possible sizes up to the cutoff length scale. Note that, at least for small values of p, the surfaces with typical sizes $\sqrt{p}^n L_p$, $n = ..0, 1, 2, ...$ correspond to the presence of all surface sizes related by a fractal scaling for a given p. A more precise formulation for what the fusion of p-adic and real physics [E2] means supports the view that topological field quanta allow a discrete scaling symmetry identifiable as scalings by powers of $\sqrt{p}$.

9.2.2   3-surfaces have outer boundaries

In length scales larger than hadronic length scale 3-surface with size L means roughly a condensate of smaller scale 3-surfaces on a piece of Minkowski space of size L. It is quite essential that these surfaces have finite size and therefore have outer boundary. The finite size of the 3-surfaces follows from the minimization of the Kähler action and from the compactness of CP_2. The argument goes as follows.

The matter inside a 3-surface creates gauge fields. In particular, the minimization of the absolute value of Kähler action in a region with definite sign of action density implies that matter serves as a source of either Kähler magnetic or Kähler electric fields. For instance, the Kähler electric field created

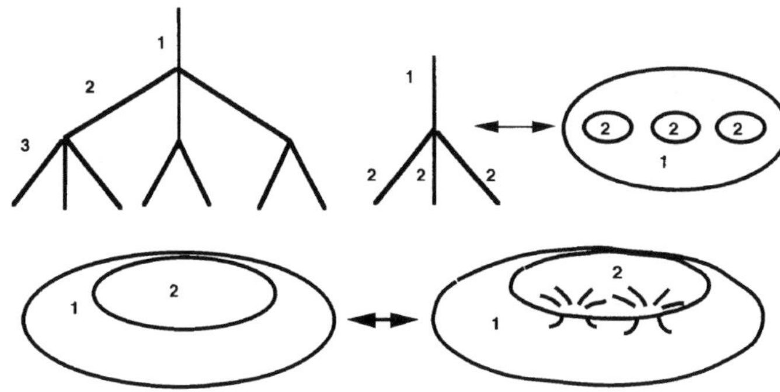

Figure 9.2: Hierarchical, fractal like structure of topological condensate predicted by RGI hypothesis: 2-dim. visualization

by a constant mass distribution increases without bound. The smooth imbeddings of the gauge fields are however not possible globally and space-time decomposes into topological field quanta and their boundaries correspond to edges of space time. The elimination of the edges leads to a 3-space consisting of disjoint components. Simple examples are provided by a cylindrically symmetric imbedding of a constant magnetic field and the Kähler electric field created by a constant mass distribution, which fail for certain critical radii.

One can understand at general level how the compactness of CP_2 enters into the game. The point is that the gauge potentials associated with the induced gauge fields are bounded functions of CP_2 coordinates. For instance, for a geodesic sphere S^2 of CP_2 gauge potentials are just proportional to $A = sin(\Theta)d\Phi$. For a generic gauge field the gauge potential is not bounded (as an example consider gauge potentials of the Coulomb field or Kähler electric field created by a constant charge distribution or by a constant magnetic field). Therefore for certain values of CP_2 coordinates the representation of the gauge potential as an induced gauge potential fails. The failure takes place at some 3-surface of X^4. One can continue the embedding by changing the values of vacuum quantum numbers but certain CP_2 coordinates possess discontinuous or even infinite derivatives on the boundary so that undesirable edges of space time result. The manner to get rid of edges is to allow boundary for X^3 so that a region, where the the representation of the gauge potential as induced gauge potential works defines a natural unit of space-time, which might be called topological field quantum. In the sequel this phenomenon will be considered in more detail.

An obvious question is what happens to the gauge fluxes of long range gauge fields near the boundaries of the topological field quantum. Same question applies also to the gravitational flux associated with the Newtonian potential at the non-relativistic limit. One possibility is the appearance of neutralizing vacuum gauge charges and negative gravitational masses near the boundaries of the field quantum, perhaps related to vacuum polarization: this alternative must be realized for the particles of vapor phase. Second possibility is topological condensation on a larger topological field quantum so that gauge and gravitational fluxes flow to the larger topological field quantum via # contacts. The larger field quantum in turn must feed its gauge fluxes in a similar manner to larger field quantum so that the hierarchical structure of topological condensate is implied by the compactness of CP_2 and gauge flux conservation. Criticality implies only that 3-surfaces of arbitrarily large size are possible and therefore the number of the condensate levels and corresponding length scales $L(n)$ is infinite. Without criticality there would be some upper bound for 3-surfaces and only vapor phase would be possible.

The # contacts feeding the gauge fluxes from level p_n to level p_{n+1} are located near the boundaries of topological field quanta: otherwise long range gauge fields would not be possible inside the topological field quanta. A more quantitative hypothesis is that # contacts are located in the boundary layer having thickness of order L_{p_n}. If topological field quantum at level n has the minimum size of order L_{p_n} then the # contacts neutralize the physical gauge charges on the average.

9.2.3 Topological field quantization

Topological field quantization is a very general phenomenon differentiating between the TGD based and Maxwellian field concepts and results from the compactness of CP_2 only, being independent of any dynamical assumptions.

Topological field quantization occurs for surfaces representable as maps from M^4 to CP_2 and means that space time surface decomposes into regions characterized by certain vacuum quantum numbers characterizing the dependence of the phase angles Ψ and Φ associated with the two complex coordinates ξ_1 and ξ_2 of CP_2. There are two frequency type vacuum quantum numbers ω_1 and ω_2 characterizing the time dependence, two wave vector like quantum numbers k_1, k_2 characterizing the z-dependence and two integer valued vacuum quantum numbers n_1, n_2 characterizing the angle dependence of these phase angles. Topological field quantization fixes unique M^4 and CP_2 coordinates inside the field quantum and is analogous to a choice of a quantization axis.

Topological field quanta

Before considering the general form of the surfaces representable as maps $M^4 \to CP_2$ some comments about CP_2 coordinates are needed:

a) The so called Eguchi-Hanson coordinates for CP_2 are given $(r, u, \Psi, \Phi) \in [0, \infty] \times [-1, 1] \times [0, 4\pi] \times [0, 2\pi]$ (see Appendix). Ψ and and Φ are angle like coordinates closely related to the phases of the two complex coordinates of CP_2 and are the interesting variables in the sequel.

b) There are following types of coordinate singularities.

i) For $r = 0$ all values of Ψ and Φ correspond to same point of CP_2.

ii) For $r = \infty$ all values of Ψ correspond to same point of CP_2. For $u = 1$ and $u = -1$ also all values of Φ correspond to same point of CP_2.

Consider now the space-time surface representable as a graph of a map $M^4 \to CP_2$. The general form of the angle coordinates Ψ and Φ as functions of M^4 cylindrical coordinates (t, z, ρ, ϕ) is given by the expression

$$
\begin{aligned}
\Phi &= \omega_1 t + k_1 z + n_1 \phi + \text{Fourier expansion} \ , \\
\Psi &= \omega_2 t + k_2 z + n_2 \phi + \text{Fourier expansion} \ .
\end{aligned}
\tag{9.2.1}
$$

There always exists a rest frame, where k_1 or k_2 vanishes. The Fourier expansion is single valued in ϕ and finite in z and t. The vacuum quantum numbers ω_1 and ω_2 are frequency type vacuum quantum numbers to be referred as "electric" quantum numbers. The quantum numbers (n_1, n_2) are integer valued and will be referred to as "magnetic" quantum numbers.

The values of the vacuum quantum numbers can change at the boundaries of the regions of space-time determined by the conditions

i) $r = 0$ and $(r = \infty, u = \pm 1)$: here all vacuum quantum numbers can change

ii) $r = \infty$: here only ω_2, n_2 and k_2 can change.

Also the choice of CP_2 coordinates and M^4 coordinates can in principle change: different CP_2 coordinates are related by color rotation and different M^4 coordinates by Lorentz transformation.

In general, the boundaries of the regions correspond to edges of space-time in the sense that CP_2 coordinates possess discontinuous or infinite derivatives at the boundaries of the field quanta. A natural manner to get rid of the edges is to consider 3- surfaces consisting of a single region only so that single region of this kind, topological field quantum, is a natural unit of 3-space. There is however an important exception to this. The join along boundaries interaction very probably means the gluing of two topological field quanta together along their boundaries and provides a manner to construct coherent quantum systems from smaller units.

The sizes of the topological field quanta are indeed finite so that the boundary of 3-space (quite essential for the ideas described before) is an unavoidable consequence of the compactness of CP_2 and the minimization of the Kähler action. The dependence of the size of the 3-surface on the vacuum quantum numbers is in accordance with the proposed interpretation: at the limit of large vacuum quantum numbers the size of the topological field quantum becomes macroscopic and at small vacuum quantum number limit the size of the surface becomes small.

Very complicated hierarchical structures predicted by the RGI are in principle possible since topological field quanta can suffer topological condensation on larger field quanta. Field quanta can become nested and both spatial and temporal structures (nesting in time like direction) are possible.

The vacuum quantum numbers associated with vacuum extremals

Vacuum extremals define a reasonable starting point for TGD based model for gravitational interactions. For vacuum extremals classical em and Z^0 fields are proportional to each other (see the Appendix of the book):

$$
\begin{aligned}
Z^0 &= 2e^0 \wedge e^3 = \frac{r}{F^2}(k+u)\frac{\partial r}{\partial u}du \wedge d\Phi = (k+u)du \wedge d\Phi \ , \\
r &= \sqrt{\frac{X}{1-X}} \ , \quad X = D|k+u| \ , \\
\gamma &= -\frac{p}{2}Z^0 \ .
\end{aligned}
\tag{9.2.2}
$$

For a vanishing value of Weinberg angle ($p = sin^2(\theta_w) = 0$) em field vanishes and only Z^0 field remains as a long range gauge field.

The study of the imbeddings of the Schwartshild metric as vacuum extremals (gravitational mass is non-vanishing but inertial mass vanishes) shows that astrophysical length scales correspond to large vacuum quantum number limit of TGD. Any mass vacuum extremal is necessarily accompanied by long ranged electro-weak and color fields and from the requirement that the corresponding force is weaker than the gravitational force one obtains that the value of the parameter ω_1 is of the order of $1/R \sim 10^{-4}\sqrt{G}$.

A simple example about the decomposition of space-time into topological field quanta is obtained by considering the cylindrically symmetric imbedding of a constant magnetic field in the z-direction as a vacuum extremal. Electromagnetic field can be written as $F^{em}_{\rho\phi} = B_0\rho$ and using the general results from the Appendix of the book one can write

$$
\begin{aligned}
u = u(\rho) \ , &\quad \Phi = n_1\phi \ , \\
r = \sqrt{\tfrac{X}{1-X}} \ , &\quad X = D|k+u| \ , \\
A^{em}_\phi = \frac{B_0\rho^2}{2} &= -\frac{p}{2}n_1(k+u)\partial_\rho u \ .
\end{aligned}
\tag{9.2.3}
$$

Assuming that $(r,u) = (0,0)$ holds true at z-axis, the equation for em gauge potential A^{em} fixes the relationship between ρ and u as

$$
u = -k \pm \sqrt{k^2 - \frac{2B_0\rho^3}{3n_1p}} \ .
\tag{9.2.4}
$$

The finite value range $0 \le u \le 1$ implies that the imbedding fails for certain values of ρ. Also the requirement that u is real implies an upper bound for ρ: the larger the value of n_1 the larger the critical radius. Imbedding can fail also for $X < 0$ and $X > 1$ corresponding to critical values of u equal $u_0 = -k$ and $D|(k+u_1)| = 1$.

9.3　Gauge charges in TGD

The concept of gauge charge has been a source of chronic headache for TGD. There are several questions waiting for definite answer. How to define gauge charge? What is the microscopic physics behind the 'anomalous' gauge charges implied by long range gravitational fields. Are the gauge charges quantized in elementary particle level or does the concept of anomalous gauge charge make sense? How gauge charges relate to the classical gauge fluxes.

9.3.1 Definition of the gauge charges in TGD

In TGD gauge fields are not primary dynamical variables but induced from the spinor connection of CP_2. There are two manners to define gauge charge: group theoretical and classical.

Group theoretical definition of the gauge charges

In the purely group theoretical approach one can associate a non-vanishing gauge charge to a 3-surface of finite size and the quantization of the gauge charge follows automatically. This definition should work at CP_2 length scale, when particles are described as 3-surfaces of size $R \sim 10^4 \times \sqrt{G}$ and the classical space-time mediating long range interactions makes no sense. Gauge interactions are mediated by a gauge boson exchange, which in TGD has topological description. Gauge boson exchanges are in a well defined sense bridges along which also the classical fields can propagate. A naive geometric argument however suggests that the exchange of CP_2 type extremals gives rise to an extremely weak gauge interaction with cross sections characterized by the geometric size of the CP_2 type extremal. The solution of the paradox is that elementary particles are actually generated in topological condensation when the light-like causal determinant (elementary particle horizon) associated with the wormhole contact becomes a carrier of partonic quantum charges creating long range fields at the space-time sheet of size at least of order Compton length of the particle.

Classical definition of the gauge charges

The classical definition of the gauge charge is as a gauge flux over a closed 2-surface. The classical quantization of the gauge charges is perhaps possible for some subset of mutually commuting charges and would be implied by the absolute minimization of the Kähler action. For a closed 3-surface gauge flux vanishes and one might argue that this is the case for finite size 3-surfaces with a boundary since the boundary conditions might generate a gauge charge near the boundary cancelling the gauge charge created by the particles condensed on the 3-surface.

This picture seems to be in conflict with the quantum view. The resolution of the paradox is simple. In length scales longer than the size of space-time sheet gauge and gravitational interactions must be described in terms of particle exchanges whereas the description in terms of classical fields works only in length scales shorter than the size of the space-time sheet.

Color gauge charges

In the case of gluons it is not so clear whether the gauge charges correspond to classical gauge fluxes since the classical gluon fields (projections of the $SU(3)$ Killing vector fields to space-time surface) do not strictly speaking correspond to genuine gauge fields. In any case, gluon field can be defined and the components of the gluon field are of the form $g_{\mu\nu}^A \propto H^A J_{\mu\nu}$, where H^A is a Hamiltonian of the color isometry and J denotes the induced Kähler form. The holonomy group of the classical color field is always Abelian which by quantum classical correspondence suggests a weak form of confinement in the sense that the allowed quantum states corresponds to the states of color multiplets having vanishing color isospin and hyper charge. The explanation of Centauro events and Pomeron in terms of quantum coherent topological evaporation of the valence quarks [F5] suggests that only color singlet states can evaporate.

There are two possibilities.

a) The quarks in the vapor phase have color charges but the total color for the evaporated quarks vanishes. This statement makes physically sense since massless gluon exchange implies a long range interaction also in the vapor phase. In this case the identification of the gauge charge as gauge flux is not possible.

b) Valence quark 3-surfaces are joined together via join along boundaries bonds so that it is this structure, which evaporates and has indeed vanishing total color gauge charge so that the identification of the color charges as gauge fluxes is possible. There is a temptation to accept this alternative since join along boundaries bonds can be identified as color flux tubes implying a string like structure for mesons. The evaporation of single sea quark or gluon is not possible in this picture. This raises the question about the definition of the quark gluon plasma: perhaps quark gluon plasma corresponds to a state in which the join along boundaries bonds between quarks are broken.

9.3.2   Questions related to gravitational interactions

Why gravitational interaction is so weak?

There are two explanations for the weakness of the gravitational interactions and although these explanations look quite different they could be consistent with each other.

a) The exchange of CP_2 type extremal would in a well defined sense create the bridge for the gravitational perturbation to propagate and also explain the extreme weakness of the gravitational interaction. As far as gravitation is considered, the evaporated particle would behave in very much the same manner as the condensate particle.

b) p-Adic fractality leads to a quantitative argument explaining the extreme weakness of the gravitational force and also predicting a hierarchy of strong gravities analogous to the force mediated by spin 2 mesons [E1]. The idea is simple: the geodesic distance $d(X^4)$ between interacting masses along space-time sheet is much longer than the distance $d(M^4)$ in M^4 and related by a very large scaling factor to the latter. This implies that strong gravitational coupling proportional to L_p^2 is scaled down to a weak coupling when $1/d^2(X^4)$ in gravitational force is expressed in terms of experimentally measured distance $d(M^4)$. A possible interpretation is that $d(X^4)$ corresponds to the length of topologically condensed gravitonic CP_2 type extremal performing random zitterbewegung. One could say that a kind of coast of Great Britain effect makes gravitational interaction weak. L_p^2 would in turn correspond to the area of the space-time sheet of the particle emitting gravitons.

Clearly, the first argument would explain the weakness of gravitational interaction described as an exchange of CP_2 type extremals whereas second argument would do the same when gravitation is described as a classical long range force.

How gravitational and inertial masses are related?

The general definition of gravitational mass is straightforward in terms of the M^4 projection of Einstein tensor and the definition involves no assumption about the space-time surface [C3]. Gravitational mass is in general not conserved. Gravitational four-momentum is in general non-vanishing and non-conserved for all kinds of vacuum extremals, in particular CP_2 type vacuum extremals, so that Equivalence Principle in strong form cannot hold true for them.

The assumption that gravitational energy corresponds to the absolute value of the conserved inertial (Poincare) energy is attractive but it is not clear whether it is really a general prediction of TGD. The identification of absolute value of inertial rest energy as the average value of non-conserved gravitational rest energy emerges very naturally in TGD based picture about particle massivation. This identification indeed allows the sign of inertial energy to depend on the time orientation of space-time sheet. At configuration space level the space-time surface with positive/negative time orientation can be assigned to the boundary of a union of future/past directed light cones.

Gravitational four-momentum of topologically evaporated particle

The interpretation of the gravitational mass as a gauge flux associated with gravitational potential makes sense in the non-relativistic limit. The problem is what happens to the particle's gravitational mass in the topological evaporation. Does gravitational mass of evaporated particle vanish? If this is the case, can gravitational energy be non-vanishing?

The answer to these questions emerges from the study of CP_2 type extremals. Einstein's tensor for CP_2 type extremal is proportional to the metric tensor so that the non-conserved gravitational momentum is non-vanishing and light-like being parallel to the light-like random curve defining M^4 projection of the CP_2 type extremal. Hence gravitational rest mass of cP_2 type extremal vanishes in the vapor phase but not gravitational energy.

Since wormhole contacts can be identified as pieces of CP_2 type extremals carry gravitational four-momentum, the net gravitational four-momentum of topologically evaporated space-time sheet carrying hierarchy of smaller space-time sheets is the sum of the gravitational four-momenta associated with wormhole contacts and non-vanishing.

In length scales shorter than the size of the topologically evaporated space-time sheet classical description of gravitational interaction makes sense whereas in longer length scales description in terms of graviton exchanges is the proper description.

What about $Diff^4$ invariant deformation of Poincare algebra?

The situation is complicated by the fact it is $Diff^4$ invariant Poincare algebra might correspond to a Lorentz invariant deformation of the ordinary Poincare algebra [de7], which corresponds to the quantum mechanical four-momentum. A hypothesis worth of considering is whether the $Diff$ invariant four-momentum is equivalent with the gravitational four-momentum. An interesting possibility is that only the $Diff$ invariant rest mass rather than four-momentum vanishes in the vapor phase. It is needless to emphasize that this alternative has rather exciting implications in the (improbable) case that topological evaporation is possible for macroscopic objects.

9.3.3 The problem of the anomalous gauge charges

The concept of anomalous gauge charges, was introduced in the earlier versions of the book. The experience from the study of the extremals of the Kähler action shows that at astrophysical length scales gauges charges are apart from numerical constant equal to the mass of the system using Planck mass as unit:

$$Q = \epsilon_1 \frac{M}{m_{proton}} \ . \tag{9.3.1}$$

The condition $\frac{\epsilon_1}{\sqrt{\alpha_K}} < 10^{-19}$ holds true in astrophysical length scales since gauge forces must be weaker than gravitational interaction in astrophysical length scales. Any mass distribution which can be modelled gravitationally in terms of vacuum extremal long range em and Z^0 gauge fields.

The path to what seems to be the correct interpretation of this result was rather tortuous. This is understandable since the conclusion that theory predicts an entire fractal hierarchy of scaled up variants of standard model physics with arbitrarily long ranges of weak and strong interactions and having interpretation in terms of dark matter hierarchy labelled by p-adic length scales and by the values of a dynamical quantized Planck constant, is not not something which would pop first in mind.

9.3.4 The concept of the # contact, particle massivation, and weakening of Equivalence Principle

There are two views about # contacts. The first view is purely classical and developed first. The quantal view dictated by quantum classical correspondence emerged much later and leads to a detailed understanding of mechanisms of particle massivation (both p-adic thermodynamics and Higgs mechanism are involved in TGD framework), of how inertial and gravitational masses and related, and of breaking, or rather weakening, of Equivalence Principle.

Classical and quantum views about # contacts

If the net gauge charge of a given condensate level is non-vanishing, there must exist some mechanism feeding the gauge charge to the lower condensate levels. The only possibility is provided by the # contacts obtained by drilling small holes on the surfaces X_1^3 and X_2^3 and connecting the holes with at tube $D_1 \times S^2$. The assumption that classical gauge charges are quantized and conserved forces also the quantization of the gauge fluxes associated with the # contacts.

The conserved gauge flux corresponds at quantum level to the partonic quantum numbers associated with either of the two light-like elementary particle horizon associated with the wormhole contact and are thus assignable to fermionic oscillator operators associated with second quantized induced spinor fields. Gauge flux is conserved only if the partonic quantum numbers sum up to zero. In this picture # contacts become special kind of particles.

 # contacts as particles

contacts are expected to be very tiny, the size is presumably of the order of CP_2 radius and the simplest model is as a piece of CP_2 type extremal. This would mean that # contact can carry gravitational four-momentum but has a vanishing gravitational rest mass. If it possesses inertial mass it can be associated with the partons at the two elementary particle horizons carrying the partonic quantum numbers.

Wormhole contacts are bosons and one can associate to them color, em and Z^0 gauge fluxes $-Q_i$ so that contacts behave as charged particles and matter-contact and contact-contact interaction energies are non-vanishing. The gauge fluxes associated with the contacts can be identified as the gauge charges associated with either light-like elementary particle horizon in absence of possible renormalization and non-conservation effects.

Bose-Einstein condensates of charged # contacts are described by a complex order parameter quantum mechanically and the lightness of the wormholes implies BE condensation to the lowest energy state. Also coherent states are possible. These states could be present in all length scales and especially interesting are the applications of the concept to bio-systems [J3]. For instance, I have proposed that # contacts carrying quantum numbers of neutrino-antineutrino pair would be involved with cognitive representations [M6].

contacts, non-conservation of gauge charges and gravitational four-momentum, and Higgs mechanism

Gravitational # contacts are necessary and if gravitational energy can be regarded in the Newtonian limit as a gauge charge, the contacts feed the gravitational energy regarded as a gauge flux to the lower condensate levels. The non-conservation of gravitational gauge flux means that # contacts can carry gravitational four-momentum and since CP_2 type vacuum extremals are the natural candidates for # contacts, the natural hypothesis is that the non-conserved light-like gravitational four-momentum of # contacts is responsible for the non-conservation of gravitational four-momentum flux. The non-conservation of the light-like gravitational four-momentum of CP_2 type extremals would in turn be responsible for the non-conservation of the net gravitational four-momentum.

contacts could be also carriers of inertial mass which must be conserved in absence of four-momentum exchange between environment and wormhole contact. Therefore Equivalence Principle cannot hold true in a strict sense. Equivalence Principle would be satisfied in a weak sense if the inertial four-momentum is equal to the average four-momentum associated with the zitterbewegung motion and corresponds to the center of mass motion for the # contact.

The non-conservation of weak gauge currents for CP_2 type extremals implies a non-conservation of weak charges and the finite range of weak forces. If wormhole contacts correspond to pieces of CP_2 type vacuum extremal, electro-weak gauge currents are not conserved classically unlike color and Kähler current. The non-conservation of weak isospin corresponds to the presence of pairs of right/left handed fermion and left/right handed antifermion at wormhole contacts. These wormhole contacts are excellent candidates for the TGD counterpart of Higgs boson providing the most natural mechanism for the massivation of weak bosons. The dominant contribution to fermion mass would be due to p-adic thermodynamics [F3]. If weak form of Equivalence Principle holds true, inertial mass would result simply as the average of non-conserved light-like gravitational four-momentum.

There would be two contributions to the mass of the elementary particle.

a) Part of the inertial mass is generated in the topological condensation of CP_2 type extremal representing elementary particle involving only single light like elementary particle horizon, say fermion, and would correspond naturally to the contribution to the mass modellable using p-adic thermodynamics. The contribution from primary topological condensation is negligible if the radius of the zitterbewegung orbit is larger than the size of the space-time sheet containing the topologically condensed boson so that the motion is along a light-like geodesic in a good approximation. For gauge bosons this contribution should be very small or vanishing. Systems like superconductors where also photons and even gravitons can become massive [D3] might form an exception in this respect.

b) The space-time sheet representing massless state suffered secondary topological condensation at a larger space-time sheet and viewed as a particle can develop an additional contribution to the mass via Higgs mechanism since the wormhole contacts cannot be regarded as moving along light like geodesics of M^4 in the length and time scale involved. # contacts carrying a net weak isospin would have interpretation as TGD counterparts of neutral Higgs bosons and the formation of coherent state involving a superposition of states with varying number of wormhole contacts would correspond to the generation of a vacuum expectation value of Higgs field. The inertial mass of the wormhole contact must be small, presumably its order of magnitude is given by $1/L_p$, where L_p is the characteristic p-adic length scale associated with a given condensate level.

Spatial distribution of # contacts

The existence of the # contacts allows also an elegant solution to the boundary conditions associated with the extremals of the Kähler action. If the space-time surface becomes vacuum near its boundaries the boundary conditions (dictated by the variational principle rather than being posed separately as in string models) are satisfied identically. Furthermore, one can obviously regard the hierarchical structure of the topological condensate more or less as a consequence of the boundary conditions.

Concerning the spatial distribution of the # contacts there are obvious constraints. Consider first the condensate blocks containing a large number of elementary particles:

a) Contacts behave as classical gauge fluxes and cause a screening of the classical gauge charges of the ordinary matter on both space-time sheets. The work with classical TGD led to the conclusion that long range classical Z^0 and/or electromagnetic gauge fields are necessary in order to have long range gravitational fields. This suggests that # contacts feeding the gauge flux to lower condensate levels are located near the boundaries of 3-surface. There is indeed a good mathematical reason for this: near the boundaries the imbedding of the gauge fields created by the interior gauge charges becomes impossible and the only possibility to satisfy the boundary conditions is that the gauge fluxes flow to lower condensate level and the surface becomes vacuum extremal near the boundary.

b) It must however emphasized that a random distribution of the # contacts inside entire condensate block is not totally excluded. The point is that the exchange of photon and graviton 3-surfaces can give rise to long range force even in the absence of classical fields. As far as photon/graviton exchange is considered, # contacts behave as extremely small dipoles. It is the results of classical TGD, which support the idea of classical long range fields. For gauge bosons the density of boundary # contacts should be very small in length scales, where the matter is essentially neutral. For gravitational # contacts the situation is different. One might well argue that there is some upper bound for the gravitational flux associated with single # contact given by, say, Planck mass or CP_2 mass so that the number of gravitational contacts would be proportional to the mass of the system.

c) An important question is how # contacts are created and destroyed. The creation of charged # contact leads to the appearance of a radial gauge field in condensate and this seems to be impossible since it involves a radical instantaneous change in the field line topology. The simplest and the least exotic manner to solve the difficulty is to assume that # contacts are created in particle-antiparticle annihilation. The particle and antiparticle belonging to different space-time sheets can join along their boundaries via the emission of radiation so that both boundary components disappear and # contact is created. This conforms with the identification of # contacts as CP_2 type vacuum extremals condensed at two space-time sheets simultaneously.

Quantum numbers of vapor phase particles

For a long time the attempts to characterize vapor phase particles remained a mixture of arguments involving classical gauge flux thinking and group theoretical quantum arguments which tended to be contradictory. The recent picture is free of these inconsistencies and allows understand how the quantum numbers of vapor phase particles are determined.

Gauge charges of vapor phase particles correspond to the net gauge charges of topologically condensed elementary particles plus those associated with # contacts and representing non-conservation: only weak isospin I_L^3 receives the latter contribution. The conserved inertial four-momentum is identifiable as the sum of average gravitational four-momenta assignable to the elementary particle horizons. The exchanges of space-time sheets, in particular CP_2 type extremals representing elementary particles provides a description for the interactions of vapor phase particles.

9.4   The new space time picture and some of its consequences

The previous considerations suggest that TGD space-time has a hierarchical, fractal like structure consisting of an infinite number of condensate levels n characterized by length scale $L(n) < L(n+1)$ identifiable as a typical size for 3-surface at level n. Spin glass analogy suggests that the label n corresponds to preferred primes characterizing p-adic length scales and to values of Planck constant labelling levels of dark matter hierarchy. p-Adic fractality means that for each p there is actually a length scale hierarchy coming in powers of $\sqrt{p}$. An infinite hierarchy of copies of standard model physics is an unavoidable prediction if quantum classical correspondence is taken seriously and can be identified as dark matter hierarchy.

9.4.1 Topological condensation and formation of bound states

It is tempting to identify the physical counterpart of the topological condensate in the length scale L as a bound state with size L. If this assumption is accepted then one ends up to the rather beautiful general scenario for the hierarchical structure of the 3-space. Quarks (3-surface of size of CP_2 length, so called CP_2 type extremals to be discussed later) condense around the hadronic 3-surfaces, hadrons condense around a piece of Minkowski space with size of order $10^{-14} - 10^{-15}$ meters to form nuclei, nuclei and electrons condense to form atoms of size of the order 10^{-10} meters or larger, atoms condense to form molecules, etc.

Generalizing the previous ideas, one ends up to a rather exciting possibility for a topological description of the macroscopic states of matter. Consider solids as an example. Solids correspond to a regular lattice of atomic or molecular 3-surfaces condensed to background 3-space. There are two kinds of forces binding the structure together.

i) There are interactions mediated via the the fields of the background 3-space and these correspond to the ordinary electric forces.

ii) There is interaction resulting from the "contacts" between the boundaries of the neighboring atoms (for a two-dimensional visualization see Fig. 9.4.1). Join along boundaries bond means mathematically a tube $D^2 \times D^1$ connecting the boundaries together or equivalently, topological condensation for the boundaries. This interaction is completely new and has as its counterpart the forces generated by the electron exchange between atoms believed to explain the binding between the atoms of certain solids. It is however clear that something quite new is introduced so that the conventional belief that Schrödinger equation in a flat 3-space alone explains these interactions would not be correct in TGD context. That the approach based on Schrödinger equation have not lead to contradictions can be understood also: what join along boundaries bond makes is to select among possible solutions of Schrödinger equation those realized in Nature by forcing the Schrödinger amplitude to the bridges connecting different structural units.

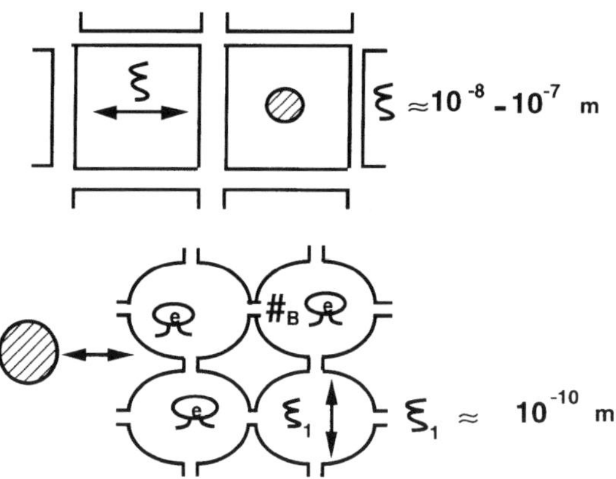

Figure 9.3: How one could understand the solid state topologically in terms of the join along boundaries interaction: 2-dim. visualization

The topological description of the liquid state goes along similar lines. Now however the contacts between neighboring atoms are not so rigid the reason being that thermal noise continually splits these contacts. A completely new element is the emergence of the vacuum quantum numbers and should lead to effects differentiating between TGD and more conventional approaches.

9.4.2 3-topology and chemistry

The practical models for chemical systems rely on the assumption that a chemical element has a well defined geometric shape. If this assumption is made then Schrödinger equation in electronic degrees of freedom combined with symmetry considerations gives satisfactory results. The general belief is that the complete Scrödinger equation treating quantum mechanically also the positions of the atoms predicts also the geometric structure of the chemical compounds. Unfortunately, in practice it is not possible to check numerically the correctness of this belief.

The "join along boundaries" interaction is a second standard phenomenological concept in the chemistry. What happens that reactants join along a part of their boundaries together to form a transition state (or a final state) and the reaction takes place in the new geometry. The chemistry of the biological systems relies heavily on this concept. For example, the catalytic action of the enzymes is often understood on the basis of key and lock principle: enzyme acts on the protein only provided the surfaces of the protein and enzyme fit together like lock and key. Usually it is believed that the association of a geometric form to chemical compounds and the "join along boundaries" mechanism provide an easy short hand description, which is in principle derivable from the complete Schrödinger equations. TGD suggests that this is not be the case.

What is exciting that this kind of idea leads to a completely fresh approach to the understanding of bio-systems: the basic principles of the underlying the biochemistry could be formulated in terms of the 3-topology. The biological information processing could involve the manipulation of the 3-topology or more precisely: the manipulation of the boundary topology if "join along boundaries" is indeed the basic mechanism. It should noticed also that the emergence of the vacuum quantum numbers is characteristic for TGD and provides a possible means for realizing the Universe as Computer idea in biological systems.

9.4.3 3-topology and super-conductivity

The #-contacts (wormholes) feeding the gauge fluxes from a given sheet of the 3-space to a larger one, can be regarded as particles carrying classical charges defined by the gauge fluxes. These particles must be light, which suggests that #-contacts can form Bose-Einstein condensate or coherent state identifiable in terms of Higgs vacuum expectation value. This BE condensate provides a possible explanation of so called Comorosan effect [la10] observed in organic molecules. A related effect is the formation of exotic atoms, when some valence electrons drop from the atomic space-time sheet to a larger space-time sheet. This process is accompanied by the generation of # contacts. The process leads to the effective lowering of the valence of the original atom and thus to "electronic alchemy". The claimed peculiar properties of so called ORMES [jf14] could have explanation as exotic atoms as suggested in [J1, J2, J3].

I have also suggested that the basic mechanism of super-conductivity somehow involves quantum coherent states of wormhole contacts. This might be the case although not quite in the original sense. There are two poorly understood problems involved with super-conductivity.

a) Super-conductor is often modelled as a coherent state of Cooper pairs. The conceptual problem is that the electric charge of this state is not well-defined and this is definitely in conflict with the conservation of electromagnetic charge.

b) The massivation of photons is a second poorly understood basic aspect of super-conductivity. The obvious question is whether this process could be interpreted in terms of a vacuum expectation value of a charged Higgs field and whether the charge of the Higgs field resolve the paradox otherwise created by the non-conservation of electromagnetic charge.

The obvious guess is that superconductor corresponds to superposition of quantum states with a well-defined total em charge such that electronic electromagnetically charge of some electronic Cooper pairs has been transferred to neutral wormhole contacts having quantum numbers of charged left/right handed positron and neutral right/left handed neutrino so that some Cooper pairs themselves have been transmuted to neutrino Cooper pairs.

In ordinary phase a space-time sheet carrying N Cooper pairs would feed em charge to a larger space-time sheet by $2N$ wormhole contacts consisting of e^+e^- parton pair. Super-conducting phase would correspond to a superposition of states for which $2M \leq 2N$ wormhole contacts have become electromagnetically charged and $2M$ electrons have transformed to neutrinos. Coherent state would thus correspond to a superposition of states with $M \leq N$ neutrino pairs, $N - M$ Cooper pairs, and

$2M$ charged wormhole contacts.

The presence of exotic W bosons mediating weak interactions in the scale of the space-time sheet would make possible this kind of states (which involved entanglement between wormhole contacts and Cooper pairs). The model would require that neutrinos and electrons in the superconducting phase have nearly identical masses and thus correspond to $p = M_{127}$, the largest Mersenne prime which corresponds to non-super-astronomical p-adic length scale. This conforms also with the absence of electro-weak symmetry breaking below the p-adic length scale characterizing the size of the Cooper pair. Also the quantum model for hearing [M6] requires that exotic neutrinos with mass very near to electron mass are involved. The TGD based model for atomic nucleus [F8] in turn predicts that quarks with mass near to electron mass appear at the ends of the color bonds connecting nucleons.

9.4.4   Macroscopic bodies as a topology of 3-space

The natural generalization of the foregoing ideas is that even the macroscopic bodies of the everyday world correspond to 3-surfaces, which have suffered topological condensation to the background 3-space. The outer surfaces of the macroscopic bodies would correspond to the boundaries of a particular space-time sheet. When macroscopic bodies touch each other, a partial join along boundaries would take place. We would live in the middle of a wild science fiction without realizing it!

Paradoxically, this new interaction is extremely familiar for us. The surface of the Earth corresponds to a boundary of a rather big 3-surface. At smaller length scales we see flowers, trees and all kinds of things and also these are 3-surfaces, which have joined along their boundaries to the surface of the Earth. Our biological bodies correspond to 3-surfaces having boundaries. We have however the special ability to cut this contact rather easily and to move quite freely although the gravitational force acting in the background 3- space takes care that the join along the boundaries with the surface of Earth is the usual state of affairs. When I touch the surface of the table by finger, a join along boundaries interaction takes place: we even recognize different objects just by touching them. We also smell and taste and at the microscopic level these senses are based on the join along boundaries interaction. Despite all this it has not been explicitly recognized that the formation of the join along boundaries bond might be a fundamental physical interaction!

What is also amusing that the implicit assumptions of any physical model of the macroscopic world is based on the assumptions about the geometric form of the physical objects and also the join along boundaries interaction is introduced implicitly into the description. For example, in order to describe solid state one draws lattice: one draws atoms in this lattice and bonds between the atoms. A second example is provided by the description of mechanical system consisting of rigid bodies.

In present picture this description is obtained by projecting the boundary of the 3- space to flat space E^3 : matter in the conventional sense corresponds to the shadow of the boundary-topology of 3-surface (for a 2-dimensional illustration see Fig.9.4.4). The fact that this kind of description is so obvious masks the fact that it is far from trivial whether one can actually deduce this kind of description starting from wave mechanics or QED.

What is so exciting that we can deduce the rough features of the topology of the surrounding 3-space just by looking it in various scales! Single glance shows that this topology is extremely complicated and contains boundaries everywhere and in all scales. In any case, it is in principle possible to make a map of a 3-surface in H both by observation of the form of macroscopic bodies and by measuring the ordinary physical observables like electromagnetic fields. Note that the fractal properties of the world are in accordance with the prediction of RGI hypothesis that topological condensate has hierarchical structure containing 3-surfaces of all possible sizes.

To summarize, topological condensation seems to provide a purely topological description for the generation of structures. The concept of matter in topologically trivial, almost flat 3-space is replaced with an empty but topologically highly nontrivial 3-space. The idea leads to a concrete program of actually finding out what is the topology of a given form of matter and understanding the physical properties matter in terms of this topology! And it would surprising if this kind of understanding would not increase our abilities to control and manipulate the properties of the matter.

Topological field quantum as a coherent quantum system

There are several arguments suggesting that topological field quanta are good candidates for coherent quantum systems and that join along boundaries provides basic means for constructing larger quantum

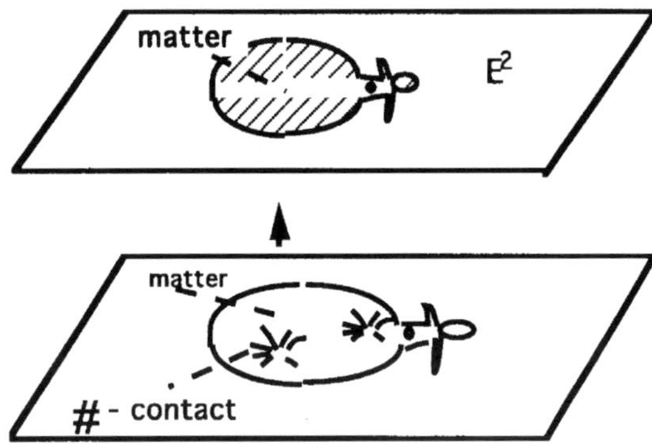

Figure 9.4: 3-dimensional matter as projection of the boundary of 3-surface to E^3: 2-dim. visualization

systems from smaller units.

a) The choice of the coordinates inside a given field quantum is analogous to the choice of the quantization axis. This suggests that the topological field quanta might provide a topological description of certain aspects of quantum phenomena. The choice of the quantization axis could indeed correspond to that taking place in quantum measurement. The fact that the quantization axes associated with different connected 3-surfaces need not be the same is in accordance with the idea that quantum coherence is possible for a connected 3-surface only. An exception is provided by a system consisting of several topological field quanta connected by "bridges" (join along boundaries bonds), for which quantization axis are same and which therefore can be be regarded as a coherent quantum system. As an example consider a spinning particle in a constant magnetic field. To describe the situation one must construct the imbedding of the magnetic field on the particle 3-surface by requiring that the resulting 4-surfaces corresponds to a preferred extremals of Kähler action. The simplest manner to achieve this is to assume that the quantization axis defining the vacuum quantum numbers n_1 and n_2 is in the direction of the magnetic field so that one say that the external magnetic field fixes the quantization axis.

b) 3-surfaces consisting of several field quanta are in general unstable in accordance with that fact that the formation of macroscopic quantum systems is also a rare phenomenon. The argument goes as follows.

i) CP_2 coordinates tend to have discontinuous or have even infinite derivatives at the boundaries of the topological field quanta if one poses some rather sensible physical requirements like the requirement that the 3-surface provides an imbedding for the Kähler electric field created by the mass distribution. As a consequence, Einstein tensor contains delta function type singularities and this is not nice. The best manner to avoid the edges is to allow boundaries.

ii) The boundaries of a 3-surface consisting of several field quanta are in general carriers of surface Kähler (Z^0) charge as the following argument shows. The embedding of the Kähler electric field associated with a given matter distribution has certain critical radius, which corresponds to the boundary of a field quantum. In general, one cannot continue the imbedding to a neighboring field quantum without allowing infinite derivatives of CP_2 coordinates.

iii) The 3-surface consisting of several field quanta is not stable unless the condition $u = cos(\Theta) = \pm 1$ is satisfied on $r = \infty$ surfaces. The point is that the excitations of Φ coordinate in general imply discontinuity of 3-surface at the boundary unless they are strongly correlated for neighboring field quanta.

c) The gluing of topological field quanta is probably possible by the join along boundaries bonds. The tube $D^2 \times D^1$ or the "bridge" between the two topological field quanta corresponds to a topological field quantum. The most probable "hot spots", where the gluing is possible correspond to parts of the surface, where the normal component of the Kähler electric field is vanishing. Now however the stability of the join along boundaries bond is not obvious. It can also happen that the directions of the induced Kähler fields are same on some portions of the boundaries and in this case the gluing by joing along boundaries bond serving as a Kähler electric flux tube is possible: in this case the stability of the bond is obvious. The color electric flux tubes between valence quarks provide a good example of this.

Topological description of supra phases

The topological construction recipe of a supra phase could be following. Take volumes of ordinary phase with a size of order of coherence length ξ, topologically condense them to the background 3-space and construct "bridges" between the boundaries of these structures. Supra phase is destroyed if the bridges are cut either thermally or by external magnetic field: the introduction of an external magnetic field indeed destroys the bridge since it implies that the quantum numbers n_1 and n_2 become in general non-vanishing inside the field quanta and bridge so that the order parameter ψ becomes discontinuous on the boundaries of bridges.

In the ground state of the supra phase the order parameter describing the supra phase is covariantly constant. Since the topology of the join along boundaries condensate is extremely complicated, the first homotopy group of the condensate is nontrivial. This means that one in general cannot find a global gauge transformation gauging the gauge potential associated with a vanishing magnetic field away. This implies that the phase increment of the order parameter along a closed homotopically nontrivial loop is in general nontrivial. These increments obviously contain information coded into the order parameter about the topology of the join along boundaries condensate.

The BE condensate of the charged # contacts, giving rise to pseudo super conductivity, playd a key role in the earlier TGD inspired model of brain as a macroscopic quantum system. In the model discussed in this chapter the coherent state of Cooper pairs is replaced with an entangled state involving product states of $2M$ charged wormhole contacts, $N - M$ electronic Cooper pairs, and M neutrino cooper pairs. One can also ask whether the vacuum quantum numbers might provide a realization for the idea about Universe as Computer. Biological information processing might be based on the manipulation of the vacuum quantum numbers. These ideas will be developed in some detail in the last part of the book.

Topological description of dissipation

The previous topological ideas lead to a general ideas about how structures are generated at macroscopic level. There is however a standard approach to the description of the generation of structures [ka3] and in this approach dissipative mechanisms play central role. The basic idea is that dissipation takes care that an open system ends up to some asymptotic state, which need not be thermal equilibrium but can be a complicated dynamic, non-equilibrium state.

The topological definition of the quantum coherence suggests that these approaches are in fact very closely related. Dissipation means certainly a loss of quantum coherence since for a coherent quantum system density matrix develops unitarily so that dissipation is impossible. Quantum coherence is lost at a given level of condensation hierarchy if the condensate consists of several 3-surfaces interacting through standard interactions only. The formation of the join along boundaries bonds however creates quantum coherence. Therefore the breaking of the join along boundaries bonds provides a good candidate for a fundamental dissipation mechanism.

To make the idea more concrete consider as an example liquid flow, assuming that there is a velocity gradient in a direction orthogonal to the velocity. What one wants to understand is the friction or how the energy of the liquid is dissipated. Liquid molecules have typically join along boundaries contacts (tube $D^2 \times D^1$) with the neighboring molecules (and due to thermal motion these contacts are continuously splitting and rejoining. The average age of a typical contact is much smaller than the time scale associated with the motion of a liquid so that the contact between two neighboring molecules suffers several thermally induced splittings and re-joinings, when the neighboring molecule pass by. A natural assumption is that the contact between two neighboring molecules is like a rubber band: energy is needed to stretch it. Assume that contact is formed between neighboring atoms moving with certain relative velocity so that the contact gets longer and splits after certain average time. The energy needed to stretch the contact longer is taken from the energy of the translational motion so that the relative motion becomes slightly slower.

As a second example, consider the understanding of the finite conductivity in metals. The neighboring atoms in the metal form a lattice and there are contacts between the neighboring atoms. These contacts are not completely stable but suffer splittings now and then. The large conductivity of the metal results from these contacts since they provide for the conduction electrons the bridges to move from one atomic 3-surface to a another one. The finiteness of the conductivity results from the fact that now and then a bridge between two neighboring atoms is broken. In the last part of the book it will be found that this kind of argument leads to a correct order of magnitude estimate for the metallic conductivity using a TGD inspired modification of the Drude model.

The concept of topological condensate affords also a second new point of view concerning the description of dissipation. The standard description of dissipation is in terms of inelastic collisions of particles. This description generalizes: particles at the condensate level n correspond to topological field quanta of level $n-1$ with typical size $L(n-1)$. In inelastic collisions of these particles join along boundaries contacts are created and split and part of kinetic energy is transferred to the kinetic energy of topological field quanta of level $n-2$ condensed on level $n-1$ field quanta. This mechanism makes possible the gradual transfer of the kinetic energy to the atomic length scales, where the collisions of ordinary particles take care of the further dissipation. Some potential applications of this picture are provided by hydrodynamics: ordinary hydrodynamics generalizes to a hierarchy of hydrodynamics, one for each condensate level plus a model for the energy and momentum transfer between two subsequent levels.

9.5 Topological condensation and color confinement

In this section a simple semiclassical model of color confinement is constructed as an application of the previous ideas. Also a view about color confinement being based on the same mechanism as the generation of macroscopic and macro-temporal quantum coherence (crucial for the TGD inspired theory of consciousness [K2] is discussed. These two arguments are separated by a temporal distance longer than decade and their different style reflects the development of my own thinking about TGD.

9.5.1 Explanation of color confinement using quantum classical correspondence

One can understand color confinement from the properties of the Kähler action by applying quantum classical correspondence.

a) The classical color gauge field is proportional to $H^A J_{\alpha\beta}$, where H^A is color Hamiltonian. This implies that the color holonomy group is Abelian. This suggests strongly that the physical states correspond states of color multiplets having vanishing color hyper charge and isospin. This would mean a weak form of color confinement.

b) The proportionality of the gluon field to the induced Kähler field, approximately satisfying free Maxwell equations, implies that the direction of the classical color field in M^4 is not random and that gluon field behaves in this sense as a massless field giving rise to long range interactions. The approximate canonical invariance of the Maxwell phase, which corresponds to the exact canonical gauge invariance of the configuration space geometry, is realized as approximately local $U(1)$ transformations which become constant color rotations below a cutoff scale identifiable as the size of space-time sheet carrying color charge.

c) The fact that the classical color field is proportional to a color Hamiltonian and Kähler field implies that the direction of the gluon field in the color algebra is random above the cutoff length scale so that color cannot propagate in length scales longer than the cutoff scale. Since color gauge currents are conserved for CP_2 type extremals representing wormhole contacts, color gauge flux is conserved in wormhole contacts which are therefore color neutral as particles so that colored variant of Higgs mechanism is not possible. The finite range of color interaction therefore leaves only the possibility that the net color charge of the elementary particles topologically condensed at the hadronic space-time sheet vanishes.

9.5.2   Hadrons as color magnetic/electric flux tubes

In this model quarks and gluons correspond to small M^4 type surfaces containing topologically condensed CP_2 type extremals and these surfaces are in turn condensed on a larger hadronic M^4 type surface. Valence quarks (at least) are connected by color electric or magnetic flux tubes (join along boundaries bonds) to form color singlets.

At elementary particle level, topological condensation means the condensation of the CP_2 type extremals around M^4 type surfaces. The condensed CP_2 type extremals perform zitterbewegung with a velocity of light although cm is at rest. The 3-space surrounding the condensed elementary particle has a finite size of the order of Compton radius (natural guess at this stage). At length scales $r << r_c$ (r_c denotes the Compton radius of the particle), condensed particles look essentially like massless particles whereas at length scales $r >> r_C$ they look like pieces of M^4 condensed to the background and moving with a velocity smaller than light velocity. CP_2 type extremals can be regarded as Kähler magnetic monopoles, whose magnetic flux runs in the internal degrees of freedom so that no long range $1/r^2$ magnetic field is generated. The fact that elementary particles are in a well defined sense Kähler magnetic monopoles supports criticality hypothesis: the strong Kähler coupling phase for the electric charges must be identical with the weak coupling phase for the magnetic monopoles and therefore Kähler action must correspond to a fixed point of the coupling constant evolution (this does not exclude the p-adic coupling constant evolution with respect to the zero modes of the Kähler metric).

The construction of the configuration space geometry and of quantum TGD lead to the conclusion that the description of the non-perturbative aspects of the color interaction must be based on the flux variables defined by the induced Kähler form. These variables include as a special case the generalized classical color fluxes. Since the low energy limit of TGD is expected to be more or less equivalent with the standard model, one can ask whether color confinement is signalled also by the divergence of the color coupling strength at low energies. p-Adic length scale hypothesis makes it possible to quantitative understand the confinement scale.

There are good reasons to expect that the quantum average space-time associated with a hadron could be regarded as an orbit of a 3-surface obtained by connecting the 3-surfaces (of size smaller than hadronic size) associated with the valence quarks with color electric flux tubes to get a color singlet state. Color singletness results from the randomness of the direction of the color field above hadronic length scales implying that the average radial color gauge flux emanating from the hadron vanish. This structure in turn has suffered a topological condensation on a larger hadronic 3-surface. The cutting of one or more color electric flux tube leads automatically to a generation of compensating color charges so that only color singlets can be created in the decays of the hadron. Also the topological evaporation of only color singlet objects is possible.

Color magnetic or electric flux tubes or both?

Both color magnetic and electric flux tubes have been used to model hadrons in TGD framework as well as in QCD, and one might wonder which of these options is the correct one. For absolute minimization of Kähler action Kähler electric fields are favored so that color electric flux tubes would be in a preferred position as models of hadron. For the more general variational principle discussed in [E2] the absolute value of Kähler action for space-time region with a definite sign of action density is either minimized or maximized (these options define dual dynamics and are consistent with the fact that 3-surfaces rather than 4-surfaces are fundamental dynamical objects). Therefore both Kähler magnetic and electric flux tubes are possible so that both color electric and magnetic models can be said to be correct.

The simplest model for the color flux tube connecting two quarks is based on the following picture.

a) The CP_2 type extremals with quark quantum numbers are topologically condensed at M^4 type 3-surfaces with size smaller than the hadronic size. These 2-surfaces are in turn condensed on the hadronic 3-surface. Quark like 3-surfaces are connected by join along boundaries contacts, which are color flux tubes connecting the boundaries of the quark 3-surfaces. These color flux tubes are the counterparts of the hadronic string.

b) The color magnetic/electric flux tube is a deformation of a vacuum extremal of type $M^2 \times D^2$ ("spring"), where D^2 is a disk orthogonal to M^2. This surface indeed looks like a tube of cross section D^2. The disk has an area of order $1/T$, where T is hadronic string tension.

c) The quarks at the ends of the flux tube serve as sources of approximately constant Kähler magnetic/electric fields (giving rise to chromo-electic fields of confining type), which generate the hadronic string tension. Since color field is proportional to Kähler field, also the Kähler charge of quark and gluon is of order $q \simeq 1$. The proportionality of the induced Kähler field and classical color field implies that hadrons can be regarded as chromo-electric flux tubes. Also QCD [gb1, ga4] affords this kind of descriptions for color confinement.

A model for color electric flux tube

Consider now in a more detail the model for the Kähler electric flux tube understood as a preferred extremal of the Kähler action. Since the actual situation is rather complicated it is useful to consider a simplified situation that is solution of the field equations with essentially constant Kähler electric field in the axal direction inside a cylinder of M^4.

The join along boundaries contact (color electric flux tube) corresponds to a surface of representable as a map from $M^4 = M^2 \times D^2$ to the homologically nontrivial geodesic sphere of type II. Here D^2 is a disk corresponding to the transversal section of the color flux tube and has size not much smaller than a typical hadronic length. One expects the Kähler action to be lowered by the generation of the Kähler electric fields. Field equations for the small deformations reduce in the lowest order to free Maxwell equations

$$D_\beta J^{\alpha\beta} \;=\; 0 \;.\tag{9.5.1}$$

Topologically condensed valence quarks at the ends of the flux tube serve as sources for the Kähler electric field.

The solution ansatz describing a constant Kähler electric field is obtained as a map from $M^4 = M^2 \times D^2$ to the geodesic sphere of type II:

$$\begin{aligned}
cos(\Theta) &= u(z) \;, \\
\Phi &= \omega t \;.
\end{aligned}\tag{9.5.2}$$

The interesting components of the induced metric and induced Kähler form are given by the expressions

$$\begin{aligned}
g_{tt} &= 1 - \frac{R^2\omega^2}{4}(1 - u^2) \;, \\
g_{zz} &= -1 - \frac{R^2}{4}\frac{u_{,z}^2}{(1 - u^2)} \;, \\
J_{tz} &= \frac{u_{,z}\omega}{4} \;.
\end{aligned}\tag{9.5.3}$$

Field equations are obtained from the conservation of four-momentum and the conservation condition for the z-component of momentum gives

$$u_{,z}^2(g_{zz}^3 g_{tt})^{-1/2} \;=\; \frac{E}{\omega^2} \;,\tag{9.5.4}$$

where E can be interpreted as the constant field strength.

The lowest order solution is obtained by approximating the induced metric with a flat metric so that one has

$$\Theta \;=\; arccos(\frac{Ez}{\omega}) \;.\tag{9.5.5}$$

The solution obtained is well defined only for the values of z having absolute value smaller that $2\pi/E$ and the g_{zz} component of the induced metric becomes infinite at the critical values of z. One might think that the appearance of the singularity is an artefact of the approximation used but this is not the case. The closer examination of the field equations shows that the singularity is unavoidable and results from the compactness of CP_2 (vector potential is proportional to $u = cos(\Theta)$) and that one cannot continue the solution in any manner for larger values of z. The nice thing is that boundary conditions are satisfied due to the singularity of the metric in the direction of the Kähler electric field. The result implies that the length of the string, and therefore the size of the hadron, is of order

$$L \sim \frac{2\pi\omega}{E} \quad .$$

The hadronic string tension is generated dynamically. One can obtain an estimate for the string tension by noticing that the situation is to a good approximation one-dimensional. This means that the Kähler electric field of the point charge is constant. Since the Kähler charges of quarks serve as sources of the Kähler field the order of magnitude for the Kähler electric field is given from Gauss theorem

$$E \;\; = \;\; \frac{q}{S} \;\; . \tag{9.5.6}$$

where $q \simeq 1$ is the Kähler charge of quark and S is the transverse area of the string. The order of magnitude estimate $q \simeq 1$ follows from the requirement that the color charges for quarks have this order of magnitude and from the fact that classical gluon field is proportional to the Kähler field. Hadronic string tension is obtained by integrating the energy momentum density over the transversal degrees of freedom

$$T \;\; \simeq \;\; \frac{1}{8\pi\alpha_K} \int E^2 dS \simeq \frac{1}{8\pi\alpha_K} \frac{q^2}{S} \;\; . \tag{9.5.7}$$

This implies that the transversal size of the hadronic string is of the order of $S \simeq 1/GeV^2$. For ground state hadrons the length of the string is therefore of same order as the transversal size of the string. Despite this, hadrons are string like objects in a well defined sense: their topology is $D^1 \times S^2$ instead of $D^1 \times D^2$.

As already found, the imbedding of the constant Kähler electric electric field associated with the flux tube becomes singular for values $z = \pm 2\pi\omega/E$ of the coordinate variable z in the direction of E (ω is the frequency associated with the solution). The study of the spherically symmetric extremal revealed that the parameter ω has value of order $10^{-4}m_{Pl}$ in long length scales. For the hadronic space-time sheet ω must of the order $\omega \sim 1/L$, where L is a typical hadronic length in order to get reasonable length for the string like object.

9.5.3　Color confinement and generation of macro-temporal quantum coherence

How macroscopic quantum coherence is possible in macroscopic time scales? This pressing problem of quantum consciousness theories involves both the question what coherence and de-coherence really mean and what really happens in quantum jump, as well as the question how the de-coherence times in living matter could be much longer than predicted by standard physics. Color confinement is the pressing problem of particle physics apparently put under the rug during last two decades. There might be a close connection between these seemingly totally un-related puzzles as the following little argument tends to show.

Classical argument: the time spent in color bound states is very long

The TGD based solution to the problem how to achieve macro-temporal quantum coherence relies on the new physics predicted by quantum TGD. The decisive factor is the gigantic almost degeneracy of states due to the fact that CP_2 canonical transformations, which effectively act as $U(1)$ gauge transformations, are approximate symmetries of the Kähler action broken only by the classical gravitation.

The argument goes as follows.

a) The increment of the psychological time in single quantum jump is estimated to be about CP_2 time, that is about 10^4 Planck times. During this time interval quantum coherence is destroyed in zero mode degrees freedom representing macroscopic degrees of freedom as well as in all degrees of freedom in which there is no bound state entanglement. This time interval is extremely brief as compared to the actual de-coherence times, which standard quantum theory allows to estimate.

b) The formation of bound states can save the situation since bound state entanglement is not reduced during state preparation phase of the quantum jump consisting of self measurements. The transformation of the zero modes (macroscopic classical degrees of freedom in which localization occurs in each quantum jump) to quantum fluctuating degrees of freedom, when join along boundaries bonds are formed between two space-time sheets representing binding systems accompanies the formation of bound states. The reason is that only over all center of mass zero modes remain zero modes. This means that the generation of macroscopic quantum fluctuating degrees of freedom and the formation of bound states accompany each other.

c) When bound state entanglement is generated, state function reduction and state preparation cease to occur in these degrees of freedom and one has macro-temporal quantum coherence. The sequence of quantum jumps effectively binds to a single quantum jump just like elementary particles bind to form atom behaving effectively as single elementary particle. The lifetime of the bound state defines the de-coherence time.

d) This does not yet explain why the lifetimes of the bound states, or more precisely, why the time spent in bound states, is much longer than predicted by the standard physics. New physics is required for this, and spin glass degeneracy provides it. What happens is following. When a bound state is formed, the space-time sheets representing the free particles are connected by join along boundaries bonds. By quantum spin glass degeneracy the number of bound states is huge as compared to the number of free states, since there is extremely large number of join along boundaries bond configurations and differing only by the classical gravitational energy. Accordingly, the time spent in bound states, and thus also de-coherence time, is much longer than that predicted by standard physics.

How could one understand color confinement in this picture? The idea is simple: when quarks form color bound states, they are connected by color flux tubes (this is the aspect of confinement which goes outside QCD). Also color flux tubes possess huge spin glass degeneracy. Free quark states do not possess this degeneracy since join along boundaries bonds are absent. Thus the time spent in free states in which color flux tubes are absent is negligible to the time time spent in color bound states so that the states consisting of free quarks are unobservable. If this picture is correct, the divergence of the color coupling strength in confinement length scale reflects mathematically the fact that number of bound states is overwhelmingly large as compared that for the free states.

Color confinement from unitarity and spin glass degeneracy

A more precise phrasing of the idea about the connection between spin glass degeneracy and color confinement relies on unitarity conditions and the assumptions $T_{MN} \simeq T$ and $T_{Mr} \simeq T_r$. Here capital subscripts refer to degenerate hadronic states and small letter subscripts to free many-quark states. In this idealization hadronic degenerate states are stable against decay to free many-quark states with only single exception. The exceptional state should act as a doorway making possible the transition to quark-gluon plasma phase.

The S-matrix can be written as sum of unit matrix and reaction matrix T: $S = 1 + iT$.

a) The unitarity conditions $SS^\dagger = 1$ read in terms of T-matrix as

$$i(T - T^\dagger) = TT^\dagger \ . \tag{9.5.8}$$

For diagonal elements one has

$$2 \times Im(T_{mm}) \ = \ \sum_r |T_{mr}|^2 \geq 0 \ . \tag{9.5.9}$$

What is essential that the right hand side is non-negative and closely related to the total rate of transitions. If this rate is high also the imaginary part at the left hand side of the equation is large and therefore also the rate for the diagonal transition. For instance, in the case of low energy strong interactions this implies that the total reaction rates are high but transitions occur mostly in the forward direction. In this case the mere large number of final many-hadron states implies that most transitions occur in the forward direction.

In the recent case one must consider both free many quark states and their bound states. Let us use capitals M, N as labels for bound states and small letters m, n as labels for free states.

a) The diagonal unitarity conditions can be written for both of these states as

$$
\begin{aligned}
2Im(T_{mm}) &= \sum_r |T_{mr}|^2 + \sum_R |T_{mR}|^2 \geq 0 \ , \\
2Im(T_{MM}) &= \sum_R |T_{MR}|^2 + \sum_r |T_{Mr}|^2 \geq 0 \ .
\end{aligned}
\tag{9.5.10}
$$

In both cases there is a large number of the degenerate states involved at the right hand side so that one expects that the right hand side has a large value. For bound states the number of degenerate states is much higher due to the additional degeneracy brought in by the join along boundaries bonds (color flux tubes). Thus the lifetime and de-coherence time should be considerably longer than expected on basis of standard physics.

b) For the non-diagonal transitions from bound states to free states one has

$$
i(T_{Mm} - \overline{T}_{mM}) = \sum_r T_{Mr}\overline{T}_{mr} + \sum_R T_{MR}\overline{T}_{mR} \ .
\tag{9.5.11}
$$

The right hand side is not positive definite and since a large number of amplitudes between widely different free and bound states of quarks are involved, one expects that a destructive interference occurs. This is consistent with a small value of the non-diagonal amplitudes T_{Mm} and with the long lifetime of bound states.

c) What happens for non-diagonal transitions between degenerate states? The unitarity conditions read as

$$
\begin{aligned}
i(T_{mn} - \overline{T}_{nm}) &= \sum_r T_{mr}\overline{T}_{nr} + \sum_r T_{mR}\overline{T}_{nR} \ , \\
i(T_{MN} - \overline{T}_{NM}) &= \sum_R T_{MR}\overline{T}_{NR} + \sum_r T_{Mr}\overline{T}_{Nr} \ .
\end{aligned}
\tag{9.5.12}
$$

The right hand side is not anymore positive definite and there is a very large number of summands present. Hence a destructive interference could occur and the amplitude would be very strongly restricted in the forward direction. This need not however be true in the case of degenerate states since they are expected to be very similar to each other.

d) One can indeed play with the idealization that the transition amplitudes between degenerate states are identical $T_{MN} = T$ and that the amplitudes T_{Mr} are independent of M and given by $T_{Mr} = T_r$.

In this case T-matrix would have the form $T = t \times X$, where X is a matrix for which all elements are equal to one. t can be written as $|t|exp(i\phi)$. T-matrix is maximally degenerate and the diagonalized form T^D of T-matrix has only a single non-vanishing element equal to Nt, N the number of degenerate states. t must satisfy the unitarity condition $|t| = 2 \times sin(\phi)/N$. S-matrix would reduce to an almost unit matrix for the diagonalized bound states.

What about the stability of the bound states in this case? The decay amplitudes for bound states corresponding to the vanishing eigen values of T are given by $T^D(M, r) = \sum c_M T_{Mr} = \sum_M c_M \times T_r = 0$ by the orthogonality of these states with the state with a non-vanishing eigen value. Thus the lifetimes of all bound states expect the one with the non-vanishing eigen value of T are infinitely long in this idealization.

9.6 TGD based view about dark matter

A rather unexpected support for the macroscopic quantum coherence comes from the work of D. Da Rocha and Laurent Nottale who have proposed that Schrödinger equation with Planck constant $\hbar$

replaced with what might be called gravitational Planck constant $\hbar_{gr} = \frac{GmM}{v_0}$ ($\hbar = c = 1$). v_0 is a velocity parameter having the value $v_0 = 144.7 \pm .7$ km/s giving $v_0/c = 4.6 \times 10^{-4}$. This is rather near to the peak orbital velocity of stars in galactic halos. Also subharmonics and harmonics of v_0 seem to appear. The support for the hypothesis coming from empirical data is impressive.

Nottale and Da Rocha believe that their Schrödinger equation results from a fractal hydrodynamics. Many-sheeted space-time however suggests astrophysical systems are not only quantum systems at larger space-time sheets but correspond to a gigantic value of gravitational Planck constant. The gravitational (ordinary) Schrödinger equation would provide a solution of the black hole collapse (IR catastrophe) problem encountered at the classical level. The resolution of the problem inspired by TGD inspired theory of living matter is that it is the dark matter at larger space-time sheets which is quantum coherent in the required time scale.

I have proposed already earlier the possibility that Planck constant is quantized and the spectrum is given in terms of logarithms of Beraha numbers: the lowest Beraha number B_3 is completely exceptional in that it predicts infinite value of Planck constant. The inverse of the gravitational Planck constant could correspond a gravitational perturbation of this as $1/\hbar_{gr} = v_0/GMm$. The general philosophy would be that when the quantum system would become non-perturbative, a phase transition increasing the value of $\hbar$ occurs to preserve the perturbative character and at the transition $n = 4 \to 3$ only the small perturbative correction to $1/\hbar(3) = 0$ remains. This would apply to QCD and to atoms with $Z > 137$ as well.

TGD predicts correctly the value of the parameter v_0 assuming that cosmic strings and their decay remnants are responsible for the dark matter. The harmonics of v_0 can be understood as corresponding to perturbations replacing cosmic strings with their n-branched coverings so that tension becomes n^2-fold: much like the replacement of a closed orbit with an orbit closing only after n turns. $1/n$-subharmonic would result when a magnetic flux tube split into n disjoint magnetic flux tubes.

The rather amazing coincidences between basic bio-rhythms and the periods associated with the states of orbits in solar system suggest that the frequencies defined by the energy levels of the gravitational Schrödinger equation might entrain with various biological frequencies such as the cyclotron frequencies associated with the magnetic flux tubes. For instance, the period associated with $n = 1$ orbit in the case of Sun is 24 hours within experimental accuracy for v_0.

9.6.1 Dark matter as macroscopic quantum phase with gigantic Planck constant

D. Da Rocha and Laurent Nottale, the developer of Scale Relativity, have ended up with an highly interesting quantum theory like model for the evolution of astrophysical systems [if1] (I am grateful for Victor Christianto for informing me about the article). The model is simply Schrödinger equation with Planck constant $\hbar$ replaced with what might be called gravitational Planck constant

$$\hbar \quad \to \quad \hbar_{gr} = \frac{GmM}{v_0} \quad . \tag{9.6.1}$$

Here I have used units $\hbar = c = 1$. v_0 is a velocity parameter having the value $v_0 = 144.7 \pm .7$ km/s giving $v_0/c = 4.6 \times 10^{-4}$. The peak orbital velocity of stars in galactic halos is 142 ± 2 km/s whereas the average velocity is 156 ± 2 km/s. Also subharmonics and harmonics of v_0 seem to appear.

The model makes fascinating predictions which hold true. For instance, the radii of planetary orbits fit nicely with the prediction of the hydrogen atom like model. The inner solar system (planets up to Mars) corresponds to v_0 and outer solar system to $v_0/5$. The predictions for the distribution of major axis and eccentrities have been tested successfully also for exo-planets. Also the periods of 3 planets around pulsar PSR B1257+12 fit with the predictions with a relative accuracy of few hours/per several months. Also predictions for the distribution of stars in the regions where morphogenesis occurs follow from the Schödinger equation.

What is important is that there are no free parameters besides v_0. In [if1] a wide variety of astrophysical data is discussed and it seem that the model works and has already now made predictions which have been later verified. A rather detailed model for the formation of solar system making quantitatively correct predictions follows from the study of inclinations and eccentricities predicted by the Bohr rules: the model proposed seems to differ from that of Nottale which makes predictions for the probability distribution of eccentricities and inclinations.

I have proposed already earlier [O4] the possibility that Planck constant is quantized and the spectrum is given in terms of logarithms of Beraha numbers: the lowest Beraha number B_3 is completely exceptional in that it predicts infinite value of Planck constant. The inverse of the gravitational Planck constant could correspond a gravitational perturbation of this as $1/\hbar_{gr} = v_0/GMm$. The general philosophy would be that when the quantum system would become non-perturbative, a phase transition increasing the value of $\hbar$ occurs to preserve the perturbative character and at the transition $n = 4 \to 3$ only the small perturbative correction to $1/\hbar(3) = 0$ remains. This would apply to QCD and to atoms with $Z > 137$ as well.

TGD predicts correctly the value of the parameter v_0 assuming that cosmic strings and their decay remnants are responsible for the dark matter. The harmonics of v_0 can be understood as corresponding to perturbations replacing cosmic strings with their n-branched coverings so that tension becomes n^2-fold: much like the replacement of a closed orbit with an orbit closing only after n turns. $1/n$-subharmonic would result when a magnetic flux tube split into n disjoint magnetic flux tubes.

The most interesting predictions from the point of view of living matter are following.

a) The dark matter is still there and forms quantum coherent structures of astrophysical size. In particular, the (Z^0) magnetic flux tubes associated with the planetary orbits define this kind of structures. The enormous value of h_{gr} makes the characteristic time scales of these quantum coherent states extremely long and implies macro-temporal quantum coherence in human and even longer time scales.

b) The rather amazing coincidences between basic bio-rhythms and the periods associated with the orbits in solar system suggest that the frequencies defined by the energy levels of the gravitational Schrödinger equation might entrain with various biological frequencies such as the cyclotron frequencies associated with the magnetic flux tubes. For instance, the period associated with $n = 1$ orbit in the case of Sun is 24 hours within experimental accuracy for v_0. Second example is the mysterious 5 second time scale associated with the Comorosan effect [la10, la11].

9.6.2   How the scaling of $\hbar$ affects physics?

It is relatively easy to deduce the basic implications of the scaling of $\hbar$.

a) If the rate for the process is non-vanishing classically, it is not affected in the lowest order. For instance, scattering cross sections for say electron-electron scattering and e^+e^- annihilation are not affected in the lowest order since the increase of Compton length compensates for the reduction of α_{em}. Photon-photon scattering cross section, which vanishes classically and is proportional to $\alpha_{em}^4 \hbar^2/E^2$, scales down as $1/\hbar^2$.

b) Higher order corrections coming as powers of the gauge coupling strength α are reduced since $\alpha = g^2/4\pi\hbar$ is reduced. Since one has $\hbar_s/\hbar = \alpha Q_1 Q_2/v_0$, $\alpha Q_1 Q_2$ is effectively replaced with a universal coupling strength v_0. In the case of QCD the paradoxical sounding implication is that α_s would become very small.

c) The binding energy scale $E \propto \alpha_{em}^2 m_e$ of atoms scales as $1/\hbar^2$ so that a partially dark matter for which protons have large value of $\hbar$ does not interact appreciably with the visible light. Scaled down spectrum of binding energies would be the experimental signature of dark matter. The resulting atomic spectrum is universal and binding energy scale $\alpha^2 m_e$ is replaced with $v_0^2 m_e$ which corresponds to $\sim .115$ eV and wavelength of $\simeq 10.78$ μm, a typical size of cell. Bohr radius is 12.2 nm for dark hydrogen atom whereas the thickness of cell membrane is about 10 nm. It would be amazing if living matter would exhibit scaled down atomic spectra with this universal energy scale.

9.6.3   Simulating big bang in laboratory

An important steps in the development of ideas were stimulated by the findings made during period 2002-2005 in Relativist Heavy Ion Collider (RHIC) in Brookhaven compared with the finding of America and for full reason.

a) The first was finding of longitudinal Lorentz invariance at single particle level suggesting a collective behavior. This was around 2002.

b) The collective behavior which was later interpreted in terms of color glass condensate meaning the presence of a blob of liquid like phase decaying later to quark gluon plasma since it was found that the density of what was expected to be quark gluon plasma was about ten times higher than expected.

c) The last finding is that this object seems to absorb partons like black hole and behaves like evaporating black hole.

In my personal Theory Universe the history went as follows.

a) I proposed 2002 a model for Gold-Gold collision as a mini big bang identified as a scaled down variant of TGD inspired cosmology. This makes sense because in TGD based critical cosmology the initial state has vanishing mass per comoving volume instead of being infinite as in radiation dominated cosmology. Any phase transition involving a generation of a new space-time sheet might proceed in this universal manner.

b) Cosmic string soup in the primordial stage is replaced by a tangle of color flux tubes containing the color glass condensate. CGC is made macroscopic quantum phase by conformal confinement (the conformal weights of partons are complex and relate to zeros of zeta) and only the net conformal weight is real in this phase). Flux tubes correspond to flow lines of incompressible liquid flow and non-perturbative phase with a very large $\hbar$ is in question. Gravitational constant is replaced by strong gravitational constant defined by the relevant p-adic length scale squared since color flux tubes are analogs of hadronic strings. Presumably L_p, $p = M_1 07 = 2^1 07 - 1$, is the p-adic length scale since Mersenne prime M_{107} labels the space-time sheet at which partons feed their color gauge fluxes. Temperature during this phase could correspond to Hagedorn temperature for strings and is determined by string tension. Density would be maximal.

c) Next phase is critical phase in which the notion of space-time in ordinary sense makes sense and 3-space is flat since there is no length scale in critical system (so that curvature vanishes). During this critical phase phase a transition to quark gluon plasma occurs. The duration of this phase fixes all relevant parameters such as temperature (which is the analog of Hagedorn temperature corresponding since critical density is maximal density of gravitational mass in TGD Universe).

d) The next phase is radiation dominated quark gluon plasma phase and then follows hadronization to matter dominated phase provided cosmological picture still applies.

Since black hole formation and evaporation is very much like formation big crunch followed by big bang, the picture is more or less equivalent with the picture in which black hole like object consisting of string like objects (mass is determined by string length just as it is determined by the radius for black holes) is formed and then evaporates. Black hole temperature corresponds to Hagedorn temperature and to the duration of critical period of the mini cosmology.

9.6.4   Living matter as dark matter

The most important gift of RIIIC was that several theoretical notions and ideas emerged during last years, and applying in hugely different length and time scales by p-adic fractality, integrate nicely.

a) Dark matter is identified as a macroscopic quantum phase with large $\hbar$ for which particles have complex conformal weights and by conformal confinement behaves like single coherent whole. Dark matter controls living matter and this explains the weird looking findings about Bohr rules for planetary orbits.

b) Living matter would be also matter with large value of $\hbar$ and form conformally confined blobs behaving like single units with extremely quantal properties, including free will of course! Dark matter would be responsible for the mysterious vital force.

c) Any system for which some interaction becomes so strong that perturbation theory does not work gives rise to this kind of system in a phase transition in which $\hbar$ increases to not lose perturbativity gives rise to this kind of "super-quantal" matter.

d) Physically $\hbar$ means a larger unit for quantum numbers and this requires that single particle states form larger particle like units. This kind of collective states with weak mutual interactions are of course very natural in strongly interacting systems. At the level of quantum jumps quantum jumps integrate effectively to single quantum jump and longer moments of consciousness result. Conformal confinement guarantees all this. Entire hierarchy of size scales for conformally confined blobs is predicted corresponding to values of $\hbar$ related to Beraha numbers but there would be only single value corresponding to very large $\hbar$ for given values of system parameters (gravitational masses, charges,...). The value of $\hbar$ determines the characteristic time and length scales associated with the conscious living system. One could say that the claim that quantum mechanics in its recent form is not enough for understanding living matter is correct: dynamical $\hbar$ is needed.

e) The picture might have implications also for the understanding of condensed matter. For instance, liquids might be liquids because they contain dark some matter at magnetic/Z^0 magnetic flux tubes (darkness follows from the large value of $\hbar$). A rather detailed model for dark super nuclei consisting of a blob of dark nuclei and for dark atoms emerges if the criticality condition $NZ\alpha_{em}/v_0 \simeq 1$ is assumed to fix the number of nuclei of dark super-atom. The basic implication is that the energy levels of dark super atoms depend only weakly on the atomic charge.

f) The facts that that dark atoms have Bohr radii in the range .2-.8 mm, which corresponds to the size scales for the basic neuronal moduli of cortex, and the wavelength associated with the ionization energies is of order 9 cm (the size of brain hemisphere) plus many other co-incidences suggest that dark matter could play key role in grey matter.

9.6.5 Anti-matter and dark matter

The usual view about matter anti-matter asymmetry is that during early cosmology matter-antimatter asymmetry characterized by the relative density difference of order $r = 10^{-9}$ was somehow generated and that the observed matter corresponds to what remained in the annihilation of quarks and leptons to bosons. A possible mechanism inducing the CP asymmetry is based on the CP breaking phase of CKM matrix.

The TGD based view about energy [D3, D5] forces the conclusion that all conserved quantum numbers including the conserved inertial energy have vanishing densities in cosmological length scales. Therefore fermion numbers associated with matter and antimatter must compensate each other. Therefore the standard option is definitely excluded in TGD framework.

An early TGD based scenario explains matter antimatter asymmetry by assuming that antimatter is in vapor phase. This requires that matter and antimatter have slightly different topological evaporation rates with the relative difference of rates characterized by the parameter r. A more general scenario assumes that matter and antimatter reside at different space-time sheets. The reader can easily guess the next step. The strict non-observability of antimatter finds an elegant explanation if anti-matter is dark matter.

9.6.6 Are long ranged classical electro-weak and color gauge fields created by dark matter?

The various model for the screening of electro-weak charges discussed during years cannot banish the unpleasant feeling that the screening cannot be complete enough to eliminate large parity breaking effects in atomic length scales The really elegant manner to avoid various catastrophes caused by long range electro-weak gauge fields could emerge only after a considerable increase in the understanding of the mathematical structure of TGD and the emergence of a view about what dark matter is.

p-Adic length scale hypothesis suggests the possibility that both electro-weak gauge bosons and gluons can appear as effectively massless particles in several length scales and there indeed exists evidence that neutrinos appear in several scaled variants [ha20] (for TGD based model see [F3]).

This inspires the working hypothesis that long range classical electro-weak gauge and gluon fields arc space-time correlates for light or massless dark electro-weak gauge bosons and gluons, which are massless.

a) Ordinary quarks and leptons would be essentially identical with their standard model counterparts with electro-weak charges screened in electro-weak length scale so that the problems related to the smallness of atomic parity breaking would be trivially resolved.

b) TGD suggests an explanation of dark matter as a macroscopically quantum coherent phase residing at larger space-time sheets [J6]. The particles of dark matter carry complex conformal weights but the net conformal weights for blocks of dark matter would be real. This implies conformal confinement and macroscopic quantum coherence. TGD suggests that $\hbar$ is dynamical and possesses a spectrum expressible in terms of generalized Beraha numbers $B_r = 4cos^2(\pi/r)$, where $r > 3$ is a rational number [O5, J6]. Just above $r = 3$ arbitrarily large values of $\hbar$ and thus also macroscopic quantum phases are possible.

c) For this scenario to make sense it is essential that p-adic thermodynamics predicts for dark quarks and leptons essentially the same masses as for their ordinary counterparts [F3]. Only the electro-weak boson masses which are determined by a different mechanism than the dominating contribution to fermion masses [F2, F3] would be small or vanishing.

d) The hypothesis that classical long ranged electro-weak gauge fields serve as classical space-time correlates for dark electro-weak gauge bosons, which are massless, could explain the special properties of bio-matter, in particular the chiral selection as resulting from the coupling to dark Z^0 quanta. Long range weak forces present in TGD counterpart of Higgs=0 phase should allow to understand the differences between biochemistry and the chemistry of dead matter.

e) For ordinary condensed matter quarks and leptons Z^0 charge would be screened in electro-weak length scale whereas in dark matter phase Z^0 electric flux would be feeded to say $k = k_Z = 169$ space-time sheet corresponding to neutrino Compton length and having size of cell. In condensed matter blobs of size larger than neutrino Compton length (about 5 μm if $k = 169$ determines the p-adic length scale of condensed matter neutrinos) the situation could be different.

The facts that parity breaking effects are strong in living matter and the p-adic length scales $k = 151, 157, 163, 167$ spanning the range between 10 nm (cell membrane thickness) and 2.5 μm cell size correspond to Gaussian Mersennes [K2], suggest that dark neutrinos are condensed at these space-time sheets. An interesting possibility is that dark matter phase quite generally condenses at the space-time sheets corresponding to Gaussian primes, in particular Gaussian Mersennes, which are much more abundant than ordinary Mersennes. Perhaps the fact that conformal weights are complex corresponds to the complex character of the primes involved. Only a fraction of the condensed matter consisting of regions of size $L(k)$ need to be in the dark phase.

Nuclear length scale $L(113)$ corresponds to Gaussian Mersenne: this raises the possibility that strong force binding nucleons can increase the value of $\hbar$ so that perturbation theory applies. By the general considerations of [J6] this would make nucleons atom sized structures so that ordinary nuclear nucleons cannot be dark. Nuclear scattering cross sections would not be however changed in the lowest order perturbation theory. In [F8, J6] a model of cold fusion based on the assumption that nuclear protons transform to dark protons is discussed.

The basic prediction of TGD based model of dark matter as a phase with a large value of Planck constant is the scaling up of various quantal length and time scales. A simple quantitative model for condensed matter with large value of $\hbar$ predicts that $\hbar$ is by a factor $\sim 2^{11}$ determined by the ratio of CP_2 length to Planck length larger than in ordinary phase meaning that the size of dark neutrons would be of order atomic size. In this kind of situation single order parameter would characterize the behavior of dark neutrinos and neutrons and the proposed model could apply as such also in this case.

Dark photon many particle states behave like laser beams decaying to ordinary photons by decoherence meaning a transformation of dark photons to ordinary ones. Also dark electro-weak bosons and gluons would be massless or have small masses determined by the p-adic length scale in question. The decay products of dark electro-weak gauge bosons would be ordinary electro-weak bosons decaying rapidly via virtual electro-weak gauge boson states to ordinary leptons. Topological light rays ("massless extremals") for which all classical gauge fields are massless are natural space-time correlates for the dark boson laser beams. Obviously this means that the basic difference between the chemistries of living and non-living matter would be the absence of electro-weak symmetry breaking in living matter (which does not mean that elementary fermions would be massless).

In conformally confined phase phase Fermi statistics allows neutrinos to have same energy if their conformal weights are different so that a kind "fermionic Bose-Einstein condensate" would be in question. If both nuclear neutrons and neutrinos are in dark phase, it is possible to achieve a rather complete local cancellation of Z^0 charge density. On the other hand, the large parity breaking effects in living matter dramatically manifesting themselves in bio-catalysis, suggest that Bose-Einstein condensates of dark electro-weak gauge bosons could appear already in molecular length scales.

9.7 Coupling constant evolution at space-time level

It is not yet possible to deduce the length scale evolution gauge coupling constants from Quantum TGD proper. Quantum classical correspondence however encourages the hope that it might be possible to achieve some understanding of the coupling constant evolution by using the classical theory.

This turns out to be the case and the earlier speculative picture about gauge coupling constants associated with a given space-time sheet as RG invariants finds support. It remains an open question whether gravitational coupling constant is RG invariant inside give space-time sheet. The discrete p-adic coupling constant evolution replacing in TGD framework the ordinary RG evolution allows also formulation at space-time level as also does the evolution of $\hbar$ associated with the phase resolution.

9.7.1   Overview

The evolution of gauge couplings at single space-time sheet

The renormalization group equations of gauge coupling constants g_i follow from the following idea. The basic observation is that gauge currents have vanishing covariant divergences whereas ordinary divergence does not vanish except in the Abelian case. The classical gauge currents are however proportional to $1/g_i^2$ and if g_i^2 is allowed to depend on the space-time point, the divergences of currents can be made vanishing and the resulting flow equations are essentially renormalization group equations. The physical motivation for the hypothesis is that gauge charges are assumed to be conserved in perturbative QFT. The space-time dependence of coupling constants takes care of the conservation of charges.

A surprisingly detailed view about RG evolution emerges.

a) The UV fixed points of RG evolution correspond to CP_2 type extremals (elementary particles).

b) The Abelianity of the induced Kähler field means that Kähler coupling strength is RG invariant which has indeed been the basic postulate of quantum TGD. The only possible interpretation is that the coupling constant evolution in sense of QFT:s corresponds to the discrete p-adic coupling constant evolution.

c) IR fixed points correspond to space-time sheets with a 2-dimensional CP_2 projection for which the induced gauge fields are Abelian so that covariant divergence reduces to ordinary divergence. Examples are cosmic strings (, which could be also seen as UV fixed points), vacuum extremals, solutions of a sub-theory defined by $M^4 \times S^2$, S^2 a homologically non-trivial geodesic sphere, and "massless extremals".

d) At the light-like boundaries of the space-time sheet gauge couplings are predicted to be constant by conformal invariance and by effective two-dimensionality implying Abelianity: note that the 4-dimensionality of the space-time surface is absolutely essential here.

e) In fact, all known extremals of Kähler action correspond to RG fixed points since gauge currents are light-like so that coupling constants are constant at a given space-time sheet. This is consistent with the earlier hypothesis that gauge couplings are renormalization group invariants and coupling constant evolution reduces to a discrete p-adic evolution. As a consequence also Weinberg angle, being determined by a ratio of $SU(2)$ and $U(1)$ couplings, is predicted to be RG invariant. A natural condition fixing its value would be the requirement that the net vacuum em charge of the space-time sheet vanishes. This would state that em charge is not screened like weak charges.

f) When the flow determined by the gauge current is not integrable in the sense that flow lines are identifiable as coordinate curves, the situation changes. If gauge currents are divergenceless for all solutions of field equations, one can assume that gauge couplings are constant at a given space-time sheet and thus continuous also in this case. Otherwise a natural guess is that the coupling constants obtained by integrating the renormalization group equations are continuous in the relevant p-adic topology below the p-adic length scale. Thus the effective p-adic topology would emerge directly from the hydrodynamics defined by gauge currents.

RG evolution of gravitational constant at single space-time sheet

Similar considerations apply in the case of gravitational and cosmological constants.

a) In this case the conservation of gravitational mass determines the RG equation (gravitational energy and momentum are not conserved in general).

b) The assumption that coupling cosmological Λ constant is proportional to $1/L_p^2$ (L_p denotes the relevant p-adic length scale) explains the mysterious smallness of the cosmological constant and leads to a RG equation which is of the same form as in the case of gauge couplings.

c) Asymptotic cosmologies for which gravitational four momentum is conserved correspond to the fixed points of coupling constant evolution now but there are much more general solutions satisfying the constraint that gravitational mass is conserved.

d) It seems that gravitational constant cannot be RG invariant in the general case and that effective p-adicity can be avoided only by a smoothing out procedure replacing the mass current with its average over a four-volume 4-volume of size of order p-adic length scale.

p-Adic evolution of gauge couplings

If RG invariance at given space-time sheet holds true, the question arises whether it is possible to understand p-adic coupling constant evolution at space-time level.

a) Simple considerations lead to the idea that M^4 scalings of the intersections of 3-surfaces defined by the intersections of space-time surfaces with light-cone boundary induce transformations of space-time surface identifiable as RG transformations. If sufficiently small they leave gauge charges invariant: this seems to be the case for known extremals which form scaling invariant families. When the scaling corresponds to a ratio p_2/p_1, $p_2 > p_1$, bifurcation would become possible replacing p_1-adic effective topology with p_2-adic one.

b) Stability considerations determine whether p_2-adic topology is actually realized and could explain why primes near powers of 2 are favored. The renormalization of coupling constant would be dictated by the requirement that Q_i/g_i^2 remains invariant.

p-Adic evolution in angular resolution and dynamical $\hbar$

For a given p-adic topology algebraic extensions of p-adic numbers define also a hierarchy ordered by the dimension of the extension and this hierarchy naturally corresponds to an increasing angular resolution so that RG flow would be associated also with it.

a) A characterization of angular scalings consistent with the identification of $\hbar$ as a characterizer of the topological condensation of 3-surface X^3 to a larger 3-surface Y^3 is that angular scalings correspond to the transformations $\Phi \to r\Phi$, $r = m/n$ in the case of X^3 and $\Phi \to \Phi$ in case of Y^3 so that X^3 becomes analogous to an m-fold covering of Y^3. Rational coverings could also correspond to m-fold scalings for X^3 and n-fold scalings for Y^3.

b) The formation of these stable multiple coverings could be seen as an analog for a transition in chaos via a process in which a closed Bohr orbit regarded as a particle itself becomes an orbit closing only after m turns. TGD predicts a hierarchy of higher level zero energy states representing S-matrix of lower level as entanglement coefficients. Particles identified as "tracks" of particles at orbits closing after m turns might serve as space-time correlates for this kind of states. There is a direct connection with the fractional quantum numbers, anyon physics and quantum groups.

c) The simplest generalization from the p-adic length scale evolution consistent with the proposed role of Beraha numbers $B_n = 4cos^2(\pi/n)$ is that bifurcations can occur for integer values of r=m and change the value of $\hbar$. The interpretation would be that single 2π rotation in δM_+^4 corresponds to the angular resolution with respect to the angular coordinate ϕ of space-time surface varying in the range $(0, 2\pi)$ and is given by $\Delta\phi = 2\pi/m$.

d) For $n = 3$ corresponding to the minimal resolution of $\Delta\phi = 2\pi/3$ $\hbar$ would be infinite. The evidence for a gigantic but finite value of "gravitational" Planck constant [J6] would mean that the simplest formula

$$\frac{1}{\hbar(n)} = \frac{log(B_n)}{log(4)}$$

for $\hbar$ fails for $n = 3$.

The first cure of the problem would be a replacement of the formula for $\hbar(n)$ by a difference equation

$$\frac{1}{\hbar(n)} - \frac{1}{\hbar(n-1)} = \frac{log(B_n)}{log(4)} - \frac{log(B_{n-1})}{log(4)}$$

having interpretation as RGE difference equation and allowing additive constant in the expression of $1/\hbar(n)$ and thus yielding finite value for $\hbar(3)$.

A more elegant resolution of the problem is that for a given n characterizing von Neumann inclusion there is spectrum of values for $\hbar(r = n/m)$ expressible in terms of $B_r = 4cos^2(\pi/r)$ as

$$\frac{1}{\hbar(n/m)} = \frac{log(B_{n/m})}{log(4)}$$

such that $m/n < 3$ holds true. This would reflect the presence of an additional degree of freedom related to the Jones inclusion. m could characterize the scaling of Φ for X^3 and n the scaling of

Φ for Y^3. A simple TGD inspired model for dark atoms and dark condensed matter [J6] predicts $\hbar/\hbar_0 = 1/v_0 \simeq 2^{11}$. This would correspond to $r \simeq .3077$.

9.7.2 The evolution of gauge and gravitational couplings at space-time level

The question is whether the RG evolution of all coupling constant parameters could have interpretation as flows at space-time level. This seems to be the case.

Renormalization group flow as a conservation of gauge current in the interior of space-time sheet

The induced gauge potentials relate to the gauge potentials A_i of perturbative gauge theory by the scaling $g_i \to g_i A_i$. Hence the gauge currents correspond to the scaled currents

$$
\begin{aligned}
J_i^\mu &= \frac{1}{g_i^2} \times J_{i,0}^\mu \ , \\
J_{i,0}^\mu &= (D_\nu F^{\mu\nu})_i \sqrt{g} \ .
\end{aligned}
\tag{9.7.1}
$$

The simplest guess for the coupling constant evolution associated with g_i^2 is that the covariant gauge current J_i^a is conserved in ordinary sense (its is identically conserved in covariant sense). This gives meaning to the perturbative approach in which gauge charges are indeed conserved. Thus one would have:

$$
\partial_\mu J_i^\mu = 0 \ .
\tag{9.7.2}
$$

or

$$
J_{i,0}^\mu \partial_\mu log(g_i^2) = \partial_\mu J_{i,0}^\mu \ .
\tag{9.7.3}
$$

Note that the non-constancy of the Weinberg angle gives an additional term to the em current given by

$$
\frac{1}{2} Z_0^{\mu\nu} \partial_\nu p \ .
\tag{9.7.4}
$$

This equation can be solved along the flow lines of the gauge current. When the flow is integrable:

$$
J_{i,0}^\mu = \phi \partial^\mu t \ ,
$$

one obtains

$$
\frac{dlog(g_i^2)}{dt} = \frac{\partial_\mu J_{i,0}^\mu}{\phi} = \nabla(log(\phi)) \cdot \nabla t + \nabla^2 t \ .
\tag{9.7.5}
$$

When this flow is not integrable coupling constants become discontinuous functions with respect to the real topology but can be continuous or even smooth with respect to some p-adic topology and the previous discussion applies as such.

The ordinary divergence of the gauge current takes the role of beta function. RG evolution is trivial in the Abelian case since in this case ordinary divergence vanishes identically. This implies that Kähler coupling strength is indeed renormalization group invariant which has been the basic hypothesis of quantum TGD.

The natural boundary conditions to the coupling constant evolution state the vanishing of the normal components of the gauge currents at boundaries

$$J_i^n \;=\; \frac{D_\beta F_i^{n\beta}\sqrt{g_4}}{g_i^2} = 0 \;. \tag{9.7.6}$$

and guarantee that the flow approaches asymptotically the boundaries. These conditions can become trivial if the four-metric at the boundary component becomes singular (effectively 2-dimensional) so that $D_\beta F_i^{n\beta}$ can approach to finite or even infinite value. This might happen in case of color gauge coupling strength if it approaches infinity near the boundary. Otherwise the conditions says nothing about coupling constants at the boundary.

Is the renormalization group evolution at the light-like boundaries trivial?

One can ask whether it is possible to define coupling constant evolution also for the gauge fields induced at light-like boundary components. The technical problems are caused by the vanishing of the determinant of the induced metric and the non-existence of contravariant metric but it is quite conceivable that the restriction to the 2-dimensional sections makes sense if one defines a contravariant metric as the inverse of the induced metric in the 2-D section.

Since CP_2 projection is 2-dimensional, RG equations suggest that coupling constants are constants on the 2-dimensional sections and that conformal invariance in the light-like direction implies constancy over the entire boundary component. Since boundary components are identifiable as parton like objects, the result would look highly satisfactory.

If the right hand side of Eq. 9.7.5 vanishes at the boundary of space-time surface g_i^2 approaches to a finite value. When the left hand side is finite and t becomes infinite as boundary is approached g_i^2 increases without limit. This happens for a finite value of t when the right hand side diverges. Classical color gauge fields are proportional to $H_A J$, where H_A are the Hamiltonians of the color isometries and J denotes the induced Kähler form. The non-triviality of renormalization group evolution is solely due to the presence of Hamiltonians. QCD suggests that α_s diverges at the outer boundary or that at least approaches to a very large value at the outer boundaries of the hadronic 4-surface.

Fixed points of coupling constant evolution

Consider now the fixed points of the coupling constant evolution.

a) The first class of fixed points corresponds to CP_2 type extremals. In this case however also gauge currents vanish so that the RG equation says nothing.

b) The second class of fixed points of the coupling constant evolution corresponds to space-time regions in which gauge fields become Abelian. This is the case for all space-time surfaces with 2-dimensional CP_2 projection: this includes vacuum extremals, massless extremals, solutions for which CP_2 projection corresponds to a homologically non-trivial geodesic sphere, and cosmic strings. This supports the view that these extremals correspond to asymptotic self-organization patters.

Are all gauge couplings RG invariants within a given space-time sheet

No extremals for which the gauge currents would have non-vanishing ordinary divergence are known at this moment (gauge currents are light-like always). Therefore one cannot exclude the possibility that all gauge coupling constants rather than only Kähler coupling strength are renormalization group invariants in TGD framework, so that the hypothesis that RG evolution reduces to a discrete p-adic coupling constant evolution would be correct.

This implies that also Weinberg angle, being determined by the ratio of $SU(2)$ and $U(1)$ couplings, is constant inside a given space-time sheet. Its value in this case is determined most naturally by the requirement that the net vacuum em charge of the space-time sheet vanishes.

The fixed point property as an implication of Abelianity is obviously in conflict with the standard picture about gauge coupling evolution and supports the view that this evolution corresponds to a discrete p-adic gauge coupling evolution.

RG equation for gravitational coupling constant

In the case of gravitational coupling constant the renormalization group equation must be formulated the current representing the contribution of Einstein tensor to the gravitational mass being defined by Einstein tensor as

$$G^\alpha \;\; = \;\; \frac{1}{16\pi G} \times G^{\alpha\beta}\partial_\beta a\sqrt{g} \; , \tag{9.7.7}$$

where a refers the proper time of future light cone (or possibly to some other preferred time coordinate determined by dynamics). In the case of cosmological constant the corresponding contribution is

$$g^\alpha \;\; = \;\; \frac{\Lambda}{16\pi G} \times g^{\alpha\beta}\partial_\beta a\sqrt{g} \; . \tag{9.7.8}$$

A natural hypothesis is that the variation of G guarantees the conservation of gravitational mass. This does not mean that gravitational energy or four-momentum would be conserved or that conservation of gravitational mass would hold true except at a given space-time sheet. One can also assume that the two contributions to the gravitational mass are not independent. This means that there is a constraint between cosmological and gravitational constants. There are two options.

a) One has

$$\Lambda \;\; = \;\; \frac{x}{G} \; . \tag{9.7.9}$$

where x is renormalization group invariant of no other length scales are involved. The RG equation would in this case read as

$$\left(G^\alpha - \frac{2x}{G}g^\alpha\right) D_\alpha log(G) = D_\alpha\left(G^\alpha + \frac{x}{G}g^\alpha\right) \; . \tag{9.7.10}$$

b) On the other hand, if p-adic length scale hypothesis is accepted, one has

$$\Lambda \;\; = \;\; \frac{x}{L_p^2} \; , \tag{9.7.11}$$

where L_p is a p-adic length scale of order of cosmic time a: $L_p \sim a$ [D5]. This would mean that Λ is RG invariant. This option resolves the mysterious smallness of the cosmological constant so that it is the most plausible option in TGD framework.

The RG equations in this case is given by

$$\left(G^\alpha + \frac{x}{L_p^2}g^\alpha\right) D_\alpha log(G) = D_\alpha\left(G^\alpha + \frac{x}{L_p^2}g^\alpha\right) \; . \tag{9.7.12}$$

and of the same general form as in the case of gauge couplings, which also supports option b).

Vacuum extremals which correspond to asymptotic cosmologies with cosmological constant satisfying

$$D_\alpha\left(G^\alpha + \frac{x}{L_p^2}g^\alpha\right) \;\; = \;\; 0 \tag{9.7.13}$$

represent examples of the fixed points of the coupling constant evolution with conserved gravitational four-momentum. Obviously much weaker conditions guarantee fixed point property.

For Schwartschild metric having imbedding as a vacuum extremal Einstein tensor vanishes so that the RG equations would say nothing about G for option a). For Reissner-Nordstöm metric also having embedding as a vacuum extremal Einstein tensor corresponds to the energy momentum tensor of Abelian gauge field and the length scale evolution of G would be non-trivial in both cases.

9.7.3   p-Adic coupling constant evolution

p-Adic coupling constant evolution associated with length scale resolution at space-time level

If gauge couplings are indeed RG invariants inside a given space-time sheet, gauge couplings must be regarded as being characterized by the p-adic prime associated with the space-time sheet. The question is whether it is possible to understand also the p-adic coupling constant evolution at space-time level.

A natural view about p-adic length scale evolution is as an existence of a dynamical symmetry mapping the preferred extremal space-time sheet of Kähler action characterized by a p-adic prime p_1 to a space-time sheet characterized by p-adic prime $p_2 > p_1$ sufficiently near to p_1. The simplest guess is that the symmetry transformation corresponds to a scaling of M^4 coordinates in the intersection X^3 of the space-time surface with light-cone boundary $\delta M_+^4 \times CP_2$ by a scaling factor p_2/p_1, which in turn induces a transformation of $X^4(X^3)$, which in general does not reduce to M^4 scaling outside X^3 since scalings are not symmetries of the Kähler action.

This transformation induces a change of the vacuum gauge charges: $Q_i \to Q_i + \Delta Q_i$, and the renormalization group evolution boils down to the condition

$$\frac{Q_i + \Delta Q_i}{g_i^2 + \Delta g_i^2} = \frac{Q_i}{g_i^2} \ . \tag{9.7.14}$$

The problem is that this transformation has a continuous variant so that p-adic length scale evolution could reduce to continuous one.

A possible resolution of the problem is based on the observation that the values of the gauge charges depend on the initial values of the time derivatives of the imbedding space coordinates. RG invariance at space-time level suggests that small scalings leave the gauge charge and thus also coupling constant invariant. As a matter fact, this seems to be the case for all known extremals since they form scaling invariant families. The scalings by p_2/p_1 for some $p_2 > p_1$ would correspond to critical points in which bi-furcations occur in the sense that two space-time surfaces $X^4(X^3)$ satisfying the minimization conditions for Kähler action and with different gauge charges appear.

The new space-time surface emerging in the bifurcation would obey effective p_2-adic topology in some length scale range instead of p_1-adic topology. Stability considerations would dictate whether $p_1 \to p_2$ transition occurs and could also explain why primes $p \simeq 2^k$, k integer, are favored. This kind of bifurcations or even multi-furcations are certainly possible by the breaking of the classical determinism.

The space-time realization of the RG evolution associated with the phase resolution

The algebraic extensions of a given p-adic number field define a hierarchy ordered by the dimension of the extension assigned to the RG evolution with respect to the phase resolution. The evolution of $\hbar$ inducing evolutions of other coupling constants have been assigned to this coupling constant evolution and an explicit formula in terms of Beraha numbers $B_n = 4cos^2(\pi/n)$ for the RG evolution has been proposed [O5, J6].

In this case the simplest candidates for the geometric transformations of space-time surface are rational scalings of the cyclic angular S^2 coordinate of $\delta M_+^4 = R_+ \times S^2$ given by $\Phi \to r\Phi$, $r = m/n$ replacing in the general case the space-time sheet with its n-fold covering acting on X^3 and inducing a transformation of $X^4(X^3)$. Single closed curve around origin in $X^4(X^3)$ would correspond to an $m2\pi$ rotation in M^4 and I have proposed that anyonic systems with fractional spin and other charges could correspond to this kind of space-time surfaces [O3, G2].

A more precise characterization consistent with the identification of $\hbar$ as a characterizer of the topological condensation of 3-surface X^3 to a larger 3-surface Y^3 is that angular scalings correspond to the transformations $\Phi \to r\Phi$, $r = m/n$ in the case of X^4 and $\Phi \to \Phi$ in case of Y^4 so that X^2 becomes analogous to an m-fold covering of Y^3. Rational coverings could also correspond to m-fold scalings for X^4 and n-fold scalings for Y^3.

The formation of these stable multiple coverings could be seen as an analog for a transition in chaos via a process in which a closed Bohr orbit regarded as a particle itself becomes an orbit closing only after m turns. TGD predicts a hierarchy of higher level zero energy states representing S-matrix of lower level as entanglement coefficients. Particles identified as "tracks" of particles at orbits closing after m turns [G2] would be natural space-time correlates for this kind of states.

The simplest generalization from the p-adic length scale evolution consistent with the proposed role of Beraha numbers is that bifurcations can occur for integer values of $r = m$ and change the value of $\hbar$. The interpretation would be that single 2π rotation in δM_+^4 corresponds to the angular resolution with respect to the angular coordinate ϕ of space-time surface varying in the range $(0, 2\pi)$ and is given by $\Delta\phi = 2\pi/m$. On the other hand, the evidence for a gigantic but finite value of "gravitational" Planck constant [J6] suggests that large values of $\hbar$ corresponding to $3 < n < 4$ and defining a "generalized" Beraha number are possible. For $n = 3$ corresponding to the minimal resolution of $\Delta\phi = 2\pi/3$ $\hbar$ would be infinite. This would allow to keep the formula for $\hbar(n)$ in its original form by replacing n with a rational number. This would mean that also rational values of r correspond to bifurcations in the range $3 < r < 4$ at least. An open question is whether the generalization of n to rational number somehow generalizes the notion of index $M : N = B_n$ of Jones inclusion.

If this picture and the explanation for the cosmological variation of the fine structure constant characterizing ordinary matter based on the relative variation of $\hbar$ of order $\Delta\hbar/\hbar \sim 10^{-6}$ [D6] are both correct, ordinary condensed matter phase would correspond to 3-surfaces X^3 condensed on larger surface Y^3 with m in the range 100-200.

9.7.4   About electro-weak coupling constant evolution

The classical space-time correlates for electro-weak coupling constant evolution deserve a separate discussion.

How to determine the value of Weinberg angle for a given space-time sheet?

The general picture about the massivation of electro-weak bosons and electro-weak gauge bosons based on the notion of induced gauge field allows to determined Weinberg angle from the condition that electromagnetic vacuum charge for a given space-time sheet vanishes.

The basic idea is that electro-weak vacuum charge densities are generated and screen weak charges transforming $1/r$ Coulomb potentials to exponentially screened ones. The massivation of fermions occurs by a different mechanism in TGD and they can be massive even in the case that electro-weak bosons are massless.

In gauge theories the screening of weak charges occurs in differential manner. In TGD framework RG invariance inside a given space-time sheet and p-adic coupling constant evolution support the view that this screening occurs in discrete manner in the sense that the weak fields would behave like massless fields inside a given space-time sheet but the net weak charges of the space-time sheets cause the screening of the weak charges and massivation in average sense. The masslessness of photons means that the vacuum em charge for a given space-time sheet vanishes. This condition allows to determine the value of Weinberg angle for a given space-time sheet.

Smoothed out position dependent Weinberg angle from the vanishing of vacuum density of em charge

A practical variant about the condition determining Weinberg angle for a given space-time sheet is obtained by a smoothing out procedure in which the distribution of discrete values of Weinberg angle is replaced with a continuous distribution interpreted as a constant below the typical size scale of space-time sheets involved.

The condition that the em charge density defined by the covariant divergence of electro-weak current vanishes, gives a differential equation allowing to solve for Weinberg angle. Using M_+^4 proper time a as a preferred time coordinate (identifiable as cosmic time and playing key role in the construction of configuration space geometry and quantum TGD [B2, B3])) this condition can be made general coordinate invariant. One can hope that with a proper choice of boundary conditions (fixed actually the the minimization of Kähler action) Weinberg angle can always have a physical value. Since gauge current is defined as the covariant divergence of gauge field the condition involves for $D > 2$ besides the ordinary divergence also a term proportional to $W_{+,\nu}W_-^{\mu\nu} - W_{-,\nu}W_+^{\mu\nu}$.

  1. Simple special cases

For vacuum extremals ordinary em current vanishes for $p = sin^2(\theta_W) = 0$. In this case the 2-dimensionality of CP_2 projection guarantees that ordinary divergence equals to the covariant one. Hence $p = 0$ guarantees trivially the vanishing of em charge density also now but there are also other solutions.

For solutions with CP_2 projection belong to a homologically non-trivial geodesic sphere of CP_2 the condition determining the Weinberg angle reduces to the vanishing of the divergence of pJ^{0i} whereas the vanishing of γ would imply a non-physical value of p.

2. General solution of the conditions

The explicit expressions for classical em and Z^0 are given by

$$
\begin{aligned}
\gamma &= 3J - pR_{03} \ , \quad p \equiv sin^2(\theta_W) \ , \\
Z^0 &= 2R_{03} \ .
\end{aligned}
\tag{9.7.15}
$$

CP_2 Kähler form J and spinor curvature component R_{03} are given in terms of vierbein by

$$
\begin{aligned}
J &= 2\left[e_1 \wedge e_2 + e_0 \wedge e_3\right] \ , \\
R_{03} &= 2e_1 \wedge e_2 + 4e_0 \wedge e_3 \ .
\end{aligned}
\tag{9.7.16}
$$

The general form of the condition determining Weinberg angle is given by

$$
\begin{aligned}
E_Z \cdot \nabla p + (\nabla \cdot E_Z)p &= F \ , \\
F &= -6\nabla \cdot E_K - 2F_1 \ .
\end{aligned}
\tag{9.7.17}
$$

Here E_Z corresponds R_{03} term in em field and E_K to Kähler electric field and F_1 corresponds to the $W_{+,\nu}W_-^{\mu\nu} - W_{-,\nu}W_+^{\mu\nu}$ term. It is assumed that $1/e^2$ factor multiplying em current is constant. If this is not the case, the replacement $F \to F + 2E_{em}\nabla 2log(e^2)$ must be made on the right hand side.

These differential equations are of the same form as renormalization group equations and continuous solutions exist if one can introduce a coordinate system in which the flow lines of Kähler electric field correspond to one coordinate. This is possible if Z^0 electric field is of the form

$$
E_Z = \phi dt \ .
\tag{9.7.18}
$$

This implies the integrability condition $dE_Z = d\phi \wedge dt$ implying

$$
dE_Z \wedge E_Z = 0 \ .
\tag{9.7.19}
$$

By introducing space-time coordinates (x, t) (t does not refer to time now) the equation can be written in the form

$$
\frac{dp}{dt} + \frac{\nabla \cdot E_Z}{\phi}p = \frac{F}{\phi} \ .
\tag{9.7.20}
$$

solutions can be written as

$$
\begin{aligned}
p &= p_0 + p_1 \ , \\
\frac{dp_0}{dt} + \frac{\nabla \cdot E_Z}{\phi}p_0 &= 0 \ , \\
\frac{dp_1}{dt} + \frac{\nabla \cdot E_Z}{\phi}p_1 &= \frac{F}{\phi} \ .
\end{aligned}
\tag{9.7.21}
$$

p_0 and p_1 are given by

$$p_0(x,t) = p_{00}(x) + exp\left(-\int_0^t du \frac{\nabla \cdot E_Z(x,u)}{\phi}\right),$$

$$p_1(x,t) = p_0(x,t) \int_0^t du \frac{F}{p_0\phi}(x,u) . \tag{9.7.22}$$

Whether $p_{00}(x) = constant$ is consistent with field equations is an open question.

3. What happens when the integrability condition fails?

The failure of the integrability condition has interpretation as failure of the smoothing out procedure. A natural guess is that in this case the coupling constant is continuous or perhaps even smooth with respect to p-adic topology below the p-adic length scale for some prime p. Non-integrability would provide a rather satisfactory differential-topological understanding of how effective p-adic topology emerges.

3. Questions related to the physical interpretation

This picture raises several interesting questions related to the physical interpretation.

a) What is the TGD counterpart of Higgs=0 phase? The dimension of CP_2 projection is is analogous to temperature and one can argue that massivation is analogous to a loss of correlations due to the increase of D bringing in additional degrees of freedom. Massless extremals having $D = 2$ all induced gauge fields are massless so that they are excellent candidates for Higgs=0 phase. Does this mean that already $D = 3$ space-time sheets correspond to a massive phase?

b) Why electro-weak length scale corresponding to Mersenne prime M_{89} is preferred [F3]? Are there also other length scales in which electro-weak massivation occurs and thus scaled copies of electro-weak bosons? These questions reduce to the questions about the stability of the proposed bifurcations.

c) The basic problem of TGD based model of condensed matter is to explain why classical long range gauge fields do not give rise to large parity breaking effects in atomic length scale but do so in cell length length scale at least in the case of living matter (bio-catalysis). The proposal has been that particles feed electro-weak and em gauge fluxes to different space-time sheets. Could it be that blocks of bio-matter with size larger than cell the space-time sheets at which em and weak charges are feeded can be in Higgs=0 phase whereas for smaller blocks screening occurs already at quark and lepton level.

This would be consistent with the fact that the dimension D of CP_2 projection tends to decrease with the size of the space-time sheet: the larger the space-time sheet, the nearer it is to a vacuum extremal. Robertson-Walker cosmologies are exact vacuum extremals carrying however non-vanishing gravitational 4-momentum densities. By previous argument W and Z masses are identical in this kind of phase if the vanishing of vacuum em field is used to fix p. The weakening of correlations caused by classical non-determinism might imply massivation.

d) Do long ranged non-screened vacuum Z^0 and W gauge fields have some quantum counterparts as quantum-classical correspondence would suggest? Does dark matter identified as a phase with large value of $\hbar$ [J6] correspond to a phase in which electro-weak symmetry breaking is absent in the bosonic sector?

This phase would differ from the ordinary one in that the weak charges of leptons and quarks are not screened in electro-weak length scale but that their masses are very nearly the same as in Higgs=0 phase since the dominant contribution to the masses of elementary fermions is not given by a coupling to Higgs type particle but determined by p-adic thermodynamics [F2, F3].

Does bio-matter involve this kind of phase at larger space-time sheets as chirality selection suggests [F9]? Does this phase of condensed matter emerge only above length scale defined by the cell size or cell membrane thickness?

Chapter 10

Elementary Particle Vacuum Functionals

10.1 Introduction

One of the basic ideas of TGD approach is genus-generation correspondence: boundary components of the 3-surface should be carriers of elementary particle numbers and the observed particle families should correspond to various boundary topologies. A more general hypothesis is that the 2-surfaces in question sections of 3-D lightlike causal determinants, say those associated with wormhole contacts carrying parton quantum numbers

10.1.1 First series of questions

The most attractive feature of this idea is universality: if the generalized string model vertices are identified as particle vertices, different particle families are predicted to behave identically with respect to the known interactions in accordance with observational facts.

Before one can accept this identification, one should however answer several questions:

a) Also elementary bosons are predicted to possess family degeneracy: why the higher boson families have not been observed? Why only $g = 0$, "spherical", bosons seem to be the bosons produced in particle accelerators? Are $g > 0$ bosons very massive or are their couplings to fermions very small?

b) Topological reactions changing the genus of boundary component are possible (some of the handles of 2-surfaces suffers pinch or new handle is created): why however different lepton numbers are conserved in such a good approximation?

c) Why the number of the observed elementary particle families seems to be three [ha2]?

10.1.2 Second series of questions

The questions above are obvious if one accepts string model picture about particle vertices. 25 years with TGD however leads to question the string model based interpretation of particle vertices and stimulates a slightly different series of questions.

a) What really happens in particle vertices? Is the generalization of string model diagrams the proper description of particle reactions in TGD framework? Or should one assume that vertices are direct generalizations of ordinary Feynmann diagrams so that the Feynmann diagrams correspond to singular 4-manifolds and vertices to non-singular 3-manifolds at which the ends of space-time sheets representing particles meet? The elegant treatment of fermion number and other conserved quantum numbers in the vertices and construction of the vertices themselves [C5] provides a considerable support for this view. In this framework string model type vertices would be interpreted in terms of a propagation of the particle through several paths simultaneously as in double-slit experiment.

b) The new picture about vertices predicts a profound difference between fermions and bosons: the lowest bosonic vacuum wave functionals must be completely delocalized with respect to the genus to guarantee that the gauge couplings to the fermions are universal. Why this delocalization does not occur for fermions as the successful calculation of elementary particle masses strongly suggests

[TGDpad]? Why would bosonic families correspond to a hierarchy of delocalized states having $g < 3$ with a phase phase factor $expi2\pi ng/3$, $n = 0, 1, 2$ characterizing the particle family. Why would fermions correspond to states localized to $g \leq 2$? What makes bells ringing is that for topologically delocalized bosons the finiteness of the vertices would require an effective reduction of the number of particle families to a finite number N. For instance, one can consider a decomposition of the lattice $\{g \geq 0\}$ to disjoint sublattices with a complete bosonic delocalization inside each lattice.

c) Why the number of genera is just three? $g \leq 2$ Riemann surfaces are always hyper-elliptic (have global Z_2 conformal symmetry) unlike $g > 2$ surfaces. Why the complete bosonic de-localization of the light families should be restricted inside the hyper-elliptic sector? Could the reason be that $g > 2$ elementary particle vacuum functionals vanish for hyper-elliptic surfaces so that states localized to $g \leq 2$ surfaces are not transformed to $g > 2$ surfaces? Does the Z_2 symmetry make these states light?

d) There is also a second intriguing observation. Configuration space Clifford algebra is a direct integral over von Neumann algebras known as hyperfinite factors of type II_1 [ea3, O5]. The hierarchy of Jones inclusions for von Neumann algebras is characterized by a quantum phase $q = exp(i\pi/N)$, $N \geq 3$. $N = 3$ corresponds to the simplest algebraic extension of rationals and is TGD framework physically completely unique as compared to $N > 3$ since the value of the inverse of $\hbar$ vanishes for $N = 3$ apart from small gravitational corrections [O5]. The huge value of Planck constant means maximal quantum coherence time natural for elementary particles.

Is the number of light particle families three because elementary particles correspond to the lowest level in the hierarchy of Jones inclusions and to the maximally quantal situation perhaps correlating with the hyper-elliptic symmetry? Could the lattice $\{g \geq 0\}$ decompose into a union of disjoint de-localization sub-lattices with $n = 3, 4, 5...$ elements corresponding to $q = exp(i\pi/n)$?

10.1.3   The notion of elementary particle vacuum functional

In order to provide answers to either series of questions one must know something about the dependence of the elementary particle state functionals on the geometric properties of the boundary component and in the sequel an attempt to construct what might be called elementary particle vacuum functionals, is made. Irrespective of what identification of interaction vertices is adopted, the arguments involved with the construction involve only the string model type vertices so that the previous discussion seems to apply more or less as such.

The basic assumptions underlying the construction are the following ones:

a) Elementary particle vacuum functionals depend on the geometric properties of the two-surface X^2 representing elementary particle.

b) Vacuum functionals possess extended Diff invariance: all 2-surfaces on the orbit of the 2-surface X^2 correspond to the same value of the vacuum functional. This condition is satisfied if vacuum functionals have as their argument, not X^2 as such, but some 2- surface Y^2 belonging to the unique orbit of X^2 (determined by the principle selecting preferred extremal of the Kähler action as a generalized Bohr orbit [B1]) and determined in $Diff^3$ invariant manner.

c) Vacuum functionals possess conformal invariance and therefore for a given genus depend on a finite number of variables specifying the conformal equivalence class of Y^2.

d) Vacuum functionals satisfy the cluster decomposition property: when the surface Y^2 degenerates to a union of two disjoint surfaces (particle decay in string model inspired picture), vacuum functional decomposes into a product of the vacuum functionals associated with disjoint surfaces.

e) Elementary particle vacuum functionals are stable against the two-particle decay $g \rightarrow g_1 + g_2$ and one particle decay $g \rightarrow g - 1$.

In the following the construction will be described in more detail.
i) Some basic concepts related to the description of the space of the conformal equivalence classes of Riemann surfaces are introduced and the concept of hyper-ellipticity is introduced. Since theta functions will play a central role in the construction of the vacuum functionals, also their basic properties are discussed.
ii) After these preliminaries the construction of elementary particle vacuum functionals is carried out.
iii) Possible explanations for the experimental absence of the higher fermion families are considered.

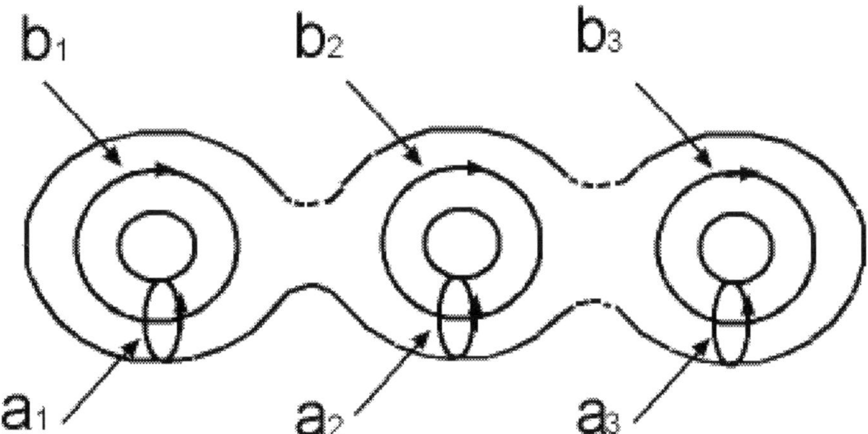

Figure 10.1: Definition of the canonical homology basis

10.2 Basic facts about Riemann surfaces

In the following some basic aspects about Riemann surfaces will be summarized. The basic topological concepts, in particular the concept of the mapping class group, are introduced, and the Teichmueller parameters are defined as conformal invariants of the Riemann surface, which in fact specify the conformal equivalence class of the Riemann surface completely.

10.2.1 Mapping class group

The first homology group $H_1(X^2)$ of a Riemann surface of genus g contains $2g$ generators [cf2, cf1, ce4]: this is easy to understand geometrically since each handle contributes two homology generators. The so called canonical homology basis can be identified as in Fig. 10.2.1.

One can define the so called intersection number $J(a, b)$ for two elements a and b of the homology group as the number of intersection points for the curves a and b counting the orientation. Since $J(a, b)$ depends on the homology classes of a and b only, it defines an antisymmetric quadratic form in $H_1(X^2)$. In the canonical homology basis the non-vanishing elements of the intersection matrix are:

$$J(a_i, b_j) \;\; = \;\; -J(b_j, a_i) = \delta_{i,j} \;\; . \tag{10.2.1}$$

J clearly defines symplectic structure in the homology group.

The dual to the canonical homology basis consists of the harmonic one-forms $\alpha_i, \beta_i, i = 1, .., g$ on X^2. These 1-forms satisfy the defining conditions

$$\begin{aligned} \int_{a_i} \alpha_j = \delta_{i,j} \quad & \int_{b_i} \alpha_j = 0 \;\; , \\ \int_{a_i} \beta_j = 0 \quad & \int_{b_i} \beta_j = \delta_{i,j} \;\; . \end{aligned} \tag{10.2.2}$$

The following identity helps to understand the basic properties of the Teichmueller parameters

$$\int_{X^2} \theta \wedge \eta \;=\; \sum_{i=1,..,g} [\int_{a_i} \theta \int_{b_i} \eta - \int_{b_i} \theta \int_{a_i} \eta] \;. \tag{10.2.3}$$

The existence of topologically nontrivial diffeomorphisms, when X^2 has genus $g > 0$, plays an important role in the sequel. Denoting by $Diff$ the group of the diffeomorphisms of X^2 and by $Diff_0$ the normal subgroup of the diffeomorphisms homotopic to identity, one can define the mapping class group M as the coset group

$$M \;=\; Diff/Diff_0 \;. \tag{10.2.4}$$

The generators of M are so called Dehn twists along closed curves a of X^2. Dehn twist is defined by excising a small tubular neighborhood of a, twisting one boundary of the resulting tube by 2π and gluing the tube back into the surface: see Fig. 10.2.1.

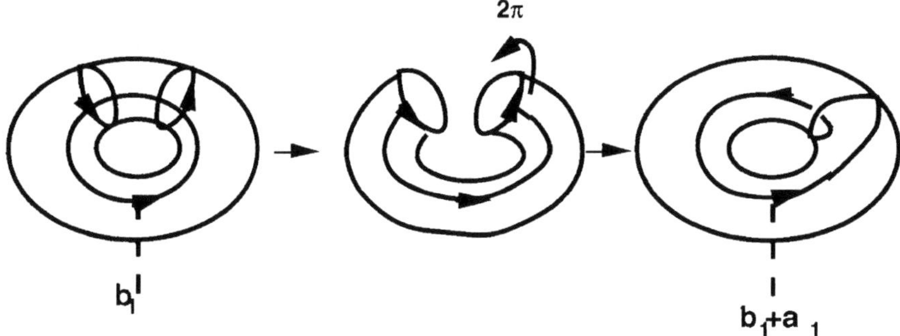

Figure 10.2: Definition of the Dehn twist

It can be shown that a minimal set of generators is defined by the following curves

$$a_1, b_1, a_1^{-1}a_2^{-1}, a_2, b_2, a_2^{-1}a_3^{-11}, ..., a_g, b_g \;. \tag{10.2.5}$$

The action of these transformations in the homology group can be regarded as a symplectic linear transformation preserving the symplectic form defined by the intersection matrix. Therefore the matrix representing the action of $Diff$ on $H_1(X^2)$ is $2g \times 2g$ matrix M with integer entries leaving J invariant: $MJM^T = J$. Mapping class group is often referred also as a symplectic modular group and denoted by $Sp(2g, Z)$. The matrix representing the action of M in the canonical homology basis decomposes into four $g \times g$ blocks A, B, C and D

$$M = \begin{pmatrix} A & B \\ C & D \end{pmatrix} \;, \tag{10.2.6}$$

where A and D operate in the subspaces spanned by the homology generators a_i and b_i respectively and C and D map these spaces to each other. The notation $D = [A, B; C, D]$ will be used in the sequel: in this notation the representation of the symplectic form J is $J = [0, 1; -1, 0]$.

10.2.2 Teichmueller parameters

The induced metric on the two-surface X^2 defines a unique complex structure. Locally the metric can always be written in the form

$$ds^2 \;=\; e^{2\phi} dz d\bar{z} \;. \tag{10.2.7}$$

where z is local complex coordinate. When one covers X^2 by coordinate patches, where the line element has the above described form, the transition functions between coordinate patches are holomorphic and therefore define a complex structure.

The conformal transformations ξ of X^2 are defined as the transformations leaving invariant the angles between the vectors of X^2 tangent space invariant: the angle between the vectors X and Y at point x is same as the angle between the images of the vectors under Jacobian map at the image point $\xi(x)$. These transformations need not be globally defined and in each coordinate patch they correspond to holomorphic (anti-holomorphic) mappings as is clear from the diagonal form of the metric in the local complex coordinates. A distinction should be made between local conformal transformations and globally defined conformal transformations, which will be referred to as conformal symmetries: for instance, for hyper-elliptic surfaces the group of the conformal symmetries contains two-element group Z_2.

Using the complex structure one can decompose one-forms to linear combinations of one-forms of type $(1,0)$ $(f(z,\bar{z})dz)$ and $(0,1)$ $(f(z,\bar{z})d\bar{z})$. $(1,0)$ form ω is holomorphic if the function f is holomorphic: $\omega = f(z)dz$ on each coordinate patch.

There are g independent holomorphic one forms ω_i known also as Abelian differentials of the first kind [cf2, cf1, ce4] and one can fix their normalization by the condition

$$\int_{a_i} \omega_j \;=\; \delta_{ij} \;. \tag{10.2.8}$$

This condition completely specifies ω_i.

Teichmueller parameters Ω_{ij} are defined as the values of the forms ω_i for the homology generators b_j

$$\Omega_{ij} \;=\; \int_{b_j} \omega_i \;. \tag{10.2.9}$$

The basic properties of Teichmueller parameters are the following:
i) The $g \times g$ matrix Ω is symmetric: this is seen by applying the formula (10.2.3) for $\theta = \omega_i$ and $\eta = \omega_j$.
ii) The imaginary part of Ω is positive: $Im(\Omega) > 0$. This is seen by the application of the same formula for $\theta = \eta$. The space of the matrices satisfying these conditions is known as Siegel upper half plane.
iii) The space of Teichmueller parameters can be regarded as a coset space $Sp(2g,R)/U(g)$ [cf1]: the action of $Sp(2g,R)$ is of the same form as the action of $Sp(2g,Z)$ and $U(g) \subset Sp(2g,R)$ is the isotropy group of a given point of Teichmueller space.
iv) Teichmueller parameters are conformal invariants as is clear from the holomorphy of the defining one-forms.
v) Teichmueller parameters specify completely the conformal structure of Riemann surface [ce4].

Although Teichmueller parameters fix the conformal structure of the 2-surface completely, they are not in one-to-one correspondence with the conformal equivalence classes of the two-surfaces:
i) The dimension for the space of the conformal equivalence classes is $D = 3g - 3$, when $g > 1$ and smaller than the dimension of Teichmueller space given by $d = (g \times g + g)/2$ for $g > 3$: all Teichmueller matrices do not correspond to a Riemann surface. In TGD approach this does not produce any problems as will be found later.
ii) The action of the topologically nontrivial diffeomorphisms on Teichmueller parameters is nontrivial and can be deduced from the action of the diffeomorphisms on the homology ($Sp(2g,Z)$ transformation) and from the defining condition $\int_{a_i} \omega_j = \delta_{i,j}$: diffeomorphisms correspond to elements $[A,B;C,D]$ of $Sp(2g,Z)$ and act as generalized Möbius transformations

$$\Omega \to (A\Omega + B)(C\Omega + D)^{-1} \;. \tag{10.2.10}$$

All Teichmueller parameters related by $Sp(2g, Z)$ transformations correspond to the same Riemann surface.

iii) The definition of the Teichmueller parameters is not unique since the definition of the canonical homology basis involves an arbitrary numbering of the homology basis. The permutation S of the handles is represented by same $g \times g$ orthogonal matrix both in the basis $\{a_i\}$ and $\{b_i\}$ and induces a similarity transformation in the space of the Teichmueller parameters

$$\Omega \to S\Omega S^{-1} \ . \tag{10.2.11}$$

Clearly, the Teichmueller matrices related by a similarity transformations correspond to the same conformal equivalence class. It is easy to show that handle permutations in fact correspond to $Sp(2g, Z)$ transformations.

10.2.3   Hyper-ellipticity

The motivation for considering hyper-elliptic surfaces comes from the fact, that $g > 2$ elementary particle vacuum functionals turn out to be vanishing for hyper-elliptic surfaces and this in turn will be later used to provide a possible explanation the non-observability of $g > 2$ particles.

Hyper-elliptic surface X can be defined abstractly as two-fold branched cover of the sphere having the group Z_2 as the group of conformal symmetries (see [cf1, ce3, ce4]. Thus there exists a map $\pi : X \to S^2$ so that the inverse image $\pi^{-1}(z)$ for a given point z of S^2 contains two points except at a finite number (say p) of points z_i (branch points) for which the inverse image contains only one point. Z_2 acts as conformal symmetries permuting the two points in $\pi^{-1}(z)$ and branch points are fixed points of the involution.

The concept can be generalized [ce3]: g-hyper-elliptic surface can be defined as a 2-fold covering of genus g surface with a finite number of branch points. One can consider also p-fold coverings instead of 2-fold coverings: a common feature of these Riemann surfaces is the existence of a discrete group of conformal symmetries.

A concrete representation for the hyper-elliptic surfaces [cf1] is obtained by studying the surface of C^2 determined by the algebraic equation

$$w^2 - P_n(z) \ = \ 0 \ , \tag{10.2.12}$$

where w and z are complex variables and $P_n(z)$ is a complex polynomial. One can solve w from the above equation

$$w_\pm \ = \ \pm\sqrt{P_n(z)} \ , \tag{10.2.13}$$

where the square root is determined so that it has a cut along the positive real axis. What happens that w has in general two roots (two-fold covering property), which coincide at the roots z_i of $P_n(z)$ and if n is odd, also at $z = \infty$: these points correspond to branch points of the hyper-elliptic surface and their number r is always even: $r = 2k$. w is discontinuous at the cuts associated with the square root in general joining two roots of $P_n(z)$ or if n is odd, also some root of P_n and the point $z = \infty$. The representation of the hyper-elliptic surface is obtained by identifying the two branches of w along the cuts. From the construction it is clear that the surface obtained in this manner has genus $k - 1$. Also it is clear that Z_2 permutes the different roots $w_\pm$ with each other and that $r = 2k$ branch points correspond to fixed points of the involution.

The following facts about the hyper-elliptic surfaces [cf1, ce4] turn out to be important in the sequel:

i) All $g < 3$ surfaces are hyper-elliptic.

ii) $g \geq 3$ hyper-elliptic surfaces are not in general hyper-elliptic and form a set of codimension 2 in the space of the conformal equivalence classes [cf1].

10.2.4   Theta functions

An extensive and detailed account of the theta functions and their applications can be found in the book of Mumford [cf1]. Theta functions appear also in the loop calculations of string model [cf2]. In the following the so called Riemann theta function and theta functions with half integer characteristics will be defined as sections (not strictly speaking functions) of the so called Jacobian variety.

For a given Teichmueller matrix Ω, Jacobian variety is defined as the $2g$-dimensional torus obtained by identifying the points z of C^g (vectors with g complex components) under the equivalence

$$z \sim z + \Omega m + n \ , \tag{10.2.14}$$

where m and n are points of Z^g (vectors with g integer valued components) and Ω acts in Z^g by matrix multiplication.

The definition of Riemann theta function reads as

$$\Theta(z|\Omega) \ = \ \sum_n exp(i\pi n \cdot \Omega \cdot n + i2\pi n \cdot z) \ . \tag{10.2.15}$$

Here $\cdot$ denotes standard inner product in C^g. Theta functions with half integer characteristics are defined in the following manner. Let a and b denote vectors of C^g with half integer components (component either vanishes or equals to $1/2$). Theta function with characteristics $[a,b]$ is defined through the following formula

$$\Theta[a,b](z|\Omega) \ = \ \sum_n exp\left[i\pi(n+a) \cdot \Omega \cdot (n+a) + i2\pi(n+a) \cdot (z+b)\right] \ . \tag{10.2.16}$$

A brief calculation shows that the following identity is satisfied

$$\Theta[a,b](z|\Omega) \ = \ exp(i\pi a \cdot \Omega \cdot a + i2\pi a \cdot b) \times \Theta(z + \Omega a + b|\Omega) \tag{10.2.17}$$

Theta functions are not strictly speaking functions in the Jacobian variety but rather sections in an appropriate bundle as can be seen from the identities

$$\begin{aligned}
\Theta[a,b](z+m|\Omega) &= exp(i2\pi a \cdot m)\Theta[a,b](z\Omega) \ , \\
\Theta[a,b](z+\Omega m|\Omega) &= exp(\alpha)\Theta[a,b](z|\Omega) \ , \\
exp(\alpha) &= exp(-i2\pi b \cdot m)exp(-i\pi m \cdot \Omega \cdot m - 2\pi m \cdot z) \ .
\end{aligned} \tag{10.2.18}$$

The number of theta functions is 2^{2g} and same as the number of nonequivalent spinor structures defined on two-surfaces. This is not an accident [cf2]: theta functions with given characteristics turn out to be in a close relation to the functional determinants associated with the Dirac operators defined on the two-surface. It is useful to divide the theta functions to even and odd theta functions according to whether the inner product $4a \cdot b$ is even or odd integer. The numbers of even and odd theta functions are $2^{g-1}(2^g+1)$ and $2^{g-1}(2^g-1)$ respectively.

The values of the theta functions at the origin of the Jacobian variety understood as functions of Teichmueller parameters turn out to be of special interest in the following and the following notation will be used:

$$\Theta[a,b](\Omega) \ \equiv \ \Theta[a,b](0|\Omega) \ , \tag{10.2.19}$$

$\Theta[a, b](\Omega)$ will be referred to as theta functions in the sequel. From the defining properties of odd theta functions it can be found that they are odd functions of z and therefore vanish at the origin of the Jacobian variety so that only even theta functions will be of interest in the sequel.

An important result is that also some *even* theta functions vanish for $g > 2$ hyper-elliptic surfaces : in fact one can characterize $g > 2$ hyper-elliptic surfaces by the vanishing properties of the theta functions [cf1, ce4]. The vanishing property derives from conformal symmetry (Z_2 in the case of hyper-elliptic surfaces) and the vanishing phenomenon is rather general [ce3]: theta functions tend to vanish for Riemann surfaces possessing discrete conformal symmetries. It is not clear (to the author) whether the presence of a conformal symmetry is in fact equivalent with the vanishing of some theta functions. As already noticed, spinor structures and the theta functions with half integer characteristics are in one-to-one correspondence and the vanishing of theta function with given half integer characteristics is equivalent with the vanishing of the Dirac determinant associated with the corresponding spinor structure or equivalently: with the existence of a zero mode for the Dirac operator [cf2]. For odd characteristics zero mode exists always: for even characteristics zero modes exist, when the surface is hyper-elliptic or possesses more general conformal symmetries.

10.3   Elementary particle vacuum functionals

The basic assumption is that elementary particle families correspond to various elementary particle vacuum functionals associated with the 2-dimensional boundary components of the 3-surface. These functionals need not be localized to a single boundary topology. Neither need their dependence on the boundary component be local. An important role in the following considerations is played by the fact that the minimization requirement of the Kähler action associates a unique 3-surface to each boundary component, the "Bohr orbit" of the boundary and this surface provides a considerable (and necessarily needed) flexibility in the definition of the elementary particle vacuum functionals. There are several natural constraints to be satisfied by elementary particle vacuum functionals.

10.3.1   Extended Diff invariance and Lorentz invariance

Extended Diff invariance is completely analogous to the extension of 3-dimensional Diff invariance to four-dimensional Diff invariance in the interior of the 3-surface. Vacuum functional must be invariant not only under diffeomorphisms of the boundary component but also under the diffeomorphisms of the 3- dimensional "orbit" Y^3 of the boundary component. In other words: the value of the vacuum functional must be same for any time slice on the orbit the boundary component. This is guaranteed if vacuum functional is functional of some two-surface Y^2 belonging to the orbit and defined in $Diff^3$ invariant manner.

An additional natural requirement is Poincare invariance. In the original formulation of the theory only Lorentz transformations of the light cone were exact symmetries of the theory. In this framework the definition of Y^2 as the intersection of the orbit with the hyperboloid $\sqrt{m_{kl}m^k m^l} = a$ is $Diff^3$ and Lorentz invariant.

Interaction vertices as generalization of stringy vertices

For stringy diagrams Poincare invariance of conformal equivalence class and general coordinate invariance are far from being a trivial issues. Vertices are now not completely unique since there is an infinite number of singular 3-manifolds which can be identified as vertices even if one assumes space-likeness. One should be able to select a unique singular 3-manifold to fix the conformal equivalence class.

One might hope that Lorentz invariant invariant and general coordinate invariant definition of Y^2 results by introducing light cone proper time a as a height function specifying uniquely the point at which 3-surface is singular (stringy diagrams help to visualize what is involved), and by restricting the singular 3-surface to be the intersection of $a = constant$ hyperboloid of M^4 containing the singular point with the space-time surface. There would be non-uniqueness of the conformal equivalence class due to the choice of the origin of the light cone but the decomposition of the configuration space of 3-surfaces to a union of configuration spaces characterized by unions of future and past light cones could resolve this difficulty.

Interaction vertices as generalization of ordinary ones

If the interaction vertices are identified as intersections for the ends of space-time sheets representing particles, the conformal equivalence class is naturally identified as the one associated with the intersection of the boundary component or light like causal determinant with the vertex. Poincare invariance of the conformal equivalence class and generalized general coordinate invariance follow trivially in this case.

10.3.2 Conformal invariance

Conformal invariance implies that vacuum functionals depend on the conformal equivalence class of the surface Y^2 only. What makes this idea so attractive is that for a given genus g configuration space becomes effectively finite-dimensional. A second nice feature is that instead of trying to find coordinates for the space of the conformal equivalence classes one can construct vacuum functionals as functions of the Teichmueller parameters.

That one can construct this kind of functions as suitable functions of the Teichmueller parameters is not trivial. The essential point is that the boundary components can be regarded as submanifolds of $M_+^4 \times CP_2$: as a consequence vacuum functional can be regarded as a composite function:

2-surface $\rightarrow$ Teichmueller matrix Ω determined by the induced metric $\rightarrow \Omega_{vac}(\Omega)$

Therefore the fact that there are Teichmueller parameters which do not correspond to any Riemann surface, doesn't produce any trouble. It should be noticed that the situation differs from that in the Polyakov formulation of string models, where one doesn't assume that the metric of the two-surface is induced metric (although classical equations of motion imply this).

10.3.3 Diff invariance

Since several values of the Teichmueller parameters correspond to the same conformal equivalence class, one must pose additional conditions on the functions of the Teichmueller parameters in order to obtain single valued functions of the conformal equivalence class.

The first requirement of this kind is the invariance under topologically nontrivial Diff transformations inducing $Sp(2g, Z)$ transformation $(A, B; C, D)$ in the homology basis. The action of these transformations on Teichmueller parameters is deduced by requiring that holomorphic one-forms satisfy the defining conditions in the transformed homology basis. It turns out that the action of the topologically nontrivial diffeomorphism on Teichmueller parameters can be regarded as a generalized Möbius transformation:

$$\Omega \rightarrow (A\Omega + B)(C\Omega + D)^{-1} \ . \tag{10.3.1}$$

Vacuum functional must be invariant under these transformations. It should be noticed that the situation differs from that encountered in the string models. In TGD the integration measure over the configuration space is Diff invariant: in string models the integration measure is the integration measure of the Teichmueller space and this is not invariant under $Sp(2g, Z)$ but transforms like a density: as a consequence the counterpart of the vacuum functional must be also modular covariant since it is the product of vacuum functional and integration measure, which must be modular invariant.

It is possible to show that the quantities

$$(\Theta[a, b]/\Theta[c, d])^4 \ . \tag{10.3.2}$$

and their complex conjugates are $Sp(2g, Z)$ invariants [cf1] and therefore can be regarded as basic building blocks of the vacuum functionals.

Teichmueller parameters are not uniquely determined since one can always perform a permutation of the g handles of the Riemann surface inducing a redefinition of the canonical homology basis (permutation of g generators). These transformations act as similarities of the Teichmueller matrix:

$$\Omega \to S\Omega S^{-1} \ , \tag{10.3.3}$$

where S is the $g \times g$ matrix representing the permutation of the homology generators understood as orthonormal vectors in the g- dimensional vector space. Therefore the Teichmueller parameters related by these similarity transformations correspond to the same conformal equivalence class of the Riemann surfaces and vacuum functionals must be invariant under these similarities.

It is easy to find out that these similarities permute the components of the theta characteristics: $[a, b] \to [S(a), S(b)]$. Therefore the invariance requirement states that the handles of the Riemann surface behave like bosons: the vacuum functional constructed from the theta functions is invariant under the permutations of the theta characteristics. In fact, this requirement brings in nothing new. Handle permutations can be regarded as $Sp(2g, Z)$ transformations so that the modular invariance alone guarantees invariance under handle permutations.

10.3.4   Cluster decomposition property

Consider next the behavior of the vacuum functional in the limit, when boundary component with genus g splits to two separate boundary components of genera g_1 and g_2 respectively. The splitting into two separate boundary components corresponds to the reduction of the Teichmueller matrix Ω^g to a direct sum of $g_1 \times g_1$ and $g_2 \times g_2$ matrices $(g_1 + g_2 = g)$:

$$\Omega^g = \Omega^{g_1} \oplus \Omega^{g_2} \ , \tag{10.3.4}$$

when a suitable definition of the Teichmueller parameters is adopted. The splitting can also take place without a reduction to a direct sum: the Teichmueller parameters obtained via $Sp(2g, Z)$ transformation from $\Omega^g = \Omega^{g_1} \oplus \Omega^{g_2}$ do not possess direct sum property in general.

The physical interpretation is obvious: the non-diagonal elements of the Teichmueller matrix describe the geometric interaction between handles and at this limit the interaction between the handles belonging to the separate surfaces vanishes. On the physical grounds it is natural to require that vacuum functionals satisfy cluster decomposition property at this limit: that is they reduce to the product of appropriate vacuum functionals associated with the composite surfaces.

Theta functions satisfy cluster decomposition property [cf1, cf2]. Theta characteristics reduce to the direct sums of the theta characteristics associated with g_1 and g_2 $(a = a_1 \oplus a_2, \ b = b_1 \oplus b_2)$ and the dependence on the Teichmueller parameters is essentially exponential so that the cluster decomposition property indeed results:

$$\Theta[a, b](\Omega^g) \ = \ \Theta[a_1, b_1](\Omega^{g_1}) \Theta[a_2, b_2](\Omega^{g_2}) \ . \tag{10.3.5}$$

Cluster decomposition property holds also true for the products of theta functions. This property is also satisfied by suitable homogenous polynomials of thetas. In particular, the following quantity playing central role in the construction of the vacuum functional obeys this property

$$Q_0 \ = \ \sum_{[a,b]} \Theta[a, b]^4 \bar{\Theta}[a, b]^4 \ , \tag{10.3.6}$$

where the summation is over all even theta characteristics (recall that odd theta functions vanish at the origin of C^g).

Together with the $Sp(2g, Z)$ invariance the requirement of cluster decomposition property implies that the vacuum functional must be representable in the form

$$\Omega_{vac} \ = \ P_{M,N}(\Theta^4, \bar{\Theta}^4)/Q_{MN}(\Theta^4, \bar{\Theta}^4) \tag{10.3.7}$$

where the homogenous polynomials $P_{M,N}$ and $Q_{M,N}$ have same degrees (M and N as polynomials of $\Theta[a, b]^4$ and $\bar{\Theta}[a, b]^4$.

10.3.5   Finiteness requirement

Vacuum functional should be finite. Finiteness requirement is satisfied provided the numerator $Q_{M,N}$ of the vacuum functional is real and positive definite. The simplest quantity of this type is the quantity Q_0 defined previously and its various powers. $Sp(2g, Z)$ invariance and finiteness requirement are satisfied provided vacuum functionals are of the following general form

$$\Omega_{vac} \;=\; \frac{P_{N,N}(\Theta^4, \bar{\Theta}^4)}{Q_0^N} \;, \tag{10.3.8}$$

where $P_{N,N}$ is homogenous polynomial of degree N with respect to $\Theta[a,b]^4$ and $\bar{\Theta}[a,b]^4$. In addition $P_{N,N}$ is invariant under the permutations of the theta characteristics and satisfies cluster decomposition property.

10.3.6   Stability against the decay $g \to g_1 + g_2$

Elementary particle vacuum functionals must be stable against the genus conserving decays $g \to g_1 + g_2$. This decay corresponds to the limit at which Teichmueller matrix reduces to a direct sum of the matrices associated with g_1 and g_2 (note however the presence of $Sp(2g, Z)$ degeneracy). In accordance with the topological description of the particle reactions one expects that this decay doesn't occur if the vacuum functional in question vanishes at this limit.

In general the theta functions are non-vanishing at this limit and vanish provided the theta characteristics reduce to a direct sum of the odd theta characteristics. For $g < 2$ surfaces this condition is trivial and gives no constraints on the form of the vacuum functional. For $g = 2$ surfaces the theta function $\Theta(a,b)$, with $a = b = (1/2, 1/2)$ satisfies the stability criterion identically (odd theta functions vanish identically), when Teichmueller parameters separate into a direct sum. One can however perform $Sp(2g, Z)$ transformations giving new points of Teichmueller space describing the decay. Since these transformations transform theta characteristics in a nontrivial manner to each other and since all even theta characteristics belong to same $Sp(2g, Z)$ orbit [cf1, cf2], the conclusion is that stability condition is satisfied provided $g = 2$ vacuum functional is proportional to the product of fourth powers of all even theta functions multiplied by its complex conjugate.

If $g > 2$ there always exists some theta functions, which vanish at this limit and the minimal vacuum functional satisfying this stability condition is of the same form as in $g = 2$ case, that is proportional to the product of the fourth powers of all even Theta functions multiplied by its complex conjugate:

$$\Omega_{vac} \;=\; \prod_{[a,b]} \Theta[a,b]^4 \bar{\Theta}[a,b]^4 / Q_0^N \;, \tag{10.3.9}$$

where N is the number of even theta functions. The results obtained imply that genus-generation correspondence is one to one for $g > 1$ for the minimal vacuum functionals. Of course, the multiplication of the minimal vacuum functionals with functionals satisfying all criteria except stability criterion gives new elementary particle vacuum functionals: a possible physical identification of these vacuum functionals is most naturally as some kind of excited states.

One of the questions posed in the beginning was related to the experimental absence of $g > 0$, possibly massless, elementary bosons. The proposed stability criterion suggests a nice explanation. The point is that elementary particles are stable against decays $g \to g_1 + g_2$ but not with respect to the decay $g \to g + \; sphere$. As a consequence the direct emission of $g > 0$ gauge bosons is impossible unlike the emission of $g = 0$ bosons: for instance the decay muon $\to$ electron $+(g = 1)$ photon is forbidden.

10.3.7   Stability against the decay $g \to g - 1$

This stability criterion states that the vacuum functional is stable against single particle decay $g \to g-1$ and, if satisfied, implies that vacuum functional vanishes, when the genus of the surface is smaller than g. In stringy framework this criterion is equivalent to a separate conservation of various lepton numbers: for instance, the spontaneous transformation of muon to electron is forbidden. Notice that this condition doesn't imply that that the vacuum functional is localized to a single genus: rather the vacuum functional of genus g vanishes for all surfaces with genus smaller than g. This hierarchical structure should have a close relationship to Cabibbo-Kobayashi-Maskawa mixing of the quarks.

The stability criterion implies that the vacuum functional must vanish at the limit, when one of the handles of the Riemann surface suffers a pinch. To deduce the behavior of the theta functions at this limit, one must find the behavior of Teichmueller parameters, when i:th handle suffers a pinch. Pinch implies that a suitable representative of the homology generator a_i or b_i contracts to a point.

Consider first the case, when a_i contracts to a point. The normalization of the holomorphic one-form ω_i must be preserved so that that ω_i must behaves as $1/z$, where z is the complex coordinate vanishing at pinch. Since the homology generator b_i goes through the pinch it seems obvious that the imaginary part of the Teichmueller parameter $\Omega_{ii} = \int_{b_i} \omega_i$ diverges at this limit (this conclusion is made also in [cf1]): $Im(\Omega_{ii}) \to \infty$.

Of course, this criterion doesn't cover all possible manners the pinch can occur: pinch might take place also, when the components of the Teichmueller matrix remain finite. In the case of torus topology one finds that $Sp(2g, Z)$ element $(A, B; C, D)$ takes $Im(\Omega) = \infty$ to the point C/D of real axis. This suggests that pinch occurs always at the boundary of the Teichmueller space: the imaginary part of Ω_{ij} either vanishes or some matrix element of $Im(\Omega)$ diverges.

Consider next the situation, when b_i contracts to a point. From the definition of the Teichmueller parameters it is clear that the matrix elements Ω_{kl}, with $k, l \neq i$ suffer no change. The matrix element Ω_{ki} obviously vanishes at this limit. The conclusion is that i:th row of Teichmueller matrix vanishes at this limit. This result is obtained also by deriving the $Sp(2g, Z)$ transformation permuting a_i and b_i with each other: in case of torus this transformation reads $\Omega \to -1/\Omega$.

Consider now the behavior of the theta functions, when pinch occurs. Consider first the limit, when $Im(\Omega_{ii})$ diverges. Using the general definition of $\Theta[a, b]$ it is easy to find out that all theta functions for which the i:th component a_i of the theta characteristic is non-vanishing (that is $a_i = 1/2$) are proportional to the exponent $exp(-\pi\Omega_{ii}/4)$ and therefore vanish at the limit. The theta functions with $a_i = 0$ reduce to $g-1$ dimensional theta functions with theta characteristic obtained by dropping i:th components of a_i and b_i and replacing Teichmueller matrix with Teichmueller matrix obtained by dropping i:th row and column. The conclusion is that all theta functions of type $\Theta(a, b)$ with $a = (1/2, 1/2, ..., 1/2)$ satisfy the stability criterion in this case.

What happens for the $Sp(2g, Z)$ transformed points on the real axis? The transformation formula for theta function is given by [cf1, cf2]

$$\Theta[a, b]((A\Omega + B)(C\Omega + D)^{-1}) \quad = \quad exp(i\phi)det(C\Omega + D)^{1/2}\Theta[c, d](\Omega) \ ,$$

$$(10.3.10)$$

where

$$\begin{pmatrix} c \\ d \end{pmatrix} = \begin{pmatrix} A & B \\ C & D \end{pmatrix}\left(\begin{pmatrix} a \\ b \end{pmatrix} - \begin{pmatrix} (CD^T)_d/2 \\ (AB^T)_d/2 \end{pmatrix}\right) \ .$$

$$(10.3.11)$$

Here ϕ is a phase factor irrelevant for the recent purposes and the index d refers to the diagonal part of the matrix in question.

The first thing to notice is the appearance of the diverging square root factor, which however disappears from the vacuum functionals (P and Q have same degree with respect to thetas). The essential point is that theta characteristics transform to each other: as already noticed all even theta characteristics belong to the same $Sp(2g, Z)$ orbit. Therefore the theta functions vanishing at $Im(\Omega_{ii}) = \infty$ do not vanish at the transformed points. It is however clear that for a given Teichmueller parametrization of pinch some theta functions vanish always.

Similar considerations in the case $\Omega_{ik} = 0$, i fixed, show that all theta functions with $b = (1/2, ..., 1/2)$ vanish identically at the pinch. Also it is clear that for $Sp(2g, Z)$ transformed points one can always find some vanishing theta functions. The overall conclusion is that the elementary particle vacuum functionals obtained by using $g \to g_1 + g_2$ stability criterion satisfy also $g \to g - 1$ stability criterion since they are proportional to the product of all even theta functions. Therefore the only nontrivial consequence of $g \to g - 1$ criterion is that also $g = 1$ vacuum functionals are of the same general form as $g > 1$ vacuum functionals.

A second manner to deduce the same result is by restricting the consideration to the hyper-elliptic surfaces and using the representation of the theta functions in terms of the roots of the polynomial appearing in the definition of the hyper-elliptic surface [cf1]. When the genus of the surface is smaller than three (the interesting case), this representation is all what is needed since all surfaces of genus $g < 3$ are hyper-elliptic.

Since hyper-elliptic surfaces can be regarded as surfaces obtained by gluing two compactified complex planes along the cuts connecting various roots of the defining polynomial it is obvious that the process $g \to g - 1$ corresponds to the limit, when two roots of the defining polynomial coincide. This limit corresponds either to disappearance of a cut or the fusion of two cuts to a single cut. Theta functions are expressible as the products of differences of various roots (Thomae's formula [cf1])

$$\Theta[a, b]^4 \propto \prod_{i<j\in T} (z_i - z_j) \prod_{k<l\in CT} (z_k - z_l) \ , \tag{10.3.12}$$

where T denotes some subset of $\{1, 2, ..., 2g\}$ containing $g+1$ elements and CT its complement. Hence the product of all even theta functions vanishes, when two roots coincide. Furthermore, stability criterion is satisfied only by the product of the theta functions.

Lowest dimensional vacuum functionals are worth of more detailed consideration.

i) $g = 0$ particle family corresponds to a constant vacuum functional: by continuity this vacuum functional is constant for all topologies.

ii) For $g = 1$ the degree of P and Q as polynomials of the theta functions is 24: the critical number of transversal degrees of freedom in bosonic string model! Probably this result is not an accident.

ii) For $g = 2$ the corresponding degree is 80 since there are 10 even genus 2 theta functions.

There are large numbers of vacuum functionals satisfying the relevant criteria, which do not satisfy the proposed stability criteria. These vacuum functionals correspond either to many particle states or to unstable single particle states.

10.3.8   Continuation of the vacuum functionals to higher genus topologies

From continuity it follows that vacuum functionals cannot be localized to single boundary topology. Besides continuity and the requirements listed above, a natural requirement is that the continuation of the vacuum functional from the sector g to the sector $g + k$ reduces to the product of the original vacuum functional associated with genus g and $g = 0$ vacuum functional at the limit when the surface with genus $g + k$ decays to surfaces with genus g and k: this requirement should guarantee the conservation of separate lepton numbers although different boundary topologies suffer mixing in the vacuum functional. These requirements are satisfied provided the continuation is constructed using the following rule:

Perform the replacement

$$\Theta[a, b]^4 \quad \to \quad \sum_{c,d} \Theta[a \oplus c, b \oplus d]^4 \tag{10.3.13}$$

for each fourth power of the theta function. Here c and d are Theta characteristics associated with a surface with genus k. The same replacement is performed for the complex conjugates of the theta function. It is straightforward to check that the continuations of elementary particle vacuum functionals indeed satisfy the cluster decomposition property and are continuous.

To summarize, the construction has provided hoped for answers to some questions stated in the beginning: stability requirements explain the separate conservation of lepton numbers and the experimental absence of $g > 0$ elementary bosons. What has not not been explained is the experimental absence of $g > 2$ fermion families. The vanishing of the $g > 2$ elementary particle vacuum functionals for the hyper-elliptic surfaces however suggest a possible explanation: under some conditions on the surface X^2 the surfaces Y^2 are hyper-elliptic or possess some conformal symmetry so that elementary particle vacuum functionals vanish for them. This conjecture indeed might make sense since the surfaces Y^2 are determined by the asymptotic dynamics and one might hope that the surfaces Y^2 are analogous to the final states of a dissipative system.

10.4 Why $g > 2$ elementary particles are absent from spectrum?

The decay properties of the intermediate gauge bosons [ha2] are consistent with the assumption that the number of the light neutrinos is $N = 3$. Also cosmological considerations pose upper bounds on the number of the light neutrino families and $N = 3$ seems to be favored [ha3]. It must be however emphasized that p-adic considerations [F5] encourage the consideration the existence of higher genera with neutrino masses such that they are not produced in the laboratory at present energies. In any case, for TGD approach the finite number of light fermion families is a potential difficulty since genus-generation correspondence suggests that the number of the fermion (and possibly also boson) families is infinite. Therefore one had better to find a good argument showing that the number of the observed neutrino families, or more generally, of the observed elementary particle families, is small also in the world described by TGD.

It will be later found that also TGD inspired cosmology requires that the number of the effectively massless fermion families must be small after Planck time. This suggests that boundary topologies with handle number $g > 2$ are unstable and/or very massive so that they, if present in the spectrum, disappear from it after Planck time, which correspond to the value of the light cone proper time $a \simeq 10^{-11}$ seconds.

In accordance with the spirit of TGD approach it is natural to wonder whether some geometric property differentiating between $g > 2$ and $g < 3$ boundary topologies might explain why only $g < 3$ boundary components are observable. One can indeed find a good candidate for this kind of property: namely hyper-ellipticity, which states that Riemann surface is a two-fold branched covering of sphere possessing two-element group Z_2 as conformal automorphisms. All $g < 3$ Riemann surfaces are hyper-elliptic unlike $g > 2$ Riemann surfaces, which in general do not posses this property. Thus it is natural to consider the possibility that hyper-ellipticity or more general conformal symmetries might explain why only $g < 2$ topologies correspond to the observed elementary particles.

As regards to the present problem the crucial observation is that some even theta functions vanish for the hyper-elliptic surfaces with genus $g > 2$ [cf1]. What is essential is that these surfaces have the group Z_2 as conformal symmetries. Indeed, the vanishing phenomenon is more general. Theta functions tend to vanish for $g > 2$ two-surfaces possessing discrete group of conformal symmetries [ce3]: for instance, instead of sphere one can consider branched coverings of higher genus surfaces.

From the general expression of the elementary particle vacuum functional it is clear that elementary particle vacuum functionals vanish, when Y^2 is hyper-elliptic surface with genus $g > 2$ and one might hope that this is enough to explain why the number of elementary particle families is three.

10.4.1 Hyper-ellipticity implies the separation of $g \leq 2$ and $g > 2$ sectors to separate worlds

If the vertices are defined as intersections of space-time sheets of elementary particles and if elementary particle vacuum functionals are required to have Z_2 symmetry, the localization of elementary particle vacuum functionals to $g \leq 2$ topologies occurs automatically. Even if one allows as limiting case vertices for which 2-manifolds are pinched to topologies intermediate between $g > 2$ and $g \leq 2$ topologies, Z_2 symmetry present for both topological interpretations implies the vanishing of this kind of vertices. This applies also in the case of stringy vertices so that also particle propagation would respect the effective number of particle families. $g > 2$ and $g \leq 2$ topologies would behave much like their own worlds in this approach. This is enough to explain the experimental findings if one can understand why the $g > 2$ particle families are absent as incoming and outgoing states or are very heavy.

10.4.2 What about $g > 2$ vacuum functionals which do not vanish for hyper-elliptic surfaces?

The vanishing of all $g \geq 2$ vacuum functionals for hyper-elliptic surfaces cannot hold true generally. There must exist vacuum functionals which do satisfy this condition. This suggest that elementary particle vacuum functionals for $g > 2$ states have interpretation as bound states of g handles and that the more general states which do not vanish for hyper-elliptic surfaces correspond to many-particle states composed of bound states $g \leq 2$ handles and cannot thus appear as incoming and outgoing states. Thus $g > 2$ elementary particles would decouple from $g \leq 2$ states.

10.4.3 Should higher elementary particle families be heavy?

TGD predicts an entire hierarchy of scaled up variants of standard model physics for which particles do not appear in the vertices containing the known elementary particles and thus behave like dark matter [A1, O5]. Also $g > 2$ elementary particles would behave like dark matter and in principle there is no absolute need for them to be heavy.

The safest option would be that $g > 2$ elementary particles are heavy and the breaking of Z_2 symmetry for $g \geq 2$ states could guarantee this. p-Adic considerations lead to a general mass formula for elementary particles such that the mass of the particle is proportional to $\frac{1}{\sqrt{p}}$ [TGDpad]. Also the dependence of the mass on particle genus is completely fixed by this formula. What remains however open is what determines the p-adic prime associated with a particle with given quantum numbers. Of course, it could quite well occur that p is much smaller for $g > 2$ genera than for $g \leq 2$ genera.

Chapter 11

Massless States and Particle Massivation

11.1 Introduction

The massless sector of the TGD and particle massivation is studied in this chapter. The identification of the spectrum of light particles reduces to two tasks: the construction of massless states and the identification of the states which remain light in p-adic thermodynamics.

11.1.1 Physical states as representations of super-canonical and Super Kac-Moody algebras

Physical states belong to the representation of super-canonical algebra and Super Kac-Moody algebra of $E^2 \times SU(3) \times U(2)_{ew}$ associated with the 2-D surfaces X^2 defined by the intersections of 3-D light like causal determinants (CDs) with 7-D CDs $X^7 = X_l^3 \times CP_2$, where X_l^3 is union of future and past directed light conc boundaries. These 2-surfaces have interpretation as partons, and the effective 2-dimensionality means that the machinery of 2-D conformal field theories can be applied in the state construction.

As explained in [B4], a generalization of the coset construction is in question in the sense that Virasoro generators are differences of those for Super Kac-Moody algebra and super-canonical algebra. The crucial condition is that the action of the super-canonical conformal algebra reduces to an X^2-conformal transformation. This follows from the fact that super-canonical conformal transformations act as S^2-local radial scalings inducing only a conformal scaling of the induced metric of X^2 ($\delta M_+^4 = S^2 \times R_+$). By the effective 2-dimensionality Super Kac-Moody and super-canonical algebras define equivalent tangent space basis for the configuration space, or alternatively, they correspond to basis for non-zero modes and zero modes identifiable as quantum and classical degrees of freedom in 1-1 correspondence with each other. Therefore the corresponding Virasoro algebras have the same central extension implying that difference algebra has vanishing central extension.

The recipe is simple. Construct first a state with a non-positive conformal weight using super-canonical generators, and then apply Super-Kac Moody generators to compensate this conformal weight to get a state with vanishing conformal weight and zero mass.

The conformal weights of super-canonical algebra generators are complex and expressible in terms of zeros of Riemann Zeta. The anti-commutators and commutators of the super-canonical generators in turn are expressible in terms of Riemann Zeta. The conformal weights of physical states must be real. This implies conformal confinement to which color confinement might reduce: what would happen that partons can have complex super-canonical conformal weights but particles have real conformal weights. The zeros of Zeta and conformal confinement represent essentially non-stringy aspects of TGD being due to the fact that the basic objects are effectively 2-dimensional rather than 1-dimensional.

The hierarchy of p-adic primes is dual to the hierarchy of zeros of Zeta, and it is possible to express the operators O_p creating states for given p-adic prime p as superpositions of products of operators O_z such that the net conformal weight is constant for each term in the superposition. The ansatz for the coefficients of the superposition inspired by the conformal invariance and an interpretation as a formal partition function fixes completely the p-adic coupling constant evolution and the infinite degeneracy due to the zeros of Zeta effectively disappears for a given p.

The fact that also certain linear combinations of zeros are possible, implies the possibility of multi-p fractality meaning that X^2 is characterized by several primes. A possible interpretation is in terms of waves in a given p-adic length scale L_p to which ripples in smaller p-adic length scale are superposed. To these ripples in turn waves in still smaller p-adic length scale are superposed, etc.. The structure of the state construction repeats the structure for the construction of infinite primes.

For spinor harmonics of CP_2 the correlation between color and electro-weak quantum numbers is not correct. Super-canonical generators provide a natural mechanism allowing to cure the problem. Boson states are identified as bi-local bilinears of fermions and anti-fermions in X^2 characterized by charge matrices and conformally invariant correlation function. $BF\overline{F}$ coupling constants can be identified in terms of normalization factors of the boson states. The small value of gravitational coupling can be understood as resulting by a fractal mechanism reducing its value from the square of p-adic length L_p, and a concrete physical interpretation for the expression of gravitational constant in terms of CP_2 length derived from number theoretic arguments emerges. The presence of primes $2, 3, ...23, p$ in the expression of the gravitational constant can be interpreted in terms of multi-p p-adic fractality involving these primes.

If the primes $p = 2, 3, ...23$ are present, the question whether besides p-adic length scales $L_p \propto \sqrt{p}$ also their multiples $\sqrt{\prod_i q_i} L_p$, where $\{q_i\}$ forms a subset of $\{2, 3, ..., 23\}$ define fundamental length scales. The implication would be small-p p-adic fractality for these small primes with each p-adic length scale L_p taking the role of CP_2 length, and there indeed is some evidence for this kind of fractality.

11.1.2 Particle massivation

Particle massivation can be regarded as a change of the vanishing parton conformal weights describable as a thermal mixing with higher conformal weights. The interpretation is made possible by the fact that the general mass formula for particle does not fix completely the conformal weights of the partons. The values of the partonic conformal weights can be only deduced from the poles of S-matrix assuming that an approximate decomposition to propagators and vertices occurs. The physical mechanism causing the thermal massivation is hydrodynamical mixing by the braiding flow defined by the normal components of energy momentum tensor of the induced Kähler field at light like 3-D CDs describing the orbits of partons.

This mechanism cannot explain the massivation of electro-weak gauge bosons, which could be caused either by TGD variant of Higgs mechanism (TGD indeed predicts a candidate for a Higgs field) or by the fact that the charge matrices of W boson and left handed component of Z^0 are not covariantly constant, which together with the hydrodynamical mixing leads to a loss of correlations. The group theoretical properties of charge matrices seem to be correct.

The massivation induced by the ergodic hydrodynamical flow can be described in terms of p-adic thermodynamics. The underlying philosophy is that real number based TGD can be algebraically continued to various p-adic number fields. This poses extremely strong conditions on the parameters of the theory. Instead of energy, the Super Kac-Moody Virasoro generator L_0 (essentially mass squared) is thermalized. This guarantees Lorentz invariance. p-Adic thermodynamics forces to conclude that CP_2 radius is essentially the p-adic length scale $R \sim L$ and thus of order $R \simeq 10^4 \sqrt{G}$ and therefore 10^4 times larger than the naive guess. Hence p-adic thermodynamics describes the mixing of states with vanishing conformal weights with their Super Kac-Moody Virasoro excitations having masses of order 10^{-4} Planck mass.

p-Adic temperature is quantized by purely number theoretical constraints (Boltzmann weight $exp(-E/kT)$ is replaced with p^{L_0/T_p}, $1/T_p$ integer) and fermions correspond to $T_p = 1$ whereas $T_p = 1/2$ seems to be the only reasonable choice for bosons. That mass squared, rather than energy, is a fundamental quantity at CP_2 length scale is also suggested by a simple dimensional argument (Planck mass squared is proportional to $\hbar$ so that it should correspond to a generator of some Lie-algebra (Virasoro generator L_0!)).

There is also modular contribution to the mass squared which can be estimated using elementary particle vacuum functionals in the conformal modular degrees of freedom of the partonic 2-surface. This contribution can be identified as a contribution coming from a thermodynamics in super-canonical

Virasoro algebra which generates excitations of the ground states with negative conformal weight. This contribution will be discussed in the next section.

The predictions of the general theory are consistent with the earlier mass calculations, and the earlier ad hoc parameters disappear. In particular, optimal lowest order predictions for the charged lepton masses are obtained and photon, gluon and graviton appear as essentially massless particles. The negative conformal weight created by super-canonical generators can have arbitrarily large magnitude so that an infinite hierarchy of exotic massless states is predicted. These states receive mass by the proposed mechanism and they are expected to be unstable but it remains to be shown that they do not appear in the spectrum of light particles. Since X^2 can have an arbitrarily large size and can even correspond to black hole horizon, the emergence of this complex structure of states is completely natural.

11.2 Generalization of the stringy mass formula

Basically configuration space Dirac equation corresponds to Super Virasoro conditions. There are several configuration space Dirac operators. The two kinds of causal determinants give rise to super-canonical and Super Kac-Moody Dirac operators. In both cases quark-lepton degeneracy gives rise to two different Dirac operators corresponding to N-S (quarks) and Ramond (leptons) representations of Super Virasoro algebras. Dirac operators are identifiable as super-generators G of the light Super Virasoro algebra and Dirac equation states that Super Virasoro algebra acts as gauge transformations.

a) Supercanonical Dirac operators $D(SC)$ are associated with 7-D CDs $X^7 = X_l^3 \times CP_2$, X_l^3 light like surface. The interpretation is in terms of creation of pairs of space-time sheets branching from X^3 and with vanishing total classical energy. In the "vibrational" part of the $L_0(SC)$ appears the quadratic Casimir operator of Lorentz group, which can have complex values in accordance with the fact that super-canonical conformal weights are in general complex. The corresponding excitations have thus spin, color, electro-weak quantum numbers, and fermion numbers.

b) The light like 3-surfaces $Y_l^3 \subset X^4$ defining light like CDs event horizons and light like boundaries of space-time sheets define Super-Kac Moody Dirac operators. 7–3 duality reduces these operators as well as super-canonical Dirac operators to the intersections X_i^2 of Y_l^3 and X^3. These operators correspond to $P \times SU(3) \times U(2)_{ew}$ localized with respect to Y_l^3 in the general case whereas the restriction of X_l^3 to surfaces $\delta M_{\pm}^4$ reduce from Poincare group P to M^4, or more precisely two $E^2 \subset M^4$: this option is indeed the one consistent with the mass calculations. The counterpart of the stringy mass formula would result from the vanishing of $L_0(SKM)$.

The Dirac equations associated with these Dirac equations two kinds of Dirac operators can be regarded as independent, and the correlation comes from the assumption that super-conformal algebra generates super-canonical spectral flows acting as local gauge symmetries and braiding operations at global level. 7–3 duality indeed realizes this correlation concretely. In this section only the general mass formula is discussed. A more detailed discussion of super Virasoro conditions and configuration space Dirac operators can be found in the chapter [B4].

11.2.1 The analog of coset construction for super generators

The analog of Olive-Goddard-Kent coset construction for super Kac-Moody and super-canonical algebras provides a highly attractive manner to obtain a Virasoro algebra with a vanishing central extension term. The coset construction for groups G and H relies on $H \subset G$ relation. The commutators of the Virasoro generators L_G with generators L_H satisfy $[L_G, L_H] = [L_H, L_H]$ by the subgroup property in Sugawara construction. This relationship generalizes to the anticommutators of super generators: $\{G_G, G_H\} = \{G_H, G_H\}$.

The following arguments support the view that Kac Moody algebra can be effectively regarded as a sub-algebra of the super-canonical algebra.

a) By 7–3 duality both canonical and Kac-Moody algebras define two basis which are in 1-1 correspondence with each other. They need not however both correspond to the tangent space basis for non-zero modes of the configuration space. Rather, Kac-Moody algebra could define zero modes in one-one correspondence with super-canonical degrees of freedom contributing to the configuration space metric.

b) Essential for the coset construction is that super-canonical conformal algebra acts effectively as Super Kac-Moody conformal algebra on Super Kac-Moody algebra so that one has $[L_G - L_H, L_H] = 0$. This is indeed true in the geometric sense if the induced metric of the partonic 2-surface X^2 suffers only an X^2-conformal scaling in the super-canonical conformal transformations realized as S^2-local radial scalings of the light like coordinate r_M of $M_+^4 = S^2 \times R_+$. This point will be discussed in detail in the next section.

c) One can wonder to what aspects of the coset construction are preserved in the recent case. Super-canonical basis is highly degenerate and this suggests that one can symbolically write $SC = SKM + degeneracy$, where "degeneracy" refers to generators which do not affect the configuration metric at all. If this is the case, Kac-Moody algebra can be regarded as a subalgebra of super-canonical algebra. The geometric argument suggests that the action of the super-canonical degeneracy algebra on super Kac-Moody conformal algebra is trivial.

The coset construction giving $L_n = L_n(SKM) - L_n(SC)$ means at the level of super generators G_n the expression

$$
\begin{aligned}
G_n^+ &= G_n^+(SKM) - G_n^+(SC) \;, \\
G_n^- &= G_n^-(SKM) - G_n^-(SC) \;.
\end{aligned}
\tag{11.2.1}
$$

Here $\pm$ refers to the positive and negative energy space-time sheets meeting at X^2 since "hermitian conjugation" (ket $\leftrightarrow$ bra) corresponds geometrically to a replacement of the space-time sheet with a positive time orientation with that having a negative time orientation. Note also that the super generators possess fermion number so that creation operators are replaced by annihilation operators in the conjugation.

If the conditions

$$
\{G^+(SKM), G^-(SC)\} = \{G^+(SKM), G^-(SKM)\}
\tag{11.2.2}
$$

hold true, the anti-commutators give just the differences $L(SKM) - L(SC)$. If 7–3 duality corresponds to a coordinate change in CH, the central extension charges for SKM and SC Super Virasoro algebras are same and resulting Super Virasoro algebra has vanishing central extension terms.

This construction works perfectly for quark sector. Leptonic super-canonical generators defined assuming the presence of operators D_+ and D_-^{-1} however anti-commute to products of configuration space Hamiltonians rather than components of Kähler form. Hence the formula for super generators $G(SC)$ does not work. The conclusion can be avoided if there exists also a leptonic representation CH gamma matrices anti-commuting to Kähler form. As already found, this kind of representation is obtained by replacing the operators D^+ and D_1^{-1} in the definitions of the super generators by projectors to the space of spinor modes with non-vanishing generalized eigen value of $D_\pm$. Also quark like super generators anti-commuting to a function algebra exist with this assumption.

The Ramond-NS degeneracy is a further important factor which must be taken into account. Possible sparticles are associated with Ramond type representation and generated by the super generators defined by the covariantly constant right handed neutrino. The helicity for neutrino is fixed by requiring that the neutrino spinor is proportional to the operator $n^k \gamma_k$, where n^k corresponds to the radial light like vector for $M_\pm^4$. If the super generators containing D and D^{-1} correspond to a Ramond type representation then covariantly right handed neutrino spinor defines an identically vanishing super generator and spartners with degenerate masses are absent. In NS sector the conformal weights of the generators generating ground states are $\pm 1/2$ so that there is no problem with the ground state degeneracy. The conclusion would be that quark/lepton like super generators G correspond to Ramond/N-S type representation. Note that leptonic super generators $G(SC)$ must carry half odd integer conformal weight.

11.2.2　General mass formula

The vanishing condition for $L_0(SKM) - L_0(SC)$ leads to a general mass formula in quark and lepton sectors separately. One of the blessings of 7–3 duality is that one can treat different 2-surfaces X_i^2 as almost independent degrees of freedom. In the case of translations this does not seem to be true since independent translations lead the surfaces X^2 outside $X_l^3 \times CP_2$. Therefore one must consider two options.

a) If one neglects the correlation between the translations and assigns to each X_i^2 independent translational degrees of freedom a separate mass formula for each X_i^2 would result:

$$M_i^2 \;=\; -\sum_i L_{0i}(SKM) + \sum_i L_{0i}(SC) \ . \tag{11.2.3}$$

Here $L_{0i}(SKM)$ contains a CP_2 cm term giving the CP_2 contribution to the mass squared known once the spinorial partial waves associated with super generators used to construct the state are known.

b) Perhaps the only internally consistent option is based on the assignment of the mass squared with the total cm. This would give

$$M^2 \;=\; \left(\sum_i p_i\right)^2 = \sum_i M_i^2 + 2\sum_{i\neq j} p_i \cdot p_j = -\sum_i L_{0i}(SKM) + \sum_i L_{0i}(SC) \ . $$
$$\tag{11.2.4}$$

Here $L_{0i}(SKM)$ contains a CP_2 cm term giving the CP_2 contribution to the mass squared known once the spinorial partial waves associated with super generators used to construct the state are known. $L_0(SC)$ term contains only leptonic or quark oscillator operators unless one allows both the lepto-quark type gamma matrices involving both D^+ and D_-^{-1} and leptonic gamma matrices involving instead of $D_\pm^{\pm 1}$ the projector P to the spinor modes with a non-vanishing eigenvalue of D.

The decomposition of the net four momentum to a sum of individual momenta can be regarded as subjective unless there is a manner to measure the individual masses. It might be that there is no unique assignment of momenta to individual partons and that this non-uniqueness is part of the gauge symmetry implied by 7–3 duality.

Mass formula before massivation

If interactions can be neglected one could assume that the free particle mass formulas

$$M_i^2 \;=\; L_{0i}(SKM) + h_i(vac) \ , \tag{11.2.5}$$

where $h_i(vac)$ corresponds to vacuum conformal weight are satisfied and what remains is the condition

$$2\sum_{i\neq j} p_i \cdot p_j = \sum_i L_{0i}(SC) - \sum_i h_i(vac) \;=\; 0 \ . \tag{11.2.6}$$

In the case of single Kac Moody algebra this condition would be replaced with

$$2\sum_{i\neq j} p_i \cdot p_j \;=\; 0 \ . \tag{11.2.7}$$

The presence of the super-canonical algebra obviously allows much more flexibility but it seems that continuum mass spectrum is not possible.

The conformal weights of the super-canonical algebra related closely to the zeros of Riemann Zeta are complex and for the physical states the sum of the imaginary parts of the conformal weights must vanish: this is satisfied if the zeros and their conjugates appear as conformal weights of generators creating the state. The presence of super-canonical half integer contribution from super-canonical sector could explain the negative values of the ground state conformal weights forced by p-adic mass calculations.

The above formula does not contain all that is needed: also a contribution from modular degrees of freedom associated with the complex structures of X_i^2 is necessary in order to understand the dependence of the mass of fermion families on the genus of X_i^2 [F3].

Hadron physics offers the most obvious application for the mass formula. The often used metaphor for the hadronic collision as a mini big bang would have a precise meaning in TGD framework, and the effective 2-dimensionality would provide a precise realization for the parton model of hadrons. The presence of both algebras could be essential for understanding the relationship between hadron and quark masses, and the presence of super-canonical spin could allow insights to the problem of proton spin.

Particle massivation

There is an objection against the proposed mass formula. The individual contributions to the mass squared eigenvalues are not fixed uniquely. The proposed identification of parton masses is only the simplest possibility, and one could add to the individual parton conformal weights contributions compensated by the super-canonical conformal weights. This might however be a blessing rather than a curse since it suggests a manner to understand particle massivation at the fundamental level.

a) A natural requirement is that parton masses are consistent with the poles of the S-matrix elements [C2]. This assumption is quite general and certainly makes sense if the S-matrix elements allow a decomposition to vertices and propagators for a tree diagram.

b) The time evolution for the quantum states of individual partons, which are always on mass shell (that is generalized eigen states of the modified Dirac operator) is a unitary process, and corresponds to a braiding for an N-puncture system defined by the product of N local operators creating the parton state. The basic requirement is that the flow contains information about the presence of other partons and thus about the normal derivatives of the imbedding space coordinates at X_l^3. Hence the S-matrix indeed contains information about the interior of the space-time surface and the effective two-dimensionality is indeed only effective. The condition that the S-matrix elements remain unchanged in conformal transformations obviously poses explicit conditions on the normal derivatives and can be regarded as conditions stating the vanishing of the corresponding beta function.

The best candidate for the flow is as the hydrodynamic flow defined by the discontinuity of the energy momentum tensor associated with the Kähler action at X_l^3 representing what might be regarded as a hydrodynamical shock wave. This flow is in general not integrable in the sense that one could assign a global coordinate varying along the flow lines. By identifying the points of X_i^2 and X_f^2 having the same value of the complex coordinate z, the flow $X_i^2 \rightarrow X_f^2$ defines a map $w : (x, y) \rightarrow (u(x, y), v(x, y))$ mixing the points of X_i^2. The inner products of the states created by the local operator $\Psi_n(x, y)$ creating one-parton state at X_i^2 with the state created by the transformed operator $\Psi_n(u(x, y), v(x, y))$ define correlation functions, which vanish above some length scale determining the mass of the parton. Massivation occurs if this map fails to be a conformal transformation.

b) By quantum classical correspondence this braiding can be also regarded as a braiding for the points of X_i^2, which correspond to the super-canonical conformal weights just like the points of the celestial sphere correspond to momenta. If the number of operators creating the parton state is larger than two, the minimal number three of threads in the braid is present and the conformal weights become "off mass shell" conformal weights. The massivation however can in principle occur always. In [O3, C5] the proposal was made that the bound state conformal weights could correspond to the zeros of poly-zetas: obviously this is a very strong prediction. Riemann Zeta would correspond to Higgs zero phase with the minimum of 2 operators with conjugate super-canonical conformal weights creating the state.

c) The successful description of particle massivation in terms of p-adic thermodynamics for the Virasoro generator L_0 (with p^2 not included) of the partons encourages to think that the change of the parton conformal weight could be understood as a generation of a thermal conformal weight by the flow induced by an ergodic braiding flow. This interpretation would allow to circumvent the problem created by the fact that the thermalization for mass squared is not consistent with the Lorentz invariance.

What is the flow defining the braiding?

The basic condition on the braiding is that it contains information about the interior of X^4 and thus about interactions with other partons. Second constraint is that the braiding flow is trivial for massless particles such as photon for which the space-time correlate should correspond to X_l^3 carrying a light-like energy momentum current.

The components $X^{n\alpha}$ of some tensor field with α restricted to X_l^3 define the most natural candidate for the braiding flow. The existence of the preferred light-like normal coordinate x^n constant at X_l^3 (in the case of δM_+^4 the light like coordinates would be $x^\pm = t \pm r$) is essential to achieve general coordinate invariance.

The discontinuity of the normal component $T^{n\alpha}$ of the energy momentum tensor associated with the Kähler action is a good candidate. At light like CDs the discontinuity of $T^{n\alpha}$ could be non-vanishing if allows light like CD to carry a shock wave also in imbedding space degrees of freedom as suggested by the super-symmetry. The discontinuity of $T^{n\alpha}$ would have an interpretation as a shock wave like hydrodynamic flow at the boundary. For massless particles the energy momentum current would have only a longitudinal component, the braiding would be trivial and particle would remain massless. The appearance of the energy momentum tensor in the definition of the S-matrix conforms with the hydrodynamic character of field equations and with the fact that the theory must be also a quantum theory of gravitation.

This guess is supported by the modified Dirac equation. By multiplying the modified Dirac equation at X_l^3 for shock waves localized at X_l^3 with the o_t defined by the light like gamma matrix along X_l^3 and doing the anti-commutations with the modified Dirac operator D, one finds

$$T^{\alpha n} D_\alpha \Psi = 0 \ . \tag{11.2.8}$$

The equation states that Ψ is covariantly constant along the flow lines of the flow defined by $T^{\alpha n}$. The equation can be written as

$$\begin{aligned} D_t \Psi + v^i D_i \Psi &= 0 \ , \\ v_i &= \frac{T^{ni}}{T^{nt}} \ . \end{aligned} \tag{11.2.9}$$

This differs from a standard flow equation for a quantity Ψ moving along field lines only by the fact that ordinary derivatives ∂_α are replaced by covariant derivatives D_α. This means that Ψ suffers a braiding transformation in spin and electro-weak degrees of freedom. Obviously, this equation states super-conformal invariance in the sense that it is not possible to poses the values of Ψ arbitrarily in entire X_l^3 but only at X^2. One could regard TGD as anyonic shydrodynamics at the space-time level. The usual dispersion characterizing Schödinger equation emerges only at the level of imbedding space when one assigns wave equations to propagators defined by S-matrix elements.

Since the induced spinor connection is continuous at X_l^3, the discontinuity for this equation reads as

$$\begin{aligned} w^i D_i \Psi &= 0 \ , \\ w_i &= \Delta \left[\frac{T^{ni}}{T^{nt}}\right] \ , \end{aligned} \tag{11.2.10}$$

thus defines an adiabatic series of braiding flows.

The naive guess is that unitarity requires that the ordinary inner product for scalar functions is preserved in the flow so that the flow defined by v^i is volume preserving. This would require that the vector field $v^i = T^{ni}/kT^{nt}$, which is analogous to a velocity field, has a vanishing diverge for some choice of the function k and thus defines a symplectic flow of X^2 generated by some Hamiltonian. This would pose an additional condition to the solution of field equations at X_l^3.

The correspondence with Higgs mechanism

A priori one can imagine three sources of mass.

a) The mixing caused by the flow reduces correlations unless it induces a conformal transformation leaving the physical state invariant. This mechanism could obviously relate to the massivation described in terms of p-adic thermodynamics.

b) The lacking covariant constancy of the charge matrices appearing in the definition of electro-weak gauge bosons except photon can induce a loss of correlations, and is a good candidate for TGD counterpart of Higgs coupling. Also fermionic states involve this kind of dependences inducing a Higgs type contribution to the mass squared.

c) In contrast to the first guess, the holonomy transformation associated with braiding does not lead to a loss of correlations. The 2-dimensionality of X^2 implies that the holonomy group defining the braiding is Abelian. In spin degrees of freedom spinor connection vanishes by the flatness of M^4 so that the braiding cannot make neither graviton nor gluons massive. The vanishing of photon mass requires that the braiding in elecro-weak spin degrees of freedom is generated by an element of the holonomy algebra commuting with the electromagnetic charge. This however means that for charge eigen states the braiding brings in only a phase factor and would not lead to a loss of correlations.

These observations lead to a view about how TGD description relates to the standard description in terms of the phases defined by the vacuum expectation values of the Higgs field.

The model for the modular contribution to the mass squared to be discussed suggests that the p-adic prime p characterizes the size and effective p-adic topology of X^2. This raises the question why the states created by super-canonical operators O_q, $q \neq p$ do not appear in the spectrum. The intuitive guess is that the states corresponding to "wrong" p-adic topology are very massive. For instance, for $q \neq p$ the states could have inverse p-adic temperature $T_q = n > 1$ possible if roots of q are allowed in the extension of R_q used. Various phases would be characterized by the primes labelling various operator basis O_q.

One can consider also an alternative interpretation based on the notion of conformal homotopy class implying that p-adic prime p characterizes X_l^3 rather than X^2.

a) Homotopy theory could generalize in the sense that there are several conformal equivalence classes of space-time sheets connecting given partonic 2-surfaces X_i^2 at various 7-D CDs representing initial and final states. The derivatives of the imbedding space coordinates with respect to the light like coordinate t at X_i^2 are same for these surfaces. These conformal homotopy classes could define the TGD counterparts of the "phases" of gauge theories labelled by vacuum expectation values of Higgs fields.

b) In the trivial conformal equivalence class braiding flow reduces to a mere conformal transformation: photon, graviton, and gluons remain massless. The thermal massivation of photon might be describable in terms of a non-trivial conformal homotopy class.

c) The success of the p-adic mass calculations would suggest that p-adic primes label conformal homotopy classes and thus phases and characterize the mass scale as well as the effective p-adic topology for which p-adic non-determinism corresponds to the classical non-determinism of Kähler action. p-Adic prime would characterize X_l^3 rather than X^2.

There are however objections against this interpretation. First, the notion of conformal homotopy class is not consistent with the notion of generalized Feynman diagram requiring that even topologically different surfaces X_l^3 are equivalent. Secondly, the model for the modular contribution to the mass squared forces to assign p-adic prime p to X^2 rather than X_l^3.

11.3 The relation between super-canonical and Super Kac-Moody algebras

The understanding of the relationship of Super Kac-Moody and super-canonical conformal symmetries is essential if one wants to justify the generalized coset construction. The basic problems are following.

a) How SKM algebra is represented and how its central extension emerges? How SKM algebra acts on SC algebra?

b) The action of super-canonical conformal algebra on super Kac-Moody algebra coincides with own action on it: $[L_{SC} - L_{SKM}, L_{SKM}] = 0$. How this commutation relation can be understood in the recent situation which is more general than that associated with the coset construction? Can SKM algebra in some sense be regarded as a sub-space or even sub-algebra of SC algebra as in the case of coset construction? Could $H \subset G$ relation correspond to the fact that isometries of H form a subgroup of canonical transformation acting also as conformal transformations?

11.3.1 Conformal symmetries and modular invariance

The full SKM invariance means that the super-conformal fields depend only on the conformal moduli of 2-surface characterizing the conformal equivalence class of the 2-surface. This means that all induced metrics differing by a mere Weyl scaling have same moduli. This symmetry is extremely powerful since the space of moduli is finite-dimensional and means that the entire infinite-dimensional space of deformations of parton 2-surface X^2 degenerates to a finite-dimensional moduli spaces under conformal equivalence. Obviously, the configurations of given parton correspond to a fiber space having moduli space as a base space. Super-canonical degrees of freedom could break conformal invariance in some appropriate sense.

Conformal and SKM symmetries leave moduli invariant

Conformal transformations and super Kac Moody symmetries must leave the moduli invariant. This means that they induce a mere Weyl scaling of the induced metric of X^2 and thus preserve its non-diagonal character $ds^2 = g_{z\bar{z}}dzd\bar{z}$. This is indeed true if
a) the Super Kac Moody symmetries are holomorphic isometries of $X^7 = \delta M^4_{\pm} \times CP_2$ made local with respect to the complex coordinate z of X^2, and
b) the complex coordinates of X^7 are holomorphic functions of z.
   Using complex coordinates for X^7 the infinitesimal generators can be written in the form

$$J^{An} \quad = \quad z^n j^{Ak} D_k + \bar{z}^n j^{A\bar{k}} D_{\bar{k}} \ . \tag{11.3.1}$$

The intuitive picture is that it should be possible to choose X^2 freely. It is however not always possible to choose the coordinate z of X^2 in such a manner that X^7 coordinates are holomorphic functions of z since a consistency of inherent complex structure of X^2 with that induced from X^7 is required. Geometrically this is like meeting of two points in the space of moduli.
   Lorentz boosts produce new inequivalent choices of S^2 with their own complex coordinate: this set of complex structures is parameterized by the hyperboloid of future light cone (Lobatchevski space or mass shell), but even this is not enough. The most plausible manner to circumvent the problem is that only the maxima of Kähler function correspond to the holomorphic situation so that super-canonical algebra representing quantum fluctuations would induce conformal anomaly.

The isometries of δM^4_+ are in one-one correspondence with conformal transformations

For CP_2 factor the isometries reduce to $SU(3)$ group acting also as canonical transformations. For $\delta M^4_+ = S^2 \times R_+$ one might expect that isometries reduce to Lorentz group containing rotation group of $SO(3)$ as conformal isometries. If r_M corresponds to a macroscopic length scale, then X^2 has a finite sized S^2 projection which spans a rather small solid angle so that group $SO(3)$ reduces in a good approximation to the group $E^2 \times SO(2)$ of translations and rotations of plane.
   This expectation is however wrong! The light-likeness of δM^4_+ allows a dramatic generalization of the notion of isometry. The point is that the conformal transformations of S^2 induce a conformal factor $|df/dw|^2$ to the metric of δM^4_+ and the local radial scaling $r_M \to r_M/|df/dw|$ compensates it. Hence the group of conformal isometries consists of conformal transformations of S^2 with compensating radial scalings. This compensation of two kinds of conformal transformations is the deep geometric phenomenon which translates to the condition $L_{SC} - L_{SKM} = 0$ in the sub-space of physical states. Note that an analogous phenomenon occurs also for the light-like CDs X^3_l with respect to the metrically 2-dimensional induced metric.
   The X^2-local radial scalings $r_M \to r_M(z, \bar{z})$ respect the conditions $g_{zz} = g_{\bar{z}\bar{z}} = 0$ so that a mere Weyl scaling leaving moduli invariant results. By multiplying the conformal isometries of δM^4_+ by z^n (z is used as a complex coordinate for X^2 and w as a complex coordinate for S^2) a conformal localization of conformal isometries would result. Kind of double conformal transformations would be in question. Note however that this requires that X^7 coordinates are holomorphic functions of X^2 coordinate. These transformations deform X^2 unlike the conformal transformations of X^2. For X^3_l similar local scalings of the light like coordinate leave the moduli invariant but lead out of X^7.

Canonical transformations break the conformal invariance

In general, infinitesimal canonical transformations induce non-vanishing components $g_{zz}, g_{\bar{z}\bar{z}}$ of the induced metric and can thus change the moduli of X^2. Thus the quantum fluctuations represented by super-canonical algebra and contributing to the configuration space metric are in general moduli changing. It would be interesting to know explicitly the conditions (the number of which is the dimension of moduli space for a given genus), which guarantee that the infinitesimal canonical transformation is moduli preserving.

What is the proper interpretation of 7–3 duality?

That KM algebra does not affect moduli is quite interesting and raises the question whether KM transformations correspond to zero modes or not. If the answer is affirmative, Kac Moody algebra cannot provide as such an alternative tangent space basis for the configuration space. For a trivial central extension Kac-Moody transformations act like gauge transformations and this would suggest that Kac-Moody transformations indeed act in zero modes and perhaps parameterize them.

This expectation is in conflict with the naive form of 7–3 duality stating that SC and and SKM algebras are two equivalent basis for the tangent space of the configuration space and related by a general coordinate transformation of the configuration space. A more general statement would be that these algebras represent two equivalent manners to characterize the tangent space of the configuration space. Only this is needed by the argument suggesting that SC and SKM conformal algebras have the same central extension. In TGD inspired quantum measurement theory zero modes are interpreted as non-quantum fluctuating classical degrees of freedom and a 1-1 correlation between quantum numbers and zero modes is assumed. This guarantees that the position of the pointer of an ideal measurement apparatus correlates in one-one manner with the measured state of the quantum system. Thus 7–3 duality would correspond to the basic assumption of quantum measurement theory.

11.3.2 How SKM algebra is represented?

Super Kac-Moody algebra involves 5 sectors corresponding to the isometries and holonomies of $\delta M_{\pm}^4 \times SU(3)$. The Kac-Moody generators of isometries $SO(3) \times SU(3)$ are representable as imbedding space transformations whereas holonomies are represented as fermionic bilinears. The value of the central extension parameter k is naturally same for all sectors.

The representation of $SO(3) \times SU(3)$ SKM algebra

The simple guess is that SKM generators act on super-canonical generators by acting on imbedding space coordinates directly as infinitesimal generators and in this manner induce the transformation of super-canonical generators defined as flux integrals over surfaces X^2_i. Their action on the modes of the imbedding space spinor fields appearing in the state construction would be completely analogous.

The central extension represents however an important delicacy. Central extension does not occur in the action on super-canonical generators. In the case of spinor harmonics of H, the central extension is realized by adding to the isometry generator and spin rotation term an integer multiple kH_A of the Hamiltonian of X^7-canonical transformation

$$ j^{Ak}D_k \quad \rightarrow \quad j^{Ak}D_k + (D^l j^{Ak} - D^k j^{Al})\frac{\Sigma_{kl}}{2} + kH_A \ . \tag{11.3.2}$$

A similar modification works also for the commutators of SKM generators.

The Hamiltonians of the isometries, which do not belong to the Cartan sub-algebra, allow a complexification. For instance, $H_{\pm} = sin(\theta)exp(\pm i\phi)$ corresponds to $J_{\pm}$ for $SO(3)$). kH_A generalizes to $kz^n H_A^+$ and $k\bar{z}^n H_A^-$. For diagonal generators Hamiltonians remain real (for instance, $H_3 = cos(\theta)$) so that one has $H_A^+ = z^n, H_A^- = \bar{z}^n H_A$.

This prescription predicts that the central extensions are same for SC and SKM algebra which gives hopes that the Super Virasoro algebras have the same central extension as suggested by the possibility to compensate the local radial scalings in super canonical algebra by conformal transformations of X^2 in SKM sector.

It deserves to be noticed that no conformal spinor fields are needed in this approach, and that the introduction of this kind of spinor fields would imply difficulties.

a) Since anticommutations are expressed as a normal ordering prescription for $\Psi(z)\Psi(w)$, Majorana condition would appear naturally and lead to difficulties with the separate conservation of lepton and quark numbers.

b) The anti-commutator for positive and negative energy components of Ψ corresponding to positive and negative powers of z would be proportional to $1/(z-w)$ and have essentially one-dimensional character. On the other hand, 2-dimensional delta function commutator for Ψ and $\Psi^\dagger$ is implied by the construction of super-canonical generators. Effective 1-dimensionality would reduce TGD to a string model like theory and this would be highly undesirable.

The representation of SKM algebra associated with holonomies

One might hope that the construction of the super Kac-Moody algebra associated with the holonomies of $\delta M_\pm^4 \times CP_2$, in particular those forming electro-weak gauge group, could run roughly along the same lines as when conformal fields are used.

a) Construct fermionic generators F_r^A, where A is an index labelling electro-weak charges and r either a half odd integer or integer, satisfying the anti-commutation relations

$$\{F_r^A, F_s^A\} \;=\; k\delta(r+s) \ . \tag{11.3.3}$$

k has the same value as the coupling to to Hamiltonian in the symplectic extension in the case of color and $SU(2)$ or E^2 generators. In standard conformal field theory representation in terms of Majorana spinors fermionic generators with positive and negative values of r are components of the same field Ψ. In the recent case they must correspond to components of Ψ and $\Psi^\dagger$. Hence the anti-commutations must be replaced with

$$\begin{aligned}
\{F_r^A, F_s^{A\dagger}\} &\;=\; k\delta(r-s) \ , \\
F_{-r}^A &\;=\; F_r^{A\dagger} \ .
\end{aligned} \tag{11.3.4}$$

b) Construct bosonic generators as bi-linears of fermionic generators

$$T_n^A \;=\; \frac{1}{\sqrt{k}} \sum_m f_{BC}^A F_{n-m}^B F_m^C \tag{11.3.5}$$

The commutation relations for the Kac-Moody generators follow from the fermionic anti-commutation relations and central extension term involving two fermionic contractions is indeed proportional to k:

$$[T_m^A, T_n^B] \;=\; f_C^{AB} T_{m+n}^C + k\delta_{A,B}\delta(m+n) \ . \tag{11.3.6}$$

c) The basic experimental input from the p-adic mass calculations is that the number of sectors of the Super Virasoro algebra is five. If the bosonic Kac Moody sectors correspond to $SO(3) \times SU(3)$ and $SU(2)_L \times U(1)$, there are only 4 sectors. The construction must be however performed also for the holonomy algebra $U(1)$ of $\delta M_+^4 = S^2 \times R_+$. The covariantly constant sigma matrices of $E^3 \subset M^4$ assignable to the rest frame defined by S^2 in $\delta M_+^4 = S^2 \times R_+$ extend this representation to a representation of $SO(3)$. There are thus 5 sectors altogether.

The construction reduces to the construction of fermionic generators but there are some important differences in comparison to the standard construction.

a) The combination of super generators F_r^A to form a conformal fermion field is neither possible nor needed. The super generators carrying lepton number can be expressed as matrix elements of electro-weak charges between covariantly constant right handed neutrino spinor field and second quantized induced spinor field:

$$\begin{aligned}
F_r^A &\;\propto\; \sqrt{k} \int_{X^2} \overline{\nu}_R o z^r Q_A \Psi \ , \quad r \geq 0 \\
o &\;=\; t^\alpha \Gamma_\alpha \ .
\end{aligned} \tag{11.3.7}$$

t^α denotes the light like tangent vector field of X_l^3 at X^2. The integer k is Kac-Moody central extension parameter. r can be either half odd integer (NS) or integer (Ramond) valued. It is essential that Γ_α

contains CP_2 gamma matrices: otherwise Q_A would annihilate $\overline{\nu}_R$. This means that CP_2 coordinates depend on the light like coordinate.

b) In the recent case use of negative powers of z would imply singularity and F^A_{-n} is defined as the Hermitian conjugate of F^A_n.

$$F^A_{-r} \quad \propto \quad \sqrt{k} \int_{X^2} \overline{\Psi} \overline{z}^r Q_A o \nu_R \ , \quad r \geq 0 \tag{11.3.8}$$

This means the replacement of z^n by $\overline{z}^n$ and performing Hermitian conjugation. The inner product should give a Kronecker delta $\delta(m+n)$ essentially due the conservation of angular momentum associated with the phase of z^n. The values of the conformal weight are now *positive* for both signs of n. There seems to be an interesting correspondence with the zeros of Riemann Zeta: super-canonical conformal weights correspond to combinations of non-trivial zeros of Zeta and Kac Moody conformal weights to $-z/2$, $z = -2n$, $n = 1, 2, \ldots$. a trivial zero of Zeta.

c) Quarks cannot contribute to Super Kac-Moody algebra. The generators should be of the form

$$F^A_n \quad \propto \quad \sqrt{k} \int \overline{\nu}_R z^n Q_A \Psi \ , \tag{11.3.9}$$

and would vanish identically since right handed neutrino spinor carries vanishing electro weak charges. The situation could be the same in the super-canonical sector since bilinears involving ν_R and second quantized quark field are somewhat questionable.

d) The $n = 0$ fermionic generators of Ramond sector responsible for a potential particle-sparticle degeneracy vanish identically in all sectors of Super Kac Moody algebra. First of all, all electro-weak charge matrices except electromagnetic charge matrix carry anomalous conformal weight $h_c = 2$. Electromagnetic charge matrix is covariantly constant and hence $n = 0$ Ramond generator vanishes since the modes of Ψ correspond to eigen values $\lambda \neq 0$ of D whereas ν_R corresponds to $\lambda = 0$. Therefore the $n = 0$ generators responsible for the sparticles in electro-weak degrees of freedom vanish identically. In $SO(3)$ degrees of freedom conformal weight of isometry generators vanishes and infinite degeneracy results unless these degrees of freedom are frozen.

11.3.3 How SKM algebra acts on the super-canonical algebra

The proposed representation of SKM algebra allows to define its action on super-canonical algebra. The action on SKM generator $z^n j_A$ on SC Hamiltonian H_B gives a term of type $z^n \{H^+_A, H_B\} + \overline{z}^n \{H^-_A, H_B\}$ For fermionic generators the action on j^k_B appearing in the SC generator gives $z^n [j^+_A, j_B]^k + \overline{z}^n [j^-_A, j^k_B]$. The anti-commutators of SC and SKM super-generators are defined in a similar manner.

The function resulting from the action of SKM generator is a real function of the configuration space coordinates and thus can be identified configuration space Hamiltonian. It is however not at all clear whether it belongs to the super-canonical algebra induced by the Hamiltonians of X^7 and thus induces configuration space isometries.

One can imagine two options.

a) With a proper interpretation of the powers z^n the closure in fact occurs.

b) The extension of the super-canonical algebra to include all the commutators resulting in this manner is accepted. This option is however clumsy since the computation of Poisson brackets would not reduce to the imbedding space level anymore.

Arguments in favor that the closure occurs

Modular invariance means that same z-coordinate charts can be used for a huge number of 2-surfaces. In fact, the finite-dimensional moduli space parameterizes these coordinate charts. This raises the idea that there are some canonical choices of X^2 complex coordinate, and that the coordinate z can be identified as a complex coordinate w of X^7 such that w coordinatizes some 2-surface Y^2 when the radial coordinate r_M and two remaining complex coordinates are kept constant. In this kind of situation the configuration space Hamiltonians resulting as commutators of SKM and SC algebra elements would have interpretation as Hamiltonians induced from the function algebra of H and hence belonging to the super-canonical algebra.

a) The challenge is to identify uniquely the complex imbedding space coordinate w. Quite generally, any 2-dimensional complex variety Y^2 in X^7 with a decomposition $\delta M_+^4 = S^2 \times R_+$ defines a good candidate for Y^2. Lorentz transformations leaving the dip of light cone boundary invariant produce a 3-dimensional family of different choices of $S^2 \times R_+$ decompositions, and some of them might be consistent with the complex structure.

A good guess is that Y^2 should have same topology as X^2. The metric of X^2 would differ by a Weyl scaling from the metric of Y^2. Kac-Moody symmetries and corresponding conformal transformations would give rise to huge family of Weyl scaled variants of X^2 and could define the family of surfaces X^2 associated with a particular Y^2.

b) The requirement that X^2 is a complex sub-manifold of X^7 is not in accordance with the idea that X^2 can be any sub-manifold of X^7. The difficulty is circumvented if only the maxima of Kähler function correspond to holomorphic sub-manifolds and if quantum fluctuations represented by super-canonical algebra break super-conformal invariance and induce central extension as a symplectic extension. This picture would provide precise geometric interpretation for the generation of the conformal anomaly.

c) To make things concrete, one can consider some special cases.
i) CP_2 projection of X^2 must be 2-dimensional in order that it contributes to the configuration space metric so that Y^2 might be a complex sub-manifold of CP_2 in some special cases.
ii) If the projection of X^2 to the sphere S^2 of $S^2 \times R_+$ is 2-dimensional (for the causal determinants associated with CP_2 type extremals this might not be the case) then the projection of X^2 to S^2 defines a multiple covering of X^2. The complex coordinate of X^2 is good if the induced metric for this coordinate has $g_{zz} = g_{\overline{z}\overline{z}} = 0$.

Y^2 as an Abelian sub-manifold of octonionic imbedding space?

This proposed manner to defined the action of SKM algebra in SC algebra relates in an interesting manner to various conjectures of TGD.

a) The number theoretic vision [E2] suggests that quaternionic sub-manifolds of the imbedding space endowed with an octonionic tangent space structure should be somehow relevant for TGD. In particular, Abelian 2-surfaces Y^2 of H for which tangent space correspond at each point to an Abelian sub-algebra identifiable as complex numbers should be of special importance. Also now the equivalence of the inherent complex structure of X^2 and that inherited from the octonionic imbedding space, is natural. Hence good guess is that Abelian 2-surfaces Y^2 correspond to the holomorphic sub-manifolds of X^7 defining also special coordinate systems of X^7 and X^2. Furthermore, of only the maxima of the Kähler function have Abelian character, X^2 can be arbitrary.

Space-time surfaces X^4 might allow a quaternionic coordinate using a projection to quaternionic sub-manifolds Y^4 of H to which the maxima of Kähler function would correspond. Also the Abelian variant of octonionic imbedding space [E2] based on 8-D hyper-complex numbers and using 4-D space-time surfaces for which tangent space at each point correspond to a 4-D hyper-complex sub-algebra could provide this kind of preferred coordinates.

Y^2 and hyper-ellipticity

This general vision provides new insights to the hyper-ellipticity, which implies that elementary particle vacuum functionals in the modular degrees of freedom vanish for $g > 2$ [F1]. Since the surfaces Y^2 are special, and since the conformal and Kac-Moody symmetries leave their conformal structures invariant, the natural guess is that the surfaces Y^2, and thus also the surfaces X^2 produced from them by the action of conformal and SKM symmetries, are hyper-elliptic.

The vanishing of the elementary particle vacuum functionals for the maxima of Kähler function would make the genera $g > 2$ very different from lower genera and perhaps allow the interpretation as the absence of higher particle families interpretable as stable bound states of $g > 2$ handles.

Hyper-ellipticity might have also a rather concrete geometric interpretation. Hyper-elliptic surfaces are defined by the general polynomial equation

$$w^2 \;=\; P_{2g+2}(z) \; . \tag{11.3.10}$$

Z_2 symmetry permutes the roots of w. For instance, z could correspond to S^2 coordinate and w to a complex coordinate of the CP_2 projection of Y^2. The two roots would emerge naturally since the S^2 projection has a finite size and spans a disc with holes. Thus X^2 must represent at least a double covering of the projection and the points which correspond to the same projection are permuted by the Z_2 symmetry.

11.3.4 Does the coset representation generalize?

Since SC and SKM algebras do not correspond to Kac-Moody algebras of groups G and $H \subset G$ and it is not obvious that coset representation generalizes. It is not at all clear whether Sugawara construction is needed or even possible. Hence it seems that direct geometric approach instead of the algebraic one is the best manner to prove that the generalization exists.

The obvious condition is that the action of SC conformal symmetries can be compensated by SKM conformal symmetries since in this kind of situation the net action is at least classically trivial. The local radial scaling $r_M \to |df/dz|r_M$ acts like conformal symmetry $z \to f(z)$ of X^2, and its effect on the induced metric can be compensated by the inverse of f. Thus the requirement that $L_{SC} - L_{SKM}$ annihilates physical states has a very natural interpretation, and only conformal anomalies can spoil the situation. This condition also states that the isometries of the light cone boundary leave the induced metric invariant if the coordinate w is a holomorphic function of z. Hence all the miraculous properties of light like 3-surfaces would be utilized in the construction.

The fact that the values of the central extension parameter k identifiable as a symplectic extension parameter are same for SC and SKM algebras, gives good hopes that Virasoro central extension is same for the two representations of the conformal algebra. The parameter k appears in the expression of super generators G of SC and SKM algebras as bi-linears of Kac Moody generators and super-generators. Since the form of these expressions is same and the same value of k appears in both cases in the action of bosonic generators on fermions, there are considerable hopes that the central extension terms cancel each other in the generalized coset representation.

11.4 About the construction of single parton states

The presence of the super-canonical algebra modifies the usual string model type construction of parton states profoundly. The basic distinction is the hierarchy of "ground states" of negative super-canonical conformal weight which must be compensated by positive Kac-Moody conformal weight. The state construction involves also number theoretical notions such as Riemann Zeta and its zeros. Even the notion of infinite primes seems to relate naturally to the construction. Super-canonical conformal invariance predicts an extremely rich spectrum of (even massless) states since the 2-surfaces X^2 can correspond not only to partons but also to macroscopic 2-surfaces and even to black hole horizons. Although sparticles are absent, the question whether the p-adic massivation yields a spectrum of light particles consistent with observations remains open.

11.4.1 Riemann Zeta and configuration space metric

The proposed proof for Riemann hypothesis [n4, jf2, E8] utilizes a very simple quantum system allowing conformal invariance. The states of this quantum system can be interpreted as generalized coherent states having overlaps expressible in terms of Riemann Zeta:

$$\langle z_1 | z_2 \rangle \propto \zeta(\overline{z}_1 + z_2) \ . \tag{11.4.1}$$

The requirement that the physical states are orthogonal to the negative norm state state with a vanishing conformal weight $z = 0$, implies that allowed conformal weights correspond to the zeros of Riemann Zeta.

p-Adic existence conditions lead to the hypothesis that the conformal weights of the Poisson algebra of configuration space Hamiltonians are expressible in terms of zeros of Zeta [B2, E8, E1]. The basic requirement is that the logarithmic waves $(r_M/r_0)^z$ at δM_+^4 are orthogonal with respect to the inner product defined as an integral over δM_+^4. The orthogonality allows super-canonical conformal weights

$z = n - 1/2 - i \sum n_i y_i$ with $\sum_i n_i = N$ is odd for n even and even for n odd [B2], where $z_i = 1/2 + y_i$ are zeros of Zeta.

In the configuration space inner product the integral over δM_+^4 is replaced with that over X^2, and this implies non-orthogonality, and there are no reasons to expect that the anti-commutators of super-generators labelled by the zeros of Riemann Zeta are diagonal. The model used in the proof of Riemann hypothesis encourages to think that anti-commutators of super generators giving configuration space metric and Kähler form are proportional to Riemann Zeta. Hence one would have

$$g_{A\bar{z}_1, B z_2} \quad \propto \quad \zeta(\bar{z}_1 + z_2) \ . \tag{11.4.2}$$

For zeros $z_i = 1/2 + iy_i$ of Zeta his would give

$$g_{A\bar{z}_1, B z_2} \quad \propto \quad \zeta(1 + i(y_2 - y_1)) \ . \tag{11.4.3}$$

The vanishing of the inner product for $z_1 = 0$ and z_2 a zero of Zeta might perhaps be interpreted to state the orthogonality with respect to a super generator corresponding to a constant Hamiltonian, which would otherwise generate $N = 1$ super-symmetry giving rise to mass degenerate sparticles. The educated guess is that the proposal for the inner product generalizes as such also to the general case. In the general case the orthogonality to $z = 0$ state is not achieved anymore.

The hypothesis would mean that the inner products of states created by the configuration space super generators and Hamiltonians are expressible as products of Riemann Zeta somewhat like the n-point function of a free field theory allows a decomposition into products of propagators.

11.4.2   General construction recipe for parton states

The general vision underlying the construction is following.

a) Mass squared eigenvalues are differences of conformal weights of Super Kac-Moody and super-canonical conformal weights. p-Adic thermodynamics causes thermal mixing of zero conformal weights associated with massless states with higher conformal weights. The states can be thought of as being constructed by first operating with appropriate super-canonical generators to a vacuum state with a vanishing conformal weight yielding a state with negative conformal weight $-N$ and then applying Super Kac-Moody generators with net weight $M \geq N$ to yield a non-tachyonic state. The presence of super-canonical algebra makes un-necessary the earlier assumption about negative vacuum conformal weight $h_{vac} \geq -4$. The most general assumption is that all super-canonical generators, both bosonic and fermionic are allowed. The notion of highest weight representation of Kac-Moody algebra however suggests that it is enough to include only the super-canonical generators corresponding to the highest weights of Kac-Moody representations.

b) Either Ramond or N-S Super Virasoro conditions or possibly both are satisfied by the physical states. The modification of vanishing conformal weights of massless states obtained by applying Kac-Moody Virasoro generators on massless states can be described as a process analogous to the generation of thermal energy and mass calculations reduce to the earlier calculations. The earlier assumption that massless states are mixed with hyper-massive states contradicting Poincare invariance is not needed.

c) The rather dramatic implication is that there is an infinite number of massless states since super-canonical conformal weight N can be arbitrarily large. This need not be a catastrophe if p-adic particle massivation makes these states massive. One must also remember that the sizes of surfaces X^2 can be arbitrarily large and that stable elementary particles probably correspond to some rather simple configurations. Indeed, the states containing now neutrino-antineutrino pair operators with conformal weight -1 define a unique subset of states serving as excellent candidates for observed particles in the construction involving super-canonical algebra. With some additional constraints, say to the value of the color Casimir operator of super-canonical Hamiltonians, this could give a constraint limiting the value of N at elementary particle level.

11.4.3   Conformal confinement

Conformal confinement follows from hermiticity requirement and means that the net conformal weight of any particle is real. A stronger statement would be that this holds true for partons. It is however easy to imagine that the color quantum numbers of quarks might be mapped to the super-canonical conformal weights and that quark partons are characterized by complex super-canonical conformal weights so that color confinement would reduce to conformal confinement.

The simplest super-canonical generators carry conformal weights $h = -z$, where $z = 1/2 + iy$ is a zero of Riemann Zeta. There are reasons to believe that that also the super-canonical conformal weights $z = n - 1/2 - i \sum n_i y_i$ are possible $\sum_i n_i = N$ odd for n even and even for n odd. Conformal confinement allows only combinations of super-canonical generators for which the net conformal weight is real. "Conformal mesons" would correspond to pairs of super-canonical generators for which the imaginary parts of the conformal weights have opposite sign. Arbitrary complex combinations are possible and one could speak of "conformal baryons" too.

The existence of super-generators associated with Hamiltonians depending only on the light like radial coordinate r_M implies a huge ground state degeneracy. By forming a product of the neutrino generators labelled by z and antineutrino generator labelled by $\overline{z}$ one obtains a bosonic operator with conformal weight -1. There are also two super Kac-Moody generators with conformal weight $h = -1/2$ corresponding to the righthanded neutrino and antineutrino and their product has conformal weight $h = -1$. The product of these generators defines an operator with vanishing conformal weight in the generalized coset construction. Every zero of Riemann Zeta defines an operator creating vacuum state but these vacua are not orthogonal if one assumes that the anti-commutators of super-generators are proportional to $\zeta(1 + i(y_1 - y_2))$ being thus analogous to coherent states.

11.4.4 p-Adic length scale hierarchy and zeros of Riemann Zeta

Concerning the physical interpretation of the degeneracy due to the zeros of ζ, the crucial observation is that primes and zeros of ζ seem to be in a dual relation analogous to that between momentum and position. This suggests that p-adic length scale hierarchy corresponds to a hierarchy of operators O_p labelled by primes and expressible as superpositions of operators $O_z O_{\overline{z}}$ constructed from Hamiltonians and their super counterparts.

p-Adic length scale hierarchy as a dual to the zeros of Zeta

The operators O_p (the labels relating to color and rotational quantum numbers are dropped to simplify the formulas) must have real conformal weights and must be superpositions of products $O_z O_{\overline{z}}$:

$$O_p = \sum_z O_p(z, \overline{z}) O_z O_{\overline{z}} \ . \tag{11.4.4}$$

If the anti-commutators and Poisson brackets are assumed to be proportional to the Riemann Zeta one can write this as

$$\langle 0_p^\dagger O_q \rangle = \sum_{z,w} \zeta(\overline{z} + w) \zeta(z + \overline{w}) \overline{O}_p(z, \overline{z}) O_q(w, \overline{w}) \ . \tag{11.4.5}$$

Of course, this formula is over-simplified since the indices labelling to super-canonical quantum numbers in rotational and color degrees of freedom have been suppressed.

p-Adic operators as operator valued partition functions in the set of zeros of Zeta

There must be some principle fixing the "Fourier coefficients" $O_p(z, \overline{z})$. The first guess is motivated by the form of Riemann Zeta and the physical intuition that $log(p)$ plays a role of inverse temperature.

a) In the case of the Hamiltonians H_p^A the counterparts of plane waves would be the analogs of bosonic partition functions

$$O_p(z, \overline{z}) = Z_{B,p}(z) \times Z_{B,p}(\overline{z}) \ ,$$
$$Z_{B,p}(z) = \frac{1}{1 - p^{-z}} \tag{11.4.6}$$

appearing as building blocks of Riemann Zeta.

b) For the fermionic generators the products $Z_{F,p}(z) \times Z_{F,p}(\overline{z})$ of fermionic partition functions $Z_{F,p} = 1 + p^{-z}$ [E8] are natural.

These formulas assume that both z and $\overline{z}$ are zeros of Zeta. This is the case only if the zeros are at the critical line $Re(z) = 1/2$. For $Re(z) = x \neq 1/2$ one could replace $\overline{z}$ with $1 - z$ in the previous formula in order to get real conformal weight always but in this case $O_p(z, \overline{z})$ would not be real. Conformal invariance suggests strongly reality for all values of z and thus Riemann hypothesis.

Note that the assumption implies the adelic formula reading in the bosonic case as

$$\prod_p O_p(z, \overline{z}) = \zeta(z)\zeta(\overline{z}) . \tag{11.4.7}$$

The vanishing of the right hand side for the zeros of Zeta might have some physical interpretation.

According to the earlier idea [C5], inspired by the quantum classical correspondence, super-canonical conformal weights are analogous to punctures of complex plane and physical states are created by conformal fields in this plane. One could indeed interpret the coefficients $O_p(z, \overline{z})$ as modes of conformal fields restricted to the set of zeros of Zeta and labelled by primes. There is no need to emphasize the implications of this hypothesis: together with the evolution of the Kähler coupling strength it would fix the p-adic coupling constant evolution completely.

The construction of the scalar propagator as a partition function in super-canonical algebra discussed in [C5] inspired the hypothesis that the ordered hierarchy of sub-algebras $A_1 \subset A_2 \subset \ldots$ of super-canonical algebra, with A_n generated by the generators with conformal weights $z_i = 1/2 + y_i$, $y_1 < y_2 < \ldots y_n$, could define a hierarchy of phases. The nearer the space-time sheet would be to a vacuum extremal, the more critical it would be, and the larger the number of zeros involved would be. The increasing number of massless states would correspond to an increasing number of modes representing long range fluctuations. This vision would require a hierarchy of z-cutoffs in the expression of $O(z, \overline{z})$ giving

$$O_{p,n}(z, \overline{z}) = \frac{1}{1 - p^{-z}} \times \frac{1}{1 - p^{-\overline{z}}} \times \theta(y_n - y) . \tag{11.4.8}$$

Can the operators O_p create orthogonal states?

One might hope that the operators O_p are orthogonal:

$$G(p, q) \propto \delta(p, q) . \tag{11.4.9}$$

It is however far from obvious that the orthogonality conditions can be satisfied or that they are are even physically plausible.

a) The over-completeness of the basis labelled by the zeros of zeta would naturally correspond to an over-completeness of the basis labelled by primes. One could even think the possibility that the over-completeness of O_z implies that all primes p are not needed and that, in accordance with the p-adic length scale hypothesis, only primes near prime powers of two are enough.

b) The model for the modular contribution to the mass squared suggests that the p-adic prime p characterizes X^2 rather than X_l^3. This means that if X^2 has an effective p-adic topology corresponding to R_p, the classical non-determinism is modellable by p-adic non-determinism in some length scale range. Various operator basis $\{O_q\}$, q prime, correspond to an infinite number of different phases. One can imagine that the temperature T_p for the phase $q \neq p$ could be high. For $T_p(q) = n > 1$, which is possible if roots of p are allowed in the extension of R_p, the mass scale would be of order CP_2 mass.

c) Higgs zero phase would correspond to CP_2 type extremals for which the operators O_p for various primes reduce to same operator apart from the normalization constant since the projection of X^2 to $\delta M_{\pm}^4$ becomes zero-dimensional. This phase would be unstable but the assumption that elementary particles are near to this phase allows to identify a mechanism eliminating light exotics. It is also possible to consider a situation in which different multi-p p-adic topologies are possible. These could relate to different fractal behaviors/classical non-determinisms of the 4-surface X^4 in the vicinity of X^2.

Multi-p partonic operators

It is also possible to have generators for which the imaginary parts of conformal weights involve a sum over arbitrary many zeros of zeta. These generators must correspond to kind of multi-p-adic states. A possible interpretation is that partonic 2-surfaces X^2 decompose into regions corresponding to effective p-adic topologies labelled by different p-adic primes (recall the non-determinism of Kähler action). By connecting 2-surfaces labelled by different p-adic primes by connected sum contacts one indeed obtains this kind of states. Just by replacing 3-D topological condensation with a 2-dimensional one, these regions could correspond to closed 2-dimensional surfaces topologically condensed on 2-D surfaces ... to 2-D surface X^2 "at the bottom".

Somewhat surprisingly, the 2-dimensional visualizations of the hierarchy of space-time sheets would be almost literally correct apart from the fact that boundaries are not present and even the world that we see around us could correspond this kind of condensation hierarchy. p-Adic fractality, if taken completely seriously, forces to conclude that also black hole horizons, and even the large voids containing galaxies near their boundaries, could be regarded as complex "partons" and that state construction applies even in these scales.

An alternative interpretation is that the topology is multi-p p-adic in the sense that the waves in a given p-adic length scale L_p have ripples in some smaller p-adic length scale. To these ripples in turn waves in still smaller p-adic length scale are superposed, etc... This interpretation, which is in some sense dual to the previous one, is favored by the explanation for the expression and smallness of the gravitational constant.

Partonic operators with complex conformal weight and color confinement

The possibility of partonic generators O_p with a well-defined complex super-canonical weight u cannot excluded. These operators would have form $O_{p,u} = O_p(z, w)O_z O_w$, $z + w = u$ and could be associated with quarks and gluons. Needles to say, conformal confinement would provide extremely elegant realization of color confinement. Also multi-p partonic operators of this kind are possible.

Conformal confinement could manifest itself even in the quark gluon plasma phase detected quite recently in the collisions of heavy nuclei [hk4]. The shocking finding, compared to Columbus's discovery of America, was that, rather than behaving as a dilute gas, the plasma behaved like a liquid with strong correlations between partons, and having density 30-50 times higher than predicted by QCD calculations. Liquid character means strong correlations and one can consider the possibility that these correlations are a signature of the conformal confinement.

11.4.5   Kähler coupling strength, infinite primes, and zeros of Zeta

The proposed interpretation could provide a deeper understanding of the mysterious presence of the 10 primes $q = 2, 3, ...23$ and prime p in the expression of the Kähler coupling constant as a function of p [E1] given by

$$\frac{1}{\alpha_K} = \frac{4}{\pi}\left[log(p) + log(K^2)\right]\ ,$$

$$K^2 = 2 \times 3 \times 5 \times 7 \times 11 \times 13 \times 17 \times 19 \times 23\ .$$

a) The proposed explanation [B4, E1] has been that elementary particles might suffer a many-step topological condensation in 3-D sense with CP_2 type extremal condensed first to $p = 2$ space-time sheet, which in condenses on $p = 3$ sheet, and so on. If the largest CD in the hierarchy labelled by p carries information about all these condensation steps it could be labelled by all these primes.

b) The proposed picture encourages to imagine that the series of topological condensations is actually 2-dimensional: 2-surface with $q = 23$ is glued to X^2 characterized by p, $q = 19$ is glued to $q = 23$, and so on up to CP_2 type extremal for which causal determinant correspond to $q = 2$. Matter could be said to reside at 2-surfaces.

The construction encourages to consider a modification of the earlier geometric interpretation of infinite primes [E3]. The construction of infinite primes can be regarded as a repeated second quantization of a super-symmetric quantum field theory. The many particle states of previous level become single particle states at the next level.

This interpretation is based on particle picture. The argument allowing to understand the weakness of the gravitational coupling strength leads to a dual interpretation in the sense that multi-p fractality corresponds to waves in length scale L_p to which ripples in length scale L_{23} are superposed. To these ripples are superposed waves in length scale L_{19}, etc... Field-particle duality encourages to take seriously both interpretations. If the primes $p = 2, 3, ...23$ are present, the question whether besides p-adic length scales $L_p \propto \sqrt{p}$ also their multiples $\sqrt{\prod_i q_i} L_p$, where $\{q_i\}$ forms a subset of $\{2, 3, ..., 23\}$ define fundamental length scales. The implication would be small-p p-adic fractality for these small primes with each p-adic length scale L_p taking the role of CP_2 length, and there indeed is some evidence for this kind of fractality.

a) The simplest infinite primes are just $X \pm 1$, where X is the product of all finite primes. Slightly more complex infinite primes are characterized by a finite set of purely bosonic primes p_i with arbitrary occupation numbers n_{p_i} for each of them, and by different fermionic primes q_i with a unit fermion number and arbitrary bosonic occupation number n_{q_i}. The super generators F_p associated with a given Hamiltonian correspond to fermionic primes and Hamiltonians H_p to bosonic primes. The states created by these operators correspond to a unique infinite prime characterizing a unique decomposition of X^2 to regions characterized by primes p. Infinite primes do not explain the group theoretical labels super-canonical generators.

b) The earlier proposal was that the occupation numbers associated with the finite primes characterizing an infinite prime label space-time sheets. Now these primes label regions of single partonic 2-surface X^2.

c) There are also more general infinite primes constructible in terms of infinite integers, and they could correspond 3-surfaces X^3 containing the partons X_i^2 [E3]. The zeros of Riemann's poly-zetas $\zeta(z_1, ..., z_n)$ assigned already earlier to bound states [C5] suggest themselves as counterparts of these infinite primes. The expressibility of the conformal weights of multi-parton state as combinations of zeros of poly-zeta would give a strong constraint on masses. Infinite integers could in turn correspond to disjoint 3-surfaces associated with the same 7-D CD $\delta M_\pm^4 \times CP_2$ or space-time sheets condensed a given space-time sheet.

d) The construction of higher level infinite primes in which the infinite primes of previous level take the role of finite primes at the first step of construction in turn would have a natural geometric interpretation in terms of the hierarchical structure of the topological condensate. At each level the net fermion number of the previous structure would tell whether the structure correspond to fermion or boson in the state.

11.4.6   A brief comparison with the earlier construction of states

It is useful to sum up the basic differences between the recent construction of quantum states and its earlier version relying on stringy intuition.

a) In the earlier p-adic mass calculations not involving super-canonical algebra the existence of $h = -1/2$ super Kac-Moody fermionic generators associated with various quark spins was assumed. The required $h_{vac} \geq -4$ ground states were constructed using these operators.

Now only right handed neutrino and antineutrino define this kind of operators. Furthermore, spinor harmonics carry conformal weight giving rise to a definite positive mass squared. This is true for both super-canonical *and* Super Kac-Moody algebra in the recent case. All spinor harmonics except right handed neutrino harmonics which are constant in CP_2 degrees of freedom carry this kind of contribution to the mass squared and thus have conformal weights with $Re(h) > 0$. Hence they cannot be used to build vacuum states with negative conformal weight. In the new construction canonical and neutrino type super-canonical generators create $h_{vac} < 0$ ground states besides configuration space Hamiltonians whose quantum numbers determine the conformal weight. There is no lower bound for h_{vac} now.

b) Since right handed neutrino generators are enough to construct the configuration space metric and spinor structure, an attractive assumption is that super-canonical algebra contains only right handed neutrino-type super generators corresponding to (p, p) type representations of the color group just like Hamiltonians. All super-canonical generators carry a radial conformal weight. Also Hamiltonians and corresponding super-generators carry definite conformal weights which correspond to the eigenvalues of CP_2 d'Alembertian.

Whether the S^2 degrees of freedom associated with δM_+^4 contribute to super-canonical conformal weight is not quite obvious. M_+^4 d'Alembert operator reads in light like coordinates ($u = t + r_M, v =$

$t - r_M)$ as

$$\partial_u \partial_v - \frac{\nabla^2_{S^2}}{r^2_M} \;,$$

which would suggest that the contribution to mass squared is of form $l(l+1)/r^2_M$ for angular momentum eigen states and thus continuous. This does not seem to make sense. A second possibility would be that the contribution is just the conformal weight associated with the S^2 conformal degrees of freedom. There is however a correlation between radial and S^2 degrees of freedom for the Hamiltonians of δM^4_+ coming from the orthogonality in δM^4_+ [B2], which suggests that only radial conformal weight contributes to the super-canonical conformal weight. This assumption will be indeed made.

11.5 Color degrees of freedom

The ground states for the Super Virasoro representations correspond to spinor harmonics in $M^4 \times CP_2$ characterized by momentum and color quantum numbers. The correlation between color and electro-weak quantum numbers is wrong for the spinor harmonics and these states would be also hypermassive. The super-canonical generators allow to build color triplet states having negative vacuum conformal weights, and their values are such that p-adic massivation is consistent with the predictions of the earlier model differing from the recent one in the quark sector. In the following the construction and the properties of the color partial waves for fermions and bosons are considered. The discussion follows closely to the discussion of [bc4].

11.5.1 SKM algebra and SKMD operator

The solutions of SKMD operator are solutions of Super Virasoro conditions for the quaternion-conformal SKM algebra associated with the group $M^4 \times SU(3) \times U(2)_{ew}$. p-Adic mass calculations require $N = 5$ sectors of super-conformal algebra. These sectors correspond to the 5 tensor factors for the $M^4 \times SU(3) \times U(2)_{ew}$ decomposition of the super Kac Moody algebra to gauge symmetries of gravitation, color and electro-weak interactions. These symmetries act on the intersections $X^2 = X^3_l \cap X^7$ of 3-D light like causal determinants (CDs) X^3_l and 7-D light like CDs $X^7 = \delta M^4_+ \times CP_2$. This constraint leaves only the 2 transversal M^4 degrees of freedom since the translations in light like directions associated with X^3_l and δM^4_+ are eliminated.

The algebra differs from the standard one in that super generators $G(z)$ carry lepton (Ramond representation) and quark (N-S representation) numbers are not Hermitian as in super-string models (Majorana conditions are not satisfied). Only Ramond or N-S type Virasoro conditions are assumed to be satisfied by the physical states.

The structure of the SKM algebra

The task is to identify the SKM algebra serving as a symmetry algebra of the theory. Here physical intuition helps to guess what the result must be. First of all, super Kac Moody conformal invariance associated with the light like causal determinants X^3_l is the counterpart of the gauge invariance associated with gravitational, color and electro-weak interactions. Hence the bosonic charges of Super-Kac-Moody algebra must correspond to the charges associated with these gauge symmetries. A further guideline is the fact that p-adic mass calculations require in the crucial fermionic sector $D = 5$ sectors of super-conformal algebra.

a) There are two kinds of fundamental super-generators. The lepton and quark like super-generators $L_{r\alpha}$ and $Q_{r\alpha}$ (α is spin label) associated with the super counterpart of the conserved quark and lepton currents are Abelian charges. Lepton and quark numbers indeed contribute in TGD framework to electromagnetic and Z^0 charges the contribution corresponding to the coupling to Kähler form.

b) There are charges associated with the isometries of the imbedding space. The super generators $L^{nA\alpha}$ and $Q^{nA\alpha}$ associated with $M^4 \times SU(3)$ generate super-symmetries of the modified Dirac action. Poincare symmetries are exact apart from the cosmological breaking caused by the presence of the light cone boundary and manifested through the fact that absolute minima of Kähler action are not

Poincare invariant. Also super Kac Moody symmetries act as symmetries of the Dirac action and can be identified as conformal localization of the translations of M^4.

A physically well-motivated guess is that the subgroup $M^4 \times SU(3)$ of the isometry group of H, or to be precise, $E^2 \times SU(3))$ becomes a local symmetry group. Thus M^4 would act as the local gauge group of gravitational interactions. For the complex-conformal symmetry only the two transversal M^4 degrees of freedom generating physical states contribute 2 sectors to the super-conformal algebra.

c) Physical intuition suggest that the remaining super-confromal sector corresponds to the conserved charges associated with electro-weak symmetries. In fact, $SU(3)$ isometry charges decompose to two separately conserved parts for the solutions of the field equations for which energy momentum tensor term and Kähler current term vanish separately (Kähler current either vanishes or is light like). One might hope that this occurs completely generally, and it indeed does for all known extremals of Kähler action [D1]. This would mean that space time surfaces are "minimal surfaces" with respect to the effective metric defined by the energy momentum tensor $T^{\alpha\beta}$ associated with the Kähler action. What is important is that for Poincare charges the corresponding decomposition brings in nothing new since the Kähler current part of the charge vanishes identically.

Since $SU(3)$ rotations are represented as electro-weak rotations, and since the $U(2)$ sub-group of $SU(3)$ corresponds naturally to the holonomy group of CP_2 spinor connection, one can identify electro-weak SKM algebra as the algebra generated by $U(2)$ color charges, and identify color charges as those associated with the energy momentum tensor. Higgs field in turn would be identified as the generators in the complement of $u(2) \subset su(3)$ and indeed has correct electro-weak quantum numbers with these identifications. This would give 2 sectors of super-conformal algebra so that one has $N = 5$ sectors as required by the p-adic mass calculations.

CM degrees of freedom

Important element in the discussion are center of mass degrees of freedom parameterized by imbedding space coordinates. By the effective 2-dimensionality it is indeed possible to assign to partons momenta and color partial waves and they behave effectively as free particles. In fact, the technical problem of the earlier scenario was that it was not possible to assign symmetry transformations acting only on on the boundary components of 3-surface.

One can assign to each eigen state of momentum and color quantum numbers a plane wave in these degrees of freedom as well as color partial wave in CP_2 degrees of freedom. Thus color quantum numbers are not spin like quantum numbers in TGD framework except effectively in the length scales much longer than CP_2 length scale. The correlation between color partial waves and electro-weak quantum numbers is not physical in general: only the covariantly constant right handed neutrino has vanishing color.

SKMD equation, the mass formula, and condition determining the effective string tension

Configuration space Dirac operator, which can be expressed in terms of mutually anto-commuting super-canonical and Super Kac-Moody Dirac operators as

$$\begin{aligned}
G &\equiv G_{SC} + iG_{SKM} \ , \\
G_{SKM} &\equiv D = D_H + \sqrt{k}G
\end{aligned} \qquad (11.5.1)$$

annihilates physical states amongst other super generators. Here D_H is the counterpart of the ordinary Dirac operator of the imbedding space defined as the contraction of the ordinary gamma matrices and leptonic or quark-like super-charges associated with the conserved momenta and color charges defined by the modified Dirac equation. Its conformal weight is 0 for the supercharges in Ramond representation and $\pm 1/2$ for the supercharges in N-S representation.

The square of D determines the possible values of mass squared operator. Mass squared eigenvalues are given by

$$M^2 = m^2_{CP_2} + kL_0 \ . \qquad (11.5.2)$$

The contribution of CP_2 spinor Laplacian to the mass squared operator is in general not integer valued.

The requirement that mass squared spectrum is integer valued for color partial waves possibly representing light states fixes the possible values of k determining the effective string tension modulo integer. The value $k = 1$ is the only possible choice. The earlier choice $k_L = 1$ and $k_q = 2/3$, $k_B = 1$ gave integer conformal weights for the lowest possible color partial waves. The assumption that the total vacuum weight h_{vac} is conserved in particle vertices implied $k_B = 1$.

11.5.2 General construction of solutions of Dirac operator of H

The construction of the solutions of massless spinor and other d'Alembertians in $M_+^4 \times CP_2$ is based on the following observations.

a) d'Alembertian corresponds to a massless wave equation $M^4 \times CP_2$ and thus Kaluza-Klein picture applies, that is M_+^4 mass is generated from the momentum in CP_2 degrees of freedom. This implies mass quantization:

$$M^2 = M_n^2 \ ,$$

where M_n^2 are eigenvalues of CP_2 Laplacian. Here of course, ordinary field theory is considered. In TGD the vacuum weight changes mass squared spectrum.

b) In order to get a respectable spinor structure in CP_2 one must couple CP_2 spinors to an odd integer multiple of the Kähler potential. Leptons and quarks correspond to $n = 3$ and $n = 1$ couplings respectively. The spectrum of the electromagnetic charge comes out correctly for leptons and quarks.

c) Right handed neutrino is covariantly constant solution of CP_2 Laplacian for $n = 3$ coupling to Kähler potential whereas right handed 'electron' corresponds to the covariantly constant solution for $n = -3$. From the covariant constancy it follows that all solutions of the spinor Laplacian are obtained from these two basic solutions by multiplying with an appropriate solution of the scalar Laplacian coupled to Kähler potential with such a coupling that a correct total Kähler charge results. Left handed solutions of spinor Laplacian are obtained simply by multiplying right handed solutions with CP_2 Dirac operator: in this operation the eigenvalues of the mass squared operator are obviously preserved.

d) The remaining task is to solve scalar Laplacian coupled to an arbitrary integer multiple of Kähler potential. This can be achieved by noticing that the solutions of the massive CP_2 Laplacian can be regarded as solutions of S^5 scalar Laplacian. S^5 can indeed be regarded as a circle bundle over CP_2 and massive solutions of CP_2 Laplacian correspond to the solutions of S^5 Laplacian with $exp(is\tau)$ dependence on S^1 coordinate such that s corresponds to the coupling to the Kähler potential:

$$s = n/2 \ .$$

Thus one obtains

$$D_5^2 \quad = \quad (D_\mu - iA_\mu \partial_\tau)(D^\mu - iA^\mu \partial_\tau) + \partial_\tau^2 \qquad (11.5.3)$$

so that the eigen values of CP_2 scalar Laplacian are

$$m^2(s) \quad = \quad m_5^2 + s^2 \qquad (11.5.4)$$

for the assumed dependence on τ.

d) What remains to do, is to find the spectrum of S^5 Laplacian and this is an easy task. All solutions of S^5 Laplacian can be written as homogenous polynomial functions of C^3 complex coordinates Z^k and their complex conjugates and have a decomposition into the representations of $SU(3)$ acting in natural manner in C^3.

e) The solutions of the scalar Laplacian belong to the representations $(p, p + s)$ for $s \geq 0$ and to the representations $(p + |s|, p)$ of $SU(3)$ for $s \leq 0$. The eigenvalues $m^2(s)$ and degeneracies d are

$$m^2(s) \quad = \quad \frac{2\Lambda}{3}[p^2 + (|s| + 2)p + |s|] \ , \quad p > 0 \ ,$$

$$d \quad = \quad \frac{1}{2}(p + 1)(p + |s| + 1)(2p + |s| + 2) \ . \qquad (11.5.5)$$

Λ denotes the 'cosmological constant' of CP_2 ($R_{ij} = \Lambda s_{ij}$).

11.5.3 Solutions of the leptonic spinor Laplacian

Right handed solutions of the leptonic spinor Laplacian are obtained from the asatz of form

$$\nu_R = \Phi_{s=0}\nu_R^0 \ ,$$

where u_R is covariantly constant right handed neutrino and Φ scalar with vanishing Kähler charge. Right handed 'electron' is obtained from the ansats

$$e_R = \Phi_{s=3}e_R^0 \ ,$$

where e_R^0 is covariantly constant for $n = -3$ coupling to Kähler potential so that scalar function must have Kähler coupling $s = n/2 = 3$ a in order to get a correct Kähler charge. The d'Alembert equation reduces to

$$(D_\mu D^\mu - (1 - \epsilon)\Lambda)\Phi = -m^2\Phi \ ,$$
$$\epsilon(\nu) = 1 \ , \quad \epsilon(e) = -1 \ . \tag{11.5.6}$$

The two additional terms correspond to the curvature scalar term and $J_{kl}\Sigma^{kl}$ terms in spinor Laplacian. The latter term is proportional to Kähler coupling and of different sign for ν and e, which explains the presence of the sign factor ϵ in the formula.

Right handed neutrinos correspond to (p,p) states with $p \geq 0$ with mass spectrum

$$m^2(\nu) = \frac{m_1^2}{3}\left[p^2 + 2p\right] \ , \ p \geq 0 \ ,$$
$$m_1^2 \equiv 2\Lambda \ . \tag{11.5.7}$$

Right handed 'electrons' correspond to $(p, p+3)$ states with mass spectrum

$$m^2(e) = \frac{m_1^2}{3}\left[p^2 + 5p + 6\right] \ , \ p \geq 0 \ . \tag{11.5.8}$$

Left handed solutions are obtained by operating with CP_2 Dirac operator on right handed solutions and have the same mass spectrum and representational content as right handed leptons with one exception: the action of the Dirac operator on the covariantly constant right handed neutrino $((p = 0, p = 0)$ state) annihilates it.

11.5.4 Quark spectrum

Quarks correspond to the second conserved H-chirality of H-spinors. The construction of the color partial waves for quarks proceeds along similar lines as for leptons. The Kähler coupling corresponds to $n = 1$ (and $s = 1/2$) and right handed U type quark corresponds to a right handed neutrino. U quark type solutions are constructed as solutions of form

$$U_R = u_R\Phi_{s==1} \ ,$$

where u_R possesses the quantum numbers of covariantly constant right handed neutrino with Kähler charge $n = 3$ ($s = 3/2$). Hence Φ_s has $s = -1$. For D_R one has

$$D_R = d_r\Phi_{s=2} \ .$$

d_R has $s = -3/2$ so that one must have $s = 2$. For U_R the representations $(p+1, p)$ with triality one are obtained and $p = 0$ corresponds to color triplet. For D_R the representations $(p, p+2)$ are obtained and color triplet is missing from the spectrum ($p = 0$ corresponds to $\bar{6}$).

The CP_2 contributions to masses are given by the formula

$$
\begin{aligned}
m^2(U,p) &= \frac{m_1^2}{3}\left[p^2 + 3p + 2\right] \;,\quad p \geq 0\;, \\
m^2(D,p) &= \frac{m_1^2}{3}\left[p^2 + 4p + 4\right] \;,\quad p \geq 0\;.
\end{aligned}
\tag{11.5.9}
$$

Left handed quarks are obtained by applying Dirac operator to right handed quark states and mass formulas and color partial wave spectrum are the same as for right handed quarks.

The color contributions to p-adic mass squared are integer valued if $m_0^2/3$ is taken as a fundamental p-adic unit of mass squared. This choice has an obvious relevance for p-adic mass calculations since canonical identification does not commute with a division by integer. More precisely, the images of number xp in canonical identification has a value of order 1 when x is a non-trivial rational whereas for $x = np$ the value is n/p and extremely is small for physically interesting primes. This choice does not however affect the spectrum of massless states but can affect the spectrum of light states in case of electro-weak gauge bosons.

11.6   Gauge bosons

The construction of gauge boson states is more intricate process than in the original scenario which did not involve super-canonical algebra and the effective 2-dimensionality and leads to profound views about how to understand the values of coupling constants, in particular the value of the gravitational constant. 7–3 duality and effective two-dimensionality allow quite detailed view about the construction of bosons, and the earlier somewhat ad hoc assumptions behind p-adic mass calculations find a justification.

11.6.1   Bi-locality of boson states

The gauge boson could correspond to either a local fermion current contracted with j^{Ak} or to a bi-local operator.

i) For a local operator no normal ordering of the current would be needed since at a given space-time sheet and for TGD quantization avoiding infinite vacuum energy, the current would contain only anti-commuting creation operators of a positive (negative) energy fermion and negative (positive) energy anti-fermion. The vacuum expectation value would involve only the contractions of fermion and anti-fermion with the legs of the current and a well-defined and finite integral over X^2 would result. This is however not enough: also the norms of the boson states must be finite and locality implies an infinite norm as an integral of $\delta(0)$ over X^2. Hence local operators seem to be excluded.

ii) If photon is created by a bi-local operator, it would involve a kind of structure function in $X^2 \times X^2$ allowing visualization as a line connecting two points x and y having fermion and anti-fermion at its ends. The bi-local current would be sum of two terms

$$
\begin{aligned}
B &= \int_{X^2 \times X^2} dV_x dV_y B(x,y)\left[\overline{\Psi}(x)E(y)\Psi(y) + \overline{\Psi}(x)E(x)\Psi(y)\right]\;, \\
E &= j^{Ak}\gamma_k\;.
\end{aligned}
\tag{11.6.1}
$$

The vacuum expectation value determining the vertex would boil down to a correlation function defined as integral over $X^2 \times X^2$ for this Hamiltonian and bilinear of functions formed from positive energy fermion and anti-fermion. $B(x,y)$ could be determined by the super conformal invariance as a correlation function.

Conformal invariance suggests that correlation functions obeying simple power scaling laws as a function of the distance r between fermion and anti-fermion are associated with the boson states. The power law holds true with respect to the distance r measured in the induced metric.

The distance r between fermion and antifermion in the induced metric of X^2 is expressed to behave fractally as function of Δr_M, where r_M is the light like radial coordinate of δM_+^4. For large values of r_M $r(\Delta r_M)$ is expected to grow very slowly since X^2 becomes almost light like in this direction. For small distance the growth is expected to be very rapid by the p-adic fractality of X^2 meaning that X^2 becomes 2-D version for the coast line of Britain. The scaling behavior

$$\frac{r}{r_0} = x^\Delta \ , \quad \Delta < 1 \ ,$$

$$x = \frac{\Delta r_M}{r_{M,0}} \tag{11.6.2}$$

is expected. A good guess for $r_{M,0}$ is as a p-adic length scale: $r_{M,0} = L_p$.

11.6.2 Bosonic charge matrices, conformal invariance, and coupling constants

Bosons are represented by fermion-antifermion bilinears. The requirement that boson state has a finite norm implies that bosons are bi-local operators creating fermion antifermion states in X^2. The bilinear representing the boson can be regarded as a second quantized version of the charge matrix of the charge represented by the boson in question. Also a polarization vector contracted with M^4 gamma matrices is involved. In the case of graviton/gluon charge matrix involve momentum operators of M^4/color rotation generators of CP_2 acting in center of mass degrees of freedom acting of the second quantized spinor fields. The normalization factors of the bosonic states determine their couplings to fermion pairs, which appear as fundamental couplings: the proportionality $1/\sqrt{N} \propto g$ between the coupling g and proportionality factor is predicted.

11.6.3 How to understand the value of gravitational constant?

The great challenge is to understand the value of the gravitational constant. The physically motivated expectation is that Planck length characterizes the failure of the locality. $\sqrt{G} \sim 10^{-4} R$ deduced from the p-adic mass calculations implies that the physics determining the coupling constants is still quite local in CP_2 length scale.

The earlier number theoretic arguments allowing to express G [B4, E1] in terms of p-adic length scale, the value of Kähler action for CP_2 as $G = L_p^2 = R^2/K^2$, $K^2 = 2 \times 3.... \times 23 \times p$, or simply $R^2 = 2 \times 3... \times 23 \times G$, where R is CP_2 size.

This expression of $\sqrt{G}$ can be substituted to $\sqrt{N} = 1/\sqrt{G}$ relating G to the graviton normalization factor N. On the other hand, N can be related to the anomalous dimension Δ by assuming that X^2 has a transversal size given by p-adic length scale L_p and that gravitational charge matrix involves a momentum operator which acts on the correlation function $1/r$ appearing in the fermionic bilinear as operator ∂_{r_M}. Using the proposed power law relating r and $\Delta r_M/L_p$, one obtains the estimate $\Delta = 1/\sqrt{2 \times 3.... \times 23 \times p}$ for the anomalous dimension appearing in the fractal relationship between r and Δr_M.

The anomalous dimension Δ would be extremely small. A possible interpretation is in terms of a hierarchy of superposed fractalities labelled by p-adic primes $q = 2, 3, ..., 23, p$: somewhat like waves containing ripples containing ripples... At the longest length scale there would the fractality corresponding to L_p, which alone would give $r_M \propto r^{1/\sqrt{p}}$. Superposed to this, like ripples on wave, is 23-adic fractality, which gives $r_M \propto r^{1/\sqrt{23p}}$, and so on. This multi-p fractality would mean that the product $n = 2 \times 3.... \times 23 \times p$ appears instead of p as a parameter in the p-adic expansions of relevant physical quantities. In the absence of this fractality gravitational coupling would be of order L_p^2.

11.6.4 The ground states associated with gauge bosons

The experience about p-adic mass calculations gives some hints about the ground state conformal weights of intermediate gauge bosons.

a) If p-adic temperature is $T = 1/2$ for bosons instead of $T = 1$ for fermions, p-adic thermodynamics does not significantly contribute to boson masses except if ground states have vanishing conformal weight so that ground state degeneracy is absent and $T = 1/2$ gives completely neglible thermal contribution to the conformal weight.

b) If Higgs mechanism in the case of gauge bosons is due to the coupling to Higgs or due to the non-covariant constancy of the charge matrix, all electro-weak bosons could have vanishing conformal weights in the ground state. This would be in conflict with the assumption $h_{vac}(W) = -2$ of the earlier model following from the following argument. W boson is not a color singlet although it does not of course belong to an irreducible representation of $SU(3)$. One could argue that W charge matrix and the left handed part of Z charge matrix correspond to a $j = 1$ triplet of $SU_L(2) \subset SU(3)$ and thus has $h_c = j(j+1) = 2$ for $j = 1$. Also, because W behaves like $e\bar{\nu}$ pair the W charge matrix must have the color conformal weight $h_c = 2$ of $e\bar{\nu}$. Also the requirement that the ground state conformal weight is conserved in electro-weak vertices supports this picture. $h_c = 2$ would be compensated by the negative conformal weight of the super-canonical operator $O_p(1 + iy)O_p(1 - iy)$ where $O_p(1 \pm iy)$ is constructed from super-canonical generators using the previous recipe.

Z^0 would be superposition of states with different super-canonical ground state conformal weights. The left handed part of charge matrix proportional to I_L^3 would have ground state super-canonical conformal weight $h_{vac} = -2$ and the vectorial part of Z^0 charge matrix proportional to Q_{em} would have $h_c = 0$ and there would be no compensating super-canonical factor. Photon of course has a vanishing ground state conformal weight.

c) In the case of gluons the isometry generator $J^A = j^{Ak}D_k$ does not change the representation associated with a color Hamiltonian. The assumption that this operator carries a conformal weight $h_c = 0$ conforms with the masslessness of gluon and with the fact that also translation generators possess a vanishing conformal weight in the stringy mass formula. If quarks have a net complex conformal weight as previous considerations suggest, gluons should have a net conformal weight of form $h = iy$. Operator $O_p(h = iy)$, constructed as a sum of the operators $O_{1/2+iy_1}O_{-1/2-iy_1-iy}$ with a sum over y_1, can compensate the imaginary part of the conformal weight conformal weights of quarks. This kind of operators might serve as counterparts for the color electric flux tubes connecting the 2-surfaces associated with quarks.

11.6.5　Bosonic charge matrices

In the following more detailed forms of bosonic charge matrices are listed.

Photon and intermediate gauge bosons

Photonic charge matrix is of the form

$$q_{em} \quad = \quad a \times Id + b \times J_{kl}\Sigma^{kl} \ , \tag{11.6.3}$$

and is covariantly constant. As a consequence photon the charge matrix does not develop any contribution to the mass squared and for $T = 1/2$ photon remains in an excellent approximation massless in p-adic thermodynamics.

The charge matrix of W^- is simply the left handed isospin matric I_L^3 and if the previous arguments are correct it carries a conformal weight $h_c = 2$. In standard model Z^0 charge matrix is a linear combination of the left handed electro-weak isospin I_L^3 and electromagnetic charge Q_{em}.

$$Q_Z \quad = \quad I_L^3 - sin^2(\theta_W)Q_{em} \ . \tag{11.6.4}$$

In standard model the mixing of I_L^3 and Q_{em} corresponds to the mixing of $U(1)$ and $U(1)_L$ bosons by Weinberg angle θ_W. I_L^3 has ground state conformal weight -2 whereas for Q_{em} the weight vanishes.

Graviton and gluon

Graviton corresponds to a charge matrix, or rather charge operator acting on fermion bi-linear, defined as

$$O \quad = \quad E^{kl}\gamma_k(\partial_{1,l} - \partial_{2,l}) \tag{11.6.5}$$

with gamma matrix contracted between fermion and antifermion: here the flatness of M^4 is essential. E_{kl} is the polarization tensor of graviton satisfying obvious constraints. The derivative operators with respect to Minkowski coordinates act on fermion and anti-fermion

For the gluon the charge matrix is given by

$$
\begin{aligned}
O &= E_k \gamma^k (J_1^A - J_2^A) \ , \\
J^A &= j^{Ak} D_k \ .
\end{aligned}
\tag{11.6.6}
$$

E_k is the polarization vector of gluon.

In the case of graviton and gluon the question about the action of the isometry generator arises since the second quantized induced spinor field Ψ and the correlation function $B(x,y)$ depend on X^2 coordinates rather than imbedding space coordinates. The problem is analogous to that of interpreting the coordinate z of X^2 in the anti-commutators and commutators of Super Kac-Moody and super-canonical generators as an imbedding space coordinate. As found, the problem can be circumvented if z is identifiable in terms of a unique imbedding coordinate w for a representative 2-surface $Y^2(X^2)$ assignable to a maximum of the Kähler function whose perturbations by super-canonical algebra appear in the configuration space functional integral.

11.6.6 $\ BF\overline{F}$ couplings and the general form of bosonic configuration space spinor fields

Conformal theory alone gives no hint about how the coupling constant appears, and configuration space-integral is necessary to understand the emergence of the gauge coupling.

a) A strong hint comes from the facts that all $BF\overline{F}$ coupling constants, except possibly gravitational constant, must be proportional to the Kähler coupling g_K. The most natural manner to achieve this is to require that the bosonic configuration space spinor fields vanish at the maximum of the Kähler function where the perturbation series is developed. That bosons should correspond to small perturbations around the maximum of the Kähler function is in accordance with the assumption that quantum fields correspond to the perturbations around the extrema of the action functional. This means that one can write $B(x,y)$ in the form

$$
\begin{aligned}
B(x,y) &= \partial_I K B^I(x,y) \ , \\
\partial_I K(X^3) &= 0 \ \text{at the maximum of } \overline{K} \ .
\end{aligned}
\tag{11.6.7}
$$

Gere $\partial_I K$ denotes partial derivatives of Kähler function with respect to the configuration space coordinates X^I vanishing at the maximum of K.

b) The functional integral in the lowest order approximation is obtained by expanding $B(x,y)$ in lowest order to functional Taylor series in using the coordinates X^I

$$
B(x,y) = \partial_R K \times B^I(x,y) \times X^R \ ,
\tag{11.6.8}
$$

It is understood that also $B^I(x,y)$ allows functional power series expansion as a functional of X^3. In the lowest order approximation the norm N of the boson state is given by the functional integral

$$
\begin{aligned}
N &= \langle \int_{X^2 \times Y^2} \overline{B}(x,y) B(x,y) dV_x dV_y \rangle = A_{IJ} \times B^{IJ} \ , \\
A_{IJ} &= \partial_I \partial_R K \times \partial_J \partial_S K \times \langle X^R X^S \rangle \ , \\
B^{IJ} &= \int_{X^2 \times Y^2} \overline{B}^I(x,y) B^J(x,y) dV_x dV_y \ .
\end{aligned}
\tag{11.6.9}
$$

Here $\langle X^R X^S \rangle$ is a two point function defined by the functional integral over small perturbations around the maximum of Kähler function Specifying the coordinates to complex coordinates and using the covariant Kahler metric $G_{K\overline{L}} = \partial_K \partial_{\overline{L}} K$ as the kinetic term. Since the contravariant Kähler metric defines the propagator, the lowest order approximation gives

$$
N = G_{K\overline{L}} \times B^{K\overline{L}} \ .
\tag{11.6.10}
$$

What is nice that the symmetry considerations allow to determine the covariant metric highly uniquely and the propagator disappears from the final formula. The normalization factor $1/\sqrt{N}$ of the boson state is obviously proportional to g_K since the Kähler function K is proportional to $1/\alpha_K$.

c) Fermion boson vertex is indeed proportional to g_K. $B(x,y)$ must be expanded in a functional Taylor series up to a second order term

$$B(x,y) \;=\; \partial_R K B^I(x,y)X^R + \partial_R K \times \partial_S B^I(x,y) \times X^R X^S + \cdots \; . \tag{11.6.11}$$

The general expression of the $BF\overline{F}$ vertex is

$$
\begin{aligned}
V_{BF\overline{F}} \;&=\; \frac{1}{\sqrt{N}}\langle \int_{X^2 \times Y^2} \overline{B}(x,y)\Gamma dV_x dV_y \rangle = \frac{1}{\sqrt{N}} A \; , \\
A \;&=\; \int_{X^2 \times Y^2} \partial_I \overline{B}^I(x,y)\Gamma dV_x dV_y \; .
\end{aligned}
\tag{11.6.12}
$$

The propagator compensates the second order derivatives of Kähler function in the functional integral average, and the vertex is indeed proportional to g_K.

11.7 Exotic states

The possibility of exotic states poses a serious problem for the proposed scenario. It however turns out that if elementary particles correspond to CP_2 type extremals, all exotic massless particles can be constructed using colored generators and by color confinement cannot induce macroscopic long range interactions. The essential assumption is that the fermionic quantization for the space-time sheets having CP_2 projection of dimension $D(CP_2) < 4$ is non-conventional. This has also direct relevance for the understanding of the matter antimatter asymmetry.

11.7.1 Non-conventional quantizations and CP_2 type extremals

In TGD Universe all elementary particles have as their building blocks fermions and anti-fermions. Thus if phase conjugate photons are possible they must result by a phase conjugation performed for fermions. Taking this idea seriously leads to the identification of the phase conjugate fermions as states created from a fermionic vacuum for which the roles of creation and annihilation operators are changed.

Non-conventional quantizations

It is possible to go even further and consider the possibility that the role change can occur separately for fermions and anti-fermions. In this kind of situation both fermion and anti-fermion oscillator operators in Ψ would be creation (or annihilation) operators. These novel kinds of vacua can carry only positive/negative fermion number whereas inertial energy can vanish so that this quantization would be naturally associated with nearly vacuum extremals. On the other hand, for the conventional vacua net fermion number can vanish unlike energy. The separate conservation of quark and lepton numbers could in principle double the bits labelling the possible fermionic vacua.

Does the dimension of the CP_2 projection of space-time sheet determine the character of the fermionic vacuum

The dimension $D(CP_2)$ of CP_2 projection of the space-time surface serves as a classifier for the asymptotic solutions of field equations for which Kähler Lorentz force vanishes (as a space-time correlate for the absence of dissipation, [D1]). Elementary particles, in particular bosons, correspond to CP_2 type extremals with $D(CP_2) = 4$. For space-time sheets with a lower-dimensional CP_2 projection the quantization could be always non-conventional. If so, the states would be always many fermion states containing ordinary fermions and what might be called anti-fermion holes. An interesting question is how to experimentally distinguish between fermions and anti-fermion holes: this might be based on the possibility of fermions and anti-fermion holes to accelerate spontaneously without any external energy feed by exchanging energy. Perhaps the reader can develop a convincing argument excluding the interpretation of condensed matter as this kind of phase.

Non-conventional quantization could tremendously simplify the spectrum of elementary particles since in the ideal situation (CP_2 type extremal with 1-D M^4 projection) the projection of X^2 to $\delta M^4_\pm = S^2 \times R_+$ reduces to a point. First, the SKM and super-Kac Moody sectors corresponding to δM^4_+ generating infinite number of states with a vanishing conformal weight disappear. Secondly, all super-canonical generators with different radial conformal weights are proportional to each other and almost all exotic massless states containing these generators would disappear from the spectrum.

Even if this picture probably involves a too strong idealization, one expects that the super-canonical generators become highly linearly independent, and that the dimension of the space spanned by them is in some sense reduced dramatically for the small deformations of CP_2 type extremals. Interestingly, the non-conventional quantization would also allow to understand the generation of matter-antimatter asymmetry [D4, D5].

The problem is to assign conformal weight to the super-canonical generators O_p at the limit of CP_2 extremals. As the system approaches CP_2 type extremal, radial waves become constant. Hence one might argue that the radial conformal weights vanish for the operators O_p contain products of generators with conjugate conformal weights.

11.7.2 The problem of exotic states

Consider now the problem of exotic states. The exotic states can emerge both from super-canonical and super Kac-Moody sectors. If one assumes that conventional quantization applies only in $D = 4$ sector, that elementary particles correspond to CP_2 type extremals, and that the idealization proposed in the earlier section "Non-conventional quantizations and CP_2 type extremals" at the limit of CP_2 extremals applies, the predicted exotic massless states are generated by color Hamiltonians only since $SO(3)$ Hamiltonians with vanishing conformal weight are "frozen" to constant. This justifies the hope that new macroscopic long range forces are absent in TGD Universe.

a) Super-canonical sector.
In super-canonical sector S^2 generators are frozen to constant and fermionic generators vanish so that infinite number of generators otherwise giving rise to degeneracy of massless states is eliminated. Color generators appear as pairs of Hamiltonian and its super-partner with an "anomalous" conformal weight determined by the color representation, and due to the breaking of conformal symmetry induced by CP_2 geometry reflecting itself as a massivation of spinor harmonics. Poisson bracket action does not conserve color conformal weights. This can be understood in terms of the breaking of conformal invariance. The ground states with negative conformal weight would be generated by color Hamiltonians and their spartners having same conformal weights. Color confinement suggests that the massless particles generated from these ground states cannot give rise to macroscopic long range forces.

b) SKM generators in NS representation.
N-S sector gives rise to super generators with conformal weight $n + 1/2, n \geq 0$ since $h = -1/2$ generators are not allowed by the representation used. Therefore the dangerous $n = 0$ operators are absent.

c) Ramond sector of SKM algebra corresponding to $SO(3) \times SU(2)_L \times U(1)$ holonomies.
$n = 0$ generators are absent in holonomy degrees of freedom are absent. That the right handed neutrino is covariantly constant, is annihilated by charge matrices, and is orthogonal with $\lambda \neq 0$ modes of the modified Dirac operator D, implies that $n = 0$ fermionic generators vanish. Also the covariant constancy of em charge matrix and the anomalous conformal weight $h_c = 2$ of the left-handed electroweak charge matrices is of importance. Hence no spartners are predicted in $SO(3) \times SU(2)_L \times U(1)$ degrees of freedom.

d) Ramond sector of SKM algebra corresponding to $SO3) \times SU(3)$ isometries.
i) $n = 0$ bosonic $SO(3) \times SU(3)$ SKM generators act directly as operators $j^{Ar}D_r$ on the Hamiltonians of X^7 appearing in the definitions of configuration space Hamiltonians. In the same manner $j^{Ar}D_r$ transforms $j^{Bk}\Gamma_k$ to $j^{[A,B]k}\Gamma_k$ and does not affect the representation of H_B. Hence the KM algebra corresponding to isometries does not increase the "particle" number defined as the number of X^2 non-local operators in the state nor change the representation of $SO(3) \times SU(3)$.
ii) Fermionic $SO(3)$ generators have $h_c = 0$ but for $n = 0$ they vanish by the orthogonality of ν_R

and $\lambda > 0$ eigen modes of D. Fermionic $SU(3)$ SKM generators have an anomalous conformal weight $h_c = 1$.

The conclusion is that massless exotics are all created by color Hamiltonians and their spartners, might be enough to achieve consistency with the experimental facts since color confinement would restrict the new long range interactions to a finite range.

11.8 Particle Massivation

In TGD framework p-adic thermodynamics provides a microscopic theory of particle massivation. The idea is very simple. The mass of the particle results from a thermal mixing of the massless states with CP_2 mass excitations of super-conformal algebra. In p-adic thermodynamics the Boltzmann weight $exp(-E/T)$ does not exist in general and must be replaced with p^{L_0/T_p} which exists for Virasoro generator L_0 if the inverse of the p-adic temperature is integer valued $T_p = 1/n$. The expansion in powers of p converges extremely rapidly for physical values of p, which are rather large. Therefore the three lowest terms in expansion give practically exact results. Thermal massivation does not not necessarily lead to light states and this drops a large number of exotic states from the spectrum of light particles. The partition functions of N-S and Ramond type representations are not changed in TGD framework despite the fact that fermionic super generators carry fermion numbers and are not Hermitian. Thus the practical calculations are relatively straightforward.

The overall conclusion about p-adic mass calculations is that fermionic mass spectrum is predicted in an excellent accuracy but that the thermal masses of the intermediate gauge bosons come 20-30 per cent to large for $T_p = 1$ and are completely negligible for $T_p = 1/2$. This forces to consider very seriously the possibility that TGD can, contrary to the original expectations, somehow provide dynamical Higgs field as a fundamental field. One can indeed identify Higgs field in TGD framework an a generalized number theory framework. An alternative manner to understand the massivation of electro-weak gauge bosons is as reflecting the breaking of conformal invariance due to the lacking covariant constancy of the left handed parts of the charge matrices of electro-weak gauge bosons. In the following some aspects of the calculations are discussed: the rather extensive calculations can be found from the four chapters "p-Adic Particle Massivation:...." of [TGDpad].

11.8.1 Partition functions are not changed

One must write Super Virasoro conditions for L_n and *both* G_n and $G_n^\dagger$ rather than for L_n and G_n as in the case of the ordinary Super Virasoro algebra, and it is a priori not at all clear whether the partition functions for the Super Virasoro representations remain unchanged. This requirement is however crucial for the construction to work at all in the fermionic sector, since even the slightest changes for the degeneracies of the excited states can change light state to a state with mass of order m_0 in the p-adic thermodynamics.

Super conformal algebra

Super Virasoro algebra is generated by the bosonic the generators L_n (n is an integer valued index) and by the fermionic generators G_r, where r can be either integer (Ramond) or half odd integer (NS). G_r creates quark/lepton for $r > 0$ and antiquark/antilepton for $r < 0$. For $r = 0$, G_0 creates lepton and its Hermitian conjugate anti-lepton. The defining commutation and anti-commutation relations are the following:

$$
\begin{aligned}
[L_m, L_n] &= (m - n)L_{m+n} + \frac{c}{2}m(m^2 - 1)\delta_{m,-n} \ , \\
[L_m, G_r] &= (\frac{m}{2} - r)G_{m+r} \ , \\
[L_m, G_r^\dagger] &= (\frac{m}{2} - r)G_{m+r}^\dagger \ , \\
\{G_r, G_s^\dagger\} &= 2L_{r+s} + \frac{c}{3}(r^2 - \frac{1}{4})\delta_{m,-n} \ , \\
\{G_r, G_s\} &= 0 \ , \\
\{G_r^\dagger, G_s^\dagger\} &= 0 \ .
\end{aligned}
$$
$$(11.8.1)$$

By the inspection of these relations one finds some results of a great practical importance.

a) For the Ramond algebra G_0, G_1 and their Hermitian conjugates generate the $r \geq 0, n \geq 0$ part of the algebra via anti-commutations and commutations. Therefore all what is needed is to assume that Super Virasoro conditions are satisfied for these generators in case that G_0 and $G_0^\dagger$ annihilate the ground state. Situation changes if the states are *not* annihilated by G_0 and $G_0^\dagger$ since then one must assume the gauge conditions for both L_1, G_1 and $G_1^\dagger$ besides the mass shell conditions associated with G_0 and $G_0^\dagger$, which however do not affect the number of the Super Virasoro excitations but give mass shell condition and constraints on the state in the cm spin degrees of freedom. This will be assumed in the following. Note that for the ordinary Super Virasoro only the gauge conditions for L_1 and G_1 are needed.

b) NS algebra is generated by $G_{1/2}$ and $G_{3/2}$ and their Hermitian conjugates (note that $G_{3/2}$ cannot be expressed as the commutator of L_1 and $G_{1/2}$) so that only the gauge conditions associated with these generators are needed. For the ordinary Super Virasoro only the conditions for $G_{1/2}$ and $G_{3/2}$ are needed.

Conditions guaranteing that partition functions are not changed

The conditions guaranteing the invariance of the partition functions in the transition to the modified algebra must be such that they reduce the number of the excitations and gauge conditions for a given conformal weight to the same number as in the case of the ordinary Super Virasoro.

a) The requirement that physical states are invariant under $G \leftrightarrow G^\dagger$ corresponds to the charge conjugation symmetry and is very natural. As a consequence, the gauge conditions for G and $G^\dagger$ are not independent and their number reduces by a factor of one half and is the same as in the case of the ordinary Super Virasoro.

b) As far as the number of the thermal excitations for a given conformal weight is considered, the only remaining problem are the operators $G_n G_n^\dagger$, which for the ordinary Super Virasoro reduce to $G_n G_n = L_{2n}$ and do not therefore correspond to independent degrees of freedom. In present case this situation is achieved only if one requires

$$(G_n G_n^\dagger - G_n^\dagger G_n)|phys\rangle \quad - \quad 0 \ . \tag{11.8.2}$$

It is not clear whether this condition must be posed separately or whether it actually follows from the representation of the Super Virasoro algebra automatically.

Partition function for Ramond algebra

Under the assumptions just stated, the partition function for the Ramond states not satisfying any gauge conditions

$$Z(t) \quad = \quad 1 + 2t + 4t^2 + 8t^3 + 14t^4 + ... \ , \tag{11.8.3}$$

which is identical to that associated with the ordinary Ramond type Super Virasoro.

For a Super Virasoro representation with $N = 5$ sectors, of main interest in TGD, one has

$$\begin{aligned} Z_N(t) \quad &= \quad Z^{N=5}(t) = \sum D(n) t^n \\ &= \quad 1 + 10t + 60t^2 + 280t^3 + ... \ . \end{aligned} \tag{11.8.4}$$

The degeneracies for the states satisfying gauge conditions are given by

$$d(n) \quad = \quad D(n) - 2D(n-1) \ . \tag{11.8.5}$$

corresponding to the gauge conditions for L_1 and G_1. Applying this formula one obtains for $N = 5$ sectors

$$d(0) = 1 \ , \quad d(1) = 8 \ , \quad d(2) = 40 \ , \quad d(3) = 160 \ . \tag{11.8.6}$$

The lowest order contribution to the p-adic mass squared is determined by the ratio

$$r(n) = \frac{D(n+1)}{D(n)} \ ,$$

where the value of n depends on the effective vacuum weight of the ground state fermion. Light state is obtained only provided the ratio is integer. The remarkable result is that for lowest lying states the ratio is integer and given by

$$r(1) = 8 \ , \quad r(2) = 5 \ , \quad r(3) = 4 \ . \tag{11.8.7}$$

It turns out that $r(2) = 5$ gives the best possible lowest order prediction for the charged lepton masses and in this manner one ends up with the condition $h_{vac} = -3$ for the tachyonic vacuum weight of Super Virasoro.

Partition function for NS algebra

For NS representations the calculation of the degeneracies of the physical states reduces to the calculation of the partition function for a single particle Super Virasoro

$$Z_{NS}(t) \quad = \quad \sum_n z(n/2) t^{n/2} \ . \tag{11.8.8}$$

Here $z(n/2)$ gives the number of Super Virasoro generators having conformal weight $n/2$. For a state with N active sectors (the sectors with a non-vanishing weight for a given ground state) the degeneracies can be read from the N-particle partition function expressible as

$$Z_N(t) \quad = \quad Z^N(t) \ . \tag{11.8.9}$$

Single particle partition function is given by the expression

$$Z(t) \quad = \quad 1 + t^{1/2} + t + 2t^{3/2} + 3t^2 + 4t^{5/2} + 5t^3 + ... \ . \tag{11.8.10}$$

Using this representation it is an easy task to calculate the degeneracies for the operators of conformal weight Δ acting on a state having N active sectors.

One can also derive explicit formulas for the degeneracies and calculation gives

$$\begin{aligned}
&D(0,N) = 1 \ , &&D(1/2,N) = N \ , \\
&D(1,N) = \frac{N(N+1)}{2} \ , &&D(3/2,N) = \frac{N}{6}(N^2 + 3N + 8) \ , \\
&D(2,N) = \frac{N}{2}(N^2 + 2N + 3) \ , &&D(5/2,N) = 9N(N-1) \ , \\
&D(3,N) = 12N(N-1) + 2N(N-1) && \ .
\end{aligned} \tag{11.8.11}$$

as a function of the conformal weight $\Delta = 0, 1/2, ..., 3$.

The number of states satisfying Super Virasoro gauge conditions created by the operators of a conformal weight Δ, when the number of the active sectors is N, is given by

$$d(\Delta, N) \quad = \quad D(\Delta, N) - D(\Delta - 1/2, N) - D(\Delta - 3/2, N) \ . \tag{11.8.12}$$

The expression derives from the observation that the physical states satisfying gauge conditions for $G^{1/2}$, $G^{3/2}$ satisfy the conditions for all Super Virasoro generators. For $T_p = 1$ light bosons correspond to the integer values of $d(\Delta + 1, N)/d(\Delta, N)$ in case that massless states correspond to thermal excitations of conformal weight Δ: they are obtained for $\Delta = 0$ only (massless ground state). This is what is required since the thermal degeneracy of the light boson ground state would imply a corresponding factor in the energy density of the black body radiation at very high temperatures. For the physically most interesting nontrivial case with $N = 2$ two active sectors the degeneracies are

$$d(0,2) = 1 \ , \quad d(1,2) = 1 \ , \quad d(2,2) = 3 \ , \quad d(3,2) = 4 \ . \tag{11.8.13}$$

11.8.2 Fundamental length and mass scales

The basic difference between quantum TGD and super-string models is that the size of CP_2 is not of order Planck length but much larger: of order 10^4 Planck lengths. This conclusion is forced by several consistency arguments, the mass scale of electron, and by the cosmological data allowing to fix the string tension of the cosmic strings which are basic structures in TGD inspired cosmology.

The relationship between CP_2 radius and fundamental p-adic length scale

One can relate CP_2 'cosmological constant' to the p-adic mass scale: for $k_L = 1$ one has

$$m_0^2 \ = \ \frac{m_1^2}{k_L} = m_1^2 = 2\Lambda \ . \tag{11.8.14}$$

$k_L = 1$ results also by requiring that p-adic thermodynamics leaves charged leptons light and leads to optimal lowest order prediction for the charged lepton masses. Λ denotes the 'cosmological constant' of CP_2 (CP_2 satisfies Einstein equations $G^{\alpha\beta} = \Lambda g^{\alpha\beta}$ with cosmological term).

The real counterpart of the p-adic thermal expectation for the mass squared is sensitive to the choice of the unit of p-adic mass squared which is by definition mapped as such to the real unit in canonical identification. Thus an important factor in the p-adic mass calculations is the correct identification of the p-adic mass squared scale, which corresponds to the mass squared unit and hence to the unit of the p-adic numbers. This choice does not affect the spectrum of massless states but can affect the spectrum of light states in case of intermediate gauge bosons.

a) For the choice

$$M^2 \ = \ m_0^2 \leftrightarrow 1 \tag{11.8.15}$$

the spectrum of L_0 is integer valued.

b) The requirement that all sufficiently small mass squared values for the color partial waves are mapped to real integers, would fix the value of p-adic mass squared unit to

$$M^2 \ = \ \frac{m_0^2}{3} \leftrightarrow 1 \ . \tag{11.8.16}$$

For this choice the spectrum of L_0 comes in multiples of 3 and it is possible to have a first order contribution to the mass which cannot be of thermal origin (say $m^2 = p$). This indeed seems to happen for electro-weak gauge bosons.

p-Adic mass calculations [F3] allow to relate m_0 to electron mass and to Planck mass by the formula

$$\frac{m_0}{m_{Pl}} \ = \ \frac{1}{\sqrt{5 + Y_e}} \times 2^{127/2} \times \frac{m_e}{m_{Pl}} \ ,$$

$$Y_e \ = \ .7798 \ , \quad m_{Pl} = \frac{1}{\sqrt{G}} \ . \tag{11.8.17}$$

Using $m_{Pl} = 2.1665 \times 10^{-8}$ kg gives $m_0/m_{Pl} = 1.9628 \times 10^{-4}$.

This means that CP_2 radius R defined by the length $L = 2\pi R$ of CP_2 geodesic is roughly 10^4 times the Planck length. More precisely, using the relationship

$$\Lambda = \frac{3}{2R^2} = M^2 = m_0^2 \ ,$$

one obtains

$$L = 2\pi R \ = \ 2\pi \sqrt{\frac{3}{2}} \frac{1}{m_0} \simeq 3.92 \times 10^4 \sqrt{G} \ . \tag{11.8.18}$$

The result came as a surprise: the first belief was that CP_2 radius is of order Planck length. It has however turned out that the new identification solved elegantly some long standing problems of TGD.

CP_2 radius as the fundamental p-adic length scale

The identification of CP_2 radius as the fundamental p-adic length scale is forced by the Super Virasoro invariance. The pleasant surprise was that the identification of the CP_2 size as the fundamental p-adic length scale rather than Planck length solved many long standing problems of older TGD.

a) The earliest formulation predicted cosmic strings with a string tension larger than the critical value giving the angle deficit 2π in Einstein's equations and thus excluded by General Relativity. The corrected value of CP_2 radius predicts the value k/G for the cosmic string tension with k in the range $10^{-7} - 10^{-6}$ as required by the TGD inspired model for the galaxy formation solving the galactic dark matter problem.

b) In the earlier formulation there was no idea as how to derive the p-adic length scale $L \sim 10^4 \sqrt{G}$ from the basic theory. Now this problem becomes trivial and one has to predict gravitational constant in terms of the p-adic length scale. This follows in principle as a prediction of quantum TGD. In fact, one can deduce G in terms of the p-adic length scale and the action exponential associated with the CP_2 extremal and gets a correct value if α_K approaches fine structure constant at electron length scale (due to the fact that electromagnetic field equals to the Kähler field if Z^0 field vanishes).

Besides this, one obtains a precise prediction for the dependence of the Kähler coupling strength on the p-adic length scale by requiring that the gravitational coupling does not depend on the p-adic length scale. p-Adic prime p in turn has a nice physical interpretation: the critical value of α_K is same for the zero modes with given p. As already found, the construction of graviton state allows to understand the small value of the gravitational constant in terms of a de-coherence caused by multi-p fractality reducing the value of the gravitational constant from L_p^2 to G.

c) p-Adic length scale is also the length scale at which super-symmetry should be restored in standard super-symmetric theories. In TGD this scale corresponds to the transition to Euclidian field theory for CP_2 type extremals. There are strong reasons to believe that sparticles are however absent and that super-symmetry is present only in the sense that super-generators have complex conformal weights with $Re(h) = \pm 1/2$ rather than $h = 0$. The action of this super-symmetry changes the mass of the state by an amount of order CP_2 mass.

11.8.3   Fermion spectrum

The fact that $k = 1$ holds true for all particles forces to modify the earlier construction of quark states. This turns out to be possible without affecting the p-adic mass calculations whose outcome depend in an essential manner on the ground state conformal weights h_{ground} of the fermions (which can be negative).

Leptonic spectrum

For $k = 1$ the leptonic mass squared is integer valued in units of m_0^2 only for the states satisfying

$$p \ mod \ 3 \neq 2 \ .$$

Only these representations can give rise to massless states. Neutrinos correspond to (p,p) representations with $p \geq 1$ whereas charged leptons correspond to $(p,p+3)$ representations. The earlier mass calculations demonstrate that leptonic masses can be understood if the ground state conformal weight is $h_{gr} = -1$ for charged leptons and $h_{gr} = -2$ for neutrinos.

The contribution of color partial wave to conformal weight is $h_c = (p^2 + 2p)/3$, $p \geq 1$, for neutrinos and $p = 1$ gives $h_c = 1$ (octet). For charged leptons $h_c = (p^2 + 5p + 6)/3$ gives $h_c = 2$ for $p = 0$ (decuplet). In both cases super-canonical operator O_p must have a net conformal weight $h_{sc} = -3$ to produce a correct conformal weight for the ground state. The requirement of conformal confinement and p-adic considerations suggests the use of operators O_p constructed as sums of products of operators with opposite imaginary parts of super-canonical conformal weight $z = -1/2 - in_1 y_1 - in_2 y_2$. If the operators in question are color Hamiltonians in octet representation net super-canonical conformal weight $h_{sc} = -3$ results. The tensor product of two octets with conjugate super-canonical conformal weights contains both octet and decuplet so that singlets are obtained. What strengthens the hopes that the construction is not adhoc is that the same operator appears in the construction of quark states too.

Right handed neutrino remains essentially massless. $p = 0$ right handed neutrino does not however generate $N = 1$ space-time (or rather, imbedding space) super symmetry so that no sparticles are predicted. The breaking of the electro-weak symmetry at the level of the masses comes out basically from the anomalous color electro-weak correlation for the Kaluza-Klein partial waves implying that the weights for the ground states of the fermions depend on the electromagnetic charge of the fermion. Interestingly, TGD predicts leptohadron physics based on color excitations of leptons and color bound states of these excitations could correspond topologically condensed on string like objects but not fundamental string like objects.

Spectrum of quarks

Earlier arguments [F4] related to a model of CKM matrix as a rational unitary matrix suggested that the string tension parameter k is different for quarks, leptons, and bosons. The basic mass formula read as

$$M^2 = m^2_{CP_2} + kL_0 \ .$$

The values of k were $k_q = 2/3$ and $k_L = k_B = 1$. The general theory however predicts that $k = 1$ for all particles.

a) By earlier mass calculations and construction of CKM matrix the ground state conformal weights of U and D type quarks must be $h_{gr}(U) = -1$ and $h_{gr}(D) = 0$. The formulas for the eigenvalues of CP_2 spinor Laplacian imply that if m_0^2 is used as unit, color conformal weight $h_c \equiv m^2_{CP_2}$ is integer for $p \bmod = \pm 1$ for U type quark belonging to $(p+1, p)$ type representation and obeying $h_c(U) = (p^2 + 3p + 2)/3$ and for $p \bmod 3 = 1$ for D type quark belonging $(p, p+2)$ type representation and obeying $h_c(D) = (p^2 + 4p + 4)/3$. Only these states can be massless since color Hamiltonians have integer valued conformal weights.

b) In the recent case $p = 1$ states correspond to $h_c(U) = 2$ and $h_c(D) = 3$. $h_{gr}(U) = -1$ and $h_{gr}(D) = 0$ reproduce the previous results for quark masses required by the construction of CKM matrix. This requires a super-canonical operator O_p with a net conformal weight $h_{sc} = -3$ just as in the leptonic case. The facts that the values of p are minimal for spinor harmonics and the super-canonical operator is same for both quarks and leptons suggest that the construction is not had hoc.

c) It would seem that the tensor product of the spinor harmonic of quarks (as also leptons) with Hamiltonians gives rise to a large number of exotic colored states which have same thermodynamical mass as ordinary quarks (and leptons). Why these states have smaller values of p-adic prime that ordinary quarks and leptons, remains a challenge for the theory. Note that the decay widths of intermediate gauge bosons pose strong restrictions on the possible color excitations of quarks. On the other hand, the large number of fermionic color exotics can spoil the asymptotic freedom, and it is possible to have and entire p-adic length scale hierarchy of QCDs existing only in a finite length scale range without affecting the decay widths of gauge bosons.

The following table summarizes the color conformal weights and super-canonical vacuum conformal weights for the elementary particles.

	L	ν_L	U	D	W	γ, G, g
h_{vac}	-3	-3	-3	-3	-2	0
h_c	2	1	2	3	2	0

Table 1. The values of the parameters h_{vac} and h_c assuming that $k = 1$. The value of $h_{vac} \leq -h_c$ is determined from the requirement that p-adic mass calculations give best possible fit to the mass spectrum.

11.8.4 Photon, graviton and gluon

For photon, gluon and graviton the conformal weight of the $p = 0$ ground state is $h_{ground} = h_{vac} = 0$. The crucial condition is that $h = 0$ ground state is non-degenerate: otherwise one would obtain several physically more or less identical photons and this would be seen in the spectrum of black-body radiation. This occurs if one can construct several ground states not expressible in terms of the action of the Super Virasoro generators.

The only possibility to get exactly massless states in thermal sense is to have $\Delta = 0$ state with one active sector so that NS thermodynamics becomes trivial due to the absence of the thermodynamical excitations satisfying the gauge conditions. For neutral gauge bosons this is indeed achieved. For $T_p = 1/2$, which is required by the mass spectrum of intermediate gauge bosons, the thermal contribution to the mass squared is however extremely small even for W boson.

11.8.5 p-Adic thermodynamics does not explain the masses of intermediate gauge bosons

The requirement that the electron-intermediate gauge boson mass ratios are sensible, serves as a stringent test for the hypothesis that intermediate gauge boson masses result from the p-adic thermodynamics. It seems that the only possible option is that the parameter k has same value for both bosons, leptons, and quarks:

$$k_B = k_L = k_q = 1 \ .$$

In this case all gauge bosons have $D(0) = 1$ and there are good changes to obtain boson masses correctly. $k = 1$ together with $T_p = 1$ implies that the thermal masses of very many boson states are extremely heavy so that the spectrum of the boson exotics is reduced drastically. For $T_p = 1/2$ the thermal contribution to the mass squared is completely negligible.

Contrary to the original optimistic beliefs based on calculational error, it seems however impossible to predict W/e and Z/e mass ratios correctly in p-adic thermodynamics scenario. Although the errors are of order 20-30 percent, they are enough to exclude p-adic thermodynamics explanation for the massivation of gauge bosons.

a) The thermal mass squared for a boson state with N active sectors (non-vanishing vacuum weight) is determined by the partition function for the tensor product of N NS type Super Virasoro algebras. The degeneracies of the excited states as a function of N and the weight Δ of the operator creating the massless state are given in the table below.

b) Both W and Z must correspond to $N = 2$ active Super Virasoro sectors for which $D(1) = 1$ and $D(2) = 3$ so that (using the formulas of p-adic thermodynamics [TGDpad, F3]) the thermal mass squared is $m^2 = k_B(p + 5p^2)$ for $T_p = 1$. The second order contribution to the thermal mass squared is extremely small so that Weinberg angle vanishes in the thermal approximation. $k_B = 1$ gives Z/e mass-ratio which is about 22 per cent too high. The thermal prediction for W-boson mass is the same as for Z^0 mass and thus even worse since the two masses are related $M_W^2 = M_Z^2 cos^2(\theta_W)$. For $T_p = 1/2$ thermal masses are completely negligible.

c) It seems that the Achilles's heel of the p-adic thermodynamics is bosonic sector whereas the weak point of the standard model is fermionic sector. This suggests that it might be possible to combine these two approaches. $T_p = 1/2$ is certainly the only possible p-adic temperature for intermediate gauge bosons so that gauge boson masses should result by a TGD variant of the Higgs mechanism. Contrary to the long-held belief, it is indeed possible to identify a candidate for Higgs boson with correct quantum numbers also in TGD framework. The point is that in quaternion-conformal Kac Moody algebra $su(3) = u(2) + t$ Kac-Moody charges decompose to two separately conserved parts Q_g and Q_J corresponding to variations with respect to induced metric and induced Kähler form. Q_J charges in $u(2)$ sub-algebra of $su(3)$ are identifiable as electro-weak charges whereas the charges in the complement t of $u(2)$ have interpretation as Higgs field possessing correct couplings to electro-weak gauge bosons. If t generates coherent state, standard electro-weak Higgs mechanism follows as a consequence. Sigma model type description in which the coupling to Higgs bosons induces only small shifts of fermion masses, suggests itself. In fact, this kind of mechanism has been also applied in the TGD inspired model of CP breaking in case of ordinary hadrons [F5].

d) An alternative option is based on the observation that the charge matrices of W and of left handed part of Z^0 are not covariantly constant and have the correct group theoretical properties to yield breaking of conformal invariance and thus mass squared as a thermodynamical vacuum expectation value.

e) The minimum p-adic mass squared is the p-adic mass squared unit $m_0^2/3$. This corresponds in a reasonable approximation to the mass of W boson so that the mass scale would be predicted correctly. The calculation of leptonic masses however requires the use of m_0^2 as a mass squared unit for which intermediate gauge boson masses are smaller than one unit. The way out of the difficulty is based on the use of a variant of the canonical identification I acting as $I_1(r/s) = I(r)/I(s)$. This map respects under certain additional conditions various symmetries and is the only sensible possibility at the level of scattering amplitudes. This variant predicts that the real counterpart of $m^2 = (m/n)p$ is $(m/n)/p$ rather than of order CP_2 squared so that intermediate gauge boson masses can be smaller than one unit even if $O(p)$ p-adically, and allows an elegant group theoretic description of m_W/m_Z mass ratio in terms of Weinberg angle. This point is discussed in [F4, F5].

N, Δ	0	1/2	1	3/2	2	5/2	3
2	1	1	1	3	3	4	4
3	1	2	3	9	11		
4	1	3	5	19	26		
5	1	4	10	24	150		

Table 2. Degeneracies $d(\Delta, N)$ of the operators satisfying NS type gauge conditions as a function of the number N of the active sectors and of the conformal weight Δ of the operator. Only those degeneracies, which are needed in the mass calculation for bosons are listed.

11.8.6   Some probabilistic considerations

There are uniqueness problems related to the mapping of p-adic probabilities to real ones. These problems find a nice resolution from the requirement that the map respects probability conservation. The implied modification of the original mapping does not change measurably the predictions for the masses of light particles.

How unique the map of p-adic probabilities and mass squared values are mapped to real numbers is?

The mapping of p-adic thermodynamical probabilities and mass squared values to real numbers is not completely unique.

a) Canonical identification $I : \sum x_n p^n \rightarrow \sum x_n p^{-n}$ takes care of this mapping but does not respect the sum of probabilities so that the real images $I(p_n)$ of the probabilities must be normalized. This is a somewhat alarming feature.

b) The modification of the canonical identification mapping rationals by the formula $I(r/s) = I(r)/I(s)$ has appeared naturally in various applications, in particular because it respects unitarity of unitary matrices with rational elements with $r < p, s < p$. In the case of p-adic thermodynamic the formula $I(g(n)p^n/Z) \rightarrow I(g(n)p^n)/I(Z)$ would be very natural although Z need not be rational anymore. For $g(n) < p$ the real counterparts of the p-adic probabilities would sum up to one automatically for this option. One cannot deny that this option is more convincing than the original one. The generalization of this formula to map p-adic mass squared to a real one is obvious.

c) Options a) and b) differ dramatically when the $n = 0$ massless ground state has ground state degeneracy $D > 1$. For option a) the real mass is predicted to be of order CP_2 mass whereas for option b) it would be by a factor $1/D$ smaller than the minimum mass predicted by the option a). Thus option b) would predict a large number of additional exotic states. For those states which are light for option a), the two options make identical predictions as far as the significant two lowest order terms are considered. Hence this interpretation would not change the predictions of the p-adic mass calculations in this respect. Option b) is definitely more in accord with the real physics based intuitions and the main role of p-adic thermodynamics would be to guarantee the quantization of the temperature and fix practically uniquely the spectrum of the "Hamiltonian".

Under what conditions the mapping of p-adic ensemble probabilities to real probabilities respects probability conservation?

One can consider also a more general situation. Assume that one has an ensemble consisting of independent elementary events such that the number of events of type i is N_i. The probabilities are given by $p_i = N_i/N$ and $N = \sum N_i$ is the total number of elementary events. Even in the case that N is infinite as a real number it is natural to map the p-adic probabilities to their real counterparts using the rational canonical identification $I(p_i) = I(N_i)/I(N)$. Of course, N_i and N exist as well defined p-adic numbers under very stringent conditions only.

The question is under what conditions this map respects probability conservation. The answer becomes obvious by looking at the pinary expansions of N_i and N. If the integers N_i (possibly infinite as real integers) have pinary expansions having no common pinary digits, the sum of probabilities is conserved in the map. Note that this condition can assign also to a finite ensemble with finite number of a unique value of p.

This means that the selection of a basis for independent events corresponds to a decomposition of the set of integers labelling pinary digits to disjoint sets and brings in mind the selection of orthonormalized basis of quantum states in quantum theory. What is physically highly non-trivial that this "orthogonalization" alone puts strong constraints on probabilities of the allowed elementary events. One can say that the probabilities define distributions of pinary digits analogous to non-negative probability amplitudes in the space of integers labelling pinary digits, and the probabilities of independent events must be orthogonal with respect to the inner product defined by point-wise multiplication in the space of pinary digits.

p-Adic thermodynamics for which Boltzman weights $g(E)exp(-E/T)$ are replaced by $g(E)p^{E/T}$ such that one has $g(E) < p$ and E/T is integer valued, satisfies this constraint. The quantization of E/T to integer values implies quantization of both T and "energy" spectrum and forces so called super conformal invariance in TGD applications, which is indeed a basic symmetry of the theory.

There are infinitely many ways to choose the elementary events and each choice corresponds to a decomposition of the infinite set of integers n labelling the powers of p to disjoint subsets. These subsets can be also infinite. One can assign to this kind of decomposition a resolution which is the poorer the larger the subsets involved are. p-Adic thermodynamics would represent the situation in which the resolution is maximal since each set contains only single pinary digit. Note the analogy with the basis of completely localized wave functions in a lattice.

11.9   Modular contribution to the mass squared

The success of the mass calculations gives convincing support for generation-genus correspondence. The basic physical picture is following.

a) Mass squared is dominated by boundary contribution, which is sum of cm and modular contributions: $M^2 = M^2(cm) + M^2(mod)$. Here 'cm' refers to the thermal contribution from boundary and interior. Modular contribution can be assumed to depend on the genus of the boundary component only.

b) Modular contribution to mass squared can be estimated apart from an overall proportionality constant. The mass scale of the contribution is fixed by the p-adic length scale hypothesis. Elementary particle vacuum functionals are proportional to a product of all even theta functions and their conjugates, the number of even theta functions and their conjugates being $2N(g) = 2^g(2^g + 1)$. Also the thermal partition function must also be proportional to $2N(g)$:th power of some elementary partition function. This implies that thermal/ quantum maxpectation $M^2(mod)$ must be proportional to $2N(g)$. Since single handle behaves effectively as particle, the contribution must be proportional to genus g also. The success of the resulting mass formula encourages the belief that the argument is essentially correct.

The challenge is to construct theoretical framework reproducing the modular contribution to mass squared. There are two alternative manners to understand the origin modular contribution.

a) The realization that super-canonical algebra is relevant for elementary particle physics leads to the idea that two thermodynamics are involved with the calculation of the vacuum conformal weight as a thermal expectation. The first thermodynamics corresponds to Super Kac-Moody algebra and second thermodynamics to super-canonical algebra. This approach allows a first principle understanding of the origin and general form of the modular contribution without any need to introduce additional structures in modular degrees of freedom. The very fact that super-canonical algebra does not commute with the modular degrees of freedom explains the dependence of the super-canonical contribution on moduli.

b) The earlier approach was based on the idea that he modular contribution could be regarded as a quantum mechanical expectation value of the Virasoro generator L_0 for the elementary particle vacuum functional. Quantum treatment would require generalization the concepts of the moduli space and theta function to the p-adic context and finding an acceptable definition of the Virasoro generator L_0 in modular degrees of freedom. The problem with this interpretation is that it forces to introduce, not only Virasoro generator L_0, but the entire super Virasoro algebra in modular degrees of freedom. One could also consider of interpreting the contribution of modular degrees of freedom to vacuum conformal weight as being analogous to that of CP_2 Laplacian but also this would raise the challenge of constructing corresponding Dirac operator. Obviously this approach has become obsolete.

The thermodynamical treatment taking into account the constraints from that p-adicization is possible might go along following lines.

a) In the real case the basic quantity is the thermal expectation value $h(M)$ of the conformal weight as a function of moduli. The average value of the deviation $\Delta h(M) = h(M) - h(M_0)$ over moduli space $\mathcal{M}$ must be calculated using elementary particle vacuum functional as a modular invariant partition function. Modular invariance is achieved if this function is proportional to the logarithm of elementary particle vacuum functional: this reproduces the qualitative features basic formula for the modular contribution to the conformal weight. p-Adicization leads to a slight modification of this formula.

b) The challenge of algebraically continuing this calculation to the p-adic context involves several sub-tasks. The notions of moduli space $\mathcal{M}_p$ and theta function must be defined in the p-adic context. An appropriately defined logarithm of the p-adic elementary particle vacuum functional should determine $\Delta h(M)$. The average of $\Delta h(M)$ requires an integration over $\mathcal{M}_p$. The problems related to the definition of this integral could be circumvented if the integral in the real case could be reduced to an algebraic expression, or if the moduli space is discrete in which case integral could be replaced by a sum.

c) The number theoretic existence of the p-adic Θ function leads to the quantization of the moduli so that the p-adic moduli space is discretized. Accepting the sharpened form of Riemann hypothesis [E8], the quantization means that the imaginary *resp.* real parts of the moduli are proportional to integers *resp.* combinations of imaginary parts of zeros of Riemann Zeta. This quantization could occur also for the real moduli for the maxima of Kähler function. This reduces the problematic p-adic integration to a sum and the resulting sum defining $\langle \Delta h \rangle$ converges extremely rapidly for physically interesting primes so that only the few lowest terms are needed.

11.9.1 The physical origin of the genus dependent contribution to the mass squared

Different p-adic length scales are not enough to explain the charged lepton mass ratios and an additional genus dependent contribution in the fermionic mass formula is required. The general form of this contribution can be guessed by regarding elementary particle vacuum functionals in the modular degrees of freedom as an analog of partition function and the modular contribution to the conformal weight as an analog of thermal energy obtained by averaging over moduli. p-Adic length scale hypothesis determines the overall scale of the contribution.

The exact physical origin of this contribution has remained mysterious but super-canonical degrees of freedom represent a good candidate for the physical origin of this contribution. This would mean a sigh of relief since there would be no need to assign conformal weights, super-algebra, Dirac operators, Laplacians, etc.. with these degrees of freedom.

Thermodynamics in super-canonical degrees of freedom as the origin of the modular contribution to the mass squared

The following general picture is the simplest found hitherto.

a) Elementary particle vacuum functionals are defined in the space of moduli of surfaces X^2 corresponding to the maxima of Kähler function. There some restrictions on X^2. In particular, p-adic length scale poses restrictions on the size of X^2. There is an infinite hierarchy of elementary particle vacuum functionals satisfying the general constraints but only the lowest elementary particle vacuum functionals are assumed to contribute significantly to the vacuum expectation value of conformal weight determining the mass squared value.

b) The contribution of Super-Kac Moody thermodynamics to the vacuum conformal weight h coming from Virasoro excitations of the $h = 0$ massless state is estimated in the previous calculations and does not depend on moduli. The new element is that for a partonic 2-surface X^2 with given moduli, Virasoro thermodynamics is present also in super-canonical degrees of freedom.

Super-canonical thermodynamics means that, besides the ground state with $h_{gr} = -h_{SC}$ with minimal value of super-canonical conformal weight h_{SC}, also thermal excitations of this state by super-canonical Virasoro algebra having $h_{gr} = -h_{SC} - n$ are possible. For these ground states the SKM Virasoro generators creating states with net conformal weight $h = h_{SKM} - h_{SC} - n \geq 0$ have larger conformal weight so that the SKM thermal average h depends on n. It depends also on the moduli M of X^2 since the Beltrami differentials representing a tangent space basis for the moduli space $\mathcal{M}$ do not commute with the super-canonical algebra. Hence the thermally averaged SKM conformal weight h_{SKM} for given values of moduli satisfies

$$h_{SKM} \quad = \quad h(n, M) \ . \tag{11.9.1}$$

c) The average conformal weight induced by this double thermodynamics can be expressed as a super-canonical thermal average $\langle \cdot \rangle_{SC}$ of the SKM thermal average $h(n, M)$:

$$h(M) \quad = \quad \langle h(n, M) \rangle_{SC} = \sum p_n(M) h(n) \ , \tag{11.9.2}$$

where the moduli dependent probability $p_n(M)$ of the super-canonical Virasoro excitation with conformal weight n should be consistent with the p-adic thermodynamics. It is convenient to write $h(M)$ as

$$h(M) \quad = \quad h_0 + \Delta h(M) \ , \tag{11.9.3}$$

where h_0 is the minimum value of $h(M)$ in the space of moduli. The form of the elementary particle vacuum functionals suggest that h_0 corresponds to moduli with $Im(\Omega_{ij}) = 0$ and thus to singular configurations for which handles degenerate to one-dimensional lines attached to a sphere.

d) There is a further averaging of $\Delta h(M)$ over the moduli space $\mathcal{M}$ by using the modulus squared of elementary particle vacuum functional so that one has

$$h \quad = \quad h_0 + \langle \Delta h(M) \rangle_{\mathcal{M}} \ . \tag{11.9.4}$$

Modular invariance allows to pose very strong conditions on the functional form of $\Delta h(M)$. The simplest assumption guaranteing this and thermodynamical interpretation is that $\Delta h(M)$ is proportional to the logarithm of the vacuum functional Ω:

$$\Delta h(M) \quad \propto \quad -log(\frac{\Omega(M)}{\Omega_{max}}) \ . \tag{11.9.5}$$

Here Ω_{max} corresponds to the maximum of Ω for which $\Delta h(M)$ vanishes.

Justification for the general form of the mass formula

The proposed general ansatz for $\Delta h(M)$ provides a justification for the general form of the mass formula deduced by intuitive arguments.

a) The factorization of the elementary particle vacuum functional Ω into a product of $2N(g) = 2^g(2^g + 1)$ terms and the logarithmic expression for $\Delta h(M)$ imply that the thermal expectation values is a sum over thermal expectation values over $2N(g)$ terms associated with various even characteristics (a, b), where a and b are g-dimensional vectors with components equal to $1/2$ or 0 and the inner product $4a \cdot b$ is an even integer. If each term gives the same result in the averaging using Ω_{vac} as a partition function, the proportionality to $2N_g$ follows.

b) For genus $g \geq 2$ the partition function defines an average in $3g - 3$ complex-dimensional space of moduli. The analogy of $\langle \Delta h \rangle$ and thermal energy suggests that the contribution is proportional to the complex dimension $3g - 3$ of this space. For $g \leq 1$ the contribution the complex dimension of moduli space is g and the contribution would be proportional to g.

$$\begin{aligned}
\langle \Delta h \rangle &\propto g \times X(g) \text{ for } g \leq 1 \ , \\
\langle \Delta h \rangle &\propto (3g - 3) \times X(g) \text{ for } g \geq 2 \ , \\
X(g) &= 2^g(2^g + 1) \ .
\end{aligned} \tag{11.9.6}$$

If X^2 is hyper-elliptic for the maxima of Kähler function, this expression makes sense only for $g \leq 2$ since vacuum functionals vanish for hyper-elliptic surfaces.

c) The earlier argument, inspired by the interpretation of elementary particle vacuum functional as a partition function, was that each factor of the elementary particle vacuum functional gives the same contribution to $\langle \Delta h \rangle$, and that this contribution is proportional to g since each handle behaves like a particle:

$$\langle \Delta h \rangle \propto g \times X(g) \ . \tag{11.9.7}$$

The prediction following from the previous differs by a factor $(3g - 3)/g$ for $g \geq 2$. This would scale up the dominant modular contribution to the masses of the third $g = 2$ fermionic generation by a factor $\sqrt{3/2} \simeq 1.22$. One must of course remember, that these rough arguments allow $g-$ dependent numerical factors of order one so that it is not possible to exclude either argument.

11.9.2   Generalization of Θ functions and quantization of p-adic moduli

The task is to find p-adic counterparts for theta functions and elementary particle vacuum functionals. The constraints come from the p-adic existence of the exponentials appearing as the summands of the theta functions and from the convergence of the sum. The exponentials must be proportional to powers of p just as the Boltzmann weights defining the p-adic partition function. The outcome is a quantization of moduli so that integration can be replaced with a summation and the average of $\Delta h(M)$ over moduli is well defined.

It is instructive to study the problem for torus in parallel with the general case. The ordinary moduli space of torus is parameterized by single complex number τ. The points related by $SL(2, Z)$ are equivalent, which means that the transformation $\tau \to (A\tau + B)/(C\tau + D)$ produces a point equivalent with τ. These transformations are generated by the shift $\tau \to \tau + 1$ and $\tau \to -1/\tau$. One can choose the fundamental domain of moduli space to be the intersection of the slice $Re(\tau) \in [-1/2, 1/2]$ with the exterior of unit circle $|\tau| = 1$. The idea is to start directly from physics and to look whether one might some define p-adic version of elementary particle vacuum functionals in the p-adic counter part of this set or in some modular invariant subset of this set.

Elementary particle vacuum functionals are expressible in terms of theta functions using the functions $\Theta^4[a, b]\overline{\Theta}^4[a, b]$ as a building block. The general expression for the theta function reads as

$$\Theta[a, b](\Omega) = \sum_n exp(i\pi(n + a) \cdot \Omega \cdot (n + a))exp(2i\pi(n + a) \cdot b) \ . \tag{11.9.8}$$

The latter exponential phase gives only a factor $\pm i$ or ± 1 since $4a \cdot b$ is integer. For $p \bmod 4 = 3$ imaginary unit exists in an algebraic extension of p-adic numbers. In the case of torus (a, b) has the values $(0, 0)$, $(1/2, 0)$ and $(0, 1/2)$ for torus since only even characteristics are allowed.

Concerning the p-adicization of the first exponential appearing in the summands in Eq. 11.9.8, the obvious problem is that π does not exists p-adically. The introduction of the scaled variable $\hat{\tau} = \pi\tau$ resolves this problem. The second modification is the replacement of the factors $exp(X)$ with $p^{X/log(p)}$ in order to achieve a rapid p-adic convergence of the sum defining the theta function. This requires a further scaling so that one has $\Omega_p = \pi\Omega/log(p)$ is the appropriate variable and the terms in the sum are apart from the phase factor of form $p^{i(n+a)\cdot\Omega_p\cdot(n+a)}$.

If the exponents

$$p^{i(n+a)\cdot Im(\Omega_{ij,p})\cdot(n+a)} = p^{-a\cdot Im(\Omega_{ij,p})\cdot a} \times p^{-2a\cdot Im(\Omega_{ij,p})n} \times p^{-n\cdot Im(\Omega_{ij,p})\cdot n}$$

are integer powers of p, $\Theta_{[a,b]}$ exist in R_p. A milder condition is that only the building blocks $\Theta^4[a, b]\overline{\Theta}^4[a, b]$ exist in R_p. The problematic factor is the first exponent since the components of the vector a can have values $1/2$ and 0 and its existence implies a quantization of $Im(\Omega_{ij,p})$ as

$$Im(\Omega_{ij,p}) = -Kn_{ij} \ , \ n_{ij} \in Z \ , \ n_{ij} \geq 1 \ , \tag{11.9.9}$$

$K = 4$ guarantees the existence of Θ functions and $K = 1$ the existence of elementary particle vacuum functionals. Obviously the sum defining Θ converges rapidly with respect to the p-adic norm.

The problem is that the condition $Im(\Omega_{ij,p}) > 0$ is not satisfied. There is however no reason why the p-adic theta function could not be defined by changing the sign of the exponents so that one would have

$$\Theta[a, b](\Omega)_p = \sum_n p^{-i(n+a)\cdot\Omega_p\cdot(n+a)} \times exp\left[2i\pi(n + a) \cdot b\right] \ ,$$

$$Im(\Omega_{ij,p}) = Kn_{ij} \ , \ n_{ij} \geq 1 \ . \tag{11.9.10}$$

$K = 4$ guarantees the existence of Θ functions in R_p and $K = 1$ the existence of elementary vacuum functional in R_p: in this case $\Theta_{[a,b]}$ exists in appropriate algebraic extension of R_p. Note that a similar change of sign must be performed in p-adic thermodynamics for powers of p to map p-adic probabilities to real ones.

A further requirement is that the phases $p^{-iRe(\Omega_{ij,p})/4}$ exist p-adically. A weaker condition that only the phases $p^{-iRe(\Omega_{ij,p})}$ exist p-adically guarantees that elementary particle vacuum functionals exist p-adically. The condition that p^{iy} exists for certain preferred values of y for all values of prime p is encountered repeatedly in the algebraic continuation of quantum TGD to p-adic context. The sharpening of the Riemann Hypothesis [E8] stating that the partition functions $1/(1-p^z)$ appearing in the product expansion of Rieman Zeta in various p-adic number fields exist for the zeros $z = 1/2 + iy$ of Riemann Zeta, is number theoretically highly attractive.

This conjecture implies that p^{iy} is in general a product of a phase factor $exp(i2\pi m/n)$ in some algebraic extension of p-adic numbers and of a Pythagorean phase $(k + il)/\sqrt{k^2 + l^2}$, $k^2 + l^2 = n^2$. A potential problem is that this phase factor does not possess unit p-adic norm in the general case.

The explicit form for the allowed (k, l) and (l, k) pairs is given by

$$\begin{aligned} k &= 2rs \ , \\ l &= r^2 - s^2 \ , \\ n &= r^2 + s^2 \ . \end{aligned} \tag{11.9.11}$$

where r and s are relatively prime integers, not both odd. Note that (l, k) is also an allowed solution. An important point to be noticed is that the p-adic norm of Pythagorean phase is not larger than one for physically most interesting primes satisfying $p \bmod 4 = 3$ since $n \bmod 4 = 1$ holds true as a simple calculation shows. This guarantees that the phase factors of the Θ function cannot spoil the p-adic convergence of the sum defining the p-adic theta function.

The sharpening of the Riemann hypothesis, when combined with the requirement that the logarithmic radial waves $(r_M/r_0)^{iz}$ exists in some finite-dimensional extension of any p-adic number fields when r_M/r_0 is rational valued, implies that the radial conformal weights z of the super-canonical algebra correspond to the zeros of Zeta and their appropriate combinations. The quantization condition is

$$Re(\Omega_{ij,p}) \;=\; K\sum n_k y_k \;,$$

(11.9.12)

where y_k correspond to zeros of Zeta. $K = 4$ guarantees that Θ functions exist p-adically. $K = 1$ is enough to guarantee the existence of elementary particle vacuum functionals.

In the real context the quantization of moduli of torus would correspond to

$$\tau \;=\; K(\sum n_k y_k + in) \times \frac{log(p)}{\pi} \;,$$

$$|\tau| \;=\; K\sqrt{n^2 + (\sum_k n_k y_k)^2} \;,$$

$$\Phi \;=\; atan(\frac{n}{\sum_k n_k y_k}) \;.$$

(11.9.13)

$K = 1$ guarantees the existence of elementary particle vacuum functionals and $K = 4$ the existence of Theta functions. The ratio for the complex vectors defining the sides of the plane parallelogram defining torus via the identification of the parallel sides is quantized. In other words, the angles Φ between the sides and the ratios of the sides given by $|\tau|$ have quantized values.

The quantization rules for the moduli of the higher genera read as

$$\Omega_{ij} \;=\; K\left[\sum n_k(i,j)y_k + in(i,j)\right] \times \frac{log(p)}{\pi} \;,$$

(11.9.14)

If the quantization rules hold true also for the maxima of Kähler function in the real context, there are good hopes that the p-adicized expression for Δh is obtained by a simple algebraic continuation of the real formula. Thus p-adic length scale characterizes partonic surface X^2 rather than the light like causal determinant X_l^3 containing X^2. Therefore the idea that various p-adic primes label various X_l^3 connecting fixed partonic surfaces X_i^2 would not be correct.

The set of the moduli allowed by the quantization rules is not invariant under modular transformations. For instance, in the case of torus the $SL(2, Z)$ Möbius transformations $\Omega \rightarrow \Omega + n$ and $\Omega \rightarrow 1/\Omega$ lead out of the allowed moduli space. This is not however a problem if there ar no modular transformations relating quantized moduli so that they can be thought of as forming single fundamental domain containing possibly non-equivalent moduli from several fundamental domains in the conventional sense of the word.

Quite generally, the quantization of moduli means that the allowed 2-dimensional shapes form a lattice and are thus additive. It also means that the maxima of Kähler function obey a linear superposition in an extreme abstract sense. The proposed number theoretical quantization is expected to apply for any complex space allowing some preferred complex coordinates. In particular, configuration space of 2-surfaces could allow this kind of quantization in the complex coordinates naturally associated with isometries and this could allow to define configuration space integration, at least the counterpart of integration in zero mode degrees of freedom, as a summation.

11.9.3 The calculation of the modular contribution $\langle \Delta h \rangle$ to the conformal weight

The quantization of the moduli implies that the integral over moduli can be defined as a sum over moduli. The theta function $\Theta[a,b](\Omega)_p(\tau_p)$ is proportional to $p^{a \cdot a Im(\Omega_{ij,p})} = p^{K n_{ij} m(a)/4}$ for $a \cdot a = m(a)/4$, where $K = 1$ *resp.* $K = 4$ corresponds to the existence existence of elementary particle vacuum functionals *resp.* theta functions in R_p. These powers of p can be extracted from the thetas defining the vacuum functional. The numerator of the vacuum functional gives $(p^n)^{2K\sum_{a,b} m(a)}$. The numerator gives $(p^n)^{2K\sum_{a,b} m(a_0)}$, where a_0 corresponds to the minimum value of $m(a)$. $a_0 = (0,0,..,0)$ is allowed and gives $m(a_0) = 0$ so that the p-adic norm of the denominator equals to one. Hence one has

$$|\Omega_{vac}(\Omega_p)|_p \quad = \quad p^{-2nK\sum_{a,b} m(a)} \qquad\qquad (11.9.15)$$

The sum converges extremely rapidly for large values of p as function of n so that in practice only few moduli contribute.

The definition of $log(\Omega_{vac})$ poses however problems since in $log(p)$ does not exist as a p-adic number in any p-adic number field. The argument of the logarithm should have a unit p-adic norm. The simplest manner to circumvent the difficulty is to use the fact that the p-adic norm $|\Omega_p|_p$ is also a modular invariant, and assume that the contribution to conformal weight depends on moduli as

$$\Delta h_p(\Omega_p) \quad \propto \quad log(\frac{\Omega_{vac}}{|\Omega_{vac}|_p}) \ . \qquad\qquad (11.9.16)$$

The sum defining $\langle \Delta h_p \rangle$ converges extremely rapidly and gives a result of order $O(p)$ p-adically as required.

The p-adic expression for $\langle \Delta h_p \rangle$ should result from the corresponding real expression by an algebraic continuation. This encourages the conjecture that the allowed moduli are quantized for the maxima of Kähler function, so that the integral over the moduli space is replaced with a sum also in the real case, and that Δh given by the double thermodynamics as a function of moduli can be defined as in the p-adic case. The positive power of p multiplying the numerator could be interpreted as a degeneracy factor. In fact, the moduli are not primary dynamical variables in the case of the induced metric, and there must be a modular invariant weight factor telling how many 2-surfaces correspond to given values of moduli. The power of p could correspond to this factor.

Chapter 12

p-Adic Particle Massivation: Elementary Particle Masses

12.1 Introduction

In this chapter the detailed predictions of the p-adic description of particle massivation are studied.

12.1.1 Basic contributions to the particle mass squared

The formula for the p-adic mass squared contains three additive contributions. The first contribution is proportional to the thermal expectation of the Virasoro generator L_0 in in Super Virasoro degrees of freedom. The miracle is that this contribution is small for the particles with the quantum numbers of the observed light particles, when Super Virasoro has $N = 5$ sectors as it does in TGD approach. These sectors correspond to the 5 tensor factors for the $M^4 \times SU(3) \times U(2)_{ew}$ decomposition of the super Kac Moody algebra to gauge symmetries of gravitation, color and electro-weak interactions. These symmetries act on the intersections $X^2 = X_l^3 \cap X^7$ of 3-D light like causal determinants (CDs) X_l^3 and 7-D light like CDs $X^7 = \delta M_+^4 \times CP_2$. This constraint leaves only the 2 transversal degrees M^4 degrees of freedom since the translations in light like directions associated with X_l^3 and δM_+^4 are eliminated.

Second contribution comes from the modular degrees of freedom associated with the boundary component of the particle like 3-surface and the interpretation as a thermal contribution is possible but not necessary.

The third contribution consists corresponds to the interactions between partonic 2-surfaces and are important inside hadrons. Also corrections from the secondary condensation to the mass squared are expected.

The lowest order contributions to the charged lepton masses are predicted correctly and information about the yet non-calculable second order corrections is obtained by requiring that the mass ratios are reproduced exactly. Since the contribution from modular degrees of freedom dominates for higher generations the successful predictions for fermion masses give strong support for the topological explanation of the family replication phenomenon.

The prediction or quark masses is more difficult since even the deduction of even the p-adic length scale determining the masses of u, d, and s is a non-trivial task. Second difficulty is related to the topological mixing of quarks. Somewhat surprisingly, the model for U and D matrices constructed for a decade ago predicts realistic quark mass spectrum although the new mass formula is based on different assumptions and different identification of p-adic mass scales.

Graviton, photon and gluons are predicted to be exactly massless. Contrary to the long held beliefs, intermediate boson mass scale is predicted to be 20-30 per cent too high if p-adic thermodynamics with temperature $T_p = 1$ is solely responsible for gauge boson masses. $T_p = 1/2$ predicts completely negligible masses.

One can consider two remedies to the situation: either Higgs type particle exists or the lack of covariant constancy of electro-weak charge matrices can be mimicked by the the generation of vacuum expectation value of Higgs.

12.1.2   The identification of Higgs as a weakly charged wormhole contact

Quantum classical correspondence suggests that electro-weak massivation should have simple space-time description allowing also to identify Higgs boson if it exists. This description indeed exists and allows also to understand the precise relationship between gravitational and inertial masses and how Equivalence Principle is weakened in TGD framework.

The basic observation is that gauge and gravitational fluxes flow to larger space-time sheets through # (wormhole) contacts. If gravitational energy can be regarded in the Newtonian limit as a gauge charge, the contacts feed the gravitational energy regarded as a gauge flux to the lower condensate levels. The non-conservation of gravitational gauge flux means that # contacts can carry gravitational four-momentum. Since CP_2 type vacuum extremals are the natural candidates for # contacts, the natural hypothesis is that the non-vanishing light-like gravitational four-momentum of # contacts is responsible for the non-conservation of gravitational four-momentum flux. The non-conservation of the light-like gravitational four-momentum of CP_2 type extremals is in turn responsible for the non-conservation of the net gravitational four-momentum.

contacts can be also carriers of inertial four-momentum which must be conserved in absence of four-momentum exchange between environment and wormhole contact. Therefore Equivalence Principle cannot hold true in strict sense. Equivalence Principle is satisfied in a weak sense if the inertial four-momentum is equal to the average four-momentum associated with the zitterbewegung motion and corresponds to the center of mass motion for the # contact.

The non-conservation of weak gauge currents for CP_2 type extremals implies a non-conservation of weak charges and the finite range of weak forces. If wormhole contacts correspond to pieces of CP_2 type vacuum extremal, electro-weak gauge currents are not conserved classically unlike color and Kähler current. The non-conservation of weak isospin corresponds to the presence of pairs of right/left handed fermion and left/right handed antifermion at wormhole contacts. These wormhole contacts are excellent candidates for the TGD counterpart of Higgs boson providing the most natural mechanism for the massivation of weak bosons. The finding that that CP_2 parts of the induced gamma matrices connect different M^4 chiralities of induced spinor fields provided the original motivation for the belief that Higgs mechanism is realized in some manner in TGD Universe. This coupling must be crucial for the formation of weakly charged wormhole contacts.

There are two contributions to the mass of elementary particle corresponding to the primary and secondary topological condensation.

a) The dominant contribution to the fermion masses would be due to p-adic thermodynamics describing primary topological condensation. If weak form of Equivalence Principle holds true, inertial mass would result simply as the average of non-conserved light-like gravitational four-momentum. This contribution to the inertial mass is generated in the topological condensation of CP_2 type extremal representing elementary particle involving only single light like elementary particle horizon, say fermion, and by randomness of the zitterbewegung corresponds naturally to the contribution given by p-adic thermodynamics.

b) For gauge bosons the contribution from primary condensation should be very small or vanishing if the radius of zitterbewegung orbit is larger than the size of the space-time sheet containing the topologically condensed boson so that the motion is along a light-like geodesic in a good approximation. The space-time sheet representing massless state suffered secondary topologically condensation at a larger space-time sheet and viewed as a particle can develop mass via Higgs mechanism since wormhole contacts cannot be regarded as moving along light like geodesics in the length and time scale involved. # contacts carrying net left handed weak isospin have interpretation as TGD counterparts of neutral Higgs bosons and the formation of a coherent state involving superposition of states with varying number of wormhole contacts corresponds to the generation of a vacuum expectation value of Higgs field.

12.1.3   Exotic states

The physical consequences of the exotic light leptons, quarks, and bosons are considered in the chapter devoted to the New Physics.

a) The existence of fermionic families suggests the existence of higher bosonic families too. The new view about particle decay as a branching of 4-surface however makes sense only if ordinary bosons are quantum superpositions of all three genera with equal probabilities and phases. The exotic bosons should correspond to non-constant Z_3-valued wave functions $exp(i2\pi ng/3))$ $g = 0, 1, 2$, $n = 1, 2$. Only three fermion families are predicted if $g > 2$ topologies for partonic 2-surfaces correspond to free many-handle states rather than bound states as for $g < 3$ topologies: who this could happen is discussed in [F1].

b) Also p-adically scaled up copies of various particles are possible as well as scaled-up/scaled-down versions of QCD associated with both quarks [F8] and colored leptons [F7]. There is now quite a lot of evidence that neutrino masses depend on environment [ha20]: this dependence could have an explanation in terms of topological condensation occurring in several p-adic length scales.

c) The original expectation was that the spectrum has $N = 1$ space-time super-symmetry and only the understanding of the super-conformal symmetries led to the conclusion that sparticles are absent. This does not affect the mass calculations in any manner and dramatically reduces the number of exotic states. The sigh of relief remained short-lasting. It became clear that since the super-canonical algebra generates ground states with an arbitrarily large negative conformal weight, which is compensated by the super Kac-Moody conformal weight, the existence of an infinite number of massless states is unavoidable. These states are expected to become massive by p-adic thermodynamics. The observed particles can be distinguished from the spectrum in a natural manner as particles having the lowest possible super-canonical vacuum weight but it is an open question whether all these super-canonical states are indeed sufficiently massive or, more or less equivalently (Uncertainty Principle), unstable enough and look like free many-particle states of a 2-dimensional system.

d) If elementary particles correspond to CP_2 type extremals, all massless exotic massless particles can be constructed using colored generators and by color confinement cannot induce macroscopic long range interactions [F2]. The essential assumption is that the fermionic quantization for the space-time sheets having CP_2 projection of dimension $D(CP_2) < 4$ is non-conventional. This has also direct relevance for the understanding of the matter antimatter asymmetry [D4, D5].

e) As already noticed electro-weak doublet Higgs particle would be present in the spectrum and be identifiable as wormhole contact, contrary to the long held beliefs. Also $q - \overline{q}$ bound states of M_{89} hadron physics such that quark and anti-quark have parallel spins and relative angular momentum $L = 1$ could mimic scalar mesons. The effective couplings of these states to leptons and quarks could mimic the couplings of Higgs boson to some degree. Scalar bound states of heavy quarks are also present in ordinary hadron physics.

To sum up, the results of the calculations provide a considerable support for TGD and the notion p-adicization. What remains still to be understood are various sources of the second order corrections to the masses and the values of some integer valued small parameters fixed completely by the empirical constraints. The detailed analysis and application of the results to derive information about hadron masses is left to the next chapter.

12.2 Various contributions to the particle masses

In the sequel various contributions to the mass squared are discussed.

12.2.1 General mass squared formula

The thermal independence of Super Virasoro and modular degrees of freedom implies that mass squared for elementary particle is the sum of Super Virasoro, modular and renormalization correction contributions:

$$M^2 = M^2(color) + M^2(SV) + M^2(mod) + M^2(ren) \ . \tag{12.2.1}$$

At this stage the small second order renormalization correction is not yet calculable but can be deduced from the known particle masses by comparing them to the predictions of the theory.

12.2.2 Color contribution to the mass squared

The mass squared contains a non-thermal color contribution to the ground state conformal weight coming from the mass squared of CP_2 spinor harmonic. The color contribution is an integer multiple of $m_0^2/3$, where $m_0^2 = 2\Lambda$ denotes the 'cosmological constant' of CP_2 (CP_2 satisfies Einstein equations $G^{\alpha\beta} = \Lambda g^{\alpha\beta}$).

The color contribution to the p-adic mass squared is integer valued only if $m_0^2/3$ is taken as a fundamental p-adic unit of mass squared. This choice has an obvious relevance for p-adic mass calculations since the simplest form of the canonical identification does not commute with a division by integer. More precisely, the image of number xp in canonical identification has a value of order 1 when x is a non-trivial rational number whereas for $x = np$ the value is n/p and extremely is small for physically interesting primes.

The choice of the p-adic mass squared unit are no effects on zeroth order contribution which must vanish for light states: this requirement eliminates quark and lepton states for which the CP_2 contribution to the mass squared is not integer valued using m_0^2 as a unit. There can be a dramatic effect on the first order contribution. The mass squared $m^2 = p$ using $m_0^2/3$ means that the particle is light. The mass squared becomes $m^2 = p/3$ when m_0^2 is used as a unit and the particle has mass of order 10^{-4} Planck masses. In the case of W and Z^0 bosons this problem is actually encountered. For light states using $m_0^2/3$ as a unit only the second order contribution to the mass squared is affected by this choice.

12.2.3   Modular contribution to the mass of elementary particle

The general form of the modular contribution is derivable from p-adic partition function for conformally invariant degrees of freedom associated with the boundary components. The general form of the vacuum functionals as modular invariant functions of Teichmuller parameters was derived in [F1] and the square of the elementary particle vacuum functional can be identified as a partition function. Even theta functions serve as basic building blocks and the functionals are proportional to the product of all even theta functions and their complex conjugates. The number of theta functions for genus $g > 0$ is given by

$$N(g) \;\; = \;\; 2^{g-1}(2^g + 1) \; . \tag{12.2.2}$$

One has $N(1) = 3$ for muon and $N(2) = 10$ for τ.

a) Single theta function is analogous to a partition function. This implies that the modular contribution to the mass squared must be proportional to $2N(g)$. The factor two follows from the presence of both theta functions and their conjugates in the partition function.

b) The factorization properties of the vacuum functionals imply that handles behave effectively as particles. For example, at the limit, when the surface splits into two pieces with g_1 and $g - g_1$ handles, the partition function reduces to a product of g_1 and $g - g_1$ partition functions. This implies that the contribution to the mass squared is proportional to the genus of the surface. Altogether one has

$$M^2(mod, g) \;\; = \;\; 2k(mod)N(g)g\frac{m_0^2}{p} \; ,$$
$$k(mod) \;\; = \;\; 1 \; . \tag{12.2.3}$$

Here $k(mod)$ is some integer valued constant (in order to avoid ultra heavy mass) to be determined. $k(mod) = 1$ turns out to be the correct choice for this parameter.

Summarizing, the real counterpart of the modular contribution to the mass of a particle belonging to $g + 1$:th generation reads as

$$M^2(mod) \;\; = \;\; 0 \; for \; e, \nu_e, u, d \; ,$$
$$M^2(mod) \;\; = \;\; 9\frac{m_0^2}{p(X))} \; for \; X = \mu, \nu_\mu, c, s \; ,$$
$$M^2(mod) \;\; = \;\; 60\frac{m_0^2}{p(X)} \; for \; X = \tau, \nu_\tau, t, b \; . \tag{12.2.4}$$

The requirement that hadronic mass spectrum and CKM matrix are sensible however forces the modular contribution to be the same for quarks, leptons and bosons. The higher order modular contributions to the mass squared are completely negligible if the degeneracy of massless state is $D(0, mod, g) = 1$ in the modular degrees of freedom as is in fact required by $k(mod) = 1$.

12.2.4 Thermal contribution to the mass squared

One can deduce the value of the thermal mass squared in order $O(p^2)$ (an excellent approximation) using the general mass formula given by p-adic thermodynamics. Assuming maximal p-adic temperature $T_p = 1$ one has

$$
\begin{aligned}
M^2 &= k(sp + Xp^2 + O(p^3)) \ , \\
s_\Delta &= \frac{D(\Delta + 1)}{D(\Delta)} \ , \\
X_\Delta &= 2\frac{D(\Delta + 2)}{D(\Delta)} - \frac{D^2(\Delta + 1)}{D^2(\Delta)} \ , \\
k &= 1 \ .
\end{aligned}
\tag{12.2.5}
$$

Δ is the conformal weight of the operator creating massless state from the ground state.

The ratios $r_n = D(n+1)/D(n)$ allowing to deduce the values of s and X have been deduced from p-adic thermodynamics in [F2]. Light state is obtained only provided $r(\Delta)$ is an integer. The remarkable result is that for lowest lying states this is the case. For instance, for Ramond representations the values of r_n are given by

$$
(r_0, r_1, r_2, r_3) = (8, 5, 4, \frac{55}{16}) \ .
\tag{12.2.6}
$$

The values of s and X are

$$
\begin{aligned}
(s_0, s_1, s_2) &= (8, 5, 4) \ , \\
(X_0, X_1, X_2) &= (16, 15, 11 + 1/2)) \ .
\end{aligned}
\tag{12.2.7}
$$

The result means that second order contribution is extremely small for quarks and charged leptons having $\Delta < 2$. For neutrinos having $\Delta = 2$ the second order contribution is non-vanishing.

12.2.5 Second order renormalization contribution

Second order renormalization contribution comes from the loops contributing to the propagators in p-adic QFT and cannot be calculated at this stage. This contribution is expected to be same for all particle families with given electro-weak quantum numbers and information about its value can be deduced from the experimental values of the particle masses. Since the value of the second order contribution Xp^2 is bounded from above $(Xp^2)_R \leq 1/p$ it is not possible to reproduce particle masses by a suitable choice of Y.

As far as the comparison of the predictions to the experimental numbers is considered, the basic problem is whether the measured mass for a particle corresponds to the primarily condensed particle and hence to the predicted mass or whether it corresponds to the mass of the secondarily condensed particle. One could argue that one must compare the mass predictions in some common p-adic length scale determined by the experimental arrangement. For instance, in case of muon and electron this could mean considering both masses in the p-adic length scale associated with the electron. In a good approximation the mass change generated by the secondary condensation can be assumed to correspond to the mass renormalization predicted by QED. It turns out that in the proposed scenario mass renormalization correction must be taken into account to achieve better than few percent accuracy for the mass ratios and that the sign and order of magnitude for the corrections are correct.

12.2.6 General mass formula for Ramond representations

By taking the modular contribution from the boundaries into account the general p-adic mass formulas for the Ramond type states read for states for which the color contribution to the conformal weight is integer valued as

$$\frac{m^2(\Delta = 0)}{m_0^2} = (8 + n(g))p + Yp^2 \ ,$$

$$\frac{m^2(\Delta = 1)}{m_0^2} = (5 + n(g)p + Yp^2 \ ,$$

$$\frac{m^2(\Delta = 2)}{m_0^2} = (4 + n(g))p + (Y + \frac{23}{2})p^2 \ ,$$

$$n(g) = 2g \cdot 2^{g-1}(2^g + 1) \ . \tag{12.2.8}$$

Here Δ denotes the conformal weight of the operators creating massless states from the ground state and g denotes the genus of the boundary component. The values of $n(g)$ for the three lowest generations are $n(0) = 0$, $n(1) = 9$ and $n(2) = 60$. The value of second order thermal contribution is nontrivial for neutrinos only. The value of the rational number Y can, which corresponds to the renormalization correction to the mass, can be determined using experimental inputs.

Using m_0^2 as a unit, the expression for the mass of a Ramond type state reads in terms of the electron mass as

$$M(\Delta, g, p)_R = K(\Delta, g, p)\sqrt{\frac{M_{127}}{p}} m_e$$

$$K(0, g, p) = \sqrt{\frac{n(g) + 8 + Y_R}{X}}$$

$$K(1, g, p) = \sqrt{\frac{n(g) + 5 + Y_R}{X}}$$

$$K(2, g, p) = \sqrt{\frac{n(g) + 4 + Y_R}{X}} \ ,$$

$$X = \sqrt{5 + Y(e)_R} \ . \tag{12.2.9}$$

Y can be assumed to depend on the electromagnetic charge and color representation of the state and is therefore same for all fermion families. Mathematica provides modules for calculating the real counterpart of the second order contribution and for finding realistic values of Y.

12.2.7   General mass formulas for NS representations

Using $m_0^2/3$ as a unit, the expression for the mass of a light NS type state for $T_p = 1$ ad $k_B = 1$ reads in terms of the electron mass as

$$M(\Delta, g, p, N)_R = K(\Delta, g, p, N)\sqrt{\frac{M_{127}}{p}} m_e$$

$$K(0, g, p, 1) = \sqrt{\frac{n(g) + Y_R}{X}} \ ,$$

$$K(0, g, p, 2) = \sqrt{\frac{n(g) + 1 + Y_R}{X}} \ ,$$

$$K(1, g, p, 3) = \sqrt{\frac{n(g) + 3 + Y_R}{X}} \ ,$$

$$K(2, g, p, 4) = \sqrt{\frac{n(g) + 5 + Y_R}{X}} \ ,$$

$$K(2, g, p, 5) = \sqrt{\frac{n(g) + 10 + Y_R}{X}} \ ,$$

$$X = \sqrt{5 + Y(e)_R} \ . \tag{12.2.10}$$

Here N is the number of the 'active' NS sectors (sectors for which the conformal weight of the massless state is non-vanishing). Y denotes the renormalization correction to the boson mass and in general depends on the electro-weak and color quantum numbers of the boson.

The thermal contribution to the mass of W boson is too large by roughly a factor $\sqrt{3}$ for $T_p = 1$. Hence $T_p = 1/2$ must hold true for gauge bosons and their masses must have a non-thermal origin perhaps analogous to Higgs mechanism. Alternatively, the non-covariant constancy of charge matrices could induce the boson mass [F2].

It is interesting to notice that the minimum mass squared for gauge boson corresponds to the p-adic mass unit $M^2 = m_0^2 p/3$ and this just what is needed in the case of W boson. This forces to ask whether $m_0^2/3$ is the correct choice for the mass squared unit so that non-thermally induced W mass would be the minimal $m_W^2 = p$ in the lowest order. This choice would mean the replacement

$$Y_R \to \frac{(3Y)_R}{3}$$

in the preceding formulas and would affect only neutrino mass in the fermionic sector. $m_0^2/3$ option is excluded by charged lepton mass calculation. This point will be discussed later.

12.2.8 Primary condensation levels from p-adic length scale hypothesis

p-Adic length scale hypothesis states that the primary condensation levels correspond to primes near prime powers of two $p \simeq 2^k$, k integer with prime values preferred. Black hole-elementary particle analogy [E5] suggests a generalization of this hypothesis by allowing k to be a power of prime. The general number theoretical vision discussed in [E1] provides a first principle justification for p-adic length scale hypothesis in its most general form. The best fit for the neutrino mass squared differences is obtained for $k = 13^2 = 169$ so that the generalization of the hypothesis might be necessary.

A particle primarily condensed on the level k can suffer secondary condensation on a level with the same value of k: for instance, electron ($k = 127$) suffers secondary condensation on $k = 127$ level. u, d, s quarks ($k = 107$) suffer secondary condensation on nuclear space-time sheet having $k = 113$). All quarks feed their color gauge fluxes at $k = 107$ space-time sheet. There is no deep reason forbidding the condensation of p on p. Primary and secondary condensation levels could also correspond to different but nearly identical values of p with the same value of k.

12.3 Fermion masses

In the earlier model the coefficient of $M^2 = kL_0$ had to be assumed to be different for various particle states. $k = 1$ was assumed for bosons and leptons and $k = 2/3$ for quarks. The fact that $k = 1$ holds true for all particles in the model including also super-canonical invariance forces to modify the earlier construction of quark states. This turns out to be possible without affecting the earlier p-adic mass calculations whose outcome depend in an essential manner on the ground state conformal weights h_{gr} of the fermions (h_{gr} can be negative). The structure of lepton and quark states in color degrees of freedom was discussed in [F2].

12.3.1 Charged lepton mass ratios

The overall mass scale for lepton and quark masses is determined by the condensation level given by prime $p \simeq 2^k$, k prime by length scale hypothesis. For charged leptons k must correspond to $k = 127$ for electron, $k = 113$ for muon and $k = 107$ for τ. For muon $p = 2^{113} - 1 - 4 * 378$ is assumed (smallest prime below 2^{113} allowing $\sqrt{2}$ but not $\sqrt{3}$). So called Gaussian primes are to complex integers what primes are for the ordinary integers and the Gaussian counterparts of the Mersenne primes are Gaussian primes of form $(1 \pm i)^k - 1$. Rather interestingly, $k = 113$ corresponds to a Gaussian Mersenne so that all charged leptons correspond to generalized Mersenne primes.

For $k = 1$ the leptonic mass squared is integer valued in units of m_0^2 only for the states satisfying

$$p \ mod \ 3 \neq 2 \ .$$

Only these representations can give rise to massless states. Neutrinos correspond to (p, p) representations with $p \geq 1$ whereas charged leptons correspond to $(p, p+3)$ representations. The earlier mass calculations demonstrate that leptonic masses can be understood if the ground state conformal weight is $h_{gr} = -1$ for charged leptons and $h_{gr} = -2$ for neutrinos.

The contribution of color partial wave to conformal weight is $h_c = (p^2 + 2p)/3$, $p \geq 1$, for neutrinos and $p = 1$ gives $h_c = 1$ (octet). For charged leptons $h_c = (p^2 + 5p + 6)/3$ gives $h_c = 2$ for $p = 0$ (decuplet). In both cases super-canonical operator O_p must have a net conformal weight $h_{sc} = -3$ to produce a correct conformal weight for the ground state. The requirement of conformal confinement and p-adic considerations suggests the use of operators O_p constructed as sums of products of operators with opposite imaginary parts of super-canonical conformal weight $z = -1/2 - in_1y_1 - in_2y_2$. If the operators in question are color Hamiltonians in octet representation net super-canonical conformal weight $h_{sc} = -3$ results. The tensor product of two octets with conjugate super-canonical conformal weights contains both octet and decuplet so that singlets are obtained. What strengthens the hopes that the construction is not adhoc is that the same operator appears in the construction of quark states too.

Using CP_2 mass scale m_0^2 [F2] as a p-adic unit, the mass formulas for the charged leptons read as

$$
\begin{aligned}
M^2(L) &= A(\nu)\frac{m_0^2}{p(L)} \ , \\
A(e) &= 5 + X(p(e)) \ , \\
A(\mu) &= 14 + X(p(\mu)) \ , \\
A(\tau) &= 65 + X(p(\tau)) \ .
\end{aligned}
\tag{12.3.1}
$$

$X(\cdot)$ corresponds to the yet unknown second order corrections to the mass squared. The following table lists the lower and upper bounds for the charged lepton mass ratios obtained by taking second order contribution to zero or allowing it to have maximum possible value. The values of lepton masses are $m_e = .510999$ MeV, $m_\mu = 105.76583$ MeV, $m_\tau = 1775$ MeV.

$$
\begin{aligned}
\frac{m(\mu)_+}{m(\mu)} &= \sqrt{\frac{15}{5}}2^7\frac{m_e}{(\mu)} \simeq 1.0722 \ , \\[2mm]
\frac{m(\mu)_-}{m(\mu)} &= \sqrt{\frac{14}{6}}2^7\frac{m_e}{m_{(\mu)}} \simeq 0.9456 \ , \\[2mm]
\frac{m(\tau)_+}{r(\tau)} &= \sqrt{\frac{66}{5}}2^{10}\frac{m_e}{m(\tau)} \simeq 1.0710 \ , \\[2mm]
\frac{m(\tau)_-}{m(\tau)} &= \sqrt{\frac{65}{6}}2^{10}\frac{m_e}{m(\tau)} \simeq .9703 \ .
\end{aligned}
\tag{12.3.2}
$$

For the maximal value of CP_2 mass the predictions for the mass ratio are systematically too large by a few per cent. From the formulas above it is clear that the second order corrections to mass squared can be such that correct masses result.

τ mass is least sensitive to $X(p(e)) \equiv Y_e$ and the maximum value of $Y_e \equiv Y_{e,max}$ consistent with τ mass corresponds to $Y_{e,max} = .7357$ and $Y_\tau = 1$. This means that the CP_2 mass is at least a fraction .9337 of its maximal value. If Y_L is same for all charged leptons and has the maximal value $Y_{e,max} = .7357$, the predictions for the mass ratios are

$$
\begin{aligned}
\frac{m(\mu)_{pr}}{m(\mu)} &= \sqrt{\frac{14 + Y_{e,max}}{5 + Y_{e,max}}} \times 2^7 \frac{m_e}{m(\mu)} \simeq .9922 \ , \\[2mm]
\frac{m(\tau)_{pr}}{m(\tau)} &= \sqrt{\frac{65 + Y_{e,max}}{5 + Y_e(max}} \times 2^{10} \frac{m_e}{m(\tau)} \simeq .9980 \ .
\end{aligned}
\tag{12.3.3}
$$

The error is .8 per cent *resp.* .2 per cent for muon *resp.* τ.

The argument leading to estimate for the modular contribution to the mass squared [F2] leaves two options for the coefficient of the modular contribution for $g = 2$ fermions: the value of coefficient is either $X = g$ for $g \leq 1$, $X = 3g - 3$ for $g \geq 2$ or $X = g$ always. For $g = 2$ the predictions are $X = 2$ and $X = 3$ in the two cases. The option $X = 3$ allows slightly larger maximal value of Y_e equal to $Y_{e,max}^{1)} = Y_{e,max} + (5 + Y_{e,max})/66$.

12.3.2   Neutrino masses

The estimation of neutrino masses is difficult at this stage since the prediction of the primary condensation level is not yet possible and neutrino mixing cannot yet be predicted from the basic principles. The cosmological bounds for neutrino masses however help to put upper bounds on the masses. If one takes seriously the LSND data on neutrino mass measurement of [ha10, ha13] and the explanation of the atmospheric ν-deficit in terms of $\nu_\mu - \nu_\tau$ mixing [ha11, ha7] one can deduce that the most plausible condensation level of μ and τ neutrinos is $k = 167$ or $k = 13^2 = 169$ allowed by the more general form of the p-adic length scale hypothesis suggested by the blackhole-elementary particle analogy. One can also deduce information about the mixing matrix associated with the neutrinos so that mass predictions become rather precise. In particular, the mass splitting of μ and τ neutrinos is predicted correctly if one assumes that the mixing matrix is a rational unitary matrix.

Super Virasoro contribution

Using $m_0^2/3$ as a p-adic unit, the expression for the Super Virasoro contribution to the mass squared of neutrinos is given by the formula

$$
\begin{aligned}
M^2(SV) &= (s + (3Yp)_R/3)\frac{m_0^2}{p} \ , \\
s &= 4 \text{ or } 5 \ , \\
Y &= \frac{23}{2} + Y_1 \ ,
\end{aligned}
\tag{12.3.4}
$$

where m_0^2 is universal mass scale. One can consider two possible identifications of neutrinos corresponding to $s(\nu) = 4$ with $\Delta = 2$ and $s(\nu) = 5$ with $\Delta = 1$. The requirement that CKM matrix is sensible forces the asymmetric scenario in which quarks and, by symmetry, also leptons correspond to lowest possible excitation so that one must have $s(\nu) = 4$. Y_1 represents second order contribution to the neutrino mass coming from renormalization effects coming from self energy diagrams involving intermediate gauge bosons. Physical intuition suggest that this contribution is very small so that the precise measurement of the neutrino masses should give an excellent test for the theory.

With the above described assumptions and for $s = 4$, one has the following mass formula for neutrinos

$$
\begin{aligned}
M^2(\nu) &= A(\nu)\frac{m_0^2}{p(\nu))} \ , \\
A(\nu_e) &= 4 + \frac{(3Y(p(\nu_e)))_R}{3} \ , \\
A(\nu_\mu) &= 13 + \frac{(3Y(p(\nu_\mu)))_R}{3} \ , \\
A(\nu_\tau) &= 64 + \frac{(3Y(p(\nu_\tau)))_R}{3} \ , \\
3Y &\simeq \frac{1}{2} \ .
\end{aligned}
\tag{12.3.5}
$$

The predictions must be consistent with the recent upper bounds [ha9] of order $10 \ eV$, $270 \ keV$ and $0.3 \ MeV$ for ν_e, ν_μ and ν_τ respectively. The recently reported results of LSND measurement [ha13] for $\nu_e - > \nu_\mu$ mixing gives string limits for $\Delta m^2(\nu_e, \nu_\mu)$ and the parameter $sin^2(2\theta)$ characterizing the mixing: the limits are given in the figure 30 of [ha13]. The results suggests that the masses of both electron and muon neutrinos are below 5 eV and that mass squared difference $\Delta m^2 = m^2(\nu_\mu) - m^2(\nu_e)$ is between $.25 - 25 \ eV^2$. The simplest possibility is that ν_μ and ν_e have common condensation level (in analogy with d and s quarks). There are three candidates for the primary condensation level: namely $k = 163, 167$ and $k = 169$. The p-adic prime associated with the primary condensation level is assumed to be the nearest prime below 2^k allowing p-adic $\sqrt{2}$ but not $\sqrt{3}$ and satisfying $p \ mod \ 4 = 3$. The following table gives the values of various parameters and unmixed neutrino masses in various cases of interest.

k	p	$(3Y)_R/3$	$m(\nu_e)/eV$	$m(\nu_\mu)/eV$	$m(\nu_\tau)/eV$
163	$2^{163} - 4*144 - 1$	1.36	1.78	3.16	6.98
167	$2^{167} - 4*144 - 1$	.34	.45	.79	1.75
169	$2^{169} - 4*210 - 1$	.17	.22	.40	.87

Could neutrino topologically condense also in other p-adic length scales than $k = 169$?

One must keep mind open for the possibility that there are several p-adic length scales at which neutrinos can condense topologically. In fact, the quantum model for hearing [M6] requires that both $k = 169$ and $k = 151$ correspond to p-adic length scales at which neutrinos can condense topologically. Rather interestingly, the ratio for the mass scales of $k = 151$ and $k = 169$ neutrinos equals to 512 and is same as the ratio of the mass scales of the ordinary $k = 107$ hadron physics and $k = 89$ hadronic physics predicted by TGD.

In fact, all intermediate p-adic length scales $k = 151, 157, 163, 167$ could correspond to metastable neutrino states. The point is that these p-adic lengths scales are number theoretically completely exceptional in the sense that there exist Gaussian Mersenne $2^k \pm i$ (prime in the ring of complex integers) for all these values of k. Since charged leptons, atomic nuclei ($k = 113$) , hadrons and intermediate gauge bosons correspond to ordinary or Gaussian Mersennes, it would not be surprising if the biologically important Gaussian Mersennes would correspond to length scales giving rise to metastable neutrino states. Of course, one can keep mind open for the possibility that $k = 167$ rather than $k = 13^2 = 169$ is the length scale defining the stable neutrino physics.

Neutrino mixing

Consider next the neutrino mixing. A quite general form of the neutrino mixing matrix D given by the table below will be considered.

	ν_e	ν_μ	ν_τ
ν_e	c_1	s_1c_3	s_1s_3
ν_μ	$-s_1c_2$	$c_1c_2c_3 - s_2s_3exp(i\delta)$	$c_1c_2s_3 + s_2c_3exp(i\delta)$
ν_τ	$-s_1s_2$	$c_1s_2c_3 + c_2s_3exp(i\delta)$	$c_1s_2s_3 - c_2c_3exp(i\delta)$

Physical intuition suggests that the angle δ related to CP breaking is small and will be assumed to be vanishing. Topological mixing is active only in modular degrees of freedom and one obtains for the first order terms of mixed masses the expressions

$$
\begin{aligned}
s(\nu_e) &= 4 + 9|U_{12}|^2 + 60|U_{13}|^2 = 4 + n_1 \ , \\
s(\nu_\mu) &= 4 + 9|U_{22}|^2 + 60|U_{23}|^2 = 4 + n_2 \ , \\
s(\nu_\tau) &= 4 + 9|U_{32}|^2 + 60|U_{33}|^2 = 4 + n_3 \ .
\end{aligned}
$$

$$(12.3.6)$$

The requirement that resulting masses are not ultraheavy implies that $s(\nu)$ must be small integers. The condition $n_1 + n_2 + n_3 = 69$ follows from unitarity. The simplest possibility is that the mixing matrix is a rational unitary matrix. The same ansatz was used successfully to deduce information about the mixing matrices of quarks. If neutrinos are condensed on the same condensation level, rationality implies that $\nu_\mu - \nu_\tau$ mass squared difference must come from the first order contribution to the mass squared and is therefore quantized and bounded from below.

The first piece of information is the atmospheric ν_μ/ν_e ratio, which is roughly by a factor 2 smaller than predicted by standard model [ha11]. A possible explanation is the CKM mixing of muon neutrino with τ-neutrino, whereas the mixing with electron neutrino is excluded as an explanation. The latest results from Kamiokande [ha11] are in accordance with the mixing $m^2(\nu_\tau) - m^2(\nu_\mu) \simeq 1.6 \cdot 10^{-2}\ eV^2$ and mixing angle $sin^2(2\theta) = 1.0$: also the zenith angle dependence of the ratio is in accordance with the mixing interpretation. If mixing matrix is assumed to be rational then only $k = 169$ condensation level is allowed for ν_μ and ν_τ. For this level $\nu_\mu - \nu_\tau$ mass squared difference turns out to be $\Delta m^2 \simeq 10^{-2}\ eV^2$ for $\Delta s \equiv s(\nu_\tau) - s(\nu_\mu) = 1$, which is the only acceptable possibility and predicts $\nu_\mu - \nu_\tau$ mass squared difference correctly within experimental uncertainties! The fact that the predictions for mass squared differences are practically exact, provides a precision test for the rationality assumption.

What is measured in LSND experiment is the probability $P(t, E)$ that ν_μ transforms to ν_e in time t after its production in muon decay as a function of energy E of ν_μ. In the limit that ν_τ and ν_μ masses are identical, the expression of $P(t, E)$ is given by

$$
\begin{aligned}
P(t, E) &= sin^2(2\theta) sin^2(\frac{\Delta Et}{2})\ , \\
sin^2(2\theta) &= 4c_1^2 s_1^2 c_2^2\ ,
\end{aligned}
\tag{12.3.7}
$$

where ΔE is energy difference of ν_μ and ν_e neutrinos and t denotes time. LSND experiment gives stringent conditions on the value of $sin^2(2\theta)$ as the figure 30 of [ha13] shows. In particular, it seems that $sin^2(2\theta)$ must be considerably below 10^{-1} and this implies that s_1^2 must be small enough.

The study of the mass formulas shows that the only possibility to satisfy the constraints for the mass squared and $sin^2(2\theta)$ given by LSND experiment is to assume that the mixing of the electron neutrino with the tau neutrino is much larger than its mixing with the muon neutrino. This means that s_3 is quite near to unity. At the limit $s_3 = 1$ one obtains the following (nonrational) solution of the mass squared conditions for $n_3 = n_2 + 1$ (forced by the atmospheric neutrino data)

$$
\begin{aligned}
s_1^2 &= \frac{69 - 2n_2 - 1}{60}\ , \\
c_2^2 &= \frac{n_2 - 9}{2n_2 - 17}\ , \\
sin^2(2\theta) &= \frac{4(n_2 \quad 9)}{51}\frac{(34 - n_2)(n_2 - 4)}{30^2}\ , \\
s(\nu_\mu) - s(\nu_e) &= 3n_2 - 68\ .
\end{aligned}
\tag{12.3.8}
$$

The study of the LSND data shows that there is only one acceptable solution to the conditions obtained by assuming maximal mass squared difference for ν_e and ν_μ

$$
\begin{aligned}
n_1 &= 2\quad n_2 = 33\quad n_3 = 34\ , \\
s_1^2 &= \frac{1}{30}\quad c_2^2 = \frac{24}{49}\ , \\
sin^2(2\theta) &= \frac{24}{49}\frac{2}{15}\frac{29}{30} \simeq .0631\ , \\
s(\nu_\mu) - s(\nu_e)) &= 31 \leftrightarrow .32\ eV^2\ .
\end{aligned}
\tag{12.3.9}
$$

That c_2^2 is near $1/2$ is not surprise taking into account the almost mass degeneracy of ν_{mu} and ν_τ. From the figure 30 of [ha13] it is clear that this solution belongs to 90 per cent likelihood region of LSND experiment but $sin^2(2\theta)$ is about two times larger than the value allowed by Bugey reactor experiment. The study of various constraints given in [ha13] shows that the solution is consistent with bounds from all other experiments. If one assumes that $k > 169$ for ν_e $\nu_\mu - \nu_e$ mass difference increases, implying slightly poorer consistency with LSND data.

There are reasons to hope that the actual rational solution can be regarded as a small deformation of this solution obtained by assuming that c_3 is non-vanishing. $s_1^2 = \frac{69-2n_2-1}{60-51c_3^2}$ increases in the deformation by $O(c_3^2)$ term but if c_3 is positive the value of $c_2^2 \simeq \frac{24-102c_1^0 c_2^0 s_2^0 c_3}{49} \sim \frac{24-61c_3}{49}$ decreases by $O(c_3)$ term so that it should be possible to reduce the value of $sin^2(2\theta)$. Consistency with Bugey reactor experiment requires $.030 \le sin^2(2\theta) < .033$. $sin^2(2\theta) = .032$ is achieved for $s_1^2 \simeq .035, s_2^2 \simeq .51$ and $c_3^2 \simeq .068$. The construction of U and D matrices for quarks shows that very stringent number theoretic conditions are obtained and as in case of quarks it might be necessary to allow complex CP breaking phase in the mixing matrix. One might even hope that the solution to the conditions is unique.

For the minimal rational mixing one has $s(\nu_e) = 5$, $s(\nu_\mu) = 36$ and $s(\nu_\tau) = 37$ if unmixed ν_e corresponds to $s = 4$. For $s = 5$ first order contributions are shifted by one unit. The masses ($s = 4$ case) and mass squared differences are given by the following table.

k	$m(\nu_e)$	$m(\nu_\mu)$	$m(\nu_\tau)$	$\Delta m^2(\nu_\mu - \nu_e)$	$\Delta m^2(\nu_\tau - \nu_\mu)$
169	.27 eV	.66 eV	.67 eV	.32 eV^2	.01 eV^2

Predictions for neutrino masses and mass squared splittings for $k = 169$ case.

Evidence for the dynamical mass scale of neutrinos

In recent years (I am writing this towards the end of year 2004 and much later than previous lines) a great progress has been made in the understanding of neutrino masses and neutrino mixing. The pleasant news from TGD perspective is that there is a strong evidence that neutrino masses depend on environment [ha20]. In TGD framework this translates to the statement that neutrinos can suffer topological condensation in several p-adic length scales. Not only in the p-adic length scales suggested by the number theoretical considerations but also in longer length scales, as will be found.

The experiments giving information about mass squared differences can be divided into three categories [ha20].

i) There along baseline experiments, which include solar neutrino experiments [ha16], KamLAND [ha19], K2K [ha17], and SuperK [ha18] as well as earlier studies of solar neutrinos. These experiments see evidence for the neutrino mixing and involve significant propagation through dense matter. For the solar neutrinos and KamLAND the mass splittings are estimated to be of order $O(8 \times 10^{-5})$ eV^2 or more cautiously 8×10^{-5} eV$^2 < \delta m^2 < 2 \times 10^{-3}$ eV^2. For K2K and atmospheric neutrinos the mass splittings are of order $O(2 \times 10^{-3})eV^2$ or more cautiously $\delta m^2 > 10^{-3}eV^2$. Thus the scale of mass splitting seems to be smaller for neutrinos in matter than in air, which would suggest that neutrinos able to propagate through a dense matter travel at space-time sheets corresponding to a larger p-adic length scale than in air.

ii) There are null short baseline experiments including CHOOZ, Bugey, and Palo Verde reactor experiments, and the higher energy CDHS, JARME, CHORUS, and NOMAD experiments, which involve muonic neutrinos (for references see [ha20]. No evidence for neutrino oscillations have been seen in these experiments.

iii) The results of LSND experiment[ha13] are consistent with oscillations with a mass splitting greater than $3 \times 10^{-2}eV^2$. LSND has been generally been interpreted as necessitating a mixing with sterile neutrino. If neutrino mass scale is dynamical, situation however changes.

If one assumes that the p-adic length scale for the space-time sheets at which neutrinos can propagate is different for matter and air, the situation changes. According to [ha20] a mass 3×10^{-2} eV in air could explain the atmospheric results whereas mass of of order .1 eV and $.07eV^2 < \delta m^2 < .26eV^2$ would explain the LSND result. These limits are of the same order as the order of magnitude predicted by $k = 169$ topological condensation.

Assuming that the scale of the mass splitting is proportional to the p-adic mass scale squared, one can consider candidates for the topological condensation levels involved.

a) Suppose that $k = 169 = 13^2$ is indeed the condensation level for LSND neutrinos. $k = 173$ would predict $m_{\nu_e} \sim 7 \times 10^{-2}$ eV and $\delta m^2 \sim .02$ eV^2. This could correspond to the masses of neutrinos propagating through air. For $k = 179$ one has $m_{\nu_e} \sim .8 \times 10^{-2}$ eV and $\delta m^2 \sim 3 \times 10^{-4}$ eV^2 which could be associated with solar neutrinos and KamLAND neutrinos.

b) The primes $k = 157, 163, 167$ associated with Gaussian Mersennes would give $\delta m^2(157) = 2^6 \delta m^2(163) = 2^{10} \delta m^2(167) = 2^{12} \delta m^2(169)$ and mass scales $m(157) \sim 22.8$ eV, $m(163) \sim 3.6$ eV, $m(167) \sim .54$ eV. These mass scales are unrealistic or propagating neutrinos. The interpretation consistent with TGD inspired model of condensed matter in which neutrinos screen the classical Z^0 force generated by nucleons would be that condensed matter neutrinos are confined inside these space-time sheets whereas the neutrinos able to propagate through condensed matter travel along $k > 167$ space-time sheets.

Comments

Some comments on the proposed scenario are in order: some of the are written much later than the previous text.

a) Mass predictions are consistent with the bound $\Delta m(\nu_\mu, \nu_e) < 2 \ eV^2$ coming from the requirement that neutrino mixing does not spoil the so called r-process producing heavy elements in Super Novae [ha12].

b) TGD neutrinos cannot solve the dark matter problem: the total neutrino mass required by the cold+hot dark matter models would be about 5 eV. In [D4] a model of galaxies based on string like objects of galaxy size and providing a more exotic source of dark matter, is discussed.

c) One could also consider the explanation of LSND data in terms of the interaction of ν_μ and nucleon via the exchange of $g = 1$ W boson. The fraction of the reactions $\bar{\nu}_\mu + p \rightarrow e^+ + n$ is at low neutrino energies $P \sim \frac{m_W^4 (g=0)}{m_W^4 (g=1)} sin^2(\theta_c)$, where θ_c denotes Cabibbo angle. Even if the condensation level of $W(g = 1)$ is $k = 89$, the ratio is by a factor of order .05 too small to explain the average $\nu_\mu \rightarrow \nu_e$ transformation probability $P \simeq .003$ extracted from LSND data.

d) The predicted masses exclude MSW and vacuum oscillation solutions to the solar neutrino problem unless one assumes that several condensation levels and thus mass scales are possible for neutrinos. This is indeed suggested by the previous considerations.

12.3.3 Quark masses

The prediction or quark masses is more difficult due the facts that the deduction of even the p-adic length scale determining the masses of these quarks is a non-trivial task, and the original identification was indeed wrong. Second difficulty is related to the topological mixing of quarks. The new scenario leads to a unique identification of masses with top quark mass as an empirical input and the thermodynamical model of topological mixing as a new theoretical input. Also CKM matrix is predicted highly uniquely.

Basic mass formulas

By the earlier mass calculations and construction of CKM matrix the ground state conformal weights of U and D type quarks must be $h_{gr}(U) = -1$ and $h_{gr}(D) = 0$. The formulas for the eigenvalues of CP_2 spinor Laplacian imply that if m_0^2 is used as a unit, color conformal weight $h_c \equiv m_{CP_2}^2$ is integer for $p \ mod \ = \pm 1$ for U type quark belonging to $(p+1, p)$ type representation and obeying $h_c(U) = (p^2 + 3p + 2)/3$ and for $p \ mod \ 3 = 1$ for D type quark belonging $(p, p+2)$ type representation and obeying $h_c(D) = (p^2 + 4p + 4)/3$. Only these states can be massless since color Hamiltonians have integer valued conformal weights.

In the recent case the minimal $p = 1$ states correspond to $h_c(U) = 2$ and $h_c(D) = 3$. $h_{gr}(U) = -1$ and $h_{gr}(D) = 0$ reproduce the previous results for quark masses required by the construction of CKM matrix. This requires super-canonical operators O_p with a net conformal weight $h_{sc} = -3$ just as in the leptonic case. The facts that the values of p are minimal for spinor harmonics and the super-canonical operator is same for both quarks and leptons suggest that the construction is not had hoc.

Consider now the mass squared values for quarks. For $h(D) = 0$ and $h(U) = -1$ and using $m_0^2/3$ as a unit the expression for the thermal contribution to the mass squared of quark is given by the formula

$$
\begin{aligned}
M^2 &= (s + X)\frac{m_0^2}{p} \ , \\
s(U) &= 5 \ , \ s(D) = 8 \ , \\
X &\equiv \frac{(3Yp)_R}{3} \ ,
\end{aligned}
\tag{12.3.10}
$$

where the second order contribution Y corresponds to renormalization effects coming and depending on the isospin of the quark. When m_0^2 is used as a unit X is replaced by $X = (Y_p)_R$.

With the above described assumptions one has the following mass formula for quarks

$$M^2(q) = A(q)\frac{m_0^2}{p(q)} \ ,$$

$$A(u) = 5 + X_U(p(u)) \ , \quad A(c) = 14 + X_U(p(c)) \ , \quad A(t) = 65 + X_U(p(t)) \ ,$$
$$A(d) = 8 + X_D(p(d)) \ , \quad A(s) = 17 + X_D(p(s)) \ , \quad A(b) = 68 + X_D(p(b)) \ .$$

$$(12.3.11)$$

p-Adic length scale hypothesis allows to identify the p-adic primes labelling quarks whereas topological mixing of U and D quarks allows to deduce topological mixing matrices U and D and CKM matrix V and precise values of the masses apart from effects like color magnetic spin orbit splitting, color Coulombic energy, etc..

Integers n_{q_i} satisfying $\sum_i n(U_i) = \sum_i n(D_i) = 69$ characterize the masses of the quarks and also the topological mixing to high degree. The reason that modular contributions remain integers is that in the p-adic context non-trivial rationals would give CP_2 mass scale for the real counterpart of the mass squared. In the absence of mixing the values of integers are $n_d = n_u = 0$, $n_s = n_c = 9$, $n_b = n_t = 60$.

The fact that CKM matrix V expressible as a product $V = U^\dagger D$ of topological mixing matrices is near to a direct sum of 2×2 unit matrix and 1×1 unit matrix motivates the approximation $n_b \simeq n_t$. The large masses of top quark and of $t\bar{t}$ meson encourage to consider a scenario in which $n_t = n_b = n \leq 60$ holds true.

The model for topological mixing matrices and CKM matrix predicts U and D matrices highly uniquely and allows to understand quark and hadron masses in surprisingly detailed level.

a) $n_d = n_u = 60$ is not allowed by number theoretical conditions for U and D matrices and by the basic facts about CKM matrix but $n_t = n_b = 59$ allows almost maximal masses for b and t. This is not yet a complete hit. The unitarity of the mixing matrices and the construction of CKM matrix to be discussed in the next section forces the assignments

$$(n_d, n_s, n_b) = (5, 5, 59) \ , \quad (n_u, n_c, n_t) = (5, 6, 58) \ . \tag{12.3.12}$$

fixing completely the quark masses apart from few per cent renormalization effects of hadronic mass scale in topological condensation which seem to be present and will be discussed in [F4]. Note that top quark mass is still rather near to its maximal value.

b) The constraint that quark contribution to pion mass does not exceed pion mass implies the constraint $n(d) \leq 6$ and $n(u) \leq 6$ in accordance with the predictions of the model of topological mixing. $u - d$ mass difference does not affect $\pi^+ - \pi^0$ mass difference and the quark contribution to $m(\pi)$ is predicted to be $\sqrt{(n_d + n_u + 13)/24} \times 136.9$ MeV for the maximal value of CP_2 mass (second order p-adic contribution to electron mass squared vanishes).

The p-adic length scales associated with quarks

The identification of p-adic length scales associated with the quarks has turned to be a highly non-trivial problem. The reasons are that for light quarks it is difficult to deduce information about quark masses for hadron masses and that the unknown details of the topological mixing (unknown until the advent of the thermodynamical model [F4]) made possible several p-adic length scales for quarks. Two natural constraints have however emerged from the recent work.

a) Quark contribution to the hadron mass cannot be larger than color contribution and for quarks having $k_q \neq 107$ quark contribution to mass is added to color contribution to the mass. For quarks with same value of k conformal weight rather than mass is additive whereas for quarks with different value of k masses are additive. An important implication is that for diagonal mesons $M = q\bar{q}$ having $k(q) \neq 107$ the condition $m(M) \geq \sqrt{2}m_q$ must hold true. This gives strong constraints on quark masses.

b) The realization that scaled up variants of quarks explain elegantly the masses of light hadrons allows to understand large mass splittings of light hadrons without the introduction of strong isospin-isospin interaction.

The new model for quark masses is based on the following identifications of the p-adic length scales.

a) The nuclear p-adic length scale $L(k)$, $k = 113$, corresponds to the p-adic length scale determining the masses of u, d, and s quarks. Note that $k = 113$ corresponds to a so called Gaussian Mersenne. The interpretation is that quark massivation occurs at nuclear space-time sheet at which quarks feed their em fluxes. At $k = 107$ space-time sheet, where quarks feed their color gauge fluxes, the quark masses are vanishing in the first p-adic order. This could be due to the fact that the p-adic temperature is $T_p = 1/2$ at this space-time sheet so that the thermal contribution to the mass squared is negligible. This would reflect the fact that color interactions do not involve any counterpart of Higgs mechanism.

p-Adic mass calculations turn out to work remarkably well for massive quarks. The reason could be that M_{107} hadron physics means that *all* quarks feed their color gauge fluxes to $k = 107$ space-time sheets so that color contribution to the masses becomes negligible for heavy quarks as compared to Super-Kac Moody and modular contributions corresponding to em gauge flux feeded to $k > 107$ space-time sheets in case of heavy quarks. Note that Z^0 gauge flux is feeded to space-time sheets at which neutrinos reside and screen the flux and their size corresponds to the neutrino mass scale. This picture might throw some light to the question of whether and how it might be possible to demonstrate the existence of M_{89} hadron physics.

One might argue that $k = 107$ is not allowed as a condensation level in accordance with the idea that color and electro-weak gauge fluxes cannot be fed at the space-time space time sheet since the classical color and electro-weak fields are functionally independent. The identification of η' meson as a bound state of scaled up $k = 107$ quarks is not however consistent with this idea unless one assumes that $k = 107$ space-time sheets in question are separate.

b) The requirement that the masses of diagonal pseudoscalar mesons of type $M = q\bar{q}$ are larger but as near as possible to the quark contribution $\sqrt{2}m_q$ to the mass, fixes the p-adic primes $p \simeq 2^k$ associated with c, b and t quark (these values of k are "nominal" since k seems to be dynamical). c quark corresponds to the p-adic length scale $k = 104 = 2^3 \times 13$. b quark corresponds to $k = 103$ for $n(b) = 5$. By the same criterion t quark must correspond to the secondary length scale $k = 3 \times 31$. It should be noticed that $k = 31$ is double Mersenne prime in the sense that $31 = 2^5 - 1$ holds true. In fact, the primes, $5, M_5 = 31$, and M_{31} form the analog of Combinatorial Hierarchy consisting of primes $2, M_2 = 3, M_3 = 7, M_7 = 127, M_{127} = 2^{127} - 1, ...$ possibly containing also larger primes than the known ones. Note that in all cases color interaction energy is measured using $k = 107$ mass scale as unit as in case of light hadrons.

A more precise definition of the model requires a detailed consideration of the renormalization of hadron masses in topological condensation which reduces the mass scale by few per cent. Some input from hadronic mass spectrum is also needed to fix the renormalization. This point will be discussed in [F4]. The following table summarizes quark masses as predicted by this model.

q	d	u	s	c	b	t
n_q	4	5	6	6	59	58
$k(q)$	113	113	113	104	103	93
$m(q)/GeV$	.090	.079	.090	1.873	6.537	202.8

Table 1. The masses of quarks predicted assuming $(n_d, n_s, n_b) = (5, 5, 59)$ and $(n_u, n_c, n_t) = (5, 6, 58)$ and maximal CP_2 mass scale consistent with η' meson mass.

12.3.4 Photon, graviton and gluon

The only possibility to get massless states is to have $\Delta = 0$ state with one active sector so that NS thermodynamics becomes trivial due to the absence of the thermodynamical excitations satisfying the gauge conditions. The model for the Weinberg mixing suggests that a second order non-thermal contribution to the mass squared is present in case of photon. Also in case of the gluon this contribution is expected to be present. For graviton the contribution should be extremely small or even higher order in p.

12.3.5 p-Adic thermodynamics does not explain intermediate gauge boson masses

For photon, gluon and graviton the conformal weight of the $p = 0$ ground state is $h_{gr} = h_{vac}$. The crucial condition is that $h = 0$ ground state is non-degenerate: otherwise one would obtain several physically more or less identical photons and this would be seen in the spectrum of black body radiation. This occurs if one can construct several ground states not expressible in terms of the action of the Super Virasoro generators.

The requirement that the electron-intermediate gauge boson mass ratios are sensible, serves as a stringent test for the hypothesis that intermediate gauge boson masses result from the p-adic thermodynamics. Contrary to the original beliefs, it seems however impossible to predict W/e and Z/e mass ratios correctly in p-adic thermodynamics scenario. Although the errors are of order ten percent, they seem to be enough to exclude p-adic thermodynamics explanation for the massivation of gauge bosons.

a) The thermal mass squared for a boson state with N active sectors (non-vanishing vacuum weight) is determined by the partition function for the tensor product of N NS type Super Virasoro algebras. The degeneracies of the excited states as a function of N and the weight Δ of the operator creating the massless state are given in the table below.

b) Both W and Z must correspond to $N = 2$ active sectors for which $D(1) = 1$ and $D(2) = 3$ so that (using the formulas of p-adic thermodynamics) the thermal mass squared is $m^2 = (p + 5p^2)$ for $T_p = 1$. The second order contribution to the thermal mass squared is extremely small so that Weinberg angle vanishes in the thermal approximation. Z/e mass-ratio is predicted to be about 22 per cent too high. The thermal prediction for W-boson mass is the same as for Z^0 mass and thus even worse since the two masses should be related $M_W^2 = M_Z^2 cos^2(\theta_W)$. For $T_p = 1/2$ thermal masses are completely negligible. Of course, thermodynamics does not provide any obvious counterpart for the group theory based explanation of Z/W mass ratio.

c) An important question concerns the mass squared unit. $m_0^3/3$ seems to correspond to m_W^2 but lepton masses favor m_0^2.

d) It seems that the Achilles's heel of the p-adic thermodynamics is bosonic sector whereas the weak point of the standard model is fermionic sector. This suggests that these two approaches should be combined.

N,Δ	0	1/2	1	3/2	2	5/2	3
2	1	1	1	3	3	4	4
3	1	2	3	9	11		
4	1	3	5	19	26		
5	1	4	10	24	150		

Table 6. Degeneracies $d(\Delta, N)$ of the operators satisfying NS type gauge conditions as a function of the number N of the active sectors and of the conformal weight Δ of the operator. Only those degeneracies, which are needed in the mass calculation of light bosons are listed.

12.3.6 Higgs mechanism for electro-weak gauge bosons

Contrary to the original beliefs, it seems that p-adic thermodynamics alone cannot explain boson masses. The problem is that the lower limit for W mass is 20-30 per cent higher than the experimental value of the W mass squared for $T_p = 1$. For $T_p = 1/2$ thermal masses are completely negligible. Rather ironically, bosonic masses are the Achilles's heel of p-adic thermodynamics whereas fermionic masses are the Achilles's heel of standard model. One can consider two solutions to the problem.

a) The first option is based on the combination of p-adic thermodynamics and Higgs mechanism. The identification of Higgs boson as a weakly charged wormhole contact provides a strong support for this approach. The vacuum expectation value of Higgs field would correspond to the generation of a coherent state of wormhole contacts.

b) Second option is based on the idea that the non-covariant constancy of electro-weak charge matrices except that of photon is responsible for the Higgs mechanism.

Group theoretical evidence for the the existence of Higgs boson

The hint for the existence of Higgs boson comes from the identification of electro-weak super-Kac Moody algebra. The identification relies on the observation that for the known solutions of field equations, field equations in CP_2 degrees of freedom separate into two separate equations corresponding to the variations with respect to induced metric and induced Kähler form. This implies that $SU(3)$ isometry charges decompose to sums of charges which are conserved separately. Color charges correspond to the conserved current associated with the variations of the induced metric. Since $U(2)_{ew}$ can be identified as the $U(2)$ subgroup of $SU(3)$ one can identify the $U(2)$ charges associated with the variations of the induced Kähler form. Thus one has beautiful identification of the basic Kac-Moody algebra structures associated with color and electro-weak sectors.

One should however find interpretation also for the $SU(3)$ charges Q_J associated with the complement of $U(2)$ in $SU(3)$ algebra associated with the variations with respect to the induced Kähler form. These Lie-algebra generators provide standard coordinatization of CP_2 in symmetric space decomposition $su(3) = h+t = u(2)+t$. The interpretation of $h = u(2)$ as electro-weak charges means symmetry breaking so that the charges in t do not form usual Kac Moody algebra acting as pure gauge symmetries but create genuine physical states. This is completely analogous to what happens at the level of CP_2 geometry: h corresponds to gauge degrees of freedom and t to CP_2 coordinates and thus to genuine dynamical degrees of freedom. The charges in t transform as two electro-weak isospin doublets with a non-vanishing color hypercharge. Thus the couplings to electro-weak gauge bosons and fermions are the same as associated with the ordinary electro-weak Higgs field. The natural conclusion is that the particles described by Kac Moody generators in t are nothing but the Higgs particles.

This suggests that Higgs mechanism could correspond to the generation of coherent states in the neutral degrees of freedom associated with t. In fact, the direction of electromagnetic charge can be defined by requiring that the vacuum expectation value of t is annihilated by em charge. What is interesting is that the components of the vacuum expectation value associated with the coherent state could be proportional to the values of the CP_2 coordinates of a space-time sheet at which particles are topologically condensed. Thus the coherent state associated with the Higgs field might provide a representation for slowly varying CP_2 coordinates of the space-time surface (in p-adic context this representation would be cognitive representation!).

This mechanism is not in conflict with the thermal massivation of fermions if the vacuum expectation is of order $O(p^2)$: the coupling to the Higgs would only induce relatively small shifts of the fermionic masses. This model predicts a strict upper bound for the induced mass consistent with the observed masses which are below the maximum mass resulting from p-adic thermodynamics with temperature $T_p = 1$.

Quantum classical correspondence requires that Higgs boson should have a concrete space-time correlate. This correlate can be identified as a wormhole contact carrying lefthanded weak charge discussed in the introduction and first chapter of this part of book.

Higgs mechanism and non-covariant constancy of electro-weak charge matrix

The second option is based on the wisdom gained about decade after the first p-adic mass calculations. It seems to be possible to understand the origin of particle massivation in terms of a hydrodynamic flow defined by the normal components $T^{n\alpha}$ of energy momentum tensor at the light like causal determinants X_l^3 representing the orbits of 2-dimensional partons. The modes of induced spinor fields representing spinorial shock waves at X_l^3 are parallelly translated along the flow lines. This induces both mixing and braiding transformation.

The thermal contribution to the mass squared has interpretation as being due to the ergodicity of the flow implying the loss of correlations in a relevant p-adic length scale. This loss of correlations in the case of CP_2 type extremals would be due to the random motion with light velocity assignable to the light-like M^4 projection of CP_2 type extremal, and as explained in the introduction, leads to an elegant picture explaining Higgs mechanism and also the relationship between gravitational and inertial masses.

One can ask whether Higgs like contribution could be also due to the lack of covariant constancy of all electro-weak charge matrices except that associated with photon. Indeed, by the arguments of the previous chapter W charge matrix and left handed part of Z^0 charge matrix have super-canonical conformal weight -2 whereas photon charge matrix and vectorial part of Z^0 charge matrix are covariantly constant and have a vanishing conformal weight. Hence the coupling to Higgs might not be necessary after all, and the vacuum expectation of Higgs would only reflect the electro-weak symmetry breaking induced by the CP_2 geometry.

Why Higgs is not detected?

Both options could explain naturally the failure to detect Higgs. Various contributions to Higgs production are discussed in [hd1, hd2]. The basic contributions to the production of Higgs bosons in p-p collisions at LHC corresponds to gluon fusion, associated production, and vector boson fusion. Various production cross sections for $p - p$ collisions at cm energy of $\sqrt{s} = 14$ TeV are given in [hd1], see also the figures of [hd2]. The dominating contribution corresponds to the triangle diagram $gg \to q\bar{q} \to H$. Since the coupling of quarks to Higgs in TGD can be much smaller than in standard model, this contribution can be very small in TGD framework. Also the rates for direct annihilations $q\bar{q} \to H$ are small for the same reason.

The rates for the vector boson fusion and associated production are in the lowest order same in TGD as in the standard model. Vector boson fusion corresponds to the scattering of quarks via the exchange of W/Z boson coupling to Higgs ($q\bar{q} \to q\bar{q}H$). The rate for this process is roughly one hundred times lower than for the associated production. Associated production corresponds to the diagram $q\bar{q} \to W \to W + H$. The rate is below the rate of the vector boson fusion if Higgs mass is above ~ 100 GeV: on basis recent searches the Higgs mass is known to be in the range $114.4 - 237$ GeV [hd2].

It seems safe to conclude that TGD predicts Higgs particle. The fact that the rate of Higgs production can be about 100 times lower than in standard model and even this could easily explain the unsuccessful search for Higgs.

How p-adic approach can be consistent with the group theoretical understanding of m_Z/m_W mass ratio?

One cannot bypass the fact that p-adic approach should be consistent with the group theoretical description of m_Z/m_W ratio. The mapping of p-adic mass squared to its real counterpart by the standard form of the canonical identification does not respect algebraic structures, which forms the core of group theoretical thinking. The problem is much more general. For instance, the unitarity of CKM matrix is not respected by canonical identification in its standard form. Also the sensible correspondence between real and p-adic coupling constants requires that algebraic structures should be preserved by the canonical identification. One could also argue that the map of p-adic probabilities to their real counterparts should respect probability conservation.

The relationship between p-adic and real scattering amplitudes discussed in [F5] encourages the conclusion that canonical identification Id at the level of the scattering amplitudes should be replaced with its variant in which rational r/s is mapped to $Id(r)/Id(p)$ rather than $Id(r/p)$, where $r(p$ is expanded in powers of $1/p$. This map is unique, when r and p are chosen to be mutually prime. The basic reason is that this correspondence respects the unitarity of CKM matrix appearing in S-matrix elements under some additional rather natural conditions.

This form of canonical identification can be modified further in p-adic thermodynamics where p-adic probabilities as a function of conformal weight n are of form $P(n) = g(n)p^n/Z$ by defining the identification as $I(P(n)) = I(g(n)p^n)/I(Z)$ so that the sum of real counterparts of p-adic probabilities equals to one without further normalization for $g(n) < p$. This number theoretic condition is extremely restrictive since the p-adic Boltzmann weights for independent events cannot have common pinary digits in their pinary expansion. In the recent case the condition $g(n) < p$ fails for sufficiently high values of n since $g(n)$ increases exponentially. The assumption about a physical cutoff in n has no physical implications since the values of p are so large. The cutoff is of order $n \sim k$ for $p \simeq 2^k$ and has interpretation in terms of the p-adic length scale L_k assignable to the elementary particle horizon.

If this variant of canonical identification is applied to map the mass squared values to their real counterparts some essential changes occur although the predictions in the fermionic sector are not changed in any essential manner [F5]. The crucial implication is that the algebraic relationship between W and Z masses might be described in this framework if $sin^2(\theta_w)$ corresponds to a ratio of small integers. The basic implications of the modified view about canonical identification are following.

i) Second order contribution to the mass squared is always negligible unless it is of form $(p-k)p^2$, where k is a reasonable fraction of p or $k << p$ holds true. The large second order contributions to charged lepton masses must be of this form. Second order thermodynamical contribution to the neutrino mass squared is negligible. Also intermediate gauge boson masses could correspond to the second order contribution

$$m_Z^2 = (p-k_Z)p^2 \ , \qquad m_W^2 = (p-k_W)p \ .$$
$$p - k_Z = \left[cos^2(\theta_W)p\right] \ , \quad p - k_W = \left[cos^4(\theta_W)p\right] \ . \tag{12.3.13}$$

Here $[x]$ denotes integer part of x. The formulas are tailored to explain also why Z/e ratio is by a factor $cos^2(\theta_W)$ smaller than for $m_Z^2 = p$. What does not look nice is that $sin^2(\theta_W)$ does not appear as a genuine rational number in the formula as its coupling constant character would suggest.

ii) First order contribution to the mass squared can be of form $(m/n) \times p$ since the real counterpart is $(m/n) \times (1/p)$ rather than being of order CP_2 mass squared for the ordinary form of canonical identification. The fact that intermediate gauge boson mass squared is smaller than the natural unit requires a ground state degeneracy or that the source of mass is not thermodynamical. The simplest manner to understand gauge boson masses is to assume

$$m_Z^2 = cos^2(\theta_W)p \ , \quad m_W^2 = cos^4(\theta_W)p \ . \tag{12.3.14}$$

Now $sin^2(\theta_W)$ is a genuine rational number. The basic assumption of the successful model for topological mixing of quarks [F4] is that the modular contribution to the masses is of form np. This assumption loses its original justification for this option and some other justification for its should be found. The first guess is that the conditions on mass squared plus probability conservation might not be consistent with unitarity unless the modular contribution to the mass squared remains integer valued in the mixing (note that all integer values are not possible [F4]). Direct numerical experimentation however shows that that this is not the case.

12.4   Appendix

The appendix has become somewhat obsolete because of the dramatic simplifications in the construction of states. I have however decided to still keep it.

12.4.1   Gauge invariant states in color sector

The construction of states satisfying Super Virasoro and Kac Moody gauge conditions for various values of conformal weight is essential ingredient in the calculation of degeneracies for various values of mass squared operator in order to estimate thermal mass expectation value. If one has obtained the multiplicities of various representations with weight n then it is easy to calculate the multiplicities for the states satisfying Super Virasoro conditions and Kac Moody conditions. Kac Moody conditions are implied by Super Virasoro conditions since $T^{a,n} \propto [L^n, T^{a,0}]$ holds true so that only Super Virasoro conditions need to be taken into account. If the gauge conditions associated with $G^{1/2}$ and $G^{3/2}$ in N-S representation induce surjective maps to the levels $n-1/2, n-1$ and $n-2$ then the multiplicity of gauge invariant representation is given by $m = m(n) - m(n-1/2) - m(n-3/2)$. In Ramond sector the gauge conditions for L^1 and G^1 guarantee the remaining gauge conditions and one has $m = m(n) - 2m(n-1)$ under similar assumptions.

The construction of the gauge invariant states relies on the following observations.

a) The states at each level n (conformal weight) of color Kac Moody algebra can be classified into irreducible representations of color group. The states are created by the monomials $O(F)$ of the 'fermionic' generators $F^{A,k}$, which can be regarded as an element of Grassmann algebra generated by $F^{A,k}$. The monomials of $F^{A,1/2}$ satisfy the gauge conditions of the bosonic Kac Moody identically.

b) The operators $O(F)$ creating nonzero norm sates can be classified into irreducible representations of the color group. The basic building blocks are the representations defined by N:th order monomials of generators F^{Ak} with k fixed. These representations are completely antisymmetrized tensor products of $N = 0, 1, ..., 8$ octets and representation content is same for all values of k. The representation content can be coded into multiplicity vector $m(N;k)$, $k = 1, 8, 10, ...$

c) Once the representation contents for antisymmetrized tensor products are known in terms of multiplicity vectors, the representation contents for tensor products of N_1, k_1 and N_2, k_2 can be determined by standard tensor product construction since anticommutativity does not produce no effects for $k_1 \neq k_2$. One can express the multiplicity vector for the tensor product $(N_1, k_1) \otimes (N_2, k_2)$ in terms of the multiplicity vector $D(k_1, k_2, k_3)$ for the tensor product of irreducible representations $k_1, k_2 = 1, 8, 10, ...$

$$m((N_1, k_1) \otimes (N_2, k_2; k) \quad = \quad m(N_1; k_1) D(k_1, k_2, k_3) m(N_2; k_2) \ . \tag{12.4.1}$$

d) It is useful to calculate total multiplicity vector $m(n;k)$ for each conformal weight n by considering all possible states having this conformal weight. The multiplicity vector is just the sum of multiplicity vectors of various tensor products satisfying $\sum N_i k_i = N$:

$$m(n;k) \quad = \quad \sum_{S=N} m((N_1, k_1) \otimes .... \otimes (N_r, k_r; k)) \ ,$$
$$S \quad \equiv \quad \sum N_i k_i \ . \tag{12.4.2}$$

The multiplicity vectors $m(n;k)$ are basic objects in the systematic construction of tensor products of several Super Virasoro algebras.

Multiplicity vectors for antisymmetric tensor products

Consider first the construction of N-fold antisymmetric tensor products of octets F^{Ak}, k fixed. The tensor products are obviously analogous to the antisymmetric tensors of 8-dimensional space. The completely antisymmetric 8-dimensional permutation symbol $\epsilon_{A_1,....,A_8}$ transforms as color singlet and induces duality operation in the set of antisymmetric representations: the antisymmetric representations N are mapped to representations $8 - N$. This implies that the representation contents are same for $N = 0$ and 8, $N = 1$ and 7, $N = 2$ and $N = 6$, $N = 3$ and $N = 5$ respectively. $N = 4$ is self dual. It is relatively easy to determine the representation content of the lowest completely antisymmetric representations and the results can be summarized conveniently as multiplicity vectors defined as

$$\bar{m} \quad \equiv \quad (m(1), m(8), m(10), m(\bar{10}), m(27), m(28), m(\bar{28}),$$
$$m(64), m(81), m(\bar{81}), m(125), ...)$$
$$\tag{12.4.3}$$

The multiplicity vectors are given by the following formulas

$$\begin{aligned}
\bar{m}(F) = \bar{m}(F^7) \quad &= \quad (0, 1) \ , \\
\bar{m}(F^2) = \bar{m}(F^6) \quad &= \quad (0, 1, 1, 1) \ , \\
\bar{m}(F^3) = \bar{m}(F^5) \quad &= \quad (1, 1, 1, 1, 1) \ , \\
\bar{m}(F^4) \quad &= \quad (0, 2, 0, 0, 2) \ , \tag{12.4.4}
\end{aligned}$$

where F^N denotes N:th tensor power of F^{Ak}.

Multiplicity vectors for general states

The next task is to calculate multiplicity vectors for various conformal weights. The task is straightforward application of Young Tableaux. The representation contents for various conformal weights for N-S algebra are given by

$$
\begin{aligned}
n \;=\;& 0 : 1 \\
n \;=\;& 1/2 : 1/2 \\
n \;=\;& 1 : (1/2)^2 \\
n \;=\;& 3/2 : 3/2 \oplus (1/2)^3 \\
n \;=\;& 2 : (3/2) \otimes (1/2) \oplus (1/2)^4 \\
n \;=\;& 5/2 : 5/2 \oplus (3/2) \otimes (1/2)^2 \oplus (1/2)^5 \\
n \;=\;& 3 : (5/2) \otimes (1/2) \oplus (3/2)^2 \oplus (3/2) \otimes (1/2)^3 \oplus (1/2)^6 \\
n \;=\;& 7/2 : 7/2 \oplus (5/2) \otimes (1/2)^2 \oplus (3/2)^2 \otimes (1/2) \oplus 3/2 \otimes (1/2)^4 \oplus (1/2)^7 \\
n \;=\;& 4 : (7/2) \otimes (1/2) \oplus (5/2) \otimes (3/2) \oplus (5/2) \otimes (1/2)^3 \oplus (3/2)^2 \otimes (1/2)^2 ... \\
& \oplus\; (3/2) \otimes (1/2)^5 \oplus (1/2)^8 \\
n \;=\;& 9/2 : 9/2 \oplus (7/2) \otimes (1/2)^2 ... \\
& \oplus\; (5/2) \otimes (3/2) \otimes (1/2) \oplus 5/2 \otimes (1/2)^4 \oplus (3/2)^3 .... \\
& \oplus\; (3/2)^2 \otimes (1/2)^3 \oplus (3/2) \otimes (1/2)^6
\end{aligned}
$$

(12.4.5)

Multiplicity vectors obtained as sums of multiplicity vectors associated with summands in the direct sum composition and are given by the following table

n	1	8	10	$\overline{10}$	27	28	$\overline{28}$	35	$\overline{35}$	64	81	$\overline{81}$
0	1											
1/2		1										
1		1	1	1								
3/2	1	2	1	1	1							
2	1	4	1	1	3							
5/2	2	6	3	3	4							
3	2	10	6	6	6			2	2	1		
7/2	4	16	8	8	12			4	4	2		
4	8	24	12	12	21	1	1	7	7	4		
9/2	10	36	21	21	32	1	1	12	12	8	1	1

Table 7. Multiplicity vectors for various conformal weights for N-S type Super Virasoro algebra.

Similar arguments can be used to deduce the multiplicity vectors in case of Ramond type Super Virasoro algebra.

n	1	8	10	$\overline{10}$	27	28	$\overline{28}$	35	$\overline{35}$	64	80	$\overline{80}$	81	$\overline{81}$	125
0	1														
1		1													
2		1	1	1											
3	2	4	2	2	2										
4	2	10	4	4	6			1	1						
5	6	20	10	10	14			4	4	1					
6	12	40	22	22	32	1	1	10	10	6					
7	17	68	36	36	55	1	1	20	20	11			1	1	
8	33	124	70	70	113	5	5	44	44	29			5	5	1
9	70	276	170	170	276	16	16	122	122	94	1	1	22	22	6

Table 8. Multiplicity vectors for various conformal weights for Ramond type Super Virasoro algebra.

Multiplicity vectors for conformally invariant states

Multiplicity vectors for gauge invariant states are obtained from the formulas $m(n) \to m(n) - m(n - 1/2) - m(n - 3/2)$ and $m(n) \to m(n) - 2m(n-1)$. The inspection of the above tables gives following tables for the multiplicity vectors of gauge invariant states needed in the mass calculations to find the possible ground states.

n	1	8	10	$\overline{10}$	27	28	$\overline{28}$	35	$\overline{35}$	64	81	$\overline{81}$
0	1											
1/2		1										
1		1	1	1								
3/2		1		2	1							
2	1	1		1	2							
5/2	1	1	1	1	1							
3	1	2	2	1	1			2	2	1		
7/2		2	1	3	2			2	2	1		
4	2	2	1	3	2	1	1	3	3	2		
9/2		2	3	5	1			3	3	3	1	1

Table 9. Multiplicity vectors for the conformal weights of gauge invariant states for N-S type Super Virasoro algebra.

n	1	8	10	$\overline{10}$	27	28	$\overline{28}$	35	$\overline{35}$	64	80	$\overline{80}$	81	$\overline{81}$	125
0	1														
1		1													
2			1	1											
3	2	2			2										
4		2			2			1	1						
5	2		2	2	2			2	2	1					
6			2	2	4	1	1	2	2	4					
7													1	1	
8					3	3	3	4	4	7			3	3	1
9	4	28	30	30	50	6	6	34	34	36	1	1	12	12	4

Table 10. Multiplicity vectors for various conformal weights of gauge invariant states of Ramond type Super Virasoro algebra.

12.4.2   Number theoretic auxiliary results

The ground state degeneracies for fermions and bosons need not to be identical to their ideal values $D = 64$ and $D = 16$ and it is of interest to find under what conditions the degeneracy can be said to be near to its ideal value. This amounts to calculating the p-adic inverse of the D in general case. Mathematica provides means for calculating modular inverses as well as modular powers (also fractional assuming that they exists). Despite this it is useful show how the real counterpart of a fractional p-adic number can be deduced.

The calculation of the modular inverse goes as follows.

a) The problem is to find the lowest order term in p-adic expansion of the inverse y of p-adic number $x \in 1, ...p-1$. The remaining terms in expansion in powers of p can be found iteratively. The equation to be solved is

$$yx \quad = \quad 1 \ mod \ p \ , \tag{12.4.6}$$

for a given value of x, which gives $y = mp + 1$.

b) One can express p in the form

$$p \quad = \quad Nx + r \ . \tag{12.4.7}$$

The evaluation of N and $r \in \{1, .., x-1\}$ is a straightforward exercise in modulo arithmetics. The defining equation for y can be written as

$$yx \;=\; m(Nx+r)+1 = mNx + mr + 1 \; . \tag{12.4.8}$$

From this one must have

$$mr + 1 \;=\; kx \; , \tag{12.4.9}$$

and any pair (m,k) satisfying this condition gives solution to y:

$$y \;=\; mN + k \; . \tag{12.4.10}$$

y must be chosen to be the smallest possible one.

Consider as examples two practical cases.

a) $p = M_n = 2^n - 1$ and $x = 15 = 2^4 - 1$. One obtains r by substituting repeatedly $2^4 = 1 \; mod \; x$ to the expression of M_n. M_n can be written in the form $M_n = 15(2^{n-4} + 2^{n-8} + ...) + r$ and the previous condition reads $mr + 1 = 15k$.

i) For M_{89} one has $r = 1$ and $(m,k) = (14,1)$ giving $y = 14(2^{n-4} + 2^{n-8} + ...) + 1$. For the real counterpart of $Xp^2/2D$ one has the approximate expression $(7X \; mod \; 16)/15$ and approximately N-S mass formula for small quantum numbers results.

ii) For M_{127} and M_{107} one has $r = 7$ and $7m + 1 = 15k$ gives $(m,k) = 2,1)$ and $y = 2(2^{n-4} + ...) + 1$. For $Xp^2/2D$ one has $X mod 16/15$: the factor 7 is absent.

b)$p = M^n$ and $x = 63 = 64 - 1$. One obtains r by substituting repeatedly $2^6 - 1 \; mod \; x$ to the expression of M_n. One has $r = 1$ for $n = 127$, $r = 31$ for $n = 107$ and $n = 89$. For the real counterpart R of Xp^2/D one has $R = (62X \; mod \; 64)/(63M_n)$ and $y = (60X \; mod \; 64)/(63M_n)$ for $n = 127$ and $107, 89$ respectively so that mass formulas change somewhat and in n-dependent manner if one has $D = 63$ instead of $D = 64$.

b) 1/5 factor appears in mass formulas for leptons and the previous argument leads to the expression $p^2/5 = (2^{126} - 2^{124} + 2^{122} - ..)p^2$. From this formula the real counterpart of, say 1/5, is in a good approximation 4/5. It must be emphasized that Mathematica provides the number theoretical modules for calculating the real counterparts for numbers of form rp, r rational number.

Acknowledgements

I am grateful for Tony Smith for pointing me the puzzling aspects related to the determination of the top quark mass.

Chapter 13

p-Adic Particle Massivation: Hadron Masses

13.1 Introduction

In this chapter the results of the calculation of elementary particle masses will be used to construct a model predicting hadron masses. The new elements are a revised identification for the p-adic length scales of quarks and the realization that number theoretical constraints on topological mixing can be realized by assuming that topological mixing leads to a thermodynamical equilibrium. This gives an upper bound of 1200 for the number of different U and D matrices and the input from top quark mass and $\pi^+ - \pi^0$ mass difference implies that physical U and D matrices can be constructed as small perturbations of matrices expressible as a direct sum of essentially unique 2×2 and 1×1 matrices.

The assumption about the presence of scaled up variants of light quarks in light hadrons leads to a surprisingly successful model for pseudo scalar meson masses in terms of only quark masses. This conforms with the idea that pseudo scalar mesons are Goldstone bosons in the sense that color Coulombic and magnetic contributions to the mass cancel each other. Also the mass differences between baryons containing different numbers of strange quarks can be understood if s quark appears as three scaled up versions. The earlier model for the purely hadronic contributions to hadron masses simplifies dramatically and only the color Coulombic and magnetic contributions to color conformal weight are needed.

13.1.1 Construction of U and D matrices

The basic constraint on the topological mixing that the modular contributions to the conformal weight defining the mass squared remain integer valued in the proper units: if this condition does not hold true, the order of magnitude for the real counterpart of the p-adic mass squared corresponds to 10^{-4} Planck masses.

Number theory gives strong constraints on CKM matrix. p-Adicization requires that U and D matrix elements are algebraic numbers. A strong constraint would be that the mixing probabilities are rational numbers implying that matrices defined by the moduli of U and D involve only square roots of rationals. The phases of matrix elements should belong to a finite extension of complex rationals.

Little can be said about the details of the dynamics of topological mixing. Nothing however prevents for constructing a thermodynamical model for the mixing. A thermodynamical model for U and D matrices maximizing the entropy defined by the mixing probabilities subject to the constraints fixing the values of n_{q_i} and the sums of row/column probabilities to one gives a thermodynamical ensemble with two quantized temperatures and two quantized chemical potentials. The resulting polynomial equations allow at most 1200 different solutions so that the number of U and D matrices is relatively small. The fact that matrix elements are algebraic numbers guarantees that the matrices are continuable to p-adic number fields as required.

The detailed study of quark mass spectrum leads to a tentative identification $(n_d, n_s, n_b) = (5, 5, 59)$ and $(n_u, n_c, n_t) = (4, 6, 58)$ of the modular contributions of conformal weights of quarks: note that in absence of mixing the contributions would be $(0, 9, 60)$ for both U and D type quarks. That b and t quark masses are nearly maximal and thus mix very little with lighter quarks is forced by the masses of t quark and $t\bar{t}$ meson. The values of n_{q_i} for light quarks follow by considering $\pi^+ - \pi^0$ mass difference.

One might consider the possibility that n_{q_i} for slightly dynamical and can vary in light mesons in order to guarantee that $u\bar{u}$, $d\bar{d}$ and $s\bar{s}$ give identical modular contributions to the conformal weight in states which are linear combinations of quark pairs. It turns out that unitarity does not allow the choices $(n_1 = 4, n_2 < 9)$, and that the choice $(n_d, n_s) = (5, 5)$, $(n_u, n_c) = (5, 6)$ is the unique choice producing a realistic CKM matrix. The requirement that quark contribution to pseudo scalar meson mass is smaller than meson mass is possible to satisfy and gives a constraint on CP_2 mass scale consistent with the prediction of leptonic masses when second order p-adic contribution to lepton mass is allowed to be non-vanishing.

The small mixing with b and t quarks is natural since the modular conformal weight of unmixed state having spectrum $\{0, 9, 60\}$ is analogous to energy so that Boltzmann weight for $n(g = 3)$ thermal excitation is small for $g = 1, 2$ ground states.

The maximally entropic solutions can be found numerically by using the fact that only the probabilities p_{11} and p_{21} can be varied freely. The solutions are unique in the accuracy used, which suggests that the system allows only single thermodynamical phase.

The matrices U and D associated with the probability matrices can be deduced straightforwardly in the standard gauge. The U and D matrices derived from the probabilities determined by the entropy maximization turn out to be unitary for most values of n_1 and n_2. This is a highly non-trivial result and means that mass and probability constraints together with entropy maximization define a sub-manifold of $SU(3)$ regarded as a sub-manifold in 9-D complex space. The choice $(n_u, n_c) = (4, n)$, $n < 9$, does not allow unitary U whereas $(n_u, n_c) = (5, 6)$ does. This choice is still consistent with top quark mass and together with $n_d = n_s = 5$ it leads to a rather reasonable CKM matrix with a value of CP breaking invariant within experimental limits. The elements V_{23} and V_{32}, $i = 1, 2$ are however roughly twice larger than their experimental values deduced assuming standard model. V_{31} is too large by a factor 1.6. The possibility of scaled up variants of light quarks could lead to too small experimental estimates for these matrix elements. The whole parameter space has not been scanned so that better candidates for CKM matrices might well exist.

13.1.2   Observations crucial for the model of hadron masses

The existence of scaled up variants of quarks is suggested by various anomalies such as Aleph anomaly [hh9] and the strange bumpy structure of the distribution of the mass of the top quark candidate. The real surprise was that the assumption about scaled up variants of u, d, s, and even c quarks in light hadrons leads to an excellent fit of meson masses with quark contribution explaining almost all of meson mass. The interpretation is that color Coulombic and color magnetic interaction conformal weights (rather than interaction energies) cancel each other in a approximation for pseudoscalar mesons in accordance with the idea that pseudo scalar mesons are massless as far as color interactions are considered. Also in case of baryons the assumption that s quark appears in three different scaled up versions (which are Λ, $\{\Sigma, \Xi\}$, and Ω) allows to understand the mass differences between baryons with different s quark content.

What remains to be understood are the contributions of color Coulombic and magnetic interactions to the mass squared. There are several delicate points to be taken into account.

a) An essential element of the new understanding is that conformal weight (mass squared is additive) for quarks with the same p-adic length scale. For instance, the masses of heavy $q\bar{q}$ mesons are equal to $\sqrt{2} \times m(q)$ rather than $2m(q)$. Since $k = 107$ for hadronic space-time sheet, for quarks with $k(q) \neq 107$, additivity holds true for the quark and color contributions for mass rather than mass squared.

b) The QCD based formula for the color magnetic interaction energy fails completely since the dependence of color magnetic spin-spin splittings on quark mass scale is nearer to logarithmic dependence on p-adic length scale than being of form $1/m(q_i)m(q_j) \propto L(k_i)L(k_j)$. This finding supports the decade old idea that the proper notion is not color interaction energy but color conformal weight.

13.1.3   A model for hadron

These findings suggest that the following model for hadrons deserves a testing. Certainly this model gives nice predictions for hadron masses and even the large color Coulombic contribution to baryon masses can be deduced from $\rho - \pi$ mass splitting in a good approximation.

a) Hadron can be characterized in terms of $k \geq 113$ partonic 2-surfaces $X^2(q_i)$ connected by join along boundaries bonds (JABs) to $k = 107$ 2-surface $X^2(H)$ corresponding to hadron and that each quark corresponds to JAB mediating classical color flux from $k = k(q)$ parton to $k = 107$ parton. Color flux tubes between quarks are replaced with pairs of flux tubes from $X^2(q_1)) \to X^2(H) \to X^2(q_2)$ mediating color Coulombic and magnetic interactions between quarks. In contrast to the standard model, mesons are characterized by two flux tubes rather than only one flux tube.

b) The contribution of color interactions to the conformal weight of hadron depends on the values of $k(q_i)$ since the interaction strengths $s_c(i,j)$ depend on $k(q_i)$ and $k(q_j)$. This makes possible the almost cancellation color magnetic and Coulombic conformal weights for scalar mesons.

13.2   Quark masses

The prediction or quark masses is more difficult due the facts that the deduction of even the p-adic length scale determining the masses of these quarks is a non-trivial task, and the original identification was indeed wrong. Second difficulty is related to the topological mixing of quarks. The new scenario leads to a unique identification of masses with top quark mass as an empirical input and the thermodynamical model of topological mixing as a new theoretical input. Also CKM matrix is predicted highly uniquely.

13.2.1   Basic mass formulas

By the earlier mass calculations and construction of CKM matrix the ground state conformal weights of U and D type quarks must be $h_{gr}(U) = -1$ and $h_{gr}(D) = 0$. The formulas for the eigenvalues of CP_2 spinor Laplacian imply that if m_0^2 is used as a unit, color conformal weight $h_c \equiv m_{CP_2}^2$ is integer for $p \bmod = \pm 1$ for U type quark belonging to $(p+1, p)$ type representation and obeying $h_c(U) = (p^2 + 3p + 2)/3$ and for $p \bmod 3 = 1$ for D type quark belonging $(p, p+2)$ type representation and obeying $h_c(D) = (p^2 + 4p + 4)/3$. Only these states can be massless since color Hamiltonians have integer valued conformal weights.

In the recent case the minimal $p = 1$ states correspond to $h_c(U) = 2$ and $h_c(D) = 3$. $h_{gr}(U) = -1$ and $h_{gr}(D) = 0$ reproduce the previous results for quark masses required by the construction of CKM matrix. This requires super-canonical operators O_p with a net conformal weight $h_{sc} = -3$ just as in the leptonic case. The facts that the values of p are minimal for spinor harmonics and the super-canonical operator is same for both quarks and leptons suggest that the construction is not had hoc.

Consider now the mass squared values for quarks. For $h(D) = 0$ and $h(U) = -1$ and using $m_0^2/3$ as a unit the expression for the thermal contribution to the mass squared of quark is given by the formula

$$
\begin{aligned}
M^2 &= (s + X)\frac{m_0^2}{p} \ , \\
s(U) &= 5 \ , \ s(D) = 8 \ , \\
X &\equiv \frac{(3Yp)_R}{3} \ ,
\end{aligned}
\tag{13.2.1}
$$

where the second order contribution Y corresponds to renormalization effects coming and depending on the isospin of the quark.

With the above described assumptions one has the following mass formula for quarks

$$M^2(q) = A(q)\frac{m_0^2}{p(q)} \ ,$$

$$A(u) = 5 + X_U(p(u)) \ , \quad A(c) = 14 + X_U(p(c)) \ , \quad A(t) = 65 + X_U(p(t)) \ ,$$
$$A(d) = 8 + X_D(p(d)) \ , \quad A(s) = 17 + X_D(p(s)) \ , \quad A(b) = 68 + X_D(p(b)) \ .$$

$$(13.2.2)$$

p-Adic length scale hypothesis allows to identify the p-adic primes labelling quarks whereas topological mixing of U and D quarks allows to deduce topological mixing matrices U and D and CKM matrix V and precise values of the masses apart from effects like color magnetic spin orbit splitting, color Coulombic energy, etc..

Integers n_{q_i} satisfying $\sum_i n(U_i) = \sum_i n(D_i) = 69$ characterize the masses of the quarks and also the topological mixing to high degree. The reason that modular contributions remain integers is that in the p-adic context non-trivial rationals would give CP_2 mass scale for the real counterpart of the mass squared. In the absence of mixing the values of integers are $n_d = n_u = 0$, $n_s = n_c = 9$, $n_b = n_t = 60$.

The fact that CKM matrix V expressible as a product $V = U^\dagger D$ of topological mixing matrices is near to a direct sum of 2×2 unit matrix and 1×1 unit matrix motivates the approximation $n_b \simeq n_t$.

The model for topological mixing matrices and CKM matrix predicts U and D matrices highly uniquely and allows to understand quark and hadron masses in surprisingly detailed level.

The large masses of top quark and of $t\bar{t}$ meson encourage to consider a scenario in which $n_t = n_b = n \le 60$ holds true.

a) $n_d = n_u = 60$ is not allowed by number theoretical conditions for U and D matrices and by the basic facts about CKM matrix but $n_t = n_b = 59$ allows almost maximal masses for b and t. This is not yet a complete hit. The unitarity of the mixing matrices and the construction of CKM matrix to be discussed in the next section forces the assignments

$$(n_d, n_s, n_b) = (5, 5, 59) \ , \quad (n_u, n_c, n_t) = (5, 6, 58) \ . \tag{13.2.3}$$

fixing completely the quark masses apart from few per cent renormalization effects of hadronic mass scale in topological condensation which seem to be present and will be discussed later. Note that top quark mass is still rather near to its maximal value.

b) The constraint that quark contribution to pion mass does not exceed pion mass implies the constraint $n(d) \le 6$ and $n(u) \le 6$ in accordance with the predictions of the model of topological mixing. It is important to notices that $u - d$ mass difference does not affect $\pi^+ - \pi^0$ mass difference and the quark contribution to $m(\pi)$ is predicted to be $\sqrt{(n_d + n_u + 13)/24} \times 136.9$ MeV for the maximal value of CP_2 mass (second order p-adic contribution to electron mass squared vanishes).

13.2.2 The p-adic length scales associated with quarks

The identification of p-adic length scales associated with the quarks has turned a highly non-trivial problem. The reasons are that for light quarks it is difficult to deduce information about quark masses for hadron masses and that the unknown topological mixing makes possible several p-adic length scales for quarks. Two natural constraints have however emerged from the recent work.

a) Quark contribution to the hadron mass cannot be larger than color contribution and for quarks having $k_q \ne 107$ quark contribution to mass is added to color contribution to the mass. For quarks with same value of k conformal weight rather than mass is additive whereas for quarks with different value of k masses are additive. An important implication is that for diagonal mesons $M = q\bar{q}$ having $k(q) \ne 107$ the condition $m(M) \ge \sqrt{2}m_q$ must hold true. This gives strong constraints on quark masses.

b) The realization that scaled up variants of quarks explain elegantly the masses of light hadrons allows to understand large mass splittings of light hadrons without the introduction of strong isospin-isospin interaction.

The new model for quark masses is based on the following identifications of the p-adic length scales.

a) The nuclear p-adic length scale $L(k)$, $k = 113$, corresponds to the p-adic length scale determining the masses of u, d, and s quarks. Note that $k = 113$ corresponds to a so called Gaussian Mersenne. The interpretation is that quark massivation occurs at nuclear space-time sheet at which quarks feed their em fluxes. At $k = 107$ space-time sheet, where quarks feed their color gauge fluxes, the quark masses are vanishing in the first p-adic order. This could be due to the fact that the p-adic temperature is $T_p = 1/2$ at this space-time sheet so that the thermal contribution to the mass squared is negligible. This would reflect the fact that color interactions do not involve any counterpart of Higgs mechanism.

p-Adic mass calculations turn out to work remarkably well for massive quarks. The reason could be that M_{107} hadron physics means that *all* quarks feed their color gauge fluxes to $k = 107$ space-time sheets so that color contribution to the masses becomes negligible for heavy quarks as compared to Super-Kac Moody and modular contributions corresponding to em gauge flux feeded to $k > 107$ space-time sheets in case of heavy quarks. Note that Z^0 gauge flux is feeded to space-time sheets at which neutrinos reside and screen the flux and their size corresponds to the neutrino mass scale. This picture might throw some light to the question of whether and how it might be possible to demonstrate the existence of M_{89} hadron physics.

One might argue that $k = 107$ is not allowed as a condensation level in accordance with the idea that color and electro-weak gauge fluxes cannot be fed at the space-time space time sheet since the classical color and electro-weak fields are functionally independent. The identification of η' meson as a bound state of scaled up $k = 107$ quarks is not however consistent with this idea unless one assumes that $k = 107$ space-time sheets in question are separate.

b) The requirement that the masses of diagonal pseudoscalar mesons of type $M = q\bar{q}$ are larger but as near as possible to the quark contribution $\sqrt{2}m_q$ to the mass, fixes the p-adic primes $p \simeq 2^k$ associated with c, b and t quark (these values of k are "nominal" since k seems to be dynamical). c quark corresponds to the p-adic length scale $k = 104 = 2^3 \times 13$. b quark corresponds to $k = 103$ for $n(b) = 5$. By the same criterion t quark must correspond to the secondary length scale $k = 3 \times 31$. It should be noticed that $k = 31$ is double Mersenne prime in the sense that $31 = 2^5 - 1$ holds true. In fact, the primes, $5, M_5 = 31$, and M_{31} form the analog of Combinatorial Hierarchy consisting of primes $2, M_2 = 3, M_3 = 7, M_7 = 127, M_{127} = 2^{127} - 1, ...$ possibly containing also larger primes than the known ones. Note that in all cases color interaction energy is measured using $k = 107$ mass scale as unit as in case of light hadrons.

A more precise definition of the model requires a detailed consideration of the renormalization of hadron masses in topological condensation which reduces the mass scale by few per cent. Some input from hadronic mass spectrum is also needed to fix the renormalization. This point will be discussed later in this chapter. The following table summarizes quark masses as predicted by this model.

q	d	u	s	c	b	t
n_q	4	5	6	6	59	58
$k(q)$	113	113	113	104	103	93
$m(q)/GeV$	.090	.079	.090	1.873	6.537	202.8

Table 1. The masses of quarks predicted assuming $(n_d, n_s, n_b) = (5, 5, 59)$ and $(n_u, n_c, n_t) = (5, 6, 58)$ and maximal CP_2 mass scale consistent with η' meson mass.

13.2.3 Are scaled up variants of quarks also there?

The following arguments suggest that p-adically scaled up variants of quarks might appear not only at very high energies but even in low energy hadron physics.

Aleph anomaly and scaled up copy of b quark

The prediction for the b quark mass is consistent with the explanation of the Aleph anomaly [hh9] inspired by the finding that neutrinos seem to condense at several p-adic length scales [ha20]. If b quark condenses at $k(b) = 97$ level, the predicted mass is $m(b, 97) = 52.3$ GeV for $n_b = 59$ for the maximal CP_2 mass consistent with η' mass. If the the mass of the particle candidate is defined experimentally as one half of the mass of resonance, b quark mass is actually by a factor $\sqrt{2}$ higher and scaled up b corresponds to $k(b) = 96 = 2^5 \times 3$. The prediction is consistent with the estimate 55 GeV for the mass of the Aleph particle and gives additional support for the model of topological mixing. Also the decay characteristics of Aleph particle are consistent with the interpretation as a scaled up b quark.

Tony Smith has emphasized the fact that the distribution for the mass of the top quark candidate has a clear structure suggesting the existence of several states, which he interprets as excited states of top quark [hc1]. According to the figures 13.2.3 and 13.2.3 representing published FermiLab data, this structure is indeed clearly visible.

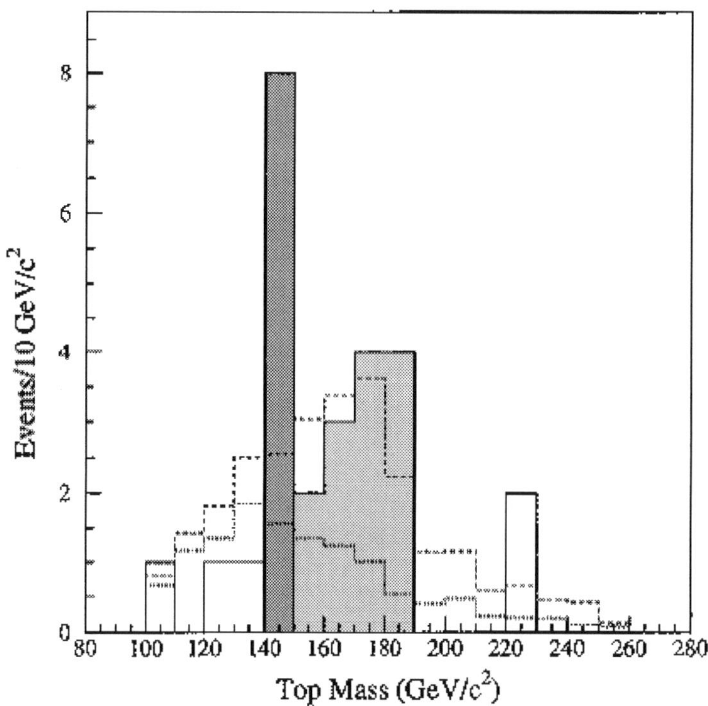

Figure 13.1: Fermilab semileptonic histogram for the distribution of the mass of top quark candidate (FERMILAB-PUB-94/097-E).

For $k(t) = 3 \times 31 = 93$, which corresponds to the tertiary p-adic length scale associated with the Mersenne prime $2^{31} - 1$ (also $31 = 2^5 - 1$ is Mersenne prime!), the mass is scaled up by a $\sqrt{2}$ factor to $m(t) = 203$ GeV for the option suggested by η' mass. This option will be discussed in more detail later. The already suggested interpretation is in terms of coherent summation of conformal weights predicting that quark contribution to $t\bar{t}$ meson mass is $\sqrt{2}m(t) = 2 \times 143$ GeV for the maximal mass scale consistent with η' meson mass. There is evidence for a sharp peak in the mass distribution of the top quark in 140-150 GeV range (Fig. 13.2.3).

Scaled up variants of d, s, u, c in top quark mass scale

The fact that all neutrinos seem to appear as scaled up versions in several scales, encourages to look whether also u, d, s, and c could appear as scaled up variants transforming to the more stable variants by a stepwise increase of the size scale involving the emission of electro-weak gauge bosons. In the following the scenario in which t and b quarks mix minimally is considered.

q	$m(92)/GeV$	$m(91)/GeV$	$m(90)/GeV$
u	114	162	229
d	130	184	261
c	119	169	240
s	130	184	261

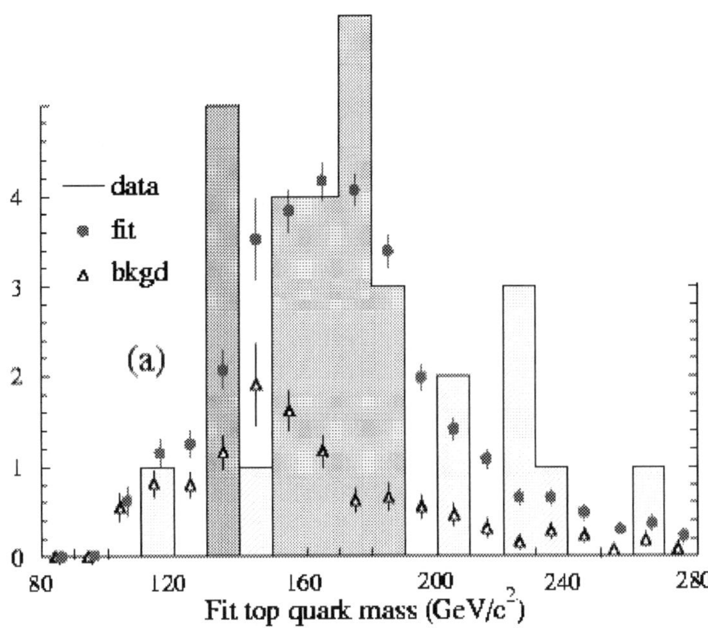

Figure 13.2: Fermilab D0 semileptonic histogram for the distribution of the mass of top quark candidate (hep-ex/9703008, April 26, 1994

Table 2. The masses of $k = 92, 91$ and $k = 90$ scaled up variants of u,d,c,s quarks assuming same integers n_{q_i} as for ordinary quarks in the scenario $(n_d, n_s, n_b) = (5, 5, 59)$ and $(n_u, n_c, n_t) = (5, 6, 58)$ and maximal CP_2 mass consistent with the η' mass.

a) For $k = 92$, the masses would be $m(q, 92) = 114$, 130, 119, 130 GeV in the order q= u,d,c,s so that all these quarks might appear in the critical region where the top quark mass has been wandering. There is also a peak slightly below 120 GeV, which could correspond to scaled up u quark.

b) For $k = 91$ copies would have masses $m(q, 91) = 162$, 184, 169, 184 GeV in the order q= u,d,c,s. $m(c(91)) = 169$ GeV is rather near to the official top quark mass. Note that $c \to bW$ decay rate is however reduced by a factor $|V(c, b)|^2 \sim 10^{-3}$ from $t \to bW$ decay rate. There is also a small peak also around 265 GeV which could relate to $m(c(90)) = 240$ GeV. Note that the mass of the u quark of M_{89} hadron physics would be $m(u_{89}) = 323$ GeV.

Note that it is possible to distinguish between scaled up quarks of M_{107} hadron physics and the quarks of M_{89} hadron physics since the unique signature of M_{89} hadron physics would be the increase of the scale of color Coulombic and magnetic energies by a factor of 512.

It must be added that the detailed identifications are sensitive to the exact value of the CP_2 mass scale. The possibility that a few per cent mass scale renormalization occurs in the topological condensation of hadrons brings in additional fuzziness. Therefore the preceding considerations cannot be taken too literally.

Fractally scaled up copies of light quarks and low mass hadrons?

One can of course ask, whether the fractally scaled up quarks could appear also in low lying hadrons. The arguments to be developed in detail later suggest that u, d, and s quark masses could be dynamical in the sense that several fractally scaled up copies can appear in low mass hadrons and explain the mass differences between hadrons.

In this picture the mass splittings of low lying hadrons with different flavors would result from fractally scaled up excitations of s and also u and d quarks in case of mesons. This notion would also throw light into the paradoxical presence of two kinds of quark masses: constituent quark masses and current quark masses having much smaller values than constituent quarks masses. That color spin-spin splittings are of same order of magnitude for all mesons supports the view that color gauge fluxes are feeded to $k = 107$ space-time sheet.

The alert reader has probably already asked whether also proton mass could be understood in terms of scaled up copies of u and d quarks. This does not seem to be the case, and an argument predicting with 23 per cent error proton mass scale from $\rho - \pi$ and $\Delta - N$ color magnetic splittings emerges.

To sum up, it seems quite possible that the scaled up quarks predicted by TGD have been observed for decade ago in FermiLab about that the prevailing dogmas has led to their neglect as statistical fluctuations. Even more, scaled up variants of s quarks might have been in front of our eyes for half century! Phenomenon is an existing phenomenon only if it is an understood phenomenon.

13.3 Topological mixing of quarks

The requirement that hadronic mass spectrum is physical requires mixing of U and D type boundary topologies. In this section quark masses and the mixing of the boundary topologies are considered on the general level and CKM matrix is derived using the existing empirical information plus the constraints on the quark masses to be derived from the hadronic mass spectrum in the later sections.

13.3.1 Mixing of the boundary topologies

In TGD the different mixings of the boundary topologies for U and D type quarks provide the fundamental mechanism for CKM mixing and also CP breaking. In the determination of CKM matrix one can use following conditions.

a) Mass squared expectation values in order $O(p)$ for the topologically mixed states must be integers and the study of the hadron mass spectrum leads to very stringent conditions on the values of these integers. Physical values for these integers imply essentially correct value for Cabibbo angle provided U and D matrices differ only slightly from the mixing matrices mixing only the two lowest generations.

b) The matrices U and D describing the mixing of U and D type boundary topologies are unitary in the p-adic sense. The requirement that the moduli squared of the matrix elements are rational numbers, is very attractive since it suggests equivalence of p-adic and real probability concepts and therefore could solve some conceptual problems related to the transition from the p-adic to real regime. It must be however immediately added that rationality assumption for the probabilities defined by S-matrix turns out to be non-physical. It turns out that the mixing scenario reproducing a physical CKM matrix is consistent with the rationality of the moduli squared of the matrix elements of U and D matrices but not with the rationality of the matrix elements themselves. The phase angles appearing in U and D matrix can be rational and in this case they correspond to Pythagorean triangles. In principle the rationality of the CKM matrix is possible.

c) The requirements that Cabibbo angle has correct value and that the elements $V(t, d)$ and $V(u, b)$ of the CKM matrix have small values not larger than 10^{-2} fixes the integers n_i characterizing quark masses to a very high degree and in a good approximation one can estimate the angle parameters analytically. remains open at this stage. The requirement of a realistic CKM matrix leads to a scenario for the values of n_i, which seems to be essentially unique.

The mass squared constraints give for the D matrix the following conditions

$$
\begin{aligned}
9|D_{12}|^2 + 60|D_{13}|^2 &= n_1(D) \equiv n_d \ , \\
9|D_{22}|^2 + 60|D_{23}|^2 &= n_2(D) \equiv n_s \ , \\
9|D_{32}|^2 + 60|D_{33}|^2 &= n_3(D) \equiv n_b = 69 - n_2(D) - n_1(D) \ .
\end{aligned}
$$

$$(13.3.1)$$

The third condition is not independent since the sum of the conditions is identically true by unitarity.

For U matrix one has similar conditions:

$$\begin{aligned}
9|U_{12}|^2 + 60|U_{13}|^2 &= n_1(U) \equiv n_u \ , \\
9|U_{22}|^2 + 60|U_{23}|^2 &= n_2(U) \equiv n_c \ , \\
9|U_{32}|^2 + 60|U_{33}|^2 &= n_3(U) \equiv n_t = 69 - n_2(U) - n_1(U) \ .
\end{aligned}$$

$$(13.3.2)$$

The integers n_d, n_s and n_u, n_c characterize the masses of the physical quarks and the task is to derive the values of these integers by studying the spectrum of the hadronic masses. The second task is to find unitary mixing matrices satisfying these conditions.

The general form of U and D matrices can be deduced from the standard parametrization of the CKM matrix given by

$$V = \begin{bmatrix} c_1 & s_1 c_3 & s_1 s_3 \\ -s_1 c_2 & c_1 c_2 c_3 - s_2 s_3 exp(i\delta_{CP}) & c_1 c_2 s_3 + s_2 c_3 exp(i\delta_{CP}) \\ -s_1 s_2 & c_1 s_2 c_3 + c_2 s_3 exp(i\delta_{CP}) & c_1 s_2 s_3 - c_2 c_3 exp(i\delta_{CP}) \end{bmatrix} \qquad (13.3.3)$$

This form of the CKM matrix is always possible to achieve by multiplying each U and D type quark fields with a suitable phase factor: this induces a multiplication U and D from left by a diagonal phase factor matrix inducing the multiplication of the columns of U and D by phase factors:

$$\begin{aligned}
U &\to U \times d(\phi_1, \phi_2, \phi_3) \ , \\
D &\to D \times d(\chi_1, \chi_2, \chi_3) \ , \\
d(\phi_1, \phi_2, \phi_3) &\equiv diag(exp(i\phi_1), exp(i\phi_2), exp(i\phi_3)) \ .
\end{aligned}$$

The multiplication of the columns by the phase factors affects CKM matrix defined as

$$V = U^\dagger D \to d(-\phi_1, -\phi_2, -\phi_3) V d(\chi_1, \chi_2), \chi_3) \ . \qquad (13.3.4)$$

By a suitable choice of the phases, the first row and column of V can be made real. The multiplication of the rows of U and D from the left by the same phase factors does not affect the elements of V. One can always choose D to be of the same general form as the CKM matrix but must allow U to have nontrivial phase overall factors on the second and third row so that the most general U matrix is parameterized by six parameters.

Mass squared conditions give two independent conditions on the values of the moduli of the matrix elements of U and D. This eliminates two coordinates so that the most general D matrix can be chosen to depend on 2 parameters, which can be taken to be $r_{11} \equiv |D_{11}|$ and $r_{21} \equiv |D_{21}|$. U matrix contains also the overall phase angles associated with the second and third row and hence depends on four parameters altogether.

13.3.2   The constraints on U and D matrices from quark masses

The new view about quark masses allows a surprisingly simple model for U and D matrices predicting in the lowest order approximation that the probabilities defined by these matrices are identical and that the integers characterizing the masses of U and D type quarks are identical.

The constraints on $|U|$ and $|D|$ matrices from quark masses

The understanding of quark masses pose strong constraints on U and D matrices. The constraints are identical in the approximation that V-matrix is identity matrix and read in the case of D-matrix as

$$\begin{aligned}
n_d &= 13 = P_{12}^D \times 9 + P_{13}^D \times 60 \ , \\
n_s &= 31 = P_{22}^D \times 9 + P_{23}^D \times 60 \ .
\end{aligned} \qquad (13.3.5)$$

The conditions for b quark give nothing new. The extreme cases when only $g = 1$ or $g = 2$ contributes to n_q gives the bounds

$$\frac{15}{36} \;\leq\; P_{13}^D \leq \frac{15}{60},$$

$$\frac{22}{60} \;\leq\; P_{23}^D \leq \frac{31}{60} \;. \tag{13.3.6}$$

Unitarity conditions

The condition $D = VU$ and the fact that V is in not too far from unit matrix being in a good approximation a direct sum of 2×2 matrix and 1×1 identity matrix, imply together that U an D cannot differ much from each other. At least the probabilities defined by the moduli squared of matrix elements are near to each other.

a) Instead of trying numerically to solve U and D matrices by a direct numerical search, it is more appropriate to try to deduce estimates for the probabilities $P_{ij}^U = |U_{ij}|^2$ and $P_{ij}^D = |D_{ij}|^2$ determined by the moduli squared of the matrix elements and satisfying the unitarity conditions $\sum_j P_{ij}^X = 1$ and $\sum_i P_{ij}^X = 1$.

b) The formula $D = UV$ using the fact that V_{i3} is small for $i = 1, 2$ implies $|D_{i3}| \simeq |U_{i3}|$. By probability conservation also the condition $|D_{33}| \simeq |U_{33}|$ must hold true so that the third columns of U and D are same in a reasonable approximation.

1. Parametrization of $|U|$ and $|D|$ matrices

The following parametrization is natural for the matrices P_{ij}^X.

$$P_{12}^D = \frac{k_D}{9} \;, \qquad P_{13}^D = \frac{n_d - k_D}{9} \;,$$

$$P_{22}^D = \frac{l_D}{9} \;, \qquad P_{23}^D = \frac{n_s - l_D}{60} \;, \tag{13.3.7}$$

$$P_{32}^D = \frac{9 - k_D - l_D}{9} \;, \qquad P_{33}^D = \frac{60 - n_s - n_d - k_D - l_D}{60} \;.$$

A similar parametrization holds true for P_{ij}^U but with $n_d = n_u$ and $n_s = n_c$ but possibly different values of k_U and l_U. Since $l_D \ll n_s$ is expected to hold true, P_{23}^D is in a good approximation equal to $P_{23}^D = n_s/60 = 31/60$. Same applies to P_{23}^U.

$k_X = 2$ (k_X need not be an integer) gives a good first estimate for mixing probabilities of u and d quark. Thus only the parameter l_X remains free if $k_D = 2$ is accepted.

The approximation $P_{i3}^U = P_{i3}^D$ motivated by the near unit matrix property of V, gives the parametrization

$$P_{12}^D \;=\; P_{12}^U = \frac{k}{9} \;, \quad P_{13}^D = P_{13}^U \frac{n_d - k}{60} \;. \tag{13.3.8}$$

2. Constraints from CKM matrix in $|U| = |D|$ approximation

The condition $D_{12} = (UV)_{12}$ when fed to the condition

$$P_{12}^U \;=\; P_{12}^D \tag{13.3.9}$$

using the approximation $k_D = k_U = k$ $l_D = l_U = l$ gives

$$|U_{i2}|^2 - |U_{i1}V_{12} + U_{i2}V_{22} + U_{i3}V_{32}|^2 = 0 \;. \tag{13.3.10}$$

$i = 1, 2, 3$ In the approximation that the small V_{32} term does not contribute, this gives

$$|U_{i1}V_{12} + U_{i2}V_{22})|^2 \;=\; |U_{i2}|^2 \;. \tag{13.3.11}$$

By dividing with $|U_{i1}|^2|V_{22}|^2$ and using the approximation $|V_{22}|^2 = 1$ this gives

$$
\begin{aligned}
v_i^2 \;+\;& 2u_iv_i \times cos(\Psi_i) = 0 \;, \\
\Psi_i \;=\;& arg(V_{i2}) - arg(V_{32}) + arg(U_{i1}) - arg(U_{i2}) \;. \\
u_i \;=\;& |\frac{U_{i2}}{U_{i1}}| \;,\; v_i = |\frac{V_{i2}}{V_{22}}| \;.
\end{aligned}
\tag{13.3.12}
$$

This gives

$$
\begin{aligned}
cos(\Psi_i) \;=\;& -\frac{v_i}{2u_i} = -\frac{v_i}{2}\sqrt{\frac{9x_i}{k_i}} \;, \\
x_i \;=\;& P_{ii}^D = 1 - \frac{k_i}{9} - \frac{n(i) - k(i)}{60} \;, \\
k(1) \;=\;& k \;,\; k(2) = l \;,\; n(1) = n_d, \; n(2) \equiv n_s \;.
\end{aligned}
\tag{13.3.13}
$$

The condition $|cos(\Psi)| \leq 1$ is trivially satisfied. For $n_d = 13$ and $k = 2$ the condition gives $x = .59$ and $cos(\Psi_1) = .185$. $k = 1.45$ gives $x = .65$ and $cos(\Psi) = .226$, which is rather near to V_{12}.

13.3.3 Constraints from CKM matrix

Besides the constraints from hadron masses, there are constraints from CKM matrix $V = U^\dagger D$ on U and D matrices.

a) The fact that CKM matrix is near unit matrix implies that U and D matrix are near to each other and the assumption $n(U_i) = n(D_i)$ predicting quark masses correctly is consistent with this.

b) Cabibbo angle allows to derive the estimate for the difference $|U_{11}| - |D_{11}|$. Together with other conditions this difference fixes the scenario essentially uniquely.

c) The requirement that CP breaking invariant J has a correct order of magnitude gives a very strong constraint on D matrix. The smallness of J implies that V is nearly orthogonal matrix and same assumption can be made about U and D matrices.

d) The requirement that the moduli the first row (column) of CKM matrix are predicted correctly makes it possible to deduce for given D (U) U (D) matrix essentially uniquely. Unitarity requirement poses very strong additional constraints. It must be emphasized that the constraints from the moduli of the CKM alone are sufficient to determine U and D matrices and hence also quark masses and hadron masses to very high degree.

1. Bounds on CKM matrix elements

The most recent experimental information [hj8] concerning CKM matrix elements is summarized in table below

$\|V_{13}\| \equiv \|V_{ub}\| = (0.087 \pm 0.075)V_{cb} : 0.42 \cdot 10^{-3} < \|V_{ub}\| < 6.98 \cdot 10^{-3}$
$\|V_{23}\| \equiv \|V_{cb}\| = (41.2 \pm 4.5) \cdot 10^{-3}$
$\|V_{31}\| \equiv \|V_{td}\| = (9.6 \pm 0.9) \cdot 10^{-3}$
$\|V_{32}\| \equiv \|V_{ts}\| = (40.2 \pm 4.4) \cdot 10^{-3}$
$s_{Cab} = 0.226 \pm 0.002$

Table 3. The experimental constraints on the absolute values of the CKM matrix elements.

$$
\begin{aligned}
s_1 \;=\;& .226 \pm .002 \;, \\
s_1s_2 \;=\;& V_{31} = (9.6 \pm .9) \cdot 10^{-3} \;, \\
s_1s_3 \;=\;& V_{13} = (.087 \pm .075) \cdot V_{23} \;, \\
V_{23} \;=\;& (40.2 \pm 4.4) \cdot 10^{-3} \;.
\end{aligned}
\tag{13.3.14}
$$

The remaining parameter is $sin(\delta)$ or equivalently the CP breaking parameter J:

$$J \;=\; Im(V_{11}V_{22}\overline{V}_{12}\overline{V}_{21}) = c_1 c_2 c_3 s_2 s_3 s_1^2 sin(\delta) \;,$$

$$(13.3.15)$$

where the upper bound is for $sin(\delta) = 1$ and the previous average values of the parameters s_i, c_i (note that the poor knowledge of s_3 affects on the upper bound for J considerably). Unitary triangle [hj9] gives for the CP breaking parameter the limits

$$1.0 \times 10^{-4} \leq J \leq 1.7 \times 10^{-4} \;.$$

$$(13.3.16)$$

2. CP breaking in $M - \overline{M}$ systems as a source of information about CP breaking phase

Information about the value of $sin(\delta)$ as well as on the range of possible top quark masses comes from CP breaking in $K - \bar{K}$ and $B - \bar{B}$ systems.

The observables in $K_L \to 2\pi$ system [hj4]

$$\eta_{+-} \;=\; \frac{A(K_L \to \pi^+\pi^-)}{A(K_S \to \pi^+\pi^-)} = \epsilon + \frac{\epsilon'}{1 + \omega/\sqrt{2}} \;,$$

$$\eta_{00} \;=\; \frac{A(K_L \to \pi^0\pi^0)}{A(K_S \to \pi^0\pi^0)} = \epsilon - 2\frac{\epsilon'}{1 - \sqrt{2}\omega} \;,$$

$$\omega \;\sim\; \frac{1}{20} \;,$$

$$\epsilon \;=\; (2.27 \pm .02) \cdot 10^{-3} \cdot exp(i43.7^o) \;,$$

$$|\frac{\epsilon'}{\epsilon}| \;=\; (3.3 \pm 1.1) \cdot 10^{-3} \;.$$

$$(13.3.17)$$

The phases of ϵ and ϵ' are in good approximation identical. CP breaking in $K - \bar{K}$ mass matrix comes from the CP breaking imaginary part of $\bar{s}d \to s\bar{d}$ amplitude M_{12} (via the decay to intermediate W^+W^- pair) whereas $K^0\bar{K}^0$ mass difference Δm_K comes from the real part of this amplitude: the calculation of the real part cannot be done reliably for kaon since perturbative QCD does not work in the energy region in question. On can however relate the real part to the known mass difference between K_L and K_S: $2Re(M_{12}) = \Delta m_K$.

Using the results of [hj4]) one can express ϵ and ϵ'/ϵ in the following numerical form

$$|\epsilon| \;=\; \frac{1}{\sqrt{2}}\frac{Im(M_{12}^{sd})}{\Delta m_K} - .05 \cdot |\frac{\epsilon'}{\epsilon}| = 2J(22.2B_K \cdot X(m_t) - .28B'_K) \;,$$

$$|\frac{\epsilon'}{\epsilon}| \;=\; C \cdot J \cdot B'_K \;,$$

$$X(m_t) \;=\; \frac{H(m_t)}{H(m_t = 60 \; GeV)} \;,$$

$$H(m_t) \;=\; -\eta_1 F(x_c) + \eta_2 F(x_t)K + \eta_3 G(x_c, x_t) \;,$$

$$x_q \;=\; \frac{m(q)^2}{m_W^2} \;,$$

$$K \;=\; s_2^2 + s_2 s_3 cos(\delta) \;.$$

$$(13.3.13)$$

Here the values of QCD parameters η_i depend on top mass slightly. B'_K and B_K are strong interaction matrix elements and vary between $1/3$ and 1. The functions F and G [hj4] are given by

$$F(x) \;=\; x\left[\frac{1}{4} + \frac{9}{4}\frac{1}{1-x} - \frac{3}{2}\frac{1}{(1-x)^2}\right] + \frac{3}{2}(\frac{x}{x-1})^3 log(x) \;,$$

$$G(x,y) \;=\; xy\left[\frac{1}{x-y}\left[\frac{1}{4} + \frac{3}{2}\frac{1}{1-x} - \frac{3}{4}\frac{1}{(1-x)^2}\right] log(x) + (y \to x) - \frac{3}{4}\frac{1}{(1-x)(1-y)}\right] \;.$$

$$(13.3.12)$$

One can solve parameter B'_K by requiring that the value of ϵ'/ϵ corresponds to the experimental mean value:

$$B'_K = \frac{1}{C \times J} \frac{\epsilon'}{\epsilon} \; . \tag{13.3.13}$$

The most recent measurements by KTeV collaboration in Fermi Lab [hj12] give for the ratio $|\epsilon'/\epsilon|$ the value $|\epsilon'/\epsilon| = (28 \pm 1) \times 10^{-4}$. The proposed standard model explanation for the large value of B'_K is that s-quark has running mass about $m_s(m_c) \simeq .1$ GeV at m_c [hj14]. The explanation is marginally consistent with the TGD prediction $m(s) = 127$ MeV for the mass of s quark. Also the effects caused by the predicted higher gluon generations having masses around 33 GeV can increase the value of ϵ'/ϵ by a factor 3 in the lowest approximation since the corrections involve sum over three different one-gluon loop diagrams with gluon mass small respect to intermediate boson mass scale [F5].

A second source of information comes from $B - \bar{B}$ mass difference. At the energies in question perturbative QCD is expected to be applicable for the calculation of the mass difference and mass difference is predicted correctly if the mass of the top quark is essentially the mass of the observed top candidate [hj5].

3. U and D matrices could be nearly orthogonal matrices

The smallness of the CP breaking phase angle δ_{CP} means that V is very near to an orthogonal matrix. This raises the hope that in a suitable gauge also U and D are nearly orthogonal matrices and would be thus almost determined by single angle parameter θ_X, $X = U, D$. Cabibbo angle $s_c = sin(\theta_c) = .226$ which is not too far from $sin^2(\theta_W) \simeq .23$ and appears in V matrix rotating the rows of U to those of D. In very vague sense this angle would characterize between the difference of angle parameters characterizing U and D matrices. If U is orthogonal matrix then the decomposition

$$V = V_1 V_2 = \begin{bmatrix} c_1 & s_1 & 0 \\ -s_1 c_2 & c_1 c_2 & s_2 exp(i\delta_{CP}) \\ -s_1 s_2 & c_1 s_2 & -c_2 exp(i\delta_{CP}) \end{bmatrix} \times \begin{bmatrix} 1 & 0 & 0 \\ 0 & c_3 & s_3 \\ 0 & -s_3 & c_3 \end{bmatrix} \tag{13.3.14}$$

suggests that CP breaking can be visualized as a process in which first s and b quarks are slightly mixed to s' and b' by V_2 ($s_3 \simeq 1.4 \times 10^{-2}$) after which V_1 induces a slightly CP-breaking mixing of d and s' with b' ($s_2 \simeq .04$).

4. How the large mixing between u and c results

The prediction that u quark spends roughly $1/3$ of time in $g = 0$ state looks bizarre and it is desirable to understand this from basic principles. The basic observations are following.

a) V matrix is in good approximation direct sum of 2×2 matrix inducing relatively large rotation with $sin(\theta_c) \simeq .23$ and unit matrix. In particular, V_{i3} are very small for $i = 1, 2$. Using the formula $D = UV$ one finds that $|U_{i3}| = |D_{i3}|$ in a good approximation for $i = 1, 2$ and by unitarity also for $I = 3$. Thus the third columns of U and D are identical in a good approximation.

b) Assume that also U_{i3} and D_{i3} are small for $i = 1, 2$. A stronger assumption is that even the contribution of D_{13} and U_{13} are so small that they do not affect u and d masses. This implies

$$\begin{aligned} n_d &= 9|D_{12}|^2 + 60|D_{13}|^2 \simeq 9|D_{12}|^2 \; , \\ n_u &\simeq 9|U_{12}|^2 \; . \end{aligned} \tag{13.3.14}$$

Unitarity implies in this approximation

$$\begin{aligned} |U_{11}|^2 &\leq 1 - \frac{n_u}{9} = \frac{1}{3} \; , \\ |D_{11}|^2 &\leq 1 - \frac{n_d}{9} = \frac{5}{9} \; . \end{aligned} \tag{13.3.14}$$

c) It might be that there are also solutions for which mixing of u resp. d quark is mostly with t resp. b quarks but numerical experimentation does not favor this idea since CP breaking becomes extremely small. Since mixing presumably involves topology change, it seems obvious that topological mixing involving a creation or annihilation of two handles is improbable.

13.4   Construction of U, D, and CKM matrices

In this section it will be found that various mathematical and experimental constraints on U and D matrices determine them essentially uniquely.

13.4.1   The constraints from CKM matrix and number theoretical conditions

The requirement that U, D and V allow an algebraic continuation to finite-dimensional extensions of various p-adic number fields provides a very strong additional constraints. The mathematical problem is to understand how many unitary V matrices acting on U as $U \to D = UV$ respect the number theoretic constraints plus the constraints $n_u = n_d + 2$ and $n_c = n_d - 2$.

It is instructive to what happens in much simpler 2-dimensional case. In this case the conditions boil down to the conditions on $n(i)$ imply $|U| = |D|$ and this condition is equivalent with (say) the condition $|U_{11}| = D_{11}$. U and D can be parameterized as

$$ U = \begin{pmatrix} cos(\theta)exp(i(\psi)) & sin(\theta)exp(i\phi) \\ -sin(\theta)exp(-i\phi) & cos(\theta)exp(-i\psi) \end{pmatrix} \quad . $$

If $cos(\theta)^2$ and $sin(\theta)^2$ are rational numbers, $exp(i\theta)$ is associated with a Gaussian integer. A more general requirement is that $exp(i\theta)$ belongs to a finite-dimensional extension of rational numbers and thus corresponds to a products of a phase associated with Gaussian integer and a phase in a finite-dimensional algebraic extension of rational numbers.

Eliminating the trivial multiplicative phases gives a set of matrices U identifiable as a double coset space $X^2 = SU(2)/U(1)_R \times U(1)_L$. The value of $cos(\theta) = |U_{11}|$ serving as a coordinate for X^2 is respected by the right multiplication with V. Eliminating trivial $U(1)_R$ phase multiplication, the space of V:s reduces to $S^2 = SU(2)/U(1)_R$. The condition that $cos(\theta)$ is not changed leaves one parameter set of allowed matrices V.

The translation of these results to 3-dimensional case is rather straightforward. In the 3-dimensional case the probabilities P_{i2}, P_{i3}, $i = 1, 2$ characterize a general matrix $|U|$, and V can affect these probabilities subject to constraints on $n(I)$. When trivial phases affecting the probabilities are eliminated, the matrices U correspond naturally to points of the 4-dimensional double coset space $X^4 = SU(3)/(U(1) \times U(1))_R \times U(1) \times U(1))_L$ having dimension $D = 4$.

The two constraints on the probabilities mean that allowed solutions for given values of $n(I)$ define a 2-dimensional surface X^2 in X^4. The allowed unitary transformations V must be such that they move U along this surface. Certainly they exist since X^2 can be regarded as a local section in $SU(3) \to X^2$ bundle obtained as a restriction of $SU(3) \to X^4$ bundle. The action of V on rows of U is ordinary unitary transformation plus a 2-dimensional unitary transformation preserving the Hermitian degenerate lengths $L_i = 9|U_{i2}|^2 + 60|U_{i3}|^2 = n_i$ defining the sub-bundle $SU(3) \to X^2$. Note for $L_1 = 0$ ($L_2 = 0$) the situation becomes 2-dimensional and solutions correspond to points in S^2. Thus these points seem to represent a conical singularity of X^2.

The 2-dimensionality of the solution space means that two moduli (probabilities) of any row or column of U or D matrix characterize the matrix apart from the non-uniqueness due to the gauge choice allowing $U(1)_L \times U(1)_R$ transformation of U. Of course, discrete sign degeneracy might be present.

A highly non-trivial problem is whether the set X^2 contains rational points and what is the number of these points. For instance, Fermat's theorem says that no rational solutions to the equation $x^n + y^n - z^n = 0$ exist for $n > 2$. The fact that the degenerate situation allows infinite number of rational solutions suggest that they exist also in the general case. Note also that the additional conditions are second order polynomial equations with rational coefficients so that $SU(3, Q)$ should contain non-trivial solutions to the equations.

It is possible to write $|U|$ in a form containing minimal number of square roots:

$$ \begin{aligned} |U_{11}| &= \sqrt{n_u}\,\frac{p_1}{N_1} \;, & |U_{12}| &= \sqrt{\frac{n_u}{9}}\,\frac{r_1}{N_1} \;, & |U_{13}| &= \sqrt{\frac{n_u}{60}}\,\frac{s_1}{N_1} \;, \\ |U_{21}| &= \sqrt{n_c}\,\frac{p_2}{N_2} \;, & |U_{22}| &= \sqrt{\frac{n_c}{9}}\,\frac{r_2}{N_2} \;, & |U_{23}| &= \sqrt{\frac{n_c}{60}}\,\frac{s_2}{N_2} \;, \\ |U_{31}| &= \sqrt{n_t}\,\frac{p_3}{N_3} \;, & |U_{32}| &= \sqrt{\frac{n_t}{9}}\,\frac{r_3}{N_3} \;, & |U_{23}| &= \sqrt{\frac{n_t}{60}}\,\frac{s_3}{N_3} \;. \end{aligned} \qquad (13.4.1) $$

Completely analogous expression holds true for D. r_i, s_i and N_i are integers, and the defining equations reduce in both cases to equations generalizing those satisfied by Pythagorean triangles

$$
\begin{aligned}
r_1^2 + s_1^2 &= N_1^2 \ , \\
r_2^2 + s_2^2 &= N_2^2 \ , \\
r_3^2 + s_3^2 &= N_3^2 \ .
\end{aligned}
\tag{13.4.0}
$$

The square roots of n_i are also eliminated from the unitarity conditions which become equations with rational coefficients for the phases appearing in U and D. Hence there are good hopes that even rational solutions to the conditions exist.

13.4.2 Number theoretic conditions on U and D matrices

The most stringent requirement would be that U and D matrices are rational unitary matrices. A less stringent condition is that only the moduli squared of U and D are rational numbers. p-Adicization allows also matrices for which various phases are products of Pythagorean phases with phases in an an extension of rational numbers defining a finite-dimensional extension of p-adic numbers. The number theoretic conditions following from the rational unitarity on the moduli of the U and D matrices are not completely independent of the parametrization used. The reason is that the products of the parameters in some algebraic extension of the rationals can combine to give a rational number. The safest parametrization to use is the one based on the moduli of the U and D matrix.

Assuming rationality for the mixing matrix all moduli can be written in the form

$$
|D_{ij}| = \frac{n_{ij}}{N} \ .
\tag{13.4.1}
$$

If only moduli squared are required to be rational, the condition is replaced with a milder one:

$$
|D_{ij}| = \frac{n_{ij}}{\sqrt{N}} \ .
\tag{13.4.2}
$$

Here $\sqrt{N}$ belongs to square root allowing algebraic extension of the p-adic numbers but is not an integer itself. An even milder condition is

$$
|D_{ij}| = \sqrt{\frac{n_{ij}}{N}} \ .
\tag{13.4.3}
$$

The following arguments show that only this option is allowed. This option is also natural in light or preceding general considerations.

1. Unitary and mass conditions modulo 8

For $p_{ij} = (\sqrt{\frac{n_{ij}}{N}})^k$, $k = 1$ or 2, the requirement that the rows are unit vectors implies

$$
\begin{aligned}
\sum_j n_{i,j}^k &= N^k \ , \\
k &= 1 \text{ or } 2 \ .
\end{aligned}
\tag{13.4.3}
$$

The problem of finding vectors with integer valued components and with a given integer valued length squared m ($k = 2$ case) is a well known and well understood problem of the number theory [cd3]. The basic idea is to write the conditions modulo 8 and use the fact that the square of odd (even) integer is 1 (0 or 4) modulo 8. The result is that one must have

$$
m \in \{1, 2, 3, 5, 6\} \ ,
\tag{13.4.4}
$$

for the conditions to possess nontrivial solutions. For $m = N$ case this is the only condition needed. In $m = N^2$ case the condition implies that N must be odd.

Using this result one can write the mass squared conditions modulo 8 for $k = 2$ as

$$
\begin{aligned}
3n_{i,2}^2 + 4n_{i,3}^2 &= n_i X \;, \\
X &= 1 \text{ for } m = N^2 \;, \\
X &\in \{1, 2, 3, 5, 6\} \text{ for } m = N \;.
\end{aligned}
\tag{13.4.3}
$$

Here modulo 8 arithmetics is understood. In $m = N^2$ case one must have $n_i \in \{0, 3, 4\}$ modulo 8. These conditions are not satisfied in general. For $m = N$ conditions allow considerably more general set of solutions. By summing the equations and using probability conservation one however obtains $7N = 5N$ implying $2N = 0$ so that the non-allowed value $N = 4$ or 0 results.

For $k = 1$ no obvious conditions result on the values of n_i and only this option is allowed by mass conditions for the physical masses.

2. Rational unitarity cannot hold true for U and D matrices separately

The mixing scenario is not consistent with the assumption that the matrix elements of U and D matrix are complex rational numbers. If this were the case then matrix elements had to be proportional to a common denominator $1/N$ such that N is odd integer (otherwise the conditions stating that the unit vector property of the rows is not satisfied). The conditions

$$
\begin{aligned}
\sum_j r_{ij} &= 1 \;, \\
9r_{12} + 60r_{13} &= n_d \;, \\
9r_{22} + 60r_{23} &= n_s \;, \\
9r_{32} + 60r_{33} &= n_b \;, \\
r_{ij} &= \frac{n_{ij}}{N_i} \;,
\end{aligned}
$$

$$\tag{13.4.-1}$$

can be written modulo 8 as

$$
\begin{aligned}
\sum_j n_{ij}^k &= N^k \;, \\
n_{12}^k + 4n_{13}^k &= n_d N^k \;, \\
n_{22}^k + 4n_{23}^k &= n_s N^k \;, \\
n_{32}^k + 4n_{33}^k &= n_b N^k \;, \\
r_{ij} &= (\frac{n_{ij}}{N})^{k/2} \;, \quad k = 1 \text{ or } 2 \;.
\end{aligned}
$$

$$\tag{13.4.-5}$$

a) Consider first the case $k = 2$. For odd n $n^2 = 1$ holds true and for even n $n^2 = 4$ or 0 holds true. It is easy to see that the conditions can be satisfied only of all integers are proportional to 4 but this cannot be possible since it would be possible since n_{ij} an N cannot contain common factors. Thus at least an extension allowing square roots is needed. Quite generally from $N^2 = 1 \ mod \ 8$ the above equations give

$$
n_{q_i} \ mod \ 8 \in \{0, 3, 4, 7\} \;.
$$

This condition fails to be satisfied by in the general case.

b) For the option $k = 1$ for which only the probabilities are rational the sum of all three equations gives $5N = 5N$ so that equations are consistent.

3. Phase factors

The phase factors associated with the rows of the mixing matrix are rational provided the corresponding angles correspond to Pythagorean triangles. Combining this property with the orthogonality conditions for the rows of the U matrix, one obtains highly nontrivial conditions relating the integers characterizing the sides of the Pythagorean triangle to the integers n_{ij}. The requirement that the imaginary parts of the inner product vanish, gives the conditions

$$\frac{s_{i,2}}{s_{i3}} = \frac{n_{13}n_{i3}}{n_{12}n_{22}} \ , \quad i = 2,3 \ . \tag{13.4.-4}$$

Combining this conditions with the general representation for the sines of the Pythagorean triangle

$$sin(\phi) = \frac{2rs}{r^2 + s^2} \ \text{or} \ \frac{r^2 - s^2}{r^2 + s^2} \ , \tag{13.4.-3}$$

one obtains conditions relating the integers appearing characterizing the triangle to the integers on the right hand side.

An interesting possibility is that the lengths of the hypothenusae of the triangles associated with $s(i,2)$ $((r(i), s(i)))$ and s_{i3} $((r_1(i), s_1(i)))$ are the same and sines correspond to the products $2rs$:

$$\begin{aligned}
r^2(i) + s^2(i) &= r_1^2(i) + s_1^2(i) \ , \\
s_{i,2} &= 2r(i)s(i)/(r^2(i) + s^2(i)) \ , \\
s_{i,3} &= 2r_1(i)s_1(i)/(r_1^2(i) + s_1^2(i)) \ .
\end{aligned} \tag{13.4.-4}$$

In this case the conditions give

$$\frac{r(i)s(i)}{r_1(i)s_1(i)} = \frac{n_{13}n_{i3}}{n_{12}n_{22}} \ . \tag{13.4.-3}$$

The conditions are satisfied if one has

$$\begin{aligned}
r(i)s(i) &= n_{13}n_{i3} \ , \\
r_1(i)s_1(i) &= n_{12}n_{22} \ .
\end{aligned} \tag{13.4.-3}$$

This implies that $r(i)$ and $s(i)$ are products of the factors contained in the product $n_{13}n_{i3}$. Analogous conclusion applies to $r_1(i)$ and $s_1(i)$.

Additional number theoretic conditions are obtained from the requirement that the real parts of the inner products between first row and second and third rows vanish:

$$n_{11}n_{i1} + c_{i,2}n_{12}n_{i2} + c_{i,3}n_{13}n_{i3} = 0 \ , \quad i = 2,3 \ . \tag{13.4.-2}$$

13.4.3　The parametrization suggested by the mass squared conditions

To understand the consequences of the mass squared conditions, it is useful to use a parametrization, which is more natural for the treatment of the mass squared conditions than the standard parametrization:

$$U = \begin{bmatrix} r_{11} & r_{12} & r_{13} \\ r_{21}x_2 & r_{22}x_2exp(i\phi_{22}) & r_{23}x_2exp(i\phi_{23}) \\ r_{31}x_3 & r_{32}x_3exp(i\phi_{32}) & r_{33}x_3exp(i\phi_{33}) \end{bmatrix} \tag{13.4.-1}$$

$$\begin{aligned}
x_2 &= exp(i\phi_2) \ , \\
x_3 &= exp(i\phi_3) \ .
\end{aligned}$$

In case of D matrix, the phase factors x_2 and x_3 can be chosen to be trivial. As far as the treatment of the mass conditions and unitarity conditions for the rows is considered, one can restrict the consideration to the case, when the overall phase factors are trivial. The remaining parameters are not independent and one on can deduce the formulas relating the moduli r_{ij} as well as the phase angles ϕ_{ij} to the parameters r_{11} and r_{12}. In general, the resulting parameters are not real and unitarity is broken.

Mass squared conditions and the requirement that the rows are unit vectors:

$$9r_{i2}^2 + 60r_{i3}^2 = n_i \ , \ i = 1, 2 \ ,$$
$$\sum_k r_{ik}^2 = 1 \ , \tag{13.4.-1}$$

allows one to express r_{i2} and r_{i3} in terms of r_{i1}

$$r_{i2} = \sqrt{[-\frac{n_i}{51} + \frac{20}{17}(1 - r_{i1}^2)]} \ ,$$
$$r_{i3} = \sqrt{[\frac{n_i}{51} - \frac{3}{17}(1 - r_{i1}^2)]} \ . \tag{13.4.-1}$$

The requirement that the rows are orthogonal to each other, relates the phase angles ϕ_{ij} in terms to r_{11} and r_{21}. Using the notations $sin(\phi_{ij}) = s_{ij}$ and $cos(\phi_{ij}) = c_{ij}$, one has

$$
\begin{aligned}
c_{i2} &= \frac{a_i}{b_i} \ , & c_{i3} &= -\frac{(A_{1i} + c_{i2}A_{2i})}{A_{3i}} \ , \\
s_{i2} &= \epsilon(i)\sqrt{1 - c_{i2}^2} \ , & s_{i3} &= -\frac{A_{2i}}{A_{3i}}s_{i2} \ , \\
A_{1i} &= r_{11}r_{i1} \ , & A_{2i} &= r_{12}r_{i2} \\
A_{3i} &= r_{13}r_{i3} \ , & \epsilon(i) &= \pm 1 \ . \\
a_i &= A_{3i}^2 - A_{1i}^2 - A_{2i}^2 \ , & b_i &= 2A_{1i}A_{2i} \ ,
\end{aligned}
\tag{13.4.0}
$$

The sign factors $\epsilon(i)$ are not completely free and must be chosen so that the second and third row are orthogonal.

The mass conditions imply the following bounds for the parameters r_{i1}

$$
\begin{aligned}
m_i &\leq \ r_{i1} \leq M_i \ , \\
m_i &= \sqrt{1 - \frac{n_i}{9}} \text{ for } n_i \leq 9 \ , \\
m_i &= 0 \text{ for } n_i \geq 9 \ , \\
M_i &= \sqrt{1 - \frac{n_i}{60}} \ .
\end{aligned}
\tag{13.4.-2}
$$

The boundaries for the regions of the solution manifold in (r_{11}, r_{21}) plane can be understood as follows. For given values of r_{11} and r_{21} there are in general two solutions corresponding to the sign factor $\epsilon(i)$ appearing in the equations defining the solutions of the mass squared conditions. This means just that complex conjugation gives a new solution from a given one. These two branches become degenerate, when the phase factors become ± 1 so that (s_{i2}, s_{i3}) vanishes for $i = 2$ or $i = 3$. Thus the curves at which one has $(s_{i2} = 0, s_{i3} = 0)$ define the boundaries of the projection of the solution manifold to (r_{11}, r_{21}) plane. At the boundaries the orthogonality conditions reduce to the form

$$r_{11}r_{i1} + \epsilon(i, 2)r_{12}r_{i2} + \epsilon_{i3}r_{13}r_{i3} = 0 \ , \ i = 2 \text{ or } 3 \ ,$$
$$\epsilon_{22} = \epsilon_{32} \ , \tag{13.4.-1}$$
$$\epsilon_{23} = -\epsilon_{33}$$

where ϵ_{ij} corresponds to the value of the cosine of the phase angle in question. Consistency requires that either second or third row becomes real on the boundary of the unitarity region and that the matrices reduce to orthogonal matrices at the dip of the region allowed by unitarity.

13.4.4　Thermodynamical model for the topological mixing

What would be needed is a physical model for the topological mixing allowing to deduce U and D matrices from first principles. The physical mechanism behind the mixing is change of the topology of X^2 in the dynamical evolution defined by the light like 2-surface X_l^3 defining parton orbit. This suggests that the topology changes $g \to g \pm 1$ dominate the dynamics so that matrix elements U_{13} and D_{13} should be indeed small so that the weird looking result $P_{11}^U \simeq 1/3$ follows from the requirement $n_u = 6$. This model however suggests that the matrix elements U_{23} and D_{23} could be large unlike in the original model for U and D matrices.

Solution of thermodynamical model

A possible approach to the construction of mixing matrices is based on the idea that the interactions causing the mixing lead to a thermal equilibrium so that the entropies for the ensemble defined by the probabilities p_{ij}^U and p_{ij}^D matrix is maximized (the subscripts U and D are dropped in the sequel).

a) The elements in the three rows of the mixing matrix represent probabilities for three states of the system with energies $(E_{i1}, E_{i2}, E_{i3}) = (0, 9, 60)$ and average energy is fixed to $\langle E \rangle = 69$.

b) There are usual constraints from probability conservation for each row plus two independent constraints from columns. The latter constraints can be regarded as a constraint on a second quantity equal to 1 for each column and brings in variable analogous to chemical potential besides temperature.

The constraint from mass squared for the third row follows from these constraints. The independent constraints can be chosen to be the following ones

$$\sum_j p_{ij} - 1 = 0 \ , \quad i = 1,2,3 \qquad\qquad \sum_i p_{ij} - 1 = 0, \quad j = 1,2 \ ,$$

$$9p_{i2} + 60p_{i3} - n_{q_i} = 0 \ , \quad i = 1,2 \ . \tag{13.4.0}$$

The obvious notations $(q_1, q_2) = (d, s)$ and $(q_1, 2_2) = (u, c)$ are introduced. The conditions on mass squared are completely analogous to the conditions fixing the energy of the ensemble and thus its temperature, and thermodynamical intuition suggests that the probabilities p_{ij} decrease exponentially as function of E_j in the absence of additional constraints coming from the probability conservation for the columns and meaning presence of chemical potential.

The variational principle maximizing entropy in presence of these constraints can be expressed as

$$
\begin{aligned}
L \ &= \ S + S_c \\
S \ &= \ \sum_{i,j} p_{ij} \times log(p_{ij}) \\
S_c \ &= \ \sum_i \lambda_i \left(\sum_j p_{ij} - 1 \right) + \sum_{j=1,2} \mu_j \left(\sum_i p_{ij} - 1 \right) + \sum_{i=1,2} \sigma_i (9p_{i2} + 60p_{i3} - n_{q_i}) \ .
\end{aligned}
\tag{13.4.-2}
$$

The variational equation is

$$\partial_{p_{ij}} L \ = \ 0 \ , \tag{13.4.-1}$$

and gives the probabilities as

$$
\begin{aligned}
p_{11} &= \tfrac{1}{Z_1} \ , & p_{12} &= \tfrac{x x_1^3}{Z_1} & p_{13} &= \tfrac{y x_1^{20}}{Z_1} \ , \\
p_{21} &= \tfrac{1}{Z_2} \ , & p_{22} &= \tfrac{x x_2^3}{Z_2} & p_{13} &= \tfrac{y x_2^{20}}{Z_2} \ , \\
p_{31} &= \tfrac{1}{Z_3} \ , & p_{32} &= \tfrac{x}{Z_3} & p_{33} &= \tfrac{y}{Z_3} \ ,
\end{aligned}
\tag{13.4.-1}
$$

Here the parameters x, y, x_1, x_2 are defined as

$$x = exp(-\mu_2) \ , \qquad y = exp(-\mu_3) \ ,$$
$$x_1 = exp(-3\sigma_1) \ , \qquad x_2 = exp(-3\sigma_2) \ .$$

$$(13.4.\text{-}1)$$

whereas the row partition functions Z_i are defined as

$$Z_1 = 1 + xx_1^3 + yx_1^{20} \ , \quad Z_2 = 1 + xx_2^3 + yx_2^{20} \ , \quad Z_3 = 1 + x + y \ . \tag{13.4.0}$$

Note that the parameters λ_i have been eliminated. There are four parameters $\mu_2, \mu_3, \sigma_2, \sigma_3$ and 2 conditions from columns and 2 mass conditions so that the number of solutions is discrete and only finite number of U and D matrices are possible in the thermodynamical approximation.

Mass squared conditions

The mass squared conditions read as

$$9xx_1^3 + 60yx_1^{20} = n(q_1)Z_1 \ , \quad 9xx_2^3 + 60yx_2^{20} = n(q_2)Z_2 \ . \tag{13.4.1}$$

These equations allow to solve y as a simple linear function of x

$$y = \frac{n(q_1) - xx_1^3(9 - n(q_1))}{(60 - n(q_1))x_1^{20}} \equiv kx + l \ , \quad y = \frac{n(q_2) - xx_2^3(9 - n(q_2))}{(60 - n(q_2)x_2^{20}} \ . \tag{13.4.2}$$

The identification of the two expressions for y allows to solve x_1 in terms of x_2 using equation of form $x_1^{20} - bx_1^3 + c = 0$:

$$\begin{aligned}
&\left[60 - n(q_2)x_2^{20}\right]\left[n(q_1) - xx_1^3(9 - n(q_1))\right] \\
= \ &\left[60 - n(q_1))x_1^{20}\right]\left[n(q_2) - xx_2^3(9 - n(q_2))\right] \ .
\end{aligned} \tag{13.4.2}$$

In the most general case the equation allows 20 roots $x_1 = x_2(x_1)$.

Probability conservation

Probability conditions give additional information. By solving $1/Z_3$ from the first column gives

$$Z_1 Z_2 Z_3 - Z_1 Z_2 - Z_2 Z_3 - Z_1 Z_3 \ = \ 0 \ , \tag{13.4.3}$$

$$\tag{13.4.4}$$

This equation is a polynomial equation for in x_1 and x_2 with degree 20 and together with Eq. 13.4.2 having same degree determines and (x_1, x_2) the possible values of x_1 and x_2 as function of x. The number of real positive roots is at most $20^2 = 400$.

Probability conservation for the second column gives

$$x\left[(1 - x_1^3)Z_2 + (1 - x_2^3)Z_1\right] + (1 - x)Z_1 Z_2 = 0 \ . \tag{13.4.5}$$

The row partition functions Z_i are linear functions of x and y and mass squared conditions give $y = kx + l$ (see Eq. 13.4.2) so that a third order polynomial equation for x results and gives the roots as functions of control parameters x_1 and x_2. Either 1 or 3 real roots are obtained for x. The values of x_1 and x_2 are determined by the probability constraint Eq. 13.4.4 for the first column and Eq. 13.4.2 relating x_1 and x_2.

The analogy with spontaneous magnetization

Physically the situation is analogous to a spontaneous symmetry breaking with y representing the external magnetizing field and x linear magnetization or vice versa. x_1 and x_2 are control parameters characterizing the interaction between spins. For single real root for x no spontaneous magnetization occurs but for 3 real roots there are two directions of spontaneous magnetization plus unstable state. In the recent case the roots must be positive. Since the maximal number of roots for (x_1, x_2) is 400, the maximal number of real roots is 1200. The trivial solution to the conditions is $p_{11} = 1, p_{22} = 1$, $p_{33} = 1$ with $x = y = 0$ represents corresponds to the absence of external magnetizing field and of magnetization.

Catastrophe theoretic description of the system

In the catastrophe theoretic approach one can see that situation as a cusp catastrophe with x as a behavior variable and x_1, x_2 in the role of control variables. In the standard parametrization of the cusp catastrophe [aa4] the conditions correspond to the equation

$$x^3 - a - bx \; = \; 0 \; ,$$

$$(13.4.5)$$

In the recent case a more general polynomial $P_3(x)$ easily transformable to the standard form is in question. The coefficients of the polynomial $P_3(x) = Dx^3 + Cx^2 + Bx + A$ are

$$
\begin{aligned}
A &= Q(x_1)Q(x_2) \; , \\
B &= P(x_1)Q(x_2) + P(x_2)Q(x_1) + R(x_2) + R(x_1) \; , \\
C &= P(x_1)R(x_2) + P(x_2)R(x_1) - R(x_1)Q(x_2) - R(x_2)Q(x_1) \; , \\
D &= R(x_1)R(x_2) \; , \\
P(u) &- 1 \quad u^3 \; , \quad Q(u) = 1 + lu^{20} \; , \quad R(u) = u^3 + ku^{20} \; .
\end{aligned}
$$

$$(13.4.2)$$

The trivial scaling transformation $A \to A/D = \hat{A}$, $B \to B/D = \hat{B}$, $C \to C/D = \hat{C}$ and the shift $x \to x + \hat{C}/3$ casts the equation in the standard form and gives

$$
\begin{aligned}
a &= -\hat{A} + \frac{\hat{C}^3}{9} \; , \\
b &= -\hat{B} + \frac{\hat{C}^2}{3} \; .
\end{aligned}
$$

$$(13.4.1)$$

The curve

$$a \; = \; \pm 2\left(\frac{b}{3}\right)^{3/2} \; , b \geq 0 \qquad (13.4.2)$$

represents the bifurcation set for the solutions. For $b \geq 0$, $|a| \leq \left(\frac{b}{3}\right)^{3/2}$ three roots are obtained for x. $a = b = 0$ corresponds to the dip of the cusp. Three solutions result under the conditions

$$\frac{\hat{C}^2}{3} \geq 3\hat{B} \ ,$$

$$(-\hat{B} + \frac{\hat{C}^2}{3})^3 \leq \frac{(-\hat{A} + \frac{\hat{C}^3}{9})^2}{4} \ ,$$

$$\hat{A} = \frac{Q(x_1)Q(x_2)}{R(x_1)R(x_2)} \ ,$$

$$\hat{B} = \frac{P(x_1)Q(x_2) + P(x_2)Q(x_1) + R(x_2) + R(x_1)}{R(x_1)R(x_2)} \ ,$$

$$\hat{C} = \frac{P(x_1)R(x_2) + P(x_2)R(x_1) - R(x_1)Q(x_2) - R(x_2)Q(x_1)}{R(x_1)R(x_2)} \ ,$$

$$P(u) = 1 - u^3 \ , \quad Q(u) = 1 + lu^{20} \ , \quad R(u) = u^3 + ku^{20} \ . \tag{13.4.-2}$$

The boundaries of the regions are defined by polynomial equations for x_1 and x_2. . The two mass squared conditions and the probability conservation for the first row select a discrete set of parameter combinations.

One might ask whether U and D matrices could correspond to different solutions of these equations for same values of n_{q_i}. This cannot be the case since this would predict too large $u - d$ mass difference. Orthogonalization conditions for the rows should determine the phases more or less uniquely and could force CP breaking. The requirement that probabilities are rational valued implies that x_1, x_2, x and y are rational and poses very strong additional conditions to the solutions. The roots should correspond to very special solutions possessing symmetries so that the solutions of polynomial equations give probabilities as rational numbers. Note however that the solutions of polynomial equations with integer coefficients are in question and the solutions are algebraic numbers: this is enough as far as the p-adicization of the theory is considered.

Maximization of entropy solving constraint equations explicitly

The mass squared conditions allow to express the probabilities p_{ij} in terms of p_{11} and p_{21} (for instance) and this allows a rather concise representation for the solution to the maximization the entropy of topological mixing. The key formulas are following.

$$p_{31} = 1 - p_{11} - p_{12} \ ,$$

$$p_{i2} = -\frac{n_i}{51} + \frac{20}{17}(1 - p_{i1}) \ , \quad i = 1,2 \ ,$$

$$p_{i3} = \frac{n_i}{51} - \frac{3}{17}(1 - p_{i1}) \ , \quad i = 1,2 \ . \tag{13.4.-3}$$

Expressing entropy directly in terms of p_{11} and p_{21}, the conditions for the maximization of entropy imply the equations

$$log(p_{ij})X^{ij} = 0 \ , \quad log(p_{ij})Y^{ij} = 0 \ , \tag{13.4.-2}$$

where a summation over repeated indices is carried out. The matrices $X^{ij} = \partial_{p_{11}} p_{ij}$ and $Y^{ij} = \partial_{p_{21}} p_{ij}$ are given by

$$X = \begin{pmatrix} 1 & -\frac{20}{17} & \frac{3}{17} \\ 0 & 0 & 0 \\ -1 & \frac{20}{17} & -\frac{3}{17} \end{pmatrix}$$

$$Y = \begin{pmatrix} 0 & 0 & 0 \\ 1 & -\frac{20}{17} & \frac{3}{17} \\ -1 & \frac{20}{17} & -\frac{3}{17} \end{pmatrix}$$

$$\tag{13.4.-3}$$

The equations can be transformed into the form

$$\prod_{ij} p_{ij}^{X_{ij}} = 1 \ , \quad \prod_{ij} p_{ij}^{Y_{ij}} = 1 \ . \tag{13.4.-2}$$

When written explicitly, these equations read as

$$\frac{p_{11}}{1-p_{11}-p_{21}} \times \left(\frac{-n_1+60(1-p_{11})}{-n_3+60(p_{11}+p_{21})}\right)^{-20/17} \times \left(\frac{n_1-9(1-p_{11})}{n_3-9(p_{11}+p_{21})}\right)^{3/17} = 1 \ ,$$

$$\frac{p_{21}}{1-p_{11}-p_{21}} \times \left(\frac{-n_2+60(1-p_{21})}{-n_3+60(p_{11}+p_{21})}\right)^{-20/17} \times \left(\frac{n_2-9(1-p_{21})}{n_3-9(p_{11}+p_{21})}\right)^{3/17} = 1 \ .$$

$$\tag{13.4.-2}$$

The equations can be cast into polynomial equations in p_{11} and p_{21} by taking 17:th power of both equations. This gives polynomial equations of degree $d = 17 + 20 + 3 = 40$. The total number of solutions to the equations is at most $40 \times 40 = 1600$. The previous estimate gave upper bound $3 \times 20 \times 20 = 1200$ for the number of solution. It might be that some symmetry is involved and reduces the upper bound by a factor $3/4$.

The solutions can be sought using gradient dynamics in which system in (p_{11}, p_{21}) plane drifts in the force field defined by the gradient ∇S of the entropy $S = -\sum_{ij} p_{ij} log(p_i j)$ and ends up to the maximum of S, $S = -\sum_{ij} p_{ij} log(p_i j)$.

$$\frac{dp_{11}}{dt} = \partial_{p_{11}} S = -X^{ij} log(p_{ij}) \ ,$$
$$\frac{dp_{21}}{dt} = \partial_{p_{21}} S = -Y^{ij} log(p_{ij}) \ ,$$

$$\tag{13.4.-2}$$

The conditions that the probabilities are positive give the constraints

$$1 - \frac{n_1}{9} \ \leq \ p_{11} \leq 1 - \frac{n_1}{60} \ ,$$
$$1 - \frac{n_2}{9} \ \leq \ p_{21} \leq 1 - \frac{n_2}{60} \ ,$$
$$0 \ \leq \ p_{21} \leq 1 - p_{11} \ ,$$
$$\frac{69 - n_1 - n_2}{60} - p_{11} \ \leq \ p_{21} \leq \frac{69 - n_1 - n_2}{9} - p_{11}$$

$$\tag{13.4.-5}$$

on the region containing the solutions.

13.4.5 U and D matrices from the knowledge of top quark mass alone?

As already found, a possible resolution to the problems created by top quark is based on the additivity of mass squared so that top quark mass would be about 230 GeV, which indeed corresponds to a peak in mass distribution of top candidate, whereas $t\bar{t}$ meson mass would be 163 GeV. This requires that top quark mass changes very little in topological mixing. It is easy to see that the mass constraints imply that for $n_t = n_b = 60$ the smallness of V_{i3} and $V(3i)$ matrix elements implies that both U and D must be direct sums of 2×2 matrix and 1×1 unit matrix and that V matrix would have also similar decomposition. Therefore $n_b = n_t = 59$ seems to be the only number theoretically acceptable option. The comparison with the predictions with pion mass led to a unique identification $(n_d, n_b, n_b) = (5, 5, 59),(n_u, n_c, n_t) = (4, 6, 59)$.

U and D matrices as perturbations of matrices mixing only the first two genera

This picture suggests that U and D matrices could be seen as small perturbations of very simple U and D matrices satisfying $|U| = |D|$ corresponding to $n = 60$ and having $(n_d, n_b, n_b) = (4, 5, 60)$, $(n_u, n_c, n_t) = (4, 5, 60)$ predicting V matrix characterized by Cabibbo angle alone. For instance, CP breaking parameter would characterize this perturbation. The perturbed matrices should obey thermodynamical constraints and it could be possible to linearize the thermodynamical conditions and in this manner to predict realistic mixing matrices from first principles. The existence of small perturbations yielding acceptable matrices implies also that these matrices be near a point at which two different matrices resulting as a solution to the thermodynamical conditions coincide.

D matrix can be deduced from U matrix since $9|D_{12}|^2 \simeq n_d$ fixes the value of the relative phase of the two terms in the expression of D_{12}.

$$
\begin{aligned}
|D_{12}|^2 &= |U_{11}V_{12} + U_{12}V_{22}|^2 \\
&= |U_{11}|^2|V_{12}|^2 + |U_{12}|^2|V_{22}|^2 \\
&+ 2|U_{11}||V_{12}||U_{12}||V_{22}|cos(\Psi) = \frac{n_d}{9} \ , \\
\Psi &= arg(U_{11}) + arg(V_{12}) - arg(U_{12}) - arg(V_{22}) \ .
\end{aligned}
$$

$$(13.4.\text{-}8)$$

Using the values of the moduli of U_{ij} and the approximation $|V_{22}| = 1$ this gives for $cos(\Psi)$

$$
\begin{aligned}
cos(\Psi) &= \frac{A}{B} \ , \\
A &= \frac{n_d - n_u}{9} - \frac{9 - n_u}{9}|V_{12}|^2 \ , \\
B &= \frac{2}{9|V_{12}|}\sqrt{n_u(9 - n_u)} \ .
\end{aligned}
$$

$$(13.4.\text{-}9)$$

The experimentation with different values of n_d and n_u shows that $n_u = 6, n_d = 4$ gives $cos(\Psi) = -1.123$. Of course, $n_u = 6, n_d = 4$ option is not even allowed by $n_t = 60$. For $n_d = 4, n_u = 5$ one has $cos(\Psi) = -0.5958$. $n_d = 5, n_u = 6$ corresponding to the perturbed solution gives $cos(\Psi) = -0.6014$.

Hence the initial situation could be $(n_u = 5, n_s = 4, n_b = 60)$, $(n_d = 4, n_s = 5, n_t = 60)$ and the physical U and D matrices result from U and D matrices by a small perturbation as one unit of t (b) mass squared is transferred to u (s) quark and produces symmetry breaking as $(n_d = 5, n_s = t, n_b = 59)$, $(n_u = 6, n_c = 4, n_t = 59)$.

The unperturbed matrices $|U|$ and $|D|$ would be identical with $|U|$ given by

$$|U_{11}| = |U_{22}| = \tfrac{2}{3} \ , \quad |U_{12}| = |U_{21}| = \tfrac{\sqrt{5}}{3} \ , \tag{13.4.-8}$$

The thermodynamical model allows solutions reducing to a direct sum of 2×2 and 1×1 matrices, and since $|U|$ matrix is fixed completely by the mass constraints, it is trivially consistent with the thermodynamical model.

Direct search of U and D matrices

The general formulas for p^U and p^D in terms of the probabilities p_{11} and p_{21} allow straightforward search for the probability matrices having maximum entropy just by scanning the (p_{11}, p_{21}) plane constrained by the conditions that all probabilities are positive and smaller than 1. In the physically interesting case the solution is sought near a solution for which the non-vanishing probabilities are $p_{11} = p_{22} = (9 - n_1)/9$, $p_{12} = p_{21} = n_1/9$, $p_{33} = 1$, $n_1 = 4$ or 5. The inequalities allow to consider only the values $p_{11} \geq (9 - n_1)/9$.

  1. Probability matrices p^U and p^D

The direct search leads to maximally entropic p^D matrix with $(n_d, n_s) = (5, 5)$:

$$
p^D = \begin{pmatrix} 0.4982 & 0.4923 & 0.0095 \\ 0.4981 & 0.4924 & 0.0095 \\ 0.0037 & 0.0153 & 0.9810 \end{pmatrix} \ , \quad
p_0^D = \begin{pmatrix} 0.5556 & 0.4444 & 0 \\ 0.4444 & 0.5556 & 0 \\ 0 & 0 & 1 \end{pmatrix} \ .
$$

$$(13.4.\text{-}8)$$

p_0^D represents the unperturbed matrix p_0^D with $n(d = 4), n_s = 5$ and is included for the purpose of comparison. The entropy $S(p^D) = 1.5603$ is larger than the entropy $S(p_0^D) = 1.3739$. A possible interpretation is in terms of the spontaneous symmetry breaking induced by entropy maximization in presence of constraints.

A maximally entropic p^U matrix with $(n_u, n_c) = (5, 6)$ is given by

$$p^U = \begin{pmatrix} 0.5137 & 0.4741 & 0.0122 \\ 0.4775 & 0.4970 & 0.0254 \\ 0.0088 & 0.0289 & 0.9623 \end{pmatrix}$$

$$(13.4.\text{-}8)$$

The value of entropy is $S(p^U) = 1.7246$. There could be also other maxima of entropy but in the range covering almost completely the allowed range of the parameters and in the accuracy used only single maximum appears.

The probabilities p_{ii}^D resp. p_{ii}^U satisfy the constraint $p(i, i) \geq .492$ resp. $p_{ii} \geq .497$ so that the earlier proposal for the solution of proton spin crisis must be given up and the solution discussed in [D2] remains the proposal in TGD framework.

2. Near orthogonality of U and D matrices

An interesting question whether U and D matrices can be transformed to approximately orthogonal matrices by a suitable $(U(1) \times U(1))_L \times (U(1) \times U(1))_R$ transformation and whether CP breaking phase appearing in CKM matrix could reflect the small breaking of orthogonality. If this expectation is correct, it should be possible to construct from $|U|$ ($|D|$) an approximately orthogonal matrix by multiplying the matrix elements $|U_{ij}|$, $i, j \in \{2, 3\}$ by appropriate sign factors. A convenient manner to achieve this is to multiply $|U|$ ($|D|$) in an element wise manner $((A \circ B)_{ij} = A_{ij} B_{ij})$ by a sign factor matrix S.

a) In the case of $|U|$ the matrix $U = S \circ |U|$, $S(2, 2) = S(2, 3) = S(3, 2) = -1$, $S_{ij} = 1$ otherwise, is approximately orthogonal as the fact that the matrix $U^T U$ given by

$$U^T U = \begin{pmatrix} 1.0000 & 0.0006 & -0.0075 \\ 0.0006 & 1.0000 & -0.0038 \\ -0.0075 & -0.0038 & 1.0000 \end{pmatrix}$$

is near unit matrix, demonstrates.

b) For D matrix there are two nearly orthogonal variants. For $D = S \circ |D|$, $S(2, 2) = S(2, 3) = S(3, 2) = -1$, $S_{ij} = 1$ otherwise, one has

$$D^T D = \begin{pmatrix} 1.0000 & -0.0075 & 0.0604 \\ -0.0075 & 1.0000 & 0.0143 \\ 0.0604 & 0.0143 & 1.0000 \end{pmatrix} .$$

The choice $D = S \circ D$, $S(2, 2) = S(2, 3) = S(3, 3) = -1$, $S_{ij} = 1$ otherwise, is slightly better

$$D^T D = \begin{pmatrix} 1.0000 & -0.0075 & 0.0604 \\ -0.0075 & 1.0000 & 0.0143 \\ 0.0601 & 0.0143 & 1.0000 \end{pmatrix} .$$

3. The matrices U and D in the standard gauge

Entropy maximization indeed yields probability matrices associated with unitary matrices. 8 phase factors are possible for the matrix elements but only 4 are relevant as far as the unitarity conditions are considered. The vanishing of the inner products between row vectors, gives 6 conditions altogether so that the system seems to be over-determined. The values of the parameters s_1, s_2, s_3 and phase angle δ in the "standard gauge" can be solved in terms of r_{11} and r_{21}.

The requirement that the norms of the parameters c_i are not larger than unity poses non-trivial constraints on the probability matrices. This should should be the case since the number of unitarity conditions is 9 whereas probability conservation for columns and rows gives only 5 conditions so that not every probability matrix can define unitary matrix. It would seem that that the constraints are satisfied only if the the 2 mass squared conditions and 2 conditions from the entropy maximization are equivalent with 4 unitarity conditions so that the number of conditions becomes 5+4=9. Therefore entropy maximization and mass squared conditions would force the points of complex 9-dimensional space defined by 3×3 matrices to a 9-dimensional surface representing group $U(3)$ so that these conditions would have a group theoretic meaning.

The formulas

$$
\begin{aligned}
r_{i2} &= \sqrt{\left[-\frac{n_i}{51} + \frac{20}{17}(1 - r_{i1}^2)\right]} \ , \\
r_{i3} &= \sqrt{\left[\frac{n_i}{51} - \frac{3}{17}(1 - r_{i1}^2)\right]} \ .
\end{aligned}
\tag{13.4.-8}
$$

and

$$
U = \begin{bmatrix}
c_1 & s_1 c_3 & s_1 s_3 \\
-s_1 c_2 & c_1 c_2 c_3 - s_2 s_3 exp(i\delta) & c_1 c_2 s_3 + s_2 c_3 exp(i\delta) \\
-s_1 s_2 & c_1 s_2 c_3 + c_2 s_3 exp(i\delta) & c_1 s_2 s_3 - c_2 c_3 exp(i\delta)
\end{bmatrix}
\tag{13.4.-7}
$$

give

$$
\begin{aligned}
c_1 = r_{11} \ , \qquad &c_2 = \frac{r_{21}}{\sqrt{1 - r_{11}^2}} \ , \\
s_3 = \frac{r_{13}}{\sqrt{1 - r_{11}^2}} \ , \qquad &cos(\delta) = \frac{c_1^2 c_2^2 c_3^2 + s_2^2 s_3^2 - r_{22}^2}{2 c_1 c_2 c_3 s_2 s_3} \ .
\end{aligned}
\tag{13.4.-6}
$$

Preliminary calculations show that for $n_1 = n_2 = 5$ case the matrix of moduli allows a continuation to a unitary matrix but that for $n_1 = 4, n_2 = 6$ the value of $cos(\delta)$ is larger than one. This would suggest that unitarity indeed gives additional constraints on the integers n_i. The unitary (in the numerical accuracy used) $(n_d, n_s) = (5, 5)$ D matrix is given by

$$
D = \begin{pmatrix}
0.7059 & 0.7016 & 0.0975 \\
-0.7057 & 0.7017 - 0.0106i & 0.0599 + 0.0766i \\
-0.0608 & 0.0005 + 0.1235i & 0.4366 - 0.8890i
\end{pmatrix} \ .
$$

The unitarity of this matrix supports the view that for certain integers n_i the mass squared conditions and entropy maximization reduce to group theoretic conditions. The numerical experimentation shows that the necessary condition for the unitarity is $n_1 > 4$ for $n_2 < 9$ whereas for $n_2 \geq 9$ the unitarity is achieved also for $n_1 = 4$.

Direct search for CKM matrices

The standard gauge in which the first row and first column of unitary matrix are real provides a convenient representation for the topological mixing matrices: it is convenient to refer to these representations as U_0 and D_0. The possibility to multiply the rows of U_0 and D_0 by phase factors $(U(1) \times U(1))_R$ transformations) provides 2 independent phases affecting the values of $|V|$. The phases $exp(i\phi_j)$, $j = 2,3$ multiplying the second and third row of D_0 can be estimated from the matrix elements of $|V|$, say from the elements $|V_{11}| = cos(\theta_c) \equiv v_{11}$, $sin\theta_c = .226 \pm .002$ and $|V_{31}| = (9.6 \pm .9) \cdot 10^{-3} \equiv v_{31}$. Hence the model would predict two parameters of the CKM matrix, say s_3 and δ_{CP}, in its standard representation.

The fact that the existing empirical bounds on the matrix elements of V are based on the standard model physics raises the question about how seriously they should be taken. The possible existence of fractally scaled up versions of light quarks could effectively reduce the matrix elements for the electro-weak decays $b \to c + W$, $b \to u + W$ resp. $t \to s + W$, $t \to d + W$ since the decays involving scaled up versions of light quarks can be counted as decays $W \to bc$ resp. $W \to tb$. This would favor too small experimental estimates for the matrix elements V_{i3} and V_{3i}, $i = 1, 2$. In particular, the matrix element $V_{31} = V_{td}$ could be larger than the accepted value.

Various constraints do not leave much freedom to choose the parameters n_{q_i}. The preliminary numerical experimentation shows that the choice $(n_d, n_s) = (5, 5)$ and $(n_u, n_c) = (5, 6)$ yields realistic

U and D matrices. In particular, the conditions $|U(1,1)| > .7$ and $|D(1,1)| > .7$ hold true and mean that the original proposal for the solution of spin puzzle of proton must be given up. In [D2] an alternative proposal based on more recent findings is discussed. Only for this choice reasonably realistic CKM matrices have been found. For $n_t = 58$ the mass of $t\bar{t}$ meson mass is reduced by one percent from 2×163 GeV for $n(5) = 59$ so that $n_t = 58$ is still acceptable if the additivity of conformal weight rather than mass is accepted for diagonal mesons.

a) The requirement that the parameters $|V_{11}|$ (or equivalently, Cabibbo angle) and $|V_{31}|$ are produced correctly, yields CKM matrices for which CP breaking parameter J is roughly one half of its accepted value. The matrix elements $V_{23} \equiv V_{cb}$, $V_{32} \equiv V_{tc}$, and $V_{13} \equiv V_{ub}$ are roughly twice their accepted value. This suggests that the condition on V_{31} should be loosened.

b) The following tables summarize the results of the search requiring that
i) the value of the Cabibbo angle s_{Cab} is within the experimental limits $s_{Cab} = .223 \pm .002$,
ii) $V_{31} = (9.6 \pm .9) \cdot 10^{-3}$, is allowed to have value at most twice its upper bound,
iii) V_{13} whose upper bound is determined by probability conservation, is within the experimental limits $.42 \cdot 10^{-3} < |V_{ub}| < 6.98 \cdot 10^{-3}$ whereas $V_{23} \simeq 4 \times 10^{-3}$ should come out as a prediction,
iv) the CP breaking parameter satisfies the condition $|(J - J_0)/J_0| < .6$, where $J_0 = 10^{-4}$ represents the lower bound for J (the experimental bounds for J are $J \times 10^4 \in (1 - 1.7)$).

The pairs of the phase angles (ϕ_1, ϕ_2) defining the phases $(exp(i\phi_1), exp(i\phi_2))$ are listed below

$$
class\ 1:\ \begin{matrix} \phi_1 & 0.1005 & 0.1005 & 4.8129 & 4.8129 \\ \phi_2 & 0.0754 & 1.4828 & 4.7878 & 6.1952 \end{matrix}
$$

$$
class\ 2:\ \begin{matrix} \phi_1 & 0.1005 & 0.1005 & 4.8129 & 4.8129 \\ \phi_2 & 2.3122 & 5.5292 & 0.7414 & 3.9584 \end{matrix} \tag{13.4.-6}
$$

The phase angle pairs correspond to two different classes of U, D, and V matrices. The U, D and V matrices inside each class are identical at least up to 11 digits(!). Very probably the phase angle pairs are related by some kind of symmetry.

The values of the fitted parameters for the two classes are given by

$$
\begin{array}{ccccc}
 & |V_{11}| & |V_{31}| & |V_{13}| & J/10^{-4} \\
class\ 1 & 0.9740 & 0.0157 & 0.0069 & .93953 \\
class\ 2 & 0.9740 & 0.0164 & 0.0067 & 1.0267
\end{array}
$$

V_{31} is predicted to be about 1.6 times larger than the experimental upper bound and for both classes V_{23} and V_{32} are roughly too times too large. Otherwise the fit is consistent with the experimental limits for class 2. For class 1 the CP breaking parameter is 7 per cent below the experimental lower bound. In fact, the value of J is fixed already by the constraints on V_{31} and V_{11} and reduces by a factor of one half if V_{31} is required to be within its experimental limits.

U, D and $|V|$ matrices for class 1 are given by

$$
U = \begin{bmatrix} 0.7167 & 0.6885 & 0.1105 \\ -0.6910 & 0.7047 - 0.0210i & 0.0909 + 0.1310i \\ -0.0938 & 0.0696 + 0.1550i & 0.1747 - 0.9653i \end{bmatrix}
$$

$$
D = \begin{bmatrix} 0.7059 & 0.7016 & 0.0975 \\ -0.6347 - 0.3085i & 0.6358 + 0.2972i & 0.0203 + 0.0951i \\ -0.0587 - 0.0159i & -0.0317 + 0.1194i & 0.6534 - 0.7444i \end{bmatrix}
$$

$$
|V| = \begin{bmatrix} 0.9740 & 0.2265 & 0.0069 \\ 0.2261 & 0.9703 & 0.0862 \\ 0.0157 & 0.0850 & 0.9963 \end{bmatrix}
$$

$$\tag{13.4.-8}$$

U, D and $|V|$ matrices for class 2 are given by

$$U = \begin{bmatrix} 0.7167 & 0.6885 & 0.1105 \\ -0.6910 & 0.7047 - 0.0210i & 0.0909 + 0.1310i \\ -0.0938 & 0.0696 + 0.1550i & 0.1747 - 0.9653i \end{bmatrix}$$

$$D = \begin{bmatrix} 0.7059 & 0.7016 & 0.0975 \\ -0.6347 - 0.3085i & 0.6358 + 0.2972i & 0.0203 + 0.0951i \\ -0.0589 - 0.0151i & -0.0302 + 0.1198i & 0.6440 - 0.7525i \end{bmatrix}$$

$$|V| = \begin{bmatrix} 0.9740 & 0.2265 & 0.0067 \\ 0.2260 & 0.9704 & 0.0851 \\ 0.0164 & 0.0838 & 0.9963 \end{bmatrix}$$

$$(13.4.\text{-}10)$$

What raises worries is that the values of $|V_{23}| = |V_{cb}|$ and $|V_{32}| = |V_{ts}|$ are roughly twice their experimental estimates. This, as well as the discrepancy related to V_{31}, might be understood in terms of the electro-weak decays of b and t to scaled up quarks causing a reduction of the branching ratios $b \to c + W$, $t \to s + W$ and $t \to t + d$. The attempts to find more successful integer combinations n_i has failed hitherto. The model for pseudoscalar meson masses, the predicted relatively small masses of light quarks, and the explanation for $t\bar{t}$ meson mass supports this mixing scenario.

13.5 Hadron masses

Besides the quark contributions already discussed, hadron mass squared can contain several other contributions and the task is to find a model allowing to identify and estimate these contributions. There are several guidelines for the numerical experimentation.

a) Conformal weight, that is mass squared, is additive for quarks corresponding to the same p-adic prime. For instance, in case of $q\bar{q}$ mesons the mass would be $\sqrt{2}m(q)$ and the contribution of $k = 113$ u, d, s quarks to nucleon mass would be $< \sqrt{3} \times 100$ MeV and thus surprisingly small. For cd meson quark masses would be additive.

b) Old fashioned quark model explains reasonably well hadron masses in terms of constituent quark masses. Effective 2-dimensionality of partons suggests an interpretation for the constituent quark as a composite structure formed by the current quark identified as a partonic 2-surface X^2 characterized by $k(q)$ and by join along boundaries bond, kind of a gluonic "rubber band" characterized by $k = 107$ and connecting X^2 to the $k = 107$ hadronic 2-surface $X^2(H)$ representing hadron. $X^2(q_i)$ could be perhaps regarded as a hole in $k = k(q)$ 3-surface. The 2-dimensional visualization for a 3-dimensional topological condensation would become much more than a mere visualization. This view about hadrons brings in mind unavoidably the surreal 2-dimensional structures formed by organs like retina. Of course, effective 2-dimensionality allows to characterize the entire Universe as an extremely complex fractal 2-surface.

The large mass of the constituent quark would be due to the color Coulombic and spin-spin interaction conformal weights of join along boundaries bond. Quark mass and the mass due to the color interaction conformal weight would be additive unless $k = 107$ for the quark (it seems that for η' this is indeed the case!). Classical color gauge fluxes would flow between $k = 107$ and $k \neq 107$ space-time sheets along the bonds. Color dynamics would take place at $k = 107$ space-time sheet in the sense that color gauge flux between quarks q_1 and q_2 flows first from $X^2((k(q_1))$ to the hadronic 2-surface $X^2(k = 107)$ and then back to $X^2(k(q_2))$. The induced Kähler field is always accompanied by a classical color gauge field and the classical color gauge flux would represent non-perturbative aspects of color interactions at space-time level.

c) A crucial observation is that the mass of η meson is rather precisely 4 times the pion mass whereas the mass of its spin excited companion ω is very nearly the same as the mass of ρ meson. This suggests that u, d quarks correspond to $k = 109$ inside η but to $k = 113$ inside ω. This inspires the idea that the p-adic mass scale of quarks is dynamical and sensitive to small perturbations as the fact that for CP_2 type extremals the operators corresponding to different p-adic primes reduce to one and same operator forces to suspect. If k characterizes the length scale associated with the elementary particle horizon as $\sqrt{k}$ multiple of CP_2 length scale, quark mass would be characterized by the size of elementary particle horizon sensitive to the dynamics in hadronic mass scale.

The physical states would result as small perturbations of this degenerate ground state and the value of $k(q)$ would be sensitive to the perturbation. A rather nice fit for meson and baryon masses results by assuming that the p-adic length scale of the quark is dynamical.

d) In the case of pseudoscalar mesons the scaled up versions of light quarks, which are not identifiable as constituent quarks, turn out to explain almost all of the pseudo scalar meson mass, and this inspires a new formulation for the old vision about pseudoscalar mesons as Goldstone bosons. Pseudoscalar mesons are Goldstone bosons in the sense that the color Coulombic and spin-spin interaction energies cancel in a good approximation so that quarks at $k \neq 107$ space-time sheets are responsible for most of the meson mass. The assumption that only $k(s)$ is dynamical for light baryons is enough to understand the mass differences between baryons having different numbers of strange quarks.

d) Color magnetic spin-spin interaction energies are indeed surprisingly constant among baryons. Also for mesons spin-spin interaction energies vary much less than the scaling of quark masses would predict on basis of QCD formula. This motivates the replacement of the interaction energy with interaction conformal weight in the case of color interactions. The interaction conformal weight is assignable to $k = 107$ space-time sheet, and the fact that spin-spin splittings of also heavy hadrons can be measured in few hundred MeVs, supports this identification. The mild dependence of color Coulombic conformal weight and spin-spin interaction conformal weight on hadron would be due to their dependence on the primes $k(q_i)$ and $k = 107$ characterizing space-time sheets connected by the the color bonds $q_i \to 107$ and $107 \to q_j$.

e) The values for the parameters s^c_{ij} and S_{ij} characterizing color Coulombic and color magnetic interaction conformal weights can be deduced from the mass squared differences for hadrons and assuming definite values for the parameters $k(q_i)$ characterizing quark masses. It seems that no other sources to meson mass (or at least pion mass) are needed. In the case of baryons the understanding of nucleon mass requires an additional contribution exactly equal to color interaction conformal weight of quarks. This contribution might relate to gluons or sea partons and its size might have explanation in terms of equipartition theorem generalized to p-adic thermodynamics.

13.5.1   Mass fit for hadrons

The values of $s(H)$ characterizing the first order contribution to the hadronic mass squared using p-adic mass scale m_{107} as a unit depend slightly on the exact value chosen for the CP_2 mass. The parameter Y_e characterizing the second order contribution to the electron mass satisfies $Y_e \leq Y_{e,max} = .7357$ coming from τ mass [F3]. Y_e manifests itself as a reduction factor $X = \sqrt{5/(5 + Y_e)}$ in the mass unit m_{107}. The basic requirement is that quark contribution to pseudoscalar meson masses is maximal but smaller than meson mass. η' meson mass turns out to determine the value of the scale factor X. Unfortunately, the value is so small that it implies $Y_e > 1$. This finding came as a cold shower in the relaxed state of mind after the calculations were finished. The reason was that the approximation $m_e = 0.5$ MeV used instead of $m_e = .511$ MeV masked the problem.

The only way out of the problem seems to be the assumption that the measured hadron masses are effective masses for hadrons which have suffered topological condensation to a larger space-time sheet, most naturally the electronic space-time sheet characterized by $k = 127$. The generalization of the QED formula for mass renormalization by replacing em charge with Kähler charge (one unit for quarks and three units for leptons), predicts correctly the order of magnitude for the about 6.4 per cent reduction of m_{107}.

The p-adic length scale associated with hadrons is naturally $k = 107$ since color gauge fluxes are feeded to this space-time sheet. The natural unit for mass squared is m_0^2 rather than $m_0^2/3$ since the expressions for U, D and V matrices use m_0^2. Also charged lepton masses can be predicted correctly only if m_0^2 is the fundamental mass scale. Multiplication by 64 expresses the mass squared in the nuclear p-adic length scale $k = 113$ associated with u, d, and s quarks.

baryon	m_{exp}/MeV	s	n
p	938.2796	22	5
n	939.5731	22	9
Δ^{++}	1231	38	0
Δ^{+}	1234.8	38	15
Δ^{0}	1233.6	38	11
Δ^{-}	≤ 1237.5	38	26
Λ	1115.60	31	13
Σ^{+}	1189.37	35	30
Σ^{0}	1192.37	35	42
Σ^{-}	1197.35	35	61
Σ^{*+}	1385	48	7
Ξ^{0}	1314.9	43	23
Ξ^{-}	1321.29	43	50
Ξ^{*0}	1531.8	58	54
Ξ^{*-}	1534.97	59	6
Ω^{-}	1672.2	70	9
Λ_c	2282.2	130	41
Λ_b	5425	738	15

Table 4. Mass fit for baryons using based on the renormalized mass unit and the requirement that η' mass is larger than quark contribution to it. Note that for Ξ^* doublet the values of $s(B)$ are different.

meson	m_{exp}/MeV	s	X
π^{0}	134.9645	0	29
π^{+}	139.5688	0	31
ρ^{0}	772	14	60
ρ^{+}	770	14	55
ω	783	15	24
η	548.9	7	35
K^{+}	493.707	6	7
K^{0}	497.7	6	13
K^{*+}	891.77	19	60
K^{*0}	896.05	20	8
η'	957.6	23	0
Φ	1019	26	2
D^{+}	1869.4	87	42
D^{0}	1864.7	87	14
D^{*+}	2010.1	101	22
D^{*0}	2007.2	101	3
F	2021	102	29
η_c	2980	222	48
Ψ	3100	241	3
B	5270	696	41
Y	9460	2244	51

Table 5. Mass fit for mesons using the same model as for baryons. Note that K^{*+} and K^{*0} have different values of s.

The tables above give the parameters $(s(H), n)$ for hadrons in the fit

$$M^2(H) = (s(H) + \frac{n(H)}{64}) \times m_{107}^2 \ ,$$

$$m_{107} = X \times \frac{m_e}{\sqrt{5 + Y_e}} \sqrt{\frac{M_{127}}{M_{107}}} \ ,$$

$$Y = .7357 \ , \ X = 0.8720 \ .$$

$$(13.5.-2)$$

The approximate representation $M^2 \simeq (s + \frac{n}{64})M_0^2$ is motivated by the fact that $(\frac{np}{64})_R \simeq \frac{n}{64}$ holds in excellent approximation for Mersenne primes and provides an excellent fit of baryon masses. The choice of $1/64$ as a unit for higher order corrections turned out to be a lucky choice since the mass squared scale $k = 113$ associated with nuclei is related by $1/64$-factor to $k = 107$ mass scale.

13.5.2 Fractally scaled up variants of quarks and masses of light hadrons

Fractally scaled up versions of light quarks allow a rather simple model for hadron masses if few per cent renormalization of the hadronic mass scale in topological condensation is accepted.

Renormalization of CP_2 mass in topological condensation

CP_2 mass defines the overall elementary particle mass scale. Electron mass determines this mass only in certain limits. The vanishing second order contribution to electron mass gives an upper bound for CP_2 mass whereas maximal second order contribution corresponds to a minimal CP_2 mass reduced by a factor $\sqrt{5/6} = .9129$ from its maximal value. Leptonic masses tend to be predicted to be few per cent too large [F3] if the second order contribution from p-adic thermodynamics to the electron mass vanishes, which suggests that second order contribution is important. The bound $Y_e \leq .7357$ can be derived from the requirement that it is possible to reproduce τ mass in p-adic thermodynamics. Also m_0^2 is implied as a fundamental mass squared unit by τ mass.

In the old fashioned $SU(3)$ based quark model η meson is regarded as a combination $u\bar{u} + d\bar{d} - 2s\bar{s}$. The basic observation is that η mass is rather precisely 4 times the mass of π whereas the mass of ω is very near to ρ mass. This suggests that η results by a fractal scaling of quark masses obtained by the replacement $k(q) = 113 \to 109$ for the quarks appearing in η. This inspires the idea that mesonic quarks are scaled up variants of light quarks and pseudoscalar mesons are almost Goldstone bosons in the sense that quark contribution to the mass is as large as possible but smaller than meson mass.

Taking this idea seriously, the masses of light hadrons provide valuable information about the value of CP_2 mass. The following arguments were developed using the approximation $m_e = .5$ MeV instead of better approximation $m_e = .511$ MeV.

a) The requirement that the quark contribution to η mass is smaller than η mass implies that CP_2 mass is reduced by a factor $X = .9403$ from its maximal value. This corresponds to $Y_e = .6551$ in the mass formula $m_e^2 = (5 + Y_e)m_0^2/p$.

b) η' corresponds to $k = 107$ naturally as will be found shortly. The contribution of $k = 107$ quark to the meson mass can be larger than meson mass itself if color contribution to the conformal weight adds to the conformal weight of quarks and is negative. One can however argue that color conformal weight cannot be negative since in the case that it does not add with quark conformal weight negative p-adic conformal weight raises hadron mass scale to CP_2 mass scale. It turns out that the constraint from η' mass implies other constraints.

c) The requirements $m(\Upsilon) \geq \sqrt{2}m_b$ and $m(\eta_c) \geq \sqrt{2}m_c$ give also constraints.

d) A possible further constraint would come from the requirement that hadrons inside a given electro-weak multiplet have same mass in the lowest p-adic order. Since the mass of baryon contains contributions corresponding to several values of p-adic prime, it can be argued that this condition can be applied only to the color contribution to the conformal weight.

e) The requirements that the predicted $\Lambda_c - n$ and $\Lambda_b - n$ mass differences are as near as possible to their experimental values but smaller than them is also a strong constraint. The constraint realizes the idea that spin $1/2$ baryons are fermionic analogs of Goldstone bosons.

All this looks fine. Unfortunately, the more precise calculation using $m_e = .511$ MeV produces however a cold shower. The requirement that it is possible to reproduce τ mass in p-adic thermodynamics gives the bound $Y_e \leq .7357$. The scaling of X by factor $.511/.5$ induces even in the case of η a large change of Y_e to $Y_e > .9$ inconsistent with the constraint $Y_e \leq .7357$. Even worse, the constraint from η' implies $Y_e > 1$. Giving up the constraint that quark contribution to meson mass is smaller than meson mass, does not help since $Y_e < .7357$ predicts too large b quark mass for $m_e = .511$ ($m_b > m(\Upsilon)/\sqrt{2}$) and the exponential sensitivity of mass to the value of k does not help. It seems that an incorrect calculation yielded the correct answer!

A possible remedy is based on the renormalization of hadron masses in topological condensation. Suppose that the measured hadron masses are not bare masses but masses of hadrons topologically condensed at some larger space-time sheet. $k(e) = 127$ is a good guess for the p-adic prime characterizing this space-time sheet. Topological condensation is expected to induce a renormalization correction to the hadron mass.

1. QED renormalization does not help

Consider first QED based model for mass renormalization. By a naive generalization from QED, the lowest order expression for the QED based mass renormalization in the condensation of a charged fermion with a unit charge with primary condensation level $p_1 = M_{107}$ at the level $p_2 = M_{127}$ now would be given by

$$\frac{\Delta m_q}{m_q} = -\frac{3\alpha_{em}}{2\pi}Q_q^2 \times log(\frac{\sqrt{p_2}}{\sqrt{p_1}}) = -\frac{3\alpha_{em}}{2\pi} \times 10 \times log(2)Q_q \ .$$

$$(13.5.\text{-}2)$$

Here Q_q denotes the em charge of the quark in the hadron. The simplest identification is $p_1 = M_{107}$ since it its hadron which suffers topological condensation. For $Q_u = 2/3$ one obtains $\Delta m_u/m_u \simeq .023$ meaning for $m_u = 100$ MeV order or magnitude of about 2 MeV. For d the renormalization would be $\sim .5$ per cent meaning mass reduction of $\sim .5$ MeV. For proton net renormalization would be roughly 4.5 MeV and for neutron 3 MeV. The predicted ~ 1.5 MeV difference in the mass renormalization might relate to the proton-neutron mass difference of about 1.3 MeV. The condensation at the nuclear $k = 113$ length scale would give $\Delta m_u/m_u \simeq .007$.

An alternative option is that p_1 corresponds to the p-adic length scale k_q of quark

$$\frac{\Delta m_q}{m_q} = -\frac{3\alpha_{em}}{2\pi}Q_q^2 \times (k_q - 127)\frac{log(2)}{2} \ . \qquad (13.5.\text{-}1)$$

For $k = 107$ quarks composing η' the renormalization of quark masses would be by a factor 8 larger than for pion and proton and mean renormalization by about 20 MeV. For η with $k_q = 109$ QED renormalization would allow to solve the problems using $Y_e = Y_{e,max}$. In the case of η' problems cannot be solved but one might argue that $k = 107$ for quarks inside η' allows to give up the constraint $m(\eta') \geq \sqrt{2}m(q_{107})$.

For b quark in Υ with $k = 103$ the renormalization would be ~ 20 MeV. For top quark with $k = 93$ the renormalization would be ~ 600 MeV. The renormalization is too small to resolve the problems in the case of heavy mesons.

An even more serious objection against both scenarios is that the renormalizations for different charge states inside electro-weak multiplets increase with the mass scale of the multiplet in contrast with the observed few MeV mass scale for the splittings.

2. Z^0 renormalization does not help

Quarks feed their Z^0 charges to the space-time sheet at which neutrinos topologically condense. A good guess is that space-time sheet corresponds to $k = 169$ (in fact, several p-adic mass scales are suggested for neutrinos). This would mean that the QED type renormalization of mass would be replaced by a mass renormalization which is by a factor $(169 - 107)/(127 - 107) \simeq 3$ larger. Also now the renormalization is different for u and d quark and treats left and right handed quarks differently. The predicted mass splittings inside electro-weak multiplets and large parity breaking effects increasing with the mass scale exclude this option.

3. Renormalization of Kähler charge

A renormalization of the mass scale is required. This suggests a renormalization effect associated with Kähler charge equal to $Q_K = 1$ for all quarks and $Q_K = 3$ for all leptons. Assuming $p_2 = M_{107}$ collectively, this gives

$$\frac{\Delta m_q}{m_q} = -\frac{3\alpha_K(107)}{2\pi} \times 10 \times log(2) \ . \tag{13.5.0}$$

The renormalization must affect also the color Coulombic and magnetic contribution and sea contributions to the mass. This is achieved if a dynamical equilibrium forces the gluonic contribution to follow the renormalization of quark masses. This might result if gluons, if the renormalization occurs also for gluonic contribution to the conformal weight which themselves are quark-antiquark bound states at the fundamental level.

In TGD framework one can seriously consider the possibility of generation of a cloud of matter surrounding condensed particle and having negative inertial energy and leading to the reduction of the inertial mass. Indeed, in cosmological scales the net density of the inertial energy in TGD Universe vanishes in the idealization based on Robertson-Walker metric whereas gravitational mass density is non-vanishing since negative inertial energy corresponds to a positive gravitational energy [D3, D5].

The generalization of the QED renormalization formula might provide a first guess for the model for how the generation of negative inertial energy occurs in the topological condensation. The condition that the inertial mass of nucleon is renormalized to zero in the length scale of a large void about 10^8 ly, would give the estimate $\Delta k = 50$ for the required renormalization and the renormalization in $k = 107 \rightarrow 127$ condensation would be $\Delta m/m = 40$ per cent. Of course, non-linear effects become important in long length scales, and the failure of the estimate to be too large by almost factor of 10 is not so serious.

The sign and order of magnitude of the renormalization effect are same as in QED case since in electron length scale Kähler coupling strength is very near to fine structure constant. By a naive extrapolation of QED formula the renormalization would be 5.2 per cent. Rather large effects would thus be in question. The renormalization of hadron mass scale can be described effectively as a reduction of the p-adic mass scale of electron appearing in the mass scale of hadron.

The detailed fit of hadron masses shows that the renormalization of hadron mass scale by 6.4 per cent with $Y_e = Y_{max} = .7357$ guarantees that η' mass is larger than the quark contribution to it, and a satisfactory fit of hadron masses results. The condition $Y_\tau \simeq 1$ is satisfied if the second order contribution to the τ mass squared is a p-adic integer of form $(p-n)p^2$, $n << p = M_{107}$. The possible logarithmic dependence of renormalization on the mass scale of quark does not affect appreciably the fit.

The definition of the model for hadron masses

The defining assumptions of the model of hadron masses are following.

a) The numerical construction of U and D matrices using the thermodynamical model for the topological mixing justifies the assumptions $n_d = n_s = 5$ and $n_u = 5, n_c = 6$.

b) The maximal value $Y_e = .7357$ for the second order contribution to electron mass is assumed.

c) η' mass is exactly equal to the contribution of $k = 107$ quarks equal to $\sqrt{m_d(107)^2 + m_u(107)^2}$. The assumption that η' contains only u and d quarks is of course questionable but the assumption that also $s\bar{s}$ and $c\bar{c}$ contributions are present with more or less equal weights however gives in a good approximation the same mass.

These conditions imply the renormalization of $m_e = .511$ MeV by a factor .9139 to $m_e^R = .4670$ MeV and this defines the renormalized hadronic unit of mass as

$$m_{107} = 2^{10} \times \sqrt{\frac{1}{5 + Y_{max}}} \times m_e^R \ .$$

Similar renormalization is applied also to the quark mass scales $m(k_q)$.

The following table gives quark masses as predicted by this model.

q	d	u	s	c	b	t
n_q	4	5	6	6	59	58
$k(q)$	113	113	113	104	103	93
$m(q)/GeV$	.090	.079	.090	1.873	6.537	202.8

Table 6. The masses of quarks predicted assuming $(n_d, n_s, n_b) = (5, 5, 59)$ and $(n_u, n_c, n_t) = (5, 6, 58)$ and maximal CP_2 mass scale consistent with η' meson mass.

The small value of strange quark mass has important physical consequences. The parameter ϵ'/ϵ characterizing direct CP breaking is proportional to $1/(m_d + m_s)^2$ so that the relatively small mass of about 90 MeV of s quark favors large direct CP breaking. Even 127 MeV corresponding to $k = 112$ represents relatively small s quark mass if constituent quark mass scale is used. Pseudoscalar masses give strong evidence that s can appear as severally scaled up variants so that current quark mass for s could correspond to $k = 113$ and 97 MeV mass. The predicted 2 exotic $k = 97$ gluons with masses ~ 31 GeV (from the data about $p\bar{p}$ collisions [F5]) would in the first approximation effectively multiply the gluon loop contribution to the direct CP breaking by a factor of 3. This together with the small s quark mass gives good hopes of understanding the value of ϵ'/ϵ.

Meson masses

The following table summarizes the predictions for meson masses assuming dynamical p-adic length scale.

Meson	scaled quarks	$m(M)/MeV$	m_{exp}/MeV
π		120	140
K^0	s(109)	487	498
K^+	s(109)	471	494
η	u(109),d(109),s(109)	509	549
η'	u(107),d(107),s(107),c(107)	958	958
$\eta' = G(1)G(1)$		847	958
η_c	c(104)	2649	2980
D	c(105),d(109)	1649	1864
Υ	b(103)	9245	9460
B	b(104),d(107)	5132	5270

Table 7. Summary of the model for the masses of mesons containing scaled up u,d, and s quarks. The model assumes the maximal value of CP_2 mass allowed by η' mass and the condition $Y_e = Y_{max} = .7357$.

Some comments about the model are in order.

a) The quark contribution to pion mass is predicted to be 120 MeV. This is consistent with idea that color contribution to pion mass is p-adically second order at $k = 107$ mass scale.

b) Just for fun one might also look what mass results if η' corresponds to a bound state of $g = 1$ gluons with modular contribution to mass squared equal to $n(g = 1) = 9 = s(g = 1)$. The maximal CP_2 mass scale consistent with eta mass predicts the 2-gluon state to have mass equal to 847 MeV. η' has an anomalously large coupling of B mesons eta $\eta'K$ and $\eta'X$ [he1], which indicate an anomalously large coupling to gluons [he2]. This suggests a considerable mixing η' with gluon-gluon bound state. Note that this model predicts identical couplings to various quark pairs as does also the model assuming that $\eta' - \Phi$ system is singlet with respect to flavor $SU(3)$ (having no fundamental status in TGD).

c) The quark contribution to $t\bar{t}$ meson mass is 2×143 GeV whereas top quark mass is predicted to be 199 GeV. There is narrow peak in the mass distribution of the top candidate in the interval 140-150 MeV (see Fig. 13.2.3) and it could actually correspond to one half or $t\bar{t}$ meson mass resulting from a erratic assumption about additivity of mass. Tony Smith [hc1] predicts the top quark mass in its ground state to be about 140 GeV. The reduction of pion mass to 120 MeV suggests the possibility that there is a mass scale dependent contribution to the meson mass of order $\Delta m_\pi / m_\pi \simeq 20/140$. For $t\bar{t}$ meson this would predict $m(t\bar{t}) \simeq 2 \times 163$ GeV.

d) A slight mixing with heavier quark pairs could increase the slightly too small masses of η_c and Υ mesons.

It must be admitted that the possibility of renormalization of the mass scale and freedom to choose p-adic length scales of quarks brings in subjective elements. The rather realistic predictions of the meson masses force to take the picture seriously. The success of the fit requires that spin-spin splitting cancels the color Coulombic contribution in a good approximation for pseudoscalar mesons (as also spin 1/2 baryons). This would be in accordance with the Goldstone boson interpretation of pseudoscalar mesons in the sense that color contribution to the mass from $k = 107$ space-time sheet vanishes in the lowest p-adic order. The spin-spin splittings predicted by QCD formulas predicting interaction energy rather than interaction conformal weight would however be quite too small for mesons containing scaled up quarks.

Baryons

Also $B - n$ mass differences can be understood of baryon are assumed to contain scaled versions of strange and heavy quarks. The deduction of precise values of $k(q)$ is however not quite straightforward since the color magnetic contribution to the mass affects the situation. A working hypothesis worth of study is that ground state contribution is same for all baryons and that for spin 1/2 baryons quark contribution to the mass is as near as possible to the real mass but smaller than it. This would realize the idea that the color contribution to the mass of spin 1/2 baryons is small.

a) $\Lambda - n$ mass difference is 176 MeV and $k(s) = 111$ would predict $m(s(111)) = 202$ MeV and mass difference $m(\lambda) - m(n) = m(s) + \sqrt{2}m(d) - \sqrt{3}m(u)$ equal to 151 MeV. Note that the spin-spin interaction energy is same if u and d quark form the paired quark system which is in $J = 0$ or $J = 1$ state so that the mass difference indeed can be regarded as quark mass difference.

b) $\Sigma - n$ mass difference is 257 MeV and predicted to be 226 MeV if Σ contains $s(110)$ instead of $s(111)$. If Ξ contains two $s(109)$ quarks the mass difference comes out as 441 MeV to be compared with the experimental value 381 MeV. For $\Xi = s(110) + s(111)$ mass differences is 369 MeV.

c) Even single hadron, such as Ω, could contain several scaled up variants of s quark. $s(108) + 2s(110)$ decomposition would give mass difference 713 MeV to be compared with the real mass difference 734 MeV.

d) For Λ_c the mass is 2282 MeV. For $k(c) = 105$ instead of $k(c) = 104$ the predicted $\Lambda_c - n$ mass difference is 1337 MeV whereas the experimental difference is 1344 MeV.

e) For Λ_b the mass is 5425 MeV. For $k(b) = 104$ instead of $k(b) = 103$ the predicted $\Lambda_b - n$ mass difference is 4467 MeV. The experimental difference is 4485 MeV.

$Baryon$	s content	$\Delta m/MeV$	$\Delta m_{exp}/MeV$
Λ	s(111)	151	176
Σ	s(110)	226	257
Ξ	s(110)+s(111)	369	381
Ω	s(108)+2s(110)	713	734
Λ_c	c(105),d(112),u(112)	1337	1344
Λ_b	b(105),u(106),d(106)	4467	4485

Table 8. Summary of the model for the masses of baryons containing strange quarks. Δm denotes the predicted $B - n$ mass difference $m(B) - m(n)$. The subscript 'exp' refer to experimental value of the quantity in question.

A model for hadrons

The spin-spin splittings predicted by QCD inspired formula for color magnetic energies would be quite too small for both mesons and baryons since spin-spin interaction energy scales as $1/m(q_1)m(q_2)$ and is very small for the fractal copies of quarks. For baryons spin-spin splittings predicted by QCD type formula would be reasonable if baryonic s-quarks behave as $k = 113$ quarks. For mesons the formula using $k = 113$ quark masses fails in the case of $\eta' - \Phi$ splitting completely and also in the case of K and η splittings come out to be wrong. This supports the view that the notion of interaction conformal weights depending only weakly on quark masses is more appropriate than interaction energy. Group theoretical considerations imply that the formulas for conformal weight differences (expressible in terms of mass squared differences) have same form as QCD based formulas for mass differences.

These findings suggest that following model for hadrons deserves a testing.

a) Hadron can be characterized in terms of $k \geq 113$ partonic 2-surfaces $X^2(q_i)$ connected by join along boundaries bonds (JABs) to $k = 107$ surface $X^2(H)$ corresponding to hadron and that each quark corresponds to JAB mediating classical color flux from $k = k(q)$ parton to $k = 107$ parton. Color flux tubes between quarks are replaced with pairs of flux tubes from $X^2(q_1)) \to X^2(H) \to X^2(q_2)$ mediating color Coulombic and magnetic interactions between quarks.

b) The contribution of color interactions to the conformal weight of hadron depends on the the the values of $k(q)$ since the interaction strengths $s_c(i,j)$ depend on $k(q_i)$ and $k(q_j)$. This makes possible the almost cancellation color magnetic and Coulombic conformal weights for scalar mesons.

c) It seems that also the integers n_{q_i} can be slightly dynamical. The model for the topological mixing matrices allows at most 1200 solutions for both U and D so that this is indeed possible. The p-adic reason for dynamical character is that otherwise the quantum average conformal weights for states like π, η and η' expressible as superposition of $q_i\overline{q_i}$ would not be integer valued and the mass of the state would be of order 10^{-4} Planck masses.

13.5.3   Color magnetic spin-spin splitting

Color magnetic hyperfine splitting makes it possible to understand the $\rho-\pi$, K^*-K, $\Delta-N$, etc. mass differences [fa6]. That the order of magnitude for the splittings remains same over the entire spectrum of hadrons serves as a support for the idea that color fluxes are feeded to $k = 107$ space-time sheet. This would suggest that color coupling strength does not run for the physical states and runs only for the intermediate states appearing in parton description of the hadron reactions. A possible manner to see the situation in terms of intermediate states feeding color gauge flux to space-time sheets with $k > 107$ so that the additive color Coulombic interaction conformal weights $s(q_i, q_j)$ would depend only on the integers $k(q_i), k(q_j)$. It will be found that the dependence is roughly of form $1/(k(q_i) + k(q_j))$, which brings in mind a logarithmic dependence of α_s on p-adic length scales involved.

There are two approaches to the problem of estimating spin-spin splitting: the first one is based on spin-spin interaction energy and the second one on spin-spin interaction conformal weight.

The model based on spin-spin interaction energy fails

Classical model would apply real number based physics to estimate the splittings and calculate color magnetic interaction energies.Standard QCD approach predicts that the color magnetic interaction energy is of form

$$\Delta E \;=\; S \sum_{pairs} \frac{\bar{s}_i \cdot \bar{s}_j}{m_i m_j r_{ij}^3} \; . \tag{13.5.1}$$

The mass differences for hadrons allow to deduce information about the nature of color magnetic interaction and make some conclusions about the applicability this model.

a) For mesons the spin-spin splitting various from 630 MeV for $\rho - \pi$ system to 120 MeV $\Psi - \eta_c$ excludes the classical model predicting that the splitting should be proportional to $1/m(q_1)m(q_2)$ (variation by a factor $2^{113-106} = 128$ instead of 5 would be predicted if the size of the hadron remains same). Also the predicted ratio of $K^* - K$ splitting to $\rho - \pi$ splitting would be $1/4$ rather than .63. The ratio of $\eta - \omega$ splitting to $\rho - \pi$ splitting would be $1/16$ rather than .34. The ratio of $\Phi - \eta'$ splitting to $\rho - \pi$ splitting would be $1/32 \simeq .03$ instead of .11.

The inspection of the spin-spin interaction energies would suggest that the interaction energy scales $E(i,j)$ obey roughly the formula

$$E(i,j) \sim \frac{5}{2} \times \frac{1}{(\Delta k(q_1)+\Delta k(q_2))} =$$

$$5 \times \frac{1}{log_2\{[L(113)/L(k(q_1)] \times [L(113)/L(k(q_2)]\}}$$

$$\Delta k(q) = 113 - k(q)$$

rather than being proportional to $2^{-k(q_1)-k(q_1)}$. The hypothesis that p-adic length scale $L(k)$ of order CP_2 length scale range corresponds to the size of elementary particle horizon associated with wormhole contacts feeding gauge fluxes of the CP_2 type extremal representing particle to the larger space-time sheet with $p \simeq 2^k$ might allow to understand this dependence.

b) $\Delta - N$, $\Sigma^* - \Sigma$, and $\Xi^* - \Xi$ mass differences are 291 MeV, 194 MeV, 220 MeV. If strange quark inside Σ corresponds to $k = 110$, the ratio of $\Sigma^* - \Sigma$ splitting to $\Delta - N$ splitting is predicted to be by a factor 1.17 larger than experimental ratio. $\Xi^* - \Xi$ splitting assuming $k(s) = 109$ the ratio would be .19 and quite too small. Assuming that s, u, d quarks have more or less same mass, the model would predict reasonably well the ratios of the splittings. Either the idea about scaled up variants of s is wrong or the notion of interaction energy must be replaced with interaction conformal weight in order to calculate the effects of color interactions to hadron masses.

The model based on spin-spin interaction conformal weight

The model based on the notion of interaction conformal weight generalizes the formula for color magnetic interaction energy to the p-adic context so that color magnetic interaction contributes directly to the conformal weight rather than rest mass. The effect is so large that it must be p-adically first order (the maximal contribution in second order to hadron mass would be however only 224 MeV) and the generalization of the mass splitting formula is rather obvious:

$$\Delta s \;=\; \sum_{pairs} S_{ij} \bar{s}_i \cdot \bar{s}_j \; . \tag{13.5.2}$$

The coefficients S_{ij} depend must be such that integer valued Δs results and CP_2 masses are avoided: this makes the model highly predictive. Coefficients can depend both on quark pair and on hadron since the size of hadron need not be constant. In any case, very limited range of possibilities remains for the coefficients.

The contribution from spin-spin splitting to the conformal weight adds to the color Coulombic contribution s_c. In the case of scalar mesons cancels s_c and spin splitting seem to cancel each other so that quark masses determine pseudo-scalar meson mass in a good approximation. This can be understood if the color flux carrying JAB connecting quark to $k = 107$ hadronic space-time sheet is also characterized by a value of $k \geq 113$. This fixes practically completely the model in case of mesons. If the interaction strengths $s_c(i,j)$ characterizing color Coulombic interaction conformal weight between two quarks depends only on the flux tube pair connecting the quarks via $k = 107$ space-time sheet via the integers $k(q_i)$, the model contains only very few parameters.

The general formula for the spin-spin splitting allowing to determine the parameters S_{ij} from the masses of a pair $H^\star - H$ of hadrons (say $\rho - \pi$ or $\Delta - N$) reads as

$$\begin{aligned} s_c + \Delta s(H^*) - s_c - \Delta s(H) \;&=\; [m(H^*) - m(H,q)]^2 - [m(H) - m(H,q)]^2 \\ &=\; \Delta s(H^*) - \Delta s(H) \; . \end{aligned} \tag{13.5.2}$$

The formula can be written also in the form

$$\Delta s(H^*) - \Delta s(H) = s(H^*) - s(H) - 2\left[\sqrt{s(H^*)} - \sqrt{s(H)}\right]\sqrt{s(q,H)} \; . \tag{13.5.3}$$

The values of $s(q, H)$ can be determined from the contribution of the quarks to the mass of the hadron. The values of $s(H)$ can be picked up from the table for hadrons.

Spin-spin interaction conformal weights for baryons

Consider now the determination of S_{ij} in the case of baryons. The general splitting pattern for baryons resulting from color Coulombic, and spin-spin interactions is given by the following table. The following equations summarize spin-spin splittings for baryons in a form of a table.

baryon	J	J_{12}	Δs^{spin}
N	$\frac{1}{2}$	0	$-\frac{3}{4}S_{d,d}$
Δ	$\frac{3}{2}$	1	$\frac{3}{4}S_{d,d}$
Λ	$\frac{1}{2}$	0	$-\frac{3}{4}S_{d,d}$
Σ	$\frac{1}{2}$	0	$-\frac{3}{4}S_{d,d}$
Σ^*	$\frac{1}{2}$	0	$\frac{1}{4}S_{d,d}+\frac{1}{2}S_{d,s}$
Ξ	$\frac{1}{2}$	0	$-\frac{3}{4}S_{s,s}$
Ξ^*	$\frac{1}{2}$	0	$\frac{1}{4}S_{s,s}+\frac{1}{2}S_{d,s}$
Ω	$\frac{3}{2}$	1	$\frac{3}{4}S_{s,s}$

$$(13.5.3)$$

Spin-spin splittings are deduced from the formulas

$$\Delta s^{spin} = S_{q_1,q)}\left(\frac{J_{12}(J_{12}+1)}{2}-\frac{3}{4}\right) ,$$
$$+ \frac{1}{4}(S_{q_1,q_3}+S_{q_2,q_3})(J(J+1)-J_{12}(J_{12}+1)-\frac{3}{4}) ,$$

$$(13.5.2)$$

where J_{12} is the angular momentum eigenvalue of the 'first two quarks', whose value is fixed by the requirement that magnetic moments are of correct sign.

The masses determine the values of the parameters uniquely if one assumes that color binding energy is constant as indeed suggested by the very notion of M_{107} hadron physics. The requirement is that the mass difference squared for $\Delta - N$, $\Sigma^* - \Sigma$, and $\Xi^* - \Xi$ come out correctly.

Consider now the values of S_{ij} for the models assuming $k = 113$ light quarks and dynamical $k(s)$.

a) For $N - \Delta$ system the equation is

$$S_{d_{113},d_{113}} = \frac{2}{3} \times \left[s(\Delta) - s(N) - 2 \times \left[\sqrt{s(\Delta)} - \sqrt{s(N)}\right]\right] \times \sqrt{s(q,N)} .$$

Using $(s(\Delta), s(N)) = (38, 22)$ the equation gives the estimate $S_{d,d} = 9$.

b) For $\Sigma^* - \Sigma$ system the basic equation can be written as

$$S_{d_{113},s_{110}} = -\frac{1}{2}S(d_{113},d_{113}) + s(\Sigma^*) - s(\Sigma) - 2 \times \left[\sqrt{s(\Sigma^*)} - \sqrt{s(\Sigma)}\right] \times \sqrt{s(q,\Sigma)} .$$

Note the identification $S(q,\Sigma) = \left[\sqrt{s(s_{110})} + \sqrt{2s(d_{113})}\right]^2$. Using $(s(\Sigma^*), s(\Sigma)) = (48, 35)$, the equation gives the estimate $S(d_{113}, s_{110}) = 6$.

c) In the case of $\Xi^* - \Xi$ system the equation is

$$S_{s_{110},s_{111}} = -\frac{1}{2}S_{d_{113},s_{110}} + s(\Xi^*) - s(\Xi) - 2 \times \left[\sqrt{s(\Xi^*)} - \sqrt{s(\Xi)}\right] \times \sqrt{s(q,\Xi)} .$$

In this case the quark contribution to the conformal weight is $S(q,\Xi) = \left[\sqrt{2s(s_{109})} + \sqrt{s(d_{113})}\right]^2$. Using $s(\Xi^*) = 58$ and $s(\Xi) = 43$ this gives the estimate $S(s_{110}, s_{111}) = 12$.

The resulting values of the parameters characterizing baryonic spin-spin splittings are summarized for $k(s) = 110, 109$ for Σ, Ξ.

$S_{d_{113},d_{113}}$	$S_{d_{113},s_{110}}$	$S_{s_{110},s_{111}}$
10	7	12

$$(13.5.3)$$

The mass squared unit used is m_0^2 and $k = 107$ defines the p-adic length scale used. It must be added that the interaction strengths seem to be rather sensitive to the exact values of the first order contributions to the hadron masses so that the number cannot be taken too literally.

Spin-spin interaction conformal weights for mesons

The values of mesonic interaction strengths $S_{i,j}$ can in principle deduced from the observed mass splittings. The following equations summarize the spin-spin splitting pattern for mesons in a form of table.

$meson$	Δs^{spin}
π	$-\frac{3}{4}S_{d,d}$
ρ	$\frac{1}{4}S_{d,d}$
η	$-\frac{3}{4}S_{d,d}$
ω	$\frac{1}{4}S_{d,d}$
$K^{\pm}, K^0(CP = 1)$	$-\frac{3}{4}S_{d,s}$
$K^0(CP = -1)$	$-\frac{3}{4}S_{d,s}$
$K^{*,\pm}, K^{*,0}(CP = 1)$	$\frac{1}{4}S_{d,s}$
$K^{*,0}(CP = -1)$	$\frac{1}{4}S_{d,s}$
η'	$-\frac{3}{4}S_{s,s}$
Φ	$\frac{1}{4}S_{s,s}$
η_c	$-\frac{3}{4}S_{c,c}$
Ψ	$\frac{1}{4}S_{c,c}$
$D^{\pm}, D^0(CP = 1)$	$-\frac{3}{4}S_{d,c}$
$D^0(CP = -1)$	$-\frac{3}{4}S_{d,c}$
$D^{*,\pm}, D^{*0}(CP = 1)$	$\frac{1}{4}S_{d,c}$
$D^{*0}(CP = -1)$	$\frac{1}{4}S_{d,c}$

$$(13.5.3)$$

Consider the spin-spin interaction for mesons.

a) For $\rho - \pi$ system one has

$$S_{d_{113},d_{113}} = s(\rho) - s(\pi) - 2 \times \left[\sqrt{s(\rho)} - \sqrt{s(\pi)} \right] \times \sqrt{s(q,\pi)} \ .$$

Using $s(\rho) = 14$ and $s(\pi) = 0$ gives $S(d_{113}, d_{113}) = 13$.

b)$\omega - \eta$ system is exceptional since only the quarks of η are scaled up from $k = 113$ to $k = 109$ so that the formula

$$\frac{1}{4}S_{d_{113},d_{113}} + \frac{3}{4}S_{d_{109},d_{109}} = s(\omega) - s(\eta) + s(q,\omega) - s(q,\eta)$$

$$-2 \times \left[\sqrt{s(\rho)}\sqrt{s(q,\omega)} - \sqrt{s(\pi)}\sqrt{s(q,\eta)} \right]$$

must be used. This gives $S_{q_{109},q_{109}} = 6$ for $q = u, d, s$.

c) $K^\star - K$-splitting gives $S_{d_{113},s_{109}} = 12$. Note that in this case $s(q, K) = (\sqrt{s(s_{109})} + \sqrt{s(d_{113})})^2$.

d) $\Phi - \eta'$ splitting gives $S_{q_{107},q_{107}} = 2$, when u quark mass is used as quark mass.

e) $D^* - D$ mass splitting gives $S_{d_{113},c_{105}} = 11$.

f) $\Psi - \eta_c$ mass difference gives $S_{c_{104},c_{104}} = 16$.

The results for the spin-spin interaction strengths S_{ij} are summarized in the table below. q_{109} refers to u, d, and s quarks.

$S_{d_{113},d_{113}}$	$S_{q_{109},q_{109}}$	$S_{q_{107},q_{107}}$	$S_{d_{113},s_{109}}$	$S_{d_{113},c_{105}}$	$S_{c_{104},c_{104}}$
13	6	2	12	11	16

$$(13.5.3)$$

Note that interaction strengths tend to be slightly stronger for mesons than for baryons: for instance, the value of $S_{d_{113},d_{113}}$ is 10 for baryons and 13 for mesons. For scaled up quarks the value of interaction strength tends to decrease and is smaller for non-diagonal than diagonal interactions. Since the values of $k(q_i)$ maximize the quark contribution to hadron masses, the interaction strength produce a satisfactory mass fit for hadrons with errors of few per cent.

Prediction for the nucleon mass scale

The large ground state contribution to the nucleon mass could be seen as a source of worries forcing also to ask why not also u and d quarks could not appear as scaled up variants in nucleons.

a) The fact that quark contribution to nucleon mass is about $\sqrt{3} \times m(d) \simeq 166$ MeV means that about 774 MeV (82 per cent) must correspond to color interaction conformal weight.

b) If the color interaction conformal weight is same for each join along boundaries bond connecting $k = 113$ quark surface $X^2(q)$ to $k = 107$ surface $X^2(H)$ the contribution to the color Coulombic conformal weight of proton should be 3/2 times the color Coulombic conformal weight of pion containing 2 color bonds (not the difference with the usual model).

c) For pion the color Coulombic conformal weight is cancelled by spin-spin interaction contribution which gives $s_c(\pi) = 3 \times 12/4 = 9$. The number $N_c(B)$ of color flux tubes is 3 for baryons. If the number of flux tubes is $N_c(M) = 2$ for mesons, this gives $s_c(N) \simeq 3 \times 9/2 \simeq 14$. Adding to this spin-spin interaction conformal weight $-3/4 S_{B,d,d} = -3 \times 9/4 \simeq -8$ gives $s(N) = 13$ so that the contribution to the proton mass is

$$m_{gr} = \sqrt{13/2} \times m_{107} \simeq 769\sqrt{2}\ MeV\ ,\quad m_{107} = .511 \times 2^{10}\tfrac{1}{\sqrt{5+Y_e}}\ MeV\ ,$$
$$Y_e = .7798\ . \tag{13.5.4}$$

By adding to this the quark mass contribution $\sqrt{3} \times m(d) \simeq 166$ MeV gives $m(N) = 720.6$ MeV which is 23 per cent too small.

Interestingly, if $s_{gr} = m_{gr}^2$ is replaced by $2s_{gr}$, the predictions are $m_{gr} = 769$ MeV and $m(N) = 935.5$ MeV which is about 4 MeV lower than neutron mass $m(n) = 939.6$ MeV. The error is .4 per cent. It seems that, besides the total color Coulombic and magnetic contribution of quarks to the nucleon conformal weight, there exists a second identical contribution. The p-adic variant of the equipartition theorem of thermodynamics suggests that the additional conformal weight could correspond to sea partons.

13.5.4 Some critical comments

The number theoretical model for quark masses and topological mixing matrices and CKM matrix as well as the simple model for hadron masses give strong support for the belief that the general vision is correct. One must bear in mind that the scenario need not be final so that the basic objections deserve an explicit articulation.

Is the canonical identification the only manner to map mass squared values to their real counterparts

In p-adic thermodynamics p-adic particle mass squared is mapped to its real counterpart by the canonical identification. If the $O(p)$ contribution corresponds to non-trivial rational number, the real mass is of order CP_2 mass. This allows to eliminate a large number of exotics. In particular, it implies that the modular contribution to the mass squared must be of form np rather than $(r/s)p$. This assumption is absolutely crucial in the model of topological mixing matrices and CKM matrix.

One can however question the use of the standard form of the canonical identification to map p-adic mass squared to its real counterpart. The requirement that p-adic and real S-matrix elements (in particular coupling constants) are related in a realistic manner, forces a modification of the canonical identification. Instead of a direct identification of real and p-adic rationals, the p-adic rationals in R_p are mapped to real rationals (or vice versa) using a variant of the canonical identification $I_{R \to R_p}$ in which the expansion of rational number $q = r/s = \sum r_n p^n / \sum s_n p^n$ is replaced with the rational number $q_1 = r_1/s_1 = \sum r_n p^{-n} / \sum s_n p^{-n}$ interpreted as a p-adic number:

$$q = \frac{r}{s} = \frac{\sum_n r_n p^n}{\sum_m s_n p^n} \rightarrow q_1 = \frac{\sum_n r_n p^{-n}}{\sum_m s_n p^{-n}} = \frac{I(r)}{I(s)} \quad . \tag{13.5.5}$$

The nice feature of this variant of the canonical identification is that it respects quantitative behavior of amplitudes, respects symmetries, and maps unitary matrices to unitary matrices if the matrix elements correspond to rationals (or generalized rationals in algebraic extension of rationals) if the p-adic integers involved are smaller than p. At the limit of infinitely large p this is always satisfied.

Quite generally, the thermodynamical contribution to the particle mass squared is in the lowest p-adic order of form rp/s, where r is the number of excitations with conformal weight 1 and s the number of massless excitations with vanishing conformal weight. The real counterpart of mass squared for the ordinary canonical identification is of order CP_2 mass by $r/s = R + r_1 p + ...$ with $R < p$ near to p. Hence the states for which massless state is degenerate become ultra heavy if r is not divisible by s. For the new variant of canonical identification these states would be light.

Even worse, the new form does not require the modular contribution to the p-adic mass squared to be of form np. Some other justification for this assumption would be needed. The first guess is that the conditions on mass squared plus probability conservation might not be consistent with unitarity unless the modular contribution to the mass squared remains integer valued in the mixing (note that all integer values are not possible). Direct numerical experimentation however shows that that this is not the case.

The predicted integer valued contributions to the mass squared are minimal in the case of u and d quarks and very nearly maximal in the case t and b quarks. This suggests a possible way out of the difficulty. Perhaps the rational valued p-adic mass squared of u and d quarks are minimal and those of b and t quarks maximal or nearly maximal. This might also allow to improve the prediction for the CKM matrix.

The objection against the use of the new variant of canonical identification is that the predictions of p-adic thermodynamics for mass squared are not rational numbers but infinite power series. p-Adic thermodynamics itself however defines a unique representation of probabilities as ratios of generalized Boltzmann weights and partition function and thus the variant of canonical identification might indeed generalize. If this representation generalizes to the sum of modular and Virasoro contributions, then the new form of canonical identification becomes very attractive. Also an elegant model for the masses of intermediate gauge bosons results if $O(p)$ contribution to mass squared is allowed to be a rational number.

Uncertainties related to the CP_2 length scale

The uncertainties related to the CP_2 length scale mean that one cannot take the detailed model for hadron masses too literally. The fundamental mass scale is in the range $[\sqrt{5/6}, 1]m_0$, where m_0 is the CP_2 mass scale reproducing electron mass in the case that second order contribution to the electron mass vanishes. The variation of mass scale changes also the identification of p-adic length scales associated with quarks in some light baryons (such as Ω).

The reduction of the hadronic mass scale in topological condensation by a mechanism in which negative (inertial) energy space-time sheets are generated in topological condensation conforms with the general vision about TGD inspired cosmology but lacks an experimental verification.

Chapter 14

p-Adic Particle Massivation: New Physics

14.1 Introduction

Concerning new physics the basic predictions are following. Also bosons have counterpart of family replication but the mixing is maximal. The covariant constancy of the right handed neutrino generating a potential super-symmetry implies that no sparticles are predicted: super-generators with vanishing conformal weight simply vanish. TGD predicts a rich spectrum of massless states for which ground states of negative super-canonical conformal weight are created by colored super-generators. By color confinement these states do not however give rise to macroscopic long range forces. A hierarchy QCD:s is possible. Whether or not TGD predicts Higgs boson remained for a long time uncertain but the identification of the Higgs as wormhole contact carrying left handed weak isospin resolved the question finally [D2].

14.1.1 How the ideas have evolved?

During the years it has become clear that there is rich spectrum of anomalies which might be explained by this new physics.

a) The observation of anomalous e^+e^- pairs in heavy ion collisions [hh4, hh3, hh6, hh7], which seem to originate from unknown pseudo-scalars and scalars of mass of order MeV, together with aesthetic arguments, suggest that lepto-hadrons exist [F7]. The decay width of Z^0 boson however seems to exclude new light fermions. A further grave counter argument against lepto-hadrons is that they should have been observed in e^+e^- annihilation at energies above few MeV.

A non-elegant solution of the problems is to give up the cherished idea of asymptotic freedom, which is indeed lost in the simplest scenario for the light exotic states. The elegant solution is based on the predicted dark matter hierarchy [F6]. The existence of hierarchy of QCD:s and electro-weak physics with different weak length scales allows to circumvent the constraints from decay widths since ordinary weak bosons cannot decay directly to dark particles. The selection rules for the couplings between particles in this hierarchy have also an elegant number theoretic characterization providing a deeper view about p-adicity and its connection with the notion of infinite primes [F6].

b) In the previous chapter it was found that hadron masses can be understood within one per cent errors and even isospin splittings can be understood. For a long time the only exception seemed to be top quark, whose mass for $k(top) = 89$ ($k(top) = 97$) was predicted to be about five times larger (3 times smaller) than the mass of the observed top candidate [hc2] (the p-adic prime p associated with the primary condensation level of elementary particle satisfies $p \simeq 2^k$, k prime, by the p-adic length scale hypothesis). As I returned almost decade later to update this chapter, it was immediately clear what had went wrong. I had not realized that the prime $k = 47$ defines secondary p-adic length scale $L(2, 47)$ for which top mass is predicted to be about 166 MeV and thus only slightly smaller than the recent consensus value of 179 GeV (for the details see Appendix).

c) What caused the decade long confusion was that the mass of the top candidate happened to be quite close to the estimates for the masses of u and d quarks of M_{89} hadron physics obtained using a simple extrapolation. A precise calculation however demonstrated that the mass scale was wrong by a factor of almost two and u_{89} mass is predicted to be $m(u_{89}) \simeq 262$ GeV. The predicted (perhaps too strong an expression) new copy of hadron physics associated with M_{89} could cause the effects shifting the top mass. The scalar mesons of M_{89} hadron physics could mimic the effects caused by Higgs particle and induce mass shifts of M_{89} hadrons as sigma mesons could induce them in ordinary hadron physics.

d) There was also an ALEPH result [hh9] suggesting the existence of a new particle with mass about $55 - 60$ GeV, the mass 59 GeV of the top corresponding to $k = 97$! The conclusion was that the top candidate discovered at that time could be fake and the 59 GeV particle is the real top. The problem was however that what I thought to be a fake top behaved like a real top and real top behaved unlike the real top. Decade later the solution of the problem was immediate: the mysterious ALEPH particle could a copy of b quark topologically condensed at $k = 97$ space-time sheet. Also neutrinos seem to topologically condense in several p-adic length scales as widely varying results for their mass squared differences strongly suggest [ha20, F3].

e) For a decade ago I regarded it as obvious that particle decays correspond to fusions and decays of 2-D boundary components and thus to smooth 4-manifold topologies: call this option a). During the last year (2004) it became clear that for mathematical reasons particle decays must correspond to branchings of partonic 2-surfaces representing time=constant sections of light like 3-surfaces, "causal determinants" (option b)). These 3-surfaces represent direct generalizations of Feynman diagrams for which lines are also branched in vertices (think of branched soap bubbles as 2-D analogy). The beauty of this option is that vertices correspond to smooth 2-surfaces and the construction of vertices becomes almost ridiculously simple even when compared to the construction of vertices in string models.

The difference between the two options is that option a) predicts flavor changing color currents so that the U and D matrices defining CKM matrix become direct observables. The physics of $M - \overline{M}$ systems, in particular the physics of $K - \overline{K}$ system, is extremely sensitive to these new effects. The prediction that $d\overline{s} \to s\overline{d}$ transition could occur by an exchange of $g = 1$ gluon and would thus induce a gigantic $K - \overline{K}$ mass difference kills option a).

14.1.2 Outline of the topics of the chapter

No attempt to discuss systematically the spectrum of various exotic bosons and fermions, basically due to the grounds states created by color super-canonical generators, will be made.

1. Color coupling constant evolution without asymptotic freedom or dark matter hierarchy?

QCD coupling constant evolution is discussed and it is found that asymptotic freedom could be lost making possible existence of several scaled up versions of QCD existing only in a finite length scale range. The basic counter arguments against lepto-hadron hypothesis are considered and it is found that the loss of asymptotic freedom could allow lepto-hadron physics.

The discovery of dark matter hierarchy about fifteen years after these argument were developed resolves the problems in much more elegant manner. TGD predicts an infinite hierarchy of electro-weak and color physics physics for which particles couple directly only via gravitons. De-coherence phase transitions can however induce processes allowing the decay of particles of a given physics to particles of another physics.

2. Summary of new physics effects

Various new physics effects are discussed.
a) There is a discussion of higher electro-weak boson and gluon families and the possible physical evidence for them, and an argument forcing the identification of partonic vertices as branchings of partonic 2-surfaces is developed.
b) ALEPH anomaly is interpreted in terms of a fractal copy of b-quark corresponding to k=197.
c) The possible signatures of M_{89} hadron physics in e^+e^- annihilation experiments are discussed using a naive scaling of ordinary hadron physics.
d) It is found that the newly born concept of Pomeron of Regge theory could be identified as the sea of perturbative QCD.

3. Cosmic primes and Mersenne primes

p-Adic length scale hypothesis suggests the existence of a scaled up copy of hadron physics associated with each Mersenne prime $M_n = 2^n - 1, n$ prime: M_{107} corresponds to ordinary hadron physics. There is some evidence for exotic hadrons. Centauro events and the peculiar events associated with $E > 10^5$ GeV radiation from Cygnus X-3 could be understood as due to the decay of gamma rays to M_{89} hadron pair in the atmosphere. The decay $\pi_n \to \gamma\gamma$ produces a peak in the spectrum of the cosmic gamma rays at energy $\frac{m(\pi_n)}{2}$ and there is evidence for the peaks at energies $E_{89} \simeq 34$ GeV and $E_{31} \simeq 3.5 \cdot 10^{10}$ GeV. The absence of the peak at $E_{61} \simeq 1.5 \cdot 10^6$ GeV can be understood as due to the strong absorption caused by the e^+e^- pair creation with photons of the cosmic microwave background. Cosmic string decays $cosmic\ string \to M_2\ hadrons\ \to M_3\ hadrons\ ..\to M_{107}\ hadrons$ is a new source of cosmic rays. The mechanism could explain the change of the slope in the hadronic cosmic ray spectrum at M_{61} pion rest energy $3 \cdot 10^6$ GeV. The cosmic ray radiation at energies near 10^9 GeV apparently consisting of protons and nuclei not lighter than Fe might be actually dominated by gamma rays: at these energies γ and p induced showers have same muon content and the decays of gamma rays to M_{89} and M_{61} hadrons in the atmosphere can mimic the presence of heavy nuclei in the cosmic radiation.

4. Anomalously large direct CP breaking in $K - \overline{K}$ system and exotic gluons

The recently observed anomalously large direct CP breaking in $K_L \to \pi\pi$ decays is explained in terms of loop corrections due to the predicted 2 exotic gluons having masses around 33.6 GeV. It will be also found that the TGD version of the chiral field theory believed to provide a phenomenological low energy description of QCD differs from its standard model version in that quark masses are replaced in TGD framework with shifts of quark masses induced by the vacuum expectation values of the scalar meson fields. This conforms with the TGD view about Higgs mechanism as causing only small mass shifts.

14.2 General vision about real and p-adic coupling constant evolution

The unification of super-canonical and Super Kac-Moody symmetries allows new view about p-adic aspects of the theory forcing a considerable modification and refinement of the almost decade old first picture about color coupling constant evolution.

Perhaps the most important questions about coupling constant evolution relate to the basic hypothesis about preferred role of primes $p \simeq 2^k$, k an integer. Why integer values of k are favored, why prime values are even more preferred, and why Mersenne primes $M_n = 2^n - 1$ and Gaussian Mersennes seem to be at the top of the hierarchy?

Second bundle of questions relates to the color coupling constant evolution. Do Mersenne primes really define a hierarchy of fixed points of color coupling constant evolution for a hierarchy of asymptotically non-free QCD type theories both in quark and lepton sector of the theory? How the transitions $M_n \to M_{n(next)}$ occur? What are the space-time correlates for the coupling constant evolution and for for these transitions and how space-time description relates to the usual description in terms of parton loops? How the condition that p-adic coupling constant evolution reflects the real coupling constant evolution can be satisfied and how strong conditions it poses on the coupling constant evolution?

14.2.1 Fusion of p-adic and real physics to single coherent whole by algebraic continuation

The development of the TGD inspired theory of consciousness theory and the vision about physics as a generalized number theory led to a general philosophy which provides powerful conceptual tools in attempts to answer the questions stated above.

Physics as a generalized number theory

The basic ideas behind physics as a generalized number theory approach deserve a brief summary.

1. The interpretation of p-adicity

Various p-adic versions of quantum TGD are interpreted as kind of cognitive representations of the real theory. p-Adic space-time sheets appear also at space-time level. Also real space-time sheets decompose into regions obeying effective p-adic topologies with the value of p characterizing the non-determinism of Kähler action in a particular region. This explains why p-adic thermodynamics predicts particle masses. Also algebraic extensions of the number fields R_p are important.

2. The construction of real and p-adic physics by algebraic continuation of rational number based physics

Real number based quantum TGD can be algebraically continued to various number fields [E1]. It seems that the generalization of the notion of number obtained by gluing reals and various p-adic number fields along common rationals might be crucial in this respect. In the spirit of manifold theory also more general gluing maps than gluing along common rationals can be imagined and canonical identification and its variants seem to be natural as far as probabilities are considered.

The algebraic continuation applies to all mathematical structures involved. Two continuations share some set of points which can be regarded as common to the number field involved. Examples of the structures involved are real and p-adic variants of imbedding space, of configuration space of 3-surfaces, and of Hilbert space of quantum states.

The continuation from rationals to various number fields abstracts the basic facts about the relationship between physical world and theories about it. Rational points correspond to physical data or numerical predictions of mathematical theories. Physical world represents algebraic continuation to reals and various p-adic continuations a hierarchy of increasingly refined theories.

3. Number theoretical existence

Number theoretical existence requirement becomes a leading guide line in the construction of the theory. p-Adic mass calculations and the construction of topological mixing matrices U, D and CKM matrix $V = U^\dagger D$ provide an example of a successful application of the number theoretical existence requirements [F4]. Coupling constant evolution represents second obvious application yet to be developed. Here the challenges relate to the realization of non-algebraic functions like logarithm appearing typically in the formulas.

Two rather general number theoretical conjectures are inspired by the physics as a generalized number theory vision [E8, E1, E2, E3].

i) The ratios of logarithms of rationals are rationals. In particular, $log_2(q) = log(q)/log(2)$ is always rational so that pits are rational multiples of bits. Among other things this makes possible to construct rational valued iterated 2-based logarithms $log_2(...(log_2(q))...)$ expected to appear in running coupling constants.

ii) The numbers p^{iy} exist in finite dimensional extensions of rational numbers for each value of prime p and each zero $z = 1/2 + iy$ of Riemann Zeta. The obvious implication is that the exponents $q^{i\sum n_k y_k}$ satisfy the same condition. The construction of p-adic variant of Teichmueller parameters and moduli space provides an application of the conjecture [F2]. The implication is that two-dimensional shapes obey linear superposition and that the maxima of Kähler function correspond to these quantized 2-dimensional shapes.

As far as coupling strength evolution is considered the question whether one should require g^2 or $\alpha = g^2/4\pi$ or some other combination to be rational or in an finite-dimensional algebraic extension of R_p is of fundamental importance. The existence of the exponent of Kähler action for CP_2 support the view that g_K^2, and presumably all coupling constants should be proportional to π^2. Unless an infinite-dimensional extension of p-adic numbers defined by powers of π (possibly making sense) is allowed, a combination of $\alpha/2\pi$ or something equivalent with it should appear in Feynman diagrams.

Is it possible to introduce infinite-dimensional extension of p-adic numbers containing π?

The assumption that only finite-dimensional extensions of p-adic numbers are possible, is only a convenient working hypothesis. π appears in the basic formulas of geometry, in particular in the geometry of CP_2. π appears also in the Feynmann rules of quantum field theories and expressions for reaction rates, and the idea about the algebraic continuation of real integrals as a manner to define various momentum space integrals p-adically is very attractive. These observations strongly encourage to consider the possibility of an infinite-dimensional extension of p-adic numbers containing both positive and negative powers of π with additional constraints coming from conditions like $exp(i\pi/2) = i$.

The definition of p-adic norm should obey the usual conditions, in particular the requirement that the norm of product is product of norms. One can imagine two alternative definitions of the p-adic norm.

a) The first definition is as $N_p(x) = |det(x)|_p$, where $det(x)$ is the determinant of the linear map of the infinite-dimensional linear space spanned by powers of π defined by $x = \sum_{k=m_x}^{n_x} x_k \pi^k$. This definition is straightforward generalization of the usual definition and guarantees that norm is indeed algebraic homomorphism respecting product.

b) The second definition is as the limit of the $N_p(x) = lim_{N\to\infty}|(det(x)_N^{1/N})|_p$, truncated to the subspace defined by basis $\{\pi^{-N}, ...\pi^N\}$. The motivation for this definition is that first definition tends to give vanishing of infinite norm. This norm does not however define an algebraic homomorphism.

The linear map is represented by a matrix for which non-vanishing elements form a band parallel to the diagonal having width $n_x - m_x + 1$ with each vertical entry in band equal to the column vector $(x_{m_x}, ..., x_{n_x})^T$ defined by the components of x. Diagonal entries are equal to x_0. The determinant is a sum over all downward direct paths along the this band with the product of components x_i along the path assigned with a given path. The paths are not allowed to visit the same horizontal point twice so that an analog of a functional integral associated with a self-avoiding random walk constrained inside the diagonal band in a discrete lattice along x-axis is in question. Quantum fluctuations are restricted to the interval $[m_x, n_x]$ surrounding the classical path.

The condition that $x_n \pi^n$ approaches zero p-adically is natural and requires that the p-adic norm of x_n approaches zero. The stronger condition

$$... < |x_{n+1}|_p < |x_n|_p < ... , \tag{14.2.1}$$

simplifies dramatically the calculation of the determinant since $x_{m_x} \pi^{m_x}$ can be factored out and $|det(x)|_p$ becomes a norm of the determinant of a lower triangular matrix with units at diagonal multiplied by an infinite power of $|x_{m_x}|_p$.

In case a) the norm is simply $|x_{m_x}|_p^{N\to\infty}$ quite generally and diverges for $|x_{m_x}| > 1$, vanishes for $|x_{m_x}| < 1$ and equals to one for $|x_{m_x}| = 1$. In case b) the norm is $|x_{m_x}|$.

Trigonometric functions $cos(\pi q)$ and $sin(\pi q)$ allow to test the sensibility of the proposed alternative definitions. For instance, it is possible to check whether the norm of $sin(n\pi)$ vanishes for allowed values of n.

Option a): $x = cos(\pi q)$ would correspond to a lower triangular matrix with units along the diagonal so that the norm would be equal to 1 irrespective of the value of q. This is not consistent with $cos(\pi/2) = 0$. This is not a catastrophe, since $q = 1/2$ has p-adic norm $(1/2)_p \geq 1$ so that the series is not p-adically converging and does not satisfy the condition posed above. The minimum requirement is $q = rp$, r rational with unit p-adic norm. By the product decomposition $sin(\pi q)$ for $|q| < 1$ has a vanishing norm. Thus the condition $|x_n|_p \leq p^{-n}$ guarantees the consistency with the basic trigonometric formulas.

Option b: The p-adic norm of $cos(\pi q)$ is equal to 1 whereas the p-adic norm of $sin(\pi q)$ is equal to $|q|_p$ from product decomposition. In particular, the norm is non-vanishing for $x = np\pi$ so that an inconsistency results.

If the conjecture that $log(p) = x_p/\pi$, where x_p belongs to some finite-dimensional extension holds true, then also the powers of logarithms of rationals would belong to the extension.

What effective p-adic topology really means?

The need to characterize elementary particle p-adically leads to the question what p-adic effective topology really means. p-Adic mass calculations leave actually a lot of room concerning the answer to this question.

a) The naivest option is that each space-time sheet corresponds to single p-adic prime. A more general possibility is that the boundary components of space-time sheet correspond to different p-adic primes. This view is not favored by the view that each particle corresponds to a collection of p-adic primes each characterizing one particular interaction that the particle in question participates.

b) A more abstract possibility is that a given space-time sheet or boundary component can correspond to several p-adic primes. Indeed, a power series in powers of given integer n gives rise to a well-defined power series with respect to all prime factors of n and effective multi-p-adicity could emerge at the level of field equations in this manner.

One could say that space-time sheet or boundary component corresponds to several p-adic primes through its effective p-adic topology in a hologram like manner. This option is the most flexible one as far as physical interpretation is considered. It is also supported by the number theoretical considerations predicting the value of gravitational coupling constant [E3].

An attractive hypothesis is that only space-time sheets characterized by integers n_i having common prime factors can be connected by join along boundaries bonds and can interact by particle exchanges and that each prime p in the decomposition corresponds to a particular interaction mediated by an elementary boson characterized by this prime.

The physics of quarks and hadrons provides an immediate test for this interpretation. The surprising and poorly understood conclusion from the p-adic mass calculations was that the p-adic primes characterizing light quarks u,d,s satisfy $k_q < 107$, where $k = 107$ characterizes hadronic space-time sheet [F4].

a) The interpretation of $k = 107$ space-time sheet as a hadronic space-time sheet implies that quarks topologically condense at this space-time sheet so that $k = 107$ cannot belong to the collection of primes characterizing quark.

b) Quark space-time sheets must satisfy $k_q < 107$ unless $\hbar$ is large for the hadronic space-time sheet so that one has $k_{eff} = 107 + 22 = 129$. This predicts two kinds of hadrons. Low energy hadrons consists of u, d, and s quarks with $k_q < 107$ so that hadronic space-time sheet must correspond to $k_{eff} = 129$ and large value of $\hbar$. One can speak of confined phase. This allows also $k = 127$ light variants of quarks appearing in the model of atomic nucleus [F8]. The hadrons consisting of c,t,b and the p-adically scaled up variants of u,d,s having $k_q > 107$, $\hbar$ has its ordinary value in accordance with the idea about asymptotic freedom and the view that the states in question correspond to short-lived resonances.

Do infinite primes code for q-adic effective space-time topologies?

Besides the hierarchy of space-time sheets, TGD predicts, or at least suggests, several hierarchies such as the hierarchy of infinite primes [E3], hierarchy of Jones inclusions [O5], hierarchy of dark matters with increasing values of $\hbar$ [F9, J6], the hierarchy of extensions of given p-adic number field, and the hierarchy of selves and quantum jumps with increasing duration with respect to geometric time. There are good reasons to expect that these hierarchies are closely related.

1. Some facts about infinite primes

The hierarchy of infinite primes can be interpreted in terms of an infinite hierarchy of second quantized super-symmetric arithmetic quantum field theories allowing a generalization to quaternionic or perhaps even octonionic context [E3]. Infinite primes, integers, and rationals have decomposition to primes of lower level.

Infinite prime has fermionic and bosonic parts having no common primes. Fermionic part is finite and corresponds to an integer containing and bosonic part is an integer multiplying the product of all primes with fermionic prime divided away. The infinite prime at the first level of hierarchy corresponds in a well defined sense a rational number $q = m/n$ defined by bosonic and fermionic integers m and n having no common prime factors.

2. Do infinite primes code for effective q-adic space-time topologies?

The most obvious question concerns the space-time interpretation of this rational number. Also the question arises about the possible relation with the integers characterizing space-time sheets having interpretation in terms of multi-p-adicity. On can assign to any rational number $q = m/n$ so called q-adic topology. This topology is not consistent with number field property like p-adic topologies. Hence the rational number q assignable to infinite prime could correspond to an effective q-adic topology.

If this interpretation is correct, arithmetic fermion and boson numbers could be coded into effective q-adic topology of the space-time sheets characterizing the non-determinism of Kähler action in the relevant length scale range. For instance, the power series of $q > 1$ in positive powers with integer coefficients in the range $[0, q)$ define q-adically converging series, which also converges with respect to the prime factors of m and can be regarded as a p-adic power series. The power series of q in negative powers define in similar converging series with respect to the prime factors of n.

I have proposed earlier that the integers defining infinite rationals and thus also the integers m and n characterizing finite rational could correspond at space-time level to particles with positive *resp.* negative time orientation with positive *resp.* negative energies. Phase conjugate laser beams would represent one example of negative energy states. With this interpretation super-symmetry exchanging the roles of m and n and thus the role of fermionic and bosonic lower level primes would correspond to a time reversal.

a) The first interpretation is that there is single q-adic space-time sheet and that positive and negative energy states correspond to primes associated with m and n respectively. Positive (negative) energy space-time sheets would thus correspond to p-adicity ($1/p$-adicity) for the field modes describing the states.

b) Second interpretation is that particle (in extremely general sense that entire universe can be regarded as a particle) corresponds to a pair of positive and negative energy space-time sheets labelled by m and n characterizing the p-adic topologies consistent with $m-$ and n-adicities. This looks natural since Universe has necessary vanishing net quantum numbers. Unless one allows the non-uniqueness due to $m/n = mr/nr$, positive and negative energy space-time sheets can be connected only by $\#$ contacts so that positive and negative energy space-time sheets cannot interact via the formation of $\#_B$ contacts and would be therefore dark matter with respect to each other.

Positive energy particles and negative energy antiparticles would also have different mass scales. If the rate for the creation of $\#$ contacts and their CP conjugates are slightly different, say due to the presence of electric components of gauge fields, matter antimatter asymmetry could be generated primordially.

These interpretations generalize to higher levels of the hierarchy. There is a homomorphism from infinite rationals to finite rationals. One can assign to a product of infinite primes the product of the corresponding rationals at the lower level and to a sum of products of infinite primes the sum of the corresponding rationals at the lower level and continue the process until one ends up with a finite rational. Same applies to infinite rationals. The resulting rational $q = m/n$ is finite and defines q-adic effective topology, which is consistent with all the effective p-adic topologies corresponding to the primes appearing in factorizations of m and n. This homomorphism is of course not 1-1.

If this picture is correct, effective p-adic topologies would appear at all levels but would be dictated by the infinite-p p-adic topology which itself could refine infinite-P p-adic topology [E3] coding information too subtle to be catched by ordinary physical measurements [O4].

Obviously, one could assign to each elementary particle infinite prime, integer, or even rational to this a rational number $q = m/n$. q would associate with the particle q-adic topology consistent with a collection of p-adic topologies corresponding to the prime factors of m and n and characterizing the interactions that the particle can participate directly. In a very precise sense particles would represent both infinite and finite numbers.

14.2.2 Effectively 2-dimensional partons and the unification of super-canonical and super Kac-Moody algebras

The important new element is the realization that 2-dimensional surfaces represent the data needed to characterize physical states at both classical and quantum level. This means that a large set of space-time surfaces, which can be even topologically non-equivalent, are physically equivalent. One manner to state the equivalence is that generalized Feynman diagrams with an arbitrary number of loops are equivalent with tree diagrams and that space-time Feynman diagrams are more like representations for computations or analytic continuations [C5].

Coupling constant evolution is replaced with p-adic coupling constant evolution, which means an existence of series of fixed points of coupling constant evolution labelled by p-adic number fields and their finite-dimensional extensions. The notions of dressed particle and loops must however have and indeed have TGD counterparts [F2, C1]. The basic point is that the states labelled by super-canonical conformal weights and having a natural grading do not form an orthonormal basis since the inner product involving integration over the light-like 7-surface $\delta M_+^4 \times CP_2$ is replaced with a 2-dimensional integral. The Gram-Schmidt orthogonalization procedure carried out in the order determined by the grading dresses bare particles and the S-matrix elements between these dressed particles have a decomposition to terms resembling the sum over ordinary Feynman diagrams.

The connection with p-adicity comes from the fact that the operators labelled by super-canonical conformal weights characterized in terms zeros of Riemann Zeta [E8, B4, F2] have operators O_p labelled by primes as a dual basis. This correspondence generalizes to the appropriate combinations of zeros of Zeta also appearing as conformal weights and multi-p-adic operators $O_{p_1,p_2\ldots p_n}$ characterized by several primes.

This implies that n-vertices are functions of n primes or n multi-primes. A representative example is the three-vertex characterized by coupling constant $g(p_1,p_2,p_3)$. Each leg of the vertex obeys its own coupling constant evolution.

The description of not only partons but also of hadrons in terms of 2-surfaces leads to a simple model for hadrons already applied to develop p-adic mass calculation to a more refined level. The important deviation from the stringy picture is that in n-vertex n 4-surfaces are glued together along their ends. This means that four-surfaces are branched and thus singular as 4-manifolds whereas vertices are non-singular as 3-manifolds. The TGD counterparts of stringy diagrams serve as correlates for the propagation of a particle along different paths with double slit experiment providing a basic application.

14.2.3   How p-adic and real coupling constant evolutions are related to each other?

The real and p-adic coupling constant evolutions should be consistent with each other. This means that the coupling constants $g(p_1,p_2,p_3)$ as functions of p-adic primes characterizing particles of the vertex should have the same qualitative behavior as real and p-adic functions. Hence the p-adic norms of complex rational valued (or those in algebraic extension) amplitudes must give a good estimate for the behavior of the real vertex. Hence a restriction of a continuous correspondence between p-adics and reals to rationals is highly suggestive. The restriction of the canonical identification to rationals would define this kind of correspondence but this correspondence respects neither symmetries nor unitarity in its basic form. Some kind of compromize between correspondence via common rationals and canonical identification should be found.

The compromise might be achieved by using a modification of canonical identification $I_{R_p \to R}$. Generalized numbers would be regarded in this picture as a generalized manifold obtained by gluing different number fields together along rationals. Instead of a direct identification of real and p-adic rationals, the p-adic rationals in R_p are mapped to real rationals (or vice versa) using a variant of the canonical identification $I_{R \to R_p}$ in which the expansion of rational number $q = r/s = \sum r_n p^n / \sum s_n p^n$ is replaced with the rational number $q_1 = r_1/s_1 = \sum r_n p^{-n} / \sum s_n p^{-n}$ interpreted as a p-adic number:

$$ q \;=\; \frac{r}{s} = \frac{\sum_n r_n p^n}{\sum_m s_n p^n} \to q_1 = \frac{\sum_n r_n p^{-n}}{\sum_m s_n p^{-n}} \tag{14.2.2}$$

This variant of canonical identification is not equivalent with the original one using the infinite expansion of q in powers of p since canonical identification does not commute with product and division. The variant is however unique in the recent context when r and s in $q = r/s$ have no common factors. For integers $n < p$ it reduces to direct correspondence. R_{p1} and R_{p2} are glued together along common rationals by an the composite map $I_{R \to R_{p_2}} I_{R_{p_1} \to R}$.

Instead of a re-interpretation of the p-adic number $g(p_1,p_2,p_3)$ as a real number or vice versa would be continued by using this variant of canonical identification. The nice feature of the map would be that continuity would be respected to high degree and something which is small in real sense would be small also in p-adic sense.

How to achieve consistency with the unitarity of topological mixing matrices and of CKM matrix?

It is easy to invent an objection against the proposed relationship between p-adic and real coupling constants. Topological mixing matrices U, D and CKM matrix $V = U^\dagger D$ define an important part of the electro-weak coupling constant structure and appear also in coupling constants. The problem is that canonical identification does not respect unitarity and does not commute with the matrix multiplication in the general case unlike gluing along common rationals. Even if matrices U and D which contain only ratios of integers smaller than p are constructed, the construction of V might be problematic since the products of two rationals can give a rational $q = r/s$ for which r or s or both are larger than p.

One might hope that the objection could be circumvented if the ratios of the integers of the algebraic extension defining the matrix elements of CKM matrix are such that the integer components of algebraic integers are smaller than p in U and D and even the products of integers in $U^\dagger D$ satisfy this condition so that modulo p arithmetics is avoided.

In the standard parametrization all matrix elements of the unitarity matrix can be expressed in terms of real and imaginary parts of complex phases ($p \mod 4 = 3$ guarantees that $\sqrt{-1}$ is not an ordinary p-adic number involving infinite expansion in powers of p). These phases are expressible as products of Pythagorean phases and phases in some algebraic extension of rationals.

i) Pythagorean phases defined as complex rationals $[r^2 - s^2 + i2rs]/(r^2 + s^2)$ are an obvious source of potential trouble. However, if the products of complex integers appearing in the numerators and denominators of the phases have real and imaginary parts smaller than p it seems to be possible to avoid difficulties in the definition of $V = U^\dagger D$.

ii) Pythagorean phases are not periodic phases. Algebraic extensions allow to introduce periodic phases of type $exp(i\pi m/n)$ expressible in terms of p-adic numbers in a finite-dimensional algebraic extension involving various roots of rationals. Also in this case the product $U^\dagger D$ poses conditions on the size of integers appearing in the numerators and denominators of the rationals involved.

If the expectation that topological mixing matrices and CKM matrix characterize the dynamics at the level $p \simeq 2^k$, $k = 107$, is correct, number theoretical constraints are not expected to bring much new to what is already predicted. Situation changes if these matrices appear already at the level k. For $k = 89$ hadron physics the restrictions would be even stronger and might force much simpler U, D and CKM matrices.

k-adicity constraint would have even stronger implications for S-matrix and could give very powerful constraints to the S-matrix of color interactions. Quite generally, the constraints would imply a p-adic hierarchy of increasingly complex S-matrices: kind of a physical realization for number theoretic emergence. The work with CKM matrix has shown how powerful the number theoretical constraints are, and there are no reasons to doubt that this could not be the case also more generally since in the lowest order the construction would be carried out in finite (Galois) fields $G(p, k)$.

How generally the hybrid of canonical identification and identification via common rationals can apply?

The proposed gluing procedure, if applied universally, has non-trivial implications which need not be consistent with all previous ideas.

a) The basic objection against the new kind of identification is that it does not commute with symmetries. Therefore its application at imbedding space and space-time level is questionable.

b) The mapping of p-adic probabilities by canonical identification to their real counterparts requires a separate normalization of the resulting probabilities. Also the new variant of canonical identification requires this since it does not commute with the sum.

c) The direct correspondence of reals and p-adics by common rationals at space-time level implies that the intersections of cognitive space-time sheets with real space-time sheet have literally infinite size (p-adically infinitesimal corresponds to infinite in real sense for rational) and consist of discrete points in general. If the new gluing procedure is adopted also at space-time level, it would considerably de-dramatize the radical idea that the size for the space-time correlates of cognition is literally infinite and cognition is a literally cosmic phenomenon.

Of course, the new kind of correspondence could be also seen as a manner to construct cognitive representations by mapping rational points to rational points in the real sense and thus as a formation of cognitive representations at space-time level mapping points close to each other in real sense to points close to each other p-adically but arbitrarily far away in real sense. The image would be a completely chaotic looking set of points in the wrong topology and would realize the idea of Bohm about hidden order in a very concrete manner. This kind of mapping might be used to code visual information using the value of p as a part of the code key.

b) In p-adic thermodynamics p-adic particle mass squared is mapped to its real counterpart by canonical identification. The objection against the use of the new variant of canonical identification is that the predictions of p-adic thermodynamics for mass squared are not rational numbers but infinite power series. p-Adic thermodynamics itself however defines a unique representation of probabilities as ratios of generalized Boltzmann weights and partition function and thus the variant of canonical identification indeed generalizes and at the same time raises worries about the fate of the earlier predictions of the p-adic thermodynamics.

Quite generally, the thermodynamical contribution to the particle mass squared is in the lowest p-adic order of form rp/s, where r is the number of excitations with conformal weight 1 and s the number of massless excitations with vanishing conformal weight. The real counterpart of mass squared for the ordinary canonical identification is of order CP_2 mass by $r/s = R + r_1 p + ...$ with $R < p$ near to p. Hence the states for which massless state is degenerate become ultra heavy if r is not divisible by s. For the new variant of canonical identification these states would be light. It is not actually clear how many states of this kind the generalized construction unifying super-canonical and super Kac-Moody algebras predicts.

A less dramatic implication would be that the second order contribution to the mass squared from p-adic thermodynamics is always very small unless the integer characterizing it is a considerable fraction of p. When ordinary canonical identification is used, the second order term of form rp^2/s can give term of form Rp^2, $R < p$ of order p. This occurs only in the case of left handed neutrinos.

The assumption that the second order term to the mass squared coming from other than thermodynamical sources gives a significant contribution is made in the most recent calculations of leptonic masses [F3]. It poses constraints on CP_2 mass which in turn are used as a guideline in the construction of a model for hadrons [F4]. This kind of contribution is possible also now and corresponds to a contribution Rp^2, $R < p$ near p.

The new variant of the canonical correspondence resolves the long standing problems related to the calculation of Z and W masses. The mass squared for intermediate gauge bosons is smaller than one unit when m_0^2 is used as a fundamental mass squared unit. The standard form of the canonical identification requires $M^2 = (m/n)p^2$ whereas in the new approach $M^2 = (m/n)p$ is allowed. Second difficult problem has been the p-adic description of the group theoretical model for m_W^2/m_Z^2 ratio. In the new framework this is not a problem anymore [F3] since canonical identification respects the ratios of small integers.

On the other hand, the basic assumption of the successful model for topological mixing of quarks [F4] is that the modular contribution to the masses is of form np. This assumption loses its original justification for this option and some other justification is needed. The first guess is that the conditions on mass squared plus probability conservation might not be consistent with unitarity unless the modular contribution to the mass squared remains integer valued in the mixing (note that all integer values are not possible [F4]). Direct numerical experimentation however shows that that this is not the case.

14.2.4 p-Adic coupling constant evolution and preferred primes

For almost decade after the discovery of p-adic length scale hypothesis, the justification for the hypothesis that primes $p \simeq 2^k$, k integer, in particular prime or power of prime, is still only heuristic. The second hypothesis has been that Mersenne primes correspond to the infrared fixed points of color coupling strength and thus to maxima of α_s and that each Mersenne prime can in principle give rise to its own hadron or leptohadron physics.

All coupling constant strengths are predicted to be proportional to Kähler coupling strength $\alpha_K(p)$. This raises the questions how color coupling strength can be so strong as compared to electro-weak coupling strength and how the evolution of color coupling strength can be consistent with the evolution of Kähler coupling strength. The number theoretic vision about coupling constant evolution allows to improve considerably the previous insights. In particular, p-adic length scale hypothesis inspires the hypothesis that electro-weak coupling constants evolve in the scale L_p whereas color coupling constant evolves in the length scale L_k explaining why strong interactions are strong.

Mersenne primes and color coupling constant evolution

There are motivations for the working hypothesis stating that Mersenne primes correspond to the maxima of color coupling constant strength. The proposed interpretation for the p-adic length scale hypothesis and the view about hadrons and partons as 2-surfaces at space-time sheets labelled by different values of p-adic primes and connected by join along boundaries bonds suggests how to understand the evolution.

Space-time sheets, or more precisely, 2-surfaces labelled by Mersenne primes could be favored because of a resonant interaction between k-adic and p-adic length scales resulting from the appearance of $1/log_2(p+1)$ factor in coupling constants modelling the effect of the presence of p-adic environment to the k-adic dynamics. This means that Mersenne gluons (presumably Mersenne gauge bosons in general) are emitted with much higher probability than gluons corresponding to other p-adic length scales. The join along boundaries bonds connecting quark like 2-surfaces to the hadronic 2-surfaces at M_n space-time sheets can be interpreted as kind of space-like gluons serving as correlates for the bound state formation. Since the maxima of color coupling constant correspond to critical points of coupling constant evolution M_n space-time sheets are expected to form larger structures by join along boundaries bonds. The formation of nuclei could be a counterpart for this. Also hadronization in strong interactions would correspond to this process.

Since $k = 113$ quarks are possible for $k = 107$ hadron physics, it seems that quarks can have join along boundaries bonds directed to M_n space-times with $n < k$. This suggests that neighboring Mersenne primes compete for join along boundaries bonds of quarks. For instance, when the p-adic length scale characterizing quark of M_{107} hadron physics begins to approach M_{89} quarks tend to feed their gauge flux to M_{89} space-time sheet and M_{89} hadron physics takes over and color coupling strength begins to increase. This would be the space-time correlate for the loss of asymptotic freedom.

The usual description of coupling constant evolution is in terms of bosonic and fermionic loops. Loop contributions of various signs compete and for too many fermions asymptotic freedom is spoiled and coupling constant begins to increase. The TGD counterparts of loops would result from Gram-Schmidt orthonormalization procedure.

The question whether the field theoretic description is consistent with the space-time description. For instance, could the couplings of the fermions of M_{89} physics to M_{107} gluons be seen as the reason for the loss of asymptotic freedom of M_{107} hadron physics? For instance, the slight evidence from the mass distribution of top quark candidate (see figures 1 and 2 in [F4]) for the scaled up versions of light quarks of M_{107} hadron physics in the mass scale of top quark, might be interpreted in this manner. The picture brings strongly in mind biological evolution as a competition.

p-Adic length scale hypothesis and the appearance of p-adic length scales L_k and L_p in pairs

p-Adic length scale hypothesis puts primes $p \simeq 2^k$ to a special position. When k is prime it would seem that there are actually two p-adic length scales involved. The p-adic length scale determined by k and p. L_k could characterize the size of the elementary horizon associated with the CP_2 type extremal topological condensed at space-time sheet with size L_p having Compton length. The scales would thus correspond to the classical and quantum sizes of elementary particle in ordinary quantum theory. The correspondence would imply automatically p-adic length scales hypothesis also in the generalized form if elementary particle horizons with integer valued surface area are allowed. In this case p-adity would be associated with some prime factor of integer k.

Coupling constant evolution and the evolution of Kähler coupling strength

The expression for gravitational constant in terms of p-adic length scale and exponent of Kähler action [E6] and the assumption that gravitational constant is renormalization group invariant (or that the exponent of Kähler action for CP_2 type extremal is rational number) lead to the hypothesis

$$\frac{1}{\alpha_K} = klog(K^2) \ , \quad K^2 = p \times 2 \times 3 \times 5 \times 7 \times 11 \times 13 \times 17 \times 19 \times 23 \ ,$$

where $k = 4/\pi$ or $k = 137/107$ holds true for the two alternative options discussed in [E8, E6]. $k = 137/107$ is favored by number theoretic arguments requiring the proportionality $g_K^2 \propto \pi^2$.

For M_{127} the value of α_K is very near to the fine structure constant which supports the view that electro-weak coupling constants are obtained from α_K by a multiplication with combinatorial factors. The identification $\alpha = \alpha_K$ is not excluded but would make the evolution of gravitational coupling constant non-trivial. If gravitational constant is renormalization group invariant, α_K increases much faster than predicted by QED. For instance, $1/\alpha_K(89) \simeq 103$ holds true to be compared with $1/\alpha_{em}(89) \simeq 128$ predicted by standard model. A logarithmic multiplicative correction to $1/\alpha_{em}$ of form

$$\tfrac{1}{\alpha_{em}} = \tfrac{1}{\alpha_K} \times \left(\tfrac{127}{k}\right)^D \ , \quad D \simeq .5949$$

is a possible parametrization for a realistic evolution in the interval $[M_{127}, M_{89}]$ at least. The parametrization is suggestive of anomalous dimension $D_A = 2D$ differing slightly from unity (note that $k^{1/2}$ corresponds to p-adic length scale). The anomalous dimension would characterize the effect of the dynamics in scale L_k to the dynamics in length scale L_p.

Do the length scales L_p *resp.* L_k correspond to electro-weak *resp.* color coupling constant evolutions?

The two different dynamics associated with exponentially differing length scales L_p and L_k should have some concrete physical manifestation. Particle massivation by p-adic thermodynamics occurs in the length scale L_p. Hence electro-weak interactions involving the exchange of massive particles could allow description as the exchange of partonic space-time sheets of size L_p. On the other hand, color interactions involving the exchange of massless particles would reduce to the exchange of CP_2 type extremals of size L_k.

The basic prediction would be that, as far orders of magnitudes of coupling constant strengths are considered, electro-weak coupling constant strengths would be of order $\alpha_K(p \simeq 2^k)$ and color coupling strength of order $\alpha_K(k)$. This would provide an elegant explanation for why color interactions are strong as compared to electro-weak interactions.

The loss of asymptotic freedom in the average sense would be caused even in the absence of loops by the evolution of α_K alone unless loop correction manage to affect dramatically the coupling constant evolution. TGD counterparts of loop corrections would induce QCD type coupling constant evolution from short to long length scales in each interval $[M_n, M_{n(next)}]$. For $k = 89$ and $n = 3$ the formula $\alpha_s = n\alpha_K$ gives $\alpha_s(89) = 0.0988$ for $k = 137/107$, which looks a rather reasonable estimate for the value of color coupling strength at M_{89}. Hence $\alpha_s = 3\alpha_K(M_n)$ would provide a first guess as regards to the initial value of α_s in coupling constant evolution $M_n \rightarrow M_{n(next)} > M_n$.

Coupling constant evolution from M_{89} towards M_{107} should decrease the value of α_s first and begin to increase it as the number of fermions in the loops is reduced. An attractive hypothesis is that modular arithmetics is involved: at $k = 107$ the value of $\alpha_s(k)$ becomes proportional to k and is reduced suddenly so that gluons in this length scale do not propagate. The requirement that p-adic $\alpha_s(k)$ becomes proportional to k at $k = 107$ fixes the beta function and means severe constraints on particle spectrum contributing to the loops unless some purely geometric mechanism is responsible for the behavior of the coupling constant.

Mersenne primes as IR fixed points of color dynamics?

The presence of scaled up versions of quarks in hadrons suggests that the size of wormhole contacts associated with CP_2 type extremals is very sensitive to the dynamics in length scale L_p since for CP_2 extremals all p-adic operators O_p reduce to essentially one and same operator [F2] (the situation generalizes the situation in the vicinity of the dip of the cusp catastrophe, where three roots of the cusp coincide). Hence it is possible to imagine a resonant interaction between the length scales defined by the primes k and $p \simeq 2^k$.

The fact that Mersenne primes are in a special role would suggest that the numbers $log_2(p + n)$, $p \simeq 2^k$, $n << p$, which are rational by the general conjecture appear as multiplicative factors in color coupling constant and represent the modulating effect of the dynamics in length scale L_p. What makes Mersenne primes special is that for $p = M_k = 2^k - 1$ $log_2(p + 1)$ has k-adic norm $1/k$. For primes p near 2^k $log_2(p + n)$ with $n > 1$ are required to obtain something proportional to k. The sudden reduction of the k-adic norm of $log_2(p+1)$ might correspond to the phase transition to confining phase at length scale M_n.

The value of color the color coupling strength should increase quasi-continuously as M_n is approached from shorter length scales. This is achieved if the effect of the p-adic electro-weak environment to the k-adic color interactions is represented by a multiplicative factor $log_2(p + 1)$, $p \simeq 2^k$, of k-adic color coupling strength. p-adic $\alpha_s(k)$ would increases and be at maximum proportional to $k - 1$. At $p = M_k$ its value becomes k so that its p-adic norm is suddenly reduced by a factor $1/k$. The amplitude for emitting gluon becomes very small, which can be interpreted by saying that gluons do not propagate in length scale L_k.

One could try to express the renormalization of $\alpha_s(k)$ in terms of an anomalous dimension D_s describing the modulating effect of the dynamics in the length scale L_p on color dynamics. The generalization of the formula for the electromagnetic coupling constant evolution would give

$$\frac{1}{\alpha_s(k)} = \frac{3}{\alpha_K(k)} \times \left(\frac{M_{89}}{p}\right)^{D_s} = \frac{3}{\alpha_K(k)} \times 2^{(89-k)D_s(k)} \ ,$$

A simplified model results when $D_s(k)$ is constant. In the recent case the anomalous dimension D_s would be negative. Even for small values of $|D_s|$ an exponential reduction factor would result and lead to a dramatic increase of α_s. The condition that α_s becomes proportional to k/n, $0 < n \leq k-1$, p-adically $k = 107$ at M_{107} so that its norm is reduced by a factor of $1/k$ gives only the condition $D_s = -log_2(107/n)/18$. The requirement that coupling constant increases gives the condition

$$|D_s| > \frac{1}{18log(2)} log \left[\frac{\alpha_K(89)}{\alpha_K(207)} \right] = .0115 \ .$$

This picture would suggests that the possible (not necessary as it has become clear) loss of the asymptotic freedom at the upper end is due to the inherent growth of Kähler coupling strength.

Comparison with QCD formulas for coupling constant evolution

TGD predicts a large number of exotics created by both bosonic and fermionic super-canonical generators and existing as massless states in the appropriate p-adic length scale. It is impossible to say anything definite about the p-adic length scales associated with these states. Besides fermionic and scalar loops also the evolution of Kähler coupling strength tends to spoil asymptotic freedom. It is however possible to compare the QCD type formula for coupling constant evolution with the proposed formula inspired by the proportionality $\alpha_s \propto \alpha_K$.

In QCD the beta function corresponds to the sum over fermion and gluon loops for gluon propagator

$$\begin{aligned}
\alpha_s(Q^2) &\simeq \frac{\alpha(Q_0^2)}{(1 + \alpha_s(Q_0^2)\frac{b_0}{4\pi}ln(Q^2/Q_0^2))} \ , \\
b_0 &= \frac{11}{6}l(J=1) - \frac{2}{3}l(J=1/2) - \frac{1}{12}l(J=0) \ , \\
l(J) &= \sum_{light \ (D,J)} l(D,J) \ .
\end{aligned} \qquad (14.2.1)$$

Here the total index $l(J)$ associated with spin J particles is sum over the indices of color representations D with spin J. Fermions are assumed to be Dirac fermions. Summation is over the colored elementary particles, which are light in the mass scale Q_0^2.

The list for the values of the indices $l(D)$ for some representations in leptonic (8 for ν and $10, \bar{10}$ for charged leptons) and quark sector ($\bar{6}$ for D type quarks) is given in table below.

N	$l(N)$	$\frac{C(N)}{C(3)}$
8	6	$\frac{9}{4}$
10	15	$\frac{9}{2}$
$\bar{10}$	15	$\frac{9}{2}$
3	1	1
$\bar{6}$	5	$\frac{15}{2}$

Table 7. The index $l(N)$ of the representation for some exotic fermion and boson representations ($l(3) = 1$). The table contains also the values of Casimir operators.

The QCD beta function $dlog(\alpha_s)/dlog(Q^2/Q_0^2)$ reads as

$$\frac{dlog(\alpha_s)}{dlog(Q^2/Q_0^2)} = -\frac{b_0}{4\pi}\frac{\alpha_s(Q^2)}{\alpha_s(Q_0^2)} \ . \qquad (14.2.2)$$

It is interesting to compare this expression with the beta function assuming that α_s differs from α_K only by a factor interpretable in terms of anomalous dimension (the variable $1/k$ corresponds to Q^2/Q_0^2))

$$\frac{dlog(\alpha_s(k))}{dlog(1/k)} = k_1 \times 2^{-(k-89)D_s} \times \alpha_s(k) - log(2)D_s \ ,$$

$$\alpha_s(k) = 3\alpha_K(k)2^{(k-89)D_s} \ , \quad k_1 = 137/107 \ . \tag{14.2.3}$$

The first term reflects the evolution of the Kähler coupling strength. The condition that coupling constant evolutions are exactly identical would give

$$D_s(k) \quad = \quad \frac{1}{log(2)} \times \left[3k_1\alpha_K(k) + \frac{b_0}{4\pi}\frac{\alpha_s(k)}{\alpha_s(89)} \right] \ . \tag{14.2.4}$$

It deserves to be noticed that the anomalous dimension $D_s(k)$ is rational number if the conjecture $log(p) = x/\pi$, x rational, holds true. The correspondence

$$D_s(89) = \frac{b_0}{4\pi log(2)}$$

is expected to hold true approximately. Gluons contribute $11/4\pi$ and each quark contributes $-(2/3)/4\pi$ to the right hand side and this contribution is considerably larger than the contribution from the evolution of Kähler coupling strength. For standard model particle spectrum the estimate for $D_s(89)$ would be $D_s(89) = \frac{7}{4\pi log(2)}$.

Contrary to the original beliefs, the loss of the asymptotic freedom is not catastrophe but allows to circumvent the counterarguments against lepto-hadronic physics (note however that the hierarchy of dark physics allows much more elegant and probably correct solution to the problem). The key idea is that the range of Q^2 for colored lepton pairs produced significantly in the decay of a virtual γ or Z^0, is rather narrow.

The new feature of QCD coupling constant evolution is the presence of $g > 0$ gluons: this means that the standard bound for the number of fermion generations resulting from the requirement of asymptotic freedom is lost, and the fate of the asymptotic freedom depends only on the spectrum of colored states with fixed value of g. All $g > 0$ bosons are massive and the modular contribution to the mass squared is same as for fermions and dominates. Coupling constant evolution is sensitive to the masses of the exotics and therefore to the topological mixing angles of the exotic fermions and bosons.

14.2.5   What happens in the transition to non-perturbative QCD?

What happens mathematically in the transition to non-perturbative QCD has remained more or less a mystery. The number theoretical considerations of [C5, O3] inspired the idea that Planck constant is dynamical and has a spectrum given in terms of Beraha numbers. The strange finding that the orbits of planets seem to obey Bohr quantization rules with a gigantic value of Planck constant inspired the hypothesis that the increase of Planck constant provides a unique mechanism allowing strongly interacting system to stay in perturbative phase [D6]. The resulting model allows to understand dark matter as a macroscopic quantum phase in astrophysical length and time scales, and strongly suggest a connection with dark matter and biology. The phase transition increasing Planck constant could provide a model for the transition to confining phase in QCD.

Beraha numbers and spectrum of Planck constant

The infinite-dimensional Clifford algebra of the configuration space ("the world of classical worlds") gamma matrices defines so called von Neumann algebra with a hierarchy of type II_1 sub-factors. So called Beraha numbers

$$B_n \quad = \quad 4cos^2(\frac{\pi}{n}), \ n \geq 3 \tag{14.2.5}$$

relate very closely to these factors as also to braid groups and quantum groups. Roughly, B_n corresponds to the renormalized dimension d of 4-component spinors of $D = 4$ dimensional space whose dimension becomes also renormalized. The formula $d_n = B_n = 2^{D_n/2}$ relating the dimension of spinors to the space-time dimension gives for the renormalized space-time dimension the expression $D_n = 2log(B_n)/log(2)$ approaching $D = 4$ at the limit $n \to \infty$. Note that the spectrum of fractal space-time dimension would have upper limit $D = 4$. In fact, there is also a continuum of dimensions $D \geq 4$ for the dimensions of sub-factors of type II_1. Obviously, the dimensions behave like energy spectrum of a quantum mechanical systems. That $D = 4$ is the limiting value of bound state dimensions suggests strongly a connection with the fact that the infinities of quantum field theory appear for $D \geq 4$.

The TGD based model for topological quantum computation [O3] based on the braiding of magnetic flux tubes led to the idea that this braiding could be seen as space-time correlate for a spectral flow for conformal weights at the level of configuration space spinor fields so that the connection with type II_1 factors emerges naturally.

In [O4] I developed general ideas related to type II_1 factors of von Neumann algebras and their connection with the physics predicted by quantum TGD. The speculation was that Beraha numbers define an entire spectrum of values of $\hbar$ or equivalently spectrum of values of Kähler coupling strength $\alpha_K \equiv g_K^2/4\pi\hbar$. The values of $\hbar$ would be given by

$$\hbar(n) \quad = \quad \frac{log(B_\infty)}{log(B_n)} \times \hbar(\infty) = \frac{log(4)}{log(4cos^2(\pi/n))} \times \hbar(\infty) \ , \quad n \geq 3 \ . \tag{14.2.6}$$

The proposed interpretation was that the spectrum corresponds to renormalization group evolution fixed points of α_K related to the angular/phase resolution whereas the p-adic length scale evolution of the Kähler coupling constant corresponds to length scale resolution. Small values of n would correspond to a poor angular/phase resolution.

The spectrum has remarkable features.

a) The ratio $\hbar(4)/\hbar(\infty) = 2$ means that in the range $n \geq 4$ $\hbar$ varies only by a factor of 2. The measured value of $\hbar$ is in the range $n \geq 4$: probably rather near to $\hbar(\infty)$. The cosmic evolution of $\hbar(n)$ induced by a gradual increase of the angular resolution might explain the reported increase of the fine structure constant $\alpha = e^2/4\pi\hbar c$ during cosmic evolution. The smallness of the increase implies that the recent value of n must be rather large so that $\hbar \simeq \hbar(\infty)$ should be a good approximation and it might be impossible to distinguish it from $\hbar(\infty)$ experimentally. Of course, the detection of varying values of fine structure constant in accordance with the prediction would be a victory for the proposed admittedly heuristic theory. It is known that different measurement methods give slightly different values for the fine structure constant so that it might be a good idea to check whether the variation could be understood in terms of Beraha numbers.

b) $n = 3$ is a complete exception since one has $\hbar(3) = \infty$ so that Kähler coupling strength and presumably also fine structure constant vanishes. This makes sense only if the Kähler action of the space-time sheet is vanishing in this phase. In fact, the requirement that the vacuum functional defined by the exponent of the Kähler function is non-vanishing for the entire universe, requires that Kähler per volume vanishes so that this condition is quite sensible and corresponds to a scale invariant situation. Vacuum extremals are a basic example of a phase with a vanishing Kähler action and correspond to a situation in which the energy densities of positive and negative energy matter cancel each other in the length scale considered. Robertson-Walker cosmologies are basic cosmological example in this respect [D5].

In hadronic physics hadronic space-time sheets could correspond to the space-time sheets with a gigantic value of $\hbar = \hbar_s$ and define the non-perturbative phase of QCD.

Does the transition to non-perturbative phase correspond to a change in the value of $\hbar$?

Nature is populated by systems for which perturbative quantum theory does not work. Examples are atoms with $Z_1 Z_2 e^2/4\pi\hbar > 1$ for which the binding energy becomes larger than rest mass, non-perturbative QCD resulting for $Q_{s,1} Q_{s,2} g_s^2/4\pi\hbar > 1$, and gravitational systems satisfying $GM_1 M_2/4\pi\hbar > 1$. Quite generally, the condition guaranteing troubles is of the form $Q_1 Q_2 g^2/4\pi\hbar > 1$. There is no general mathematical approach for solving the quantum physics of these systems but it is believed that a phase transition to a new phase of some kind occurs.

The gravitational Schrödinger equation forces to ask whether Nature herself takes care of the problem so that this phase transition would involve a change of the value of the Planck constant to guarantee that the perturbative approach works. The values of $\hbar$ would vary in a stepwise manner from $\hbar(\infty)$ to $\hbar(4) = \hbar(\infty)/2$ corresponding to $B(4)$ and the last step would be a transition to a phase which differs only slightly from the phase $1/\hbar(3) = 0$ would occur and correspond to

$$\hbar \quad \to \quad \frac{Q_1 Q_2 g^2}{v} \tag{14.2.7}$$

inducing

$$\frac{Q_1 Q_2 g^2}{4\pi\hbar} \to \frac{v}{4\pi} \quad . \tag{14.2.8}$$

The simplest (and of course ad hoc) assumption making sense in TGD Universe is that v is a harmonic or subharmonic of v_0 appearing in the gravitational Schrödinger equation and allowing identification in terms of the ratio of Planck and CP_2 lengths. For instance, for the Kepler problem the spectrum of binding energies would be universal (independent of the values of charges) and given by $E_n = v^2 m/2n^2$ with v playing the role of small coupling. Bohr radius would be $g^2 Q_2/v^2$ for $Q_2 \gg Q_1$.

Hadronic space-time sheets and non-perturbative QCD

The mechanism is applied in [D6] to explain the strange finding that the orbits of planets seem to correspond to the Bohr orbits with a gigantic value of $\hbar$. The phase transition changing the value of $\hbar$ would be

$$\hbar \quad \to \quad \hbar_{gr} = \frac{GmM}{v_0} \quad . \tag{14.2.9}$$

Here I have used units $\hbar = c = 1$. v_0 is a velocity parameter having the value $v_0 = 144.7 \pm .7$ km/s giving $v_0/c = 4.6 \times 10^{-4}$. The parameter v_0 becomes the new coupling constant and corresponds essentially to the ratio G/R^2 of Planck length squared and CP_2 length squared and is thus a universal small coupling constant in the non-perturbative phase in which $GmM > 1$ holds true. Also harmonics and sub-harmonics of v_0 are possible.

In the case of non-perturbative QCD the transition

$$\hbar \quad \to \quad \hbar_s = \frac{g_s^2}{v_0} \tag{14.2.10}$$

would occur for $g_s^2 \sim 1$. $\hbar_s$ would characterize the dynamics of the hadronic space-time sheets with $k = 107$ and more generally space-time sheets characterized by Mersenne primes. These space-time sheets would be maximally critical with especially strong quantum fluctuations in the length scale L_p coupling strongly to the color dynamics in the length scale L_k.

Various hadronic physics in the hierarchy of hadronic physics correspond to Mersenne primes M_n and color coupling strength becomes extremely small below $\Lambda(M_n)$. Each hadronic physics could be said to exists only in some length scale interval. The fact that bosons determine coupling constant evolution near $\Lambda(M_n)$ plus fractality considerations suggest that the scales $\Lambda(M_n)$ characterizing the transition to non-perturbative phase are universal and of form

$$\Lambda(M_n) \quad = \quad \frac{k}{\sqrt{M_n} l} \quad ,$$
$$l \quad \simeq \quad 1.288 \times 10^4 \sqrt{G} \quad . \tag{14.2.10}$$

For lepto-hadrons the hypothesis implies $\Lambda(M_{127}) = 2^{-10}\Lambda_{QCD} \simeq 0.3~MeV$ for $\Lambda_{QCD} \simeq 310~MeV$. With same assumption one has

$$\Lambda(M_{89}) = \sqrt{\frac{M_{107}}{M_{89}}}\Lambda(QCD) \simeq 158.7~GeV \quad ,$$

which means that the observed top quark candidate could belong to M_{89} Physics. If $g = 0$ quark (u or d is in question) one might expect that this quark is observed in the decay of scaled up ρ_{89} or $\omega(89)$ meson (the mass ratio of ω and ρ is 1.02 so that the prediction for the effective $u(89)$ mass does not change appreciably).

Could only the hadronic space-time sheet correspond to large $\hbar$

The ideas about dark matter are still developing and leave room for several options. One basic question is whether the space-time sheets of hadrons or both hadrons and quarks (or valence quarks) are scaled up in the transition to large $\hbar$ phase. The assumption that only the $\hbar$ associated with hadronic space-time sheet is scaled up seems to be the most elegant one. In [F8] it will be found that this assumption need not hold true for heavier nuclei with $Z \geq 12$ for which also quarks are scaled up.

The surprising and poorly understood conclusion from the p-adic mass calculations was that the p-adic primes characterizing light quarks u,d,s satisfy $k_q < 107$, where $k = 107$ characterizes hadronic space-time sheet [F4].

a) The assignment of a collection of primes to elementary particle as characterizer of p-adic primes characterizing the particles coupling directly to it, is inspired by the notion of infinite primes [E3], and discussed in [F6]. The idea is that only particles characterized by integers having common prime factors can interact by exchange of elementary bosons whose p-adic length scales are characterized by the common primes. The interpretation of $k = 107$ space-time sheet as a hadronic space-time sheet implies that quarks topologically condense at this space-time sheet so that $k = 107$ cannot belong to the collection of primes characterizing quark.

b) Since hadron is expected to be larger than quark, quark space-time sheets should satisfy $k_q < 107$ unless $\hbar$ is large for the hadronic space-time sheet so that one has $k_{eff} = 107 + 22 = 129$: this follows from the fact that the scaling factor $1/v_0$ satisfies in a reasonable approximation $v_0 \simeq 2^{11}$. This would predict two kinds of hadrons. Low energy hadrons consists of u, d, and s quarks with $k_q < 107$ so that hadronic space-time sheet must correspond to $k_{eff} = 129$ and large value of $\hbar$. One can speak of confined phase. This allows also $k = 127$ light variants of quarks appearing in the model of atomic nucleus. The hadrons consisting of c,t,b and the p-adically scaled up variants of u,d,s having $k_q > 107$, $\hbar$ has its ordinary value in accordance with the idea about asymptotic freedom and the view that the states in question correspond to short-lived resonances.

This picture is very elegant but would mean that it would be light *hadron* rather than quark which should have large $\hbar$ and scaled up Compton length. This does not affect appreciably the model of atomic nucleus since the crucial length scales $L(127)$ and $L(129)$ are still present.

14.3 Various New Physics effects

Besides lepto-hadrons and M_{89} hadrons TGD suggests also other new physics effects such as higher generations for bosons and fractally scaled up versions of quarks. The basic challenge is to decide on experimental grounds whether partonic vertices correspond to fusions or branchings and the physics of $M\overline{M}$ systems allows to do this. More exotic effects are related to the new concept of space time: for example the concept of topological evaporation (formation of Baby Universes in elementary particle length scale) suggests an explanation for the Pomeron. Also exotic p-adic Super Virasoro representations for which the CP_2 mass scale is replaced effectively divided by a power of p can be considered as possible associated with non-perturbative aspects of hadronic physics.

14.3.1 Higher boson families

TGD predicts that also bosons, with gravitons included, should have at least 3 light generations. Whether the higher genera are light is an open question: an argument suggesting that these generations might be very heavy, perhaps having a mass of order 10^{-4} Planck masses, was developed in [F1].

New view about interaction vertices and bosons

There are two options for the identification of particle vertices as topological vertices.

1. Option a)

The original assumption was that one can assign also to bosons a partonic 2-surface X^2 with more or less well defined genus g. The hypothesis is consistent with the view that particle reactions are described by smooth 4-surfaces with vertices being singular 3-surfaces intermediate between two three-topologies. The basic objection against this option is that it can induce too high rates for flavor changing currents. In particular $g > 0$ gluons could induce these currents. Second counter argument is that stable $n > 4$-particle vertices are not possible.

2. Option b)

According to the new vision (option b)), particle decays correspond to branchings of the partonic 2-surfaces in the same sense as the vertices of the ordinary Feynman diagrams do correspond to branchings of lines. The basic mathematical justification for this vision is the enormous simplification caused by the fact that vertices correspond to non-singular 2-manifolds. This option allows also $n > 3$-vertices as stable vertices.

A consistency with the experimental facts is achieved if the observed gauge bosons have each value of $g(X^2)$ with the same probability. Hence the general boson state would correspond to a phase $exp(in2\pi g/3)$, $n = 0, 1, 2$, in the discrete space of 3 lowest topologies $g = 0, 1, 2$. The observed bosons would correspond to $n = 0$ state and exotic higher states to $n = 1, 2$.

The nice feature of this option is that no flavor changing neutral electro-weak or color currents are predicted. This conforms with the fact that CKM mixing can be understood as electro-weak phenomenon described most naturally by causal determinants X_l^3 (appearing as lines of generalized Feynman diagram) connecting fermionic 2-surfaces of different genus.

Consider now objections against this scenario.

a) Since the modular contribution does not depend on the gradient of the elementary particle vacuum functional but only on its logarithm, all three boson states should have mass squared which is the average of the mass squared values $M^2(g)$ associated with three generations. The fact that modular contribution to the mass squared is due to the super-canonical thermodynamics allows to circumvent this objection. If the super-canonical p-adic temperature is small, say $T_p = 1/2$, then the modular contribution to the mass squared is completely negligible also for $g > 0$ and photon, graviton, and gluons could remain massless. The wiggling of the elementary particle vacuum functionals at the boundaries of the moduli spaces $\mathcal{M}_g$ corresponding to 2-surfaces intermediate between different 2-topologies (say pinched torus and self-touching sphere) caused by the change of overall phase might relate to the higher p-adic temperature T_p for exotic bosons.

b) If photon states had a 3-fold degeneracy, the energy density of black body radiation would be three times higher than it is. This problem is avoided if the the super-canonical temperature for $n = 1, 2$ states is higher than for $n = 0$ states, and same as for fermions, say $T_p = 1$. In this case two mass degenerate bosons would be predicted with mass squared being the average over the three genera. In this kind of situation the factor $1/3$ could make the real mass squared very large, or order CP_2 mass squared, unless the sum of the modular contributions to the mass squared values $M_{mod}^2(g) \propto n(g)$ is divisible by 3. This would make also photon, graviton, and gluons massive. Fortunately, $n(g)$ is divisible by 3 as is clear form n(0)=0,n(1)=9,$n(2) = 60$.

Masses of exotic bosons

For option a) ordinary bosons are accompanied by $g > 0$ massive partners. For option b) both ordinary bosons and their exotic partners have suffered maximal topological mixing. The general expressions for the masses for the electro-weak gauge bosons and gluons are given in the following table for both options. Mass squared are given in order $O(1/p)$: in this approximation Z^0 and W have identical masses.

	boson	γ, g, G	W, Z
a)	$\dfrac{M_R}{m_0\sqrt{\frac{M_{127}}{p}}}$	$\sqrt{n(g)}$	$\sqrt{n(g) + 1}$
b)	$\dfrac{M_R}{m_0\sqrt{\frac{M_{127}}{p}}}$	$0, \sqrt{23}, \sqrt{23}$	$1, \sqrt{24}, \sqrt{24}$

Table 1. Masses of ordinary boson generations when a) mixing between generations is absent, b) when the mixing is maximal as in the proposed model. Genus dependent contribution is given by

$n(0) = 0$, $n(1) = 9$, $n(2) = 60$. The mass scale m_0 is defined in terms of electron mass as $m_0 = \frac{m_e}{\sqrt{5+Y}}$, $Y \geq .0317$.

When applied at $k = 89$ condensation level the formulas give lower bounds for the masses of the higher generations of electro-weak boson families listed in the following table.

	boson	$M(\gamma/g/G)$	$M(W/Z)$
a)	$g = 0$	0	80.1
	$g = 1$	350.6	369.6
	$g = 2$	905.2	912.7
b)	$n = 0$	0	80.1
	$n = 1, 2$	560.5	572.5

Table 2. Lowest order predictions for the masses of various boson families for option a) and b) corresponding to the absence of mixing and maximal mixing with g-dependent phase. The primary condensation level is assumed to be $k = 89$. Electro-weak boson masses are given in approximation $cos^2(\theta_W) = 1$ ($m_W^2 = cos^2(\theta_W)M_Z^2$).

Possible evidence for exotic gluons

The widening of pionic p_T distributions in $p\bar{p}$ collisions at cm energies higher than $63\ GeV$ might signal the production of exotic gluons. It is known that this widening takes place in stepwise manner rather than smoothly between laboratory energies $2\ TeV$ and $7\ TeV$ [hi11].

a) Option a) $k = 97$ $g > 0$ gluon would have masses about $21\ GeV$ and $55GeV$. The average cm energy for quark pair is indeed about $21\ GeV$ at $\sqrt{s} = 63\ GeV$ so that the annihilation to $g = 1$ gluon followed by a spherically symmetric decay of gluon to two jets could explain the widening of p_T distributions. The production of $g = 2$ gluons via quark antiquark annihilation should give rise to a stepwise widening of meson distributions at cm energy roughly equal to $3m_g(g = 2) = 165\ GeV$.

b) For option b) the masses of exotic $k = 97$ gluons would be 33.6 GeV and quark pair annihilating to an exotic gluon should have energy higher than the average energy.

Possible evidence for exotic electro-weak gauge bosons

Exotic electro-weak gauge bosons imply new, possibly flavor changing, neutral and charged current interactions. If higher electro-weak genera are primarily condensed at level $k = 89$ and if the corresponding gauge couplings have same value as ordinary electro-weak couplings, the Fermi constant associated with the new interactions is given by

$$G_F(g) \;=\; r(g)G_F \;,$$
$$r \;=\; \frac{m_W^2}{m_W^2(g, k(g))}G_F \;. \tag{14.3.0}$$

For $k(g) = 89$ one has roughly $r(1) \sim 1/9$ and $r(2) \sim 1/60$ for option a) (no mixing) and $r(1) \sim 1/24$ for option b) (complete mixing). It should be possible to deduce upper bound for the prime $k(g)$ characterizing the primary condensation level from existing experimental data ($k(g) = 83$ is the next possibility).

1. Do new interactions between quarks become effective in $p\bar{p}$ collisions above $\sqrt{s} \sim 1.8\ TeV$?

A first piece of evidence for higher genera of electro-weak bosons comes from $x_t = 2p_t/\sqrt{s}$ scaling result of CDF [hc6]. For parton model inclusive jet cross section scales exactly and the ratio of production cross sections in $p\bar{p}$ collisions at two different cm energies and same value of x_t should be equal to one. QCD coupling constant evolution introduces scaling violations and inclusive jet cross section should decrease with cm energy. For cm energies $\sqrt{s} = 546\ GeV \equiv E_0$ and $\sqrt{s} = 1800\ GeV$ QCD predicts roughly $\sigma(546\ GeV)/\sigma(1800\ GeV) \sim 2$.

Experimentally the ratio was however found to be about 1.5, which suggests that new interactions between quarks become effective at $\sqrt{s} = 1.8\ TeV$. The masses of exotic W and Z bosons are 572.5 for option b) and slightly above E_0. Also exotic gluons could contribute to the inclusive jet production rate but this contribution is present already at E_0. The production and decay of vector mesons of M_{89} hadron physics via one photon intermediate states should give only small contribution to inclusive jet cross section.

2. Excess of di-jet rate at large transverse momentum as evidence for exotic electro-weak gauge bosons?

Further slight evidence for higher generation electro-weak gauge bosons (or color exotic bosons) comes from the excess of the di-jet rate at large transverse momentum observed in $p\bar{p}$ collisions at CDF [hh1, hc5]. Explanation for the enhanced di-jet rate in terms of new vector bosons with mass of the order of TeV has been already proposed [hh10]. The large masses of intermediate gauge bosons (about $350\ GeV$ for $g=1$ and $870\ GeV$ for $g=2$ for option a) and $572.5\ GeV$ for option b) assuming $k=89$ condensation level) imply that the angle distributions of scattered quarks are more central and spherically symmetric at the limit of infinite mass. Therefore the exchange of exotic electro-weak gauge bosons could explain the enhancement of di-jet rate for p_T in the range $200-400$ GeV. The exchange of $g>0$ gluons having $k=97$ (with masses about $21\ GeV$ and $55 GeV$) should affect only the normalization of p_T- distribution but not its shape.

3. α_s anomaly

Both option a) and b) predict new Z^0 decay channels. For example, the decays of Z^0 to quark-antiquark pair and exotic gluon are possible and provide a test for TGD picture. These channels give order α_s correction to the total hadronic decay width of Z^0. This correction could explain why the value of α_s deduced from the total hadronic decay width of Z^0 is larger than the low energy value of α_s ('α_s anomaly').

4. Evidence for flavor changing neutral current interactions?

The neutral current interaction between leptons and quarks mediated by $g=1$ Z boson (option a)) could be seen in ep collisions as anomalous events of type $e+p \rightarrow \mu + X$. H1 group [hh8] has reported an event with high transverse momenta observed in Hera. The energies of colliding positron and proton are $E_e = 27.5\ GeV$ and $E_p = 820\ GeV$ respectively. The event is characterized by

a) large transverse momentum of single muon μ^+: $P_T(\mu) = 23.2 \pm^{+7}_{-5}\ GeV$.

b) small $\delta \equiv \sum E(1 - cos(\theta))$, where E and θ denote the energy and angle of any detected particle with respect to the direction of proton beam in laboratory frame.

c) large transverse momentum of the total hadronic system $p_T(H) = 42.1 \pm 4.2\ GeV$.

d) large missing transverse momentum $P_T(miss) = 18.7 \pm 4.8^{+5}_{-7}\ GeV$, which could belong to the undetected particle in the direction of of e^+ beam.

Option b)

A possible explanation of the events in the framework of option b) is as an exchange of photon between e and q followed by an emission of W boson by q decaying into $\mu \bar{\nu}_\mu$ pair. Large missing leptonic and hadronic transverse momenta could be assigned to the neutrinos.

Option a)

Option a) allows to consider the interpretation of the event in [hh8] is as neutral flavor changing process. The process $e^+ + q_1 \rightarrow \mu^+ + q_2$ mediated by Z_1 exchange does however explain neither the small value of δ nor the missing transverse momentum. The scattering favors large transverse momenta since boson propagator is essentially constant rather than being concentrated in forward direction. A possible explanation [hh8] for the smallness of δ is that photon in the direction of e^+ beam is radiated from initial state positron and therefore remains undetected. Photon does not contribute to missing transverse momentum balance. Events of this origin show topologies of ordinary neutral current events and the detector resolutions for $p_T(miss)$ and δ have been determined from a sample of events satisfying neutral current pre-selection criteria with $p_T > 25\ GeV$ for scattered positron [hh8]. The fraction of neutral current events showing a value $\delta < 20\ GeV$ is smaller than 1 per cent. The probability that missing transverse momentum is due to detector resolution is small.

An alternative reconstruction of the event is following. The muon produced in $e^+ + q_1 \rightarrow \mu_V^+ + q_2$ is virtual and emits on mass shell Z^0 boson, $\mu_V \rightarrow \mu + Z^0$, which then decays to neutrino pair. Neutrinos give large invisible contribution to transverse momentum and δ.

a) Assuming that q_1 carries fraction x, $x = 1/3$ for definiteness, of proton momentum one has $E(q_1) = p(q_1) = E_p/3$. The natural assumption is that hadronic transverse momentum is due to the scattering of quark with positron: $p_T(H) \sim p_T(q_2) \sim p_T(\mu_V)$ giving $p_T(q_2) = p_T(\mu_V) \simeq 42\ GeV$.

b) Assuming that approximately massless scattered quark q_2 gives rise to the observed two hadronic clusters, whose transverse momenta and polar angles are $p_T(1) = 25.3\ GeV$, $\theta_1 = 22.3$ degrees and $p_T(2) = 15.2\ GeV$, $\theta_2 = 15.2$ degrees respectively, one can deduce total longitudinal momentum of the quark after scattering as the sum of the longitudinal momenta of quark clusters and equal to $p_L(q_2) = p_L(H) \simeq 113\ GeV$. The energy and longitudinal momentum of the virtual muon can be deduced from energy momentum balance and one has $E(\mu_V) \simeq 180\ GeV$ and $p_L(\mu_V) \simeq 133\ GeV$. The virtual mass of virtual muon is $m(\mu_V) \simeq 114\ GeV$ so that the decay $\mu_V \to Z^0 + \mu$ is kinematically possible.

c) The energy and polar angle of Z^0 can be deduced from the transverse momentum and polar angle of the final state muon using energy momentum balance: $E_Z = 148\ GeV$, $\theta(Z) \simeq 19$ degrees.

14.3.2 The physics of $M - \overline{M}$ systems forces the identification of vertices as branchings of partonic 2-surfaces

For option b) gluons are superpositions of $g = 0, 1, 2$ states with identical probabilities and vertices correspond to branchings of partonic 2-surfaces. Exotic gluons do not induce mixing of quark families and genus changing transitions correspond to light like 3-surfaces connecting partonic 2-surfaces with different genera. CKM mixing is induced by this topological mixing. The basic testable predictions relate to the physics of $M\overline{M}$ systems and are due to the contribution of exotic gluons and large direct CP breaking effects in $K - \overline{K}$ favor this option.

For option a) vertices correspond to fusions rather than branchings of the partonic 2-surfaces. The prediction that quarks can exchange handle number by exchanging $g > 0$ gluons (to be denoted by G_g in the sequel) could be in conflict with the experimental facts.

1. CP breaking in $K - \overline{K}$ as a basic test

CP breaking physics in kaon-antikaon and other neutral pseudoscalar meson systems is very sensitive to the new physics. What makes the situation especially interesting, is the recently reported high precision value for the parameter ϵ'/ϵ describing direct CP breaking in kaon-antikaon system [hj13]. The value is almost by an order of magnitude larger than the standard model expectation. $K - \overline{K}$ mass difference predicted by perturbative standard model is 30 per cent smaller than the the experimental value and one cannot exclude the possibility that new physics instead of/besides non-perturbative QCD might be involved.

In standard model the low energy effective action is determined by box and penguin diagrams. $\Delta S = 2$ piece of the effective weak Lagrangian, which describes processes like $s\bar{d} \to d\bar{s}$, determines the value of the $K - \overline{K}$ mass difference Δm_K and since this piece determines $K \to \overline{K}$ amplitude it also contributes to the parameter ϵ characterizing indirect CP breaking. $\Delta S = 2$ part of the weak effective action corresponds to box diagrams involving two W boson exchanges.

2. Δm_K kills option a

For option a) box diagrams involving Z and $g > 0$ exchanges are allowed provided exchanges correspond to exchange of both Z and $g > 0$ gluon. The most obvious objection is that the exchanges of $g > 0$ gluons make strong $\Delta S > 0$ decays of mesons possible: $K_S \to \pi\pi$ is a good example of this kind of decay. The enhancement of the decay rate would be of order $(\alpha_s(g = 1)/\alpha_{em})^2(m_W/m_G(g = 1)^2 \sim 10^3$. Also other $\Delta S = 1$ decay rates would be enhanced by this factor. The real killer prediction is a gigantic value of Δm_K for kaon-antikaon system resulting from the possibility of $\bar{s}d \to \bar{d}s$ decay by single $g = 1$ gluon exchange. This prediction alone excludes option a).

3. Option b) could explain direct CP breaking

For option b) box diagrams are not affected in the lowest order by exotic gluons. The standard model contributions to Δm_K and indirect CP breaking are correct for the observed value of the top quark mass which results if top corresponds to a secondary p-adic length scale $L(2, k)$ associated with $k = 47$ (Appendix). Higher order gluonic contribution could increase the value of Δm_K predicted to be about 30 per cent too small by the standard model.

In standard model penguin diagrams contribute to $\Delta S = 1$ piece of the weak Lagrangian, which determines the direct CP breaking characterized by the parameter ϵ'/ϵ. Penguin diagrams, which describe processes like $s\bar{d} \to d\bar{d}$, are characterized by effective vertices dsB, where B denotes photon, gluon or Z boson. dsB vertices give the dominant contribution to direct CP breaking in standard model. The new penguin diagrams are obtained from ordinary penguin diagrams by replacing ordinary gluons with exotic gluons.

For option b) the contributions predicted by the standard model are multiplied by a factor 3 in the approximation that exotic gluon mass is negligible in the mass scale of intermediate gauge boson. These diagrams affect the value of the parameter ϵ'/ϵ characterizing direct CP breaking in $K - \bar{K}$ system found experimentally to be almost order of magnitude larger than standard model expectation [hj13].

14.3.3   A new twist in the spin puzzle of proton

The so called proton spin crisis or spin puzzle of proton was an outcome of the experimental finding that the quarks contribute only 13-17 per cent of proton spin [hb2, hb1] whereas the simplest valence quark model predicts that quarks contribute about 75 per cent to the spin of proton with the remaining 25 per cent being due to the orbital motion of quarks. Besides the orbital motion of valence quarks also gluons could contribute to the spin of proton. Also polarized sea quarks can be considered as a source of proton spin.

Quite recently, the spin crisis got a new twist [hb4]. One of the few absolute predictions of perturbative QCD (pQCD) is that at the limit, when the momentum fraction of quark approaches unity, quark spin should be parallel to the proton spin. This is due to the helicity conservation predicted by pQCD in the lowest order. The findings are consistent with this expectation in the case of protonic u quarks but not in the case of protonic d quark. The discovery is of a special interest from the point of view of TGD since it might have an explanation involving the notions of many-sheeted space-time, of color-magnetic flux tubes, the predicted super-canonical "vacuum" spin, and also the concept of quantum parallel dissipation.

The experimental findings

In the experiment performed in Jefferson Lab [hb4] neutron spin asymmetries A_1^n and polarized structure functions $g_{1,2}^n$ were deduced for three kinematic configurations in the deep inelastic region from e-^{3}He scattering using 5.7 GeV longitudinally polarized electron beam and a polarized ^3He target. A_1^n and $g_{1,2}^n$ were deduced for $x = .33, .47$, and $.60$ and $Q^2 = 2.7, 3.5$ and 4.8 (GeV/c)^2. A_1^n and g_1^n at $x = .33$ are consistent with the world data. At $x = .47$ A_1^n crosses zero and is significantly positive at $x = 0.60$. This finding agrees with the next-to-leading order QCD analysis of previous world data without the helicity conservation constraint. The trend of the data agrees with the predictions of the constituent quark model but disagrees with the leading order pQCD assuming hadron helicity conservation.

By isospin symmetry one can translate the result to the case of proton by the replacement $u \leftrightarrow d$. By using world proton data, the polarized quark distribution functions were deduced for proton using isospin symmetry between neuron and proton. It was found that $\Delta u/u$ agrees with the predictions of various models while $\Delta d/d$ disagrees with the leading-order pQCD.

Let us denote by $q(x) = q^\uparrow + q^\downarrow(x)$ the spin independent quark distribution function. The difference $\Delta q(x) = q^\uparrow - q^\downarrow(x)$ measures the contribution of quark q to the spin of hadron. The measurement allowed to deduce estimates for the ratios $(\Delta q(x) + \Delta \bar{q}(x))/(q(x) + \bar{q}(x))$.

The conclusion of [hb4] is that for proton one has

$$\frac{\Delta u(x) + \Delta \overline{u}(x)}{u(x) + \overline{u}(x)} \simeq .737 \pm .007 \ , \ \text{for} \ x = .6 \ .$$

This is consistent with the pQCD prediction. For d quark the experiment gives

$$\frac{\Delta d(x) + \Delta \overline{d}(x)}{d(x) + \overline{d}(x)} \simeq -.324 \pm .083 \ \text{for} \ x = .6 \ .$$

The interpretation is that d quark with momentum fraction $x > .6$ in proton spends a considerable fraction of time in a state in which its spin is opposite to the spin of proton so that the helicity conservation predicted by first order pQCD fails. This prediction is of special importance as one of the few absolute predictions of pQCD.

The finding is consistent with the relativistic $SU(6)$ symmetry broken by spin-spin interaction and the QCD based model interpolated from data but giving up helicity conservation [hb4]. $SU(6)$ is however not a fundamental symmetry so that its success is probably accidental.

It has been also proposed that the spin crisis might be illusory [hb3] and due to the fact that the vector sum of quark spins is not a Lorentz invariant quantity so that the sum of quark spins in infinite-momentum frame where quark distribution functions are defined is not same as, and could thus be smaller than, the spin sum in the rest frame. The correction due to the transverse momentum of the quark brings in a non-negative numerical correction factor which is in the range $(0, 1)$. The negative sign of $\Delta d/d$ is not consistent with this proposal.

TGD based model for the findings

The TGD based explanation for the finding involves the following elements.

a) TGD predicts the possibility of vacuum spin due to the super-canonical symmetry. Valence quarks can be modelled as a star like formation of magnetic flux tubes emanating from a vertex with the conservation of color magnetic flux forcing the valence quarks to form a single coherent structure. A good guess is that the super-canonical spin corresponds classically to the rotation of the the star like structure.

b) By parity conservation only even values of super-canonical spin J are allowed and the simplest assumption is that the valence quark state is a superposition of ordinary $J = 0$ states predicted by pQCD and $J = 2$ state in which all quarks have spin which is in a direction opposite to the direction of the proton spin. The state of $J = 1/2$ baryon is thus replaced by a new one:

$$
\begin{aligned}
|B, \frac{1}{2}, \uparrow\rangle &= a|B, 1/2, \frac{1}{2}\rangle|J = J_z = 0\rangle + b|B, \frac{3}{2}, -\frac{3}{2}\rangle|J = J_z = 2\rangle \ , \\
|B, 1/2, \frac{1}{2}\rangle &= \sum_{q_1,q_2,q_3} c_{q_1,q_2,q_3} q_1^{\uparrow} q_2^{\uparrow} q_3^{\downarrow} \ , \\
|B, \frac{3}{2}, -\frac{3}{2}\rangle &= d_{q_1,q_2,q_3} q_1^{\downarrow} q_2^{\downarrow} q_3^{\downarrow} \ .
\end{aligned}
\tag{14.3.-1}
$$

$|B, 1/2, \frac{1}{2}\rangle$ is in a good approximation the baryon state as predicted by pQCD. The coefficients c_{q_1,q_2,q_3} and d_{q_1,q_2,q_3} depend on momentum fractions of quarks and the states are normalized so that $|a|^2 + |b|^2 = 1$ is satisfied: the notation $p = |a|^2$ will be used in the sequel. The quark parts of $J = 0$ and $J = 2$ have quantum numbers of proton and Δ resonance. $J = 2$ part need not however have the quark distribution functions of Δ.

c) The introduction of $J = 0$ and $J = 2$ ground states with a simultaneous use of quark distribution functions makes sense if one allows quantum parallel dissipation. Although the system is coherent in the super-canonical degrees of freedom which correspond to the hadron size scale, there is a decoherence in quark degrees of freedom which correspond to a shorter p-adic length scale and smaller space-time sheets.

d) Consider now the detailed structure of the $J = 2$ state in the case of proton. If the d quark is at the rotation axis, the rotating part of the triangular flux tube structure resembles a string containing u-quarks at its ends and forming a di-quark like structure. Di-quark structure is taken to mean correlations between u-quarks in the sense that they have nearly the same value of x so that $x < 1/2$ holds true for them whereas the d-quark behaving more like a free quark can have $x > 1/2$.

A stronger assumption is that di-quark behaves like a single colored hadron with a small value of x and only the d-quark behaves as a free quark able to have large values of x. Certainly this would be achieved if u quarks reside at their own string like space-time sheet having $J = 2$.

From these assumptions it follows that if u quark has $x > 1/2$, the state effectively reduces to a state predicted by pQCD and $u(x) \to 1$ for $x \to 1$ is predicted. For the d quark the situation is different and introducing distribution functions $q^{J}(x)$ for $J = 0, 2$ separately, one can write the spin asymmetry at the limit $x \to 1$ as

$$
\begin{aligned}
A_d &\equiv \frac{\Delta d(x) + \Delta \overline{d}(x)}{d(x) + \overline{d}(x)} = \frac{p(\Delta d_0 + \Delta \overline{d}_0) + (1 - p)(\Delta d_2 + \Delta \overline{d}_2)}{p(d_0 + \overline{d}_0) + (1 - p)(d_2 + \overline{d}_2)} \ , \\
p &= |a|^2 \ .
\end{aligned}
\tag{14.3.-1}
$$

Helicity conservation gives $\Delta d_0/d_0 \to 1$ at the limit $x \to 1$ and one has trivially $\Delta d_2/d_2 = -1$. Taking the ratio

$$y = \frac{d_2}{d_0}$$

as a parameter, one can write

$$A_d \quad \to \quad \frac{p - (1-p)y}{p + (1-p)y} \tag{14.3.0}$$

at the limit $x \to 1$. This allows to deduce the value of the parameter y once the value of p is known:

$$y \quad = \quad \frac{p}{1-p} \times \frac{1 - A_d}{1 + A_d} \quad . \tag{14.3.1}$$

From the requirement that quarks contribute a fraction $\Sigma = \sum_q \Delta q \in (13, 17)$ per cent to proton spin, one can deduce the value of p using

$$\frac{p \times \frac{1}{2} - (1-p) \times \frac{3}{2}}{\frac{1}{2}} \quad = \quad \Sigma \tag{14.3.2}$$

giving $p = (3 + \Sigma)/4 \simeq .75$.

Eq. 14.3.1 allows estimate the value of y. In the range $\Sigma \in (.13, .30)$ defined by the lower and upper bounds for the contribution of quarks to the proton spin, $A_d = -.32$ gives $y \in (6.98, 9.15)$. $d_2(x)$ would be more strongly concentrated at high values of x than $d_0(x)$. This conforms with the assumption that u quarks tend to carry a small fraction of proton momentum in $J = 2$ state for which uu can be regarded as a string like di-quark state.

A further input to the model comes from the ratio of neutron and proton F_2 structure functions expressible in terms of quark distribution functions of proton as

$$R^{np} \quad \equiv \quad \frac{F_2^n}{F_2 p} = \frac{u(x) + 4d(x)}{4u(x) + d(x)} \quad . \tag{14.3.3}$$

According to [hb4] $R^{np}(x)$ is a straight line starting with $R^{np}(x \to 0) \simeq 1$ and dropping below $1/2$ as $x \to 1$. The behavior for small x can be understood in terms of sea quark dominance. The pQCD prediction for R^{np} is $R^{np} \to 3/7$ for $x \to 1$, which corresponds to $d/u \to z = 1/5$. TGD prediction for R^{np} for $x \to 1$

$$R^{np} \quad \equiv \quad \frac{F_2^n}{F_2^p} = \frac{pu_0 + 4(pd_0 + (1-p)d_2)}{4pu_0 + pd_0 + (1-p)d_2}$$

$$= \quad \frac{p + 4z(p + (1-p)y)}{4p + z(p + (1-p)y)} \quad . \tag{14.3.3}$$

In the range $\Sigma \in (.13, .30)$ which corresponds to $y \in (6.98, 9.15)$ for $A_d = -.32$ $R^{np} = 1/2$ gives $z \simeq .1$, which is 20 per cent of pQCD prediction. 80 percent of d-quarks with large x predicted to be in $J = 0$ state by pQCD would be in $J = 2$ state.

14.3.4   Fractally scaled up versions of quarks

The strange anomalies of neutrino oscillations [ha20] suggesting that neutrino mass scale depends on environment can be understood if neutrinos can suffer topological condensation in several p-adic length scales [F3]. The obvious question whether this could occur also in the case of quarks led to a very fruitful developments leading to the understanding of hadronic mass spectrum in terms of scaled up variants of quarks. Also the mass distribution of top quark candidate exhibits structure which could be interpreted in terms of heavy variants of light quarks. The ALEPH anomaly [hh9], which

I first erratically explained in terms of a light top quark has a nice explanation in terms of b quark condensed at $k = 97$ level and having mass ~ 55 GeV. These points are discussed in detail in [F4].

The emergence of ALEPH results [hh9] meant a an important twist in the development of ideas related to the identification of top quark. In the LEP 1.5 run with $E_{cm} = 130 - 140\ GeV$, ALEPH found 14 e^+e^- annihilation events, which pass their 4-jet criteria whereas 7.1 events are expected from standard model physics. Pairs of dijets with vanishing mass difference are in question and dijets could result from the decay of a new particle with mass about 55 GeV.

The data do not allow to conclude whether the new particle candidate is a fermion or boson. Top quark pairs produced in e^+e^- annihilation could produce 4-jets via gluon emission but this mechanism does not lead to an enhancement of 4-jet fraction. No $b\bar{b}b\bar{b}$ jets have been observed and only one event containing b has been identified so that the interpretation in terms of top quark is not possible unless there exists some new decay channel, which dominates in decays and leads to hadronic jets not initiated by b quarks. For option b), which seems to be the only sensible option, this kind of decay channels are absent.

Super symmetrized standard model suggests the interpretation in terms of super partners of quarks or/and gauge bosons [hf2]. It seems now safe to conclude that TGD does not predict sparticles. If the exotic particles are gluons their presence does not affect Z^0 and W decay widths. If the condensation level of gluons is $k = 97$ and mixing is absent the gluon masses are given by $m_g(0) = 0$, $m_g(1) = 19.2\ GeV$ and $m_g(2) = 49.5\ GeV$ for option a) and assuming $k = 97$ and hadronic mass renormalization. It is however very difficult to understand how a pair of $g = 2$ gluons could be created in e^+e^- annihilation. Moreover, for option b), which seems to be the only sensible option, the gluon masses are $m_g(0) = 0$, $m_g(1) = m_g(2) = 30.6\ GeV$ for $k = 97$. In this case also other values of k are possible since strong decays of quarks are not possible.

The strong variations in the order of magnitude of mass squared differences between neutrino families [ha20] can be understood if they can suffer a topological condensation in several p-adic length scales. One can ask whether also t and b quark could do the same. In absence of mixing effects the masses of $k = 97$ t and b quarks would be given by $m_t \simeq 48.7$ GeV and $m_b \simeq 52.3$ GeV taking into account the hadronic mass renormalization. Topological mixing reduces the masses somewhat. The fact that b quarks are not observed in the final state leaves only $b(97)$ as a realistic option. Since Z^0 boson mass is ~ 94 GeV, $b(97)$ does not appreciably affect Z^0 boson decay width. The observed anomalies concentrate at cm energy about 105 GeV. This energy is 15 percent smaller than the total mass of top pair. The discrepancy could be understood as resulting from the binding energy of the $b(97)\bar{b}(97)$ bound states. Binding energy should be a fraction of order $\alpha_s \simeq .1$ of the total energy and about ten per cent so that consistency is achieved.

14.3.5 What M_{89} Hadron Physics would look like?

TGD suggests the existence of the scaled up copies of hadron physics corresponding to the Mersenne primes $M_n = 89, 61, 31, ..$ at least in the sense that α_s has maximum at these length scales. The assumption of QCD:s decoupling completely from each other seems more unrealistic.

The requirement of unitarity forces the existence of Higgs particle in gauge theories. The failure of the p-adic mass calculations to predict intermediate gauge boson masses correctly forces to give up the idea that boson masses are of purely thermodynamical origin. A possible TGD counterpart for the Higgs fields is as the fields defined by the Kac-Moody generators associated with the complement of the $u(2)$ algebra of $su(3)$ associated with the conserved charges Q_J defined by the variation of the modified Dirac action with respect to the induced Kähler form. As found in the [F3], the small coupling of the Higgs to fermionic masses resolves the paradoxical situation created by the failure to detect Higgs boson. Also the fact that left handed electro-weak charge matrices are not covariantly constant could explain Higgs vacuum expectation value without an introduction of an elementary scalar field.

One could of course, consider also other explanations for Higgs. The only scalar mesons with masses in intermediate boson mass scale allowed by TGD are bound states of quark and antiquark of M_{89} hadron physics such that quark and antiquark have parallel spins and relative angular momentum $L = 1$. The effective couplings of these states to leptons and quarks could mimic the couplings of Higgs boson to some degree. Scalar bound states of heavy quarks are also present in ordinary hadron physics. The coherent states formed by these particles could mimic the effects caused by a fundamental Higgs field.

M_{89} would be obtained in the first approximation by scaling the ordinary hadron physics by the ratio $\sqrt{\frac{M_{89}}{M_{107}}}$. This implies that QCD Λ, string tension, etc. get scaled by the appropriate power of this factor. If one estimates the u_{89} mass as $m(u_{89}) = m(\rho_{89})/2$ one obtains the TGD prediction for its mass as $m(u_{89}) = 512m(\rho_{107})/2 \simeq 197\ GeV$. Defining $u(89)$ mass by scaling the mass of ordinary u quark defined as one third of proton mass one obtains u_{89} mass about $160\ GeV$. This estimate for u_{89} mass happened to be within experimental uncertainties equal to the mass of the top candidate discovered just when the mass calculations were carried out and led to a tentative identification of the top candidate as u_{89}.

The fact that top candidate turned out to have production and decay characteristics of the real top forced to give up this hypothesis. Also the study of CKM matrix led to the cautious conclusion that only the mass of the experimental top candidate is consistent with CP breaking observed in $K - \bar{K}$ and $B - \bar{B}$ system (Appendix). Even more, the direct calculation of the u_{89} mass from p-adic thermodynamics gives $m(u_{89}) \simeq 262$ GeV and demonstrates the the idea about identifying top quark as u_{89} quark was a result of sloppy order of magnitude thinking. The relatively high mass however leaves open the possibility that M_{89} physics exists.

M_{89} physics means the emergence of a new condensate level in the hadronic physics. One can visualize M_{89} hadrons as very tiny objects possibly condensed on the quarks and gluons of M_{107} hadron physics. The New Physics begins to reveal itself, when the collision energy is so high that M_{89} hadrons inside quarks and gluons can exist as on mass shell particles (M_{89} hadron inside M_{107} hadron is comparable to a bee of size of one cm in a room of size about 5 meters!).

The new Physics at the energies not much above the energy scale of top is essentially the counterpart of ordinary hadron physics at cm energies of the order of ρ/ω meson mass. Therefore M_{89} meson resonances and their interactions described rather satisfactorily by the old fashioned string model with string tension scaled by factor 2^{18} should describe the situation. The electro-weak interactions should be in turn describable using generalization of current algebra ideas, such as PCAC and vector dominance model. If M_{89} hadrons condense on quarks and gluons this physics must be convoluted with the distribution functions of M_{89} hadrons inside quarks and gluons. The resonance structures are partially smeared out by the convolution process.

M_{89} vector mesons should be observed as resonances in e^+e^- annihilation and charged M_{89} pion should be pair produced at e^+e^- collision energies achievable in near future at LEP. Gamma pairs form unique signature of neutral leptopion. The following table gives the naive scaling estimate for the masses of lowest lying M_{89} hadrons.

meson	m/GeV	baryon	m/GeV
π^0	69.1	p	480.4
π^+	71.5	n	481.0
K^+	252.8	Λ	571.2
K^0	254.8	Σ^+	609.0
η	281.0	Σ^0	610.4
η'	490.5	Σ^-	610.5
ρ	394.2	Ξ^0	673.2
ω	400.9	Ξ^-	676.5
K^*	456.7	Ω^-	856.2
Φ	522		

Table 10. Masses of low lying hadrons for M_{89} hadron physics obtained by scaling ordinary hadron masses by a factor of 512.

Consider next the estimation of the production and decay rates for $\rho(89)\ /\omega(89)$ and more generally M_{89} mesons. In e^+e^- annihilation vector boson resonances are produced via the decay of virtual photon or Z^0. Since low energies are in question at M_{89} level the scaled up version of vector dominance model described in the nice book of Feynman [db1] should give a satisfactory description for the production of M_{89} mesons via resonance mechanism. The idea is to introduce direct coupling $F_V = m_V^2/g_V$ of photon (or gauge boson) to vector boson (ρ, ω, ϕ). The diagrams describing the production of mesons via decay of vector boson contain vector boson propagator $\frac{1}{p^2 - m_V^2 + im_V\Delta}$ and the production rate is enhanced by a factor $R = 4\pi m_V^2/(\Delta^2 g_V^2)$ in the resonance: the factor should be same in M_{89}

physics as in ordinary hadron physics. The ratio $r = \alpha_{em} R/\alpha_s$ gives a rough measure for the ratio of the rates of production for $u(89)$ and ordinary top quark. A rough estimate for what is to be expected is obtained by scaling the results of ordinary hadron physics. The table below gives the estimates for the quantity r and one has $r = 15.1$ for ω.

meson	$m/512\ MeV$	$\Delta/512\ MeV$	$g_V^2/4\pi$	r
ρ	770	150	2.27	0.52
ω	783	10	18.3	15.1
Φ	1019	4.2	13.3	230.8

Table 11. Scaled up resonance production parameters for ρ, ω and Φ. The last column of the table gives the value of the quantity $r = \alpha_{em} R/\alpha_s$, which should give a measure for the ratio of production rate of $u(_89)$ and of the production of ordinary top quark pair.

Centauro type events [hi11] might find nice explanation in terms of M_{89} hadron physics. If electroweak decay channels dominate over hadronic decay channels for M_{89} mesons this might lead to anomalously small abundance of ordinary pions in Centauro events. In particular, neutral M_{89} pions are expected to decay dominantly to photon pairs and since monoenergetic gamma pairs are used as a signature of pions the observed abundance of ordinary pions becomes small. Evidence for M_{89} pions comes from anomalous gamma pairs detected in the decays of Z^0 bosons [hh2] with total energy of about 60 GeV. The pairs might be related to the decay of M_{89} exotic pion predicted to have mass $m_{89} \simeq 2^9 m_\pi \simeq 67.5\ GeV$.

The resonance production of M_{89} vector mesons via the graph $q\bar{q} \to \gamma(virt) \to M(M_{89})$ and their decay to dijets gives small contribution to dijet production rate.

At high enough cm energies, presumably of order $\sqrt{s} \sim 10\ TeV$ in $p\bar{p}$ collisions the jets of M_{89} hadron physics should begin to manifest themselves. The unique signature of M_{89} jets is that the p_T spectrum for the hadrons of the jet, which is of form $exp(-kp_T^2/\Lambda(89))$, is by factor 512 wider than the p_T spectrum of hadrons for ordinary jets.

Following list gives some of the unique signatures of New Physics.

a) At higher energies exotic pions are produced abundantly and might be detectable via annihilation to monoenergetic photon pair. π^0 of the New Physics should have mass 69.1 GeV and $\gamma\gamma$ annihilation width $512 \cdot 7.63\ eV = 3.9\ MeV$ (obtained by scaling from that for ordinary pion). The width for the decay by W emission from either quark of $\pi^0(89)$ (the second is assumed to act as spectator) is of order $G_F^2 m(u(89))^5/(192\pi^3)$ and of order 2.5 MeV.

b) The scaling of mass splittings inside isopin multiplets with the scale factor 512 as compared to ordinary hadron physics is a unique signature of M_{89} hadrons.

c) The scaled up versions of ρ and ω meson should be found at nearby energies. Kaon (and s quark) of the New Physics should be seen as a decay product of $\Phi(522\ GeV) \to K + \bar{K}$: from table 14.3.5 one finds that that Φ should have rather small hadronic width $\Delta \simeq 2.2\ GeV$ so that the parameter measuring its production rate to the production rate of ordinary quark is as high as $r \simeq 230.8$ at resonance.

14.3.6   Topological evaporation and the concept of Pomeron

Topological evaporation provides a possible explanation for the mysterious concept of Pomeron originally introduced to describe hadronic diffractive scattering as the exchange of Pomeron Regge trajectory [hg1]. No hadrons belonging to Pomeron trajectory were however found and via the advent of QCD Pomeron was almost forgotten. Pomeron has recently experienced reincarnation [hg3, hg4, hg2]. In Hera [hg3] $e - p$ collisions, where proton scatters essentially elastically whereas jets in the direction of incoming virtual photon emitted by electron are observed. These events can be understood by assuming that proton emits color singlet particle carrying small fraction of proton's momentum. This particle in turn collides with virtual photon (antiproton) whereas proton scatters essentially elastically.

The identification of the color singlet particle as Pomeron looks natural since Pomeron emission describes nicely diffractive scattering of hadrons. Analogous hard diffractive scattering events in pX diffractive scattering with $X = \bar{p}$ [hg4] or $X = p$ [hg2] have also been observed. What happens is that proton scatters essentially elastically and emitted Pomeron collides with X and suffers hard scattering so that large rapidity gap jets in the direction of X are observed. These results suggest that Pomeron is real and consists of ordinary partons.

The TGD based identification of Pomeron is very economical: Pomeron corresponds to sea partons, when valence quarks are in vapour phase. In TGD inspired phenomenology events involving Pomeron correspond to pX collisions, where incoming X collides with proton, when valence quarks have suffered coherent simultaneous (by color confinement) evaporation into vapor phase. System X sees only the sea left behind in evaporation and scatters from it whereas valence quarks continue without noticing X and condense later to form quasi-elastically scattered proton. If X suffers hard scattering from the sea the peculiar hard diffractive scattering events are observed. The fraction of these events is equal to the fraction f of time spent by valence quarks in vapor phase.

Dimensional argument can be used to derive a rough order of magnitude estimate for f as $f \sim 1/\alpha = 1/137 \sim 10^{-2}$ for f: f is of same order of magnitude as the fraction (about 5 per cent) of peculiar events from all deep inelastic scattering events in Hera. The time spent in condensate is by dimensional arguments of the order of the p-adic legth scale $L(M_{107})$, not far from proton Compton length. Time dilation effects at high collision energies guarantee that valence quarks indeed stay in vapor phase during the collision. The identification of Pomeron as sea explains also why Pomeron Regge trajectory does not correspond to actual on mass shell particles.

The existing detailed knowledge about the properties of sea structure functions provides a stringent test for the TGD scenario. According to [hg4] Pomeron structure function seems to consist of soft $((1-x)^5$), hard $((1-x)$) and super-hard component (delta function like component at $x = 1$). The peculiar super hard component finds explanation in TGD based picture. The structure function $q_P(x, z)$ of parton in Pomeron contains the longitudinal momentum fraction z of the Pomeron as a parameter and $q_P(x, z)$ is obtained by scaling from the sea structure function $q(x)$ for proton $q_P(x, z) = q(zx)$. The value of structure function at $x = 1$ is non-vanishing: $q_P(x = 1, z) = q(z)$ and this explains the necessity to introduce super hard delta function component in the fit of [hg4].

14.4　Cosmic Rays and Mersenne Primes

TGD suggests the existence of a scaled up copy of hadron physics associated with each Mersenne prime $M_n = 2^n - 1$, n prime: M_{107} corresponds to ordinary hadron physics. There is some evidence for exotic hadrons. Centauro events and the peculiar events associated with $E > 10^5$ GeV radiation from Cygnus X-3 could be understood as due to the decay of gamma rays to M_{89} hadron pair in the atmosphere. The decay $\pi_n \to \gamma\gamma$ produces a peak in the spectrum of the cosmic gamma rays at energy $\frac{m(\pi_n)}{2}$ and there is evidence for the peaks at energies $E_{89} \simeq 34$ GeV and $E_{31} \simeq 3.5 \cdot 10^{10}$ GeV. The absence of the peak at $E_{61} \simeq 1.5 \cdot 10^6$ GeV can be understood as due to the strong absorption caused by the e^+e^- pair creation with photons of the cosmic microwave background.

Cosmic string decays *cosmic string* $\to M_2$ *hadrons* $\to M_3$ *hadrons* $..\to M_{107}$ *hadrons* is a new source of cosmic rays. The mechanism could explain the change of the slope in the hadronic cosmic ray spectrum at M_{61} pion rest energy $3 \cdot 10^6$ GeV. The cosmic ray radiation at energies near 10^9 GeV apparently consisting of protons and nuclei not lighter than Fe might be actually dominated by gamma rays: at these energies γ and p induced showers have same muon content and the decays of gamma rays to M_{89} and M_{61} hadrons in the atmosphere can mimic the presence of heavy nuclei in the cosmic radiation.

14.4.1　Mersenne primes and mass scales

p-Adic mass calculations lead to quite detailed predictions for elementary particle masses. In particular, there are reasons to believe that the most important fundamental elementary particle mass scales correspond to Mersenne primes $M_n = 2^n - 1$, $n = 2, 3, 7, 13, 17, 19, ...$

$$m_n^2 = \frac{m_0^2}{M_n} ,$$

$$m_0 \simeq 1.41 \cdot \frac{10^{-4}}{\sqrt{G}} , \tag{14.4.0}$$

where $\sqrt{G}$ is Planck length. The known elementary particle mass scales were identified as mass scales associated identified with Mersenne primes $M_{127} \simeq 10^{38}$ (leptons), M_{107} (hadrons) and M_{89}

(intermediate gauge bosons). Of course, also other p-adic length scales are possible and it is quite possible that not all Mersenne primes are realized.

Theory predicts also some higher mass scales corresponding to the Mersenne primes M_n for $n = 89, 61, 31, 19, 17, 13, 7, 3$ and suggests the existence of a scaled up copy of hadron physics with each of these mass scales. In particular, masses should be related by simple scalings to the masses of the ordinary hadrons.

An attractive hypothesis is that the color interactions of the particles of level M_n can be described using the ordinary QCD scaled up to the level M_n so that that masses and the confinement mass scale Λ is scaled up by the factor $\sqrt{M_n/M_{107}}$.

$$\Lambda_n = \sqrt{\frac{M_n}{M_{107}}} \Lambda \ . \tag{14.4.1}$$

In particular, the masses of the exotic pions associated with M_n are given by

$$m(\pi_n) = \sqrt{\frac{M_n}{M_{107}}} m_\pi \ . \tag{14.4.2}$$

Here $m_\pi \simeq 135 \ MeV$ is the mass of the ordinary pion.

The interactions between the different level hadrons are mediated by the emission of electro-weak gauge bosons and by gluons with cm energies larger than the energy defined by the confinement scale of level with smaller p. The decay of the exotic hadrons at level M_{n_k} to exotic hadrons at level $M_{n_{k+1}}$ must take place by a transition sequence leading from the effective M_{n_k}-adic space-time topology to effective $M_{n_{k+1}}$-adic topology. All intermediate p-adic topologies might be involved.

14.4.2 Cosmic strings and cosmic rays

Cosmic strings

Cosmic strings (not quite the same thing in TGD as in GUTs) are basic objects in TGD inspired cosmology [D5, D4].

a) In TGD inspired galaxy model galaxies are regarded as mass concentrations around cosmic strings and the energy of the string corresponds to the dark energy whereas the particles condensed at cosmic strings and magnetic flux tubes resulting from them during cosmic expansion correspond to dark matter [D5, D4]. The galactic nuclei, often regarded as candidates for black holes, are the most probable seats for decaying highly entangled cosmic strings.

b) Galaxies are known to organize to form larger linear structures. This can be understood if the highly entangled galactic strings organize around long strings like pearls in necklace. Long strings could correspond to galactic jets and their gravitational field could explain the constant velocity spectrum of distant stars in the galactic halo.

c) In [D5, D4, D6] it is suggested that decaying cosmic strings might provide a common explanation for the energy production of quasars, galactic jets and gamma ray bursters and that the visible matter in galaxies could be regarded as decay products of cosmic strings. The magnetic and Z^0 magnetic flux tubes resulting during the cosmic expansion from cosmic strings allow to assign at least part of gamma ray bursts to neutron stars. Hot spots (with temperature even as high as $T \sim \frac{10^{-3,5}}{\sqrt{G}}$) in the cosmic string emitting ultra high energy cosmic rays might be created under the violent conditions prevailing in the galactic nucleus.

The decay of the cosmic strings provides a possible mechanism for the production of the exotic hadrons and in particular, exotic pions. In [hi16] the idea that cosmic strings might produce gamma rays by decaying first into 'X' particles with mass of order $10^{15} \ GeV$ and then to gamma rays, was proposed. As authors notice this model has some potential difficulties resulting from the direct production of gamma rays in the source region and the presence of intensive electromagnetic fields near the source. These difficulties are overcome if cosmic strings decay first into exotic hadrons of type M_{n_0}, $n_0 \geq 3$ of energy of order $2^{-n_0+2} 10^{25} \ GeV$, which in turn decay to exotic hadrons corresponding to M_k, $k > n_0$ via ordinary color interaction, and so on so that a sequence of M_k:s starting some value of n_0 in $n = 2, 3, 7, 13, 17, 19, 31, 61, 89, 107$ is obtained. The value of n remains open at this stage and depends on the temperature of the hot spot and much smaller temperatures than the $T \sim m_0$ are possible: favored temperatures are the temperatures $T_n \sim m_n$ at which M_n hadrons become unstable against thermal decay.

Decays of cosmic strings as producer of high energy cosmic gamma rays

In [hi17] the gamma ray signatures from ordinary cosmic strings were considered and a dynamical QCD based model for the decay of cosmic string was developed. In this model the final state particles were assumed to be ordinary hadrons and final state interactions were neglected. In present case the string decays first to M_{n_0} hadrons and the time scale of for color interaction between M_{n_0} hadrons is extremely short (given by the length scale defined by the inverse of π_{n_0} mass) as compared to the time time scale in case of ordinary hadrons. Therefore the interactions between the final state particles must be taken into account and there are good reasons to expect that thermal equilibrium sets on and much simpler thermodynamic description of the process becomes possible.

A possible description for the decaying part of the highly tangled cosmic string is as a 'fireball' containing various M_{n_0} ($n \geq 3$) partons in thermal equilibrium at Hagedorn temperature T_{n_0} of order $T_{n_0} \sim m_{n_0} = 2^{-2+n_0} \frac{10^{-4}}{k\sqrt{G}}$, $k \simeq 1.288$. The experimental discoveries made in RHIC suggest [hk3] that high energy nuclear collisions create instead of quark gluon plasma a liquid like phase involving gluonic BE condensate christened as color glass condensate. Also black hole like behavior is suggested by the experiments.

RHIC findings inspire a TGD based model for this phase as a macroscopic quantum phase condensed on a highly tangled color magnetic string at Hagedorn temperature. The model the notions of conformal confinement and dynamical but quantized $\hbar$ [J6]. This phase has no direct gauge interactions with ordinary matter and is identified in TGD framework as a particular instance of dark matter. Quite generally, quantum coherent dark matter would reside at magnetic flux tubes idealizable as string like objects with string tension determined by the p-adic length scale and thus outside the "ordinary" space-time.

a) The tangled cosmic string begins to cool down and when the temperature becomes smaller than $m(\pi_{n_0})$ mass it has decayed to M_{n_1} matter which in turn continues to decay to M_{n_2} matter. The decay to M_{n_1} matter could occur via a sequence $n_0 \rightarrow n_0 - 1 \rightarrow ...n_1$ of phase transitions corresponding to the intermediate p-adic length scales $p \simeq 2^k$, $n_1 \geq k > n_0$. Of course, all intermediate p-adic length scales are in principle possible so that the process would be practically continuous and analogous to p-adic length scale evolution with $p \simeq 2^k$ representing more stable intermediate states.

b) The first possibility is that virtual hadrons decay to virtual hadrons in the transition $k \rightarrow k-1$. The alternative option is that the density of final state hadrons is so high that they fuse to form a single highly entangled hadronic string at Hagedorn temperature T_{k-1} so that the process would resemble an evaporation of a hadronic black hole staying in quark plasma phase without freezing to hadrons in the intermediate states. This entangled string would contain partons as "color glass condensate".

c) The process continues until all particles have decayed to ordinary hadrons. Part of the M_n low energy thermal pions decay to gamma ray pairs and produce a characteristic peak in cosmic gamma ray spectrum at energies $E_n = \frac{m(\pi_n)}{2}$ (possibly red-shifted by the expansion of the Universe). The decay of the cosmic string generates also ultra high energy hadronic cosmic rays, say protons. Since the creation of ordinary hadron with ultra high energy is certainly a rare process there are good hopes of avoiding the problems related to the direct production of protons by cosmic strings (these protons produce two high flux of low energy gamma rays, when interacting with cosmic microwave background [hi16]).

Exotic cosmic ray events and exotic hadrons

One signature of the exotic hadrons is related to the interaction of the ultra high energy gamma rays with the atmosphere. What can happen is that gamma rays in the presence of an atmospheric nucleus decay to virtual exotic quark pair associated with M_{n_k}, which in turn produces a cascade of exotic hadrons associated with M_{n_k} through the ordinary scaled up color interaction. These hadrons in turn decay $M_{n_{k+1}}$ type hadrons via mechanisms to be discussed later. At the last step ordinary hadrons are produced. The collision creates in the atmospheric nucleus the analog of quark gluon plasma which forms a second kind of fireball decaying to ordinary hadrons. RHIC experiments have already

discovered these fireballs and identified them as color glass condensates [hk3]. It must be emphasized
that it is far from clear whether QCD really predicts this phase.

These showers differ from ordinary gamma ray showers in several respects.

a) Exotic hadrons can have small momenta and the decay products can have isotropic angular
distribution so that the shower created by gamma rays looks like that created by a massive particle.

b) The muon content is expected to be similar to that of a typical hadronic shower generated by
proton and larger than the muon content of ordinary gamma ray shower [hi1].

c) Due to the kinematics of the reactions of type $\gamma + p \rightarrow H_{M_n} + ... + p$ the only possibility at the
available gamma ray energies is that M_{89} hadrons are produced at gamma ray energies above $10\ TeV$.
The masses of these hadrons are predicted to be above $70\ GeV$ and this suggests that these hadrons
might be identified incorrectly as heavy nuclei (heavier than ^{56}Fe). These signatures will be discussed
in more detail in the sequel in relation to Centauro type events, Cygnus X-3 events and other exotic
cosmic ray events. For a good review for these events and models form them see the review article
[hi19].

Some cosmic ray events [hi4, hi3] have total laboratory energy as high as 3000 TeV which suggests
that the shower contains hadron like particles, which are more penetrating than ordinary hadrons.
One might argue that exotic hadrons interact only electro-weakly (color is confined in the length
scale associated with M_n) with the atmosphere one might argue that they are more penetrating than
the ordinary hadrons. The observed highly penetrating fireballs could also correspond color glass
condensates of partons identifiable as conformally confined phase and having no direct interactions
with ordinary matter or even to BE condensate like states of conformally confined dark hadrons
possessing complex conformal weights.

Exotic mesons can also decay to lepton pairs and neutral exotic pions produce gamma pairs. These
gamma pairs in principle provide a signature for the presence of exotic pions in the cosmic ray shower.
If M_{89} proton is sufficiently long-lived enough they might be detectable. The properties of Centauro
type events however suggest that M_{89} protons are short lived.

14.4.3   Peaks in cosmic gamma ray spectrum

The decay of the M_n pions at rest to two gamma rays produces gamma rays with energy $E_n = \frac{m(\pi_n)}{2}$.
Therefore the cosmic gamma ray spectrum might show detectable signatures at these energies.

There is indeed some evidence for this kind of signatures in cosmic gamma ray background.

a) There are indications that the energy density of the cosmic gamma ray spectrum has peak at
energy near $33.5\ GeV$ ([hi13], see Fig. 14.4.3). A possible identification is as gamma rays produced by
the decay of M_{89} pions. The energy distribution would be induced from the non-relativistic thermal
distribution with temperature near $m(\pi_{89})$.

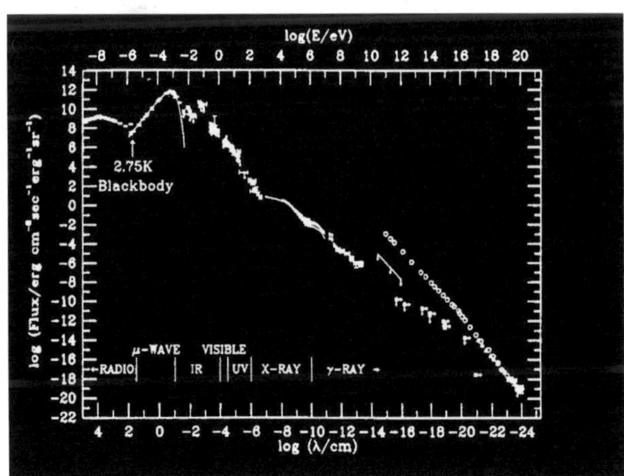

Figure 14.1: There are some indications that cosmic gamma ray flux contains a peak in the energy
interval $10^{10} - 10^{11}\ eV$. Figure is taken from [hi13].

b) M_{61} corresponds to gamma ray threshold energy of $1.7 \cdot 10^6 \ GeV$. There is no visible signature at this energy but there is a good explanation for this. The e^+e^- pair production of the gamma rays with energy in certain energy range above $10^6 \ GeV$ with the photons of the cosmic microwave background implies strong reduction of the gamma ray flux by 2-3 orders of magnitude [hi17]. According to [hi17] the cutoff red-shift is of order $z - 1 \simeq e^{-5}$ at this energy and corresponds to an upper bound of order 10^8 light years (the size of the large voids) for the distance of the source to be observable. The energy of the gamma rays coming from M_{61} pions happens to belong to the region with strongest absorption.

c) M_{31} corresponds to energy of the order of $1.7 \cdot 10^{10} \ GeV$ and jump in cosmic ray energy density is expected. As figure 14.4.3 shows, the cosmic ray spectrum contains indeed an bump at this energy [hi10, hi16, hi17]: the energy flux has a peak in short energy interval above $1.7 \cdot 10^{10} \ GeV$. The simplest possibility is that the bump results from the decay of thermal M_{31} pions created in the decay of cosmic string. The effect is partially masked by the annihilation of gamma rays and photons of the cosmic radio wave background to e^+e^- pairs above the energy $5 \cdot 10^9 \ GeV$ and the greatest effect comes at $3 \cdot 10^{10} \ GeV$ [hi17] (the mass of the exotic pion!).

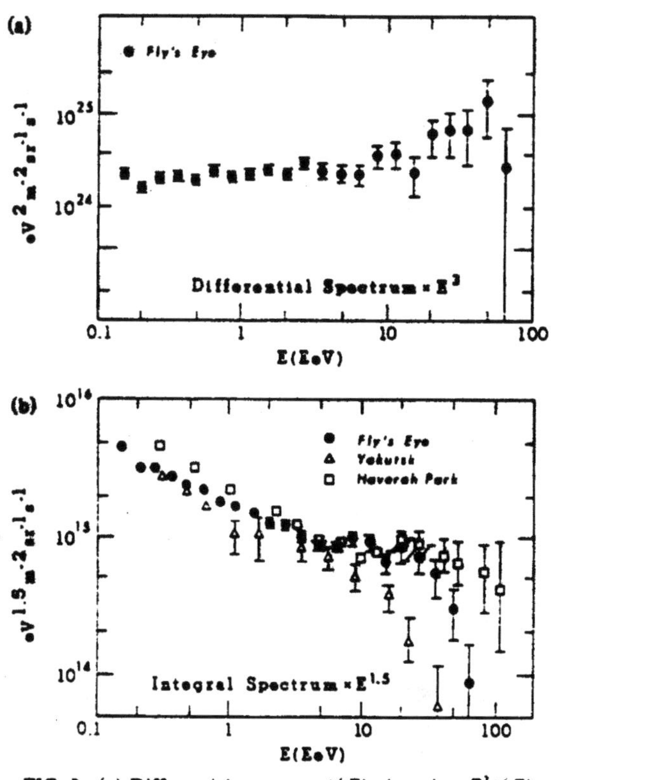

FIG. 2. (a) Differential spectrum $j(E)$ plotted as $E^3 j(E)$. A power-law best fit of the form $j(E) = aE^{-\gamma}$ yields $a = 109.6 \pm 2.2$ EeV^{-1} km^{-2} sr^{-1} yr^{-1} and $\gamma = 2.94 \pm 0.02$ for events at $E < 10$ EeV. Between 10 and 50 EeV we obtain $a = 34 \pm 17$ EeV^{-1} km^{-2} sr^{-1} yr^{-1} and $\gamma = 2.42 \pm 0.27$. The lack of events above 50 EeV indicates that the flattened slope does not continue. (b) Integral spectrum $I(>E)$ plotted as $E^{1.5}I(>E)$. Data from both Haverah Park and Yakutsk (Refs. 10, 12, and 13) experiments are also shown.

Figure 14.2: A bump in cosmic ray spectrum as a possible indication for M_{31} hadron physics.

An alternative explanation for the bump is based on the assumption that cosmic rays are predominantly protons at these energies [hi9]. The proton component of the cosmic ray spectrum is predicted to effectively terminate at energy about $7 \cdot 10^{10} \ GeV$ due to pion production from cosmic microwave background. The experimental situation is unclear at this moment. Haverah Park detector claims the detection of 4 events with energies above 10^{11} GeV whereas Fly's Eye detector reports no events [hi14].

d) The theory predicts further peaks at $m_{19} = 64 \cdot m_{31} \simeq 2 \cdot 10^{12}$ GeV, $m_{17} = 2m_{19}$,... It might well be possible in not so far future to verify whether cosmic gamma ray flux contains these peaks.

14.4.4 Centauro type events, Cygnus X-3 and M_{89} hadrons

The results reported by Brazil-Japan Emulsion Chamber Collaboration [hi4, hi6] on multiple production of hadrons induced by cosmic rays with energies $E_{lab} > 10^5$ GeV provide evidence for new Physics. The distributions for the transverse momentum p_T and longitudinal momentum fraction x for pions were found to differ from the distributions extrapolated from lower energies. The widening of the transversal momentum distributions has also been observed at accelerator energies (ISR above $\sqrt{s} = 63$ GeV and CERN SPS-$p\bar{p}$ Collider at $\sqrt{s} = 540$ GeV). Furthermore, exotic events called Geminion, Centauro, Chiron with emission of $n_B \leq 100$ hundred baryons but practically no pions were detected. There are also peculiar events associated with the radiation coming from Cygnus X-3. A recent summary about peculiar events is given in the review article [hi19].

Mirim, Acu and Quacu

The exotic cosmic ray events are described in the review article of [hi4]. In [hi4] the multiple production of pions is classified into 3 jet types called Mirim, Acu and Quacu. Although the transverse momentum distributions for pions observed at low energies are universal, Acu and Quacu jets are characterized by wider transverse momentum distributions with larger value of average transverse momentum p_T than in low energy pionization: this widening is in accordance with accelerator results. The distributions for the longitudinal momentum fraction x scale but differ from the low energy situation for Acu and Quacu jets.

In [hi4, hi7, hi11] a description of these events in terms of 'fireballs' decaying into ordinary hadrons were considered. The p_T distribution associated with Mirim is just the ordinary low energy transverse momentum distribution whereas the distributions associated with Acu and Quacu are wider. The masses of the fireballs were assumed to be discrete and were found to be $M_0 \sim 2 - 3$ GeV (Mirim), $M_1 \sim 15 - 30$ GeV (Acu), $M_2 \sim 100 - 300$ GeV (Quacu). It should be noticed that the upper bounds for the masses associated with Acu and Quacu fireballs are roughly by a factor of two smaller than the masses associated with M_{89} pion and M_{89} proton. The temperatures were found to be in range $0.4 - 10$ GeV for Acu and Quacu fireball and to be substantially larger than the ordinary Hagedorn temperature $T_H \simeq 0.16$ GeV.

Chirons, Centauros, anti-Centauros, and Geminions

For the second class of events consisting of Chirons, Centauros and Geminions observed at laboratory energies $100 - 1000$ TeV pion production is strongly suppressed (gamma pairs resulting from the decay of neutral pions are almost absent) [hi4]. The primary event takes place few hundred meters above the detector and decay products are known to be hadrons and mostly baryons: about 15 (100) for Mini-Centauros (Centauros). This excludes the possibility that exotic hadrons decay in emulsion chamber and implies also that the decay mechanism of the primary particle is such that very few mesons are produced.

The fireball hypothesis has been applied also to Centauro type events assuming that fireballs corresponds to a different phase than in the case of Mirim, Acu and Quacu [hi4, hi11]. The fireball masses associated with Mini-Centauro and Centauro are according to the estimate of [hi4] $M_{mini} = 35$ GeV and $M_{Centauro} = 23$ GeV. These masses are almost exactly one half of the masses of the M_{89} pion (70 GeV) and proton (470 GeV) respectively!

$$
\begin{aligned}
M_{Mini} &\simeq \frac{m(\pi_{89})}{2} \ , \\
M_{Centauro} &\simeq \frac{m(p_{89})}{2} \ .
\end{aligned}
\tag{14.4.2}
$$

This suggests that the decay of cosmic gamma ray to M_{89} quark pair which in turn hadronizes to (possibly virtual) M_{89} hadrons induced by the interaction with the nucleon of atmosphere is the origin of Mini-Centauro/Centauro events.

The basic difference between the decaying fireballs in Acu/Quacu events and Centauro type events is that Acu/Quacu decays produce neutral pions unlike Centauros.

The appearance of the factor of 1/2 in the mass estimates needs an explanation. One explanation is systematic error in the evaluation of hadronic energy: for instance, the gamma inelasticity k_γ telling which fraction of hadronic energy is transformed to electromagnetic energy might be actually smaller than believed by a factor of order two. An alternative explanation is related to the decay mechanism of M_{89} particle: if the decay takes place via a decay to two off mass shell M_{89} hadrons decaying in turn to hadrons then the average rest energy of the fireball is indeed one half of the mass of the decaying on mass shell particle. The reason for the necessity of off mass shell intermediate states is perhaps the stability of the on mass shell exotic hadrons against the direct decay to ordinary hadrons.

Anti-Centauros are much like Centauros except that neutral pions are over-abundant [hi19]. The speculative model [hi18] relies on the notion of chiral condensates consisting of neutral pions in the case of Centauros and charged pions in the case of anti-Centauros.

The case of Cygnus X-3

There are peculiar events associated with the cosmic rays coming from Cygnus X-3 at gamma ray energies above 10^5 GeV [hi12]. The primary particle must be massless particle and is most probably ordinary gamma ray. The structure of the shower however suggests that the decaying particle is very massive! Furthermore, the muon content of the shower is larger than that associated with gamma ray shower. A possible explanation is that the gamma rays coming from Cygnus X-3 with energy above the threshold 10^4 GeV produce M_{89} hadrons, which in turn create the cosmic ray shower through the decay to M_{89} hadrons and the decay of these to the ordinary M_{107} hadrons: this indeed means that the gamma rays behave like massive particles in the atmosphere.

14.4.5   TGD based explanation of the exotic events

The TGD based model for exotic events involve p-adic length scale hierarchy, many-sheeted space-time, and TGD inspired view about dark matter. A decisive empirical input comes from RHIC events suggesting that quark gluon plasma is actually a liquid like "macroscopic" quantum phase identifiable as a particular instance of dark matter.

General considerations

The mass estimates for the fireballs and the absence of neutral pions suggest that Mini-Centauro/Centauro type events correspond to the decay of M_{89} hadrons (pion/proton) to ordinary hadrons. The general model for the exotic events would be following.

a) Cosmic gamma ray decays first into M_{89} quark pair via electromagnetic interaction with the nucleon of the atmosphere. Pairs of Centauros/anti-Centauros and quark-gluon-plasma blobs explaing Mirim/Qcu/Quacu events would be naturally created in these collisions.

b) The quark pair in turn hadronizes to M_{89} hadrons decaying to virtual $k > 89$ hadrons which in turn end up via a sequential decay process to ordinary hadrons. This process is kinematically possible if the condition $E_{tot} > 2M^2/m_p$, is satisfied (M is the mass of the exotic hadron). For example, the energy of the gamma ray must be larger than $500\,TeV$ for exotic proton pair production. For the exotic pion the corresponding lower bound is about $10\,TeV$. The energies of the exotic events are indeed above $100\,TeV$ in accordance with these bounds. The average total energy is about $E_{tot} = 1740\,TeV$ for Centauros and $E_{tot} \simeq 903\,TeV$ for Mini-Centauros [hi11]. The mechanism implies that two M_{89} fireballs are produced. 'Binocular' events (Geminions) consisting of two widely separated fireballs have indeed been observed [hi4].

c) If anti-Centauros result via the same mechanism there must be a mechanism explaining why the production of neural pions varies from event to event. One proposal is that the difference is due to a formation of pion condensates consisting of neural *resp.* charged pions in the two situations [hi18]. This hypothesis would unify Centauro events with anti-Centauro events in which the production of neutral pions is abnormally high [hi19].

d) Mirim/Acu/Quacu events could correspond to the decay of a high temperature quark-gluon plasma blob, or rather color glass condensate, to hadrons (recall that the estimated plasma temperatures are much lower than for Centauros). The collision of M_{89} hadron possibly generated in the interaction of the cosmic gamma ray with ordinary nucleon could induce both the decay of M_{89} hadron to virtual hadrons and generate quark-gluon plasma blob in the atmospheric target nucleus. Hagedorn temperature $T(k)$, $89 < k \leq 107$ is a good guess for the temperature of this plasma blob. RHIC findings [hk3] suggest that the blob corresponds to highly tangled hadronic string containing partons in conformally confined phase identifiable as dark matter and decaying by de-coherence to ordinary hadrons [J6].

Connection with TGD based model for RHIC events

The counterparts of Centauros and other exotic events have not been observed in accelerator experiments. More than a decade after writing the first version of the model for Centauros came however data from RHIC experiment [hk3], which seems to provide a connection between laboratory and cosmic ray data. In RHIC collisions of very energetic Gold nuclei are studied. The collisions were expected to create a quark gluon plasma freezing to ordinary hadrons. The surprise was that the resulting state behaves like an ideal liquid and has also black hole like properties [hk3].

TGD based model for RHIC findings

The TGD based model [D6, J6] for RHIC findings is following.

a) The state in question corresponds to a highly entangled hadronic string at Hagedorn temperature defining the analog of black hole and decaying by evaporation. The gravitational constant defined by Planck length is effectively replaced by a hadronic gravitational constant defined by the hadronic length scale. p-Adic length scale hypothesis predicts entire hierarchy of Hagedorn temperatures.

b) Bose-Einstein condensate of gluons referred to as color glass condensate has been proposed as an explanation for the liquid like behavior of the quark-gluon phase. TGD based explanation for the liquid like state is that that quarks and gluons are not in the ordinary state but possess complex conformal weights closely related to the zeros of Riemann Zeta (of the basic distinctions between TGD and super string models) such that the net conformal weight of the system is real. This gives rise to what might be called conformal confinement forcing the system to behave like single coherent unit. Dark matter is identified as conformally confined matter and interacts with the ordinary matter only via a coherent emission of "laser beams" of dark bosons which must de-cohere into ordinary bosons before the interaction.

c) A further assumption is that the state corresponds to a large value of $\hbar$ increasing the length and time scales of quantum coherence since typical length and time scales are proportional to $\hbar$. In lowest order in $\hbar$ (classical limit) the physics does not change but higher order corrections are reduced since gauge coupling strengths are reduced. For the situation involving non-perturbative effects (typically binding energies) the change of $\hbar$ induces more dramatic effects.

d) An even stronger assumption is that electro-weak gauge symmetry is restored in dark matter phase in the sense that electro-weak gauge bosons are massless but fermions possess essentially the masses predicted by p-adic thermodynamics. The hypothesis is motivated by the proposed resolution of the interpretational problems related to the existence of long range classical electro-weak gauge fields: in condensed matter length scales these fields would be space-time correlates for massless dark electro-weak gauge bosons.

For M_{89} hadrons the restoration of the electro-weak symmetry is especially natural since in TGD framework classical induce fields are massless for known non-vacuum extremals. The finite size of the space-time sheet carrying these fields brings in the length scale determining the boson mass when the space-time sheet in question looks point like in the length scale resolution used.

A more precise model for exotic events

A more detailed formulation necessitates a rough model for the transformation of M_{89} hadrons to M_{107} hadrons.

a) On mass shell exotic hadrons can be assumed to be stable against direct decay to ordinary hadrons so that their decay must take place via a sequential decay to off mass shell exotic hadrons characterized by $107 > k > 89$, which eventually decay to ordinary hadrons. The simplest decay mode is the decay to two virtual exotic hadrons with average mass, which is one half of the mass of the decaying exotic hadron in accordance with observations.

b) M_{89} hadron decays to virtual hadrons with $p \simeq 2^k > M_{89}$ dominate over electro-weak decays since the characteristic time scale is defined by $\Lambda(QCD, M_{89}) = 512\Lambda(QCD, 107)$. This means that most of the energy in the process goes to virtual $k > 89$ virtual mesons. Neutral $k > 89$ virtual pions, if created, can decay to gamma pairs so that the problem of understanding the absence of neutral pions remains.

c) M_{89} hadronic space-time sheet suffers a topological phase transition to M_{107} hadronic space-time sheet via several steps $k = 89 \rightarrow k_1 > 89.. \rightarrow k_n = 107$. In the process the size of hadronic surface suffers a $2^9 = 512$-fold expansion meaning the increase of volume by a factor for $2^{27} \sim 10^9/8$ so that a small scale Big Bang is really in question! The expansion brings in mind liquid-vapor phase transition but the freezing to hadrons (due to the properties of color coupling constant evolution) makes the transition more like a liquid-solid phase transition.

As noticed all p-adic length scales in the range involved could be present but $p \simeq 2^k$ would define more stable intermediate states. A possible experimental signature for the sequence of the phase transitions labelled by $89 \leq k \leq 107$ is a bumpy structure of the detected hadronic cascades with a maximum of 17 maxima. This kind of structure with a constant distance between maxima and 11 maxima has been indeed observed for some cascades (see Fig. 8 of [hi19]).

A good guess for the critical temperature of the Big Bang like phase transition to occur is $T_{cr}(89) = km_{89}$, where k is some numerical factor. TGD inspired model for the early cosmology provides a universal hydrodynamics model for this period as a mini Big Bang, or rather "a soft whisper amplified to a relatively big bang", containing the duration of the period as the only parameter [D5].

d) If the decay process is fast enough, the density of virtual hadrons in the final state becomes so high that they form single highly tangled cosmic string in Hagedorn temperature $T(k)$. An entire sequence of $T(k) = km_k$, $107 > k > 89$ of phase transition temperatures could be involved without intermediate freezing to hadrons. Since the transformation of $k = 89$ hadrons to $k = 107$ hadrons would be essentially a decay process, the distribution of decay products is isotropic in the center of mass frame of $k = 89$ hadron (Centauros/anti-Centauros). The same conclusion holds true for the decay of quark gluon plasma (Mirim/Qcu/Quacu).

How to understand the anomalous production of pions?

Both Centauros and anti-Centauros can be understood if the transformation of M_{89} hadrons to ordinary hadrons generates "mis-aligned" pionic BE condensates. $U(2)_{ew}$ symmetry is restored for M_{89} hadrons and there is no preferred isospin direction for the order parameter of M_{89} pionic BE condensate. This BE condensate is however excluded by energetic considerations. The sequence of phase transitions leading to M_{107} hadrons involving intermediate p-adic length scales could however generate this kind of BE condensate.

If an overcooling occurs in the sense that electro-weak symmetry is not lost, the first intermediate pion condensate can correspond to π_+, π_- or π_0. Charged π condensates would be created in pairs with opposite charges. In this kind of situation the number of gamma rays produced in the decay to ordinary hadrons would vary from event to event.

The presence of pionic BE condensates favors the decay to M_{107} hadrons via hadronic intermediate states rather than via the cooling of partonic phase condensed on single tangled string whose length grows. This and the idea that $U(2)_{ew}$ symmetry could be exact for the dark matter phase, encourages to consider also the possibility that M_{89} hadron decays to a state consisting of dark M_{107} hadrons having complex conformal weights and forming a BE condensate like state behaving like single coherent unit and interacting with the ordinary matter only via emission of dark gauge boson BE condensates de-cohering to ordinary gauge bosons.

Dark pionic BE condensates with various charges could be present. These dark π condensates would decay coherently to pairs of dark ew boson "laser beams", which can interact with the ordinary matter only after they have de-cohered to ordinary ew gauge bosons and remain undetected if the de-coherence time for dark bosons is long enough, probably not so. Dark hadron option could thus explain also the abnormally long penetration lengths.

14.4.6   Cosmic ray spectrum and exotic hadrons

The hierarchy of M_n hadron physics provides also a mechanism producing ultra high energy cosmic gamma rays and hadrons.

Do gamma rays dominate the spectrum at ultrahigh energies?

A possible piece of evidence for M_{89} hadrons is related to the analysis [hi15] of the cosmic ray composition near 10^9 GeV. The analysis was based on the assumption that the spectrum consists of nuclei. The assumptions and conclusions of the analysis can be criticized:

a) There is argument [hi2], which states that the interaction of protons having energy above 10^9 GeV with the cosmic microwave background implies pion pair creation and a rapid loss of proton energy so that the contribution of protons should be strongly suppressed in the cosmic ray spectrum above $E = 7 \cdot 10^{10}$ GeV. If protons dominate, cosmic ray spectrum should effectively terminate at energy of order $7 \cdot 10^{10}$ GeV: some events above $E = 10^{11}$ GeV have been however detected [hi14].

b) It is not obvious whether one can distinguish between protons and gamma rays at these energies since the muon content of the photon and proton showers are near to each other at these energies [hi16]. Therefore the particles identified as protons might well be gamma rays.

c) The spectrum can be fitted assuming that cosmic ray spectrum has two components. Light component ('protons') can be identified as protons and He nuclei. The heavy component ('Fe') corresponds to Fe and heavier nuclei. The nuclei between He and Fe seem to be peculiarly absent. Furthermore, there are also indications that spectrum contains only light nuclei in the range $3 \cdot 10^7 - 10^{11}$ GeV [hi5].

An alternative interpretation suggested also in [hi16] is that cosmic ray flux is dominated by gamma rays at these energies. 'Protons' correspond to gamma rays interacting ordinarily with matter. 'Fe nuclei' correspond to the fraction of gamma rays decaying first into M_{89} exotic quark pair producing corresponding exotic hadrons, which then decay to ordinary hadrons and produce showers resembling ordinary heavy nucleus shower.

Hadronic component of the cosmic ray spectrum

The properties of the hadronic cosmic ray spectrum above $4 \cdot 10^5$ GeV are not well understood.

a) It has turned out difficult to invent acceleration mechanisms producing hadronic cosmic rays having energies above 10^5 GeV [hi15].

b) The spectrum contains a 'knee' (power $E^{-2.7}$ changes to about E^{-3} at the knee), which is at the energy $3 \cdot 10^6$ GeV [hi15] and equals to the mass of M_{61} pion. It is difficult to understand how the knee is generated although several explanations have been proposed (these are reviewed shortly in [hi15]).

A possible solution of the problems is that part of the hadronic cosmic rays are generated in the decay of string like objects rather than by some acceleration mechanism. Assume that M_{n_k} hadron is created in the decay cascade. Since $M_{n_{k+m}}$, $m = 1, 2, ..$ hadrons can have rest masses above M_{n_k} threshold mass, one can consider the possibility that M_{n_k} hadron decays sequentially to ordinary M_{107} hadron with arbitrary large rest mass (even larger than M_{n_k} pion mass) and that this ordinary hadron in turn produces some very energetic low mass hadrons, say proton and antiproton, identifiable as cosmic rays. The most efficient producers of hadrons are M_{n_k} pions since these are produced most abundantly in the decay of $M_{n_{k+1}}$ hadrons. M_{n_k} pion at rest cannot however decay to ordinary hadrons with energy above M_{n_k} pion mass. Therefore the slope of the cosmic ray energy flux should become steeper above M_{n_k}, in particular M_{61}, threshold.

The problem of relic quarks and hierarchy of QCD:s

Baryon and lepton numbers are conserved separately in TGD and one of the basic problems of the gauge theories with conserved baryon number is the problem of relic quarks. Hadronization starts in temperature of the order of quark mass and since hadronization is basically many quark process it continues until the expansion rate of the Universe becomes larger than the rate of the hadronization. As a consequence, the number density of relic quarks is much larger than the upper bound $n_{relic} < \rho_B/m_q = 10^{-9} n_\gamma m_p/m_q$ obtained from the requirement that the contribution of relic quarks to mass density is smaller than the baryonic mass density. There is also an experimental upper bound $n_{relic} < 10^{-28} n_\gamma$.

The assumption about the existence of QCD:s with a hierarchy of increasing scales $\Lambda_{QCD}(M_n)$ implies that the length scale $L(n) \sim 1/\sqrt{\Lambda_{QCD}(M_n)}$ below which quarks are free, decreases with increasing cosmic temperature and therefore the problem of the relic quarks disappears.

14.5 The anomalously large direct CP violation in $K \to 2\pi$ decay

KTeV collaboration in Fermilab [hj13] has measured the parameter $|\epsilon'/\epsilon|$ characterizing the size of the direct CP violation in the decays of kaons to two pions. The value of the parameter was found to be $|\epsilon'/\epsilon| = (2.8 \pm .1)10^{-3}$ and is almost by an order of magnitude larger than the naive standard model expectations based on the hypothesis that direct CP breaking is induced by CKM matrix. In [hj14] it was shown that the value of the parameter could be understood without introducing any new physics if the value of running strange quark mass at m_c is about $m_s(m_c) = .1$ GeV and $m_d << m_s$ holds true.

14.5.1 How to solve the problems in TGD framework

Problems

Also in TGD framework the situation looks confusing.

a) The TGD based prediction for the value of the CP breaking parameter for CKM matrices satisfying the constraints coming from p-adicity is within the experimental constraints $1.0 \times 10^{-4} \leq J \leq 1.7 \times 10^{-4}$ coming from the standard model so that J produces no problems (see [F4] or Appendix for the CKM matrix as predicted by TGD).

b) The dominating contributions of the chiral field theory to $Re(\epsilon'/\epsilon)$ are proportional to $1/(m_s + m_d)^2$. The predictions of p-adic thermodynamics for s and d quark masses for $k(d) = k(s) = 113$ are $m_d(113) = m_s(113) = 90$ MeV and if this mass can be interpreted as $m_s(m_c) \simeq 0.1$ GeV, the prediction is too small by a factor $1/4$. Even worse, if m_s corresponds to the scaled up mass $m_s(109) \simeq 360$ MeV of the s quark inside kaon, the situation changes completely and ϵ'/ϵ is too small by a factor $\sim 1/4.5^2 \simeq .05$.

c) TGD predicts that family replication phenomenon has also a bosonic counterpart and two exotic gluons are predicted with couplings depending on whether 3-vertices correspond to fusions (option a)) or branchings (option b)) of partonic 2-surfaces. Option a) is excluded since it predicts a gigantic value of Δm in $K - \overline{K}$ system.

Option b) predicts 2 heavy gluon generations with mass of order 32 GeV for $k_g = 97$. Flavor changing color currents are absent and in the standard diagrammatic expression for CP breaking parameter gluon propagators are replaced by a sum of ordinary massless and two exotic gluon massive gluon propagators. The fact that the matrix elements relevant for the estimation of the CP breaking parameter are estimated at momentum transfer of order $\mu = m_W$, implies that gluon masses do not significantly change the contribution of higher gluons to the amplitude. Hence the penguin amplitudes are simply multiplied by a factor 3 and the ratio ϵ'/ϵ is multiplied by a factor 3.

The model based on exotic gluons and current quarks

The model is based on the idea that a transition $s_{109} \to s_{113}$ occurs before electro-weak process and has an interpretation as a transformation of constituent quark to current quark. Suppose that $m_s = 100$ MeV and $m_d << m_s$ is a manner to produce correct value of ϵ'/ϵ in standard model and look what modifications of this picture TGD implies.

Assuming exotic gluons ϵ'/ϵ would be still by a factor .15 too small for $m_s(109)$. For $m_s(113)$ ϵ'/ϵ would be a fraction $3.7/4$ of the correct value since $m_d = m_s$ rather than $m_d << m_s$ holds true. This however requires that the amplitudes for the transition $s(109) \to s(113)$ and its reversal are near to unity.

The question is why $s(109) \to s(113)$ constituent-current transformation should occur in electro-weak interactions and why it occurs with amplitude $A \sim 1$. Of course it could also be that also d

quark is transformed to a very low mass variant with mass about 4 MeV predicted by chiral field theory. This would correspond to $k = 125$. As a result the amplitude would be multiplied by a factor 4 and $A = 1/2$ would becomes possible.

For some reason the join along boundaries bonds feeding em gauge flux of s quark to $k = 109$ space-time sheet would be transferred to nuclear space-time sheets with $k = 113$ before the electro-weak scattering process responsible for the CP breaking. Note that the value of strange quark mass about 176 MeV deduced from τ lepton decay rate corresponds to $m_s(111)$ in a good approximation. Also this indicates that various scaled up variants of quark masses can appear in the electro-weak dynamics as intermediate states.

The assumption for the proportionality $\epsilon'/\epsilon \propto 1/(m_s + m_d)^2$ derivable from chiral field theory can be criticized. Finding a justification for this assumption seems to be a non-trivial challenge since it is not at all clear that chiral field theory based on $SU(3)$ flavor symmetry makes sense in TGD context.

In the following the basic notions and chiral field theory approach are discussed in more detail to clarify what is involved.

14.5.2 Basic notations and concepts

Until 1963 CP symmetry was believed to be an exact symmetry of Nature. In this year it was however observed by Christensen, Cronin, Fitch and Turlay that CP symmetry is violated in hadronic decays of neutral kaons. In order to interpret the experimental evidence one must consider the strong Hamiltonian eigen states K^0 and its CP conjugate $\bar{K}^0$ as a mixture of physical short lived K_S decaying predominantly to two pions and long-lived K_L decaying mostly semi-leptonically and into 3 pion states. Two- and three pion final states have odd and even CP parity. In absence of CP breaking one would identify K_S and K_L as the CP even and CP odd states

$$
\begin{aligned}
K_1 &= (K^0 + \bar{K}^0)/\sqrt{2} \; , \\
K_2 &= (K^0 - \bar{K}^0)/\sqrt{2} \; .
\end{aligned}
\tag{14.5.1}
$$

What was observed in 1963 was that K_L decays also to two-pion final states.

There are two mechanisms of CP violation. The indirect mechanism involves a slight mixing of K^1 and K^2 characterized by a complex mixing parameter $\bar{\epsilon}$

$$
\begin{aligned}
K_S &= \frac{K_1 + \bar{\epsilon} K_2}{1 + |\bar{\epsilon}|^2} \; , \\
K_L &= \frac{K_2 + \bar{\epsilon} K_1}{1 + |\bar{\epsilon}|^2} \; .
\end{aligned}
\tag{14.5.2}
$$

Direct mechanism involves the direct decay of K_2 to two pion state and is induced by the weak interaction Lagrangian L_W directly. Both mechanisms can be parameterized in terms of the small ratios

$$
\begin{aligned}
\eta_{00} &= \frac{\langle \pi^0 \pi^0 | L_W | K_L \rangle}{\langle \pi^0 \pi^0 | L_W | K_S \rangle} \; , \\
\eta_{+-} &= \frac{\langle \pi^+ \pi^- | L_W | K_L \rangle}{\langle \pi^+ \pi^- | L_W | K_S \rangle} \; .
\end{aligned}
\tag{14.5.3}
$$

Here L_W represents the $\Delta S = 1$ part of the weak Lagrangian. The equations for η parameters can be also written as

$$
\begin{aligned}
\eta_{00} &= \epsilon - \frac{2\epsilon'}{1 - \omega\sqrt{2}} \simeq \epsilon - 2\epsilon' \; , \\
\eta_{+-} &= \epsilon - \frac{2\epsilon'}{1 + \omega/\sqrt{2}} \simeq \epsilon + \epsilon' \; .
\end{aligned}
\tag{14.5.4}
$$

Parameter $\bar{\epsilon}$ is simply related to ϵ. The parameter ω measures the ratio

$$|\omega| = \frac{|\langle (\pi\pi)_{I=2}|L_W|K_S\rangle|}{|\langle (\pi\pi)_{I=0}|L_W|K_S\rangle|} \simeq 1/22.2 \ . \tag{14.5.5}$$

$I = 0$ and $I = 2$ denote the isospin states of final state pions.

The CP violating parameters are expressible in terms of $K_{S,L}$ decay amplitudes as

$$
\begin{aligned}
\epsilon &= \frac{\langle (\pi\pi)_{I=0}|L_W|K_L\rangle}{\langle (\pi\pi)_{I=0}|L_W|K_S\rangle} \ , \\
\epsilon' &= \frac{\epsilon}{\sqrt{2}}\left[\frac{\langle (\pi\pi)_{I=2}|L_W|K_L\rangle}{\langle (\pi\pi)_{I=0}|L_W|K_L\rangle} - \frac{\langle (\pi\pi)_{I=2}|L_W|K_S\rangle}{\langle (\pi\pi)_{I=0}|L_W|K_S\rangle}\right] \ .
\end{aligned}
\tag{14.5.6}
$$

By Watson's theorem one can write the generic amplitudes for K^0 and $\bar{K}^0$ decay as

$$
\begin{aligned}
\langle (\pi\pi)_I|L_W|K^0\rangle &= -iA_I exp(i\delta_I) \ , \\
\langle (\pi\pi)_I|L_W|\bar{K}^0\rangle &= -iA_I^* exp(i\delta_I) \ ,
\end{aligned}
\tag{14.5.7}
$$

where the phases δ_I arise from the pion finals state interactions. In good approximation ($|\bar{\epsilon}ImA_0| \ll |ReA_0|$, $|\bar{\epsilon}|^2 \ll 1$) one can write

$$
\begin{aligned}
\epsilon' &= exp(i(\pi/2 + \delta_2 - \delta_1)) \times \frac{\omega}{\sqrt{2}} \times (\frac{ImA_2}{ReA_2} - \frac{ImA_0}{ReA_0}) \ , \\
\omega &= \frac{ReA_2}{ReA_0} \ .
\end{aligned}
\tag{14.5.8}
$$

With the approximations used one obtains a relationship

$$\epsilon' = \bar{\epsilon} + i\frac{ImA_0}{ReA_0} \ . \tag{14.5.9}$$

One can find a more detailed representation of the subject in various review articles [hj4, cf7]. The standard model of CP breaking is based on the presence of complex phases in CKM matrix.

The value of the parameter ϵ describing indirect CP violation is well established and given by

$$|\epsilon| = (2.266 \pm .017) \times 10^{-3} \ .$$

The phases of ϵ and ϵ' are in good approximation identical so that their signs are same. The value of $Re(\epsilon'/\epsilon)$ was finally established by KTeV collaboration at Fermi Lab to be

$$Re(\frac{\epsilon'}{\epsilon}) = (2.8 \pm .01) \times 10^{-3} \ .$$

The measurement is consistent with the result of the CERN experiment NA31, which has also found a non-vanishing value for this parameter.

There are several theories of CP violation. The so called milliweak theory predicts vanishing value of ϵ'. The model based on the presence of CP breaking phases in three-generation CKM matrix predicts non-vanishing value for the parameter. Also Higgs particles can effect the value of the parameter in standard model. Standard model predicts this parameter to be nonzero but the expectation has been that the value is roughly ten times smaller than the measured value.

A possible explanation of the effect which does not introduce new physics is based on the hypothesis that the mass of s quark is smaller than the mass of d quark: the running mass $m_s(2\ GeV) \simeq .1$ GeV is needed to explain the anomaly if CP breaking parameter J is taken to be in the range $(1 - 1.7) \times 10^{-4}$ claimed in [hj9] to follow from unitarity. There is however experimental evidence from τ decays for $m_s(m(\tau)) = (172 \pm 31)\ MeV$. This suggests that some new short length scale physics is involved.

Standard model prediction for $Re(\epsilon'/\epsilon)$ [hj14] can be summarized in a handy formula

$$
\begin{aligned}
Re(\frac{\epsilon'}{\epsilon}) &= J \times \left[-1.35 + R_s \left(A_6 B_6^{1/2} + A_8 B_8^{3/2} \right) \right] \ , \\
A_6 &= 1.1 |r_Z^8| \ , \\
A_8 &= 1.0 - .67 |r_Z^8| \ .
\end{aligned}
\tag{14.5.10}
$$

The subscript Z refers to renormalization mass m_Z. The parameter R_s is given by

$$
R_s \simeq \left[\frac{150 \ MeV}{m_s(m_c) + m_d(m_c)} \right]^2 \ .
\tag{14.5.11}
$$

The dominating contributions to $Re(\frac{\epsilon'}{\epsilon})$ come from second (A_6) and third terms (A_8). These terms correspond to gluonic and electro-weak penguin diagram contributions to the CP breaking decays and of opposite sign. Clearly, the sum of the two terms is roughly one third of the gluonic term alone.

14.5.3 Separation of short and long distance physics using operator product expansion

The calculation of CP breaking parameters involves physics in very wide energy scale. The strategy is to derive low energy effective action by functionally integrating over the short distance effects coming from energies larger than m_c. This leads to Wilson expansion for the low energy electro-weak effective Lagrangian

$$
L_{low,W} = -\sum_i C_i(\mu, m_c, m_b, m_t, m_W, ...) Q_i(\mu) \ .
\tag{14.5.12}
$$

The coefficients C_i of the operators Q_i in the low energy effective action for light quarks (u, d, s) are functionals of various parameters characterizing short distance physics. The coefficients $C_i(\mu)$ in Wilson expansion of electro-weak effective action can be written as

$$
C_i(\mu) = \frac{G_F}{\sqrt{2}} V_{ud} V_{us}^* \left[x_i(\mu) + \tau y_i(\mu) \right] \ .
\tag{14.5.13}
$$

Here x_i and y_i are Wilson coefficients. V_{ij} denotes CKM matrix and τ is defined as $\tau = V_{td} V_{ts}^* / V_{ud} V_{us}^*$. $V_{td} V_{ts}^*$ is identical with CP breaking invariant J in standard parametrization. Coefficients $y_i(\mu)$ summarize short distance CP breaking physics and in order to determine CP breaking one needs to consider only the coefficients y_i.

Long distance physics is the difficult part of the calculation since it involves calculation of matrix elements of the quark operators Q_i between initial kaon state and final two-pion state. There are several approaches to the problem. Chiral field theory [hj3] is phenomenological approach and relies on the idea that low energy effective action for quarks can be expressed in a good approximation using meson fields. Lattice QCD is believed to provide a more fundamental direct method for the calculation of the correlation functions of Q_i.

Short distance physics

In present initial states are kaons and μ denotes the momentum exchange for a typical diagram associated with the scattering of $d\bar{s}$ quark to final state consisting of of light quarks. μ is taken to be of order m_W and by using renormalization group equations one can deduce the values of the coefficients $C_i(\mu)$ at energy scales, typically of order 1 GeV.

The basic standard diagrams contributing to the $\Delta S = 1$ and $\Delta S = 2$ processes are given by the figure below.

The quark operators Q_i appearing in the expansion can be classified. In present case the list of relevant operators correspond to various terms possible in four-fermion Fermi interaction and are given by the following list.

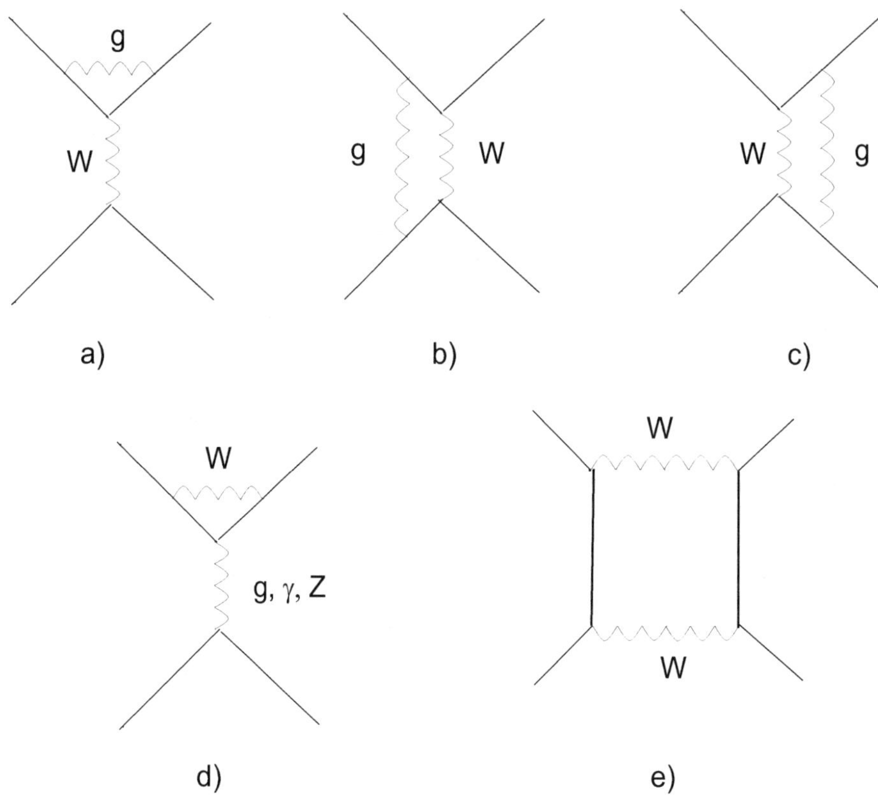

Figure 14.3: Standard model contributions to the matching of the quark operators in the effective flavor-changing Lagrangian

$$Q_1 = (\bar{s}_\alpha u_\beta)_{V-A}(\bar{u}_\beta d_\alpha)_{V-A} \ , \tag{14.5.14}$$

$$Q_2 = (\bar{s}u)_{V-A}(\bar{u}d)_{V-A} \ , \tag{14.5.15}$$

$$Q_{3,5} = (\bar{s}d)_{V-A}\sum_q (\bar{q}q)_{V\mp A} \ ,$$

$$Q_{4,6} = (\bar{s}_\alpha d_\beta)_{V-A}\sum_q (\bar{q}_\beta q_\alpha)_{V\mp A} \ ,$$

$$Q_{7,9} = \frac{3}{2}(\bar{s}d)_{V-A}\sum_q \hat{e}_q(\bar{q}q)_{V\pm A} \ ,$$

$$Q_{8,10} = \frac{3}{2}(\bar{s}_\alpha d_\beta)_{V-A}\sum_q \hat{e}_q(\bar{q}_\beta q_\alpha)_{V\pm A} \ . \tag{14.5.16}$$

α, β denote color indices and $\hat{e}_q$ denote quark charges. $V \pm A$ refers to the Dirac structure $\gamma_\mu(1 \pm \gamma_5)$. Q_2 is induced by mere W exchange whereas gluonic loop corrections to Q_2 induce Q_1. QCD through penguin loop induces the penguin operators Q_{3-6}. Electro-weak loops, in which penguin gluon is replaced with electro-weak gauge boson, induce $Q_{7,9}$ and part of Q_3. The operators $Q_{8,10}$ are induced by the QCD renormalization of the electro-weak loop operators $Q_{7,9}$.

As far as the calculation of ϵ'/ϵ is considered, the dominating contributions come from the penguin diagrams, which are proportional to the vertices $s\bar{d}V$, where V is either gluon or electro-weak gauge boson and to the propagator denominator of V with momentum squared equal to momentum exchange between initial state quarks, which equals to $(p_i - p_j)^2 = \mu^2$. For option b) the standard gluon contribution is replaced with a sum over contributions of ordinary and exotic gluons. For option a) situation is more complicated since $g > 0$ gluons can change the genus of the fermion.

The operators Q_6 and Q_8 give the dominating contributions to ϵ'/ϵ and these contributions are competing. Q_6 and Q_8 differ only by the fact that in Q_8 penguin gluon is replaced with penguin electro-weak boson γ or Z^0. For neutral kaon initial state electro-weak penguin diagram is proportional to the product $e_q e_{\bar{q}} = -e_q^2$ of the virtual quark whereas in case of gluons the factor $Tr(T^a T^a) > 0$ appears. Therefore the contributions associated with Q_6 and Q_8 are of opposite sign and mutually competing.

Detailed calculations lead to the formula already described:

$$
\begin{aligned}
Re(\frac{\epsilon'}{\epsilon}) &= J \times \left[-1.35 + R_s \left(A_6 B_6^{1/2} + A_8 B_8^{3/2} \right) \right] \ , \\
A_6 &= 1.1 |r_Z^8| \ , \\
A_8 &= 1.0 - .67 |r_Z^8| \ .
\end{aligned}
\tag{14.5.17}
$$

for $Re(\epsilon'/\epsilon)$. The coefficients B_6 and B_8 code the long distance physics and their values do not differ too much from $B_6 = B_8 = 1$. Clearly, the sum of Q_6 and Q_8 contributions is roughly one third of the Q_6 contribution alone. From the general structure of Feynman diagrams it ss clear that for option b) the effect caused by the introduction of exotic gluons is in a good approximation a simple scaling of the Q_6 contribution by a factor 3 in the approximation that gluon masses are negligible as compared to W mass, and that this new contribution can enhance direct CP breaking dramatically.

Chiral field theory approach

The basic problem is to calculate electro-weak matrix elements of the quark effective action between hadronic states. These matrix elements reduce to vacuum expectation values of various quark bilinears appearing in four-fermion Fermi interaction Lagrangian. This problem is very difficult since non-perturbative QCD is involved in an essential manner. An attempt to circumvent this problem [hj3] is based on the hypothesis that low energy effective action for quarks is essentially equivalent with the low energy effective action, where pseudoscalar meson fields as dynamical fields and scalar, vector and axial vector meson fields occur as external fields not subject to variations. Quark masses are identified as vacuum expectation values of the external scalar meson field. The approximate symmetry of the chiral field theory is flavor $SU(3)_L \times SU(3)_R$ which is exact symmetry at the limit of massless quarks. This symmetry can be realized if mesons are represented by an element U of $SU(3)$ regarded as a dynamical field: the two $SU(3)$:s act on U from left and right respectively. For small perturbations around ground state mesons correspond to various Lie-algebra generators of $SU(3)$. Chiral field develops vacuum expectation value. If vacuum expectation is not proportional to unit matrix it corresponds to the presence of coherent states associated with the neutral components of the pseudo scalar meson field.

The basic formulation of the chiral field theory approach is described in [hj3] whereas its application to the calculation of ϵ'/ϵ is described in [cf7]. The strong part of the chiral action [hj3] is given by the formula

$$
\begin{aligned}
L_S &= \frac{f^2}{4} \left[Tr\{D_\mu U^\dagger D^\mu U\} + 2B_0 Tr\{(s - ip)U\} + 2B_0^* Tr\{(s + ip)U^\dagger\} \right] \\
&+ \frac{1}{12} H_0 D_\mu \theta D^\mu \theta \ .
\end{aligned}
\tag{14.5.18}
$$

D_μ denotes the covariant derivative defined by the couplings to the left and right handed gauge bosons L_μ and R_μ defined as superpositions $R_\mu = v_\mu + a_\mu$ and $L_\mu = v_\mu - a_\mu$ of the vector and axial vector mesons fields v and a. Action contains three coupling constant parameters: f, B_0 and H_0, which is present because the presence of color instantons can lead to a non-vanishing value of the θ parameter in QCD. In lowest order f is pion decay constant f_π and B_0 sets the scale in the formula $M_M^2 = B_0(\sum_i m(q_i))$ inspired by broken $SU(3)$ symmetry and resulting as a prediction of the model. The components for the non-vanishing vacuum expectation value for the external scalar field are identified as quark masses. The generation of vacuum expectation value of s implies that quark condensates are developed:

$$\langle \bar{q}_i q_j \rangle = B_0 f^2 \delta_{i,j} \ ,$$

$$B_0 f^2 = \frac{f_\pi^2 m_\pi^2}{(m_u + m_d)} = \frac{f_K^2 m_K^2}{(m_s + m_d)} \ . \tag{14.5.19}$$

Note that the strong part of the chiral Lagrangian is invariant under the overall scaling of quark masses.

The weak part of the chiral action corresponds to the sigma model counterpart of the most general electro-weak four-fermion action. The recipe for constructing this action is described in more detail in [cf7] and can be summarized as rules associating with various fermionic bi-linears appearing in the generalized Fermi action corresponding terms of the weak part of the chiral action. In particular, the following rules hold true:

$$\bar{q}_L^j \gamma^\mu q_L^i \ \rightarrow \ -i f_\pi^2 (U^\dagger D_\mu U)_{ij} \ ,$$

$$\bar{q}_R^j \gamma^\mu q_R^i \ \rightarrow \ -i f_\pi^2 (U D_\mu U^\dagger)_{ij} \ ,$$

$$\bar{q}_L^j \gamma^\mu q_R^i \ \rightarrow \ -2 B_0 \left[\frac{f^2}{4} U + \text{higher order terms} \right]_{ij} \ ,$$

$$\bar{q}_R^j \gamma^\mu q_L^i \ \rightarrow \ -2 B_0 \left[\frac{f^2}{4} U^\dagger + \text{higher order terms} \right]_{ij} \ . \tag{14.5.20}$$

The chiral counterparts of the left and right handed currents are proportional to BM and depend on the ratios of quark masses only. The terms giving dominating contribution to the $\Delta S = 1$ part of the weak effective action involve the chiral counterparts of terms $\bar{q}_L^j q_R^i$ breaking chiral invariance. The chiral counterparts of these terms are proportional to B and, in accordance with expectations, fail to be invariant under the overall scaling of quark masses. The higher order contributions to these terms are important for the calculations of direct CP breaking effects but are not written explicitly here because they are not needed in the estimate for how the predictions of the standard model are modified in TGD framework. The terms breaking chiral symmetry give rise to ϵ'/ϵ a contribution, which is proportional to $1/(m_s + m_d)^2$.

The $\Delta S = 2$ part of effective quark action is involved with $K^0 \rightarrow \bar{K}^0$ transitions and the corresponding quark operator is given by

$$Q_{S2} = (\bar{s}_L \gamma^\mu d_L)(\bar{s}_L \gamma^\mu d_L) \ . \tag{14.5.21}$$

The chiral counterpart of this operator is obviously invariant under overall scaling of quark masses.

Does chiral theory approach make sense in TGD framework?

The TGD based model for the large direct CP breaking based on exotic gluons and on the transformation of s_{109} to s_{113} has been already discussed. The open question is whether the $1/(m_s + m_d)^2$ proportionality of the CP breaking amplitude can be justified in TGD context where it is not at all clear that chiral theory approach makes sense.

In standard model framework chiral field theory provides a phenomenological description of the low energy hadron physics and makes possible the calculation of various hadronic matrix elements needed to derive the predictions for CP breaking effect.

Chiral field theory limit however involves some questionable assumptions about the relationship between QCD and low energy hadron physics.

a) $SU(3)$ symmetry is assumed and allows description of light mesons in terms of $SU(3)$ valued chiral field U possessing $SU(3)_R \times SU(3)_R$ symmetry broken only by quark mass matrix. In TGD framework $SU(3)$ symmetry is purely phenomenological symmetry since the fundamental gauge group is the gauge group of the standard model.

b) The generation of quark masses is described as effective spontaneous symmetry breaking caused by the vacuum expectation value of $SU(3)$ Lie-algebra valued external scalar field s. Quark masses are identified as the components of the diagonal vacuum expectation value of this field. Physically the scalar field corresponds to scalar meson field so that quark masses would result from the coupling of the quarks to coherent states of scalar mesons. This cannot be a correct physical description in TGD framework, where p-adic thermodynamics gives rise to quark masses. Of course, the presence of the scalar field can give rise to a small shift in the values of the quark masses. Also Higgs field could be in question.

c) The coupling of the field s to chiral field U implies in the standard model context that the mass squared values of mesons are proportional to the sums of masses of the mesonic quarks: for instance, $M_\pi^2 = B_0(m_u + m_d)$ and $M_K^2 = B_0(m_s + m_d)$, where B_0 is one of the basic coupling constants of the chiral field theory. This formula is not consistent with the p-adic mass calculations, where quark mass squared is additive for quarks with the same value of k_q and quark mass for different values of k_q. Indeed, the formulas $M_\pi^2 = m_u^2 + m_d^2$ and $M_K^2 = (m_s + m_d)^2$ are true. The chiral field formula predicts $m_s/m_d \simeq 24$ requiring $m_u = m_d \simeq 13$ MeV ($k = 121$) for $m_s(113) = 320$ MeV whereas TGD predicts $m_s(109)/m_d(107) = 4$. For $m_s \simeq 100$ MeV the prediction is $m_d \simeq 4.2$ MeV. This looks suspiciously small.

To sum up, although the basic assumptions of chiral field theory limit look too specific in TGD framework, its predictions for low energy hadron physics are well-tested and TGD could be consistent with them. If this the case, the assumption about $s_{109} \to s_{107}$ transition allows a correct prediction of direct CP breaking amplitude using chiral field theory limit.

14.6 Appendix

14.6.1 Effective Feynman rules and the effect of top quark mass on the effective action

The effective low energy field theory relevant for $K - \bar{K}$ systems is in the standard model context summarized elegantly using the Feynman rules of effective field theory deriving from box and penguin diagrams. The rules in t'Hooft-Feynman gauge are summarized in excellent review article of Buras and Fleischer [hj10]. For box diagrams the rules are following:

$$
\begin{aligned}
Box(\Delta S = 2) &= \lambda_i^2 \frac{G_F^2}{16\pi^2} M_W^2 S_0(x_i)(\bar{s}d)_{V-A}(\bar{s}d)_{V-A} \ , \\
Box(T_3 = -1/2) &= \lambda_i \frac{G_F}{\sqrt{2}} \frac{\alpha}{sin^2(\theta_W)} B_0(x_i)(\bar{s}d)_{V-A}(\bar{\mu}\mu)_{V-A} \ , \\
Box(T_3 = 1/2) &= \lambda_i \frac{G_F}{\sqrt{2}} \frac{\alpha}{sin^2(\theta_W)} \left[-4B_0(x_i)\right] (\bar{s}d)_{V-A}(\bar{\nu}\nu)_{V-A} \ , \\
\lambda_i &= V_{is}^* V_{id} \ .
\end{aligned}
\tag{14.6.1}
$$

The box vertices listed here describe the decays $K_0 \to \bar{K}_0$ and contribute to $K_0 \to \bar{\mu}\mu$ and $K_0 \to \bar{\nu}\nu$ decays. $(\bar{q}_1 q_2)_{V-A}$ is shorthand notation for the left handed weak current involving gamma matrices and the products of fermionic bi-linears actually involve contraction of the gamma matrix indices.

Penguin diagrams can be characterized by the effective vertices $\bar{s}dB$, where B is photon, Z boson or gluon, which is treated as usual in effective field theory

$$
\begin{aligned}
\bar{s}Zd &= i\lambda_i \frac{G_F}{\sqrt{2}} \frac{g_Z}{2\pi^2} M_Z^2 g_Z C_0(x_i)\bar{s}\gamma^\mu(1 - \gamma_5)d \ , \\
\bar{s}\gamma d &= -i\lambda_i \frac{G_F}{\sqrt{2}} \frac{e}{8\pi^2} D_0(x_i)\bar{s}(q^2\gamma^\mu - q^\mu q^\nu \gamma_\nu)(1 - \gamma_5)d \ , \\
\bar{s}G^a d &= -i\lambda_i \frac{G_F}{\sqrt{2}} \frac{g_s}{8\pi^2} E_0(x_i)\bar{s}(q^2\gamma^\mu - q^\mu q^\nu \gamma_\nu)(1 - \gamma_5)T^a d \ .
\end{aligned}
\tag{14.6.2}
$$

The vertices above correspond to the exchange of Z, photon and gluon between the quarks. Boson propagator and second vertex is constructed using the standard Feynman rules. The counterparts of the sdB vertices are easily constructed for $g > 0$ gluons. The orthogonality of single hadron states requires that flavor is conserved for $g > 0$ exchanges.

The functions $B_0, C_0, \ldots$ characterize the low energy effective action at mass scale $\mu = m_W$. The subscript '0' refers to the values of these functions without QCD corrections, which are taken into account using renormalization group equations to deduced the functions at mass scale of order 1 GeV. The functions are listed below:

$$
\begin{aligned}
B_0(x_t) &= \frac{1}{4}\left[\frac{x_t}{y_t} + \frac{x_t log(x_t)}{y_t^2}\right] \ , \\[2mm]
C_0(x_t) &= \frac{x_t}{8}\left[-\frac{x_t - 6}{y_t} + \frac{3x_t + 2}{y_t^2}log(x_t)\right] \ , \\[2mm]
D_0(x_t) &= -\frac{4}{9}log(x_t) - \frac{25x_t^2 - 19x_t^3}{36y_t^3} + \frac{x_t^2(-6 - 2x_t + 5x_t^2)}{18y_t^3}log(x_t) \ , \\[2mm]
E_0(x_t) &= -\frac{2}{3}log(x_t) + \frac{x_t^2(15 - 16x_t - 4x_t^2)}{6y_t^4}log(x_t) + \frac{x_t(18 - 11x_t - x_t^2)}{12y_t^3} \ , \\[2mm]
S_0(x_t) &= \frac{4x_t - 11x_t^2 + x_t^3}{4y_t^2} - \frac{3x_t^3 log(x_t)}{2y_t^3} \ , \\[2mm]
S_0(x_c, x_t) &= x_c\left[log(\frac{x_t}{x_c}) - \frac{3x_t}{4y_t} - \frac{3x_t^2 log(x_t)}{4y_t^2}\right] \ , \\[2mm]
& x_c = (\tfrac{m_c}{m_W})^2 \quad x_t = (\tfrac{m_t}{m_W})^2 \ , \quad y_t = 1 - x_t \ .
\end{aligned}
\qquad (14.6.3)
$$

Although x_t, being the interesting parameter, appears as the only argument of these functions, also the contributions coming from light quarks propagating in the loops are included. For comparison purposes it is useful to give the explicit relations between electro-weak coupling parameters and G_F.

$$
\begin{aligned}
\frac{G_F}{\sqrt{2}} &= \frac{g_W^2}{8m_W^2} \ , \\[2mm]
g_W &= \frac{e}{sin(\theta_W)} \ , \\[2mm]
g_Z &= \frac{e}{sin(\theta_W)cost(\theta_W)} \ .
\end{aligned}
\qquad (14.6.4)
$$

The following table summarizes the effect of the change of the top quark mass on the functions $B_0, C_0, \ldots$ What is given are the ratios $r(f) = f(55)/f(175)$ of the functions $B_0, C_0, \ldots$ evaluated for top quark masses 55 GeV and 175 GeV respectively.

$$
\begin{array}{ccccccc}
f & B_0(x_t) & C_0(x_t) & D_0(x_t) & E_0(x_t) & S_0(x_t) & S_0(x_c, x_t) \\
r & .51 & .09 & -.70 & 3.44 & .15 & .81
\end{array}
\qquad (14.6.5)
$$

These results leave allow only the identification of the experimental candidate as a realistic candidate for top quark.

a) The function B_0 is reduced only by a factor of $1/2$ and there are no new physics contributions to B_0 in the lowest order. The function C_0 characterizing Z penguin diagrams is reduced by an order of magnitude. The coefficient $C_0(x_t) - 4B_0(x_t)$ characterizes the dominating contribution to $K \to \mu^+\mu^-$ decay in standard model and the decay amplitude is reduced by a factor .27 so that this decay would provide a stringent test selecting between 55 GeV top quark and 175 GeV top quark. Unfortunately, the predicted $K \to \mu^+\mu^-$ rate is still by several orders of magnitude below the experimental upper bound.

b) The function $S_0(x_t)$ characterizing $B - \bar{B}$ and $K - \bar{K}$ mass differences is reduced almost by an order of magnitude. Note that in case of Δm_K the ratio $r(tt/ct)$ of the WW box diagram amplitudes with two top quarks and c and t in internal fermion lines is $r(tt/ct) \sim 738$ for $m_t = 175$ GeV and $r(tt/ct) \sim 138$ for $m_t = 55$ GeV (the moduli of the factors coming from CKM matrix are taken into account). Thus $m_t = 175$ GeV is the only sensible choice.

14.6.2 U and D matrices from the knowledge of top quark mass alone?

As already found, a possible resolution to the problems created by top quark is based on the additivity of mass squared so that top quark mass would be about 230 GeV, which indeed corresponds to a peak in mass distribution of top candidate, whereas $t\bar{t}$ meson mass would be 163 GeV. This requires that top quark mass changes very little in topological mixing. It is easy to see that the mass constraints imply that for $n_t = n_b = 60$ the smallness of V_{i3} and $V(3i)$ matrix elements implies that both U and D must be direct sums of 2×2 matrix and 1×1 unit matrix and that V matrix would have also similar decomposition. Therefore $n_b = n_t = 59$ seems to be the only number theoretically acceptable option. The comparison with the predictions with pion mass led to a unique identification $(n_d, n_b, n_b) = (5, 5, 59), (n_u, n_c, n_t) = (4, 6, 59)$.

U and D matrices as perturbations of matrices mixing only the first two genera

This picture suggests that U and D matrices could be seen as small perturbations of very simple U and D matrices satisfying $|U| = |D|$ corresponding to $n = 60$ and having $(n_d, n_b, n_b) = (4, 5, 60)$, $(n_u, n_c, n_t) = (4, 5, 60)$ predicting V matrix characterized by Cabibbo angle alone. For instance, CP breaking parameter would characterize this perturbation. The perturbed matrices should obey thermodynamical constraints and it could be possible to linearize the thermodynamical conditions and in this manner to predict realistic mixing matrices from first principles. The existence of small perturbations yielding acceptable matrices implies also that these matrices be near a point at which two different matrices resulting as a solution to the thermodynamical conditions coincide.

D matrix can be deduced from U matrix since $9|D_{12}|^2 \simeq n_d$ fixes the value of the relative phase of the two terms in the expression of D_{12}.

$$
\begin{aligned}
|D_{12}|^2 &= |U_{11}V_{12} + U_{12}V_{22}|^2 \\
&= |U_{11}|^2|V_{12}|^2 + |U_{12}|^2|V_{22}|^2 \\
&+ 2|U_{11}||V_{12}||U_{12}||V_{22}|cos(\Psi) = \frac{n_d}{9} \;, \\
\Psi &= arg(U_{11}) + arg(V_{12}) - arg(U_{12}) - arg(V_{22}) \;.
\end{aligned}
$$

$$(14.6.6)$$

Using the values of the moduli of U_{ij} and the approximation $|V_{22}| = 1$ this gives for $cos(\Psi)$

$$
\begin{aligned}
\cos(\Psi) &= \frac{A}{B} \;, \\
A &= \frac{n_d - n_u}{9} - \frac{9 - n_u}{9}|V_{12}|^2 \;, \\
B &= \frac{2}{9|V_{12}|}\sqrt{n_u(9 - n_u)} \;.
\end{aligned}
$$

$$(14.6.7)$$

The experimentation with different values of n_d and n_u shows that $n_u = 6, n_d = 4$ gives $cos(\Psi) = -1.123$. Of course, $n_u = 6, n_d = 4$ option is not even allowed by $n_t = 60$. For $n_d = 4, n_u = 5$ one has $cos(\Psi) = -0.5958$. $n_d = 5, n_u = 6$ corresponding to the perturbed solution gives $cos(\Psi) = -0.6014$.

Hence the initial situation could be $(n_u = 5, n_s = 4, n_b = 60)$, $(n_d = 4, n_s = 5, n_t = 60)$ and the physical U and D matrices result from U and D matrices by a small perturbation as one unit of t (b) mass squared is transferred to u (s) quark and produces symmetry breaking as $(n_d = 5, n_s = t, n_b = 59)$, $(n_u = 6, n_c = 4, n_t = 59)$.

The unperturbed matrices $|U|$ and $|D|$ would be identical with $|U|$ given by

$$
|U_{11}| = |U_{22}| = \tfrac{2}{3} \;, \quad |U_{12}| = |U_{21}| = \tfrac{\sqrt{5}}{3} \;,
$$

$$(14.6.8)$$

The thermodynamical model allows solutions reducing to a direct sum of 2×2 and 1×1 matrices, and since $|U|$ matrix is fixed completely by the mass constraints, it is trivially consistent with the thermodynamical model.

Direct search of U and D matrices

The general formulas for p^U and p^D in terms of the probabilities p_{11} and p_{21} allow straightforward search for the probability matrices having maximum entropy just by scanning the (p_{11}, p_{21}) plane constrained by the conditions that all probabilities are positive and smaller than 1. In the physically interesting case the solution is sought near a solution for which the non-vanishing probabilities are $p_{11} = p_{22} = (9 - n_1)/9$, $p_{12} = p_{21} = n_1/9$, $p_{33} = 1$, $n_1 = 4$ or 5. The inequalities allow to consider only the values $p_{11} \geq (9 - n_1)/9$.

1. Probability matrices p^U and p^D

The direct search leads to maximally entropic p^D matrix with $(n_d, n_s) = (5, 5)$:

$$p^D = \begin{pmatrix} 0.4982 & 0.4923 & 0.0095 \\ 0.4981 & 0.4924 & 0.0095 \\ 0.0037 & 0.0153 & 0.9810 \end{pmatrix} , \quad p_0^D = \begin{pmatrix} 0.5556 & 0.4444 & 0 \\ 0.4444 & 0.5556 & 0 \\ 0 & 0 & 1 \end{pmatrix} .$$

$$(14.6.9)$$

p_0^D represents the unperturbed matrix p_0^D with $n(d = 4), n_s = 5$ and is included for the purpose of comparison. The entropy $S(p^D) = 1.5603$ is larger than the entropy $S(p_0^D) = 1.3739$. A possible interpretation is in terms of the spontaneous symmetry breaking induced by entropy maximization in presence of constraints.

A maximally entropic p^U matrix with $(n_u, n_c) = (5, 6)$ is given by

$$p^U = \begin{pmatrix} 0.5137 & 0.4741 & 0.0122 \\ 0.4775 & 0.4970 & 0.0254 \\ 0.0088 & 0.0289 & 0.9623 \end{pmatrix}$$

$$(14.6.10)$$

The value of entropy is $S(p^U) = 1.7246$. There could be also other maxima of entropy but in the range covering almost completely the allowed range of the parameters and in the accuracy used only single maximum appears.

The probabilities p_{ii}^D resp. p_{ii}^U satisfy the constraint $p(i, i) \geq .492$ resp. $p_{ii} \geq .497$ so that the earlier proposal for the solution of proton spin crisis must be given up and the solution discussed in [D2] remains the proposal in TGD framework.

2. Near orthogonality of U and D matrices

An interesting question whether U and D matrices can be transformed to approximately orthogonal matrices by a suitable $(U(1) \times U(1))_L \times (U(1) \times U(1))_R$ transformation and whether CP breaking phase appearing in CKM matrix could reflect the small breaking of orthogonality. If this expectation is correct, it should be possible to construct from $|U|$ $(|D|)$ an approximately orthogonal matrix by multiplying the matrix elements $|U_{ij}|$, $i, j \in \{2, 3\}$ by appropriate sign factors. A convenient manner to achieve this is to multiply $|U|$ $(|D|)$ in an element wise manner $((A \circ B)_{ij} = A_{ij} B_{ij})$ by a sign factor matrix S.

a) In the case of $|U|$ the matrix $U = S \circ |U|$, $S(2, 2) = S(2, 3) = S(3, 2) = -1$, $S_{ij} = 1$ otherwise, is approximately orthogonal as the fact that the matrix $U^T U$ given by

$$U^T U = \begin{pmatrix} 1.0000 & 0.0006 & -0.0075 \\ 0.0006 & 1.0000 & -0.0038 \\ -0.0075 & -0.0038 & 1.0000 \end{pmatrix}$$

is near unit matrix, demonstrates.

b) For D matrix there are two nearly orthogonal variants. For $D = S \circ |D|$, $S(2, 2) = S(2, 3) = S(3, 2) = -1$, $S_{ij} = 1$ otherwise, one has

$$D^T D = \begin{pmatrix} 1.0000 & -0.0075 & 0.0604 \\ -0.0075 & 1.0000 & 0.0143 \\ 0.0604 & 0.0143 & 1.0000 \end{pmatrix} .$$

The choice $D = S \circ D$, $S(2,2) = S(2,3) = S(3,3) = -1$, $S_{ij} = 1$ otherwise, is slightly better

$$D^T D = \begin{pmatrix} 1.0000 & -0.0075 & 0.0604 \\ -0.0075 & 1.0000 & 0.0143 \\ 0.0601 & 0.0143 & 1.0000 \end{pmatrix} .$$

3. The matrices U and D in the standard gauge

Entropy maximization indeed yields probability matrices associated with unitary matrices. 8 phase factors are possible for the matrix elements but only 4 are relevant as far as the unitarity conditions are considered. The vanishing of the inner products between row vectors, gives 6 conditions altogether so that the system seems to be over-determined. The values of the parameters s_1, s_2, s_3 and phase angle δ in the "standard gauge" can be solved in terms of r_{11} and r_{21}.

The requirement that the norms of the parameters c_i are not larger than unity poses non-trivial constraints on the probability matrices. This should should be the case since the number of unitarity conditions is 9 whereas probability conservation for columns and rows gives only 5 conditions so that not every probability matrix can define unitary matrix. It would seem that that the constraints are satisfied only if the the 2 mass squared conditions and 2 conditions from the entropy maximization are equivalent with 4 unitarity conditions so that the number of conditions becomes 5+4=9. Therefore entropy maximization and mass squared conditions would force the points of complex 9-dimensional space defined by 3×3 matrices to a 9-dimensional surface representing group $U(3)$ so that these conditions would have a group theoretic meaning.

The formulas

$$r_{i2} = \sqrt{[-\frac{n_i}{51} + \frac{20}{17}(1 - r_{i1}^2)]} \ ,$$

$$r_{i3} = \sqrt{[\frac{n_i}{51} - \frac{3}{17}(1 - r_{i1}^2)]} \ . \tag{14.6.11}$$

and

$$U = \begin{bmatrix} c_1 & s_1 c_3 & s_1 s_3 \\ -s_1 c_2 & c_1 c_2 c_3 - s_2 s_3 exp(i\delta) & c_1 c_2 s_3 + s_2 c_3 exp(i\delta) \\ -s_1 s_2 & c_1 s_2 c_3 + c_2 s_3 exp(i\delta) & c_1 s_2 s_3 - c_2 c_3 exp(i\delta) \end{bmatrix} \tag{14.6.12}$$

give

$$c_1 = r_{11} \ , \qquad c_2 = \frac{r_{21}}{\sqrt{1-r_{11}^2}} \ ,$$

$$s_3 = \frac{r_{13}}{\sqrt{1-r_{11}^2}} \ , \qquad cos(\delta) = \frac{c_1^2 c_2^2 c_3^2 + s_2^2 s_3^2 - r_{22}^2}{2 c_1 c_2 c_3 s_2 s_3} \ . \tag{14.6.13}$$

Preliminary calculations show that for $n_1 = n_2 = 5$ case the matrix of moduli allows a continuation to a unitary matrix but that for $n_1 = 4, n_2 = 6$ the value of $cos(\delta)$ is larger than one. This would suggest that unitarity indeed gives additional constraints on the integers n_i. The unitary (in the numerical accuracy used) $(n_d, n_s) = (5, 5)$ D matrix is given by

$$D = \begin{pmatrix} 0.7059 & 0.7016 & 0.0975 \\ -0.7057 & 0.7017 - 0.0106i & 0.0599 + 0.0766i \\ -0.0608 & 0.0005 + 0.1235i & 0.4366 - 0.8890i \end{pmatrix} .$$

The unitarity of this matrix supports the view that for certain integers n_i the mass squared conditions and entropy maximization reduce to group theoretic conditions. The numerical experimentation shows that the necessary condition for the unitarity is $n_1 > 4$ for $n_2 < 9$ whereas for $n_2 \geq 9$ the unitarity is achieved also for $n_1 = 4$.

Direct search for CKM matrices

The standard gauge in which the first row and first column of unitary matrix are real provides a convenient representation for the topological mixing matrices: it is convenient to refer to these representations as U_0 and D_0. The possibility to multiply the rows of U_0 and D_0 by phase factors $(U(1) \times U(1))_R$ transformations) provides 2 independent phases affecting the values of $|V|$. The phases $exp(i\phi_j)$, $j = 2, 3$ multiplying the second and third row of D_0 can be estimated from the matrix elements of $|V|$, say from the elements $|V_{11}| = cos(\theta_c) \equiv v_{11}$, $sin\theta_c = .226 \pm .002$ and $|V_{31}| = (9.6 \pm .9) \cdot 10^{-3} \equiv v_{31}$. Hence the model would predict two parameters of the CKM matrix, say s_3 and δ_{CP}, in its standard representation.

The fact that the existing empirical bounds on the matrix elements of V are based on the standard model physics raises the question about how seriously they should be taken. The possible existence of fractally scaled up versions of light quarks could effectively reduce the matrix elements for the electro-weak decays $b \to c + W$, $b \to u + W$ resp. $t \to s + W$, $t \to d + W$ since the decays involving scaled up versions of light quarks can be counted as decays $W \to bc$ resp. $W \to tb$. This would favor too small experimental estimates for the matrix elements V_{i3} and V_{3i}, $i = 1, 2$. In particular, the matrix element $V_{31} = V_{td}$ could be larger than the accepted value.

Various constraints do not leave much freedom to choose the parameters n_{q_i}. The preliminary numerical experimentation shows that the choice $(n_d, n_s) = (5, 5)$ and $(n_u, n_c) = (5, 6)$ yields realistic U and D matrices. In particular, the conditions $|U(1, 1)| > .7$ and $|D(1, 1)| > .7$ hold true and mean that the original proposal for the solution of spin puzzle of proton must be given up. In [D2] an alternative proposal based on more recent findings is discussed. Only for this choice reasonably realistic CKM matrices have been found. For $n_t = 58$ the mass of $t\bar{t}$ meson mass is reduced by one percent from 2×163 GeV for $n(5) = 59$ so that $n_t = 58$ is still acceptable if the additivity of conformal weight rather than mass is accepted for diagonal mesons.

a) The requirement that the parameters $|V_{11}|$ (or equivalently, Cabibbo angle) and $|V_{31}|$ are produced correctly, yields CKM matrices for which CP breaking parameter J is roughly one half of its accepted value. The matrix elements $V_{23} \equiv V_{cb}$, $V_{32} \equiv V_{tc}$, and $V_{13} \equiv V_{ub}$ are roughly twice their accepted value. This suggests that the condition on V_{31} should be loosened.

b) The following tables summarize the results of the search requiring that
i) the value of the Cabibbo angle s_{Cab} is within the experimental limits $s_{Cab} = .223 \pm .002$,
ii) $V_{31} = (9.6 \pm .9) \cdot 10^{-3}$, is allowed to have value at most twice its upper bound,
iii) V_{13} whose upper bound is determined by probability conservation, is within the experimental limits $.42 \cdot 10^{-3} < |V_{ub}| < 6.98 \cdot 10^{-3}$ whereas $V_{23} \simeq 4 \times 10^{-3}$ should come out as a prediction,
iv) the CP breaking parameter satisfies the condition $|(J - J_0)/J_0| < .6$, where $J_0 = 10^{-4}$ represents the lower bound for J (the experimental bounds for J are $J \times 10^4 \in (1 - 1.7)$).
The pairs of the phase angles (ϕ_1, ϕ_2) defining the phases $(exp(i\phi_1), exp(i\phi_2))$ are listed below

$$
\begin{array}{llllll}
class\ 1: & \phi_1 & 0.1005 & 0.1005 & 4.8129 & 4.8129 \\
& \phi_2 & 0.0754 & 1.4828 & 4.7878 & 6.1952 \\
\\
class\ 2: & \phi_1 & 0.1005 & 0.1005 & 4.8129 & 4.8129 \\
& \phi_2 & 2.3122 & 5.5292 & 0.7414 & 3.9584
\end{array}
\qquad (14.6.14)
$$

The phase angle pairs correspond to two different classes of U, D, and V matrices. The U, D and V matrices inside each class are identical at least up to 11 digits(!). Very probably the phase angle pairs are related by some kind of symmetry.

The values of the fitted parameters for the two classes are given by

	$|V_{11}|$	$|V_{31}|$	$|V_{13}|$	$J/10^{-4}$
class 1	0.9740	0.0157	0.0069	.93953
class 2	0.9740	0.0164	0.0067	1.0267

V_{31} is predicted to be about 1.6 times larger than the experimental upper bound and for both classes V_{23} and V_{32} are roughly too times too large. Otherwise the fit is consistent with the experimental limits for class 2. For class 1 the CP breaking parameter is 7 per cent below the experimental lower bound. In fact, the value of J is fixed already by the constraints on V_{31} and V_{11} and reduces by a factor of one half if V_{31} is required to be within its experimental limits.

U, D and $|V|$ matrices for class 1 are given by

$$U = \begin{bmatrix} 0.7167 & 0.6885 & 0.1105 \\ -0.6910 & 0.7047 - 0.0210i & 0.0909 + 0.1310i \\ -0.0938 & 0.0696 + 0.1550i & 0.1747 - 0.9653i \end{bmatrix}$$

$$D = \begin{bmatrix} 0.7059 & 0.7016 & 0.0975 \\ -0.6347 - 0.3085i & 0.6358 + 0.2972i & 0.0203 + 0.0951i \\ -0.0587 - 0.0159i & -0.0317 + 0.1194i & 0.6534 - 0.7444i \end{bmatrix}$$

$$|V| = \begin{bmatrix} 0.9740 & 0.2265 & 0.0069 \\ 0.2261 & 0.9703 & 0.0862 \\ 0.0157 & 0.0850 & 0.9963 \end{bmatrix}$$

$$(14.6.15)$$

U, D and $|V|$ matrices for class 2 are given by

$$U = \begin{bmatrix} 0.7167 & 0.6885 & 0.1105 \\ -0.6910 & 0.7047 - 0.0210i & 0.0909 + 0.1310i \\ -0.0938 & 0.0696 + 0.1550i & 0.1747 - 0.9653i \end{bmatrix}$$

$$D = \begin{bmatrix} 0.7059 & 0.7016 & 0.0975 \\ -0.6347 - 0.3085i & 0.6358 + 0.2972i & 0.0203 + 0.0951i \\ -0.0589 - 0.0151i & -0.0302 + 0.1198i & 0.6440 - 0.7525i \end{bmatrix}$$

$$|V| = \begin{bmatrix} 0.9740 & 0.2265 & 0.0067 \\ 0.2260 & 0.9704 & 0.0851 \\ 0.0164 & 0.0838 & 0.9963 \end{bmatrix}$$

$$(14.6.16)$$

What raises worries is that the values of $|V_{23}| = |V_{cb}|$ and $|V_{32}| = |V_{ts}|$ are roughly twice their experimental estimates. This, as well as the discrepancy related to V_{31}, might be understood in terms of the electro-weak decays of b and t to scaled up quarks causing a reduction of the branching ratios $b \to c + W$, $t \to s + W$ and $t \to t + d$. The attempts to find more successful integer combinations n_i has failed hitherto. The model for pseudoscalar meson masses, the predicted relatively small masses of light quarks, and the explanation for $t\bar{t}$ meson mass supports this mixing scenario.

Part V

APPLICATIONS TO COSMOLOGY AND ASTROPHYSICS

Chapter 15

The Relationship Between TGD and GRT

15.1   Introduction

There seems to be a fundamental obstacle against the existence of a Poincare invariant theory of gravitation:

a) The conservation laws of energy and momentum should be exact in this kind of a theory and broken only by the presence of the light cone boundary in the case of TGD.

b) GRT space-time seems to be an empirically well established concept and for GRT space-time these conservation laws are not exact: even the concept of the four-momentum is somewhat nebulous. In fact, it will be found that for the space-time surfaces satisfying Einstein's equations gravitational energy momentum is not conserved and non-conservation becomes large for small values of cosmic time.

One can imagine several solutions to this problem without giving up Einstein's equations.

a) Although Poincare invariance is an exact symmetry at the classical level even for $H = M_+^4 \times CP_2$, the situation could be different at quantum level in this case since the presence of the light cone boundary breaks Poincare symmetry geometrically and one must define momentum generators in Diff^4 invariant manner. Therefore it might be possible to explain the non-conservation of the four-momentum quantum mechanically. Quantum effects are expected to be important for small values of cosmic time only so that in the following this possibility will not be considered although it is in principle quite interesting. It has however turned out that the proper formulation of theory must be based on $H = M^4 x CP_2$ as imbedding space so that Poincare invariance is satisfied also at quantum level.

b) In TGD, space-time is many-sheeted and the apparent non-conservation of the energy at a given space-time sheet could result from the exchange of matter between various space-time sheets. Also there exists what might be called vapor phase, which consists of disjoint particle like 3-surfaces. The non-conservation of energy for GRT space-time could be also caused by the energy flow between GRT space-time (condensate) and vapor phase.

c) It took 25 years to discover the third option. When space-times are 4-surfaces, four-momentum currents are vector currents rather than components of a tensor and the sign of energy depends on the time orientation of the space-time sheet. This led to the identification of phase conjugate photons as negative energy photons and forced to ask whether also phase conjugate fermions are possible and whether matter anti-matter asymmetry could be resolved by assuming that antimatter reside at negative energy space-time sheets. This option has also elegant formulation at the level of configuration space of 3-surfaces.

Gravitational mass would be identified as the absolute value of the inertial mass so that its density corresponds to the difference of inertial mass densities for matter and antimatter and is not conserved. Similar interpretation applies to the gravitational counterparts of gauge charges coupling to the long range electro-weak and color gauge fields. The resulting Eastern view about cosmology is maximally predictive, consistent with the crossing symmetry of elementary particle physics, and also conforms with the materialistic illusion about unique objective reality at the limit when the interaction between negative and positive energy matter can be neglected.

d) Even this interpretation was not quite correct yet and it took still two years to end up with what looks the final one. The observation that inertial four-momentum vanishes for CP_2 type extremals but that gravitational four-momentum is light-like and has randomly varying direction led to the realization that the non-conservation of the gravitational four-momentum occurs already at elementary particle level and that inertial four-momentum is the time average of the gravitational four-momentum and corresponds to the average center of mass motion for the CP_2 type extremal [D2]. At elementary particle level Equivalence Principle holds true in average sense. This picture leads naturally to p-adic thermodynamics and allows to identify Higgs bosons as weakly charged wormhole contacts which are pieces of CP_2 type extremals [F3].

The topics discussed in the chapter are following.

a) The relationship between TGD and GRT is discussed in light of the new view about energy. One of the most fascinating outcomes is the resolution of the most gigantic failure in the art of order magnitude estimates. The naive estimate for the cosmological constant predicted also by TGD is by a factor 10^{120} larger than its value deduced from the accelerated expansion of the Universe. The resolution comes naturally from the p-adic fractality predicting that cosmological constant is reduced by a factor of 2 in a step wise manner in phase transitions occurring at times $T(k) \propto 2^{k/2}$, which correspond to p-adic time scales. On the average $\Lambda(k)$ behaves as $1/a^2$, where a is the light-cone proper time. This predicts correctly the observed value of Λ.

b) The notion of many-sheeted space time interpreted as a hierarchy of smoothed out space-times produced by Nature itself rather than only renormalization group theorist is discussed. The dynamics of what might be called gravitational charges is discussed.

c) The theory is applied to the vacuum extremal embeddings of Reissner-Nordström and Schwartschild metric.

d) A model for the final state of a star, which indicates that Z^0 force, presumably created by dark matter, might have an important role in the dynamics of the compact objects. During year 2003, more than decade after the formulation of the model, the discovery of the connection between supernovas and gamma ray bursts [ij6] provided strong support for the predicted axial magnetic and Z^0 magnetic flux tube structures predicted by the model for the final state of a rotating star. Two years later the interpretation of the predicted long range weak forces as being caused by dark matter emerged.

15.2 How do General Relativity and TGD relate?

Mora that two and half decades with TGD have taught that the most important and most difficult problems are those related to the interpretation of the theory. The most difficult lesson has however been that the world view represented by so called well-established theories can be totally wrong, or perhaps better to say not-even-wrong, even when they work amazingly well apart from easily forgotten anomalies.

From the beginning there was an unpleasant feeling that the proposed relationship between TGD and General Relativity might not be elegant enough to be correct. In the sequel I summarize the most important problems, their resolution in terms of new notion of energy, and then discuss in more detailed level the basic predictions.

15.2.1 The problems

There were several closely related problems.

a) How the principle selecting preferred extremals of Kähler action as generalized Bohr orbits can be consistent with Einstein's equations? The energy tensor of Kähler action is certainly not proportional to the Einstein tensor.

b) Equivalence Principle encourages the identification of the inertial energy with gravitational energy. But how the exact conservation of inertial energy can be consistent with the non-conservation of gravitational energy defined by the Einstein tensor? Basically the problem is about relationship between inertial and gravitational energy. Poincare invariance does not yet resolve completely the conceptual problems related to the definition of energy.

c) Robertson-Walker cosmologies are imbeddable into $H = M^4_+ \times CP_2$ but they are vacua with respect to the inertial energy. This forced to give up the very elegant idea that the energy momentum tensor associated with Kähler action would correspond to the classical energy.

Also General Relativity has its problems.

a) The problem of cosmological constant is the most acute problem of General Relativity and also of string models and of M-theory. Recent experimental findings support the view that cosmological constant might be non-vanishing. In TGD Einstein's equations emerge as structural equations and the finite size of space-time sheets allows a non-vanishing cosmological constant also in TGD. My personal strong belief has been that cosmological constant vanishes and the justification has been that the action with a cosmological constant becomes infinite. Unfortunately, the situation changes if Einstein's equations are not derivable from action but are structural equations. Note also that if space-time sheets have a finite temporal extension the volume term in action becomes finite.

b) Despite huge amount of work done during last decades (during the GUT era the problem was regarded as being solved!) matter antimatter asymmetry remains still an unresolved problem of cosmology.

15.2.2   The new view about energy as a solution of the problems

The solution to these problems did not emerge by staying inside the framework of respected physics, and the frame of mind making possible the liberation from the jail of old ideas evolved only gradually. Basically this occurred during last one and half decades when I played with ideas related with quantum consciousness, quantum biology, and also with over unity energy production claimed by the advocates of "free energy" (see the section B-10.6 of bibliography).

The liberation process was initiated by a simple observation which led to the identification of what I believe to be the basic mechanism of conscious information processing and functioning of living matter [TGDconsc, K1]. Since space-time is 4-surface, the sign of energy depends on the time orientation. The identification of phase conjugate photons as negative energy photons allows readily to understand their mysterious properties and communication and control of geometric past in terms of negative energy signals becomes possible as well as remote metabolism by emitting negative energy particles received by a system acting as energy reservoir.

It took few years until I found myself asking whether phase conjugation might make sense for also fermions, and whether one could resolve the puzzle of matter antimatter asymmetry by assuming that anti-fermions carry negative energies. The idea that it might be possible to generate material objects from vacuum remained for a long time too high a threshold for taking this idea seriously. The pleasant realization was that these ideas are immediately testable by just looking whether pairs of photons and their phase conjugates can annihilate to pairs of electrons and negative energy positrons. Needless to say, the technological implications for energy and information technologies are something which no one has dared to dream of.

This bottle neck idea led rather rapidly to quite a dramatic revision of the vision about the world around us and resolved elegantly the interpretational problems.

a) The most predictive variant of TGD assumes that the net conserved quantum numbers of the Universe vanish. Both quantum and classical worlds are vacua and these vacua are replaced with new ones in each quantum jump. Crossing symmetry guarantees that the new picture is consistent with elementary particle physics. In fact, I used years ago crossing symmetry to derive S-matrix and high energy limit and indeed interpreted the zero energy system as kind of cognitive representation for the real world system.

b) The interpretation necessitates negative energies. In TGD Universe all elementary particles have as their building blocks fermions and anti-fermions. Thus if phase conjugate photons are possible they must result by a phase conjugation performed for fermions. Taking this idea seriously leads to the identification of the phase conjugate fermions as states created from a fermionic vacuum for which the roles of creation and annihilation operators are changed.

One can go even further and consider the possibility that this role change can occur separately for fermions and anti-fermions: in this kind of situation both fermion and anti-fermion oscillator operators would be creation operators or annihilation operators. These novel kinds of vacua can carry only positive/negative fermion number and inertial energy can vanish so that they would be naturally associated with nearly vacuum extremals. For the conventional vacua net fermion number can vanish whereas energy cannot vanish. The separate conservation of quark and lepton numbers in principle doubles the bits labelling the possible vacua.

c) The Eastern vision about particle reactions as creation of positive and negative energy particles from vacuum encourages also a more refined view about particle massivation. In TGD Universe total energy vanishes. This would mean that massive gauge bosons can be created from vacuum with appreciable probably only as pairs of positive and negative energy bosons. The Compton wave length of intermediate boson would define the p-adic length scale giving the size of the pair and imply also its instability. In the case of fermions, say electron, the situation cannot be quite so simple. p-Adic mass calculations rely on the idea that massless states mix slightly with states having CP_2 mass scale and the dominant massless contribution explains the stability of electron. Intermediate gauge bosons have a genuine coupling to the TGD counterpart of Higgs field which could explain the mass of the bosons whereas for fermions this contribution is small (there is also a second mechanism based on the lacking covariant constancy of intermediate boson charge matrices). Higgs would thus be the basic cause for the inability to create massive gauge boson pairs with too large separation. Analogous picture applies to color confinement: it is not possible to create colored particles from vacuum without negative energy companion outside the confinement length.

d) The energy momentum tensor of the Kähler action corresponds to the net energy density of matter and antimatter. For vacuum extremals these energies cancel each other. Thus the gigantic vacuum degeneracy of Kähler action finally finds a precise physical meaning. In particular, this holds true in cosmological length scales.

e) Equivalence principle holds true in the sense that gravitational mass is the *absolute* value of inertial mass and the energy density appearing in Einstein's equations corresponds to the *difference* of the energy densities associated with matter and antimatter and is thus positive. Since gravitational energy density is the difference of matter and antimatter energy densities, it is conserved at the classical limit when the annihilation of matter and antimatter is negligible. Einstein's equations are structural equations, kind of equations of state and gravitational constant and cosmological constant can vary.

This would resolve the puzzle created by the finding made already during the first years of TGD. The mystery was that for space-time surfaces which are surfaces in $M_+^4 \times S^2$, where S^2 is homologically nontrivial geodesic sphere, Kähler energy density satisfies approximately Einstein's equation but with gravitational coupling G_{eg}, which is about 10^7 times stronger than Newton's constant. Similar conclusion holds true for cosmic strings.

The correct interpretation is that the density of inertial energy is in good approximation a fraction 10^{-7} about the density of gravitational energy so that there is slight matter antimatter asymmetry. For vacuum extremals G_{eg} is formally infinite and one could speak of strong gravity. The earlier interpretation was in terms of electro-gravity based on hypothesis that gravitation couples more strongly to classical fields and recent interpretation allows to give up this hypothesis. The absolute minimization Kähler action is the fundamental classical dynamics serving as a kind of symbolic representation of quantum dynamics, and the dynamics of Einstein tensor is determined by it.

f) p-Adic length scale hypothesis plus the detailed study of membrane like vacuum extremals lead to the hypothesis that cosmological constant depends on p-adic length scale $\Lambda/R^2 \propto 1/R^2 L^2(k) \propto 2^{-k}$. Amazingly, the recent value of the cosmological constant suggested by the accelerated expansion of the Universe comes out as a correct prediction!

Cosmological expansion at a particular space-time sheet becomes a TGD counterpart for a sequence of periods of increasingly slow inflation which a reduction of Λ by a factor of 2 at each time when the size of space-time sheet exceeds a p-adic length scale. It must be however emphasized that Kähler action determines the classical dynamics and it is by no means clear that exponential expansion is involved. What certainly occurs is liberation of gravitational energy, which means that the difference of inertial energy densities for matter and antimatter is reduced in a phase transition like manner. Maybe the interpretation in terms of annihilation of matter and antimatter is appropriate. Perhaps particles with masses of order p-adic length scale become non-relativistic and annihilate to lighter particles, most naturally those corresponding to the next p-adic length scale.

One must consider also the possibility that the sign of Λ is negative. This would imply that at each step when Λ is reduced by a factor of two, cosmic expansion suffers kind of a jerk and then starts to slow down. The requirement that the gravitational energy is positive however favors positive value of Λ.

15.2.3 Basic predictions at quantitative level

In the following I consider basic quantitative predictions following from the proposed scenario by studying some of the simplest extremals of Kähler action.

Membrane like objects and the p-adic evolution of cosmological constant

The first argument in favor of a non-vanishing cosmological constant comes from the requirement that gravitational energy is larger or equal than the absolute value of the inertial energy. If cosmological constant vanishes, one might argue that the gravitational energy of any 3-surface of finite size vanishes. This would require that also the inertial energy vanishes. The way out of dilemma is to assume that energy density contains also the contribution corresponding to cosmological constant which is proportional to volume. Actually this argument must be taken with a big grain of salt as following more convincing argument demonstrates.

The density of gravitational energy should be non-negative always and this forces to introduce cosmological constant as the following argument shows.

Consider membrane like vacuum extremals of form $X^4 = M^1 \times X^2 \times S^2$, where S^1 is circle in CP_2, M^1 corresponds to time axis of M^4 and X^2 is arbitrary two-surface. Inertial energy density vanishes for this surface so that matter and antimatter energies cancel each other locally. Einstein tensor is given by

$$
G^{\alpha\beta} \quad = \quad R^{\alpha\beta}(X^2) - \frac{1}{2}g^{\alpha\beta}R(X^2) \ . \tag{15.2.1}
$$

One can integrate the contribution of the Einstein tensor to the total energy (one can multiply $G^{\alpha\beta}$ the vector field generating time translation in M^4_+ to get a vector field) to get given by

$$
16\pi G \times E_{gr} \quad = \quad -\frac{1}{2}L(S^1) \times \int_{X^2} R(X^2)d^2x = \frac{1}{2}L(S^1) \times 2\pi(g-1) \ . \tag{15.2.2}
$$

Note that $\xi(X^2) = 1 - g$ is the Euler characteristic and g denotes the number of handles of the two surface. The gravitational energy would be negative for spherical topology without a compensating cosmological contribution from the action $S = \hat{\Lambda}\int \sqrt{(g)}d^4x$, $\hat{\Lambda} \equiv 8\pi G\Lambda$:

$$
E_{vac} \quad = \quad \hat{\Lambda}L(S^1) \times S(X^2) \ . \tag{15.2.3}
$$

Here $S(X^2)$ denotes the area of the two-surface.

Cosmological constant Λ can depend on the p-adic length scale characterizing the size of X^2. In fact, it seems that this must be the case. First of all, for very small radii the contribution of the Einstein tensor implies negative gravitational energy irrespective of the value of Λ. One can however assume that Λ scales as $1/L_{p,n}^2$ where $L_{p,n} = p^{(n-1)/2}L_p$ is the n-ary p-adic length scale characterizing the size of X^2. For $n = 1$ one obtains the hierarchy $L_p = \frac{\sqrt{p}}{\sqrt{2}}L_2$ of p-adic length scale with $p \simeq 2^k$, k integer, defining the physically favored p-adic length scales. This would mean that the contribution of the cosmological constant term does not grow with surface area in long length scales. For p-adic length scales below L_2 which is of order CP_2 size, the scales $n = -1, -2, ..$ become possible. For $n = -1$ the scaling $\Lambda(k) \propto 2^k/L^2(2)$ takes care that the the the sum of energies is independent of p-adic length scale.

This hypothesis implies that the value of cosmological constant is in cosmological length scales related by a p-adic scaling to its value at the 2-adic length scale $L(2) \sim R$:

$$
\frac{\Lambda(k)}{\Lambda(k=2)} = \frac{a^2(2)}{a^2(k)} = 2^{-k+1} \ . \tag{15.2.4}
$$

Here $a(k)$ is the p-adic time scale defined by the p-adic length scale $L(k)$ identified as Minkowski light cone proper time. $a(k)$ increases by a factor 2 in each phase transition reducing the value of the cosmological constant. The implication is that cosmological constant decreases as $1/a^2$ one the average. This provides an elegant solution to the mysterious smallness of cosmological constant assuming just the naive value at the primordial stage.

As known, the naive estimate $\Lambda_0 \sim 1/G$ implies that the cosmological constant is about 10^{120} times larger than the experimental upper bound for it. This is no doubt the most dramatic discrepancy between theory and experiment known in the history of physics. The detailed discussion of cosmic strings leads to the conclusion that the value of $\Lambda(2)$ is of order

$$\Lambda(2) \quad \sim \quad \frac{1}{R^2} \sim 10^{-7}\Lambda_0 \ . \tag{15.2.5}$$

This alone does not help much!

The p-adic scaling implies that the recent value of $\Lambda(a)$ is about

$$\Lambda(a) \quad \sim \quad \sim 10^{-7}\Lambda(2)\frac{a^2(2)}{a^2(k)} = 10^{-7}\Lambda(2)2^{-k+1} \ . \tag{15.2.6}$$

The requirement $\Lambda(a)/\Lambda(2) \simeq 10^{-120}$ gives $2^{-k+1} \simeq 10^{-113} \sim (2^{-10})^{113/3}$ gives $k \sim 378$. This gives together with $L(k = 151) = 10$ nm gives $L(k) = 2^{(k-151)/2} \times L(151) \sim 2^{113}L(151) \sim 10^{26}$ m which corresponds to 10^{11} light years and to a good estimate for the recent age of the universe. Thus this simple argument resolves elegantly the hugest discrepancy in the history of theoretical physics (only in cosmology one can tolerate so cosmic discrepancies!).

Cosmic strings and the value of cosmological constant

Cosmic strings are objects of form $X^4 = X^2 \times S_I^2$, where S_I^2 is the homologically non-trivial geodesic sphere of CP_2. Non-vacuum extremals of Kähler carrying ultra-strong Kähler magnetic field are in question. For simplicity one can assume that X^2 is a flat piece M^2 of M_+^4 so that the Einstein tensor has expression

$$G^{\alpha\beta} = \mathbf{R}(S_I^2) - \frac{1}{2}\mathbf{g}R(S_I^2) \ . \tag{15.2.7}$$

where $R(S_I^2)$ denotes the curvature scalar. The energy density $G^{tt} = -(1/2)R(S_I^2)$ is negative because of spherical spherical topology. Obviously cosmic strings correspond to the p-adic length scale $L(2)$ naturally so that it should be possible to deduce a lower bound for $\Lambda(2)$ from the requirement that gravitational energy density is non-negative.

There are actually two constraints involved.

a) The first constraint comes from the sum of the energies of matter and and antimatter expressible as the conserved energy of Kähler field classically. A priori one cannot exclude the possibility that for cosmic strings the quantization in fermionic sector is such that both fermions and anti-fermions are either positive or negative. This requires that cosmic strings created in the big bang or possibly later appear in pairs which opposite vacuum energies. Since very simple objects are in question this might be the case although arbitrary small difference of induced metrics might be amplified into infinitely large difference in zero point energies.

The energy momentum tensor for Kähler field and Einstein tensor can be written as

$$\begin{aligned}
\mathbf{T}_K &= \frac{1}{2 \times 16\pi\alpha_K R^4}(\mathbf{g}(M^2) - \mathbf{g}(S^2)) \ , \\
\mathbf{G} &= -\frac{4}{R^2}\mathbf{g}(M^2) \ .
\end{aligned} \tag{15.2.8}$$

Note that the radius of S_I^2 is $R/4$ for the conventions used.

$\mathbf{T}_K$ can be written as sum of Einstein tensor and metric tensor as

$$\mathbf{T}_K = \frac{1}{16\pi G_{eg}}\mathbf{G} + \hat{\Lambda}_{eg}\mathbf{g} \ ,$$

$$G_{eg} = 8R^2\alpha_K \ ,$$

$$\hat{\Lambda}_{eg} = \frac{1}{2R^4 \times 16\pi\alpha_K} \ . \tag{15.2.9}$$

Note that the ratio $G_{eg}/G = \frac{8R^2}{G}\alpha_K$ is much larger than one.

The study of simple imbeddings of Maxwell field to $M^4 \times S_I^2$, demonstrates that Einstein's equations hold true approximately also more generally in the non-relativistic approximation and the value of G_{eq} is as above. This raises the possibility that the dynamics of Kähler action could be reflected in phenomena of electro-gravitation and that near vacuum extremals could be seen as a limit at which electro-gravitational coupling strength becomes very large.

There could be thus two "gravitations": one for the sum and one for the difference of energies of matter and antimatter. The consistency with the existing theory requires an approximate separate conservation of these energies meaning that the annihilation of matter and antimatter can be neglected. TGD based models for the asymptotic cosmology and the asymptotic state of star rely on this assumption.

b) Gravitational energy tensor can be written as a sum of Einstein tensor and metric tensor

$$\mathbf{T}_{gr} = \frac{1}{16\pi G}\mathbf{G} + \Lambda(2)\mathbf{g} \ . \tag{15.2.10}$$

c) The basic constraint comes from the requirement that irrespective of the option chosen for the fermionic quantization one has

$$E_{gr} \geq E_K \ , \tag{15.2.11}$$

where one has

$$E_{gr} = -\frac{1}{16\ G} + \hat{\Lambda}(2)\frac{\pi R^2}{4} \ ,$$

$$E_K = -\frac{1}{16G_{ge}} + \hat{\Lambda}_{gc}\frac{\pi R^2}{4} = \frac{1}{4 \times 16\alpha_K R^2} \ . \tag{15.2.12}$$

The lower bound means that the two terms in gravitational energy nearly cancel each other since the right hand side is of order 10^{-7} as compared to the left hand side. One has

$$\hat{\Lambda}(2) \geq \frac{1}{16R^4}\left[\frac{16R^2}{G} + \frac{1}{\alpha_K}\right] \ . \tag{15.2.13}$$

For the minimal value of $\Lambda(2)$ gravitational and inertial energies are identical. This corresponds to the situation in which negative energy contribution to the inertial energy vanishes. This situation can result in two manners. Either antimatter is absent or fermionic quantization is performed in the standard manner. The latter option fixes the value of $\hat{\Lambda}(2)$ completely.

The cancellation of the total inertial energy can be achieved if strings have opposite time directions and the zero energy fermionic ground states for the two strings are phase conjugates of each other. Strings with conventional fermionic vacua can emit positive energy fermions during later stages of evolution: this reduces the energy density of the string. If the stringy fermionic vacuum is unconventional, the string can generate fermion anti-fermion pairs during later phases of cosmic evolution such that the negative energy member of the pair remains inside string and reduces its energy density (note that fermion number increases!). These TGD arguments generalize to phase conjugate strings. The reduction of the energy density means a thickening of the string since flux quantization condition must be satisfied. Energy density behaves as $1/S$ during this process.

More detailed considerations [D4, D5] show that only the non-conventional vacua are consistent with the galactic rotation curves favoring $T_{gr} \sim 1/R^2$ rather than $T_{gr} \sim 1/G$ predicted by the mechanism based on the conventional stringy vacua.

Do absolute minima of Kähler action satisfy the inertial counterpart of Einstein's equations?

Quantum classical correspondence suggests that the extremals of Kähler action are able to represent "symbolically" various aspects of the dynamics of quantum jumps, also dissipation. In particular, absolute minima should asymptotically serve as space-time correlates of non-dissipating self-organization patterns. This requires that Lorentz Kähler force vanishes. This ansatz leads to a construction of very general solution families for the field equations.

The vanishing of Lorentz Kähler force implies that the energy momentum tensor $T_K^{\alpha\beta}$ associated with Kähler action has a vanishing covariant divergence. This raises the question whether it could be expressible as a linear combination of induced metric and Einstein tensor:

$$T_K^{\alpha\beta} \;=\; \frac{1}{16\pi G_{eg}} G^{\alpha\beta} + \hat{\Lambda}_{eg} g^{\alpha\beta} \; . \tag{15.2.14}$$

Since inertial energy momentum is in question, the counterparts of Newton's constant and cosmological constant must differ from their gravitational counterparts as they indeed do according to the previous considerations.

The strongly non-deterministic dynamics of vacuum extremals (with respect to inertial energy only) brings strongly in mind 4-dimensional topological quantum field theories and one can ask whether the time evolution could in some sense define a topological quantum field theory providing information about the topology of the vacuum extremal. The vacuum extremal property could pose constraints on the four-topology and also define non-trivial cobordism theory for 3-topologies.

Boundary conditions in presence of cosmological constant

When space-time sheet has boundaries the conservation of gravitational energy gives additional conditions. One must assign energy momentum tensor $T_\delta^{\alpha\beta}$ to the boundary. The boundary counterpart of Einstein's equations can be written as

$$D_\beta T_{\delta)}^{\alpha\beta} \;=\; X^\alpha \; , \tag{15.2.15}$$

where X^α corresponds to the variation of the boundary term resulting from the variation of the Einstein action $(\kappa R + \hat{\Lambda})\sqrt{g}$ in the interior. Only the first and second derivatives of the interior action contribute to the boundary term so that cosmological constant term does not contribute to the boundary term. One has

$$X^\alpha \;=\; \kappa\{ - \left[\frac{\partial(R\sqrt{g})}{\partial g_{\alpha\beta|\gamma|n}}\right]_{|\gamma} + \frac{\partial(R\sqrt{g})}{\partial g_{\alpha\beta|n}} \} \; . \tag{15.2.16}$$

The requirement that gravitational energy does not leak out from the space-time sheet gives the boundary conditions

$$\kappa G^{n\alpha} + \hat{\Lambda} g^{n\alpha} \;=\; X^\alpha \; . \tag{15.2.17}$$

These conditions refer only to the metric and it should be relatively easy to satisfy these conditions in the case of vacuum extremals of Kähler action. In case of flat 4-surfaces it is obviously not possible to satisfy the boundary conditions.

CP_2 type extremals and Einstein's equations

One might argue that Einstein's equations provide a phenomenological description of many-particle system and it is not possible to apply them to CP_2 extremals. One can still look what one obtains in this manner. Since vacuum extremals are in question inertial energy density vanishes and one has $T_+ = -T_-$. If CP_2 extremals carry elementary particle quantum numbers, such as fermion number, this is possible only if bosonic energy cancels fermionic energy.

Since CP_2 a is constant curvature space, Einstein tensor is proportional to the metric tensor $G^{\alpha\beta} = -\frac{8}{R^2}g^{\alpha\beta}$. The contribution to the gravitational energy is negative. If one assumes that cosmological constant is same as for cosmic strings and equals to $\hat{\Lambda}(2)$, one obtains an additional positive and dominating contribution to the gravitational energy.

15.2.4 Non-conservation of gravitational four-momentum

TGD predicts non-conservation of gravitational four-momentum. At elementary particle level the non-conservation of the gravitational momentum corresponds to the fact that the light-like gravitational momentum of CP_2 type extremal changes its direction during zitterbewegung. The natural hypothesis is that the conserved inertial four-momentum is the average of the gravitational momentum so that massivation results automatically and the rest energy corresponds to the center of mass motion. The randomness of the light-like curve associated with the CP_2 type extremal explains why p-adic thermodynamics is the proper description for the massivation.

As discussed in the chapter "General Ideas about Topological Condensation and Evaporation", an additional contribution to the mass comes from the wormhole contacts in secondary topological condensation which are also pieces of CP_2 type extremals and perform zitterbewegung. p-Adic length scale characterizing the particle corresponds naturally to the scale below which the motion is in a good approximation along light-like geodesic of M^4.

The standard belief is that a compact 3-manifold must have a vanishing gravitational mass. In TGD framework this would mean that the gravitational mass of the vapor phase particles vanishes. As already seen, this is not the case in TGD without further assumptions about the dynamics of topological evaporation. Although the evaporated particle does not interact via classical long range gravitational fields, it can interact via the emission and absorption of gravitons. This should indeed be the case if the particle has a vanishing gravitational mass. But since gravitons form join bonds with the external world, the system would not actually be like an isolated universe. If particle is topologically evaporated in strict four-dimensional sense, it must satisfy empty space Einstein equations with a cosmological term, and be a vacuum extremal so that it does not carry either matter or antimatter and is rather uninteresting physically.

If the space-time sheets of 3-surface has no contacts with larger space-time sheets, it is like an isolated universe, and its inertial mass should vanish and be small if the number of these contacts is small. By the same argument also Lorentz-, color-, fermionic, etc. quantum numbers of the evaporated space-time sheet should vanish. This does not mean that the space-time sheet could not be in motion and possess cm momentum and angular momentum. The point is that these quantum numbers are compensated by the correspond quantum numbers in internal degrees of freedom. Situation changes in the case of non-conserved quantum numbers and the quantum numbers associated with the representations of super-canonical symmetries (discussed in the chapters of the first part of the book) could be non-vanishing.

The inertial mass possibly possessed by a topologically evaporated particle is left at the boundaries of holes resulting in the evaporation. Topologically evaporated particle can indeed develop negative inertial energy density at its boundaries guaranteing the vanishing of its inertial mass since space-time sheets with negative time orientation possess negative energy.

One can evaluate the divergence of the gravitational energy momentum tensor satisfying Einstein equations with a cosmological term by calculating the divergence of the four momentum current $T^{A\alpha}$ (A labels different momentum components)

$$D_\alpha T^{A\alpha} = \left[\frac{1}{16\pi G}G^{\alpha\beta} + \hat{\Lambda}(2)g^{\alpha\beta}\right]H^k_{\alpha\beta}j^A_k \ . \tag{15.2.18}$$

Here j^{Ak} denotes the Killing vector of translation. This equation will be applied later in the context of Robertson-Walker cosmologies.

A situation of a special physical interest corresponds to the separate conservation of inertial and gravitational masses. In this case the following equations of motion derivable from Einstein action $S = \int(kR + \hat{\Lambda}(2))\sqrt{g}d^4x$, $k = 1/16\pi G$, hold true

$$\left[kG^{\alpha\beta} + \hat{\Lambda}(2)g^{\alpha\beta}\right]H^k_{\alpha\beta} = 0 \ . \tag{15.2.19}$$

At cosmological length scales the cosmological term is negligible, and one can replace the equation with

$$G^{\alpha\beta} H^k_{\alpha\beta} \;=\; 0 \;. \tag{15.2.20}$$

This action was my first attempt to try to construct "sub-manifold gravity". This equation is expected to provide a reasonable when the energies associated with matter and antimatter are separately conserved.

For Schwartshild metric the divergence indeed vanishes identically if $\hat\Lambda(2)$ is vanishing (or sufficiently small as in the recent Universe). One can say that stationary situations in which gravitational mass is conserved are those in which gravitational mass appears as lumps. Thus asymptotic self-organization patterns necessarily correspond to gravitational lumping of the matter. In the case of general Reissner-Nordström metric energy flow turns out to be non-vanishing although the metric itself is stationary in the sense of GRT. It will be later found that the Robertson-Walker cosmology satisfying stationarity conditions is essentially unique and provides a natural candidate for the asymptotic cosmology.

The earlier interpretation for the non-conservation of gravitational four-momentum was in terms of a loss of inertial four-momentum due to topological evaporation and flow between different space-time sheets. This interpretation must be given up. If the flow occurs along join along boundaries bonds then the fluxes of energy momentum currents through the cross sections of join along boundaries bonds give the net flow. Not only the flow of energy-momentum but also the flow of fermion numbers, charges, etc., can occur between various space-time sheets. In particular, the flow of entropy is possible. An interesting possibility is that the vapor phase could serve as a "cosmic paper basket" allowing to circumvent the consequences of the second law of thermodynamics: the creation of order through gravitational interactions would become possible in topological condensate through the transfer of entropy to the vapor phase.

15.3   TGD and GRT descriptions of space-time

The absolute minimization of Kähler action does not imply Einstein's equation and the original attempt to understand GRT was as a long length scale limit of TGD for which GRT space-time is kind of quantum average space-time obtained by smoothing out topological non-homogenities representing particles and various structures and describing their presence using various kinds of currents. An improved picture is based on the new view about energy, the idea that many-sheeted space-time in some sense takes care of the smoothing out procedure itself by providing a hierarchy of smoothed out dynamics, and the conviction that Einstein's equations coupled to matter provide a statistical rather than a fundamental description. In particular, quantization of EYM action does not make sense in this conceptual framework.

15.3.1   Many-sheeted space-time defines a hierarchy of smoothed out space-times

The notion of quantum average space-time obtained by smoothing out details below the scale of resolution was inspired by renormalization philosophy and for long time I regarded it as a fictive concept. The rough idea was that quantum average effective space-times correspond to the absolute minima of the Kähler action associated with the maxima of the Kähler function. Therefore the dynamics of the quantum average effective space-time is fixed and the stationarity requirement for the effective action should only select some physically preferred maxima of the Kähler function. The topologically trivial space time of classical GRT cannot directly correspond to the topologically highly nontrivial TGD space-time but should be obtained only as an idealized, length scale dependent and essentially macroscopic concept. This allows the possibility that also the dynamics of the effective smoothed out space-times is determined by the effective action.

The space-time in length scale L is obtained by smoothing out all topological details (particles) and by describing their presence using various densities such as energy momentum tensor $T^{\alpha\beta}_\#$ and Yang Mills current densities $J^{\alpha}_{a\#}$ serving as sources of classical electro-weak and color gauge fields (see Fig. 15.3.1). It is important to notice that the smoothing out procedure eliminates elementary particle type boundary components in all length scales: this suggests that the size of a typical elementary particle boundary component sets lower limit for the scale, where the smoothing out procedure applies.

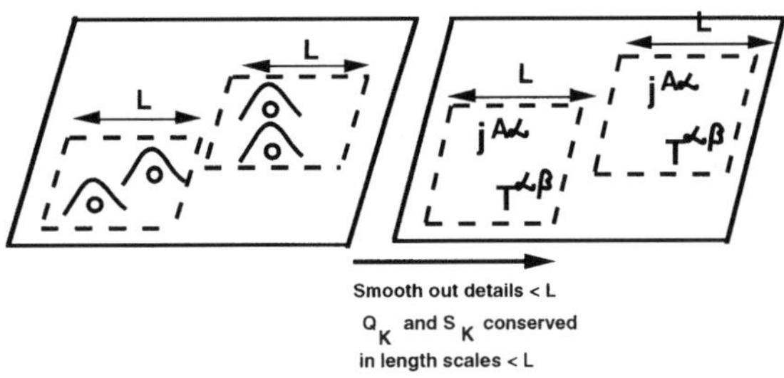

Figure 15.1: Intuitive definition of length scale dependent space-time

During development of the many-sheeted space-time concept it has become obvious that the notions of classical space-time and of smoothing out of details are not only activities of a theoretician, but that the many-sheeted space-time itself can be said to perform renormalization theory.

a) Classical space-time is much more than a fiction produced by the stationary phase approximation. The localization in the so called zero modes, which corresponds to state function reduction in TGD, which occurs in each quantum jump (the delicacies due to macro-temporal quantum coherence will not be discussed here) means that the superposition of space-time surfaces in the final state of quantum jump, consists of space-time surfaces equivalent from the point of view of observer.

b) The notion of many-sheeted space-time predicts a hierarchy of space-time sheets labelled by p-adic primes $p \simeq 2^k$, k integer with primes and prime powers being in preferred role. The space-time sheets at a given level of hierarchy play a role of particles topologically condensed at larger space-time sheets. Hence the physics at larger space-time sheets is quite concretely a smoothed out version of the physics at smaller space-time sheets. Many-sheeted space-time itself performs renormalization group theory, and p-adic primes characterizing the sizes of the space-time sheets correspond to the fixed points of the renormalization group evolution.

c) There are good reasons to expect that the absolute minimum value for the Kähler action vanishes for large enough space-time sheets, and that space-time sheets result as small deformations of the vacuum extremals at the long length scale limit. The equations derived from Einstein-Hilbert action for the induced metric can be posed as an additional constraint on stationary vacuum extremals for which the gravitational four momentum current is conserved.

An important difference to the standard view is that energy momentum tensor is dictated by the Einstein tensor (plus metric) rather than vice versa. Since the dynamics of the induced EYM fields is dictated by the absolute minimization of Kähler action, EYM equations cannot in general be satisfied without the introduction of particle currents. This conforms with the view that Einstein's equations relate to a statistical description of matter in terms of both particle densities and classical fields. The imbeddability to $H = M_+^4 \times CP_2$ means a rich spectrum of predictions not made by GRT. TGD inspired cosmology and TGD based model for the final state of the star are good examples of these predictions, and are consistent with experimental facts.

15.3.2 The dynamics of "gravitational" charges

The idea is that the dynamics of topologically condensed matter is about the evolution of "gravitational" counterparts of various currents defined as differences of the currents associated with positive and negative energy matter. Interestingly, this is what one could expect on the basis of quantum classical correspondence. According to TGD inspired theory of consciousness, quantum jump is the basic building block of conscious experience [TGDconsc] and only changes can be experienced consciously, and crossing symmetry allows to regard positive and negative energy particles as the initial and final states of particle reaction. In terms of TGD inspired theory of consciousness, "inertial" corresponds to the "objective" existence (quantum states) and "gravitational" to the "subjective" existence (quantum jumps).

a) The space-time sheet is a preferred extremal of Kähler action (absolute minimum or a more general preferred extremal [E2]) and classical electro-weak and color fields are determined by this dynamics. The energy momentum tensor of Kähler action characterizes the net four-momentum density. The dynamics of the induced spinor fields is the super-symmetric counterpart of Kähler action and thus corresponds to the "inertial sector" of the theory. The fundamental role of the induced spinor fields in quantum theory suggests the same. That induced Kähler field and induced spinor field do not directly correspond to change or difference of any quantum number and are not directly observable would conform with their "inertial" character.

b) The divergences of classical YM fields are non-vanishing in general and define non-conserved vacuum gauge currents. Einstein's equations make sense only if the divergences of the energy momentum tensor of matter and those of classical YM fields cancel each other. The goal is achieved if the YM currents associated with the topologically condensed matter are equal to the vacuum YM currents. Also the net divergences of other contributions to the energy momentum tensor must vanish and this gives to the analogs of hydrodynamical equations.

c) The question is whether the YM currents correspond to the sum or difference of currents associated with positive and negative energy matter. Internal consistency would suggest that the dynamics is for the differences of various charges for matter and antimatter so that the currents would correspond to "gravitational" counterparts of gauge charges. This would be also consistent with the facts that only the covariant divergence of these currents is vanishing and that they are not genuinely conserved. Fortunately, the expressions for the currents would be exactly the same as in the standard approach assuming opposite charges for negative energy matter.

For instance, the presence of long range $W^\pm$ fields would code for the occurrence of quantum jumps in which electro-weak quantum numbers are changed. These quantum jumps would occur in all time and length scales. Classical gauge fields would be in a well-defined sense "gravitational" counterparts of the quantum fields describing corresponding elementary particles. This interpretation would apply also in the case of electromagnetic current. Indeed, classical em current corresponds to the covariant divergence of em field and if classical $W^\pm$ fields are present em current fails to be conserved.

d) This interpretation does not allow to understand why long range gauge fields are present since standard model predicts that gauge fields created by ordinary particles vanish above weak length scale. The identification of these fields is as fields generated by dark matter, this notion understood in a very general sense [A1, A2]. Weak fields behave like massless fields below the p-adic length scale determining the size scale of the space-time sheets of weak bosons and are vacuum screened above this length scale in the sense that they do not feed appreciable gauge fluxes to larger space-time sheets: the weak screening occurs in # contacts modellable using CP_2 type extremals. The causal horizon associated with a # contact behaves as a bound pair of dark partons being partially responsible for these long range forces. Color confinement has the same effect on the color gauge fields as weak vacuum screening.

As always, one can invent also an objection. To Einstein's tensor it is indeed possible to assign non-conserved four-momentum and color currents but no electro-weak currents, and one might argue that this interpretation cannot make sense for classical electro-weak gauge charges which allow only definition as gauge fluxes.

e) Usually matter currents identified as YM currents would determine YM fields. Now vacuum YM currents dictate the behavior of the matter currents.

A more formal prescription goes as follows:

a) The effective action describes the presence of matter and antimatter using YM currents, particle densities, etc.. and can be taken to be Einstein Yang Mills action for the induced gauge fields plus matter currents. Energy momentum tensor can contain also a hydrodynamics part: typical example comes from cosmology. Not only classical electromagnetic and Z^0 fields but also $W^\pm$ and color gauge fields can appear as classical fields in macroscopic length scales.

b) For YM fields the interaction term is just the external YM current multiplied with YM gauge potential. For the metric the corresponding term is the trace of energy momentum tensor. This assumption implies the addition of the various interaction terms to the EYM action density

$$
\begin{aligned}
L_{EYM} \;\rightarrow\;& L_{EYM} \\
&+\; \sum_a Tr(J^\alpha_{a\#} A_{a\alpha}) \\
&+\; T^{\alpha\beta}_{\#} g_{\alpha\beta} \;.
\end{aligned}
\tag{15.3.1}
$$

The YM currents associated with topologically condensed matter are denoted by $J^\alpha_{a\#}$, where 'a' refers to a specific component of YM field. The corresponding energy momentum tensor is denoted by $T^{\alpha\beta}_{\#}$.

c) L can be decomposed into a sum of terms corresponding to color and electro-weak interactions and gravitational interaction:

$$
L_{EYM} \;=\; L_{ew} + L_c + L_{gr} \;.
\tag{15.3.2}
$$

The couplings associated with the various actions depend on the length scale considered. For Kähler function, the space-time associated with a given 3- surface is fixed by the requirement that it corresponds to an absolute minimum of the Kähler action. For the maxima of Kähler function also the extremization with respect to X^3 degrees of freedom is performed. This procedure is the same as applied usually, when finding the physically interesting solutions of the effective action. It must be emphasized that the spin glass nature of the TGD Universe might imply complications.

d) Field equations are solved by taking YM equations and Einstein's equations as *definitions* of the source densities associated with various particle types. The remaining field equations are then those describing the motion of the topologically condensed matter and are identically satisfied.

Consider now the general features of this scenario.

a) That the absolute minimization of Kähler action defines the fundamental dynamics means an enormous simplification at the level of principle. For instance, for the topologically condensed Kähler charged cosmic strings the exterior space-time can be regarded as an extremal of the Kähler action with the Kähler electric charge of the cosmic string serving as a source of the radial Kähler field. This leads to a model for the formation of galaxies, which treats Kähler charged cosmic strings as line sources of Kähler electric field rather than line sources of gravitational field. The predicted gravitational field is analogous to the gravitational field of a straight string in Newtonian approximation so that Einstein tensor vanishes in this approximation, which in turn conforms with the fact that matter resides at the boundaries of the void and is necessary in order to neutralized the total Kähler charge. GRT based model predicts flat metric with an angular defect.

b) For sufficiently long length scales the average density of inertial energy vanishes so that the vacuum extremals of the Kähler action should become important as the minima of the Kähler action also. The proposed model for inhomogeneous cosmology leads to the conclusion that the length scale, where vacuum extremals of Kähler action apply is the length scale, at which matter is in average Kähler neutral (so that net inertial 4-momentum density vanishes). This length scale is estimated to be about 10^8 light years at present.

c) The study of the imbeddings of various metrics shows that massive objects are in general accompanied by long range electro-weak fields. Electromagnetic and Z^0 fields are the most plausible candidates, which means that dark gluons and electro-weak bosons with arbitrary small mass scales but not coupling to ordinary matter are predicted.

15.4   Imbedding of the Reissner-Nordström metric

In the following the imbedding of electromagnetically neutral Reissner-Nordström metric to $M^4_+ \times CP_2$ will be studied. The imbedding generalizes to an imbedding of any spherically symmetric metric. The imbeddings as vacuum extremals reduces to an imbedding into 6-dimensional $M^4 \times Y^2$, Y^2 Lagrange manifold (vanishing induced Kähler form). Any vacuum extremal defines a solution of Einstein's equations if energy momentum tensor is defined by Einstein's equations. Non-vacuum imbeddings of Reissner-Nordström solutions would correspond to homologically non-trivial geodesic sphere of CP_2, and it is implausible that non-vacuum imbeddings could be extremals. Whether the imbeddings of the metrics believed to describe rotating objects in GRT Universe are possible at all, is not known but it might well be that the dimension of the imbedding space is too low to allow them.

15.4.1   Two basic types of imbeddings

One can construct a large number of imbeddings for Reissner-Nordström metric. These imbeddings need not extremals except when they are represent vacua.

a) X^4 could be a sub-manifold of $M^4 \times S_i^2$, $i = I, II$, where S_i^2 is one of the geodesic spheres of CP_2. For $i = II$ the imbeddings are extremals but this is not the case for $i = I$. The properties of these imbeddings are essentially those associated with the spherically symmetric stationary extremals of the Kähler action. Long range electromagnetic and Z^0 fields assignable to dark matter [A1, A2, F6] are present but the corresponding forces are by a factor 10^{-4} weaker than gravitational force, when the parameter ωR is of order one.

b) The vacuum extremals of the Kähler action are the physically most interesting candidates for the imbeddings and represent a situation in which the inertial energies of matter and antimatter cancel each other. For these imbeddings electro-weak fields are in general non-vanishing. Em neutrality is possible to achieve only for $p = sin^2(\theta_W) = 0$. Long ranged W^+ and W^- fields can be present and they induce a small mixing between charged dark lepton and corresponding neutrino spinors.

15.4.2   The condition guaranteing the vanishing of em, Z^0, or Kähler fields

In order to obtain imbedding with vanishing em, Z^0, or Kähler field, one must pose the condition guaranteing the vanishing of corresponding field (see the Appendix of the book). For extremals of Kähler action em Z^0 fields are always simultaneously present unless Weinberg angle vanishes. In practice only the condition guaranteing vanishing of Kähler field is thus interesting.

Using coordinates $(r, u = cos(\Theta), \Psi, \Phi)$ for CP_2 the surfaces in question can be expressed as

$$
\begin{aligned}
r &= \sqrt{\frac{X}{1-X}} \ , \\
X &= D|k + u|^\epsilon \ , \\
u &\equiv cos(\Theta) \ , \quad D = \frac{r_0^2}{1 + r_0^2} \times \frac{1}{C} \ , \quad C = |k + cos(\Theta_0)|^\epsilon \ .
\end{aligned}
\tag{15.4.1}
$$

Here C and D are integration constants. The value of the parameter ϵ characterizes which field vanishes:

$$
a) \ \ \epsilon = \tfrac{3+p}{3+2p} \ , \quad b) \ \ \epsilon = \tfrac{1}{2} \ , \quad c) \ \ \epsilon = 1 \ ,
$$

$$
p = sin^2(\Theta_W) \ .
\tag{15.4.2}
$$

Here a/b/c corresponds to the vanishing of em/Z^0/Kähler field.

$0 \le X \le 1$ is required by the reality of r. $r = 0$ would correspond to $X = 0$ giving $u = -k$ achieved only for $|k| \le 1$ and $r = \infty$ to $X = 1$ giving $|u + k| = [(1 + r_0^2)/r_0^2)]^\epsilon$ achieved only for

$$
sign(u + k) \times [\frac{1 + r_0^2}{r_0^2}]^\epsilon \le k + 1 \ ,
$$

where $sign(x)$ denotes the sign of x.

These imbeddings obviously possess a 2-dimensional CP_2 projection. The generation of long range vacuum weak and color electric fields is a purely TGD based phenomenon related to the fact that gauge fields are not primary dynamical variables.

For future purposes it is convenient to list the explicit expressions of relevant gauge field when em or Kähler field vanishes.

a) Using coordinates $(u = cos(\Theta), \Phi)$ the expressions for the Kähler form and Z^0 field for space-time surfaces with vanishing em field read as

$$
\begin{aligned}
J &= -\frac{p}{3 + 2p} X du \wedge d\Phi \ , \quad X = D|k + u|^{\frac{3+p}{3+2p}} \\
Z^0 &= -\frac{6}{p} J \ .
\end{aligned}
$$

$$
\tag{15.4.3}
$$

b) For vacuum extremals ($\epsilon = 1$) classical em and Z^0 fields are proportional to each other:

$$
\begin{aligned}
Z^0 &= 2e^0 \wedge e^3 = \frac{r}{F^2}(k+u)\frac{\partial r}{\partial u}du \wedge d\Phi = (k+u)du \wedge d\Phi \ , \\
r &= \sqrt{\frac{X}{1-X}} \ , \quad X = D|k+u| \ , \\
\gamma &= -\frac{p}{2}Z^0 \ .
\end{aligned}
\tag{15.4.4}
$$

For a vanishing value of Weinberg angle ($p = 0$) em field vanishes and only Z^0 field remains as a long range gauge field. Vacuum extremals for which long range Z^0 field vanishes but em field is non-vanishing are not possible. The only reasonable physical interpretation seems to be in terms of a hierarchy of electro-weak physics with arbitrarily light weak boson mass scales.

The effective form of the CP_2 metric is given by

$$
\begin{aligned}
ds^2_{eff} &= (s_{rr}(\frac{dr}{d\Theta})^2 + s_{\Theta\Theta})d\Theta^2 + (s_{\Phi\Phi} + 2ks_{\Phi\Psi})d\Phi^2 = \frac{R^2}{4}[s^{eff}_{\Theta\Theta}d\Theta^2 + s^{eff}_{\Phi\Phi}d\Phi^2] \ , \\
s^{eff}_{\Theta\Theta} &= X \times \left[\frac{\epsilon^2(1-u^2)}{(k+u)^2} \times \frac{1}{1-X} + 1 - X\right] \ , \\
s^{eff}_{\Phi\Phi} &= X \times \left[(1-X)(k+u)^2 + 1 - u^2\right] \ .
\end{aligned}
\tag{15.4.5}
$$

This expression is useful in the construction of electromagnetically neutral imbedding of, say Schwartchild metric. For $k \neq 1$ $u = \pm 1$ corresponds in general to circle rather than single point as is clear from the fact that $s^{eff}_{\Phi\Phi}$ is non-vanishing at $u = \pm 1$ so that u and Φ parameterize a piece of cylinder.

15.4.3 Imbedding of Reissner-Nordström metric

The imbedding of R-N metric to be discussed generalizes with minor modifications to an imbedding of a spherically symmetric star model characterized by a mass density $\rho(r_M)$ and pressure $p(r_M)$ since the corresponding line element can be written in the form $ds^2 = A(r_M)dt^2 - B(r_M)dr^2_M - r^2_M d\Omega^2$ [ia2]. For vacuum extremal a solution of field equations results.

Denote the coordinates of M^4_+ by (m^0, r_M, θ, ϕ) and those of X^4 by (t, r_M, θ, ϕ). The expression for Reissner-Nordström metric reads as

$$
\begin{aligned}
ds^2 &= Adt^2 - Bdr^2_M - r^2_M d\Omega^2 \ , \\
A &= 1 - \frac{a}{r_M} - \frac{b}{r^2_M} \ , \quad B = \frac{1}{A} \ , \\
a &= 2GM \ , \quad b = G\pi q^2 \ .
\end{aligned}
\tag{15.4.6}
$$

The imbedding is given by the expression

$$
\begin{aligned}
\Phi &= \omega_1 t + f(r_M) \ , \\
\Psi &= k\Phi = \omega_2 t + kf(r_M) \ , \\
m^0 &= \lambda t + h(r_M) \ , \\
\lambda &= \sqrt{1 + \frac{R^2\omega_1^2}{4}s^{eff}_{\Phi\Phi}(\infty)} \ , \quad k = \frac{\omega_2}{\omega_1} \ .
\end{aligned}
\tag{15.4.7}
$$

The components of s^{eff} are given by Eq. 15.4.5 and general form of imbedding by Eqs. 15.4.1 and 15.4.2.

The functions $f(r_M)$ and $h(r_M)$ are determined by the condition

$$
\lambda\partial_{r_M}h = \frac{R^2}{4}s^{eff}_{\Phi\Phi}\omega_1\partial_{r_M}f
\tag{15.4.8}
$$

resulting from the requirement $g_{tr_M} = 0$ and from the expression for $g_{r_M r_M} = -B$:

$$
\begin{aligned}
h &= \int dr_M \sqrt{Y} \ , \quad Y = \frac{Y_1}{Y_2} \ , \\
Y_1 &= -B + 1 + \frac{R^2}{4} s_{\Theta\Theta}^{eff} \frac{(\partial_{r_M} u)^2}{(1 - u^2)} \ , \\
Y_2 &= 1 - \frac{4\lambda^2 \ s_{\Theta\Theta}^{eff}}{R^2 \omega_1^2 \ s_{\Phi\Phi}^{eff}} \ .
\end{aligned}
\tag{15.4.9}
$$

The condition $Y > 0$ at the limit $r \to \infty$ gives non-trivial conditions. Y_1 is positive at large values of r_M and this gives

$$
Y_1 = -B + 1 + s_{\Theta\Theta}^{eff} \frac{(\partial_{r_M} u)^2}{(1 - u^2)} \geq 0
$$

for the allowed values of r_M. Y_1 can change sign at some critical radius above Schwartschild radius $r_S = 2GM$ since B becomes infinite at r_S: this can be avoided only provided one has $u \to 1$ at $r_M \to r_S$. Y_2 must preserve its sign and this is possible if the value of $R\omega_1$ is sufficiently large. Below $r = r_S$ Y_1 has positive and also Y_2 can be positive down to some critical radius. At $r = r_S$ Y_1 has infinite discontinuity in case that Y_1 approaches finite value from above and CP_2 coordinates are continuous. It is easy to see that square root singularity of Θ as a function of $r_M - r_S$ is in question so that the function h is continuous so that the solution is well-defined.

The dependence of $u \equiv cos(\Theta)$ on radial coordinate r_M is determined by the expression for $g_{tt} = A$ giving the condition

$$
A = \lambda^2 - \frac{R^2 \omega_1^2}{4} s_{\Phi\Phi}^{eff} \omega_1^2 \ .
\tag{15.4.10}
$$

The asymptotic behavior of the coordinate $u = cos(\Theta)$ is of form

$$
u \ \simeq \ u_\infty + \frac{K}{r_M} \ ,
\tag{15.4.11}
$$

u_∞ is fixed by the condition $A(\infty) = 1$:

$$
\begin{aligned}
\lambda^2 &- \frac{R^2 \omega_1^2}{4} s_{\Phi\Phi}^{eff}(\infty) = 1 \ , \\
s_{\Phi\Phi}^{eff} &= X \times \left[(1 - X)(k + u)^2 + 1 - u^2\right] \ , \quad X = D|k + u|^\epsilon \ .
\end{aligned}
\tag{15.4.12}
$$

The value of K is given by

$$
K = \frac{8GM}{R^2 \omega_1^2} \left[\frac{\partial s_{\Phi\Phi}^{eff}}{\partial u}(\infty)\right]^{-1} \ .
\tag{15.4.13}
$$

The values of K and u_∞ depend on parameters $\lambda, R\omega_1, k, D$.

For definiteness one can assume that the value of u at infinity is non-negative:

$$
u_\infty \ \geq \ 0 \ .
\tag{15.4.14}
$$

There are two different solution types depending on the sign of the parameter K.

a) For $K < 0$ u decreases and approaches to $u_{min} \geq 0$ as r_M decreases.

b) For $K > 0$ u increases and approaches to $u_{max} \leq 1$. The requirement that the solution can be continued below Schwartshild radius allows only this option. Below Schwartshild radius u must transform to a solution of type a).

Imbeddability breaks for a critical value of the radial coordinate

The imbeddability breaks for some critical value of the coordinate r_M. The extremal value of u and the radius r_c below which the imbedding fails corresponds to the maximum possible value of $s_{\Phi\Phi}^{eff}$. This value corresponds either to $u = 0, 1$ or to a vanishing derivative of $s_{\Phi\Phi}^{eff}$

$$\frac{\partial s_{\Phi\Phi}^{eff}}{\partial u} = 0 \ . \tag{15.4.15}$$

For $\epsilon = 1$ corresponding to vacuum extremals s_{eff} is a fourth order polynomial as a function of u depending on external parameters. One has

$$s_{\Phi\Phi}^{eff} = D|k + u| \times \left[(1 - D|k + u|)(k + u)^2 + 1 - u^2\right] \ . \tag{15.4.16}$$

s_{eff} becomes negative for very large values of u. Hence a restriction of the standard form of the dual of the cusp catastrophe to the range $u \in (0, 1)$ results. Depending on the values of external parameters there are either 2 maxima or single maximum. For $k = 1$ the positive extremum correspond to $u = 1/|D|$.

In the case of the Schwartshild metric this gives for the critical radius the expression

$$
\begin{aligned}
r_c &= \frac{r_S}{\delta} \ , \\
\delta &= 1 - \lambda^2 + \frac{R^2 \omega_1^2}{4} s_{\Phi\Phi}^{eff}(max) \ , \\
r_S &= 2GM \ .
\end{aligned}
\tag{15.4.17}
$$

The existing evidence for black hole like objects suggests that it would be better to have $\delta >> 1$ in order to get imbeddings of the Schwartschild metric containing also horizon and part of the interior region. A sufficiently large value of $R\omega_1$ indeed allows to have arbitrarily small value of r_c. There the experimental evidence for the existence of black hole like objects leads to no problems.

The imbeddings of Schwartschild metric possess electro-weak and color charges

Both vacuum and non-vacuum imbeddings of Reissner-Nodrström and Scwartshild metric necessarily possess some non-vanishing electro-weak and color charges. Consider first vacuum extremals. Z^0 electric field Z_{tr}^0 is proportional to ω_1

$$Z_{tr_M}^0 = \omega_1(k + u)\partial_{r_M} u \ . \tag{15.4.18}$$

The gauge flux through a sphere with radius r_M depends on r_M so that Z^0 vacuum charge density is necessarily present.

The condition $\theta \propto \sqrt{r - r_S}$ allowing to continue the imbedding below $r_M < r_S$ implies that gauge fluxes, which are proportional to $sin(\Theta)\partial_{r_M}\Theta$, are finite at $r = r_S$ so that the renormalizations of gauge couplings remain finite at least down to Schwartschild radius.

At large distances the gauge flux approaches to

$$
\begin{aligned}
Q_Z(\infty) &= \frac{1}{g_Z} \int_{r_M \to \infty} Z_{tr_M}^0 r_M^2 d\Omega \ , \\
&= \frac{4\pi}{g_Z}\omega_1(k + u_\infty)K = \frac{4\pi}{g_Z}(k + u_\infty)\frac{8GM}{R^2\omega_1}\left[\frac{\partial s_{\Phi\Phi}^{eff}}{\partial u}(\infty)\right]^{-1}
\end{aligned}
\tag{15.4.19}
$$

at the limit $r_M \to \infty$. Z^0 charge is proportional to the gravitational mass. The gauge flux grows at small distances in accordance with the general wisdom about the coupling constant evolution of $U(1)$ gauge field.

The requirement that Z^0 force is weaker than gravitational force expressed as the condition

$$\frac{Q_Z^2}{GM^2} \ll 1$$

implies

$$\frac{32\pi}{R\omega_1 g_Z}(k + u_\infty)\left[\frac{\partial s^{eff}_{\Phi\Phi}}{\partial u}(\infty)\right]^{-1} \ll \frac{R}{\sqrt{G}} \ . \tag{15.4.20}$$

It seems that a sufficiently large value of $R\omega_1$ allows arbitrarily small values for both the Z^0 charge and the critical radius r_c. In the earliest scenario, which was based on the assumption that CP_2 radius is of order Planck length the situation was different. It is clear that the larger radius of CP_2 makes it possible to avoid too strong classical electro-weak forces.

The non-extremal imbedding to S^2_I studied in detail here is Kähler charged and therefore also Z^0 charged since the condition $Z^0 = 6J/p$ holds true by electromagnetic neutrality. The value of the Kähler charge for non-vacuum imbedding depends on the distance from the origin

$$\begin{aligned}
Q_K(r_M) &= \frac{1}{g_K}\int_{r_M=const} J_{tr_M} r_M^2 d\Omega \ , \\
J_{tr_M} &= -\frac{p}{2(3+p)}\omega_1|k + u|^{\frac{3+p}{3+2p}}\partial_{r_M}u \ ,
\end{aligned}$$

$$\tag{15.4.21}$$

The expression for the charge differs only in minor details from that for Z^0 charge for vacuum extremals. Essentially similar conclusions about the behavior of the gauge charges hold true also in the case of vacuum extremals and the expressions differ only by the value of the parameter ϵ characterizing whether em, Z^0, of Kähler field vanishes.

Equivalence Principle and critical radius

When one considers Equivalence Principle, one must keep in mind that the Kähler charged imbeddings of Reissner Nodström and Scwartschild metrics are *not* extremals. Since the general properties of imbeddings to S^2_I and S^2_{II}, the Kähler charged imbedding allow provide information about general properties of the imbedding.

1. New view about Equivalence Principle

a) In the case of vacuum extremals the Kähler mass vanishes. The interpretation is that net inertial energy density is vanishing but the density of gravitational energy is non-vanishing and non-conserved in general. For Reissner Nordström solution it would be natural to interpret the gravitational mass as the gravitational energy of the classical gauge fields. Schwartschild metric possesses necessarily a vacuum density of some color and electro-weak charges implying that the charge grows at short distances. The net contribution to the gravitational energy momentum tensor however vanishes.

b) One can consider Equivalence Principle also in the case of Kähler charged imbeddings if one believes that the imbedding is a reasonable approximation an extremal. The interpretation of the gravitational energy density as the difference of energy densities of matter and antimatter does not require nor even allow the interpretation of the mass associated with the electro-weak fields of R-N solution field as gravitational mass. The Kähler mass of the solution should be however smaller than its gravitational mass. This does not pose any conditions on the critical radius since the density of Kähler charge can change sign inside the critical radius (meaning that antimatter dominates inside the critical radius). Thus no constraints results.

2. Standard view about Equivalence Principle

If the energies of both matter and antimatter where positive, the strongest form of Equivalence Principle would require that the Kähler mass of the solution equals to its gravitational mass. It is difficult to see how this could be implied by any deep principle, say by the absolute minimization.

This requirement poses a lower limit to the critical radius since the Kähler energy outside the critical radius should be smaller than the gravitational mass of the system. In the lowest order approximation this energy is given by the expression

$$
\begin{aligned}
\frac{E_K}{M} &= \frac{1}{8\pi\alpha_K M} \int_{r_M \geq r_c} \lambda E_K^2 dV \\
&= \frac{\lambda Q_K^2}{GM^2}\frac{r_S}{r_c} \ .
\end{aligned}
\tag{15.4.22}
$$

The requirement that electro-weak interactions are much weak than gravitational interaction imply the condition $Q_K^2/GM^2 \ll 1$ so that the ratio can be equal to 1 as Equivalence Principle requires only if $r_S/r_c \gg 1$ holds true.

Gravitational energy is not conserved for vacuum imbedding of Reissner-Nordström metric

The inertial energy associated with Kähler action inside a ball of given radius is not conserved for Reissner-Nordström metric imbedded as a non-vacuum extremal. This follows from the dependence $m^0 = \lambda t + h(r_M)$ implying that energy current has a radial component and from the non-vanishing of $T^{r_M r_M}$. The non-conservation is not due to the outflow of energy but due to the fact that in the case of Kähler charged imbedding field equations are not satisfied. The basic reason is that the contraction of the energy momentum tensor with the second fundamental form is non-vanishing.

For vacuum extremals it is gravitational energy which fails to be conserved. For instance, for the imbedding of Reissner-Nordström this happens. Only at the limit of Schwartshild metric gravitational energy is conserved. The vacuum extremals which are extremals of Einstein-Hilbert action for the induced metric conserve gravitational four momenta and color charges and are excellent candidates for models of the asymptotic state of star.

The non-stationarity of the vacuum extremal imbedding ($m^0 = \lambda t + h(r_M)$) of R-N metric leads to the following expression for the rate of the change of gravitational energy per time inside a sphere of radius r

$$
\begin{aligned}
\frac{dE_{vap}/dt}{E(r_M)} &= \frac{dE(r_M)/dt}{E(r_M)} + X \ , \\
X &= \frac{\int T^{r_M r_M} \partial_{r_M} m^0 \sqrt{g} d\Omega}{E(r_M)} \ , \\
E(r_M) &= \int T^{tt} \partial_0 m^0 \sqrt{g} dV \ .
\end{aligned}
\tag{15.4.23}
$$

The latter term depending on $T^{r_M r_M}$ takes into account the flow of gravitational energy through boundaries of the sphere and is in general non-vanishing for Reissner-Nordström metric.

Since the proposed solution ansatz works also in the more general case of a stationary spherically symmetric star model, characterized by the pressure $p(r_M)$ and the energy density $\rho(r_M)$, one can write a general order of magnitude estimate for the gravitational energy transfer associated with the boundary of the sphere approximating $h(r_M)$ with the corresponding function for the Schwartshild metric for large values of r_M as

$$
X \simeq -\partial_{r_M} h(r_M)\frac{4\pi p r_M^2}{M} \ .
\tag{15.4.24}
$$

The explicit expression for $\partial_{r_M} h(r_M)$ is given by

$$
\begin{aligned}
\partial_{r_M} h(r_M) &\simeq \frac{1}{\lambda}\sqrt{\frac{Y_1}{Y_2}} \;, \\
Y_1 &= -B + 1 + \frac{R^2}{4}s^{eff}_{\Theta\Theta}\frac{(\partial_{r_M} u)^2}{(1-u^2)} \;, \\
Y_2 &= 1 - \frac{4\lambda^2}{R^2\omega_1^2}\frac{s^{eff}_{\Theta\Theta}}{s^{eff}_{\Phi\Phi}} \;.
\end{aligned}
\tag{15.4.25}
$$

Here B is determined by the Einstein equations defining the star model and can be approximated with its value for Schwartshild metric.

At the surface of the Sun ($r_M \simeq 6\cdot 10^8\ m$, particle density $n \simeq 10^{21}/m^3$, $T \simeq 0.5\ eV$, $M \simeq 10^{57} m_p$ and pressure $p \simeq nT$) the order of magnitude of this term is about $X/E \simeq 2\pi K 10^{-13}/year$. For $K \sim 1$ (obtained if the radius of CP_2 is of order Planck length) the loss would be of the same order of magnitude as the inertial energy loss associated with the solar wind: $K \sim 1/k$, $10^4 < k << 10^8$, however implies that the loss is roughly four orders of magnitudes smaller.

It should be noticed that in the case of matter dominated cosmology shows that the rate for the reduction of the gravitational energy is of the order of $(dE/da)/E \simeq 1/a \simeq 10^{-11}/year$, which is of the same order as the fusion energy production of Sun. Thus it would seem that the rate for the change of gravitational energy in cosmological length scales is same as that for the inertial energy in the solar length scale. This could be understood if the ratio of inertial energy densities of matter and negative energy antimatter inside Sun and possibly everywhere remains constant not too far from unity.

15.4.4 Anomalous time dilation effects due to warping as basic distinction between TGD and GRT

TGD predicts the possibility of large anomalous time dilation effects due to the warping of space-time surfaces, and the experimental findings of Russian physicist Chernobrov about anomalous changes in the rate of flow of time [jb4, jb7] provide indirect support for this prediction.

Anomalous time dilation effect due to the warping

Consider first the ordinary gravitational time dilation predicted by GRT. For simplicity consider a stationary spherically symmetric metric $ds^2 = g_{tt}dt^2 - g_{rr}dr^2 - r^2 d\Omega^2$ in spherical coordinates. The time dilation is characterized by the difference $\Delta = \sqrt{g_{tt}} - 1$. In the weak field approximation one has $g_{tt} = 1 + 2\Phi_{gr}$, where Φ_{gr} is gravitational potential. The ordinary time dilation is given by $\Delta = \sqrt{g_{tt}} - 1 \simeq 2\Phi_{gr}$. At the Earth's surface the gravitational potential of the Earth is about $\Phi_{gr} = GM/R_E \simeq 10^{-9}$.

Consider next the situation for space-time surfaces. There exists an infinite number of warped imbeddings of M^4 to $M^4 \times CP_2$ given by $s^k = s^k(m^0)$, which are metrically equivalent with the canonical imbedding with CP_2 coordinates constant. New M^4 time coordinate is related by a diffeomorphism to the standard one. By restricting the imbedding to $M^4 \times S^1$, where S^1 a geodesic circle with radius $R/2$ (using the chosen convention for the definition CP_2 radius), the time component of the induced metric is $g_{m^0m^0} = 1 - R^2\omega^2/4$. The identification of M^4 coordinates as the preferred natural standard coordinate frame allows to overcome the difficulties related to the identification of the preferred time coordinate in general relativity in the case the metric does not approach asymptotically flat metric. For this choice an anomalous time dilation $\sqrt{1 - R^2\omega^2/4}$ due to the warping results even when gravitational fields are absent. Moreover, the dilation can be large.

The study of the imbeddings of Scwhartshild metric as vacuum extremals demonstrates that this vacuum warping is also seen as the degeneracy of the imbeddings of stationary spherically symmetric metrics. If m^0 is used as a time coordinate, anomalous time dilation is obtained also at $r_M \to \infty$ and is given by

$$
\sqrt{g_{m^0 m^0}} = \tfrac{1}{\lambda} \;.
\tag{15.4.26}
$$

This time dilation is seen only if the clocks to be compared are at different space-time sheets. The anomalous time dilation can be quite large since the order of magnitude for the parameter ωR is naturally of order one for the imbeddings of R-N metrics.

Mechanisms producing anomalous time dilation

Anomalous time dilation could result in many manners.

a) An adiabatic variation of the parameters λ and ω_1 of the space-time sheet containing the clock could be induced by some physical mechanism. For instance, X_c^4 could move "over" a large space-time sheet X^4 and gradually form $\#$ and $\#_B$ contacts with it. Topological light rays (MEs) define a good candidate for X^4. The parameter values λ and ω could change quasi-continuously if X_c^4 gradually generates CP_2 sized wormhole contacts or join along boundaries bonds connecting it to X^4. This process would not affect the gravitational flux fed to X_c^4.

For instance, if X^4 is at rest with respect to Earth, this motion would result from the rotation of Earth and the effect should appear periodically from day to day. If it is at rest with respect to Sun, the effect should appear once a year.

The generation of vacuum extremals X_{vac}^4 (not gravitational vacua), which is in principle possible even by intentional action since conservation laws are not broken, could induce anomalous time dilation by this mechanism.

b) A phase transition increasing the value of $\hbar$ increases the size of the space-time sheet in the same proportion. This transition could quite well affect also the parameter λ. If this phase transition occurs for the space-time sheet X_c^4 at which the clock feeds its gravitational flux, this mechanism could provide a feasible manner to induce an anomalous time dilation.

c) The system containing the clock could suffer a temporary topological condensation to a smaller space-time sheet and thus feed its gravitational flux to this space-time sheet. This would require coherently occurring splitting of $\#$ contacts and their regeneration. It is not possible to say anything definite about the probability of this kind of process except that it does not look very feasible.

The findings of Chernobrov

The findings claimed by Russian researcher V. Chernobrov support anomalous time dilation effect [jb4, jb7]. Chernobrov has studied anomalies in the rate of time flow defined operationally by comparing the readings of clocks enclosed inside a spherical volume with the readings of clocks outside this volume. The experimental apparatus involves a complex Russian doll like structure of electromagnets.

Chernobrov reports a slowing down of time by about 30 seconds per hour inside his experimental apparatus [jb4] so that the average dilation factor during hour would be about $\Delta = 1/120$. If the dilation is present all the time, the anomalous contribution to the gravitational potential would be by a factor $\sim 10^7$ larger than that of Earth's gravitational potential and huge gravitational perturbations would be required to produce this kind of effect.

The slowing of the time flow is reported to occur gradually whereas the increase for the rate of time flow is reported to occur discontinuously. Time dilation effects were observed in connection with the cycles of moon, diurnal fluctuations, and even the presence of operator.

Consider now the explanation of the basic qualitative findings of Chernobrov.

a) The gradual slowing of the time flow suggests that the parameter values of λ and ω change adiabatically. This favors option a) since the formation of $\#$ contacts occurs with some finite rate.

b) Also the sudden increase of the rate of time flow is consistent with option a) since the splitting of $\#$ contacts occurs immediately when the sheets X_c^4 and X^4 are not "over" each other.

c) The occurrence of the effect in connection with the cycles of moon, diurnal fluctuations, and in the presence of operator support this interpretation. The last observation would support the view that intentional generation of almost vacuum space-time sheets is indeed possible.

Vacuum extremals as means of generating time dilation effects intentionally?

Field equations allow a gigantic family of vacuum extremals: any 4-surface having CP_2 projection, which belongs to a 2-dimensional Lagrange manifold with a vanishing induced Kähler form, is a vacuum extremal. Canonical transformations and diffeomorphisms of CP_2 produce new vacuum extremals. Vacuum extremals carry non-vanishing classical electro-weak and color fields which are reduced to some $U(1)$ subgroup of the full gauge group and also classical gravitational field. Although the vacuum extremals are not absolute minima, their small deformations could define such. These vacuum extremals, call them X^4_{vac} for brevity, could be generated by intentional action. In the first quantum jump the p-adic variant of the vacuum extremal representing an intention to create X^4_{vac} would appear and in some subsequent quantum jump it would be transformed to a real space-time sheet.

The creation of these almost vacuum extremals could generate time dilation effects. The material system would gradually generate CP_2 sized wormhole contacts and/or join along boundaries connecting its space-time sheet to X^4_{vac} and this could change the values of the vacuum parameters λ, ω.

Could warping have something to do with condensed matter physics?

Warping predicts the reduction of the effective light velocity. There is a report [je1] of an experimental study of a condensed-matter system (graphene, a single atomic layer of carbon, in which electron transport is essentially governed by massless Dirac's equation. According to the report, the charge carriers in graphene mimic relativistic particles with zero rest mass and have an effective 'speed of light' $c_1 = c/300 = 10^6$ m/s.

The study reveals a variety of unusual phenomena that are characteristic of two-dimensional Dirac fermions. Graphene's conductivity never falls below a minimum value corresponding to the quantum unit of conductance, even when the concentrations of charge carriers tend to zero. The integer quantum Hall effect in graphene is anomalous in that it occurs at half-integer filling factors. The cyclotron mass m_c of massless current carriers in graphene is defined in terms of the energy of current carrier as $E = m_c c_1^2$.

The authors believe that these phenomena can be understood in the framework of the ordinary QED and this might be the case. One can however wonder whether the massless Dirac equation for the 2-dimensional system could correspond to the modified Dirac equation for the induced spinor fields and whether the reduction of the maximal signal velocity to c_1 could have the warping of the space-time sheet as a space-time correlate. In the idealization that the CP_2 projection of the space-time surface is a geodesic circle of CP_2, and using M^4 coordinates for space-time surface, so that one would have $\Phi = \omega t$ for S^2 coordinate Φ, one would have $g_{tt} = c_1^2 = 1 - R^2\omega^2/4 = 10^{-4}/9$.

15.5   A model for the final state of the star

As found, the energy production by fusion inside stars is of the same order of magnitude as the rate of change for the gravitational energy associated with the recent matter dominated cosmology. This suggests that the ratio of inertial energy densities of matter and negative energy antimatter remains constant at least in the solar length scale. Since no energy is produced in the final state of the star, the stationary solutions for which inertial energies are conserved separately for matter and antimatter provide a natural model for the final state of the star.

Besides stationarity, there is also a second new element, namely color and electro-weak long ranged forces coupling to the dark matter. For instance, for Kähler charged extremals one necessary has classical Z^0 force even when classical em force can vanish. For Schwartshild solution this force becomes very strong at small values of the radial distance. Therefore the presence of the Z^0 force, and presumably also other classical electro-weak forces, are expected play crucial role in the dynamics of the compact objects. The most plausible physical interpretation is in terms of dark matter.

The topics to be discussed in the following are:

a) Spherically symmetric stationary model for the final state of the star. It is found that the model cannot be completely realistic since the stationarity assumption fails at the origin and at the surface of the star.

b) Generalization of the model to what could be called dynamo model in order to achieve stationarity.

c) The possible consequences of long range weak and color forces associated with dark matter, in particular the Z^0 force, concerning the dynamics of the compact objects.

The original discussion was based on a different view about energy and motivated the study of Kähler charged solutions with the stationarity property. These 4-surfaces are *not* extremals of the Kähler action. The replacement of the stationary solutions with vacuum extremals requires however only the replacement of the geodesic sphere S^2_I with S^2_{II} implying that both em and Z^0 fields are unavoidably present (or even $W^{\pm}$ fields, depending on vacuum extremal). A serious limitation of the model is that it is single-sheeted. Indeed, the fact that the rotation axis and magnetic axis of super novae are different can be seen as a signal of many-sheeted-ness: the dominantly em and Z^0 fields would reside at different space-time sheets and would correspond to ordinary and dark matter. Of course, entire hierarchy of space-time sheets are expected to be present.

15.5.1   Spherically symmetric model

The simplest model for the final state of the star that one can imagine is obtained by assuming time translation invariance plus spherical symmetry and imbeddability to $M^4 \times S_i^2$, where S_i^2, $i = I, II$ is the geodesic sphere of CP_2. For the homologically non-trivial sphere S_I^2 the solution is *not* an extremal whereas S_{II}^2 gives an extremal with a vanishing density of inertial energy. In the original discussion cosmological constant was assumed to vanish. There are excellent reasons to assume that this constant is so small that it does not have any appreciable effects in the scale of the star and can thus be neglected. The nice feature of this kind of model is that symmetry assumptions plus stationarity requirement fix almost completely the model: no assumptions about the equation of state for the matter inside the star are needed.

The solution ansatz giving rise to vacuum extremal corresponds to a surface $X^4 \subset M_+^4 \times S_{II}^2$, where S_{II}^2 is the homologically trivial geodesic sphere of CP_2. The solution ansatz has the same general form as the imbedding of spherically symmetric metric.

$$
\begin{aligned}
m^0 &= \lambda t + h(r) \;, \\
\Theta &= \Theta(r) \;, \\
\Phi &= \omega t + k(r) \;.
\end{aligned}
\tag{15.5.1}
$$

The requirement that g_{tr} vanishes, implies a relationship between the functions $h(r)$ and $k(r)$. One might think that the simplest model is obtained, when the functions $h(r)$ and $k(r)$ vanish identically. One doesn't however obtain physically acceptable solutions in this manner: this is seen by expressing the g_{rr} component of the metric in terms of the mass function

$$
-g_{rr} = 1 + \frac{R^2}{4}(\partial_r\Theta)^2 = \frac{1}{1 - \frac{2GM(r)}{r}} \;.
$$

At the radii of order star radius (larger than Schwartshild radius $r_S = 2GM$) the gradient of Θ must be of the order of $1/R$ and this is inconsistent with the finite range of possible values for Θ.

As already shown the field equations $G^{\alpha\beta}D_\beta\partial_\alpha h^k = 0$ are obtained by varying the integral of the curvature scalar over the space time surface. Field equations reduce to conservation conditions for suitably chosen conserved current: for instance the relevant components of the gravitational 4-momentum and gravitational color currents and express the conservation of gravitational four-momentum current and corresponding color currents.

The expression for the induced metric is given by

$$
\begin{aligned}
ds^2 &= B dt^2 - A dr^2 - r^2 d\Omega^2 \;, \\
B &= \lambda^2 - \frac{R^2\omega^2}{4}\sin^2\Theta \;, \\
A &= 1 + \frac{R^2}{4}(\partial_r\Theta)^2 + \frac{R^2}{4}\sin^2\Theta(\partial_r k)^2 - (\partial_r h)^2 \;.
\end{aligned}
\tag{15.5.2}
$$

The vanishing of the g_{tr} component of the metric implies the condition

$$
\lambda\partial_r h - \frac{R^2}{4}\sin^2\Theta\,\omega\partial_r k = 0 \;.
\tag{15.5.3}
$$

The expressions for the components of Einstein tensor for spherically symmetric stationary metric are given by

$$
\begin{aligned}
G^{rr} &= \frac{1}{A^2}\left(-\frac{\partial_r B}{Br} + \frac{(A-1)}{r^2}\right) \ , \\
G^{\theta\theta} &= \frac{1}{r^2}\left[-\frac{\partial_r^2 B}{2BA} + \frac{1}{2Ar}\left(\frac{\partial_r A}{A} - \frac{\partial_r B}{B}\right)\right. \\
&\quad + \left.\frac{\partial_r B}{4AB}\left(\frac{\partial_r A}{A} + \frac{\partial_r B}{B}\right)\right] \ , \\
G^{tt} &= \frac{1}{AB}\left(-\frac{\partial_r A}{Ar} + \frac{(1-A)}{r^2}\right) \ .
\end{aligned}
\tag{15.5.4}
$$

A solution of the field equations with one-dimensional CP_2 projection and vanishing gauge fields is obtained by specifying the solution ansatz in the following manner

$$
\begin{aligned}
\Theta &= \frac{\pi}{2} \ , \\
h(r) &= hr \ , \\
k(r) &= kr \ .
\end{aligned}
\tag{15.5.5}
$$

The requirement that g_{rt} vanishes gives the condition

$$
h\lambda = R^2 \omega k/4 \ .
$$

The functions A and B are in this case just constants. Since A differs from unity, the resulting metric is however non-flat and the non-vanishing components of the Einstein tensor are given by the expressions

$$
\begin{aligned}
G^{tt} &= \frac{(1-A)}{ABr^2} \ , \\
G^{rr} &= -\frac{(1-A)}{A^2 r^2} \ .
\end{aligned}
\tag{15.5.6}
$$

Field equations can be written as conservation conditions, say for the components of gravitational 4-momentum and the conserved "gravitational" color charges associated with the symmetry $\Phi \to \Phi + \varepsilon$. Quite generally, "gravitational" isometry currents have only time and radial components and radial component represent radial flow to or from the origin. Since the time component is time independent, the field equations state that radial flow is constant so that radial component of the current must behave as $1/r^2$. This is guaranteed provided the condition

$$
\partial_r(G^{rr}\sqrt{g}) = 0
\tag{15.5.7}
$$

holds true: this is indeed the case since G^{rr} is proportional to $1/r^2$.

The radial flow of gravitational energy is non-vanishing due to $m^0 = \lambda tt + h(r)$ behavior and given by the expression

$$
J^r = \frac{G^{rr}h}{16\pi G} \ .
$$

The conservation condition for J^r fails to be satisfied at origin, which acts as a source or a sink for the gravitational energy. Conservation law fails also at the surface of the star.

One can consider several interpretations.

a) Gravitational mass could be genuinely non-conserved at these locations. Dark particles *resp.* antiparticles with positive *resp.* negative energy would be created in the center of star such that the net density of inertial energy remains zero. Positive and negative energy particles would flow along their own space-time sheets to the outer surface and annihilate there so that there would be no net growth of the gravitational mass. The simplest possibility is that $\#$ contacts, which correspond to bound states of parton and negative energy antiparton [F6], split to give rise to particles of opposite inertial energy. At the outer surface $\#$ contacts fuse together again.

b) Second option assumes the conservation of gravitational four-momentum. At the surface the non-conservation could result from the flow of the gravitational 4-momentum to a larger space-time sheet via join along boundaries bonds. No net flow of inertial energy would be involved since positive and negative energy flows must cancel each other. For example, for a physically acceptable solution the opposite inertial energies of matter and antimatter might flow radially from or towards the z-axis, flow to say north pole at the surface of the object and return back along z-axis. The gravitatioanl energy could also flow at origin to a second space-time sheet and return back at the surface of the star.

The (gravitational) mass function of the solution is given by the expression

$$M(r) \;=\; \frac{\lambda}{16\pi G}\frac{(A-1)}{AB}\sqrt{AB}r \;. \tag{15.5.8}$$

Mass is positive for Minkowskian metric with $B > 0$ and Euclidian metric but negative for the interior black hole metric $(A < 0, B < 0)$. Mass is proportional to the radius of the star and in order to obtain an object with about Schwartshild radius one must assume that the parameter k is of the order of $1/R$.

Concerning the physical interpretation of the $\Theta = \pi/2$ solution following remarks are in order:

a) Various gauge fields vanish since CP_2 projection is actually a geodesic circle. The interpretation is that various gauge charges vanish separately for matter and antimatter. Note that one-dimensional CP_2 projection conforms with the similar property of Robertson Walker cosmologies.

b) Both gravitational, color and weak forcec vanish inside the star and the motion along radial geodesics takes place with constant velocity. This is consistent with the radial flow of gravitational energy.

c) Solution ansatz allows generalizations. For example, the following modification is stationary with respect to energy: $m^0 = \lambda t$, $\Phi = \omega_1 t + k_1 r$, $\Psi = \omega_2 t + k_2 r$, $u = constant < \infty$, $\Theta = \pi/2$. By choosing the values of the parameters suitably all the field equations are satisfied but stationarity is not achieved.

d) The solution should allow gluing to the Schwartshild metric at $\Theta = \pi/2$. As found, for the imbedding of Schwartshild metric the $\Theta = 0$ correspond to the Schwartchild radius so that $\Theta = \pi/2$ would most naturally correspond to $r_M < r_S$. Since radial gauge fluxcs arc non-vanishing and finite at Schwartshild radius, they must be non-vanishing $\Theta = \pi/2$ surface too, so that the star would carry surface charges and behave somewhat like a conducting sphere.

15.5.2 Dynamo model

The previous considerations have shown that the spherically symmetric solution is probably not physically realistic as such and it seems also clear that spherical symmetry must be given up and be replaced with a symmetry with respect to rotations around z-axis in order to obtain more realistic solutions. Since realistic stars rotate and have strong magnetic fields it is natural to ask whether rotation and magnetic fields might provide remedy for the pathological features of the solution. The rotation of the gauge charged matter (in "gravitational" sense) indeed creates classical gauge magnetic fields, which become very strong near the surface of the star, where the condition $\Theta \simeq \pi/2$ holds. If matter is approximately gauge neutral in the interior, the gauge fields should vanish to a very good approximation in the interior and the previous solution should be a good approximation to the actual situation. The rotating star could therefore be regarded as a rotating electro-weak conductor. Both Z^0 and em fields are present for a vanishing Kähler field and the ratio of field strengths is $\gamma/Z^0 = -\sin^2(\theta_W)/2 \simeq -1/8$ (see Appendix) so that Z^0 field dominates.

The generation of strong em and Z^0 electric and magnetic fields suggests a mechanism guaranteing the stability of the solution: star behaves like a dynamo. For solutions with a 2-dimensional CP_2 projection em and Z^0 electric and magnetic fields are automatically orthogonal. For $\Theta \simeq \pi/2$ they are very strong and dominate over gravitation and centrifugal force. Therefore the stability of the surface region naturally results from the cancellation of the electric and magnetic em and Z^0 forces $(\bar{E} + \bar{v} \times \bar{B} = 0)$, which takes place, when the velocity field of the matter is suitably chosen. This condition is completely analogous to the vanishing of Kähler Lorentz 4-force which seems to be a general property of the solutions of field equations [D1] and there are reasons to hope non-vacuum extremals describing rotating star can be found.

Conditions for the vanishing of the induced Kähler field

Although the situation becomes too complicated in order to allow the finding of exact solutions describing rotating star, one can identify some general properties of the solution ansatz describing the rotating configuration with Kähler electric and magnetic fields. In order to study the general properties of the solution ansatz in its most general form the explicit expressions for the line element and Kähler form of CP_2 given by the expression

$$
\begin{aligned}
ds^2 &= \frac{dr^2}{F^2} + \frac{r^2}{4F^2}(d\Psi + cos(\Theta)d\Phi)^2 + \frac{r^2}{4F}(d\Theta^2 + sin^2(\Theta)d\Phi^2) \ , \\
J &= \frac{r}{2F^2}dr \wedge (d\Psi + cos(\Theta)d\Phi) + \frac{r^2}{2F}sin\Theta d\Theta \wedge d\Phi \ , \\
F &= 1 + r^2 \ ,
\end{aligned}
\tag{15.5.9}
$$

0
are needed.

The vanishing of Kähler field is can be guaranteed by the conditions (not the most general ones, canonical transformations generate new solutions)

$$
\begin{aligned}
\Phi &= q\Psi \ , \\
\frac{dr}{d\theta} &= -qrF\frac{sin(\theta)}{1 + qcos(\theta)} \ .
\end{aligned}
\tag{15.5.10}
$$

Note that this ansatz excludes the case $qcos(\Theta) = -1$ for which only $W^\pm$ fields are non-vanishing. For this ansatz the expressions for em and Z^0 fields (see Appendix for general formulas) are

$$
\gamma = -sin^2(\Theta_W)R_{03} \ , \qquad Z^0 = 2R_{03} \ ,
$$
$$
R_{03} = -qr^2 F sin(\theta)d\Theta \wedge d\Psi \ .
\tag{15.5.11}
$$

Here R_{03} denotes a component of spinor curvature.

Topological quantum numbers

The crucial point is that the expansions for the angle coordinates Φ and Ψ using spherical coordinates contain linear terms in t, r and ϕ

$$
\begin{aligned}
\Phi &= n_1\phi + \omega_1 t + k_1 \ , \\
\Psi &= n_2\phi + \omega_2 t + k_2 \ .
\end{aligned}
\tag{15.5.12}
$$

The functions k_1 and k_2 corresponds to Fourier expansion in terms of the plane waves $exp(in\phi)$ with coefficients depending on the coordinates (t, r, θ).

The terms depending linearly on ϕ imply a nontrivial topological structure for the gauge fields not present for the ordinary Maxwell fields. What happens is that space-time divides into regions, which correspond to different values of the topological quantum numbers (n_1, n_2). In the boundaries of these regions the values of the coordinates u and Θ must be such that different values of Φ and Ψ correspond to same point of CP_2. From the expression of the line element one finds that for Ψ the point $u = 0$ and the sphere $u = \infty$ corresponds to these kinds of points. For Φ the surfaces $u = 0$ and $u = \infty$, $\Theta = 0$ correspond to these kinds of surfaces. The form of Φ and Ψ implies that both electric and magnetic gauge fields are nontrivial and rather closely related as is clear from the expression for the Kähler form. Therefore the non-triviality of the winding numbers n_1 and n_2 is what seems to be the crucial, purely TGD based feature of rotating gauge field structures.

Stationary, axially symmetric ansatz with a non-vanishing Kähler field

To make the discussion more concrete, let us assume that the induced metric is invariant with respect to rotations around z-axis and time translations. This is achieved if CP_2 coordinates (apart from linear dependence on ϕ) depend on the coordinates r_M and θ only.

$$
\begin{aligned}
r &= r(r_M, \theta) \\
\Theta &= \Theta(r_M, \theta) \ , \\
k_i &= k_i(r_M, \theta) \ , \quad i = 1, 2 \ .
\end{aligned}
\tag{15.5.13}
$$

This kind of ansatz is clearly consistent: field equations reduce to four equations since second fundamental form is orthogonal to the four-surface and there are four free functions of r and θ: one has effectively two dimensional field theory. Since the general solution ansatz for field equations relies on the vanishing of the Lorentz Kähler force central for the dynamo mechanism, it is of interest to study the general properties of the solution ansatz with a non-vanishing Kähler field. This ansatz can give as special cases space-time sheets carrying Z^0 and em fields with magnetic fields having different rotation axis.

In order to further simplify the discussion let us assume that X^4 corresponds to a sub-manifold of $M^4 \times S_I^2$. For instance, the ansatz

$$
\begin{aligned}
r &= \infty \ , \\
\Theta &= \Theta(r_M, \theta) \ , \\
\Phi &= n\phi + \omega t + k(r_M, \theta) \ .
\end{aligned}
\tag{15.5.14}
$$

is consistent with this assumption. A simpler ansatz is obtained by assuming $k(r_M, \theta) = 0$. This ansatz has the following properties.

a) Induced Kähler (and Z^0-) electric and magnetic fields are automatically orthogonal since CP_2 projection is two-dimensional. In fact, the orthogonality holds to an excellent approximation also for the values of u different but near to $u = \infty$ since the resulting additional components of the Kähler field are extremely small. Kähler electric and magnetic fields are given by

$$
\begin{aligned}
E_{r_M} &= J_{r_M t} = -\partial_{r_M} cos(\Theta)\omega/2 \ , \\
E_\theta &= -\partial_\theta cos(\Theta)\omega/2 \ , \\
B_\theta &= -\partial_{r_M} cos(\Theta)n/2 \ , \\
B_{r_M} &= -\partial_\theta cos(\Theta)n/2 \ .
\end{aligned}
$$

$$
\tag{15.5.15}
$$

The field strengths are related by

$$
\begin{aligned}
E &= vB \ , \\
v &= \frac{\omega}{n}\sqrt{-\frac{g_{\phi\phi}}{g_{tt}}} \simeq \frac{\omega}{n}\rho \ ,
\end{aligned}
\tag{15.5.16}
$$

where ρ denotes radial distance from the rotation axis. v can be interpreted as a velocity type parameter. The requirement that $v < 1$ gives a lower bound for the value of n: $n > \omega r_0$, where r_0 denotes the radius of the star: the condition implies that n must be larger than the mass of the star using Planck mass as unit. Somewhat counter intuitively, small rotation velocities seem to correspond to large values of n.

b) Kähler electric and magnetic fields indeed provide a possible mechanism guaranteeing the stability of the star at the surface, where Z^0 forces dominate over gravitation and centrifugal force. Star behaves like a dynamo: matter rotates with a velocity guaranteeing the vanishing of the Z^0 force. It should be noticed that no upper bound for the rotation velocity except that resulting from causality is obtained

$(\Omega < 1/r_0)$. Therefore this mechanism might explain the observed very large rotation velocities (for instance in Super Nova SN1987A), which are hard to understand in GRT based models [ia14].

c) The ansatz indeed describes a rotating object. First, the dynamo mechanism for the stability necessitates the presence of rotation and determines rotation velocity also. Secondly, the presence of Kähler magnetic field can be understood as being created by the rotation of gauge charges. Thirdly, the $g_{t\phi}$ component of the induced metric and therefore the angular momentum density $J_z^t \propto G^{t\phi} r^2 sin^2\theta$ is non-vanishing. A rough order of magnitude estimate for the angular momentum gives $J \simeq M\sqrt{G}n$. In order to obtain angular momentum of order $MR \simeq GM^2$ the order of magnitude for the parameter n must be $n \simeq M\sqrt{G}$ or the mass of the star using Planck mass as unit or: notice that also the Kähler charge of the star is of the same order of magnitude.

d) The gluing of the solution to Schwartshild solution realized as a vacuum extremal is possible at a surface $\Theta = 0$, which corresponds to Scwartschild radius, since at this surface different values of Φ correspond to same point of CP_2. The gluing condition gives additional constraint $u = \infty$ at $r_M = r_S$.

e) The experience with the radially symmetric solution ansatz suggests that Θ is very nearly constant $\Theta \simeq \pi/2$ in the interior and varies considerably only at the surface of the star where Θ must go to zero in order to allow gluing to Schwartshild metric at $r = r_S$. A possible picture is therefore the following. On z-axis there is a Z^0 charged vortex creating radial Z^0 electric field and Z^0 magnetic field in the direction of the vortex. In order to obtain cyclic energy flow matter velocity near the surface of the star must have besides the rotational component a component in θ direction (Z^0 force vanishes in this direction).

f) An interesting possibility is that the vortex actually corresponds to a Kähler charged cosmic string which has gradually lost its enormous inertial mass by a generating pairs of positive and negative energy particles, such that positive energy particles have left the string and participated in the formation of the star. The weakening of the magnetic field would have forced a gradual thickening of the cosmic string to an ordinary magnetic flux tube. This 'stars as pearls in necklace' picture would be consistent with the idea that cosmic strings serve as seeds of galaxy and star formation. Both negative and positive energy strings should be present in order to guarantee vanishing of net inertial energy and one can wonder whether the axis of Z^0 and em magnetic fields correspond to these two kinds of strings.

Does Sun have a solid surface?

The model for the asymptotic state of star predicts that mass at given space-time sheet is concentrated in a spherical shell so that star would have a multi-sheeted onion-like structure. This brings in mind the model for the formation of planetary systems in which spherical layers of quantum coherent dark matter serve as templates for the formation of visible matter which eventually condensed to planets [D6, J6]. It would not be surprising if also younger stars and also planets would possess similar structure. This picture is in conflict with the simplest model of Sun as a gas sphere.

Recently new satellites have begun to provide information about what lurks beneath the photosphere. The pictures produced by Lockheed Martin's Trace Satellite and YOHKOH, TRACE and SOHO satellite programs are publicly available in the web. SERTS program for the spectral analysis suggest a new picture challenging the simple gas sphere picture [if5]. The visual inspection of the pictures combined with spectral analysis has led Michael Moshina to suggests that Sun has a solid, conductive spherical surface layer consisting of calcium ferrite. The article of Moshina [if5] provides impressive pictures, which in my humble non-specialist opinion support this view. Of course, I have not worked personally with the analysis of these pictures so that I do not have the competence to decide how compelling the conclusions of Moshina are. In any case, I think that his web article [if5] deserves a summary.

Before SERTS people were familiar with hydrogen, helium, and calcium emissions from Sun. The careful analysis of SERTS spectrum however suggest the presence of a layer or layers containing ferrite and other heavy metals. Besides ferrite SERTS found silicon, magnesium, manganese, chromium, aluminum, and neon in solar emissions. Also elevated levels of sulphur and nickel were observed during more active cycles of Sun. In the gas sphere model these elements are expected to be present only in minor amounts. As many as 57 different types of emissions from 10 different kinds of elements had to be considered to construct a picture about the surface of the Sun.

Moshina has visually analyzed the pictures constructed from the surface of Sun using light at wave lengths corresponding to three lines of ferrite ions (171, 195, 284 Angstroms). On basis of his analysis he concludes that the spectrum originates from rigid and fixed surface structures, which can survive for days. A further analysis shows that these rigid structure rotate uniformly.

The existence of rigid structures idealizable as spherical shells in the first approximation would conform with the model for the final state of star extrapolated to a qualitative picture about the structure younger stars.

15.5.3 Z^0 force and dynamics of compact objects

The fact that long ranged color fields and weak fields, in particular Z^0 electric fields, could become strong under certain conditions and in fact dominate over gravitation might have interesting consequences in the physics of compact objects. Besides the dynamo mechanism guaranteing the stability of the compact object the following ideas come immediately into mind.

a) In GRT based models Super Nova explosion is explained in terms of the pressure of the collapsed matter. Numerical simulations however fail to produce the explosion [ia14] and it might even be that GRT based models in fact predict the collapse to black hole. Z^0 and em electric fields created by dark matter plus the existence of particles with Z^0 charge, which are suggested by TGD based model of nucleus and condensed matter to be present already in ordinary condensed matter, might provide a natural mechanism preventing the formation of the black hole (also excluded by the failure of complete imbeddability). When matter collapses to a sufficiently small volume the value of Θ approaches $\pi/2$ in the surface region and very strong repulsive radial Z^0 force is generated and could indeed lead to the explosion. Very light exotic variants of Higgs bosons identified as wormhole contacts having lefthanded weak charge provides a possible mechanism generating Z^0 charge.

b) The strong Z^0 fields at the surface of the star might provide energy source and acceleration mechanism for very high energy cosmic rays and a mechanism producing very high energy X-rays. These rays would be dark matter particles but could transform to ordinary matter by the mechanism discussed in [F9, J6]. For instance, one can imagine the ejection of a particle beam from the surface of a compact object: particles in dark matter phase gain very high energies in the Z^0 electric field and emit brehmstrahlung in the direction of their motion: most intense emission appears in the region very near the surface of the star, where the Z^0 electric field is strongest. This kind of mechanism might provide alternative explanation for the pulsars. In standard explanations the emission takes place in the direction of the magnetic axis, which does not coincide with the rotation axis. In present case the emission point could be anywhere on the surface of the star and magnetic and rotation axes might well coincide as they do in the simplest model. What one has to do is to invent a mechanism creating the surface instability pushing the matter from the surface of the star to the Kähler electric field.

c) The topological character of the magnetic structures might have applications also in the physics of the ordinary stars. It is known that solar magnetic fields correspond to definite isolated structures [il1]. Since electromagnetic fields must be accompanied by Kähler fields it is tempting to assume that these structures indeed correspond to the structures predicted by TGD. At the surface of the Sun the value of Θ near $\pi/2$ are possible and therefore Z^0 force can be very strong inside the magnetic structures.

15.5.4 Correlation between gamma ray bursts and supernovae and dynamo model for the final state of the star

The correlation between gamma ray bursts and supernovae is certainly the cosmological discovery of the year 2003 [ij6, ij7].

a) The first indications for supernova gamma ray burst connection came 1998 when a supernova was seen few days after the gamma ray burst in the same region of sky. In this case the intensity of the burst was however by four orders of magnitude weaker than for the typical gamma ray bursts so that the idea about the correlation was not taken seriously. On 29 March, observers recorded a burst christened as GRB030329. On 6 April, theorists at the Technion Institute of Technology in Israel and CERN in Geneva predicted that there would be signs of a supernova in the visible light and infrared spectra on 8 April [ij6]. On cue, two days later, observers picked up the telltale spectrum of a type Ic supernova in the same region of sky, triggered as the collapsing star lost hydrogen from its surface. It has now become clear that a large class of gamma ray bursts correlate with supernovae of type

Ib and Ic [ij8], and that they could thus be powered by the mere core collapse leading to supernova. Recall that supernovae of type II involve hydrogen lines unlike those of type I. Supernovae of type Ib shows Helium lines, and Ic shows neither hydrogen nor helium but intermediate mass elements instead. Supernovae of type Ib and Ic are thought to result as core collapse of massive stars.

b) One of the most enigmatic findings were the "mystery spots" accompanying supernova SN1987A at a distance of few light weeks at the symmetry axis at opposite sides of the supernova [ib6]. Their luminosity was nearly 5 per cent of the maximal one. SN1987A was also accompanied by an expanding axi-symmetric remnant surrounded by three concentric rings.

c) The latest finding [ij5] is that the radiation associated with the gamma ray bursts is maximally polarized. The polarization degree is the incredible 80 ± 20 per cent, which tells that it must be generated in an extremely strong magnetic field rather than in a simple explosion. The magnetic field must have a strong component parallel to the eye sight direction.

According to the updated model discussed in detail in [D4], cosmic strings transform in topological condensation to magnetic flux tubes about which they represent a limiting case. Primordial magnetic flux tubes forming ferro-magnet like structures become seeds for gravitational condensation leading to the formation of stars and galaxies. The TGD based model for the asymptotic state of a rotating star as dynamo leads to the identification of the predicted magnetic flux tube at the rotation axis of the star as Z^0 magnetic flux tube of primordial origin and assignable to dark matter. Besides Z^0 magnetic flux tube structure also magnetic flux tube structure exists at different space-time sheet but is in general not parallel to the Z^0 magnetic structure. This structure cannot have primordial origin (the magnetic field of star can even flip its polarity).

The flow of matter along Z^0 magnetic (rotation) axis generates synchrotron radiation, which escapes as a precisely targeted beam along magnetic axis and leaves the star. The identification is as the rotating light beam associated with ordinary neutron stars. During the core collapse leading to the supernova this beam becomes gamma ray burst. The mechanism is very much analogous to the squeezing of the tooth paste from the tube.

TGD based models of nuclei [F8] and condensed matter [F9] suggests that the nuclei of dense condensed matter develop anomalous color and weak charges coupling to dark weak bosons having Compton length L_w of order atomic size. Also lighter copies of weak bosons can be important in living matter. This weak charge is vacuum screened above L_w and by dark particles below it. Dark neutrinos, which according to TGD based explanation of tritium beta decay anomaly [F8] should have the same mass scale as ordinary neutrinos, are good candidates for screening dark particles. The Z^0 charge unbalance caused by the ejection of screening dark neutrinos hinders the gravitational collapse. The strong radial compression amplifies the tooth paste effect in this kind of situation so that there are hopes to understand the observed incredibly high polarization of 80 ± 20 per cent [ij5].

15.5.5 Z^0 force and Super Nova explosion

The mechanism behind Super Nova explosion is not completely understood. The general picture is roughly the following.

a) The formation of iron means the end of the nuclear processes. The inner parts of the star contract and the degeneracy pressure of the non-relativistic electrons ($E_F \propto \rho^{2/3}$) increases and compensates the gravitational force. The equilibrium state is not stable. When the mass of the iron core approaches Chandrasekhar mass $1.4 M_{Sun}$ electrons become relativistic. The milder dependence of the electron Fermi energy on density $E_F \propto \rho^{1/3}$ at the relativistic limit leads to the loss of stability. The high Fermi energy of the electrons allows also the reactions $p + e^- \to n + \nu_e$ implying decrease of the electronic pressure and neutronization of nuclear matter in the core. Gravitational collapse starts.

b) Collapse stops, when the density of the core reaches the density of the nuclear matter. The degeneracy pressure of the neutrons stops contraction, a shock wave is created and the shock wave and neutrino radiation blow the outer regions of the star away so that Super Nova explosion results.

The problem of this scenario is that numerical simulations do not lead to a strong enough Super Nova explosion and the star tends to collapse into a black hole. A repulsive long ranged Z^0 force predicted by TGD based model of atomic nuclei [F8] generating an additional pressure provides a possible mechanism hindering the collapse and leading to the explosion.

a) The TGD based model for nuclei [F8] and condensed matter [F9] suggests that the nuclei of dense condensed matter develop anomalous color and weak charges coupling to dark weak bosons having Compton length L_w of order atomic size. Weak charge is due to the charged color bonds between nucleons: for instance, tetraneutron can be understood as an alpha particle containing two negatively charged color bonds [F8]. This weak charge is vacuum screened above L_w and by dark particles below it. The charged bonds could exist and also generated between nucleons of different nuclei during collapse.

Dark neutrinos, which according to the TGD based explanation of tritium beta decay anomaly [F8] should have the same mass scale as ordinary neutrinos, are good candidates for screening dark weak force partially below length scale L_w. In equilibrium color force compensates the partially screened Z^0 force in the bonds. For the ordinary condensed matter densities vacuum screening effectively eliminates the force between neighboring nuclei, and the force makes it visible only via low compressibility. The gravitational collapse could be hindered by the strong additional pressure created by the repulsive L_w-ranged weak interaction between nucleons becoming manifest in the resulting dense phase.

b) In the initial state Z^0 charge is screened by dark neutrinos below L_w so that the repulsive Z^0 force is weaker than gravitational force and attractive color force associated with the bonds. Neutronization reactions $p + e^- \rightarrow n + \nu_e$ trigger the collapse. During collapse density increases so that dark neutrinos are not able to screen the anomalous Z^0 charge density. The dark neutrino radiation escaping from the star can also reduce the Z^0 screening. The resulting repulsive weak force implies a rapid increase of pressure with increasing density and thus a very low compressibility as it is proposed to imply also in the case of ordinary condensed matter[F9]. The repulsive weak force thus stops the collapse to black-hole.

c) The study of the spherically symmetric star models as 4-surfaces imbedded in $M_+^4 \times CP_2$ shows that the extreme nonlinearity of Kähler action implies that Z^0 force dominates over gravitation near the surface of the star.

15.6   Has strong gravimagnetism been observed?

Physicists M. Tajmar and C. J. Matos and their collaborators working in ESA (European Satellite Agency) have made an amazing claim of having detected strong gravimagnetism with gravimagnetic field having a magnitude which is about 20 orders of magnitude higher than predicted by General Relativity [ik3]. If the findings are replicable they mean a revolution in the science of gravity and, as one might hope, force a long-waited serious reconsideration of the basic assumptions of the dominating super-string approach.

15.6.1   Gravimagnetic Thomson field and rotating superconductors

The starting point of the theory of Tajmar and Matos [ik2] is the so called Thomson magnetic moment generated in rotating charged super-conductors adding a constant contribution to the exponentially damped Meissner contribution to the magnetic field. This contribution can be understood as being due to the massivation of photons in super-conductors. The modified Maxwell equations are obtained by just adding scalar potential mass term to Gauss law and vector potential mass term to the equation related the curl of the magnetic field to the em current.

The expression for the Thomson magnetic field is given by

$$B = 2\omega_R n_s \times \lambda_\gamma^2 \,, \qquad (15.6.1)$$

where ω_R is the angular velocity of superconductor, n_s is charge density of super-conducting particles and $\lambda_\gamma = \hbar/m_\gamma$ is the wave length of a massive photon at rest. In the case of ordinary superconductor one has $\lambda_\gamma = \sqrt{m^*/q^* n_s}$, where $m^* \simeq 2m_e$ and $q^* = -2e$ are the mass and charge of Cooper pair. Hence one has

$$B = -2\frac{m^*}{2e}\omega_R \,. \qquad (15.6.2)$$

Magnetic field extends also outside the super-conductor and by measuring it with a sufficient accuracy outside the super-conductor one can determine the value of the electron mass. Instead of the theoretical value $m^*/2m_e = .999992$ which is smaller than one due to the binding energy of the Cooper pair the value $m^*/2m_e = 1.000084$ was found by Tate [ik1]. This inspired the theoretical work generalizing the notion of Thomson field to gravimagnetism and the attemnpt to explain the anomaly in terms of the effects caused by the gravimagnetic field.

Note that in the case of ordinary matter the equations would lead to an inconsistency at the limit $m_\gamma = 0$ since the value of Thomson magnetic field would become infinite. The resolution of the problem proposed in [ik2] is based on the replacement of rotation frequency ω with electron's spin precession frequency $\omega_L = -eB/2m$ so that the consistency equation becomes $B = -B = 0$ for a unique choice $1/\lambda_\gamma^2 = -\frac{q}{m}n$. One could also consider the replacement of ω with electron's cyclotron frequency $\omega_c = 2\omega_L$. To my opinion there is no need to assume that the modified Maxwell's equations hold true in the case of ordinary matter.

Gravimagnetic field

The perturbative approach to the Einstein equations leads to equations which are essentially identical with Maxwell's equations. The g_{tt} component of the metric plays the role of scalar potential and the components g_{ti} define gravitational vector potential. Also the generalization to the super-conducting situation in which graviphotons develop a mass is straightforward. Just add the scalar potential mass term to the counterpart of Gauss law and vector potential mass term to the equation relating the curl of the gravitomagnetic field to the gravitational mass current.

In the case of a rotating superconductor Thomson magnetic moment is replaced with its gravito-magnetic counterpart

$$B_{gr} = -2\omega_R \rho_m \lambda_g^2 \ . \tag{15.6.3}$$

Obviously this formula would give rise to huge gravimagnetic fields in ordinary matter approaching infinite values at the limit of vanishing gravitational mass. Needless to say, these kind of fields have not been observed.

Equivalence Principle however suggests that the gravimagnetic field must be assigned with the rotating coordinate frame of the super-conductor. Equivalence principle would state that seing the things in a rotating reference frame is equivalent of being in a gravimagnetic field $B_{gr} = -2\omega_R$ in the rest frame. This fixes the graviphoton mass to

$$\frac{1}{\lambda_{gr}^2} = (\frac{m_{gr}}{\hbar})^2 = G\rho_m \ . \tag{15.6.4}$$

For a typical condensed matter density parameterized as $\rho_m = Nm_p/a^3$, $a = 10^{-10}$ m this gives the order of magnitude estimate $m_{gr} \sim N^{1/2}10^{-21}/a$ so that graviton mass would be extremely small.

If this is all what is involved, gravimagnetic field should have no special effects. In [ik2] it is however proposed that in superconductors a small breaking of Equivalence Principle occurs. The basic assumptions are following.

a) Super-conducting phase and the entire system obey separately their gravitational analogs of Maxwell field equations.

b) The ad hoc assumption is that for super-conducting phase the sign of the gravimagnetic field is opposite to that for the ordinary matter. If purely kinematic effect were in question so that graviphotons were pure gauge degrees of freedom, the value of m_{gr}^2 should should be proportional to ρ_m^* and $\rho_m - \rho_m^*$ respectively.

c) Graviphoton mass is same for both ordinary and super-conducting matter and corresponds to the net density ρ_m of matter. This is essential for obtaining the breaking of Equivalence Principle.

With these assumptions the gravimagnetic field giving rise to acceleration field detected in the rest system would be given by

$$B_{gr}^* = \frac{\rho_m^*}{\rho} \times 2\omega \tag{15.6.5}$$

This is claimed to give rise to a genuine acceleration field

$$g^* = -\frac{\rho_m^*}{\rho}a \tag{15.6.6}$$

where a is the radial acceleration due to the rotational motion.

Explanation for the too high value of measured electron mass in terms of gravimagnetic field

A possible explanation of the anomalous value of the measured electron mass [ik1] is in terms of gravimagnetic field affecting the flux Bohr quantization condition for electrons by adding to the electromagnetic vector potential term q^*A_{em} gravitational vector potential m^*A_{gr}. By requiring that the quantization condition

$$\oint (m^*v + q^*A_{em} + m^*A_{gr})dl = 0 \tag{15.6.7}$$

is satisfied for the superconducting ring, one obtains

$$B = -\frac{2m}{e}\omega - \frac{m}{e}B_{gr} . \tag{15.6.8}$$

This means that the magnetic field is slightly stronger than predicted and it has been known that this is indeed the case experimentally.

The higher value of the magnetic field could explain the slightly too high value of electron mass as determined from the magnetic field. This gives

$$B_{gr} = \frac{\Delta m_e}{m_e} \times 2\omega = \frac{\Delta m_e}{m_e} \times em_e \times B . \tag{15.6.9}$$

The measurement implies $\Delta m_e/m_e = 9.2 \times 10^{-5}$. The model discussed in [ik2] predicts $\Delta m_e/m_e \sim \rho^*/\rho$. The prediction is about 23 time smaller than the experimental result.

15.6.2 Is the large gravimagnetic field possible in TGD framework?

TGD allows top consider several alternative solutions for the claimed effect.

a) TGD predicts the possibility of classical electro-weak fields at larger space-time sheets If these couple to Cooper pairs generate exotic weak charge at super-conducting space-time sheets the Bohr quantization conditions modify the value of the magnetic field. Exotic weak charge would however mean also exotic electronic em charge so that this option is excluded. It would also require that the Z^0 charge of test bodies used to measure the acceleration field is proportional to their gravitational mass.

b) TGD suggests a hierarchy of strong gravities analogous to those generated by spin 2 mesons. These gravitons behave like massless particles below the appropriate Compton length. This Compton length can be arbitrarily long at higher levels of dark matter hierarchy. Electrons do not however couple to these gravitons so that this option soes not seem to work.

c) The rotation of the super-conductor would correspond quite concretely to a rotation of the corresponding space-time sheet. Also the space-time sheet defining the magnetic and gravito-magnetic body of the system could participate to the rotation. Since the rotation affects the shape of the space-time surface the breaking of the Equivalence Principle is unavoidable. This predicts gravitational analogs of the effects found in rotating magnetic systems [jf2], for instance the radial gravitational field $E^{gr} = vB^{gr}$ but does not seem to be enough to understand what is involved.

d) As already noticed, the failure of Equivalence Principle could be understood if the gravimagnetic fields of super-conducting and ordinary matter do not interfere. Many-sheeted space-time suggest the lack of interference is caused by the the fact that super-conducting and ordinary matter reside at different space-time sheets. If the measurement of the gravimagnetic field (or rather its change causing Faraday effect and tangential acceleration) is carried out at either space-time sheet a breaking of Equivalence Principle is observed. In the similar manner magnetic field is affected via Bohr rules for angular momentum and gives rise to the desired effect. In this framework the model of [ik2] looks rather plausible one.

e) Induced field concept implies extremely tight correlations between induced classical gauge fields and induced gravitational field and one expects that the magnetic field associated with the rotating super-conductor gives rise to a gravimagnetic field so that ordinary Meissner and Thompson effects would force their gravitational counterparts. This is not at all obvious in standard physics framework. It turns out however that the predicted gravimagnetic field is far two small in the simplest model.

f) The dependence of the mass of graviphoton on magnetic penetration length involves Planck constant. TGD predicts a hierarchy of Planck constants $\hbar(k) = \lambda^k \hbar_0$. This means that for given value of m_γ and m_{gr} there is a hierarchy of increasing photon graviton rest Compton length defining the penetration depths for superconductors. It turns out that if graviphotons are dark, one can indeed understand the huge value of gravimagnetic field in TGD framework.

Key observations

Two observations are essential for what follows.

a) It would seem that the the superposition of the of the gravitomagnetic fields of the superconducting and ordinary ordinary matter gives the net gravitomagnetic field which would not give any anomalous effects. The breaking of Equivalence Principle could be understood if these fields do not interfere in the experimental situation. In TGD framework the separation of ordinary matter and Cooper pairs at separate space-time sheets can explain the absence of the interference.

b) In order to obtain large enough an effect one must assume that ρ_m^* contains an additional contribution besides Cooper pairs. In TGD framework the presence of also other particles besides Cooper pairs at super-conducting space-time sheets could increase the value of ρ_m^* by a factor of order 23. For instance, the space-time sheet could contain $23m_e/m_p$ protons $23m_e/Am_p$ heavier atoms per electron.

Could gravimagnetism and breaking of Equivalence Principle be forced by the induced field concept?

The safest starting point seems to be that the separation of super-conducting phase to its own space-time sheet induces the reduction $\rho_m \to \rho_m - \rho_m^*$ at the space-time of ordinary matter. If the photograviton is not modified correspondingly one has in inertial frame effective B_{gr} obtained by the replacement $\rho - \rho_m \to -\rho_m^*$ in the defining formula as one goes to rest system. This gives also the sign of the gravimagnetic field correctly. The task is to find whether the notion of induced gauge field is consistent with or can even predict the generation of B_{gr}.

TGD predicts an extremely tight correlation between various kinds of classical fields so that the Thomson magnetic field associated with super-conductor is expected to be accompanied by a gravimagnetic field which need not have a value consistent with the Equivalence Principle.

The assumption that the space-time sheet in question has a CP_2 projection belonging to either Lagrange manifold Y^2 of CP_2, say homologically trivial geodesic sphere, or to a non-trivial geodesic sphere allows to model the situation in a simple manner. For the homologically trivial sphere field equations are identically satisfied by the vacuum extremal property. The treatments are essentially identical so that the consideration is restricted to the non-vacuum case for definiteness.

For the simplest cylindrically symmetric situations CP_2 coordinates can be expressed as ($\Theta = f(\rho), \Phi = \omega t - n\phi$) cylindrical coordinates for M^4. Kähler magnetic field would be $g_K B^K_{\rho\phi} = sin(\Theta)\rho n$ and gravimagnetic vector potential would have the non-vanishing component $A^{gr}_\phi = g_{t\phi} = -R^2 \sin^2(\Theta)n\omega$. This would give $B^{gr}_{\rho\phi} = -2R^2\omega sin(\Theta)g_K B^K_{\rho\phi}$. For the ratio $2m_e B_{gr}/g_K B^K$ one would obtain

$$\frac{2m_e B^{gr}}{g_K B^K} = -2R^2\omega m_e sin(\Theta) \ . \tag{15.6.10}$$

Equivalence Principle would require $2m_e B_{gr}/eB_{em} = 1$ so that one would have

$$2R^2\omega m_e sin(\Theta) = -x \ . \tag{15.6.11}$$

where one has $eB_{em} \equiv xg_K B_K$. The value of x is completely fixed for homologically non-trivial geodesic sphere. In the appendix of [TGDview] it is shown to be $x = 3 - 2sin^2(\theta_W)$, where $sin^2(\theta_W) \simeq$ .23 denotes Weinberg angle.

The condition is impossible to satisfy in precise sense since $sin(\Theta)$ cannot be constant so that at least a small breaking of Equivalence Principle is unavoidable but probably does not offer an explanation of the effect. The assumption that that B^K is constant implies $sin(\Theta) = \alpha + \beta\rho^2$ and implies that also B_{gr} is constant in the lowest order approximation.

The condition above would require an extremely large value of the parameter ω of order $\omega \sim 1/m_e R^2 \sim 10^{19}/R$. This would imply that the induced metric has Euclidian signature by $g_{tt} < 1 - R^2\omega^2 sin^2(\Theta) < 0$. It would also imply a huge Kähler electric field $g_K E^K = g_K B^K_{\rho\phi}\omega/n$. The situation is obviously same also in the case that the value gravitomagnetic field has the value needed to explain the experimental findings of Tate.

Could the large gravimagnetic field correspond to dark graviphotons?

A possible way out of the difficulty is based on the assumption that the graviphotons are dark and have a large value of Planck constant increasing in turn the value of λ_{gr} and thus gravitomagnetic field.

The TGD based model for the hierarchy of Planck constants associated with the dark matter hierarchy assumes that the various values $\hbar = \lambda^k \hbar_0$, $\lambda \simeq 2^{11}$ correspond at the level of imbedding space to a book like structure. Different algebraic extensions of rational numbers and p-adic numbers correspond to different values of Planck constant and copies of imbedding space. The metrics for these different copies differ by a scaling in M^4 degrees of freedom and are glued together by along a subset of rationals such that the distances of the glued points from the common origin of glued copies of M^4 are identical. The configuration space of 3-surfaces decomposes into sectors labelled by unions of future and past light cones and the dips of these light cones define the preferred origins.

The key observation is that the role of Planck constant in the d'Alembertian at the level of the imbdding space is to multiply M^4 part of the d'Alembertian but leave CP_2 part unaffected. This is in accordance with the fact that induced spinor connection corresponds to gauge couplings not involving $\hbar$ and also with the fact that the scaling of CP_2 spinor connection does not make sense. At the level of induced spinor fields Planck constant in turn corresponds to the scaling factor of the M^4 part of the induced metric.

Hence it is natural to assume that contravariant M^4 metric scales as $\hbar^2(k) \propto \lambda^{2k}$. $\lambda \simeq 2^{11}$ as a function of $\hbar$ whereas CP_2 metric is not affected. This would mean that M^4 contribution to the induced covariant metric scales as λ^{-2k}: this implies that Kähler action can be seen as a function of the value of Planck constant and thus codes for the higher level corrections in powers of $\hbar$. This allows to have TGD to predict a series of higher order corrections in powers of $\hbar$ although the perturbation theory defined by the configuration space functional integral could reduce to the lowest order approximation as the general number theoretic and integrability arguments inspired by symmetric space property of the zero mode constant sectors of the configuration space suggest.

Using scaled coordinates in which M^4 metric is represented by a unit tensor, this geometrization of the dynamics of Planck constant would mean an effective scaling $R^2 \to \lambda^{2k} R^2$ for the CP_2 radius R increasing the contribution of the CP_2 metric to the induced metric. This is just what is needed to preserve the Minkowskian signature of the induced metric in ultrastrong gravimagnetic fields.

For a dynamical Planck constant the expression for ω would become $\omega \sim 1/m_e\lambda^{2k}R^2 \sim 10^{19}/\lambda^{2k}R$. The requirement that the signature of the induced metric is Minkowskian gives $g_{tt} = 1 - R^2\lambda^{2k}\omega^2 > 0$. This boils down to the conditions

$$\lambda^k > \frac{x}{Rm_e} \sim 10^{19}x \ ,$$
$$\omega \leq \frac{m_e}{x^2} \ , \tag{15.6.12}$$

where x is the numerical constant defined earlier. Using $\lambda \simeq 2^{11}$ this would give $k \geq 6$. The hierarchy of dark matter levels associated with living matter contains $k = 7$ levels relevant to human consciousness and $k = 7$ corresponds to a characteristic time scale of about 50 years [M3].

One can deduce an estimate for the dark graviphoton mass by assuming the value for λ_{gr} implied by B_{gr} necessary to explain the anomaly observed by Tate [ik1]. This would give

$$m_{gr} \sim \sqrt{\frac{N}{10}} \times \frac{1}{10^2 a} \ , \tag{15.6.13}$$

where gravitational mass density has been parameterized as $\rho_m = N m_p / a^2$, $a = 10^{-10}$ m. Note that the rest energy is above the thermal threshold at room temperature. The mass corresponds to an ordinary Compton length of order 10 nm, size scale for Cooper pairs and cell membrane thickness which emerges as a fundamental length scale characterizing Cooper pairs in the TGD based model for high T_c super-conductor [J1, J2, J3]. TGD inspired model of living matter predicts that also the magnetic structures corresponding to scaled up variants of cell membrane having sizes scaled up by λ^k are fundamental [M3].

A possible physical interpretation for the origin of the ordinary graviphoton mass would be that the confinement of longitudinal graviton inside a magnetic flux tube of thickness $L(151) = 10$ nm gives its a non-vanishing effective rest mass due to the confinement in transversal degrees of freedom which is same for all scaled up variants. The effect would be completely analogous to the generation effective photon mass in waveguide. For $k = 6$ level the Compton length of the dark graviphoton would be about 10^{11} m, the size scale of the solar system, so that a genuine long range interaction would be in question.

The gravitational mass of the photon associated with super-conductivity would be enormous for ordinary value of Planck constant. For the ordinary value of Planck constant the mass would be around $m_\gamma \simeq 10^{-3} m_e$ and for $k = 6$ one would have $m_\gamma \simeq 10^{16} m_e$. This weird implication suggests that super-conducting photons are ordinary. In this case the wavelength would be of order one nanometer and of the same order of magnitude as the wavelength of ordinary graviphoton, which supports the interpretation that transversal confinement to magnetic flux tubes gives rise to the mass in both cases.

The model ties together both electron mass scale, the order of magnitude for the size of Cooper pairs, cell membrane thickness crucial for high T_c super-conductivity, and the size scale of the solar system which in the case finite space-time sheets would give natural estimate for the much lower mass scale of the ordinary graviphoton. This raises the hope that the model might have at least something to do with reality. The model also suggest that dark gravimagnetism might be of importance in living systems.

Chapter 16

Cosmic Strings

16.1   Introduction

Cosmic strings belong to the basic extremals of the Kähler action. The string tension of the cosmic strings is $T \simeq .2 \times 10^{-6}/G$ and slightly smaller than the string tension of the GUT strings and this makes them very interesting cosmologically. Indeed, string like objects play fundamental role in TGD inspired cosmology and also provide TGD based models for the galaxy formation, galactic dark matter, and for the generation of the large voids. Therefore the study of the properties of cosmic strings deserves a separate chapter.

The physical interpretation of strings depends on the principle assumed to select the preferred extremals as generalized Bohr orbits. There are some objections against absolute minimization (call this option $P1$) and the number theoretically favored principle selects preferred extremals as surfaces minimizing ($P2$) or maximizing ($P3$) the absolute value of Kähler action separately for each space-time region where the action has a definite sign [E2]. $P2$ and $P3$ dual of each other are symmetric under exchange of electric and magnetic fields unlike absolute minimization. Both $P2$ and $P3$ can be acceptable options since it is 3-surface which is fundamental notion whereas space-time surface is a derived concept. $P2$ would represent a conservative view about future and past minimizing gradients whereas $P3$ would tend to exaggerate gradients and give more dramatic view about future and past. It would depend on situation which of the principles gives best space-time correlates for the quantum theory.

There are two kinds of strings: free and topological condensed ones.

a) Free cosmic strings are not absolute minima of the Kähler action (the action has wrong sign). $P3$ would favor cosmic strings and also their electric duals expected to exist.

b) In long enough length scales Kähler action per volume must vanish so that the idealization of cosmology as a vacuum extremal becomes possible and there must be some mechanism compensating the positive action of the free cosmic strings and topological condensation provides this kind mechanism. It however became possible to understand what really occurs in the topological condensation only after a more precise view about the relationship between inertial and gravitational masses had emerged.

16.1.1   The relationship between inertial and gravitational masses

Concerning the understanding of the topological condensation of cosmic strings the decisive breakthrough came through the understanding of the relationship between inertial and gravitational masses.

a) TGD predicts that inertial four-momentum is conserved whereas an empirical fact is that gravitational four-momentum is not conserved in cosmological length scales. The solution of the paradox came through the realization that the inertial energies of matter and antimatter have opposite signs if one requires that that the vacuum energy associated with second quantized free induced spinor fields vanishes. The most elegant and most predictive theory results if all quantum states of the Universe have vanishing net energy. Gravitational four-momentum can in turn be defined as a difference of inertial momenta associated with matter and antimatter. As it has turned out there is also non-conservation of gravitational momentum at elementary particle length scales and inertial four-momentum can be identified as a time average of gravitational four-momentum at elementary particle level.

b) The study of the simplest extremals of Kähler action demonstrates that cosmological constant in TGD Universe is non-vanishing. What is new that p-adic fractality predicts that Λ scales as $1/L^2(k)$ as a function of the p-adic scale characterizing the space-time sheet implying a series of phase transitions reducing Λ. The recent value of the cosmological constant comes out correctly. The gravitational energy density assignable to the cosmological constant is identifiable as that associated with topologically condensed cosmic strings and magnetic flux tubes to which they are gradually transformed during cosmological evolution.

16.1.2 Topological condensation of cosmic strings

Free cosmic strings are expected to decay rapidly to ordinary particles. Topological condensation of the cosmic strings with associated generation of the radial Kähler electric field provides a mechanism for the generation of negative action. The requirement that the Kähler electric field leads to the cancellation of the action of the cosmic string implies that cosmic string carries enormous Kähler charge (roughly one unit of Kähler charge per CP_2 radius).

The basic question whether the exterior region of the topologically condensed cosmic string can be modelled using only General Relativity.

a) In Kähler neutral case, a very simple imbedding of the straight string exterior metric exists. This model cannot however guarantee the cancellation of the positive Kähler magnetic action. The idea is obvious. If the string generates a sufficiently strong Kähler charge, the negative Kähler electric action cancels the positive Kähler magnetic action of the string.

b) The string creates a gravitational potential consisting of terms assignable to the gravitational mass of string itself and with the Kähler electric field created by it. The problem is that the gravitational constant is predicted to be roughly 10^7 times too large. The new view about energy resolves this problem: Kähler energy represents the density of inertial energy and is much smaller than gravitational energy. The cancellation of Kähler action implies that the gravitational mass of string is by a factor about 10^7 larger than the inertial mass of the string. Factor oneis expected in the case that string carries no matter or antimatter besides field energy.

A possible resolution of the problem comes from the observation that cosmic strings must be created from vacuum as pairs of strings with opposite time orientation and inertial energies. For tightly bound, perhaps even coiled, pairs of cosmic strings radial Kähler fields have opposite directions and tend to cancel each other. There is however a slight matter antimatter asymmetry so that the induced Kähler field is not a pure dipole field and can cancel the Kähler magnetic action.

The pairing must be rather precise so that strings form a coiled structure geometrically analogous to a DNA double helix. This structure could contract to a point at the limit $a \to 0$ for M_+^4 option. One could say that two cosmic strings, whose inertial energies are of opposite sign and of equal magnitude are created from vacuum at or shortly after the moment of big bang. For M^4 option the evolution before the moment of big bang could be preceded by a gravitational collapse and one would have a mirror pair of evolutions. Even a sequence of them making up a temporal mirror hall is possible if the reduction of cosmological constant during cosmic evolution initiates gravitational collapse of the space-time sheet containing the strings at some moment. p-Adic fractality and the fact that the evolution of stars ends with a collapse support this view.

c) The two cosmic strings, which are phase conjugates of each other in the sense that fermionic vacua are related by Hermitian conjugation, must reduce the ratio of their inertial and gravitational masses. The simplest options correspond to the growth of gravitational mass and reduction of inertial mass. Depending on the character of the stringy fermionic vacua, it is possible to imagine two different models for how this could occur.

i) If the phase conjugate strings correspond to conventional fermionic vacua and are condensed at space-time sheets X_i^4, $i = I, II$, assumed to correspond to non-conventional vacua phase conjugates to each other, the vacuum polarization of X_i^4 could generate gravitational mass of order $10^7 m_I$ but leave inertial mass unaltered. The gravitational string tension of strings as seen from large distance would be given by $T_{gr} \sim 1/G$ and inertial string tension would remain essentially $T_I \sim 1/R^2$, and the two fundamental scales of TGD would correspond to gravitational and inertial masses. The gravitational string tension of cosmic strings explaining the observed galactic rotation curves corresponds to T_I rather than $T_{gr} \sim 1/G$. Hence this option is not favored.

ii) If strings correspond to non-conventional fermionic vacua they could reduce the magnitudes of their inertial masses by the counterpart of Hawking radiation. In this case gravitational mass could remain as such. This option is favored by the galactic rotation curves.

d) The phase conjugacy of two cosmic strings has deep implications for the cosmic and ultimately also for biological evolution (magnetic flux tubes paly a fundamental role in TGD inspired biology and cosmic strings are limiting cases of them). The point is that the arrows of geometric time are opposite for the strings and also for positive energy matter and negative energy antimatter. This implies a competition between two dissipative time developments proceeding in different directions of geometric time and looking self-organization and even self-assembly from the point of view of a conscious observer living in an opposite direction of geometric time. This resolves paradoxes created by gravitational self-organization contra second law of thermodynamics.

16.1.3 Cosmic strings and generation of structures

p-Adic fractality and simple quantitative observations lead to the hypothesis that pairs of cosmic strings are responsible for the evolution of astrophysical structures in a very wide length scale range. Large voids with size of order 10^8 light years can be seen as structures containing knotted and linked cosmic string pairs wound around the boundaries of the void. Galaxies correspond to same structure with smaller size and linked around the supra-galactic strings. This conforms with the finding that galaxies tend to be grouped along linear structures. Simple quantitative estimates show that even stars and planets could be seen as structures formed around cosmic strings of appropriate size. Thus Universe could be seen as fractal cosmic necklace consisting of cosmic strings linked like pearls around longer cosmic strings linked like...

The observed quantization of the cosmic recession velocity [ig2] supports the proposed view. The space-time sheet containing closed cosmic string pairs is closed solid torus like structure. If photons from a distant astrophysical object are not able to escape this space-time sheet they can be detected after having traversed several times around the closed loop and the red shift is proportional to the number of traversals. In case of larger void the order of magnitude for the quantization is predicted correctly.

16.1.4 Correlation between super-novae and cosmic strings

During year 2003 two important findings related to cosmic strings were made.

a) A correlation between supernovae and gamma ray bursts was observed.

b) Evidence that some unknown particles of mass $m \simeq 2m_e$ and decaying to gamma rays and/or electron positron pairs annihilating immediately serve as signatures of dark matter. These findings challenge the identification of cosmic strings and/or their decay products as dark matter, and also the idea that gamma ray bursts correspond to cosmic fire crackers formed by the decaying ends of cosmic strings. This forces the updating of the more than decade old rough vision about topologically condensed cosmic strings and about gamma ray bursts described in this chapter (old version is left essentially untouched in order to demonstrate how important the experimental input is for the evolution of ideas).

According to the updated model, cosmic strings transform in topological condensation to magnetic flux tubes about which they represent a limiting case. Primordial magnetic flux tubes forming ferromagnet like structures become seeds for gravitational condensation leading to the formation of stars and galaxies. The TGD based model for the asymptotic state of a rotating star as dynamo [D3] leads to the identification of the predicted magnetic flux tube at the rotation axis of the star as Z^0 magnetic flux tube of primordial origin (flux tube carries also em field but could carry only Z^0 charge). Besides Z^0 magnetic flux tube structure also magnetic flux tube structure exists at different space-time sheet but is in general not parallel to the Z^0 magnetic structure. This structure cannot have primordial origin (the magnetic field of star can even flip its polarity).

The flow of matter along Z^0 magnetic (rotation) axis generates synchrotron radiation, which escapes as a precisely targeted beam along magnetic axis and leaves the star. The identification is as the rotating light beam associated with ordinary neutron stars. During the core collapse leading to the supernova this beam becomes gamma ray burst. The mechanism is very much analogous to the squeezing of the tooth paste from the tube.

TGD based models of nuclei [F8] and condensed matter [F9] suggests that the nuclei of dense condensed matter develop anomalous color and weak charges coupling to dark weak bosons having Compton length L_w of order atomic size. Also lighter copies of weak bosons can be important in living matter. This weak charge is vacuum screened above L_w and by dark particles below it. Dark neutrinos, which according to TGD based explanation of tritium beta decay anomaly [F8] should have the same mass scale as ordinary neutrinos, are good candidates for screening dark particles. The Z^0 charge unbalance caused by the ejection of screening dark neutrinos hinders the gravitational collapse. The strong radial compression amplifies the tooth paste effect in this kind of situation so that there are hopes to understand the observed incredibly high polarization of 80 ± 20 per cent [ij5].

TGD suggests the identification of particles of mass $m \simeq 2m_e$ accompanying dark matter as lepto-pions [F7] formed by color excited leptons, and topologically condensed at magnetic flux tubes having thickness of about lepto-pion Compton length. Lepto-pions would serve as signatures of dark matter whereas dark matter itself would correspond to the magnetic energy of topologically condensed cosmic strings transformed to magnetic flux tubes.

16.2   General vision about topological condensation of cosmic strings

In this section the basic properties of free cosmic strings are discussed and a general vision about topological condensation of cosmic strings is proposed. In the later sections the vision is developed at a more quantitative level

16.2.1   Free cosmic strings

The free cosmic strings correspond to four-surfaces of type $X^2 \times S^2$, where S^2 is the homologically nontrivial geodesic sphere of CP_2 [Appendix of the book] and X^2 is minimal surface in M_+^4. In this section, a co-moving cosmic string solution inside the light cone $M_+^4(m)$ associated with a given m point of M_+^4 will be constructed.

Recall that the line element of the light cone in co-moving coordinates inside the light cone is given by

$$ds^2 \;\; = \;\; da^2 - a^2(\frac{dr^2}{1+r^2} + r^2 d\Omega^2) \; . \tag{16.2.1}$$

Outside the light cone the line element is given

$$ds^2 \;\; = \;\; -da^2 - a^2(-\frac{dr^2}{1-r^2} + r^2 d\Omega^2) \; , \tag{16.2.2}$$

and is obtained from the line element inside the light cone by replacements $a \to ia$ and $r \to -ir$.

Using coordinates $(a = \sqrt{(m^0)^2 - r_M^2}, ar = r_M)$ for X^2 the orbit of the cosmic string is given by

$$\theta \;\; = \;\; \frac{\pi}{2} \; ,$$
$$\phi \;\; = \;\; f(r) \; . \tag{16.2.3}$$

Inside the light cone the line element of the induced metric of X^2 is given by

$$ds^2 \;\; = \;\; da^2 - a^2(\frac{1}{1+r^2} + r^2 f_{,r}^2)dr^2 \; . \tag{16.2.4}$$

The equations stating the minimal surface property of X^2 can be expressed as a differential conservation law for energy or equivalently for the component of the angular momentum in the direction orthogonal to the plane of the string. The conservation of the energy current T^α gives

$$
\begin{aligned}
T^{\alpha}_{,\alpha} &= 0 \;, \\
T^{\alpha} &= Tg^{\alpha\beta}m^{0}_{,\beta}\sqrt{g} \;, \\
T &= \frac{1}{8\alpha_K R^2} \simeq .22 \times 10^{-6}\frac{1}{G} \;.
\end{aligned}
\tag{16.2.5}
$$

The string is slightly smaller than that associated with the GUT strings.

The non-vanishing components of energy current are given by

$$
\begin{aligned}
T^{a} &= TUa \;, \\
T^{r} &= -T\frac{r}{U} \;, \\
U &= \sqrt{1 + r^2(1 + r^2)f_{,r}^2} \;.
\end{aligned}
\tag{16.2.6}
$$

The equations of motion give

$$
U = \frac{r}{\sqrt{r^2 - r_0^2}} \;,
\tag{16.2.7}
$$

or equivalently

$$
\phi_{,r} = \frac{r_0}{r\sqrt{(r^2 - r_0^2)(1 + r^2)}} \;,
\tag{16.2.8}
$$

where r_0 is an integration constant to be determined later. Outside the light cone the solution has the form

$$
\phi_{,r} = \frac{r_0}{\sqrt{r^2 + r_0^2}\,r\sqrt{1 - r^2}} \;.
\tag{16.2.9}
$$

In the region inside the light cone, where the conditions

$$
r_0 \;\; << \;\; r << 1
\tag{16.2.10}
$$

hold, the solution has the form

$$
\begin{aligned}
\phi(r) &\simeq \phi_0 + \frac{v}{r} \;, \\
v &= \frac{r_0}{\sqrt{1 + r_0^2}} \;,
\end{aligned}
\tag{16.2.11}
$$

corresponding to the linearized equations of motion

$$
f_{,rr} + \frac{2f_{,r}}{r} = 0 \;,
\tag{16.2.12}
$$

obtained most nicely from the angular momentum conservation condition.

In co-moving coordinates (in general the co-moving coordinates of sub-light-cone M^4_+!) the string is stationary. In Minkowski coordinates string rotates with an angular velocity inversely proportional to the distance from the origin

$$
\omega \simeq \frac{v}{r_M}
\tag{16.2.13}
$$

so that the orbital velocity of the string becomes essentially constant in this region. For very large values of r the orbital velocity of the string vanishes as $1/r$. Outside the light cone the variable r is in the role of time and for a given value of the time variable r strings are straight and one can regard the string as a rigidly rotating straight string in this region.

Inside the light cone, the solution becomes ill defined for the values of r smaller than the critical value r_0. Although the derivative $\phi_{,r}$ becomes infinite at this limit, the limiting value of ϕ is finite so that strings winds through a finite angle. The normal component T^r of the energy momentum current vanishes at $r = r_0$ identically, which means that no energy flows out at the end of the string. The coordinate variable r becomes however bad at $r = r_0$ (string resembles a circle at r_0) and this conclusion must be checked using ϕ as coordinate instead of r. The result is that the normal component of the energy current indeed vanishes.

Field equations are not however satisfied at the end of the string since the normal component of the angular momentum current (in z- direction) is non-vanishing at the boundary and given by

$$J^r = Tr^2 a \ . \tag{16.2.14}$$

This means that the string loses angular momentum through its ends although the angular momentum density of the string is vanishing. The angular momentum lost at moment a is given by

$$J = \frac{Tr^2 a^2}{2} = \frac{Tr_M^2}{2} \ . \tag{16.2.15}$$

This angular momentum is of the same order of magnitude as the angular momentum of a typical galaxy [ia9].

In M^4 coordinates singularity corresponds to a disk in the plane of string growing with a constant velocity, when the coordinate m^0 is positive

$$
\begin{aligned}
r_M &= vm^0 \ , \\
v &= \frac{r_0}{\sqrt{1 + r_0^2}} \ .
\end{aligned}
\tag{16.2.16}
$$

From the expression of the energy density of the string

$$
\begin{aligned}
T^a &= \frac{ar}{\sqrt{r^2 - r_0^2}} \ , \\
T &= \frac{1}{8\alpha_K R^2} \ ,
\end{aligned}
\tag{16.2.17}
$$

it is clear that energy density diverges at the singularity. As also noticed, the string tension is by a factor of order 10^{-6} smaller than the critical string tension $T_{cr} = 1/4G$ implying angle deficit of 2π in GRT so that there is no conflict with General Relativity (unlike in the earlier scenario, in which the CP_2 radius was of order Planck length). The energy of the string portion ranging from r_0 to r_1 is given by

$$E = T\sqrt{(r_1^2 - r_0^2)a} = T\sqrt{\delta r_M^2} \ . \tag{16.2.18}$$

It should be noticed that M^4 time development of the string can be regarded as a scaling: each point of the string moves to radial direction with a constant velocity v.

One can calculate the total change of the angle ϕ from the integral

$$\Delta\phi = \sqrt{\frac{r_0^2}{1 + r_0^2}} \int_{r_0}^{\infty} dr \frac{1}{r\sqrt{(r^2 - r_0^2)(1 + r^2)}} \ . \tag{16.2.19}$$

The upper bound of this quantity is obtained at the limit $r_0 \to 0$ and equals to $\Delta\phi = \pi/2$.

Cosmic strings lead to a model for the formation of large voids containing galaxies at their boundaries. This model leads to quite a fascinating picture providing new interpretation for the second law of thermodynamics and suggesting how gravitational self assembly circumvents its implications.

16.2.2   Pairing of strings as a manner to satisfy Einstein's equations

The standard view about topologically condensation of cosmic strings would be very simple. The metric of the space-time sheet at which string condenses would be flat except at the position of string and develops angular defect. In TGD Universe this configuration is not however physical.

a) The absolute minimization of Kähler action requires that the positive Kähler magnetic action of a topologically condensed string must be somehow compensated and the only manner to achieve this is that the strings develop Kähler charges giving rise to a radial Kähler electric fields whose combined action compensates the Kähler magnetic action of the string.

b) Since the net inertial energy in the most elegant TGD Universe must cancel in the length scale of large voids, cosmic string has an enormous (Kähler) energy, In particular, the gravitational mass of the strings deduce from the induced metric associated with the Kähler electric field would be by a factor of order 10^7 too large. The only manner to avoid paradox is to allow the pairing of strings having having opposite inertial energy and radial Kähler electric fields of opposite sign. A slight matter antimatter asymmetry is necessary since the Kähler electric fields of two strings with exactly the same magnitude would sum to that of a line dipole.

The pairing must be rather precise so that strings form a coiled structure geometrically analogous to a DNA double helix. This structure could contract to a point at the limit $a \to 0$ for M_+^4 option. One could say that two cosmic strings, whose inertial energies are of opposite sign and of equal magnitude are created from vacuum at or shortly after the moment of big bang. Of course this could occur also later. For M^4 option the evolution before the moment of big bang could be preceded by a gravitational collapse and one would have a mirror pair of evolutions. Even a sequence of them making up a temporal mirror hall is possible if the reduction of cosmological constant during cosmic evolution initiates gravitational collapse of the space-time sheet containing the strings at some moment. p-Adic fractality and the fact that the evolution of stars ends with a collapse support this view.

Depending on the character of the fermionic vacuum of strings, one can imagine two mechanisms for the reduction of m_I/m_{gr} ratio. For the conventional fermionic vacuum inertial mass remains as such and gravitational mass increases. For the non-conventional vacuum m_I is reduced and m_{gr} remains unchanged. Galactic rotation curves favor $T_{gr} \sim 1/R^2$, which suggests that the reduction of m_I/m_{gr} is achieved by the reduction of m_I rather than growth of m_{gr}. Despite this both options deserve a more detailed consideration.

16.2.3   Reduction of the m_I/m_{gr} ratio as a vacuum polarization effect?

The fact that cosmic strings are not small perturbations of vacuum extremals could be seen as a support for the view that the string vacua are conventional so that both fermions and anti-fermions have positive inertial energy and $m_{gr} = m_I$ holds true for them. Let $X_{I/II}^4$ be nearly vacuum extremal space-time sheets of opposite time orientation containing strings I and II and correspond to un-conventional phase vacua which are phase conjugates of each other. Since $X_{I/II}^4$ can carry only a positive/negative fermion number, X_I^4 and X_{II}^4 must be created from the vacuum as a phase conjugate pair with opposite fermion numbers. Matter antimatter asymmetry would result from the vacuum polarization of $X_{I/II}^4$ in the immediate vicinity of strings generating fermion/antifermion number and gravitational mass of magnitude $\sim 10^7 m_I$. As found, the increase of T_{gr} to $T_{gr} \sim 1/G$ is not favored by the galactic rotation curves.

16.2.4   Generation of ordinary matter via Hawking radiation?

The second option is that the string vacua are non-conventional so that fermions and anti-fermions have opposite energies inside them. In this case the strings can reduce their inertial masses by the analog of Hawking radiation involving the generation of fermion anti-fermion pairs, whose second member remains inside string. Assume that the vacuum of the space-time sheet X^4 containing phase conjugate strings I and II, corresponds to the vacuum of either string. String I ($E > 0$) emits positive energy fermions and develops inside it a density of negative energy anti-fermions (fermion number increases) so that the inertial mass density is reduced. String II ($E < 0$) emits negative energy anti-fermions and develops inside it a density of positive energy fermions. The signs of the fermion number and Kähler charge of the string correlate.

This "Hawking radiation" could generate at least part of the visible matter. The splitting of cosmic strings followed by a "burning" of the string ends provides a second manner to generate visible matter. Fermions and negative energy anti-fermions dominate the energy density at the space-time sheet containing strings and strings themselves contain dominantly negative energy fermions and positive energy anti-fermions. This explains the effective absence of antimatter.

The conservation of magnetic flux implies that the reduction of Kähler energy leads to the thickening of the strings and they become Kähler magnetic flux tubes. As far as net fermion and anti-fermion numbers are considered the exterior region contains positive energy fermions and negative energy anti-fermions (for both of which the sign of fermion number is same!) whereas the string space-time sheets contain negative energy fermions and positive energy anti-fermions. Hence positive energy anti-fermions are effectively absent and in this sense the Universe looks matter-antimatter asymmetric.

Before continuing it must however be emphasized that non-conventional quantization cannot be the only one. The reason is that both kinds of creation operators create states with the same sign of fermion number so that this kind of space-time sheets cannot carry bosons with a vanishing fermion number. The dimension $D(CP_2)$ of CP_2 projection of the space-time surface serves as a classifier for the asymptotic solutions of field equations for which Kähler Lorentz force vanishes (as a space-time correlate for the absence of dissipation, [D1]). Elementary particles, in particular bosons, correspond to CP_2 type extremals with $D(CP_2) = 4$.

For space-time surfaces with a lower-dimensional CP_2 projection the quantization could be always non-conventional. If so, the states would be always many fermion states containing ordinary fermions and what might be called anti-fermion holes. An interesting question is how to experimentally distinguish between fermions and anti-fermion holes: this might be based on the possibility of fermions and anti-fermion holes to accelerate spontaneously without any external energy feed by exchanging energy. Perhaps the reader can develop a convincing argument excluding the interpretation of condensed matter as this kind of phase.

A more detailed picture

A little bit more quantitative picture is obtained by looking more concretely the second quantization of fermions for non-conventional vacuum.

a) Unlike in ordinary quantization both fermionic and anti-fermionic oscillator operators appearing in the second quantized spinor field are either creation - or annihilation operators. There are two fermionic vacua with vanishing vacuum energy and these vacua are phase conjugates of each other.

$$
\begin{aligned}
|vac_+\rangle &= \prod_k a^\dagger_{-k} \prod_k b^\dagger_{+k}|0\rangle \ , \\
|vac_-\rangle &= \prod_k a_{-k} \prod_k b_{+k}|0\rangle \ .
\end{aligned}
\qquad (16.2.20)
$$

Here $>$ and < 0 refer to positive and negative energy state respectively.

b) One can assume that string I with positive inertial energy corresponds to $|vac_+\rangle$ and string II to $|vac_-\rangle$. For definiteness one can assume that the larger space-time sheet X^4 corresponds to vacuum $|vac_+\rangle$. The total vacuum is the tensor product of these vacua.

$$
\begin{aligned}
|vac\rangle &= |vac_I\rangle \otimes |vac_0\rangle \otimes |vac_{II}\rangle \ , \\
|vac_I\rangle &= |vac_0\rangle = |vac_+\rangle \ , \quad |vac_{II}\rangle = |vac_-\rangle \ .
\end{aligned}
$$

$$(16.2.21)$$

Obviously the selection breaks matter-antimatter symmetry.

c) The generation of Hawking radiation can be modelled by a Hamiltonian, which corresponds to the gravitational energy so that its matrix elements are non-vanishing between states with the same inertial energy. The free part of the Hamiltonian is enough to generate transitions and is of the general form

$$
\begin{aligned}
H &= H_- + H_+ + H_a + H_b \ , \\
H_- &= \sum h_k a_{-k} b_{+k} \ , \quad H_+ = (H_-)^\dagger = \sum \overline{h}_k a^\dagger_{-k} b^\dagger_{+k} \ , \\
H_a &= (H_a)^\dagger = \sum \alpha_k a^\dagger_{-k} a_{-k} \ , \quad H_b = (H_b)^\dagger = \sum \beta_k b^\dagger_{+k} b_{+k} \ .
\end{aligned}
\tag{16.2.22}
$$

H_- and its conjugate create or annihilate zero energy fermion anti-fermion pairs from vacuum and have vanishing vacuum expectation value. H_a and H_b have a non-vanishing vacuum expectation value. It is assumed that the oscillator operators act in the same manner in the state spaces associated with the stringy space-time sheets and the larger space-time sheet containing them. The assumption is consistent with the universality of oscillator operators meaning that same oscillator operators make sense even in different number fields.

d) To see that Hawking radiation can generate the ordinary matter consider the action of operator H to the vacuum formed as the product of three vacua. In this case the members of the oscillator operator pairs appearing in H can act on different vacua and thus induce Hawking radiation. Both H_b and H_- are needed in order to generate non-vanishing positive energy fermion- and negative energy anti-fermion numbers in X^4.

i) The observations

$$
H_-|vac_-\rangle = 0 \ , \quad H_+|vac_+\rangle = 0
\tag{16.2.23}
$$

help to figure out what parts of H can reduce the absolute value of the inertial energy of a given string.

ii) The negative inertial energy of string II can be increased by acting by creation operator $b^\dagger_+$. H_+ and H_b contain these operators. H_b acts on it like b_- and creates negative energy anti-fermion. H_+ acts on $|vac_0\rangle$ like $a^\dagger_-$ and annihilates it or a positive energy fermion radiated from string I.

iii) The positive inertial energy of string I can be reduced by acting by annihilation operator b_+. H_- and H_b contain these operators. H_- creates to $|vac_0$ like a_- and creates positive energy fermion. H_b acts on it like $b^\dagger_+$ and annihilates the ground state or annihilates fermion radiated from string II. What is interesting is that the strings could reduce the absolute values of their inertial energies without generating matter at the larger space-time sheet.

The new view about second law

Quantum classical correspondence suggests negative and positive energy strings tend to dissipate backwards in opposite directions of the geometric time in their geometric degrees of freedom. This would mean a continual competition between ordinary dissipation of tightly paired positive energy strings. Time reversed dissipation of negative energy strings looks from the point of view of systems consisting of positive energy matter self-organization and even self assembly. The matter at the space-time sheet containing strings in turn consists of positive energy matter and negative energy antimatter and also here same competition would prevail.

This tension suggests a general manner to understand the paradoxical aspects of the cosmic and biological evolution.

a) The first paradox is that the initial state of cosmic evolution seems to correspond to a maximally entropic state. Entropic state could relate to space-time sheets with negative time orientation but there would be also negentropic state corresponding to the positive energy matter. The dissipative evolution of matter at space-time sheets with positive time orientation (space-time sheets of positive energy cosmic strings and those containing pairs of cosmic strings) would obey second law and evolution of space-time sheets with negative time orientation (in particular, negative energy cosmic strings) its geometric time reversal. Second law would hold true in the standard sense as long as one can neglect the interaction with negative energy antimatter and strings. TGD inspired theory of consciousness predicts p-adic evolution and this would mean that negentropic tendency would win. Perhaps matter antimatter asymmetry reflects this.

b) The presence of the cosmic strings with negative energy and time orientation could explain why gravitational interaction leads to a self-assembly of systems in cosmic time scales. The formation of supernovae, black holes and the possible eventual concentration of positive energy matter at the negative energy cosmic strings could reflect the self assembly aspect due to the presence of negative energy strings. An analog of biological self assembly identified as the geometric time reversal for ordinary entropy generating evolution would be in question.

c) In the standard physics framework the emergence of life requires extreme fine tuning of the parameters playing the role of constants of Nature and the initial state of the Universe should be fixed with extreme accuracy in order to predict correctly the emergence of life. In the proposed framework situation is different. The competition between dissipations occurring in reverse time directions means that the analog of homeostasis fundamental for the functioning of living matter is realized at the level of cosmic evolution. The signalling in both directions of geometric time makes the system essentially four-dimensional with feedback loops realized as geometric time loops so that the evolution of the system would be comparable to the carving of a four-dimensional statue rather than approach to chaos.

Creation of matter from vacuum by annihilation of laser waves and their phase conjugates?

The possibility of negative energy anti-fermions suggests a new energy technology. Photons and their phase conjugates with opposite energies could only annihilate to a pair of positive energy fermion and negative energy anti-fermion. Vacuum could effectively serve as an unlimited source of positive energy and make creation of matter from nothing literally possible. The idea could be tested by allowing laser beams and their phase conjugates to interact and by looking whether fermions pop out via two-photon annihilation. Fermion-anti-fermion pairs with arbitrarily large fermion masses could be generated by utilizing photons of arbitrarily low energy. The energies of the final state fermions are completely fixed from conservation laws so that it should be relatively easy to check whether the process really occurs. Generalized Feynman rules predict the cross section for the process and it should behave as $\sigma \propto \alpha^2/m^2$, where m is the mass of the fermion so that annihilation to electrons is the best candidate for study. Bio-systems might have already invented intentional generation of matter in this manner. Certainly the possible new energy technology should be applied with some caution in order to not to build a new quasar!

Speaking more seriously, the creation and annihilation of pairs of positive and negative energy space-time sheets would have a change of gravitational mass as its signature so that the process could be detected. In particular, the annihilation of fermions and anti-fermions to pairs of phase conjugate photons would reduce gravitational mass and give rise to antigravity effects.

16.2.5   Cosmic strings and cosmological constant

The study of Einstein's equations for cosmic strings forces the conclusion that cosmological constant is non-vanishing in TGD Universe. p-Adic length scale hypothesis leads to a scenario for the evolution of cosmological constant predicting its recent value correctly. Cosmic strings allow to interpret cosmological constant as characterizing the density of gravitational energy associated with the space-time sheets carrying pairs of positive and negative energy cosmic strings and having a vanishing net Kähler charge. Therefore, contrary to the original beliefs, the models for dark matter based on cosmological constant and cosmic strings are actually equivalent.

Cosmological constant for the internal dynamics of strings

In order to understand why cosmological constant is unavoidable at the level of gravitational dynamics in the interior of cosmic strings, one can simplify things by assuming that X^2 in $X^4 = X^2 \times S_I^2$ is a flat piece M^2 of M_+^4 so that the Einstein tensor has expression

$$G^{\alpha\beta} = \mathbf{R}(S_I^2) - \frac{1}{2}\mathbf{g}R(S_I^2) \ . \tag{16.2.24}$$

where $R(S_I^2)$ denotes the curvature scalar. The energy density $G^{tt} = -(1/2)R(S_I^2)$ is negative because of spherical spherical topology. Obviously cosmic strings correspond to the p-adic length scale $L(2)$ naturally so that it should be possible to deduce a lower bound for $\Lambda(2)$ from the requirement that gravitational energy density is non-negative.

There are actually two constraints involved.

a) The first constraint comes from the sum of the energies of matter and antimatter expressible as the energy of Kähler field. One cannot exclude the possibility that for cosmic strings the quantization in fermionic sector is such that the energies both fermions and anti-fermions are either positive or negative. This requires that cosmic strings created in the big bang or possibly later appear in pairs which opposite vacuum energies.

The energy momentum tensor for Kähler field and Einstein tensor can be written as

$$
\begin{aligned}
\mathbf{T}_K &= \frac{1}{2 \times 16\pi\alpha_K R^4}(\mathbf{g}(M^2) - \mathbf{g}(S^2)) \ , \\
\mathbf{G} &= -\frac{4}{R^2}\mathbf{g}(M^2) \ .
\end{aligned}
\tag{16.2.25}
$$

Note that the radius of S_I^2 is $R/4$ for the conventions used.

$\mathbf{T}_K$ can be written as sum of Einstein tensor and metric tensor as

$$
\begin{aligned}
\mathbf{T}_K &= \frac{1}{16\pi G_{eg}}\mathbf{G} + \hat{\Lambda}_{eg}\mathbf{g} \ , \\
G_{eg} &= 8R^2\alpha_K \ , \\
\hat{\Lambda}_{eg} &= \frac{1}{2R^4 \times 16\pi\alpha_K} \ .
\end{aligned}
\tag{16.2.26}
$$

Note that the ratio $G_{eg}/G = \frac{8R^2}{G}\alpha_K$ is much larger than one.

The study of simple imbeddings of Maxwell field to $M^4 \times S_I^2$, demonstrates that Einstein's equations hold true approximately also more generally in the non-relativistic approximation and the value of G_{eq} is as above. This raises the possibility that the dynamics of Kähler action could be reflected in phenomena of electro-gravitation and that near vacuum extremals could be seen as a limit at which electro-gravitational coupling strength becomes very large.

There would be thus two "gravitations": one for the sum and one for the difference of energies of matter and antimatter. The consistency with the existing theory requires an approximate separate conservation of these energies meaning that the annihilation of matter and antimatter can be neglected. TGD based models for the asymptotic cosmology and the asymptotic state of star rely on this assumption.

b) Gravitational energy tensor can be written as a sum of Einstein tensor and metric tensor

$$
\mathbf{T}_{gr} = \frac{1}{16\pi G}\mathbf{G} + \Lambda(2)\mathbf{g} \ .
\tag{16.2.27}
$$

c) The basic constraint comes from the requirement that irrespective of the option chosen for the fermionic quantization one has

$$
E_{gr} \geq E_K \ ,
\tag{16.2.28}
$$

where one has

$$
\begin{aligned}
E_{gr} &= -\frac{1}{16\,G} + \hat{\Lambda}(2)\frac{\pi R^2}{4} \ , \\
E_K &= -\frac{1}{16G_{ge}} + \hat{\Lambda}_{ge}\frac{\pi R^2}{4} = \frac{1}{4 \times 16\alpha_K R^2} \ .
\end{aligned}
\tag{16.2.29}
$$

The lower bound means that the two terms in gravitational energy nearly cancel each other since the right hand side is of order 10^{-7} as compared to the left hand side. One has

$$\hat{\Lambda}(2) \;\geq\; \frac{1}{16R^4}\left[\frac{16R^2}{G} + \frac{1}{\alpha_K}\right] \; . \tag{16.2.30}$$

For the minimal value of $\Lambda(2)$ gravitational and inertial energies are identical. This corresponds to the situation in which negative energy contribution to the inertial energy vanishes. This situation can result in two manners. Either antimatter is absent or fermionic quantization is performed in the standard manner. The latter option fixes the value of $\hat{\Lambda}(2)$ completely.

Strings with conventional fermionic vacua can emit positive energy fermions during later stages of evolution: this reduces the energy density of the string. In the case that the stringy fermionic vacuum is non-conventional, the string can generate fermion anti-fermion pairs during later phases of cosmic evolution such that the negative energy member of the pair remains inside string and reduces its energy density (note that fermion number increases!). The reduction of the energy density means a thickening of the string since flux quantization condition must be satisfied. Energy density behaves as $1/S$ during this process.

p-Adic evolution of cosmological constant

The evolution of the cosmological constant Λ is different at each space-time sheet, and the value of Λ is determined by the p-adic length scale size of the space-time sheet according to the formula $\Lambda(k) = \Lambda(2) \times (L(2)/L(k))^2$ derived in the [D5] from the requirement that gravitational energy as difference of inertial energies and matter and antimatter (or vice versa) is non-negative. Since cosmological expansion forces both strings to become very thin as the initial moment is approached, $\Lambda(k)$ must decrease during cosmological evolution of the void and increase again during the possibly occurring contraction phase. The reduction of the value of Λ below critical value might initiate the phase of contraction. The value of $\Lambda(k)$ at the space-time sheet at which strings are condensed is smaller than for cosmic strings since the thickness of the string characterizes the value of Λ.

In standard physics context piecewise constant cosmological constant would be naturally replaced by a cosmological constant behaving like $1/a^2$ as a function of cosmic time. p-Adic prediction is consistent with the recent study [ic1] according to which cosmological constant has not changed during the last 8 billion years: the conclusion comes from the reshifts of supernovae of type I_a. If p-adic length scales $L(k) = p \simeq 2^k$, k any positive integer, are allowed, the finding gives the lower bound $T_N > \sqrt{(2)}/(\sqrt{2} - 1)) \times 8 = 27.3$ billion years for the recent age of the universe.

Now Brad Shaefer from Lousiana University has studied the red shifts of gamma ray bursters up to a red shift $z = 6.3$, which corresponds to a distance of 13 billion light years [id6], and claims that the fit to the data is not consistent with the time independence of the cosmological constant. In TGD framework this would mean that a phase transition changing the value of the cosmological constant must have occurred during last 13 billion years.

Cosmological constant for the Kähler neutral space-time sheets carrying pairs of strings

There is a strong temptation to relate the presence of the cosmological constant in the model for topological condensation in terms of constant density of $U(1)$ gauge charge (not Kähler charge for vacuum extremals).

a) In case of Kähler charge this would lead to a modification of the previous ansatz for the space-time sheet containing strings of opposite inertial energies by adding to $u = cos(\Theta)$ a term which is linear in coordinate ρ: $u \to u + k_1\frac{\rho}{\rho_0}$. This ansatz does not however satisfy field equations even approximately.

b) One can however consider vacuum extremals with similar structure imbedded to $M^4 \times Y^2$, Y^2 could be S^2_{II} or a more general Lagrange sub-manifold of CP_2 so that field equations reduce to boundary conditions stating that gravitational four-momentum does not leak out from boundaries. Also in this case cosmological constant would correspond to a presence of constant $U(1)$ gauge charge density.

c) The ordinary gravitational potential energy of a co-moving volume decreases as $E_{gr} \propto M^2(L(k))/L(k)$ as a function of the p-adic scale, whereas cosmological constant varying as $1/L(k)^2$ contributes to the gravitational energy a positive term $E_\Lambda \propto L(k)$ implying $V_{gr} \propto L(k)$ and $E_{tot} \propto L(k)$. The prediction that the energy of the space-time sheet is proportional to its size is just what has been observed to hold true for galactic dark matter distribution. This suggest that the dark matter is due to Kähler neutral space-time sheets containing pairs of cosmic strings and cosmological constant characterizes the density of the dark matter decreasing as $1/T^2(k)$ and on the average as $1/a^2$ as a function of cosmic time. Thus the original idea about cosmic strings as the source of dark matter would be basically correct.

16.3 Topologically condensed cosmic strings

The purpose of this section is to represent in more detail the calculations behind the vision discussed in the previous section. As already noticed, free cosmic strings as such cannot correspond to the absolute minima of the action since their action is large and positive. The fact that vacuum functional favors surfaces with a large Kähler action however implies that the most probable configurations must have a small total action. For the free string of type $X^2 \times S_I^2$ the generation of the Kähler electric action is however impossible: S_I^2 doesn't allow Kähler electric fields with non-vanishing charge (an elementary topological fact). Therefore one must consider the topological condensation of free cosmic strings. This is achieved by connecting $S^2 \times D^2$ to the background space-time by small wormholes. Alternatively, one can drill small holes in cosmic string and background and connect the boundaries of these holes by join along boundaries bonds.

In this kind of situation it is possible for the string to have also a net Kähler electric charge since the field lines can flow to the exterior space-time. Charge carriers could be topologically condensed elementary particles both inside and around the string. A breaking of matter antimatter symmetry at the length scale of the string is required. The simplest field configuration corresponds to a radial Kähler electric field. The basic constraints on the model of condensation are following.

a) The contribution of the Kähler electric field to action cancels the Kähler magnetic action of strings.

b) Einstein's equations are satisfied. This requires that strings appear in parallel, possibly coiled pairs of positive and negative energy strings so that Kähler electric fields interfere destructively. Matter antimatter symmetry is broken in the sense that the fields do not have quite the same strength so that the interference does not lead to a pure dipole field.

It is relatively simple to modify the ansatz of the radial Kähler electric field to produce a non-vanishing cosmological constant and having interpretation in terms of constant density of Kähler charge. Field equations do not however allow this kind of modification. For vacuum extremals for which the density of inertial mass vanishes the situation is different. A space-time sheet containing topologically condensed space-time sheets containing pairs of strings and having a vanishing net Kähler charge can be modelled as a vacuum extremal with cosmological constant characterizing the gravitational energy density of strings.

16.3.1 Topological condensation of a neutral cosmic string

The topological condensation of neutral free cosmic string does not produce stable state since the value of the Kähler action is enormous and of wrong sign. It is however useful to build a model of exterior space-time as a solution of Einstein's equations also in this case. For a straight string this solution is flat except at the position of the string. What happens is that the 2-dimensional plane orthogonal to the string becomes a conical surface. The angular defect is given by

$$\Delta\phi \;=\; \frac{T}{T_{max}} \times 2\pi \;\;,\;\; T_{max} = \frac{1}{4G} \;\;. \tag{16.3.1}$$

Here the string tension T refers to the gravitational mass density of the string and this is not necessarily identical with the inertial mass density. Obviously $T_{max} = 1/4G$ represents an upper bound for the gravitational mass density of the string.

The metric can be written as

$$ds^2 \;=\; dt^2 - dz^2 - \frac{d\rho^2}{k_1^2} - \rho^2 d\phi^2 \;\;,$$

$$k_1^2 \;=\; 1 - 4GT \;\;. \tag{16.3.2}$$

The imbeddings of this metric as an induced metric are easy to find. The simplest imbedding is obtained by considering a map $M^4 \to S^1$, where S^1 is a geodesic circle of CP_2. Denoting by Φ the angle coordinate of S^1, one has

$$
\begin{aligned}
\Phi &= k\rho \ , \\
1 + \frac{R^2 k^2}{4} &= \frac{1}{k_1^2} \ .
\end{aligned}
\tag{16.3.3}
$$

The geodesic lines associated with this metric are easy to find in Cartesian coordinates. In M^4 coordinates the geodesics are slightly curved, which is nothing but the lense effect [ib4]. To see what happens consider geodesic lines in the plane; cut from the plane a sector corresponding to the deficit angle and bend it to form a cone; after this operation project the geodesic lines on the cone to the plane again to see how the geodesics look like in M^4 coordinates. The observation of this bending is possible if the coordinates used by the observers are actually M^4 coordinates rather than space-time coordinates.

16.3.2 Exterior space-time of a static Kähler charged string

In order to get rid of unessential complications it is useful to study first the detailed properties of the radial Kähler field associated with the topologically condensed, straight Kähler charged string as an extremal of the Kähler action, which in the approximation that gravitational effects are neglected, satisfies Maxwell's equations. A natural constraint for the solution is that the metric reduces to the metric associated with the neutral string at the limit of the vanishing Kähler charge. Despite the idealizations involved, the calculation makes sense since it gives a good grasp of a more realistic situation of s non-straight string, which is at rest in the co-moving coordinates but rotates in M^4 coordinates.

The ansatz leading to a radial Kähler field in the background space is given by the following expression in cylindrical coordinates for M_+^4

$$
\begin{aligned}
cos(\Theta) &= u(\rho) \ , \\
\Phi &= \omega t + k_1 \rho \ .
\end{aligned}
\tag{16.3.4}
$$

It deserves to be noticed that a more general ansatz $\Phi = \omega t + k_1 \rho + n\phi$ would give rise also to Kähler magnetic field but this ansatz is not consistent with field equations.

The interesting components of the induced metric and Kähler field in the cylindrical coordinates are given by the expression

$$
\begin{aligned}
J_{\rho t} &= \frac{\omega}{4} \partial_\rho u \ , \\
g_{tt} &= 1 - K(1 - u^2) \ , \\
g_{\rho\rho} &= -1 - k_1^2 - \frac{K}{\omega^2}(\partial_\rho u)^2 \frac{1}{(1 - u^2)} \ , \\
g_{\rho t} &= -\frac{K}{\omega} k_1 (1 - u^2) \partial_\rho u \ , \\
K &= \frac{R^2 \omega^2}{4} \ , \\
k_1^2 &= -1 + \frac{1}{1 - \frac{T}{4G}} \ .
\end{aligned}
\tag{16.3.5}
$$

The field equations are equivalent to the conservation of four momentum say in x-direction and one obtains in this manner the following field equation

$$
\begin{aligned}
\partial_\rho(g^{\rho\rho} X) &= -\frac{2X}{\rho} \ , \\
X &= (\partial_\rho u)^2 \frac{1}{(g_{tt} g_{\rho\rho} - (g_{\rho t})^2)^{1/2}} \ .
\end{aligned}
\tag{16.3.6}
$$

In the approximation $g^{\rho\rho} = 1$ the solution of this equation can be written as

$$X \;=\; \frac{X_0}{\rho^2} \; . \tag{16.3.7}$$

This gives extremely nonlinear first order differential equation for u.

In the lowest order approximation the surrounding space-time is just the flat M^4. In this approximation one obtains just source free Maxwell equation in M^4 in the region exterior to the string

$$
\begin{aligned}
J^2_{\rho t} &= \frac{C^2}{\rho^2} \; , \\[6pt]
u(\rho) &= -k \ln\left(\frac{\rho}{\rho_0}\right) \; , \\[6pt]
k &= -\frac{C}{\omega} \; .
\end{aligned}
\tag{16.3.8}
$$

The result is expected since u corresponds essentially to the gauge potential of the electric field created by a line charge. This result seems to exclude the presence of constant Kähler charge density giving naturally rise to a cosmological constant since in this case one should have $u = k_1\rho + ...$ not consistent with approximate flatness of the metric. Thus it would seem that cosmological constant is possible only for vacuum or nearly vacuum extremals.

Lowest order solution ceases to be well defined at $u = \pm 1$. These values of u correspond finite values of the radial coordinate ρ

$$
\begin{aligned}
\rho_\pm &= \rho_0 \exp\left(\pm\frac{1}{k}\right) \; , \\[6pt]
\frac{\rho_+}{\rho_-} &= \exp\left(\frac{2}{k}\right) \; .
\end{aligned}
\tag{16.3.9}
$$

One might think that this is due to the failure of the lowest order approximation and that situation changes for exact solution so that $u = 1$ corresponds to the limit, when the radial coordinate approaches infinity. This seems not to be the case however.

Notice that the boundary conditions are satisfied at the singular value of ρ. This is due to the divergence of the radial component of metric, which implies that the normal components of the energy momentum current $T^{\alpha\beta}$ and Kähler field $J^{\alpha\beta}$ vanish. Also this suggests that the singularity is actual: space time simply ceases to exist at the larger critical radius!

16.3.3   Newtonian limit and the new view about energy

The relevant components of the induced metric are given by the expression

$$
\begin{aligned}
g_{tt} &= 1 - K(2U - U^2) \; , \quad g_{\rho t} = -\frac{Kk_1}{\omega}(2U - U^2) \; , \\[6pt]
g_{\rho\rho} &= -1 - k_1^2 - \frac{Kk^2}{\omega^2}\frac{1}{\rho^2(1 - U^2)} \; , \\[6pt]
U &= k \ln\left(\frac{\rho}{\rho_0}\right) \; , \quad \ln\left(\frac{\rho_+}{\rho_0}\right) = \frac{1}{k} \; , \quad K = \frac{R^2\omega^2}{4} \; .
\end{aligned}
\tag{16.3.10}
$$

A correct sign for the logarithmic part of the gravitational potential is obtained if k is positive.

Note that the radial component of the metric approaches its flat space value rapidly in the region, where the logarithmic term is small. Near the critical radius ρ_+ the radial component of the metric begins to grow so that one obtains a large lens effect[ib4]. This behavior differs radically from that obtained for the neutral strings.

The following argument leads to which I believe is a correct interpretation of the metric.

a) One apply Newtonian approximation in which one has

$$g_{tt} - 1 = 2\Phi_{gr} \ , \qquad \nabla^2 \Phi_{gr} = -4\pi \rho_{gr} \ . \tag{16.3.11}$$

$\Phi_g r$ contains two terms. Logarithmic term identifiable corresponds to the gravitational mass of the cosmic string. The square of the logarithm corresponds to the gravitational mass associated with the Kähler electric field.

b) By applying these definitions a simple calculation gives that the gravitational mass density associated with the square of the logarithm is

$$\rho_{gr} = \frac{A_{gr}}{\rho^2} \ , \qquad A_{gr} \equiv \frac{1}{4\pi} \frac{k^2 K}{G} \ . \tag{16.3.12}$$

c) The inertial mass density of the Kähler electric field is given by

$$\rho_I = \frac{1}{16\pi\alpha_K} \frac{E_K^2}{2} = \frac{A_I}{\rho^2} \ , \qquad A_I \equiv \frac{1}{16\pi\alpha_K} \frac{k^2 \omega^2}{2} \ . \tag{16.3.13}$$

d) From this one obtains for the ratio of inertial and gravitational mass densities of the Kähler electric field the following expression

$$x \equiv \frac{\rho_I}{\rho_{gr}} = \frac{A_I}{A_{gr}} = \frac{G}{8\alpha_K R^2} \simeq .22 \times 10^{-6} \ . \tag{16.3.14}$$

Here the value of R and α_K deduced from p-adic mass calculations have been used. Note that the parameter $k\omega$ disappears from the ratio.

If gravitational energy and inertial energy are identical and inertial energies are positive, the identification $G = 8\alpha_K R^2$ would be forced by Equivalence Principle. This identification is not however consistent with the rest of TGD (p-adic mass calculations for instance). The assumption of a physical value for R is in turn in conflict with Equivalence Principle in it standard form, and the only manner to resolve the paradox is to accept the new view about energy.

a) The assumption that the inertial energies of matter and antimatter are of opposite signs and gravitational energy density is sum of absolute values of these energy densities allows a consistent interpretation of the result. Cosmic strings must be created from vacuum as pairs. The Kähler electric field makes sense only if the system actually consists of a pair of parallel strings with opposite time orientations and sufficiently near to each other. Strings have opposite signs of inertial energies, and they generate Kähler charge of opposite sign so that their radial Kähler electric fields must point in opposite directions so that destructive interference occurs. The magnitudes of the fields must be slightly different in order that the field would not reduce to a dipole field and this means matter antimatter asymmetry. This reduces the Kähler field from its full value to a value corresponding to the ratio ρ_I/ρ_{gr}. The gravitational masses of both strings are identical and equal to absolute values of inertial masses so that Equivalence Principle holds true. The Kähler electric field energy can be interpreted as the inertial energy lost by the strings by generalized Hawking radiation.

b) One can also deduce the gravitational energy density $e_{gr} = |e_S|$ of the free cosmic string using the expression

$$e_S = \frac{1}{16\pi\alpha_K} \int_{S_I^2} J_{\mu\nu} J^{\mu\nu} d^2x \ .$$

This gives $e_S = T = 1/32\alpha_K R^2$.

c) For a *single* topologically condensed string creating a Kähler electric field the inertial and gravitational string tensions and angle deficit δ would have expressions

$$
\begin{aligned}
T_I &= e_S = \frac{1}{32\alpha_K R^2} \ , \\
T_{gr} &= \frac{T_I}{x} = \frac{1}{4G} \ , \\
\delta &= \frac{1}{4GT_{gr}} 2\pi = 2\pi \ .
\end{aligned}
\tag{16.3.15}
$$

The gravitational string tension would be too large by a factor 10^7 and the angle defect $1 - T_{gr}/4GT_{gr}$ would be a full 2π. The conclusion is that the inertial string tension T_I must be reduced in the topological condensation by a factor $x \simeq 10^{-7}$ from the value applying to free cosmic strings by a generation of the TGD counterpart of Hawking radiation leading by flux conservaiton also to the thickening of the cosmic string to a Kähler magnetic flux tube with a transversal area $S/S_0 \sim 1/x$, where S_0 is the area of S_I^2. The angle defect would reduce to $2\pi - \delta \simeq 10^{-7} \times 2\pi$.

c) In the case of positive energy cosmic string Hawking radiation means generation of pairs of negative energy anti fermion and positive energy anti-fermion. Negative energy fermions remains inside the string and positive energy anti-fermion ends to the topological condensate. The phase conjugate of this process applies to the negative energy string. This process reduces the inertial string tension T_I by a factor ϵ and one has $\delta = \epsilon 2\pi$.

The maximum value of of the reduction factor can be estimated by the requirement that the gravitational mass is conserved during the reduction of inertial mass of string so that the gravitational string tension equals to the inertial string tension of the free string: $T_{gr}(\epsilon) = T_I(free)$. This gives

$$\epsilon_{max} = \frac{G}{8\alpha_K R^2} \simeq .22 \times 10^{-6} \ . \tag{16.3.16}$$

$\delta = \epsilon_{max} \times 2\pi \simeq 10^{-7} \times 2\pi$ would be the maximal angle deficit.

The value of the Kähler coupling strength α_K used in the estimate corresponds to the p-adic length scale of electron. According to the proposed model for the p-adic evolution of Kähler coupling constant the Kähler coupling strength behaves as

$$\frac{\alpha_K(p)}{\alpha_K(q)} = \frac{log(qK^2)}{log(pK^2)} \ , \quad K^2 = 2 \times 3 \times 5 \times ....23 \ .$$

(K^2 is the product of primes not larger than 23). Substituting for q the smallest possible p-adic prime $q = 2$ and for p the p-adic prime $M_{127} = 2^{127} - 1$ characterizing electron, one obtains $\epsilon_{max} \simeq 4 \times 10^{-8}$.

d) Physical intuition suggests that cosmic strings are gradually transformed to Kähler magnetic flux tubes of finite thickness during cosmic evolution. Since magnetic flux is conserved the Kähler magnetic field scales as $B_K \propto 1/S$, where S is the transverse area of string. Inertial string tension T_I scales also as $T_I \propto 1/S$. This would mean that the minimal reduction of the inertial mass by the fraction ϵ_{max} would increase the transverse area of the string roughly by factor $1/\epsilon_{max} \simeq 2.5 \times 10^7$ for $q = 2$.

16.3.4 The cancellation of the Kähler magnetic action by Kähler electric action

By requiring that the action associated with the Kähler electric field cancels the action of the *two* topologically condensed cosmic strings involved (approximating metric with a flat metric) one obtains the approximate condition that the Kähler electric energy e_E per unit length of string is $n = 2$ times the Kähler magnetic e_B per unit length of string: $e_E = 2e_B$. One has

$$e_E = \frac{1}{2 \times 16\pi\alpha_K} \int_{\rho_l}^{\rho_u} E_K^2 \rho d\rho 2\pi = \frac{k^2\omega^2}{16\alpha_K} log(\rho_u/\rho_l) \ ,$$

$$e_B = \frac{\epsilon}{32\alpha_K R^2} \ . \tag{16.3.17}$$

The precise choice of the upper and lower bounds ρ_u and ρ_l for the radial integration will be considered below. By using the parametrization

$$log(\frac{\rho_u}{\rho_l}) = r \times log(\frac{\rho_+}{\rho_-}) \equiv \frac{2r}{k} \ , \quad r < 1 \ ,$$

and equating $e_E = ne_B$, $n = 2$, one obtains

$$k = \frac{n}{8r} \times \frac{2\epsilon}{\alpha_K \omega^2 R^2} \ . \tag{16.3.18}$$

One can deduce an estimate for the value of the parameter k also by using the defining equation for Newton's gravitational potential. $\Phi_{gr} = Kkln(\rho/\rho_+)$ gives $Kk = 2GT_{gr} = \epsilon/2$ implying

$$k = \frac{2\epsilon}{R^2\omega^2} \ . \tag{16.3.19}$$

This estimate is not quite consistent with the previous estimate for the choice $\rho_u = \rho_+$ and $\rho_l = \rho_-$. In this case consistency would require that Kähler electric action compensates for the action of 8 cosmic strings rather than two! The good news is that for $n = 1$ (single string) the inconsistency would be even worse.

The discrepancy might be due to a calculational error. On the other hand, the selection of the upper and lower bounds of the radial integration is an important factor. One might argue that since ρ_+ and ρ_- are singularities of the radial component of the metric, the actual bounds must reside between them. At ρ_0 the gravitational potential has minimum and this would make it natural choice for the radius of the cosmic string giving $\rho_l = \rho_0$. If one chooses the upper bound such that one has $log(\rho_u/\rho_0) = 1/2k$ instead of $1/k$, consistency results with $n = 2$.

It is of interest to see find the value of the parameter $R^2\omega^2$ in real life situation by using the expression

$$\frac{\rho_u}{\rho_0} = exp(\frac{1}{2k}) = exp(\frac{R^2\omega^2}{4\epsilon}) \ .$$

Taking ρ_u to correspond to the size of a larger void: $\rho_u = 10^8$ ly and assuming $\rho_0 = R$, one would obtain $R^2\omega^2 \sim 4.8 \times 10^{-5}$ so that the induced metric does not deviate very much from the flat metric. The value of ρ_+ would be completely super-astrophysical being related by a factor of order 10^{53} to $\rho_u = 10^8$ ly.

16.3.5   Exterior space-time of a co-moving Kähler charged string

Since cosmic strings are at rest in the co-moving coordinates, it is natural to try to find the exterior space-time as a small deformation of M_+^4. The previous considerations suggest that in a good approximation exterior space-time can be constructed by finding the imbedding of the Kähler electric field for a string at rest. In co-moving coordinates the rotation of the string correspond to a deformation of a straight static string from the straight string $\phi \rightarrow \phi_0 + v/r$, where v has interpretation as the velocity of a rotating string, and for large values of r the treatment as a straight string is expected to be a good approximation.

Taking z-axis to be in the direction of string, Gauss law gives $n_\rho E^\rho \sqrt{g} = C$ (where n^α is unit normal in radial direction). In co-moving cylindrical coordinates

$$
\begin{aligned}
ds^2 \ &- \ da^2 - a^2(F_1 d\rho^2 + F_2 dz^2 + 2F_3 d\rho dz + \rho^2 d\phi^2) \ , \\
F_1 \ &= \ 1 - \frac{\rho^2}{1 + \rho^2 + z^2} \ , \\
F_2 \ &= \ 1 - \frac{z^2}{1 + \rho^2 + z^2} \ , \\
F_3 \ &= \ -\frac{\rho z}{1 + \rho^2 + z^2} \ .
\end{aligned}
\tag{16.3.20}
$$

Gauss law gives for the gauge potential Φ_K

$$F_2 \partial_\rho \Phi_1 - F_3 \partial_z \Phi_K \ = \ -\frac{C}{\rho\sqrt{F_1}} \ . \tag{16.3.21}$$

This equation fixes the behavior of Φ_K along the flow lines of flow generated by the vector field $(J^\rho, J^z) = (F_2, -F_3)$ of (ρ, z) plane. The flow lines are given conical sections

$$(\frac{z}{z_0})^2 - \frac{(1+\rho^2)}{(1+\rho_0^2)} = 0 , \tag{16.3.22}$$

from which z can be solved. Φ_K can be integrated as

$$\Phi_K(\rho, z(\rho)) = -C \int_{flowline} \frac{1}{2\rho\sqrt{F_1 F_2}} d\rho , \tag{16.3.23}$$

where the flow line can be assumed to begin from the surface of a cylinder of radius, say, ρ_0. Not surprisingly, the contribution from large values of ρ to the potential gives logarithmic term.

A simple imbedding of Kähler gauge potential to a geodesic sphere of CP_2 is obtained from the following ansatz

$$\begin{aligned}
\Phi_K &= u\partial_a\Phi , \\
\Phi &= \omega a + k_2\rho , \\
u &= \frac{\Phi_K}{\omega} ,
\end{aligned} \tag{16.3.24}$$

where k_2 term takes care that the imbedding approaches to the neutral string solution at the appropriate limit. The imbeddability fails at points $u = \pm 1$ and just as in previous case one obtains a family of imbeddings parameterized by ω.

16.3.6   A model of cosmic evolution inside large void

In the original model the interior of the void contained single straight "big" string and the oppositely Kähler charged string responsible for the formation of galaxies was at the boundaries of the void. This model is not consistent with the Equivalence Principle.

The model of for the pairs of cosmic strings inspires a model for the evolution of large voids of recent size about $L = 10^8$ ly as regions containing pairs of closed and strongly knotted and linked cosmic strings located near the boundaries of the void. The gravitational mass of a cosmic string of length L has the same order of magnitude as the gravitational mass of ordinary matter contained by the void (roughly one proton per cubic meter) so that the number of string pairs cannot be too large. The interior of the void is modelled as a vacuum extremal with cosmological constant behaving according to the prediction of TGD, and one must consider the possibility that the void starts collapsing when the cosmological constant becomes sufficiently small.

Conservation of gravitational mass and vanishing of Kähler action

Gravitational mass is assumed to be conserved in reasonable approximation after the generation of a pair of cosmic strings. It is however quite possible that zero inertial energy pairs of cosmic strings and magnetic flux tubes are generated from vacuum during the whole cosmological evolution so that gravitational mass can increase. Also cosmologies within cosmologies can be generated in this manner.

For free cosmic string gravitational energy is equal to inertial energy. Kähler energy which corresponds to inertial energy corresponds to a fraction of order $\epsilon \leq 10^{-7}$ about gravitational energy. $\epsilon \simeq 10^{-7}$ holds only true for when the space-time sheet is sub-manifold of $M^4 \times S_I^2$. In a more general case ϵ is smaller and approaches zero at the limit of vacuum extremal. The parameter $\epsilon_1 \sim 10^{-9}$ giving the number of protons per single photon characterizes matter antimatter asymmetry and is not very far from ϵ. In fact, taking into account the dependence of the Kähler coupling strength on p-adic length scale, the value of ϵ for short p-adic length scales becomes $\epsilon \sim 10^{-8}$.

Critical cosmology as a period of phase transition

Critical cosmology could be interpreted as the period during which the newly created pair of strings radiates positive and negative energy particles and establishes its Kähler field. In cosmology this period would be followed by a radiation dominated period. The change of the cosmology to exactly critical cosmology with a 2-dimensional CP_2 projection instead of 1-dimensional projection and possessing $SO(4)$ instead of $SO(3,1)$ as the group of space-time isometries should correlate with this process. The density of gravitational energy approaches infinity towards the end of the critical period and behaves like $1/a^2$ at the limit $a \to 0$ so that the gravitational energy per co-moving volume vanishes at this limit.

Pairs of cosmic strings and magnetic flux tubes with vanishing net energy can be created also later during cosmic evolution and the critical sub-cosmology provides a universal model for this phase transition involving only single parameter characterizing the duration of the critical phase. This might make it possible to test the model in laboratory.

TGD inspired theory of consciousness [TGDconsc] leads to a model for intentional action in which the generation of pairs of positive and negative energy space-time sheets is in a key role. Thus the signature of intentional action would be the generation of gravitational mass. In the chapter [D6] the strange finding that planets seem to be in accelerated motion guaranteing that the reduction of orbit radii compensates for their increase by cosmic expansion. With some tongue in cheek but with an inspiration coming from the self hierarchy predicted by TGD inspired theory of consciousness, the acceleration was explained in terms of intentional action realized as quantum jumps in astrophysical length scales and guaranteing that Sun does not participate to the cosmic expansion. A possible concrete realization of this process could be in terms of creation of pairs of Kähler magnetic flux tubes whose net inertial energy vanishes. Their gravitational energy would gradually increase the gravitational mass of Sun so that also the planetary motion would be accelerated.

Later stages of the evolution

During the later states of evolution strings are transformed to Kähler magnetic flux tubes. It is also possible that new pairs of strings or flux tubes are created so that Universe could quite literally grow gravitational mass. It would seem that the coiling of strings preserved during evolution is the best manner to guarantee Einstein's equations. The space-time sheet at which the strings are condensed can approach to a vacuum extremal so that the parameter ϵ in the above conditions can become arbitrarily small. Note that the transverse area S/S_0 scales as $S/S_0 = 1/\epsilon$ and can become arbitrarily large. For a Kähler magnetic flux tube of thickness $d = 1$ μm this would give $\epsilon \sim 10^{-48}$ so that the surrounding space-time sheet would be extremely near to a vacuum extremal and positive and negative energy antimatter would compensate each other in very high precision.

Also pairs of positive and negative energy magnetic flux with sufficiently weak field strength can be created from vacuum. If the initial transverse area of the Kähler magnetic flux tube is S_i, the Kähler electric field energy per unit length of string would be

$$e_E \quad = \quad \epsilon \frac{S_0}{S_i} \frac{e_S}{2} \quad , \tag{16.3.25}$$

where e_E is the Kähler magnetic energy per unit length for a free cosmic string. This assuming that the gravitational energy is the inertial energy of the initial flux tube containing no fermions nor anti-fermions.

Could large voids collapse?

The predicted weakening of the cosmological constant in the region between the strings might imply that the gravitational attraction eventually wins and the system starts to collapse. An indefinite growth of the gravitational energy of a co-moving volume would require a continual creation of pairs of positive and negative energy fermions from vacuum. As found the assumption that cosmological constant behaves as $\Lambda = 1/a^2$ on the average implies that mass per volume is proportional to the p-adic length $L(k)$ scale characterizing it. This requires a continual generation matter and antimatter, presumably in form of pairs of cosmic strings or magnetic flux tubes, whose density is responsible for the cosmological constant.

Unless pairs of cosmic strings or magnetic flux tubes with zero net energy, the system must eventually start to collapse back and could pulsate between minimal and maximal radii. During the contraction phase the magnetic flux tubes representing cosmic strings would be forced to return back to the cosmic string configuration.

At first this proposal seems to be in conflict with the fact that TGD does not allow imbeddings of cosmologies with over-critical matter density. TGD however allows imbeddings of time reversals of sub-critical sub-cosmologies (such as large voids) and one can imagine of gluing the time reversals together to get a symmetric cosmology. The two halves of the cosmology would also represent evolutions of conscious life in opposite directions.

16.4   Topological condensation around a pair of cosmic strings

In order to understand the possible cosmological role of the topologically condensed cosmic strings, one must study the topological condensation of the ordinary matter around cosmic strings. In the following, the condensation both at elementary particle level and for astrophysical objects will be considered.

16.4.1   The motion of charged particle in the field of a pair of cosmic strings

Although Kähler field differs in character from other gauge fields one can speak about Lorentz Kähler force defined as the sum of em and Z^0 forces which for S_I^2 extremal are related to Kähler field by

$$
\begin{aligned}
\gamma &= 3J - sin^2\theta_W R_{03} \ , \\
Z^0 &= 2R_{03} \ .
\end{aligned}
\tag{16.4.1}
$$

Here R_{03} denotes a component of CP_2 spinor curvature and one has

$$
\begin{aligned}
R_{03} &= 2(2e^0 \wedge e^3 + e^1 \wedge e^2) \ , \\
J &= 2(e^0 \wedge e^3 + e^1 \wedge e^2) \ .
\end{aligned}
\tag{16.4.2}
$$

For two-dimensional CP_2 projection γ and Z^0 are proportional to J:

$$
\gamma \ - \ (3 - \frac{sin^2(\theta_W)}{2} k_Z)J \ , \ Z^0 = k_Z J \ .
\tag{16.4.3}
$$

In particular, when γ vanishes one has $Z^0 = \frac{6}{sin^2(\theta_W)}J$. Thus if particle as em and Z^0 charges q_{em} and q_Z one can call the sum of em and Z^0 forces simply Kähler force and the particle has effective Kähler charge given by

$$
Q = (3 - \frac{sin^2(\theta_W)}{2} k_Z)q_{em} + k_Z q_Z \ .
\tag{16.4.4}
$$

For ordinary particles screening of weak fields means that $k_Z = 0$. For dark matter particles coupling to light variant of electro-weak bosons k_Z is vanishing below the corresponding weak length scale.

In order to develop a more detailed picture about the topological condensation, let us consider next the motion of Kähler charged, massive particle in the field of a cosmic string (or as a matter pair of cosmic strings with opposite energies). In the microscopic description the motion is governed by Lorentz-Kähler force and field equations read

$$
\frac{dU^\alpha}{ds} = \frac{Q}{M} J^\alpha{}_\beta U^\beta \ ,
\tag{16.4.5}
$$

where U^α denotes the four-velocity of particle and Q its Kähler charge.

In the following, only the circular orbits

$$
\begin{aligned}
\rho &= constant \ , \\
\phi &= \omega_0 s \ . \\
z &= \beta s \ ,
\end{aligned}
\tag{16.4.6}
$$

will be considered.

Combining the radial equation of motion with the mass shell condition $U \cdot U = 1$ one obtains the following condition

$$\rho \omega_0^2 \;=\; [K k^2 A \times X - \frac{Q \omega k}{M}] \frac{A}{\rho} \;,$$

$$A \;=\; \sqrt{\frac{1 + \rho^2 \omega_0^2}{1 - K(1 - k^2 X^2)}} \;.$$

$$X \;=\; \ln\left(\frac{\rho}{\rho_0}\right) < 1/k \;. \tag{16.4.7}$$

Using $Kk = \epsilon/2\alpha_K$ and $\omega = 2\sqrt{K}/R$ one can transform the positivity condition for the right hand side to

$$\frac{\epsilon}{2\alpha_K} A \times X - \frac{Q \omega}{M} \geq 0 \;. \tag{16.4.8}$$

For $Q < 0$ the condition is identically satisfied: this obviously reflects matter antimatter asymmetry. For the vanishing Kähler charge the condition can be satisfied in the region $\rho > \rho_0$: one would obtain negative gravitational mass in the region $\rho < \rho_0$ by assuming that the asymptotic value of u_∞ for the spherically symmetric extremal representing topologically condensed system corresponds to the local value of u for the larger space-time sheet. This supports the view that $\rho = \rho_0$ corresponds to size of the cosmic string.

For $Q > 0$ the condition gives

$$\frac{\rho}{\rho_0} \;\geq\; exp\left[\frac{Q\omega}{M} \frac{2\alpha_K}{\epsilon A}\right] \;. \tag{16.4.9}$$

This condition is satisfied only for sufficiently large radii so that $Q > 0$ matter tends to be driven far away from the string. Thus in case of a pair of strings with opposite Kähler charges positive energy matter and negative energy antimatter are separated.

Gravitational and Kähler forces cancel each other for the value of u given by

$$u \;=\; \pm\sqrt{\frac{M^2}{Q^2 \omega^2} - \frac{K}{1 - K}} \;. \tag{16.4.10}$$

Unless the condition $1 < \frac{M^2}{Q^2 \omega^2} - \frac{K}{1-K} < 2$ hold true either gravitational or Kähler force dominates everywhere.

16.4.2 Matter distribution around a pair of cosmic strings

The distribution of stars in the vicinity of cosmic string can be modelled using kinetic model for the evolution of the distribution of stars. Assuming that stars have some average mass M and that the situation is non-relativistic the kinetic equation for the distribution of stars reads

$$\frac{dn}{dt} \;=\; \nabla \cdot (D\nabla n + \bar{w}n) \;. \tag{16.4.11}$$

The second term is the divergence of the current consisting of diffusion term and drift term caused by the Kähler force.

The drift velocity $\bar{w}$ is related to the Kähler force F_K

$$\bar{w} \;=\; b\bar{F}_K \;, \tag{16.4.12}$$

where b is the mobility of the star. Assuming that one can associate a well defined temperature parameter to the star distribution the mobility is related to the diffusion constant D by the Einstein relation $D = bT$. Kähler force is expressible in terms of Kähler gauge potential

$$\bar{F}_K \;=\; \nabla Q\Phi \;. \tag{16.4.13}$$

Here $\Phi = kT_s G\omega ln(\rho/\rho_0)$ is the gauge potential of the Kähler electric field. T_s denotes the string tension:

$$T_s \simeq .22 \times 10^{-6} \times \frac{\epsilon}{G} \;.$$

The lower bound for ϵ is about 10^{-7} from the previous considerations. Q is the average Kähler charge of the star: $Q \simeq \epsilon M\sqrt{G}$,

An order of magnitude estimate for diffusion constant is given by $D \simeq \langle v \rangle / n\sigma$, where $\langle v \rangle = \sqrt{(T/M)}$ is the average thermal velocity of star and σ is the collision cross section for collisions with other stars.

The equilibrium distribution corresponds to the cancellation of diffusion and drift currents

$$\frac{dn}{dr} \;\simeq\; -\frac{M\sqrt{G}\omega}{T}\partial_r \Phi n \;. \tag{16.4.14}$$

In isothermal case one obtains for the distribution of stars the following expression

$$n(\rho) \;=\; n_0 exp(-\frac{M\sqrt{G}\Phi_K\omega}{T}) = n_0(\frac{r}{r_0})^\alpha \;,$$

$$\alpha \;=\; \frac{M\sqrt{G}T_s G\omega}{T} \;, \tag{16.4.15}$$

so that power law behavior results. Unfortunately, concerning the value of the temperature parameter there is nothing interesting to say.

The second alternative is based on the adiabaticity assumption

$$\frac{T}{T_0} \;=\; (\frac{n}{n_0})^{1-\gamma} \;, \tag{16.4.16}$$

where γ denotes adiabatic constant. In this case one obtains

$$n(r) \;=\; n_0(Aln(\frac{r}{r_0}))^{\frac{1}{(1-\gamma)}} \;,$$

$$A \;=\; (1-\gamma)M\sqrt{G}T_s\frac{G}{T_0} \;. \tag{16.4.17}$$

for the distribution of stars.

16.4.3　Quantization of the cosmic recession velocity

The statistical analysis of the observational data about red shift of quasars [ig2] shows that the distribution of emission line red shifts of quasars have a periodicity, which can be explained most nicely by assuming that the recession velocity v calculated from red shift is quantized so that one has, using the standard relation between the recession velocity and distance of the emitting object,

$$v \;=\; H_0(r_0 + nR) \;. \tag{16.4.18}$$

Here H_0 denotes the present value of Hubble constant. The order of magnitude for the parameter R is $R \simeq 10^8$ ly.

There is also a problem of the association between galaxies and quasars. There are indications that galaxies and quasars form correlated pairs but that the red shift of the quasar is much larger than the red shift of the galaxy [ig1]. In case that the two systems are actually different physical systems, this implies that the red shift of the quasar member is of non-cosmological origin.

Various explanations for these effects have been proposed. For example, the idea that Universe is multiply connected has been put forward [ig2]. According to this explanation the emission lines with different red shifts correspond to images of single object: the light emitted from the object can travel several times "around the world" before being detected and the distance to the observe is thus quantized: $r = r_0 + nL$, where L is the size of the non-simply connected Universe. Observations require that L is of the order of $L \simeq 10^8 - 10^9 \; ly$.

The TGD based explanation for the phenomenon is similar in spirit to this explanation (see Fig. 16.4.3). The original model for the phenomenon turned out to be inconsistent with the revised view about cosmic strings. The model however allows an obvious modification.

Original model for the quantization of red shifts

The original model was based on the idea is that null geodesic lines around the topologically condensed "big" strings ("big" means that the parameter $K = \omega^2 R^2/4$ is not too far from unity) do not leave the 3-space surrounding "big" string in the center of large void of radius of order 10^8 ly and carrying strong Kähler electric field cancelling its magnetic action: for the simplest geodesic the projection to the plane orthogonal to the string is just circle. Galaxies tend to be situated near the boundaries of the 3-space surrounding big string and the light emitted from quasar can travel several times around the string before being detected.

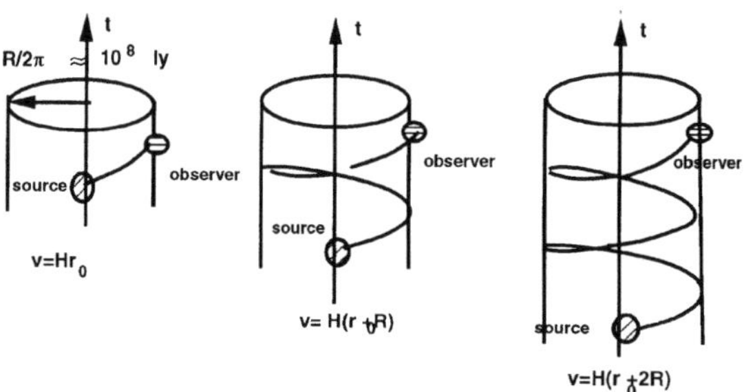

Figure 16.1: Quantization of the cosmic recession velocity.

A simplified situation is obtained, when the distance R of the emitting quasar and observer from the string is same ($R \simeq 10^8 \; ly$) and when the distance along string direction is L. In this case the projection of the light like geodesic on plane is circle and the motion in z-direction is along straight line. The distance travelled by light before its detection is given by the expression

$$ r = \sqrt{L^2 + (r_0 + n2\pi R)^2} \; . \tag{16.4.19} $$

If observer and source are in same plane one obtains the previous formula for the quantized recession velocity. The size of the parameter R, which is fixed by the hypothesis that big void regions correspond to cosmic strings is indeed in accordance with the observational constraints.

It is not at all obvious that the orbit of photon can indeed be confined inside the outer critical radius ρ_+ associated with the string having $\omega R \sim 1$: the Kähler charge cannot obviously be all that matters since photons do not couple to it. For "big" strings however $\omega R \sim 1$ holds true. This is indeed the case: the physical reason is the extremely strong gravitational field caused by the big string. To see this consider the equations of motion for an orbit with circular projection in the plane orthogonal to the string. Orbit is characterized by energy conservation condition, momentum conservation condition in the direction of string, masslessness condition and the equation of motion in radial direction (essentially Kepler law)

$$
\begin{aligned}
\frac{dt}{ds} &= E \ , \\
\frac{dz}{ds} &= p \ , \\
E^2 g_{tt} - p^2 - \rho^2 \omega_0^2 &= 0 \ , \\
\rho \omega_0^2 &= \frac{\partial_\rho g_{tt} E^2}{2} \ ,
\end{aligned}
\tag{16.4.20}
$$

The last equation forces the photon to a circular orbit if some additional consistency conditions are satisfied and obviously requires Kähler charged string. The expression for the time component of the metric is given by

$$
\begin{aligned}
g_{tt} &= 1 - \frac{R^2 \omega^2}{4}(1 - u^2) \ , \\
u &= cos(\Theta) = kln(\frac{\rho}{\rho_0}) \ , \\
k &= \frac{1}{ln(\frac{\rho_+}{\rho_0})} \ .
\end{aligned}
\tag{16.4.21}
$$

Here $u = cos\Theta$ denotes the coordinate variable of the geodesic sphere S^2 as a function of the radial coordinate approaching value $u = -1$ at the boundary of the cylindrical region surrounding big string. These conditions boil down to the following condition fixing the value for the radius ρ of the circular orbit

$$
cos(\Theta) = \frac{1}{\sqrt{K}}\sqrt{\frac{1 - \frac{p^2}{E^2} - K}{1 - \frac{Kk^2}{\omega_0^2 \rho^2}}} \ .
\tag{16.4.22}
$$

This equation has real solutions provided the argument of the square root term is positive. In addition the condition $|cos\Theta| \leq 1$ must hold true.

If the longitudinal momentum of the photon vanishes, one has

$$
cos(\Theta) = \frac{1}{\sqrt{K}}\sqrt{\frac{1 - K}{1 - \frac{Kk^2}{E^2}}} \ .
\tag{16.4.23}
$$

In the approximation $\frac{Kk^2}{E^2} \simeq 0$ this gives the bounds $1/2 < K < 1$. This condition is not consistent with the assumption that $K = R^2 \omega^2 / 4$ is a small parameter given by

$$
K = \frac{\epsilon}{2\alpha_K k} \ .
$$

The small value of K is consistent with $p/E \simeq 1$ so that most of photons momentum is in the direction of string. The result means that the original model based on "big" strings in the center of the large void and explaining the observations must be given up.

Modified model for the quantization of red shifts

The new view about the relationship between inertial and gravitational energy forces to give up the idea that the center of large void could contain "big" string. There are only coiled pairs of "big" strings of opposite inertial energies near the boundaries of the large void having $K << 1$ and creating Kähler fields with much weaker strength.

The modification of the previous model is obvious and much analogous to the topological model for the quantization. If the pairs of strings and the torus like space-time sheets containing them and winding around the boundary of the large void are closed and are able to confine photons inside them and thus acting as cosmic wave guides, the photons from a distant star can rotate several times along these space-time sheets and same quantization of the red shift would result also now.

If the proposed explanation for the quantized red shift is correct, one can in principle observe the time development of single object from quasar to galaxy by a series of images, the time difference between two successive images being of the order of 10^8 ly. These images are observed on the same line of sight, when the light comes from a distant object.

16.5 Cosmic string model for galaxies and other astrophysical objects

The new view about the relationship between gravitational and inertial energy forces to modify the original model based of galaxy based on split cosmic strings. Splitting, although possible, might not be needed since Hawking radiation might replace it as a basic mechanism generating visible matter. By p-adic fractality the mechanism generalizes and provides a universal mechanism for the generation of astrophysical structures and universe can be seen as fractal necklace containing coiled pairs of cosmic strings linked around larger structures of similar kind linked... The mysterious cosmological constant can be identified as the gravitational energy density of cosmic strings and of Kähler magnetic flux tubes.

16.5.1 Creation of pairs of cosmic strings from vacuum as a universal mechanism for the generation of structures

Astronomical observations suggest that galaxies form linear structures
[ib3]. This inspired the original TGD based model of galaxies as decay products of split cosmic strings forming kind of cosmic fire crackers. The required order of magnitude for the string tension was of order $10^{-6}/G$ the same as the string tension of the cosmic strings predicted by TGD (so that CP_2 radius would reflect itself directly in the galactic dynamics!). The model suggested also a solution of galactic dark matter problem since the net mass of a ball containing string is expected to depend linearly on the radius of the ball as indeed found.

One problem of this model was that galactic strings ought be in the plane of the galaxy. The galactic jets which one might expect to be parallel to the strings are however orthogonal to the galactic plane which suggests that visible matter condensed on certain points of a long string roughly orthogonal to the galactic plane.

The new view about the relationship between inertial and gravitational energy and the necessity of cosmological constant forces to modify this scenario.

a) The observation that galaxies are organized in linear structures can be understood if the basic structures are (possibly) coiled pairs of positive and negative energy cosmic strings winding in a spaghetti like manner along the boundaries of large voids. Part of ordinary matter would results as a Hawking radiation from these strings but the very fact these strings are mostly invisible suggests that the matter emitted by them remains in the vicinity of strings. Visible jets orthogonal to the galactic plane usually interpreted in terms of black hole emissions could correspond to the emission of Hawking radiation from these structures. Galaxies are concentrations of visible matter around these strings and they are roughly orthogonal to the plane of galaxy.

b) If the cosmological constant behaves according to the TGD based prediction, the contributions of both Einstein tensor and metric to the gravitational mass depend linearly on the size of the volume as the mass of the dark matter is found to behave. This suggests that dark matter corresponds to the cosmological constant and the challenge is to identify this matter and understand why it does not radiate.

c) The generation of positive and negative energy matter with zero net energy from vacuum does not contribute to the inertial energy but increases gravitational mass. This has occurred already during string dominated critical period during which the density of gravitational mass behaves as $\rho \propto 1/a^2$ as a function of the light cone proper time and the mass per co-moving volume is proportional to a. The fractality of TGD inspired cosmology suggests that the creation pairs of positive and negative energy cosmic strings giving rise to cosmologies inside cosmologies has occurred also later in smaller length scales. In particular, galaxies and even smaller structures could be seen as cosmologies within cosmologies. Pairs of cosmic strings and magnetic flux tubes are not visible and are thus excellent candidates for the dark matter.

If the initial inertial and gravitational mass per unit length of these objects is same as that for a free string, the order of magnitude for the gravitational energy density of dark matter per volume is predicted correctly if the length L of string inside sphere R is proportional to its radius: $L \propto R$. Hence the original model would not be totally wrong. Galaxies could be strongly knotted relatively short cosmic strings linked around the long cosmic strings like pearls in a necklace. Their shortness would mean that they do not contribute significantly to the mass of the void.

d) p-Adic fractality suggests that even smaller astrophysical structures might involve strings linked with larger strings linked with...., the cosmic necklace would be a fractal necklace. In the case of Sun a string of length $L \sim 10^{11}$ m, which is not far from the distance $AU = 1.5 \times 10^{11}$ m between Earth and Sun, would be needed whereas the radius of Sun is $\sim 7 \times 10^{8}$ meters. Thus the magnetic flux tubes resulting form these strings could wind around solar system and bind the entire system into single coherent magnetic structure. For Earth one would have $L \sim 3 \times 10^{5}$ m, which is smaller than the radius $R = 6.4 \times 10^{6}$ m of Earth. What makes this interesting is that quite recently it has been announced that Earth contains a previously unidentified core region with size of 3×10^{5} m [il2]. This picture suggest a universal mechanism for the evolution of the solar system replacing the existing Newtonian model based on the amplification of gravitational perturbations.

16.5.2　Cosmic strings and dark matter problem

Consider now the idea that the presence of cosmic strings might solve the the dark matter puzzle [ia10]. The presence of the dark matter is indicated by the velocity spectrum of the distant stars (at distance of few tens of kilo-parsecs from the center of the galaxy), which according to the recent observations [ib1, ib2] approaches to a constant depending on the galaxy in question and having the general order of magnitude $V \simeq 10^{-3}$.

One can estimate the velocity V of a distant star in galactic plane from Kepler law (the spherically symmetric model for galaxy suggests that this argument indeed applies)

$$\frac{V^2}{R} \;=\; \frac{GM(R)}{R^2} \;, \tag{16.5.1}$$

where $M(R)$ denotes the mass inside a sphere of radius R. Since the mass of the cosmic string dominates the mass inside a sphere of radius R one gets the following very rough estimate for the effective gravitational mass inside the sphere of radius R

$$M(R) \;\simeq\; n2TR \;, \tag{16.5.2}$$

where $n > 1$ accounts for the fact that straight string is not in question. From the known velocity V one obtains for the string tension the estimate

$$T \;\sim\; \frac{V^2}{4nG} \sim \frac{10^{-6}}{4nG} \sim v_D T_{free} \;. \tag{16.5.3}$$

This estimate is of the same order of magnitude as the lower bound of string tension obtained from the Jeans criterion. The result is also consistent with the assumption that, due to their gravitational binding to strings, stars rotate with the same velocity as strings.

Recall that the string tension of the TGD cosmic string is given by

$$T = \frac{1}{8\alpha_K R^2} \simeq .22 \times 10^{-6} \frac{1}{G} \;.$$

This is somewhat smaller than the required tension for $n = 1$. For a pair of phase conjugate strings the gravitational tension is twice this value. The effective string tension of the co-moving string also increases for $r \to r_0$ (see the general description of cosmic string solution) and diverges at $r = 0$. Furthermore, since the cosmic string is not straight there appears additional factor n making $M(R)$ larger than the simple estimate above.

On basis of these observations one has a strong temptation to think that the still existing cosmic strings, possibly thickened to magnetic flux tubes, correspond to galactic and extragalactic dark matter. At this stage one must leave open whether the naive argument leads to a correct form for the velocity spectrum of stars. Whether or not true this prediction would have nice features in that it would relate the velocity spectrum directly to the size and age of the galaxy since the velocity v determines the recent size of the visible galaxy (if it corresponds to the recent distance of the string end from the center of galaxy): the older the galaxy with given size the smaller the rotational velocity v. Elliptic galaxies are older than spiral galaxies: rotational velocities for the elliptic galaxies are indeed smaller than for spiral galaxies [ib2]. Furthermore, the rotational velocities increase with the size of the galaxy, when the age of the galaxy is kept constant: also this feature is in qualitative accordance with observed facts [ib1, ib2].

An interesting question is whether one could explain the angular momentum of galaxies in terms of the tidal forces acting between the galaxies [ib7] at the opposite ends of a string (having length of order 10^5 light years. The idea is following. For free cosmic string there is a flux of angular momentum of order Tar^2 (using Robertson-Walker coordinates (a, r))) through the end of the string, which produces a correct order of magnitude for the galactic angular momentum at time a given by $J \sim Ta^2r^2 = Tr_M^2$, $r_M \sim 10^5 \ ly$.

16.5.3　Estimate for the velocity parameters

The first task is to fix the value of the velocity parameter, to be denoted by V, appearing in the general solution describing one arm of the split cosmic string. In the region, where linearized equations of motion hold the orbital velocity V of the cosmic string is constant.

The radius of the singular region associated with cosmic string increases with some velocity v_D identifiable as the velocity with which the size of a typical galaxy (defined for example as the distance of spiral arm L from the center of galaxy) is about $L \simeq 10^4 - 10^5$ light years [ia7, ia8]. The condition $vT < L$, where $T \simeq 10^9 - 10^{10}$ years is the typical age of the galaxy, gives the estimate

$$v_D \quad < \quad 10^{-5} \ , \tag{16.5.4}$$

for the velocity v_D using the velocity of light as unit.

One can relate the velocity v_D to the string tension if one accepts the assumption that the relative motion of the string ends results from the shortening of strings, which in turn results from the decay of the string ends to elementary particles (some of them possibly exotics). A rough estimate for the velocity of the shortening of the string [ib4] is based on the observation that the velocity

$$v \quad \simeq \quad TG \sim 10^{-6} \tag{16.5.5}$$

seems to set the time scale for the various dynamical processes leading to the decay of strings [ib4]: for example, the shortening of loop with radius L via gravitational radiation as well as the shortening of the string connecting the monopole pair takes place with this velocity [ib4]. This velocity is considerably smaller than the typical velocity $V \simeq 10^{-3}$ [ib1, ib2] of the distant stars moving in the galactic plane, which in turn can be understood using Kepler law.

The idea that the spiral arms of the spiral galaxy correspond to cosmic strings seems to be in accordance with the observational facts. In case of Milky Way [ia8] the distance of spiral arms is about $L = 10^4 - 10^5$ light years from the center of the galaxy so that the order of magnitude for the velocity v_D is $v_D \sim 10^{-6} - 10^{-5}$. Furthermore, spiral arms are known to recede from the center of the Milky Way [ia8].

The model suggests also an explanation for the observed bar like structure connecting the ends of the spiral arms of the spiral galaxies. The gravitational field is most intense near the string end so that the density of the ordinary matter is expected to be largest near the end of the string. On the other hand, the orbit of the string end is straight line so that "bar" like structure might be formed [ia7], when the the ends of the spiral arms recede from each other.

It should be stressed that the visible form of galaxies is not so closely related with the form of strings contrary to the original expectations (we used the term "spiral string"). This is clear from the observation that the total change of angle ϕ is smaller than $\pi/2$, which means that strings are really not "spiral" like. Of course, this result holds for free strings and it might be that condensation in fact creates spiral structure somehow. A more conventional explanation is the generation of density waves with spiral structure [ia9] and the presence of strings might have something to do with this phenomenon.

16.5.4 Galaxies as split cosmic strings?

It is not clear whether the Hawking radiation from a coiled pair of cosmic strings is able to explain galactic visible matter. The reason is that the cosmic strings responsible for linear structures formed by galaxies are not visible along their entire length. One might argue that same applies also the knotted and linked galactic cosmic string pairs. If this is the case, the dark matter problem becomes visible matter problem. A possible solution of the problem is based on split cosmic strings with splitting possibly resulting in the collision of galactic strings with the long supra-galactic strings.

This scenario has indeed some attractive features (see Fig. 16.5.4).

a) The ends of the split cosmic string create strong gravitational fields and serve as seeds for the galaxy formation. Lense effect [ib4] is predicted to be a signature of the string pairs. The fact that spiral galaxies have in general two arms, has a nice topological explanation.

b) One ends up to a rather simple scenario for the evolution of the galaxy.
i) The splitting occurs most probably during the string dominated phase for $t < L \sim 10^4 \sqrt{G}$ (L is essentially CP_2 radius) and results most naturally from the collision of two strings.
ii) The split strings begin to decay by emitting particles from their ends. The decay leads to a shortening of the split strings with constant velocity v so that the ends of the split strings recede from each other. This velocity can be identified with the velocity parameter $v \sim TG$ associated with the motion of the spiral arms. A correct size for the visible part of the galaxy is predicted.
iii) Decaying cosmic string ends provide a model for the 'central engines' associated with the galactic nuclei [ia13]. The energy production by string decay turns out to be of same order of magnitude as the energy production in quasars assuming that the energy is produced in a narrow jet parallel to the string (momentum conservation favors this option). This was proposed as an explanation for the visible jets associated with the active galaxies as resulting from the interaction of the decay products with the ordinary matter. The fact that these jets are orthogonal to the galactic plane suggests Hawking radiation from supra-galactic string stimulated by the collision as an alternative explanation.
iv) Co-moving cosmic strings happen to rotate with the same velocity as distant stars (relative to the center of galaxy) are found to rotate. The gravitational binding of stars by the average gravitational field created by cosmic strings would explain the rotational velocity spectrum.

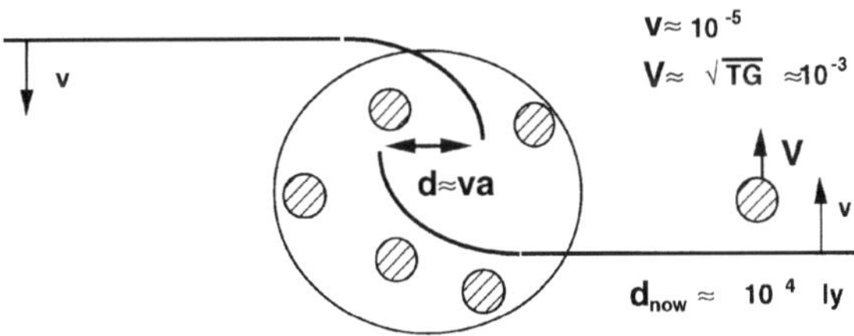

Figure 16.2: String model for galaxies.

In the following the model will be discussed in more detail to see whether it really works. The value for the velocity parameter v will be derived, Jeans criterion for the formation of the structures around a split cosmic string will be discussed, a simple toy model for a galaxy using spherically symmetric mass distribution will be constructed and the possibility that cosmic strings might provide a solution to the galactic dark matter problem will be studied.

Jeans criterion for the galaxy formation

It is not obvious that Jeans criterion for the generation of structures by gravitational interaction can be applied to galaxy formation in the recent situation differing so dramatically from Newtonian framework. One can however check what Jeans criterion would give in the case of split cosmic strings [ib4].

a) The size L of the density fluctuation leading to the formation of a structure satisfies the inequality

$$l_J \quad < \quad L < l_H \ , \tag{16.5.6}$$

where the Jeans length l_J is given by [ib4]

$$l_J \quad \simeq \quad 10 v_s t \ , \tag{16.5.7}$$

where v_s denotes the velocity of sound. Notice that the formation of structures is not possible at the radiation dominated era since Jeans length is larger than horizon: $l_H \simeq t < l_J \simeq 10t$ since the velocity of sound is of order 1.

b) When radiation and matter decouple from each other (corresponding to the value of about $a_{dec} = 10^8$ light years [ia2]), the formation of galaxies becomes possible due to the lowering of the pressure, which leads also to the lowering of the sound velocity v_s from $v_s \simeq 1$ to $v_s \simeq 10^{-5}$ (thermal velocity of hydrogen). Jeans length shortens by a factor 10^{-5} and the formation of structures becomes possible.

In accordance with the idea that the split strings act as seeds for the galaxy formation, one can identify Jeans length as the minimal distance between the ends of the split string, which leads to a formation of galaxy

$$v_D a_{dec} \quad > \quad l_J \ . \tag{16.5.8}$$

Using the values for a_{dec} and l_J one obtains lower bounds for the velocity v_D between the ends of the galactic string and for the string tension of the galactic strings (accepting the proposed relationship between v_D and string tension)

$$v_D \quad > \quad 10^{-6} \ ,$$
$$T \quad > \quad \frac{10^{-6}}{G} \ . \tag{16.5.9}$$

One obtains also a lower bound for the recent size L_{now} of the galactic nuclei assuming that the decay of galactic strings continues with velocity v_D

$$L_{now} \quad > \quad 10^4 \ ly \ . \tag{16.5.10}$$

These numbers are in accordance with the estimate obtained for the string tension of a typical galactic strings and with what is known about recent sizes of the galaxies [ia7].

Spherically symmetric model

The imbeddability requirement plays central role in TGD inspired cosmology and the galaxy model based on spherically symmetric mass ($M(r) = kr$) distribution is of some interest. This model could be regarded as a large length scale idealization of galaxy mass distribution. In case that galactic dark matter consists of the exotic decay products of the cosmic string the model might be even reasonably realistic. The line element for an energy momentum tensor characterized by energy density $\rho(r)$ and pressure $p(r)$ is given by the expression $ds^2 = A(r)dt^2 - B(r)dr^2 - r^2 d\Omega^2$ and to find an imbedding for this metric one can use the general imbedding ansatz introduced, when discussing the imbedding of Reissner- Nordström metric.

Under rather general assumptions about the mass density the time component of the metric for a spherically symmetric mass distribution $M(r)$ (the mass inside the sphere of radius r) is given by the expression $g_{tt} = 1 - 2GM(r)/r$. In present case one would obtain $g_{tt} = constant$ so that some of the underlying assumptions must fail. The following form leads to a correct gravitational force

$$g_{tt} \;\; = \;\; 1 + 2Gk ln(\frac{r}{r_0}) \; . \tag{16.5.11}$$

The gravitational force in the Newtonian limit is $2Gk/r = 2GM(r)/r^2$ and implies that Kepler law to be used later to derive velocity distribution of distant stars is indeed applicable.

The general expression for the metric component g_{tt} in terms of the imbedding ($m^0 = \lambda t, \Theta = \Theta(r), \Phi = \omega t + f(r)$)

$$g_{tt} \;\; = \;\; \lambda^2 - R^2 \omega^2 sin^2(\Theta) \; , \tag{16.5.12}$$

which gives

$$sin^2(\Theta) \;\; = \;\; \lambda^2 - 1 - \frac{2Gk}{R^2\omega^2} ln(\frac{r}{r_0}) \; . \tag{16.5.13}$$

Imbedding fails for two critical radii r_{in} $(sin^2|(\Theta) = 1)$ and r_{out} $(sin^2(\Theta) = 0)$

$$
\begin{aligned}
ln(\frac{r_{in}}{r_0}) \;\; &= \;\; \frac{(\lambda^2 - 1 - R^2\omega^2)}{2Gk} \; , \\
ln(\frac{r_{out}}{r_0}) \;\; &= \;\; \frac{(\lambda^2 - 1)}{2Gk} \; .
\end{aligned}
\tag{16.5.14}
$$

An interesting question is whether one could relate the inner critical radii to the existence of the galactic nucleus having diameter of the order of 2 parsecs (.65 light years).

16.5.5 Cylindrically symmetric model for the galactic dark matter

TGD allows also a model of the dark matter based on cylindrical symmetry. In this case the dark matter would correspond to the mass of a cosmic string orthogonal to the galactic plane and traversing through the galactic nucleus. The string tension would the one predicted by TGD. In the directions orthogonal to the plane of galaxy the motion would be free motion so that the orbits would be helical, and this should make it possible to test the model. In this kind of situation general theory of relativity would predict only an angle deficit giving rise to a lens effect. TGD predicts a Newtonian $1/\rho$ potential in a good approximation.

Spiral galaxies are accompanied by jets orthogonal to the galactic plane and a good guess is that they are associated with the cosmic strings. The two models need not exclude each other. The vision about astrophysical structures as pearls of a fractal necklace would suggest that the visible matter has resulted in the decay of cosmic strings originally linked around the cosmic string going through the galactic plane and creating $M(R) \propto R$ for the density of the visible matter in the galactic bulge. The finding that galaxies are organized along linear structures [ib3] fits nicely with this picture.

16.6 Cosmic strings and energy production in quasars

One of the basic mysteries of astrophysics are so called 'central engines' in the centers of the galaxies [ia13]. These engines are very massive, have very small size of at most few light hours, their luminosity fluctuates in hour time scale, their electromagnetic spectrum is non-thermal and they are often accompanied by two jets in opposite directions. One should also understand why some galaxies are active (have a pair of jets) and others not. A mysterious property of jets is their microstructure: main jets with length of order 10^6 light years are accompanied by short jets with length of order one light year and with directions parallel to the long jets.

In the standard model the central engine is a galactic black hole but the mechanism of the jet production is not well understood. In the following it is shown that decaying cosmic string ends provide a good candidate for the central engine. Note that in the standard picture jets are orthogonal to galactic plane whereas in the proposed model jets are parallel to the galactic plane. One could consider also the possibility that galaxies are formed in the splitting of cosmic strings orthogonal to galactic plane but this option will not be discussed here.

16.6.1   Basic properties of the decaying cosmic strings

The rate for the shortening of a split galactic cosmic string can be deduced by an order of magnitude argument

$$
\begin{aligned}
v \ &\sim \ kTG \ , \\
T \ &\simeq \ \frac{2 \times 10^{-7}}{G} \ .
\end{aligned}
\tag{16.6.1}
$$

T is the string tension of the cosmic string. k is some numerical constant not too far from unity. The numerical study of the *ordinary* cosmic strings [ib4] gives support for this order of magnitude estimate.

Taking the age of the Universe to be $a \sim 10^{11}$ years and assuming that the cosmic string is split in early phase of cosmology, the length of the portion of the decayed string is of the order

$$
L \ \sim \ kTGa \simeq 2 \times 10^4 k \ \text{light years} \ ,
\tag{16.6.2}
$$

which is of the same order of magnitude as the typical size of the visible part of the galaxy.

An estimate for the rate of the energy production by single cosmic string is given by

$$
P \ \sim \ Tv = kT^2G \sim \frac{4 \times 10^{-14}k}{G} \sim 10^{47}k \times m(proton)/sec \ .
\tag{16.6.3}
$$

The energy production in quasars is roughly 10^{14} times larger than the energy production in Sun, which is about 10^{25} W: this gives $P \sim 10^{49}m_p/sec$. In order to have same order of magnitude one should have

$$
k \ \sim \ 25 \ .
\tag{16.6.4}
$$

The required value of k looks suspiciously large and suggests that the energy flux from the decaying cosmic string could well be a jet directed to a narrow cone, which would increase the observed effective energy flux.

16.6.2   Decaying cosmic string ends as a central engine

It seems that the decaying cosmic string could explain elegantly the basic properties of the central engines. There are two alternative scenarios to be considered.

I) Galaxies are formed around the ends created in the splitting of a very long cosmic string.

II) Galaxies are formed by a decay of a piece of cosmic string. The decay of a finite piece of cosmic might explain the existence of some stellar objects accompanied by jet like structures.

In both cases the rate of the string decay gives a correct upper bound for the recent size of the visible part of the galaxies. Consider now the explanation of basic characteristics of active galaxies.

a) Visible jets are created by the energy beams

The rate of the energy production in the decay of a cosmic string is few per cent about the estimated energy production in quasars assuming spherical symmetry. A correct rate for the observed energy flux from quasars is obtained if the energy from the decay of the string is liberated in a jet. Since two string ends are involved, the visible two-jet structure is an automatic consequence. The jets emerging from the active galactic nuclei are created by the interaction of the primary jets with the ordinary matter.

b) Quasars

Quasars differ from the ordinary galaxies only in that the energy jet from the cosmic string decay meets Earth. This explains the non-thermal nature of the spectrum and the absence of the atomic lines for the most intensive quasars (they are masked by the primary radiation). The rapid variations (a time scale of an hour) in the luminosity can be understood as resulting from the motion of the cosmic string inducing changes in the direction of the jet. Also the similarity between active and inactive galaxies is an automatic consequence.

c) Active-inactive distinction

For the option I possible explanation is that the galactic black hole has absorbed all matter around the galaxy and the jets coming from the decay of the cosmic strings have nothing with which to interact. It could however happen that the two jets interact with matter in very distant regions creating two tightly correlated jets but apparently originating from very distant sources. It could also occur that string ends are inside a galactic black hole for inactive galaxies so that the decay products remain inside the black hole and no visible jets are created. For the option II inactive galaxies without any jets, one can also consider the possibility that the piece of cosmic string has already decayed completely.

d) Dark matter halo

There are two alternative explanations for the the velocity spectrum of the distant stars around the galaxy. The first, purely TGD based, explanation is that distant stars are gravitationally bound to the rotating cosmic string. Cosmic string indeed rotates with a correct velocity and, being Kähler charged, creates a genuine gravitational field unlike neutral cosmic string. The standard explanation is based on the assumption that galaxy is surrounded by a dark matter halo.

An interesting possibility is that a halo of dark matter could result from the decay of the cosmic strings, perhaps in the form of ordinary and exotic neutrino like matter predicted by TGD. The decay could produce also part of the visible matter around the galactic nucleus. The jet model suggests that most of the decay products of the cosmic string escape the visible region of the galaxy but massive and Kähler charged particles with a proper sign of charge could remain bound to the cosmic string. Dark variants of ordinary elementary particles, in particular dark neutrinos, suffer classical Z^0 force below appropriate p-adic length scale. Clearly, Kähler force favors the generation of matter antimatter asymmetry. The average density in the halo would however be perhaps too small to explain the velocity spectrum.

e) Production mechanism for ultrahigh energy cosmic rays

The decay of the cosmic string should also give rise to ultrahigh energy cosmic rays. This production mechanism would provide an alternative for the production mechanisms based on the acceleration of the charged particles [it is difficult to conceive how any acceleration mechanism could lead to the generation of ultra high energy cosmic rays].

16.6.3   How to understand the micro-jet structure?

The long jets with length of roughly 10^6 light years have microstructure consisting of micro-jets with length of order one light year. This feature could be regarded as a shortcoming of the model. A possible TGD based explanation is based on lense effect on the gravitational field of the split cosmic string (scenario I).

In option I, a lense effect, caused by the strong gravitational field of the cosmic string itself, and creating multiple images could be involved. Since charged cosmic string is in question, the situation is more complicated than for the ordinary cosmic string. For instance, photon could rotate several times around the cosmic string before leaving the galactic region. The disappearance of the effect in distant regions (of length of order light year) could be understood if the energy jet were on the wrong side of the string at large distances or the distance between the jet and cosmic string would become so large that photons would not anymore circulate around the string.

16.6.4   Gamma-ray bursts and cosmic strings

Gamma ray bursters [ij1] are now quite generally believed to have a cosmological origin. The energy flux from the gamma ray bursters (assuming spherical symmetry and cosmological origin and distance of order 10^8 ly) is about 10^{16} times the energy flux from Sun and by a factor of 10^2 larger than the total energy flux from the decaying cosmic string. The order of magnitude is same as for the energy flux of quasars. Typically the energy is produced in pulses lasting for a few seconds but also long lasting bursts consisting of a train of smaller pulses with a duration of order second are detected. It seems that the system emitting pulses is in some sense near criticality. The distribution of the gamma ray bursters is isotropic.

An interesting possibility is that decaying cosmic strings might explain also this phenomenon. The string would produce a continuous stream of energy, which fails slightly to meet Earth. Small perturbations causing the string end to oscillate (random oscillation of the direction of a flicker is a good analogy) imply that the beam of energy can meet the Earth at each period of oscillation and cause a sequence of pulses. A unique maximum intensity is predicted.

The shape of the pulse is predicted to reflect only the time development of the direction of the cosmic string rather than the actual intensity distribution of the pulse and this should make it possible to distinguish between TGD based and other explanations for the bursts. For instance, the typical bi-modality of the pulse could reflect directly to a perturbation taking string direction from the equilibrium position and bringing it back. The asymmetry of this perturbation caused by dissipative effects should explain the asymmetry of the two intensity peaks. The observed hardness-brightness correlation could be understood as following from the cosmic red shift and cosmic time dilatation increasing the observed duration of the pulse.

From the estimate that there are

$$\frac{dN}{dt} \sim 10^{-6} \; year^{-1} \; galaxy^{-1}$$

bursts per galaxy per year and taking the average duration t_P of the pulse to be

$$t_P \sim 1 \; sec \; ,$$

one obtains a *very* rough estimate for the the probability that a given galaxy acts as a gamma ray burster at a given moment as

$$P \sim t_P \times \frac{dN}{dt} \sim 10^{-13}.$$

One can estimate the solid angle Ω of the cone to which the energy of the decaying cosmic string is emitted: the probability P for galaxy being a burster, is simply the product of the probability $p(A)$ that galaxy is active multiplied with the probability $\Omega/(4\pi)$ that Earth happens to be in the solid angle Omega

$$P = \frac{p(A)\Omega}{4\pi} \sim 10^{-13} \; ,$$

which gives

$$\Omega \sim \frac{4\pi P}{p(A)} \sim \frac{10^{-12}}{p(A)} \; .$$

To proceed further an estimate for the probability of being active galaxy is needed. The value of Ω had better to be rather small since the oscillations in the direction of the cosmic string leading to fluctuations in the intensity of beam must be of the order of Ω and too large fluctuations are not expected (cosmic string is quite a heavy object!).

16.7 Light dark matter particles, gamma ray bursts, and supernovae

Both the model for dark matter identified as cosmic strings or their decay products and the model for gamma ray bursts identified as beams resulting in the fire cracker like decay of cosmic strings were constructed more than decade ago. During year 2003 came several astonishing observations, which at first seemed to be in a dramatic conflict with both the model of the dark matter and the model of gamma ray bursts.

It however turned out that these findings allow to relate, modify, and generalize as many as five models sketched at that time as the first applications of TGD. The subjects modelled were following:
i) The final state of a rotating star predicting flux tube like magnetic field along the symmetry axis [D3],
ii) Dark matter identified as cosmic strings or their decay products,
iii) Sunspots identified as the throats of magnetic flux tubes feeding magnetic flux to larger space-time sheet and behaving effectively as magnetic monopoles [D6]),
iv) Gamma ray bursts explained as cosmic firecrackers resulting from the decay of split cosmic strings to elementary particles,
v) The anomalous e^+e^- pairs produced in the collisions of heavy nuclei at energy near the Coulomb wall as decay products of lepto-pions consisting of color excited leptons [F7].

16.7.1   Correlations between gamma ray bursts and supernovae

The established correlation between gamma ray bursts and supernovae is certainly the cosmological discovery of the year 2003 [ij6, ij7].

a) The first indications for supernova gamma ray burst connection came 1998 when a supernova was seen few days after the gamma ray burst in the same region of sky. In this case the intensity of the burst was however by four orders of magnitude weaker than for the typical gamma ray bursts so that the idea about the correlation was not taken seriously. On 29 March, observers recorded a burst christened as GRB030329. On 6 April, theorists at the Technion Institute of Technology in Israel and CERN in Geneva predicted that there would be signs of a supernova in the visible light and infrared spectra on 8 April [ij6]. On cue, two days later, observers picked up the telltale spectrum of a type Ic supernova in the same region of sky, triggered as the collapsing star lost hydrogen from its surface. It has now become clear that a large class of gamma ray bursts correlate with supernovae of type Ib and Ic [ij8], and that they could thus be powered by the mere core collapse leading to supernova. Recall that supernovae of type II involve hydrogen lines unlike those of type I. Supernovae of type Ib shows Helium lines, and Ic shows neither hydrogen nor helium but intermediate mass elements instead. Supernovae of type Ib and Ic are thought to result as core collapse of massive stars.

b) One of the most enigmatic findings were the "mystery spots" accompanying supernova SN1987A at a distance of few light weeks at the symmetry axis at opposite sides of the supernova [ib6] Their luminosity was nearly 5 per cent of the maximal one. SN1987A was also accompanied by an expanding axi-symmetric remnant surrounded by three concentric rings.

c) The latest finding [ij5] is that the radiation associated with the gamma ray bursts is maximally polarized. The polarization degree is the incredible 80 ± 20 per cent, which tells that it must be generated in an extremely strong magnetic field rather than in a simple explosion. The magnetic field must have a strong component parallel to the eye sight direction.

Do topologically condensed cosmic strings become co-moving magnetic flux tubes serving as seeds for the formation of stars and galaxies

According to the model for the formation of stars and galaxies proposed already fifteen years ago, topologically condensed pieces of cosmic strings perhaps resulting in the collision of long possibly knotted cosmic strings would serve as seeds making possible formation of lumps of matter forming later stars. The assumption that the pieces of cosmic strings result in the collision of cosmic strings leading to the splitting of them to pieces with some fractal length distribution perhaps concentrated around p-adic length scales would explain why the mass $M(R)$ of galactic dark matter inside a sphere of radius R is proportional to the radius: $M(R) \propto R$.

1. Topologically condensed cosmic strings as co-stretching magnetic flux tubes

I considered already 15 years ago a model for topological condensation of cosmic strings assuming that strong radial Kähler electric fields are generated to compensate the large positive magnetic action. Cosmic strings are actually a special case of magnetic flux tube solutions of field equations. This leads to a revised vision for what happens for topologically condensed cosmic strings. This model does not exclude the presence of the radial electric fields due to the charging of the cosmic strings.

Cosmic strings, which are in the ideal situation string like objects of type $X^2 \times Y^2$, X^2 string like object in M_+^4 and Y^2 geodesic sphere of CP_2 or a piece of it, generate an M_+^4 projection which increases in thickness so that the solution becomes increasingly thicker magnetic flux tube. In the topological condensation the open ends of the string disappear and thus no decay to elementary particles can occur. Thus the topological condensation would stabilize the cosmic strings against decay.

a) The simplest assumption is that the topologically condensed piece of a magnetic flux tube of finite length co-stretches with the expanding universe so that its length increases as $L \propto a$, a light cone proper time.

b) The requirement that magnetic flux is conserved and quantized implies $B \propto 1/S$, S the transverse area of the flux tube. The condition that magnetic energy is conserved, implies $S \propto L \propto a$ and $B \propto 1/a$. This of course applies both to the magnetic and Z^0 magnetic flux tubes.

The assumption that topologically condensed pieces of cosmic strings remain co-stretching forever is questionable, and it might be that when the thickness of the flux tube reaches a critical value corresponding to a Compton length of say pion or lepto-pion, expansion stops, and the flux tube freezes to a very long hadronic or lepto-hadronic (color) magnetic flux tube (a Kähler field giving rise to em or Z^0 field gives also rise to a classical color field).

"Wormhole magnetic fields" consist of pairs of magnetic flux tubes represented by space-time sheets with opposite time orientations and thus having opposite energies. These structures have zero energy and I have proposed that they play a key role in the physics of living matter. In particular, they could be generated by intentional action by first generating a p-adic variant of the wormhole magnetic field representing the intention to generate wormhole magnetic field, and then transforming it to its real counterpart in quantum jump. One cannot exclude the possibility that cosmic strings could also be generated as zero energy pairs of cosmic strings with opposite time orientation. This would make possible to intentionally create universe from nothing. This is actually the only possibility if one poses the boundary condition that no quantum numbers flow out of the future light cone at its boundary.

2. Stars and galaxies as gravitational condensates around fragments of cosmic strings

The gravitational condensation of matter around short parallel flux tubes topologically condensed at larger space-time sheets is a natural mechanism for generating structures like galaxies and stars. The pieces of magnetic flux tubes would form expanding ferro-magnet like structure in the self-consistent magnetic field defined by the by the return flux flowing at the space-time sheet at which strings have suffered topological condensation. The contribution of the magnetic flux tubes to the total mass of the star can be small and the ordinary matter can be seen as decay products of cosmic strings as in the earlier model. Similar mechanism with different initial length of topologically condensed cosmic strings and resulting in fragmentation in the collision of say two long cosmic strings could give rise to the birth of galactic nuclei.

According to the TGD based model of primordial critical cosmology, the transition from string dominated to radiation dominated cosmology should have occurred at $a_0 \sim 10^{-10}$ s, and one could argue that the topological condensation of the magnetic flux tubes should have started at this time. With this assumption the recent thickness of the magnetic flux tubes would be $d = (a/a_0)^{1/2} \times 10^4 \sqrt{G} \sim 10^{-16}$ m for $a \sim 10^{11}$ years. This corresponds to a hadronic length scale. Quite generally, this would suggest that at light cone proper time a the fragments of long cosmic strings, which have survived the decay to elementary particles, have typical length $L \sim a$.

From the recent length of about light month associated with super nova SN1987A (identifying the mysterious light spots as ends of the flux tube), one can deduce that the length L_0 of the cosmic strings at a_0 would have been $L_0 \simeq 10^{-14}$ m, roughly the Compton length of pion. The corresponding magnetic field would be about 10^{16} Tesla and extremely strong. Fields of similar magnitude have been proposed to result in the core collapse of supernovae [ij4]. It however seems that the flux tubes of the primordial magnetic fields cannot explain the highly polarized synchrotron radiation but that the temporary extremely strong Z^0 magnetic field induced by the core collapse are responsible for the polarization.

Magnetic and Z^0 magnetic flux tubes as templates for the formation of material structures is an idea borrowed from TGD inspired theory of consciousness and of bio-systems as macroscopic quantum systems [TGDconsc]. The TGD based quantum model for bio-matter assumes that the magnetic flux tubes of Earth serve as templates for the formation of bio-matter, and also define what I have called magnetic bodies controlling pre-biotic and biotic evolution [N4]. Also the idea that magnetic flux tubes act as wave guides and make precisely targeted communications possible originates from TGD inspired theory of consciousness [K1]. Thus magnetic flux tube structures could serve as templates for and even guide the evolution of matter in all length and time scales: this is certainly in spirit with the fractality of TGD Universe.

A mechanism producing gamma ray burst and polarized synchrotron radiation

The dynamo model for the final state of a rotating star leads to a model for gamma ray bursts consistent with ultrahigh polarization of the synchrotron radiation. The model is consistent with the standard model for the radiation beams from neutron stars.

1. Generalizing the dynamo model for the final state of rotating star

TGD based dynamo model for the final state of rotating star predicts that the rotation axis star contains extremely strong magnetic or Z^0 magnetic field. The field along the axis can also be helical and B_ϕ would naturally result from the rotation of the matter. While attempting to interpret the dynamo model I proposed that the axial field might somehow relate to a cosmic string. This might be indeed the case.

What I did not realize 15 years ago that many-sheeted space-time allows both magnetic and Z^0 magnetic dynamo fields and their symmetry axes of the fields need not coincide.

a) The atomic nuclei of even ordinary condensed matter can carry anomalous weak charges due to the presence of color bonds between nucleons having at their ends exotic quarks with mass of order electron mass and carrying also weak charges [F8, F9]. If some color bonds become charged they have also net weak charges. The Z^0 repulsion due to the weak bosons with Compton length of order atomic radius can explain the low compressibility of condensed matter and give rise to the repulsive term in van der Waals equation. Weak repulsion due to exotic weak bosons is expected to become important in the extremely dense phase of matter inside star.

b) There are good justifications for the assumption that Z^0 magnetic axis is parallel to the rotation axes- Z^0 magnetic field having neutron number as its source receives a large varying contribution dictated by the flow dynamics of the star. Hence Z^0 magnetic field is expected to be very strong, at least in the situations in which currents of different dark matter particle species do not cancel each other. In particular, the ejection of dark neutrinos during the formation of supernova is expected to generate a strong Z^0 charge due to the anomalous Z^0 charges of nuclei. This induces both Z^0 electric field and Z^0 magnetic fields. Since rotation and Z^0 magnetic fields are so strongly coupled, the Z^0 magnetic and rotation axes should coincide.

c) The fact that the rotation axis of the star is rather stable is consistent with the primordial origin of the Z^0 magnetic field and suggests that Z^0 magnetic field as the primordial cause of the rotation.

c) Magnetic axis need not coincide with the rotation axis. The direction of the magnetic field of the star can be reversed (this is happening just now in case of Sun). This suggests that magnetic field does not have primordial origin and reflects the dynamics of the star.

d) TGD based variant for charged particle currents frozen to the magnetic field lines (assumed to have infinity conductivity in magnetohydrodynamics) are non-dissipative supra currents flowing along magnetic flux tubes of the magnetic and Z^0 magnetic fields. These currents in turn generate magnetic and/or Z^0 magnetic fields with field lines circulating around the rotation axes and thus make the magnetic field along symmetry axis helical.

e) Both in the case of magnetic or Z^0 magnetic field, the charged particles topologically condensed at the super-conducting flux tubes could be also spin polarized and amplify the field further.

In many-sheeted space-time topologically condensed magnetic flux tubes must feed their fluxes to larger space-time sheets so that a many-sheeted variant of the dipole field would result. The return fluxes would flow at larger space-time sheet and correspond to thicker flux tubes with weaker intensity of the magnetic flux. The regions, where the flux would be transferred between space-time sheets could correspond to join along boundaries bonds or wormhole contacts. In the latter case they would look like magnetic charges. As the in case of the sunspots, a fractal structure containing flux tubes inside flux tubes is expected [D6].

The mysterious light spots associated with SN1987A [ib6] could correspond to join along boundaries bonds or the throats of the magnetic flux tubes of or primordial Z^0 magnetic flux tubes.

3. Synchrotron radiation in strong Z^0 magnetic field as a mechanism generating strong polarization

Usually the degree of polarization for the radiation from supernovae is around few per cent [ij9].

The polarization associated with gamma ray burst GRB021206 is however incredibly high 80 ± 20 per cent and maximal polarization of the radiation [ij5]. This requires extremely strong Z^0 magnetic field. The helical Z^0 magnetic field along the rotation axis can have flux quanta of astrophysical size and is ideal for accelerating dark charges flowing along the rotation axis and for producing dark photon synchrotron radiation leaking out in the direction of the rotating magnetic axis and transforming to ordinary photons by a mechanism analogous to decoherence of laser beams [F9, J6]. Gamma ray bursts could be seen as a particular case of this radiation resulting when an especially strong dark current (say dark electron current) flows along the rotational axis in an exceptionally strong dynamically generated Z^0 magnetic field, and induces a beam of synchrotron radiation along the rotating magnetic axis.

The radiation is linearly polarized with the polarization direction and intensity defined by the vector

$$\overline{n} \times (\overline{n} \times \overline{B}^Z) = \overline{B}^Z - B_z^Z cos(\theta)\overline{n} \ ,$$

where $\overline{n}$ is the direction of the observer in the direction of the axial magnetic flux tubes and characterized by the angle θ. The direction of polarization is constant during the observation period if the symmetry axis associated with B^Z coincides with the rotation axis. It is essential that magnetic and Z^0 magnetic fields are not parallel and reside at different space-time sheets. The intensity is proportional to the square of the polarization factor given by

$$(B^Z)^2 \times \left(1 - cos^2(\alpha)cos^2(\theta)\right) \ , \quad cos(\alpha) \equiv \frac{B_z^Z}{B^Z} \ .$$

If the Z^0 magnetic field has only z-component, the intensity is proportional to $(B^Z)^2 sin^2(\theta)$ and at minimum.

4. Radial compression as a mechanism producing strong Z^0 magnetic field

A sudden compression in radial directions orthogonal to the rotation axis at the core collapse could be seen as a process analogous to the squeezing of the tooth paste tube. A strong non-dissipative supra current along the axis of magnetic field is induced because this is the route of the lowest resistance. This current in turn generates a strong magnetic field component B_ϕ^Z, and the charges accelerated in the axial direction in this field emit synchrotron radiation with a direction of polarization tangential to the magnetic field component B_ϕ^Z. If all nuclei possess anomalous Z^0 charges, the matter flow along rotation axis can generate very strong Z^0 magnetic field so that there are good hopes of explaining the anomalously high value of polarization of the synchrotron radiation.

The three expanding ring like structures associated with SN1987A [ib9] could be identified as being due to dark Z^0 currents rotating around the strong axial Z^0 magnetic field. Even the identification as torus like flux quanta of Z^0 magnetic field induced by the very strong Z^0 current along the z-axis is possible. This kind of Z^0 magnetic dark currents rotating around axial Z^0 magnetic field could be even responsible for the rings associated with planets like Saturnus and even with the ring current associated with Earth. This picture conforms with the model for the formation of solar system in which macroscopically quantum coherent dark matter serves as a template around which ordinary matter is condensed [D6, J6] as also with the explanation of tritium beta decay anomaly assuming that Earth's orbit is surrounded by dark neutrino belt [F8].

It is known that spherical and even axial symmetry is broken in case of SN1987A and this is consistent with the fact that magnetic and Z^0 magnetic axis are not parallel. Let L be the line of sight orthogonal to the the plane S of sky, and R the projection of the ring to S. Let z-axis correspond to L and x- and y-axis to the directions of the minor and major axis of R. Denote by E_z and E_y the projections of ejecta to S and xz-plane. From the figure 2 of [ij3] one can deduce that the plane of the ring forms an angle of 44 degrees with respect L. The symmetry axes of E_y resp. E_z forms an angle of 45 degrees resp. 15 degrees with respect to x-axis. From this one can conclude the polar and azimuthal angles of the symmetry axis of ejecta are $\theta = 45.4$ degrees and $\phi = 9$ degrees. A good guess is that this axis corresponds to the rotation axis and axis of Z^0 magnetic field tilted by 45.4 degrees with respect to the line of sight parallel to the magnetic axis. Mystery spots are known to be located at this axis too [ij3] so that they could indeed correspond to sunspot like throats at which Z^0 magnetic flux is transferred between space-time sheets.

Magnetic flux tubes as wave guides

Magnetic flux tubes are ideal wave guides forcing the confined radiation to propagate in a precisely targeted manner along them. Topological light rays (MEs) accompany magnetic flux tubes involved and have interpretation as space-time correlates for a radiation propagating in the waveguide defined by the magnetic flux tube. They are accompanied by coherent light generated by light like vacuum currents associated with then. Topological light rays would couple to Alfwen waves representing transversal oscillations of the magnetic flux tubes propagating also with light velocity.

The wave guide function of magnetic flux tubes suggests a generalization and modification of the model of gamma ray bursts. Gamma ray bursts would be generated by the synchrotron radiation generated in the acceleration of charges when they move along rotation axis with dynamically generated component B_ϕ^Z. Part of the resulting radiation would end up to a rotating magnetic flux tube bundle in the direction of the rotating magnetic axis. The initial channelling at the magnetic flux tubes would force synchrotron radiation to propagate to distant parts of the universe in a precisely targeted manner. This mechanism would explain the observed universal properties for the gamma ray bursts [ij10] difficult to understand in the models involving mergers, say collisions of white dwarf binaries [ij8]. As already noticed, the model is consistent with the existing model for the ordinary radiation arriving from supernovae and thought of as involving a beam rotating with the supernova.

16.7.2   Lepto-pions as a signature dark matter?

The identification of cosmic strings as the ultimate source of both visible and dark matter does not exclude the possibility that a considerable portion of topologically condensed cosmic strings have decayed to some light particles. In particular, this could be the situation in the galactic nuclei. On the other hand, if some fraction of cosmic strings evolve to magnetic flux tubes, these flux tubes identifiable as dominant part of the dark matter can carry phases of some exotic particles serving as signatures of the dark matter. Quite recent experimental findings [ic5] suggest that these exotic particles could be lepto-hadrons predicted by TGD [F7].

Lepto-hadrons

In the chapter [F7] I have discussed the TGD based explanation for the anomalous production of electron positron pairs in the collisions of heavy nuclei at energies corresponding to the height of Coulomb wall. The effect was observed for more than fifteen years ago [hh5] but after string model revolution has been forgotten by theorists like many other anomalies of particle physics. The hypothesis is that so called lepto-pions are produced in the strong, non-orthogonal, and rapidly varying electric and magnetic fields of the colliding nuclei. Lepto-hadrons are color bound states of colored excitations of leptons predicted by TGD defining an asymptotically non-free QCD. Actually an entire hierarchy of non-asymptotically free QCD:s are allowed in TGD Universe.

The idea that lepto-hadrons might have something to do with the dark matter has popped up now and then during the last decade but for some reason I have not taken it seriously. Situation changed towards the end of the year 2003. There exist now detailed maps of the dark matter in the center of galaxy and it has been found that the density of dark matter correlates strongly with the intensity of monochromatic photons with energy equal to the rest mass of electron [ic5].

The only explanation for the radiation is that some yet unidentified particle of mass very nearly equal to $2m_e$ decays to an electron positron pair or directly to gamma pair. Electron and positron are almost at rest and this implies a high rate for the annihilation to a pair of gamma rays. A natural identification for the particle in question would be as a lepto-pion. By their low mass lepto-pions, just like ordinary pions, would be produced in high abundance, in lepto-hadronic strong reactions and therefore the intensity of the monochromatic photons resulting in their decays would serve as a measure for the density of the lepto-hadronic matter. Also the presence of lepto-pionic condensates can be considered. Lepto-pions decay directly to both gamma pairs and electron-positron pairs. Indeed, galaxy is for long time known to be a source of positrons and there is no generally accepted mechanism producing them [ic5].

These findings force to take seriously either the identification
a) of the dark matter as lepto-hadrons or
b) of lepto-pions as a signature of dark matter, which itself would be basically magnetic energy associated the with cosmic strings transformed to magnetic flux tubes in topological condensation.

In fact, lepto-pions are not the only possibility. The TGD based model for tetra-neutrons [hh11] [F8] is based on the hypothesis that mesons made of scaled down versions of quarks corresponding to Mersenne prime M_{127} (ordinary quarks correspond to $k = 107$) and having masses around one MeV could correspond to the color electric flux tubes binding the neutrons to form a tetra-neutron. The same force would be also relevant for the understanding of alpha particles.

Why lepto-hadrons cannot directly correspond to dark matter?

The identification of lepto-hadrons as dark matter raises several questions leading to the conclusion that lepto-pions are most probably only a signature of dark matter.

a) Why the ratio of the lepto-hadronic mass density to the mass density of the ordinary hadrons would be so high, of order 7? Could an entire hierarchy of asymptotically non-free QCDs be responsible for the dark matter so that lepto-hadrons would explain only a small portion of the dark matter? Is even the hierarchy of QCD:s enough?

b) Under what conditions one can regard lepto-hadronic matter as a dark matter? Could short life-times of lepto-hadrons make them effectively dark matter in the sense that there would be no stable enough atom like structures consisting of say charged lepto-baryons bound electromagnetically to the ordinary nuclei or electrons? But what would be the mechanism producing lepto-hadrons in this case (nuclear collisions produce lepto-pions only under very special conditions)?

c) What would be the role of the many-sheeted space-time: could lepto-hadrons and atomic nuclei reside at different space-time sheets so that lepto-baryons could be long-lived? Could dark matter quite generally correspond to the matter at different space-time sheets and thus serve as a direct signature of the many-sheeted space-time topology? Magnetic flux tubes are excellent candidates for the the the space-time sheets accommodate the dark matter but there are good reasons to believe that magnetic energy is considerably higher than the energy of particles condensed on magnetic flux tubes so that magnetic energy is the best candidate for dark matter.

These objections strongly suggests that lepto-pions serve only as a signature of dark matter whereas dark matter itself corresponds to the magnetic energy of the magnetic flux tubes and cosmic strings.

Lepto-pions topologically condensed on magnetic flux tubes as a signature of dark matter?

Lepto-pions and other leptohadrons producing copiously lepto-pions could reside at magnetic of Z^0 magnetic flux tubes of thickness of order Compton length of lepto-pion. These strings could be seen as kind of very long lepto-hadronic strings. Also long hadronic flux tubes carrying coherent states of ordinary pions are possible and Z^0 flux tubes beaming the gamma ray bursts could correspond to them.

One could identify the lepto-hadronic magnetic flux tubes as structures generated later in the cosmic evolution, when the magnetic flux of hadronic flux tubes flow to larger space-time sheets. The transversal length scales of the flux tubes would be in ratio m_e/m_p and the magnetic field would be by a factor of about 10^{-6} weaker, about 10^{10} Tesla whereas the magnetic field of supernovae are around 10^9 Tesla. If the thickness of the magnetic flux tube at the moment of the annihilation of lepto-pion is of the order of Compton length of electron, one obtains an estimate for its thickness at the moment when the transition to the radiation dominated phase occurred.

If the strength of the magnetic field is of order $eB \sim m_e^2 \sim 10^9$ Tesla, the cyclotron frequency would be of same order as electron mass $eB/m_e \sim m_e$ and in gamma ray region. For $eB \sim m_p^2$ the field strength would be 10^{15} Tesla and cyclotron energy would be of order proton mass. Harmonics of this line might serve as a signature for the strength of the magnetic field. The monochromatic gamma lines at electron mass could also result in cyclotron transitions of electrons if the magnetic field at magnetic flux tubes that $eB = m_e^2$ holds true in high precision.

One can imagine two mechanisms of lepto-pion production.

a) The magnetic and Z^0 magnetic fields associated with the magnetic flux tubes give rise to classical color fields, which suggest that one could regard the flux tubes as macroscopic color magnetic and possibly also color electric flux tubes carrying lepto-hadrons, which produce copiously lepto-pions in their reactions.

b) In heavy ion collisions lepto-pion production is caused by the presence of the rapidly varying non-orthogonal electric and magnetic fields of colliding nuclei, whose "instanton density" $E \cdot B$ is non-vanishing (this means that the magnetic flux tube has higher than 2-dimensional CP_2 projection). The amplitude for lepto-pion production as a decay of the coherent state is proportional to the Fourier component of the "instanton density". The mechanism could be at work also now if magnetic flux tubes carry strong charges and generate radial electric fields. Lepto-pions would serve as signature for rapid changes of the magnetic and electric fields induced by rapid deformations of the magnetic flux tubes.

Solar X-ray halo and scaled up QCDs at magnetic flux tubes

Quite recently New Scientist told about an explanation proposed by Kostantin Zioukas and his colleagues [ic7] for the X-ray halo of Sun in terms of axions, one of the many candidates for the dark matter [ic6]. The X-ray halo of Sun was detected at 1940. The halo extends from the surface of Sun (free path for photons increases at the surface). The X-ray intensity decays exponentially and extends several solar radii from the surface. The energy range of X-rays is $3 - 15$ keV. The origin of the X-ray halo has remained a mystery.

The axions in the required mass range are predicted by certain higher-dimensional theories [ic7]. The axions would be produced in the solar core and because of their extremely long lifetime they would propagate to the surface of Sun and some fraction of non-relativistic axions would remain bound in the solar gravitational field where they would decay. The estimated mean distance of the proposed axion population from the solar surface is about 6.2 solar radii. Zioukas and his colleagues are able to deduce the value of the coupling constant $g_{A\gamma\gamma}$ characterizing the rate of axion decay and the interaction cross section of axion with matter from the fact that the X-ray luminosity must be proportional to $g_{A\gamma\gamma}^4$. The resulting lifetime of the axion is about 10^{21} s to be compared with the lifetime of ordinary pion about 10^{-16} s.

TGD suggests an alternative explanation based on a non-asymptotically free exotic QCD at a magnetic flux tube corresponding to a p-adic length scale $L(k)$ for which the scaled down value of pion mass corresponds to mass of about 3 keV. Assuming that pion corresponds to $k = 107$ ($k = 109$ is the second candidate) this gives $2^{(k-107)/2} \sim m_{\pi(107)}/m_{\pi(k)}$. The lower limit for the energy spectrum would favor the p-adic length scale $L(139)$ giving $m_{\pi(139)} \simeq 2.2$ keV. The lifetime of lepto-pion would be scaled up by a factor 2^{16} so that one would have $\tau \sim 10^{-11}$ s. One cannot exclude the presence of several scaled up QCDs with $k - 139, 137$ and $k = 131$ being the most favored ones in the energy range of about 3 octaves spanned by the X-ray spectrum.

In the recent case the intensity of the X-ray halo from a given spherical volume V of the halo defining the pixel is determined by the density $dn(\pi)/dl$ of the exotic pions per unit length of the magnetic flux tube and the length $l(V)$ of the magnetic flux tube inside the volume, which is expected behave as $l(V) \sim V^{1/3}$. A rough estimate is

$$ I(V) \sim \frac{dn(\pi)}{dl} \times l(V) \times \Gamma \times \langle E(\pi) \rangle \Delta\Omega \; , $$

where $\Delta\Omega = A/4\pi R^2$ is the solid angle defined spanned by the the active detection area A of the measuring instrument at a given point of the magnetic flux tube and R is the distance of Earth from Sun. In principle this allows to estimate the density of exotic pions per unit length of the magnetic flux tube.

The exponential decay of the intensity with distance from the surface of the Sun would suggest that magnetic flux tubes might be regarded as threads extending from the solar surface and returning back to it, and that the probability of a path of given length decreases exponentially with its length. If the probability for the appearance of a thread of given length is proportional to the Boltzman weight $exp(-E_B/T)$, where E_B is magnetic energy of the thread and T is temperature parameter, this indeed holds true.

The intensity of the magnetic field at the flux tubes can be estimated from the nominal value $B_E = .5 \times 10^{-4}$ Tesla of the Earth's magnetic field at the space-time sheet $k = 169$. By scaling one would obtain $B = 2^{169-139}B_E = 5 \times 10^4$ Tesla. The field is extremely strong and could be perhaps assigned to remnants of primordial cosmic strings. Note that also Z^0 magnetic field could be in question in which case dark matter coupling to scaled down copies of electro-weak bosons would be in question [F6, F9].

Do the length scale ratios for astrophysical objects reflect Compton length ratios of elementary particles?

The ratio for the size $L_l \sim 10^5$ light years of a galactic nucleus to the distance $L_h \sim 1$ light month between the light spots of super nova gives an estimate for the ratio of the lengths of the lepto-hadronic and hadronic magnetic flux tubes. This would predict $L_l/L_h \sim 10^6$ and that the ratio of transverse thicknesses $d_l/d_h = 10^3$, which is the ratio of lepto-pion Compton length scale scale to proton Compton length. This would suggests that the length scale hierarchy for astrophysical objects could represent a scaled up version of the p-adic length scale hierarchy associated with elementary particles.

Frequency cutoff for zero point frequencies as a test for many-sheeted space-time?

For a quantum system modellable in terms of harmonic oscillators (say photon field) the frequency spectrum in the thermal equilibrium obeys Planck distribution. Besides this the system exhibits zero point fluctuations whose energy density is given by $\rho_0(f) = 8\pi^2 f^3$ ($\hbar = c = 1$) in the 3-dimensional case. Zero point fluctuations appear in many models of physical phenomena such as X-ray scattering in solids, Lamb shift, Casimir effect, and the interpretation of the Aharonov Bohm effect (for references see [ic8]).

The zero point fluctuations are predicted to appear also in electronic systems, and the experimentally measured spectral density of the current noise measured by Koch [ic2] in Josephson junctions provides a direct support for this prediction. The fluctuations have been observed up to the frequency of $f = .6$ THz which corresponds to a microwave wavelength of .5 mm.

It has been proposed by Beck and Mackey [ic8] that if these fluctuations are associated with the vacuum energy, the total vacuum energy density associated with these fluctuations cannot exceed the recently measured dark energy density of the Universe: this leads to a cutoff frequency of $f_c = (1.69 \pm .05)$ THz for the measured frequency spectrum.

In TGD framework dark matter is ordinary matter at larger space-time sheets. First of all, the finite size of the space-time sheet poses an IR cutoff. p-Adic length scale hierarchy suggests that that there is also UV cutoff that corresponds to the next p-adic length scale in the hierarchy. Hence the frequencies above the UV cutoff would correspond to oscillations at smaller space-time sheets. The interpretation would be in terms of de-coherence.

Thus a given space-time sheet would contain half octave of frequencies between the frequency cutoffs $f_{low}(k) = c/L(k) \propto 2^{-k/2}$ and $f_{up}(k) = c/L(k+1)$. Cutoff frequencies would come as half octaves for k integer as predicted by the most general form of the p-adic length scale hypothesis. The stronger form of the hypothesis favors prime values of k. Note that for $k = 179$ (prime) the predicted cutoff frequency would be $f_c(179) \simeq 1.74$ THz, which happens consistent with the prediction of [ic8] deduced from the estimate for the dark matter density. This need not be an accident. According to the TGD based model explaining the finding that neutrino mass depends on the environment, neutrinos can condense on several space-time sheets and neutrinos in dense matter travel along $k = 179$ space-time sheet [F3].

The problem is that the spectral density would be same at every space-time sheet. One might however hope that the shift of the spectrum from a space-time sheet to another one manifests itself as some kind of structure at half-integer octaves of a basic frequency. By using a suitable arrangement one might be even able to eliminate some space-time sheet so that a gap would result. An interesting question is how the measurement instrument could be constructed to detect only the frequencies associated with a space-time sheet corresponding to a fixed value of k.

Chapter 17

TGD and Cosmology

17.1 Introduction

TGD Universe is quantum counterpart of a statistical system at a critical temperature. As a consequence, topological condensate is expected to possess hierarchical, fractal like structure containing topologically condensed 3-surfaces with all possible sizes. Both Kähler magnetized and Kähler electric 3-surfaces ought to be important and string like objects indeed provide a good example of Kähler magnetic structures important in TGD inspired cosmology. In particular space-time is expected to be many-sheeted even at cosmological scales and ordinary cosmology must be replaced with many-sheeted cosmology. The presence of vapor phase consisting of free cosmic strings and possibly also elementary particles is second crucial aspects of TGD inspired cosmology.

Quantum criticality of TGD Universe supports the view that many-sheeted cosmology is in some sense critical. Criticality in turn suggests fractality. Phase transitions, in particular the topological phase transitions giving rise to new space-time sheets, are (quantum) critical phenomena involving no scales. If the curvature of the 3-space does not vanish, it defines scale: hence the flatness of the cosmic time=constant section of the cosmology implied by the criticality is consistent with the scale invariance of the critical phenomena. This motivates the assumption that the new space-time sheets created in topological phase transitions are in good approximation modellable as critical Robertson-Walker cosmologies for some period of time at least.

Any one-dimensional sub-manifold of CP_2 allows global imbeddings of subcritical cosmologies whereas for a given 2-dimensional Lagrange manifold of CP_2 critical and overcritical cosmologies allow only one-parameter family of partial imbeddings. The infinite size of the horizon for the imbeddable critical cosmologies is in accordance with the presence of arbitrarily long range fluctuations at criticality and guarantees the average isotropy of the cosmology. Imbedding is possible for some critical duration of time. The parameter labelling these cosmologies is a scale factor characterizing the duration of the critical period. These cosmologies have the same optical properties as inflationary cosmologies but exponential expansion is replaced with logarithmic one. Critical cosmology can be regarded as a 'Silent Whisper amplified to Bang' rather than 'Big Bang' and transformed to hyperbolic cosmology before its imbedding fails. Split strings decay to elementary particles in this transition and give rise to seeds of galaxies. In some later stage the hyperbolic cosmology can decompose to disjoint 3-surfaces. Thus each sub-cosmology is analogous to biological growth process leading eventually to biological death.

The critical cosmologies can be used as a building blocks of a fractal cosmology containing cosmologies containing ... cosmologies. p-Adic length scale hypothesis allows a quantitative formulation of the fractality [D6]. Fractal cosmology predicts cosmos to have essentially same optical properties as inflationary scenario but avoids the prediction of unknown vacuum energy density. Fractal cosmology explains the paradoxical result that the observed density of the matter is much lower than the critical density associated with the largest space-time sheet of the fractal cosmology. Also the observation that some astrophysical objects seem to be older than the Universe, finds a nice explanation.

The key difference between inflationary and quantum critical cosmologies relates to the interpretation of the fluctuations of the microwave background. In the inflationary option fluctuations are amplified to long length scale fluctuations during inflationary expansion. In quantum critical cosmology the fluctuations be assigned to the quantum critical period accompanying macroscopic quantum fluctuations of the dark matter appearing in very long length scales during the phase transition so that no inflationary expansion is needed. Sub-critical cosmology is predicted after the inflationary period.

Absolutely essential element of the considerations (and longstanding puzzle of TGD inspired cosmology) is the conservation of energy implied by Poincare invariance which seems to be in conflict with the non-conservation of gravitational energy. It took long time to discover the natural resolution of the paradox. In TGD Universe matter and antimatter have opposite energies and gravitational four-momentum is identified as difference of the four momenta of matter and antimatter (or vice versa, so that gravitational energy is positive). The assumption that the net inertial energy density vanishes in cosmological length scales is the proper interpretation for the fact that Robertson-Walker cosmologies correspond to vacuum extremals of Kähler action.

Tightly bound, possibly coiled pairs of cosmic strings are the basic building block of TGD inspired cosmology and all al structures including large voids, galaxies, stars, and even planets can be seen as pearls in a cosmic fractal necklace consisting of cosmic strings containing smaller cosmic strings linked around them containing... During cosmological evolution the cosmic strings are transformed to magnetic flux tubes and these structures are also key players in TGD inspired quantum biology.

Negative energy virtual gravitons represented by topological quanta having negative time orientation and hence also negative energy. The absorption of negative energy gravitons by photons could explain the gradual red-shifting of the microwave background radiation. Negative energy virtual gravitons give also rise to a negative gravitational potential energy. Quite generally, negative energy virtual bosons build up the negative interaction potential energy. An important constraint to TGD inspired cosmology is the requirement that Hagedorn temperature $T_H \sim 1/R$, where R is CP_2 size, is the limiting temperature of radiation dominated phase.

In the following this scenario is described in detail.

a) Basic ingredients of TGD inspired cosmology are introduced. The implications of absolute minimization of Kähler action and of the identification of gravitational four-momentum as a difference of four-momenta of matter and negative energy antimatter are discussed. The consequences of the imbeddability requirement are analyzed. The basic properties of cosmic strings are summarized and simple model for vapor phase as consisting of critical density of cosmic strings are introduced. Additional topics are thermodynamical aspects of cosmology, in particular the new view about second law and the consequences of Hagedorn temperature. Non-conservation of gravitational momentum is considered.

c) The evolution of the fractal cosmology is described in more detail.

d) TGD inspired cosmology is compared to inflationary scenario: in particular, the TGD based explanation for the recently observed flatness of 3-space and a possible solution to the Hubble constant controversy are discussed.

e) Certain problems of the cosmology such as the questions why some stars seem to be older than the Universe, the claimed time dependence of the fine structure constant, the generation of matter antimatter asymmetry, and the problem of the fermion families, are discussed.

f) Simulating Big Bang in laboratory is the title of the last section. The motivation comes from the observation that critical cosmology could serve as a universal model for phase transitions.

17.2 Basic ingredients of TGD inspired cosmology

In this section the general principles and ingredients of the TGD inspired cosmology are discussed briefly.

17.2.1 Many-sheeted space-time defines a hierarchy of "smoothed out" space-times

The notion of quantum average space-time obtained by smoothing out details below the scale of resolution was inspired by renormalization philosophy and for long time I regarded it as a purely fictive concept. The rough idea was that quantum average effective space-times correspond to the absolute minima of the Kähler action associated with the maxima of the Kähler function. Therefore the dynamics of the quantum average effective space-time is fixed and the stationarity requirement for the effective action should only select some physically preferred maxima of the Kähler function. The topologically trivial space time of classical GRT cannot directly correspond to the topologically highly nontrivial TGD space-time but should be obtained only as an idealized, length scale dependent and essentially macroscopic concept. This allows the possibility that also the dynamics of the effective smoothed out space-times is determined by the effective action.

The space-time in length scale L is obtained by smoothing out all topological details (particles) and by describing their presence using various densities such as energy momentum tensor $T^{\alpha\beta}_{\#}$ and Yang Mills current densities $J^{\alpha}_{a\#}$ serving as sources of classical electro-weak and color gauge fields (see figure 15.3.1. It is important to notice that the smoothing out procedure eliminates elementary particle type boundary components in all length scales: this suggests that the size of a typical elementary particle boundary component sets lower limit for the scale, where the smoothing out procedure applies.

During the development of the many-sheeted space-time concept it has become obvious that the notions of classical space-time and of smoothing out of details are not only activities of a theoretician, but that the many-sheeted space-time itself can be said to perform renormalization theory.

a) Classical space-time is much more than a fiction produced by the stationary phase approximation. The localization in the so called zero modes, which corresponds to state function reduction in TGD, which occurs in each quantum jump (the delicacies due to macro-temporal quantum coherence will not be discussed here) means that the superposition of space-time surfaces in the final state of quantum jump, consists of space-time surfaces equivalent from the point of view of observer.

b) The notion of many-sheeted space-time predicts a hierarchy of space-time sheets labelled by p-adic primes $p \simeq 2^k$, k integer with primes and prime powers being in preferred role. The space-time sheets at a given level of hierarchy play a role of particles topologically condensed at larger space-time sheets. Hence the physics at larger space-time sheets is quite concretely a smoothed out version of the physics at smaller space-time sheets. Many-sheeted space-time itself performs renormalization group theory, and p-adic primes characterizing the sizes of the space-time sheets correspond to the fixed points of the renormalization group evolution. p-Adic length scale hierarchy quantifies the vision about fractal cosmology.

c) There are good reasons to expect that the absolute minimum value for the Kähler action vanishes for large enough space-time sheets, and that space-time sheets result as small deformations of the vacuum extremals at the long length scale limit. Einstein's equations can be posed as an additional constraint on vacuum extremals so that there are good hopes of satisfying Einstein's equations exactly at long length scale limit, when one can expect that positive and negative energies of matter and antimatter are conserved separately in statistical sense.

An important difference to the standard view is that energy momentum tensor is dictated by the Einstein tensor (plus metric) rather than vice versa. Since the dynamics of the induced EYM fields is dictated by the absolute minimization of Kähler action, EYM equations cannot in general be satisfied without the introduction of particle currents. This conforms with the view that Einstein's equations relate to a statistical description of matter in terms of both particle densities and classical fields. The imbeddability to $H = M^4_+ \times CP_2$ means a rich spectrum of predictions not made by GRT. TGD inspired cosmology and TGD based model for the final state of the star are good examples of these predictions, and are consistent with experimental facts.

17.2.2   The notion of energy in TGD Universe

Although TGD was born as a solution to the "energy problem" of GRT, the relationship between inertial and gravitational energy remained poorly understood, and only the development of TGD inspired theory of consciousness allowed to find a satisfactory resolution of the interpretational problems.

Negative energy space-time sheets

Negative energy space-time sheets represents an important distinction between TGD and standard physics. They are possible because energy momentum tensor is replaced by a collection of conserved currents associated with various components of four momentum. This resolves the energy problem of general relativity but, since the sign of the conserved charged depends on the time orientation of the space-time sheet, the sign of energy is not positive definite anymore.

Quantum classical correspondence implies that also elementary particles can have negative energies and this means a new kind of physics. It seems that this physics has been already discovered: the strange properties of phase conjugate laser waves can be understood if they consist of negative energy photons.

Negative energy space-time sheets have far reaching implications for TGD inspired theory of consciousness. The so called time mirror mechanism involves the reflection of negative energy signals sent to the geometric past from population inverted lasers as amplified positive energy signals propagating to the geometric future. Time mirror mechanism provides the holy grail to the understanding of the mechanisms of brain functioning and also of the workings of the living matter. There are obvious implications for communication and energy technologies since negative energy signals could make possible instantaneous remote sensing and quantum control over arbitrarily long distances so that light velocity would cease to be a restriction forcing us to be habitants of 3-space instead of space-time.

It took years before time was ripe to consider seriously the generalization of phase conjugate photons to that of phase conjugate fermions corresponding to two different fermionic vacua related by Hermitian conjugation of oscillator operators (inducing a phase conjugation for photons). It is possible to imagine two different options which both might be realized.

a) The first option is the conventional quantization for which fermions and anti-fermions have the same sign of energy at a given space-time sheet. Positive/negative energy modes correspond to creation/annihilation operators. At a given space-time sheet both fermion and anti-fermion energies are either positive or negative. This conforms with the fact that classical energy density associated with Kähler action is either positive or negative.

b) At a given space-time sheet fermions and anti-fermions have opposite energies so that both of them could correspond to creation operators at a given space-time sheet in the expansion of second quantized induced spinor field. This option looks rather attractive since it implies the vanishing of vacuum energy density in the standard quantization. In TGD framework there is however no vacuum energy density to be cancelled since the energy eigenvalues are replaced with the generalized eigenvalues of the modified Dirac operator. Furthermore, TGD based construction of S-matrix does not rely on Hamiltonian formalism.

This option leads to difficulties with the construction of gauge bosons as fermion-antifermion bound states: the bilinears in question would be bilinears in creation and annihilation operators. This difficulty can be circumvented if the quantization of fermions depends on the dimension $D(CP_2)$ of the CP_2 projection classifying the extremals of Kähler action [D1] so that the conventional quantization is applied only if the dimension is $D(CP_2) = 4$ as it indeed is for CP_2 type extremals. This option reduces also dramatically the number of massless elementary particles [F2].

This option is favored by the fact that the huge vacuum degeneracy, in particular the vacuum property of extremals representing Robertson-Walker cosmologies, cries for a natural explanation. This could perhaps be interpreted in terms of vanishing of net energy density of the topologically condensed matter: energy density would vanish only in certain length scale resolution. The alternative interpretation would be that zero energy states with the proposed unconventional fermionic quantization are in question assuming however that the sign of the net energy density correlates with the sign of the energy density associated with the Kähler action. Hence one must leave mind open for both options.

In any case, these findings led to the realization that the maximally predictive theory results when the net conserved quantities of the Universe are vanishing. The vanishing of net quantum numbers of Universe would resolve the unpleasant philosophical questions like "What is the total fermion number of the Universe" and would mean the cosmic realization of the crossing symmetry based view about particle reaction as generating of positive and negative energy particles from vacuum. One could see the entire Universe as a result of intentional actions in which intentions represented by p-adic space-time sheets are transformed to actions represented by real space-time sheets. Everyone knows the anecdotes about yogis and gurus creating material objects from nothing and very few "scientifically thinking" westerner can take these stories really seriously. Whether or not these stories are true, they might however express a deep truth about reality.

The relationship between inertial and gravitational masses

Concerning the understanding of the topological condensation of cosmic strings the decisive breakthrough came through the understanding of the relationship between inertial and gravitational four-momenta.

a) TGD predicts that inertial four-momentum is conserved whereas an empirical fact is that gravitational four-momentum is not conserved in cosmological length scales. The solution of the paradox came through the realization that the inertial energy densities of space-time sheets with opposite time orientation have opposite signs: I shall refer to these options as phase conjugates of each other. These vacua can appear as two variants: the conventional vacua for which fermions and anti-fermions have same sign of energy and vacua for which fermions and anti-fermions have opposite signs of energy. For the latter vacua the sign of the net energy is dictated by the time orientation. These vacua would be naturally associated with space-time surfaces which are small deformations of vacuum extremals.

The most elegant and most predictive theory results if all quantum states of the Universe have a vanishing net energy (in fact, all conserved quantum numbers must vanish). The inertial energy density vanishes for vacuum extremals. The fact that Kähler energy is non-negative means that the sign of inertial energy density is fixed for a given space-time sheet.

This picture suggest a general mechanism generating matter antimatter asymmetry. The generalized eigenvalues λ for the modified Dirac operator [B4] can have both signs, and the first guess is that the different signs corresponds to oscillator operators for fermions and anti-fermions. In a sharp contrast to the standard quantum field theory, the eigen modes corresponding to the different signs are not related by any symmetry. Hence it is quite possible that the number of say negative eigenvalues is much smaller than the number of positive eigenvalues, so that matter antimatter asymmetry would result for the conventional vacuum. For non-conventional fermionic vacua for which both kinds of oscillator operators in Ψ are creation or annihilation type, space-time sheet can carry only carry states with fermion number 1 or -1, depending on time orientation. Hence matter and antimatter would tend to reside on space-time sheets of opposite time orientation in consistency with the idea that the Universe is created from vacuum.

Gravitational four-momentum can be defined as a difference of inertial momenta associated with matter and antimatter or vice versa. There are two options.

a) The difference of the inertial four-momenta is multiplied by a sign factor, which is $+1$ for $|vac_+\rangle$ and -1 for $|vac_-\rangle$, to guarantee that gravitational energy is always non-negative. This option is favored for cosmological reasons and the belief that gravitational energy is positive always.

b) One could also argue that the sign of gravitational energy is different for $|vac_+\rangle$ and $|vac_-\rangle$ so that no sign factor is needed. This option could be defended: the sign of gravitational mass as determined from the asymptotic behavior of the induced metric for spherically symmetric stationary metrics can be also negative and vacuum extremals presumably allow also this kind of behavior. One can of course challenge the assumption that asymptotic behavior of the metric characterizes the value of gravitational mass completely generally. One could also assume that Einstein's equations hold true only for the asymptotic self-organization patterns just like Lorentz-Kähler four-force vanishes for them and that gravitational mass is positive for the asymptotic patterns. For pairs of phase conjugate cosmic strings the dominant contribution to gravitational mass would vanish for this option so that the option looks un-physical. Hence option a) is is assumed in the sequel.

Non-conservation of gravitational four-momentum

TGD predicts non-conservation of gravitational four-momentum. The interpretation as a non-conservation of the difference of four-momenta of positive and negative energy matter replaces the previous interpretation as a flow of inertial four-momentum between different space-time sheets and this alters the previous picture.

The standard belief is that a compact 3-manifold must have a vanishing gravitational mass. In TGD framework this would mean that the gravitational mass of the vapor phase particles vanishes. As already seen, this is not the case in TGD without further assumptions about the dynamics of topological evaporation. Although the evaporated particle does not interact via classical long range gravitational fields, it can interact via the emission and absorption of gravitons. This should indeed be the case if the particle has a vanishing gravitational mass. But since gravitons form join bonds with the external world, the system would not actually be like an isolated universe. If particle is topologically evaporated in strict four-dimensional sense, it must satisfy empty space Einstein equations with a cosmological term, and be a vacuum extremal so that it does not carry either matter or antimatter and is rather uninteresting physically.

If the space-time sheets of 3-surface has no contacts with larger space-time sheets, it is like an isolated universe, and its inertial mass should vanish and be small if the number of these contacts is small. By the same argument also Lorentz-, color-, fermionic, etc. quantum numbers of the evaporated space-time sheet should vanish. This does not mean that the space-time sheet could not be in motion and possess cm momentum and angular momentum. The point is that these quantum numbers are compensated by the correspond quantum numbers in internal degrees of freedom. Situation changes in the case of non-conserved quantum numbers and the quantum numbers associated with the representations of super-canonical symmetries (discussed in the chapters of the first part of the book) could be non-vanishing.

The inertial mass possibly possessed by a topologically evaporated particle is left at the boundaries of holes resulting in the evaporation. Topologically evaporated particle can indeed develop negative inertial energy density at its boundaries guaranteing the vanishing of its inertial mass since space-time sheets with negative time orientation possess negative energy.

One can evaluate the divergence of the gravitational energy momentum tensor satisfying Einstein equations with a cosmological term by calculating the divergence of the four momentum current $T^{A\alpha}$ (A labels different momentum components)

$$ D_\alpha T^{A\alpha} = \left[\frac{1}{16\pi G} G^{\alpha\beta} + \hat{\Lambda}(2) g^{\alpha\beta} \right] H^k_{\alpha\beta} j^A_k \ . \tag{17.2.1} $$

Here j^{Ak} denotes the Killing vector of translation. This equation will be applied later in the context of Robertson-Walker cosmologies.

A situation of a special physical interest corresponds to the separate conservation of inertial and gravitational masses. In this case the following equations of motion derivable from Einstein action $S = \int (kR + \hat{\Lambda}(2)) \sqrt{g} d^4 x$, $k = 1/16\pi G$, hold true

$$ \left[kG^{\alpha\beta} + \hat{\Lambda}(2) g^{\alpha\beta} \right] H^k_{\alpha\beta} = 0 \ . \tag{17.2.2} $$

At cosmological length scales the cosmological term is negligible, and one can replace the equation with

$$ G^{\alpha\beta} H^k_{\alpha\beta} = 0 \ . \tag{17.2.3} $$

This action was my first attempt to try to construct "sub-manifold gravity". This equation is expected to provide a reasonable when the energies associated with matter and antimatter are separately conserved.

For Schwartshild metric the divergence indeed vanishes identically if $\hat{\Lambda}(k)$ is vanishing (or sufficiently small as in the recent Universe). One can say that stationary situations in which gravitational mass is conserved are those in which gravitational mass appears as lumps. Thus asymptotic self-organization patterns necessarily correspond to gravitational lumping of the matter. In the case of general Reissner-Nordström metric energy flow turns out to be non-vanishing although the metric itself is stationary in the sense of GRT. It will be later found that the Robertson-Walker cosmology satisfying stationarity conditions is essentially unique and provides a natural candidate for the asymptotic cosmology.

The earlier interpretation for the non-conservation of gravitational four-momentum was in terms of a loss of inertial four-momentum due to topological evaporation and flow between different space-time sheets. This interpretation must be given up. If the flow occurs along join along boundaries bonds then the fluxes of energy momentum currents through the cross sections of join along boundaries bonds give the net flow. Not only the flow of energy-momentum but also the flow of fermion numbers, charges, etc., can occur between various space-time sheets. In particular, the flow of entropy is possible. An interesting possibility is that the vapor phase could serve as a "cosmic paper basket" allowing to circumvent the consequences of the second law of thermodynamics: the creation of order through gravitational interactions would become possible in topological condensate through the transfer of entropy to the vapor phase.

The notion of negative potential energy involves also new physics

The concept of negative potential energy is completely standard notion in physics. Perhaps so standard that physicists have begun to regard it as understood. The precise physical origin of the negative potential energy is however complete mystery, and one is forced to take the potential energy as a purely phenomenological concept deriving from quantum theory as an effective description.

In TGD framework topological field quantization leads to the hypothesis that quantum concepts should have geometric counterparts and also potential energy should have precise correlate at the level of description based on topological field quanta. This indeed seems to be the case. As already explained, TGD allows space-time sheets to have both positive and negative time orientations. This in turn implies that also the sign of energy can be also negative. This suggests that the generation of negative energy space-time sheets representing virtual gravitons together with energy conservation makes possible the generation of huge gravitationally induced kinetic energies and gravitational collapse. In this process inertial energy would be conserved since instead, of positive energy gravitons, the inertial energy would go to the energy of matter.

This picture has direct correlate in quantum field theory where the exchange negative energy virtual bosons gives rise to the interaction potential. The gravitational red-shift of microwave background photons is the strongest support for the non-conservation of energy in General Relativity. In TGD it could have concrete explanation in terms of absorption of negative energy virtual gravitons by photons leading to gradual reduction of their energies. This explanation is consistent with the classical geometry based explanation of the red-shift based on the stretching of electromagnetic wave lengths. This explanation is also consistent with the intuition based on Feynman diagram description of gravitational acceleration in terms of graviton exchanges.

Minkowski space or its future light cone?

If Kähler action were strictly deterministic, the only possible choice for H would be $M_+^4 \times CP_2$. Together with negative energies the classical non-determinism of the Kähler action however means that one cannot exclude the possibility that imbedding space is $M^4 \times CP_2$ meaning exact Poincare invariance. The point is that generation of pairs of positive and negative energy space-time sheet at light-like 7-surfaces $X_l^3 \times CP_2$ means emergence of new kind of causal determinants generalizing the light cone boundary $\delta M_+^4 \times CP_2$ as a fundamental causal determinant. All states of the Universe have vanishing net quantum numbers and everything in the Universe would have been pair-created from vacuum.

Of course, also M_+^4 option allows vanishing net quantum numbers naturally: just pose the natural boundary conditions that nothing flows our from the future light cone or into it. The recent proposal for TGD inspired cosmology relies in this assumption and cosmic string dominated cosmology is initiated by a process in which pairs of positive and negative energy cosmic string space-time sheets are generated from vacuum at the initial moments.

The possible difference between M^4 and M_+^4 based zero energy cosmologies might relate to the absence of the light cone boundary. In principle nothing forbids the generation of zero energy sub-cosmologies effectively belonging to M_+^4 also in the case of M^4 but in case of M_+^4 this option is unavoidable. The idea that space-time surface can be regarded as the field of quaternions locally slightly favors M^4 option. Although M^4 cosmology has strong aesthetic appeal, the discussion of this chapter is restricted to M_+^4 cosmology.

The category of light cones, the construction of the configuration space geometry, and the problem of psychological time

Light-like 7-surfaces of imbedding space are central in the construction of the geometry of the world of classical worlds. The original hypothesis was that space-times are 4-surfaces of $H = M_+^4 \times CP_2$, where M_+^4 is the future light cone of Minkowski space with the moment of big bang identified as its boundary $\delta H = \delta M_+^4 \times CP_2$: "the boundary of light-cone". The naive quantum holography would suggest that by classical determinism everything reduces to the light cone boundary. The classical non-determinism of Kähler action forces to give up this naive picture which also spoils the full Poincare invariance.

The new view about energy and time forces to conclude that space-time surfaces approach vacua at the boundary of the future light cone. The world of classical worlds, call it CH, would consist of classical universes having a vanishing inertial 4-momentum and other conserved quantities and being created from vacuum: big bang would be replaced with a "silent whisper amplified to a big bang". The net gravitational mass density can be non-vanishing since gravitational momentum is difference of inertial momenta of positive and negative energy matter: Einstein's Equivalence Principle is exact truth only at the limit when the interaction between positive and negative energy matter can be neglected.

Poincare invariant theory results if one replaces CH with the union of its copies $CH(a)$ associated with the light cones $M_+^4(a)$ with a specifying the position of the dip of $M_+^4(a)$ in M^4. Also past directed light-cones $M_-^4(a)$ are allowed. The unions of the light cones with inclusion as a basic arrow would form category analogous to the category Set with inclusion defining the arrow of time. This category formalizes the ideas that cosmology has a fractal Russian doll like structure, that the cosmologies inside cosmologies are singularity free, and that cosmology is analogous to an organic evolution and organic evolution to a mini cosmology.

The view also unifies the proposed two explanations for the arrow of psychological time [K1].

a) The mind like space-time sheets representing conscious self drift quantum jump by quantum jump towards geometric future whereas the matter like space-time sheets remain stationary. The self of the organism presumably consisting mostly of topological field quanta, would be like a passenger in a moving train seeing the changing landscape. The organism would be a mini cosmology drifting quantum jump to the geometric future. Also selves living in the reverse direction of time are possible.

b) Psychological time could correspond to a phase transition front in which intentions represented by p-adic space-time sheets transform to actions represented by real space-time sheets moving to the direction of geometric future. The motion would be due to the drift of $M_+^4(a)$. The very fact that the mini cosmology is created from vacuum, implies that space-time sheets of both negative and positive field energy are abundantly generated as realizations of intentions. The intentional resources are richest near the boundary of $M_+^4(a)$ and depleted during the ageing with respect to subjective time as asymptotic self-organization patterns are reached. Interestingly, mini cosmology can be seen as a fractally scaled up variant of quantum jump. The realization of intentions as negative energy signals (phase conjugate light) sent to the geometric past and inducing a positive energy response (say neural activity) is consistent with the TGD based models for motor action and long term memory [K1].

The spectral asymmetry, infinite primes, negative energies, and electric-magnetic duality

In the [B4] the proposal is made that quantum gravitational holography, which was basic element of TGD long before the discovery of quantum gravitational holography, could be realized in TGD in the sense that the exponent of Kähler function can be deduced from the Dirac determinant closely related to the modified Dirac operator. The information allowing to deduce the exponent of Kähler function for a given space-time sheet would be deducible from the determinants of the operators $D_+ D_-^{-1}$, where D_+ and D_- are the modified Dirac operators associated with the maximal deterministic space-time regions separated by a 3-dimensional causal determinants.

In the chapters [E3] a detailed model inspired by the notion of infinite prime is developed allowing to predict the values of Kähler coupling strength and gravitational constant. This discussion leads to new insights about the construction of the configuration space geometry and allows to sharpen the ideas about fractal cosmology.

The interpretation of the ansatz for the fermionic determinant in terms of infinite primes would be following.

a) The spectra of D_+ and D_- could correspond to the presence of the infinite primes generated from infinite vacuum primes $X \pm 1$, $X = \prod_k p_k$ (product over all finite primes). The asymmetric modes labelled by p_i and p would be coded by a fermionic infinite prime of type $X - 1$ or $X + 1$. In the arithmetic quantum field theory the asymmetric modes correspond to a difference of frequency scales proportional to $\omega_0 = \omega_+ - \omega_-$.

b) The two families of infinite primes correspond to the two possible signs of scaling momentum and "arrow of scaling" in a complete analogy with the corresponding interpretation based on time orientation in the case of ordinary Dirac operator.

c) The "arrow of scaling" would mean that for the 10 exceptional primes generated from $X + 1$ or $X - 1$ the asymmetric modes for number theoretical fermions would have different frequency scales. The "arrow of scaling" would naturally correlate with whether the sub-cosmology is expanding or contracting and thus with the arrow of the geometric and experienced time.

d) The two vacua correspond to the two possible quantizations in which creation and annihilation operators change their role and would thus characterize positive and negative energy matter predicted by quantum TGD. Phase conjugate photons behaving like negative energy photons would indeed be negative energy matter.

e) If the configuration space is identified as the union of configuration spaces $CH(a)$ ("subcosmologies") associated with all possible unions of future and past light cones $M_\pm^4(a)$ with the dip any point a of M^4 then $M_\pm^4$ naturally corresponds to $X \pm 1$. Why the ontology of TGD makes this picture natural is discussed in [O1].

f) It is possible to understand how the sub-cosmologies are created from vacuum at the moments of "big bangs" for sub-cosmologies in the fractal cosmology. Negative energy magnetic flux tubes stable in the reverse time direction are time-reflected as positive energy magnetic flux tubes which are unstable and decay to thicker magnetic flux tubes with a weaker field strength. This picture justifies the TGD inspired cosmology in which positive and negative energy sectors dissipate in different directions of geometric time. Both sectors give positive contribution to the gravitational mass and negative energy sector would give the dominating contribution to the dark gravitational energy.

17.2.3 Robertson-Walker cosmologies

Robertson-Walker cosmologies are the basic building block of standard cosmologies and sub-critical R-W cosmologies have a very natural place in TGD framework as Lorentz invariant cosmologies. Inflationary cosmologies are replaced with critical cosmologies being parameterized by a single parameter telling the duration of the critical cosmology. Over-critical cosmologies are not possible at all.

Why Robertson-Walker cosmologies?

Robertson Walker cosmology, which is a vacuum extremal of the Kähler action, is a reasonable idealization only in the length scales, where the density of the Kähler charge vanishes. Since (visible) matter and antimatter carry Kähler charges of opposite sign this means that Kähler charge density vanishes in length scales, where matter-antimatter asymmetry disappears on the average. This length scale is certainly very large in present day cosmology: in the proposed model for cosmology its present value is of the order of 10^8 light years: the size of the observed regions containing visible matter predominantly on their boundaries [ib3]. That only matter is observed can be understood from the fact that fermions reside dominantly at future oriented space-time sheets and anti-fermions on past-oriented space-time sheets.

Robertson Walker cosmology is expected to apply in the description of the condensate locally at each condensate level and it is assumed that the GRT based criteria for the formation of "structures" apply. In particular, the Jeans criterion stating that density fluctuations with size between Jeans length and horizon size can lead to the development of the "structures" will be applied.

Imbeddability requirement for RW cosmologies

Standard Robertson-Walker cosmology is characterized by the line element [ia2]

$$ds^2 \;\; = \;\; f(a)da^2 - a^2(\frac{dr^2}{1 - kr^2} + r^2 d\Omega^2) \;\; , \tag{17.2.4}$$

where the values $k = 0, \pm 1$ of k are possible.

The line element of the light cone is given by the expression

$$ds^2 \;\; = \;\; da^2 - a^2(\frac{dr^2}{1 + r^2} + r^2 d\Omega^2) \;\; . \tag{17.2.5}$$

Here the variables a and r are defined in terms of standard Minkowksi coordinates as

$$a \;\; = \;\; \sqrt{(m^0)^2 - r_M^2} \;\; ,$$
$$r_M \;\; = \;\; ar \;\; . \tag{17.2.6}$$

Light cone clearly corresponds to mass density zero cosmology with $k = -1$ and this makes the case $k = -1$ is rather special as far imbeddings are considered since any Lorentz invariant map $M^4_+ \to CP_2$ defines imbedding

$$s^k = f^k(a) \ . \tag{17.2.7}$$

Here f^k are arbitrary functions of a.

$k = -1$ requirement guarantees imbeddability if the matter density is positive as is easy to see. The matter density is given by the expression

$$\rho = \frac{3}{8\pi G a^2}\left(\frac{1}{g_{aa}} + k\right) \ . \tag{17.2.8}$$

A typical imbedding of $k = -1$ cosmology is given by

$$\begin{aligned} \phi &= f(a) \ , \\ g_{aa} &= 1 - \frac{R^2}{4}(\partial_a f)^2 \ . \end{aligned} \tag{17.2.9}$$

where ϕ can be chosen to be the angular coordinate associated with a geodesic sphere of CP_2 (any one-dimensional sub-manifold of CP_2 works equally well). The square root term is always positive by the positivity of the mass density and the imbedding is indeed well defined. Since g_{aa} is smaller than one, the matter density is necessarily positive.

Critical and over-critical cosmologies

TGD allows the imbeddings of a one-parameter family of critical over-critical cosmologies. Critical cosmologies are however not inflationary in the sense that they would involve the presence of scalar fields. Exponential expansion is replaced with a logarithmic one so that the cosmologies are in this sense exact opposites of each other. Critical cosmology has been used hitherto as a possible model for the very early cosmology. What is remarkable that this cosmology becomes vacuum at the moment of 'Big Bang' since mass density behaves as $1/a^2$ as function of the light cone proper time. Instead of 'Big Bang' one could talk about 'Small Whisper amplified to bang' gradually. This is consistent with the idea that space-time sheet begins as a vacuum space-time sheet for some moment of cosmic time. As an imbedded 4-surface this cosmology would correspond to a deformed future light cone having its tip inside the future light cone. The interpretation of the tip as a seed of a phase transition is possible. The imbedding makes sense up to some moment of cosmic time after which the cosmology becomes necessarily hyperbolic. At later time hyperbolic cosmology stops expanding and decomposes to disjoint 3-surfaces behaving as particle like objects co-moving at larger cosmological space-time sheet. These 3-surfaces topologically condense on larger space-time sheets representing new critical cosmologies.

Consider now in more detail the imbeddings of the critical and overcritical cosmologies. For $k = 0, 1$ the imbeddability requirement fixes the cosmology almost uniquely. To see this, consider as an example of $k = 0/1$ imbedding the map from the light cone to S^2, where S^2 is a geodesic sphere of CP_2 with a vanishing Kähler form (any Lagrage manifold of CP_2 would do instead of S^2). In the standard coordinates (Θ, Φ) for S^2 and Robertson-Walker coordinates (a, r, θ, ϕ) for future light cone (, which can be regarded as empty hyperbolic cosmology), the imbedding is given as

$$\begin{aligned} sin(\Theta) &= \frac{a}{a_1} \ , \\ (\partial_r \Phi)^2 &= \frac{1}{K_0}\left[\frac{1}{1 - kr^2} - \frac{1}{1 + r^2}\right] \ , \\ K_0 &= \frac{R^2}{4a_1^2} \ , \quad k = 0, 1 \ , \end{aligned} \tag{17.2.10}$$

when Robertson-Walker coordinates are used for both the future light cone and space-time surface.

The differential equation for Φ can be written as

$$\partial_r \Phi = \pm \sqrt{\frac{1}{K_0}\left[\frac{1}{1-kr^2} - \frac{1}{1+r^2}\right]} \ . \qquad (17.2.11)$$

For $k = 0$ case the solution exists for all values of r. For $k = 1$ the solution extends only to $r = 1$, which corresponds to a 4-surface $r_M = m^0/\sqrt{2}$ identifiable as a ball expanding with the velocity $v = c/\sqrt{2}$. For $r \to 1$ Φ approaches constant Φ_0 as $\Phi - \Phi_0 \propto \sqrt{1-r}$. The space-time sheets corresponding to the two signs in the previous equation can be glued together at $r = 1$ to obtain sphere S^3.

The expression of the induced metric follows from the line element of future light cone

$$ds^2 = da^2 - a^2\left(\frac{dr^2}{1-kr^2} + r^2 d\Omega^2\right) \ . \qquad (17.2.12)$$

The imbeddability requirement fixes almost uniquely the dependence of the S^2 coordinates a and r and the g_{aa} component of the metric is given by the same expression for both $k = 0$ and $k = 1$.

$$\begin{aligned}
g_{aa} &= 1 - K \ , \\
K &\equiv K_0 \frac{1}{(1-u^2)} \ , \\
u &\equiv \frac{a}{a_1} \ .
\end{aligned} \qquad (17.2.13)$$

The imbedding fails for $a \geq a_1$. For $a_1 \gg R$ the cosmology is essentially flat up to immediate vicinity of $a = a_1$. Energy density and "pressure" follow from the general equation of Einstein tensor and are given by the expressions

$$\begin{aligned}
\rho &= \frac{3}{8\pi G a^2}\left(\frac{1}{g_{aa}} + k\right) \ , \quad k = 0, 1 \ , \\
\frac{1}{g_{aa}} &= \frac{1}{1-K} \ , \\
p &= -\left(\rho + \frac{a\partial_a\rho}{3}\right) = -\frac{\rho}{3} + \frac{2}{3}K_0 u^2 \frac{1}{(1-K)(1-u^2)^2}\rho_{cr} \ , \\
u &\equiv \frac{a}{a_1} \ .
\end{aligned} \qquad (17.2.14)$$

Here the subscript 'cr' refers to $k = 0$ case. Since the time component g_{aa} of the metric approaches constant for very small values of the cosmic time, there are no horizons associated with this metric. This is clear from the formula

$$r(a) = \int_0^a \sqrt{g_{aa}}\frac{da}{a}$$

for the horizon radius.

The mass density associated with these cosmologies behaves as $\rho \propto 1/a^2$ for very small values of the M_+^4 proper time. The mass in a co-moving volume is proportional to $a/(1-K)$ and goes to zero at the limit $a \to 0$. Thus, instead of Big Bang one has 'Silent Whisper' gradually amplifying to Big Bang. The imbedding fails at the limit $a \to a_1$. At this limit energy density becomes infinite. This cosmology can be regarded as a cosmology for which co-moving strings ($\rho \propto 1/a^2$) dominate the mass density as is clear also from the fact that the "pressure" becomes negative at big bang ($p \to -\rho/3$) reflecting the presence of the string tension. The natural interpretation is that cosmic strings condense on the space-time sheet which is originally empty.

The facts that the imbedding fails and gravitational energy density diverges for $a = a_1$ necessitates a transition to a hyperbolic cosmology. For instance, a transition to radiation or matter dominated hyperbolic cosmology can occur at the limit $\theta \to \pi/2$. At this limit $\phi(r)$ must transform to a function $\phi(a)$. The fact, that vacuum extremals of Kähler action are in question, allows large flexibility for the modelling of what happens in this transition. Quantum criticality and p-adic fractality suggest the presence of an entire fractal hierarchy of space-time sheets representing critical cosmologies created at certain values of cosmic time and having as their light cone projection sub-light cone with its tip at some a=constant hyperboloid.

More general imbeddings of critical and over-critical cosmologies as vacuum extremals

In order to obtain imbeddings as more general vacuum extremals, one must pose the condition guaranteeing the vanishing of corresponding the induced Kähler form (see the Appendix of this book). Using coordinates $(r, u = cos(\Theta), \Psi, \Phi)$ for CP_2 the surfaces in question can be expressed as

$$
\begin{aligned}
r &= \sqrt{\frac{X}{1-X}} \; , \\
X &= D|k+u| \; , \\
u &\equiv cos(\Theta) \; , \quad D = \frac{r_0^2}{1+r_0^2} \times \frac{1}{C} \; , \quad C = |k + cos(\Theta_0)| \; .
\end{aligned}
\tag{17.2.15}
$$

Here C and D are integration constants.

These imbeddings generalize to imbeddings to $M^4 \times Y^2$, where Y^2 belongs to a family of Lagrange manifolds described in the Appendix of this book with induced metric

$$
\begin{aligned}
ds^2_{eff} &= \frac{R^2}{4}[s^{eff}_{\Theta\Theta} d\Theta^2 + s^{eff}_{\Phi\Phi} d\Phi^2] \; , \\
s^{eff}_{\Theta\Theta} &= X \times \left[\frac{(1-u^2)}{(k+u)^2} \times \frac{1}{1-X} + 1 - X \right] \; , \\
s^{eff}_{\Phi\Phi} &= X \times \left[(1-X)(k+u)^2 + 1 - u^2 \right] \; .
\end{aligned}
\tag{17.2.16}
$$

For $k \neq 1$ $u = \pm 1$ corresponds in general to circle rather than single point as is clear from the fact that $s^{eff}_{\Phi\Phi}$ is non-vanishing at $u = \pm 1$ so that u and Φ parameterize a piece of cylinder. The generalization of the previous imbedding is as

$$
sin(\Theta) = ka \quad \to \quad \sqrt{s^{eff}_{\Phi\Phi}} = ka \; .
\tag{17.2.17}
$$

For Φ the expression is as in the previous case and determined by the requirement that g_{rr} corresponds to $k = 0, 1$.

The time component of the metric can be expressed as

$$
g_{aa} = 1 - \frac{R^2 k^2}{4} \frac{s^{eff}_{\Theta\Theta}}{\frac{d\sqrt{s^{eff}_{\Phi\Phi}}}{d\Theta}}
\tag{17.2.18}
$$

In this case the $1/(1-k^2a^2)$ singularity of the density of gravitational mass at $\Theta = \pi/2$ is shifted to the maximum of $s^{eff}_{\Phi\Phi}$ as function of Θ defining the maximal value a_{max} of a for which the imbedding exists at all. Already for $a_0 < a_{max}$ the vanishing of g_{aa} implies the non-physicality of the imbedding since gravitational mass density becomes infinite.

The geometric properties of critical cosmology change radically in the transition to the radiation dominated cosmology: before the transition the CP_2 projection of the critical cosmology is two-dimensional. After the transition it is one-dimensional. Also the isometry group of the cosmology changes from $SO(3) \times E^3$ to $SO(3,1)$ in the transition. One could say that critical cosmology represents Galilean Universe whereas hyperbolic cosmology represents Lorentzian Universe.

String dominated cosmology

A particularly interesting cosmology is string dominated cosmology with very nearly critical mass density. Assuming that strings are co-moving the mass density of this cosmology is proportional to $1/a^2$ instead of the $1/a^3$ behavior characteristic to the standard matter dominated cosmology. The line element of this metric is very simple: the time component of the metric is simply constant smaller than 1:

$$g_{aa} = K < 1 \ . \tag{17.2.19}$$

The Hubble constant for this cosmology is given by

$$H = \frac{1}{\sqrt{K}a} \ , \tag{17.2.20}$$

and the so called acceleration parameter [ia2] k_0 proportional to the second derivative $\ddot{a}$ therefore vanishes. Mass density and pressure are given by the expression

$$\rho = \frac{3}{8\pi GKa^2}(1 - K) = -3p \ . \tag{17.2.21}$$

What makes this cosmology so interesting is the absence of the horizons. The comparison with the critical cosmology shows that these two cosmologies resemble each other very closely and both could be used as a model for the very early cosmology.

Stationary cosmology

An interesting candidate for the asymptotic cosmology is stationary cosmology for which gravitational four-momentum currents (and also gravitational color currents) are conserved. This cosmology extremizes the Einstein-Hilbert action with cosmological term given by $\int (kR + \lambda)\sqrt{g}d^4x + \lambda$ and is obtained as a sub-manifold $X^4 \subset M_+^4 \times S^1$, where S^1 is the geodesic circle of CP_2 (note that imbedding is now unique apart from isometries by variational principle).

For a vanishing cosmological constant, field equations reduce to the conservation law for the isometry associated with S^1 and read

$$\partial_a(G^{aa}\partial_a\phi\sqrt{g}) = 0 \ , \tag{17.2.22}$$

where ϕ denotes the angle coordinate associated with S^1. From this one finds for the relevant component of the metric the expression

$$g_{aa} = \frac{(1 - 2x)}{(1 - x)} \ ,$$
$$x = (\frac{C}{a})^{2/3} \ . \tag{17.2.23}$$

The mass density and "pressure" of this cosmology are given by the expressions

$$\rho = \frac{3}{8\pi Ga^2}\frac{x}{(1 - 2x)} \ ,$$
$$p = -(\rho + \frac{a\partial_a\rho}{3}) = -\frac{\rho}{9}\left[3 - \frac{2}{(1 - 2x)}\right] \ . \tag{17.2.24}$$

The asymptotic behavior of the energy density is $\rho \propto a^{-8/3}$. "Pressure" becomes negative indicating that this cosmology is dominated by the string like objects, whose string tension gives negative contribution to to the "pressure". Also this cosmology is horizon free as are all string dominated cosmologies: this is of crucial importance in TGD inspired cosmology.

It should be noticed that energy density for this cosmology becomes infinite for $x = (C/a)^{2/3} = 1/2$ implying that this cosmology doesn't make sense at very early times so that the non-conservation of gravitational energy is necessary during the early stages of the cosmology.

Non-conservation of gravitational energy in RW cosmologies

In RW cosmology the gravitational energy in a given co-moving sphere of radius r in local light cone coordinates (a, r, θ, ϕ) is given by

$$E \;=\; \int \rho g^{aa} \partial_a m^0 \sqrt{g}\, dV \; . \tag{17.2.25}$$

The rate characterizing the non-conservation of gravitational energy is determined by the parameter X defined as

$$X \;\equiv\; \frac{(dE/da)_{vap}}{E} = \frac{(dE/da + \int |g^{rr}| p \partial_r m^0 \sqrt{g}\, d\Omega)}{E} \;\;, \tag{17.2.26}$$

where p denotes the pressure and $d\Omega$ denotes angular integration over a sphere with radius r. The latter term subtracts the energy flow through the boundary of the sphere.

The generation of the pairs of positive and negative (inertial) energy space-time sheets leads to non-conservation of gravitational energy. The generation of pairs of positive and negative energy cosmic strings would be involved with the generation of a critical sub-cosmology.

For RW cosmology with subcritical mass density the calculation gives

$$X \;=\; \frac{\partial_a(\rho a^3/\sqrt{g_{aa}})}{(\rho a^3/\sqrt{g_{aa}})} + \frac{3 p g_{aa}}{\rho a} \; . \tag{17.2.27}$$

This formula applies to any infinitesimal volume. The rate doesn't depend on the details of the imbedding (recall that practically any one-dimensional sub-manifold of CP_2 defines a huge family of subcritical cosmologies). Apart from the numerical factors, the rate behaves as $1/a$ in the most physically interesting RW cosmologies. In the radiation dominated and matter dominated cosmologies one has $X = -1/a$ and $X = -1/2a$ respectively so that gravitational energy decreases in radiation and matter dominated cosmologies. For the string dominated cosmology with $k = -1$ having $g_{aa} = K$ one has $X = 2/a$ so that gravitational energy increases: this might be due to the generation of dark matter due to pairs of cosmic strings with vanishing net inertial energy.

For the cosmology with exactly critical mass density Lorentz invariance is broken and the contribution of the rate from 3-volume depends on the position of the co-moving volume. Taking the limit of infinitesimal volume one obtains for the parameter X the expression

$$\begin{aligned}
X &= X_1 + X_2 \;, \\
X_1 &= \frac{\partial_a(\rho a^3/\sqrt{g_{aa}})}{(\rho a^3/\sqrt{g_{aa}})} \;, \\
X_2 &= \frac{p g_{aa}}{\rho a} \times \frac{3 + 2r^2}{(1 + r^2)^{3/2}} \; .
\end{aligned} \tag{17.2.28}$$

Here r refers to the position of the infinitesimal volume. Simple calculation gives

$$X = X_1 + X_2 \;,$$

$$X_1 = \frac{1}{a}\left[1 + 3K_0 u^2 \frac{1}{1-K}\right] \;,$$

$$X_2 = -\frac{1}{3a}\left[1 - K - \frac{2K_0 u^2}{(1-u^2)^2}\right] \times \frac{3 + 2r^2}{(1 + r^2)^{3/2}} \;,$$

$$K = \frac{K_0}{1 - u^2} \;, \quad u = \frac{a}{a_0} \;, \quad K_0 = \frac{R^2}{4a_0^2} \; . \tag{17.2.29}$$

The positive density term X_1 corresponds to increase of gravitational energy which is gradually amplified whereas pressure term ($p < 0$) corresponds to a decrease of gravitational energy changing however its sign at the limit $a \to a_0$.

The interpretation is in terms of creation of pairs of positive and negative energy particles contributing nothing to the inertial energy. Also pairs of positive energy gravitons and negative anti-gravitons are involved. The contributions of all particle species are determined by thermal arguments so that gravitons should not play any special role as thought originally.

Pressure term is negligible at the limit $r \to \infty$ so that topological condensation occurs all the time at this limit. For $a \to 0, r \to 0$ one has $X > 0 \to 0$ so that condensation starts from zero at $r = 0$. For $a \to 0, r \to \infty$ one has $X = 1/a$ which means that topological condensation is present already at the limit $a \to 0$.

Both the existence of the finite limiting temperature and of the critical mass density imply separately finite energy per co-moving volume for the condensate at the very early stages of the cosmic evolution. In fact, the mere requirement that the energy per co-moving volume in the vapor phase remains finite and non-vanishing at the limit $a \to 0$ implies string dominance as the following argument shows.

Assuming that the mass density of the condensate behaves as $\rho \propto 1/a^{2(1+\alpha)}$ one finds from the expression

$$\rho \propto \frac{(\frac{1}{g_{aa}} - 1)}{a^2} \ ,$$

that the time component of the metric behaves as $g_{aa} \propto a^\alpha$. Unless the condition $\alpha < 1/3$ is satisfied or equivalently the condition

$$\rho \ < \ \frac{k}{a^{2+2/3}} \tag{17.2.30}$$

is satisfied, gravitational energy density is reduced. In fact, the limiting behavior corresponds to the stationary cosmology, which is not imbeddable for the small values of the cosmic time. For stationary cosmology gravitational energy density is conserved which suggests that the reduction of the density of cosmic strings is solely due to the cosmic expansion.

17.2.4 Cosmic strings and TGD inspired cosmology

The vapor phase density of cosmic strings is source of all matter and radiation in TGD based cosmology. Therefore cosmic strings deserve a separate discussion. More details can be found in [D4].

Topological condensation of cosmic strings

Free cosmic strings as such are certainly not absolute minima of Kähler action and they are expected to decay rapidly to ordinary particles. Topological condensation of the cosmic strings with associated generation of the radial Kähler electric field provides a mechanism for the generation of negative action. The requirement that the Kähler electric field leads to the cancellation of the action of the cosmic string implies that cosmic string carries enormous Kähler charge (roughly one unit of Kähler charge per CP_2 radius).

The basic question whether the exterior region of the topologically condensed cosmic string can be modelled using only General Relativity.

a) In Kähler neutral case, an extremely simple imbedding of the straight string exterior metric exists. This model cannot however guarantee the cancellation of the positive Kähler magnetic action. The idea is obvious. If the string generates a sufficiently strong Kähler charge, the negative Kähler electric action cancels the positive Kähler magnetic action of the string.

b) The string creates a gravitational potential consisting of terms assignable to the gravitational mass of string itself and with the Kähler electric field created by it. The problem is that the gravitational constant is predicted to be roughly 10^7 times too large. The new view about energy resolves this problem: Kähler energy represents the density of inertial energy and is much smaller than gravitational energy. The cancellation of Kähler action implies that the gravitational mass of string is by a factor about 10^7 larger than the inertial mass of the string. Equality is expected in the case that string carries no matter or antimatter besides field energy.

A possible resolution of the problem comes from the observation that cosmic strings must be created from vacuum as pairs of strings with opposite time orientation and inertial energies. For tightly bound, perhaps even coiled, pairs of cosmic strings radial Kähler fields have opposite directions and tend to cancel each other. There is however a slight matter antimatter asymmetry so that the induced Kähler field is not a pure dipole field and can cancel the Kähler magnetic action.

The pairing must be rather precise so that strings form a coiled structure geometrically analogous to a DNA double helix. This structure could contract to a point at the limit $a \to 0$ for M_+^4 option. One could say that two cosmic strings, whose inertial energies are of opposite sign and of equal magnitude are created from vacuum at or shortly after the moment of big bang. For M^4 option the evolution before the moment of big bang could be preceded by a gravitational collapse and one would have a mirror pair of evolutions. Even a sequence of them making up a temporal mirror hall is possible if the reduction of cosmological constant during cosmic evolution initiates gravitational collapse of the space-time sheet containing the strings at some moment. p-Adic fractality and the fact that the evolution of stars ends with a collapse support this view.

c) The two cosmic strings, which are phase conjugates of each other in the sense that fermionic vacua are related by Hermitian conjugation, must reduce the ratio of their inertial and gravitational masses. The simplest options correspond to either the growth of gravitational mass or the reduction of inertial mass.

Depending on the character of the stringy fermionic vacua, it is possible to imagine two different models for how this could occur.

i) The fact that cosmic strings are not small perturbations of vacuum extremals might be seen as favoring the view that the string vacua are conventional so that both fermions and anti-fermions have positive inertial energy and $m_{gr} = m_I$ holds true for each string. Let $X^4_{I/II}$ be nearly vacuum extremal space-time sheets of opposite time orientation containing strings I and II and correspond to un-conventional phase vacua which are phase conjugates of each other. Since $X^4_{I/II}$ can carry only a positive/negative fermion number, X^4_I and X^4_{II} must be created from the vacuum as a phase conjugate pair with opposite fermion numbers. Matter antimatter asymmetry would result from the vacuum polarization of $X^4_{I/II}$ in the immediate vicinity of strings generating fermion/antifermion number and gravitational mass of magnitude $\sim 10^7 m_I$.

The gravitational string tension of string as seen from large distance would be given by $T_{gr} \sim 1/G$ and inertial string tension would be essentially $T_I \sim 1/R^2$ and the two fundamental scales would correspond to gravitational and inertial masses. The gravitational string tension of cosmic strings explaining the observed galactic rotation curves corresponds to T_I rather than $T_{gr} \sim 1/G$. Hence it seems that this option is not favored.

ii) The second option is that the string vacua are non-conventional so that fermions and anti-fermions have opposite energies inside them. In this case the strings can reduce their inertial masses by the analog of Hawking radiation involving the generation of fermion anti-fermion pairs, whose second member remains inside string. Assume that the vacuum of the space-time sheet X^4 containing phase conjugate strings I and II, corresponds to the vacuum of either string. String I ($E > 0$) emits positive energy fermions and develops inside it a density of negative energy anti-fermions (fermion number increases). String II ($E < 0$) emits negative energy anti-fermions and develops inside it a density of positive energy fermions. The signs of the fermion number and Kähler charge of the string correlate.

This "Hawking radiation" generates at least part of visible matter. The splitting of cosmic strings followed by a "burning" of the string ends provides a second manner to generate visible matter. Fermions and negative energy anti-fermions dominate the energy density at the space-time sheet containing strings and strings themselves contain dominantly negative energy fermions and positive energy anti-fermions. This explains the effective absence of antimatter.

The conservation of magnetic flux implies that the reduction of Kähler energy leads to the thickening of the strings and they become Kähler magnetic flux tubes.

d) p-Adic fractality and simple quantitative observations lead to the hypothesis that pairs of positive and negative energy cosmic strings are responsible for the evolution of astrophysical structures in a very wide length scale range. Large voids with size of order 10^8 light years can be seen as structures containing knotted and linked cosmic string pairs wound around the boundaries of the void. Galaxies correspond to same structure with smaller size and linked around the supra-galactic strings. This conforms with the finding that galaxies tend to be grouped along linear structures [ib10]. Simple quantitative estimates of [D4] show that even stars and planets could be seen as structures formed around cosmic strings of appropriate size. Thus Universe could be seen as fractal cosmic necklace consisting of cosmic strings linked like pearls around longer cosmic strings linked like...

Cosmic strings and cosmological constant

The study of the simplest extremals of Kähler action demonstrates that cosmological constant in TGD Universe is non-vanishing [D3, D4]. What is new that p-adic length fractality predicts that Λ scales as $1/L^2(k)$ as a function of the p-adic scale characterizing the space-time sheet implying a series of phase transitions reducing Λ. The recent value of the cosmological constant comes out correctly. The gravitational energy density assignable to the cosmological constant is identifiable as that associated with topologically condensed cosmic strings and magnetic flux tubes to which they are gradually transformed during cosmological evolution.

The value of the cosmological constant Λ is different at each space-time sheet the formula $\Lambda(k) = \Lambda(2) \times (L(2)/L(k))^2$ follows from the requirement that gravitational energy as difference of inertial energies and matter and antimatter (with an appropriate sign factor) is non-negative.

The ordinary gravitational potential energy of a co-moving volume decreases as $E_{gr} \propto M^2(L(k))/L(k)$ as a function of the p-adic scale, whereas cosmological constant varying as $1/L(k)^2$ contributes to the gravitational energy a positive term $E_\Lambda \propto L(k)$ implying $V_{gr} \propto L(k)$ and $E_{tot} \propto L(k)$. The prediction that the energy of the space-time sheet is proportional to its size is just what has been observed to hold true for galactic dark matter distribution. This suggest that the dark matter is due to Kähler neutral space-time sheets containing pairs of cosmic strings and cosmological constant characterizes the density of the dark matter decreasing as $1/T^2(k)$ and on the average as $1/a^2$ as a function of cosmic time. Thus the original idea about cosmic strings as the source of dark matter would be basically correct.

Since cosmological expansion implies a transformation of cosmic strings to gradually thickening Kähler magnetic flux tubes, $\Lambda(k)$ must decrease during cosmological evolution of the void and increase again during the possibly occurring contraction phase. The reduction of the value of Λ below a critical value might initiate the phase of contraction.

Cosmic strings and matter-antimatter asymmetry

Matter-antimatter asymmetry is one of the deepest puzzles of modern physics and during the last 25 years of TGD it has become clear that its understanding might require new physical principles instead of mechanisms. The new view about energy changes dramatically the interpretation of the early periods of cosmology although the basic equations for the density of gravitational mass are not changed.

1. Pairs of cosmic strings and generation of matter antimatter asymmetry

a) The creation of the Universe from vacuum would mean generation of tightly bound, possibly coiled, pairs of cosmic strings having opposite time orientations from vacuum during the first moments of cosmology. I have used the notion of "wormhole magnetic field" about pairs of space-time sheets carrying magnetic fields in opposite directions and having opposite time orientations [J5]. This terminology would be appropriate also now.

b) The topological condensation of cosmic strings to the background space-time sheets would induce a phase transition in which positive energy antimatter falls into negative energy states by emitting the difference energy as ordinary particles of positive or negative energy depending on the sign of time orientation. Besides this process also the decay of the ends of split cosmic strings to elementary particles could serve as a source of ordinary matter. Since Kähler magnetic flux is conserved in the process of thickening of the flux tubes, the magnetic energy density per unit length for topologically condensed cosmic strings decreases like $1/S$ (S is the area of cross section) in the process, and they become gradually magnetic or Z^0 magnetic flux tubes serving as templates for various structures. The emitted particles topologically condense at the background space-time and possess vanishing average inertial energy density in the length scales where cosmology applies.

Critical cosmology naturally represents this phase transition, and the breaking of Lorentz symmetry can be interpreted as being caused by the seed of the phase transition located at $r = 0$. With the conventional interpretation for the Einstein tensor, the recent cosmology of course poses strong lower bounds on the sizes of these regions but with the new interpretation the separation of matter and antimatter corresponds to the breaking of the Cosmological Principle (Lorentz invariance) due to the presence of non-vanishing Kähler fields.

c) The Hawking radiation generating ordinary matter outside strings implies that outside region contains net number of positive energy fermions number and net number of negative energy anti-fermions and Kähler force (sum of Z^0 and em forces) is opposite for these particles so that separation could occur.

d) The photon to baryon ratio $r \sim 10^{-9}$ characterizes matter asymmetry quantitatively. The parameter ϵ characterizing the ratio of the densities of inertial energy and gravitational energy has the upper bound

$$\epsilon_{max} \;=\; \frac{G}{8\alpha_K R^2} \simeq .22 \times 10^{-6} \; . \tag{17.2.31}$$

This bound is for $\alpha_K(p = M_{127})$ corresponding to electron's p-adic length scale. The bound increase to $\epsilon_{max} \simeq 2.5 \times 10^{-8}$ for $p = 2$. Thus it would not be too difficult to relate r to the parameter ϵ which indeed measures matter antimatter asymmetry since $\epsilon = 0$ means complete symmetry.

2. Does a separation matter and antimatter occur?

The study of spherically symmetric stationary non-vacuum extremals of Kähler action [D1] representable as surfaces in $M_+^4 \times S_I^2$ led to the conclusion that the frequency ω characterizing the extremal has opposite sign for particle and anti-particle. Suppose that the asymptotic behavior of the metric of the elementary particle space-time sheet determines the gravitational mass of the particle for vacuum extremals, and that the asymptotic value u_∞ of the coordinate $u = cos(\Theta)$ at particle's space-time sheet (having much smaller size than condensate) is same as the local value of u in condensate. Under these assumptions one must conclude that gravitational mass is of opposite sign for fermions and anti-fermions at a given point of topological condensate.

This argument, which was developed before the new view about the relationship between gravitational and inertial four-momenta and could have holes in it, would suggest that the gravitational masses of the fermions and anti-fermions are of opposite sign at a given point of the space-time sheet. Hence matter and antimatter must reside in different space-time regions, most naturally at different space-time sheets. The negativity of the gravitational mass could induce a gradual drift of matter and antimatter into regions, where the gravitational mass is positive and thus generate matter antimatter asymmetry. With this interpretation Einstein's equations would hold true for asymptotic self-organization patterns just like the Lorentz Kähler four-force vanishes only for them.

3. Other aspects of matter antimatter asymmetry

The CKM matrix for quarks contains CP breaking phase factors and this could lead to different evaporation rates for baryons and anti-baryons are different (quark cannot appear as vapor phase particle since vapor phase particle must have vanishing color gauge charges). The CP breaking at the level of CKM matrix in turn might have explanation in terms of the hadronic Kähler electric fields.

Also the presence of CP breaking phase factor $exp(ikS_{CS})$, where $S_{CS} = \int_{X^4} J \wedge J + boundary\ term$ is purely topological Chern Simons term and naturally associated with the boundaries of space-time sheets with at most $D = 3$-dimensional CP_2 projection, could have something to do with the matter antimatter asymmetry.

Development of strings in the string dominated cosmology

The development of the string perturbations in the Robertson Walker cosmology has been studied [ib4] and the general conclusion seems to be that that all the details smaller than horizon are rapidly smoothed out. In present case, the horizon has an infinite size so that details in all scales should die away. To see what actually happens consider small perturbations of a static string along z-axis. Restrict the consideration to a perturbation in the y-direction. Using instead of the proper time coordinate t the "conformal time coordinate" τ defined by $d\tau = dt/a$ the equations of motion read [ib4]

$$(\partial_\tau + \frac{2\dot{a}}{a})(\dot{y}U) = \partial_z(y'U) \; ,$$
$$U = \frac{1}{\sqrt{1 + (y')^2 - \dot{y}^2}} \; . \tag{17.2.32}$$

Restrict the consideration to small perturbations for which the condition $U \simeq 1$ holds. For the string dominated cosmology the quantity $\dot{a}/a = 1/\sqrt{K}$ is constant and the equations of motion reduce to a very simple approximate form

$$\ddot{y} + \frac{2}{\sqrt{K}}\dot{y} - y'' = 0 \; . \tag{17.2.33}$$

The separable solutions of this equation are of type

$$y = g(a)(C\sin(kz) + D\cos(kz)) \; ,$$
$$g(a) = (\frac{a}{a_0})^r \; . \tag{17.2.34}$$

where r is a solution of the characteristic equation $r^2 + 2r/\sqrt{K} + k^2 = 0$:

$$r = -\frac{1}{\sqrt{K}}(1 \pm \sqrt{1 - k^2 K}) \; . \tag{17.2.35}$$

For perturbations of small wavelength $k > 1/\sqrt{K}$, an extremely rapid attenuation occurs; $1/\sqrt{K} \simeq 10^{27}$! For the long wavelength perturbations with $k << 1/\sqrt{K}$ (physical wavelength is larger than t) the attenuation is milder for the second root of above equation: attenuation takes place as $(a/a_0)^{\sqrt{K}k^2/2}$. The conclusion is that irregularities in all scales are smoothed away but that attenuation is much slower for the long wave length perturbations.

The absence of horizons in the string dominated phase has a rather interesting consequence. According to the well known Jeans criterion the size L of density fluctuations leading to the formation of structures [ib4] must satisfy the following conditions

$$l_J < L < l_H \; , \tag{17.2.36}$$

where l_H denotes the size of horizon and l_J denotes the Jeans length related to the sound velocity v_s and cosmic proper time as [ib4]

$$l_J \simeq 10 v_s t \; . \tag{17.2.37}$$

For a string dominated cosmology the size of the horizon is infinite so that no upper bound for the size of the possible structures results. These structures of course, correspond to string like objects of various sizes in the microscopic description. This suggests that primordial fluctuations create structures of arbitrary large size, which become visible at much later time, when cosmology becomes string dominated again.

17.2.5 Thermodynamical considerations

The new view about energy challenging the universal applicability of the second law of thermodynamics, the existence of 'vapor phase' consisting mainly of cosmic strings and critical temperature equal to Hagedorn temperature are basic characteristics of TGD inspired cosmology.

The new view about second law

Quantum classical correspondence suggests negative and positive energy strings tend to dissipate backwards in opposite directions of the geometric time in their geometric degrees of freedom. This would mean a continual competition between ordinary dissipation of tightly paired positive energy strings. Time reversed dissipation of negative energy strings looks from the point of view of systems consisting of positive energy matter self-organization and even self assembly. The matter at the space-time sheet containing strings in turn consists of positive energy matter and negative energy antimatter and also here same competition would prevail.

This suggests a general manner to understand the paradoxical aspects of the cosmic and biological evolution.

a) The first paradox is that the initial state of cosmic evolution seems to correspond to a maximally entropic state. Entropic state could relate to space-time sheets with negative time orientation but there would be also negentropic state corresponding to the positive energy matter. The dissipative evolution of matter at space-time sheets with positive time orientation (space-time sheets of positive energy cosmic strings and those containing pairs of cosmic strings) would obey second law and evolution of space-time sheets with negative time orientation (in particular negative energy cosmic strings) its geometric time reversal. Second law would hold true in the standard sense as long as one can neglect the interaction with negative energy antimatter and strings. TGD inspired theory of consciousness predicts p-adic evolution and this would mean that negentropic tendency would win. Perhaps matter antimatter asymmetry reflects this.

b) The presence of the cosmic strings with negative energy and time orientation could explain why gravitational interaction leads to a self-assembly of systems in cosmic time scales. The formation of supernovae, black holes and the possible eventual concentration of positive energy matter at the negative energy cosmic strings could reflect the self assembly aspect due to the presence of negative energy strings. An analog of biological self assembly identified as the geometric time reversal for ordinary entropy generating evolution would be in question.

c) In the standard physics framework the emergence of life requires extreme fine tuning of the parameters playing the role of constants of Nature and the initial state of the Universe should be fixed with extreme accuracy in order to predict correctly the emergence of life. In the proposed framework situation is different. The competition between dissipations occurring in reverse time directions means that the analog of homeostasis fundamental for the functioning of living matter is realized at the level of cosmic evolution. The signalling in both directions of geometric time makes the system essentially four-dimensional with feedback loops realized as geometric time loops so that the evolution of the system would be comparable to the carving of a four-dimensional statue rather than approach to chaos.

d) The apparent creation of order by the gravitational interactions is a mystery in the standard cosmology. A naive application of the second law of thermodynamics suggests that in GRT based cosmology the most probable end state corresponds to a black hole dominated Universe since the entropy of the black hole is much larger than the entropy of a typical star with the same mass.

TGD allows to consider two alternative solutions of this puzzle.

i) The new view about second law inspires the view that gravitational self-organization corresponds to the temporal mirror image of dissipative time evolution for space-time sheets with negative time orientation competing with thermalization. The self organizing tendency of negative energy cosmic strings competing with the opposite tendency of positive energy strings and ordinary matter could give rise to kind of gravitational homeostasis. Although black hole like structure could result as outcome of gravitational self-organization they would not be sinks of information but have complex internal information carrying structure.

ii) It is also possible that elementary particles take the role of black holes in TGD framework. CP_2 type extremals are the counterparts of the black holes in TGD. Hawking-Bekenstein area law generalizes and states that elementary particles are carriers of p-adic entropy. Thus this p-adic entropy associated with the thermodynamics of Virasoro generator L_0 could be the counterpart of black hole entropy and the decay of the free cosmic strings to elementary particles would thus generate "invisible" entropy. The upper bound for the p-adic entropy depends on p-adic condensation level as $log(p)$ so that the generation of the new space-time sheets with increasing size (and thus p) generates new entropy since the particles, which are topologically condensed on these sheets, can have entropy of order $log(p)$.

Vapor phase

The structure of $M_+^4 \times CP_2$ suggests kinematic constraints on the cosmology: for the very small values of the M_+^4 proper time a the allowed 3-surfaces are necessarily CP_2 type surfaces or string like objects rather than pieces of M^4. As a consequence, topological evaporation should take place so that the space-time resembles enormous Feynman diagram rather than continuous "classical" space-time. It indeed turns out that although the condensate could be present also in the primordial stage, the dominant contribution to the energy density is in the vapor phase during the primordial cosmology (and as it turns out, also in recent cosmology unless one takes into account the fact that at each level of condensate cosmic expansion is only local!). The properties of the critical cosmology suggest that space-time sheets representing critical sub-cosmologies are generated only after some value $a_0 \sim R$ of light cone proper time, where $R \sim 10^4$ Planck times corresponds is CP_2 time. Before this moment there is no macroscopic space-time but only vapor phase consisting of cosmic strings having purely geometric contact interactions. Thus the idea about primordial cosmology as a stage preceding the formation of space-time in the sense of General Relativity seems to be correct in TGD framework.

The key object of the TGD inspired cosmology is cosmic string with string tension $T \simeq .2 \times 10^{-6}/G$ of same order as the string tension of the GUT strings but with totally different physical and geometric interpretation. Cosmic strings play a key role in the very early string dominated cosmology, they might generate the matter antimatter asymmetry, they lead to the formation of the large voids and galaxies, they give rise to the galactic dark matter and the also dominate the mass density in the asymptotic cosmology. Vapor phase cosmic strings might be present also in the cosmology of later times and correspond closely to the vacuum energy density of inflationary cosmologies since the condensation of cosmic strings would give rise to the cosmological constant.

For critical cosmology the gravitational energy of the co-moving volume is proportional to a at the limit $a \to 0$ and vanishes so that 'Silent Whisper' amplifying to 'Big Bang' is in question. The assumption that also vapor phase gravitational energy density (that is density in imbedding space) behaves in similar manner implies the absence of initial singularities also at vapor phase level. Thus the condition

$$\rho \;\; \propto \;\; \frac{1}{a^2} \;, \tag{17.2.38}$$

and hence the string dominated primordial cosmology both in vapor phase and space-time sheets is an attractive hypothesis mathematically. The simplest hypothesis suggested by dimensional considerations is that the mass density of the vapor phase near $a = 0$ behaves as

$$\rho \;\; = \;\; n \frac{3}{8\pi G a^2} \;. \tag{17.2.39}$$

Here n is numerical factor of order one. This hypothesis fixes the total energy density of the universe and sets strong constraints on energetics of the cosmology. At later stages topological condensation of the strings reduces the mass density in vapor phase and replaces n by a decreasing function of a. A very attractive hypothesis is that the value of of n is

$$n = 1 \;. \tag{17.2.40}$$

This gravitational energy density is same as that of critical cosmology at the limit of flatness and can be interpreted as TGD counterpart for the basic hypothesis of inflationary cosmologies. In inflationary cosmologies 70 per cent of the critical mass density is in form of vacuum energy deriving from cosmological constant. In TGD the counterpart of vacuum energy is the mass density of cosmic strings in vapor phase.

One can criticize the assumption as un-necessarily strong. There is no absolute necessity for the density of gravitational four-momentum of strings in M_+^4 to be conserved and one can consider the possibility that zero inertial energy string pairs are created from vacuum everywhere inside future light cone.

Long range interactions in the vapor phase are generated only by the exchange of particle like 3-surfaces and the long range interactions mediated by the exchange of the boundary components are impossible. The exchange of CP_2 type extremals has geometric cross section and the same is expected to be true for the other exchanges of the particle like surfaces. This would mean that the interaction cross sections are determined by the size of the particle of the order of CP_2 radius: $\sigma \simeq l^2 \sim 10^8 G$. In this sense the asymptotic freedom of gauge theories would be realized in the vapor phase. It should be emphasized that this assumption might be wrong and that the gauge interactions between two particles belonging to vapor phase and condensate respectively are certainly present and topological condensation can be indeed seen as this interaction. It should be noticed that the expansion of the Universe in vapor phase is slower than in condensed phase: the ratio of the expansion rates of the universe in vapor and condensed phases is given by the velocity of light in the condensed phase $(c_\# = \sqrt{g_{aa}})$.

Also the cross sections for the purely geometric contact interactions of free cosmic strings are extremely low. This suggests that vapor phase is in essentially in temperature zero string dominated state and that the energy density of strings behaves as $1/a^2$.

Limiting temperature

Since particles are extended objects in TGD, one expects the existence of the limiting temperature T_H (Hagedorn temperature as it is called in string models) so that the primordial cosmology is in Hagedorn temperature. A special consequence is that the contribution of the light particles to the energy density becomes negligible: this is in accordance with the string dominance of the critical mass cosmology. The value of T_H is of order $T_H \sim 1/R$, where R is CP_2 radius of order $R \sim 10^4\sqrt{G}$ and thus considerably smaller than Planck temperature.

The existence of limiting temperature gives strong constraint to the value of the light cone proper time a_F when radiation dominance must have established itself in the critical cosmology which gave rise to our sub-cosmology. Before the moment of transition to hyperbolic cosmology critical cosmology is string dominated and the generation of negative energy virtual gravitons builds up gradually the huge energy density density, which can lead to gravitational collapse, splitting of the strings and establishment of thermal equilibrium with gradually rising temperature. This temperature cannot however become higher than Hagedorn temperature T_H, which serves thus as the highest possible temperature of the effectively radiation dominated cosmology following the critical period. The decay of the split strings generates elementary particles providing the seeds of galaxies.

If most strings decay to light particles then energy density is certainly of the form $1/a^4$ of radiation dominated cosmology. This is not the only manner to obtain effective radiation dominance. Part of the thermal energy goes to the kinetic energy of the vibrational motion of strings and energy density $\rho \propto 1/a^2$ cannot hold anymore. The strings of the condensate is expected to obey the scaling law $\rho \propto 1/a^4$, $p = \rho/3$ [ib4]. The simulations with string networks suggest that the energy density of the string network behaves as $\rho \propto 1/a^{2(1+v^2)}$, where v^2 is the mean square velocity of the point of the string [ib8]. Therefore, if the value of the mean square velocity approaches light velocity, effective radiation dominance results even when strings dominate [ib5]. In radiation dominated cosmology the velocity of sound is $v = 1/\sqrt{3}$. When v lowers to sound velocity one obtains stationary cosmology which is string dominated.

An estimate for a_F is obtained from the requirement that the temperature of the radiation dominated cosmology, when extrapolated from its value $T_R \simeq .3\text{eV}$ at the time about $a_R \sim 3 \times 10^7$ years for the decoupling of radiation and matter to $a = a_F$ using the scaling law $T \propto 1/a$, corresponds to Hagedorn temperature. This gives

$$a_F = a_R \frac{T_R}{T_H} \;,$$

$$T_H = \frac{n}{R} \;, \qquad a_R \sim 3 \times 10^7 \; y \;, \quad T_R = .27 \; eV \;.$$

$$(17.2.41)$$

This gives a rough estimate $a_F \sim 3 \times 10^{-10}$ seconds, which corresponds to length scale of order 7.7×10^{-2} meters. The value of a_F is quite large.

The result does not mean that radiation dominated sub-cosmologies might have not developed before $a = a_F$. In fact, entire series of critical sub-cosmologies could have developed to radiation dominated phase before the final one leading to our sub-cosmology is actually possible. The contribution of sub-cosmology i to the total energy density of recent cosmology is in the first approximation equal to the fraction $(a_F(i)/a_F)^4$. This ratio is multiplied by a ratio of numerical factors telling the number of effectively massless particle species present in the condensate if elementary particles dominate the mass density. If strings dominate the mass density (as expected) the numerical factor is absent.

For some reason the later critical cosmologies have not evolved to the radiation dominated phase. This might be due to the reduced density of cosmic strings in the vapor phase caused by the formation of the earlier cosmologies which does not allow sufficiently strong gravitational collapse to develop and implies that critical cosmology transforms directly to stationary cosmology without the intervening effectively radiation dominated phase. Indeed, condensed cosmic strings develop Kähler electric field compensating the huge positive Kähler action of free string and can survive the decay to light particles if they are not split. The density of split strings yielding light particles is presumably the proper parameter in this respect.

p-Adic length scale hypothesis allows rather predictive quantitative model for the series of sub-cosmologies [D6] predicting the number of them and allowing to estimate the moments of their birth, the durations of the critical periods and also the durations of radiation dominated phases. p-Adic length scale hypothesis allows also to estimate the maximum temperature achieved during the critical period: this temperature depends on the duration of the critical period a_1 as $T \sim n/a_1$, where n turns out to be of order 10^{30}. This means that if the duration of the critical period is long enough, transition to string dominated asymptotic cosmology occurs with the intervening decay of cosmic strings leading to the radiation dominated phase.

The existence of the limiting temperature has radical consequences concerning the properties of the very early cosmology. The contribution of a given massless particle to the energy density becomes constant. So, unless the number of the effectively massless particle families $N(a)$ increases too fast the contribution of the effectively massless particles to the energy density becomes negligible. The massive excitations of large size (string like objects) are indeed expected to become dominant in the mass density.

What about thermodynamical implications of dark matter hierarchy?

The previous discussion has not mentioned dark matter hierarchy labelled by increasing values of Planck constants and predicted macroscopic quantum coherence in arbitrarily long scales. In TGD Universe dark matter hierarchy means also a hierarchy of conscious entities with increasingly long span of memory and higher intelligence [TGDconsc, M3].

This forces to ask whether the second law is really a fundamental law and whether it could reflect a wrong view about existence resulting resulting when all these dark matter levels and information associated with conscious experiences at these levels is neglected. For instance, biological evolution difficult to understand in a universe obeying second law relies crucially on evolution as gradual progress in which sudden leaps occur as new dark matter levels emerge.

TGD inspired consciousness suggests that Second Law holds true only for the mental images of a given self (a system able to avoid bound state entanglement with environment [TGDconsc]) rather than being a universal physical law. Besides these mental images there is irreducible basic awareness of self and second law does not apply to it. Also the hierarchy of higher level conscious entities is there. In this framework second law would basically reflect the exclusion of conscious observers from the the physical model of the Universe.

17.3   TGD inspired cosmology

Quantum criticality suggests strongly quantum critical fractal cosmology containing cosmologies inside cosmologies such that each sub-cosmology is critical before transition to hyperbolic phase. The general conceptual framework represented in the previous section give rather strong constraints on fractal cosmology. There are reasons to believe that the scenario to be represented, although by no means the final formulation, contains several essential features of what might be called TGD inspired cosmology.

17.3.1   Primordial cosmology

TGD inspired cosmology has primordial phase in which only vapor phase containing only cosmic strings is present and lasting to $a \sim R$. During this period it is not possible to speak about space-time in the sense of General Relativity. The energy density and 'pressure' of cosmic strings in vapor phase (densities in $M_+^4 \times CP_2$ are assumed to be

$$\rho_V = \frac{3}{8\pi G a^2} \ ,$$
$$p = -\frac{\rho}{3} \ . \tag{17.3.1}$$

This assumption would mean that gravitational energy and various gravitational counterparts of the classical charges associated with the isometries of H are conserved during vapor phase period. This assumption guarantees consistency with the critical cosmology and by the requirement that the mass per co-moving volume vanishes at the limit $a \to 0$ so that the Universe is apparently created from nothing. The interactions between cosmic strings are pure contact interactions and extremely weak and it seems natural to assume that the temperature of the vapor phase is zero.

It must be however added that the conservation of gravitational four-momentum is an un-necessarily strong assumption and one can quite well consider the possibility that sub-cosmologies are generated from vacuum spontaneously.

17.3.2 Critical phases

The absolute minimization of Kähler action does not allow vapor phase to endure indefinitely since the Kähler magnetic action of the free cosmic string is positive and infinite at the limit of infinite duration. What happens is that space-time sheets are created and cosmic strings condense on them and generate Kähler electric fields compensating the positive Kähler magnetic action. Individual cosmic string can however stay as free cosmic string for arbitrarily long time since the finite magnetic Kähler action can be compensated by the correspondingly larger electric Kähler action. In principle cosmic strings can be created as pairs of positive and negative inertial energy cosmic strings from vacuum in vapor phase.

In accordance with this primordial phase is followed by the generation of critical cosmologies as 'Silent Whispers' amplifying to 'Big Bangs' basically by emission of ordinary matter by Hawking radiation, and possibly by gravitational heating made possible by the emission of negative energy virtual gravitons as "acceleration radiation" as matter gains strong inertial energies in gravitational fields. p-Adic length scale hypothesis allows to deduce estimates for the typical time for the creation of a critical cosmology, the duration of the critical phase, the temperature achieved during the critical phase and the duration of the hyperbolic expanding phase possibly following it and transforming to a phase in which cosmic expansion ceases and space-time surface behaves like a particle.

17.3.3 Radiation dominated phases

p-Adic length scale hypothesis suggests that the typical moments of birth $a_0(k)$ and durations $a_1(k)$ for the critical cosmologies satisfy $a_0(k) \sim L(k)$ and $a_1(k) \sim L(k)$, where k prime or power of prime, $L(k) = l \times 2^{k/2}, l = R \simeq 10^4$ Planck lengths, and n is a numerical factor. p-Adic length scale hypothesis suggest that the temperature just after the transition to the effectively radiation dominated phase is

$$T(k) = \tfrac{n}{L(k)} \ , \qquad \text{for } k > k_{cr} \ ,$$
$$T(k) = T_H \sim \tfrac{1}{R} \ , \quad \text{for } k \leq k_{cr} \ . \tag{17.3.2}$$

Here n is rather large numerical factor. Since $a_F \sim 2.7 \times 10^{-10}$ seconds which corresponds to length scale $L \simeq .08$ meters roughly to p-adic length scale $L(197) \simeq .08$ meters (which by the way corresponds to the largest p-adic length scale associated with brain, a cosmic joke?!), should correspond to the establishment of Hagedorn temperature, one has the conditions

$$k_{cr} = 197 \ ,$$
$$n \simeq 2^{197/2} \sim 10^{30} \sim \frac{m_{CP_2}^2}{m_p^2} \ .$$

Thus n is in of same order of magnitude as the ratio of the CP_2 mass squared ($m_{CP_2} \simeq 10^{-4}$ Planck masses) to proton mass squared.

Dimensional considerations suggest also that the energy density in the beginning of the radiation dominated phase (in case that it is achieved) is

$$\rho = nT^4(k) \ , \tag{17.3.3}$$

where n a numerical factor of order one. n does not count for the number of light particle species since the thermal energy of strings could give rise to the effective radiation dominance. Furthermore, if ordinary matter is created by Hawking radiation and by radiation generated by the ends of split strings, the large mass and Hagedorn temperature as a limiting temperature could make impossible the generation of particle genera higher the three lowest ones (see [F1] for the argument why $g > 2$ particle families have ultra heavy masses). Thus it seems that the infinite number of fermion families cannot lead to infinite density of thermal energy and why their presence leaves no trace in present day cosmology.

When the time parameter a_1 of the critical cosmology grows too large, it cannot anymore generate radiation dominated phase since the temperature remains too low. Previous considerations suggest that the maximum value of a_1 is roughly $a_1(max) = a_F \sim 3 \times 10^{-10}$. After this critical sub-cosmologies would transform directly to the stationary cosmologies.

Radiation dominated phase transforms to matter dominated phase and possibly decomposes to disjoint 3-surfaces with size of order horizon size at the same time. p-Adic length scale hypothesis suggests that the duration of the radiation dominated phase with respect to the proper time of the space-time sheet is or order

$$s_2 \ \equiv \ \int_{a_1}^{a_1+a_2} \sqrt{g_{aa}} da \sim L(k) \ . \tag{17.3.4}$$

In case of 'our' radiation dominated cosmology this gives correct estimate for the moment of time when transition to matter dominated phase occurs since one has $L(k) \sim a_F$ in this case.

That the decomposition to disjoint 3-surfaces occurs after the transition to matter dominated phase is suggested by simple arguments. First of all, the decomposition into regions has obviously interpretation as a formation of visible structures around hidden structures formed by pairs of cosmic strings thickened to magnetic flux tubes. Secondly, of decomposition occurs, the photons coming from distant objects 'drop' to the space-time sheets representing later critical cosmologies. This explains why the optical properties of the Universe seem to be those of a critical cosmology.

17.3.4   Matter dominated phases

The transition to the matter dominated phase followed by the decoupling of the radiation and matter makes possible the formation of structures. This is expected to involve compression of matter to dense regions and to lead to at least a temporary decomposition of the matter dominated cosmology to disjoint 3-surfaces condensed on larger space-time surfaces. The reason is that Jeans length becomes smaller than the size of the horizon. A galaxy model based on the assumption that the region around the two curved ends of a split cosmic string serve as a seed for galaxy formation has been considered in [D4]. In particular, it was found that Jeans criterion leads to a lower bound for the string tension of the galactic strings of same order of magnitude as the string tension of the cosmic strings.

If one assumes that matter dominated regions continue cosmological expansion so that the radius of region equals to the horizon size $R = a^{1/2}$, the fraction of the volume occupied by matter dominated regions grows as $\epsilon(a) = (a/a_R)\epsilon(a_R)$. In recent cosmology the regions have joined together for $\epsilon(a_R) > 10^{-3}$ which would suggest that ultimately asymptotic string dominated cosmology results. One could however argue that matter dominated cosmology does not expand. Taking into account the horizon size of about 5×10^5 light years at the time of the transition to matter dominance, this would mean that galaxies do not participate in cosmic expansion but move as particles on background cosmology.

TGD allows an entire sequence of matter dominated cosmologies associated with the radiation dominated cosmologies labelled by p-adic primes allowed by p-adic length scale hypothesis. Forgetting the delicacies related to nucleo-synthesis, the matter densities associated with these matter dominated cosmologies are scaled down like $(a_1(k)/a_F)^3$ where $a_1(k)\sin L(k)$ is the moment at which the corresponding critical cosmology was created. Thus the latest matter dominated cosmology gives the dominating contribution to the matter density.

Sooner or later matter dominated cosmology becomes string dominated. A good guess is that the transition to string dominance occurs if cosmic expansion of the space-time sheet indeed continues. To see what is involved consider the bounds on the total length of string per large void with size of order $a_* \sim 10^8$ light years. This length can be parameterized as $L = nL(void)$. The requirement that the mass density of the strings is below the critical density gives, when applied to the large void with size of $a_* \simeq 10^8$ light years at recent time a, gives

$$\frac{3}{4\pi}\frac{nT}{a_*^2} < \rho_s = \frac{3}{8\pi G a^2} \ . \tag{17.3.5}$$

Here one has $T \simeq .22 \times 10^{-6}\frac{1}{G}$. This gives roughly

$$n < 2 \times 10^6 \times (\frac{a_*}{a})^2 \ . \tag{17.3.6}$$

The second constraint is obtained from the requirement that the ratio of the string mass per void to the mass of the ordinary matter per void is not too large at present time. Using the expression

$$\rho_m \simeq \frac{3}{32\pi G}\frac{a_*}{a^3} \ ,$$

with $a_* \sim 10^8$ years (time of recombination) and the expression for the string mass per void one has

$$\frac{\rho_s(a)}{\rho_m a(a)} \quad = \quad n \times 1.8 \times 10^{-6}(\frac{a}{a_*})^3 \ . \tag{17.3.7}$$

for the ratio of the densities. For $a = 10^{10+1/2}$ ly the two conditions give

$$\begin{aligned} n \quad &< \quad 20 \ , \\ \frac{\rho_s(a)}{\rho_m} \quad &\simeq \quad n \times 18 \times \sqrt{10} \ . \end{aligned} \tag{17.3.8}$$

These equations suggest that n cannot be much larger than one and suggest the simple picture in which the Kähler charges associated with the "big" string in the interior of the large void and with the galactic strings on the boundaries of the void cancel each other. The minimal value of n is clearly $n = 4$ corresponding to a straight string in the interior of the void. It must be however emphasized that these estimates are rough.

The rate $dlog(E_{gr})/dlog(a)$ for the change of gravitational energy in co-moving volume at present moment in the matter dominated cosmology is determined by

$$\frac{(d\rho_c/da)}{\rho_c} \quad = \quad -\frac{1}{2a} \sim 10^{-11}\frac{1}{year} \ . \tag{17.3.9}$$

The rate is of the same order of magnitude as the rate of energy production in Sun [ia2] so that the rates dE_{I+}/da and dE_{I-}/da for the change of positive and negative contributions to the inertial energy would be of same order of magnitude and sum up to dE_{gr}/da.

17.3.5 Asymptotic cosmology

During asymptotic cosmology one has $dE_{gr}/da = 0$ so that matter and antimatter interact weakly and annihilation processes do not affect the density of gravitational energy. The following argument suggests that asymptotic cosmology is equivalent with the assumption that the cosmic expansion of the space-time sheets almost halts. The expression for the horizon radius for the cosmology decomposing into critical, radiation and matter dominated and asymptotic phases. The expression for the radius reads as

$$R = \int_0^a \sqrt{g_{aa}} \frac{da}{a} = R_0 + R_{as} \ ,$$

where R_0 corresponds from the cosmology before the transition to the asymptotic cosmology and R_{as} gives the contribution after that. Formally this expression is infinite since the contribution to R_{as} from the critical period is infinite. Since one has $g_{aa} \to 1$ asymptotically R_{as} is in good approximation equal to $R_{as} = log(a/a_{as})$, where a_{as} denotes the time for the transition to asymptotic cosmology. This means that the growth of the horizon radius becomes logarithmically slow: $dR(a)/da = 1/a$. A possible interpretation is that the sizes of various structures during asymptotic cosmology are almost frozen. One can however consider the possibility that the disjoint structures formed during the period of matter dominated phase expand and fuse together so that the there is basically single structure of infinite size formed by the join along boundaries condensate of various matter carrying regions.

From the known estimates [ib3] for the total length of galactic string per void one obtains estimate for the needed string tension of the galactic strings. The resulting string tension is indeed of the order of GUT string tension $T \sim 10^{-6}/G$. It will be found later that Jeans criterion gives same lower bound for the string tension of the galactic strings. The resulting contribution to the mass density is smaller than the critical mass density so that no inconsistencies result.

The simplest mechanism generating galactic strings is the splitting of long strings to pieces resulting from the collisions of the strings during very early string dominated cosmology. This mechanism implies that galaxies should form linear structures: this seems indeed to be the case [ib3].

The recent mass density of the strings pairs is considerably larger than that associated with the visible matter. This implies string dominance sooner or later. There are two possible alternatives for the string dominated cosmology.

a) Cosmology with co-moving strings.

b) Stationary cosmology, which seems a natural candidate for the asymptotic cosmology.

Consider first the co-moving string dominated cosmology. The mass density for the string dominated Robertson-Walker cosmology (necessarily smaller than critical density now) is given by the expression

$$\begin{aligned}
\rho &= \frac{3}{8\pi G a^2}(\frac{1}{K} - 1) \ , \\
H^2 &= \frac{1}{K a^2} \ ,
\end{aligned} \tag{17.3.10}$$

and is a considerable fraction of the critical mass density unless the parameter K happens to be very close to 1. Sub-criticality gives the condition

$$c_\# = \sqrt{K} > \frac{1}{\sqrt{2}} \ .$$

The requirement that the gravitational force dominates over the Kähler force implies that the value of $g_{aa} = K$ differs considerably from unity. The recent value of the quantity $K a^2$ can be evaluated from the known value of the Hubble constant. By the previous argument, the ratio of the string mass density to the matter mass density for the recent time $a \sim 10^{10+1/2}$ years is about $\rho_s/\rho_m \sim 50$. This gives the estimates for the light velocity in the condensate and the ratio of the density to the critical density

$$\begin{aligned}
c_\# &= \sqrt{K} \simeq .93 \ , \\
\Omega \equiv \frac{\rho}{\rho_{cr}} &\simeq .16 \ .
\end{aligned} \tag{17.3.11}$$

One also obtains an estimate for the time a_1 , when the transition to string dominated phase has occurred

$$
\begin{aligned}
\rho_{m0} &= \rho_m(\frac{a}{a_1})^3 = \rho_s = \rho_{s0}(\frac{a}{a_1})^2 \ , \\
a_1 &= \frac{\rho_m}{\rho_s}a \sim 6 \times 10^8 \ \text{ly} \ .
\end{aligned}
\tag{17.3.12}
$$

The fraction of the total mass density about the critical mass density is about 4 per cent an perhaps two small.

Consider next the stationary cosmology. The relevant component of the metric and mass density are given by the expressions

$$
\begin{aligned}
g_{aa} &= \frac{(1-2x)}{(1-x)} \ , \\
\rho &= \frac{3}{8\pi Ga^2}\frac{x}{(1-2x)} \ , \\
x &= (\frac{a_1}{a})^{2/3} \ .
\end{aligned}
\tag{17.3.13}
$$

Asymptotically the mass density for this cosmology behaves as
$\rho \simeq 1/a^{2(1+v^2)}$, $v^2 = 1/3$ and "pressure" ($p \simeq -1/9\rho$) is negative indicating that strings indeed dominate the mass density. The results from the numerical simulation of the GUT cosmic strings suggest the interpretation of v^2 as mean square velocity for a long string [ib8]: the relative velocities of the big strings seem rather large.

The transition to the stationary cosmology must take place at some finite time since the energy density

$$
\rho = \frac{3}{8\pi Ga^2}\frac{(\frac{a_1}{a})^{2/3}}{(1 - 2(\frac{a_1}{a})^{2/3})} \ ,
\tag{17.3.14}
$$

is negative, when the condition $a < a_1(1/2)^{-3/2}$ holds true. An estimate for the parameter a_1 is obtained by requiring that the ratio of the mass density to the recent density of the ordinary matter is of order $r \sim 200$ at time $a \sim 10^{10.5}$ ly (this requires $n = 4$, which corresponds to the lower bound for the length of cosmic string per void): $\frac{\rho}{\rho_m}(a) = r$. This gives for the parameter x, the time parameter a_1, the velocity of light in the condensate and for the fraction of the mass density about the critical mass density the following estimates:

$$
\begin{aligned}
x &= \frac{\frac{r}{4}\frac{a_*}{a}}{1 + \frac{r}{2}\frac{a_*}{a}} \simeq .16 \ , \\
a_1 &\simeq 2 \times 10^9 \ \text{ly} \ , \\
c_\# &\simeq .93 \ , \\
\Omega &\simeq .16 \ .
\end{aligned}
\tag{17.3.15}
$$

Apart from the value of the transition time, the results are essentially the same as for the string dominated cosmology. By increasing the amount of a string per void one could reduce the value of the light velocity in the condensate. The experimental lower bound on Ω is $\Omega > .016$ and the favored value is $\Omega \sim .3$. The latter value would require $n \simeq 6.8$ instead of the lower bound $n = 4$ and give $c_\# \simeq .87$

If the proposed physical interpretation for dE_{gr}/da in terms of the energy production inside the stars is correct, then stationary cosmology should be a good idealization for the cosmology provided that the rate of the energy production of stars is negligibly small as compared with the total energy density. This is expected to case, when the energy density of the string like objects begins to dominate over the ordinary matter.

String dominated and stationary cosmologies have certain common characteristic features:

a) Horizons are absent. This implies that the formation of the structures of arbitrarily large size should be possible at this stage and in certain sense the formation of these structures can be regarded as a manifestation of the structures already formed during the very early string dominated cosmology.

b) The so called acceleration parameter q_0 vanishes asymptotically for the stationary cosmology and identically for string dominated cosmology:

$$q_0(stat) \quad = \quad -\frac{\ddot{a}a}{\dot{a}^2} \simeq -\frac{1}{3}(\frac{a_1}{a})^{2/3} \to 0 \ ,$$
$$q_0(string) \quad = \quad 0 \ . \tag{17.3.16}$$

For the recent stationary cosmology the value of q_0 would be very small and negative: of order $q_0(stat) \simeq -.05$. For the matter dominated cosmology the value of this parameter is $q_0 = 1/2$ and positive $(a \simeq t^{2/3}$). The earlier attempts made to evaluate the value of this parameter from the observations are consistent with the value $q_0 = 0$ as well as with the value $q_0 = 1/2$ [ia2]. Quite recent determinations of the parameter [id1] are consistent with $q_0 \geq 0$ but exclude large negative values of q_0 typical for the inflationary scenarios with a large value of the cosmological constant. A precise measurement of q_0 provides an important test for the stationary cosmology.

17.4 Inflationary cosmology or quantum critical cosmology?

The measurements [ii4] allow to deduce information about the curvature properties of the space-time in cosmological scales. These experimental findings force the conclusion that cosmological time= constant sections are essentially flat after the decoupling of the em radiation from matter which occurred roughly one half million years after the Big Bang. The findings allowed to build a much more detailed model for the many-sheeted cosmology leading also to a considerable increase in the understanding of the general principles of TGD inspired cosmology. In the following the observational facts are discussed first and then TGD based explanation relying on the many-sheeted cosmology is briefly discussed. One ends up to a cosmological realization of quantum criticality in terms of a fractal cosmology having Russian doll like structure. The cosmologies within cosmologies are critical cosmologies before transition to hyperbolicity followed by an eventual decay to disjoint non-expanding 3-surfaces.

Critical cosmologies can be regarded as 'Silent Whispers' amplifying to Big Bangs and are generated from vacuum by the gradual condensation of cosmic strings to initially empty and flat space-time sheets. The transition to hyperbolicity involves topological condensation of the remnants of the earlier sub-cosmologies. Hyperbolic period is followed by a decay to disjoint non-expanding 3-surfaces, remnants of the sub-cosmology. There is thus a strong analogy with biological evolution involving growth, metabolism and death. Sub-cosmologies are characterized by three parameters: moment of birth and durations of the critical period and hyperbolic periods. p-Adic length scale hypothesis makes model quantitative by providing estimates for the moments of cosmic time when the phase transitions generating new critical sub-cosmologies occur and fixes the number of the phase transitions already occurred. What is especially remarkable, that the time for the generation of CMB is predicted correctly from p-adic fractality and from the absence of the second acoustic peak in the spectrum of CMB fluctuations.

The recent measurements [ii4] allow to deduce information about the curvature properties of the space-time in cosmological scales. The conclusion is that cosmological time= constant sections are essentially flat after the decoupling of the em radiation from matter which occurred roughly one half million years after the Big Bang. This forces to consider a more detailed model for the many-sheeted cosmology provided by TGD. In the following the observational facts are discussed first and then TGD based explanation relying on the many-sheeted cosmology is discussed. One ends up to a cosmological realization of quantum criticality in terms of a fractal cosmology having Russian doll like structure. The cosmologies within cosmologies are critical cosmologies before transition to hyperbolicity followed by an eventual decay to disjoint non-expanding 3-surfaces.

Critical cosmologies can be regarded as 'Silent Whispers' amplifying to Big Bangs and are generated from vacuum by the gradual condensation of cosmic strings to initially empty and flat space-time sheets. The transition to hyperbolicity involves topological condensation of the remnants of the earlier sub-cosmologies. Hyperbolic period is followed by a decay to disjoint non-expanding 3-surfaces, remnants of the sub-cosmology. There is thus a strong analogy with biological evolution involving growth, metabolism and death. Sub-cosmologies are characterized by three parameters: moment of birth and durations of the critical period and hyperbolic periods. p-Adic length scale hypothesis makes model quantitative and gives estimate for the moments of cosmic time when the phase transitions generating new critical sub-cosmologies occur and fixes the number of the phase transitions already occurred. The time for the generation of CMB is predicted correctly from p-adic fractality and from the absence of the second acoustic peak in the spectrum of CMB fluctuations.

It must be emphasized that in TGD framework critical cosmology reflects quantum criticality and the presence of two kinds of two-dimensional conformal symmetries acting at the level of imbedding space and space-time [B4]. Thus the correlation function for the fluctuations of the mass density at the surface of a sphere of fixed radius is dictated by conformal invariance and by quantal effects the naive scaling dimension predicted by the scaling invariance can be modified to an anomalous dimension. The implications of replacing scaling invariance with conformal invariance for the correlation function of density fluctuations is discussed at the general level in [ii3].

17.4.1 Comparison with inflationary cosmology

TGD differs from GRT in several respects. Many-sheeted space-time concept forces fractal cosmology containing cosmologies within cosmologies. A p-adic hierarchy of long ranged electro-weak and color physics assignable to dark matter at various space-time sheets is predicted if one interprets the unavoidable long ranged classical gauge fields as space-time correlates of corresponding quantum fields. Confinement (weak) length scales associated with these physics correspond to p-adic length scales characterizing the sizes of the space-time sheets of corresponding hadrons (weak bosons). Topological condensation involves a formation of # contacts identifiable as parton- antiparton pairs defining a particular instance of dark matter. Infinite variety of dark matters, or more precisely partially dark matters with respect to each other, is predicted.

Z^0 force competes with gravitational force and it will be found that the role of this force seems to be crucial in understanding the formation of the observed large void regions (with recent size of order 10^8 light years) containing ordinary matter predominantly on their boundaries. Einstein's equations provide only special solutions of the field equations for the length scale dependent space-time of TGD. For instance, in case of strongly Kähler charged cosmic strings it seems better to regard the strings as sources of the Kähler electric field rather than gravitational field. Vapor phase containing at least cosmic strings is the crucial element of TGD inspired cosmology.

The proposed scenario for cosmology deserves a comparison with the inflationary scenarios [ii1, ia15].

a) In inflationary cosmologies exponentially expanding phase corresponds to a symmetry non-broken phase and de-Sitter cosmology follows from the vacuum energy density for the Higgs field. The vacuum energy of the Higgs field creates "negative pressure" giving rise to the exponential expansion. The string tension of the topologically condensed cosmic strings creates the "negative pressure" in TGD context.

In TGD framework situation is diametric opposite of this since exponential expansion is replaced by a logarithmic expansion. This can be seen by solving proper time t in terms of M_+^4 proper time a from the equation

$$\frac{dt}{da} = \sqrt{g_{aa}} = \sqrt{1 - \frac{R^2 k^2}{4} \frac{1}{1 - k^2 a^2}} \ . \tag{17.4.1}$$

One obtains

$$kt = \int^{sinh(ka)} du \sqrt{cosh^2(u) - \frac{R^2 k^2}{4}} \simeq sinh(ka) \tag{17.4.2}$$

since $R^2 k^2 \ll 1$ holds true. This gives $kt \sim sinh(ka) \sim exp(ka)$ rather than $a \sim exp(Ht)$ as in inflationary scenarios so that expansion takes place with a logarithmic slowness.

b) In the inflationary scenarios the exponential expansion destroys inhomogenities and implies the isotropy of $3\,K$ radiation and the decay of the Higgs field to radiation creates entropy. In TGD string dominance implies the absence of horizons. There are no horizons associated with the vapor phase neither since it obeys light cone cosmology. Also critical, string dominated and asymptotic cosmologies are horizon free.

c) In inflationary scenarios the transition to the radiation dominated phase corresponds to the transition from the symmetric phase to a symmetry broken phase. In TGD something analogous happens. Cosmic strings are free at primordial stage but unstable against decay to elementary particles because their action has wrong sign. Some of these strings achieve stability by topologically condensing and generating large Kähler electric charge to cancel their Kähler magnetic action. Light particles of matter in turn suffer a gradual condensation around Kähler electric strings. The Kähler charge of the string induces automatically a slight matter-antimatter asymmetry in the exterior space-time. At the limit $a \to 0$ the contribution of the condensate to the energy of a given co-moving volume vanishes and in this sense condensate can be regarded as a seed of the symmetry broken phase.

d) In inflationary scenarios the critical mass density is reached from above and final state corresponds to a cosmology with a critical mass density. In TGD scenario in its simplest form, the mass density is exactly critical before the transition to the "radiation dominated phase" and overcritical mass density resides in the vapor phase. In a well defined sense vapor phase makes possible subcritical cosmology. The mysterious vacuum energy density of the inflationary cosmologies corresponds in TGD framework to the energy density of cosmic strings in vapor phase.

17.4.2 Balloon measurements of the cosmic microwave background favor flat cosmos

Inflationary scenario has been one of the dominating candidates for cosmology. The basic prediction of the inflationary cosmology is criticality of the mass density which means that cosmic time=constant sections are flat. Observations about the density of known forms of matter are not consistent with this and the only possible manner to get critical mass density is to assume that there exist some hitherto unknown form of vacuum energy density contributing roughly 70 percent to the energy density of the universe. This vacuum energy density is believed to cause the observed acceleration of the cosmic expansion.

The basic geometrical prediction of the inflationary scenario is that cosmic time=constant sections are flat Euclidian 3-spaces. This prediction has been now tested experimentally and it seems that the predictions are consistent with the observations. The test is based on the study of non-uniformities of the cosmic microwave background (CMB). CMB was created about half million years after the moment of Big Bang when opaque plasma of electrons and ions coalesced into transparent gas of neutral hydrogen and helium. Thermal photons decoupled from matter to form cosmic microwave background and have been propagating practically freely after that. The fluctuations of the temperature of the cosmic microwave background reflect the density fluctuations of the universe at the time when this transition occurred. The prediction is that the relative fluctuations of temperature are proportional to the relative fluctuations of mass density and are few parts to 10^5.

Happily, it is possible to estimate the size spectrum for the regions of unusually high and low density theoretically and compare the predictions with the experimentally determined distribution of hot and cold spots in CMB. Since the light from hot and cold spots propagates through the intervening curved space, its intrinsic geometry reflects itself in the properties of the observed spectrum of CMB fluctuations. Hence it becomes possible to experimentally determine whether the 3-space (cosmic time=constant section) is negatively curved (expansion forever), positively curved and closed (big crunch) or flat.

The acoustic properties of the plasma help in the task of determining the spectrum of CBM fluctuations. The competition between gravity and radiation pressure during radiation dominated period produced regions of slow, attenuated oscillatory contraction and expansion. The maximum size of over-dense region that could have shrunk coherently during the half million years before the plasma became transparent was limited by the velocity of sound which is $c/\sqrt{3}$ in radiation dominated plasma. This gives $R = 5/\sqrt{3} \times 10^5$ light years which is about 300 thousand light years for the maximal size of the hot spot. The observed position and size of the first acoustic peak corresponding to the largest hot spots and its observed position depends on the presence or absence of the distorting cosmic curvature. If the intervening 3-space is positively (negatively) curved the parallel rays coming from hot spot diverge and hot spots look larger (smaller) than they actually are: also distances between hot spots look larger (smaller).

To abstract cosmological details from the observations one calculates the power spectrum of the thermal fluctuations by fitting the CMB temperature map to a spherical harmonic series. The absolute square of the fitted amplitude for l:th order spherical-harmonic component is essentially the mean-square point-to-point temperature fluctuation of the CMB on angular scale about π/l radians. The observed fluctuation power spectrum as function of l has maximum at $l = 200$. This is consistent with flat intervening 3-space and inflationary scenario. The next maximum of power spectrum as function of l corresponds to the second acoustic peak (recall that acoustic oscillations are in question) with smaller size of hot spot and should be observed at $l = 500$ according to inflationary scenario. In fact this peak has not been observed. This might be due to the small statistics or due to the fact that the scale free prediction of the inflationary scenario for the spectrum of fluctuations is quite not correct but that fluctuations have cutoff at some length scale larger than the size of the size of the hot spot associated with the second acoustic peak.

In standard cosmology the result means that 3-space has remained flat for most of the time after the moment when CMB was generated. Of course, the cosmology can have changed hyperbolic after that since the small mass density of the recent day universe implies that the effects of the curvature on the optical properties of the universe are small. Inflationary scenario predicts this if one repeats the biggest blunder of Einstein's life by adding to Einstein's equations cosmological constant, which means that vacuum energy density of an unknown origin contributes about 70 per cent to the mass density of the universe. Besides this one must assume that primordial baryon density is about 50 per cent higher than standard expectation. Thus inflationary model survives the test but not gracefully.

17.4.3 Quantum critical fractal cosmology as TGD counterpart of the inflationary cosmology

In TGD framework Einstein's equations are structural equations relating the energy momentum tensor of topologically condensed matter to the geometry of the space-time surface rather than fundamental equations derivable from a variational principle. Furthermore, the solutions of Einstein's equations are only a special case of the equations characterizing the macroscopic limit of the theory. The simplest assumption is however that Einstein's equations hold true for each sheet of the many-sheeted space-time and is made in TGD inspired cosmology.

Does quantum criticality of TGD imply criticality and fractality of TGD based cosmology?

Quantum criticality of the TGD Universe supports the view that many-sheeted cosmology is in some sense critical. Criticality in turn suggests p-adic fractality. Phase transitions, in particular the topological phase transitions giving rise to new space-time sheets, are (quantum) critical phenomena involving no scales. If the curvature of the 3-space does not vanish, it defines scale: hence the flatness of the cosmic time=constant section of the cosmology implied by the criticality is consistent with the scale invariance of the critical phenomena. This motivates the assumption that the new space-time sheets created in topological phase transitions are in good approximation modellable as critical Robertson-Walker cosmologies for some period of time at least.

Neither inflationary cosmologies nor overcritical cosmologies allow global imbeddings. TGD however allows the imbedding of a one-parameter family of critical and overcritical cosmologies. Imbedding is possible for some critical duration of time. The parameter labelling these cosmologies is a scale factor characterizing the duration of the critical period. The infinite size of the horizon for the imbeddable critical cosmologies is in accordance with the presence of arbitrarily long range fluctuations at criticality and guarantees the average isotropy of the cosmology. These cosmologies have the same optical properties as inflationary cosmologies.

The critical cosmologies can be used as a building blocks of a fractal cosmology containing cosmologies containing ... cosmologies. p-Adic length scale hypothesis allows a quantitative formulation of the fractality. Fractal cosmology provides explanation for the balloon experiments and also for the paradoxal result that the observed density of the matter is much lower than the critical density associated with the largest space-time sheet of the fractal cosmology. Also the observation that some astrophysical objects seem to be older than the Universe, finds a nice explanation.

Cosmic strings and vapor phase

En essential element of TGD inspired cosmology is the presence of vapor phase consisting dominantly of cosmic strings. For the values of light cone proper time a smaller than CP_2 time R, space-time does not exist in sense as it is defined in General Relativity. Instead, very early Universe consists of a primordial soup of cosmic strings. General arguments lead to the hypothesis that the density of the cosmic strings in vapor phase in in this period is

$$\rho_V \quad = \quad \frac{3}{8\pi G a^2} \ . \tag{17.4.3}$$

The expression of the density is formally same as the critical density of flat critical cosmology (note that future light cone is hyperbolic vacuum cosmology). The topological condensation of free cosmic strings forced by the absolute minimization of Kähler action (free cosmic strings have infinite positive Kähler magnetic action) to critical space-time sheets leads to fractal hierarchy of critical cosmologies and reduces the density of vapor phase. Obviously the energy density in vapor phase is very much analogous to the vacuum energy density needed in inflationary cosmologies.

What happens when criticality becomes impossible?

Given critical sub-cosmology is created at the moment $a = a_0$ of the light cone proper time. The imbeddability of the critical cosmology fails for $a = a_1$. The question is what happens for the space-time sheet before this occurs. A natural assumption is that when the value of the cosmic time for which imbeddability fails is approached, cosmology is transformed to hyperbolic cosmology. One can imagine several scenarios but the following one involving two transitions is the most plausible one. The first step is the transition of the critical cosmology to a hyperbolic cosmology which is either matter or radiation dominated or to a stationary cosmology for which gravitational energy density is conserved. The next step is possible decomposition of $a = constant$ 3-surface of hyperbolic cosmology to disjoint non-expanding 3-surfaces topologically condensing on critical cosmology created later. This process in turn could induce the transition of the critical cosmology to hyperbolicity: when critical sub-cosmology eats the remnants of earlier sub-cosmology it could become hyperbolic itself. Of course, this is not the only mechanism. This scenario resembles to high degree the lifecycle of a biological organism involving gradual growth, metabolism and death.

1. *Transition to matter or radiation dominated phase*

The critical cosmology is transformed to a hyperbolic cosmology with sub-critical mass density. This option is very general and means that criticality is gradually shifted to increasingly longer length scales when it breaks down in short length scales. The continuity condition in the transformation to hyperbolic cosmology with $\theta - \pi/2$ and $\phi = \phi(a)$ for g_{aa} reads as

$$\frac{1}{g_{aa}^H} - 1 \quad = \quad \frac{1}{1 - K} \equiv \epsilon \ ,$$

$$K \quad \equiv \quad \frac{R^2}{4a_1^2} \frac{1}{(1 - (\frac{a}{a_1})^2)} \ . \tag{17.4.4}$$

The light cone projection of the sub-cosmology is sub-ligtcone of M_+^4. a denotes light cone proper time for this sub-light cone: its value is obviously smaller than the value of M_+^4 proper time. Upper index 'H' refers to to the metric of the hyperbolic cosmology. The value of the parameter ϵ must deviate considerably from unity and and since R/a_1 is extremely small number, the transformation to hyperbolic cosmology must happen very near to $a = a_1$: for all practical purposes this fixes the moment of transition to be $a = a_1$. Critical cosmology is also flat in excellent approximation up to $a = a_1$. The mass density of the hyperbolic cosmology behaves during the matter (radiation) dominated phase as

$$\rho \quad = \quad \frac{3}{8\pi G} \epsilon \frac{a_1^{1+n}}{a^{3+n}} \ . \tag{17.4.5}$$

Here $n = 0$ corresponds to matter dominance and $n = 1$ to radiation dominance.

2. Decomposition of $a = constant$ surface to disjoint non-expanding components

p-Adic length scale hypothesis suggests that hyperbolic sub-cosmology ceases to participate in the cosmic expansion sooner or later and that $a = constant$ 3-surface decomposes to disjoint particle like non-expanding objects topologically condensing at and comoving on the sub-cosmologies generated later. A possible mechanism causing the decomposition of a hyperbolic sub-cosmology into disjoint space-time sheets is the intersection of the sub-light cones defined by the sub-cosmologies initiated at same $a = constant$ hyperboloid. The transition to non-expanding phase has certainly occurred for stellar objects.

The disjoint 3-surfaces generated in this process are topologically condensed at (or are 'metabolized' by) younger critical cosmologies and the simplest assumption is that this condensation process changes the newer cosmology to matter dominated hyperbolic cosmology. This assumption is consistent with the fact that the mass density of the critical cosmologies is very small before the transformation to the matter dominated phase so that they cannot contain topologically condensed matter. Before the condensation process the condensation of free cosmic strings gives rise to the gradual increase of the mass density of the critical cosmology.

This picture implies that cosmic expansion occurs only above some length scale and that the long length scale optical properties of the universe are determined by the competion of sub-cosmologies in hyperbolic and critical stages since photons travel along space-time sheets of both type.

p-Adic fractality

p-Adic fractality suggests that all cosmological phase transitions giving rise to the generation of new space-time sheets should be describable using the same universal Robertson-Walker cosmology during their critical period so that cosmology would contain cosmologies containing cosmologies... like Russian doll contains Russian dolls inside it. The light cone projection of each sub-cosmology is sub-light cone. Lorentz invariance requires that the probability distribution for the position the tip of the sub-light cone is constant along $a = constant$ hyperboloid.

Sub-cosmology is characterized by three parameters a_0, a_1 and a_2. a_0 characterizes the moment of birth for sub-cosmology, a_1 characterizes in excellent approximation the value of the sub-light cone proper time for which the transition from critical to hyperbolic sub-cosmology occurs. $a_1 + a_2$ in turn characterizes the sub-light cone proper time for the decay of the hyperbolic sub-cosmology to comoving non-expanding surfaces. p-Adic length scale hypothesis allows to make educated guesses for the values of a_0, a_1 and a_2 so that TGD inspired cosmology becomes highly predictive.

Since a_0 characterizes the moment of birth for sub-cosmology, it is not expected to reflect in any manner the dynamics of earlier sub-cosmologies. In contrast to this, a_1 and a_2 characterize the internal dynamics of sub-cosmology involving gravitational time dilation effects in an essential manner and this suggests that the fundamental parameters are the values of the proper times s_1 and s_2 for sub-cosmologies to which a_1 and a_2 are related in simple manner.

More quantitatively, the proper time s of the space-time surface representing cosmology is defined as

$$s = \int_0^a \sqrt{g_{aa}} da \ .$$

The relationship between light cone proper time and proper time of the critical cosmology implies the relationship

$$
\begin{aligned}
s_1 &= \int_0^{a_1} \sqrt{1 - K} da \ , \\
K &\equiv \frac{R^2}{4a_1^2} \frac{1}{(1 - (\frac{a}{a_1})^2)} \ .
\end{aligned}
\tag{17.4.6}
$$

between a_1 and s_1. Up to to $a \simeq a_1$ the value of the parameter K is nearly vanishing so that $s \simeq a$ holds in a good approximation during the critical period. This means that the values of s_1 and a_1 are in excellent approximation identical:

$$s_1 \simeq a_1 \ .$$

The relationship between s_2 and a_1 and a_2 is

$$s_2 \quad = \quad \int_{a_1}^{a_1+a_2} \sqrt{g_{aa}}\, da \quad . \tag{17.4.7}$$

The gravitational dilation effects for hyperbolic cosmology are large and s_2 and a_2 can differ by orders of magnitude.

p-Adic length scale hypothesis states two things.

a) Each p-adic prime p corresponds to p-adic length scale $L_p = \sqrt{p} \times l$, where $l \simeq 10^4$ Planck lengths is CP_2 'radius'.

b) The primes $p \simeq 2^k$, k prime or power of prime are physically preferred so that one has

$$L_p \equiv L(k) \simeq 2^{k/2} \times l \quad .$$

p-Adic fractality allows to make educated guesses for the most plausible values of the parameters a_0, a_1 and a_2 characterizing the evolution of the sub-cosmologies.

1. Moments of birth of sub-cosmologies

It seems that the generation of new sub-cosmologies is a process having nothing to do with the internal dynamics of sub-cosmologies themselves. Therefore p-adic fractality suggests that the dips of the sub-light cones associated with the critical cosmologies are concentrated in good approximation at the hyperboloids

$$a_0(k) = x_0 L(k)$$

of the light cone M_+^4 where x_0 is some numerical constant: note that a_0 refers to the proper time of the light cone M_+^4 rather than sub-light cone. The number of primes k in the interval $[2, ..., 401]$ (see Table 2) is rather small which implies that the number of sub-cosmologies created after Big Bang is smaller than 100.

2. Moments for the transition to hyperbolicity

The natural guess is that the imbedding for the cosmology characterized by $p \simeq 2^k$ fails for $a \simeq a_1$ (in excellent approximation) when sub-cosmology also starts to metabolize the remnants of earlier sub-cosmologies. p-Adic length scale hypothesis gives the estimate

$$s_1(k) \simeq a_1(k) = x_1 L(k) \quad ,$$

where x_1 is numerical constant of order unity. The most natural interpretation is that transition to radiation or matter dominated cosmology occurs. It is natural to assume that topological condensation of 3-surfaces resulting from earlier cosmology accompanies this transition. One can also say that cosmological metabolism causes transition to hyperbolicity.

3. Moments of death for sub-cosmologies

The death of the sub-cosmology means decay to disjoint 3-surfaces. The simplest assumption is that this occurs when the age of sub-cosmology measured with respect to sub-cosmological proper time s exceeds p-adic time scale defined by the next p-adic prime in the hierarchy. Thus one has

$$s_1 + s_2 \simeq a_1 + s_2 = x_2 L(k(next))$$

giving

$$s_2 = x_2 L(k(next)) - x_1 L(k) \quad . \tag{17.4.8}$$

From this one can relate the parameter a_2 with the p-adic length scales $L(k(next))$ and $L(k)$. $L(k)$ gives the size scale of the 3-surfaces resulting when the connected space-time sheet $a_2 = constant$ decomposes to pieces. Due to gravitational time dilation s_2 can be smaller than a_2 by several orders of magnitude so that the duration of the hyperbolic period when measured using sub-light cone proper time is lengthened by gravitational time dilation and topological condensation of the remnants of sub-cosmology can take place to a critical cosmology having $k > k(next)$.

4. Temperature and energy density of the critical cosmology at the moment of transition to hyperbolicity.

p-Adic length scale hypothesis suggest that the temperature just after the transition to the effectively radiation dominated phase is

$$T(k) = \frac{n}{L(k)} \ , \qquad \text{for } k > k_{cr} \ ,$$

$$T(k) = T_H \sim \frac{1}{R} \ , \quad \text{for } k \leq k_{cr} \ .$$

(17.4.9)

Here n is rather large numerical factor. Since $a_F \sim 2.7 \times 10^{-10}$ seconds which corresponds to length scale $L \simeq .08$ meters roughly to p-adic length scale $L(197) \simeq .08$ meters (which by the way corresponds to the largest p-adic length scale associated with brain, cosmic joke?), should correspond to the establishment of Hagedorn temperature, one has the conditions

$$k_{cr} \ = \ 197 \ ,$$

$$n \ \simeq \ 2^{197/2} \sim 10^{30} \sim \frac{m_{CP_2}^2}{m_p^2} \ .$$

Thus n is in of same order of magnitude as the ratio of the CP_2 mass squared ($m_{CP_2} \simeq 10^{-4}$ Planck masses) to proton mass squared.

Dimensional considerations suggest also that the energy density in the beginning of the radiation dominated phase (in case that it is achieved) is

$$\rho = nT(k)^4 \ , \tag{17.4.10}$$

where n a numerical factor of order one. n does not count for the number of light particle species since the thermal energy of strings gives rise to the effective radiation dominance. This explains why infinite number of fermion families does not lead to infinite density of thermal energy and why their presence leaves no trace in present day cosmology.

When the time parameter a_1 of the critical cosmology becomes too high, it cannot anymore generate radiation dominated phase since the temperature remains too low. Previous considerations suggest that the maximum value of a_1 is roughly $a_1(max) = a_F \sim 3 \times 10^{-10}$. After this critical sub-cosmologies transform directly to the stationary cosmologies.

These estimates fix the the structure of the fractal cosmology to rather high degree. Note that the expanding space-time surfaces associated with the new critical cosmologies created in the phase transition can fuse since corresponding light cones can intersect. The number of the phase transitions occurred after the light cone proper time corresponding to electron Compton length is roughly forty. The tables below give the p-adic length scales in the range extending from electron Compton radius to 10^{10} light years.

k	127	131	137	139	149
$L_p/10^{-10}m$	.025	.1	.8	1.6	50

k	151	157	163	167	169
$L_p/10^{-8}m$	1	8	64	256	512

k	173	179	181	191	193
$L_p/10^{-4}m$	.2	1.6	3.2	100	200

k	197	199	211	223	227
L_p/m	.08	.16	10	640	2560

Table 1. p-Adic length scales $L_p = 2^{k-151}L_{151}$, $p \simeq 2^k$, k prime, possibly relevant to astro- and biophysics. The last 3 scales are included in order to show that twin pairs are very frequent in the biologically interesting range of length scales. The length scale $L(151)$ is take to be thickness of cell scale, which is 10^{-8} meters in good approximation.

k	227	229	233	239	241
L_p/m	$2.3E+3$	$4.6E+3$	$1.9E+4$	$1.5E+5$	$3.0E+5$
k	251	257	263	269	271
L_p/m	$.96E+7$	$7.7E+7$	$6.0E+8$	$4.8E+9$	$.9E+10$
k	277	289	293	307	311
L_p/m	$7.7E+10$	$5.0E+12$	$2.0E+13$	$2.5E+15$	$1.0E+16$
k	313	317	329	331	337
L_p/ly	2.2	$5.4E+2$	$1.0E+3$	$2.2E+3$	$8.4E+3$
k	347	349	353	359	367
L_p/ly	$2.8E+5$	$5.6E+5$	$2.2E+6$	$1.8E+7$	$2.9E+8$
k	373	379	381	391	397
L_p/ly	$2.2E+9$	$1.9E+10$	$3.8\ E+10$	$1.2E+12$	$.96E+13$

Table 2. p-Adic length scales $L_p = 2^{(k-127)/2}L_{127}$, $p \simeq 2^k$, k prime, possibly relevant to large scale astrophysics. The definition of the length scale involves an unknown factor r of order one and the requirement $L(151) \simeq 10^{-8}$ meters, the thickness of the cell membrane, implies that this factor is $r \simeq 1.1$.

17.4.4 The problem of cosmological missing mass

In inflationary cosmology the basic problem is related to the missing mass. The experimentally determined recent density of the ordinary matter is about 4 per cent of the critical mass density and it seems that ordinary sources (other than vacuum energy density) can contribute about 30 percent of the critical mass density in inflationary scenarios. In TGD framework the situation is different as following arguments show.

1. Criticality does not force missing mass in TGD framework.

There is no absolute need for vacuum energy density since the mass densities of critical cosmologies present in condensate are extremely low before the transition to the hyperbolicity. In TGD framework the observed mass density corresponds to the mass density at 'our' cosmological space-time sheet condensed to some larger space-time sheet... condensed on the largest space-time sheet present in the topological condensate now. Since vapor phase density equals to the critical density of flat critical cosmology, the net energy density of the entire topological condensate is bound to be smaller than the critical density. This is in accordance with experimental facts. In fact, vapor phase energy density corresponds closely to the vacuum energy density of inflationary scenarios. By the conservation of energy the total energy density at various space-time sheets is indeed equal to 'critical' vapor phase density apart from effects caused by different expansion rates. The possibility of negative energy virtual gravitons however makes possible for a given space-time sheet to have energy density much larger than the energy density of the vapor phase.

2. The observed optical properties of the Universe require that photons travel in critical cosmologies for a sufficiently long fraction of time.

The photons coming from a distant source must propagate along a space-time sheet of a critical cosmology for a sufficiently long fraction of time during their travel to detector. If the period of the matter dominance is too long, photons spend too long time fraction in matter dominated phase and the spectrum of anisotropies is seriously affected. This is avoided if the period between the initiation of the matter dominance and decomposition into disjoint 3-surfaces is sufficiently short. Generation of lumps of matter could in fact involve gravitational collapse leading to the decomposition of the 3-surface to pieces. Second possibility is that the topological condensation of photons is more probable on critical and essentially flat cosmologies (present always) than on matter dominated cosmologies. The large rate of topological evaporation from radiation and matter dominated cosmologies is consistent with this. An alternative explanation is that topological evaporation is only effective and caused by the reduction of the energy density by the absorption of negative energy virtual gravitons. Both effects might of course be involved.

3. The mass density of later matter dominated cosmologies should be larger than that of previous matter dominated cosmologies.

Assume that previous cosmology have made transition to non-expanding phase and behaves as co-moving matter with density $\rho(p_1)$ on the next expanding matter dominated cosmology with density $\rho(p_2)$. Under this assumption the condition

$$\rho(p_1) \equiv p\rho(1) < \rho(p_2)$$

implies

$$a_1(p_1)\epsilon(p_1)\frac{1}{a^3(p_1)} = p \times a_1(p_2)\epsilon(p_2)\frac{1}{a^3(p_2)} \ .$$

The larger the space-time sheet, the later it is created, and therefore one has $a(p_1) > a(p_2)$ as well as $a_1(p_1) < a_1(p_2)$. For large values of $a(p_1)$ and $a(p_2)$ one has $a(p_1) \sim a(p_2)$ in good approximation and one has

$$a_1(p_1)\epsilon(p_1) \quad = \quad p \times a_1(p_2)\epsilon(p_2) \ . \tag{17.4.11}$$

The parameters ϵ are of order unity in recent day cosmology.

If one assumes the relationship $s_1 \simeq a_1 = xL(k)$, one obtains

$$\frac{\epsilon(k_1)}{\epsilon(k_2)} \quad = \quad p \times 2^{(k_2-k_1)/2} \ . \tag{17.4.12}$$

It is possible to satisfy this constraint for $p < 1$.

The assumption about cosmologies inside cosmologies implies distribution of ages of the Universe and provides a natural explanation for why the observed mass density is subcritical. Cosmic strings topologically condensed at the larger space-time sheet could correspond to the missing mass. The age of the space-time sheet of an astrophysical object can be much longer than the age of the largest space-time sheet: this could explain the paradoxical observation that some stars seem to be older than the Universe.

17.4.5   TGD based explanation of the results of the balloon experiments

TGD based model explaining the results of balloon experiments relies on the notion of the fractal cosmology.

Under what conditions Universe is effectively critical?

TGD based model explaining the results of balloon experiments relies on the notion of the fractal cosmology. If $a = constant$ sections of hyperbolic cosmologies decompose to disjoint 3-surfaces after sufficiently short matter dominated period, the photons propagating along these space-time sheets must 'drop' on the critical space-time sheets so that situation stays effectively critical and model yields same predictions as inflationary cosmology. The decoupling of radiation from matter involved a topological phase transition leading to a generation of new expanding space-time sheets along which the CMB radiation could propagate.

The following argument shows under what conditions the total duration of the matter dominated periods is negligible as compared with the total duration of the critical periods. The ratio of the observed angular separation $\Delta\phi_{obs}$ between hot spots to real angular separation $\Delta\phi_r$ between them can be deduced from

$$\Delta\phi_{obs} \simeq tan(\Delta\phi_{obs}) \;=\; \frac{\sqrt{g_{\phi\phi}}\Delta\phi_r}{R(r)} \;,$$

$$R(r) \;=\; \int \sqrt{g_{aa}}\,da = \int \sqrt{g_{rr}}\frac{dr}{da}\,da \qquad (17.4.13)$$

$R(r)$ is the Euclidian distance calculated along the light like geodesic associated with photon and depends on the curvature properties of the intervening space. Flat cosmology serves as a natural reference and the ratio

$$\frac{\Delta\phi_{obs}}{\Delta\phi_{obs}(flat)} \;=\; \frac{R(r,flat)}{R(r)}$$

$$\;=\; \frac{a - a_1}{\int_{a_1}^{a} \sqrt{g_{aa}}\,da} \qquad (17.4.14)$$

measures the effect of the intervening space to the observed angular distance between hot spots of CMB. Note that the integral must be expressed in terms of the initial values of the coordinate r.

When photons travel along critical cosmology, $g_{aa} \simeq 1$ holds true and this corresponds to flat situation. For a fixed value of r one has the following approximate expressions in various cosmologies

$$a - a_1 = r - r_1 \;, \qquad \text{(critical cosmology with } g_{aa} = 1) \;,$$

$$a - a_1 \sim log(\tfrac{r}{r_1}) \;, \qquad \text{(hyperbolic cosmology with } g_{aa} = 1) \;,$$

$$\tfrac{2}{3}ka\big((\tfrac{a}{a_R})^{1/2} - (\tfrac{a_1}{a_R})^{1/2}\big)$$

$$= log(\tfrac{r}{r_1}) \;, \qquad \text{(matter dominance with } g_{aa} = k(\tfrac{a}{a_R})^{1/2}) \;.$$

$$(17.4.15)$$

From these expressions one finds that same increment of r gives rise to much smaller increment of a in hyperbolic cosmology than in critical cosmology. Thus the fractions of r spent in critical cosmology gives the dominating contribution to the integral unless this fraction happens to be especially small. From these expressions one finds that for a given distance r the red shift in approximately flat (no horizon) hyperbolic cosmology is exponentially larger than in critical cosmology. The arrival of photons along hyperbolic cosmology could thus explain why their ages when derived from the red shift seem to be larger than the age of the Universe derived assuming that photons travel along critical cosmology.

During periods of matter dominance g_{aa} behaves as $g_{aa} = k\frac{a}{a_2}$ and gives smaller contribution than critical period. Integral can be expressed as sum of critical and matter dominated contributions as

$$\int_{a_2}^{a} \sqrt{g_{aa}}\,da \;=\; \sum_i [\Delta a_0(i) + s_2(i)] \;. \qquad (17.4.16)$$

Here the durations of periods are of order $L(k_i)$ and last period gives the dominant contribution. If the last propagation has occurred along critical cosmology for a sufficiently long time, the contribution of the earlier matter dominated periods to the integral are small and the last critical period can dominate in the integral. If the last critical period corresponds to $k = 379$ preceded by $k = 373$, then the ratio for angle separations does not differ more than about 10 per cent from the value guaranteeing ideal criticality.

What the absence of the second acoustic peak implies?

The absence of the second acoustic peak (which might be also a statistical artefact) fixes the TGD based model to a very high degree.

a) By quantum criticality scale free spectrum for the size L of the density fluctuations is a natural assumption when L is above the p-adic length scale $L(k(prev))$ characterizing the size of the remnants of the previous cosmology condensing to the critical space-time sheets in the transition to hyperbolic cosmology. Below this size ($L < L(k(prev))$) the spectrum for fluctuations has however natural cutoff. This cutoff could also correspond to the length of the cosmic strings giving rise to large voids containing cosmic strings inside them in TGD based model of galaxy formation and to the recent size of large voids containing galaxies at their boundaries. The space-time sheets of large voids should have been born in the phase transition generating CMB if this picture is correct.

b) The first acoustic maximum corresponds to $l = 200$ and $L(k_R)$. The second acoustic maximum corresponds to $l = 500$ and has thus size which is $2/5$ of the size of the first hot spot. $L(k_R(prev))$ defines the lower bound for the size of the density and temperature fluctuations as the minimum size of topologically condensed space-time sheets. Therefore, if second acoustic maximum is present, the size of the corresponding hot spot must be larger than $L(k_R(prev))$. Thus the condition for the absence of the second acoustic maximum is

$$\frac{L(k_R(prev))}{L(k_R)} < \frac{2}{5} \ .$$

Thus the experimental absence of the second maximum requires that k_R and $k_R(prev)$ form twin pair ($k_R(prev) = k_R - 2$) so that one has $L(k_R(prev)) = L(k_R)/2$.

There are two candidates for the twin pairs in question: the twin pairs are $(347, 349)$ and $(359, 381)$ (see table 2 for the values of corresponding p-adic length scales). Only the first pair is consistent with the previous considerations related to p-adic fractality.

a) The pair ($k_R(prev) = 347, k_R = 349$) corresponds to the p-adic length scales $L(347) = 2.8E + 5$ ly and $L(349) = 5.6E + 5$ ly. $L(347)$ clearly corresponds to the minimum size of the first acoustic peak. Rather remarkably, the length scale $L(347)$, which corresponds also to the size of the typical spatial structures frozen in the transition to matter dominated cosmology, corresponds rather closely to the estimated time $s_R \sim 5E+5$ years for the transition to matter dominance and also to the typical size of galaxies. In consistency with the general picture, the estimate

$$s_R = s_1 + s_2 = x_2 L(349)$$

gives $s_R = 5.8E + 5$ years for $x_2 = 1$.

b) If one takes seriously the order of magnitude estimate $s = s_R = 5 \times 10^5$ light years for the age of the cosmology when CMB was created, and assumes that hyperbolic cosmology was radiation dominated before s_R, one can estimate the value of light cone proper time a at this time using the formula

$$s_R = \int_{a_1}^{a_R} \sqrt{g_{aa}} da \ ,$$

$$g_{aa} \simeq 10^{-3} \frac{a^2}{a_R^2} \ . \tag{17.4.17}$$

This gives $a_R \sim 3.3 \times 10^7$ ligh years: this corresponds to the p-adic length scale $L(359)$. Thus gravitational time dilatation implies that topological condensation does not occur to $L(353)$ next to $L(349)$ but to $L(259)$. 5 new cosmologies corresponding to $k = 353, 359, 367, 373$ and 379 should have emerged after the transition to matter dominated cosmology and could correspond to cosmological structures. Large voids are certainly this kind of structures and correspond to the p-adic length scale $L(367) \sim 2.9E + 8$ ly. The predicted age of the Universe is about $L(381) \sim 1.9E + 10$ years in this scenario.

Fluctuations of the microwave background as a support the notion of many-sheeted space-time

The fluctuations of the microwave background temperature are due to the un-isotropies of the mass density: enhanced mass density induces larger red shift visible as a local lowering of the temperature. Hence the fluctuations of the microwave temperatures spectrum provide statistical information about the deviations of the geometry of the 3-space from global homogenuity. The symmetries of the fluctuation spectrum can also provide information about the global topology of 3-space and for over-critical topologies the presence of symmetries is easily testable [ig8].

The first year Wilkinson microwave anisotropy probe observations [ig3] allow to deduce the angular correlation function. For angular separations smaller the 60 degrees the correlation function agrees well with that predicted by the inflationary scenarios and deriving essentially from the assumption of a flat 3-space (due to quantum criticality in TGD framework). For larger angular separations the correlations however vanish, which means the existence of a preferred length scale. The correlation function can be expressed as a sum of spherical harmonics. The $J = 1$ harmonic is not detectable due to the strong local perturbation masking it completely. The strength of $J = 2$ partial wave is only $1/7$ of the predicted one whereas $J = 3$ strength is about 72 per cent of the predicted. The coefficients of higher harmonics agree well with the predictions based on infinite flat 3-space.

Later some interpretational difficulties have emerged: there is evidence that the shape of spectrum might reflect local conditions. There are differences between northern and southern galactic hemispheres and largest fluctuations are in the plane of the solar system. In TGD framework these anomalies could be interpreted as evidence for the presence of galactic and solar system space-time sheets.

1. Dodecahedral cosmology?

The WMAP result means a discrepancy with the inflationary scenario and explanations based on finite closed cosmologies necessarily having $\Omega > 1$ but very near to $\Omega = 1$ have been proposed. In [ig4] Poincare dodecahedral space, which is globally homogenous space obtained by identifying the points of S^3 related by the action of dodecahedral group, or more concretely, by taking a dodecahedron in S^3 (12 faces, 20 vertices, and 30 edges) and identifying opposite faces after 36 degree rotation, was discussed. It was found to fit quadrupole and octupole strengths for $1.012 < \Omega < 1.014$ without an introduction of any other parameters than Ω.

However, according to [ic4] the quadrupole and octupole moments have a common preferred spatial axis along which the spectral power is suppressed so that dodecahedron model seems to be excluded. The analysis of [ig5] led to the same result. According to the article of Luminet [ig9], the situation is however not yet completely settled, and there is even some experimental evidence for the predicted icosahedral symmetry of the thermal fluctuations.

The possibility to imbed also a very restricted family of over-critical cosmologies raises the question whether it might be possible to develop a TGD based version of the dodecahedral cosmology. The dodecahedral property could have two interpretations in TGD framework.

a) Space-time sheet with boundaries could correspond to a fundamental dodecahedron of S^3. If temperature fluctuations are assumed to be invariant under the so called icosahedral group, which is subgroup of $SO(3)$ leaving the vertices of dodecahedron invariant as a point set, the predictions of the dodecahedral model result.

b) An alternative interpretation is that the temperature fluctuations for S^3 decomposing to 120 copies of fundamental dodecahedron are invariant under the icosahedral group.

For neither option topological lensing phenomenon is present since icosahedral symmetry is not due to the identification of points of 3-space in widely different directions but due to symmetry which is not be strict. An objection against both options is that there is no obvious justification for the G invariance of the thermal fluctuations. The only justification that one can imagine is in terms of quantum coherent dark matter.

The finding of WMAP that the ratio Ω of the mass density of the Universe to critical mass density is $\Omega = 1 + g_{aa} = 1 + \epsilon$, $\epsilon = 0.02 \pm 02$. This is consistent with critical cosmology. If only slightly overcritical cosmology is realized, there must be a very good reason for this.

The WMAP constraint implies that the value of a which corresponds to the value of cosmic time a_s which characterizes the thermal fluctuations must be such that $g_{aa} = \epsilon$ holds true. The inspection of the explicit form of g_{aa} deduced in the subsection "Critical and over-critical cosmologies" requires that a_s is extremely near to the value a_0 of cosmic time for which $g_{aa} = 0$ holds true: the deviation of a from a_0 should be of order $(R/_0)R$ and most of the thermal radiation should have been generated at this moment.

Since gravitational mass density approaches infinity at $a \to a_0$ one can imagine that the spectrum of thermal fluctuations reflects the situation at the transition to sub-criticality occurring for $\Omega = 1 + \epsilon$. Thermal fluctuations would be identifiable as long ranged quantum critical fluctuations accompanying this transition and realized as a hierarchy of space-time sheets inducing the formation of structures.

The scaling invariance of the fluctuation spectrum generalizes in TGD framework to conformal invariance. This means that the correlation function for fluctuations can have anomalous scaling dimension [ii3]. The hadron physics analogy would be the transition from hadronic phase to quark gluon plasma via a critical phase discussed in section "Simulating Big Bang in laboratory".

The transition $k = 1 \rightarrow 0 \rightarrow -1$ would involve the change in the shape of the $S^2 \subset CP_2$ angle coordinate Φ as a function $f(r)$ of radial coordinate of RW cosmology. The shape is fixed by the value of $k = 1, 0, -1$. In particular, Φ would become constant in the transition to subcriticality. $k = 1 \rightarrow 0$ phase transition would be accompanied by the increase of the maximal size of space-time sheets to infinite in accordance with the emergence of infinite quantum coherence length at criticality. Whether this could be regarded as the TGD counterpart for the exponential expansion during inflationary period is an interesting question. In the transition to subcriticality also the shape of Θ as function of a necessarily changes since $sin(\Theta(a > a_0)) > 1$ would be required otherwise.

2. Hyperbolic cosmology with finite volume?

Also hyperbolic cosmologies allow infinite number of non-simply connected variants with 3-space having finite volume. For these cosmologies the points of $a = constant$ hyperboloid are identified under some discrete subgroup G of $SO(3,1)$. Also now fundamental domain determines the resulting space and it has a finite volume.

It has been found that a hyperbolic cosmology with finite-sized 3-space based on so called Picard hyperbolic space [ig6, ig7], which in the representation of hyperbolic space H^3 as upper half space $z > 0$ with line element $ds^2 = (dx^2 + dy^2 + dz^2)/z^2$ can be modelled as the space obtained by the identifications $(x, y, z) = (x + ma, y + nb, z)$. This space can be regarded as an infinitely long trumpet in z-direction having however a finite volume. The cross section is obviously 2-torus. This metric corresponds to a foliation of H^3 represented as hyperboloid of M^4 by surfaces $m^3 = f(\rho)$, $\rho^2 = (m^1)^2 + (m^2)^2$ with f determined from the requirement that the induced metric is flat so that x, y correspond to Minkowski coordinates (m^1, m^2) and z a parameter labelling the flat 2-planes corresponds to m^3 varying from ∞ to ∞.

This model allows to explain the small intensities of the lowest partial waves as being due to constraints posed by G invariance but requires $\Omega = .95$. This is not quite consistent with $\Omega = 1.02 \pm .02$.

Also now two interpretations are possible in TGD framework. Thermal photons could originate from a space-time sheet identifiable as the fundamental domain invariant under G. Alternatively, $a = constant$ hyperboloid could have a lattice-like structure having fundamental domain as a lattice cell with thermal fluctuations invariant under G. The shape of the fundamental domain interpreted as a surface of M^4 is rather weird and one could argue that already this excludes this model.

Quantum criticality and the presence of quantum coherent dark matter in arbitrarily long length scales could explain the invariance of fluctuations. If Ω reflects the situation after the transition to subcriticality, one has $\Omega = g_{aa} - 1 = .95$. This gives $g_{aa} = 1.95$ which is in conflict with $g_{aa} < 1$ holding true for the imbeddings of all hyperbolic cosmologies. Thus Ω must correspond to the critical period and one should explain the deviation from $\Omega = 1$. A detailed model for the temperature fluctuations possibly fixed by conformal invariance alone would be needed in order to conclude whether many-sheeted space-time might allow this option.

3. Is the loss of correlations due to the finite size of the space-time sheet?

One can imagine a much more concrete explanation for the vanishing of the correlations at angles larger than 60 degrees in terms of the many-sheeted space-time. Large angular separations mean large spatial distances. Too large spatial distance, together with the fact that the size of the space-time sheet containing the two astrophysical objects was smaller than now, means that they cannot belong to the same space-time sheet if the red shift is large enough, and cannot thus correlate. The size of the space-time sheet defines the preferred scale. The preferred direction would be most naturally defined by cosmic string(s) in the length scale of the space-time sheet. For instance, closed cosmic string would define an expanding 3-space with torus topology and thus having symmetries. This option would explain also the WMAP anomalies suggesting local effects as effects due to galactic and solar space-time sheets.

Empirical support for the hyperbolic period

TGD inspired cosmology predicts that critical cosmology is followed by a hyperbolic cosmology. A natural question is whether the travel of microwave photons through the negative curvature cosmology might induce some signatures in microwave background. This is indeed the case.

The geodesics in negative curvature 3-space diverge exponentially. The divergence of the nearly parallel light-like geodesic lines is due to the negative curvature making 2-dimensional sections of 3-space analogous to saddle surfaces. The scatterings during the travel of light induce geodesic mixing so that light from regions with differing temperature mix. Hence negative curvature tends to smooth out the anisotropies of the temperature distribution.

Negative curvature has also a more dramatic signature. Gurzadyan [ii2, ii5] has developed a very refined argument involving algorithmic information theory and complexity theory to show that in the hyperbolic cosmology the hot and cold spots of the temperature distribution of the cosmic microwave radiation look elongated. The direction of elongation is random but the shape of the ellipse is characterized by the curvature of 3-space and does not depend on temperature or size of the spot. For a flat or positively curved space this kind of elongation does not occur.

The emergence of a preferred direction in a Lorentz invariant cosmology looks highly counter-intuitive. My humble understanding is that a scattering of photons from a large geometric structure must be involved somehow. The elongation should relate to what happens at the last scattering surface whose position together with the positions of observer and previous scattering surface define a plane whose normal defines the preferred direction, which would presumably correspond to the shorter axis of the ellipse. In TGD framework the transfer of photons from a larger space-time sheet to that of observer might correspond to this scattering process. Scattering surface would correspond to the boundary of the space-time sheet of the observer whereas scattering would correspond to refraction at the boundary.

The analysis of BOOMERanG, COBE and WMAP CMB maps indeed shows that the spots have elliptic shape with ellipticity parameter ~ 2 whereas the prediction for hyperbolic RW cosmology is 1.4. [ii6]. This would suggest that some additional effect is involved and TGD inspired bet have been already described.

17.5 Some problems of cosmology

In this chapter some problems, most of them common to both standard and TGD inspired cosmology, are discussed.

17.5.1 Why some stars seem to be older than the Universe?

There exists experimental evidence that some stars are older than the Universe [ih1, ih2, ih3]. A related problem is the problem of the two Hubble constants. These paradoxical results can be understood in TGD inspired cosmology. In TGD light can propagate via several routes. In the topological condensate light ray can propagate along one of the many curved space-time sheet as a small condensed particle and in the vapor phase as a small 3-surface in imbedding space $H = M_+^4 \times CP_2$, where M_+^4 is future light cone of M^4. The time needed to travel from point A to point B is shorter in the vapor phase than in any space-time surface since the geodesic length along the space-time surface in the induced metric is obviously longer than in free Minkowski space. This time depends also on the space-time sheet so that entire spectrum of effective light velocities and Hubble constants results. The failure to distinguish between vapor phase photons and photons propagating along various space-time sheets leads to the paradox as following arguments shows and possibly also to the problem of two (or in fact more than two) different Hubble constants. The possibility of the vapor phase photons or photons propagating along almost flat space-time sheets emitted by the objects outside the space-time horizon of 'our' space-time sheet explains also objects with anomalously large red shifts.

Basic facts

To understand these results one must study TGD based cosmology in more quantitative level.

a) The most general cosmological imbedding of M_+^4 to $M_+^4 \times CP_2$, is of form

$$
\begin{aligned}
s^k &= s^k(a) \ , \\
g_{aa} &= 1 - s_{kl}\frac{ds^k}{da}\frac{ds^l}{da} \ , \\
ds^2 &= g_{aa}da^2 - a^2\left(\frac{dr^2}{1+r^2} + r^2 d\Omega^2\right) \ .
\end{aligned}
\tag{17.5.1}
$$

Here s_{kl} is CP_2 metric tensor and describes always expanding cosmology with subcritical or at most critical mass density.

b) The age of the Universe defined as M_+^4 proper time a of the co-moving observer (the co-moving observer on the space-time surfaces is also co-moving in M_+^4) is larger than the age defined as the proper time $s(a)$ of the co-moving observer on space-time surface. For the matter dominated Universe one has $g_{aa} = Ka$, which gives

$$
\frac{age(cond)}{age(vapor)} = \frac{s(a)}{a} = \frac{2}{3}\sqrt{g_{aa}} \ ,
\tag{17.5.2}
$$

for the ratio of the ages.

c) The recent value of g_{aa} can estimated from the expression for the mass density in the expanding cosmology

$$
\begin{aligned}
\rho &= \frac{3}{8\pi G}\left(\frac{1}{g_{aa}} + k\right) \ , \\
k &= -1 \ .
\end{aligned}
\tag{17.5.3}
$$

$k = 0$ mass density corresponds to the critical mass density ρ_c. The mass density is believed to be a fraction of order $\epsilon = 0.1 - 0.5$ of the critical mass density and this gives estimate for $\sqrt{g_{aa}}$:

$$
\begin{aligned}
\sqrt{g_{aa}} &= \sqrt{1-\epsilon} \ , \\
\epsilon &= \frac{\rho}{\rho_c} \ .
\end{aligned}
\tag{17.5.4}
$$

$\sqrt{g_{aa}} = 2/3$ suggested by the proposed solution to the Hubble constant discrepancy gives $\epsilon = \frac{9}{4}$. $\epsilon = .1$ gives $\sqrt{g_{aa}} \simeq .95$.

d) The ratio of the condensate travel time to the vapor phase travel time for short distances is given by

$$
\frac{\tau(cond)}{\tau(vapor)} = \frac{1}{\sqrt{g_{aa}}} \ .
\tag{17.5.5}
$$

This effect is in principle observable. The effect provides also a means of measuring the mass density of the Universe.

e) The light travelling in the vapor phase can reach the observer from a region, which is the intersection of the past light cone of the observer with the boundary of M_+^4 and therefore finite region of M^4. The M^4 radius of this region in the rest frame of the observer is equal $r_M = a/2$ by elementary geometry.

f) For a null geodesic of the space-time surface representing cosmology, starting at (a_0, r) and ending at $(a, 0)$, one has

$$
\begin{aligned}
r &= sinh(X) \ , \qquad \text{(hyperbolic cosmology)} \ , \\[2mm]
r &= X \ , \qquad \text{(critical cosmology)} \ ,
\end{aligned}
\tag{17.5.6}
$$

$$
X = \int_{a_0}^{a}\frac{\sqrt{g_{aa}}}{a}da \ .
$$

If g_{aa} approaches zero for $a_0 \to 0$, as it does for the radiation dominated cosmology, the integral defining X is finite. This means that the value of $r_M(a_0)$ (M^4 distance of the object from the observer) approaches zero at this limit. All radiation from the moment of the big bang comes from the tip of the light cone. The very early cosmology with a critical mass density corresponds to $g_{aa} = 1 - K$, K a very small number, and also in this case the radiation comes from the origin.

Maximum Minkowski distance from which light can propagate

It is interesting to find the maximum value of M_+^4 distance r_M from which it is possible to receive information in various cosmologies. The radius $r_M(a_0)$ has maximum for some finite value of a_0 and this radius defines the M^4 radius of the Universe observed using the condensate photons. For a_0 corresponding to maximum the condition

$$
\begin{aligned}
\sqrt{g_{aa}} &= tanh(X) \;, \quad \text{(hyperbolic cosmology)} \;, \\
\sqrt{g_{aa}} &= X \;, \quad \text{(critical cosmology)} \;.
\end{aligned}
\tag{17.5.7}
$$

The maximum corresponds to a rather large value of a_0. Consider now various cases.

i) In case of matter dominated cosmology one has $g_{aa} = Ka$ and one has the condition

$$
u_0 = tanh(2(u - u_0)) \simeq 2(u - u_0) \;, \quad u = \sqrt{Ka} \;, \quad u_0 = \sqrt{Ka_0} \;.
\tag{17.5.8}
$$

This gives in good approximation

$$
u_0 = r = \tfrac{2}{3}u \;, \quad a_0 = \tfrac{4}{9}a \;, \quad r_M^0 = \tfrac{8}{27}ua = \tfrac{16}{81}\sqrt{Ka} \times a \;.
\tag{17.5.9}
$$

ii) In case of vapor phase and also for asymptotic cosmology in the limit of flatness one obviously has

$$
r_M^0 = a \;.
\tag{17.5.10}
$$

iii) In case of critical cosmology with $g_{aa} = 1$ one has

$$
a_0 = \tfrac{a}{e} \;, \quad r_0 = 1 \;, \quad r_M^0 = \tfrac{a}{e} \;.
\tag{17.5.11}
$$

The value of r_M^0 is clearly smallest in matter dominated cosmology.

Many-sheeted space-time allows several snapshots from the evolution of astrophysical objects

Vapor phase photons and condensate photons propagating along various space-time sheets provide in principle a possibility to obtain simultaneous information about the astrophysical object in various different phases of its development. For an object situated at distance r and observed at $(a, r = 0)$, the emission moments a_0 and $a_1 > a_0$ (in Minkowski proper time) for the condensate photon and vapor phase photon are related by the formula

$$
\frac{a}{a_1} = exp(2\sqrt{K_1}(a^{1/2} - a_0^{1/2})) \;.
\tag{17.5.12}
$$

in the matter dominated cosmology $g_{aa} = K_1 a$ ($K_1 a \sim 1$). Hence a sufficiently nearby Super Nova would provide a test for this effect. The first burst of light corresponds to vapor phase photons and subsequent bursts to the condensate photons. The time lag between the bursts provides a manner to measure the value of $\sqrt{g_{aa}}$. Unfortunately, the time lag in case of SN1987A is quite too large since the distance of order $1.5 \cdot 10^5 \; ly$. The observation of the same spectral line with two different cosmological red-shifts is second effect of this kind and might be erratically interpreted as the existence of two different objects on same line of sight.

Why some stars seem to be older than the Universe?

Red-shifts are determined by the apparent velocity of astrophysical object which is in good approximation given $v = Hr$, where H is Hubble constant which in TGD depends on space-time sheet along which photons propagate. One has $r = sinh(X)$ for hyperbolic cosmology and $r = X$ for critical cosmology, where the function X is defined by Eq. 17.5.6. For matter dominated cosmology with $g_{aa} = Ka$ and for almost flat hyperbolic cosmology with $g_{aa} = 1 - \epsilon$ one has

$$X = 2\left[(Ka)^{1/2} - (Ka_0)^{1/2}\right] < 1 \ , \quad \text{(matter dominance)} \ ,$$
$$X = \sqrt{(1-\epsilon)}log(\tfrac{a}{a_0}) \ , \qquad \text{(almost flat hyperbolic)} \ . \tag{17.5.13}$$

From this it is clear that the approximation $sinh(X) \simeq X$ makes sense in case of matter dominated cosmology and the red-shifts do not differ much from those predicted by critical cosmology.

For almost flat hyperbolic cosmology and for vapor phase situation is dramatically different since red-shifts can be exponentially larger. Therefore, if most of radiation comes along matter dominated or critical space-time sheets, then the radiation coming in vapor phase or along almost flat hyperbolic space-time sheets can give rise to huge red-shifts and stars which seem to be older than the Universe. The presence of several space-time sheets means that using common value of Hubble constant one obtains entire spectrum of ages of the Universe. Same astrophysical can also give rise to several images corresponding to the photons propagating along various space-time sheets. It might be that this mechanism might be involved with the observed multiple images of stars.

The puzzle of several Hubble constants

Each cosmic space-time has its own Hubble constant defined as

$$H \ = \ \frac{1}{a\sqrt{g_{aa}}} \ , \tag{17.5.14}$$

where the value of the light cone proper time corresponds to the light cone proper time of observer in the sub-light cone defined by the sub-cosmology. The value of Hubble constant is smallest at almost flat space-time sheets. Photons propagating along almost flat space-time sheet or in vapor phase provide a possible solution to the puzzle of two different Hubble constants if the mass density is sufficiently large. The distances derived from type Ia super-novae give $H_0^a = 54 \pm 8 \ kms^{-1}Mpc^{-1}$ to be compared with the Hubble result $H_0^b = 80 \pm 17 \ kms^{-1}Mpc^{-1}$ [ih2].

The discrepancy is resolved if the measurement of the distance is correct and made using photons propagating in vapor phase or along almost flat hyperbolic space-time sheets so that H_0^a corresponds in good approximation to the Hubble constant of M_+^4, which is by a factor

$$\frac{H_0^a}{H_0^b} \ = \ \frac{H_0(M_+^4)}{H_0(X^4)} = \sqrt{g_{aa}} = \sqrt{1-\epsilon} \sim 2/3 \tag{17.5.15}$$

smaller than the Hubble constant of the space-time surface. The needed mass density $\epsilon = 5/9$ and the ratio of the propagation velocities of light differs considerably from unity. For $\epsilon = .1$ the ratio of two Hubble constants is predicted to be .95 and some other explanation for discrepancy is needed. The model for the stationary cosmology indeed suggests that the density of matter is much below the value needed to explain the Hubble discrepancy in this manner.

For instance, for the space-time outside the Kähler charged cosmic string, discussed in [D4], one has

$$g_{tt} = 1 - \frac{R^2\omega^2}{4}(1 - u^2) \ , \quad -1 < u(\rho) < 1.$$

The model for the galaxy formation requires $exp(4\omega R) \sim 10^3$ and this gives $\frac{\omega^2 R^2}{4} \simeq .86$ implying $\sqrt{g_{tt}} \geq .37$ so that the reduction of the local light velocity can be rather large and explain the Hubble controversy.

In fact, there are quite recent results [id3], which can be interpreted as a support for the many-sheeted space-time picture with separate Hubble constant associated with each sheet. The preliminary result is that the Hubble constant determined from the nearby supernovas is larger than that determined from the faraway supernovas. The proposed interpretation is that the rate of the expansion of the Universe is increasing in the course of time. The increase could be due to the non-vanishing cosmological constant corresponding to a vacuum energy density about 40 per cent of the critical density: the origin of this vacuum energy density remains a mystery.

TGD suggests suggests that Hubble constant depends on the (p-adic) length scale associated with the space-time sheet and decreases as the length scale increases. [This could also solve the problem of the two different Hubble constants since entire spectrum of Hubble constants is predicted]. Photons from nearby supernovas have suffered a topological condensation on a smaller space-time sheet as those from faraway supernovas. Hence the Hubble constant for nearby supernovas is larger and the rate of the expansion of the Universe is found to apparently increase in the course of time.

The decrease of the Hubble constant as a function of the (p-adic) length scale characterizing a given space-time sheet would follow from the fractality of the TGD Universe implying that the mass density as a function of the p-adic length scale decreases in the long length scales. Fractality could in turn would follow from the basic hypothesis necessary to get a sensible cosmology in TGD, namely that a space-time sheet corresponding to a given p-adic length scale expands until it reaches critical size not too much larger than the p-adic length scale in question. This does not exclude the possibility that the matter topologically condensed on the space-time sheet in question continues expanding and is therefore gradually drifted to the boundaries of the space-time sheet. The presence of the large voids with galaxies on their boundaries, is consistent with this assumption. From the view point of a given space-time sheet, smaller space-time sheets behave like particles of fixed size, whose density is gradually reduced in the cosmic expansion.

17.5.2 Many-sheeted cosmology explains accelerated cosmological expansion and the claimed time dependence of the fine structure constant

There is recent evidence for the time dependence of the fine structure constant in cosmological time scales [ie2]. The spectroscopic observations of a number of absorption systems in the spectra of distant quasars indicate a smaller value of α in the past. The comparison of the ratios of the frequencies for relativistic atomic transitions depending non-linearly on α^2 gives the average value $\Delta\alpha/\alpha = -0.72 \pm .18 \times 10^{-5}$ in the red shift range $z = .5 - 3.5$.

On the other hand, the data about the isotopic abundances in Oklo natural reactor which operated at 1.8×10^9 years ago gives the upper bound $\Delta\alpha/\alpha \leq 10^{-7}$ [ie4]: this corresponds to the red shift $z = .13$. This suggests an abrupt change of the fine structure constant in the range $.13 < z_0 \leq .5$.

A further important piece of data is about type Ia super-novae in distant galaxies. These data have extended the Hubble diagram to red shifts $z \geq 1$ [id2]. The data imply an accelerated expansion of the universe in the framework of standard cosmology requiring the introduction of cosmological constant and vacuum energy density of unknown origin.

The notion of the many-sheeted cosmology might explain the apparent acceleration of the cosmological expansion. The notion of the many-sheeted space-time could also explain the apparent time variation of the fine structure constant as the following arguments tend to demonstrate.

Explanation for the acceleration of the cosmic expansion

One can imagine two alternative explanations for the acceleration of the cosmic expansion. I proposed the first explanation before the understanding the relationship between gravitational and inertial energy and it relies on many-sheeted nature of cosmology in an essential manner. Second explanation is analogous to the standard explanation in terms of quintessence with quintessence replaced with the the gravitational four-momentum density of space-time sheets containing pairs of cosmic strings and modelled by cosmological constant.

1. Acceleration is only apparent and is due to the many-sheeted cosmology

The typical sizes of space-time sheets are characterized by p-adic length scales $L_p = l \times \sqrt{p}$, $l \simeq 10^4$ Planck lengths (CP_2 length). By p-adic length scale hypothesis the favored primary p-adic length scales correspond to $p \simeq 2^k$, k prime or power of prime. Secondary and higher p-adic length scales are defined as $L_p(n) = p^{n/2}L_p$ and their number is not large. There are relatively few primary p-adic length scales between electron length scale and longest cosmological length scales of order 10 Gy (something like 45). The number of p-adic length scales between CP_2 length scale and 10 Gy provides an estimate for the number of space-time sheets in the many-sheeted cosmology and the number is relatively small, smaller than hundred. The simplest model is that the space-time sheets corresponding to $L(k)$ emerge for the first time at $t \sim L(k)$. These space-time sheets would naturally correspond to sub-cosmologies spanning light cones inside light cone. t could correspond to the light cone proper time: $t = a$ (note that in TGD framework one has $z + 1 = a_{now}/a$, where a_{now}/a is the ratio of the recent value of the light cone proper time to that at the moment of emission). t could also correspond to the proper time s for the sub-cosmology in question: $t = s$.

By p-adic fractality of the average density of the gravitational mass is lower at later space-time sheets so that gravitational attraction is weaker and hence the expansion occurs more rapidly and Hubble constant is larger. The simplest model for how light arrives from a distant source at $L(k) < t_s < L(k(next))$ is that photons propagate along a space-time sheet corresponding to $L(k)$. Obviously the effective value of the Hubble constant can increase with t_s if this is the case. Much more dramatic increase of the effective Hubble constant appears if the sub-cosmology changes from critical to subcritical cosmology as already found.

2. Acceleration expansion is due to the non-vanishing of the cosmological constant

TGD predicts that cosmological constant cannot vanish for string like objects and space-time sheets with closed 3-topology. It is quite possible that cosmological constant depends on 3-topology. Hence for space-time sheets with boundary constant could however vanish. Indeed, for non-vacuum extremals modelling space-time sheets carrying pairs of cosmic strings cosmological cosmological constant must vanish by field equations.

For vacuum extremals situation could be different since Einstein's action with cosmological constant allow a natural boundary term and boundary conditions can guarantee that no gravitational momentum flows out from boundaries. Cosmological constant could be in terms of the density of gravitational four-momentum associated with space-time sheets containing pairs of cosmic strings with a vanishing net inertial energy.

p-Adic length scale hypothesis predicts the dependence on p-adic length scale so that one has $\Lambda(k) = 1/L(k)^2$ and thus decreases effectively like $1/a^2$ as a function of light cone proper time. A correct order order of magnitude is predicted $\Lambda(k)$ in the sense that it correspond to a considerable fraction about critical density able to explain the accelerated expansion as being due to cosmological constant.

3. Is the reported cosmic jerk consistent with the gradual reduction of cosmological constant?

There is an objection against the hypothesis that cosmological constant has been gradually decreasing during the cosmic evolution. Type Ia supernovae at red shift $z \sim .45$ are fainter than expected, and the interpretation is in terms of an accelerated cosmic expansion [id4]. If a period of an accelerated expansion has been preceded by a decelerated one, one would naively expect that for older supernovae from the period of decelerating expansion, say at redshifts about $z > 1$, the effect should be opposite. The team led by Adam Riess [id5] has identified 16 type Ia supernovae at redshifts $z > 1.25$ and concluded that these supernovae are indeed brighter. The conclusion is that about about 5 billion years ago corresponding to $z \simeq .48$, the expansion of the Universe has suffered a cosmic jerk and transformed from a decelerated to an accelerated expansion.

The apparent dimming/brightening of supernovae at the period of accelerated/decelerated expansion the follows from the luminosity distance relation

$$\mathcal{F} = \frac{\mathcal{L}}{4\pi d_L^2} , \tag{17.5.16}$$

where $\mathcal{L}$ is actual luminosity and $\mathcal{F}$ measured luminosity, and from the expression for the distance d_L in flat cosmology in terms of red shift z in a flat Universe

$$
\begin{aligned}
d_L &= (1+z) \int_0^z \frac{du}{H(u)} \\
&= (1+z)H_0^{-1} \int_0^z exp\left[-\int_0^u du\,[1+q(u)]\,d(ln(1+u))\right] du \ ,
\end{aligned}
\tag{17.5.17}
$$

where one has

$$
\begin{aligned}
H(z) &= \frac{dln(a)}{ds} \ , \\
q &\equiv -\frac{d^2a/ds^2}{aH^2} = \frac{dH^{-1}}{ds} - 1 \ .
\end{aligned}
\tag{17.5.18}
$$

In TGD framework a corresponds to the lightcone proper time and s to the proper time of Robertson-Walker cosmology. Depending on the sign of the acceleration parameter q, the distance d_L is larger or smaller and accordingly the object looks dimmer or brighter.

At first the conclusion of [id5] seems to be in conflict with the TGD based cosmology for which cosmological constant would have been gradually reduced during the cosmic evolution. The conflict is however only apparent. By the fractality of TGD based cosmology the reduction of the cosmological constant as function of cosmic time means only that the minimum value of the cosmological constant for the space-time sheets in the region $a < a_0$ decreases as $\Lambda \propto 1/L(k)^2$, where $L(k)$ corresponds to the p-adic length scale associated with a_0, and approximately one has $\Lambda \propto 1/a_0^2$. The value of the cosmological constant depends only on the size of the space-time sheet but not on its temporal position, and by the fractality of TGD inspired cosmology new space-time sheets can be created at any value of cosmic time.

Hence the interpretation for the observations of [id5] is same as for the variation for the value of Hubble constant. The radiation coming from distances larger than 5 billion light years arrives along larger space-time sheets having a smaller cosmological constant and smaller acceleration (but not deceleration as concluded in [id5]). The observations would reflect more our temporal position as observers rather than the early cosmological evolution. The prediction is that also other jerks have occurred and correspond to the effective doubling of Λ and occurring at half octaves of a from which corresponding critical red shifts can be deduced.

Apparent time dependence of the fine structure constant

Consider next a possible explanation for the apparent time dependence of the fine structure constant. It is assumed that new space-time sheets with size determined by the p-adic length scale $L(k)$ emerge at values $t \sim L(k)$ of the time coordinate during the cosmological evolution. It is also assumed that the proper description of atoms involves in an essential manner the concept of classical em field. This is indeed the case in TGD framework but not for the Bether-Salpeter equation relying on correlation functions and the abstraction of the basic features of perturbative QED.

a) The basic point is that atomic nuclei need not feed their entire electric gauge fluxes to the atomic space-time sheet, which presumably corresponds to $p \simeq 2^k$, $k = 131$ or $k = 137$, but can feed a small fraction of the electric flux also to the larger space-time sheets. The simplest assumption is that each new cosmological space-time sheet receives a constant fraction of the existing nuclear gauge charge. Stability requirement suggests that also each electron feeds a negative fraction of its electric flux to the larger space-time sheet so that an overall charge neutrality is preserved. The fraction must be negative to guarantee that the nuclear and electronic charges effectively increase in magnitude when new larger space-time sheets emerge during the cosmological evolution. Negative fraction is favored also by the fact that the effective nuclear charge would otherwise approach zero in the sufficiently distant geometric future. The effect corresponds to an apparent renormalization of the fine structure constant having nothing to do with the ordinary QED renormalization or the renormalization of the fine structure constant suggested by the p-adic coupling constant evolution.

b) The experimental findings suggest that the distribution of the electric gauge fluxes between different space-time sheets could have changed in some abrupt manner during the period $.16 < z_0 < .5$. The lower bound follows from the fact that Oklo natural reactor data are consistent with the laboratory value of the effective fine structure constant. Assume that this abrupt change corresponds to the emergence of a new space-time sheet at $z = z_0$ taking a negative fraction of order $\epsilon \sim -10^{-5}$ of the nuclear and electronic gauge fluxes so that the effective nuclear and electronic charges increase correspondingly in magnitude. More generally, assume that this occurs for all values of cosmic time $t(k) \sim L(k)$ corresponding to p-adic length scales.

c) If the p-adic length scale L_p appears at $t = a \simeq L_p$ then p-adic length scales appear at $a(k_n) = 2^{(k_n - k_0)/2} a_{k_0}$. The effective fine structure constant is predicted to be constant inside intervals $[a(k_n), a(k_{n-1})]$. The minimum value for the increment of k_n is $\Delta k = k_n - k_{n-1} = 2$ and corresponds to a variation of a by single octave and to a pair of twin primes $k_n = k_{n-1} + 2$. This predicts the constancy of the effective fine structure constant after $z = z_0$ in accordance with the experimental facts. If $z_0 = a_{now}/a_0 - 1$ corresponds to the first abrupt change in the range $.13 < z_0 < .5$ then for $\Delta k = 2$ another abrupt change would occur at $z_1 = 2z_0 + 1$, $1.26 < z_1 < 3$. If each space-time sheet receives the same amount of electric flux, one has $\Delta[log(\alpha)](z_1) \simeq 2\Delta[log(\alpha)](z_0)$, which is excluded in the range considered. For $\Delta k = 4$ the next abrupt change would correspond to $z_2 = 4z_0 + 3$: $3.52 < z_2 < 5$. Unfortunately, this value of z is slightly above the range studied in [ie3]. For $\Delta k = 6$ one would have $z_3 = 8z_0 + 7$, $8 < z_3 < 11$.

d) The negative em flux which is fraction of order $\epsilon \sim -10^{-5}$ of nuclear electromagnetic charge flowing to single space-time sheet does not lead to any inconsistencies since the number of the primary p-adic length scales between atomic length scale and cosmological length scales is only 45. Therefore the total variation between $a = a_{now} \sim 10^{10}$ years and $a = 10^7$ years (this is the range probed by the cosmic microwave background) would correspond to something like five p-adic length scales for $t = a$ and the predicted net variation in the red shift interval $.13 < z < 10^3$ would not be larger than $\Delta[log(\alpha)] \sim 10^{-4}$ if each p-adic space-time sheet receives the same amount of the electric flux.

Note that this model might be seen as a topological and microscopic version of the Bekenstein's field theory model [ie1] based on the assumption that fine structure constant is a slowly varying scalar field Φ having naturally the needed linear coupling to the Maxwell action. In [ie4] it was suggested that Φ could correspond to the so called quintessence field believed to give rise to cosmological vacuum energy and that Bekenstein's model could explain the observed variation of the fine structure constant. Note that in many-sheeted cosmology charge conservation is not lost although the effective fine structure constant depends on cosmological time.

Chapter 18

TGD Inspired Speculations on Astrophysics and Consciousness

18.1   Introduction

The concept of 3-space in TGD is considerably more general than in the conventional theories. 3-space is not any more connected but can have arbitrary many disjoint components. Even macroscopic boundaries are allowed: macroscopic bodies are interpreted as 3-surfaces having outer boundary. There are strong indications that 3-space has a hierarchical fractal structure: 3-surfaces topologically condensed on 3-surfaces condensed on..., where topological condensation means that 'small' 3-surface is 'glued' to a larger 3-surface by connected sum operation.

The fundamental feature of the topological condensation is the generation of Kähler electric fields implied by the minimization of Kähler action: gravitational fields are always accompanied by long range electro-weak and color gauge fields with gauge charges, which are in astrophysical scales apart from a small but non-vanishing numerical factor equal the mass of particle using Planck mass as unit. For non-vacuum extremals also the Kähler charge can be non-vanishing. For simplest vacuum extremals with CP_2 projection a homologically trivial geodesic sphere only W boson fields are vanishing, which suggests that W generates charge entanglement in astrophysical length scales.

Topological field quantization is a central concept: the presence of gauge charges implies that 3-surface has outer boundary: the larger the charge the smaller the size of the 3-surface. This makes it possible to relate the size of the 3-surface (topological field quantum) to the gauge charges of typical particles in the condensate. The formation of macroscopic quantum systems, such as super conductors, corresponds to the formation of bonds between the boundaries of the neighboring topological field quanta. A possible astrophysical example is neutron star: join along boundaries bonds are formed between neutrons so that single giant nucleus results.

18.1.1   Dark matter as large $\hbar$ phase

D. Da Rocha and Laurent Nottale have proposed that Schrödinger equation with Planck constant $\hbar$ replaced with what might be called gravitational Planck constant $\hbar_{gr} = \frac{GmM}{v_0}$ ($\hbar = c = 1$). v_0 is a velocity parameter having the value $v_0 = 144.7 \pm .7$ km/s giving $v_0/c = 4.6 \times 10^{-4}$. This is rather near to the peak orbital velocity of stars in galactic halos. Also subharmonics and harmonics of v_0 seem to appear. The support for the hypothesis coming from empirical data is impressive.

Nottale and Da Rocha believe that their Schrödinger equation results from a fractal hydrodynamics. Many-sheeted space-time however suggests astrophysical systems are not only quantum systems at larger space-time sheets but correspond to a gigantic value of gravitational Planck constant. The gravitational (ordinary) Schrödinger equation would provide a solution of the black hole collapse (IR catastrophe) problem encountered at the classical level. The resolution of the problem inspired by TGD inspired theory of living matter is that it is the dark matter at larger space-time sheets which is quantum coherent in the required time scale.

I have proposed earlier the possibility that Planck constant is quantized and the spectrum is given in terms of logarithms of Beraha numbers: the lowest Beraha number B_3 is completely exceptional in that it predicts infinite value of Planck constant. The inverse of the gravitational Planck constant could correspond a gravitational perturbation of this as $1/\hbar_{gr} = v_0/GMm$. The general philosophy would be that when the quantum system would become non-perturbative, a phase transition increasing the value of $\hbar$ occurs to preserve the perturbative character and at the transition $n = 4 \to 3$ only the small perturbative correction to $1/\hbar(3) = 0$ remains. This would apply to QCD and to atoms with $Z > 137$ as well.

TGD predicts correctly the value of the parameter v_0 assuming that cosmic strings and their decay remnants are responsible for the dark matter. The harmonics of v_0 can be understood as corresponding to perturbations replacing cosmic strings with their n-branched coverings so that tension becomes n^2-fold: much like the replacement of a closed orbit with an orbit closing only after n turns. $1/n$-sub-harmonic would result when a magnetic flux tube split into n disjoint magnetic flux tubes.

The study of inclinations (tilt angles with respect to the Earth's orbital plane) leads to a concrete model for the quantum evolution of the planetary system. Only a stepwise breaking of the rotational symmetry and angular momentum Bohr rules plus Newton's equation (or geodesic equation) are needed, and gravitational Shrödinger equation holds true only inside flux quanta for the dark matter.

a) During pre-planetary period dark matter formed a quantum coherent state on the (Z^0) magnetic flux quanta (spherical cells or flux tubes). This made the flux quantum effectively a single rigid body with rotational degrees of freedom corresponding to a sphere or circle (full SO(3) or SO(2) symmetry).

b) In the case of spherical shells associated with inner planets the $SO(3) \to SO(2)$ symmetry breaking led to the generation of a flux tube with the inclination determined by m and j and a further symmetry breaking, kind of an astral traffic jam inside the flux tube, generated a planet moving inside flux tube. The semiclassical interpretation of the angular momentum algebra predicts the inclinations of the inner planets. The predicted (real) inclinations are 6 (7) resp. 2.6 (3.4) degrees for Mercury resp. Venus). The predicted (real) inclination of the Earth's spin axis is 24 (23.5) degrees.

c) The $v_0 \to v_0/5$ transition allowing to understand the radii of the outer planets in the model of Da Rocha and Nottale could be understood as resulting from the splitting of (Z^0) magnetic flux tube to five flux tubes representing Earth and outer planets except Pluto, whose orbital parameters indeed differ dramatically from those of other planets. The flux tube has a shape of a disk with a hole glued to the Earth's spherical flux shell.

It is important to notice that effectively a multiplication $n \to 5n$ of the principal quantum number is in question. This allows to consider also alternative explanations. Perhaps external gravitational perturbations have kicked dark matter from the orbit or Earth to $n = 5k$, $k = 2, 3, ..., 7$ orbits: the fact that the tilt angles for Earth and all outer planets except Pluto are nearly the same, supports this explanation. Or perhaps there exist at least small amounts of dark matter at all orbits but visible matter is concentrated only around orbits containing some critical amount of dark matter and these orbits satisfy $n \bmod 5 = 0$ for some reason.

d) A remnant of the dark matter is still in a macroscopic quantum state at the flux quanta. It couples to photons as a quantum coherent state but the coupling is extremely small due to the gigantic value of $\hbar_{gr}$ scaling alpha by $\hbar/\hbar_{gr}$: hence the darkness.

The rather amazing coincidences between basic bio-rhythms and the periods associated with the states of orbits in solar system suggest that the frequencies defined by the energy levels of the gravitational Schrödinger equation might entrain with various biological frequencies such as the cyclotron frequencies associated with the magnetic flux tubes. For instance, the period associated with $n = 1$ orbit in the case of Sun is 24 hours within experimental accuracy for v_0.

18.1.2   Dark matter as a source of long ranged weak and color fields

Long ranged classical electro-weak and color gauge fields are unavoidable in TGD framework. The smallness of the parity breaking effects in hadronic, nuclear, and atomic length scales does not however seem to allow long ranged electro-weak gauge fields. The problem disappears if long range classical electro-weak gauge fields are identified as space-time correlates for massless gauge fields created by dark matter. The identification explains chiral selection in living matter and unbroken $U(2)_{ew}$ invariance and free color in bio length scales become characteristics of living matter and of bio-chemistry and bio-nuclear physics. An attractive solution of the matter antimatter asymmetry is based on the identification of also antimatter as dark matter.

18.1.3   Consciousness and cosmology

Consciousness and cosmology represents a rather weird association from the point of view of materialistically inclined cosmologist. p-Adic physics of cognition however predicts that cognitive consciousness is unavoidably a cosmic phenomenon as far its space-time correlates are considered. Magnetic flux tube hierarchy provides the template for the evolution of conscious, intelligent systems in all length scales in TGD Universe, and bio-systems are predicted to possess magnetic bodies of astrophysical size. Adding to this the enormous spectrum of non-deterministic vacuum extremals (with respect to inertial energy) of field equations allowing interpretation as space-time correlates of intentional action, one has good motivations for a serious consideration of the possibility that intentionality might be realized in astrophysical length scales. There is even some evidence that Sun might act as an intentional system. These speculations are not empty since rather dramatic testable phenomena are predicted. The model explains also the anomalous acceleration of spacecrafts and the finding that the age distribution of stars in a given galaxy does not seem to depend on the age of the galaxy.

18.2   Gravitational Bohr rules and dark matter as a macroscopic quantum phase in astrophysical length scales

D. Da Rocha and Laurent Nottale, the developer of Scale Relativity, have ended up with an highly interesting quantum theory like model for the evolution of astrophysical systems [if1] (I am grateful for Victor Christianito for informing me about the article). The model is simply Schrödinger equation with Planck constant $\hbar$ replaced with what might be called gravitational Planck constant

$$\hbar \quad \rightarrow \quad \hbar_{gr} = \frac{GmM}{v_0} \ . \tag{18.2.1}$$

Here the units used are $\hbar = c = 1$. v_0 is a velocity parameter having the value $v_0 = 144.7 \pm .7$ km/s giving $v_0/c = 4.6 \times 10^{-4}$. The peak orbital velocity of stars in galactic halos is 142 ± 2 km/s whereas the average velocity is 156 ± 2 km/s. Also sub-harmonics and harmonics of v_0 seem to appear.

The model makes fascinating predictions which hold true. For instance, the radii of planetary orbits fit nicely with the prediction of the hydrogen atom like model. The inner solar system (planets up to Mars) corresponds to v_0 and outer solar system to $v_0/5$.

The predictions for the distribution of major axis and eccentrities have been tested successfully also for exoplanets. Also the periods of 3 planets around pulsar PSR B1257+12 fit with the predictions with a relative accuracy of few hours/per several months. Also predictions for the distribution of stars in the regions where morphogenesis occurs follow from the gravitational Schödinger equation.

What is important is that there are no free parameters besides v_0. In [if1] a wide variety of astrophysical data is discussed and it seem that the model works and has already now made predictions which have been later verified. In the following I shall discuss Nottale's model from the point of view of TGD.

18.2.1   TGD prediction for the parameter v_0

One of the basic questions is the origin of the parameter v_0, which according to a rich amount of experimental data discussed in [if1] seems to play a role of a constant of Nature. One of the first applications of cosmic strings in TGD sense was an explanation of the velocity spectrum of stars in the galactic halo in terms of dark matter which could consists of cosmic strings. Cosmic strings could be orthogonal to the galactic plane going through the nucleus (jets) or they could be in galactic plane in which case the strings and their decay products would explain dark matter assuming that the length of cosmic string inside a sphere of radius R is or has been roughly R [D4]. The predicted value of the string tension is determined by the CP_2 radius whose ratio to Planck length is fixed by electron mass via p-adic mass calculations. The resulting prediction for the v_0 is correct and provides a working model for the constant orbital velocity of stars in the galactic halo.

The parameter $v_0 \simeq 2^{-11}$, which has actually the dimension of velocity unless on puts $c = 1$, and also its harmonics and sub-harmonics appear in the scaling of $\hbar$. v_0 corresponds to the velocity of distant stars in the model of galactic dark matter. TGD allows to identify this parameter as the parameter

$$v_0 = 2\sqrt{TG} = \sqrt{\frac{1}{2\alpha_K}}\sqrt{\frac{G}{R^2}} \ ,$$
$$T = \frac{1}{8\alpha_K}\frac{\hbar_0}{R^2} \ . \tag{18.2.2}$$

Here T is the string tension of cosmic strings, R denotes the "radius" of CP_2 ($2R$ is the radius of geodesic sphere of CP_2). α_K is Kähler coupling strength, the basic coupling constant strength of TGD, whose evolution as a function of p-adic length scale is fixed by quantum criticality. The condition that G is invariant in the p-adic coupling constant evolution and number theoretical arguments predict

$$\alpha_K(p) = k\frac{1}{log(p)+log(K)} \ ,$$
$$K = \frac{R^2}{\hbar_0 G} = 2\times 3\times 5\times 7\times 11\times 13\times 17\times 19\times 23 \ , \quad k \simeq \pi/4 \ . \tag{18.2.3}$$

The predicted value of v_0 depends logarithmically on the p-adic length scale and for $p \simeq 2^{127} - 1$ (electron's p-adic length scale) one has $v_0 \simeq 2^{-11}$.

18.2.2   Model for planetary orbits without $v_0 \to v_0/5$ scaling

Also harmonics and sub-harmonics of v_0 appear in the model of Nottale and Da Rocha. For instance, the outer planets (Jupiter, Saturn,...) correspond to $v_0/5$ whereas inner planets correspond to v_0. Quite generally, it is found that the values seem to come as harmonics and sub-harmonics of v_0: $v_n = nv_0$ and v_0/n, and the argument [if1] is that the different values of n relate to fractality. This scaling is not necessary for the planetary orbits in TGD based model.

Effectively a multiplication $n \to 5n$ of the principal quantum number is in question in the case of outer planets. If one accepts the interpretation that visible matter has concentrated around dark matter, which is in macroscopic quantum phase around Bohr orbits, this allows to consider also the possibility that $\hbar_{gr}$ has the same value for all planets.

a) Some gravitational perturbation has kicked dark matter from the region of the asteroid belt to $n \simeq 5k$, $k = 2,..,6$, orbits. The best fit is obtained by using values of n deviating somewhat from multiples of 5 which suggests that the scaling of v_0 is not needed. Gravitational perturbations might have caused the same for the visible matter. The fact that the tilt angles of Earth and outer planets other than Pluto are nearly the same suggests that the orbits of these planets might be an outcome of some violent quantum process for dark matter preserving the orbital plane in a good approximation. Pluto might in turn have experienced some violent collision changing its orbital plane.

b) There could exist at least small amounts of dark matter at all orbits but visible matter is concentrated only around orbits containing some critical amount of dark matter.

	Exp.	Titius-Bode	Bohr$_1$	Bohr$_2$
Planet	R/R_M	R/R_M	$[n, R/R_M]$	$[n, R/R_M]$
Mercury	1	1	$[3, 1]$	
Venus	1.89	1.75	$[4, 1.8]$	
Earth	2.6	2.5	$[5, 2.8]$	
Mars	3.9	4	$[6, 4]$	
Asteroid belt	6.1-8.7	7	$[(7, 8, 9), (5.4, 7.1, 9)]$	
Jupiter	13.7	13	$[11, 13.4]$	$[2\times 5, 11.1]$
Saturn	25.0	25	$[3\times 5, 25]$	$[3\times 5, 25]$
Uranus	51.5	49	$[22, 53.8]$	$[4\times 5, 44.4]$
Neptune	78.9	97	$[27 , 81]$	$[5\times 5, 69.4]$
Pluto	105.2	97	$[31, 106.7]$	$[6\times 5, 100]$

Table 1. The table represents the experimental average orbital radii of planets, the predictions of Titius-Bode law (note the failure for Neptune), and the predictions of Bohr orbit model assuming a)that the principal quantum number n corresponds to best possible fit, b) the scaling $\lambda \to \lambda/5$ for outer planets. Option a) gives the best fit with errors being considerably smaller than the maximal error $|\Delta R|/R \simeq 1/n$ except for Uranus.

How to understand the harmonics and sub-harmonics of v_0 in TGD framework?

Also harmonics and sub-harmonics of v_0 appear in the model of Nottale and Da Rocha. In particular, the outer planets (Jupiter, Saturn,...) correspond to $v_0/5$ whereas inner planets correspond to v_0 in this model. As already found, TGD allows also an alternative explanation.

Quite generally, it is found that the values seem to come as harmonics and sub-harmonics of v_0: $v_n = nv_0$ and v_0/n, and the argument [if1] is that the different values of n relate to fractality. This quantization is a challenge for TGD since v_0 certainly defines a fundamental constant in TGD Universe.

a) Consider first the harmonics of v_0. Besides cosmic strings of type $X^2 \times S^2 \subset M^4 \times CP_2$ one can consider also deformations of these strings defining their multiple coverings so that the deformation is n-valued as a function of S^2-coordinates (Θ, Φ) and the projection to S^2 is thus an $n \to 1$ map. The solutions are higher dimensional analogs of originally closed orbits which after perturbation close only after n turns. This kind of surfaces emerge in the TGD inspired model of quantum Hall effect naturally [O3] and $n \to \infty$ limit has an interpretation as an approach to chaos [G2].

Using the coordinates (x, y, θ, ϕ) of $X^2 \times S^2$ and coordinates m^k for M^4 of the unperturbed solution the space-time surface the deformation can be expressed as

$$
\begin{aligned}
m^k &= m^k(x, y, \theta, \phi) \ , \\
(\Theta, \Phi) &= (\theta, n\phi) \ .
\end{aligned}
\tag{18.2.4}
$$

The value of the string tension would be indeed n^2-fold in the first approximation since the induced Kähler form defining the Kähler magnetic field would be $J_{\theta\phi} = nsin(\Theta)$ and one would have $v_n = nv_0$. At the limit $m^k = m^k(x, y)$ different branches for these solutions collapse together.

b) Consider next how sub-harmonics appear in TGD framework. Cosmic strings are predicted to decay to magnetic flux tube structures by absolute minimization of Kähler action. The Kähler magnetic flux $\Phi = BS$ is conserved in the process but the thickness of the M^4 projection of the cosmic string increases field strength is reduced. This means that string tension, which is proportional to B^2S, is reduced (so that also Kähler action is reduced). The fact that space-time surface is Bohr orbit in generalized sense means that the reduced string tension (magnetic energy per unit length) is quantized. The task is to guess how the quantization occurs. There are two options.

a) The simplest explanation for the reduction of v_0 is based on the decay of a flux tube resembling a disk with a hole to n identical flux tubes so that $v_0 \to v_0/n$ results for the resulting flux tubes. It turns out that this mechanism is favored and explains elegantly the value of $\hbar_{gr}$ for outer planetary system. One can also consider small-p p-adicity so that n would be prime.

b) Second explanation is more intricate. Consider a magnetic flux tube. Since magnetic flux is quantized, the magnetic field strengths are quantized in integer multiples of basic strength: $B = nB_0$ and would rather naturally correspond to the multiple coverings of the original magnetic flux tube with magnetic energy quantized in multiples of n^2. The idea is to require internal consistency in the sense that the allowed reduced field strengths are such that the spectrum associated with B_0 is contained to the spectrum associated with the quantized field strengths $B_1 > B_0$. This would allow only field strengths $B = B_S/n^2$, where B_S denotes the field strength of the fundamental cosmic string and one would have $v_n = v_0/n$. Flux conservation requires that the area of the flux tube scales as n^2.

Sub-harmonics might appear in the outer planetary system and there are indications for the higher harmonics below the inner planetary system [if1]: for instance, solar radius corresponds to $n = 1$ orbital for $v_3 = 3v_0$. This would suggest that Sun and also planets have an onion like structure with highest harmonics of v_0 and strongest string tensions appearing in the solar core and highest sub-harmonics appearing in the outer regions. If the matter results as decay remnants of cosmic strings this means that the mass density inside Sun should correlate strongly with the local value of n characterizing the multiple covering of cosmic strings.

One can ask whether the very process of the formation of the structures could have excited the higher values of n just like closed orbits in a perturbed system become closed only after n turns. The energy density of the cosmic string is about one Planck mass per $\sim 10^7$ Planck lengths so that $n > 1$ excitation increasing this density by a factor of n^2 is obviously impossible except under the primordial cosmic string dominated period of cosmology during which the net inertial energy density must have vanished. The structure of the future solar system would have been dictated already during the primordial phase of cosmology when negative energy cosmic string suffered a time reflection to positive energy cosmic strings.

Nottale equation is consistent with the TGD based model for dark matter

TGD allows two models of dark matter. The first one is spherically symmetric and the second one cylindrically symmetric. The first thing to do is to check whether these models are consistent with the gravitational Schödinger equation/Bohr quantization.

1. Spherically symmetric model for the dark matter

The following argument based on Bohr orbit quantization demonstrates that this is indeed the case for the spherically symmetric model for dark matter. The argument generalizes in a trivial manner to the cylindrically symmetric case.

a) The gravitational potential energy $V(r)$ for a mass distribution $M(r) = xTr$ (T denotes string tension) is given by

$$V(r) \;=\; Gm \int_r^{R_0} \frac{M(r)}{r^2} dr = GmxTlog(\frac{r}{R_0}) \; . \tag{18.2.5}$$

Here R_0 corresponds to a large radius so that the potential is negative as it should in the region where binding energy is negative.

b) The Newton equation $\frac{mv^2}{r} = \frac{GmxT}{r}$ for circular orbits gives

$$v \;=\; xGT \; . \tag{18.2.6}$$

c) Bohr quantization condition for angular momentum by replacing $\hbar$ with $\hbar_{gr}$ reads as $mvr = n\hbar_{gr}$ and gives

$$r_n \;=\; \frac{n\hbar_{gr}}{mv} = nr_1 \; ,$$
$$r_1 \;=\; \frac{GM}{vv_0} \; . \tag{18.2.7}$$

Here v is rather near to v_0.

d) Bound state energies are given by

$$E_n \;=\; \frac{mv^2}{2} - xTlog(\frac{r_1}{R_0}) + xTlog(n) \; . \tag{18.2.8}$$

The energies depend only weakly on the radius of the orbit.

e) The centrifugal potential $l(l+1)/r^2$ in the Schrödinger equation is negligible as compared to the potential term at large distances so that one expects that degeneracies of orbits with small values of l do not depend on the radius. This would mean that each orbit is occupied with same probability irrespective of value of its radius. If the mass distribution for the starts does not depend on r, the number of stars rotating around galactic nucleus is simply the number of orbits inside sphere of radius R and thus given by $N(R) \propto R/r_0$ so that one has $M(R) \propto R$. Hence the model is self consistent in the sense that one can regard the orbiting stars as remnants of cosmic strings and thus obeying same mass distribution.

2. Cylindrically symmetric model for the galactic dark matter

TGD allows also a model of the dark matter based on cylindrical symmetry. In this case the dark matter would correspond to the mass of a cosmic string orthogonal to the galactic plane and traversing through the galactic nucleus. The string tension would the one predicted by TGD. In the directions orthogonal to the plane of galaxy the motion would be free motion so that the orbits would be helical, and this should make it possible to test the model. The quantization of radii of the orbits would be exactly the same as in the spherically symmetric model. Also the quantization of inclinations predicted by the spherically symmetric model could serve as a sensitive test. In this kind of situation general theory of relativity would predict only an angle deficit giving rise to a lens effect. TGD predicts a Newtonian $1/\rho$ potential in a good approximation.

Spiral galaxies are accompanied by jets orthogonal to the galactic plane and a good guess is that they are associated with the cosmic strings. The two models need not exclude each other. The vision about astrophysical structures as pearls of a fractal necklace would suggest that the visible matter has resulted in the decay of cosmic strings originally linked around the cosmic string going through the galactic plane and creating $M(R) \propto R$ for the density of the visible matter in the galactic bulge. The finding that galaxies are organized along linear structures [ib3] fits nicely with this picture.

3. MOND and TGD

TGD based model explains also the MOND (Modified Newton Dynamics) model of Milgrom [ic3] for the dark matter. Instead of dark matter the model assumes a modification of Newton's laws. The model is based on the observation that the transition to a constant velocity spectrum seems in the galactic halos seems to occur at a constant value of the stellar acceleration equal to $a_0 \simeq 10^{-11}g$, where g is the gravitational acceleration at the Earth. MOND theory assumes that Newtonian laws are modified below a_0.

The explanation relies on Bohr quantization. Since the stellar radii in the halo are quantized in integer multiples of a basic radius and since also rotation velocity v_0 is constant, the values of the acceleration are quantized as $a(n) = v_0^2/r(n)$ and a_0 correspond to the radius $r(n)$ of the smallest Bohr orbit for which the velocity is still constant. For larger orbital radii the acceleration would indeed be below a_0. a_0 would correspond to the distance above which the density of the visible matter does not appreciably perturb the gravitational potential of the straight string. This of course requires that gravitational potential is that given by Newton's theory and is indeed allowed by TGD.

18.2.3 The interpretation of $\hbar_{gr}$ and pre-planetary period

$\hbar_{gr}$ could corresponds to a unit of angular momentum for quantum coherent states at magnetic flux tubes or walls containing macroscopic quantum states. Quantitative estimate demonstrates that $\hbar_{gr}$ for astrophysical objects cannot correspond to spin angular momentum. For Sun-Earth system one would have $\hbar_{gr} \simeq 10^{77}\hbar$. This amount of angular momentum realized as a mere spin would require 10^{77} particles! Hence the only possible interpretation is as a unit of orbital angular momentum. The linear dependence of $\hbar_{gr}$ on m is consistent with the additivity of angular momenta in the fusion of magnetic flux tubes to larger units if the angular momentum associated with the tubes is proportional to both m and M.

Just as the gravitational acceleration is a more natural concept than gravitational force, also $\hbar_{gr}/m = GM/v_0$ could be more natural unit than $\hbar_{gr}$. It would define a universal unit for the circulation $\oint v \cdot dl$, which is apart from $1/m$-factor equal to the phase integral $\oint p_\phi d\phi$ appearing in Bohr rules for angular momentum. The circulation could be associated with the flow associated with outer boundaries of magnetic flux tubes surrounding the orbit of mass m around the central mass $M \gg m$ and defining light like 3-D CDs analogous to black hole horizons.

The expression of $\hbar_{gr}$ depends on masses M and m and can apply only in space-time regions carrying information about the space-time sheets of M and and the orbit of m. Quantum gravitational holography suggests that the formula applies at 3-D light like causal determinant (CD) X_l^3 defined by the wormhole contacts gluing the space-time sheet X^3 of the planet to that of Sun. More generally, X_l^3 could be the space-time sheet containing the planet, most naturally the magnetic flux tube surrounding the orbit of the planet and possibly containing dark matter in super-conducting state. This would give a precise meaning for $\hbar_{gr}$ and explain why $\hbar_{gr}$ does not depend on the masses of other planets.

The simplest option consistent with the quantization rules and with the explanatory role of magnetic flux structures is perhaps the following one.

a) X_l^3 is a torus like surface around the orbit of the planet containing delocalized dark matter. The key role of magnetic flux quantization in understanding the values of v_0 suggests the interpretation of the torus as a magnetic or Z^0 magnetic flux tube. At pre-planetary period the dark matter formed a torus like quantum object. The conditions defining the radii of Bohr orbits follow from the requirement that the torus-like object is in an eigen state of angular momentum in the center of mass rotational degrees of freedom. The requirement that rotations do not leave the torus-like object invariant is obviously satisfied. Newton's law required by the quantum-classical correspondence stating that the orbit corresponds to a geodesic line in general relativistic framework gives the additional condition implying Bohr quantization.

b) A simple mechanism leading to the localization of the matter would have been the pinching of the torus causing kind of a traffic jam leading to the formation of the planet. This process could quite well have involved a flow of matter to a smaller planet space-time sheet Y_l^3 topologically condensed at X_l^3. Most of the angular momentum associated with torus like object would have transformed to that of planet and situation would have become effectively classical.

c) The conservation of magnetic flux means that the splitting of the orbital torus would generate a pair of Kähler magnetic charges. It is not clear whether this is possible dynamically and hence the torus could still be there. In fact, TGD explanation for the tritium beta decay anomaly citeTroitsk,Mainz in terms of classical Z^0 force [F8] requires the existence of this kind of torus containing neutrino cloud whose density varies along the torus. This picture suggests that the lacking $n = 1$ and $n = 2$ orbits in the region between Sun and Mercury are still in magnetic flux tube state containing mostly dark matter.

d) The fact that $\hbar_{gr}$ is proportional to m means that it could have varied continuously during the accumulation of the planetary mass without any effect in the planetary motion: this is of course nothing but a manifestation of Equivalence Principle.

e) It is interesting to look for the scaled up versions of Planck mass $m_{Pl} = \sqrt{\hbar_{gr}/\hbar} \times \sqrt{\hbar/G} = \sqrt{M_1 M_2/v_0}$ and Planck length $L_{Pl} = \sqrt{\hbar_{gr}/\hbar} \times \sqrt{\hbar/G} = G\sqrt{M_1 M_2/v_0}$. For $M_1 = M_2 = M$ this gives $m_{Pl} = M/\sqrt{v_0} \simeq 45.6 \times M$ and $L_{Pl} = r_S/2\sqrt{v_0} \simeq 22.8 \times r_S$, where r_S is Schwartshild radius. For Sun r_S is about 2.9 km so that one has $L_{Pl} \simeq 66$ km. For a few years ago it was found that Sun contains "inner-inner" core of radius about $R = 300$ km [il2] which is about $4.5 \times L_{Pl}$.

18.2.4 Inclinations for the planetary orbits and the quantum evolution of the planetary system

The inclinations of planetary orbits provide a test bed for the theory. The semiclassical quantization of angular momentum gives the directions of angular momentum from the formula

$$cos(\theta) = \frac{m}{\sqrt{j(j+1)}} \quad , \quad |m| \leq j \ . \tag{18.2.9}$$

where θ is the angle between angular momentum and quantization axis and thus also that between orbital plane and (x,y)-plane. This angle defines the angle of tilt between the orbital plane and (x,y)-plane.

$m = j = n$ gives minimal value of angle of tilt for a given value of n of the principal quantum number as

$$cos(\theta) \quad = \quad \frac{n}{\sqrt{n(n+1)}} \ . \tag{18.2.10}$$

For $n = 3, 4, 5$ (Mercury, Venus, Earth) this gives $\theta = 30.0, 26.6$, and 24.0 degrees respectively.

Only the relative tilt angles can be compared with the experimental data. Taking as usual the Earth's orbital plane as the reference the relative tilt angles give what are known as inclinations. The predicted inclinations are 6 degrees for Mercury and 2.6 degrees for Venus. The observed values [if2] are 7.0 and 3.4 degrees so that the agreement is satisfactory. If one allows half-odd integer spin the fit is improved. For $j = m = n - 1/2$ the predictions are 7.1 and 2.9 degrees for Mercury and Venus respectively. For Mars, Jupiter, Saturn, Uranus, Neptune, and Pluto the inclinations are 1.9, 1.3, 2.5, 0.8, 1.8, 17.1 degrees. For Mars and outer planets the tilt angles are predicted to have wrong sign for

$m = j$. In a good approximation the inclinations vanish for outer planets except Pluto and this would allow to determine m as $m \simeq \sqrt{5n(n+1)/6}$: the fit is not good.

The assumption that matter has condensed from a matter rotating in (x,y)-plane orthogonal to the quantization axis suggests that the directions of the planetary rotation axes are more or less the same and by angular momentum conservation have not changed appreciably. The prediction for the tilt of the rotation axis of the Earth is 24 degrees of freedom in the limit that the Earth's spin can be treated completely classically, that is for $m = j >> 1$ in the units used for the quantization of the Earth's angular momentum. What is the value of $\hbar_{gr}$ for Earth is not obvious (using the unit $\hbar_{gr} = GM^2/v_0$ the Earth's angular momentum would be much smaller than one). The tilt of the rotation axis of Earth with respect to the orbit plane is 23.5 degrees so that the agreement is again satisfactory. This prediction is essentially quantal: in purely classical theory the most natural guess for the tilt angle for planetary spins is 0 degrees.

The observation that the inner planets Mercury, Venus, and Earth have in a reasonable approximation the predicted inclinations suggest that they originate from a primordial period during which they formed spherical cells of dark matter and had thus full rotational degrees of freedom and were in eigen states of angular momentum corresponding to a full rotational symmetry. The subsequent $SO(3) \rightarrow SO(2)$ symmetry breaking leading to the formation of torus like configurations did not destroy the information about this period since the information about the value of j and m was coded by the inclination of the planetary orbit.

In contrast to this, the dark matter associated with Earth and outer planets up to Neptune formed a flattened magnetic or Z^0 magnetic flux tube resembling a disk with a hole and the subsequent symmetry breaking broke it to separate flux tubes. Earth's spherical disk was joined to the disk formed by the outer planets. The spherical disk could be still present and contain super-conducting dark matter. The presence of this "heavenly sphere" might closely relate to the fact that Earth is a living planet. The time scale $T = 2\pi R/c$ is very nearly equal to 5 minutes and defines a candidate for a bio-rhythm.

If this flux tube carried the same magnetic flux as the flux tubes associated with the inner planets, the decomposition of the disk with a hole to 5 flux tubes corresponding to Earth and to the outer planets Mars, Jupiter, Saturn and Neptune, would explain the value of v_0 correctly and also the small inclinations of outer planets. That Pluto would not originate from this structure, is consistent with its anomalously large values of inclination $i = 17.1$ degrees, small value of eccentricity $e = .248$, and anomalously large value of inclination of equator to orbit about 122 degrees as compared to 23.5 degrees in the case of Earth [if2].

18.2.5 Eccentricities and comets

Bohr-Sommerfeld quantization allows also to deduce the eccentricities of the planetary and comet orbits. One can write the quantization of energy as

$$\frac{p_r^2}{2m_1} + \frac{p_\theta^2}{2m_1 r^2} + \frac{p_\phi^2}{2m_1 r^2 sin^2(\theta)} - \frac{k}{r} = -\frac{E_1}{n^2} \ ,$$

$$E_1 = \frac{k^2}{2\hbar_{gr}^2} \times m_1 = \frac{v_0^2}{2} \times m_1 \ . \tag{18.2.11}$$

Here one has $k = GMm_1$. E_1 is the binding energy of $n = 1$ state. In the orbital plane ($\theta = \pi/2, p_\theta = 0$) the conditions are simplified. Bohr quantization gives $p_\phi = m\hbar_{gr}$ implying

$$\frac{p_r^2}{2m_1} + \frac{k^2\hbar_{gr}^2}{2m_1 r^2} - \frac{k}{r} = -\frac{E_1}{n^2} \ . \tag{18.2.12}$$

For $p_r = 0$ the formula gives maximum and minimum radii $r_\pm$ and eccentricity is given by

$$e^2 = \frac{r_+ - r_-}{r_+} = \frac{2\sqrt{1 - \frac{m^2}{n^2}}}{1 + \sqrt{1 - \frac{m^2}{n^2}}} \ . \tag{18.2.13}$$

For small values of n the eccentricities are very large except for $m = n$. For instance, for $(m = n-1, n)$ for $n = 3, 4, 5$ gives $e = (.93, .89, .86)$ to be compared with the experimental values $(.206, .007, .0167)$. Thus the planetary eccentricities with Pluto included $(e = .248)$ must vanish in the lowest order approximation and must result as a perturbation of the magnetic flux tube.

The large eccentricities of comet orbits might however have an interpretation in terms of $m < n$ states. The prediction is that comets with small eccentricities have very large orbital radius. Oort's cloud is a system weakly bound to a solar system extending up to 3 light years. This gives the upper bound $n \leq 700$ if the comets of the cloud belong to the same family as Mercury, otherwise the bound is smaller. This gives a lower bound to the eccentricity of not nearly circular orbits in the Oort cloud as $e > .32$.

18.2.6   Why the quantum coherent dark matter is not visible?

The obvious objection against quantal astrophysics is that astrophysical systems look extremely classical. Quantal dark matter in many-sheeted space-time resolves this counter argument. As already explained, the sequence of symmetry breakings of the rotational symmetry would explain nicely why astral Bohr rules work. The prediction is however that delocalized quantal dark matter is probably still present at (the boundaries of) magnetic flux tubes and spherical shells. It is however the entire structure defined by the orbit which behaves like a single extended particle so that the localization in quantum measurement does not mean a localization to a point of the orbit. Planet itself corresponds to a smaller localized space-time sheet condensed at the flux tube.

One should however understand why this dark matter with a gigantic Planck constant is not visible. The simplest explanation is that there cannot be any direct quantum interactions between ordinary and dark matter in the sense that particles with different values of Planck constant could appear in the same particle vertex. This would allow also a fractal hierarchy copies of standard model physics to exist with different p-adic mass scales.

There is also second argument. The inability to observe dark matter could mean inability to perform state function reduction localizing the dark matter. The probability for this should be proportional to the strength of the measurement interaction. For photons the strength of the interaction is characterized by the fine structure constant. In the case of dark matter the fine structure constant is replaced with

$$\alpha_{em,gr} \quad = \quad \alpha_{em} \times \frac{\hbar}{\hbar_{gr}} = \alpha_{em} \times \frac{v_0}{GMm} \quad . \tag{18.2.14}$$

For $M = m = m_{Pl} \simeq 10^{-8}$ kg the value of the fine structure constant is smaller than $\alpha_{em}v_0$ and completely negligible for astrophysical masses. However, for processes for which the lowest order classical rates are non-vanishing, rates are not affected in the lowest order since the increase of the Compton length compensates the reduction of α. Higher order corrections become however small. What makes dark matter invisible is not the smallness of α_{em} but the fact that the binding energies of say hydrogen atom proportional to $\alpha^2 m_e$ are scaled as $1/\hbar^2$ so that the spectrum is scaled down.

18.2.7   Quantum interpretation of gravitational Schrödinger equation

Schrödinger equation in astrophysical length scales with a gigantic value of Planck constant looks sheer madness idea from the standard physics point of view. In TGD Universe situation might be different.

a) In TGD inertial four-momentum (or conserved four-momentum) is not positive definite and the net four-momentum of the Universe vanishes. Already in cosmological length scales the density of inertial mass vanishes. Gravitational masses and inertial masses can be identified only at the limit when one can neglect the interaction between positive and negative energy matter. The masses appearing in the gravitational Schrödinger equation are gravitational masses and one can ask whether inertial and gravitational Planck constants are different.

b) The fractality of the many-sheeted space-time predicts that quantum effects appear in all length and time scales. In particular, dark matter is at larger space-time sheets and hence almost invisible.

c) An even more weirder looks the idea that Planck constant could have a gigantic value in astrophysical length scales being of order of magnitude of product of masses using Planck mass as a unit for $\hbar = c = 1$. This would mean that gravitation at space-time sheets of astrophysical size would have super quantal character! But even the gigantic value of Planck constant might be understood in TGD framework.

Beraha numbers and spectrum of Planck constant

The infinite-dimensional Clifford algebra of the configuration space ("the world of classical worlds") gamma matrices defines so called von Neumann algebra with a hierarchy of type II_1 sub-factors. So called Beraha numbers

$$B_n \quad = \quad 4cos^2(\frac{\pi}{n}), \quad n \geq 3 \tag{18.2.15}$$

relate very closely to these factors as also to braid groups and quantum groups. Roughly, B_n corresponds to the renormalized dimension d of 4-component spinors of $D = 4$ dimensional space whose dimension becomes also renormalized. The formula $d_n = B_n = 2^{D_n/2}$ relating the dimension of spinors to the space-time dimension gives for the renormalized space-time dimension the expression $D_n = 2log(B_n)/log(2)$ approaching $D = 4$ at the limit $n \to \infty$. Note that the spectrum of fractal space-time dimension would have upper limit $D = 4$. In fact, there is also a continuum of dimensions $D \geq 4$ for the dimensions of sub-factors of type II_1. Obviously, the dimensions behave like energy spectrum of a quantum mechanical systems. That $D = 4$ is the limiting value of bound state dimensions suggests strongly a connection with the fact that the infinities of quantum field theory appear for $D \geq 4$.

The TGD based model for topological quantum computation [O3] based on the braiding of magnetic flux tubes led to the idea that this braiding could be seen as space-time correlate for a spectral flow for conformal weights at the level of configuration space spinor fields so that the connection with type II_1 factors emerges naturally.

In [O4] I developed general ideas related to type II_1 factors of von Neumann algebras and their connection with the physics predicted by quantum TGD. The speculation was that Beraha numbers define an entire spectrum of values of $\hbar$ or equivalently spectrum of values of Kähler coupling strength $\alpha_K \equiv g_K^2/4\pi\hbar$. The values of $\hbar$ would be given by

$$\hbar(n) \quad = \quad \frac{log(B_\infty)}{log(B_n)} \times \hbar(\infty) = \frac{log(4)}{log(4cos^2(\pi/n))} \times \hbar(\infty) \ , \quad n \geq 3 \ . \tag{18.2.16}$$

The proposed interpretation was that the spectrum corresponds to renormalization group evolution fixed points of α_K related to the angular/phase resolution whereas the p-adic length scale evolution of the Kähler coupling constant corresponds to length scale resolution. Small values of n would correspond to a poor angular/phase resolution.

The spectrum has remarkable features.

a) The ratio $\hbar(4)/\hbar(\infty) = 2$ means that in the range $n \geq 4$ $\hbar$ varies only by a factor of 2. The measured value of $\hbar$ is in the range $n \geq 4$: probably rather near to $\hbar(\infty)$. The cosmic evolution of $\hbar(n)$ induced by a gradual increase of the angular resolution might explain the reported increase of the fine structure constant $\alpha = e^2/4\pi\hbar c$ during cosmic evolution. The smallness of the increase implies that the recent value of n must be rather large so that $\hbar \simeq \hbar(\infty)$ should be a good approximation and it might be impossible to distinguish it from $\hbar(\infty)$ experimentally. Of course, the detection of varying values of fine structure constant in accordance with the prediction would be a victory for the proposed admittedly heuristic theory. It is known that different measurement methods give slightly different values for the fine structure constant so that it might be a good idea to check whether the variation could be understood in terms of Beraha numbers.

b) $n = 3$ is a complete exception since one has $\hbar(3) = \infty$ so that Kähler coupling strength and presumably also fine structure constant vanishes. This makes sense only if the Kähler action of the space-time sheet is vanishing in this phase. In fact, the requirement that the vacuum functional defined by the exponent of the Kähler function is non-vanishing for the entire universe, requires that Kähler per volume vanishes so that this condition is quite sensible and corresponds to a scale invariant situation. Vacuum extremals are a basic example of a phase with a vanishing Kähler action and correspond to a situation in which the energy densities of positive and negative energy matter cancel each other in the length scale considered. Robertson-Walker cosmologies are basic cosmological example in this respect [D5].

Gravitational Planck constant as a small perturbation of $1/\hbar(3) = 0$

Although the value of $\hbar_{gr}$ in the Nottale's variant of Schrödinger equation is not strictly infinite, it is infinite for almost all numerical purposes. From the point of view of α_K $\hbar/\hbar_{gr}$ is the correct number to consider and the deviation of $\hbar/\hbar_{gr}$ from zero could be interpreted as a gravitational perturbative effect changing the value of x from zero. The modification would be given by

$$\frac{\hbar}{\hbar_{gr}} = \frac{v_0}{GMm} \tag{18.2.17}$$

would be extremely small, and would have a natural interpretation as resulting from the gravitational interaction between masses M and m.

What is interesting is that the modification is not proportional to GmM but to the small parameter v_0/GMm One could interpret the parameter as proportional to the product of Compton lengths associated with M and m using CP_2 radius R as the natural fundamental length unit.

A possible interpretation for the deviation of $\hbar/\hbar_{gr}$ from zero is as a deviation $\Delta\phi = \pi/3 - \phi$ of the angle ϕ defining Beraha number from the maximal value $\phi = \pi/3$. One would have

$$\Delta\phi = \frac{\pi}{3} - arcos(2^{\hbar/\hbar_{gr}-1}) \simeq \frac{log(2)}{\sqrt{3}} \frac{\hbar}{\hbar_{gr}} = \frac{log(2)}{\sqrt{3}} \frac{v_0}{GMm} \ . \tag{18.2.18}$$

The proposed picture would suggest that when the system size becomes very large $n = 3$ super quantum phase is approached. This requires that the extremals of Kähler action have vanishing or extremely small action. This is indeed the case for the vacuum extremals and Robertson-Walker cosmologies are the most important example of vacuum extremals cosmologically. What is interesting that inertial and gravitational Planck constants seem to lie at the opposite ends of the spectrum of Planck constants.

Gravitational Schrödinger equation as a means of avoiding gravitational collapse

Schrödinger equation provided a solution to the infrared catastrophe of the classical model of atom: the classical prediction was that electron would radiate its energy as brehmstrahlung and would be captured by the nucleus. The gravitational variant of this process would be the capture of the planet by a black hole, and more generally, a collapse of the star to a black hole. Gravitational Schrödinger equation could obviously prevent the catastrophe.

For $1/r$ gravitation potential the Bohr radius is given by $a_{gr} = GM/v_0^2 = r_S/2v_0^2$, where $r_S = 2GM$ is the Schwartchild radius of the mass creating the gravitational potential: obviously Bohr radius is much larger than the Schwartschild radius. That the gravitational Bohr radius does not depend on m conforms with Equivalence Principle, and the proportionality $\hbar_{gr} \propto Mm$ can be deduced from it. Gravitational Bohr radius is by a factor $1/2v_0^2$ larger than black hole radius so that black hole can swallow the piece of matter with a considerable rate only if it is in the ground state and also in this state the rate is proportional to the black hole volume to the volume defined by the black hole radius given by $2^3 v_0^6 \sim 10^{-20}$.

The $\hbar_{gr} \to \infty$ limit for $1/r$ gravitational potential means that the exponential factor $exp(-r/a_0)$ of the wave function becomes constant: on the other hand, also Schwartshild and Bohr radii become infinite at this limit. The gravitational Compton length associated with mass m does not depend on m and is given by GM/v_0 and the time $T = E_{gr}/\hbar_{gr}$ defined by the gravitational binding energy is twice the time taken to travel a distance defined by the radius of the orbit with velocity v_0 which suggests that signals travelling with a maximal velocity v_0 are involved with the quantum dynamics.

In the case of planetary system the proportionality $\hbar_{gr} \propto mM$ creates problems of principle since the influence of the other planets is not taken account. One might argue that the generalization of the formula should be such that M is determined by the gravitational field experienced by mass m and thus contains also the effect of other planets. The problem is that this field depends on the position of m which would mean that $\hbar_{gr}$ itself would become kind of field quantity.

Does the transition to non-perturbative phase correspond to a change in the value of $\hbar$?

Nature is populated by systems for which perturbative quantum theory does not work. Examples are atoms with $Z_1 Z_2 e^2/4\pi\hbar > 1$ for which the binding energy becomes larger than rest mass, non-perturbative QCD resulting for $Q_{s,1} Q_{s,2} g_s^2/4\pi\hbar > 1$, and gravitational systems satisfying $GM_1 M_2/4\pi\hbar > 1$. Quite generally, the condition guaranteing troubles is of the form $Q_1 Q_2 g^2/4\pi\hbar > 1$. There is no general mathematical approach for solving the quantum physics of these systems but it is believed that a phase transition to a new phase of some kind occurs.

The gravitational Schrödinger equation forces to ask whether Nature herself takes care of the problem so that this phase transition would involve a change of the value of the Planck constant to guarantee that the perturbative approach works. The values of $\hbar$ would vary in a stepwise manner from $\hbar(\infty)$ to $\hbar(4) = \hbar(\infty)/2$ corresponding to $B(4)$ and the last step would be a transition to a phase which differs only slightly from the phase $1/\hbar(3) = 0$ would occur and correspond to

$$\hbar \quad \rightarrow \quad \frac{Q_1 Q_2 g^2}{v} \tag{18.2.19}$$

inducing

$$\frac{Q_1 Q_2 g^2}{4\pi\hbar} \rightarrow \frac{v}{4\pi} \quad . \tag{18.2.20}$$

The simplest (and of course ad hoc) assumption making sense in TGD Universe is that v is a harmonic or subharmonic of v_0 appearing in the gravitational Schrödinger equation. For instance, for the Kepler problem the spectrum of binding energies would be universal (independent of the values of charges) and given by $E_n = v^2 m/2n^2$ with v playing the role of small coupling. Bohr radius would be $g^2 Q_2/v^2$ for $Q_2 \gg Q_1$.

This provides a new insight to the problems encountered in quantizing gravity. QED started from the model of atom solving the infrared catastrophe. In quantum gravity theories one has started directly from the quantum field theory level and the recent decline of the M-theory shows that we are still practically where we started. If the gravitational Schrödinger equation indeed allows quantum interpretation, one could be more modest and start from the solution of the gravitational IR catastrophe by assuming a dynamical spectrum of $1/\hbar$ fixed in the first approximation by Beraha numbers and perhaps containing perturbative additive corrections defined by the previous general formula which for $n = 3$ case would give the entire coupling. The implications would be profound: the whole program of quantum gravity would have been misled as far as the quantization of systems with $GM_1 M_2/\hbar > 1$ is considered. In practice, these systems are the most interesting ones and the prejudice that their quantization is a mere academic exercise would have been completely wrong.

An alternative formulation for the occurrence of a transition increasing the value of $\hbar$ could rely on the requirement that classical bound states have reasonable quantum counterparts. In the gravitational case one would have $r_n = n^2 \hbar_{gr}^2/GM_1^2 M$, for $M_1 \ll M$, which is extremely small distance for $\hbar_{gr} = \hbar$ and reasonable values of n. Hence, either n is so large that the system is classical or $\hbar_{gr}/\hbar$ is very large. Equivalence Principle requires the independence of r_n on M_1, which gives $\hbar = kGM_1 M_2$ giving $r_n = n^2 kGM$. The requirement that the radius is above Schwartshild radius gives $k \geq 2$. In the case of Dirac equation the solutions cease to exist for $Z \geq 137$ and which suggests that $\hbar$ is large for hypothetical atoms having $Z \geq 137$.

18.2.8 How do the magnetic flux tube structures and quantum gravitational bound states relate?

In the case of stars in galactic halo the appearance of the parameter v_0 characterizing cosmic strings as orbital rotation velocity can be understood classically. That v_0 appears also in the gravitational dynamics of planetary orbits could relate to the dark matter at magnetic flux tubes. The argument explaining the harmonics and sub-harmonics of v_0 in terms of properties of cosmic strings and magnetic flux tubes identifiable as their descendants strengthens this expectation.

The notion of magnetic body

In TGD inspired theory of consciousness the notion of magnetic body plays a key role: magnetic body is the ultimate intentional agent, experiencer, and performer of bio-control and can have astrophysical size: this does not sound so counter-intuitive if one takes seriously the idea that cognition has p-adic space-time sheets as space-time correlates and that rational points are common to real and p-adic number fields. The point is that infinitesimal in p-adic topology corresponds to infinite in real sense so that cognitive and intentional structures would have literally infinite size.

The magnetic flux tubes carrying various supra phases can be interpreted as special instance of dark energy and dark matter. This suggests a correlation between gravitational self-organization and quantum phases at the magnetic flux tubes and that the gravitational Schrödinger equation somehow relates to the ordinary Schrödinger equation satisfied by the macroscopic quantum phases at magnetic flux tubes. Interestingly, the transition to large Planck constant phase should occur when the masses of interacting is above Planck mass since gravitational self-interaction energy is $V \sim GM^2/R$. For the density of water about 10^3 kg/m^3 the volume carrying a Planck mass correspond to a cube with side 2.8×10^{-4} meters. This corresponds to a volume of a large neuron, which suggests that this phase transition might play an important role in neuronal dynamics.

Could gravitational Schrödinger equation relate to a quantum control at magnetic flux tubes?

An infinite self hierarchy is the basic prediction of TGD inspired theory of consciousness ("everything is conscious and consciousness can be only lost"). Topological quantization allows to assign to any material system a field body as the topologically quantized field pattern created by the system [N4, K1]. This field body can have an astrophysical size and would utilize the material body as a sensory receptor and motor instrument.

Magnetic flux tube and flux wall structures are natural candidates for the field bodies. Various empirical inputs have led to the hypothesis that the magnetic flux tube structures define a hierarchy of magnetic bodies, and that even Earth and larger astrophysical systems possess magnetic body which makes them conscious self-organizing living systems. In particular, life at Earth would have developed first as a self-organization of the super-conducting dark matter at magnetic flux tubes [N4].

For instance, EEG frequencies corresponds to wavelengths of order Earth size scale and the strange findings of Libet about time delays of conscious experience [ma3, ma4] find an elegant explanation in terms of time taken for signals propagate from brain to the magnetic body [K1]. Cyclotron frequencies, various cavity frequencies, and the frequencies associated with various p-adic frequency scales are in a key role in the model of bio-control performed by the magnetic body. The cyclotron frequency scale is given by $f = eB/m$ and rather low as are also cavity frequencies such as Schumann frequencies: the lowest Schumann frequency is in a good approximation given by $f = 1/2\pi R$ for Earth and equals to 7.8 Hz.

1. Quantum time scales as "bio-rhythms" in solar system?

To get some idea about the possible connection of the quantum control possibly performed by the dark matter with gravitational Schrödinger equation, it is useful to look for the values of the periods defined by the gravitational binding energies of test particles in the fields of Sun and Earth and look whether they correspond to some natural time scales. For instance, the period $T = 2GM_S n^2/v_0^3$ defined by the energy of nth planetary orbit depends only on the mass of Sun and defines thus an ideal candidate for a universal "bio-rhytm".

For Sun black hole radius is about 2.9 km. The period defined by the binding energy of lowest state in the gravitational field of Sun is given $T_S = 2GM_S/v_0^3$ and equals to 23.979 hours for $v_0/c = 4.8233 \times 10^{-4}$. Within experimental limits for v_0/c the prediction is consistent with 24 hours! The value of v_0 corresponding to exactly 24 hours would be $v_0 = 144.6578$ km/s (as a matter fact, the rotational period of Earth is 23.9345 hours). As if as the frequency defined by the lowest energy state would define a "biological" clock at Earth! Mars is now a strong candidate for a seat of life and the day in Mars lasts 24hr 37m 23s! $n = 1$ and $n = 2$ are orbitals are not realized in solar system as planets but there is evidence for the $n = 1$ orbital as being realized as a peak in the density of IR-dust [if1]. One can of course consider the possibility that these levels are populated by small dark matter planets with matter at larger space-time sheets. Bet as it may, the result supports the notion of quantum gravitational entrainment in the solar system.

The slower rhythms would become as n^2 sub-harmonics of this time scale. Earth itself corresponds to $n = 5$ state and to a rhythm of .96 hours: perhaps the choice of 1 hour to serve as a fundamental time unit is not merely accidental. The magnetic field with a typical ionic cyclotron frequency around 24 hours would be very weak: for 10 Hz cyclotron frequency in Earth's magnetic field the field strength would about 10^{-11} T. However, $T = 24$ hours corresponds with 6 per cent accuracy to the p-adic time scale $T(k = 280) = 2^{13}T(2, 127)$, where $T(2, 127)$ corresponds to the secondary p-adic time scale of .1 s associated with the Mersenne prime $M_{127} = 2^{127} - 1$ characterizing electron and defining a fundamental bio-rhytm and the duration of memetic codon [TGDgeme].

Comorosan effect [la10, la11, J5] demonstrates rather peculiar looking facts about the interaction of organic molecules with visible laser light at wavelength $\lambda = 546\ nm$. As a result of irradiation molecules seem to undergo a transition $S \to S^*$. S^* state has anomalously long lifetime and stability in solution. $S \to S^*$ transition has been detected through the interaction of S^* molecules with different biological macromolecules, like enzymes and cellular receptors. Later Comorosan found that the effect occurs also in non-living matter. The basic time scale is $\tau = 5$ seconds. p-Adic length scale hypothesis does not explain τ, and it does not correspond to any obvious astrophysical time scale and has remained a mystery.

The idea about astro-quantal dark matter as a fundamental bio-controller inspires the guess that τ could correspond to some Bohr radius R for a solar system via the correspondence $\tau = R/c$. As observed by Nottale, $n = 1$ orbit for $v_0 \to 3v_0$ corresponds in a good approximation to the solar radius and to $\tau = 2.18$ seconds. For $v_0 \to 2v_0$ $n = 1$ orbit corresponds to $\tau = AU/(4 \times 25) = 4.992$ seconds: here $R = AU$ is the astronomical unit equal to the average distance of Earth from Sun. The deviation from τ_C is only one per cent and of the same order of magnitude as the variation of the radius for the orbit due to orbital eccentricity $(a - b)/a = .0167$ [if2].

2. Earth-Moon system

For Earth serving as the central mass the Bohr radius is about 18.7 km, much smaller than Earth radius so that Moon would correspond to $n = 147.47$ for v_0 and $n = 1.02$ for the sub-harmonic $v_0/12$ of v_0. For an afficionado of cosmic jokes or a numerologist the presence of the number of months in this formula might be of some interest. Those knowing that the Mayan calendar had 11 months and that Moon is receding from Earth might rush to check whether a transition from $v/11$ to $v/12$ state has occurred after the Mayan culture ceased to exist: the increase of the orbital radius by about 3 per cent would be required! Returning to a more serious mode, an interesting question is whether light satellites of Earth consisting of dark matter at larger space-time sheets could be present. For instance, in [N4] I have discussed the possibility that the larger space-time sheets of Earth could carry some kind of intelligent life crucial for the bio-control in the Earth's length scale.

The period corresponding to the lowest energy state is from the ratio of the masses of Earth and Sun given by $M_E/M_S = (5.974/1.989) \times 10^{-6}$ given by $T_E = (M_E/M_S) \times T_S = .2595$ s. The corresponding frequency $f_E = 3.8535$ Hz frequency is at the lower end of the theta band in EEG and is by 10 per cent higher than the p-adic frequency $f(251) = 3.5355$ Hz associated with the p-adic prime $p \simeq 2^k$, $k = 251$. The corresponding wavelength is 2.02 times Earth's circumference. Note that the cyclotron frequencies of Nn, Fe, Co, Ni, and Cu are $5.5, 5.0, 5.2, 4.8$ Hz in the magnetic field of $.5 \times 10^{-4}$ Tesla, which is the nominal value of the Earth's magnetic field. In [M4] I have proposed that the cyclotron frequencies of Fe and Co could define biological rhythms important for brain functioning. For $v_0/12$ associated with Moon orbit the period would be 7.47 s: I do not know whether this corresponds to some bio-rhytm.

It is better to leave for the reader to decide whether these findings support the idea that the super conducting cold dark matter at the magnetic flux tubes could perform bio-control and whether the gravitational quantum states and ordinary quantum states associated with the magnetic flux tubes couple to each other and are synchronized.

18.2.9 p-Adic length scale hypothesis and $v_0 \to v_0/5$ transition at inner-outer border for planetary system

$v_0 \to v_0/5$ transition would allow to interpret the orbits of outer planets as $n \geq 1$ orbits. The obvious question is whether inner to outer zone as $v_0 \to v_0/5$ transition could be interpreted in terms of the p-adic length scale hierarchy.

a) The most important p-adic length scale are given by primary p-adic length scales $L(k) = 2^{(k-151)/2} \times 10$ nm and secondary p-adic length scales $L(2,k) = 2^{k-151} \times 10$ nm, k prime.

b) The p-adic scale $L(2,139) = 114$ Mkm is slightly above the orbital radius 109.4 Mkm of Venus. The p-adic length scale $L(2,137) \simeq 28.5$ Mkm is roughly one half of Mercury's orbital radius 57.9 Mkm. Thus strong form of p-adic length scale hypothesis could explain why the transition $v_0 \to v_0/5$ occurs in the region between Venus and Earth ($n = 5$ orbit for v_0 layer and $n = 1$ orbit for $v_0/5$ layer).

c) Interestingly, the *primary* p-adic length scales $L(137)$ and $L(139)$ correspond to fundamental atomic length scales which suggests that solar system be seen as a fractally scaled up "secondary" version of atomic system.

d) Planetary radii have been fitted also using Titius-Bode law predicting $r(n) = r_0 + r_1 \times 2^n$. Hence on can ask whether planets are in one-one correspondence with primary and secondary p-adic length scales $L(k)$. For the orbital radii $58, 110, 150, 228$ Mkm of Mercury, Venus, Earth, and Mars indeed correspond approximately to k= $276, 278, 279, 281$: note the special position of Earth with respect to its predecessor. For Jupiter, Saturn, Uranus, Neptune, and Pluto the radii are 52,95,191,301,395 Mkm and would correspond to p-adic length scales $L(280+2n))$, $n = 0, ..., 3$. Obviously the transition $v_0 \to v_0/5$ could occur in order to make the planet–p-adic length scale one-one correspondence possible.

e) It is interesting to look whether the p-adic length scale hierarchy applies also to the solar structure. In a good approximation solar radius .696 Mkm corresponds to $L(270)$, the lower radius .496 Mkm of the convective zone corresponds to $L(269)$, and the lower radius .174 Mkm of the radiative zone (radius of the solar core) corresponds to $L(266)$. This encourages the hypothesis that solar core has an onion like sub-structure corresponding to various p-adic length scales. In particular, $L(2,127)$ ($L(127)$ corresponds to electron) would correspond to 28 Mm. The core is believed to contain a structure with radius of about 10 km: this would correspond to $L(231)$. This picture would suggest universality of star structure in the sense that stars would differ basically by the number of the onion like shells having standard sizes.

Quite generally, in TGD Universe the formation of join along boundaries bonds is the space-time correlate for the formation of bound states. This encourages to think that (Z^0) magnetic flux tubes are involved with the formation of gravitational bound states and that for $v_0 \to v_0/k$ corresponds either to a splitting of a flux tube resembling a disk with a whole to k pieces, or to the scaling down $B \to B/k^2$ so that the magnetic energy for the flux tube thickened and stretched by the same factor k^2 would not change.

18.2.10 Further evidence for dark matter

The notion of many-sheeted space-time has been continually receiving qualitative support from various anomalies. In the following two latest anomalies are summarized briefly.

First dark matter galaxy found

The propose model for dark matter suggests an existence of dark matter planets and even dark matter galaxies. Therefore the news about finding of the first dark galaxy in New Scientist [ic9] came as a pleasant surprise. The galaxy is located at a distance of 10^7 light years. It contains 1 per cent hydrogen gas and and 99 per cent dark matter and is identified by 21 cm hydrogen line: hence the name VIRGOH21. The amount of dark matter counts as 10^8 average stars.

Anomalous chemical compositions at the surface of Sun as evidence for dark matter

Physics in Action, February 2005 contained the popular article "Chemical Controversy at the Solar Surface" by J. Bahcall in Physics in Action [if6]. The article describes the problems created by results reported in the article "The Solar Chemical Decomposition" by M. Asplund, N. Grevesse, J. Sauval [if7]. The abundances of C, N, O, Ne, Ar at the solar surface are about 30-40 per cent less than predicted by the standard solar model. If these abundances are feeded into the standard solar model as input the predictions change in the range $.45R - .73R$ of distances from solar interior (R is solar radius). In particular, sound velocity is predicted incorrectly. Interestingly, these abundances are consistent with the abundances in the gaseous medium in the neighborhood of our galaxy.

In TGD framework a possible solution of paradox comes from already old model of solar corona and solar magnetic field. Part of matter resides as dark matter at magnetic and Z^0 magnetic flux tubes of Sun (dark energy) and enters to the solar corona along these. That also gaseous medium in the neighborhood of our galaxy contains same abundances suggests that the formation of Sun has proceeded by a transformation of part of dark matter to a visible matter by leakage to space-time sheets visible to us. This is indeed what TGD inspired model for the formation of solar system based on quantal dark matter suggests.

Does Sun have a solid surface?

$n = 1$ Bohr orbit corresponds in a reasonable approximation to $L(276)/9 \simeq L(270)$ and thus to solar radius. This raises the question whether solar surface could contain spherical shell representing a topological condensate of dense matter around dark matter, kind of spherical preform of planet below the photosphere.

Recently new satellites have begun to provide information about what lurks beneath the photosphere. The pictures produced by Lockheed Martin's Trace Satellite and YOHKOH, TRACE and SOHO satellite programs are publicly available in the web. SERTS program for the spectral analysis suggest a new picture challenging the simple gas sphere picture [if5]. The visual inspection of the pictures combined with spectral analysis has led Michael Moshina to suggests that Sun has a solid, conductive spherical surface layer consisting of calcium ferrite. The article of Moshina [if5] provides impressive pictures, which in my humble non-specialist opinion support this view. Of course, I have not worked personally with the analysis of these pictures so that I do not have the competence to decide how compelling the conclusions of Moshina are. In any case, I think that his web article [if5] deserves a summary.

Before SERTS people were familiar with hydrogen, helium, and calcium emissions from Sun. The careful analysis of SERTS spectrum however suggest the presence of a layer or layers containing ferrite and other heavy metals. Besides ferrite SERTS found silicon, magnesium, manganese, chromium, aluminum, and neon in solar emissions. Also elevated levels of sulphur and nickel were observed during more active cycles of Sun. In the gas sphere model these elements are expected to be present only in minor amounts. As many as 57 different types of emissions from 10 different kinds of elements had to be considered to construct a picture about the surface of the Sun.

Moshina has visually analyzed the pictures constructed from the surface of Sun using light at wave lengths corresponding to three lines of ferrite ions (171, 195, 284 Angstroms). On basis of his analysis he concludes that the spectrum originates from rigid and fixed surface structures, which can survive for days. A further analysis shows that these rigid structure rotate uniformly.

The existence of a rigid structure idealizable as spherical shell in the first approximation could by previous observation be interpreted as a spherical shell corresponding to $n = 1$ Bohr orbit of a planet not yet formed. This structure would already contain the germs of iron core and of crust containing Silicon, Ca and and other elements.

There is also another similar piece of evidence [if4]. A new planet has been discovered orbiting around a star in a triple-star system in the constellation Cygnus. The planet is a so-called hot Jupiter but it orbits the parent star at distance of .05 AU, which much less that than allowed by current theories of planetary formation. Indeed, the so called migration theory predicts that the gravitational pull of the two stars should have stripped away the proto-planetary disk from the parent star. If an underlying dark matter structure serves as a condensation template for the visible matter, the planetary orbit is stabilized by Bohr quantization.

There is however a problem: ordinary iron and also ordinary iron topologically condensed at dark space-time sheets, becomes liquid at temperature 1811 K at atmospheric pressure. Using for the photospheric pressure p_{ph}, the ideal gas approximation $p_{ph} = n_{ph}T_{ph}$, the values of photospheric temperature $T_{ph} \sim 5800$ K and density $\rho_{ph} \sim 10^{-2}\rho_{atm}$, and idealizing photosphere as a plasma of hydrogen ions and atmosphere as a gas of O_2 molecules, one obtains $n_{ph} \sim .32n_{atm}$ giving $p_{ph} \sim 6.4p_{atm}$. This suggests that calcium ferrite cannot be solid at temperatures of order 5800 K prevailing in the photosphere (the material with highest known melting temperature is graphite with melting temperature of 3984 K at atmospheric pressure). Thus it would seem that dark calcium ferrite at the surface of the Sun cannot be just ordinary calcium ferrite at dark space-time sheets.

The alternative model of inherently dark atom developed in this chapter and applied also in [J6, L2, N2] provides a possible solution of the problem. The transition energies of dark N-molecules are N-fold as compared to those of the ordinary molecules. Therefore N-molecules can be stable and also exist in solid phase at N times higher temperatures than ordinary molecules. The N-photons emitted by dark N-molecules can decay to N ordinary photons and produce the spectral signatures of the ordinary molecules. Therefore the solid surface of the Sun would consists of dark N-matter. Quite generally, the spectral lines of molecules detected in environments where they are not thermally stable would be the signature of N-molecules and could be used to test the proposed model.

How to create dark matter in laboratory...

The creation of dark matter at laboratory is of course the crucial test. The hints for what to do come already from the findings of Tesla, which did not fit completely with Maxwell's electrodynamics (, which, using M-theory inspired jargon, had become "the only known classical theory of electromagnetism",) and were thus forgotten.

To transform visible matter to dark matter in laboratory one might try to generate conditions in which visible matter leaks to larger space-time sheets. What one could try is to generate pulsed current of electrons. For instance, current could flow to a circuit component acting as a charge reservoir. When the circuit is opened, and current cannot leave the charge reservoir, a situation analogous to a traffic jam occurs and some electrons might leak to larger space-time sheets via join along boundaries bonds generated in the process. Di-electric breakdown along larger space-time sheet would be in question. Recoil effects and zero point kinetic energy liberated as ionizing radiation would serve as a signature of the process. The production of dark matter might occur also in the usual di-electric breakdown and lead to the appearance of electrons in much larger volume after it partially re-enters original space-time sheets. The change of zero point kinetic energy would be liberated as radiation and would cause formation of plasma. Tesla detected dramatic effects of this kind in experiments utilizing sharp pulses.

..or has it already been done?

In their article "Investigation of high voltage discharges in low pressure gases through large ceramic super-conducting electrodes", Modanese and Podkletnov [jf5] report a fascinating discovery suggesting that some new form of radiation is generated in the di-electric breakdown of a capacitor at low temperature and having super-conductor as a second electrode. This radiation induces oscillatory motion of test penduli but, and this is very strange, its intensity is not reduced with distance.

The TGD based explanation [G3] would be in terms of either "topological light rays" or what I call in honor of Tesla "scalar wave pulses" (much like a capacitor moving with velocity of light predicted by TGD but not allowed by Maxwell's ED). This radiation would induce the formation of join along boundaries bonds between atomic and larger space-time sheets and part of electrons from penduli would leak to larger space-time sheets and their motion would result as a recoil effect. The radiation would have only the role of control signal and this would explain why its intensity is not weakened.

From the point of view of single sheeted space-time an over-unity device would be in question since the zero point kinetic energy would be transformed to kinetic energy. The transformation of visible matter to dark matter is in TGD Universe the basic mechanism of metabolism predicting universality of metabolic energy currencies and living matter in TGD Universe has developed a refined machinery to recycle the dropped charges back to the atomic space-time sheets to be used again. Combined with time mirror mechanism this makes, not a perpetuum mobile, but an extremely flexible mechanism of metabolism.

Conformal confinement and dark matter

The basic question is what makes dark matter macroscopic quantum coherent system having a large Planck constant. There must be something distinguishing in a profound manner between ordinary and dark matter. The findings in RHIC (Relativistic Heavy Ion Collider) made during last years [hk3] led to an identification of a candidate for this mechanism [D5]. The super-canonical conformal weights, which are closely related to the non-trivial zeros of Riemann Zeta, are in general complex for single particle states whereas the net conformal weights of physical states are real.

Super-canonical conformal weights can of course be real already at single particle level and probably are so for ordinary particles with valence quarks forming a possible exception. If this is not the case, the reality condition for the net conformal weight forces the many particle system to behave like a single coherent unit. Complex conformal weights make possible entanglement and the resulting system would be a macroscopic quantum system in these new degrees of freedom reflecting directly configuration space degrees of freedom predicted by TGD. This mechanism would work also in case of non-gravitational interactions which become so strong that $\hbar$ becomes large, and could play a key role in the understanding of living matter since dark matter at magnetic flux tubes could be responsible for the quantum control of the ordinary matter. Knotted and linked magnetic flux tubes containing conformally entangled particles are also ideal candidates for topological quantum computers [O3, J6] and DNA and RNA are excellent candidates for topological quantum computer like systems.

18.2.11   How dark matter and visible matter interact?

The hypothesis that the value of $\hbar$ is dynamical, quantized and becomes large at the verge of a transition to a non-perturbative phase in the ordinary sense of the word has fascinating implications. In particular, dark matter, would correspond to a large value of $\hbar$ and could be responsible for the properties of the living matter. In order to test the idea experimentally, a more concrete model for the interaction of ordinary matter and dark matter must be developed and here of course experimental input and the consistency with the earlier quantum model of living matter is of considerable help.

Basic implications from the scaling of $\hbar$

It is relatively easy to deduce the basic implications of the scaling of $\hbar$.

a) If the rate for the process is non-vanishing classically, it is not affected in the lowest order. For instance, scattering cross sections for say electron-electron scattering and e^+e^- annihilation are not affected in the lowest order since the increase of Compton length compensates for the reduction of α_{em}. Photon-photon scattering cross section, which vanishes classically and is proportional to $\alpha_{em}^4 \hbar^2 / E^2$, scales down as $1/\hbar^2$.

b) Higher order corrections coming as powers of the gauge coupling strength α are reduced since $\alpha = g^2/4\pi\hbar$ is reduced. Since one has $\hbar_s/\hbar = \alpha Q_1 Q_2/v_0$, $\alpha Q_1 Q_2$ is effectively replaced with a universal coupling strength $v_0 \simeq 4.6 \times 10^{-4}$. In the case of QCD the paradoxical sounding implication is that α_s becomes very small in non-perturbative phase.

Simple model for dark atoms

The simplest model for dark nuclei is as blobs of N_{cr} dark nuclei surrounded by dark electrons at distance which is scaled by a factor k^2, $k = \hbar_s/\hbar$ from that for ordinary atom. Thus the dark nucleus looks point-like from the point of view of the electron cloud. As far as electrons are considered there are two options.

i) Dark electrons could behave as independent particles with the interaction strength with the dark nucleus given by $NZ\alpha$. The critical value N_{cr} of N is determined from the condition of criticality for single electron dark nucleus interaction: $[X] = 1$, $X = N_{cr} Z \alpha$. Here $[x]$ denotes the largest integer smaller than x. Criticality implies $\hbar_s/\hbar = X/v_0 \sim 1/v_0$. Note that the condition $1 \leq X < 2$ implying $1/v_0 \leq \hbar_s/\hbar < 2/v_0$ holds true.

ii) Dark electrons are in the same state apart from the values of super-canonical conformal weights making possible to satisfy fermionic statistics. They behave like a single super-electron with mass $N_{cr} m_e$ and em charge $N_{cr} e$. In this case the criticality condition reads $[X] = 1$, $X = N_{cr}^2 Z \alpha$. The criticality implies $\hbar_s/\hbar = X/v_0 \sim 1/v_0$.

b) The binding energy scale $E \propto Z^2 \alpha_{em}^2 m_e$ of atoms scales as $1/\hbar^2$ so that a partially dark matter for which protons have large value of $\hbar$ does not interact appreciably with the visible light. Scaled down spectrum of atomic binding energies would be the experimental signature of dark atoms. The resulting binding energy spectrum is independent of the atom in the approximation $N_{cr} = 1/Z\alpha$. The binding energy scale defined by the ionization energy $E_0 = Z^2 \alpha^2 m_e/4$ as given by Bohr model is replaced with $E_0 = v_0^2 m_e/4X^2 \simeq 26.5/X^2$ meV. Different values of X allow to distinguish between different atoms since the energy scale differ by a factor $1/4$ from the maximal one. It should be noticed that the resting potential of neuron is around .64 meV (the value varies in considerable limits up to 80 meV).

c) The ionization wavelength for ordinary $\hbar$ would be $\simeq 46.5X^2$ μm, which for $X = 1$ below the maximal size of neuron about 100 μm. For $\hbar_s = \hbar/v_0$ the wavelength is given by factor $46.5X^3/v_0$ and $\simeq 9.4X^3$ cm, which happens to be the size scale of brain hemisphere for $X = 1$.

d) The Bohr radius of dark atom scales as $\hbar^2$ and is given by $a_d = (X/v_0)^2 a_0 \simeq .2X^2 < .8$ mm, ($a_0 = \hbar/\alpha_{em} m_e$). The size of basic multi-neuron modules in cortex is about 1 mm. These intriguing observations give hints about the possible role of dark atoms in the functioning of living matter and brain.

An alternative model for dark atoms

A work with attempts to understand dark matter hierarchy led to an alternative model of dark atoms in which the energy spectra of dark atoms and molecules are nearly the same as their ordinary counterparts.

a) The original model for dark atoms relies on the scaling of $\hbar$ by λ^k at the kth level of the dark matter hierarchy. Here λ is integer and $\lambda \simeq 2^{11}$ seems to define a preferred value of λ. Also the harmonics and integer valued sub-harmonics of λ are possible.

b) In the case of hydrogen atom the model predicts that the energies of hydrogen atom proportional to $1/\hbar^2$ are scaled down by $1/\lambda^2$ so that dark atoms would not be thermally stable. In practice this would exclude dark atoms and molecules as inherently dark systems. The topological condensation of ordinary atoms and molecules at λ^k-sheeted (now in the sense of "Riemann surfaces" over M^4) dark magnetic flux quanta is however possible and means scaling up of the cyclotron energy by λ^k making possible cyclotron Bose-Einstein condensates at high temperatures identifiable as dark quantum plasmas. The same scaling occurs to the energy of dark plasma oscillations so that their energies can be above thermal threshold. Dark plasmoids and plasma oscillations are indeed fundamental in the TGD based model of quantum control in living matter.

c) One must be however very cautious in drawing conclusions since the model for the dark matter is not precise enough to exclude the possibility that the notion of dark atom and molecules makes also sense. For instance, dark atoms having ordinary size and ordinary energy spectrum could be possible if the principal quantum number n is fractionized to $n \to n/\lambda$. The fractionization could make sense if the atomic space-time sheet is λ-folded and atoms become radial anyons. The corresponding Bohr orbits would close in the radial direction only after λ turns. The formation of dark atoms could be interpreted as a transition to chaos by period λ-folding in radial and angular degrees of freedom. This option would differ from the first one in that radial scaling in M^4 by a factor λ^k is replaced by a radial λ^k-folding so that the M^4 projection of dark atom has the same size as in the case of ordinary atom.

This picture is favored by the requirement that four-momenta and angular momenta remain invariant in the transition to the dark matter phase but does not conform with the first model of dark atoms which assumes that n is integer. This model was formulated before the realization of the λ^k-fold Riemann surface like structure of dark space-time sheets following from the conservation of angular momentum.

d) Since dark atom would define a λ^k-fold covering of M^4, one expects a degeneracy of states corresponding to the phase factors $exp(ikn2\pi/\lambda^k)$, $k = 0, ..., \lambda^k - 1$, where n labels the sheets of the λ^k-fold covering of M^4. The nuclei and electrons of $N \leq \lambda^k$ dark atom could form many-particle states separately and fermionic statistics becomes effectively para-statistics for the resulting N-atoms. Note that the N electrons and nuclei would be in identical states in ordinary sense of the word since Bohr orbits must be identical: kind of fermionic Bose-Einstein condensates become thus possible.

e) The quantum transitions of N-atoms for $N = \lambda^k$ would give rise to dark counterparts of the photons emitted in the ordinary atomic transitions. For $N \leq \lambda^k$ the energies of dark photons would be N times higher than the energies liberated in the ordinary transitions. The claims of Mills [jf12] about the scaling up of the binding energy of the hydrogen ground state by a square k^2 of an integer in plasma state might be understood as being due to the formation of dark $N = k^2$-atoms emitting dark photons with k^2-fold energies de-cohering to ordinary photons. The plasma phase would contain a fraction which is in dark plasma state. The chemistry of bio-molecules identified as N-molecules would definitely differ from the ordinary chemistry.

The fractionization $n \to n/\lambda$ of integer n labelling vibrational modes and cyclotron states would be unavoidable. Single particle cyclotron states having $E = \hbar(k)\omega$ of the earlier picture would in this framework correspond to single particle states having $n = \lambda^k$ or to $N = \lambda^k$-ion states. Fermionic $N = \lambda^k$-states are expected to have a special role since these configurations are analogous noble gas atoms with full shells of electrons and to magic nuclei with full cells of nucleons. Most biologically important ions are fermions and $N = \lambda^k$ states would give rise to what might be regarded as fermionic analogs of Bose-Einstein condensates. For bosonic ions there is no restriction to the occupation numbers of λ^k single particle states involved.

f) Quite generally, the presence of spectral lines of transitions of atoms or molecules which are not be stable at the temperature of environment, would serve as a signature for the presence of N-particles. Interestingly, spectral analysis demonstrates the presence water inside sunspots [if3], where the temperature varies in the range 3000-4500 K. The de-coherence of N-photons emitted by thermally stable N-water molecules with $N > 10$ would explain the finding.

g) The phase $q = exp(i2\pi/\lambda^k)$ brings unavoidably in mind the phases defining quantum groups and playing also a key role in the model of topological quantum computation [O3]. Quantum groups indeed emerge from the spinor structure in the "world of classical worlds" realized as the space of 3-surfaces in $M^4 \times CP_2$ and being closely related to von Neumann algebras known as hyper-finite factors of type II_1 [O5]. Unfortunately, the integer n characterizing the phase cannot be identified as λ. Could it be that quantum groups emerge in two different manners in TGD framework?

If so, living matter could perhaps be understood in terms of quantum deformations of the ordinary matter, which would be characterized by the quantum phases $q = exp(i2\pi/\lambda^k)$. Hence quantum groups, which have for long time suspected to have significance in elementary particle physics, might explain the mystery of living matter and predict an entire hierarchy of new forms of matter.

Are both options for dark matter realized?

For $N = \lambda^k$ molecules which dark photons emitted in the rotational and conformational transitions would be above thermal threshold. It is of course quite possible that both options are realized. The fact that also fermionic ions (such as Na^+, K^+, Cl^-) are important for living system suggests that this is the case. This would also provide a justification for the hypothesis that microtubular conformations represent bits and allow conformational dynamics to serve as metabolic controller by providing microwave dark photons with energies above thermal threshold.

As demonstrated in [L2], the notion of N-particle leads to an amazingly elegant model for the lock and key mechanism of bio-catalysis as well as the understanding of the DNA replication based on the spontaneous decay and completion of fermionic $N < \lambda^k$-particles to λ^k-particles. Optimal candidates for the N-particles are N-hydrogen atoms associated with bio-molecules appearing as letters in the "pieces of text" labelling the molecules. Lock and key would correspond to conjugate names in the sense that N_1 and N_2 for the letters in the name and its conjugate satisfy $N_1 + N_2 = \lambda^k$: as the molecules combine, a full fermion shell represented by λ^k- fermion is formed.

How dark photons transform to ordinary photons?

The transitions of dark atoms naturally correspond to coherent transitions of the entire dark electron BE condensate and thus generate N_{cr} dark photons which have complex conformal weights and are conformally confined and behave thus like laser beams. Dark photons do not interact directly with the visible matter.

The simplest guess is that the transformation of dark photon BE condensates to ordinary photons corresponds to a loss of coherence by conformal liberation in which conformal weights of photons become real. An open question is whether even ordinary laser beams could be identified as beams of dark photons. Note that the transition from dark to ordinary photons implies the reduction of wave length and thus also of coherence length by a factor v_0.

Dark $\leftrightarrow$ visible transition should have also a space-time correlate. The so called topological light rays or MEs ("massless extremals") represent a crucial deviation of TGD from Maxwell's ED and have all the properties characterizing macroscopic classical coherence. Therefore MEs are excellent candidates for the space-time correlate of BE condensate of dark photons.

MEs carry in general a superposition of harmonics of some basic frequency determined by the length of ME. A natural expectation is that the frequency of classical field corresponds to the generalized de Broglie frequency of dark photon and is thus $\hbar/\hbar_s$ times lower than for ordinary photons. In completely analogous manner de Broglie wave length is scaled up by $k = \hbar_s/\hbar$. Classically the decay of dark photons to visible photons would mean that an oscillation with frequency f inside topological light ray transforms to an oscillation of frequency f/k such that the intensity of the oscillation is scaled up by a factor k. Furthermore, the ME in question could naturally decompose into $1 < N_{cr} \leq 137$ ordinary photons in case that dark atoms are in question. Of course also MEs could decay to lower level MEs and this has an interpretation in terms of hierarchy of dark matters to be discussed next.

Hierarchy of dark matters and hierarchy of minds

The notion of dark matter is only relative concept in the sense that dark matter is invisible from the point of view of the ordinary matter. One can imagine an entire hierarchy of dark matter structures corresponding to the hierarchy of space-time sheets for which p-adic length scales differ by a factor $1/v_0 \sim 2^{11}$. The BE condensates of N_{cr} ordinary matter particles would serve as dynamical units for "dark dark matter" invisible to the dark matter. The above discussed criticality criterion can be applied at all levels of the hierarchy to determine the value of the dynamical interaction strength for which BE condensates of BE condensates are formed.

This hierarchy would give rise to a hierarchy of the values of $\hbar_n/\hbar$ coming as powers of v_0^{-n} as well as a hierarchy of wavelengths with same energy coming as powers or v_0^n. For zero point kinetic energies proportional to $\hbar^2$ this hierarchy would come in powers of v_0^{-2n}, for magnetic interaction energies proportional to $\hbar$ the hierarchy would come in powers v_0^{-n} whereas for atomics energy levels the hierarchy would come in powers of v_0^{2n} (assuming that this hierarchy makes sense).

The most interesting new physics would emerge from the interaction between length scales differing by powers of v_0 made possible by the decay of BE condensates of dark photons to ordinary photons having wavelength shorter by a factor $\sim v_0$. This interaction could provide the royal road to the quantitative understanding how living matter manages to build up extremely complex coherent interactions between different length and time scales.

In the time domain dark matter hierarchy could allow to understand how moments of consciousness organize to a hierarchy with respect to the time scales of moment of consciousness coming as 2^{11k} multiples of CP_2 time scale. Even human life span could be seen as single moment of consciousness at $k = 14^{th}$ level of the dark matter hierarchy whereas single day in human life would correspond to $k = 12$.

Realization of intentional action and hierarchy of dark matters

How long length scales are able to control the dynamics in short length scales so that the extremely complex process extending down to atomic length scales realizing my intention to write this word is possible. This question has remained without a convincing answer in the recent day biology and there strong objections against the idea that this process is planned and initiated at neuronal level.

I have proposed a concrete mechanism for the realization of intentional action in terms of time mirror mechanism involving the emission of negative energy photons and proceeding as a cascade in a reversed direction of geometric time from long to short length scales [K1]. This cascade would induce as a reaction analogous processes proceeding in the normal direction of geometric time as a response and would correspond to the neural correlates of intentional action in very general sense of the word.

The counterparts for the negative energy signals propagating to the geometric past would be phase conjugate (negative energy) laser beams identifiable as Bose-Einstein condensates of dark photons. In the time reflection these beams would transform to positive energy dark matter photons eventually decaying to ordinary photons. The space-time correlate would be MEs decaying into MEs and eventually to CP_2 type extremals representing ordinary photons.

The realization of intentional action as desires of boss expressed to lower level boss would naturally represented the decay of the phase conjugate dark laser beam to lower level laser beams decaying to lower level laser beams decaying to... . This would represent the desire for action whereas the time reflection at some level would represent the realization desire as stepwise decay to lower level laser beams and eventually to ordinary photons. The strong quantitative prediction would be that these levels correspond to a length and time scale hierarchies coming in powers of $1/v_0 \sim 2^{11}$.

Wave-length hierarchy, coherent metabolism, and proton-electron mass ratio

The fact that a given wavelength length corresponds to energies related to each other by a scaling with powers of v_0 provides a mechanism allowing to transfer energy from long to short long scales by a de-coherence occurring either in the standard or reversed direction of geometric time. De-coherence in the reversed direction of time would be associated with mysterious looking processes like self-assembly allowing thus an interpretation as a normal decay process in reversed time direction.

It is perhaps not an accident that the value of $v_0 \simeq 4.6 \times 10^{-4}$ is not too far from the ratio of $m_e/m_p \simeq 5.3 \times 10^{-4}$ giving the ratio of zero point kinetic energies of proton and electron for a given space-time sheet. This co-incidence could in principle make possible a metabolic mechanism in which dark protons and ordinary electrons co-operate in the sense that dark protons generate dark photon BE condensates with wave length λ transforming to ordinary photons with wavelength $v_0\lambda$ absorbed by ordinary electrons.

Some examples are in order to illustrate these ideas.

a) As already found, in the case of dark atoms the scaling of binding energies as $1/\hbar^2$ allows the coupling of ~ 9 cm scale of brain hemisphere with the length scale ~ 50 μm of large neuron. $N_{cr} \leq 137$ ordinary IR photons would be emitted in single burst and interacting with neuron.

b) For a non-relativistic particle in a box of size L the energy scale is given by $E_1 = \hbar^2\pi^2/2mL^2$ so that the visible photons emitted would have energy scaled up by a factor $(\hbar_s/\hbar)^2 \simeq 4 \times 10^6$. The collective dropping of N_{cr} dark protons to larger space-time sheet would liberate a laser beam of dark photons with energy equal to the liberated zero point kinetic energy. For instance, for the p-adic length scale $L(k = 159 = 3 \times 53) \simeq .63$ μm this process would generate laser beam of IR dark photons with energy $\sim .5$ eV also generated by the dropping of ordinary protons from $k = 137$ atomic space-time sheet. There would thus be an interaction between dark protons in cell length scale and ordinary protons in atomic length scale. For instance, the dropping of dark protons in cell length scale could induce driving of protons back to the atomic space-time sheet essential for the metabolism [K6]. Similar argument applies to electrons with the scale of the zero point kinetic energy about 1 keV.

c) If the energy spectrum associated with the conformational degrees of freedom of proteins, which corresponds roughly to a frequency scale of 10 GHz remains also invariant in the phase transition to dark protein state, coherent emissions of dark photons with microwave wave lengths would generate ordinary infrared photons. For instance, metabolic energy quanta of $\sim .5$ eV could result from macroscopic Bose-Einstein condensates of 58 GHz dark photons resulting from the oscillations in the conformational degrees of freedom of dark proteins. A second option is that the conformal energies are scaled by $\hbar_s/\hbar$ (ω would remain invariant). In this case these coherent excitations would generate ordinary photons with energy of about 1 keV able to drive electrons back to the atomic $k = 137$ space-time sheet.

d) Since magnetic flux tubes have a profound role in TGD inspired theory of consciousness, it is interesting to look also for the behavior of effective magnetic transition energies in the phase transition to the dark matter phase. This transition increases the scale of the magnetic interaction energy so that anomalously large magnetic spin splitting $\hbar_s eB/m$ in the external magnetic field could serve as a signature of dark atoms. The dark transition energies relate by a factor $\hbar_s/\hbar$ to the ordinary magnetic transition energies.

For instance, in the magnetic field of Earth with a nominal value $.5 \times 10^{-4}$ Tesla dark electron cyclotron frequency is 6×10^5 Hz and corresponds to ordinary microwave photon with frequency ~ 1.2 GHz and wavelength $\lambda \simeq 25$ cm. For proton the cyclotron frequency of 300 Hz would correspond to energy of ordinary photon with frequency of 6×10^5 Hz and could induce electronic cyclotron transitions and spin flips in turn generating for instance magneto-static waves.

It is easy to imagine a few step dark matter hierarchy connecting EEG frequencies of dark matter with frequencies of visible light for ordinary photons. This kind of hierarchy would give considerable concreteness for the notion of magnetic body having size scale of Earth.

A connection with bio-photons

The biologically active radiation at UV energies was first discovered by Russian researcher Gurwitz using a very elegant experimental arrangement [la1]. Gurwitz christened this radiation mitogenetic radiation since it was especially intense during the division of cell.

A direct proof for the biological activity of mitogenetic radiation consisted of a simple experiment in which either quartz or glass plate was put between two samples. The first sample contained already growing onion roots whereas the second sample contained roots which did not yet grow. In the case of quartz plate no stimulation of growth occurred unlike for glass plate. Since quartz is not transparent to UV light wher3as the ordinary glass is, the conclusion was that the stimulation of growth is due to UV light.

The phenomenon was condemned by skeptics as a pseudo science and only the modern detection technologies demonstrated its existence [la3], and mitogenetic radiation became also known as bio-photons (the TGD based model for bio-photons is discussed in [K6]). Bio-photons form a relatively featureless continuum at visible wavelengths continuing also to UV energies, and are believed to be generated by DNA or at least to couple with DNA. The emission of bio-photons is most intense from biologically active organisms and the irradiation by UV light induces an emission of mitogenetic radiation by a some kind of amplification mechanism. It has been suggested that bio-photons represent some kind of leakage of a coherent light emitted by living matter.

According to Russian researcher V. M. Injushin [la2], mitochondrios emit red light at wavelengths 620 nm and 680 nm corresponding to energies 2 eV and 1.82 eV. According to the same source, the nucleus of cell sends UV light at wavelengths 190, 280 and 330 nm corresponding to the energies 6.5, 4.4 and 3.8 eV. The interpretation as a kind of leakage of coherent light would conform with the identification in terms of BE condensates of dark photons with $\hbar_s/\hbar \simeq 2^{11}$ emitted at wavelengths varying in the range $.3 - 1.25$ mm and decaying to photons with energies visible and UV range. For instance, 1.82 eV radiation corresponds to a dark photon wave length of 1.4 mm for $v_0(eff) = 2^{-11}$. A bio-control of ordinary bio-matter at sub-cellular level performed by dark matter from the millimeter length scale could be in question. This proposal conforms with the fact that 1 mm defines the scale of the blobs of neurons serving as structural units in cortex.

The analysis of Kirlian photographs has shown that the pattern of visible light emitted by various body parts, for instance ear, code information about other body parts [la4]. These bio-holograms for which a general model is discussed in [K4] could be realized as dark photon laser beams.

In phantom DNA effect [la5] a chamber containing DNA is irradiated with a visible laser light and the DNA generates as a response coherent visible radiation at same wavelength. Strangely enough, the chamber continues to emit weak laser light even after the removal of DNA. This effect could be due to the decay of a dark photon BE condensate remaining in the chamber. Also the findings of Peter Gariaev [la6] about the effects of visible laser light on DNA, in particular the stimulated emission of radio waves in kHz-MHz frequency range might also relate to dark photons somehow.

18.2.12 Does dark matter provide a correct interpretation of long ranged classical electro-weak gauge fields?

For two decades one of the basic interpretational challenges of TGD has been to understand how the un-avoidable presence of long range classical electro-weak gauge fields can be consistent with the small parity breaking effects in atomic and nuclear length scales. Also classical color gauge fields are predicted, and I have proposed that color qualia correspond to increments of color quantum numbers [K3]. The proposed model for screening cannot banish the unpleasant feeling that the screening cannot be complete enough to eliminate large parity breaking effects in atomic length scales so that one one must keep mind open for alternatives.

p-Adic length scale hypothesis suggests the possibility that both electro-weak gauge bosons and gluons can appear as effectively massless particles in several length scales and there indeed exists evidence that neutrinos appear in several scaled variants [ha20] (for TGD based model see [F3]).

This inspires the working hypothesis that long range classical electro-weak gauge and gluon fields are correlates for light or massless dark electro-weak gauge bosons and gluons.

a) In this kind of scenario ordinary quarks and leptons could be essentially identical with their standard counterparts with electro-weak charges screened in electro-weak length scale so that the problems related to the smallness of atomic parity breaking would be trivially resolved.

b) In condensed matter blobs of size larger than neutrino Compton length (about 5 μm if $k = 169$ determines the p-adic length scale of condensed matter neutrinos) the situation could be different. Also the presence of dark matter phases with sizes and neutrino Compton lengths corresponding to the length scales $L(k)$, $k = 151, 157, 163, 167$ in the range 10 nm-2.5 μm are suggested by the number theoretic considerations (these values of k correspond to so called Gaussian Mersennes [K2]). Only a fraction of the condensed matter consisting of regions of size $L(k)$ need to be in the dark phase.

c) Dark quarks and leptons would have masses essentially identical to their standard model counterparts. Only the electro-weak boson masses which are determined by a different mechanism than the dominating contribution to fermion masses [F2, F3] would be small or vanishing.

d) The large parity breaking effects in living matter would be due to the presence of dark nuclei and leptons. Later the idea that super-fluidity corresponds to Z^0 super-conductivity will be discussed: it might be that also super-fluid phase corresponds to dark neutron phase.

The basic prediction of TGD based model of dark matter as a phase with a large value of Planck constant is the scaling up of various quantal length and time scales. A simple quantitative model for condensed matter with large value of $\hbar$ predicts that $\hbar$ is by a factor $\sim 2^{11}$ determined by the ratio of CP_2 length to Planck length larger than in ordinary phase meaning that the size of dark neutrons would be of order atomic size. In this kind of situation single order parameter would characterize the behavior of dark neutrinos and neutrons and the proposed model could apply as such also in this case.

Dark photon many particle states behave like laser beams decaying to ordinary photons by decoherence meaning a transformation of dark photons to ordinary ones. Also dark electro-weak bosons and gluons would be massless or have small masses determined by the p-adic length scale in question. The decay products of dark electro-weak gauge bosons would be ordinary electro-weak bosons decaying rapidly via virtual electro-weak gauge boson states to ordinary leptons. Topological light rays ("massless extremals") for which all classical gauge fields are massless are natural space-time correlates for the dark boson laser beams. Obviously this means that the basic difference between the chemistries of living and non-living matter would be the absence of electro-weak symmetry breaking in living matter (which does not mean that elementary fermions would be massless).

In conformally confined phase phase Fermi statistics allows neutrinos to have same energy if their conformal weights are different so that a kind "fermionic Bose-Einstein condensate" would be in question. If both nuclear neutrons and neutrinos are in dark phase, it is possible to achieve a rather complete local cancellation of Z^0 charge density.

The model for neutrino screening developed in [F9] was developed years before the ideas about the identification of the dark matter emerged. The generalization of the discussion to the case of dark matter option should be rather trivial and is left to the reader as well as generalization of the discussion of the effects of long range Z^0 force on bio-chemistry.

18.2.13   Anti-matter and dark matter

The usual view about matter anti-matter asymmetry is that during early cosmology matter-antimatter asymmetry characterized by the relative density difference of order $r = 10^{-9}$ was somehow generated and that the observed matter corresponds to what remained in the annihilation of quarks and leptons to bosons. A possible mechanism inducing the CP asymmetry is based on the CP breaking phase of CKM matrix.

The TGD based view about energy [D3, D5] forces the conclusion that all conserved quantum numbers including the conserved inertial energy have vanishing densities in cosmological length scales. Therefore fermion numbers associated with matter and antimatter must compensate each other. Therefore the standard option as such is definitely excluded in TGD framework although CKM matrix might well relate to the generation of matter antimatter asymmetry as discussed in [F6].

An early TGD based scenario explains matter antimatter asymmetry by assuming that antimatter is in topological vapor phase. This requires that matter and antimatter have slightly different topological evaporation rates with the relative difference of rates characterized by the parameter r. A more general scenario assumes that matter and antimatter reside at different space-time sheets.

The reader can easily guess the next step. The strict non-observability of antimatter finds an elegant explanation if matter and anti-matter are dark relative to each other. For instance, the masses of particles of antimatter could be scaled down so that antimatter could be practically everywhere without appreciably affecting the density of gravitational mass.

The matter-antimatter asymmetry should be generated during cosmic evolution already before the formation of nucleons during the primordial synthesis of matter and antimatter. The number theoretical model for topological condensation based on formation of # contacts between space-time sheets of opposite time orientations (and thus opposite signs for energies) leads to a more detailed view about what might happen.

contacts can be modelled as CP_2 type extremals which simultaneously topologically condense to the two space-time sheets with Minkowskian signature of induced metric. The resulting two causal horizons are carriers of elementary particle quantum numbers and are identifiable as partons. The # contacts with vanishing net quantum numbers could be generated spontaneously and the splitting of # contact would create positive particle and negative energy particle at the two space-time sheets involved. The requirement that the net quantum numbers of Universe vanish is consistent with this kind of pairing of positive and negative energy space-time sheets.

Number theoretical vision [E3, F6] leads to a vision in which elementary particles correspond to infinite primes, perhaps also integers, or even rationals which in turn can be mapped to finite rationals. To infinite primes, integers, and rationals is is possible to associate a finite rational $q = m/n$ by a homomorphism. q defines an effective q-adic topology of space-time sheet consistent with p-adic topologies defined by the primes dividing m and n ($1/p$-adic topology is homeomorphic to p-adic topology).m and n are exchanged by super-symmetry and the primes dividing m (n) correspond to space-time sheets with positive (negative) time orientation. The largest prime dividing m (n) determines the mass scale of the space-time sheet in p-adic thermodynamics. Two space-time sheets characterized by rationals having common prime factors can be connected by a $\#_B$ contact and can interact by exchange of particles characterized by divisors of m or n. Thus fundamental topological selection rules would be coded by the hierarchy of infinite primes.

A possible interpretation is that particle (in extremely general sense that even entire universe can be regarded as a particle) corresponds to a pair of positive and negative energy space-time sheets labelled by m and n characterizing the p-adic topologies consistent with $m-$ and n-adicities. This looks natural since Universe has necessary vanishing net quantum numbers. Unless one allows the non-uniqueness due to $m/n = mr/nr$, positive and negative energy space-time sheets can be connected only by $\#$ contacts so that positive and negative energy space-time sheets cannot interact via the formation of $\#_B$ contacts and would be therefore dark relative to each other. Negative energy antiparticles would also have different p-adic mass scales. If the rate for the creation of $\#$ contacts and their CP conjugates are slightly different, say due to the presence of electric components of gauge fields, matter antimatter asymmetry could be generated primordially.

18.3 Consciousness and cosmology

The words consciousness and cosmology at the same line represent a totally insane association unless one has assimilated the conceptual background provided by TGD inspired theory of consciousness, where p-adic physics of cognition means that cognitive consciousness is unavoidably cosmic phenomenon as far its space-time correlates are considered and magnetic flux tube hierarchy provides the template for the evolution of conscious, intelligent systems in all length scales.

There is a further argument supporting the view that conscious intelligence is the basic property of Universe. As found in [O4], the need to algebraically continue rational physics to all number fields was shown to lead to a generalization of the notion of space-time point to what might be regarded as realization of the monad concept of Leibniz. Mathematical points are still completely structureless in the real sense but with respect to p-adic topologies situation changes profoundly. One can assign to space-time point a free algebra, kind of "Mother of All Algebraic Structures" allowing representation of any algebraic structure. This would realize Universe as an algebraic hologram, and give good hopes of understanding of what are the space-time correlates of mathematical cognition.

18.3.1 Gravitation and consciousness

The purpose of the following arguments is to persuade the reader to consider seriously the possibility that classical gravitational interactions are space-time correlate for the subjective existence defined as sequence of quantum jumps.

p-Adic self hierarchy

Quantum jump as moment of consciousness and self as a system able to not develop bound state entanglement are basic notions of TGD inspired theory of consciousness [TGDconsc]. The contents of consciousness of self are determined as a statistical average over the ensemble formed by quantum jumps which have occurred after the last "wake-up".

Macro-temporal quantum coherence corresponds to generation of bound state entanglement with rational or extended rational entanglement probabilities stable against state function reduction and preparation processes occurring in each quantum jump. The implication is that state function reduction and preparation processes cease in appropriate degrees of freedom and de-coherence does not occur. As a consequence, the entropy of quantum jump ensemble does not increase and self stays negentropic. Since mental images correspond to sub-selves, mental images remain sharp during macro-temporal quantum coherence. One can say that moments of consciousness effectively integrate to a single long lasting moment of consciousness during macro-temporal quantum coherence. Quantum jumps represent the elementary particles of consciousness, and bound state entanglement binds both elementary particles and moments of consciousness to larger structural elements. The outcome is a fractal hierarchy of consciousness completely analogous and very intimately related to the corresponding hierarchy of matter.

Self hierarchy is the basic prediction of TGD inspired theory of consciousness differentiating it from many competing theories: everything is conscious and consciousness can be only lost. At the bottom of the self hierarchy are elementary particles and at the top of it is the entire Universe. p-Adic fractality suggests that the self hierarchy looks essentially the same at every level. p-Adic physics as physics of cognition adds additional weight to this vision since real and p-adic space-time sheets have rational points as common points and p-adically infinitesimal distances correspond to infinite real distances. On basis of these arguments one has reasons to believe that even structures of cosmological size should be conscious, intelligent, and capable of intentional action. The non-determinism of the vacuum extremals suggests the same at classical level.

Subjective time-geometric time, gravitational mass-inertial mass

Quite generally, the dynamics of topologically condensed matter is about the evolution of "gravitational" counterparts of various currents defined as differences of the currents associated with positive and negative energy matter. Interestingly, this is what one could expect on the basis of quantum classical correspondence. Quantum jump is the basic building block of conscious experience and only changes can be experienced consciously. One the other hand, crossing symmetry allows to regard positive and negative energy particles as the initial and final states of particle reaction. In terms of TGD inspired theory of consciousness, "inertial" corresponds to the "objective" existence (quantum states) and "gravitational" to the "subjective" existence (quantum jumps). Classical fields and quanta of fields (CP_2 type extremals) correspond also to the gravitational-inertial dichotomy.

a) The space-time sheet is an absolute minimum of Kähler action and classical electro-weak and color fields are determined by this dynamics. The energy momentum tensor of Kähler action characterizes the net four-momentum density. The dynamics of the induced spinor fields is the super-symmetric counterpart of Kähler action and thus corresponds to the "inertial sector" of the theory. The fundamental role of the induced spinor fields in quantum theory suggests the same. That induced Kähler field and induced spinor field do not directly correspond to change or difference of any quantum number and are not directly observable conforms with their "inertial" character.

b) The divergences of classical YM fields are non-vanishing in general and define non-conserved vacuum gauge currents. Einstein's equations make sense only if the divergences of the energy momentum tensor of matter and those of classical YM fields cancel each other. The goal is achieved if the YM currents associated with the topologically condensed matter are equal to the vacuum YM currents. Also the net divergences of other contributions to the energy momentum tensor must vanish and this gives to the analogs of hydrodynamical equations.

c) The question is whether the YM currents correspond to the sum or difference of currents associated with positive and negative energy matter. Internal consistency suggests that the dynamics is for the differences of various charges for matter and antimatter so that the currents would correspond to "gravitational" counterparts of gauge charges. This would be also consistent with the facts that only the covariant divergence of these currents is vanishing and that they are not genuinely conserved. Fortunately, the expressions for the currents would be exactly the same as in the standard approach assuming opposite charges for negative energy matter.

With this interpretation even the predicted presence of long range electro-weak and color gauge fields becomes understandable. For instance, the presence of long range $W^\pm$ fields codes for the occurrence of quantum jumps in which electro-weak quantum numbers are changed. Classical gauge fields would be in a well-defined sense "gravitational" counterparts of the quantum fields corresponding to elementary particles. This interpretation applies also in the case of electromagnetic current. Indeed, classical em current corresponds to the covariant divergence of em field and if classical $W^\pm$ fields are present em current fails to be conserved.

Vacuum extremals, gravitation, intentionality, and Universe engineering itself

Kähler action has a gigantic vacuum degeneracy: any four-surface $X^4 \subset M^4_+ \times Y^2$, $Y^2 \subset CP_2$ a Lagrange manifold with a vanishing induced Kähler form, is a vacuum extremal with respect to the inertial energy. The physical interpretation of this degeneracy remained open for a long time although it became clear that it is responsible for spin glass degeneracy and quantum criticality of TGD Universe crucial for understanding of living matter. The fact that vacuum extremals are not vacua with respect to gravitational four-momentum provides the sought for interpretation. Vacuum extremals correspond to the space-time correlate for that aspect of the Universe, which is engineered and can be affected by intentional action. Living creatures are certainly natural candidates for a phase of matter modellable in terms vacuum extremals or their small deformations.

p-Adic fractality and the fact that the evolution of cognition as cognitive growth proceeds from long to short length scales (p-adically infinitesimal corresponds to infinite in real context) forces to take very seriously the possibility that Universe is a product of conscious, intentional engineering in all length scales and could be seen as a kind of four-dimensional artwork or a living organism of cosmic size in 4-dimensional homeostatic equilibrium.

The changes of gravitational mass provide direct signature of intentional action. Intentional action leads to a generation of a pair of positive and negative energy space-time sheets with a vanishing net (conserved) inertial energy: $E_+ + E_- = 0$. The gravitational energy of the system increases by $\Delta E_{gr} = E_+ - E_- = 2E_+$. It would be possible to measure the weight of intention by a gravitational scale! Also the reverse effect is possible. The ceasing of intentional actions at the moment of biological death could relate to the claimed loss of weight at the moment of biological death. Note that the annihilation of pairs of positive energy fermions and negative energy anti-fermions to photons and their phase conjugates provides a possible source of antigravity effects and it might be possible to develop antigravity machines utilizing this effect and possibly already utilized by living matter.

It would seem that Penrose's intuition about the role of quantum gravity is both right and wrong. Penrose proposes that it is possible to reduce the non-determinism of quantum jump to non-computable dynamics of gravitational fields, perhaps even at classical level. In TGD Universe classical gravitation involves genuine classical non-determinism which serves as a symbolic representations for conscious, intentional existence which at fundamental level corresponds to the dynamics of quantum jumps.

Topological thermodynamics and TGD

The dimension of CP_2 projection of space-time surface is an important classifier for the phases of matter in TGD Universe. R. M. Kiehn has developed highly interesting differential-topological view about thermodynamics, which he calls topological thermodynamics [ab7]. The basic goal is to reduce thermodynamics to differential topology using so called Pfaff systems defined by one-forms A_k identified as various thermodynamical potentials and completely analogous to action 1-forms in symplectic mechanics. The basic classifier for thermodynamical phases is the so called topological dimension associated with the Pfaff system defined by one-form A identified as action in Kiehn's approach. The topological dimension d of Pfaff system is the number of non-vanishing terms in the sequence $\{A, dA, A \wedge dA, dA \wedge dA, ..\}$. Even differential-topological definition of entropy is achieved in this approach.

The construction of solutions of field equations in TGD framework relies on the idea that the absolute minima of Kähler action corresponds asymptotically to self-organization patterns which do not dissipate anymore and that this is reflected as the vanishing of Lorentz-Kähler 4-force. Against this essentially thermodynamical background it is not surprising that there is very close relationship to Kiehn's approach. In TGD context action A is identifiable as the induced Kähler gauge potential A whereas d corresponds to the dimension of CP_2 projection of the space-time sheet.

For vacuum vacuum extremals d is 1 or 2. For instance, Lorentz invariant Robertson-Walker cosmologies have $d = 1$, whereas critical cosmology has $d = 2$. Besides vacuum extremals also Kähler magnetic flux tubes have $d = 2$ and correspond to highly ordered ferromagnetic phase. $d = 4$ corresponds to chaos. The most interesting phase from the point of view of living systems and information processing (perhaps by topological quantum computation) has $d = 3$ and is identifiable as spin glass phase representing the critical transition region between $d = 2$ and $d = 4$ phases. Thus $d = 2$ vacuum extremals seem to provide the hardware, the matter which conscious intelligent systems manipulate and perhaps also create from it $d = 3$ systems by small perturbations.

Kiehn's detailed thermodynamical interpretation of the contact and symplectic structure associated with A allows in principle a rather detailed thermodynamical interpretation of the absolute minima of Kähler action. For instance, vacuum extremals can be interpreted as thermodynamical equilibria for which entropy function associated with various points of space-time sheet is constant. The classical non-determinism of the space-time sheets could thus be also seen as a thermal randomness. This randomness would be however absolutely essential for space-time engineering by first melting the space-time sheet to a vacuum extremal, deforming it to the desired shape, and then allowing to cool it again. $d = 3$ corresponds in Kiehn's approach to thermal non-equilibrium states far from equilibrium and identifiable as topological defects.

Topological quantum computation in cosmic length scales?

Magnetic flux tube structures are the basic building blocks of living matter in TGD inspired quantum biology [M1, M3], and the assumption has been that the creation of pairs of positive and negative energy flux tube structures is one of the basic mechanisms of intentional action. For example, the self-assembly of proteins could be identified as a reversed time evolution for negative energy magnetic flux tube structures serving as templates for the formation of living matter. TGD inspired model of quantum biology leads to model for DNA in which the DNA double strand and also other helical structures corresponds to pairs of positive and negative energy magnetic flux tubes.

Also cosmic strings form tightly coiled pairs and can form complex knotted and linked structures. The scale is different but by p-adic fractality and quantum criticality of TGD Universe these structures should be very similar. Even more, cosmic strings should evolve to magnetic and Z^0 magnetic flux tubes by gradual thickening and by generation of fermion or anti-fermion number.

The TGD inspired model for topological quantum computation (TQC) using linking and braiding of magnetic flux tubes [O3] led to the proposal that also the double strand of DNA and/or RNA might perform TQC. The two strands would correspond to positive and negative energy magnetic flux tubes with linking and knotting of RNA coding for TQC program. The braiding of the two strands would define not only a TQC program but also its geometric time reversal. The possibility of quantum parallel dissipation makes also possible generalization of topological quantum computation to that involving dissipation.

p-Adic fractality and the view about evolution of cognition as a process proceeding from long length and time scales towards shorter ones inspires the question whether the coiled pairs of cosmic strings could be seen as fractally scaled up versions of pairs of RNA strands and define cosmic topological quantum computations such that negative energy string corresponds to communication to and control of the geometric past making possible endless iteration of TQC.

18.3.2   Is solar system a genuine self-organizing quantum system?

There are two means of determining the positions of planets in the solar system [ik8, ik9, ik7, ik6]. The first method is based on optical measurements and determines the position of planets with respect to the distant stars. Already thirty years ago [ik6] came the first indications that the planetary positions determined in this manner drift from their predicted values as if planets were in accelerated motion. The second method determines the relative positions of planets using radar ranging: this method does not reveal any such acceleration.

C. J. Masreliez [ik9] has proposed that this acceleration could be due to a gradual scaling of the planetary system so that the sizes L of the planetary orbits are reduced by an over-all scale factor $L \rightarrow L/\lambda$, which implies the acceleration $\omega \rightarrow \lambda^{3/2}\omega$ in accordance with the Kepler's law $\omega \propto 1/L^{3/2}$. This scaling would exactly compensate the cosmological scaling $L \rightarrow (R(t)/R_0) \times L$ of the solar system size L, where $R(t)$ the curvature parameter of Robertson-Walker cosmology having the line element

$$ ds^2 \;\; = \;\; dt^2 - R^2(t) \left(\frac{dr^2}{1 + r^2} + r^2 d\Omega^2 \right) \;\; . \tag{18.3.1}$$

According to Masreliez, the model explains also some other anomalies in the solar system, such as angular momentum discrepancy between the lunar motion and the spin-down of the Earth [ik9]. The model also changes the rate for the estimated drift of the Moon away from the Earth so that the Moon could have very well formed together with Earth some five billion years ago.

In the TGD framework planetary acceleration could reflect non-trivial dynamics of quantum jumps so that quantum effects would be directly observable in the scale of solar system. Two times would be involved: subjective time defined by sequence of quantum jumps and geometric time. The dynamics with respect to the geometric time would be determined by the equations of motion for planetary dynamics in expanding background for the space-time sheet representing solar system: the radar measurements would probe this dynamics.

The dynamics with respect to subjective time would make itself visible since the space-time sheet would be replaced by a new one in each quantum jump and characterized by slightly different parameters after each quantum jump. The optical measurements measuring positions of stars with respect to distant stars would probe this dynamics. The basic parameter would be the size of the solar system as measured with respect to M_+^4 metric. The scaling compensating the cosmic expansion would guarantee that the observed size remains constant. For each geometric history the size would increase in the geometric future. The solar system would utilize quantum non-determinism in order to avoid decay by cosmic expansion.

The basic coordinate systems

Consider now the previous argument in more detail. The first task is to identify the coordinates appearing in the equations of motion of the planetary system. Denote the standard spherical Minkowski coordinates by (m^0, r_M, θ, ϕ). The line element reads as

$$ds^2 = d(m^0)^2 - dr_M^2 - r_M^2 d\Omega^2 \ . \tag{18.3.2}$$

Light cone coordinates are related to these coordinates by the relationship

$$a = \sqrt{m_0^2 - r_M^2} \ , \quad r = r_M/a \ . \tag{18.3.3}$$

Here a is the light cone proper time along radii from the dip of the light cone $a = constant$ surfaces are hyperboloids. The line element is given

$$ds^2 = da^2 - a^2 \left(\frac{dr^2}{1 + r^2} + r^2 d\Omega^2 \right) \tag{18.3.4}$$

and is nothing but the empty space Minkowski metric.

The Robertson-Walker metric for the space-time sheet reads as

$$ds^2 = g_{aa} da^2 - a^2 (\frac{dr^2}{1 + r^2} + r^2 d\Omega^2) \ . \tag{18.3.5}$$

The space-time sheet possessing this metric as induced metric is obtained as a map $M_+^4 \to CP_2$ having the form $s^k = s^k(a)$, where s^k denote CP_2 coordinates satisfying the constraint

$$g_{aa} = 1 - s_{kl} \partial_a s^k \partial_a s^l \ , \tag{18.3.6}$$

where s_{kl} denotes the metric tensor of CP_2.

One can introduce cosmic time as proper time coordinate t, or Hubble time as it is called, by the equation

$$\frac{dt}{da} = \sqrt{g_{aa}} \ . \tag{18.3.7}$$

For the matter-dominated cosmology one as

$$\frac{t}{t_0} = (\frac{a}{a_0})^{3/2} \ . \tag{18.3.8}$$

$t \simeq 1.5 \times 10^{10}$ ly is the value which explains the planetary acceleration in the model of Masreliez.

The basic question concerns the connection between cosmic coordinates and the radial and time coordinates (r_{PN}, t_{PN}) used in Post-Newtonian approximation. The correspondence $(t = t_{PN}, r = r_{PN})$ is the natural first approximation.

The cosmic time dilation would slow down the time scale of the planetary dynamics and cosmic expansion would lead to adiabatic expansion of the size of the solar system. This would predict the scaling $L(a)/L(a_0) = a/a_0$ for the sizes of the planetary orbits as measured using the r_M coordinate of M_+^4 metric whereas angular velocities of planets would remain constant $\omega(a)/\omega(a_0) = constant$. The solar system would gradually decay.

Quantum compensation of the cosmic expansion

In order to compensate the cosmic expansion solar system should perform quantum jumps compensating the scaling caused by the cosmic expansion so that the observed size of the solar system would remain unchanged. This requires $L(a)/L(a_0) = constant$ and the scalings

$$\frac{\omega(a)}{\omega(a_0)} = \left(\frac{a}{a_0}\right)^{3/2} = \frac{t}{t_0} \;,\quad \frac{v(a)}{v(a_0)} = \left(\frac{a}{a_0}\right)^{1/2} = \left(\frac{t}{t_0}\right)^{1/2} \tag{18.3.9}$$

for the angular velocity ω and and tangential velocity v along the orbit. The equation for the angular acceleration is $d\omega/dt = \omega/t$. This result differs by a factor of 3 from the equation $d\omega/dt = 3\omega/t$ of Masreliez [ik9]. On basis of work of Masreliez one can conclude this kind of scaling indeed explains the observed drift quite satisfactorily for $t \simeq 5$ billion years (instead of $t = 15$ billion years of [ik9]). Thus the effect would allow to see the effects of the cosmic expansion in human time scale and would make possible to determine the value of cosmic time t from the planetary dynamics.

The obvious question is why this kind of scaling would occur and a possible answer suggested by TGD inspired theory of consciousness is that the solar system is conscious self-organizing system which purposefully tends to avoid the decay caused by cosmic expansion. There must be some mechanism making the scaling possible. One can imagine at least the following mechanisms.

a) A gradual generation of Z^0 charges, which are opposite sign for Sun and planets would provide a possible mechanism of this kind. The weakness of the model is that it requires constant Z^0 charge/mass ratio.

b) Also the creation of pairs of positive and negative energy magnetic flux tubes from vacuum could generate gravitational mass. The strong magnetic and Z^0 magnetic field of solar magnetic field zone could consist pairs of flux tubes with positive and negative inertial energies. If the gravitational mass M_{gr} of Sun relates to its radius R by $M_{gr}R^3 = constant$ law holding true for non-rotating non-relativistic astrophysical objects [ia3], the growth of mass and solar radius would obey the law $dlog(M)/dlog(a) = 3/4$ and $dlog(R)/dlog(a) = -1/4$. In the context provided by general philosophy of TGD inspired theory of consciousness the extremely complex self-organizing magnetic flux tube structure of convective zone brings unavoidably in mind cortical layers of brain. The new view about dark matter as a super-quantal system gives additional impetus to the otherwise weird sounding idea about the solar system as intelligent and conscious system.

Explanation of the anomalous acceleration of space-crafts

The data gathered during one quarter of century ([ha15, ik5]) seem to suggest that spacecrafts do not obey the laws of Newtonian gravitation. What has been observed is anomalous constant acceleration of order $(8 \pm 3) \times 10^{-11} g$ ($g = 9.81 \; m/s^2$ is gravitational acceleration at the surface of Earth) for the Pioneer/10/11, Galileo and Ulysses [ik5]. The acceleration is directed towards Sun and could have an explanation in terms of $1/r^2$ long range force if the density of charge carriers of the force has $1/r$ dependence on distance from the Sun. From the data in [ha15, ik5], the anomalous acceleration of the spacecraft is of order

$$\delta a \sim .8 \times 10^{-10} g \;,$$

where $g \simeq 9.81 \; m/s^2$ is gravitational acceleration at the surface of Earth. Using the values of Jupiter distance $R_J \simeq .8 \times 10^{12}$ meters, radius of Earth $R_E \simeq 6 \times 10^6$ meters and the value Sun to Earth mass ratio $M_S/M_E \simeq .3 * 10^6$, one can relate the gravitational acceleration

$$a(R) = \frac{GM_S}{R^2} = \frac{M_S}{M_E}\frac{R_E^2}{R^2}$$

of the spacecraft at distance $R = R_J$ from the Sun to g, getting roughly $a \simeq 1.6 \times 10^{-5} g$. One has also

$$\frac{\delta a}{a} \simeq 1.3 \times 10^{-4} \;.$$

The model for solar system as a self-organizing system provides also an explanation for the anomalous acceleration $a_F = (8.744 \pm 1.33) \times 10^{-8}$ cm/s^2 of space crafts. The value of the anomalous acceleration has been found to be given by Hubble constant: $a_F = cH$. $H = 82$ km/s/Mpc gives $a_F = 8 \times 10^{-8}$ cm/s^2. It is very difficult to believe that this could be an accident. Indeed, if the planets suffer an anomalous inward acceleration compensating for the cosmic expansion, then also the space-crafts in the radial motion experience the same acceleration.

18.3.3   The independence of the age distribution of stars in galaxies on the age of galaxy as evidence for quantum coherent dark matter

Big bang cosmology is in a middle of deep crisis. Various aspects of the situation were discussed in the first Crisis in Cosmology conference held 23-25 June 2005 in Portugal. One of the most serious arguments against Big Bang cosmology is the evidence that the age distribution of stars in galaxies does not depend on the age of the galaxy. As if the cosmology were steady state cosmology in a sharp contrast to the voluminous experimental evidence suggesting an expanding cosmology [id7]. The defenders of the standard cosmology have claimed that the measurement inaccuracies are so high that one cannot draw definite conclusions about the situation.

In TGD framework, the independence of the age distribution of stars on the age of the galaxy would add a further item to the long list of paradoxes due to the erratic identification of the notions of geometric time and experienced time. The TGD based explanation of the anomaly generalizes the earlier model for the shrinking of planetary radii for which there is also evidence [ik8, ik9, ik7]. Rather unexpectedly, the finding lends support for the basic predictions of TGD inspired theory of consciousness including the existence of the infinite hierarchy of conscious entities, and allows to considerably sharpen the earlier view about the relationship between geometric and experienced time.

The connection between the notions of geometric and experienced time

Consider first the TGD based view about the connection between geometric and experienced time.

   1. Time flow as a drift of the space-time sheet of observer relative to environment

a) The space-time sheet X_o^4 associated with the conscious observer drifts towards geometric future with respect to the space-time sheet X_e^4 defined by the environment. From the perceived change of the environment the conscious observer concludes that time flows. The motion towards geometric future corresponds to some average increment τ of geometric time per quantum jump: τ can be identified as a drift velocity $v = \tau$ with respect to subjective time measured using single quantum jump as a unit. Its inverse $1/v$ gives the number of quantum jumps per geometric time defining a measure for the temporal resolution of the conscious experience. This connection explains the origin of the natural but erratic identification of the geometric time with the experienced time.

b) Drift velocity τ characterizes system-subsystem pair and can in principle vary. Geometric considerations suggests that there is a minimal drift velocity (maximal temporal resolution) corresponding to an average increment of geometric time per quantum jump or order CP_2 time $\tau_{CP_2} = R/c$. Quantum classical correspondence allows to consider the possibility that relativity principle for geometric time might have an analog at the level of subjective time.

c) The experience of the flow of subjective time is associated with the number $1/v$ of quantum jumps per unit of geometric time measuring the rate of dissipation. The dissipated power P could be identified as the time component of four-force causing a drift towards geometric future. Dimensional considerations and p-adic fractality suggests that the inverse of the drift velocity (number $1/\tau$ of quantum jumps per geometric time) would be proportional to P/M, where M is the rest mass of the sub-system:

$$\frac{1}{\tau}\frac{dt}{ds} = \frac{P}{M} \; . \tag{18.3.10}$$

Here t denotes the geometric time coordinate, most naturally M_+^4 proper time, and s the curve length along the orbit of the space-time sheet regarded as a point-like object. This predicts that for light systems such as elementary particles the number of quantum jumps per geometric time is high and small for larger systems. For $P = 0$ there would be no experience of time flow since single quantum jump would have an infinite geometric duration and the system would be in a state of Eternal Now.

   2. p-Adic time scale hierarchy for time resolutions of conscious experience

The hierarchy of p-adic length and time scales would relate naturally to the hierarchy of drift velocities and dissipation rates P. The first guess is

$$P \quad \propto \quad \frac{kM}{T_p} \quad , \tag{18.3.11}$$

where M is the rest mass of particles condensed at space-time sheet characterized by the p-adic prime p. This gives

$$\tau_p \quad = \quad \frac{T_p}{k} \quad , \tag{18.3.12}$$

which is also what p-adic fractality would suggest. The time resolution of conscious experience with respect to the geometric time would weaken as p grows in accordance with the idea that p-adic hierarchy defines a hierarchy of length and time resolutions. The higher the dissipation rate, the shorter then geometric duration of quantum jump and the slower the system drifts towards geometric future.

3. Hierarchy of quantum jumps and dark matter

TGD inspired theory of consciousness predicts a hierarchy of quantum jumps such that single quantum jump at given level would correspond to a sequence of quantum jumps at the level below it. The self at the higher level of hierarchy experiences its sub-selves as mental images, and the sub-sub-..-selves at lower levels give rise to a kind of diffuse background experience giving rise to an experience of time flow during single quantum jump.

This hierarchy of moments of consciousness of increasing subjective and geometric durations defines a direct counterpart for the hierarchy formed by elementary particles, hadrons, atoms, molecules, etc.... Indeed, the formation of bound states with rational entanglement probabilities characterized by positive entanglement negentropy identified as a p-adic variant of Shannon entropy is identified as be the physical correlate for the hierarchy or moments of consciousness [H2]. The generalization of this picture to the case of of hyper-finite type II_1 factors is also discussed in [H2].

This hierarchy of consciousness should relate closely to the hierarchy of Planck constants defined by Beraha numbers $B_n = 4sin^2(\pi/n)$ and their generalizations B_r with r rational. For large values of $\hbar$ and dark matter the average increment of geometric time per quantum jump would be longer than for small values of $\hbar$ and to ordinary matter and each quantum jump would be experienced as longer and longer lasting "Eternal Now".

4. Life as an attempt to climb as far as possible to the geometric future

That human beings and presumably all living systems experience time flow would be due to the dissipation by the component of the living system consisting of ordinary matter. The basic goal of the life cycle would be to run faster than the environment to future get as far as possible to the geometric future to be able to experience what the environment at more advanced levels of evolution looks like. Climbing up hill would be a good metaphor for life.

This would serve a good quantum physical definition for the notion of progress. One can imagine several means for achieving this goal: the increase of the p-adic prime p (size of the system), the minimization of the dissipation rate, and the transition to a higher level in the hierarchy of dark matter. The price paid would be a poorer time resolution of direct conscious experience. The diffuse background due to sub-sub-...-selves however provides the experience about time flow. Somewhat paradoxically, in quantum jumps of long geometric duration the highest level contribution to the experience would be that of Eternal Now experienced in pure form during experiences like NDEs when the contributions from the biological body are minimal.

The p-adic length scale hierarchy associated with the biological body would define a hierarchy of these goals. At a given level of hierarchy of moments of consciousness (the hierarchy of Planck constants) the geometric time would run the faster, the longer the p-adic length scale characterizing partially the level in the evolutionary hierarchy is. If the number of quantum jumps is roughly the same at various levels of p-adic hierarchy, the subjectively experienced life span defined as number of quantum jumps would be also more or less the same.

5. Life cycle as single moment of consciousness?

There is a group of questions relating to the identification of "me" in the hierarchy of conscious entities. What is the geometric duration of moment of consciousness for "me" as the highest level intentional agent affecting my life ("silent witness")? Does it correspond to some neurophysiological time scale, say some EEG period, to single day in life, or does biological life cycle correspond to a single moment of consciousness of "silent witness" so that all its experiences in shorter time scales would correspond to mental images defined by sub-sub-...-selves with a smaller value of τ (in particular, moments of sensory consciousness with a geometric time duration of order .1 seconds). Meditative practices indeed claim that we are conscious also during deep sleep but do not usually remember anything about this period.

If "I" would cease to exist as a conscious entity during deep sleep my directly experienced life history could not be longer than single day in life unless the communications with other conscious entities generate a mental image about longer life history as a cognitive construct.

This supports the view that "me" as an intentional agent responsible for the grand design of the life cycle corresponds to the "silent witness". Lower level intentional "me's" are present but their effects of lower level selves to the geometric future have a shorter time span by statistical determinism, and characterized by the corresponding p-adic time scales.

6. Time mirror mechanism for intentional action and time localization of the sensory experience and intentional inertness of the geometric past

The proposed picture does not yet resolve all imaginable paradoxes.

a) My sensory experience is more or less from single moment of geometric time rather than from the entire 4-dimensional body as would be suggested by life span a single moment of consciousness picture. It would seem that the state of the sensory space-time sheets of the geometric past is such that they do not contribute to the everyday conscious experience. Sensory memories and perhaps also cognitive memories (clearly distinguishable from sensory reality) define exceptions to this rule. If the basic character of the sensory experience is such that it takes the sensory receptor into a state in which further sensory perception is not possible for a sufficiently long period of time, this would be the case.

b) In TGD based model for sensory receptors sensory receptors are analogous to population inverted lasers which return to the ground state during sensory perception (in the capacitor model of sensory receptors sensory input generates a discharge of the capacitor [K3]). By classical non-determinism it is possible to have a situation in which the return to the ground state in a given quantum jump occurs only in a finite time interval defined naturally by an appropriate space-time sheet. Since the return to the initial state requires energy feed, there is some recovery time, perhaps even of order life span, for the sensory receptors. Quantum classical correspondence and the fact subjective past does not change in quantum jumps would suggest that no recovery occurs. The quantum jump sequence should induce the falling down of the the the domino pieces ordered along the time axis and the pieces should not get up too soon.

Unless the geometric past is inert with respect to intentional action, one can imagine a situation in which the intentional action on the geometric past can affect the fate of the entire biological body of geometric future in time scales of life span. Rather paradoxical situations can result. For instance, a musician could wake up as physicist when a young "me" in the geometric past or the highest level intentional agent makes a different choice. Paradoxes are avoided if geometric past becomes inert with respect to intentional action.

The view about how intentional action is realized allows to understand both the time localization of the contents of sensory experience and the inertness of the geometric past with respect to intentional action.

a) According to the TGD based model, intentional action is realized as a negative energy signal sent to the geometric past representing a desire about action inducing positive energy signals realizing the action. The intentional action has a fractal structure: the desire about action at highest level and longest length and time scale induces desires at lower levels proceeding down to the bio-molecular level. The temporal distances at which the signals propagate into geometric past is expected to be proportional to T_p so that lower dimensional intentional agents would have a shorter span of intentional action.

b) A negative energy signal from the geometric future induces the return of the population inverted many-sheeted laser defining a sensory receptor to the ground state inducing the positive energy signal responsible for the action. Because of the finite recovery time the subsequent intentional actions of the lower level intentional agents induce a drift of the front of sensory experience towards geometric future. One can of course imagine the possibility that the recovery time is shorter than life span in which the revived youth begins to contribute to the conscious experience: this might relate to the fact that old people can relive the youth.

c) Intentional and sensory life could be seen as an analog of p-adic length scale evolution in which shorter and shorter p-adic time scales are excited with downwards scaling mapped to the time evolution. Life cycle would be like carving a statue by starting from a rough sketch and adding details. This conforms also with the basic anatomy of quantum jump.

7. The drift velocities towards geometric future must be same for communicating intentional agents in biosphere

The assumption that the drift velocities towards geometric future are different for conscious entities able to communicate also leads to a paradoxical situations. For instance, if my drift velocity is lower than that of my friend, I soon find that my friend looks more or less like a dead statue if the hypothesis about intentional inertness is true. If my drift velocity is higher, I would learn that my friend is in a macroscopic quantum superposition of different variants of my friend: one would be musician, one would be physicist, etc... This suggests that the drift velocity is same for conscious entities able to communicate. The findings about planetary system and age distribution of stars are consistent with this assumption. The assumption becomes natural if the intentional agents correspond to dark matter in general.

Quantum explanation for the shrinking planetary radii and apparent steady state cosmology

The independence of the age distribution of stars on the age of galaxy suggests that cosmology looks like a steady state cosmology with respect to the subjective time and expanding cosmology with respect to the geometric time. The explanation for why this is the case would be same as for the shrinking of the orbital radii of planets [ik8, ik9].

1. The model

Consider next the explanation for the shrinking of the orbital radii of planets and for the independence of the age distribution of stars on the age of galaxy.

a) Assume that the Newtonian radii correspond to the radial coordinate r of the Robertson-Walker coordinate system with origin at the Sun. Assume that quantum jumps have physical effects even in astrophysical length scales, and are such that they compensate completely the increase of the distance s of the planet from Sun caused by the cosmic expansion so that $s = a \int_0^r dr/(1 + r^2) \simeq ar$ stays constant apart from the oscillatory variation caused by the non-circular motion.

b) This situation is achieved if the space-time sheet X_o^4 associated with the observer drifts with respect to the space-time sheet X_p^4 associated with the planetary system, which in turn drifts with the same velocity at the space-time sheet X_g^4 of galaxy. This implies that the change of perceived 3-D environment at X_p^4 due to the drift of X_o^4 at X_p^4 is compensated by the drift of X_p^4 at X_g^4.

c) In the same manner, the independence of the age distribution of stars on the age of galaxy can be understood if X_o^4 drifts at X_g^4 with the same velocity as X_g^4 sheets drift at the cosmological space-time sheet X_c^4. The equality of the drift velocities is consistent with the hypothesis that field/magnetic bodies of even galactic size contribute to our conscious experience.

d) Also p-adic fractality implying cosmologies with cosmologies picture suggests that the age distribution of stars does not depend on the age of galaxy.

2. The interpretation of the astrophysical and cosmic anomalies as a support for the quantum coherence of dark matter

The assumption that dark matter is in a quantum coherent state in astrophysical and even cosmic length and time scales means that the systems consisting of dark matter do not dissipate much and thus do not drift much with respect to each other.

Since the Universe consists mostly of dark matter, the shrinking of planetary radii and the constancy of the age distributions of stars in galaxies can be seen as an evidence for the quantum coherence of dark matter and for the assumption that dark matter at our magnetic bodies is what makes us intentional agents. The findings support also the view that universe is conscious even in the length and time scales of galaxies and even enjoy what meditative practices call enlightened states or cosmic consciousness and that these length scales contribute also to our consciousness.

Acknowledgements

I am grateful for Victor Christianto for informing me about the article of Nottale and Da Rocha. Also the highly useful discussions with him and Carlos Castro are acknowledged.

Appendix A

A-1 Basic properties of CP_2

A-1.1 CP_2 as a manifold

CP_2, the complex projective space of two complex dimensions, is obtained by identifying the points of complex 3-space C^3 under the projective equivalence

$$(z^1, z^2, z^3) \equiv \lambda(z^1, z^2, z^3) \ . \tag{A-1.1}$$

Here λ is any non-zero complex number. Note that CP_2 can also regarded as the coset space $SU(3)/U(2)$. The pair z^i/z^j for fixed j and $z^i \neq 0$ defines a complex coordinate chart for CP_2. As j runs from 1 to 3 one obtains an atlas of three coordinate charts covering CP_2, the charts being holomorphically related to each other (e.g. CP_2 is a complex manifold). The points $z^3 \neq 0$ form a subset of CP_2 homoeomorphic to R^4 and the points with $z^3 = 0$ a set homeomorphic to S^2. Therefore CP_2 is obtained by "adding the 2-sphere at infinity to R^4".

Besides the standard complex coordinates $\xi^i = z^i/z^3$, $i = 1, 2$ the coordinates of Eguchi and Freund [bc3] will be used and their relation to the complex coordinates is given by

$$\begin{aligned}
\xi^1 &= z + it \ , \\
\xi^2 &= x + iy \ .
\end{aligned} \tag{A-1.2}$$

These are related to the "spherical coordinates" via the equations

$$\begin{aligned}
\xi^1 &= r\exp(i\frac{(\Psi + \Phi)}{2})cos(\frac{\Theta}{2}) \ , \\
\xi^2 &= r\exp(i\frac{(\Psi - \Phi)}{2})sin(\frac{\Theta}{2}) \ .
\end{aligned} \tag{A-1.3}$$

The ranges of the variables r, Θ, Φ, Ψ are $[0, \infty], [0, \pi], [0, 4\pi], [0, 2\pi]$ respectively.

Considered as a real four-manifold CP_2 is compact and simply connected, with Euler number 3, Pontryagin number 3 and second $b = 1$.

A-1.2 Metric and Kähler structures of CP_2

In order to obtain a natural metric for CP_2, observe that CP_2 can be thought of as a set of the orbits of the isometries $z^i \rightarrow exp(i\alpha)z^i$ on the sphere S^5: $\sum z^i\bar{z}^i = R^2$. The metric of CP_2 is obtained by projecting the metric of S^5 orthogonally to the orbits of the isometries. Therefore the distance between the points of CP_2 is that between the representative orbits on S^5. The line element has the following form in the complex coordinates

$$ds^2 = g_{a\bar{b}}d\xi^a d\bar{\xi}^b \ , \tag{A-1.4}$$

where the Hermitian, in fact Kähler metric $g_{a\bar{b}}$ is defined by

$$g_{a\bar{b}} = R^2 \partial_a \partial_{\bar{b}} K \ , \tag{A-1.5}$$

where the function K, Kähler function, is defined as

$$K = log(F) \ ,$$
$$F = 1 + r^2 \ . \tag{A-1.6}$$

The representation of the metric is given by

$$\frac{ds^2}{R^2} = \frac{(dr^2 + r^2\sigma_3^2)}{F^2} + \frac{r^2(\sigma_1^2 + \sigma_2^2)}{F} \ , \tag{A-1.7}$$

where the quantities σ_i are defined as

$$r^2\sigma_1 = Im(\xi^1 d\xi^2 - \xi^2 d\xi^1) \ ,$$
$$r^2\sigma_2 = -Re(\xi^1 d\xi^2 - \xi^2 d\xi^1) \ ,$$
$$r^2\sigma_3 = -Im(\xi^1 d\bar{\xi}^1 + \xi^2 d\bar{\xi}^2) \ . \tag{A-1.8}$$

The vierbein forms, which satisfy the defining relation

$$s_{kl} = R^2 \sum_A e_k^A e_l^A \ , \tag{A-1.9}$$

are given by

$$e^0 = \frac{dr}{F} \ , \quad e^1 = \frac{r\sigma_1}{\sqrt{F}} \ ,$$
$$e^2 = \frac{r\sigma_2}{\sqrt{F}} \ , \quad e^3 = \frac{r\sigma_3}{F} \ . \tag{A-1.10}$$

The explicit representations of vierbein vectors are given by

$$e^0 = \frac{dr}{F} \ , \qquad e^1 = \frac{r(sin\Theta cos\Psi d\Phi + sin\Psi d\Theta)}{2\sqrt{F}} \ ,$$
$$e^2 = \frac{r(sin\Theta sin\Psi d\Phi - cos\Psi d\Theta)}{2\sqrt{F}} \ , \quad e^3 = \frac{r(d\Psi + cos\Theta d\Phi)}{2F} \ . \tag{A-1.11}$$

The explicit representation of the line element is given by the expression

$$ds^2/R^2 = dr^2/F^2 + (r^2/4F^2)(d\Psi + cos\Theta d\Phi)^2 + (r^2/4F)(d\Theta^2 + sin^2\Theta d\Phi^2) \ . \tag{A-1.12}$$

The vierbein connection satisfying the defining relation

$$de^A = -V_B^A \wedge e^B \ , \tag{A-1.13}$$

is given by

$$V_{01} = -\frac{e^1}{r} \ , \qquad V_{23} = \frac{e^1}{r} \ ,$$
$$V_{02} = -\frac{e^2}{r} \ , \qquad V_{31} = \frac{e^2}{r} \ ,$$
$$V_{03} = (r - \tfrac{1}{r})e^3 \ , \quad V_{12} = (2r + \tfrac{1}{r})e^3 \ . \tag{A-1.14}$$

The representation of the covariantly constant curvature tensor is given by

$$
\begin{aligned}
R_{01} &= e^0 \wedge e^1 - e^2 \wedge e^3 \;, & R_{23} &= e^0 \wedge e^1 - e^2 \wedge e^3 \;, \\
R_{02} &= e^0 \wedge e^2 - e^3 \wedge e^1 \;, & R_{31} &= -e^0 \wedge e^2 + e^3 \wedge e^1 \;, \\
R_{03} &= 4e^0 \wedge e^3 + 2e^1 \wedge e^2 \;, & R_{12} &= 2e^0 \wedge e^3 + 4e^1 \wedge e^2 \;.
\end{aligned}
\tag{A-1.15}
$$

Metric defines a real, covariantly constant, and therefore closed 2-form J

$$
J = -ig_{a\bar{b}}d\xi^a d\bar{\xi}^b \;,
\tag{A-1.16}
$$

the so called Kähler form. Kähler form J defines in CP_2 a symplectic structure because it satisfies the condition

$$
J^k_{\;r}J^{rl} = -s^{kl} \;.
\tag{A-1.17}
$$

The form J is integer valued and by its covariant constancy satisfies free Maxwell equations. Hence it can be regarded as a curvature form of a $U(1)$ gauge potential B carrying a magnetic charge of unit $1/2g$ (g denotes the gauge coupling). Locally one has therefore

$$
J = dB \;,
\tag{A-1.18}
$$

where B is the so called Kähler potential, which is not defined globally since J describes homological magnetic monopole.

It should be noticed that the magnetic flux of J through a 2-surface in CP_2 is proportional to its homology equivalence class, which is integer valued. The explicit representations of J and B are given by

$$
\begin{aligned}
B &= 2re^3 \;, \\
J &= 2(e^0 \wedge e^3 + e^1 \wedge e^2) = \frac{r}{F^2}dr \wedge (d\Psi + cos\Theta d\Phi) + \frac{r^2}{2F}sin\Theta d\Theta d\Phi \;.
\end{aligned}
\tag{A-1.19}
$$

The vierbein curvature form and Kähler form are covariantly constant and have in the complex coordinates only components of type (1,1).

Useful coordinates for CP_2 are the so called canonical coordinates in which Kähler potential and Kähler form have very simple expressions

$$
\begin{aligned}
B &= \sum_{k=1,2} P_k dQ_k \;, \\
J &= \sum_{k=1,2} dP_k \wedge dQ_k \;.
\end{aligned}
\tag{A-1.20}
$$

The relationship of the canonical coordinates to the "spherical" coordinates is given by the equations

$$
\begin{aligned}
P_1 &= -\frac{1}{1+r^2} \;, \\
P_2 &= \frac{r^2 cos\Theta}{2(1+r^2)} \;, \\
Q_1 &= \Psi \;, \\
Q_2 &= \Phi \;.
\end{aligned}
\tag{A-1.21}
$$

A-1.3   Spinors in CP_2

CP_2 doesn't allow spinor structure in the conventional sense [bc4]. However, the coupling of the spinors to a half odd multiple of the Kähler potential leads to a respectable spinor structure. Because the delicacies associated with the spinor structure of CP_2 play a fundamental role in TGD, the arguments of Hawking are repeated here.

To see how the space can fail to have an ordinary spinor structure consider the parallel transport of the vierbein in a simply connected space M. The parallel propagation around a closed curve with a base point x leads to a rotated vierbein at x: $e^A = R^A_B e^B$ and one can associate to each closed path an element of $SO(4)$.

Consider now a one-parameter family of closed curves $\gamma(v) : v \in (0,1)$ with the same base point x and $\gamma(0)$ and $\gamma(1)$ trivial paths. Clearly these paths define a sphere S^2 in M and the element $R^A_B(v)$ defines a closed path in $SO(4)$. When the sphere S^2 is contractible to a point e.g., homologically trivial, the path in $SO(4)$ is also contractible to a point and therefore represents a trivial element of the homotopy group $\Pi_1(SO(4)) = Z_2$.

For a homologically nontrivial 2-surface S^2 the associated path in $SO(4)$ can be homotopically nontrivial and therefore corresponds to a nonclosed path in the covering group Spin(4) (leading from the matrix 1 to -1 in the matrix representation). Assume this is the case.

Assume now that the space allows spinor structure. Then one can parallel propagate also spinors and by the above construction associate a closed path of Spin(4) to the surface S^2. Now, however this path corresponds to a lift of the corresponding $SO(4)$ path and cannot be closed. Thus one ends up with a contradiction.

From the preceding argument it is clear that one could compensate the non-allowed -1- factor associated with the parallel transport of the spinor around the sphere S^2 by coupling it to a gauge potential in such a way that in the parallel transport the gauge potential introduces a compensating -1-factor. For a $U(1)$ gauge potential this factor is given by the exponential $exp(i2\Phi)$, where Φ is the magnetic flux through the surface. This factor has the value -1 provided the $U(1)$ potential carries half odd multiple of Dirac charge $1/2g$. In case of CP_2 the required gauge potential is half odd multiple of the Kähler potential B defined previously. In the case of $M^4 \times CP_2$ one can in addition couple the spinor components with different chiralities independently to an odd multiple of $B/2$.

A-1.4   Geodesic sub-manifolds of CP_2

Geodesic sub-manifolds are defined as sub-manifolds having common geodesic lines with the imbedding space. As a consequence the second fundamental form of the geodesic manifold vanishes, which means that the tangent vectors h^k_α (understood as vectors of H) are covariantly constant quantities with respect to the covariant derivative taking into account that the tangent vectors are vectors both with respect to H and X^4.

In [ba2] a general characterization of the geodesic sub-manifolds for an arbitrary symmetric space G/H is given. Geodesic sub-manifolds are in 1-1-correspondence with the so called Lie triple systems of the Lie-algebra g of the group G. The Lie triple system t is defined as a subspace of g characterized by the closedness property with respect to double commutation

$$[X,[Y,Z]] \in t \ \text{ for } \ X,Y,Z \in t \ . \tag{A-1.22}$$

$SU(3)$ allows, besides geodesic lines, two nonequivalent (not isometry related) geodesic spheres. This is understood by observing that $SU(3)$ allows two nonequivalent $SU(2)$ algebras corresponding to subgroups $SO(3)$ (orthogonal 3×3 matrices) and the usual isospin group $SU(2)$. By taking any subset of two generators from these algebras, one obtains a Lie triple system and by exponentiating this system, one obtains a 2-dimensional geodesic sub-manifold of CP_2.

Standard representatives for the geodesic spheres of CP_2 are given by the equations

$$S^2_I \ : \ \xi^1 = \bar{\xi}^2 \ \text{ or equivalently } \ (\Theta = \pi/2, \Psi = 0) \ ,$$

$$S^2_{II} \ : \ \xi^1 = \xi^2 \ \text{ or equivalently } \ (\Theta = \pi/2, \Phi = 0) \ .$$

The non-equivalence of these sub-manifolds is clear from the fact that isometries act as holomorphic transformations in CP_2. The vanishing of the second fundamental form is also easy to verify. The first geodesic manifold is homologically trivial: in fact, the induced Kähler form vanishes identically for S_I^2. S_{II}^2 is homologically nontrivial and the flux of the Kähler form gives its homology equivalence class.

A-2 $\quad CP_2$ geometry and standard model symmetries

A-2.1 Identification of the electro-weak couplings

The delicacies of the spinor structure of CP_2 make it a unique candidate for space S. First, the coupling of the spinors to the $U(1)$ gauge potential defined by the Kähler structure provides the missing $U(1)$ factor in the gauge group. Secondly, it is possible to couple different H-chiralities independently to a half odd multiple of the Kähler potential. Thus the hopes of obtaining a correct spectrum for the electromagnetic charge are considerable. In the following it will be demonstrated that the couplings of the induced spinor connection are indeed those of the GWS model [fa4] and in particular that the right handed neutrinos decouple completely from the electro-weak interactions.

To begin with, recall that the space H allows to define three different chiralities for spinors. Spinors with fixed H-chirality $e = \pm 1$, CP_2-chirality l, r and M^4-chirality L, R are defined by the condition

$$
\begin{aligned}
\Gamma \Psi &= e\Psi \ , \\
e &= \pm 1 \ ,
\end{aligned}
\tag{A-2.1}
$$

where Γ denotes the matrix $\Gamma_9 = \gamma_5 \times \gamma_5$, $1 \times \gamma_5$ and $\gamma_5 \times 1$ respectively. Clearly, for a fixed H-chirality CP_2- and M^4-chiralities are correlated.

The spinors with H-chirality $e = \pm 1$ can be identified as quark and lepton like spinors respectively. The separate conservation of baryon and lepton numbers can be understood as a consequence of generalized chiral invariance if this identification is accepted. For the spinors with a definite H-chirality one can identify the vielbein group of CP_2 as the electro-weak group: $SO(4) = SU(2)_L \times SU(2)_R$.

The covariant derivatives are defined by the spinorial connection

$$
A = V + \frac{B}{2}(n_+ 1_+ + n_- 1_-) \ .
\tag{A-2.2}
$$

Here V and B denote the projections of the vielbein and Kähler gauge potentials respectively and $1_{+(-)}$ projects to the spinor H-chirality $+(-)$. The integers $n_\pm$ are odd from the requirement of a respectable spinor structure.

The explicit representation of the vielbein connection V and of B are given by the equations

$$
\begin{aligned}
V_{01} &= -\frac{e^1}{r} \ , & V_{23} &= \frac{e^1}{r} \ , \\
V_{02} &= -\frac{e^2}{r} \ , & V_{31} &= \frac{e^2}{r} \ , \\
V_{03} &= (r - \tfrac{1}{r})e^3 \ , & V_{12} &= (2r + \tfrac{1}{r})e^3 \ ,
\end{aligned}
\tag{A-2.3}
$$

and

$$
B = 2re^3 \ ,
\tag{A-2.4}
$$

respectively. The explicit representation of the vielbein is not needed here.

Let us first show that the charged part of the spinor connection couples purely left handedly. Identifying Σ_3^0 and Σ_2^1 as the diagonal (neutral) Lie-algebra generators of $SO(4)$, one finds that the charged part of the spinor connection is given by

$$
A_{ch} = 2V_{23}I_L^1 + 2V_{13}I_L^2 \ ,
\tag{A-2.5}
$$

where one have defined

$$I_L^1 = \frac{(\Sigma_{01} - \Sigma_{23})}{2} \ ,$$

$$I_L^2 = \frac{(\Sigma_{02} - \Sigma_{13})}{2} \ . \tag{A-2.6}$$

A_{ch} is clearly left handed so that one can perform the identification

$$W^\pm = \frac{2(e^1 \pm ie^2)}{r} \ , \tag{A-2.7}$$

where $W^\pm$ denotes the charged intermediate vector boson.

Consider next the identification of the neutral gauge bosons γ and Z^0 as appropriate linear combinations of the two functionally independent quantities

$$X = re^3 \ ,$$

$$Y = \frac{e^3}{r} \ , \tag{A-2.8}$$

appearing in the neutral part of the spinor connection. We show first that the mere requirement that photon couples vectorially implies the basic coupling structure of the GWS model leaving only the value of Weinberg angle undetermined.

To begin with let us define

$$\bar{\gamma} = aX + bY \ ,$$

$$\bar{Z}^0 = cX + dY \ , \tag{A-2.9}$$

where the normalization condition

$$ad - bc = 1 \ ,$$

is satisfied. The physical fields γ and Z^0 are related to $\bar{\gamma}$ and $\bar{Z}^0$ by simple normalization factors.

Expressing the neutral part of the spinor connection in term of these fields one obtains

$$\begin{aligned} A_{nc} = {}& [(c+d)2\Sigma_{03} + (2d-c)2\Sigma_{12} + d(n_+1_+ + n_-1_-)]\bar{\gamma} \\ &+ [(a-b)2\Sigma_{03} + (a-2b)2\Sigma_{12} - b(n_+1_+ + n_-1_-)]\bar{Z}^0 \ . \end{aligned} \tag{A-2.10}$$

Identifying Σ_{12} and $\Sigma_{03} = 1 \times \gamma_5\Sigma_{12}$ as vectorial and axial Lie-algebra generators, respectively, the requirement that γ couples vectorially leads to the condition

$$c = -d \ . \tag{A-2.11}$$

Using this result plus previous equations, one obtains for the neutral part of the connection the expression

$$A_{nc} = \gamma Q_{em} + Z^0(I_L^3 - sin^2\theta_W Q_{em}) \ . \tag{A-2.12}$$

Here the electromagnetic charge Q_{em} and the weak isospin are defined by

$$Q_{em} = \Sigma^{12} + \frac{(n_+1_+ + n_-1_-)}{6} \ ,$$

$$I_L^3 = \frac{(\Sigma^{12} - \Sigma^{03})}{2} \ . \tag{A-2.13}$$

The fields γ and Z^0 are defined via the relations

$$
\begin{aligned}
\gamma &= 6d\bar{\gamma} = \frac{6}{(a+b)}(aX + bY) \ , \\
Z^0 &= 4(a+b)\bar{Z}^0 = 4(X - Y) \ .
\end{aligned}
\tag{A-2.14}
$$

The value of the Weinberg angle is given by

$$
sin^2\theta_W = \frac{3b}{2(a+b)} \ ,
\tag{A-2.15}
$$

and is not fixed completely. Observe that right handed neutrinos decouple completely from the electro-weak interactions.

The determination of the value of Weinberg angle is a dynamical problem. The angle is completely fixed once the YM action is fixed by requiring that action contains no cross term of type γZ^0. Pure symmetry non-broken electro-weak YM action leads to a definite value for the Weinberg angle. One can however add a symmetry breaking term proportional to Kähler action and this changes the value of the Weinberg angle.

To evaluate the value of the Weinberg angle one can express the neutral part F_{nc} of the induced gauge field as

$$
F_{nc} = 2R_{03}\Sigma^{03} + 2R_{12}\Sigma^{12} + J(n_+1_+ + n_-1_-) \ ,
\tag{A-2.16}
$$

where one has

$$
\begin{aligned}
R_{03} &= 2(2e^0 \wedge e^3 + e^1 \wedge e^2) \ , \\
R_{12} &= 2(e^0 \wedge e^3 + 2e^1 \wedge e^2) \ , \\
J &= 2(e^0 \wedge e^3 + e^1 \wedge e^2) \ ,
\end{aligned}
\tag{A-2.17}
$$

in terms of the fields γ and Z^0 (photon and Z- boson)

$$
F_{nc} = \gamma Q_{em} + Z^0(I_L^3 - sin^2\theta_W Q_{em}) \ .
\tag{A-2.18}
$$

Evaluating the expressions above one obtains for γ and Z^0 the expressions

$$
\begin{aligned}
\gamma &= 3J - sin^2\theta_W R_{03} \ , \\
Z^0 &= 2R_{03} \ .
\end{aligned}
\tag{A-2.19}
$$

For the Kähler field one obtains

$$
J = \frac{1}{3}(\gamma + sin^2\theta_W Z^0) \ .
\tag{A-2.20}
$$

Expressing the neutral part of the symmetry broken YM action

$$
\begin{aligned}
L_{ew} &= L_{sym} + fJ^{\alpha\beta}J_{\alpha\beta} \ , \\
L_{sym} &= \frac{1}{4g^2}Tr(F^{\alpha\beta}F_{\alpha\beta}) \ ,
\end{aligned}
\tag{A-2.21}
$$

where the trace is taken in spinor representation, in terms of γ and Z^0 one obtains for the coefficient X of the γZ^0 cross term (this coefficient must vanish) the expression

$$X = -\frac{K}{2g^2} + \frac{fp}{18} \ ,$$
$$K = Tr\left[Q_{em}(I_L^3 - sin^2\theta_W Q_{em})\right] \ , \tag{A-2.22}$$

In the general case the value of the coefficient K is given by

$$K = \sum_i \left[-\frac{(18 + 2n_i^2)sin^2\theta_W}{9}\right] \ , \tag{A-2.23}$$

where the sum is over the spinor chiralities, which appear as elementary fermions and n_i is the integer describing the coupling of the spinor field to the Kähler potential. The cross term vanishes provided the value of the Weinberg angle is given by

$$sin^2\theta_W = \frac{9\sum_i 1}{(fg^2 + 2\sum_i(18 + n_i^2))} \ . \tag{A-2.24}$$

In the scenario where both leptons and quarks are elementary fermions the value of the Weinberg angle is given by

$$sin^2\theta_W = \frac{9}{(\frac{fg^2}{2} + 28)} \ . \tag{A-2.25}$$

The bare value of the Weinberg angle is 9/28 in this scenario, which is quite close to the typical value 9/24 of GUTs [fa5].

A-2.2 Discrete symmetries

The treatment of discrete symmetries C, P, and T is based on the following requirements:
a) Symmetries must be realized as purely geometric transformations.
b) Transformation properties of the field variables should be essentially the same as in the conventional quantum field theories [fa1].

The action of the reflection P on spinors of is given by

$$\Psi \ \rightarrow \ P\Psi = \gamma^0 \times \gamma^0 \Psi \ . \tag{A-2.26}$$

in the representation of the gamma matrices for which γ^0 is diagonal. It should be noticed that W and Z^0 bosons break parity symmetry as they should since their charge matrices do not commute with the matrix of P.

The guess that a complex conjugation in CP_2 is associated with T transformation of the physicist turns out to be correct. One can verify by a direct calculation that pure Dirac action is invariant under T realized according to

$$m^k \ \rightarrow \ T(M^k) \ ,$$
$$\xi^k \ \rightarrow \ \bar{\xi}^k \ ,$$
$$\Psi \ \rightarrow \ \gamma^1\gamma^3 \times 1 \times \Psi \ . \tag{A-2.27}$$

The operation bearing closest resemblance to the ordinary charge conjugation corresponds geometrically to complex conjugation in CP_2:

$$\xi^k \ \rightarrow \ \bar{\xi}^k \ ,$$
$$\Psi \ \rightarrow \ \gamma^2\gamma^0 \times 1 \times \Psi^\dagger \ . \tag{A-2.28}$$

As one might have expected symmetries CP and T are exact symmetries of the pure Dirac action.

A-3 Basic facts about induced gauge fields

Since the classical gauge fields are closely related in TGD framework, it is not possible to have space-time sheets carrying only single kind of gauge field. For instance, em fields are accompanied by Z^0 fields for extremals of Kähler action. Weak forces is however absent unless the space-time sheets contains topologically condensed exotic weakly charged particles responding to this force. Same applies to classical color forces. The fact that these long range fields are present forces to assume that there exists a hierarchy of scaled up variants of standard model physics identifiable in terms of dark matter.

Classical em fields are always accompanied by Z^0 field and some components of color gauge field. For extremals having homologically non-trivial sphere as a CP_2 projection em and Z^0 fields are the only non-vanishing electro-weak gauge fields. For homologically trivial sphere only W fields are non-vanishing. Color rotations does not affect the situation.

For vacuum extremals all electro-weak gauge fields are in general non-vanishing although the net gauge field has U(1) holonomy by 2-dimensionality of the CP_2 projection. Color gauge field has $U(1)$ holonomy for all space-time surfaces and quantum classical correspondence suggest a weak form of color confinement meaning that physical states correspond to color neutral members of color multiplets.

A-3.1 Induced gauge fields for space-times for which CP$_2$ projection is a geodesic sphere

If one requires that space-time surface is an extremal of Kähler action and has a 2-dimensional CP$_2$ projection, only vacuum extremals and space-time surfaces for which CP$_2$ projection is a geodesic sphere, are allowed. Homologically non-trivial geodesic sphere correspond to vanishing W fields and homologically non-trivial sphere to non-vanishing W fields but vanishing γ and Z^0. This can be verified by explicit examples.

$r = \infty$ surface gives rise to a homologically non-trivial geodesic sphere for which e_0 and e_3 vanish imply the vanishing of W field. For space-time sheets for which CP$_2$ projection is $r = \infty$ homologically non-trivial geodesic sphere of CP_2 one has

$$\gamma = (\frac{3}{4} - \frac{sin^2(\theta_W)}{2})Z^0 \simeq \frac{5Z^0}{8} \quad .$$

The induced W fields vanish in this case and they vanish also for all geodesic sphere obtained by $SU(3)$ rotation.

$Im(\xi^1) = Im(\xi^2) = 0$ corresponds to homologically trivial geodesic sphere. A more general representative is obtained by using for the phase angles of standard complex CP_2 coordinates constant values. In this case e^1 and e^3 vanish so that the induced em, Z^0, and Kähler fields vanish but induced W fields are non-vanishing. This holds also for surfaces obtained by color rotation. Hence one can say that for non-vacuum extremals with 2-D CP$_2$ projection color rotations and weak symmetries commute.

A-3.2 Space-time surfaces with vanishing em, Z^0, or Kähler fields

In the following the induced gauge fields are studied for general space-time surface with 2-dimensional CP_2 projection without assuming the extremal property. In fact, extremal property reduces the study of extremals with 2-dimensional CP_2 projection to the study of vacuum extremals and surfaces having geodesic sphere as a CP_2 projection and in this sense the following arguments are somewhat obsolete in their generality.

Space-times with vanishing em, Z^0, or Kähler fields

The following considerations apply to a more general situation in which the homologically trivial geodesic sphere and extremal property are not assumed. It must be emphasized that this case is possible in TGD framework only for a vanishing Kähler field.

Using spherical coordinates (r, Θ, Ψ, Φ) for CP_2, the expression of Kähler form reads as

$$J = \frac{r}{F^2} dr \wedge (d\Psi + cos(\Theta) d\Phi) + \frac{r^2}{2F} sin(\Theta) d\Theta \wedge d\Phi \ ,$$

$$F = 1 + r^2 \ .$$
$$\text{(A-3.1)}$$

The general expression of electromagnetic field reads as

$$F_{em} = (3 + 2p) \frac{r}{F^2} dr \wedge (d\Psi + cos(\Theta) d\Phi) + (3 + p) \frac{r^2}{2F} sin(\Theta) d\Theta \wedge d\Phi \ ,$$

$$p = sin^2(\Theta_W) \ ,$$
$$\text{(A-3.2)}$$

where Θ_W denotes Weinberg angle.

a) The vanishing of the electromagnetic fields is guaranteed, when the conditions

$$\Psi = k\Phi \ ,$$

$$(3 + 2p) \frac{1}{r^2 F} (d(r^2)/d\Theta)(k + cos(\Theta)) + (3 + p) sin(\Theta) = 0 \ ,$$
$$\text{(A-3.3)}$$

hold true. The conditions imply that CP_2 projection of the electromagnetically neutral space-time is 2-dimensional. Solving the differential equation one obtains

$$r = \sqrt{\frac{X}{1 - X}} \ ,$$

$$X = D \left[|\frac{(k + u)}{C}| \right]^\epsilon \ ,$$

$$u \equiv cos(\Theta) \ , \quad C = k + cos(\Theta_0) \ , \quad D = \frac{r_0^2}{1 + r_0^2} \ , \quad \epsilon = \frac{3 + p}{3 + 2p} \ ,$$
$$\text{(A-3.4)}$$

where C and D are integration constants. $0 \leq X \leq 1$ is required by the reality of r. $r = 0$ would correspond to $X = 0$ giving $u = -k$ achieved only for $|k| \leq 1$ and $r = \infty$ to $X = 1$ giving $|u + k| = [(1 + r_0^2)/r_0^2)]^{(3+2p)/(3+p)}$ achieved only for

$$sign(u + k) \times [\frac{1 + r_0^2}{r_0^2}]^{\frac{3+2p}{3+p}} \leq k + 1 \ ,$$

where $sign(x)$ denotes the sign of x.

The expressions for Kähler form and Z^0 field are given by

$$J = -\frac{p}{3 + 2p} X du \wedge d\Phi \ ,$$

$$Z^0 = -\frac{6}{p} J \ .$$
$$\text{(A-3.5)}$$

The components of the electromagnetic field generated by varying vacuum parameters are proportional to the components of the Kähler field: in particular, the magnetic field is parallel to the Kähler magnetic field. The generation of a long range Z^0 vacuum field is a purely TGD based feature not encountered in the standard gauge theories.

b) The vanishing of Z^0 fields is achieved by the replacement of the parameter ϵ with $\epsilon = 1/2$ as becomes clear by considering the condition stating that Z^0 field vanishes identically. Also the relationship $F_{em} = 3J = -\frac{3}{4} \frac{r^2}{F} du \wedge d\Phi$ is useful.

c) The vanishing Kähler field corresponds to $\epsilon = 1, p = 0$ in the formula for em neutral space-times. In this case classical em and Z^0 fields are proportional to each other:

$$Z^0 \;=\; 2e^0 \wedge e^3 = \frac{r}{F^2}(k+u)\frac{\partial r}{\partial u}du \wedge d\Phi = (k+u)du \wedge d\Phi \;,$$

$$r \;=\; \sqrt{\frac{X}{1-X}} \;,\quad X = D|k+u| \;,$$

$$\gamma \;=\; -\frac{p}{2}Z^0 \;. \tag{A-3.6}$$

For a vanishing value of Weinberg angle ($p = 0$) em field vanishes and only Z^0 field remains as a long range gauge field. Vacuum extremals for which long range Z^0 field vanishes but em field is non-vanishing are not possible.

The effective form of CP_2 metric for surfaces with 2-dimensional CP_2 projection

The effective form of the CP_2 metric for a space-time having vanishing em,Z^0, or Kähler field is of practical value in the case of vacuum extremals and is given by

$$ds^2_{eff} \;=\; (s_{rr}(\frac{dr}{d\Theta})^2 + s_{\Theta\Theta})d\Theta^2 + (s_{\Phi\Phi} + 2ks_{\Phi\Psi})d\Phi^2 = \frac{R^2}{4}[s^{eff}_{\Theta\Theta}d\Theta^2 + s^{eff}_{\Phi\Phi}d\Phi^2] \;,$$

$$s^{eff}_{\Theta\Theta} \;=\; X \times \left[\frac{\epsilon^2(1-u^2)}{(k+u)^2} \times \frac{1}{1-X} + 1 - X\right] \;,$$

$$s^{eff}_{\Phi\Phi} \;=\; X \times \left[(1-X)(k+u)^2 + 1 - u^2\right] \;, \tag{A-3.7}$$

and is useful in the construction of vacuum imbedding of, say Schwartchild metric.

Topological quantum numbers

Space-times for which either em, Z^0, or Kähler field vanishes decompose into regions characterized by six vacuum parameters: two of these quantum numbers (ω_1 and ω_2) are frequency type parameters, two (k_1 and k_2) are wave vector like quantum numbers, two of the quantum numbers (n_1 and n_2) are integers. The parameters ω_i and n_i will be referred as electric and magnetic quantum numbers. The existence of these quantum numbers is not a feature of these solutions alone but represents a much more general phenomenon differentiating in a clear cut manner between TGD and Maxwell's electrodynamics.

The simplest manner to avoid surface Kähler charges and discontinuities or infinities in the derivatives of CP_2 coordinates on the common boundary of two neighboring regions with different vacuum quantum numbers is topological field quantization, 3-space decomposes into disjoint topological field quanta, 3-surfaces having outer boundaries with possibly macroscopic size.

Under rather general conditions the coordinates Ψ and Φ can be written in the form

$$\Psi \;=\; \omega_2 m^0 + k_2 m^3 + n_2\phi + \text{Fourier expansion} \;,$$

$$\Phi \;=\; \omega_1 m^0 + k_1 m^3 + n_1\phi + \text{Fourier expansion} \;. \tag{A-3.8}$$

m^0,m^3 and ϕ denote the coordinate variables of the cylindrical M^4 coordinates) so that one has $k = \omega_2/\omega_1 = n_2/n_1 = k_2/k_1$. The regions of the space-time surface with given values of the vacuum parameters ω_i,k_i and n_i and m and C are bounded by the surfaces at which space-time surface becomes ill-defined, say by $r > 0$ or $r < \infty$ surfaces.

The space-time surface decomposes into regions characterized by different values of the vacuum parameters r_0 and Θ_0. At $r = \infty$ surfaces n_2,ω_2 and m can change since all values of Ψ correspond to the same point of CP_2: at $r = 0$ surfaces also n_1 and ω_1 can change since all values of Φ correspond to same point of CP_2, too. If $r = 0$ or $r = \infty$ is not in the allowed range space-time surface develops a boundary.

This implies what might be called topological quantization since in general it is not possible to find a smooth global imbedding for, say a constant magnetic field. Although global imbedding exists it decomposes into regions with different values of the vacuum parameters and the coordinate u in general possesses discontinuous derivative at $r = 0$ and $r = \infty$ surfaces. A possible manner to avoid edges of space-time is to allow field quantization so that 3-space (and field) decomposes into disjoint quanta, which can be regarded as structurally stable units a 3-space (and of the gauge field). This doesn't exclude partial join along boundaries for neighboring field quanta provided some additional conditions guaranteing the absence of edges are satisfied.

For instance, the vanishing of the electromagnetic fields implies that the condition

$$\Omega \; \equiv \; \frac{\omega_2}{n_2} - \frac{\omega_1}{n_1} = 0 \; , \tag{A-3.9}$$

is satisfied. In particular, the ratio ω_2/ω_1 is rational number for the electromagnetically neutral regions of space-time surface. The change of the parameter n_1 and n_2 (ω_1 and ω_2) in general generates magnetic field and therefore these integers will be referred to as magnetic (electric) quantum numbers.

A-4   p-Adic numbers and TGD

A-4.1   p-Adic number fields

p-Adic numbers (p is prime: 2,3,5,...) can be regarded as a completion of the rational numbers using a norm, which is different from the ordinary norm of real numbers [ca2]. p-Adic numbers are representable as power expansion of the prime number p of form:

$$x \; = \; \sum_{k \geq k_0} x(k)p^k, \; x(k) = 0, ..., p - 1 \; . \tag{A-4.1}$$

The norm of a p-adic number is given by

$$|x| \; = \; p^{-k_0(x)} \; . \tag{A-4.2}$$

Here $k_0(x)$ is the lowest power in the expansion of the p-adic number. The norm differs drastically from the norm of the ordinary real numbers since it depends on the lowest pinary digit of the p-adic number only. Arbitrarily high powers in the expansion are possible since the norm of the p-adic number is finite also for numbers, which are infinite with respect to the ordinary norm. A convenient representation for p-adic numbers is in the form

$$x \; = \; p^{k_0}\varepsilon(x) \; , \tag{A-4.3}$$

where $\varepsilon(x) = k + ...$ with $0 < k < p$, is p-adic number with unit norm and analogous to the phase factor $exp(i\phi)$ of a complex number.

The distance function $d(x, y) = |x - y|_p$ defined by the p-adic norm possesses a very general property called ultra-metricity:

$$d(x, z) \; \leq \; max\{d(x, y), d(y, z)\} \; . \tag{A-4.4}$$

The properties of the distance function make it possible to decompose R_p into a union of disjoint sets using the criterion that x and y belong to same class if the distance between x and y satisfies the condition

$$d(x, y) \; \leq \; D \; . \tag{A-4.5}$$

This division of the metric space into classes has following properties:

a) Distances between the members of two different classes X and Y do not depend on the choice of points x and y inside classes. One can therefore speak about distance function between classes.

b) Distances of points x and y inside single class are smaller than distances between different classes.

c) Classes form a hierarchical tree.

Notice that the concept of the ultra-metricity emerged in physics from the models for spin glasses and is believed to have also applications in biology [cc1]. The emergence of p-adic topology as the topology of the effective space-time would make ultra-metricity property basic feature of physics.

A-4.2 Canonical correspondence between p-adic and real numbers

The basic challenge encountered by p-adic physicist is how to map the predictions of the p-adic physics to real numbers. p-Adic probabilities provide a basic example in this respect. Identification via common rationals and canonical identification and its variants have turned out to play a key role in this respect.

Basic form of canonical identification

There exists a natural continuous map $I : R_p \to R_+$ from p-adic numbers to non-negative real numbers given by the "pinary" expansion of the real number for $x \in R$ and $y \in R_p$ this correspondence reads

$$
\begin{aligned}
y &= \sum_{k>N} y_k p^k \to x = \sum_{k<N} y_k p^{-k} \ , \\
y_k &\in \{0, 1, .., p-1\} \ .
\end{aligned}
\tag{A-4.6}
$$

This map is continuous as one easily finds out. There is however a little difficulty associated with the definition of the inverse map since the pinary expansion like also decimal expansion is not unique ($1 = 0.999...$) for the real numbers x, which allow pinary expansion with finite number of pinary digits

$$
\begin{aligned}
x &= \sum_{k=N_0}^{N} x_k p^{-k} \ , \\
x &= \sum_{k=N_0}^{N-1} x_k p^{-k} + (x_N - 1)p^{-N} + (p-1)p^{-N-1} \sum_{k=0,..} p^{-k} \ .
\end{aligned}
\tag{A-4.7}
$$

The p-adic images associated with these expansions are different

$$
\begin{aligned}
y_1 &= \sum_{k=N_0}^{N} x_k p^k \ , \\
y_2 &= \sum_{k=N_0}^{N-1} x_k p^k + (x_N - 1)p^N + (p-1)p^{N+1} \sum_{k=0,..} p^k \\
&= y_1 + (x_N - 1)p^N - p^{N+1} \ ,
\end{aligned}
\tag{A-4.8}
$$

so that the inverse map is either two-valued for p-adic numbers having expansion with finite pinary digits or single valued and discontinuous and non-surjective if one makes pinary expansion unique by choosing the one with finite pinary digits. The finite pinary digit expansion is a natural choice since in the numerical work one always must use a pinary cutoff on the real axis.

The topology induced by canonical identification

The topology induced by the canonical identification in the set of positive real numbers differs from the ordinary topology. The difference is easily understood by interpreting the p-adic norm as a norm in the set of the real numbers. The norm is constant in each interval $[p^k, p^{k+1})$ (see Fig. A-4.2) and is equal to the usual real norm at the points $x = p^k$: the usual linear norm is replaced with a piecewise constant norm. This means that p-adic topology is coarser than the usual real topology and the higher the value of p is, the coarser the resulting topology is above a given length scale. This hierarchical ordering of the p-adic topologies will be a central feature as far as the proposed applications of the p-adic numbers are considered.

Ordinary continuity implies p-adic continuity since the norm induced from the p-adic topology is rougher than the ordinary norm. p-Adic continuity implies ordinary continuity from right as is clear already from the properties of the p-adic norm (the graph of the norm is indeed continuous from right). This feature is one clear signature of the p-adic topology.

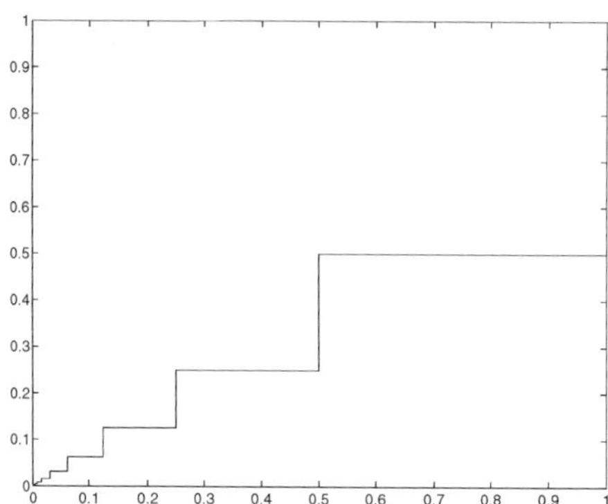

Figure A.1: The real norm induced by canonical identification from 2-adic norm.

The linear structure of the p-adic numbers induces a corresponding structure in the set of the non-negative real numbers and p-adic linearity in general differs from the ordinary concept of linearity. For example, p-adic sum is equal to real sum only provided the summands have no common pinary digits. Furthermore, the condition $x +_p y < max\{x, y\}$ holds in general for the p-adic sum of the real numbers. p-Adic multiplication is equivalent with the ordinary multiplication only provided that either of the members of the product is power of p. Moreover one has $x \times_p y < x \times y$ in general. The p-Adic negative -1_p associated with p-adic unit 1 is given by $(-1)_p = \sum_k (p-1)p^k$ and defines p-adic negative for each real number x. An interesting possibility is that p-adic linearity might replace the ordinary linearity in some strongly nonlinear systems so these systems would look simple in the p-adic topology.

These results suggest that canonical identification is involved with some deeper mathematical structure. The following inequalities hold true:

$$
\begin{aligned}
(x + y)_R &\leq x_R + y_R \ , \\
|x|_p |y|_R \leq (xy)_R &\leq x_R y_R \ ,
\end{aligned}
\tag{A-4.9}
$$

where $|x|_p$ denotes p-adic norm. These inequalities can be generalized to the case of $(R_p)^n$ (a linear vector space over the p-adic numbers).

$$
\begin{aligned}
(x + y)_R &\leq x_R + y_R \ , \\
|\lambda|_p |y|_R \leq (\lambda y)_R &\leq \lambda_R y_R \ ,
\end{aligned}
\tag{A-4.10}
$$

where the norm of the vector $x \in T_p^n$ is defined in some manner. The case of Euclidian space suggests the definition

$$
(x_R)^2 = (\sum_n x_n^2)_R \ .
\tag{A-4.11}
$$

These inequalities resemble those satisfied by the vector norm. The only difference is the failure of linearity in the sense that the norm of a scaled vector is not obtained by scaling the norm of the original vector. Ordinary situation prevails only if the scaling corresponds to a power of p.

These observations suggests that the concept of a normed space or Banach space might have a generalization and physically the generalization might apply to the description of some non-linear systems. The nonlinearity would be concentrated in the nonlinear behavior of the norm under scaling.

Modified form of the canonical identification

The original form of the canonical identification is continuous but does not respect symmetries even approximately. This led to a search of variants which would do better in this respect. The modification of the canonical identification applying to rationals only and given by

$$I_Q(q = p^k \times \frac{r}{s}) = p^k \times \frac{I(r)}{I(s)} \tag{A-4.12}$$

is uniquely defined for rationals, maps rationals to rationals, has also a symmetry under exchange of target and domain. This map reduces to a direct identification of rationals for $0 \leq r < p$ and $0 \leq s < p$. It has turned out that it is this map which most naturally appears in the applications. The map is obviously continuous locally since p-adically small modifications of r and s mean small modifications of the real counterparts.

Canonical identification is in a key role in the successful predictions of the elementary particle masses. The predictions for the light elementary particle masses are within extreme accuracy same for I and I_Q but I_Q is theoretically preferred since the real probabilities obtained from p-adic ones by I_Q sum up to one in p-adic thermodynamics.

Generalization of number concept and notion of imbedding space

TGD forces an extension of number concept: roughly a fusion of reals and various p-adic number fields along common rationals is in question. This induces a similar fusion of real and p-adic imbedding spaces. Since finite p-adic numbers correspond always to non-negative reals n-dimensional space R^n must be covered by 2^n copies of the p-adic variant R_p^n of R^n each of which projects to a copy of R_+^n (four quadrants in the case of plane). The common points of p-adic and real imbedding spaces are rational points and most p-adic points are at real infinity.

For a given p-adic space-time sheet most points are literally infinite as real points and the projection to the real imbedding space consists of a discrete set of rational points: the interpretation in terms of the unavoidable discreteness of the physical representations of cognition is natural. Purely local p-adic physics implies real p-adic fractality and thus long range correlations for the real space-time surfaces having enough common points with this projection.

p-Adic fractality means that M^4 projections for the rational points of space-time surface X^4 are related by a direct identification whereas CP_2 coordinates of X^4 at these points are related by I, I_Q or some of its variants implying long range correlates for CP_2 coordinates. Since only a discrete set of points are related in this manner, both real and p-adic field equations can be satisfied and there are no problems with symmetries. p-Adic effective topology is expected to be a good approximation only within some length scale range which means infrared and UV cutoffs. Also multi-p-fractality is possible.

Bibliography

B-1 Topology

B-1.1 Basic background material

[aa1] R. Thom (1954), Commentarii Math. Helvet., 28, 17.

[aa2] J. Milnor (1965), *Topology form Differential Point of View*. The University Press of Virginia.

[aa3] A. Wallace (1968), *Differential Topology*. W. A. Benjamin, New York.

[aa4] E. C. Zeeman (ed.)(1977), *Catastrophe Theory*, Addison-Wessley Publishing Company.

B-1.2 Topology and classical physics

[ab1] A. Lakthakia (1994), *Beltrami Fields in Chiral Media*, Series in Contemporary Chemical Physics - Vol. 2, World Scientific, Singapore.

[ab2] D. Reed (1995), in *Advanced Electromagnetism: Theories, Foundations, Applications*, edited by T. Barrett (Chap. 7), World Scientific, Singapore.

[ab3] J. Etnyre and R. Ghrist (2001), *An index for closed orbits in Beltrami field*. ArXiv:math.DS/0101095.

[ab4] O. I. Bogoyavlenskij (2003), *Exact unsteady solutions to the Navier-Stokes equations and viscous MHD equations*. Phys. Lett. A, 281-286.

[ab5] J. Etnyre and G. Ghrist (2003)*Generic hydrodynamic instability of curl eigen fields*, arXiv:hat.DS/0306310.

[ab6] G. E. Marsh (1995), *Helicity and Electromagnetic Field Topology* in *Advanced Electromagnetism*, Eds. T. W. Barrett and D. M. Grimes, Word Scientific.

[ab7] R. M. Kiehn (2004), *Non-Equilibrium and Irreversible Thermodynamics-from a Perspective of Topological Evolution*.
http://www22.pair.com/csdc/download/topthermo.pdf .

B-2 Differential geometry and CP_2 as a manifold

B-2.1 Differential and Riemannin geometry

[ba1] L. P. Eisenhart (1964), *Riemannian Geometry*. Princeton University Press.

[ba2] S. Helgason (1962), *Differential Geometry and Symmetric Spaces*. Academic Press, New York.

[ba3] M. Spivak (1970), *Differential Geometry I,II,III,IV*. Publish or Perish. Boston.

B-2.2　Calibrations

[bb1]　R. Harvey (1990), *Spinors and Calibrations*, Academic Press, New York.

[bb2]　Homepage of Jose Figueroa-O'Farrill,
http://www.maths.ed.ac.uk/~jmf/Research/susy_7.html.

[bb3]　K. Becker, M. Becker, and A. Strominger (1995), *Fivebranes, membranes, and non-perturbative string theory*, Nucl. Phys. B456, 130.

[bb4]　G. W. Gibbons and G. Papadopouls, *Calibrations and Intersecting Branes*, hep-th/9803163.

[bb5]　J. M. Figueroa-O'Farrill (1998), *Intersecting Brane Geometries*, hep-th/9806040.

[bb6]　R. C. McLean(1998), em Deformations and Calibrated Submanifolds, Comm. Anal. Geom. 6, 705-747.
http://www.math.duke.edu/preprints/1996.html

[bb7]　J. Gutowski, G. Papadopoulos and P. K. Townsend (1999), *Supersymmetry and Generalized Calibrations*, arXiv:hep-th/9905156.

B-2.3　CP_2 as a manifold

[bc1]　G. W. Gibbons and C. N. Pope (1977), CP_2 *as gravitational instanton.* Commun. Math. Phys. 55, 53.

[bc2]　S. W. Hawking and C. N. Pope (1978), *Generalized Spin Structures in Quantum Gravity.* Physics Letters Vol 73 B, no 1.

[bc3]　T. Eguchi, T. Gilkey, J. Hanson (1980), Phys. Rep. 66, 6.

[bc4]　C. N. Pope (1980), *Eigenfunctions and Spin^c Structures on* CP_2 D.A.M.T.P. preprint.

B-3　Number theory and algebraic geometry

B-3.1　Algebraic numbers

[ca1]　M. Eichler (1966), *Introduction to the theory of algebraic numbers and functions*, Academic Press, New York.

[ca2]　Z. I. Borevich and I. R. Shafarevich (1966) ,*Number Theory*, Academic Press.

[ca3]　U. Dudley (1969), *Elementary Number Theory*, W.H. Freeman and Company.

[ca4]　R. B. J. T. Allenby and E. J. Redfern (1989), *Introduction to Number Theory with Computing*, Edward Arnold.

[ca5]　J. Esmonde and M. Ram Murty (1991), *Problems in Algebraic Number Theory*, Springer-Verlag, New York.

B-3.2　Quaternions and octonions

[cb1]　J. C. Baez (2001), *The Octonions*, Bull. Amer. Math. Soc. 39 (2002), 145-205.
http://math.ucr.edu/home/baez/Octonions/octonions.html.

[cb2]　J. Daboul and R. Delborough (1999) *Matrix Representations of Octonions and Generalizations*, hep-th/9906065.

[cb3]　J. Schray and C. A. Manogue (1994) *Octonionic representations of Clifford algebras and triality*, hep-th/9407179.

[cb4] S. De Leo and K. Abdel-Khalek (1996), *Octonionic quantum mechanics and complex geometry*, Prog.Theor.Phys. 96 (1996) 823-832, hep-th/9609032.

[cb5] T. Smith (1997), *D4-D5-E6 Physics*. Homepage of Tony Smith. *http : //galaxy.cau.edu/tsmith/d4d5e6hist.html*. The homepage contains a lot of information and ideas about the possible relationship of octonions and quaternions to physics.

[cb6] S. Okubo (1997), *Angular momentum, quaternion, octonion, and Lie-super algebra Osp(1,2)*, ph/9710038.

[cb7] L. P. Horwitz (1996), *Hypercomplex quantum mechanics*, quant-ph/9602001.

[cb8] M. Brenner (1999), *Quantum octonions*, Communications in Algebra 27 (1999), 2809-2831. http://math.usask.ca/ bremner/research/publications/qo.pdf

[cb9] G. Benkart, J. M. Perez-Izquierdo, *A quantum octonion algebra*, math.QA/9801141.

B-3.3   p-Adic numbers

[cc1] G. Parisi (1992), Field Theory, Disorder and Simulations, World Scientific.

[cc2] A. Khrennikov (1994), *p-Adic Valued Distributions in Mathematical Physics*, Kluwer Academic Publishers, Dordrecht.

[cc3] L. Brekke and P. G. O. Freund (1993), *p-Adic Numbers in Physics*, Phys. Rep. vol. 233, No 1.

[cc4] S. Solla, G. Sorkin, and S. White (1986), "Configuration space analysis for optimization problems", in *Disordered Systems and Biological Organization*, E. Bienenstock et al. (Eds.), NATO ASI Series, ¡b¿ F20¡/b¿, Berlin: Springer Verlag, pp. 283-293.

[cc5] A. Yu. Khrennikov (1992), *p-Adic Probability and Statistics*, Dokl. Akad Nauk, vol 433 , No 6.

[cc6] F. Q. Gouvêa (1997), *p-adic Numbers: An Introduction*, Springer.

B-3.4   Miscellaneous topics related to number theory

[cd1] A. Y. Khinchin (1964), *Continued Fractions*, Third Edition, The Univesity of Chicago Press, Chicago. See also
P. Guerzhoy, *A Short Introduction to Continued Fractions*, http://www.math.temple.edu/~pasha/contfrac.pdf.

[cd2] *Khinchin's constant*, http://mathworld.wolfram.com/ContinuedFraction.html.

[cd3] G. Grosswald (1985), *Representations of Integers as Sums of Squares*, Springer Verlag, New York.

[cd4] A. Robinson (1974), *Non-standard Analysis*, North-Holland, Amsterdam.

B-3.5   Algebraic geometry and Riemann surfaces

[ce1] S. S. Abhyankar (1980), *Algebraic Geometry for Scientists and Engineers*, Mathematical Surveys and Monographs, No 35, American Mathematical Society.

[ce2] K. E. Smith. L. Kahanpää, P. Kekäläinen and W. Traves (2000), *Invitation to Algebraic Geometry*, Springer Verlag, New York, Inc..

[ce3] R. Accola (1975), *Riemann Surfaces, Theta functions and Abelian Automorphism Groups*. Lecture Notes in Mathematics 483. Springer Verlag.

[ce4] H. M. Farkas and I. Kra (1980), *Riemann Surfaces*. Springer Verlag.

B-3.6   Theta functions and Riemann Hypothesis

[cf1] D. Mumford(1983,1984), *Tata Lectures on Theta I,II,III*. Birkhäuser.

[cf2] L. Alvarez-Gaume, G. Moore and C. Vafa (1986), *Theta functions, Modular invariance and Strings*. Commun. Math. Phys. 106. 40-87.

[cf3] D. Zagier (1994), *Values of Zeta Functions and Their Applications*, First European Congress of Mathematics (Paris, 1992), Vol. II, Progress in Mathematics 120, Birkhauser, 497-512.

[cf4] U. Mueller and C. Schubert (2002), *A Quantum Field Theoretical Representation of Euler-Zagier Sums*, arXiv:math.QA/9908067. Int. J. Math. Sc. Vol 31, issue 3 (2002), 127-148.

[cf5] W. T. Lu and S. Sridhar (2004), *Correlations Among the Riemann Zeros: Invariance, Resurgence, Prophecy and Self-Duality*. arXiv:nlin.CD/0405058.

[cf6] S. Chowia (1965), *The Riemann Hypothesis and Hilbert's Tenth Problem*, Gordon and Breach, p. 79.

[cf7] E. Bogomolny(2003), *Quantum and Arithmetical Chaos*. arXiv:nlin.CD/0312061.

B-4   Mathematics of Quantum Theory

B-4.1   Quantum Mechanics

[da1] P. A. M. Dirac (1939), *A New Notation for Quantum Mechanics*, Proceedings of the Cambridge Philosophical Society, 35: 416-418.
J. E. Roberts (1966), *The Dirac Bra and Ket Formalism*, Journal of Mathematical Physics, 7: 1097-1104. Halvorson, Hans and Clifton, Rob (2001),

[da2] L. Schwartz (1945),*Generalisation de la Notion de Fonction, de Derivation, de Transformation de Fourier et Applications Mathematiques et Physiques*, Annales de l'Universite de Grenoble, 21: 57-74. 1951-1952,
L. Scwartz (1951-1952), *Theorie des Distributions*, Publications de l'Institut de Mathematique de l'Universite de Strasbourg, Vøls 9-10, Paris: Hermann.

[da3] I. Gelfand and M. Neumark (1943), *On the Imbedding of Normed Rings into the Ring of Operators in Hilbert Space*, Recueil Mathematique [Matematicheskii Sbornik] Nouvelle Serie, 12 [54]: 197-213. [Reprinted in C*-algebras: 1943-1993, in the series Contemporary Mathematics, 167, Providence, R.I. : American Mathematical Society, 1994.]
I. Gelfand and N. Vilenkin (1964), *Generalized Functions*, Volume 4, New York: Academic Press. [First published in Russian in 1961.]

[da4] J. von Neumann (1955), *Mathematical Foundations of Quantum Mechanics*, Princeton, NJ: Princeton University.

[da5] I. Gelfand, M. Graev, N. Vilenkin (1966) *Generalized Functions, Volume 5*, Academic Press.

[da6] . A. M. Dirac (1981),*The Principles of Quantum Mechanics*, Oxford: Oxford University Press.

B-4.2   Path integrals

[db1] R. P. Feynman (1948), *Space-Time Approach to Non-Relativistic Quantum Mechanics*, Reviews of Modern Physics, 20: 367387. [It is reprinted in (Schwinger 1958).]

[db2] R. J. Rivers (1987), *Path Integral Methods in Quantum Field Theory*, Cambridge: Cambridge University Press.

[db3] J. J. Duistermaat and G., J. Heckmann (1982), Inv. Math. 69, 259.

[db4] M. Chaichian and A. Denisev (2001), *Path Integrals in Physics*, Institute of Physics Publishing.

B-4.3 Algebraic quantum field theory

[dc1] R. Haag (1992), *Local Quantum Physics*, Springer, Berlin.

[dc2] M. Takesaki (1970), *Tomita's Theory of Modular Hilbert Algebras and Its Applications*, Lecture Notes in Mathematics 128, Springer, Berlin.

[dc3] M. Rainer (2000), *Algebraic Quantum Field Theory on Manifolds: A Haag-Kastler Setting for Quantum Geometry*, gr-qc/9911076.

[dc4] S. Doplicher, R. Haag, J. E. Roberts (1971-1974), *Local Observables and Particle Statistics I and II*. Commun. Math. Phys. 23 and 35.
S. Doplicher, J. E. Roberts (1989), Ann. Math. 130, 75-119.
S. Doplicher, J. E. Roberts (1989), Invent. Math. 98, 157-218.

[dc5] H. P. Halvorson (2001), *Locality, Localization, and the Particle Concept: Topics in the Foundations of Quantum Field Theory*, Ph. D Dissertation.
http://philsci-archive.pitt.edu/archive/00000346/00/main-new.pdf.

[dc6] R. Longo (2004), *Operators algebras and Index Theorems in Quantum Field Theory*, Andrejevski lectures in Göttingen 2004 (lecture notes).

[dc7] R. Haag and D. Kastler (1964), *An Algebraic Approach to Quantum Field Theory*, Journal of Mathematical Physics, 5: 848-861.

[dc8] B. Schroer (2001), *Lectures on Algebraic Quantum Field Theory and Operator Algebras*, http://xxx.lanl.gov/abs/math-ph/0102018.

[dc9] H. J. Borchers (2000), *On Revolutionizing QFT with Tomita's Modular Theory* http://www.lqp.uni-goettingen.de/papers/99/04/99042900.html. J. Math. Phys. 41 (2000) 3604-3673.

B-4.4 Geometric quantization

[dd1] N. M. J. Woodhouse(1997), *Geometric Quantization*, Second Edition, Oxford University.

B-4.5 Symmetries and supersymmetries

[de1] H. Weyl (1931) *Theory of Groups and Quantum Mechanics*, Princeton, NJ: Princeton University.

[de2] E. P. Wigner (1959) *Group Theory and its Application to the Quantum Mechanics of Atomic Spectra*, New York: Academic.

[de3] I. M. Gelfand, R. A. Minklos and Z. Ya. Shapiro (1963), *Representations of the rotation and Lorentz groups and their applications*, Pergamon Press.

[de4] M. A. Naimark (1964), *Linear Representations of the Lorentz Group*, Oxford: Pergamon.

[de5] M. Chaichian and R. Hagedorn (1998), *Symmetries in Quantum Mechanics*, Institute of Physics Publishing.

[de6] C. Itzykson, H. Saleur, J.-B. Zuber (Editors) (1988), *Conformal Invariance and Applications to Statistical Mechanics*, Word Scientific.

[de7] A. A. Kehagias and P. A. Meessen(1994), Phys. Lett. B 324, p. 20.

[de8] S. J. Gates M. T. Grisaru, M. Rocek, and W. Siegel (1983), *Superspace: Or One Thousand and One Lessons in Supersymmetry*, London: Benjamin-Cummings.

B-4.6   Kac-Moody algebras and loop groups

[df1] F. Gliozzi, D. Olive, and J. Scherk (1977), Phys. Lett. 65B, 282; Nucl. Phys. B122, 253.

[df2] P. Goddard and D. Olive (1986), *Kac-Moody and Virasoro algebras in relation to Quantum Physics* . Preprint DAMTP 86-5.

[df3] P. Goddard, A. Kent, D. Olive (1986), Commun. Math. Phys 103, 105-119.

[df4] P. Goddard and D. Olive (1986), *The Vertex Operator Construction for Non-Simply-Laced Kac-Moody Algebras I,II* in *Topological and Geometrical Methods in Field Theory*, Eds. J. Hietarinta and J. Westerholm. Word Scientific.

[df5] E. Witten (1987), *Coadjoint orbits of the Virasoro Group*, PUPT-1061 preprint.

[df6] D. S. Freed (1985), *The Geometry of Loop Groups*(Thesis). Berkeley: University of California.

B-4.7   Anomalies, current algebras and groups

[dg1] J. Mickelson (1989), *Current Algebras and Groups*. Plenum Press, New York.

[dg2] J. Mickelson (2002), *Gerbes, (Twisted) K-Theory, and the Supersymmetric WZW Model*, hep-th/0206139.

[dg3] J. L. Manes (1986), *Anomalies in Quantum Field Theory and Differential Geometry* Ph.D. Thesis LBL-22304.

[dg4] L. D. Faddeev (1984), *Operator Anomaly for Gauss Law*. Phys. Lett. Vol 145 B, no 1, 2.

[dg5] R. Jackiw (1983), in *Gauge Theories of Eighties*, Conference Proceedings, Äkäslompolo, Finland (1982) Lecture Notes in Physics, Springer Verlag.

B-4.8   Category theory and quantum physics

[dh1] B. Mitchell (1965), *Theory of Categories*. Academic Press.

[dh2] H. Schubert (1972), *Categories*. Springer-Verlag, New York, Heidelberg.

[dh3] C. J. Isham and J. Butterfield (1999), *Some Possible Roles for Topos Theory in Quantum Theory and Quantum Gravity*. arXiv:gr-gc/9910005 .

B-5   Von Neumann algebras and related topics

B-5.1   Von Neumann algebras

[ea1] F. J. Murray, J. von Neumann (1936), *On Rings of Operators*, Ann. Math. 37,116-229.
F. J. Murray, J. von Neumann (1937), *On Rings of Operators II*, Trans. Am. Math. Soc. 41, 208-248.
F. J. Murray, J. von Neumann (1940), *On Rings of Operators III*, Ann. Math. 41,94-161.
F. J. Murray, J. von Neumann (1943), *On Rings of Operators IV*, Ann. Math. 44,716-808.

[ea2] J. von Neumann (1937), *Quantum Mechanics of Infinite Systems*, first published in (Redei and Stöltzner 2001, 249-268). [A mimeographed version of a lecture given at Pauli's seminar held at the Institute for Advanced Study in 1937, John von Neumann Archive, Library of Congress, Washington, D.C.]

[ea3] M. Redei and M. Stöltzner (eds) (2001), *John von Neumann and the Foundations of Quantum Physics*, Vol. 8, Dordrecht: Kluwer Academic Publishers. V. S. Sunder (1987), *An Invitation to von Neumann Algebras*, New York: Springer-Verlag.

[ea4] J. Dixmier (1981), *Von Neumann Algebras*, Amsterdam: North-Holland Publishing Company. [First published in French in 1957: Les Algebres d'Operateurs dans l'Espace Hilbertien, Paris: Gauthier-Villars.]

[ea5] F. M. Goodman, P. de la Harpe and V. F. R. Jones (1989), *Coxeter graphs and towers of algebras*, Springer-Verlag.

[ea6] V. F. R. Jones(2000), *The planar algebra of a bipartite graph*, in "Knots in Hellas'98", World Sci. Publishing (2000), 94-117.

[ea7] A. Connes (1973), *Une classification des facteurs de type III*, Ann. Sci. Ecole Norm. Sup. 6, 133-2

[ea8] V. F. R. Jones (1983), *Index for Subfactors*, Invent. Math. (72),1-25.

[ea9] Y. Ito and I. Nakamura(1996) *Hilbert schemes and simple singularities*, Proc. of EuroConference on Algebraic Geometry, Warwick 1996, ed. by K. Hulek et al., CUP, (1999),pp. 151–233, http://www.math.sci.hokudai.ac.jp/ nakamura/ADEHilb.pdf.

B-5.2 Braid groups and topological quantum field theories

[eb1] C. N. Yang, M. L. Ge (1989), *Braid Group, Knot Theory, and Statistical Mechanics*, World Scientific.

[eb2] C. N. Yang and M. L. Ge (1989), *Braid Group, Knot Theory, and Statistical Mechanics*, World Scientific.

[eb3] V. F. R. Jones (1983), *Braid groups, Hecke algebras and type II_1 factors*, Geometric methods in operator algebras, Proc. of the US-Japan Seminar, Kyoto, July 1983.

[eb4] V. F. R. Jones, *Hecke algebra representations of braid groups and link polynomial*, Ann. Math., 126(1987), 335-388.

[eb5] V. F. R. Jones (1987), *Hecke algebra representations of braid groups and link polynomial*, Ann. Math., 126, 335-388.

B-5.3 Quantum groups

[ec1] M. Chaichian and A. Demichev (1996), *Introduction to Quantum Groups*, Singapore: World Scientific.

[ec2] S. Sawin (1995), *Links, Quantum Groups, and TQFT's*, q-alg/9506002.

[ec3] E. Witten 1989), *Quantum field theory and the Jones polynomial*, Comm. Math. Phys. 121 , 351-399.

[ec4] C. Kassel (1995), *Quantum Groups*, Springer Verlag.

[ec5] V. Jones (2003), *In and around the origin of quantum groups*, arXiv:math.OA/0309199.

[ec6] H. Saleur (1990), *Zeroes of chromatic polynomials: a new approach to the Beraha conjecture using quantum groups*, Comm. Math. Phys. 132, 657.

B-5.4 Noncommutative geometry

[ed1] A. Connes (1994), *Non-commutative Geometry*, San Diego: Academic Press.

[ed2] A. Connes and D. Kreimer (1998), *Hofp algebras, renormalization, and non-commutative geome-try*, Quantum Field Theory; Perspective and Prospective (C. DeWitt-Morette and J.-B.- Zueber, eds.), Kluwer, Dortrecth, 1999, 59-109. CMP 2000:05, Commun. Math. Phys. 199, 203.242(1998). MR 99h:81137.
Ibid (1999), *Renormalization in quantum field theory and the Riemann-Hilbert problem I: the Hopf algebra structure of graphs and the main theorem*, arXiv:hep-th/9912092.
Ibid (2000), *Renormalization in quantum field theory and the Riemann-Hilbert problem II: the β function, diffeomorphisms and the renormalization group*, arXiv:hep-th/0003188.

[ed3] A. Connes (2005), *Quantum Fields and Motives*, hep-th/0504085.

[ed4] M. Kontsevich (1999), *Operads and Motives in Deformation Quantization*, arXiv: math.QA/9904055.

[ed5] P. Cartier (2001), *A Mad Day's Work: From Grothendienck to Connes and Kontsevich: the Evolution of Concepts of Space and Symmetry*, Bulletin of the American Mathematical Society, Vol 38, No 4, pp. 389-408.

B-5.5   Physical applications

[ee1] N. H. V. Temperley and E. H. Lieb (1971), *Relations between the percolation and colouring problem and other graph-theoretical problems associated with regular planar lattices:some exact results for the percolation problem*, Proc. Roy. Soc. London 322 (1971), 251-280.

[ee2] C. Gomez, M. Ruiz-Altaba, G. Sierra (1996), *Quantum Groups and Two-Dimensional Physics*, Cambridge University Press.

[ee3] O. Bratteli and D. W. Robinson (1979-1981), *Operator Algebras and Quantum Statistical Mechanics*, Volumes 1-2, New York: Springer-Verlag.

[ee4] R. E. Cutkosky (1960), J. Math. Phys. 1:429-433.

[ee5] F. Wilzek (1990), *Fractional Statistics and Anyon Super-Conductivity*, World Scientific.

[ee6] S. M. Girvin (1999), *Quantum Hall Effect, Novel Excitations and Broken Symmetries*, cond-mat/9907002.

[ee7] R. B. Laughlin (1990), Phys. Rev. Lett. 50, 1395.

[ee8] Y. Zhang (2004), *Hopf Algebraic Structures in the Cutting Rules*, hep-th/0408042.

[ee9] D. J. Broadhurst and D. Kreimer (1996), *Association of multiple zeta values with positive knots via Feynman diagrams up to 9 loops*, arXiv: hep-th/9609128 .

B-5.6   Topological quantum computation

[ef1] M. Freedman, H. Larsen, and Z. Wang (2002), *A modular functor which is universal for quantum computation*, Found. Comput. Math. 1, no 2, 183-204. Comm. Math. Phys. 227, no 3, 605-622. quant-ph/0001108.
M. H. Freedman (2001), *Quantum Computation and the localization of Modular Functors*, Found. Comput. Math. 1, no 2, 183-204.
M. H. Freedman (1998), *P/NP, and the quantum field computer*, Proc. Natl. Acad. Sci. USA 95, no. 1, 98-101.
A. Kitaev (1997), Annals of Physics, vol 303, p.2. See also *Fault tolerant quantum computation by anyons*, quant-ph/9707021.
L. H. Kauffmann and S, J. Lomonaco Jr. (2004), *Braiding operations are universal quantum gates*, arxiv.org/quant-ph/0401090.
Paul Parsons (2004) , *Dancing the Quantum Dream*, New Scientist 24. January. www.newscientist.com/hottopics.

[ef2] M. Freedman, A. Kitaev, M. Larson, Z. Wang (200?), www.arxiv.of/quant-ph/0101025.

[ef3] M. Freedman, H. Larsen, and Z. Wang (2002), *A modular functor which is universal for quantum computation*, Found. Comput. Math. 1, no 2, 183-204. Comm. Math. Phys. 227, no 3, 605-622. quant-ph/0001108.

[ef4] L. H. Kauffmann and S, J. Lomonaco Jr. (2004), *Braiding operations are universal quantum gates*, arxiv.org/quant-ph/0401090.

B-6   Quantum field theories

[fa1] J. Björken, S. Drell (1965), *Relativistic Quantum Fields*. Mc-Graw-Hill, New York.

[fa2] R. P. Feynman (1972),*Photon-Hadron Interactions*, W.A. Benjamin Inc.

[fa3] C. Iztykson, J. B. Zuber (1980), *Quantum Field Theory*. New York: Mc-Graw-Hill.

[fa4] K. Huang (1982), *Quarks,Leptons & Gauge Fields*. World Scientific.

[fa5] A. Zee (1982), *The Unity of Forces in the Universe* World Science Press, Singapore.

[fa6] F. E.C lose (1979), *An Introduction to Quarks and Partons*, Academic Press.

[fa7] L. B. Okun (1982), *Leptons and Quarks*, North-Holland Publishing Company.

[fa8] K. Huang (1992) *Quarks, Leptons and Gauge Fields*, World Scientific.

[fa9] J. Wess and J. Bagger (1983),*Supersymmetry and Supergravity*, Princeton, NJ: Princeton University.

[fa10] S. M. Christensen (1984), *Quantum Theory of Gravity*, Adam Hilger Ltd.

[fa11] S. Weinberg (1995), *Quantum Theory of Fields*, New York: Cambridge University Press.

B-7   Conformal field theories, strings, and superstrings

B-7.1   Books about conformal field theories and strings

[ga1] J. H. Schwartz (ed) (1985), *Super strings. The first 15 years of Superstring Theory*. World Scientific.

[ga2] M. B. Green, J. H Schwartz, and E. Witten (1987), *Superstring Theory*. Cambridge University Press.

[ga3] M. Kaku (1991),*Strings, Conformal Fields and Topology*, Springer Verlag.

[ga4] A. M. Polyakov (1987), *Gauge Fields and Strings*. Harwood Academic Publishers.

B-7.2   Articles about strings and superstrings

[gb1] P. Olesen (1985), Phys. Lett. B 160, 144.

[gb2] M. M. Bowick and S. G. Rajeev (1987), *The holomorphic geometry of closed bosonic string theory and $Diff(S^1)/S^1$*, Nucl. Phys. B293, 348-384.

[gb3] S. B. Giddings (1989), *Strings at Hagedorn temperature*, Phys. Lett. B vol 226, no 1, 2.

[gb4] P. S. Aspinwall, B. R. Greene, and D. R. Morrison (1993), *Calabi-Yau Moduli Space, Mirror Manifolds, and Space-time Topology Change in String Theory*, hep-th/9309097.

[gb5] J. M. Maldacena (1997), *The Large N Limit of Superconformal Field Theories and Supergravity*, hep-th/9711200.

[1] Gavedcki K. Gawedzki (1999), *Conformal field theory: a case study*, hep-th/9904145.

[gb7] T. Ericson and J. Rafelski (2002), *The tale of the Hagedorn temperature*, Cern Courier, vol 43, No 7,
http://www.cerncourier.com/main/toc/43/7.

[gb8] W. Broniowski (2002), *Two Hagedorn temperatures*, hep-ph/0006020.

[gb9] H. Hori *et al* (2003), *Mirror Symmetry*, American Mathematical Society.

[gb10] M. Marino (2004), *Chern-Simons Theory and Topological Strings*, arXiv:hep-the/0406005.

B-8   Particle Physics

B-8.1   Neutrinos

[ha1] R. Davis Jr. *et al*(1988), Phys. Rev. Lett., 20, 1205.

[ha2] Decamp *et al* (Aleph Collaboration)(1989), CERN-EP/89-141, preprint.

[ha3] D. Denegri *et* (1989), *The number of neutrino species.* CERN EP/89-72, preprint.

[ha4] K. S. Hirata *et al* (1989), Phys. Rev. Lett., 63, 16

[ha5] A. I. Abazov *et al*(1991) , Phys. Rev. Lett. 67, 24, 3332.

[ha6] P. Anselman *et al*(1992) , Phys. Lett. B

[ha7] L. Borodovsky *et al* (1992), Phys. Rev. Lett. 68, p. 274.

[ha8] Ch. Weinheimer *et al* (1993), Phys. Lett. 300B, 210. 285, 376.

[ha9] A. D. Dolgov and I.Z. Rothstein (1993), Phys. Rev. Lett. vol 71, No 4.

[ha10] W. C. Louis (1994), in Proceedings of the XVI Conference on Neutrino Physics and Astrophysics, Eilat, Israel.

[ha11] Y. Fukuda *et al* (1994), Phys. Lett. B 335, p. 237.

[ha12] Q. -Z. Qian and G. M. Fuller (1995), Phys. Rev. D 51, 1479.

[ha13] C. Athanassopoulos *et al* (LSND collaboration) (1996),*Evidence for Neutrino Oscillations from Muon Decay at Rest*, nucl-ex/9605001.

[ha14] V. M. Lobashev *et al*(1996), in *Neutrino 96* (Ed. K. Enqvist, K. Huitu, J. Maalampi). World Scientific, Singapore.

[ha15] C. Seife (1998), New Scientist, No 2151, Sept. 12, p. 4.

[ha16] Q. R Ahmad *et al*(2002), Phys. Rev. Lett. 89 011301, nucl-ex/0204008.

[ha17] R. J. Wilkes (K2K)(2002), ECONF C020805, TTH02, hep-ex/0212035.

[ha18] M. B. Smy (Super-Kamiokande)(2003), Nucl. Phys. Proc. Suppl. 118, 25. hep-ex/0208004.

[ha19] K. Eguchi *et al* (KamLAND) (2003), Phys. Rev. Lett. 90, 021802, hep-ex/0212021.

[ha20] D. B. Kaplan, A. E. Nelson and N. Weiner (2004), *Neutrino Oscillations as a Probe of Dark Energy*,hep-ph/0401099.

B-8.2   Proton spin crisis

[hb1] M. J. Alguard *et al* (1978), Phys. Rev. Lett. 41, 70; G. Baum *et al* (1983), Phys. Rev.Lett. 51, 1135.

[hb2] J. Ashman *et al* (1988), Phys. Lett. B 206, 364;
J. Ashman *et al* (1989), Nucl. Phys. B 328, 1.

[hb3] Bo-Ciang Ma (2000), *The spin structure of the proton*, RIKEN Review No. 28.

[hb4] X. Zheng *et al* (2004), The Jefferson Lab Hall A Collaboration, *Precision Measurement of the Neutron Spin Asymmetries and Spin-Dependent Structure Functions in the Valence Quark Region*, arXiv:nucl-ex/0405006 .

B-8.3   Top quark

[hc1] T. Smith(2003), *Truth Quark, Higgs, and Vacua*,
http://www.innerx.net/personal/tsmith/TQvacua.html .

[hc2] F. Abe *et al* (1944), Phys. Rev. Lett. 73, 1.

[hc3] F. Abe *et al* (1995), Phys. Rev. Lett. 74, 14, p.2626.

[hc4] F. Abe *et al* (1995), Phys. Rev. Lett. 74,3,p.343

[hc5] F. Abe$_4$ *et al*(1996), hep-ex/9601008.

[hc6] F. Abe$_5$ *et al* (1993), Phys. Rev. Lett. 70, 1376.

B-8.4   Higgs boson

[hd1] U. Egede (1998) *The search for a standard model Higgs at the LHC and electron identification using transition radiation in the ATLAS tracker*, thesis
http://www.quark.lu.se/ atlas/thesis/egede/thesis-node10.html.

[hd2] D. Wackeroth (2004), *NLO QCD Predictions for Hadronic Higgs Production with Heavy Quarks*, FermiLab Theory Seminar, Batavia, July 2004,
http://ubpheno.physics.buffalo.edu/ dow/fermilab2004.pdf.

B-8.5   Bound states of gluons

[he1] B. H. Behrens *et al* (1998), Phys. Rev. Lett. 80, 3710;
T. E. Browder *et al* (1998), Phys. Rev. Lett. 81, 1786.

[he2] D. Atwood and A. Somi (1997), JLAB-THY-97;
ibid (1997), Phys. Rev. Lett. 79, 5206. See also
H. W. Pfaff (2000), η' *Meson Production*.
http://pcweb.physik.uni-giessen.de/disto/papers/etaprime.htm.

B-8.6   Supersymmetry

[hf1] J. Ellis, J. L. Lopez and D. V. Nanopoulos(1995), hep-ph/9512288.

[hf2] G. R. Farrar (1995), *New Signatures of Squarks*, hep-ph/9512306.

[hf3] P. Fayet (1977), Phys. Lett. 70 B, 4, p.461.

[hf4] P. Fayet (1986), Phys. Lett. 175 B, 4, p. 471.

[hf5] S. Ambrosanio *et al*(1996), hep-ph/9602239.

B-8.7 Pomeron

[hg1] N. M. Queen, G. Violini (1974), *Dispersion Theory in High Energy Physics*, The Macmillan Press Limited.

[hg2] A. M. Smith *et al*(1985), Phys. Lett. B 163, p. 267.

[hg3] M. Derrick*et al* (1993), Phys. Lett B 315, p. 481.

[hg4] A. Brandt *et al* (1992), Phys. Lett. B 297, p. 417.

[hg6] P. E. Schlein (1994), Phys. Lett. B, p. 136.

B-8.8 Laboratory evidence for particles which should not exist

[hh1] A. Bhatti, for the CDF and D0 collaborations (1995), FNAL-CONF-95/192-E, Presented at Topical Workshop on Proton-Antiproton Collider Physics, Batavia, IL, 9-13, May 1995.

[hh2] L3 collaboration (1992), *High mass photon pairs in $l^+l^-\gamma\gamma$ events at LEP*, Phys. Lett. B., vol 295, No 3, 4.

[hh3] A. T. Goshaw *et al*(1979), Phys. Rev. Lett. 43, 1065.

[hh4] T. Akesson *et al*(1987), Phys. Lett. B192, 463, T. Akesson *et al*(1987), Phys. Rev. D36, 2615.

[hh5] J. Schweppe *et al*(1983), Phys. Rev. Lett. 51, 2261.

[hh6] P. V. Chliapnikov *et al*(1984), Phys. Lett. B 141, 276.

[hh7] S. Barshay (1992) , Mod. Phys. Lett. A, Vol 7, No 20, p. 1843.

[hh8] H1 Collaboration (1994) ,*Observation of an $e^+p \rightarrow \mu^+X$ Event with High Transverse Momenta at HERA*, DESY 94-248 preprint.

[hh9] H. Waschmuth (CERN, for the Aleph collaboration) (1996), *Results from e^+e^- collisions at 130, 136 and 140 GeV center of mass energies in the ALEPH Experiment http : //alephwww.cern.ch/ALPUB/pub/pub$_9$6.html*.

[hh10] G. Altarelli *et al* (1996), CERN-TH/96-20, hep-ph/9601324.

[hh11] F. M. Marquez *et al* (2003), Phys. Rev. C65, 044006. See also E. Samuel, *Ghost in the Atom*, New Scientist, vol 176, issue 2366 - 26 October 2002, page 30.

B-8.9 Evidence for exotic particles from cosmic ray spectrum

[hi1] J. Wdowczyk (1965), Proc. 9th Int. Conf. Cosmic Rays, Vol 2 , p. 691.

[hi2] K. Greisen (1966), Phys. Rev. Lett. 16, 748.

[hi3] Pamir Collaboration (1979), Proc. 16:th Intern. Cosmic Ray Conf. Vol. 7, p. 279.

[hi4] C. M. G. Lattes. Y. Fujimoto and S. Hasegava (1980), Phys. Rep., Vol. 65, No 3.

[hi5] J. Linsley and A. A. Watson (1981), Phys. Rev. Lett. 436, 459.

[hi6] Brazil-Japan Collaboration on Chacaltaya Emulsion Chamber Experiments (J. A. Chinellato et al), in Proton-Antiproton Collider Physics, 1981, Madison, Wis. (New York, N.Y. 1982)

[hi7] P. Ammiraju, E. Recami, W. A. Rodriguez Jr. (1983), *Chirons, Geminions, Centauros, Decays into Pions:a Phenomenological and Theoretical Analysis*, Il Nuovo Cimento, vol. 78 A, No 2, p. 173.

[hi8] M. Samorski and W. Stamm (1983), Astrophys. J. Lett. 268 L17.

[hi9] C. T. Hill and D. N. Schramm(1984), Phys. Rev. D 31, p. 564.

[hi10] R. Baltrusaitis *et al* (1984), Astrophys. J. 281, L9.
     R. Baltrusaitis *et al* (1985), Phys. Rev. Lett. 54, 1875.

[hi11] K. Goulianos (1987), Comm. Nucl. Part. Phys., Vol. 17, No. 4, p. 195.

[hi12] J. M. Bonnet-Bidaud and G. Chardin (1988), *Cygnus X-3, a critical review*, Phys. Rep. Vol 170, No 6.

[hi13] M. T. Ressell and M. S. Turner (1990), Comments Astrophys. 14, 323.

[hi14] P. Sokolsky, P. Sommers and B. R. Dawson (1992), *Extremely High Energy Cosmic Rays*, Phys. Rep. Vol. 217, No 5.

[hi15] T. K. Gaiser *et al* (1993), *Cosmic ray composition around* 10^{18} *eV* , Phys. Rev. D, vol 47, no 5.

[hi16] X. Chi, C. Dahanayake, J. Wdowczyk abd A. W. Wolfendale (1993), *Cosmic rays and cosmic strings* and *Gamma rays from collapsing cosmic strings*, Astroparticle Physics 1, 129-131 and 239-243.

[hi17] J. H. MacGibbon and R. H. Brandenberger (1993), *Gamma-ray signatures for ordinary cosmic strings*, Phys. Rev. D, Vol 47, No 6, p. 2883.

[hi18] J. D. Bjorken (1997), Acta Phys. Polonica B. V. 28. p. 2773.

[hi19] D. Gladysz-Dziadus (2003), *Are Centauros Exotic Signals of Quark-Gluon Plasma.* http://www1.jinr.ru/Archive/Pepan/v-34-3/v-34-3-3.pdf.

B-8.10   CP breaking in meson-antimeson systms

[hj1] J. Ellis, M. K. Gaillard, D. V. Nanopoulos (1976), Nucl. Phys B109, p. 213.

[hj2] J. F. Donoghue and B. R. Holstein (1983), Phys. Rev. D, Vol 29, No 9, p. 2088.

[hj3] J. Gasser and H. Luytweyler (1985), *Chiral perturbation theory: expansions the the mass of the strange quark*, Nucl. Phys. B 250, p. 465.

[hj4] E. A. Paschos and U. Turke (1989), Phys. Rep., Vol. 178, No 4.

[hj5] A. Ali and D. London (1995), Zeischrift fur Phys. C 65, p. 431.

[hj6] A. Bazarko (1996), http://xxx.lanl.gov/abs/hep-ex/9609005.

[hj7] D. Schaile (1994), Fortshcr. Phys. 42, 5.

[hj8] A. J. Buras (1994), Phys. Lett. B. 333, No 3,4, p. 476.

[hj9] G. Buchalla, A. J. Buras and M. Lautenbacher (1996), Rev. Mod. Phys. 68, p. 1125.

[hj10] A. J. Buras and R. Fleischer (1997), *Quark mixing, CP violation and Rare Decays After the Top Quark Discovery.* hep-ph/9704376.

[hj11] S. Bertolini, Marco Fabbrichesi and Jan. O. Eeg (1998). *Estimating* ϵ'/ϵ. *A review.* hep-ph/9802405.

[hj12] http://fnphyx-www.fnal.gov/experiments/ktev/epsprime/epsprime.html.

[hj13] http://fnphyx-www.fnal.gov/experiments/ktev/epsprime/epsprime.html.

[hj14] Y. Keum, U. Nierste and A. I. Sanda (1999), *A short look at* ϵ'/ϵ. hep-ph/9903230.

B-8.11   Possible new hadronic physics

[hk1] B. B. Back *et al*(2002), Nucl. Phys. A698, 416 (2002).

[hk2] B. B. Back *et al*(2002), Phys. Rev. Lett. Vol. 89, No 22, 25 November. See also http://www.scienceblog.com/community/modules.php?name=News&file=article&sid= 357.

[hk3] T. Ludham and L. McLerran (2003), *What Have We Learned From the Relativistic Heavy Ion Collider?*, Physics Today, October issue. http://www.physicstoday.org/vol-56/iss-10/p48.html. E. S. Reich (2005), *Black hole like phenomenon created by collider*, New Scientist 19, issue 2491.

[hk4] A. Gefter (2004), *Liquid Universe*, a popular article about the unexpected properties of recently discovered quark gluon plasma. New Scientist, Vol 184, No 2469 (16 October).

[hk5] E. S. Reich (2005), *Black hole like phenomenon created by collider*, New Scientist 19, issue 2491.

[hk6] H. Nastase (2005), *The RHIC fireball as a dual black hole*, hep-th/0501068.

B-9   Gravitation, cosmology, and astrophysics

B-9.1   Books related to gravitation, cosmology, and astrophysics

[ia1] S. Chandrasekhar (1961), *Hydrodynamic and Hydromagnetic Stability*, Oxford University Press.

[ia2] S. Weinberg (1967), *Gravitation and Cosmology*. Wiley, New York.

[ia3] L. D. Landau and E. M. Lifshitz (1970), *Statistical Physics*, Pergamon Press.

[ia4] C. W. Misner, K. S. Thorne, and J. A. Wheeler (1973), *Gravitation*. Freeman and Company, New York.

[ia5] S. A. Kaplan and V. N. Tsytovich (1973), *Plasma Astrophysics*. Pergamon Press.

[ia6] V. L. Ginzburg (1979), *Theoretical Physics and Astrophysics*. Pergamon Press.

[ia7] S. M. Fall and D. Lynden-Bell (1981), *The Structure and Evolution of Normal Galaxies*. Cambridge University Press, Cambridge.

[ia8] D. Mihalas and J. Binney (1981), *Galactic Astronomy: Structure and Kinematics*. Freeman & Company.

[ia9] W. C. Saslaw (1985), *Gravitational Physics of Stellar and Galactic Systems*. Cambridge University Press.

[ia10] N. Cohen (1988), *Gravity's Lens. Views of the New Cosmology*. John Wiley & Sons ,Inc..

[ia11] H. Zirin (1988), *Astrophysics of The Sun*, Cambridge University Press.

[ia12] S. P. Maran (Editor) (1992), *The Astronomy and Astrophysics Encyclopedia*, van Nostrand Reinhold , New York.

[ia13] J.-P. Luminet (1992), *Black Holes*, Cambridge University Press.

[ia14] S. L. Shapiro and S. A. Teukolsky (1986), *Black Holes, White Dwarfs and Neutron Stars*. John Wiley & Sons, Inc.

[ia15] L. F. Abbotand So-Young Pi (1986), *Inflationary Cosmology*. World Scientific.

B-9.2 String like structures in cosmology

[ib1] J. Einasto, A.Kaasik, E. Saar (1974), Nature 250 , 309.

[ib2] J. E. Gallagher and S. M. Faber (1979) Rev. of Astrophys., 17, 135.

[ib3] Ya. B. Zeldovich, J. Einasto and S. F. Shandarin (1982), *Giant Voids in the Universe*. Nature, Vol. 300, 2.

[ib4] A. Vilenkin (1985), *Cosmic Strings and Domain Walls*. Phys.Rep. 121, No 5 North-Holland, Amsterdam.

[ib5] T. W. B. Kibble (1985) Nucl. Phys. B 252, 227.

[ib6] W. Meikle *et al* (1987), Nature 329, 262.
P. Nisenson *et al* (1999), Astrophysics Journal 320, L15.

[ib7] R. H. Brandenberger and E. P. S. Shellard (1988), *Angular momentum and mass function of galaxies seeded by cosmic strings*. Preprint BROWN-HET -663.

[ib8] F. R. Bouchet and D. P. Bennet (1989), preprint PUPT-89-1128.

[ib9] P. Jacobsen *et al* (1993), Astrophysics Journal 369, L63.

[ib10] J. Einasto *et al* (1997), Nature, vol 385.

B-9.3 Cosmological constant, dark matter, and dark energy

[ic1] Y. Wang and M. Tegmark (2005). Phys. Rev. Lett. 92, 241302.
New Light on Dark Energy, http://physicsweb.org/articles/news/8/6/14.
S. Battersby (2005), *Dark energy: was Einstein right all along?*, New Scientist, 3 December, Vol. 188, No 2528.

[ic2] R. H. Koch, D. van Harlingen, and J. Clarke (1982), Phys. Rev. B 26, 74.

[ic3] M. Milgrom (1983), *A modification of the Newtonian dynamics as a possible alternative to the hidden mass hypothesis*, ApJ, 270, 365. See also http://www.astro.umd.edu/~ssm/mond/astronow.html.

[ic4] M. Tegmark *et al* (2003), *Cosmological Parameters from SDSS and WMAP*, arXiv:astro-ph/0310723.

[ic5] C. Boehm, D. Hooper, J. Silk, M. Casse(2003), *MeV Dark Matter: Has It Been Detected?*, arXiv:astro-ph/0309686. See also *Astronomers claim dark matter breakthrough*, New Scientist, 6 October, 2003. http://www.newscientist.com/news/news.jsp?id=ns99994214.

[ic6] J. Hogan, *Sun's halo linked to dark matter particles*, New Scientist, vol 182, issue 2443 - 17 April 2004, page 8.

[ic7] K. Zioutas *et al* (2004), *Quiet Sun X-rays as a Signature for New Particles*, arXiv: astro-ph/0403176.

[ic8] C. Beck and M. C. Mackey (2004), *Has Dark Energy been Measured in Lab?*, astro-ph/0406504.

[ic9] S. Clark, *First Dark Galaxy Found*, New Scientist 26 February 2005, vol 185, No 2488.

B-9.4   Accelerated cosmic expansion

[id1] S. Perlmutter *et al* (1996), astro-ph/9602122.

[id2] S. Perlmutter *et al* (1997), Ap. J. 483, 565.

[id3] Science, vol. 279, no. 5351 (January 30. 1998), p. 651. Article about the preliminary results announced by S. Perlmutter and his team.

[id4] A. G. Riess (2000), PASP, 112, 1284.

[id5] A. G. Riess *et al* (2004), *Type Ia Supernova Discoveries at $z > 1$ from the Hubble Space Telescope: Evidence for Past Deceleration and Constraints on Dark Energy Evolution.* arXiv: astro-ph/0402512 .

[id6] B. Schaefer (2006), *The hubble diagram to $z = 6.3$ with swift gamma ray bursts.* Abstract from the American Astronomical Society meeting in Washington, D.C., for talk 157.06 on 11 January 2006. http://www.phys.lsu.edu/GRBHD/. See also http://cosmicvariance.com/2006/01/11/evolving-dark-energy/.

[id7] M. Chown (2005) , *End of the Beginning*, New Scientist 2. July 2005, vol 187, No 2506. http://www.newscientist.com/article.ns?id=mg18625061.800.

B-9.5   Possibly varying fine structure constant

[ie1] J. Bekenstein (1982), Phys. Rev. D 25, 1527.

[ie2] J. K. Webb *et al* (2001), *Further Evidence for Cosmological Evolution of the Fine Structure Constant*, arXiv:astro-ph/0012539.
T. Chiva and K. Kohri (2001), *Quintessence Cosmology and Varying α*, arXiv:hep-ph/0111086.

[ie3] J. K. Webb, M. T. Murphy, V. V. Flambaum, V. A. Dzuba, J. D. Barrow, C. W. Churchill, J. X. Prochaska and A. M. Wolfe (2001), *Further Evidence for Cosmological Evolution of the Fine Structure Constant*, Phys. Rev. Lett., Vol. 87, No. 9, Paper No. 091301; August 27, 2001. arxiv.org/abs/astro-ph/0012539.

[ie4] T. Chiva and K. Kohri (2001), *Quintessence Cosmology and Varying α*, arXiv:hep-ph/0111086.

[ie5] R. Srianand, H. Chand, P. Petitjean and B. Aracil (2004), *Limits on the Time Variation of the Electromagnetic Fine-Structure Constant in the Low Energy Limit from Absorption Lines in the Spectra of Distant Quasars*, Phys. Rev. Lett., Vol. 92, Paper No. 121302; March 26, 2004. astro-ph/0402177.

[ie6] J. D. Barrow and J. K. Webb (2005), *Inconstant Constants Do the inner workings of nature change with time?*. Scientific American, June issue. http://www.sciam.com/article.cfm?chanID=sa006&colID=1&articleID= 0005BFE6-2965-128A-A96583414B7F0000.

B-9.6   Dark matter in astrophysical length scales

[if1] D. Da Roacha and L. Nottale (2003), *Gravitational Structure Formation in Scale Relativity*, astro-ph/0310036.

[if2] http://hyperphysics.phy-astr.gsu.edu/hbase/solar/soldata2.html.

[if3] *Some sunspot facts*, http://www.sunblock99.org.uk/sb99/people/KMacpher/properties.html.

[if4] N. Dume (2005), *New Exoplanet Defies Theory*, Physics Web, http://physicsweb.org/articles/news/9/7/6/1.

[if5] M. Moshina (2005), *The surface ferrite layer of Sun*,
http://www.thesurfaceofthesun.com/TheSurfaceOfTheSun.pdf.

[if6] J. Bahcall (2005), *Chemical Controversy at the Solar Surface*, Physics in Action, February 2005, vol 18, No 2.

[if7] M. Asplund, N. Grevesse, J. Sauval (2004) *The Solar Chemical Decomposition*, astro-ph/0410214.

B-9.7 Cosmic topology

[ig1] Y. Q. Chu and X. F. Zhu (1983). Astrophys. J., 271, 507.

[ig2] L. Z. Fang and H. Sato (1985), *Is the Periodicity in the Distribution of Quasar Red Shifts an Evidence of Multiple Connectedness of the Universe?*, Gen. Rel. and Grav. Vol 17 , No 11.

[ig3] C. L. Bennett *et al* (2003), *First Year Wilkinson Microwave Anisotropy Probe (WMAP1) Observations: Preliminary Maps and Basic Results*, Astrophys. J. Suppl. 148, 1-27.

[ig4] J.-P Luminet *et al* (2003), *Dodecahedral Space Topology as an Explanation for Weak Wide Angle Correlations in the Cosmic Microwave Background*, arXiv:astro-ph/0310253 .

[ig5] N. J. Cornish, D. N. Spergel, G. D. Starkman, and E. Komatsu. *Constraining the Topology of the Universe*, astro-ph/0310233.

[ig6] R. Aurich, S. Lustig, F. Steiner and H. Then (2004), *Hyperbolic Universes with a Horned Topology and the CMB Anisotropy*, Class. Quant. Grav. 21, 4901-4925. astro-ph/0403597.

[ig7] S. Battersby (2004), *Big Bang glow hints at funnel-shaped Universe*, New Scientist 15 April. Popular summary of the work of group of Steiner about Picard hyperbolic cosmology.
http://www.newscientist.com/article.ns?id=dn4879

[ig8] M. J. Reboucas and G. I. Gomero (2004), *Cosmic Topology: A Brief Overview*, arXiv:astro-ph/04102324 .

[ig9] J.-P. Luminet (2005), *A cosmic hall of mirrors*, Physics World, September.

B-9.8 Stars older than the Universe

[ih1] M. J. Pierce *et al* (1994), Nature 371, 385-389.

[ih2] W. L. Freedman *et al*(1994), Nature 371, 757-762.

[ih3] A. Saha *et al* (1995), Astrophys. J. Astrophys. J. 438, 826.

B-9.9 Cosmic microwave background

[ii1] A. H. Guth (1981), Phys. Rev. D 23, 347.

[ii2] V. G. Gurzadyan and A. A. Kocharayan (1992), Astronomy and Astrophysics, vol. B21, p. 19.

[ii3] I. Antoniadis, P. O. Mazur, and E. Mottola (1996), *Conformal Invariance and Cosmic Background Radiation*, astro-ph/9611208.

[ii4] B. Schwarzchild (2000), *The most recent balloon measurements on fluctuations in microwave background*, Physics Today, July.

[ii5] V. G. Gurzadyan (2003), *Kolmogorov Complexity, Cosmic Background Radiation, and Irreversibility*, arXiv:astro-ph/0312523.

[ii6] V. G. Gurzadyan *et al* (2004), *WMAP conforming the ellipticity in BOOMERanG and COBE CMB maps*, arXiv:astro-ph/0402399.
For a popular article see
A. Gefter (2005), *The riddle of time*, New Scientists, vol 188, no 2521, 15 October.

B-9.10   Gamma ray bursts

[ij1] I. G. Mitrofanov (1997), astro-ph/9707342.

[ij2] H. Muir (2000), *Trailblazer*, New Scientist, 23, September, No 2257.

[ij3] L. Wang *et al*(2002), *The Axially Symmetric Ejecta of Supernova 1987A*, Astrophys.J. 579, 671-677. arXiv:astro-ph/0205337.

[ij4] S. Akiyama and J. G. Wheeler (2002), *Magnetic Fields in Supernovae*, arXiv:astro-ph/0211458.

[ij5] W. Coburn and S. E. Boggs (2003), *Polarization of Prompt γ-ray emission from the γ-ray burst of 6 December 2002* , arXiv:astro-ph/0305377. Nature, 2003, May 22, 423, 415-417.
See also *Twisted secrets of gamma-ray bursts*, New Scientist, vol. 178 issue 2398 - 07 June 2003, page 25.

[ij6] S. Dado, A. Dar, A. De Rujula (2003), *The Supernova associated with GRB030329*, arXiv:/astro-ph/0304106.

[ij7] J. Hjorth *et al*(2003), *A very energetic supernova associated with the gamma-ray burst of 29 March 2003*, Nature, 423, 847.

[ij8] A. Middleditch (2003), *A White Dwarf Merger Paradigm for Supernovae and Gamma-Ray Bursts*, arXiv:astro-ph/0311484.

[ij9] L. Wang (2003), *Supernova Explosions: Lessons from Spectro-Polarimetry*, arXiv:astro-ph/0311299.

[ij10] D. A. Frail (2003), *Gamma-Ray Bursts: Jets and Energetics*, arXiv:astro-ph/0311301.

B-9.11   Experimental evidence for gravitational anomalies and long ranged weak forces in astrophysical length scales

[ik1] J. Tate *et al* (1989), Phys. Rev. Lett. 62 (8), 845-848.
Ibid. (1990), Phys. Rev. B 42(13), 7885-7893.

[ik2] M. Tajmar and C. J. de Matos (2006), *Local Photon and Graviton Mass and Its Consequences*, arXiv.org gr-gc 0603032.

[ik3] M. Tajmar *et al* (2006), *Experimental Detection of Gravimagnetic London Moment*,arXiv.org gr-gc 0603033. See also the popular article "Towards new tests of general relativity", at http://www.esa.int/SPECIALS/GSP/SEM0L6OVGJE_0.html.

[ik4] M. Allais (1959), *Should the Laws of Gravitation be modified?: Part I -Abnormalities in the Motion of a Paraconical Pendulum with Anisotropic Support*, Aero/Space Engineering, Sept 1959.
M. Allais (1959), *Should the Laws of Gravitation be modified?: Part II -Experiments in Connection with the Abnormalities the Motion of a Paraconical Pendulum with an Isotropic Support*, Aero/Space Engineering, Oct. 1959.
G. T. Gilles (1990), American J. of Physics, 58, 530 (review of work of Allais).

[ik5] J. D. Anderson *et al*(1998), Phys. Rev.Lett. Vol. 81, No 14,p. 2858.

[ik6] C. Oesterwinter and C. J. Cohen (1972), Cel. Mech. 5, 317.

[ik7] Y. B. Kolesnik (2000), *Applied Historical Astronomy, 24th meeting of the IAU*, Joint Discussion 6, Manchester, England.
Ibid (2001a), Journees 2000 *Systemes de reference spatio-temporels*, J2000, a fundamental epoch for origins of reference systems and astronomomical models, Paris.

[ik8] C. J. Masreliez (2001), *Do the planets accelerate*.
http://www.estfound.org.

[ik9] C. J. Masreliez (2001), *Expanding Space-Time Theory*,
       http://www.estfound.org.

[ik10] H. Mueller, *Global Scaling*,
        http://www.dr-nawrocki.de/globalscalingengl2.html .

B-9.12 Miscellaneous

[il1] C. J. Durrant (1988), *The Atmosphere of the Sun.* IOP Publishing Ltd..

[il2] BBC NEWS Science/Nature (2002), *Quakes reveal 'core within
       a core',* Wednesday, 2 October,
       http://news.bbc.co.uk/1/hi/sci/tech/2290551.stm .

B-10   Possible New Physics in condensed matter scale

B-10.1   Findings challenging standard view about chemistry

[ja1] W. D. Knight *et al* (1984), Phys.Rev. Lett. 52, 2141.

[ja2] P. Ball (2005), *A new kind of alchemy*, New Scientist, 16 April issue.
       http://www.newscientist.com/channel/fundamentals/mg18624951.800.

[ja3] A. W. Castleman *et al* (2005), *Al Cluster Superatoms as Halogens in Polyhalides and as Alkaline
       Earths in Iodide Salts*, Science, 307, 231-235.

B-10.2   Possible new physics related to time's arrow and light velocity

[jb1] D. M. Pepper (1982), *Nonlinear Optical Phase Conjugation*, in *Optical Engineering*, vol. 21, no.
       2, March/April.

[jb2] http://www.usc.edu/dept/ee/People/Faculty/feinberg.html.

[jb3] Akimov A.E., Kovalchuk G.U., Medvedev V.G., Oleinik V.K., Pugach A.F. "Predvaritelnyye
       rezultaty astronomicheskikh nabludenii po metodike N.A.Kozyreva.", Kiev, 1992, GAO AN
       Ukrainy, preprint # GAO-92-5R. (russian) ("Preliminary results of astronomical observations
       using N. A. Kozyrev's method.").

[jb4] V. Chernobrov (1996), *Experiments on the Change of Direction and Rate of Time Motion*, Proc.
       of III Int. Conf. On Problems of Space, Time, and Gravitation, Russian Academi of Sciences, St.
       Petersburg, Russia.

[jb5] Obolensky, A. (1988), Electronics and Wireless World.

[jb6] R. Y. Chiao (1998), *Tunneling Times and Super-luminality: a Tutorial.* Dept. of Physics, Univ.
       of California, Berkeley, CA 94720-7300, U.S.A. Noberber 6.
       R. Y. Chiao and A. M. Steinberg (1997), in Progress in Optics XXXVII, E. Wolf, ed., (Elsevier,
       Amsterdam), p. 345.

[jb7] D. Reed (2003), *Conceptual Hurdles to New Millennium Physics*, Explore!, vol 12, No 1.

B-10.3   Sonoluminescence,sonofusion, cold fusion

[jc1] B. R. Barber *et al* (1994), Phys. Rev. Lett. , Vol 72 , No 9, p, 1380.

[jc2] Jed Rothwell(1996).
Some recent developments in cold fusion,
http://ourworld.compuserve.com/homepages/JedRothwell/brieftec.htm.
Report on The Second International Low Energy Nuclear Reactions Conference Holiday Inn, College Station, Texas, September 13-14, 1996.
http://ourworld.compuserve.com/homepages/JedRothwell/ilenrc2s.htm,
Review of the Sixth International Conference on Cold Fusion (ICCF6),
http://ourworld.compuserve.com/homepages/JedRothwell/iccf6rev.htm

[jc3] E. Storms (1996), *Review of cold fusion effect.*
http://www.jse.com/storms/1.html.

B-10.4   Anomalous physics of water

[jd1] M. Chaplin (2005), *Water Structure and Behavior,*
http://www.lsbu.ac.uk/water/index.html.
For 41 anomalies see http://www.lsbu.ac.uk/water/anmlies.html.
For the icosahedral clustering see http://www.lsbu.ac.uk/water/clusters.html.
J. K. Borchardt(2003), *The chemical formula H2O - a misnomer,* The Alchemist 8 Aug (2003).
R. A. Cowley (2004), *Neutron-scattering experiments and quantum entanglement,* Physica B 350 (2004) 243-245.
R. Moreh, R. C. Block, Y. Danon, and M. Neumann (2005), *Search for anomalous scattering of keV neutrons from H2O-D2O mixtures,* Phys. Rev. Lett. 94, 185301.

[jd2] S. L. Glashow (1999), *Can Science Save the World?,*
http://www.hypothesis.it/nobel/nobel99/eng/pro/pro_2_1_1.htm.

[jd3] M. Chaplin (2000), *Molecular Vibration and Absorption,*
Online book. Selected Science Educators, London Southbank University,
http://www.lsbu.ac.uk/water/vibrat.html.

[jd4] C. Smith (2001), *Learning From Water, A Possible Quantum Computing Medium,* talk in CASYS'2001, 5th international conference on Computing Anticipating Systems held in Liege, Belgium, August 13-18. Abstract book published by Chaos.

B-10.5   Possible other new physics

[je1] K. S. Novoselov *et al* (2005), *Two-dimensional gas of massless Dirac fermions in graphene,* Letter Nature 438, 197-200 . (10 November).

[je2] M. Chown (2004), *Quantum Rebel,* popular article about the experiment of Shahriar Afshar. New Scientist, Vol 183, No 2457.

[je3] S. E. Shnoll *et al* (1998), *Realization of discrete states during fluctuations in macroscopic processes,* Uspekhi Fisicheskikh Nauk, Vol. 41, No. 10, pp. 1025-1035.

B-10.6   Rotating magnetic systems, over unity devices, etc.

[jf1] Hayasaka H., Takeuchi S. (1989), *Anomalous weight reduction on a gyroscopes right rotation around the vertical axis of the earth,*
Phys. Rev. Lett., # 25, p.2701.

[jf2] V. V. Roshchin and S. M. Godin (2001), *An Experimental Investigation of the Physical Effects in a Dynamic Magnetic System,* New Energy Technologies Issue #1 July-August 2001.

[jf3] P. LaViolette (2001), *How the Searl Effect Works: an Analysis of the Magnetic Energy Converter* , Neue Wasserstofftechnologien un Raumantriebe, Vortäge der Kongresses vom 23.-24. Juni, Weinfelden. Jupiter Verlag.

[jf4] E. Podkletnov and R. Nieminen (1992), Physica C 203 441. E. Podkletnov, "Weak gravitational shielding properties of composite bulk YBa2Cu3O7-x super-conductor below 70 K under electromagnetic field", report MSU-chem 95, improved version (cond-mat/9701074).

[jf5] E. Podkletnov and G. Modanese (2002), *Investigation of high voltage discharges in low pressure gases through large ceramic super-conducting electrodes,* http://xxx.lanl.gov/abs/physics/0209051.

[jf6] J.-L. Naudin *The Quest for Over-Unity.* The page contains material about over-unity devices, in particular magnetic motors. http://jnaudin.free.fr/.

[jf7] J. Naudin (2005), *Free Energy Atomic Hydrogen: the MAHG project,* http://jlnlabs.imars.com/mahg/tests/index.htm.

[jf8] I. Langmuir (1915), em The Dissociation of Hydrogen Into Atoms, Journal of American Chemical Society 37, 417.

[jf9] P. Kanarev (2002) *Water is New Source of Energy,* Krasnodar.

[jf10] S. Meyer (1996) *Water Fuel Cell,* International News Release, Issue No. 11A-Rv. S. Meyer (1990) *Method for Production of A Fuel Gas,* US Patent 4,936,961.

[jf11] J. B. Bateman (1978), *A Biologically Active Combination of Modulated Magnetic and Microwave Fields: the Priore Machine,* in "Office of Naval Research," London Report R-5- 78, Aug. 1978.

[jf12] R. Mills *et al*(2003), *Spectroscopic and NMR identification of novel hybrid ions in fractional quantum energy states formed by an exothermic reaction of atomic hydrogen with certain catalysts.* http://www.blacklightpower.com/techpapers.html .

[jf13] T. B. Bahler and G. Fazi (2002) *Force on an Asymmetric Capacitor,* http://membres.lycos.fr/jlnlabs/arl_fac/index.html.

[jf14] 'Monatomic' homepage. http://monatomic.earth.com/.

B-11   Physics, information, and self-organization

[ka1] H. Haken (1988), *Information and Self-Organization,* Springer Verlag, Berlin.

[ka2] H. Haken (1978), *Synergetics,* Springer Verlag.

[ka3] P. Glandsdorff and I. Prigogine (1991), *Thermodynamic Theory of Structure, Stability and Fluctuations,* Wiley, New York 1991.

B-11.1   Possible new physics related to living matter

[la1] A. Gurwitsch (1923), *Die Natur des Specifischen Erregurs der Zeliteilung,* Roux, Archiv: 100; 11. D. Downing (2001), *Daylight Robber - The importance of sunlight to health,* chapter 8. Online book at http://www.bio-immuno-development.com/books/daylight/924.htm.

[la2] V. M. Inyushin and P. R. Chekorov (1975), *Biostimulation through laser radiation and bioplasma,* Alma-Ata, Kazakh SSR.Translated into english in 1976.

[la3] F. A. Popp, B.Ruth, W. Bahr, J. Böhm, P. Grass (1981), G. Grolig, M.Rattemeyer, H. G. Schmidt and P. Wulle: *Emission of Visible and Ultraviolet Radiation by Active Biological Systems.* Collective Phenomena(Gordon and Breach), 3, 187-214.
F. A. Popp, K. H. Li, and Q. Gu (eds.) (1992), *Recent Advances in Bio-photon Research and its Applications.* World Scientific, Singapore-New Jersey.
F.- A. Popp: *Photon-storage in biological systems, in: Electromagnetic Bio-Information*, pp.123-149. Eds. F. A. Popp, G. Becker, W. L. Knig, and W.Peschka. Urban & Schwarzenberg, Mnchen-Baltimore.
F.-A. Popp (2001), *About the Coherence of Bio-photons,*
http://www.datadiwan.de/iib/ib0201e_.htm .
F.-A. Popp and J.-J. Chang (2001), *Photon Sucking and the Basis of Biological Organization,*
http://www.datadiwan.de/iib/ib0201e3.htm .
F.-A. Popp and Y. Yan (2001), *Delayed Luminescence of Biological Systems in Terms of States,*
http://www.datadiwan.de/iib/pub2001-07.htm .

[la4] M. Shaduri. & G.Tshitshinadze (1999), *On the problem of application of Bioenergography in medicine.* Georgian Engineering News 2, 109-112.
See also http://www.bioholography.org/ .

[la5] P. Gariaev *et al* (2000), "The DNA-wave-biocomputer", CASYS'2000, Fourth International Conference on Computing Anticipatory Systems, Liege, 2000. Abstract Book, Ed. M. Dubois.

[la6] P. P. Gariaev *et al*(2002), *The spectroscopy of bio-photons in non-local genetic regulation*, Journal of Non-Locality and Remote Mental Interactions, Vol 1, Nr 3.
http://www.emergentmind.org/gariaevI3.htm .

[la7] M. W. Ho (1993), *The Rainbow and the Worm*, World Scientific, Singapore.
M. W. Ho (1994), *Coherent Energy, Liquid Crystallinity and Acupuncture,*
http://www.consciousness.arizona.edu/quantum/Archives/Uploads/
mifdex.cgi?msgindex.mif.
M. W. Ho and P. T. Saunders(1994), *Liquid Crystalline Mesophase in living organisms*, in *Bio-electrodynamics and Biocommunication* (M. W Ho, F. A. Popp and U. Warnke, eds), World Scientific, Singapore.

[la8] N. Cherry (2000), *Conference report on effects of ELF fields on brain,*
http://www.tassie.net.au/emfacts/icnirp.txt .

[la9] W. A. Tiller, W. E. Dibble Jr., and M. J. Kohane(2001), *Conscious Acts of Creation: the Emergence of a New Physics*, p.17.
Pavior Publishing. http://www.pavior.com/.

[la10] S. Comorosan (1975), *On a possible biological spectroscopy*, Bull. of Math. Biol., Vol 37, p. 419.

[la11] S. Comorosan, M. Hristea, P. Murogoki (1980), *On a new symmetry in biological systems*, Bull. of Math. Biol., Vol 42, p. 107.

B-11.2 Consciousness and new physics

[ma1] D. J. Benor (2001), *Spiritual Healing: scientific validation of a healing revolution*, Vol. I. Vision publications, Southfield MI.

[ma2] M. Germine (2002), *Scientific Validation of Planetary Consciousness,*
Journal of Non-Locality and Remote Mental Interactions Vol.I Nr. 3.
http://www.emergentmind.org/germineI3.htm .

[ma3] S. Klein (2002), *Libet's Research on Timing of Conscious Intention to Act: A Commentary* of Stanley Klein, Consciousness and Cognition 11, 273-279.
http://cornea.berkeley.edu/pubs/ccog_2002_0580-Klein-Commentary.pdf.

[ma4] B. Libet, E. W. Wright Jr., B. Feinstein, and D. K. Pearl (1979), *Subjective referral of the timing for a conscious sensory experience* Brain, 102, 193-224.

B-12   Own publications

[n1] Pitkänen, M. (1983) International Journal of Theor. Phys. ,22, 575.

[n2] Pitkänen, M. (1988) Annalen der Physik. 45, 37, 235-248.

[n3] Pitkänen, M. (1983) International Journal of Theor. Phys. ,22, 575.

[n4] M. Pitkänen (2002), *A Strategy for Proving Riemann Hypothesis*, matharXiv.org/0111262.

[n5] M. Pitkänen (2003), *A Strategy for Proving Riemann Hypothesis*, Acta Math. Univ. Comeniae, vol. 72.

[n6] M. Pitkänen (2003), *Time, Space-Time, and Consciousness, Quantum Model for Sensory Receptor*, and *Quantum Model for Nerve Pulse, EEG, and ZEG*. Journal of nonlocality and remote mental interactions, issue 3.
http://www.emergentmind.org/journal.htm.

[n7] Articles about TGD inspired theory of consciousness in previous issues of Journal of Non-Locality and Remote Mental Interactions.
http://www.emergentmind.org.

B-13   Online Books About TGD

[TGDview] M. Pitkänen (2006), *Topological Geometrodynamics: Overview*.
http://www.physics.helsinki.fi/~matpitka/tgdview/tgdview.html.

[A1] The chapter *An Overview about the Evolution of Quantum TGD* of [TGDview].
http://www.physics.helsinki.fi/~matpitka/tgdview/tgdview.html#tgdevoI.

[A2] The chapter *An Overview about Quantum TGD* of [TGDview].
http://www.physics.helsinki.fi/~matpitka/tgdview/tgdview.html#tgdevoII.

[A3] The chapter *TGD and M-Theory* of [TGDview].
http://www.physics.helsinki.fi/~matpitka/tgdview/tgdview.html#MTGD.

[A4] M. Pitkänen (2006), *Quantum Physics as Infinite-Dimensional Geometry* of [TGDview].
http://www.physics.helsinki.fi/~matpitka/tgdview.html#tgdgeom.

[A5] The chapter *Construction of Quantum Theory* of [TGDview].
http://www.physics.helsinki.fi/~matpitka/tgdview/tgdview.html#quthe.

[A6] The chapter *Physics as a Generalized Number Theory* of [TGDview].
http://www.physics.helsinki.fi/~matpitka/tgdview/tgdview.html#tgdnumber.

[A7] The chapter *Cosmology and Astrophysics in Many-Sheeted Space-Time* of [TGDview].
http://www.physics.helsinki.fi/~matpitka/tgdview/tgdview.html#tgdclass.

[A8] The chapter *Elementary Particle Vacuum Functionals* of [TGDview].
http://www.physics.helsinki.fi/~matpitka/tgdview/tgdview.html#elvafu.

[A9] The chapter *Massless States and Particle Massivation* of [TGDview].
http://www.physics.helsinki.fi/~matpitka/tgdview/tgdview.html#mless.

[TGDgeom] M. Pitkänen (2006), *Quantum Physics as Infinite-Dimensional Geometry*.
http://www.physics.helsinki.fi/~matpitka/tgdgeom/tgdgeom.html.

[B1] The chapter *Identification of the Configuration Space Kähler Function* of [TGDgeom].
http://www.physics.helsinki.fi/~matpitka/tgdgeom/tgdgeom.html#kahler.

[B2] The chapter *Construction of Configuration Space Kähler Geometry from Symmetry Principles: Part I* of [TGDgeom].
http://www.physics.helsinki.fi/~matpitka/tgdgeom/tgdgeom.html#compl1.

[B3] The chapter *Construction of Configuration Space Kähler Geometry from Symmetry Principles: Part II* of [TGDgeom].
http://www.physics.helsinki.fi/~matpitka/tgdgeom/tgdgeom.html#compl2.

[B4] The chapter *Configuration Space Spinor Structure* of [TGDgeom].
http://www.physics.helsinki.fi/~matpitka/tgdgeom/tgdgeom.html#cspin.

[TGDquant] M. Pitkänen (2006), *Quantum TGD*.
http://www.physics.helsinki.fi/~matpitka/tgdquant/tgdquant.html.

[C1] The chapter *Construction of Quantum Theory* of [TGDquant].
http://www.physics.helsinki.fi/~matpitka/tgdquant/tgdquant.html#quthe.

[C2] The chapter *Construction of S-matrix* of [TGDquant].
http://www.physics.helsinki.fi/~matpitka/tgdquant/tgdquant.html#smatrix.

[C3] The chapter *Is it Possible to Understand Coupling Constant Evolution at Space-Time Level?* of [TGDquant].
http://www.physics.helsinki.fi/~matpitka/tgdquant/tgdquant.html#rgflow.

[C4] The chapter *Is it Possible to Understand Coupling Constant Evolution at Space-Time Level?* of [TGDquant].
http://www.physics.helsinki.fi/~matpitka/tgdquant/tgdquant.html#limit.

[C5] The chapter *Equivalence of Loop Diagrams with Tree Diagrams and Cancellation of Infinities in Quantum TGD* of [TGDquant].
http://www.physics.helsinki.fi/~matpitka/tgdquant/tgdquant.html#bialgebra.

[C6] The chapter *Was von Neumann Right After All* of [TGDquant].
http://www.physics.helsinki.fi/~matpitka/tgdquant/tgdquant.html#vNeumann.

[TGDclass] M. Pitkänen (2006), *Physics in Many-Sheeted Space-Time*.
http://www.physics.helsinki.fi/~matpitka/tgdclass/tgdclass.html.

[D1] The chapter *Basic Extremals of Kähler Action* of [TGDclass].
http://www.physics.helsinki.fi/~matpitka/tgdclass/tgdclass.html#class.

[D2] The chapter *General Ideas about Topological Condensation and Evaporation* of [TGDclass].
http://www.physics.helsinki.fi/~matpitka/tgdclass/tgdclass.html#topcond.

[D3] The chapter *The Relationship Between TGD and GRT* of [TGDclass].
http://www.physics.helsinki.fi/~matpitka/tgdclass/tgdclass.html#tgdgrt.

[D4] The chapter *Cosmic Strings* of [TGDclass].
http://www.physics.helsinki.fi/~matpitka/tgdclass/tgdclass.html#cstrings.

[D5] The chapter *TGD and Cosmology* of [TGDclass].
http://www.physics.helsinki.fi/~matpitka/tgdclass/tgdclass.html#cosmo.

[D6] The chapter *TGD and Astrophysics* of [TGDclass].
http://www.physics.helsinki.fi/~matpitka/tgdclass/tgdclass.html#astro.

[D7] The chapter *Macroscopic Quantum Phenomena and CP_2 Geometry* of [TGDclass].
http://www.physics.helsinki.fi/~matpitka/tgdclass/tgdclass.html#super.

[D8] The chapter *Hydrodynamics and CP_2 Geometry* of [TGDclass].
http://www.physics.helsinki.fi/~matpitka/tgdclass/tgdclass.html#hydro.

[TGDnumber] M. Pitkänen (2006), *TGD as a Generalized Number Theory*.
http://www.physics.helsinki.fi/~matpitka/tgdnumber/tgdnumber.html.

[E1] The chapter *TGD as a Generalized Number Theory: p-Adicization Program* of [TGDnumber].
      http://www.physics.helsinki.fi/∼matpitka/tgdnumber/tgdnumber.html#visiona.

[E2] The chapter *TGD as a Generalized Number Theory: Quaternions, Octonions, and their Hyper Counterparts* of [TGDnumber].
      http://www.physics.helsinki.fi/∼matpitka/tgdnumber/tgdnumber.html#visionb.

[E3] The chapter *TGD as a Generalized Number Theory: Infinite Primes* of [TGDnumber].
      http://www.physics.helsinki.fi/∼matpitka/tgdnumber/tgdnumber.html#visionc.

[E4] The chapter *p-Adic Numbers and Generalization of Number Concept* of [TGDnumber].
      http://www.physics.helsinki.fi/∼matpitka/tgdnumber/tgdnumber.html#padmat.

[E5] The chapter *p-Adic Physics: Physical Ideas* of [TGDnumber].
      http://www.physics.helsinki.fi/∼matpitka/tgdnumber/tgdnumber.html#phblocks.

[E6] The chapter *Fusion of p-Adic and Real Variants of Quantum TGD to a More General Theory* of [TGDnumber].
      http://www.physics.helsinki.fi/∼matpitka/tgdnumber/tgdnumber.html#mblocks.

[E7] The chapter *Category Theory, Quantum TGD, and TGD Inspired Theory of Consciousness* of [TGDnumber].
      http://www.physics.helsinki.fi/∼matpitka/tgdnumber/tgdnumber.html#categoryc.

[E8] The chapter *Riemann Hypothesis and Physics* of [TGDnumber].
      http://www.physics.helsinki.fi/∼matpitka/tgdnumber/tgdnumber.html#riema.

[E9] The chapter *Topological Quantum Computation in TGD Universe* of [TGDnumber].
      http://www.physics.helsinki.fi/∼matpitka/tgdnumber/tgdnumber.html#tqc.

[E10] The chapter *Intentionality, Cognition, and Physics as Number theory or Space-Time Point as Platonia* of [TGDnumber].
      http://www.physics.helsinki.fi/∼matpitka/tgdnumber/tgdnumber.html#intcognc.

[TGDpad] M. Pitkänen (2006), *p-Adic length Scale Hypothesis and Dark Matter Hierarchy*
      http://www.physics.helsinki.fi/∼matpitka/paddark/paddark.html.

[F1] The chapter *Elementary Particle Vacuum Functionals* of [TGDpad].
      http://www.physics.helsinki.fi/∼matpitka/paddark/paddark.html#elvafu.

[F2] The chapter *Massless States and Particle Massivation* of [TGDpad].
      http://www.physics.helsinki.fi/∼matpitka/paddark/paddark.html#mless.

[F3] The chapter *p-Adic Particle Massivation: Hadron Masses* of [TGDpad].
      http://www.physics.helsinki.fi/∼matpitka/paddark/paddark.html#padmass2.

[F4] The chapter *p-Adic Particle Massivation: Hadron Masses* of [TGDpad].
      http://www.physics.helsinki.fi/∼matpitka/paddark/paddark.html#padmass3.

[F5] The chapter *p-Adic Particle Massivation: New Physics* of [TGDpad].
      http://www.physics.helsinki.fi/∼matpitka/paddark/paddark.html#padmass4.

[F6] The chapter *Topological Condensation and Evaporation* of [TGDpad].
      http://www.physics.helsinki.fi/∼matpitka/paddark/paddark.html#padaelem.

[F7] The chapter *The Recent Status of Leptohadron Hypothesis* of [TGDpad].
      http://www.physics.helsinki.fi/∼matpitka/paddark/paddark.html#leptc.

[F8] The chapter *TGD and Nuclear Physics* of [TGDpad].
      http://www.physics.helsinki.fi/∼matpitka/paddark/paddark.html#padnucl.

[F9] The chapter *Dark Nuclear Physics and Living Matter* of [TGDpad].
      http://www.physics.helsinki.fi/∼matpitka/paddark/paddark.html#exonuclear.

[F10] The chapter *Super-Conductivity in Many-Sheeted Space-Time* of [TGDpad].
http://www.physics.helsinki.fi/~matpitka/paddark/paddark.html#supercond.

[TGDfree] M. Pitkänen (2006), *TGD and Fringe Physics*.
http://www.physics.helsinki.fi/~matpitka/freenergy/freenergy.html.

[G1] The chapter *Anomalies Related to the Classical Z^0 Force and Gravitation* of [TGDfree].
http://www.physics.helsinki.fi/~matpitka/freenergy/freenergy.html#Zanom.

[G2] The chapter *The Notion of Free Energy and Many-Sheeted Space-Time Concept* of [TGDfree].
http://www.physics.helsinki.fi/~matpitka/freenergy/freenergy.html#freenergy.

[G3] The chapter *Did Tesla Discover the Mechanism Changing the Arrow of Time?* of [TGDfree].
http://www.physics.helsinki.fi/~matpitka/freenergy/freenergy.html#tesla.

[G4] The chapter *Ufos, Aliens, and the New Physics* of [TGDfree].
http://www.physics.helsinki.fi/~matpitka/freenergy/freenergy.html#mantleufo.

[TGDconsc] M. Pitkänen (2006), *TGD Inspired Theory of Consciousness*.
http://www.physics.helsinki.fi/~matpitka/tgdconsc/tgdconsc.html.

[H1] The chapter *Matter, Mind, Quantum* of [TGDconsc].
http://www.physics.helsinki.fi/~matpitka/tgdconsc/tgdconsc.html#conscic.

[H2] The chapter *Negentropy Maximization Principle* of [TGDconsc].
http://www.physics.helsinki.fi/~matpitka/tgdconsc/tgdconsc.html#nmpc.

[H3] The chapter *Self and Binding* of [TGDconsc].
http://www.physics.helsinki.fi/~matpitka/tgdconsc/tgdconsc.html#selfbindc.

[H4] The chapter *Quantum Model for Sensory Representations*
of [TGDconsc].
http://www.physics.helsinki.fi/~matpitka/tgdconsc/tgdconsc.html#expc.

[H5] The chapter *Time and Consciousness* of [TGDconsc].
http://www.physics.helsinki.fi/~matpitka/tgdconsc/tgdconsc.html#timesc.

[H6] The chapter *Quantum Model of Memory* of [TGDconsc].
http://www.physics.helsinki.fi/~matpitka/tgdconsc/tgdconsc.html#memoryc.

[H7] The chapter *Conscious Information and Intelligence* of [TGDconsc].
http://www.physics.helsinki.fi/~matpitka/tgdconsc/tgdconsc.html#intsysc.

[H8] The chapter *p-Adic Physics as Physics of Cognition and Intention* of [TGDconsc].
http://www.physics.helsinki.fi/~matpitka/tgdconsc/tgdconsc.html#cognic.

[H9] The chapter *Quantum Model for Paranormal Phenomena*
of [TGDconsc].
http://www.physics.helsinki.fi/~matpitka/tgdconsc/tgdconsc.html#parac.

[H10] The chapter *TGD Based Model for OBEs* of [TGDconsc].
http://www.physics.helsinki.fi/~matpitka/tgdconsc/tgdconsc.html#OBE.

[TGDselforg] M. Pitkänen (2006), *Bio-Systems as Self-Organizing Quantum Systems*.
http://www.physics.helsinki.fi/~matpitka/bioselforg/bioselforg.html.

[I1] The chapter *Quantum Theory of Self-Organization* of [TGDselforg].
http://www.physics.helsinki.fi/~matpitka/bioselforg/bioselforg.html#selforgac.

[I2] The chapter *Possible Role of p-Adic Numbers in Bio-Systems* of [TGDselforg].
http://www.physics.helsinki.fi/~matpitka/bioselforg/bioselforg.html#biopadc.

[I3] The chapter *Biological Realization of Self Hierarchy* of [TGDselforg].
http://www.physics.helsinki.fi/~matpitka/bioselforg/bioselforg.html#bioselfc.

[I4] The chapter *Quantum Control and Coordination in Bio-systems: Part I* of [TGDselforg].
http://www.physics.helsinki.fi/~matpitka/bioselforg/bioselforg.html#qcococI.

[I5] The chapter *Quantum Control and Coordination in Bio-Systems: Part II* of [TGDselforg].
http://www.physics.helsinki.fi/~matpitka/bioselforg/bioselforg.html#qcococII.

[TGDware] M. Pitkänen (2006), *Quantum Hardware of Living Matter*.
http://www.physics.helsinki.fi/~matpitka/bioware/bioware.html.

[J1] The chapter *Bio-Systems as Super-Conductors: part I* of [TGDware].
http://www.physics.helsinki.fi/~matpitka/bioware/bioware.html#superc1.

[J2] The chapter *Bio-Systems as Super-Conductors: part II* of [TGDware].
http://www.physics.helsinki.fi/~matpitka/bioware/bioware.html#superc2.

[J3] The chapter *Bio-Systems as Super-Conductors: part III* of [TGDware].
http://www.physics.helsinki.fi/~matpitka/bioware/bioware.html#superc3.

[J4] The chapter *Quantum Antenna Hypothesis* of [TGDware].
http://www.physics.helsinki.fi/~matpitka/bioware/bioware.html#tubuc.

[J5] The chapter *Wormhole Magnetic Fields* of [TGDware].
http://www.physics.helsinki.fi/~matpitka/bioware/bioware.html#wormc.

[J6] The chapter *Coherent Dark Matter and Bio-Systems as Macroscopic Quantum Systems* of
[TGDware].
http://www.physics.helsinki.fi/~matpitka/bioware/bioware.html#darkbio.

[J7] The chapter *About the New Physics Behind Qualia* of [TGDware].
http://www.physics.helsinki.fi/~matpitka/bioware/bioware.html#newphys.

[TGDholo] M. Pitkänen (2006), *Bio-Systems as Conscious Holograms*.
http://www.physics.helsinki.fi/~matpitka/hologram/hologram.html.

[K1] The chapter *Time, Spacetime and Consciousness* of [TGDholo].
http://www.physics.helsinki.fi/~matpitka/hologram/hologram.html#time.

[K2] The chapter *Macro-Temporal Quantum Coherence and Spin Glass Degeneracy* of [TGDholo].
http://www.physics.helsinki.fi/~matpitka/hologram/hologram.html#macro.

[K3] The chapter *General Theory of Qualia* of [TGDholo].
http://www.physics.helsinki.fi/~matpitka/hologram/hologram.html#qualia.

[K4] The chapter *Bio-Systems as Conscious Holograms* of [TGDholo].
http://www.physics.helsinki.fi/~matpitka/hologram/hologram.html#hologram.

[K5] The chapter *Homeopathy in Many-Sheeted Space-Time* of [TGDholo].
http://www.physics.helsinki.fi/~matpitka/hologram/hologram.html#homeoc.

[K6] The chapter *Macroscopic Quantum Coherence and Quantum Metabolism as Different Sides of
the Same Coin* of [TGDholo].
http://www.physics.helsinki.fi/~matpitka/hologram/hologram.html#metab.

[TGDgeme] M. Pitkänen (2006), *Mathematical Aspects of Consciousness Theory*.
http://www.physics.helsinki.fi/~matpitka/genememe/genememe.html.

[L1] The chapter *Genes and Memes* of [TGDgeme].
http://www.physics.helsinki.fi/~matpitka/genememe/genememe.html#genememec.

[L2] The chapter *Many-Sheeted DNA* of [TGDgeme].
http://www.physics.helsinki.fi/~matpitka/genememe/genememe.html#genecodec.

[L3] The chapter *Could Genetic Code Be Understood Number Theoretically?* of [TGDgeme].
http://www.physics.helsinki.fi/~matpitka/genememe/genememe.html#genenumber.

[L4] The chapter *Pre-Biotic Evolution in Many-Sheeted Space-Time* of [TGDgeme].
http://www.physics.helsinki.fi/~matpitka/genememe/genememe.html#prebio.

[TGDeeg] M. Pitkänen (2006), *TGD and EEG*.
http://www.physics.helsinki.fi/~matpitka/tgdeeg/tgdeeg/tgdeeg.html.

[M1] The chapter *Magnetic Sensory Canvas Hypothesis* of [TGDeeg].
http://www.physics.helsinki.fi/~matpitka/tgdeeg/tgdeeg/tgdeeg.html#mec.

[M2] The chapter *Quantum Model for Nerve Pulse* of [TGDeeg].
http://www.physics.helsinki.fi/~matpitka/tgdeeg/tgdeeg/tgdeeg.html#pulse.

[M3] The chapter *Dark Matter Hierarchy and Hierarchy of EEGs* of [TGDeeg].
http://www.physics.helsinki.fi/~matpitka/tgdeeg/tgdeeg/tgdeeg.html#eegdark.

[M4] The chapter *Quantum Model for EEG: Part I* of [TGDeeg].
http://www.physics.helsinki.fi/~matpitka/tgdeeg/tgdeeg/tgdeeg.html#eegI.

[M5] The chapter *Quantum Model of EEG: Part II* of [TGDeeg].
http://www.physics.helsinki.fi/~matpitka/tgdeeg/tgdeeg/tgdeeg.html#eegII.

[M6] The chapter *Quantum Model for Hearing* of [TGDeeg].
http://www.physics.helsinki.fi/~matpitka/tgdeeg/tgdeeg/tgdeeg.html#hearing.

[TGDmagn] M. Pitkänen (2006), *Magnetospheric Consciousness*.
http://www.physics.helsinki.fi/~matpitka/magnconsc/magnconsc.html.

[N1] The chapter *Magnetospheric Sensory Representations* of [TGDmagn].
http://www.physics.helsinki.fi/~matpitka/magnconsc/magnconsc.html#srepres.

[N2] The chapter *Crop Circles and Life at Parallel Space-Time Sheets* of [TGDmagn].
http://www.physics.helsinki.fi/~matpitka/magnconsc/magnconsc.html#crop1.

[N3] The chapter *Crop Circles and Life at Parallel Space-Time Sheets* of [TGDmagn].
http://www.physics.helsinki.fi/~matpitka/magnconsc/magnconsc.html#crop2.

[N4] The chapter *Pre-Biotic Evolution in Many-Sheeted Space-Time* of [TGDmagn].
http://www.physics.helsinki.fi/~matpitka/magnconsc/magnconsc.html#prebio.

[N5] The chapter *Semi-trance, Mental Illness, and Altered States of Consciousness* of [TGDmagn].
http://www.physics.helsinki.fi/~matpitka/magnconsc/magnconsc.html#semitrancec.

[N6] The chapter *Semitrance, Language, and Development of Civilization* of [TGDmagn].
http://www.physics.helsinki.fi/~matpitka/magnconsc/magnconsc.html#langsoc.

[TGDmathc] M. Pitkänen (2006), *Mathematical Aspects of Consciousness Theory*.
http://www.physics.helsinki.fi/~matpitka/magnconsc/mathconsc.html.

[O1] The chapter *Category Theory, Quantum TGD, and TGD Inspired Theory of Consciousness* of
[TGDmathc].
http://www.physics.helsinki.fi/~matpitka/mathconsc/mathconsc.html#categoryc.

[O2] The chapter *Infinite Primes and Consciousness* of [TGDmathc].
http://www.physics.helsinki.fi/~matpitka/mathconsc/mathconsc.html#infpc.

[O3] The chapter *Topological Quantum Computation in TGD Universe* of [TGDmathc].
http://www.physics.helsinki.fi/~matpitka/mathconsc/mathconsc.html#tqc.

[O4] The chapter *Intentionality, Cognition, and Physics as Number theory or Space-Time Point as
Platonia* of [TGDmathc].
http://www.physics.helsinki.fi/~matpitka/mathconsc/mathconsc.html#intcognc.

[O5] The chapter *Was von Neumann Right After All* of [TGDmathc].
http://www.physics.helsinki.fi/~matpitka/mathconsc/mathconsc.html#vNeumann.

Index

9 780955 117084